*Edited by
Richard Dronskowski,
Shinichi Kikkawa, and
Andreas Stein*

**Handbook of
Solid State Chemistry**

Edited by
*Richard Dronskowski,
Shinichi Kikkawa, and
Andreas Stein*

Handbook of Solid State Chemistry

Volume 1: Materials and Structure of Solids

WILEY-VCH Verlag GmbH & Co. KGaA

Editors

Richard Dronskowski
RWTH Aachen
Institute of Inorganic Chemistry
Landoltweg 1
52056 Aachen
Germany

Shinichi Kikkawa
Hokkaido University
Faculty of Engineering
N13 W8, Kita-ku
060-8628 Sapporo
Japan

Andreas Stein
University of Minnesota
Department of Chemistry
207 Pleasant St. SE
Minneapolis, MN 55455
USA

Cover Credit: Sven Lidin, Arndt Simon and Franck Tessier

All books published by **Wiley-VCH** are carefully produced. Nevertheless, authors, editors, and publisher do not warrant the information contained in these books, including this book, to be free of errors. Readers are advised to keep in mind that statements, data, illustrations, procedural details or other items may inadvertently be inaccurate.

Library of Congress Card No.: applied for

British Library Cataloguing-in-Publication Data
A catalogue record for this book is available from the British Library.

Bibliographic information published by the Deutsche Nationalbibliothek
The Deutsche Nationalbibliothek lists this publication in the Deutsche Nationalbibliografie; detailed bibliographic data are available on the Internet at http://dnb.d-nb.de.

© 2017 Wiley-VCH Verlag GmbH & Co. KGaA, Boschstr. 12, 69469 Weinheim, Germany

All rights reserved (including those of translation into other languages). No part of this book may be reproduced in any form – by photoprinting, microfilm, or any other means – nor transmitted or translated into a machine language without written permission from the publishers. Registered names, trademarks, etc. used in this book, even when not specifically marked as such, are not to be considered unprotected by law.

Print ISBN: 978-3-527-32587-0
oBook ISBN: 978-3-527-69103-6

Cover Design   Formgeber
Typesetting   Thomson Digital, Noida, India
Printing and Binding   Markono Print Media Pte Ltd, Singapore

Printed on acid-free paper

Preface

When you do great science, you do not have to make a lot of fuss. This oft-forgotten saying from the twentieth century has served these editors pretty well, so the foreword to this definitive six-volume *Handbook of Solid-State Chemistry* in the early twenty-first century will be brief. After all, is there any real need to highlight the paramount successes of solid-state chemistry in the last half century? – Successes that have led to novel magnets, solid-state lighting, dielectrics, phase-change materials, batteries, superconducting compounds, and a lot more? Probably not, but we should stress that many of these exciting matters were derived from curiosity-driven research — work that many practitioners of our beloved branch of chemistry truly appreciate, and this is exactly why they do it. Our objects of study may be immensely important for various applications but, first of all, they are interesting to us; that is, how chemistry defines and challenges itself. Let us also not forget that solid-state chemistry is a neighbor to physics, crystallography, materials science, and other fields, so there is plenty of room at the border, to paraphrase another important quote from a courageous physicist.

Given the incredibly rich heritage of solid-state chemistry, it is probably hard for a newcomer (a young doctoral student, for example) to see the forest for all the trees. In other words, there is a real need to cover solid-state chemistry in its entirety, but only if it is conveniently grouped into digestible categories. Because such an endeavor is not possible in introductory textbooks, this is what we have tried to put together here. The compendium starts with an overview of materials and of the structure of solids. Not too surprisingly, the next volume deals with synthetic techniques, followed by another volume on various ways of (structural) characterization. Being a timely handbook, the fourth volume touches upon nano and hybrid materials, while volume V introduces the reader to the theoretical description of the solid state. Finally, the sixth volume reaches into the real world by focusing on functional materials. Should we have considered more volumes? Yes, probably, but life is short, dear friends.

This handbook would have been impossible to compile for three authors, let alone a single one. Instead, the editors take enormous pride in saying that they managed to motivate more than a hundred first-class scientists living across the globe, each of them specializing in (and sometimes even shaping) a subfield of solid-state chemistry or a related discipline, and all of these wonderful colleagues did their very best to make our dream come true. Thanks to all of you; we sincerely appreciate your contributions. Thank you, once again, on behalf of solid-state chemistry. The editors also would like to thank Wiley-VCH, in particular Dr. Waltraud Wüst and also Dr. Frank Otmar Weinreich, for spiritually (and practically) accompanying us over a few years, and for reminding us here and there that there must be a final deadline. That being said, it is up to the reader to judge whether the tremendous effort was justified. We sincerely hope that this is the case.

A toast to our wonderful science! Long live solid-state chemistry!

Richard Dronskowski
RWTH Aachen, Aachen, Germany

Shinichi Kikkawa
Hokkaido University, Sapporo, Japan

Andreas Stein
University of Minnesota, Minneapolis, USA

Contents

Volume 1: Materials and Structure of Solids

1 Intermetallic Compounds and Alloy Bonding Theory Derived from Quantum Mechanical One-Electron Models *1*
Stephen Lee and Daniel C. Fredrickson
1.1 Introduction *1*
1.1.1 The H_2 Molecule *2*
1.2 Tight-Binding Theories *4*
1.2.1 Simple and Extended Hückel Theory *4*
1.2.1.1 DOS, Projected DOS, COOP, and COHP *7*
1.2.1.2 Applications of COOP and COHP Analysis *11*
1.2.2 Site Preferences and Mulliken Populations *17*
1.2.3 Moment Analyses *20*
1.2.3.1 Rigid Band Model *20*
1.2.3.2 Higher Moments *21*
1.2.3.3 The Structural Energy Difference Theorem, μ_2-Scaling, and Higher Moments *25*
1.2.3.4 Tight Binding-Based Structure Maps *27*
1.3 Nearly Free Electron Model *29*
1.3.1 Pseudopotentials *33*
1.4 Magnetism *35*
1.4.1 Ferromagnetism *35*
1.4.1.1 Fe, Co, and Ni *37*
1.4.2 Symmetry-Breaking and Antiferromagnetism *38*
1.5 Synergism of the Jones and Tight-Binding Bonding Models *40*
1.5.1 Cu_5Zn_8 *40*
1.5.2 Li_8Ag_5 *42*
1.5.3 Quasicrystals and Quasicrystalline Approximants *45*
1.5.3.1 $Li_{52.0}Al_{88.7}Cu_{19.3}$ *46*

1.6	Emerging Directions in Tight-Binding Theory *48*
1.6.1	Localized Bonding Models *48*
1.6.1.1	Wannier Functions *48*
1.6.1.2	The 18-Electron Rule *51*
1.6.1.3	The Nowotny Chimney Ladder Phases *53*
1.6.1.4	Other Transition Metal – Main Group Metalloid Phases *55*
1.6.2	Elucidating Atomic Size Effects *57*
1.6.2.1	Steric Strain *57*
1.6.2.2	μ_2-Hückel Chemical Pressure Analysis *58*
1.6.3	Electronegativity Differences *60*
1.6.3.1	Brewer's Extension of the Lewis Concept to Transition Metals *60*
1.6.3.2	The μ_3 Acidity Model and the Lewis Theory *62*
1.7	Conclusions *65*
	References *67*

2	**Quasicrystal Approximants** *73*
	Sven Lidin
2.1	What Is an Approximant? *73*
2.2	How Can the Structure of an Aperiodic Crystal be Solved, Refined, and Understood? *73*
2.3	The Quasicrystals and Their Approximants *75*
2.4	Octagonal Approximants *75*
2.5	Decagonal Approximants *76*
2.6	Dodecagonal Approximants *80*
2.7	Icosahedral Approximants *80*
2.8	Conclusion *89*
	References *89*

3	**Medium-Range Order in Oxide Glasses** *93*
	Hellmut Eckert
3.1	Introduction *93*
3.2	General Concepts of Structural Order in Glasses *95*
3.3	Experimental Characterization of Medium-Range Order *99*
3.3.1	X-Ray and Neutron Diffraction Techniques *99*
3.3.2	Vibrational Spectroscopy *101*
3.3.3	Solid-State NMR *103*
3.3.3.1	Magic-Angle Spinning NMR *104*
3.3.3.2	Dipolar NMR Methods *105*
3.4	Medium Range Order in Covalent Network Glasses *109*
3.4.1	Glassy SiO_2, GeO_2, and the Binary System SiO_2–GeO_2 *109*
3.4.1.1	T—O—T Bond Angle Distributions *109*
3.4.1.2	Ring Size Distributions *112*
3.4.1.3	Network Connectivity in SiO_2–GeO_2 Glasses *113*
3.4.2	Boron Oxide, B_2O_3–SiO_2, and B_2O_3–GeO_2 Glasses *113*

3.4.3	P_2O_5, P_2O_5-GeO_2, and Ternary P_2O_5-Based Network Glasses	*115*
3.5	Medium Range Order in Modified Oxide Glasses	*118*
3.5.1	Medium-Range Order Effects in Binary Model Glasses	*118*
3.5.1.1	Cation Next-Nearest Neighbor Environments and Spatial Distributions Studied by Neutron Diffraction	*118*
3.5.1.2	Cation Next-Nearest Environments and Spatial Distributions Studied by Dipolar NMR	*120*
3.5.1.3	Ring-Size Distributions and "Super-Structural Units"	*124*
3.5.2	Connectivities and Cation Distributions in Glasses with Multiple Network Formers	*126*
3.5.2.1	Borophosphate Glasses and Related Systems	*127*
3.5.2.2	Nanosegregation and Phase Separation Effects in Borosilicate Glasses:	*129*
3.6	Conclusions and Recommendations for Further Work	*131*
	Acknowledgments	*132*
	References	*133*
4	**Suboxides and Other Low-Valent Species**	*139*
	Arndt Simon	
4.1	Materials and Structure of Solids	*139*
4.1.1	Bulk Materials	*139*
4.1.1.1	Suboxides and Other Low-Valent Species	*139*
4.2	Epilogue	*157*
	References	*159*
5	**Introduction to the Crystal Chemistry of Transition Metal Oxides**	*161*
	J.E. Greedan	
5.1	Introduction	*161*
5.2	Crystal Chemistry of TMO: Some Basics	*163*
5.2.1	Close Packing and Hole Filling	*163*
5.2.2	Ionic Radii, Radius Ratios, and Site Preferences	*166*
5.2.3	Electronic Influences on Site Symmetry	*168*
5.2.4	Representation of Crystal Structures of TMO	*168*
5.3	Most Important Structure Types for TMO	*169*
5.3.1	NaCl or Rock Salt	*169*
5.3.1.1	Binary Oxides: MO	*169*
5.3.1.2	Ternary and More Complex Oxides with the Ordered-Rock Salt Structure (ORS)	*172*
5.3.2	Corundum or Al_2O_3	*177*
5.3.2.1	Binary Oxides	*178*
5.3.2.2	Ilmenite and Other Ordered Corundum-Like Structures	*179*
5.3.3	Rutile and Related Structures	*181*
5.3.3.1	Binary Oxides	*181*
5.3.3.2	Site-Ordered Rutiles: the Trirutile Structure and Columbite	*183*

5.3.3.3	Structures Compositionally Related to Rutile: Hollandites and Others *184*
5.3.3.4	Other Forms of TiO_2 and MnO_2 *186*
5.3.4	Fluorite, CaF_2, and Related Structures *187*
5.3.4.1	Binary Oxides *187*
5.3.4.2	Ordered Defect Fluorite Strutures; -Bixbyite, Pyrochlore, and Weberite *189*
5.3.4.3	Other Defect Fluorite Structures *196*
5.3.5	Perovskite and Related Structures *196*
5.3.5.1	ABO_3 *196*
5.3.5.2	Perovskite Related Structures *201*
5.3.6	Spinels *212*
5.4	Conclusion *214*
	References *215*

6	**Perovskite Structure Compounds** *221*
	Yuichi Shimakawa
6.1	Chemical Variety of Perovskites *222*
6.2	Crystal Structure of Perovskite *223*
6.3	Ion Order in Perovskites *229*
6.3.1	B-Site Cation Order *229*
6.3.2	A-Site Cation Order *232*
6.3.3	A- and B-Site Cation Order *236*
6.3.4	X Anion Vacancy Order *236*
6.4	Perovskite-Related Compounds *241*
6.4.1	Hexagonal Perovskite Compounds *241*
6.4.2	Post-Perovskite Compounds *242*
6.4.3	Antiperovskite Structure *243*
6.4.4	Perovskite-Related Layered Structure Compounds *243*
	References *247*

7	**Nitrides of Non-Main Group Elements** *251*
	P. Höhn and R. Niewa
7.1	Introduction *251*
7.2	Preparative Aspects *252*
7.3	Structural Chemistry of Nitrides *257*
7.3.1	Binary Nitrides of Transition and Rare Earth Metals *258*
7.3.1.1	Group 3, Sc, Y, La, and Rare Earth Metals, U, Th *259*
7.3.1.2	Group 4, Ti, Zr, Hf *262*
7.3.1.3	Group 5, V, Nb, Ta *264*
7.3.1.4	Group 6, Cr, Mo, W *267*
7.3.1.5	Group 7, Mn, Re *268*
7.3.1.6	Group 8, Fe, Ru, Os *269*
7.3.1.7	Group 9, Co, Rh, Ir *271*
7.3.1.8	Group 10, Ni, Pd, Pt *271*

7.3.1.9	Group 11, Cu, Ag, Au	*271*
7.3.1.10	Group 12, Zn, Cd, Hg	*272*
7.3.2	Multinary Derivatives of Metal-Rich Compounds	*272*
7.3.2.1	γ′-Fe$_4$N Derivatives	*273*
7.3.2.2	ε-Fe$_3$N Derivatives	*278*
7.3.2.3	β-Mn Derivatives	*279*
7.3.2.4	η-Carbide Derivatives	*280*
7.3.2.5	MAX Phases $M_{n+1}AN_n$ and Related Compounds M_3AN	*281*
7.3.2.6	Z-Phases $MM'N$	*283*
7.3.2.7	$RE_2M_{17}N_y$ and Related Phases	*283*
7.3.2.8	Related Systems Based on Close-Packed Arrangements of M	*285*
7.3.3	Multinary Nitrides and Nitridometalates	*286*
7.3.3.1	$CN(M) = 2$	*286*
7.3.3.2	$CN(M) = 2 + 3$	*298*
7.3.3.3	$CN(M) = 3$	*300*
7.3.3.4	$CN(M) = 3 + 4$	*305*
7.3.3.5	$CN(M) = 4$	*309*
7.3.3.6	$CN(M) = 5$	*323*
7.3.3.7	$CN(M) = 6$	*324*
7.3.3.8	Further Unexplored Systems	*328*
7.3.4	Metallide Nitrides	*328*
7.4	Conclusion and Future Trends	*330*
	Abbreviations	*333*
	References	*333*
8	**Fluorite-Type Transition Metal Oxynitrides**	
	Franck Tessier	
8.1	Introduction	*361*
8.2	Fluorite-Type Structures	*362*
8.2.1	From Diamond to Fluorite Structures	*362*
8.2.1.1	Antifluorite Structure	*363*
8.2.2	Defect Fluorites	*364*
8.2.2.1	Zirconium-Based Oxynitrides	*365*
8.2.2.2	Bixbyite and Antibixbyite Structures	*365*
8.2.2.3	Scheelite Structure	*366*
8.2.2.4	Pyrochlore Structure	*369*
8.3	Synthesis	*370*
8.4	Defect-Fluorite Solid Solutions	*371*
8.4.1	R–W–O–N System	*371*
8.4.2	R–Ta–O–N System	*375*
8.5	Conclusion	*378*
	Acknowledgments	*378*
	References	*378*

9	Mechanochemical Synthesis, Vacancy-Ordered Structures and Low-Dimensional Properties of Transition Metal Chalcogenides *383*
	Yutaka Ueda and Tsukio Ohtani
9.1	Introduction *383*
9.2	Synthesis of Metal Chalcogenides by Mechanical Alloying *384*
9.2.1	Synthesis of Copper Chalcogenides by Ball Milling *385*
9.2.2	Sonochemical Synthesis of Cu and Ag Chalcogenides *389*
9.2.3	Solid-State Reaction: Formation of Cu_3Se_2 by Making α-CuSe and α-Cu_2Se Pellets Contact *394*
9.2.4	Mechanically Induced Self-Propagating Reactions (MSR) of CdSe and ZnSe: Premilling Effects *399*
9.3	Vacancy-Ordered Structure, Order–Disorder Transition and Site Preference in the $3d$ Transition Metal Chalcogenides *402*
9.3.1	Vacancy-Ordered Structures in Binary MX–MX_2 System *402*
9.3.2	Phase Diagram and Order–Disorder Transition in Binary MSe–MSe_2 System *404*
9.3.3	Phase Diagram and Site Preference in Pseudobinary M_3X_4–M'_3X_4 System *406*
9.3.4	The Site Preference and Magnetic Properties *412*
9.4	Quasi-One-Dimensional Chalcogenides *417*
9.4.1	Charge Density Wave (CDW) Instability in Quasi-One-Dimensional Compound $NbSe_3$ *417*
9.4.2	CDW and Superconductivity in Quasi-One-Dimensional Compound $In_xNb_3Te_4$ *418*
9.4.3	Successive Phase Transitions in Quasi-One-Dimensional Sulfides ACu_7S_4 (A = Tl, K, Rb) *423*
	References *430*

10	Metal Borides: Versatile Structures and Properties *435*
	Barbara Albert and Kathrin Hofmann
10.1	Introduction *435*
10.2	Synthesis of Borides *439*
10.2.1	Powder Preparation *439*
10.2.2	Single-Crystal Growth *439*
10.2.3	Nanomaterials *440*
10.3	Metal-Rich Borides: Interstitials or More? *440*
10.3.1	General *440*
10.3.2	M_3B to MB_2: From Isolated Boron Atoms to Chains and Sheets *441*
10.3.3	τ-Borides *442*
10.3.4	Diborides and Related Compounds *442*
10.4	Boron-Rich Solids *446*
10.4.1	General *446*
10.4.2	Octahedra and Bipyramids in Borides: MB_4, MB_6, and Na_3B_{20} *446*
10.4.3	Dodecaborides and Borides with Icosahedra *448*
10.4.4	Carbon-Like Frameworks of Boron Atoms in Borides *449*

10.5	Higher Borides *450*	
10.6	Conclusion *450*	
	References *451*	

11 Metal Pnictides: Structures and Thermoelectric Properties *455*
Abdeljalil Assoud and Holger Kleinke
- 11.1   Introduction *455*
- 11.2   Binary Antimonides *457*
- 11.2.1 Main Group Antimonides *457*
- 11.2.2 Early Transition Metal Antimonides *459*
- 11.2.3 Late Transition Metal Antimonides *462*
- 11.3   Ternary Antimonides *465*
- 11.3.1 Ternary Antimonides without Sb−Sb Bonds *465*
- 11.3.2 Ternary Antimonides with Sb−Sb Bonds *469*
- 11.4   Conclusions *471*
- References *472*

12 Metal Hydrides *477*
Yaoqing Zhang, Maarten C. Verbraeken, Cédric Tassel, and Hiroshi Kageyama
- 12.1   Introduction *477*
- 12.2   Hydrogen Chemistry *478*
- 12.2.1 Hydrogen Species *478*
- 12.2.2 Hydride Compounds *479*
- 12.3   Saline Hydrides *480*
- 12.3.1 Thermodynamics of Hydride Compounds *480*
- 12.3.2 Crystal Structures *483*
- 12.3.3 High-Pressure Behavior *489*
- 12.3.4 Hydride Ion Mobility in Saline Hydrides *491*
- 12.4   Ternary Ionic Hydrides *494*
- 12.5   Metallic Hydrides *496*
- 12.5.1 Crystal Structures and Compositions *498*
- 12.5.2 Bonding, Stoichiometry, and Properties *500*
- 12.6   Oxyhydrides: The Hydride Anion in Oxide Structures *501*
- 12.6.1 p- and f-Block Oxyhydrides *503*
- 12.6.1.1 Rare Earth Oxyhydride *503*
- 12.6.1.2 Antiperovskite *503*
- 12.6.1.3 Suboxide Hydrides *504*
- 12.6.1.4 Mayenite *505*
- 12.6.2 Low-Temperature Synthesis of Transition Metal Oxyhydrides *506*
- 12.6.2.1 Cobalt Oxyhydrides *507*
- 12.6.2.2 Titanium Oxyhydrides *509*
- 12.6.2.3 Vanadium Oxyhydrides *512*
- 12.6.3 High-Pressure Synthesis of Oxyhydrides *513*
- Acknowledgement *515*
- References *515*

13 Local Atomic Order in Intermetallics and Alloys *521*
Frank Haarmann

13.1 Motivation *521*
13.2 Synthesis *522*
13.3 Hume-Rothery Phases *522*
13.4 Laves Phases *523*
13.5 Zintl Phases *524*
13.6 Heusler Phases *526*
13.7 Thermoelectric Materials *527*
13.8 Clathrates *528*
13.9 Phase Change Materials (PCM) *529*
13.10 Disorder in IPs *531*
13.11 Vacancies in $Cu_{1-x}Al_2$ *532*
13.12 Preferred Site Occupation *533*
13.13 Substitution by Similar Type of Atoms *535*
References *537*

14 Layered Double Hydroxides: Structure–Property Relationships *541*
Shan He, Jingbin Han, Mingfei Shao, Ruizheng Liang, Min Wei, David G. Evans, and Xue Duan

14.1 Two-Dimensional Structure of Layered Double Hydroxides (LDHs) *541*
14.1.1 Introduction of LDHs *541*
14.1.2 Basic Structural Features *542*
14.1.2.1 Brucite-Like Layers *542*
14.1.2.2 Host Layer Structure of LDHs *543*
14.1.2.3 Interlayer Structure of LDHs *543*
14.1.3 Theoretical Studies of LDHs Structure *544*
14.1.3.1 Molecular Dynamics Simulations of LDHs *544*
14.1.3.2 Density Functional Theory Calculations of LDHs *545*
14.2 Properties and Applications of LDHs *546*
14.2.1 Confinement Effect *546*
14.2.2 Catalytic Properties of LDHs *548*
14.2.2.1 Solid Base Catalysis *549*
14.2.2.2 Photocatalysis *549*
14.2.2.3 Metal-Based Catalysts *550*
14.2.2.4 Active Species-Intercalated Nanocatalysts *552*
14.2.3 Electrochemical Energy Storage Properties *553*
14.2.3.1 Supercapacitors *554*
14.2.3.2 Li-Ion Batteries *556*
14.2.3.3 Electrochemical Catalysis *556*
14.2.4 Gas Barrier Properties *558*
14.2.4.1 Gas Barrier Films with Brick–Mortar Structure *558*
14.2.4.2 Gas Barrier Films with Brick–Mortar–Sand Structure *558*

14.2.4.3	Gas Barrier Films with Self-Healing Properties	*560*
14.2.5	Biological and Medical Properties	*561*
14.2.5.1	Synthesis of LDH-Based Biomaterials	*561*
14.2.5.2	Advantages of LDH-Based Materials in Bioapplications	*562*
14.2.5.3	Applications of LDH-Based Materials in Drug Delivery and Treatment	*562*
	References	*564*
15	**Structural Diversity in Complex Layered Oxides**	*571*
	S. Uma	
15.1	Introduction	*571*
15.2	Types of Layered Oxides	*572*
15.2.1	Origin of Interest and Basic Features	*572*
15.2.2	Layered Oxides Containing Metal-Oxygen Octahedra	*573*
15.2.2.1	Layered Perovskites	*573*
15.2.2.2	Layered Oxides Possessing Rocksalt Superstructures	*581*
15.2.2.3	Layered Oxides Made up of Edge- and/or Corner-Shared Metal-Oxygen Octahedra	*584*
15.2.2.4	Miscellaneous Layered Oxides	*586*
15.3	Interlayer Reactions	*587*
15.3.1	Ion Exchange	*587*
15.3.2	Intercalation	*589*
15.4	Potential Applications and Future Possibilities	*590*
	Acknowledgment	*590*
	References	*591*
16	**Magnetoresistance Materials**	*595*
	Ichiro Terasaki	
16.1	Classical Magnetoresistance	*595*
16.2	Classification of Magnetoresistance	*597*
16.3	Giant Magnetoresistance Materials	*599*
16.4	Colossal Magnetoresistance Materials	*604*
16.5	Dilute Magnetic Semiconductors	*607*
16.6	Quantum Transport Phenomena	*609*
16.7	Concluding Remarks	*610*
	References	*611*
17	**Magnetic Frustration in Spinels, Spin Ice Compounds, $A_3B_5O_{12}$ Garnet, and Multiferroic Materials**	*617*
	Hongyang Zhao, Hideo Kimura, Zhenxiang Cheng, and Tingting Jia	
17.1	Frustration: An Introduction	*617*
17.2	Several Concepts in Frustrated Magnets	*619*
17.2.1	Geometric Frustration in Simple Atomic Systems: The Curved-Space Approach	*620*

| XVI | Contents

17.2.2	Fluctuations of the Spins in a Spin Liquid can be Classical or Quantum *622*
17.2.2.1	Classical Spin Liquid *622*
17.2.2.2	Quantum Spin Liquid (QSL) *623*
17.2.3	Spin Ice *624*
17.3	Interesting Magnetic Frustrated Materials *625*
17.3.1	Highly Frustrated Magnetism in Spinels *626*
17.3.2	Spin Ice Compounds: $Ho_2Ti_2O_7$ and $Dy_2Ti_2O_7$ *629*
17.3.3	$A_3B_5O_{12}$ Garnet Structure *631*
17.3.4	Magnetic Frustration in Multiferroic Materials – Aurivillius Bilayered Multiferroic: $Bi_5Ti_3FeO_{15}$ *634*
17.4	Conclusions *638*
	Acknowledgments *639*
	References *639*

18   Structures and Properties of Dielectrics and Ferroelectrics *643*
Mitsuru Itoh
18.1	Crystal Symmetry and the Properties *643*
18.2	Second-Order Phase Transition in Ferroelectric *647*
18.3	First-Order Phase Transition in Ferrelectric *648*
18.4	Successive Phase Transition and Free Energy of Ferroelectric *649*
18.5	Generalization of the Formula to a Real Ferroelectric *650*
18.6	Antiferroelectric and Ferrielectric *651*
18.7	Order–Disorder-Type Ferroelectric: Rochelle Salt – Potassium Sodium Tartrate ($NaKC_4H_4O_6 \cdot 4H_2O$) *653*
18.8	Hydrogen-Bonded System: Potassium Dihydrogen Phosphor (KDP) *653*
18.9	Perovskites and Related Compounds *654*
18.10	Quantum Paraelectricity *657*
18.11	Relaxor Ferroelectric *659*
18.12	Piezoelectric: $Pb(Zr_{1-x}Ti_x)O_3$ (PZT) *661*
18.13	Organic Ferroelectric *662*
	References *663*

19   Defect Chemistry and Its Relevance for Ionic Conduction and Reactivity *665*
Joachim Maier
19.1	Introduction: On the Relevance of Defect Chemistry *665*
19.2	Ionic and Electronic Defects as Acid–Base and Redox Particles in Equilibrium *669*
19.3	Transport and Transfer *682*
19.4	Defect Chemistry at Boundaries and Heterogeneous Electrolytes *686*
19.5	Reaction and Catalysis *689*
19.6	Electrochemical Applications *693*
19.6.1	Ion Conductors as Solid Electrolytes *693*

19.6.1.1 Fuel Cells 693
19.6.1.2 Batteries 695
19.6.1.3 Chemical Sensors 695
19.6.2 Mixed Conductors as Electrodes or Absorbers 695
19.6.2.1 Fuel Cells 695
19.6.2.2 Batteries 696
19.6.2.3 Sensors, Chemical Storage and Permeation Devices 696
19.6.3 Electronic Conductors as Current Collectors or Taguchi Sensors 696
19.6.3.1 Fuel Cells and Batteries 696
19.6.3.2 Sensors 696
19.7 Conclusions 697
References 697

20 Molecular Magnets 703
J.V. Yakhmi
20.1 Introduction 703
20.2 Strategies of Design and Synthesis of Molecular Magnets 704
20.3 Purely Organic Magnets 705
20.4 Ferrimagnetic Building-Blocks 706
20.5 Molecular Magnetic Sponges and Hydrogels 709
20.6 Polycyanometallates 710
20.6.1 Prussian Blue and its Hexacyanate Analogues 710
20.6.2 Negative Magnetization and Pole Reversal 713
20.6.3 Magnetocaloric Effect 713
20.6.4 Hepta/Octacyanates 713
20.7 High Spin (HS)–Low Spin (LS) transitions. Spin Crossover (SCO) 714
20.8 Design of Molecular Magnets Using AZIDO Bridge to Mediate Spin–Spin Couplings 715
20.9 Single Molecule Magnets (SMM) and Single Chain Magnets (SCM) 716
20.10 Molecular Spintronics 718
20.11 Multifunctional Molecular Magnetic Materials 719
20.12 Biogenic Magnetism 720
20.12.1 Magnetotactic Bacteria 720
20.12.2 Magnetoreception 721
20.13 Other Recent Developments, Current scenario, and Future Perspectives 723
Acknowledgments 724
References 724

21 Ge–Sb–Te Phase-Change Materials 735
Volker L. Deringer and Matthias Wuttig
21.1 Introduction 735
21.2 Crystalline Ge–Sb–Te Phases: Stable and Metastable 737
21.2.1 GeTe and Sb_2Te_3 737

21.2.2	The Stable Quasibinary Phases	*738*
21.2.3	Metastable (Rock Salt-Type) Phases	*739*
21.3	The Amorphous Phases	*739*
21.3.1	What Makes Amorphous PCMs So Special?	*739*
21.3.2	Experimental Probes	*741*
21.3.3	First-Principles Simulations	*742*
21.4	Materials Properties and Property Contrast	*742*
21.4.1	Resonant Bonding in PCMs	*742*
21.4.2	Origin and Nature of Vacancies	*744*
21.4.3	How Disorder Determines Electronic Properties	*744*
21.5	From Phase-Change Data Storage to a Versatile Materials Platform	*745*
	Acknowledgments	*746*
	References	*746*

Index *751*

Volume 2: Synthesis

1 High-Temperature Methods *1*
Rainer Pöttgen and Oliver Janka

2 High-Pressure Methods in Solid-State Chemistry *23*
Hubert Huppertz, Gunter Heymann, Ulrich Schwarz, and Marcus R. Schwarz

3 High-Pressure Perovskite: Synthesis, Structure, and Phase Relation *49*
Yoshiyuki Inaguma

4 Solvothermal Methods *107*
Nobuhiro Kumada

5 High-Throughput Synthesis Under Hydrothermal Conditions *123*
Nobuaki Aoki, Gimyeong Seong, Tsutomu Aida, Daisuke Hojo, Seiichi Takami, and Tadafumi Adschiri

6 Particle-Mediated Crystal Growth *155*
R. Lee Penn

7 Sol–Gel Synthesis of Solid-State Materials *179*
Guido Kickelbick and Patrick Wenderoth

8 Templated Synthesis for Nanostructured Materials *201*
Yoshiyuki Kuroda and Kazuyuki Kuroda

9	**Bio-Inspired Synthesis and Application of Functional Inorganic Materials by Polymer-Controlled Crystallization** *233* Lei Liu and Shu-Hong Yu	
10	**Reactive Fluxes** *275*	
11	**Glass Formation and Crystallization** *287* T. Komatsu	
12	**Glass-Forming Ability, Recent Trends, and Synthesis Methods of Metallic Glasses** *319* Hidemi Kato, Takeshi Wada, Rui Yamada, and Junji Saida	
13	**Crystal Growth Via the Gas Phase by Chemical Vapor Transport Reactions** *351* Michael Binnewies, Robert Glaum, Marcus Schmidt, and Peer Schmidt	
14	**Thermodynamic and Kinetic Aspects of Crystal Growth** *375* Detlef Klimm	
15	**Chemical Vapor Deposition** *399* Takashi Goto and Hirokazu Katsui	
16	**Growth of Wide Bandgap Semiconductors by Halide Vapor Phase Epitaxy** *429* Yuichi Oshima, Encarnación G. Víllora, and Kiyoshi Shimamura	
17	**Growth of Silicon Nanowires** *467* Fengji Li and Sam Zhang	
18	**Chemical Patterning on Surfaces and in Bulk Gels** *539* Olaf Karthaus	
19	**Microcontact Printing** *563* Kiyoshi Yase	
20	**Nanolithography Based on Surface Plasmon** *573* Kosei Ueno and Hiroaki Misawa	

Index *589*

Volume 3: Characterization

1 **Single-Crystal X-Ray Diffraction** *1*
   Ulli Englert

2 **Laboratory and Synchrotron Powder Diffraction** *29*
   R. E. Dinnebier, M. Etter, and T. Runcevski

3 **Neutron Diffraction** *77*
   Martin Meven and Georg Roth

4 **Modulated Crystal Structures** *109*
   Sander van Smaalen

5 **Characterization of Quasicrystals** *131*
   Walter Steurer

6 **Transmission Electron Microscopy** *155*
   Krumeich Frank

7 **Scanning Probe Microscopy** *183*
   Marek Nowicki and Klaus Wandelt

8 **Solid-State NMR Spectroscopy: Introduction for Solid-State Chemists** *245*
   Christoph S. Zehe, Renée Siegel, and Jürgen Senker

9 **Modern Electron Paramagnetic Resonance Techniques and Their Applications to Magnetic Systems** *279*
   Andrej Zorko, Matej Pregelj, and Denis Arčon

10 **Photoelectron Spectroscopy** *311*
   Stephan Breuer and Klaus Wandelt

11 **Recent Developments in Soft X-Ray Absorption Spectroscopy** *361*
   Alexander Moewes

12 **Vibrational Spectroscopy** *393*
   Götz Eckold and Helmut Schober

13 **Mößbauer Spectroscopy** *443*
   Hermann Raphael

14 **Macroscopic Magnetic Behavior: Spontaneous Magnetic Ordering** *485*
   Heiko Lueken and Manfred Speldrich

15	**Dielectric Properties** *523* Rainer Waser and Susanne Hoffmann-Eifert
16	**Mechanical Properties** *561* Volker Schnabel, Moritz to Baben, Denis Music, William J. Clegg, and Jochen M. Schneider
17	**Calorimetry** *589* Hitoshi Kawaji

Index *615*

Volume 4: Nano and Hybrid Materials

1	**Self-Assembly of Molecular Metal Oxide Nanoclusters** *1* Laia Vilà-Nadal and Leroy Cronin
2	**Inorganic Nanotubes and Fullerene–Like Nanoparticles from Layered (2D) Compounds** *21* L. Yadgarov, R. Popovitz-Biro, and R. Tenne
3	**Layered Materials: Oxides and Hydroxides** *53* Ida Shintaro
4	**Organoclays and Polymer-Clay Nanocomposites** *79* M.A. Vicente and A. Gil
5	**Zeolite and Zeolite-Like Materials** *97* Watcharop Chaikittisilp and Tatsuya Okubo
6	**Ordered Mesoporous Materials** *121* Michal Kruk
7	**Porous Coordination Polymers/Metal–Organic Frameworks** *141* Ohtani Ryo and Kitagawa Susumu
8	**Metal–Organic Frameworks: An Emerging Class of Solid-State Materials** *165* Joseph E. Mondloch, Rachel C. Klet, Ashlee J. Howarth, Joseph T. Hupp, and Omar K. Farha
9	**Sol–Gel Processing of Porous Materials** *195* Kazuki Nakanishi, Kazuyoshi Kanamori, Yasuaki Tokudome, George Hasegawa, and Yang Zhu

10	**Macroporous Materials Synthesized by Colloidal Crystal Templating** *243* Jinbo Hu and Andreas Stein	
11	**Optical Properties of Hybrid Organic–Inorganic Materials and their Applications – Part I: Luminescence and Photochromism** *275* Stephane Parola, Beatriz Julián-López, Luís D. Carlos, and Clément Sanchez	
12	**Optical Properties of Hybrid Organic–inorganic Materials and their Applications – Part II: Nonlinear Optics and Plasmonics** *317* Stephane Parola, Beatriz Julián-López, Luís D. Carlos, and Clément Sanchez	
13	**Bioactive Glasses** *357* Hirotaka Maeda and Toshihiro Kasuga	
14	**Materials for Tissue Engineering** *383* María Vallet-Regí and Antonio J. Salinas	
	Index *411*	

Volume 5: Theoretical Description

1	**Density Functional Theory** *1* Michael Springborg and Yi Dong
2	**Eliminating Core Electrons in Electronic Structure Calculations: Pseudopotentials and PAW Potentials** *29* Stefan Goedecker and Santanu Saha
3	**Periodic Local Møller–Plesset Perturbation Theory of Second Order for Solids** *59* Denis Usvyat, Lorenzo Maschio, and Martin Schütz
4	**Resonating Valence Bonds in Chemistry and Solid State** *87* Evgeny A. Plekhanov and Andrei L. Tchougréeff
5	**Many Body Perturbation Theory, Dynamical Mean Field Theory and All That** *119* Silke Biermann and Alexander Lichtenstein
6	**Semiempirical Molecular Orbital Methods** *159* Thomas Bredow and Karl Jug
7	**Tight-Binding Density Functional Theory: DFTB** *203* Gotthard Seifert

8	**DFT Calculations for Real Solids** *227*	
	Karlheinz Schwarz and Peter Blaha	
9	**Spin Polarization** *261*	
	Dong-Kyun Seo	
10	**Magnetic Properties from the Perspectives of Electronic Hamiltonian: Spin Exchange Parameters, Spin Orientation, and Spin-Half Misconception** *285*	
	Myung-Hwan Whangbo and Hongjun Xiang	
11	**Basic Properties of Well-Known Intermetallics and Some New Complex Magnetic Intermetallics** *345*	
	Peter Entel	
12	**Chemical Bonding in Solids** *405*	
	Gordon J. Miller, Yuemei Zhang, and Frank R. Wagner	
13	**Lattice Dynamics and Thermochemistry of Solid-State Materials from First-Principles Quantum-Chemical Calculations** *491*	
	Ralf Peter Stoffel and Richard Dronskowski	
14	**Predicting the Structure and Chemistry of Low-Dimensional Materials** *527*	
	Xiaohu Yu, Artem R. Oganov, Zhenhai Wang, Gabriele Saleh, Vinit Sharma, Qiang Zhu, Qinggao Wang, Xiang-Feng Zhou, Ivan A. Popov, Alexander I. Boldyrev, Vladimir S. Baturin, and Sergey V. Lepeshkin	
15	**The Pressing Role of Theory in Studies of Compressed Matter** *571*	
	Eva Zurek	
16	**First-Principles Computation of NMR Parameters in Solid-State Chemistry** *607*	
	Jérôme Cuny, Régis Gautier, and Jean-François Halet	
17	**Quantum Mechanical/Molecular Mechanical (QM/MM) Approaches** *647*	
	C. Richard A. Catlow, John Buckeridge, Matthew R. Farrow, Andrew J. Logsdail, and Alexey A. Sokol	
18	**Modeling Crystal Nucleation and Growth and Polymorphic Transitions** *681*	
	Dirk Zahn	
	Index *701*	

Volume 6: Functional Materials

1. **Electrical Energy Storage: Batteries** *1*
   Eric McCalla

2. **Electrical Energy Storage: Supercapacitors** *25*
   Enbo Zhao, Wentian Gu, and Gleb Yushin

3. **Dye-Sensitized Solar Cells** *61*
   Anna Nikolskaia and Oleg Shevaleevskiy

4. **Electronics and Bioelectronic Interfaces** *75*
   Seong-Min Kim, Sungjun Park, Won-June Lee, and Myung-Han Yoon

5. **Designing Thermoelectric Materials Using 2D Layers** *93*
   Sage R. Bauers and David C. Johnson

6. **Magnetically Responsive Photonic Nanostructures for Display Applications** *123*
   Mingsheng Wang and Yadong Yin

7. **Functional Materials: For Sensing/Diagnostics** *151*
   Rujuta D. Munje, Shalini Prasad, and Edward Graef

8. **Superhard Materials** *175*
   Ralf Riedel, Leonore Wiehl, Andreas Zerr, Pavel Zinin, and Peter Kroll

9. **Self-healing Materials** *201*
   Martin D. Hager

10. **Functional Surfaces for Biomaterials** *227*
    Akiko Nagai, Naohiro Horiuchi, Miho Nakamura, Norio Wada, and Kimihiro Yamashita

11. **Functional Materials for Gas Storage. Part I: Carbon Dioxide and Toxic Compounds** *249*
    L. Reguera and E. Reguera

12. **Functional Materials for Gas Storage. Part II: Hydrogen and Methane** *281*
    L. Reguera and E. Reguera

13. **Supported Catalysts** *313*
    Isao Ogino, Pedro Serna, and Bruce C. Gates

14	**Hydrogenation by Metals** *339*	
	Xin Jin and Raghunath V. Chaudhari	
15	**Catalysis/Selective Oxidation by Metal Oxides** *393*	
	Wataru Ueda	
16	**Activity of Zeolitic Catalysts** *417*	
	Xiangju Meng, Liang Wang, and Feng-Shou Xiao	
17	**Nanocatalysis: Catalysis with Nanoscale Materials** *443*	
	Tewodros Asefa and Xiaoxi Huang	
18	**Heterogeneous Asymmetric Catalysis** *479*	
	Ágnes Mastalir and Mihály Bartók	
19	**Catalysis by Metal Carbides and Nitrides** *511*	
	Connor Nash, Matt Yung, Yuan Chen, Sarah Carl, Levi Thompson, and Josh Schaidle	
20	**Combinatorial Approaches for Bulk Solid-State Synthesis of Oxides** *553*	
	Paul J. McGinn	

Index *573*

1
Intermetallic Compounds and Alloy Bonding Theory Derived from Quantum Mechanical One-Electron Models

Stephen Lee[1] and Daniel C. Fredrickson[2]

[1]Department of Chemistry and Chemical Biology, Cornell University, Ithaca, NY, 14853, USA
[2]Department of Chemistry, University of Wisconsin-Madison, Madison, WI, 53706, USA

1.1 Introduction

Quantum theory serves dual roles in intermetallic chemistry. On the one hand, there is the ever-advancing frontier of more sophisticated and accurate quantum calculations, each carried out for their own specific systems. But on the other hand, quantum theory lies, in a general way, at the very core of our overall understanding of intermetallic bonding as a whole.

An analogy can be made to organic chemistry. While each successive generation of quantum chemists has more and more powerful tools to calculate the electronic structure of organic molecules, it is the Lewis structure, the two-electron two-center bond, electron arrow pushing, electrophilicity, nucleophilicity, frontier orbitals, and the Woodward–Hoffmann rules that remain virtually unchanged as the core bonding concepts of the practicing organic chemist.

The question then is what kind of quantum models, by analogy, lie at the heart of intermetallic chemistry? Even if this question requires a subjective response, its answer is important. Historically, our understanding of the intermetallic bond comes from two very different theories: the tight-binding and the nearly free electron models [1]. They in turn have led to a variety of key concepts: the projected density of states (projected DOS), Mulliken populations, crystal orbital overlap populations (COOP), Wannier functions, crystal orbital Hamilton populations (COHP), pseudopotentials, the structural energy difference theorem, and the Brillouin and Jones zones among them [2–15].

The goal of this chapter is to set forth these two very different theories side-by-side and to show how these two models, sometimes individually and sometimes in concert, continue to develop today, illuminating such diverse and current areas as structural diversity, the Hume-Rothery electron phases,

Handbook of Solid State Chemistry, First Edition. Edited by Richard Dronskowski, Shinichi Kikkawa, and Andreas Stein.
© 2017 Wiley-VCH Verlag GmbH & Co. KGaA. Published 2017 by Wiley-VCH Verlag GmbH & Co. KGaA.

coherent magnetic states, Zintl phases, a qualitative understanding of phase diagrams, Nowotny chimney ladder phases, half-Heusler compounds, transition metal – main group electron counting rules, Laves phases, and quasi-crystals [14,16–27].

1.1.1
The H_2 Molecule

At the core of any discussion of intermetallic bonding is the chemical bond itself. While our main focus in this chapter will be on metallic (and even magnetic) systems, much can be gained from looking again at the traditional covalent bond first [28,29]. Along these lines, it is nothing short of remarkable how much can be learned from the mere H_2 molecule. Figure 1.1 shows the electron density of two hydrogen atoms, first as isolated atoms and finally in the H_2 molecule itself. The expected concentration of electrons in the space between the H nuclei is apparent in this picture. So is another striking feature: a pronounced contraction of the atomic orbitals (AOs) upon forming the molecule. The original hydrogen atom AOs are much more diffuse, while in H_2 both the electrons adhere more closely to the individual hydrogen nuclei themselves, even while spending more time between the nuclei.

We rationalize these results in the following manner. First, we recognize that solutions to the Schrödinger equation are always compromises between minimization of the kinetic and the potential energies. Potential energy, taken by itself,

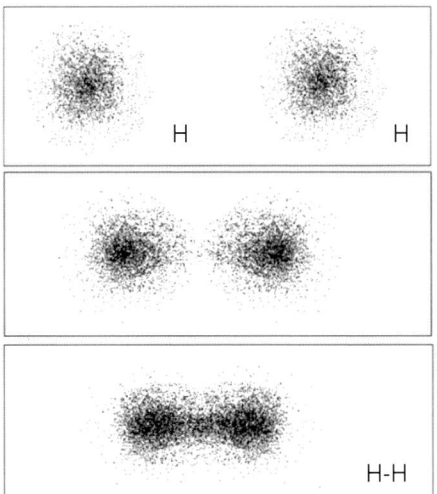

Figure 1.1 Change in the hydrogen AO electron clouds in the formation of the H_2 molecule. While most elementary texts emphasize the increased electron density in the shared space between the two H nuclei, equally significant is the orbital contraction around each nucleus. (Reproduced with permission from Ref. [28]. Copyright 1962, American Physical Society.)

would require the ground-state electron to lie entirely at the nucleus, while the kinetic energy, as the sole factor, would lead to a ground state with zero curvature, an entirely flat wave function. The actual ground state, based on exponential decay functions, is a compromise between these two extrema. The balance in this compromise in the final wave function is always in keeping with the virial theorem $E_{potential} = -2E_{kinetic}$.

Great weight is placed on showing our introductory chemistry students that the two hydrogen atomic orbitals meld together in the H_2 MO diagram (Figure 1.2). The key point of the classic MO diagram is the in-phase character of the bonding MO. This in-phase character lessens orbital curvature and loosens the role that kinetic energy plays in the balance of the total energy. With the kinetic energy constraints weakened, AO contraction occurs, lowering the potential energy and concomitantly raising the kinetic energy; the requirements of the virial theorem are met.

The lesson here is that quite generally, the covalent bond requires not just orbital overlap but AO contraction as well. And in turn, the propensity for atomic orbital contraction is a dominant feature in the covalent bond. Compare the bonding enthalpies of the Group 1 diatomics: H_2: 432 kJ/mol; Li_2: 97.7 kJ/mol; Na_2: 89.2 kJ/mol; K_2: 70.5 kJ/mol; Rb_2: 67.0 kJ/mol; and Cs_2: 60.0 kJ/mol [30]. In hydrogen, the nonmetal, the principal AO is the 1s orbital. The 1s orbital, alone among all s AOs, is the only orbital with no core orbitals to prevent its further contraction. Its diatomic covalent bond is then stronger than any of the other first column elements.

The other members of the periodic table's first column, the alkali metals, have core s orbitals that restrain any contraction that covalent-bonding might induce in a valence orbital (due to the orthogonality requirements). The gradual reduction in the propensity for AO contraction mirrors the weakening of the covalent bonds on going from Li_2 to Cs_2.

The above analysis suggests that fundamental differences should arise elsewhere in the periodic table. Wherever, in the periodic table, a new type of AO enters the valence set, for example, the first time a p, d, or f orbital is seen, there

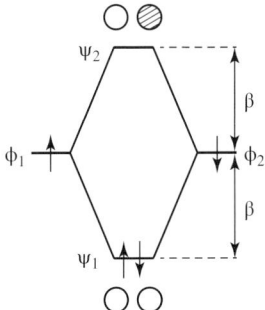

Figure 1.2 Schematic of the H_2 molecular orbital diagram.

will be no core orbitals of like angular momentum to prevent covalent bond valence–orbital contraction (orbitals with different angular momentum quantum numbers require no readjustments to be orthogonal with one another).

The unusual bonding characteristics of the 2p elements are well known in organic chemistry (we cite in particular the ability of these elements to form multiple and hydrogen bonds). The corresponding uniqueness of the 3d elements, as we will discuss later in this chapter, has another manifestation: magnetic order.

1.2 Tight-Binding Theories

1.2.1 Simple and Extended Hückel Theory

We begin our analysis of intermetallic bonding itself with a brief discussion of key tight-binding concepts. We use a simple semiempirical derivative of Density Functional Theory (DFT). In DFT, the total energy of a system is written as a functional of the electron density, $E = E[\rho(\mathbf{r})]$, in which the $\rho(\mathbf{r})$ is built (via the Kohn–Sham Ansatz) from independent-electron wave functions, $\psi_j(\mathbf{r})$ [31]. The ground state of the system is obtained by minimizing the energy with respect to each of the occupied one-electron functions (subject to the constraint of normalization), which corresponds to solving the Kohn–Sham equation:

$$\left(-\frac{\hbar^2}{2m}\nabla^2 + V_{\text{local}}(\mathbf{r}) + V_{\text{Hartree}}(\mathbf{r}) + V_{\text{XC}}(\mathbf{r})\right)\psi_j(\mathbf{r}) = E_j\psi_j(\mathbf{r}), \quad (1.1)$$

where V_{local} gives the potential energy the electron experiences due to the nuclei, while V_{Hartree} and V_{XC} are the Hartree and exchange–correlation potentials, respectively. As V_{Hartree} and V_{XC} are both functionals of the ground state $\rho(\mathbf{r})$, the Kohn–Sham equation must be solved through a series of iterations to give self-consistency between the set of ψ_j's and the $\rho(\mathbf{r})$ derived from them.

The Kohn–Sham equation can be used as a basis (in retrospect) for the formulation of the semiempirical extended and simple Hückel methods. We first approximate the potential experienced by the electrons as being set by semiempirical parameters, rather than determined self-consistently, such that the Kohn–Sham equation is reduced to a one-electron Schrödinger equation:

$$\left(-\frac{\hbar^2}{2m}\nabla^2 + V_{\text{semiempirical}}(\mathbf{r})\right)\psi_j(\mathbf{r}) = E_j\psi_j(\mathbf{r}). \quad (1.2)$$

The next assumption we make regards the form of the wave function: We consider each ψ_j to be a linear combination of atomic orbitals (LCAO), ϕ_i's, where the set is limited to valence orbitals deemed essential for understanding the bonding of the system. By writing the wave function as $\psi_j = \sum_i c_i^{(j)} \phi_i$, the energy of ψ_j becomes

$$\begin{aligned}\langle E_j\rangle &= \iiint \psi_j(\mathbf{r})^*\left(-\frac{\hbar^2}{2m}\nabla^2 + V_{\text{semiempirical}}(\mathbf{r})\right)\psi_j(\mathbf{r})d\mathbf{r} \\ &= \iiint \left(\sum_i c_i^{(j)*}\phi_i^*\right)\left(-\frac{\hbar^2}{2m}\nabla^2 + V_{\text{semiempirical}}(\mathbf{r})\right)\left(\sum_{i'} c_{i'}^{(j)}\phi_{i'}\right)d\mathbf{r} \\ &= \sum_i \sum_{i'} c_i^{(j)*}c_{i'}^{(j)}\iiint \phi_i^*\left(-\frac{\hbar^2}{2m}\nabla^2 + V_{\text{semiempirical}}(\mathbf{r})\right)\phi_{i'}d\mathbf{r} \\ &= \sum_i \sum_{i'} c_i^{(j)*}c_{i'}^{(j)}H_{ii'} \\ &= \mathbf{c}^{(j)*\dagger}\mathbf{H}\mathbf{c}^{(j)},\end{aligned}\qquad(1.3)$$

where $H_{ii'}$ is the Hamiltonian matrix element representing the interaction between atomic orbitals ϕ_i and $\phi_{i'}$, $\mathbf{H}$ is the Hamiltonian matrix expressed in the atomic orbital representation, and $\mathbf{c}^{(j)}$ is the column vector of atomic orbital coefficients corresponding to ψ_j. The solutions to the one-electron Schrödinger equation are then found by solving the eigenvalue problem:

$$\mathbf{H}\mathbf{c}^{(j)} = E_j\mathbf{S}\mathbf{c}^{(j)},\qquad(1.4)$$

where the inclusion of the overlap matrix $\mathbf{S}$ (with $S_{ii'} = \iiint \phi_i^*\phi_{i'}d\mathbf{r}$) recognizes that the atomic orbitals do not form an orthogonal basis set. This formulation of the tight-binding approach is known as the extended Hückel (eH) method [32–34]. Replacing the $\mathbf{S}$ matrix in the above equation by the identity matrix ($\mathbf{I}$) can open new paths to analyzing the electronic structure of compounds. This approximation can be seen as an orthogonalization of the basis set that is incorporated implicitly through changes to the parameters, and leads to the *simple* Hückel (sH) approach.

The key semiempirical parameters in the above equations are in the forms of the atomic orbitals and the Hamiltonian matrix elements used. In the eH and sH methods, the basis functions for the valence s and p orbitals are usually chosen to be Slater-type orbitals (STOs):

$$\phi_i(r,\theta,\varphi) = Nr^{n-1}e^{-\zeta r}Y_{l,m}(\theta,\varphi)\qquad(1.5)$$

with N being a normalization factor, n being the principal quantum number for the atomic orbital, and $Y_{l,m}$ being the proper spherical harmonic to represent the angular character of the orbital. These orbitals have a single parameter ζ, which tunes how quickly the orbital decays as a function of distance. Due to their relatively complex radial character, the d orbitals are generally built from a combination of two STOs with differing zeta values (double-ζ functions):

$$\phi_i(r,\theta,\varphi) = c_1\left(N_1r^{n-1}e^{-\zeta_1 r}Y_{l,m}(\theta,\varphi)\right) + c_2\left(N_2r^{n-1}e^{-\zeta_2 r}Y_{l,m}(\theta,\varphi)\right)\qquad(1.6)$$

for which three parameters are necessary (ζ_1, ζ_2, and the c_1/c_2 ratio).

The remaining parameters in the method are the form the $H_{ii'}$ matrix elements take, which in turn implicitly define the semiempirical potential $V_{\text{semiempirical}}(\mathbf{r})$ experienced by the electrons. The diagonal elements H_{ii} can be simply interpreted as the on-site energies of the atomic orbitals. The off-diagonal elements $H_{ii'}$ with $i \neq i'$ are treated as being proportional to the overlap between the two orbitals using the Wolfsberg–Helmholz approximation [35]:

$$H_{ii'} = K \left(\frac{H_{ii} + H_{i'i'}}{2} \right) S_{ii'}, \qquad (1.7)$$

where K is a constant traditionally set to 1.75 (and sometimes modified with a weight to counteract an effect in eH calculations known as counterintuitive orbital mixing) [36].

The extended and simple Hückel (eH and sH) methods then rely on only a handful of parameters for each element: the on-site energy for each atomic orbital (H_{ii}), and the ζ parameters for each STO orbital used to represent the atomic orbitals (and c_1/c_2 ratios for any double-ζ functions).

This tight-binding formulation is convenient not only in the small number of parameters used but also in the chemical nature of these parameters. The H_{ii} values can be interpreted as capturing the electronegativity of the element: The Mulliken definition of electronegativity is simply the average of the ionization energy and electron affinity of an element. These are represented, very approximately, by the $|H_{ii}|$ parameters, with a particular focus on the ionization energies. The ζ values, on the other hand, dictate the radial character of the atomic orbitals, and thus provide a measure of the sizes of the atoms.

Originally, the parameters of this method were chosen to best-reproduce the atomic properties given by higher level calculations on free atoms, creating a standard set that allowed calculations to be carried out on molecules for which *ab initio* calculations would be prohibitive. The increases in computational power and wide-spread availability of DFT, however, have diminished the advantages of eH in terms of speed relative to the disadvantages it introduces in its approximations.

The use of these approaches has thus evolved to harness their conceptual simplicity and transparency, rather than their speed. The flexibility of the Hückel approaches (and other tight-binding methods) and their deep relationship to the Kohn–Sham equation means that with the proper parameterization they can *quantitatively* reproduce the band energies of DFT calculations for a variety of solid-state structures [37–43]. As is illustrated in Figure 1.3 for the $ScNi_2$ ($MgCu_2$-type) structure, fitting programs such as *eHtuner* can be used to refine the parameters of an eH or sH calculation against DFT results to obtain root-mean-squared deviations in the energies of the occupied bands of less than 0.2 eV [41]. These can then be seen as a means of interpreting DFT results in terms of an effective Hückel model, in a manner analogous to the way a Curie–Weiss law is used to interpret magnetization data.

This ability to reproduce DFT results for intermetallic phases with models derived from orbital interactions is deeply reassuring as it places these compounds firmly in the realm of chemistry, where molecular orbital theory has proven to be a powerful aid for relating bonding, structure, and reactivity. As we will see in the remainder of this chapter, the Hückel approach can bring parallel insights to intermetallic systems, both by providing a simpler substitute for the

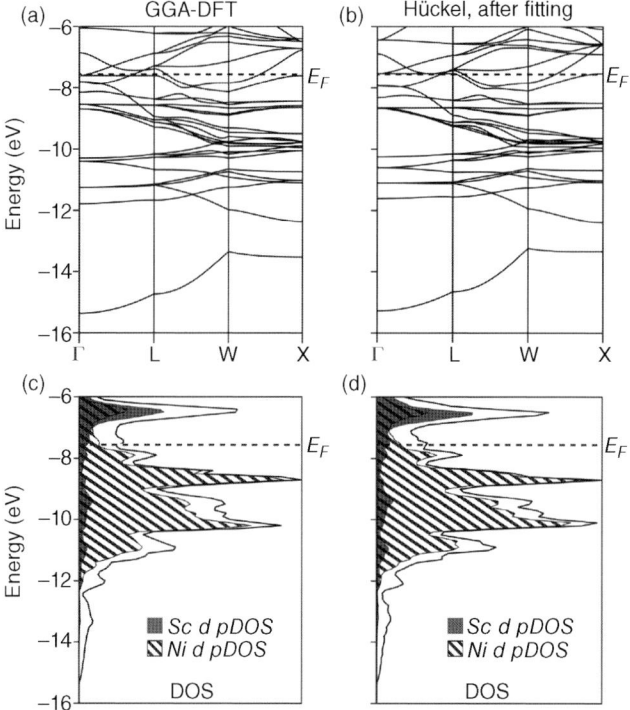

Figure 1.3 Comparison of the GGA-DFT electronic structure of ScNi$_2$ with that calculated using a best-fit Hückel model. (a and b) Band structures calculated with the two methods. (c and d) Density of state (DOS) distributions with contributions from the Sc and Ni d orbitals shaded. (Reproduced with permission from Ref. [41]. Copyright 2012, RCS.)

complex output of a DFT calculation, and by serving as a platform for the development of prototypes for new approaches to bonding analysis. These prototypes can later be extended to first-principles calculations.

1.2.1.1 DOS, Projected DOS, COOP, and COHP

Even after the DFT results for an intermetallic structure are translated into a Hückel-based model, developing a bonding picture is still a challenging process. A general principle in the construction of MO diagrams is that the number of final MOs should equal the number of atomic orbitals in the system. For an intermetallic phase, with its periodic repetition of unit cells, this leads to a limitless number of crystal orbitals to analyze, such that the discrete MO energy levels of a molecule broaden into continuous energy bands. One approach to characterizing this complex situation is to focus on the distribution of states over the energy scale, rather than discussing any state in particular. Tools for viewing the electronic structure in this way are given by the density of states (DOS) [6], projected DOS [6], the crystal orbital overlap population (COOP) [44,45], and crystal orbital Hamilton population (COHP) distributions [46].

The electronic DOS curve forms the core of all of these approaches. It can be thought of simply as a histogram of the number of crystal orbitals as a function of energy. Sharp peaks in the DOS distribution typically arise from relatively localized components of the electronic structure that exhibit little interaction between neighboring unit cells, such as lone pairs, transition metal d orbitals, or (as a particularly extreme case) lanthanide 4f orbitals. Broader DOS features generally correspond to functions with more overlap between neighboring unit cells. In particular, intermetallic compounds with strongly delocalized s–p electrons exhibit a DOS tail at low energies with a nearly parabolic shape [10]. These states approximate the lowest states of a free electron gas.

On its own, however, the DOS distribution for a phase is usually not very informative, as illustrated in the corresponding curve for the NaTl-type Zintl Phase LiGa (Figure 1.4). Moving from the lowest energy upward, the DOS distribution appears like a hilly landscape, with some hills defined by gentle slopes, while others are sharper. One feature to note is the Fermi energy (E_F), separating filled and empty states, which lies in a minimum between these mounds. Such minima are often interpreted as a partial bandgap, referred to as a pseudogap, which coincides with a favorable electron count. In this case, the presence of a pseudogap at the E_F can be easily rationalized in terms of LiGa adhering to the Zintl concept: Each Li atom transfers its single electron to the Ga sublattice, while the newly formed Ga^{1-} anions link together into a diamond network through covalent Ga—Ga bonds. The result is an electron-precise compound whose Ga sublattice is isoelectronic to elemental Si, a semiconductor [14,47–49].

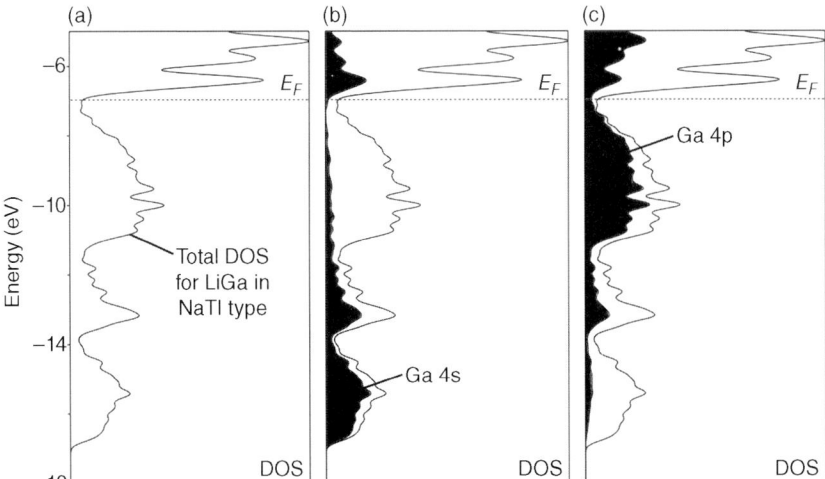

Figure 1.4 The DFT-calibrated simple Hückel density of states (DOS) distribution of LiGa in the NaTl type. (a) The total DOS. (b) The total DOS overlaid with the projected DOS for the Ga 4s orbitals. (c) The corresponding projected DOS plot for the Ga 4p. For each plot, a dotted line gives the Fermi energy (E_F), the energy separating filled from empty orbitals.

Deeper insights into the electronic structure can be obtained by examining how much each of the features can be attributed to valence orbitals of the Li and Ga atoms. This decomposition is accomplished with the *projected DOS* distribution. Whereas in the standard DOS, each state is included with an equal weight, the projected DOS weights each state in proportion to the probability of finding an electron in that state on a given atom or group of orbitals. For example, in Figure 1.4b and c we present the projected DOS curves for the Ga 4s and 4p orbitals of LiGa. The DOS distribution below the E_F can be divided roughly into three main peaks, centered at about −15, −13, and −9 eV. From the projected DOS curves, the lowest peak is predominantly composed of Ga 4s character, as can be understood from this orbital being the lowest energy valence orbital in the system. The next peak up represents a transition from Ga 4s-based states to Ga 4p-based ones, with the projected DOS for the two orbitals being about equal to each other. Finally, the highest occupied peak of states is dominated by the Ga 4p orbitals.

The sum of these two contributions accounts for most of the total DOS curve; in other words, there is little character from the Li atoms present in the occupied levels. This is consistent with the Zintl formalism's assignment of the phase as Li^+Ga^-, where all of the valence electrons are associated with the Ga atoms. Above the E_F, however, the situation is different: the sum of the Ga 4s and 4p contributions for these unoccupied levels is now much smaller than the total DOS. The remainder arises from the Li 2s and 2p levels.

From the projected DOS curve, we can quickly tell how the character for a given atomic orbital is distributed in the band structure of a material. However, we do not learn very much about the bonding basis of this character. Two tools for investigating chemical bonding are the *crystal orbital overlap population (COOP)* and *crystal orbital Hamilton population (COHP)* curves.

In Figure 1.5, we illustrate the first of these tools, again using LiGa as an example. We begin by showing again the DOS curve for this phase (Figure 1.5a) for reference, and then present alongside it (Figure 1.5b and c) the COOP curves for the Li–Ga and Ga–Ga interactions. Like the projected DOS curves shown earlier, these COOP functions are constructed from a weighted DOS curve, but with the weights in this case based on the overlap, the bonding, between the adjacent AOs. States are weighted by the fraction of their components associated with specific bonding interactions between atoms or atomic orbitals.

Positive values of the COOP indicate that the corresponding states, on the whole, contribute in a favorable way to the bonding interactions. Such is the case for most states in the energy range shown for the Li—Ga interactions, and for all states below the E_F for the Ga—Ga interactions. Above the E_F, the COOP for the Ga—Ga interaction switches and is strongly negative. These negative values suggest that the occupation of these states would reduce the net electron population of the Ga—Ga bonding interactions; these high-energy states are Ga—Ga antibonding.

Like the pseudogap we pointed out in the DOS curve earlier, the placement of the E_F at the transition between Ga—Ga bonding and Ga—Ga antibonding states

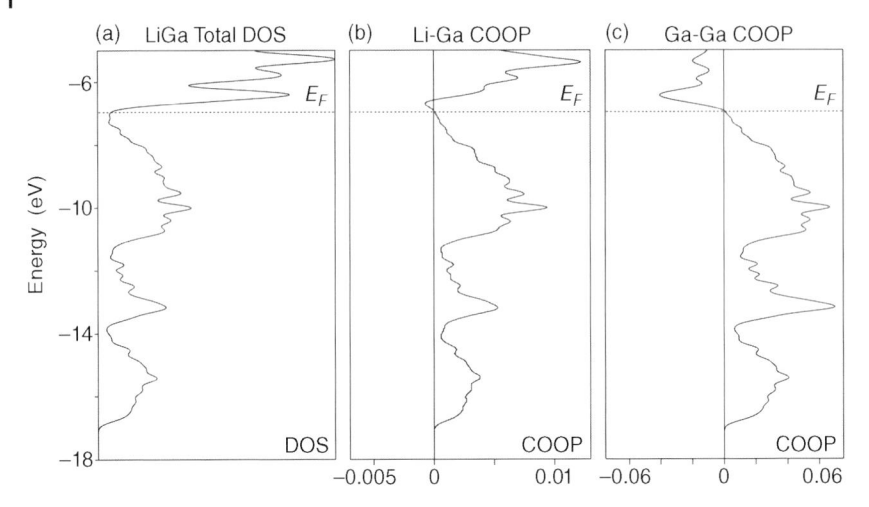

Figure 1.5 The crystal orbital overlap population (COOP) analysis applied to LiGa. (a) The DOS for LiGa shown for reference. (b) The COOP curve for the Li–Ga interactions. (c) The Ga–Ga COOP curve.

nicely affirms the notion that LiGa is a Zintl phase: The separate levels based on covalent bonds between Ga sp^3-hybrid orbitals should be filled, while their antibonding counterparts should be empty. It is common to refer to interactions whose bonding is maximized at the E_F in this way as *optimized*. The Li–Ga interactions, which still show a number of bonding states above the E_F, are then considered to be unoptimized.

So far, the COOP function coupled with the Zintl concept has yielded a clear albeit qualitative picture of the bonding in LiGa. Using the COOP alone, however, it is difficult to make more quantitative conclusions about the interaction strengths. Consider the relative scales in the COOP values for the Li–Ga and Ga–Ga interactions. The maximum COOP for the Ga–Ga curve is more than six times greater than that for the Li–Ga curve, but it would be premature to conclude from this that the strongest Ga–Ga interaction is simply six times that for the Li–Ga interaction: We first need to know how much a Ga–Ga overlap counts energetically compared to the same amount of Li–Ga overlap.

The COHP analysis is designed to allow a more direct comparison between different interaction types. Rather than using the number of electrons in an interaction as the weighting factor for each state in the function, it uses the portion of the state's energy due to that interaction as the weight. In this way, a COHP curve reflects the quantitative energetic contribution of an interaction. As is evident from the COHP curves for Li–Ga in Figure 1.6, the results of a COHP analysis are qualitatively very similar to those from a COOP one. The main difference is that the horizontal axis has become inverted as positive orbital overlap (seen in the COOP) gives rise to a negative change in the energy (the effect shown in the COHP). To make the connection between COOP and

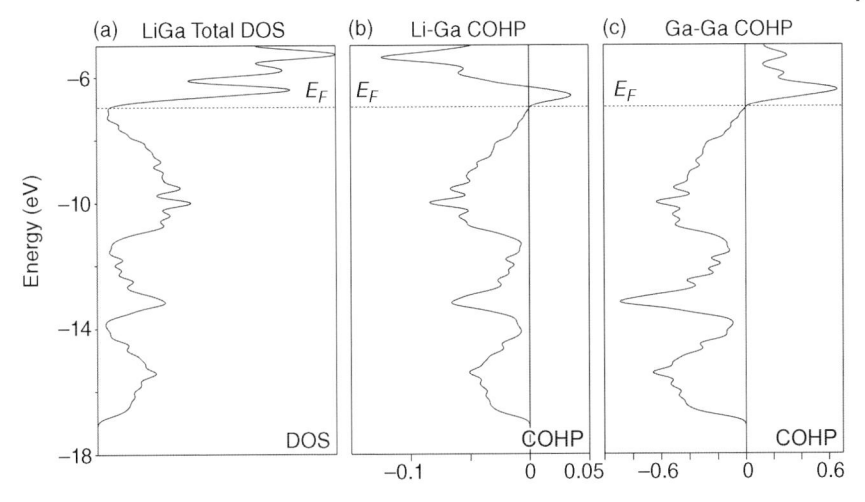

Figure 1.6 Crystal orbital Hamilton population (COHP) analysis of LiGa. (a) The total DOS for LiGa. (b) The Li–Ga COHP curve. (c) The Ga–Ga COHP curve.

COHP plots more direct, COHP data are most often plotted as −COHP versus E.

A more profound difference between the COOP and COHP results is in the scales along the x-axis. Between different COHP curves, these values can now be directly compared in terms of the amounts each interaction contributes to the total energy. In this case, the results largely confirm that the Ga—Ga interactions are stronger than the Li—Ga ones. However, the magnitude of the difference is now larger, indicating that energetically Ga—Ga overlap counts more than Li—Ga overlap.

1.2.1.2 Applications of COOP and COHP Analysis

COOP analysis was introduced by Roald Hoffmann and his students and spread quickly among the intermetallic community [6,45,50–59]. In an early instance, Hoffmann and Zheng examined the $ThCr_2Si_2$ structure type, which (while largely unknown among chemists in general) is perhaps the most common of all structure types [60].

Hoffmann and Zheng, in examining this structure type, noted a striking periodic trend: when late transition metal elements occupy the Wyckoff position 4d Cr site, short bonds between the 4e Si sites are found. However, if an early transition element occupies this same 4d site, longer 4e—4e bonds are observed (in the following discussion, 4d will refer to the Wyckoff position and not an AO).

Their analysis, resolving the bonding issues underlying this structural trend, is in hindsight simple and clear. Hoffmann and Zheng first cleaved the 4e–4e contacts in question to create an isolated layer of the structure based on the 4d and 4e sites. They then derived the DOS and COOP curves of Figure 1.7, with Mn and P, respectively, at the 4d and 4e sites. The two lowest energy bands in this

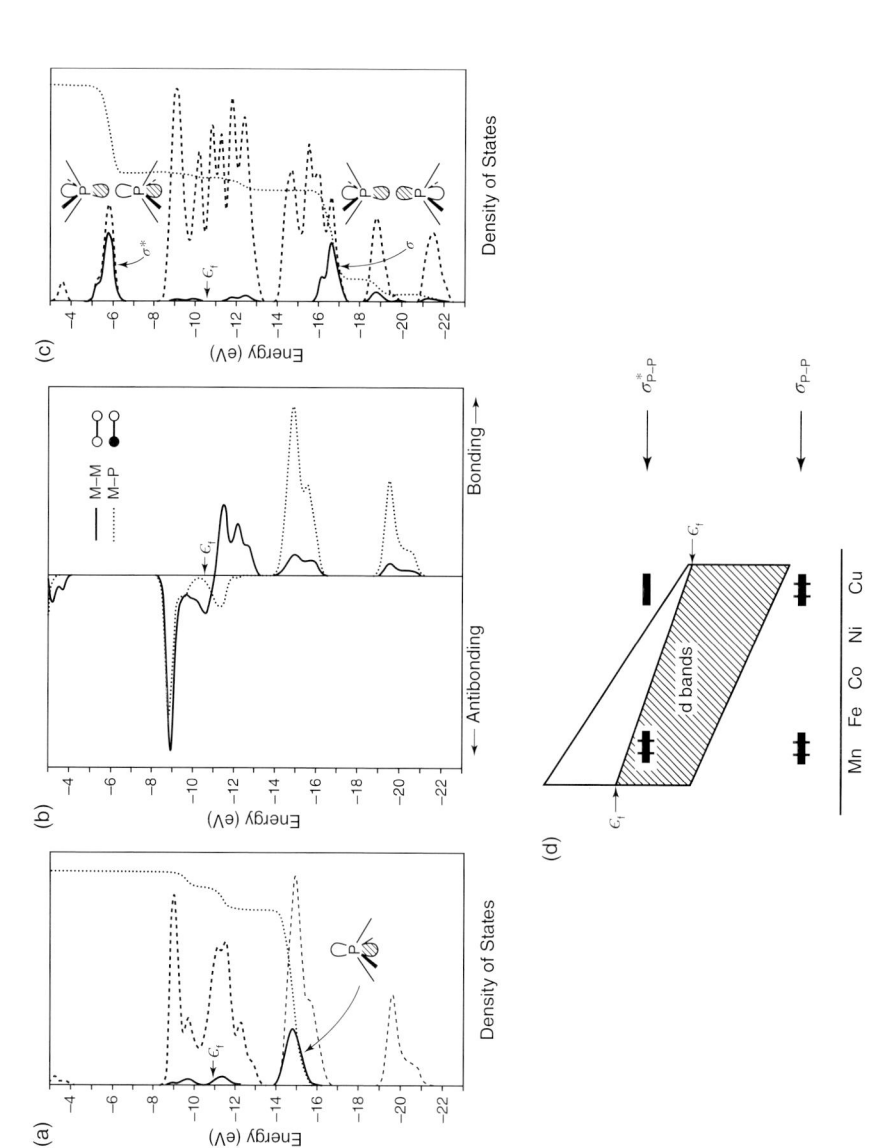

Figure 1.7 Hoffmann and Zheng's application of COOP to explaining structural trends in the ThCr$_2$Si$_2$ type. (a and b) DOS (with projected fragment MO DOS) and COOP curves for an isolated sheet (composed of Wyckoff 4d and 4e sites) of the ThCr$_2$Si$_2$ structure type populated respectively by Mn and P atoms. (c) DOS showing interlayer P–P interactions in the 3D crystal structure. (d) Schematic showing relative positions of interlayer P–P bonding and antibonding bands with respect to transition metal d-band. For early transition metals, the P–P antibonding states lie below the E_F, leading to long P–P distances. For late transition metals, these antibonding states are depopulated, resulting in shorter P–P contacts. (Reproduced with permission from Ref. [52]. Copyright 1985, American Chemical Society.)

figure are primarily 4e-based followed by, at higher energy, a transition metal d orbital band between −14 and −8 eV.

Hoffmann and Zheng subsequently joined these 4d–4e layers together into the full 3D $ThCr_2Si_2$ structure type. As Figure 1.7 shows, if strong interlayer 4e–4e contacts are adopted, the initially low-energy 4e bands split into very low-energy bonding and higher energy antibonding bands. These authors deduced that it is the relative energetic placement of the antibonding 4e–4e band with respect to the 4d-site-based valence d orbital band that lies at the heart of the observed periodic trend.

As Figure 1.7 shows, while for early transition metal elements such as Mn, the antibonding 4e–4e band is occupied, for later transition metal elements, such as Cu, it remains vacant. If the antibonding 4e–4e orbitals are occupied, then intersheet 4e—4e bonds are weak and 4e—4e bonds are long. It is for this reason that for those $ThCr_2Si_2$-type structures made up of early transition metal elements (such as Mn), the 4e–4e contacts are long, but for compounds containing late transition metal elements they are short.

A systematic and comprehensive review of the multiple ways COOP and COHP have been used by the intermetallic community could fill many pages. A few varied examples, however, may give a hint as to its broad applicability.

Hughbanks and coworkers used eH calculations to elucidate the structural diversity of early transition metal chalcogenides [61–64]. For instance, Kim *et al.* examined Ta-S phases in which columns of face-sharing Ta@Ta$_{10}$ pentagonal antiprisms are condensed with varying degrees of ligation by sulfur atoms, which cap the triangular faces of the column [65]. Phases with the stoichiometries Ta_2S and $Ta_3S_{\sim 2}$ were known, and the authors were able to extrapolate from these phases the plausible stability of a Ta_6S_5 compound.

Through a comparison of the DOS curves calculated for these phases to that of elemental Ta, Kim *et al.* saw a clear driving force for the incorporation of sulfur (Figure 1.8a). As sulfur content is gradually added to the Ta–S compound, a pseudogap at the E_F becomes increasingly apparent, culminating in a pronounced bandgap for Ta_6S_5. The addition of sulfur thus appears to confer a special electronic stability to the compounds.

Using COOP analysis, the authors then traced the formation of this gap to Ta—Ta interactions (Figure 1.8b). In elemental Ta, the Ta—Ta bonding is suboptimal, as the placement of the E_F leaves some states with Ta—Ta bonding character unfilled. With the introduction of S, a near perfect match is obtained between the E_F and the onset of Ta—Ta antibonding states. The gaps/pseudogaps at the E_F energies for the Ta–S phases coincide with optimization of the Ta–Ta interactions.

A curious feature noted by Kim *et al.* in these COOP curves is that the Ta—Ta bonding remains optimized for each of the Ta–S phases, despite the expectation that each S atom added to the structure should oxidize the Ta sublattice by two electrons. This was interpreted in terms of the cleavage of Ta–Ta interactions between neighboring $(Ta@Ta_{10/2})_\infty$ columns as S atoms are introduced, which perfectly accommodates the loss of these electrons.

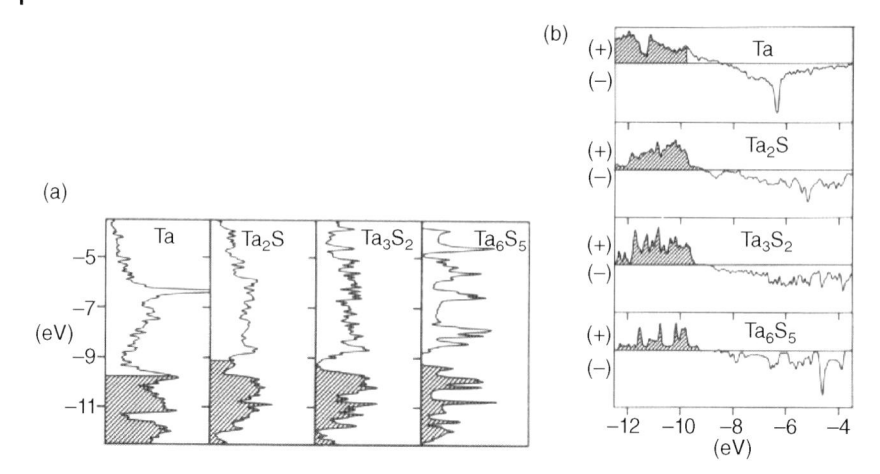

Figure 1.8 Comparison of the electronic structures of elemental Ta, Ta$_2$S, Ta$_3$S$_2$, and a proposed Ta$_6$S$_5$ phase. (a) DOS distributions. (b) Ta–Ta COOP curves. The regions of the curves corresponding to occupied states are shaded. (Reproduced with permission from Ref. [65]. Copyright 1991, American Chemical Society.)

COOP and COHP have been used to good effect with electropositive metal and main group metalloid containing compounds, within the general umbrella of Zintl phases [14,59,66–73]. Fässler *et al.*, for example, investigated the bonding in K$_5$Pb$_{24}$. K$_5$Pb$_{24}$ crystallizes in an ordered variant of the α-Mn structure [74]. The K atoms occupy a selection of sites that are best optimized for relatively electropositive atoms [75], with the more electronegative Pb atoms taking up the remainder of the sites to create a dense network of Pb–Pb contacts.

The DOS distribution (Figure 1.9) is dominated by Pb character, consistent with the Zintl-like charge assignment of $(K^+)_5(Pb_{24})^{-5}$. As was found for the Ga in LiGa, the s character of the Pb is concentrated at the lower end of the energy range. This time the s character is concentrated in a separate block of states, reflecting the core-like nature of the s electrons of heavy p block elements. The p block states occur higher in energy, with the E_F passing through them in a narrow pseudogap. The filling of the band structure up to a DOS minimum again suggests a system that has reached a favorable electron count.

A COOP analysis of the Pb–Pb interactions indicates that the origin of this pseudogap is less straightforward than in the previous cases we have discussed. The COOP curve calculated over all Pb–Pb contacts shows a small amount of antibonding at the E_F, hinting that the electron count is a little high relative to the optimal count for the phase.

Moving to the COOPs for individual Pb–Pb contacts within the structure provides some explanation. The crossing between bonding and antibonding for the different symmetry-distinct Pb–Pb interactions occurs at different electron counts. While the crossing for the crystal structure's contact "b" aligns well with the E_F, that for contact "a" appears almost 2 eV lower. This mismatch in the ideal E_F's for these interactions means that no electron count will be able to simultaneously optimize all interactions. The actual electron count has the advantage

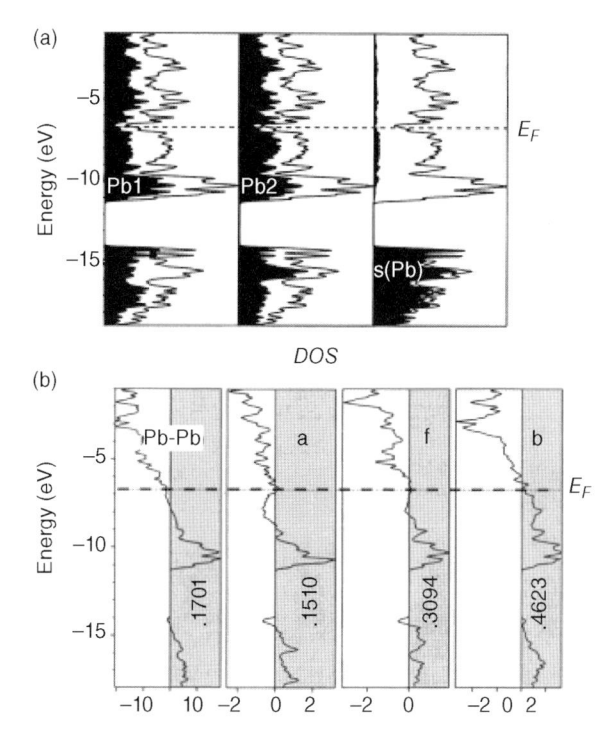

Figure 1.9 eH bonding analysis of K_5Pb_{24}. (a) DOS distributions with contributions from the individual Pb sites and Pb 6s orbitals shaded in black. (b) COOP curves for all Pb–Pb contacts, followed by the corresponding curves for individual contacts. Numbers alongside the COOP curves give the integrated values of the COOP up to the E_F (Mulliken overlap populations). (Reproduced with permission from Ref. [74]. Copyright 1999, WILEY-VCH Verlag GmbH, Weinheim.)

that it optimizes the "b" contacts, while only slightly populating the antibonding realm for the average of the Pb–Pb interactions.

Unlike the COOP analysis, the major applications of the COHP analysis have been through its implementation for DFT calculations, particularly those performed with the LMTO-ASA-TB (Linear Muffin Tin Orbital-Atomic Sphere Approximation-Tight-Binding) program [76–81]. The interpretation of the results from this approach is very similar to those obtained via Hückel-based approaches. While COOP and COHP should, in principle, arrive at similar findings, an actual comparison of DFT-derived eH COOP curves with LMTO COHP has never been systematically carried out in practice.

An example of the use of COHP is provided by the Corbett and Miller groups' analysis of the compound $K_{23}Au_{12}Sn_9$. In this compound $(Au_{12}Sn_9)^{23-}$ clusters (resembling fragments of a $(AuSn_{4/2})_\infty$ layer of vertex-sharing square pyramidal units) are immersed in a matrix of K cations. The LMTO band structure and DOS distributions (Figure 1.10) indicate that the phase is one electron/cluster

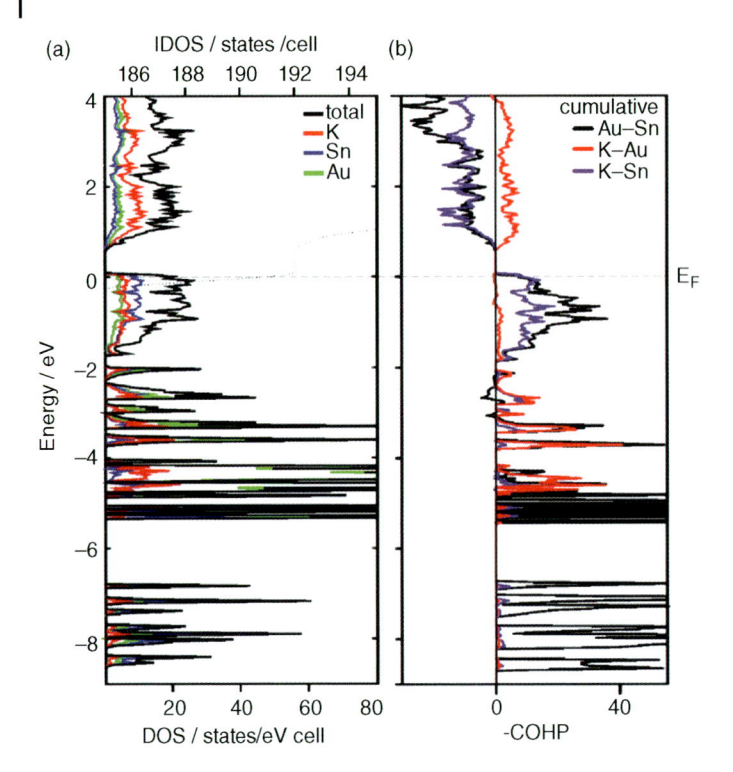

Figure 1.10 Electronic structure analysis of $K_{23}Au_{12}Sn_9$ by the groups of Miller and Corbett. (a) DOS and projected DOS distributions calculated using the ASA-LMTO-TB method, an implementation of DFT. (b) The COHP curves obtained in this analysis. (Reproduced with permission from Ref. [82]. Copyright 2010, American Chemical Society.)

short of a ca. 0.5 eV bandgap, suggesting that the compound is slightly electron deficient.

COHP data (shown as −COHP versus E, as described above, to make the connection to the COOP more intuitive) yields a clear picture of which interactions are and are not responsible for the bandgap (Figure 1.10b). Both the Au–Sn and K–Sn COHP curves display a sharp transition between strong bonding and antibonding character at the gap. The K–Au interactions, on the other hand, show moderate bonding above the gap; these interactions would require many more electrons before their energetic contribution to the structure would be maximized. The presence of these high-energy bonding states can be partially understood from the high energies of the K 4s, K 4p, and Au 6p orbitals; bonding states between such atomic orbitals would be expected to be very high-lying in the band structure.

In this section, we have shared a sample of the large pool of applications of COOP and COHP to the electronic structure of intermetallic compounds. As we have seen, these methods provide a general sense of which interactions are strongly bonding or optimized at the E_F, and give hints about whether the bonding in a compound may be improved through the introduction or removal of a small number of electrons. Such information sets up one of the key questions that will frequently recur in this chapter: How does the structure of an intermetallic compound determine the electron counts at which its bonding interactions are optimized?

1.2.2
Site Preferences and Mulliken Populations

One common – almost defining – characteristic of intermetallic compounds is the segregation of their component elements to distinct sites within a crystal structure. A central line of inquiry into these compounds is thus the elucidation of the elemental site preferences among the fixed set of atomic sites in a given crystal structure. The question of which is most favorable of the numerous possible ordering patterns of elements over the sites of a crystal structure is known as the *coloring problem* [83]. One of the earliest explorations of this problem for intermetallics involved the resolution of the atomic ordering of CaB_2C_2. The initial proposed structure for this compound had the ordering of the B and C atoms as shown in Figure 1.11a. Based on eH calculations and symmetry arguments, however, Burdett et al. proposed the alternative structure shown in Figure 1.11b [84]. This prediction has since been definitively confirmed experimentally [85,86]. An earlier application of the method, combined with moments analysis, applied to 1:1 transition metal intermetallic compounds with different atomically ordered bcc arrangements (the CsCl and the CuTi structures) may also be noted [83].

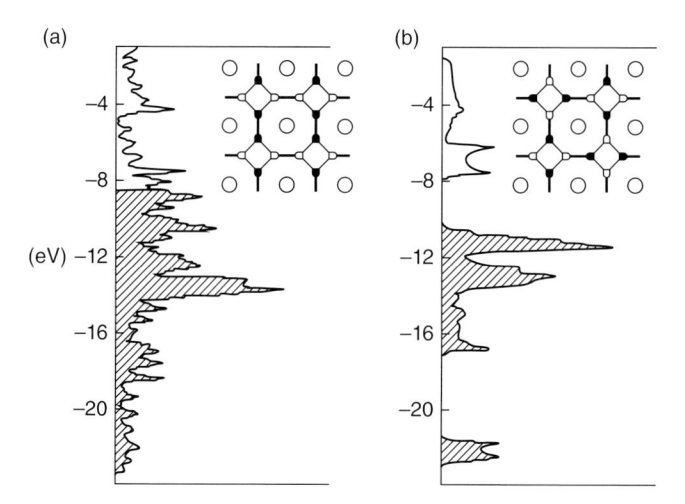

Figure 1.11 (a) Early experimentally proposed structure for the B and C ordering in $Ca_2B_2C_2$. (b) Proposed alternative structure for this phase based on eH calculations. *Large circles:* Ca atoms at height 0.5; *small filled circles:* C atoms at height 0; *small open circles:* B atoms at height 0. Even visual inspection of the two DOS curves shows that the alternative in part (b) is lower in energy. (Reproduced with permission from Ref. [84]. Copyright 1986, American Chemical Society.)

The extended Hückel method has provided one of the most elegant treatments of the coloring problem through the concept of topological charge stabilization [87,88]. At the core of this approach is the notion that calculations on a hypothetical structure composed of a single element contain indications about the relative energetics of placing a diverse set of elements at each and any of the hypothetical structure's sites. Elemental substitutions in general can affect both the on-site and pairwise bonding terms in the total energy, and researchers have taken different approaches to explaining and predicting site preferences depending on which of these is deemed to play a larger role.

In their work examining site preferences in ternary transition metal sulfides and phosphides, Franzen and coworkers considered the bonding terms to dominate [89,90]. Based on the calculation of the Mulliken overlap populations (the values obtained by integrating COOP curves up to the E_F) and the related Pauling bond order function, they were able to explain the role of bond strength in transition metal site preferences. One example is given by the phase $Zr_{2.7}Hf_{11.3}P_9$, a complex phosphide Kleinke and Franzen discovered with 15 symmetry-distinct transition metal positions with mixed Zr/Hf occupancy [90]. Using eH calculations on a $Hf_{14}P_9$ model, they were able to draw a strong correlation (with one outlier) between the fraction of Hf on each of these sites and the Hf—Hf and Hf—P bond strengths. Their rationale for this trend is deeply founded in chemical intuition: The greater diffuseness of the Hf d orbitals over that of the Zr d orbitals will lead to stronger bonds. It is thus advantageous to put the Hf atoms where the potential for bonding is greatest.

Considerations based on the on-site energies have also been very fruitful. Here, one uses the Mulliken populations of the atoms (which can loosely be thought of as the integration of the filled portion of an atom's projected DOS). One then calculates the Mulliken populations for a network where all atomic sites are assumed to have the same eH parameters. The sites with the largest Mulliken populations are predicted to be occupied by the compound's most electronegative elements.

C.-S. Lee and Miller, in their investigation of Bergman type 1/1 quasi-crystal approximants derived from the R-phase, have vividly demonstrated how this simple approach to site preference can form the basis for both prediction and new synthetic work (Figure 1.12) [91]. To understand the observed coloring patterns in these compounds, they performed an eH calculation on an all-Al model of the Bergman phase $Mg_{32}(Al,Zn)_{49}$, and plotted the deviation of the site Mulliken populations from the average (*relative Mulliken populations*) as a function of the valence electron concentration (VEC, Figure 1.12b). In the resulting graph, the sites are clearly divided into two separate groups: a relatively electron-poor group (positive values of the relative Mulliken population) consisting of sites A1–A3 and an electron-rich group (negative values) containing sites M1–M4. Based on our above considerations, the latter group is predicted to have a strong affinity for the most electronegative atoms in the system. The observed site occupancies for $Mg_{32}(Al,Zn)_{49}$ match this expectation well: The A sites are exclusively occupied by Mg, while most of the M sites are taken by the Zn or Al atoms.

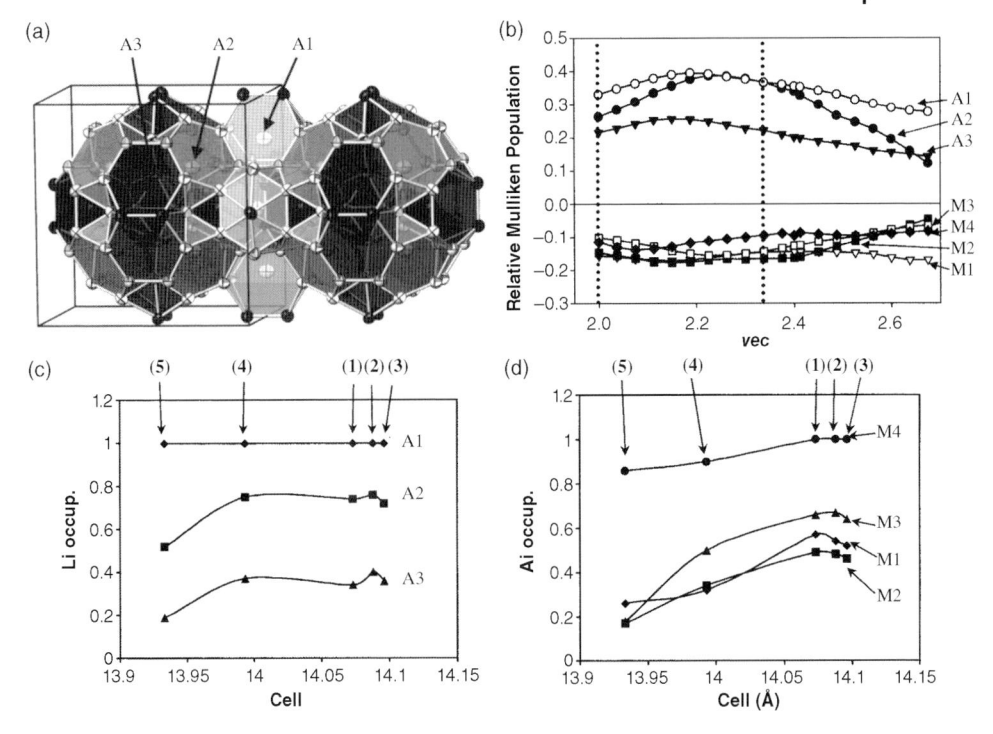

Figure 1.12 Explanation of site preferences in Li-Mg-Zn-Al Bergman phases using relative Mulliken populations. (a) The crystal structure of the Bergman phase with the positions of the A1, A2, and A3 sites indicated. (b) Relative Mulliken populations for the symmetry distinct atomic sites in an all-Al model of the phase, plotted as functions of the valence electron concentration (VEC). (c) Experimentally determined Li occupations on the A (Mg/Li) sites for phases with different compositions. (d) The Al occupancies for the M (Al/Zn) sites in the same phases. (Reproduced with permission from Ref. [92]. Copyright 2001, American Chemical Society.)

Finer details can also be discerned within the relative Mulliken populations for the A and M groups. Most strikingly, for much of the VEC range of interest, the Mulliken populations of the three A sites fall in the order A3>A2>A1. Lee and Miller designed an exquisite experiment to test the predictions made by these more subtle differences: They synthesized a series of Li-substituted versions of the $Mg_{32}(Al,Zn)_{49}$ phase to explore how Li and Mg would order on the A sites. Their structure refinements reveal that the content of Li on the A sites increases from A3 (about 20–35% Li) to A2 (about 55–75% Li) to A1 (100% Li). This sequence is in full accord with the predictions of the Mulliken populations.

Explaining the site preferences within the M sites was more challenging. However, Lee and Miller found that if the A1 Li atoms are considered as simple noninteracting Li^+ ions, this situation becomes clearer. Removing the A1 atoms from the all-Al model leads to a greater differentiation among the M sites, with the M4 site showing the highest electron population. This reasoning can be used to account for the affinity of the M4 site for Al over Zn, as the Al 3s and 3p valence orbitals are lower in energy than their Zn 4s and 4p counterparts.

1.2.3
Moment Analyses

In parallel with the development of projected DOS, COOP, and COHP, tight-binding theory has established a number of concepts derived from the DOS itself. Among these are the moment theories, where the tight-binding DOS's role as a distribution function is explored in detail [93,94]. Distribution functions are characterized by a series of measures, of which the best known are the mean and the standard deviation but this extends further to the less familiar statistical concepts of the skewness and the kurtosis (if mean and standard deviation measure respectively average and breadth, skewness and kurtosis measure respectively asymmetry and central peakedness) [95].

All of these measures can be derived from the power moments of the distribution, μ_n, where

$$\mu_n = \int E^n \rho(E) dE,$$

and $\rho(E)$ is the tight-binding electronic DOS. In the case of sH theory, μ_1, the mean, is just the average of the valence atomic orbital energies and is therefore a constant.

1.2.3.1 Rigid Band Model

The simplest of all moments models is the rigid band approximation, where the tight-binding DOS is represented solely by its first and second moments [96,97]. Following tradition, this second moment is viewed in the context of a rectangular-shaped DOS (Figure 1.13). The rectangular band is characterized by a single parameter, the bandwidth W.

The essential point of rigid band theory is that the variation in the value of W is less important than the band-filling of this rectangular band. W can be considered to be a constant, independent of structure. The very bottom and top of this rectangular band correspond to respectively the most bonding and antibonding states. The electronic energy relies solely on electron filling, and is a parabola with respect to this variable.

Figure 1.13 illustrates the essential correctness of this simple rigid band model for the group 1–12 metals. The net binding energies in these elements start small for the earliest metals, reaching a maximum energy near the half-filling of the valence s and d orbitals followed by a gradual decline as increasingly antibonding orbitals are filled. Placed in the context of a moment analysis, the rigid band model tells us that both the mean and (for a rectangular DOS) the width of the DOS can be considered fixed. This corresponds to assuming that both the first and second moments (μ_1 and μ_2) are structurally independent.

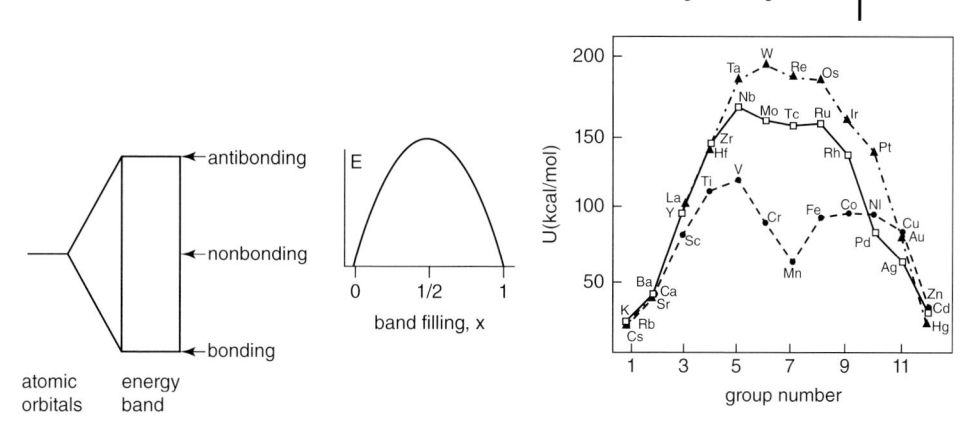

Figure 1.13 The rigid band model. Band shape and width are considered to be independent of electron count resulting in the binding energy acquiring a parabolic shape vis-à-vis the electron concentration. The right panel illustrates the approximate correctness of this model for the transition elements. Adapted with permission from Refs. [9,96]. The deviation for Cr, Mn, Fe, and Co in the first row transition metals is most likely related to atomic-based exchange energies. The overall parabolic shape corresponds to the structure-independent μ_2 term of a moment energy expansion.

1.2.3.2 Higher Moments

The structural energy difference theorem, discussed in the next section, defines and supports the earlier rigid band analysis, as to the lesser role μ_1 and μ_2 play in the differences in crystal structure energies [10]. In tight-binding theory, the key structural energy parameters can be fully derived from the higher moments, although the third and fourth moments, quantities related to the DOS skewness and kurtosis, are particularly significant [9,98–101].

A few energetic results demonstrate the utility of tight-binding and moments analysis. The first is the evolution of crystal structure across the transition metal elements [102]. Plotted in Figure 1.14 are the differences in energy of the three main elemental metallic structure types: face-centered cubic (fcc), body-centered cubic (bcc), and hexagonal closest-packed (hcp). As comparison of this figure with experimentally observed crystal structures (Table 1.1) shows, the higher moments of the electronic density of states correctly anticipate the evolution of all the transition metal structures, with the exception of La, the magnetically coherent elements Mn and Fe, and the late transition metal elements (groups 9 and higher) for which valence p orbital effects must be included [103].

In a similar vein, Haydock, Johannes, and Heine used similar tight-binding methods for transition metal Laves phases (Figure 1.15 and Table 1.2) [104,105]. In this early work, the $MgCu_2$ structure type was calculated to be the most stable for a valence electron concentration (VEC) of 10.2 to 14.6 electrons/per ZT_2 formula unit; the $MgZn_2$ structure type was lowest in energy from 14.6 to 20.0; and $MgCu_2$ again was found as the preferred structure for VEC from 20.8 to 23.1, after which the $MgNi_2$ structure dominated. A comparison of these predicted zones with experimental data shows these energetic predictions are largely

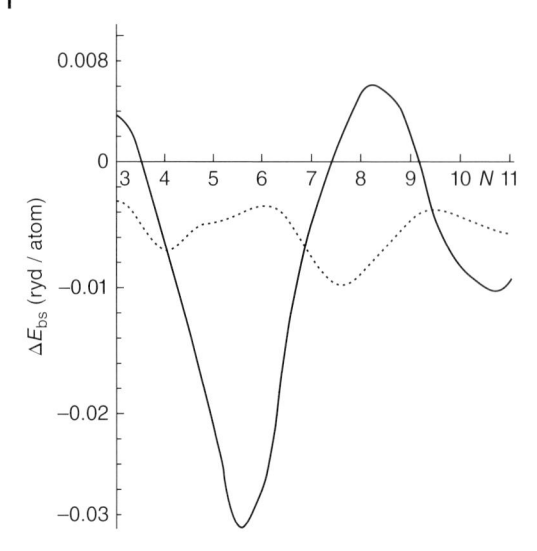

Figure 1.14 Differences in tight-binding energies for transition metals in common elemental structure types. *Solid line:* bcc – fcc energy differences as a function of d-band filling. *Dotted line:* The corresponding hcp – fcc energy difference curve. This early work predates the use of second moment constancy and valence s and p orbitals in tight-binding transition metal calculations. In this figure, the structure whose energy difference curve is lowest at a given electron count is the most favored structure. (Reproduced with permission from Ref. [102]. Copyright 1970, IOP.)

correct, with a deviation in the central $MgCu_2$ to $MgZn_2$ energetic crossing point. This result is particularly remarkable as in these early calculations only d orbitals were treated, and no steric factors were allowed.

Later calculations have refined this earlier work. With the explicit assumption of μ_2 constancy and the inclusion of valence s and p orbitals, it is possible to rationalize the differences in energy between the primarily noble metal Hume-Rothery fcc, CsCl, γ-brass, β-Mn, and ε-, ζ-, and η-hcp phases [103]. The electron concentrations of the known binary alloys found in these structure types are given in Figure 1.16a. There is a strong correspondence between the experimental electron concentration domains and the tight-binding energies of Figure 1.17a.

Table 1.1 STP transition metal element crystal structure types.

Period	3	4	5	6	7	8	9	10
3d	Sc	Ti	V	Cr	—	—	Co	Ni
4d	Y	Zr	Nb	Mo	Tc	Ru	Rh	Pd
5d	—	Hf	Ta	W	Re	Os	Ir	Pt
Structure	hcp	hcp	bcc	bcc	hcp	hcp	fcc	fcc

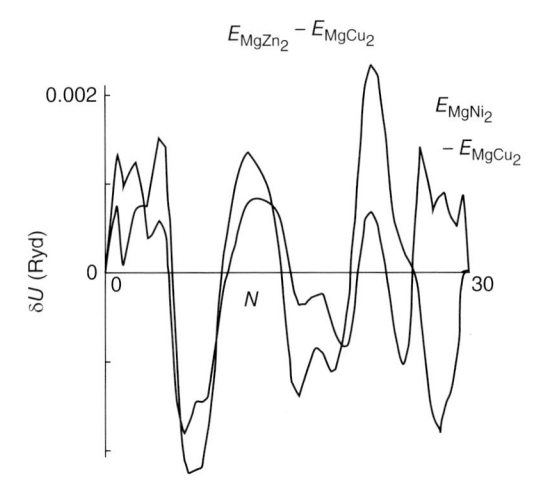

Figure 1.15 For pure transition metal compounds, the differences in tight-binding energies of the MgCu$_2$, MgZn$_2$, and MgNi$_2$ Laves structure types as a function of valence electron concentration. The curves follow the early convention that the structure whose difference curve is lowest in energy is energetically the most stable. Reproduced with permission from Refs. [104,105].

Table 1.2 Valence electron concentration (VEC, e^-/ZT_2) for pure transition metal Laves phases.

Compound	VEC	Compound	VEC	Compound	VEC	Compound	VEC
MgCu$_2$ type							
HfV$_2$	14	ZrV$_2$	14	TaV$_2$	15	ZrCr$_2$	16
TiCr$_2$	16	HfCr$_2$	16	ZrMo$_2$	16	HfMo$_2$	16
ZrW$_2$	16	HfW$_2$	16	TaV$_{1.5}$Mn$_{0.5}$	16	TaCr$_2$	17
MgZn$_2$ type							
ScMn$_2$	17	TiVFe	17	ScTc$_2$	17	YTc$_2$	17
ScRe$_2$	17	YRe$_2$	17	TiMn$_2$	18	ZrMn$_2$	18
HfMn$_2$	18	TaTiCo	18	ZrTc$_2$	18	ZrRe$_2$	18
HfRe$_2$	18	ZrVNi	19	HfVNi	19	TaVCo	19
NbMn$_2$	19	TaMn$_2$	19	ScFe$_2$	19	ScRu$_2$	19
YRu$_2$	19	ScOs$_2$	19	YOs$_2$	19	NbVNi	20
TaVNi	20	TaCrCo	20	TiFe$_2$	20	ZrRu$_2$	20
ZrOs$_2$	20	HfOs$_2$	20	NbCrNi	21	TaCrNi	21
NbFe$_2$	21	TaFe$_2$	21				
MgCu$_2$ type							
ScCo$_2$	21	YCo$_2$	21	YRh$_2$	21	ScIr$_2$	21
YIr$_2$	21	ZrV$_{0.5}$Ni$_{1.5}$	21.5	TiCo$_2$	22	ZrCo$_2$	22
HfCo$_2$	22	ZrIr$_2$	22	TaFeNi	23	NbCo$_2$	23
TaCo$_2$	23	ScNi$_2$	23	YNi$_2$	23	YPt$_2$	23
MgNi$_2$ type							
Ta$_{0.8}$Co$_{2.2}$	23.8	NbCo$_3$	24				

Thermodynamically stable Laves phases taken from the analysis of Haydock *et al.* [104,105].

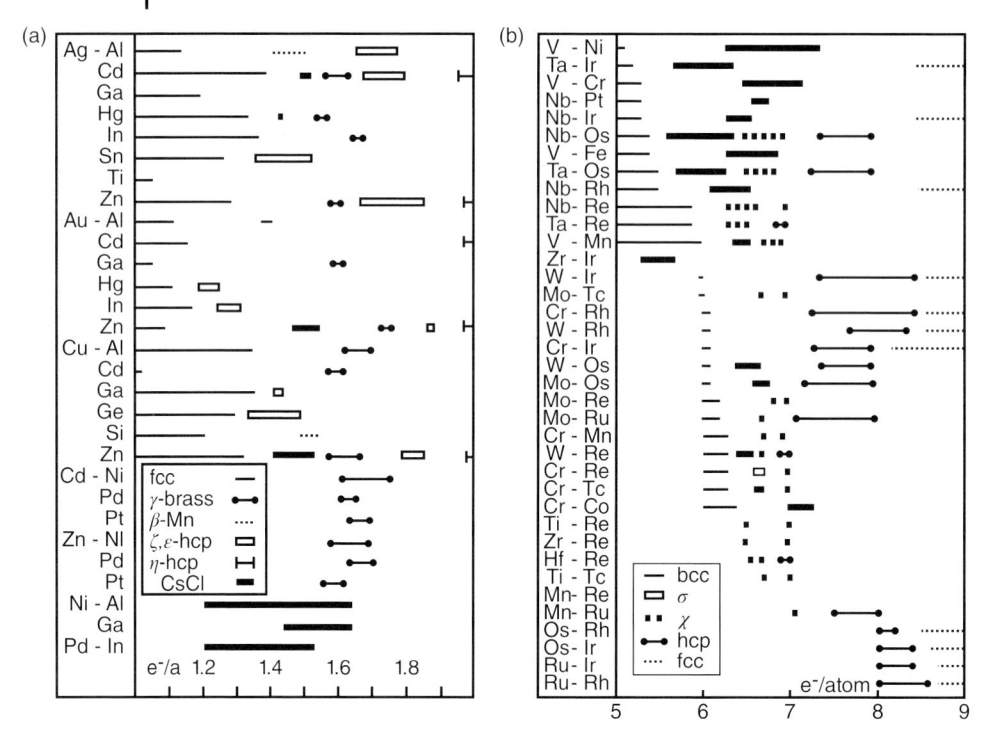

Figure 1.16 The low temperature stability regions for groups 10–14 (a) and 5–9 (b) binary alloys. Data are extrapolated from binary phase diagrams. Adapted with permission from Refs. [106–108].

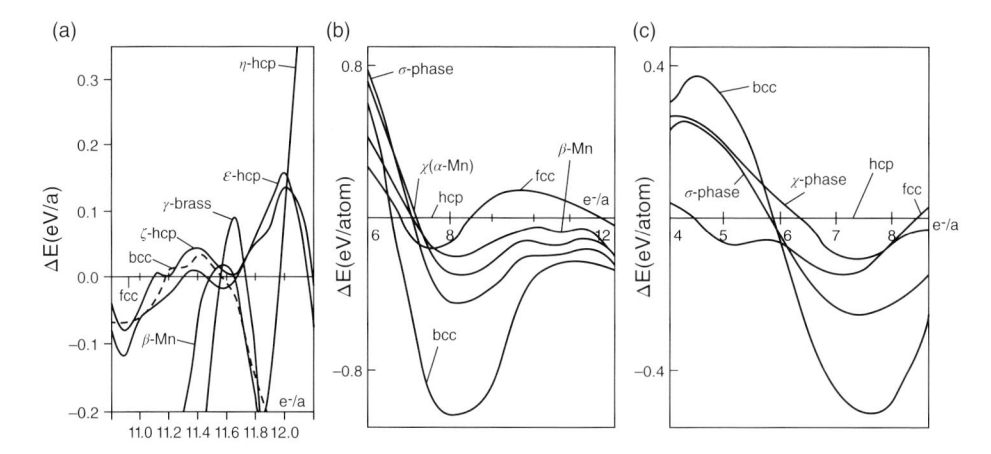

Figure 1.17 Differences in tight-binding energy for both (a) Hume-Rothery noble metal and (b)–(c) transition metal electron concentration phases. For the transition metal series, tight-binding calculations are shown (b) with and (c) without the inclusion of valence p orbitals. The theoretical results are in good agreement with experiment. Early transition metals have pronounced less p character in their valence orbitals. The α-Mn structure is referred to by its metallurgical name as the χ-phase. In this figure, the curve that is highest corresponds to the most stable alternative. Reproduced with permission from Ref. [103]. Copyright 1991, American Chemical Society.

In Figures 1.16b and 1.17b, the corresponding data for the transition metal-based electron concentration phases are shown, including the χ and σ phases found between 6–8 e^-/a. For the theoretical predictions in Figure 1.17, two results are presented (parts b and c), one with and one without p orbital contributions. A strong correspondence between experimental electron concentration rules and calculated energies can again be observed. As discussed for the transition metal structures, the former graph applies to later transition metal elements, the latter to the earlier groups.

1.2.3.3 The Structural Energy Difference Theorem, μ_2-Scaling, and Higher Moments

In the above sections, we have discussed how setting the μ_2 values for a series of structures to a constant and common value allows us to directly investigate factors that govern their relative stabilities. The structural energy difference theorem offers a way to think about this constant μ_2, opening paths to new insights from Hückel calculations [10]. This theorem considers the total energy of a structure to be the sum of an attractive and a repulsive term (i.e., $E_{\text{total}} = E_{\text{repulsive}} + E_{\text{attractive}}$), and states that we can measure the relative energies of two structures by comparing their attractive terms when the structures are scaled so that their repulsive terms are equal. By reference to common characteristics of repulsive energy, it can be deduced that μ_2 is proportional to the repulsive energy term of this theorem [101,109,110].

Adding such a μ_2 component term to the energy in a simple Hückel calculation yields the μ_2-Hückel method, where the energy is given by

$$E_{\mu_2\text{-Hückel}} = E_{\text{repulsive}} + E_{\text{attractive}} = \gamma \cdot \mu_2 + E_{\text{Hückel}}. \tag{1.8}$$

Here, γ is a constant that can be adjusted so that the energy is minimized at the correct unit cell volume for a chosen phase. This equation has a simple interpretation. The $E_{\mu_2\text{-Hückel}}$ is decomposable into pairwise interactions, so that the total energy is reducible as follows:

$$\begin{aligned} E_{\mu_2\text{-Hückel}} &= \gamma \cdot \mu_2 + E_{\text{Hückel}} \\ &= \gamma \cdot \sum_i \sum_{i'} H_{ii'} H_{i'i} + \sum_n o_n \sum_i \sum_{i'} c_i^{(n)*} c_{i'}^{(n)} H_{ii'} \\ &= \sum_i \sum_{i'} \left(\gamma \cdot H_{ii'}^2 + \left(\sum_n o_n c_i^{(n)*} c_{i'}^{(n)} \right) \cdot H_{ii'} \right). \end{aligned}$$

The total energy now becomes a sum over contributions from individual pairs of orbitals (when $i \neq i'$) and on-site contributions (when $i = i'$). Each interaction consists of an attractive component proportional to $H_{ii'}$, which is balanced against a repulsive component proportional to $H_{ii'}^2$, a form that recalls common models for interatomic interactions such as the Lennard-Jones 6–12 potential [101].

The supplementation of the Hückel energy with the repulsive $\gamma \cdot \mu_2$ helps resolve a long-standing issue with the sH and eH models: On their own they predict that the stabilization due to a bond will increase as the interatomic distances shrink, even to the point where the two atoms merge into a single atom. With the μ_2 term in place, however, the DFT-calibrated Hückel calculations have the potential to not only reproduce the band energies and density of states curves of a DFT result, but also, in some cases, to rationalize the distance dependence of interactions.

In Figure 1.18, we illustrate this potential with a comparison of the DFT and best-fit Hückel results for the CaCu$_5$-type phase SrAg$_5$ [111]. As in our previous example above, the band structures calculated with the two approaches show a high degree of similarity (Figure 1.18a and b). Below these band diagrams, we also show contour plots giving the total energies of the two models as functions of V and c/a. In both cases, a single minimum is visible at a c/a ratio just above 0.8. Qualitative agreement is also seen in the shape of the minimum: in both pictures, traveling from the minimum to lower V comes with a much steeper climb than moving to higher V.

Within tight-binding theory, the ability to capture aspects of the distance dependence of interactions plays a useful role in modern developments of the field [101,110–113]. In the past, these ideas have been applied to the energy isosbestic-like points seen in DFT calculations of transition metal-based CsCl-type

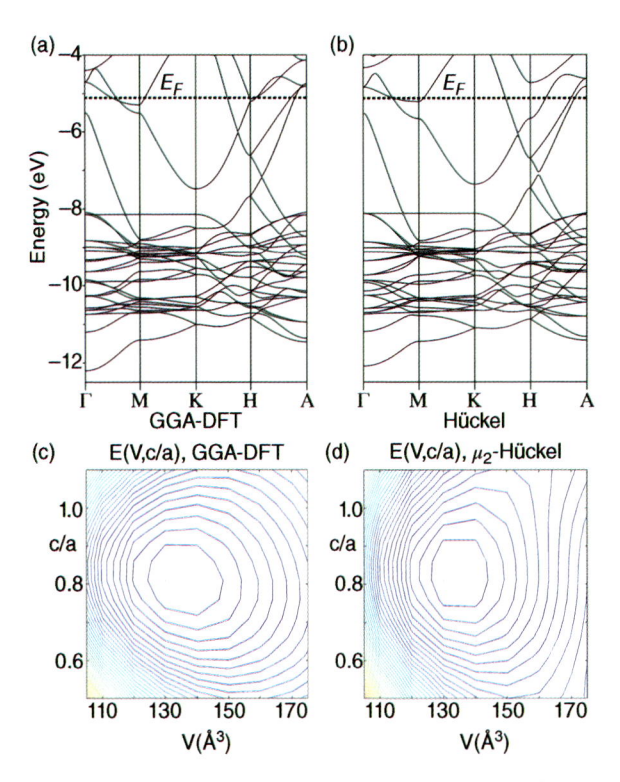

Figure 1.18 Validation of a μ_2-Hückel model of SrAg$_5$ (CaCu$_5$-type) against DFT results. (a and b) Electronic band structures calculated with GGA-DFT and a best-fit simple Hückel model. (c and d) Contour maps of the total energies of SrAg$_5$ obtained from GGA-DFT and the μ_2-Hückel model, as functions of unit cell volume (V) and c/a ratio. (Reproduced with permission from Ref. [111]. Copyright 2011, American Chemical Society.)

phases [110]. Similarly, it has been applied to the growth of a second energy minimum in further DFT calculations on the same phases (traceable to a Jahn–Teller effect in k-space) [112].

Within the context of μ_2-scaling, higher moments can also be assigned specific energetic roles [83,98,99]. First, as is illustrated in Figure 1.19, each higher moment oscillates between stabilizing or destabilizing effects as we scan across the different band fillings. These energy contributions look much like a Fourier series, with higher moments having ever-decreasing importance. In addition, higher moments can be directly tied to atomistic features of the compound: the ionicity of the binary components, bond angles, and the number of triangles, squares, pentagons, and hexagons within the crystal structure [98,101].

Moments analyses based on these ideas have been applied to a variety of Zintl-like main group metalloid structures. On the one hand, researchers have wished to take advantage of the higher moments energy difference curves. On the other hand, researchers have been hampered by the complex geometries of intermetallic structures, their vast number of bond angles, an inability to deal with steric factors, and a lack of methods for dealing with elements whose electronegativities lead to only partial electron transfer [114–120]. To date, special attention has been paid to electron concentration rules for main group Laves structures. In Zintl compounds with fixed Zintl electropositive/electronegative atom ratios, some progress has been achieved, as is shown in Figure 1.20 [114,116].

1.2.3.4 Tight Binding-Based Structure Maps

Structure maps, the sorting of crystal structure types based on a few key parameters, go to the heart of intermetallic structure systematics [10]. From early on, tight-binding theory has played a role in structure map development. Figure 1.21 shows a tight-binding treatment of 1 : 1 T–E compounds, where T is a transition metal element and E a main group atom [121].

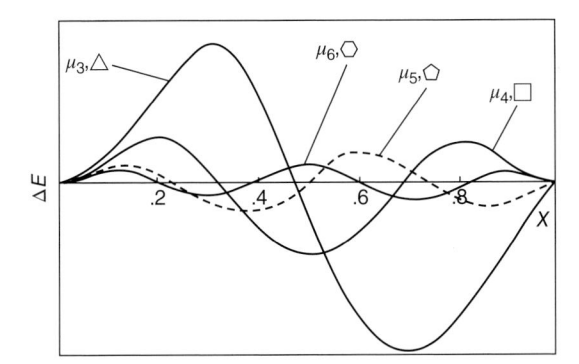

Figure 1.19 μ_2-scaled sH differences in energies between structures composed of triangles, hexagons, pentagons, or squares of bonded atoms plotted as a function of the band filling, x. Reproduced with permission from Ref. [101]. In this figure, the curve that is highest corresponds to the most stable alternative.

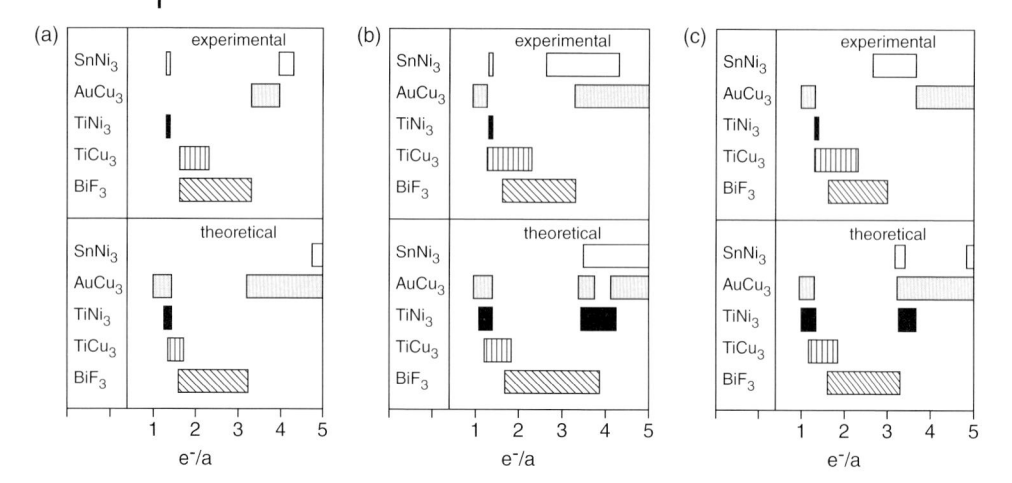

Figure 1.20 Comparison of experimental and predicted stability ranges for structure types adopted by Zintl ZE_3 compounds (Z = electropositive metal atom, E = electronegative one). The E elements are main group metalloid atoms taken from (a) rows 13 and 14, (b) rows 13–16, and (c) rows 15 and 16 of the periodic table. Theoretical fields are based on tight-binding calculations with valence AOs whose principal quantum numbers and exponential decay parameters correspond to the modeled elements. (Reproduced with permission from Ref. [116]. Copyright 1995, American Chemical Society.)

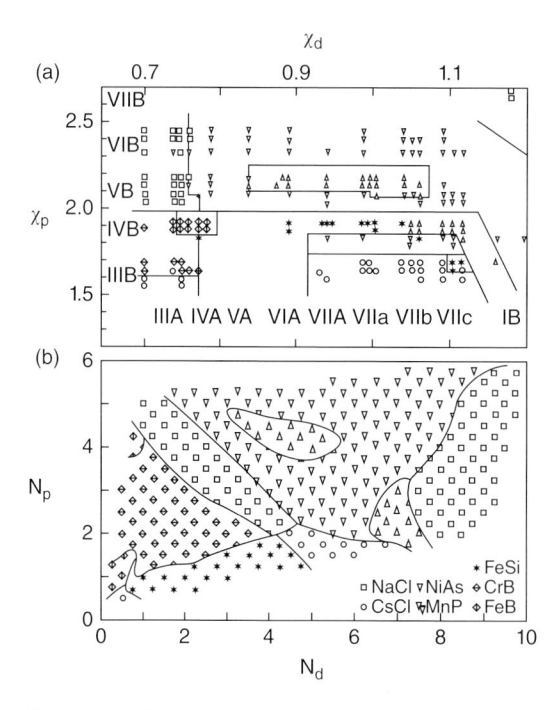

Figure 1.21 (a) Experimental structure map for 1:1 T–E compounds (T = transition metal, E = main group element), where χ_p, χ_d are given elemental parameters (related to electronegativities). (Reproduced with permission from Ref. [122]. Copyright 1984, Elsevier.) (b) Structure map based on tight-binding calculations where axes correspond to the respective number of valence electrons of the T and E atoms. (Reproduced with permission from Ref. [121]. Copyright 1984, American Physical Society.)

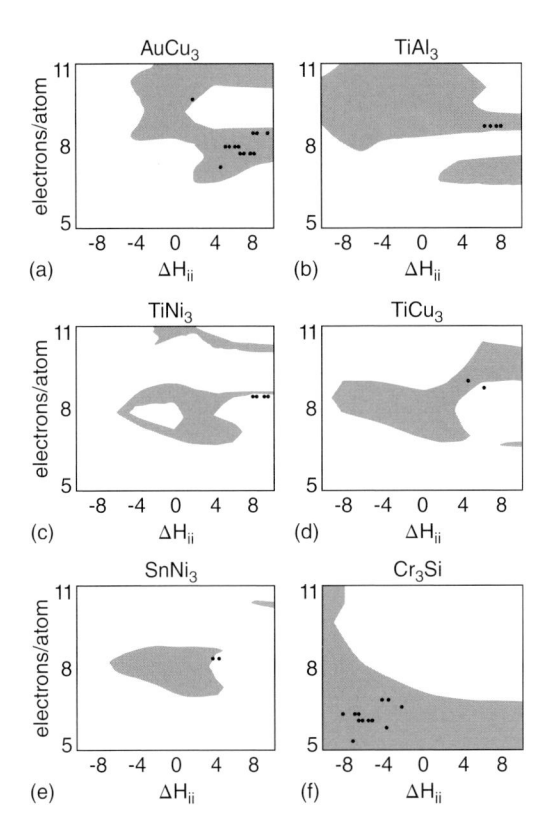

Figure 1.22 Direct comparison of 1:3 ordered transition metal compounds with tight-binding derived energies. Gray regions indicate where the indicated structure type is within 0.05 eV/atom of the most stable structure type. In these figures, a second quantum mechanical variable, the difference in d orbital AO energies (ΔH_{ii}, a measure of electronegativity differences) forms the horizontal axis. (Reproduced with permission from Ref. [123]. Copyright 2005, Elsevier.)

In these early calculations, tight-binding calculations were used to determine qualitative features of the experimentally observed structure map. Such qualitative treatments still play a useful role today [10]. In certain instances, it has proven possible to directly map tight-binding energetics onto experimentally observed structures. An example of such work is shown for noncoherently magnetic 1:3 transition metal species in Figure 1.22.

1.3
Nearly Free Electron Model

Even if today the practicing researcher rarely undertakes nearly free electron band calculations themselves, concepts based on this model continue to play a central role in our understanding of intermetallic bonding. Knowledge of nearly free electron theory remains essential [1]. Unlike tight-binding calculations that begin with an AO basis set, nearly free electron theory starts with a free-electron plane wave basis. Electron density is then treated as a perturbing factor, mixing these plane wave states. This perturbing electron density is well known to experimental intermetallic chemists, its Fourier transform being the sum of the X-ray diffraction structure factors. Once corrected for Debye–Waller factors and X-ray machine collection factors, it may be viewed as the X-ray diffraction pattern!

The essential point of the theory is the comparison of the Fermi sphere, the reciprocal space sphere whose interior encompasses all unperturbed filled electronic plane wave states, with the Jones zone, a reciprocal space polyhedron derived solely from the strongest X-ray diffraction peaks [2,11,124,125]. The key feature of the former is its volume, a parameter determined solely from the number of valence electrons per atom and the number of atoms per unit cell. The method for generating the latter is illustrated by the Jones zone of the bcc structure in Figure 1.23. As this figure shows, the strongest k-vectors, for bcc the $\{1\,1\,0\}$ reflections, define the Jones polyhedron faces.

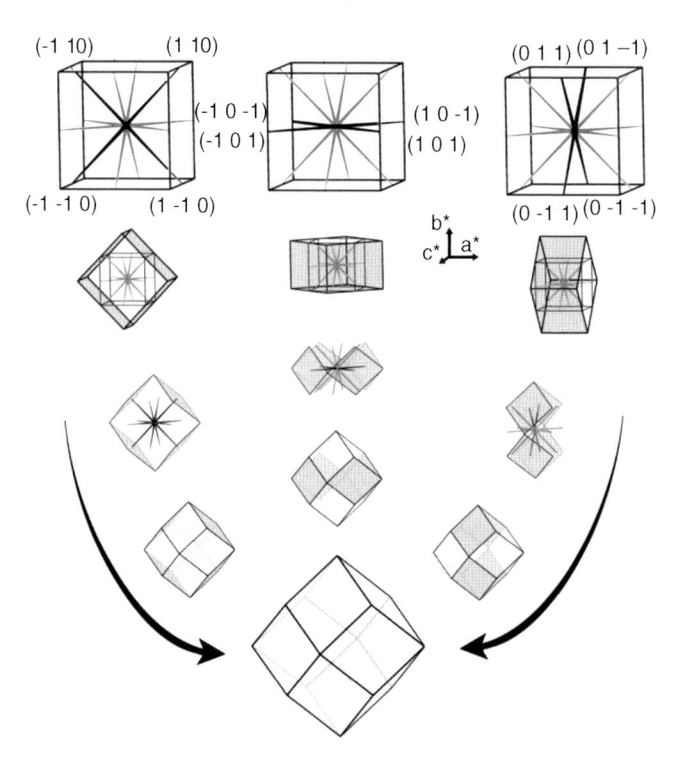

Figure 1.23 The bcc Jones zone. The strongest diffracting reflections in bcc, the $\{1\,1\,0\}$ reflections, define the faces of the Jones zone, a reciprocal space polyhedron. Reflections are shown first (on top) as vectors, second (in middle) as planes or truncated planes, and finally (on bottom) as polyhedral faces. For bcc, a Bravais lattice, the Jones zone is similar in shape to bcc's first Brillouin zone. Importantly, the Jones zone establishes lower and upper bounds for optimal e^-/a values.

Especially pertinent for intermetallic compounds derived from mixtures of group 10–14 elements is the direct comparison of the Jones zone with the Fermi sphere itself. Figure 1.24 shows this in two different ways [126,127]. To the right of the figure are the Jones zone and the free electron Fermi sphere plotted on a standard scale: It can be seen that the volumes of the two constructs approach

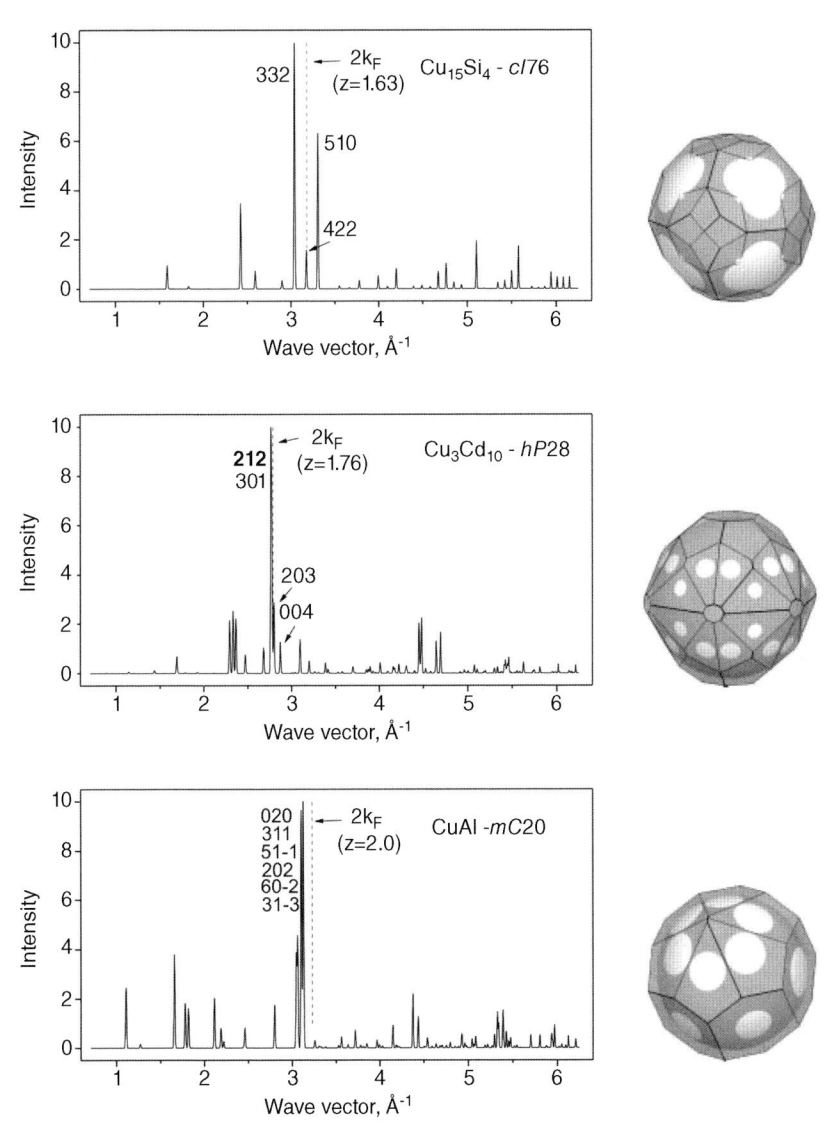

Figure 1.24 Comparison of $Cu_{15}Si_4$, Cu_3Cd_{10}, and CuAl powder diffraction patterns, Fermi sphere, and Jones zone. The diameter of the Fermi sphere is determined solely from the number of valence electrons/atom and the number of atoms per unit cell. The diameter of the Fermi sphere, $2k_F$ is indicated with a dashed line. (Reproduced with permission from Ref. [126]. Copyright 2007, De Gruyter.) Of note is that the dashed line, indicating the Fermi sphere radius, is always to the immediate right of one of the strongest diffraction peaks: The Fermi sphere therefore intersects and bypasses the Jones zone. The Jones planes therefore specify a lower bound for optimal electron concentrations.

one another. To the left of this same figure, the diameter of the Fermi sphere (indicated with a dashed line) is shown to correspond closely to the lengths in reciprocal space of the strongest diffraction peaks.

These agreements are not accidental. In the nearly free electron model, strong diffraction peaks allow strong mixing between plane wave states whose k-vectors have the same magnitude but which point in different directions. If these strong peaks are of the same length as the $2k_F$ Fermi sphere diameter, plane waves near the surface of the Fermi sphere mix and separate into two sets: one set lowers in energy, the other rises. The former is filled with electrons; the latter remains unoccupied. Both a pseudogap in the electron energy DOS and the energetic stability for the system as a whole ensue.

Jones theory has the added virtue that bounds for both upper and lower electron concentrations can be easily obtained. The lower bound is defined by the electron filling where the Fermi sphere is exactly circumscribed by the Jones zone, while the upper bound corresponds to the electron filling in which the Jones zone, and just the Jones zone, is fully occupied. For example for bcc, a Bravais lattice, the Jones zone allows exactly two e^-/a. The sphere exactly circumscribed by the Jones zone has a volume 0.74 times this value. Jones theory therefore establishes lower and upper bounds of respectively 1.48 ($1.48 = 2 \times 0.74$) and 2 e^-/a.

One great success of the Jones theory of lower and upper e^-/a bounds has been in complex cubic metal crystals [2,124,126–129]. For the most complex of the Hume-Rothery electron concentration phases, γ-brass with 52 atoms in its cubic unit cell, the Jones zone is composed of 36 faces (Figure 1.25). The Jones zone's lower and upper boundaries are 1.54–1.73 e^-/a. These values are in excellent agreement with the known γ-brass binary alloy phase boundaries. Table 1.3 shows this agreement between these bounds and the actual experimentally observed range of these phases [2].

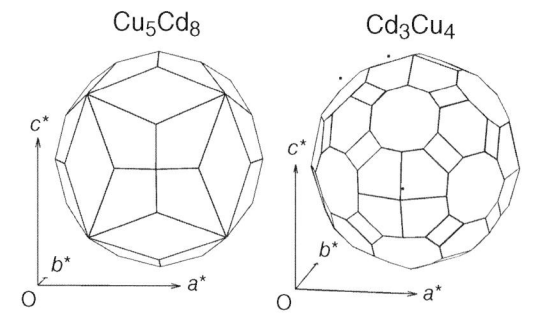

Figure 1.25 Jones zones for the γ-brass (Cu$_5$Zn$_8$-type) Cu$_5$Cd$_8$ and Cd$_3$Cu$_4$ (with respectively 52 and 1124 atoms/unit cell). γ-brass Jones zone faces: {3 3 0} and {4 1 1} reflections; Cd$_3$Cu$_4$: {8 8 0}, {11 3 3}, {7 7 7}, and {10 4 4}. The Jones zone lower and upper bounds, respectively, 1.54–1.73 and 1.35–1.52 e^-/a, are among the most effective ways of understanding the observed e^-/a values [2,106–108,129].

Table 1.3 Noble metal γ-brass e^-/a ranges [2].

Alloy	Lower e^-/a boundary	Upper e^-/a boundary
Cu–Zn	1.58	1.66
Cu–Cd	1.60	1.67
Cu–Sn	1.67	1.67
Cu–Al	1.60	1.67
Ag–Zn	1.58	1.66
Ag–Cd	1.60	1.67
Ni–Zn	1.58	1.66
Li–Ag	1.00	1.00
Jones zone	1.54	1.73

Less well known is that the Jones model aids in the understanding of very large cubic crystal structures [130–132]. We illustrate here the Jones zone for the Samson phase, Cd_3Cu_4, with an extraordinary 1124 atoms in its cubic unit cell [133–135]. This zone proves to have 68 faces (Figure 1.26), and lower and upper bounds of respectively 1.35 and 1.52 e^-/a [129]. These values can be compared with these phases' binary phase boundaries. Extrapolating known-phase boundaries to their low-temperature limit, we estimate phase boundaries of 40–46% atomic cadmium composition and a 1.40–1.46 e^-/a phase range, in agreement with the Jones lower and upper limits. Recent work has shown the accuracy of these bounds for all phases in the Cu–Cd and Au–Cd binary phase diagrams [127,129]. In addition, the Jones zone concept has been applied to high-pressure metals and intermetallic phases [136,137].

1.3.1
Pseudopotentials

At the divide between the early nearly free-electron models and current DFT-based derived k-space interactions lies the concept of the pseudopotential [4,5,138]. In this section, we explore the way in which pseudopotentials allow for the qualitative understanding of the distorted crystal structures of elemental gallium, indium, and mercury, as well as the elongated c-axes in hcp Zn and Cd.

In pseudopotential theory, the original nearly free-electron model interaction between different plane waves is improved by the addition of a pseudopotential $v(q)$:

$$V(\mathbf{r}) = \sum_{q}{}' S(\mathbf{q})v(q)e^{i\mathbf{q}\mathbf{r}},$$

where $S(\mathbf{q})$ are the X-ray k-space structure factors and $v(q)$ can be obtained directly from calculation and verified from experimental observations.

At its most useful, the pseudopotential is independent of the crystal structure and depends only on the element in question. An example of such a calculated pseudopotential along with experimental verification of its form is given in Figure 1.26 for the element indium. The pseudopotential combines with the structure factors and exchange- and correlation-dependent terms $\chi(q)$ and $\beta(q)$ in determining the structure-dependent energy of an atomic arrangement:

$$U = \sum_q' |S(\mathbf{q})|^2 v^2(q)\beta(q)\chi(q) = \sum_q' |S(\mathbf{q})|^2 \Phi_{bs}(q).$$

The crux of the pseudopotential theory is that the $\chi(q)$ and $\beta(q)$ functions are fairly smooth and nodeless. Only the pseudopotential carries a node, whose position is used to define q_0 (Figure 1.26). If an elemental crystal structure were to have its principal Jones reflections separated by this q_0 value, there would be no interaction between plane waves on opposite sides of the Jones zone and hence no energetic stability conferred on the system [5].

The key factor is therefore to compare a given structure's Jones zone reflection $2k$ values with an element's q_0 value. If the Jones zone reflections have nearly the same value as q_0, the structure can be expected to distort, moving some planes to shorter and others to longer reciprocal space distances and hence increasing overall the structural energetic stabilization energy.

In Figure 1.27, the q_0 values for a variety of elements are directly compared with the $2k$ values of the principal Jones reflections for the fcc, ideal $c/a = 1.63$ hcp, and bcc structures. Two points can be garnered from this comparison. First, the q_0 values for Hg, In, and Ga lie centered on the Jones zones of all three structure types. As a consequence, Hg and In have distorted fcc structures, while Ga adopts a distorted hcp structure. Second, among divalent hcp-adopting metallic elements, Be and Mg have q_0 values far away from ideal hcp's $2k$ values. In Zn and Cd, by contrast, the q_0 and $2k$ values are near one another. The former therefore have c/a ratios within a few percent of the ideal value of 1.63,

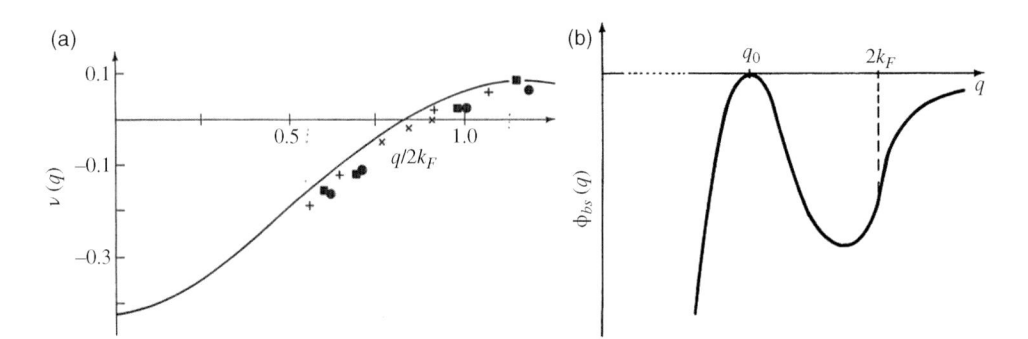

Figure 1.26 (a) Indium pseudopotential in Ry fitted from filled circles: InP, filled square: InAs, + signs: InSb, × signs: elemental In; all scaled to volume of elemental In. (b) Schematic for the energy stabilization function $\Phi(q)$. Reproduced with permission from Ref. [5].

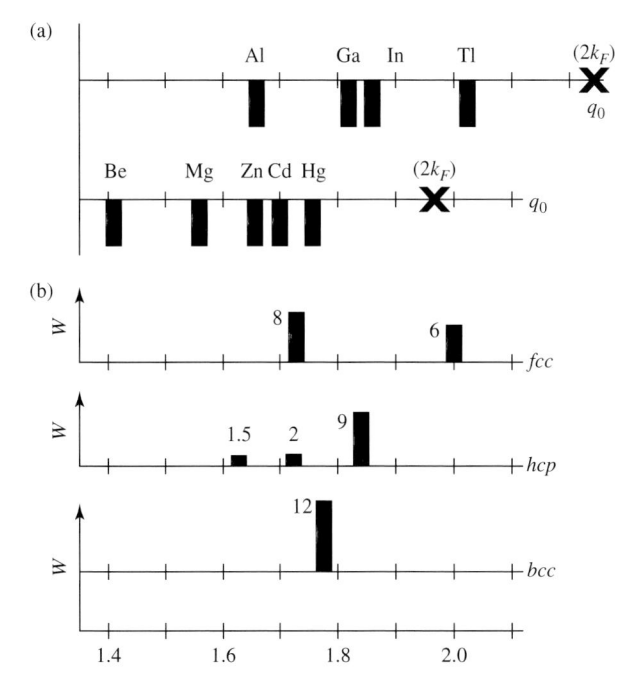

Figure 1.27 (a) For metals with valency two or three, the k-space position of the pseudopotential node, q_0. (b) Location of the principal Jones reflections. Integer values noted in bottom portion of figure are the multiplicities of the respective Jones faces. Reproduced with permission from Ref. [5].

while the latter have elongated cells with c/a ratios of respectively 1.86 and 1.89 [139].

It should be noted that the pseudopotential theory does not solely rely on the position of the q_0 nodes. Were that to be true, the nearly free electron Jones model itself would be invalidated. Instead, within the context of pseudopotentials, this model as applied to Hume-Rothery noble metal phases has been entirely substantiated [140].

1.4 Magnetism

1.4.1 Ferromagnetism

While transition metal itinerant ferromagnetic theory has traditionally evolved from ideas unrelated to intermetallic bonding theory, this is not to say that such bonding models cannot be usefully applied to the understanding of metallic ferromagnetism [141,144]. As Dronskowski and Landrum have explained, just as much of our understanding of the covalent bond can be derived from considering the H$_2$ molecule, itinerant ferromagnetism can be extrapolated from examination of the simplest of paramagnetic molecules, the triplet ground-state O$_2$ dimer [12,145–147].

For this molecule in its ground state, Landrum and Dronskowski compared the filled up-spin $2\pi(\pi^*)$ orbitals with their corresponding unfilled down-spin orbitals. They found the difference in energy between the up- and down-spin orbitals to be 1.85 eV and, as Figure 1.28 demonstrates, that the former orbitals are distinctly more contracted than the latter.

They account for these differences with the concept of the Fermi exchange hole. Electrons are fermions. Independent of any classical forces, identical fermions repel one another and therefore electrons with identical spin repel one another. Around each electron, an atomic-sized region exists where the electron density of other electrons with like-spin is depleted relative to what would have been expected were the electrons not to have been identical.

Having pushed other identical spin electrons aside, identical spin electrons shield each other more poorly than opposite spin electrons: In O$_2$, there are more up-spin α-electrons than down-spin β-electrons and therefore the Z_{eff} for the α-electrons is greater than for the β-electrons: α-electrons are thus more spatially contracted, and hence have lower π^* energy.

We continue this approach to O$_2$'s heavier congeners: S$_2$, and Se$_2$. For these systems, Landrum and Dronskowski found that the α and β valence π^* energy splitting falls from 1.85 to 0.81 to 0.63 eV with increasingly smaller differences in sizes of up- and down-spin valence π^* orbitals (Figure 1.28).

The second row main group molecule oxygen, as its valence p orbital is the first of the p series, exhibits a much greater ability to disproportionate in the size of its π orbitals, a fact we previously discussed for H$_2$, the first diatomic in which the s AOs are utilized. That the 2p orbitals are the first instance of orbitals with p angular momentum properties is of paramount importance.

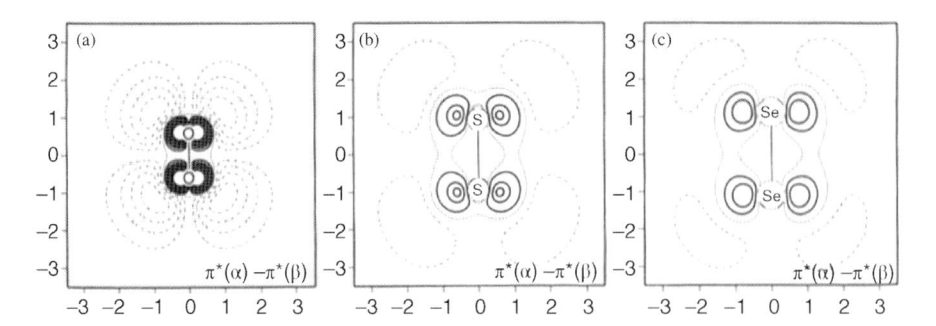

Figure 1.28 Difference maps between α- and β-spin molecular orbitals for O$_2$ (a), S$_2$ (b), and Se$_2$ (c) dimers. The darker contrast for the O$_2$ system is due to the greater AO contraction experienced by the α-spin MOs of this 2p valence element. (Reproduced with permission from Ref. [145]. Copyright 2000, John Wiley & Sons.)

1.4.1.1 Fe, Co, and Ni

We now turn to the ferromagnetic elements, iron, cobalt, and nickel. In them, we see the confluence of two issues. First, these elements are all first row transition elements, the first elements with d orbital characteristics. Their valence d orbital is 3d, an orbital unconstrained by lower d orbitals in its ability to contract. As such these elements are among those whose electronic structure will have the greatest ability to disproportionate in the size and energies of their α- and β-spins. Second, in the rigid band approximation, Fe, Co, and Ni correspond to the very last of the first row transition elements, and hence are the elements where, in a simple one-electron picture, antibonding states are maximally filled.

Figure 1.29 presents the DOS, and COHP curve for nonmagnetic body-centered cubic iron (Fe is body-centered cubic and magnetic, while as discussed in the next section, Co and Ni are face-centered cubic). The band structure is a remarkable one. Not only are there many filled antibonding states in elemental Fe without magnetic ordering, but the Fermi level actually also lies in the middle of a very large spike in the DOS of such antibonding states. Until raised to very high temperatures, nonmagnetic body-centered cubic iron is a most unstable compound.

These results can be contrasted with calculations on ferromagnetic iron. As Figure 1.29 shows, the shift between up-spin (α) and down-spin (β) orbitals is around 2 eV, a shift similar to that seen in molecular oxygen. This 2 eV difference corresponds to a much greater and smaller Z_{eff} for respectively α- and β-electrons.

The effect of the different Z_{eff} values is profound. With their greater Z_{eff}, α-electrons are much more contracted and much less capable at ordinary Fe—Fe bond distances of overlapping with one another. The antibonding character of the large spike in the nonmagnetic DOS and COHP is therefore greatly reduced. The opposite is found for the β-electrons. Their Z_{eff} is reduced in magnitude, the β orbitals expand, and their antibonding peak is much the greater.

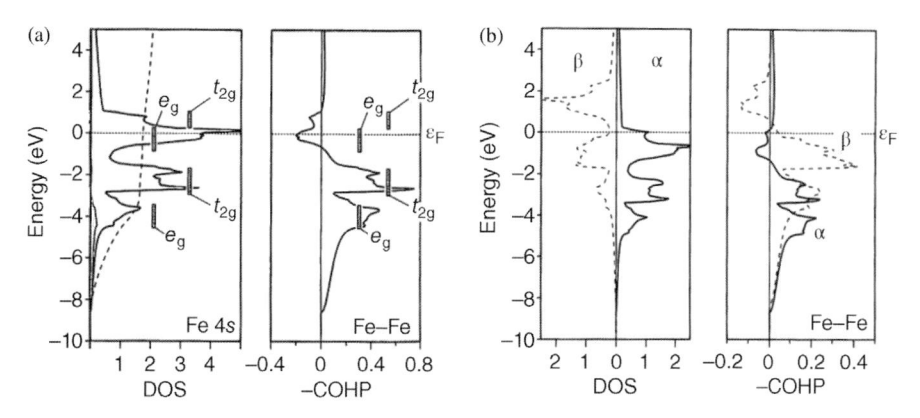

Figure 1.29 (a) DOS and COHP for nonmagnetic Fe. The e_g and t_{2g} indicate the classic O_h point group-derived splittings of these bcc orbitals. (b) DOS and COHP for ferromagnetic Fe. (Reproduced with permission from Ref. [145]. Copyright 2000, John Wiley & Sons.)

With the 2 eV energy difference, the large spike of antibonding α orbitals becomes completely filled, while the corresponding antibonding β spike remains empty. It is this filling change coupled with the different antibonding propensities of the up- and down-spin electrons that stabilizes the iron ferromagnetic state. Landrum and Dronskowski show similar effects for cobalt and nickel.

Alone among the transition elements, the late first row transition metals iron, cobalt, and nickel are ferromagnetic. Their ferromagnetism is an outcome of (i) the greater ability of the first row d-block to disproportionate in the size of their α and β electrons, and (ii) the antibonding character of a fairly narrow band straddling the Fermi energy.

The work of Landrum and Dronskowski elicited vehement objections from some members of the solid-state physics community upon its initial publication [146]. It is therefore relevant that similar analyses for more complex intermetallic ferromagnets have since been successfully carried out, by a number of groups, on substantially different classes of intermetallic compounds (including pyrite derivatives, borides, half-Heusler compounds, and phosphides) [12,148–150]. It should be noted that the Landrum–Dronskowski model need not have implications for all itinerant ferromagnetic systems; instead, it examines in detail one of nature's "solutions" for the avoidance of filled antibonding states.

1.4.2
Symmetry-Breaking and Antiferromagnetism

Itinerant ferromagnetism provides one outlet for the reduction of antibonding character of late transition metal species, but other avenues exist as well. Among the foremost of these are structural distortion (well known from the Peierls distortion and charge density waves) and the onset of antiferromagnetism [151–158]. These mechanisms are instances of spatial and spin symmetry-breaking, respectively.

For structural distortions, we center our attention on the bcc structure that has increasingly populated antibonding states as the VEC progresses from 6 to 8 e^-/a. For nonmagnetic transition metal compounds, the structural distortion from body-centered cubic to the body-centered tetragonal (bct) structure is known to occur, bct being the structure where the two atoms in the former bcc unit cell continue to reside at the corner and body center, but where the c-axis elongates so that $a = b < c$ (Table 1.4). When the structure distorts to the point that $c = \sqrt{2}a$, the bct structure becomes equivalent to fcc.

Figure 1.30 contrasts two different energetic measures for the bcc and bct structures [112]. The first, labeled μ_2, are the energy results of a higher moment calculation between the two structure types. The second, labeled k_{star}, plots the energy difference that results solely from the redistribution of electron filling between k-points that formerly in bcc were fixed to have identical energy orbitals but where in the lower symmetry bct structure are now permitted to have different energies. By the coincidence of the two energy curves, we conclude that it is the orbital mixing allowed by the spatial symmetry-breaking from bcc to bct that

Table 1.4 Atomically ordered, magnetically unordered, low-temperature binary bcc, bct, and fcc compounds. (Reproduced with permission from Ref. [112]. Copyright 2002, American Chemical Society.)

\multicolumn{6}{c}{CsCl type (c/a = 1.00)}					

Compound	e^-/a	Compound	e^-/a	Compound	e^-/a
TiTc	5.50	HfTc	5.50	VTc	6.00
HfRu	6.00	TiOs	6.00	ZrOs	6.00
HfOs	6.00	HfRh	6.50		

		HgMn type (c/a variable)			
Compound (c/a)	e^-/a	Compound (c/a)	e^-/a	Compound (c/a)	e^-/a
TaRu (1.02)	6.50	NbRu (1.12)	6.50	TiRh (1.14)	6.50
TiIr (1.18)	6.50				

		$AuCu_3$ type (c/a = 1.41)			
Compound	e^-/a	Compound	e^-/a	Compound	e^-/a
$NbRu_3$	7.25	$TiRh_3$	7.50	$HfRh_3$	7.75
$TiIr_3$	7.75	$ZrIr_3$	7.50	$HfIr_3$	7.75
VRh_3	8.00	$TaRh_3$	8.00	VIr_3	8.00
$NbIr_3$	8.00	$TaIr_3$	8.00		

is responsible for bct stability at higher electron counts and that this symmetry breaking term is largest near 7.8 e^-/a.

As Table 1.4 shows, a valence electron count of 6.5 e^-/a, corresponding to a d electron count of roughly 5.5 e^-/a, initiates the crossover from bcc to bct type, in accordance with Figure 1.30. At greater values, as further calculations show, this bct phase crosses over to the fcc type.

It may be noted that such cubic to tetragonal space group spatial distortions do not preclude the possibility of further alleviation of antibonding character through itinerant ferromagnetism. Just such a case has been studied – that of the ferromagnet, τ-MnAl [12,159]. The methods described in the previous section appear to give a good account for this complex system.

With these as preambles, we can now turn to an even more fascinating variant of symmetry-breaking: the antiferromagnetic state. Just two transition metal elements, Cr and Mn, in their elemental form, are known to be ground-state antiferromagnets. The former is found in the bcc structure, while the latter adopts the complex cubic α-Mn structure. We restrict our discussion here to the simpler bcc Cr system.

Figure 1.31 shows a band calculation for nonmagnetic bcc Cr [160]. Highlighted with gray ovals in this figure are three unavoided crossings, all lying near the nonmagnetic Cr Fermi energy. We contrast these results with antiferromagnetic bcc Cr, a compound in which the bcc atom positions are maintained at the cell corner and center, but are now taken to be of opposite directional spin. As

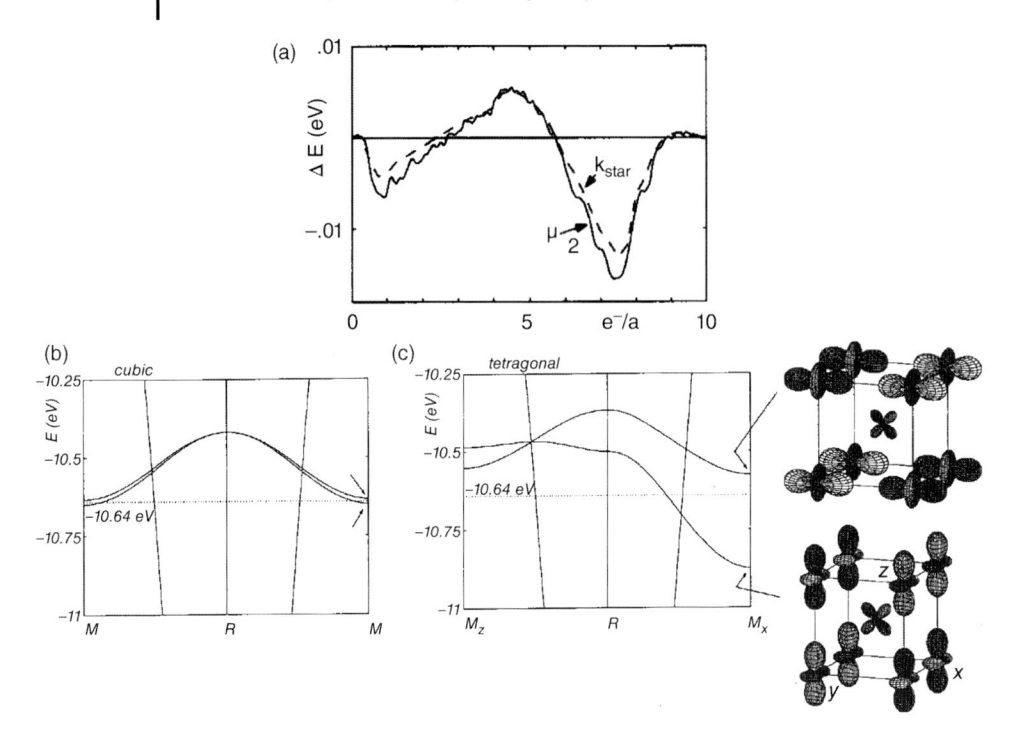

Figure 1.30 The O_h to D_{4h} (Jahn–Teller type) symmetry breaking that occurs maximally near 8 e^-/a in the bcc structure. The distortion, shown here with a lengthening of the c/a ratio to 1.02, stabilizes the bct structure type. (a) Overall energetics for this distortion. k_{star} and μ_2 refer respectively to the energy effect of electron count reorganization caused by the symmetry breaking and the full tight-binding band calculation. (b and c) Key crystal orbitals responsible for the 8 e^-/a symmetry breaking. In the lower energy crystal orbital, the z-direction antibond becomes weaker, while the more xy-directed bonding interactions become stronger. (Reproduced with permission from Ref. [112]. Copyright 2002, American Chemical Society.)

Figure 1.31 shows, this symmetry lowering turns the three previously unavoided crossings into avoided ones. The lower and upper energy branches become respectively filled and unfilled, stabilizing the system. COHP calculations, as shown in this figure, confirm these bonding results.

1.5
Synergism of the Jones and Tight-Binding Bonding Models

1.5.1
Cu_5Zn_8

With the advent of high-quality band calculations, the researcher's reliance on the tight-binding and nearly free-electron theory may appear reduced. The concepts based on these theories, however, are still of vital importance. In this

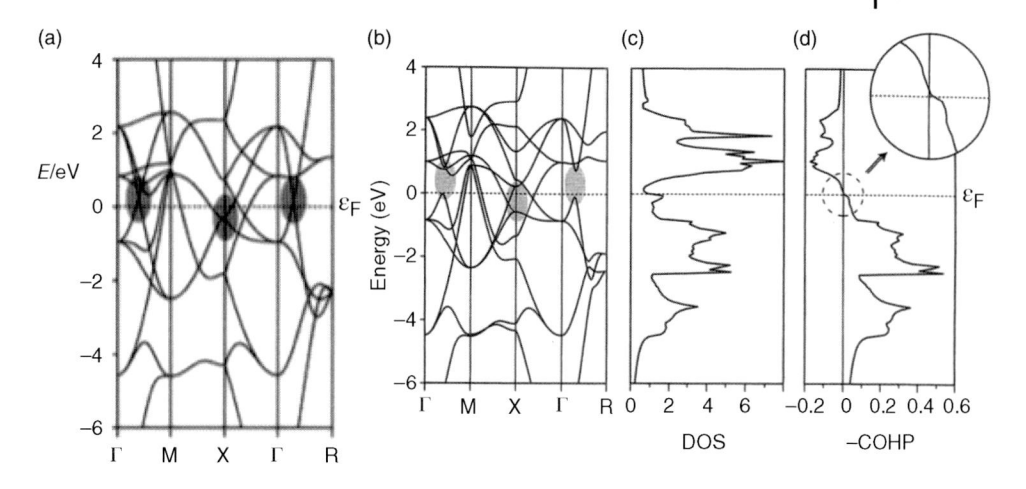

Figure 1.31 (a) Non-spin polarized bcc Cr band structure. Gray ovals indicate regions of symmetry-based unavoided crossings. (b–d) Band structure, DOS, and COHP, respectively, for antiferromagnetic bcc Cr. Gray ovals indicate symmetry-breaking avoided crossings. (Reproduced with permission from Ref. [160]. Copyright 2002, John Wiley & Sons.)

section, we explore ways in which high-quality band calculations allow us to strengthen, renew, and unite these earlier concepts.

We begin with the γ-brass structure for which, seemingly independent of one another, both the tight-binding and Jones model nearly free electron approach are so illuminating. To understand the connection between these models, we recast a DFT-parameter-fitted tight-binding calculation into the format of Jones theory [40]. The procedure has two steps: (i) We find the linear combination of tight-binding crystal orbitals that best combine to make orbitals maximally similar to pure plane wave states and (ii) we weight this linear combination by the calculated tight-binding energy eigenvalues to determine the tight-binding-based energy corresponding to the various crystal structure perturbed plane wave states.

The results are illustrated in Figure 1.32. First, in Figure 1.32a, we illustrate the strong fit between properly parametrized eH theory and LDA–DFT band calculations. Second, in Figure 1.32b, we consider plane wave approximants as determined by the linear combination of eH orbitals. The red curve in this panel corresponds to the plane wave energies in the $\{k_x\ k_x\ 0\}$ direction, derived from the eH Hamiltonian. As can be seen, and as predicted by the Jones model, there is a large energy splitting at $\{3\ 3\ 0\}$.

The advantage of this analysis is that it allows us to examine, from the tight-binding bonding AO perspective, what it is about the $\{3\ 3\ 0\}$ point that creates such a large splitting. The upper and lower energy states are illustrated in Figure 1.32c. As the figure shows, valence s and p orbitals assemble into plane wave-looking states. But while the lower energy state is entirely bonding, the high energy state has a mixture of bonding and antibonding interactions. The

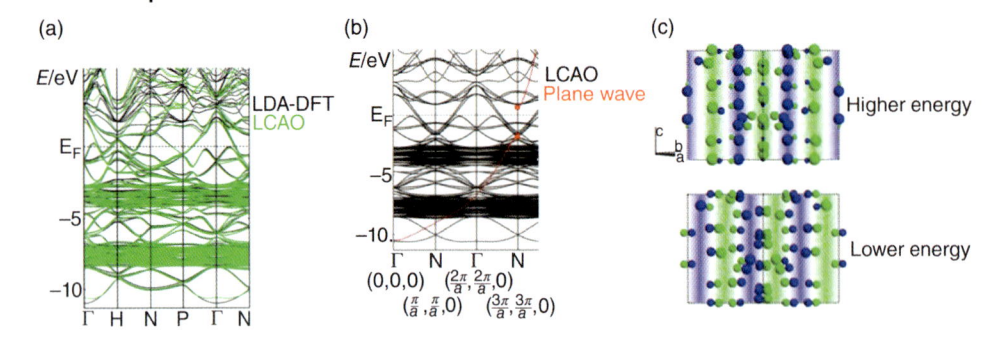

Figure 1.32 (a) Direct comparison of LDA-DFT and eH-fitted band structure of γ-brass. For all filled orbitals, the energy matches are strong. (b) eH-fitted (LCAO) band structure and plane wave energies derived from this LCAO Hamiltonian. The Jones splitting along {3 3 0}, $(3\pi/a, 3\pi/a, 0)$, expressed as LCAOs is shown in red. (c) The lower/upper energy Jones branches in LCAO format, illustrating their respective net bonding/nonbonding character – a synergism between the nearly free electron and tight-binding-derived concepts. (Reproduced with permission from Ref. [40]. Copyright 2011, American Chemical Society.)

tight-binding and Jones models give an overlapping, synergistic, and convergent picture of the noble metal chemical bond.

1.5.2
Li$_8$Ag$_5$

The synergism between the nearly free electron and tight-binding models can be further explored in the γ-brass structure, Li$_8$Ag$_5$, the lone example of this structure type with just 1 e^-/a [161–163]. As will prove important later, Li$_8$Ag$_5$ is never found in a perfectly ordered form: Instead, it exists over a wide range of Ag/Li compositions, with the Ag-rich composition near that of Li$_{33}$Ag$_{19}$.

We consider first the diffraction image of the perfectly ordered compound Li$_8$Ag$_5$, electron density being the perturbative element in the nearly free-electron model. Figure 1.33 shows two powder patterns: one for Li$_8$Ag$_5$ itself and the other where all atoms are kept at the previous positions but where all atoms are given equalized scattering factors. The latter diffraction image shows dominant reflections {3 3 0} and {4 1 1}, the reflections responsible for the standard 36-faced γ-brass Jones zone.

The former tells a quite different story. Outside the standard Jones model, there is a significant {2 1 1} peak [164]. It is of a reciprocal length much too short to be connected to the Jones model k-space mixing. As we demonstrate below through the application of tight-binding analysis, this peak expresses the disproportionation between the light Li and heavy Ag sites. Recast into the nearly free electron plane wave picture, the {2 1 1} peak allows for the transfer of electron density from the Li-containing regions to the silver-rich portions of the structure.

The crystallographic positions preferred by Ag and Li are illustrated in Figure 1.34. If Li$_8$Ag$_5$ wave functions are centered on Ag atoms, it is because the

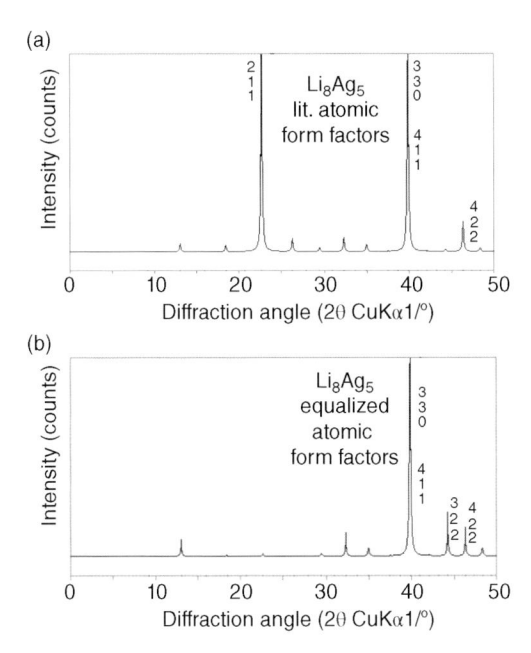

1.5 Synergism of the Jones and Tight-Binding Bonding Models

Figure 1.33 (a) Powder pattern for $Li_{33}Ag_{19}$ idealized by full single-element occupancies at each crystallographic site, as Li_8Ag_5. (b) Calculated powder pattern where all atoms have been given the same atomic form factor. For the sake of convenience, literature Ag atomic form factors were used at all sites in this latter calculation. The difference in the {2 1 1} peak height between the two patterns suggests that this reflection allows the electron transfer from the electropositive Li to the more electronegative Ag sites.

electropositive Li atom has transferred electron density to the more electronegative Ag sites [14,48,49]. Figure 1.35 shows the DOS and COOP for just this Ag-containing network. The COOP crossover from bonding to antibonding states is found at 260 e^-/unit cell.

A chief virtue of our analysis is its ability to interpret these results. With 20 Ag atoms/unit cell and 200 e^-/unit cell taken up by their 4d orbitals, there are 60 e^-/unit cell to occupy Ag valence s and p bonding orbitals. As Figure 1.34 shows, in a covalent bonding scheme, these Ag atoms can be thought of as either sp^2 or sp^3 hybridized, with exactly three bonds/Ag atom. These bonds account for the 60 e^-/unit cell needed to optimize the Ag–Ag interactions, as revealed by the COOP curve.

We compare these electronic structure results with the actual compositional phase boundaries of the Ag–Li γ-brass phase, a phase that is always deficient in Ag atoms. The above calculations on nondefective $Li_{32}Ag_{20}$ are never in agreement with any observed Li–Ag γ-brass phase. The nondefective structure has 30 bonding orbitals and requires 60 s and p valence electrons per unit cell. Nondefective $Li_{32}Ag_{20}$ has only 52 e^-/unit cell.

We can, however, change the number of Ag—Ag σ bonds by placing defects into the Ag network. As each Ag atom is removed, three σ bonds are lost.

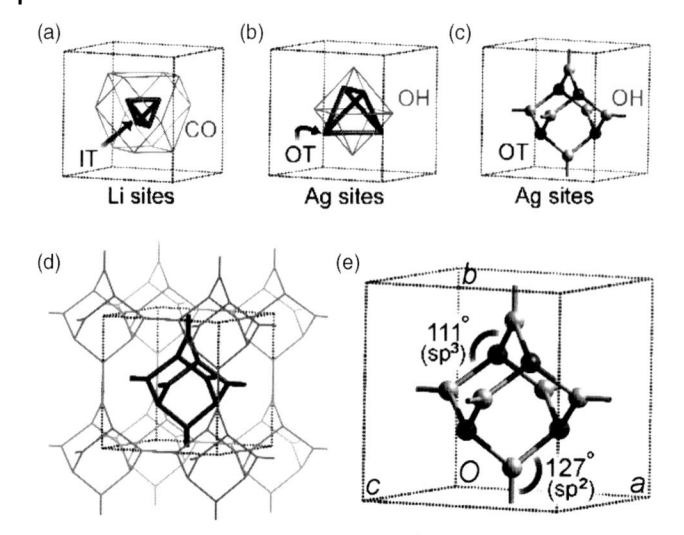

Figure 1.34 The γ-brass structure adopted by Li$_8$Ag$_5$. (a and b) The four γ-brass sites, traditionally named inner tetrahedra IT, cubooctahedra CO, outer tetrahedra OT, and octahedra OH, the former two occupied by lithium, the latter two predominantly by silver. Zintl analysis solely considers the Ag-site framework. (c–e) Ag atoms form two interconnected nets of adamantane-like clusters. OT and OH have respectively trigonal pyramidal (sp^3-like) and trigonal planar sp^2-like environments.

Li$_{33}$Ag$_{19}$, with just 1 out of 20 silver atoms removed, requires only 54 ($54 = 60 - 3 \times 2$) electrons, a value much closer to the designated 52 electrons.

In closing the two-electron gap between the 54 e^-/unit cell in Li$_{33}$Ag$_{19}$ and the actual 52 e^-/unit cell, we follow previous single-crystal and powder diffraction studies and consider the IT and CO sites fully occupied by Li atoms, but the OT and CO sites, while remaining fully occupied have mixed Li and Ag occupation.

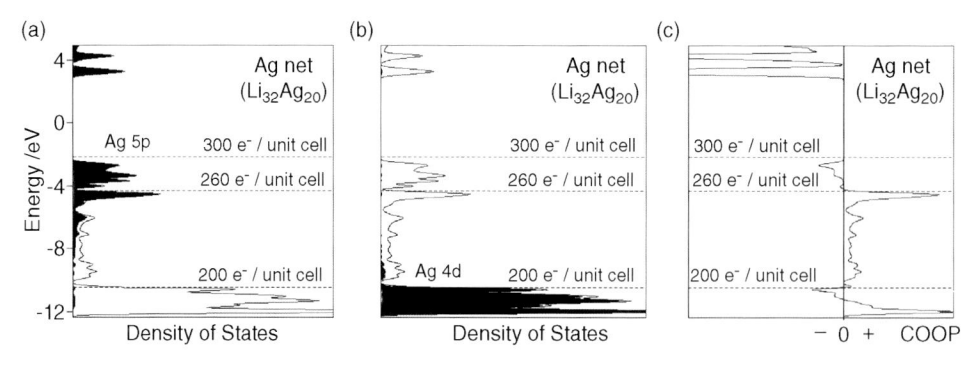

Figure 1.35 (a–c) DOS and COOP for the Ag framework in Li$_{32}$Ag$_{20}$ with valence 4d, 5s, and 5p bands. COOP bonding–antibonding crossover at 260 e^-/unit cell corresponds to filling of Ag 4d and bonding sp^2 and sp^3 bands. Between 260 and 300 e^-/unit cell are Ag lone pair states derived from sp^2 and sp^3 hybridization.

The composition would then be $Li_{32+x}Ag_{20-x}$, with exactly 52 e^-/unit cell and a bonding requirement of $60 - 6x$ e^-/unit cell. Solving for x, we find $x = 4/3$, giving a nominal composition of $Li_{33.33}Ag_{18.67}$ and a phase with 64.1% Li and 35.9% Ag. These values match the Rietveld refined Ag-rich phase reported by Noritake et al. with composition 64.3% Li and 35.7% Ag [163]. This excellent agreement could only be obtained by the synergistic application of the tight-binding and nearly free-electron models.

1.5.3
Quasicrystals and Quasicrystalline Approximants

Quasicrystals and quasicrystalline approximants are among the most complicated of all intermetallic structures. The understanding of their structures provides a stringent test as to the applicability of a synergistic use of the tight-binding and nearly free electron concepts. Here, we restrict our discussion to just one of these complex phases, the Bergman and Pauling 1/1 quasicrystalline approximant structure type, the R-phase [165,166]. We begin our discussion of these phases with the lone member in this family of structures that is solely comprised of group 10–12 elements, $Zn_{11}Au_{15}Cd_{23}$ [129]. In the absence of early alkali or transition elements, a Jones analysis is warranted.

The reported Jones zone for Bergman 1/1 quasicrystalline approximants is the 84-faced O_h polyhedron formed by the {5 4 3}, {7 1 0}, {5 3 4}, {1 7 0}, and {5 5 0} reflections (Figure 1.36) [13,125,167]. Surprisingly, this Jones zone has a point group O_h, even though 1/1 Bergman phases crystallize in $Im\bar{3}$ or T_h^5, have Laue class T_h, and have a crystal symmetry dominated by near I_h symmetry.

The diffraction data from $Zn_{11}Au_{15}Cd_{23}$ (Table 1.5) suggests a different Jones zone. The strongest reflections for this and, in fact, for most other Bergman phases are {5 0 3}, {6 0 0}, and {5 3 2}. As Figure 1.36 shows, these reflections lead to a T_h point group with a near I_h symmetry Jones zone.

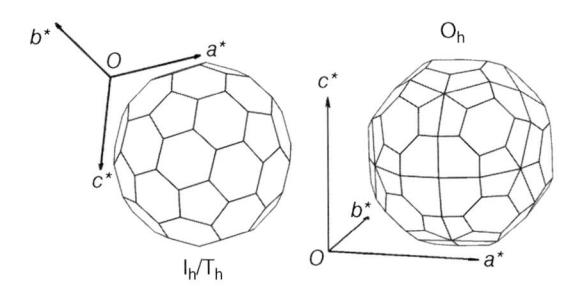

Figure 1.36 Literature and suggested Jones zones for $Li_{52.0}Al_{88.7}Cu_{19.3}$. Literature Jones zone and suggested Jones zones have respectively O_h and I_h/T_h point group symmetries. The literature Jones zone faces: {5 4 3}, {7 1 0}, {5 3 4}, {1 7 0}, and {5 5 0}. Suggested faces: {5 0 3}, {6 0 0}, and {5 3 2}. The suggested Jones zone conforms to the near I_h symmetry of this quasicrystalline approximant phase and uses only strong diffraction reflections.

Table 1.5 Comparison of the X-ray scattering factors of 1/1 Bergman-type quasicrystalline approximant phases' I_h/T_h and O_h Jones zones.

{h k l} Peak	Jones zone	Multiplicity reflections	I_{hkl} all Al $Li_{52.0}Al_{88.7}Cu_{19.3}$ [a]	I_{hkl} experimental $Li_{52.0}Al_{88.7}Cu_{19.3}$ [b]	I_{hkl} experimental $Zn_{11}Au_{15}Cd_{23}$ [c]
{5 0 3}	I_h/T_h	12	100.0%	22.9%	38.3%
{6 0 0}	I_h/T_h	6	93.0%	00.0%	100.0%
{5 3 2}	I_h/T_h	24	67.1%	60.5%	79.0%
{5 4 3}	O_h	24	8.7%	25.6%	0.1%
{7 1 0}	O_h	12	7.9%	22.4%	0.0%
{5 3 4}	O_h	24	6.0%	10.8%	6.7%
{1 7 0}	O_h	12	0.9%	4.6%	0.6%
{5 5 0}	O_h	12	0.3%	0.0%	1.7%

a) 100% hypothetical Al occupation at each atomic site. In the all-Al system, the {5 0 3}, {6 0 0}, and {5 3 2} reflections are the three strongest reflections. The reduction in the {5 4 3}, {7 1 0}, and {5 3 4} intensities relative to those in the middle column suggests that these reflections allow disproportionation from the more electropositive to the more electronegative atomic sites – and hence are not suitable in the makeup of the Jones zone.
b) Single-crystal atomic site occupations [169].
c) Single-crystal atomic site occupations [129].

The I_h/T_h Jones zone volume is 247.0 e^-/unit cell. The Jones zone circumscribes a sphere that contains 207.4 e^-/unit cell. In Jones theory, these two values form respectively an upper and a lower bound for the number of electrons/unit cell. $Zn_{11}Au_{15}Cd_{23}$ has 246.7 e^-/unit cell, within the expected boundaries [168]. The close agreement between the Jones-predicted upper bound, the volume of the Jones zone, and the actual number of valence electrons suggests a sizable pseudogap in the density of states. (In Jones theory, insulators, such as diamond, have electron concentrations which exactly fill the Jones zone.)

1.5.3.1 $Li_{52.0}Al_{88.7}Cu_{19.3}$

As Table 1.5 shows, strong intensity reflections do exist outside the Jones reflections in many 1/1 Bergmann quasicrystalline approximant phases. Diffraction and other data of $Li_{52.0}Al_{88.7}Cu_{19.3}$ further clarify this point (Figure 1.37) [169,170]. Just as was observed in $Li_{33}Ag_{19}$, there are a number of strong reflections outside the I_h/T_h Jones zone, but the intensity of these peaks appears largely due to the different diffracting intensities of the Li atoms. The suggestion is that, just like in this earlier case, such diffraction peaks allow for electron transfer between the light Li and the other heavier atoms.

DFT-parameter-fitted tight-binding calculations on the Al–Cu network of $Li_{52.0}Al_{88.7}Cu_{19.3}$ support this view, although some care has to be taken vis-à-vis the imperfect crystallographic ordering in the system (see Figure 1.38 caption for details). In the extreme that the Li atoms entirely pass their valence electrons into bands dominated by the Cu and Al atomic orbitals, we find a COOP crossover point of 337 e^-/unit cell. This value is in good agreement with the value

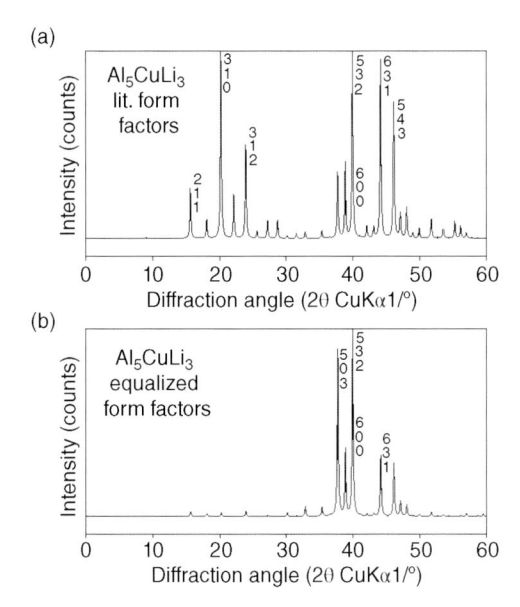

Figure 1.37 (a) Calculated $Li_{54.0}Al_{89.2}Cu_{19.0}$ powder pattern using literature site occupancies. (b) Calculated powder pattern where all atoms have been given the same atomic form factor. For the sake of convenience, literature Al atomic form factors were used at all sites in this latter calculation. The difference between the two powder patterns suggests that the lower angle reflections allow for electron transfer from the lighter and more electropositive Li to the heavier and more electronegative Cu and Al sites.

reported from single-crystal X-ray work of 337.4 e^-/unit cell [168,169]. It would appear relevant to study other alkali-metal based 1/1 Bergman phases, such as those listed in Table 1.6, by similar means [92,171–174].

Table 1.6 Known 1/1 quasicrystalline approximant (R-phase) compounds containing a mixture of Li and elements from groups 10–14 [91,92,166,169,175,176].

Compound	e^-/unit cell
$Li_{52.0}Si_{72.0}Ni_{36.0}$	340.0
$Li_{52.0}Si_{51.1}Cu_{56.9}$	313.3
$Li_{52.0}Ga_{84.0}Cu_{24.0}$	328.0
$Li_{52.0}Al_{88.7}Cu_{19.3}$	337.4
$Li_{52.0}Al_{78.2}Zn_{29.8}$	346.2
$Li_{52.0}Al_{62.2}Zn_{45.8}$	330.2
$Li_{52.0}Al_{56.6}Zn_{51.4}$	324.6
$Li_{52.0}Al_{51.4}Zn_{56.6}$	319.4

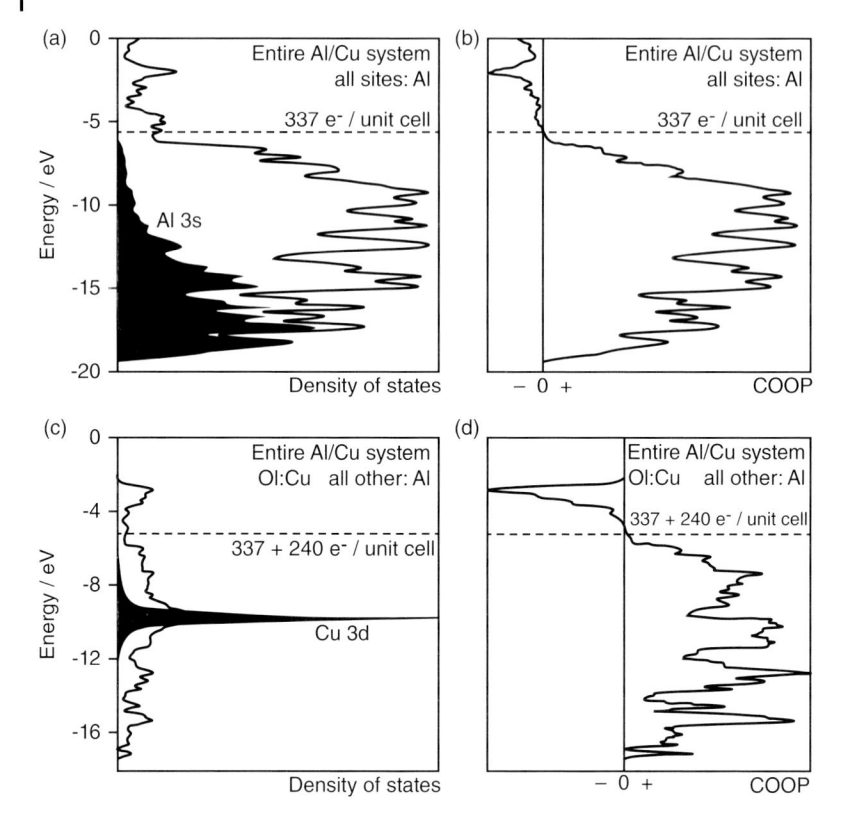

Figure 1.38 (a) DOS and (b) COOP for $Li_{52.0}Al_{88.7}Cu_{19.3}$ with Al placed at all Al-predominant sites. (c) DOS and (d) COOP show an alternate band calculation where the lone site with substantial Cu presence (the 24g OI or outer icosahedral site, with 46% Cu presence, which caps the vertices of the 24g II or inner icosahedral site of this structure) is treated as a fully Cu occupied site. Removing from the electron count the 240 e^- occupying Cu 3d orbitals, the bonding/antibonding crossover remains at 337 e^-/unit cell, the same value as found in the upper panel. These values are in agreement with the single crystal-derived electron count of 337.4 e^-/unit cell.

1.6
Emerging Directions in Tight-Binding Theory

1.6.1
Localized Bonding Models

1.6.1.1 Wannier Functions

When chemists approach intermetallic phases, it is natural for them to seek out localized schemes associated with specific features of a crystal structure, a pursuit that has yielded fruitful connections to molecular systems [61,177]. One could, for example, determine the atomic electron configurations of the atoms of transition metal intermetallic compounds (Figure 1.39) [178]. Or one could look for localized orbitals. It is this latter approach that is our primary focus here. The most famous of all localized crystal orbitals are the Wannier functions.

1.6 Emerging Directions in Tight-Binding Theory

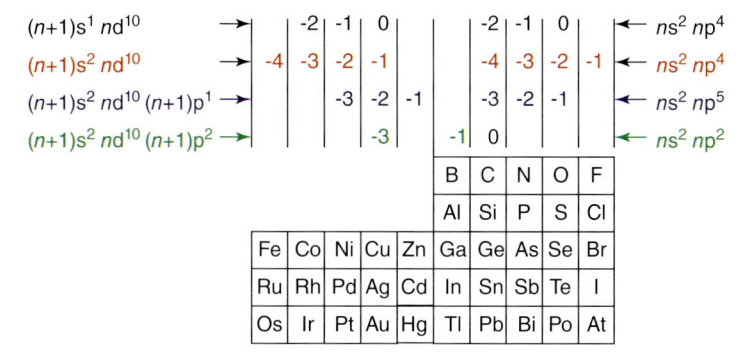

$(n+1)s^1\,nd^{10}$ → $\quad$ -2 | -1 | 0 $\quad\quad$ -2 | -1 | 0 $\quad$ ← $ns^2\,np^4$
$(n+1)s^2\,nd^{10}$ → -4 | -3 | -2 | -1 $\quad$ -4 | -3 | -2 | -1 ← $ns^2\,np^4$
$(n+1)s^2\,nd^{10}\,(n+1)p^1$ → $\quad\quad$ -3 | -2 | -1 $\quad\quad$ -3 | -2 | -1 ← $ns^2\,np^5$
$(n+1)s^2\,nd^{10}\,(n+1)p^2$ → $\quad\quad\quad\quad$ -3 $\quad\quad\quad$ -1 | 0 $\quad\quad$ ← $ns^2\,np^2$

					B	C	N	O	F
					Al	Si	P	S	Cl
Fe	Co	Ni	Cu	Zn	Ga	Ge	As	Se	Br
Ru	Rh	Pd	Ag	Cd	In	Sn	Sb	Te	I
Os	Ir	Pt	Au	Hg	Tl	Pb	Bi	Po	At

Figure 1.39 Oxidation states and valence electron configurations of the anions associated with late transition metal and main group elements. The colors black, red, blue, and green indicate the tendency to form chains, discrete anions, dumbbells, and diamond-like frameworks, respectively. (Reproduced with permission from Ref. [178]. Copyright 2011, John Wiley & Sons.)

The Wannier function may be understood by noting the simple quantum mechanical theorem that the expectation value of the energy for a system of occupied E eigenstates is unaffected when we take linear combinations of these states (while preserving the orthogonality of the functions) [9]. Alternatively, one can note that the occupied MOs and crystal orbitals must be antisymmetrized with respect to the exchange of electrons, as through Slater determinants, and a property of determinants is that one can linearly recombine the filled orbitals without changing the multielectron wave function [179].

For example, the energy and multielectron wave function that we calculate for a methane molecule will be the same whether we use the standard $a_1 + t_2$ set of bonding MOs for the eight electrons of its filled octet, or use them to construct localized functions corresponding to the individual C sp^3-H $1s$ bonds. In terms of the energy of the system, the four localized bonds are a perfectly valid way of interpreting the wave functions of the system.

In a similar way, linear combinations of the crystal orbitals can be taken to produce functions that are localized to specific portions of a unit cell, combinations that are referred to as Wannier functions. Perhaps the most common implementation of this idea is given by the *maximally localized Wannier function* [180,181], in which the coefficients of the crystal orbitals contributing to the Wannier function are chosen to optimize localization to a chosen point in space, say the center of a bond. Through this and other methods [61,182,183], recognizable bonding functions can be constructed from the crystal orbitals [184–186].

Yee and Hughbank illustrate this approach vividly in their extended Hückel analysis of MoS_2 and other transition metal sulfides [61]. Their investigation of MoS_2 using Wannier functions revealed the compound's Mo triangles were stabilized by three-center two-electron bonds (Figure 1.40). Analogous bonding motifs were also identified in $LiNbO_2$, $Hx_x(Ta/Nb)S_2$, and ZrS.

Another example is given by theoretical analysis of Hooper and Zurek of metal polyhydrides under pressure using DFT [187]. In their proposed high-pressure

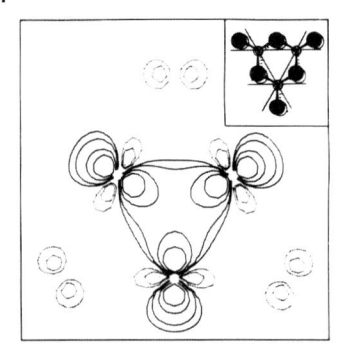

Figure 1.40 Wannier function revealing a three-center two-electron bond at Mo_3 triangles in MoS_2. (Reproduced with permission from Ref. [61]. Copyright 1991, American Chemical Society.)

RbH_5, they noted the presence of linear H_3 units. Through Wannier analysis of the H-based bands just below the E_F of this phase, they beautifully reproduced the bonding and nonbonding MOs of the three-center-four-electron bond of an H_3^- anion, demonstrating conclusively the nature of the bonding in these units (Figure 1.41).

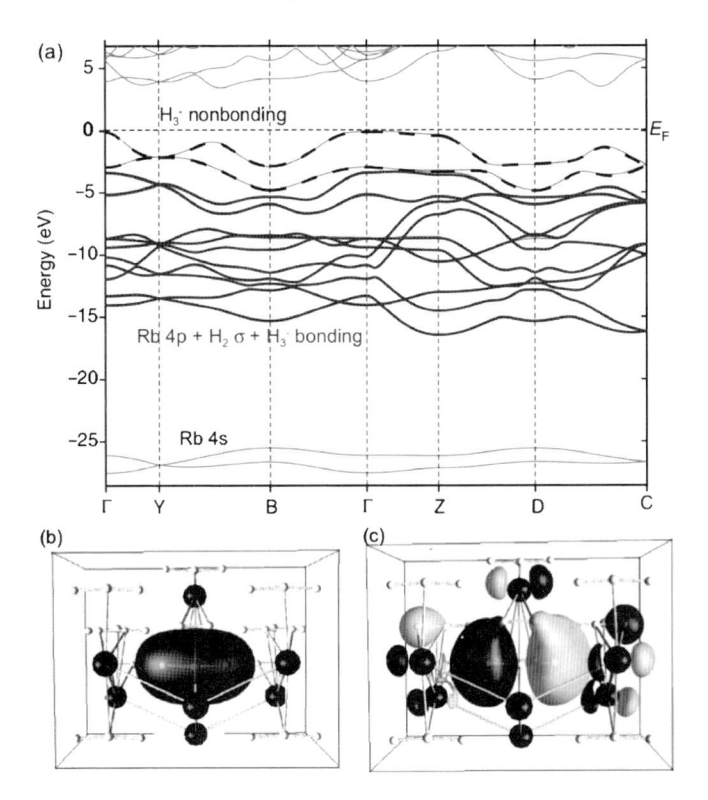

Figure 1.41 Interpretation of (a) the occupied bands of RbH_5 in terms of the (b) bonding and (c) nonbonding MOs of the three-center four-electron bond in an H_3^- anion. (Reproduced with permission from Ref. [187]. Copyright 2012, John Wiley & Sons.)

1.6.1.2 The 18-Electron Rule

The Wannier approach can be further developed to extract local MO diagrams for structural motifs in intermetallic phases, allowing us to directly draw analogies between molecular and intermetallic chemistry. Consider the 18-electron rule. This simple rule lies at the heart of transition metal organometallic chemistry, and yet when it comes to intermetallic combinations of transition metal and main group metalloid elements, the overall concept has been much less explored. The electronic DOS pseudogaps in a handful of phases – 18-electron half-Heusler phases [188,189], $NiSi_2$ (fluorite type), $Fe_8Al_{17.4}Si_{7.6}$ [190], and $Co_3Al_4Si_2$ [190] – have been connected to the filling of 18-electron configurations on transition metal sites, but sparingly little else. Where applied, however, such analyses have been of use: for example, in the half-Heusler phases (Table 1.7), the 18-electron species are insulators; it is away from this magic electron count that interesting electronic properties arise [188,189].

The problem is clear: Intermetallic compounds tend to have a range of experimentally observed electron counts, and their structures are complex. In an organometallic compound, the chemist can easily consider the transition metals and ligands as separate electron sources and define a specific electron counting role to each subunit separately. Equivalent procedures for intermetallic chemistry are more involved and harder to come by.

Using the Wannier approach discussed in the previous section, however, a systematic process, analogous to the schemes used to develop the 18-electron rule, may be derived for intermetallic compounds. This approach, *reversed approximation molecular orbital (raMO) analysis*, presupposes that a chemist can envision a local MO diagram applicable to the system in question [191]. From the choice of this local orbital set, a model Hamiltonian is formed whose eigenvectors are the linear combination of the occupied crystal orbitals that most resemble the initially chosen local orbital set. Accurate energies for these locally derived raMOs can then be obtained by calculating their energy expectation values with the original Hamiltonian matrix for the full compound.

As an example, consider the $CrGa_4$ structure – an ideal first system to consider as the principal bonds are all between the transition metal (T) and the main group metalloid "ligand" atoms (E), and hence the subsidiary complications found in many T–E intermetallic compounds just do not arise (Figure 1.42). The local orbital starting points are the nine transition metal s, p, and d atomic

Table 1.7 Representative semi-Heusler compounds ABX and their valence electron concentration per formula unit (VEC). (Reproduced with permission from Ref. [188]. Copyright 2000, Elsevier Science B.V.)

VEC	Compounds				
17	TiIrSn	ZrIrSn	HfIrSn	TiCoSn	TiFeSb
18	TiNiSn	ZrNiSn	NbCoSn	TiCoSb	VFeSb
19		TiNiSb		VCoSb	
22		MnNiSb		MnPtSb	

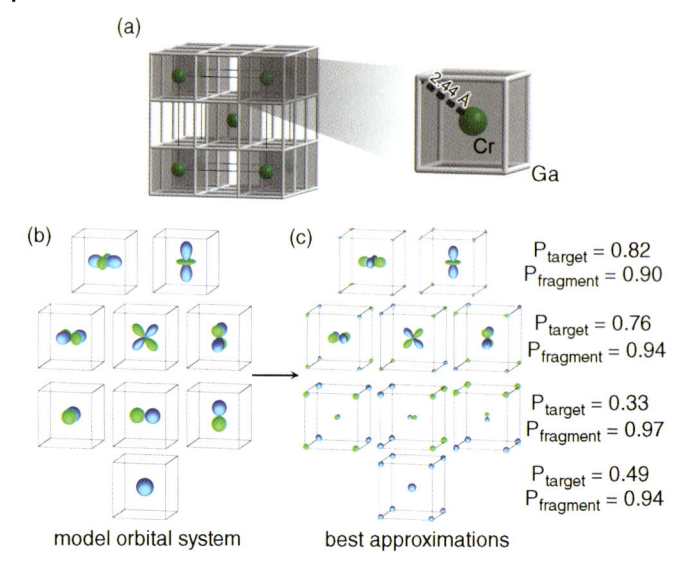

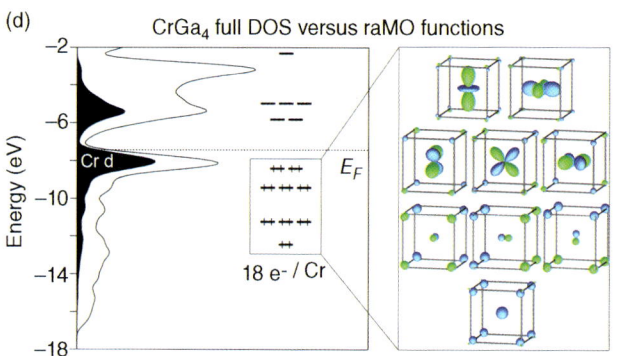

Figure 1.42 The role of 18-electron configurations in CrGa$_4$, revealed through the raMO method. (a) The crystal structure of CrGa$_4$, with the Cr@Ga$_8$ coordination environments emphasized. (b) The set of local orbitals used to interpret the electronic structure of the phase: a Cr atom's valence s, p, and d orbitals. (c) The raMO reconstructions of these local orbitals. (d) Alignment of the raMO energies with the full DOS for the compound. (Reproduced with permission from Ref. [191]. Copyright 2014, American Chemical Society.)

orbitals (AOs). The starting AOs themselves cannot be the final optimized orbitals: As the DOS curve of Figure 1.42d shows, even the more localized *d* AOs are splayed all across the energy range, and are significant contributors to many unfilled antibonding states.

But as Figure 1.42 shows, nine localized orthogonal raMOs very much like the nine transition metal AOs can be derived from the electronic structure of the compound. Based on our effective tight-binding Hamiltonian, the energies of these nine raMOs all lie below the Fermi energy. An 18-electron configuration based on the nine raMOs can then be assigned to Cr atom, in agreement with the number of valence electrons/Cr atom deduced from CrGa$_4$'s stoichiometry ($18 = 6 + (4 \times 3)$).

1.6.1.3 The Nowotny Chimney Ladder Phases

One of the most intriguing families of transition metal – main group metalloid binary compounds – are the Nowotny chimney ladder (NCL) phases [192]. In the NCLs, the T atoms trace out fourfold helices as they join together in a β-Sn-type sublattice. The E atoms form a separate set of helices that run along the inside of the T helices to give a net composition of TE$_{2-x}$ (Figure 1.43).

As the figure shows, while the β-Sn-type T sublattice is constant, the main group metalloid E helix is highly variable. One near constant within the structure, however, is the electron count. In Table 1.8, the electron counts for these phases are tabulated on a per T atom basis. While for the earlier transition

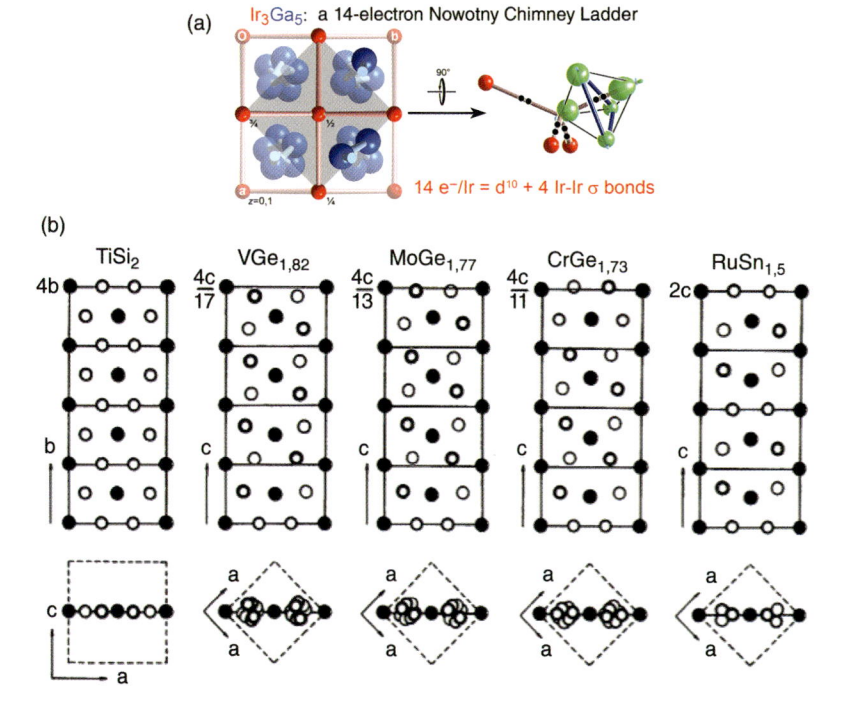

Figure 1.43 The Nowotny Chimney Ladders (NCLs), and the orbital origin of their 14-electron rule. (a) *Left*: View of the Ir$_3$Ga$_5$ structure, Ir and Ga, in respectively red and blue. Reproduced with permission from Ref. [195]. (b) Comparative side view of a number of NCL structure types. Filled and open circles, respectively, the T and E atoms. While the T atoms remain in a β-Sn-like arrangement, the E atoms, while always in a helix, adopt a wide variety of helicities. Reproduced with permission from Ref. [18]. (a) *Right*: The β-Sn-like network leads to each T atom having four T neighbors arranged in a flattened tetrahedral geometry. raMO analysis reveals the covalent sharing of electron pairs in multicenter functions along each of these T–T contacts. Filled 18-electron configurations are then obtainable on the T atoms with only $18-4 = 14$ electrons/T atom.

Table 1.8 Nowotny chimney ladder phases. (Reproduced with permission from Ref. [18]. Copyright 1972, John Wiley & Sons, NY.)

Compound	Pearson symbol	e^-/T atom
$TiSi_2$	$oF24$	12.00
$V_{17}Ge_{31}$	$tP192$	12.29
$Cr_{11}Ge_{19}$	$tP120$	12.91
Mo_9Ge_{16}	$tP100$	13.11
$Mo_{13}Ge_{23}$	$tP144$	13.10
$Mn_{11}Si_{19}$	$tP120$	13.91
$Mn_{15}Si_{26}$	$tP164$	13.93
Tc_4Si_7	$tP44$	14.00
Ru_2Sn_3	$tP20$	14.00
$Rh_{10}Ga_{17}$	$tP108$	14.10
$Rh_{17}Ge_{22}$	$tP156$	14.18
Ir_3Ga_5	$tP32$	14.00
Ir_4Ge_5	$tP36$	14.00

metals (groups 4–6), the number of electrons per T atom increases as we move from left to right, the situation changes for later T atoms. Once the transition metal belongs to column 7 or a higher column, this electron count stays fixed at 14 e^-/T atom [193,194].

A raMO analysis of these compounds provides a simple explanation for this electronic preference for 14 electrons/T atom and simultaneously accounts for the respectively rigid and flexible forms of the T and E sublattices across these compounds [195]. When using the full spd set of valence orbitals of a single T atom as the starting point of a raMO analysis, electron pairs are found to be associated with each local orbital in a manner similar to that that we showed earlier for $CrGa_4$. These results indicate that the T atoms do achieve closed-shell 18-electron configurations. However, some of these functions show not only T—E bonding, but T—T bonding as well. In particular, the T s and p raMOs each shows strong bonding contributions from their four T neighbors, arranged at the corners of a flattened β-Sn type tetrahedron (Figure 1.43).

These raMOs can be combined in an sp^3 fashion to create bonding functions directed along each of the T—T contacts. Each of these functions contains an electron pair that is covalently shared along the T—T contact, and that counts toward the 18-electron configurations for both T atoms involved. As each T atom has four such contacts, it can achieve an 18-electron configuration with only 14 electrons of its own, in the same way that C in CH_4 has an octet despite bringing only four valence electrons to the molecule. As all of the NCLs share this diamond-like T sublattice, they should all have closed shells at 14 electrons/T atom.

A look at one of these T—T bonding functions, however, indicates that these are not two-center two-electron bonds in the traditional sense (Figure 1.43a, right). Bridging lobes are also present on E atoms that encircle the T—T contact, so that the function is not only T—T bonding but also T—E bonding. Properly speaking, then, this represents a multicenter bonding function that shares the same nodal properties as a classical T—T σ bond. These interactions are termed *isolobal T—T bonds*.

1.6.1.4 Other Transition Metal – Main Group Metalloid Phases

The raMO analysis can be applied to a wide range of T—E compounds. The crux in its application is the selection of the initial local MO diagram. With the complexity of intermetallic compounds, an iterative procedure may be called for: A first set of chosen local orbitals can lead to choices for a second iteration. Also, the wave functions for the electrons unaccounted for in the mapping to the first set of local orbitals can be used as a basis set for a second raMO analysis. In this way, the complicated electronic structure of an intermetallic is broken down into its component bonding subsystems. A graphical example of this procedure is shown in Figure 1.44 [196].

A simple electron counting rule has emerged from this study. Each T atom requires $18-n+m$ electrons to achieve a closed-shell electron count, where n is the number of isolobal T—T bonds that it participates in, and m is the number of electrons on E-based orbitals per T atom that are orthogonal to the T sublattice orbitals. The values of n and m are determined through the raMO analysis itself, as is tabulated for more than 30 structure types in Table 1.9.

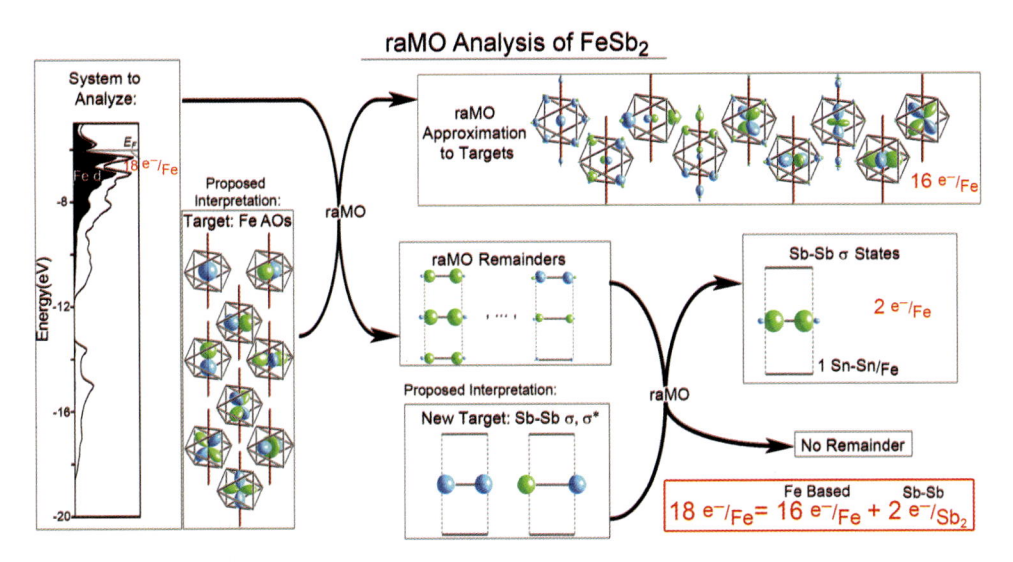

Figure 1.44 Schematic illustration of the raMO approach, using the marcasite-type phase FeSb$_2$ as an example. The occupied crystal orbitals of FeSb$_2$ are first used as a basis set for reproducing the local Fe spd orbitals. The raMO functions that result can be divided into Fe-based states (with each Fe being involved in two Fe—Fe bonds) and remainder states, orthogonal to the Fe-centered raMOs. A mapping of the remainder states onto Sb–Sb dimer functions allocates the remaining electrons to Sb–Sb σ bonds. (Reproduced with permission from Ref. [196]. Copyright 2015, American Chemical Society.)

Table 1.9 Binary intermetallic phases following the 18-n rule. (Reproduced with permission from Ref. [196]. Copyright 2015, American Chemical Society.)

Structure type	Phase	n	m	18-n+m	VEC	ΔVEC histogram for all phases in the ICSD					
						< −3	−2	−1	0	1	2	3
$PtHg_4$	$CrGa_4$	0	0	18	18	0	0	0	1	1	0	0
CaF_2	$NiSi_2$	0	0	18	18	0	2	5	2	0	0	0
Fe_3C	$IrIn_3$	0	0	18	18	0	0	0	1	1	0	0
$CoGa_3$	$FeGa_3$	1	0	17	17	0	0	1	4	5	0	0
Co_2Al_5	Co_2Al_5	1.5	0	16.5	16.5	1	1	0	2	0	0	0
Ru_2B_3	Pt_2Sn_3	2	0	16	16	2	0	1	1	0	0	0
β-$FeSi_2$	$FeSi_2$	2	0	16	16	0	0	0	2	0	0	0
NiAs	NiAs	3	0	15	15	16	4	6	15	7	0	0
Ni_2Al_3	Ni_2Al_3	3	0	15	14.5	1	0	11	0	0	0	0
Rh_4Si_5	Rh_4Si_5	3.5	0	14.5	14	0	0	2	0	0	0	0
$CrSi_2$	$CrSi_2$	4	0	14	14	0	2	5	1	0	0	0
Ir_3Ga_5	Ir_3Ga_5	4	0	14	14	0	0	0	1	0	0	0
Mo_3Al_8	Mo_3Al_8	4	0	14	14	0	0	0	1	0	0	0
MnP	PdGe	4	0	14	14	3	0	7	10	0	0	0
$TiSi_2$	$RuGa_2$	4	0	14	14	0	3	1	2	0	0	0
Ru_2Sn_3	Ru_2Sn_3	4	0	14	14	0	0	0	1	0	0	0
$TiAl_3$	$NbAl_3$	4	0	14	14	0	0	3	6	0	0	0
$MoSi_2$	$MoSi_2$	4	0	14	14	0	0	0	7	1	0	0
CoSn	CoSn	4	0	14	13	0	1	4	0	0	0	0
$ZrAl_3$	$ZrAl_3$	5	0	13	13	0	0	0	3	0	0	0
CsCl	CoAl	6	0	12	12	3	1	7	14	7	9	0
FeSi	FeSi	6	0	12	12	2	2	3	6	6	0	0
$AuCu_3$	$ScAl_3$	6	0	12	12	0	0	0	5	0	0	1
$ZrSi_2$	$ZrSi_2$	6	0	12	12	0	0	3	4	0	0	0
Mn_2Hg_5	Mn_2Hg_5	6	0	12	12	0	0	2	1	1	0	2
$FeS_2(oP6)$	$NiSb_2$	0	2	20	20	1	7	1	9	4	0	0
	$FeSb_2$	2	2	18	18	0	1	0	7	1	9	4
$CoAs_3$	$CoAs_3$	0	6	24	24	0	0	0	7	0	0	0
$PdSn_2$	$PdSn_2$	1	1	18	18	0	0	1	1	0	0	0
	$CoGe_2$	2	1	17	17	0	0	0	1	1	0	0
$CoSb_2$	$CoSb_2$	1	2	19	19	0	0	0	23	0	0	0
Ir_3Ge_7	Ir_3Sn_7	1	1.33	18.33	18.33	0	0	18	4	0	0	0
	Os_3Sn_7	2	1.33	17.33	17.33	0	0	0	18	4	0	0
$CuAl_2$	$CuAl_2$	2	2	17	18	7	3	6	2	1	0	0
Re_3Ge_7	Re_3Ge_7	2.67	1.33	16.67	16.33	0	0	0	1	0	0	0
CrP_2	$OsGe_2$	3	1	16	16	0	0	4	2	0	0	0
Mg_2Ni	$MoSn_2$	6	2	14	14	0	0	0	2	1	0	0

As can be seen from this table, for electron-precise compounds such as the Ni$_2$Al$_3$ structure type, the predicted electron filling is in complete agreement with the observed electron concentration. But in systems where stability is controlled by a number of factors, the calculated electron concentration lies to the upper end of the experimental distribution.

1.6.2
Elucidating Atomic Size Effects

1.6.2.1 Steric Strain

While this chapter has so far underplayed the importance of atomic steric factors, such steric factors are crucial in the understanding of intermetallic structures. For alloy phases, the importance of atomic size effects was recognized early in the steric-based Hume-Rothery 15% rule [16,197]. As crystal structures of intermetallic compounds emerged from the first X-ray diffraction studies, the role of sterics was clearly demonstrated by the Laves phases [198].

A classic geometrical analysis of the Laves phase structures assuming that the atoms behave as hard spheres quickly runs into issues. Consider the cubic MgCu$_2$ Laves phase. Assuming that the spheres on the Cu 16d sites (denoted as N in Figure 1.45) pack such that they touch each other, the remaining spaces for the spheres on the Mg 8a sites (M in Figure 1.45) have a definite size. If we denote the diameters of the N and M atoms as D_N and D_M, respectively, the M atoms make no contacts to their neighbors for $D_M/D_N < \sqrt{3/2}$. At $D_M/D_N = \sqrt{3/2}$, M spheres make contact with each other, but M–N contacts

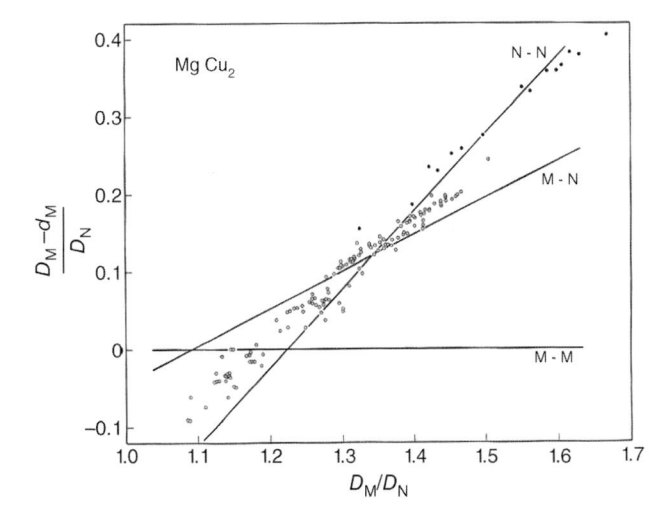

Figure 1.45 Analysis of steric factors in the MgCu$_2$ structure type based on atomic diameters (D_M and D_N for the Mg- and Cu-type sites, respectively) and bond distances (d_M and d_N). Filled circles represent binary compounds whose component elements differ in their Pauling electronegativity by a value greater than or equal to 1. (Reproduced with permission from Ref. [18]. Copyright 1972, John Wiley & Sons, NY.)

would not occur until $D_M/D_N = 1.345$, at which point the M site spheres would be overlapping.

Pearson developed an insightful way of analyzing such situations, as is shown in Figure 1.45 [18]. Here, the tabulated atomic diameters for the elements are compared with the diameters deduced from the M–M and N–N interatomic distances of the individual Laves crystal structures, d_M and d_N. The difference between D and d values provides a measure of the strain induced by the packing of atoms in the structures. Where $D_M - d_M$ or $D_N - d_N$ is positive, the bond is stretched; where negative, compressed.

In this figure, the unitless ratio $(D_M - d_M)/D_N$ is plotted against the ratio D_M/D_N. Each instance of a MgCu$_2$ Laves phase results in a single data point on the graph. The key to the figure is the interplay between these values and the three straight lines, labeled $N - N$, $M - N$, and $M - M$. Where the specific crystal structure data point lies above or below the specific bond-type line, this bond type is respectively compressed or stretched out.

As a look at the points in Figure 1.45 demonstrates, the actual MgCu$_2$ structures exhibit a compromise between the N—N and M—N bond types, with a mild compression of one type leading to an elongation of the other. The one exception, understandably, is where the difference of N and M electronegativity is large (filled circles); in such cases, the N–N contact has the dominant role, as can be expected from a Zintl analysis.

1.6.2.2 μ_2-Hückel Chemical Pressure Analysis

It is difficult to extend analyses like the one given above to more complex intermetallic systems. A typical intermetallic compound can have many more than two atomic sites, a host of atomic site parameters, more than two elements, and many bond types. In much of intermetallic chemistry, even the concept of an ideal metal–metal bond length is ill-defined. As Pearson has noted, bcc and fcc may have respectively 68 and 74% packing efficiencies, but for those elements that have phase transitions between the two structure types, densities generally shift by less than 1% [18].

Quantum mechanics does not offer simple terms to add to the energy for the costs of one atom encroaching on the territory of its neighbors. The electronic energy is given together as a lump sum, and it is difficult to determine if a given bond is either stretched out or compressed relative to an energetic ideal.

It is here that the μ_2-Hückel model discussed previously in this paper can play a role. The simplicity of the μ_2-Hückel model makes it possible to resolve this equilibrium into its competing components. The decomposition of the total energy into a sum of on-site and pairwise interactions means that the macroscopic pressure $(-\partial E_{\mu_2-\text{Hückel}}/\partial V)$ can be written as an average over the microscopic pressures acting along each interatomic contact [111].

As these local pressures arise entirely from strains within a crystal structure, and not from an external force, we equate them with the notion of chemical pressure (CP) often used in thinking about how atomic size effects influence the properties of a material. This approach, and its adaptation to DFT calculations [199–201], has revealed how the frustration between packing constraints and electronic interactions can drive the formation of a range of complex intermetallic structures.

A model system for this analysis is the Ca–Ag system, where a variant of the $CaCu_5$ structure, Ca_2Ag_7, is found but the much simpler $CaCu_5$ type remains unknown [202,203]. In Figure 1.46a, we begin with a simple plot of the large coordination polyhedron of the Ca atoms in this structure, at the optimal cell volume for $CaAg_5$ were it to adopt the $CaCu_5$ structure.

We superimpose spheres on the atomic positions whose radii are proportional to the magnitude of the pressure experienced by those atoms. The color of each sphere gives the sign of the pressure. White spheres, as found for the Ag atoms, correspond to positive pressures, which would be soothed by the expansion of the structure. Black spheres, on the other hand, as found in the Ca sites, represent negative pressures, which pull their surroundings toward them (much like a black hole).

A clearer picture of the interatomic interactions underlying this conflict can be obtained by turning from the net pressure on each atom to the distributions of CPs around them. To visualize these distributions, the CPs around each atom are projected onto low-order spherical harmonics, and then represented as radial plots. In these plots, a CP surface is drawn on each atom, whose distance from the nucleus along any direction is proportional to the pressure along that direction. For $CaAg_5$, the large sphere at the Ca site becomes slightly elongated vertically, indicating its interactions with the Ag atoms above and below are particularly strained. The CP features on the Ag sublattice undergo a much more pronounced change: Instead of small white spheres, the surfaces on the Ag atoms now have dumbbells of large white lobes pointing along the Ag–Ag contacts in blue. The Ca atoms are too small for its coordination environment, and pull the surrounding Ag atoms toward it. The Ag–Ag contacts, however, are already too short. From this scheme, it is not surprising that no Ca analog to $CaCu_5$ exists in the Ca–Ag phase diagram.

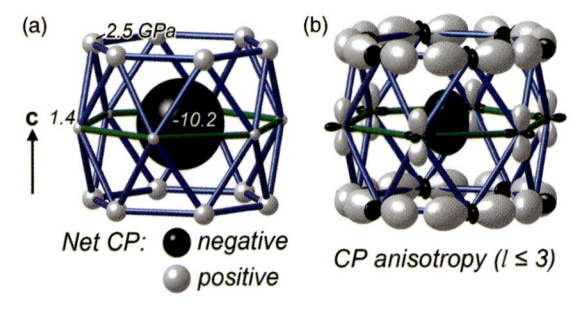

Figure 1.46 μ_2-Chemical pressures (CPs) plotted for the Ca@Ag$_{18}$ polyhedra in the expected but unobserved $CaAg_5$ phase with the $CaCu_5$ structure type. (a) Net atomic chemical pressures plotted as spheres with radii proportional to pressure. (b) Plots of the directional dependence of the CPs around each atom. l refers to the range of spherical harmonics used in this representation. (Reproduced with permission from Ref. [111]. Copyright 2011, American Chemical Society.)

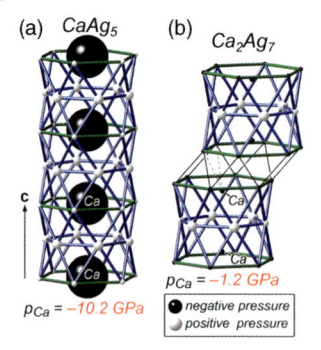

Figure 1.47 Comparison of the net atomic CPs calculated for (a) a hypothetical CaCu$_5$-type CaAg$_5$ phase and (b) its experimentally observed superstructure variant Ca$_2$Ag$_7$. (Reproduced with permission from Ref. [111]. Copyright 2011, American Chemical Society.)

We now consider the more complex Ca$_2$Ag$_7$, in which slabs of the CaCu$_5$ type are separated by the deletion of layers of atoms. First is shown the net atomic CPs for a larger, columnar section of a hypothetical CaCu$_5$-type CaAg$_5$, where the fusion of the Ca-centered polyhedra through shared hexagonal faces along c is visible (Figure 1.47). The Ca atoms appear again with large black spheres indicative of their being too small for their environments. On going to the observed structure of Ca$_2$Ag$_7$, layers of Ag atoms are removed. The net CPs on the Ca atoms have become substantially reduced, indicating that the size of the Ca atoms is a good fit for this particular electronic context.

The essential feature of the μ_2-Hückel method that makes this approach possible is the ease with which the total energy and pressure can be spatially decomposed into contributions from individual bonds. Since the first applications of the μ_2-Hückel CP approach, it has become clear that a similar decomposition can be made of the results of DFT calculations (now mapping contributions to a grid rather than bond centers). The DFT adaptation of the CP approach allows for similar insights to be obtained without the need of parameterizing a semi-empirical model. This development provides one example of how Hückel-type models can serve as a testing ground for new theoretical concepts, which can then be implemented with higher levels of theory [199–201].

1.6.3
Electronegativity Differences

1.6.3.1 Brewer's Extension of the Lewis Concept to Transition Metals

In addition to the electron-counting and atomic size effects discussed in the previous two sections, electronegativity differences often play a decisive role in the crystal structures and stability of intermetallic phases. Compare, for example, the binary phase diagram of Li–Hg with that of In–Hg.

Li and In have somewhat similar metallic radii (respectively 157 and 167 pm), can both be thought of as univalent, and have (for metals) low melting points (respectively 179° and 157 °C). But as Figure 1.48 shows, the Li–Hg and In–Hg

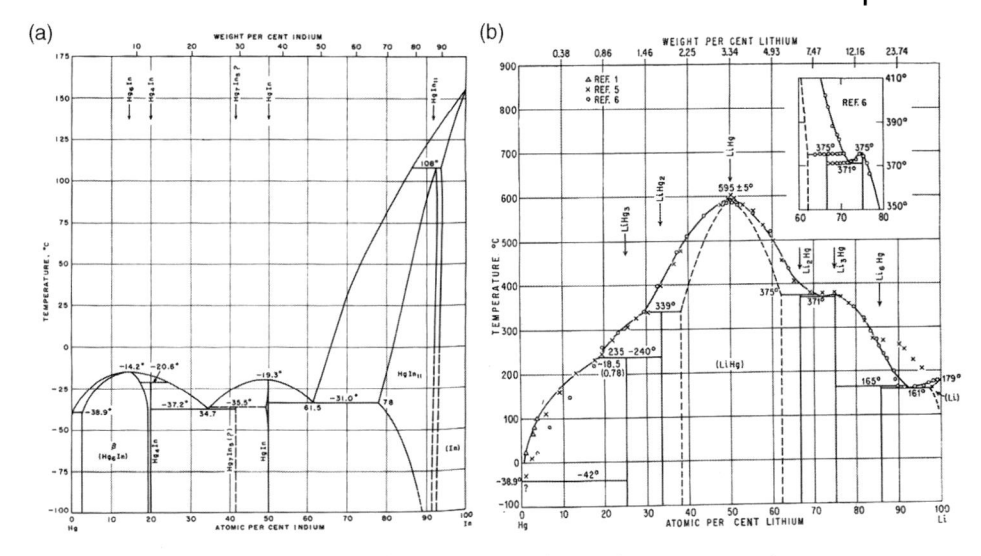

Figure 1.48 The binary phase diagrams of the Hg–In (a) and Hg–Li (b) systems. Note that the melting point of the 1:1 LiHg phase is much higher than that of its InHg counterpart [106–108].

phase diagrams show little in common. Both phase diagrams reveal a stable 1:1 compound, but while InHg's melting point is −19°, that of LiHg is 595 °C.

In hindsight, the difference between the Hg–In and Hg–Li phase diagrams is one of electronegativities [197]. Li is an electropositive metal (Pauling $\chi = 1.0$), while Hg and In are much more electronegative metals (χ of respectively 1.9 and 1.7). LiHg has a much higher melting point as there is electron transfer between the two elements.

Such electronegativity-based effects take place not just with the noble and main group metals but with transition metal (T) elements as well. Pairs of elements taken from opposite sides of the d-block (with thus large electronegativity differences) tend to form highly stable intermetallic phases. For example, the Sc–Cu system contains three distinct intermetallic compounds, each of which appear to be line phases that are stable from low temperatures up until the liquidus is reached above temperatures of about 900 °C [204].

For combinations of elements that are closer to each other in the d-block, the driving force for the formation of intermetallic phases versus random alloys is much less strong. The phase diagrams of combinations such as Cr–Fe, Cu–Au, Mn–Co are characterized by large homogeneity ranges for the elemental phases just below the liquidus, with any intermetallics appearing as low-temperature phases. Even such famous phases as $AuCu_3$ and the σ-phase CrFe, which serve as the prototypes of their respective structure types, become unstable relative to disordered alloys at high temperatures.

Brewer and Wengert offered an insightful chemical interpretation for the enhanced thermodynamic favorability of compounds between late and early transition metals [205]. They pointed out that for early T elements, many of the atoms' d orbitals will be empty, creating an opportunity for them to act as electron pair acceptors. Atoms of the late T elements, however, will have most of their d orbitals filled, which could then serve as electron pair donors. The formation of intermetallic phases between these two types of elements can then be envisioned as a Lewis acid–base reaction. In such a reaction, the empty high-lying d orbitals of the early transition metal interact with the filled low-lying d orbitals of the late transition metal to form a polarized bonding orbital.

1.6.3.2 The μ_3 Acidity Model and the Lewis Theory

Brewer's elegant concept of acid–base reactions between metallic elements can be further developed in the context of tight-binding theory. One avenue is given by the method of moments, and in particular by the third moment, μ_3. The profound influence of μ_3 on the shape of the DOS distribution is illustrated in Figure 1.49 with three DOS curves based on different values for μ_3: one negative, one zero, and one positive (with the lower order moments standardized to $\mu_0 = 1$, $\mu_1 = 0$, and $\mu_2 = 1$).

As this figure shows, μ_3 indicates the skewness of the DOS: negative μ_3 values have a tail strongly dipping toward the bonding region, while the opposite is seen for positive μ_3 values. For the half-filled band, where μ_3 is negative, antibonding orbitals are filled, but where it is positive, bonding orbitals remain unfilled.

Stacey and Fredrickson quantified this relationship by exploring the question of which μ_3 and μ_4 values minimize the total energy of the system at any given electron count. They found that the match between the μ_3 value and the fractional band-filling (BF) to which it is best suited (BF_{ideal}) is given by a simple equation: $BF_{ideal} = \frac{1}{2} + \frac{\mu_3}{2\sqrt{\mu_3^2+4}}$. For standardized DOS curves defined by μ_3 and μ_4, this equation determines the location of the global minimum of the $E(\mu_3, \mu_4)$

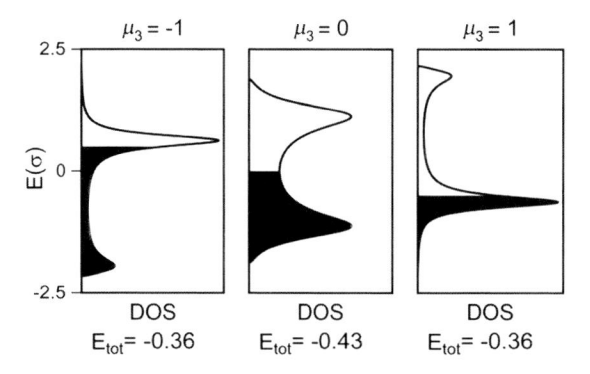

Figure 1.49 The role of μ_3 in determining the shape of a DOS curve. DOS curves are shown with $\mu_3 < 0$, =0, and >0. The shaded portion of the DOS give the occupied states for a half-filled system. The generation of the DOS curves from the moments has been truncated to μ_4 and smaller terms, to help clarify the role of these earlier moments. (Reproduced with permission from Ref. [206]. Copyright 2014, American Chemical Society.)

function for each electron count (for μ_4, the minimum invariably occurs for the value where the kurtosis, which equals $\mu_4 - \mu_3^2 - 1$, is zero).

Importantly, this μ_3 construct can be usefully applied to transition metal intermetallic phase formation. For example, in Figure 1.50a, the μ_3 and μ_4-derived DOS curves for d-only models of Sc and Ir are compared [101,206]. The shapes of the curves are almost identical (as are the d-blocks for the other elemental transition metals, regardless of structure type), with their μ_3 values being best-adapted for an electron count of about 4.4 electrons/atom ($BF_{ideal} = 0.44$). The actual electron counts of Sc and Ir deviate from this ideal substantially: Sc, being on the early side of the d-block, has only 1.49 electrons in its d orbitals (according to the DFT-calibrated Hückel model), making it short of the ideal by almost three electrons/atom; Ir, on the other hand, is a late transition metal with 7.3 electrons in its d-block, about three electrons/atom in excess of the μ_3 ideal.

This identification of Ir and Sc as electron-rich and electron-poor, respectively, has parallels with the Lewis acid–base picture in line with Brewer's model. Just as Lewis bases react to donate electron pairs, elemental Ir would benefit from losing electrons to further approach its μ_3-ideal electron count. Sc, however, would prefer to gain electrons, much as a Lewis acid accepts electron pairs. In this way, Ir and other elements with electron counts higher than the μ_3 ideal can be classified as μ_3-bases, while Sc and others whose electron counts fall short of the μ_3 ideal can be termed μ_3-acids. The large range of valence electron counts among the transition metals, combined with a μ_3 value that is relatively inflexible for elemental transition metal structures, gives rise to a virtual continuum of acidities across the d-block (Figure 1.51).

The notion of μ_3-acidic and μ_3-basic elements suggests the possibility of neutralization reactions between them. Such a process is demonstrated in Figure 1.50, where upon going from the elements to the intermetallic phase ScIr, the DOS distributions for the Sc and Ir atoms change dramatically. The Sc DOS

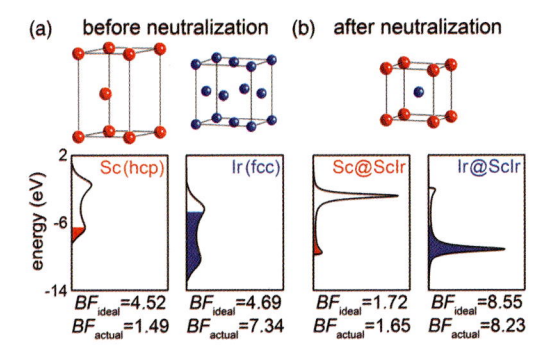

Figure 1.50 The formation of ScIr from its component elements as a μ_3-acid/base reaction. (a) The DOS curves of d-only models of elemental Sc and Ir, with occupied states shaded. (b) The Sc- and Ir-projected DOS of the corresponding model for ScIr. Note the much closer match between the actual and predicted band-fillings (BFs) in the intermetallic phase. (Reproduced with permission from Ref. [206]. Copyright 2014, American Chemical Society.)

Sc	Ti	V	Cr	Mn	Fe	Co	Ni	Cu
3.04	2.11	0.52	-0.09	-1.22	-2.27	-3.12	-3.89	-4.80
Y	Zr	Nb	Mo	Tc	Ru	Rh	Pd	Ag
2.52	2.02	0.92	-0.17	-0.94	-1.66	-3.11	-4.06	-4.86
Lu	Hf	Ta	W	Re	Os	Ir	Pt	Au
2.69	2.40	1.29	0.15	-0.44	-1.68	-2.64	-3.71	-4.60

Figure 1.51 μ_3-acidity elements of the d-block. Acidities are defined as the μ_3-ideal electron count minus the actual electron count for the d orbitals in the elemental phase of each element. Positive numbers correspond to μ_3-acidic character, while negative numbers indicate μ_3-basicity. (Reproduced with permission from Ref. [206]. Copyright 2014, American Chemical Society.)

curve becomes skewed so that only a small number of states are in the low-energy region, as is appropriate to its low electron count. The Ir DOS distribution skews in accord with its higher electron count. These changes in the shape of the DOS (much more than electron transfers between the elements) give a much tighter consistency between the ideal and actual electron counts for both elements. We term this effect μ_3-*neutralization*. One result is that the E_F now lies in a DOS pseudogap.

As a demonstration of the importance of this interaction, Stacey and Fredrickson calculated a stability map, based on the CsCl structure type, for the degree of incomplete neutralization as a function of the acid and base strengths of the component elements for first-row transition metals [207,208]. The resulting map (Figure 1.52) shows a deep valley of high neutralization along the diagonal,

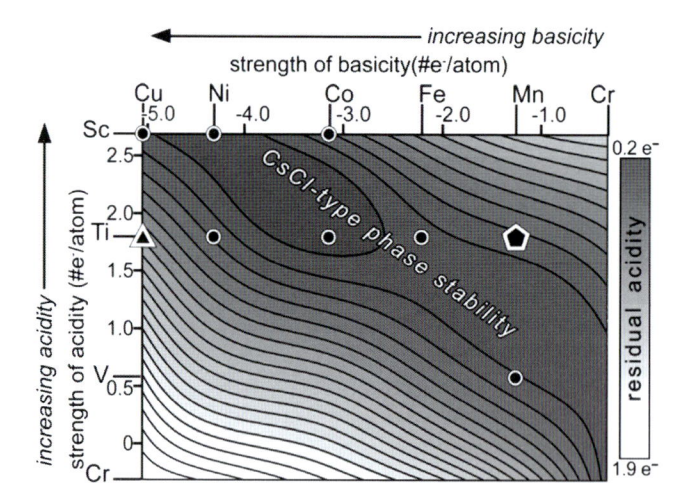

Figure 1.52 Smoothed contour map of residual acidities following reaction of μ_3 acids and μ_3 bases with varying strengths to form CsCl-type phases. Darker regions correspond to higher neutralization. The positions of experimentally observed CsCl-type phases are indicated with circles, while two more complex phases at the edges of the CsCl-type stability range are indicated with a triangle and pentagon. Reproduced with permission from Ref. [208]. Copyright 2013, American Chemical Society.

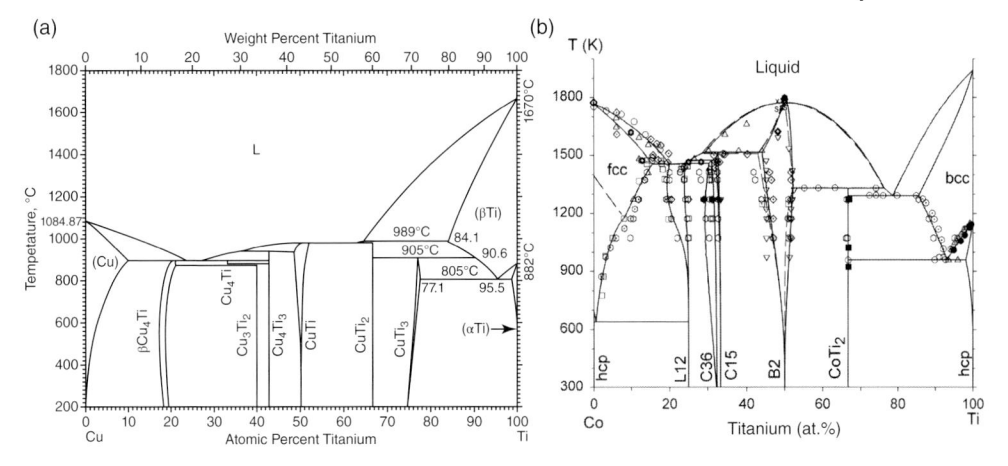

Figure 1.53 The binary phase diagrams of the Cu–Ti (a) and Co–Ti (b) phase diagrams. Note that the melting point of the 1:1 CoTi phase is much higher than that of its CuTi counterpart [209,210].

where the acid–base strengths are well matched between the two reactants, but steep ridges away from the diagonal where there is much less effective neutralization. The position of this valley correlates well with the experimentally observed CsCl phases (circles).

In Figure 1.53, we compare the binary phase diagrams of Ti–Cu and Ti–Co. By electronegativity arguments, one might presuppose that TiCu would have a greater range of thermodynamic stability than say TiCo. The position of TiCu on the structure stability map (Figure 1.52) suggests, however, that the Cu is too strong of a base for the moderate acid Ti. As the CsCl structure stability map and the phase diagrams make clear, TiCo, in actuality, has a greater range of thermodynamic stability than TiCu. Even without directly looking at entropic factors, the Lewis acid/base neutralization structure stability map is of qualitative utility. Of further interest are the more complex structures adopted in the Ti–Cu and the Ti–Mn phase diagrams (respectively the CuTi and $Ti_{21}Mn_{25}$ structures) based on more complete Lewis acid/base neutralization [208].

1.7
Conclusions

One might suppose that to understand the intermetallic compound and alloy bond would require knowledge of the newest directions of first-principles-derived quantum mechanics. QTAIMAC (quantum theory of atoms in molecules and crystals) might then be taken as a likely starting point for discussion [211,212]. Or one might take that the appropriate start lies in classic concepts, like the organic chemists' Lewis structures and that from these classic concepts one should develop suitable quantum mechanical theories. In molecular chemistry, the modern valence bond theory, discussed by Shaik and Hiberty, or the natural bond orbital (NBO) theory of Weinhold and Landis would come to mind, while for intermetallic chemists the analogues might be ELF (electron localization function) or the new extension of NBO to periodic systems [179,213–216]. Or one might believe the best approach is to make a simple Ansatz, and deduce from this Ansatz new theoretical constructs as Fukui, Woodward, and Hoffmann did with frontier orbitals and their symmetry- and MO-based rules of organic reactivity [217].

The difference between the approaches is not one of calculational accuracy. Today most quantum mechanical methods can be improved to a sufficiently high level of accuracy to satisfy all but the most demanding of chemists. It is more a philosophical one: Whether the most productive mode of thinking lies in the fashioning of new theoretical constructs of experimentally predictive value, which eventually live outside the calculations themselves, or whether a scientist's thinking is most correct solely within the confines of the quantum mechanical calculation itself.

The question is a complex one for intermetallic chemists. Our arsenal of phenomenological tools is not nearly as developed as the organic chemist's counterpart. For intermetallic chemistry, few bonding concepts stem from outside quantum mechanical thinking: Zintl phases, the Hume-Rothery rules, the Schubert two-component theory of metals, and the Engel–Brewers and Miedema's theories for alloy phase formation come to mind but little else [11,48,218,219]. And, of these, it is the Zintl concept and Hume-Rothery's insights that have survived best the combined test of time and investigation [13,14].

For the intermetallic chemist, as opposed to the organic chemist, bonding concepts and stability criteria tend to emanate from theoretical analysis itself. The Peierls distortion, the Mott–Jones zone, the Stoner criterion for ferromagnetism, the Cooper pair, Whangbo and Canadell's hidden Fermi nesting, Pettifor's structural energy difference theorem, Heine–Cohen–Weaire pseudopotential theory, the Landrum–Dronskowski ferromagnetism model, and the Fukui–Woodward–Hoffmann idea of maximal in-phase AO overlap are all pertinent examples. All these concepts now live outside the realm of the quantum calculations that initially engendered them.

This chapter has focused on the intermetallic bonding concepts among this short list. For us, QTAIMAC results today continue to live largely in the confines of QTAIMAC calculations. ELF is especially insightful in well-defined areas, such as intermetallic compounds with localized lone pairs and two-center two-electron bonds but can, at least for us, devolve into bewildering complexity in delocalized intermetallic systems [213,220–222]. But the one-electron models, nearly free-electron and tight-binding derived theories, the same models that can most readily account for the Peierls distortion, the Mott–Jones zone, the dominance of pseudopotential nodes, and the Pettifor structural energy difference theorem continue to provide an initial platform for the testing of new theoretical concepts.

Young chemists might be told today that one-electron ideas are inaccurate and dangerous. The point of this chapter has been to demonstrate the opposite – with the help of high-level *ab initio* calculations, today, it is more possible than ever to carry out effective one-electron modeling (effective here taken in both senses of the word: on the one hand, something useful, and on the other hand an accurate semiempirical rendition of a multielectron calculation). The future of our understanding of atomic size differences, electronegativity effects, valency, thermodynamic phase stability, and electron concentration rules – the heart of any discussion of intermetallic bonding – has never looked brighter.

References

1. Ashcroft, N.W. and Mermin, N.D. (1976) *Solid State Physics*, Holt, New York.
2. Mott, N.F. and Jones, H. (1936) *The Theory of the Properties of Metals and Alloys*, Clarendon Press, Oxford.
3. Wannier, G.H. (1959) *Elements of Solid State Theory*, Cambridge Press, Cambridge.
4. Harrison, W.A. (1966) *Pseudopotentials in the Theory of Metals*, Benjamin, New York.
5. Heine, V. (1969) *The Physics of Metals, Vol. 1: Electrons* (ed. J.M. Ziman), Cambridge University Press, Cambridge, pp. 1–61.
6. Hoffmann, R. (1988) *Solids and Surfaces: A Chemist's View of Bonding in Extended Structures*, Wiley-VCH Verlag GmbH, Weinheim, Germany.
7. Nesper, R. (1991) *Angew. Chem., Int. Ed.*, **30**, 789–817.
8. Sutton, A.P. (1993) *Electronic Structure of Materials*, Clarendon Press, Oxford.
9. Burdett, J.K. (1995) *Chemical Bonding in Solids*, Oxford University Press, New York.
10. Pettifor, D.G. (1995) *Bonding and Structure of Molecules and Solids*, Clarendon Press, Oxford.
11. Pettifor, D.G. (2002) in *Materials Science and Engineering: Its Nucleation and Growth* (ed. M. McLean), Maney Publishing, Leeds, pp. 297–335.
12. Dronskowski, R. (2005) *Computational Chemistry of Solid State Materials*, Wiley-VCH Verlag GmbH, Weinheim, Germany.
13. Mizutani, U. (2011) *Hume-Rothery Rules for Structurally Complex Alloy Phases*, CRC Press, Boca Raton, FL.
14. Fässler T.F. (ed.) (2011) *Zintl Phases: Principles and Recent Developments*, Structure and Bonding Series, vol. **139**, Springer, Heidelberg.
15. Alemany, P. and Canadell, E. (2014) *The Chemical Bond: Chemical Bonding Across the Periodic Table* (eds G. Frenking and S. Shaik), Wiley-VCH Verlag GmbH, Weinheim, Germany, pp. 445–475.
16. Hume-Rothery, W. and Raynor, G.V. (1954) *The Structure of Metals and Alloys*, 3rd edn, Institute of Metals., London.
17. Hume-Rothery, W. (1963) *Electrons, Atoms, Metals, and Alloys*, Dover, New York.
18. Pearson, W.B. (1972) *The Crystal Chemistry and Physics of Metals and Alloys*, Wiley Series on the Science and Technology of Materials, Wiley-Interscience, New York.
19. Janot, C. (1994) *Quasicrystals: A Primer*, Clarendon Press, Oxford.
20. Sauthoff, G. (1995) *Intermetallics*, Wiley-VCH Verlag GmbH, Weinheim, Germany.
21. Kauzlarich, S.M. (ed.) (1996) *Chemistry, Structure, and Bonding of Zintl Phases and Ions*, Wiley VCH Verlag GmbH, Weinheim, Germany.
22. Westbrook, J.H. and Fleischer, R.L. (eds) (2000) *Intermetallic Compounds: Principles and Practices*, John Wiley & Sons, Inc., New York.

23 Shevchenko, V.Y. (2011) *Search in Chemistry, Biology and Physics of the Nanostate*, Lema, Saint Petersburg, Russia.

24 Trebin, H.-R. (ed.) (2003) *Quasicrystals*, Wiley-VCH Verlag GmbH, Weinheim, Germany.

25 Ferro, R. and Saccone, A. (2008) *Intermetallic Chemistry*, Elsevier, Amsterdam.

26 DuBois, J.M. and Belin-Ferré, E. (eds) (2011) *Complex Metallic Alloys: Fundamentals and Applications*, Wiley-VCH Verlag GmbH, Weinheim, Germany.

27 Pöttgen, R. and Johrendt, D. (2014) *Intermetallics: Synthesis, Structure and Function*, de Gruyter, Berlin.

28 Ruedenberg, K. (1962) *Rev. Mod. Phys.*, **34**, 326–376.

29 Baird, N.C. (1986) *J. Chem. Educ.*, **63**, 660–664.

30 Lombardi, E. and Jansen, L. (1986) *Phys. Rev. A*, **33**, 2907–2912.

31 Martin, R.M. (2004) *Electronic Structure: Basic Theory and Practical Methods*, Cambridge University Press, Cambridge, UK.

32 Hoffmann, R. (1963) *J. Am. Chem. Soc.*, **39**, 1397–1412.

33 Whangbo, M.H., Hoffmann, R., and Woodward, R.B. (1979) *Proc. R. Soc. Lond. A*, **366**, 23–46.

34 Lohr, L.L. and Pyykkö, P. (1979) *Chem. Phys. Lett.*, **62**, 333–338.

35 Wolfsberg, M. and Helmholz, L. (1952) *J. Chem. Phys.*, **20**, 837–843.

36 Ammeter, J.H., Buergi, H.B., Thibeault, J.C., and Hoffmann, R. (1978) *J. Am. Chem. Soc.*, **100**, 3686–3692.

37 Cerda, J. and Soria, F. (2000) *Phys. Rev. B*, **61**, 7965–7971.

38 Roy, L.E. and Hughbanks, T. (2006) *J. Phys. Chem. B*, **110**, 20290–20296.

39 Roy, L.E. and Hughbanks, T. (2007) *J. Solid State. Chem.*, **180**, 818–823.

40 Berger, R.F., Walters, P.L., Lee, S., and Hoffmann, R. (2011) *Chem. Rev.*, **111**, 4522–4545.

41 Stacey, T.E. and Fredrickson, D.C. (2012) *Dalton Trans.*, **41**, 7801–7813.

42 Stacey, T.E., Borg, C.K.H., Zavalij, P.J., and Rodriguez, E.E. (2014) *Dalton Trans.*, **43**, 14612–14624.

43 Fang, S. K., Defo, R., Shirodkar, S.N., Lieu, S., Tritsaris, G.A., and Kaxiras, E. (2015) *Phys. Rev. B*, **92**, 205108.

44 Hughbanks, T. and Hoffmann, R. (1983) *J. Am. Chem. Soc.*, **105**, 1150–1162.

45 Hughbanks, T. and Hoffmann, R. (1983) *J. Am. Chem. Soc.*, **105**, 3528–3537.

46 Dronskowski, R. and Blöchl, P.E. (1993) *J. Phys. Chem.*, **97**, 8617–8624.

47 Zintl, E. (1939) *Angew. Chem.*, **52**, 1–6.

48 Schäfer, H., Eisenmann, B., and Müller, W. (1973) *Angew. Chem.*, **85**, 742–760.

49 Nesper, R. (2014) *Z. Anorg. Allg. Chem.*, **640**, 2639–2648.

50 Wijeyesekera, S.D. and Hoffmann, R. (1984) *Organometallics*, **3**, 949–961.

51 Kertesz, M. and Hoffmann, R. (1984) *J. Am. Chem. Soc.*, **106**, 3453–3460.

52 Hoffmann, R. and Zheng, C. (1985) *J. Phys. Chem.*, **89**, 4175–4181.

53 Tremel, W., Hoffmann, R., and Silvestre, J. (1986) *J. Am. Chem. Soc.*, **108**, 5174–5187.

54 Li, J. and Hoffmann, R.Z. (1986) *Z. Naturforsch. B*, **41**, 1399–1415.

55 Hoffmann, R., Li, J., and Wheeler, R.A. (1987) *J. Am. Chem. Soc.*, **109**, 6600–6602.

56 Kleinke, H. and Franzen, H.F. (1996) *Angew. Chem., Int. Ed.*, **35**, 1934–1936.

57 Todorov, E. and Sevov, S.C. (1997) *Inorg. Chem.*, **36**, 4298–4302.

58 Xu, Z. and Guloy, A.M. (1997) *J. Am. Chem. Soc.*, **119**, 10541–10542.

59 Kaskel, S. and Corbett, J.D. (2003) *Inorg. Chem.*, **42**, 1835–1841.

60 Villars, P. and Calvert, L.D. (1991) *Pearson's Handbook of Crystallographic Data for Intermetallic Phases*, 2nd edn, ASM International, Materials Park, OH.

61 Yee, K.A. and Hughbanks, T. (1991) *Inorg. Chem.*, **30**, 2321–2328.

62 Nanjundaswamy, K.S. and Hughbanks, T. (1992) *J. Solid State Chem.*, **98**, 278–290.

63 Abdon, R.L. and Hughbanks, T. (1994) *Angew. Chem., Int. Ed.*, **33**, 2328–2330.

64 Abdon, R.L. and Hughbanks, T. (1995) *J. Am. Chem. Soc.*, **117**, 10035–10040.

65 Kim, S.J., Nanjundaswamy, K.S., and Hughbanks, T. (1991) *Inorg. Chem.*, **39**, 159–164.
66 Bobev, S. and Sevov, S.C. (2002) *J. Alloys Compd.*, **338**, 87–92.
67 Mao, J.-G., Goodey, J., and Guloy, A.M. (2004) *Inorg. Chem.*, **43**, 282–289.
68 Lupu, C., Downie, C., Guloy, A.M., Albright, T.A., and Mao, J.-G. (2004) *J. Am. Chem. Soc.*, **126**, 4386–4397.
69 Kim, S.-J., Kraus, F., and Fässler, T.F. (2009) *J. Am. Chem. Soc.*, **131**, 1469–1478.
70 Dai, J.-C., Gupta, S., Gourdon, O., Kim, H.-J., and Corbett, J.D. (2009) *J. Am. Chem. Soc.*, **131**, 8677–8682.
71 Stoyko, S.S., Khatun, M., and Mar, A. (2012) *Inorg. Chem.*, **51**, 8517–9521.
72 Palasyuk, A., Grin, Y., and Miller, G.J. (2014) *J. Am. Chem. Soc.*, **136**, 3108–3117.
73 Henze, A. and Fässler, T.F. (2016) *Inorg. Chem.*, **5**, 822–827.
74 Fässler, T.F., Kronseder, C., and Wörle, M. (1999) *Z. Anorg. Allg. Chem.*, **625**, 15–23.
75 Fredrickson, D.C., Lee, S., and Hoffmann, R. (2007) *Angew. Chem., Int. Ed.*, **46**, 1958–1976.
76 Boucher, F. and Rousseau, R. (1998) *Inorg. Chem.*, **37**, 2351–2357.
77 Xia, S.-Q. and Bobev, S. (2006) *Inorg. Chem.*, **45**, 7126–7132.
78 You, T.-S., Grin, Y., and Miller, G.J. (2007) *Inorg. Chem.*, **48**, 8801–8811.
79 Zelinska, O.Y. and Mar, A. (2008) *Inorg. Chem.*, **47**, 297–305.
80 Fokwa, B.P.T. and Hermus, M. (2011) *Inorg. Chem.*, **50**, 3332–3341.
81 Smetana, V., Lin, Q., Pratt, D.K., Kreyssig, A., Ramazaoglu, M., Corbett, J.D., Goldman, A.I., and Miller, G.J. (2012) *Angew. Chem., Int. Ed.*, **51**, 12699–12702.
82 Li, B., Kim, S.-J., Miller, G.J., and Corbett, J.D. (2010) *Inorg. Chem.*, **49**, 1503–1509.
83 Burdett, J.K., Lee, S., and McLarnan, T. (1985) *J. Am. Chem. Soc.*, **107**, 3083–3089.
84 Burdett, J.K., Canadell, E., and Hughbanks, T. (1986) *J. Am. Chem. Soc.*, **108**, 3971–3976.
85 Albert, B. and Schmitt, K. (1999) *Inorg. Chem.*, **38**, 6159–6163.
86 Hofmann, K. and Albert, B. (1991) *J. Am. Chem. Soc.*, **113**, 8216–8220.
87 Longuet-Higgins, H.C., Rector, C.W., and Platt, J.R. (1950) *J. Chem. Phys.*, **18**, 1174–1181.
88 Gimarc, B.M. (1983) *J. Am. Chem. Soc.*, **105**, 1979–1984.
89 Franzen, H.F. and Köckerling, M. (1995) *Prog. Solid State Chem.*, **23**, 265–289.
90 Kleinke, H. and Franzen, H.F. (1998) *J. Solid State Chem.*, **136**, 221–226.
91 Lee, C.-S. and Miller, G.J. (2001) *Inorg. Chem.*, **40**, 338–345.
92 Miller, G.J. and Lee, C.-S. (2001) *Eur. J. Inorg. Chem.*, **40**, 338–345.
93 Cyrot-Lackmann, F. (1968) *J. Phys. Chem. Solids*, **29**, 1235–1243.
94 Ducastelle, F. and Cyrot-Lackmann, F. (1970) *J. Phys. Chem. Solids*, **31**, 1295–1306.
95 Kendall, M.G. (1944) *The Advanced Theory of Statistics Vol. I*, Lippencott, Philadelphia, PA.
96 Friedel, J. (1969) *The Physics of Metals*, vol. **1** (ed. J.M. Ziman), Cambridge University Press, Cambridge, pp. 340–403.
97 Burdett, J.K. (1997) *Chemical Bonds: A Dialog*, John Wiley & Sons, Inc., New York.
98 Burdett, J.K. and Lee, S. (1985) *J. Am. Chem. Soc.*, **107**, 3050–3063.
99 Burdett, J.K. and Lee, S. (1985) *J. Am. Chem. Soc.*, **107**, 3063–3082.
100 Pettifor, D.G. and Podloucky, R. (1986) *J. Phys. C*, **19**, 315–330.
101 Lee, S. (1996) *Annu. Rev. Phys. Chem.*, **47**, 397–419.
102 Pettifor, D.G. (1970) *J. Phys. C*, **3**, 367–377.
103 Hoistad, L.M. and Lee, S. (1991) *J. Am. Chem. Soc.*, **113**, 8216–8220.
104 Haydock, R. and Johannes, R.L. (1975) *J. Phy. F Metal Phys.*, **5**, 2055–2067.
105 Johannes, R.L., Haydock, R., and Heine, V. (1976) *Phys. Rev. Lett.*, **36**, 372–376.
106 Hansen, M. (1958) *Constitution of Binary Alloys*, McGraw-Hill, New York.
107 Elliot, R.P. (1965) *Constitution of Binary Alloys: First Supplement*, McGraw-Hill, New York.

108 Shunk, F.A. (1969) *Constitution of Binary Alloys: Second Supplement*, McGraw-Hill, New York.
109 Rose, J.H., Ferrante, J., and Smith, J.R. (1981) *Phys. Rev. Lett.*, **47**, 675–678.
110 Todorov, E., Evans, M., Lee, S., and Rousseau, R. (2001) *Chem. Eur. J.*, **7**, 2652–2662.
111 Fredrickson, D.C. (2011) *J. Am. Chem. Soc.*, **133**, 10070–10073.
112 Lee, S. and Hoffmann, R. (2002) *J. Am. Chem. Soc.*, **124**, 4811–4823.
113 Harris, N.A., Hadler, A.B., and Fredrickson, D.C. (2011) *Z. Anorg. Allg. Chem.*, **637**, 1961–1974.
114 Lee, S. (1991) *J. Am. Chem. Soc.*, **113**, 101–105.
115 Nesper, R. and Miller, G. (1993) *J. Alloys Compd.*, **197**, 109–121.
116 Hoistad, L.M. (1995) *Inorg. Chem.*, **34**, 2711–2717.
117 Häussermann, U. and Nesper, R. (1975) *J. Alloys Compd.*, **5**, 2055–2067.
118 Häussermann, U., Svensson, C., and Lidin, S. (1998) *J. Am. Chem. Soc.*, **120**, 3867–3880.
119 Tse, J.S., Frapper, G., Ker, A., Rousseau, R., and Klug, D.D. (1986) *Phys. Rev. Lett.*, **82**, 4472–4475.
120 Amerioun, S., Simak, S., and Häussermann, U. (2003) *Inorg. Chem.*, **42**, 1467–1474.
121 Pettifor, D.G. and Podloucky, R. (1984) *Phys. Rev. Lett.*, **53**, 1080–1083.
122 Pettifor, D.G. (1984) *Solid State Commun.*, **51**, 31–34.
123 Clark, P.M., Lee, S., and Fredrickson, D.C. (2005) *J. Solid State Chem.*, **178**, 1269–1283.
124 Jones, H. (1934) *Proc. R. Soc. A*, **144**, 225–234.
125 de Laissardière, G.T., Nguyen-Manh, D., and Mayou, D. (2005) *Progress in Materials Science*, **50**, 679–788.
126 Degtyareva, V.F. and Smirnova, H.S. (2007) *Z. Kristallogr.*, **222**, 718–721.
127 Degtyareva, V.F. and Afonikova, N.S. (2015) *Solid State Sci.*, **49**, 61–67.
128 Sato, H. and Toth, R.S. (1962) *Phys. Rev. Lett.*, **8**, 239–241.
129 Lee, S., Henderson, R., Kaminsky, C., Nelson, Z., Nguyen, J., Settje, N.F., Schmidt, T.S., and Feng, J. (2013) *Chem. Eur. J.*, **19**, 10244–10270.
130 Samson, S. (1962) *Nature*, **195**, 259–262.
131 Weber, T., Dshemuchadse, J., Kobas, M., Conrad, M., Harbrecht, B., and Steurer, W. (2009) *Acta Crystallogr. B*, **65**, 308–317.
132 Conrad, M., Harbrecht, B., Weber, T., Jung, D.Y., and Steurer, W. (2009) *Acta Crystallogr. B*, **65**, 318–325.
133 Samson, S. (1967) *Acta Crystallogr.*, **23**, 586–600.
134 Andersson, S. (1980) *Acta Crystallogr. B*, **36**, 2513.
135 Kreiner, G. and Schapers, M. (1997) *J. Alloys Compd.*, **259**, 83–114.
136 Feng, J., Hoffmann, R., and Ashcroft, N.W. (2010) *J. Chem. Phys.*, **132**, 114106.
137 Degtyareva, V.F. and Degtyareva, O. (2009) *New J. Phys.*, **11**, 063037.
138 Heine, V. and Weaire, D. (eds) (1970) Pseudopotential theory of cohesion and structure, in *Solid State Physics* (eds H. Ehrenreich, F. Seitz, and D. Turnbull), vol. **24**, Academic Press, New York.
139 Brennan, T.D. and Burdett, J.K. (1993) *Inorg. Chem.*, **32**, 746–749.
140 Stroud, D. and Ashcroft, N.W. (1971) *J. Phys. F Metal Phys.*, **1**, 113–124.
141 Stoner, E.C. (1939) *Proc. R. Soc. Lond. A*, **169**, 339–371.
142 Herring, C. (1966) *Magnetism, Vol. IV* (eds G.T. Rado and H. Suhl), Academic Press, New York, pp. 1–407.
143 Janak, J.F. (1977) *Phys. Rev. B*, **16**, 255–262.
144 Gubanov, V.A., Liechtenstein, A.I., and Postnikov, A.V. (1992) *Magnetism and the Electronic Structure of Crystals*, Springer, Berlin.
145 Landrum, G.A. and Dronskowski, R. (2000) *Angew. Chem., Int. Ed.*, **39**, 1560–1585.
146 Stollhoff, G. (2000) *Angew. Chem., Int. Ed.*, **39**, 4471–4475.
147 Seo, D.-K. and Kim, S.-H. (2008) *J. Comput. Chem.*, **29**, 2172–2176.
148 Ramesha, K., Seshadri, R., Ederer, C., He, T., and Subramanian, M. (2004) *Phys. Rev. B*, **70**, 214409.

149 Offernes, L., Ravindran, P., Seim, C.W., and Kjekshus, A. (2008) *J. Alloys Compd.*, **458**, 47–60.

150 Kovnir, K., Thompson, C.M., Zhou, H.D., Wiebe, C.R., and Shatruk, M. (2010) *Chem. Mater.*, **22**, 1704–1713.

151 Peierls, R.E. (1955) *Quantum Theory of Solids*, Clarendon Press, Oxford.

152 Bertaut, E.F. (1968) *Acta Crystallogr. A*, **24**, 217–231.

153 Wilson, J.A., DiSalvo, F.J., and Mahajan, S. (1974) *Phys. Rev. Lett.*, **32**, 882–885.

154 Monceau, P., Ong, N.P., Portis, A.M., Meerschaut, A., and Rouxel, J. (1976) *Phys. Rev. Lett.*, **37**, 602–606.

155 Burdett, J.K. and Lee, S. (1983) *J. Am. Chem. Soc.*, **105**, 1079–1083.

156 Whangbo, M.H. and Canadell, E. (1992) *J. Am. Chem. Soc.*, **114**, 9587–9600.

157 Whangbo, M.H., Seo, D.-K., and Canadell, E. (1996) *NATO ASI Series B*, **354**, 285–302.

158 Papoian, G.A. and Hoffmann, R. (2000) *Angew. Chem., Int. Ed.*, **39**, 2408–2448.

159 Kurtulus, Y. and Dronskowski, R. (2003) *J. Solid State Chem.*, **176**, 390–399.

160 Decker, A., Landrum, G.A., and Dronskowski, R. (2002) *Z. Anorg. Allg. Chem.*, **628**, 303–309.

161 Freeth, W.E. and Raynor, G.V. (1954) *J. Inst. Met.*, **82**, 569–574.

162 Arnberg, L. and Westman, S. (1972) *Acta Chem. Scand.*, **26**, 1748–1750.

163 Noritake, T., Aoki, M., Towata, S., Takeuchi, T., and Mizutani, U. (2007) *Acta Crystallogr. B*, **63**, 726–734.

164 Mizutani, U., Asahi, R., Sato, H., Noritake, T., and Takeuchi, T. (2008) *J. Phys. Condens. Matter*, **20**, 275228.

165 Bergman, G., Waugh, J.L.T., and Pauling, L. (1957) *Acta Crystallogr.*, **10**, 254–259.

166 Doering, W., Seelentag, W., Buchholz, W., and Schuster, H.U. (1979) *Z. Naturforsch.*, **34**, 1715–1718.

167 Mizutani, U., Takeuchi, T., and Sato, H. (2004) *Prog. Mater. Sci.*, **49**, 227–261.

168 Raghavan, V. (2010) *J. Phase Equilibria Diffus.*, **31**, 288–290.

169 Audier, M., Pannetier, J., Leblanc, M., Janot, C., Lang, J.-M., and Dubost, B. (1988) *Physica B.*, **153**, 136–142.

170 Sato, H., Takeuchi, T., and Mizutani, U. (2004) *Phys. Rev. B*, **70**, 024210.

171 Belin, C. and Ling, R.G. (1983) *J. Solid State Chem.*, **48**, 40–48.

172 King, R. B. (1991) *Inorg. Chim. Acta*, **181**, 217–225.

173 Belin, C. and Tillard-Charbonnel, M. (1993) *Prog. Solid State Chem.*, **22**, 59–109.

174 Lee, C.-S. and Miller, G.J. (2000) *J. Am. Chem. Soc.*, **122**, 4937–4947.

175 Tillard-Charbonnel, M. and Belin, C. (1991) *J. Solid State Chem.*, **90**, 270–278.

176 Tillard, M., Belin, C., Spina, L., and Jia, Y.Z. (2005) *Solid State Sci.*, **7**, 1125–1134.

177 Todorov, E. and Sevov, S.C. (2005) *Inorg. Chem.*, **44**, 5361–5369.

178 Whangbo, M.-H., Lee, C., and Köhler, J. (2011) *Eur. J. Inorg. Chem.*, **2011**, 3841–3847.

179 Shaik, S. and Hiberty, P.C. (2008) *A Chemist's Guide to Valence Bond Theory*, John Wiley & Sons, Inc., Hoboken, NJ.

180 Mostofi, A.A., Yates, J.R., Lee, Y.-S., Souza, I., Vanderbilt, D., and Marzari, N. (2008) *Comput. Phys. Commun.*, **178**, 685–699.

181 Marzari, N., Mostofi, A.A., Yates, J.R., Souza, I., and Vanderbilt, D. (2012) *Rev. Mod. Phys.*, **84**, 1419–1475.

182 Zurek, E., Jepsen, O., and Andersen, O.K. (2005) *ChemPhysChem*, **6**, 1934–1942.

183 Yin, Z.P. and Pickett, W.E. (2010) *Phys. Rev. B*, **81**, 174534.

184 Zurek, E., Jepsen, O., and Andersen, O.K. (2010) *Inorg. Chem.*, **49**, 1384–1396.

185 Rudenko, A.N. and Katsnelson, M.I. (2014) *Phys. Rev. B*, **89**, 201408.

186 Botana, A.S., Quan, Y., and Pickett, W.E. (2015) *Phys. Rev. B*, **92**, 155134.

187 Hooper, J. and Zurek, E. (2012) *Chem. Eur. J.*, **18**, 5013–5021.

188 Jung, D., Koo, H.-J., and Whangbo, M.-H. (2000) *J. Mol. Struct.*, **527**, 113–119.

189 Kandpal, H.C., Felser, C., and Seshadri, R. (2006) *J. Phys. D Appl. Phys.*, **39**, 776–785.

190 Fredrickson, R.T. and Fredrickson, D.C. (2013) *Inorg. Chem.*, **52**, 3178–3189.

191 Yannello, V.J., Kilduff, B.J., and Fredrickson, D.C. (2014) *Inorg. Chem.*, **53**, 2730–2741.

192 Nowotny, H. (1970) *Chemistry of Extended Defects in Nonmetallic Solids* (eds L. Eyring and M. O'Keeffe), North-

Holland Publishing Company, Amsterdam, pp. 223–237.

193 Jeitschko, W and Parthé, E. (1967) *Acta Crystallogr.*, **22**, 417–430.

194 Pearson, W.B. (1970) *Acta Crystallogr. B*, **26**, 1044–1046.

195 Yannello, V.J. and Fredrickson, D.C. (2014) *Inorg. Chem.*, **53**, 10627–10631.

196 Yannello, V.J. and Fredrickson, D.C. (2015) *Inorg. Chem.*, **54**, 11385–11398.

197 Hume-Rothery, W. (1965) *Phase Stability in Metals and Alloys*, McGraw-Hill, New York.

198 Laves, F. and Witte, H. (1935) *Metallwirtschaft, Metallwissenschaft, Metalltechnik*, **14**, 645–649.

199 Fredrickson, D.C. (2012) *J. Am. Chem. Soc.*, **134**, 5991–5999.

200 Engelkemier, J., Berns, V.M., and Fredrickson, D.C. (2013) *J. Chem. Theory Comput.*, **9**, 3170–3180.

201 Berns, V.M., Engelkemier, J., Guo, Y., Kilduff, B.J., and Fredrickson, D.C. (2014) *J. Chem. Theory Comput.*, **10**, 3380–3392.

202 Baren, M.R. (1990) *Binary Alloy Phase Diagrams* (eds T.B. Massalski, H. Okamoto, P.R. Subramanian, and L. Kacprzak), ASM International, Materials Park, OH, pp. 20–21.

203 Snyder, G.J and Simon, A. (1995) *J. Alloys Compd.*, **223**, 65–69.

204 Okamoto, H. (2013) *J. Phase Equilibria Diffus.*, **34**, 493–505.

205 Brewer, L. and Wengert, P.R. (1973) *Metall. Trans.*, **4**, 83–104.

206 Guo, Y., Stacey, T.E., and Fredrickson, D.C. (2014) *Inorg. Chem.*, **53**, 5280–5293.

207 Stacey, T.E. and Fredrickson, D.C. (2012) *Inorg. Chem.*, **51**, 4250–4264.

208 Stacey, T.E. and Fredrickson, D.C. (2013) *Inorg. Chem.*, **52**, 8349–8359.

209 Davydov, A.V., Kattner, U.R., Josell, D., Blendell, J.E., Waterstrat, R.M., Shapiro, A.J., and Boettinger, W.J. (2009) *Metall. Mater. Trans. A*, **32**, 2175–2186.

210 Okamoto, H. (2002) *J. Phase Equilibria Diffus.*, **26**, 549–500.

211 Bader, R.F.W. (1990) *Atoms in Molecules: A Quantum Theory*, Clarendon Press, Oxford.

212 Gatti, C. (2005) *Z. Kristallogr.*, **220**, 399–457.

213 Savin, A., Nesper, R., Weggert, S., and Fässler, T.F. (1997) *Angew. Chem., Int. Ed.*, **36**, 1808–1832.

214 Weinhold, F. and Landis, C.R. (2005) *Valency and Bonding*, Cambridge Press, Cambridge.

215 Galeev, T.R., Dunnington, B.D., Schmidt, J.R., and Boldyrev, A.I. (2013) *Phys. Chem. Chem. Phys.*, **15**, 5022–5029.

216 Dolyniuk, J.-A., He, H., Ivanov, A.S., Boldyrev, A.I., Bobev, S., and Kovnir, K. (2015) *Inorg. Chem.*, **54**, 8608–8616.

217 Fleming, I. (1978) *Frontier Orbitals and Organic Chemical Reactions*, John Wiley & Sons, Inc., London.

218 Schubert, K. (1964) *Kristallstrukturen Zweikomponentiger Phasen*, Springer, Berlin.

219 Williams, A.R., Gelatt, C.D., and Moruzzi, V.L. (1980) *Phys. Rev. Lett.*, **44**, 429–433.

220 Gieck, C., Schreyer, M., Fässler, T.F., Raif, F., and Claus, P. (2006) *Eur. J. Inorg. Chem.*, **2006**, 3482–3488.

221 Hlukhyy, V. and Fässler, T.F. (2010) *Z. Anorg. Allg. Chem.*, **636**, 100–107.

222 Stalder, E.D., Wörle, M., and Nesper, R. (2010) *Inorg. Chim. Acta*, **363**, 4355–4360.

2
Quasicrystal Approximants

Sven Lidin

Lund University, CAS, Getingevägen 60, 22100 Lund, Sweden

2.1
What Is an Approximant?

Aperiodic crystals were recognized as early as the 1920s [1], but it was through the application of higher dimensional crystallographic methods [2,3] that the relationship between aperiodic structures and approximants became clear. An aperiodic crystal is an object that displays the prerequisite long-range order of crystallinity, as evidenced in an "essentially discrete diffraction diagram" [4] but lacks periodicity. Structures that are closely related to aperiodic structures having similar compositions and diffraction patterns are called approximants. Since approximants are normal periodic crystals, the analysis of their structures is in principle straightforward, although complications arise since they tend to be complex with large unit cells and many independent atomic positions. Approximants are useful in the analysis of their aperiodic companions and provide simple models for understanding the physical properties of aperiodic crystals.

The term approximant is used in two, slightly different, contexts. The first is simply that used above, that is, phases that are chemically and structurally similar to aperiodic structures. This loose use of the term is quite convenient and also allows its application to systems that do not follow the second, more rigorous, definition that precisely defines the relationship between an aperiodic crystal and its rational approximants in mathematical terms. To understand this second definition, it is first necessary to explain how higher dimensional crystallography can be used to handle aperiodic crystals.

2.2
How Can the Structure of an Aperiodic Crystal be Solved, Refined, and Understood?

Modern higher dimensional crystallography for aperiodic structures was first used by P.M. de Wolff who introduced this formalism to describe the modulated

Handbook of Solid State Chemistry, First Edition. Edited by Richard Dronskowski, Shinichi Kikkawa, and Andreas Stein.
© 2017 Wiley-VCH Verlag GmbH & Co. KGaA. Published 2017 by Wiley-VCH Verlag GmbH & Co. KGaA.

structure of γ-Na$_2$CO$_3$ [2]. The diffraction pattern of this compound is characterized by a set of strong main reflections that may be indexed using the normal three vectors **a***, **b***, and **c*** and additionally a set of satellites with positions that are incommensurate with the basic lattice. All satellite positions may however be descried by an additional diffraction vector **q** so that the position of any diffraction spot in the pattern may be given as $h\mathbf{a}^* + k\mathbf{b}^* + l\mathbf{c}^* + m\mathbf{q}$. The vector **q** is not independent of the vectors **a***, **b***, and **c*** but when expressed in terms of these $\mathbf{q} = \alpha \mathbf{a}^* + \beta \mathbf{b}^* + \gamma \mathbf{c}^*$, at least one of the three coefficients α, β, and γ is an irrational number. Formally, this makes **q** linearly independent of the lattice defined by integer multiples of the basis vectors and it may be considered as orthogonal to these. This is the basis of the 3+1 dimensional indexing introduced by P.M. de Wolff for γ-Na$_2$CO$_3$. The intensity of all reflections $hklm$ carry information of how the modulation affects position and occupancy of the atoms in the structure and refining an aperiodic structure amounts to modeling atomic modulation functions that generate satellites that mimic the experimental intensity distribution, and the intersection between real space and higher dimensional space determines how the atomic modulation functions should be sampled for a particular atomic position. Thus, the 3+1 dimensional indexing of reciprocal space corresponds to a 3+1 dimensional direct space where the electron density of atoms is no longer a discrete object, but instead a one-dimensional hyperatom, an occupational domain, where the modulation of the atomic position is periodic in the additional dimension (internal space). In Figure 2.1, the atomic surface of an antimony atom in ζ-Zn$_{3-\delta}$Sb$_2$ [5] is shown.

Note the periodicity along $x1$ (abscissa) and along internal space $x4$ (ordinate). Neither of these two directions represents the real structure in three-dimensional space. The direction $x1$ represents a fictitious, average structure devoid of modulation. Real space is instead given by the section through this representation shown by the red line determined by the direction and magnitude of the modulation **q**-vector. For this particular structure, the **q**-vector along with **a*** is

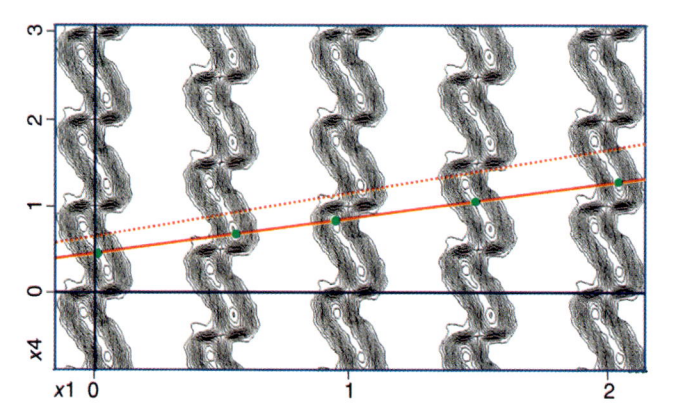

Figure 2.1 A section of the electron density of an antimony atom in incommensurate ζ-Zn$_{3-\delta}$Sb$_2$.

given by $\mathbf{q} = \alpha 00$ ($\alpha \cdot 0.385$). The point of intersection between the $x4$-axis and real space is arbitrary, and although a parallel shift of the red line leads to a locally different atomic configuration, those structures generated are globally equivalent. The atomic positions along this direction (shown as green disks) are given by the intersections of real space (red line) with the maxima of the atomic surfaces.

For an incommensurate $\mathbf{q}$-vector, the angle between the $x1$-axis and real space is irrational and the sequence of intercepts between real space and the maxima of the electron density is an aperiodic function. For commensurate values of α, the sequence of intercepts becomes periodic. In the figure, the dotted red line represents a structure where $\alpha = 1/2$ and the resulting real space structure is a twofold superstructure. This superstructure is an approximant to the incommensurate structure.

The research field of aperiodic structures remained relatively obscure until the discovery of icosahedral quasicrystals in rapidly quenched Al–Mn samples [6,7], which made a huge impact. It was quickly realized that these quasicrystals must be structurally closely related to the approximant μ-AlMn phase [8].

2.3
The Quasicrystals and Their Approximants

The metallic quasicrystals known today may be conveniently classified into axial and icosahedral. Additionally, quasicrystals do not necessarily have noncrystallographic rotational symmetry, and cubic quasicrystals have been reported [9]. The axial quasicrystals include those with pentagonal, octagonal, decagonal, and dodecagonal symmetry. Steurer has discussed the possibility of heptagonal quasicrystals and what their approximants would be. Additionally, dodecagonal quasicrystals have been identified in mesoporous silica [10] and in soft matter [11–13] and for these, approximants have also been identified. In this chapter, we will concentrate on the metallic systems.

2.4
Octagonal Approximants

The reports of octagonal quasicrystals are few, the phases first reported were o-$V_{15}Ni_{10}Si$ and o-$Cr_5Ni_3Si_2$ [14] followed by o-$Mn_{82}Si_{15}Al_3$ [15], o-Mn_4Si [16], and o-Mn-Fe-Si [17]. The approximants for all these phases have the β-Mn structure type. This is a widespread structure type with 76 entries in Pearson Crystal Database [18], including three elemental structures: Mn [19], Co [20], and Re [21]. A colored version of this is the structure type Mg_3Ru_2 [22,23] with three entries in the database. A simulation of crystallization from a monoatomic liquid was shown to yield patches of twinned β-Mn intergrown with an aperiodic structure that mimics the diffraction pattern of the octagonal phase well [24].

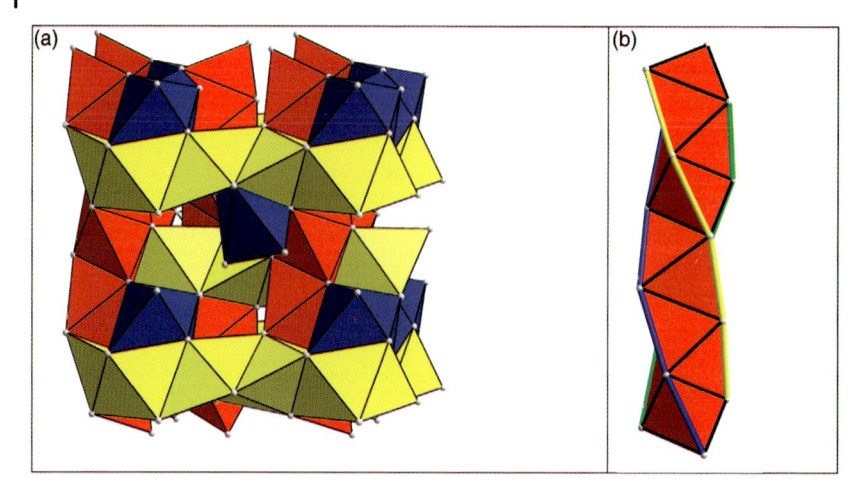

Figure 2.2 (a) The β-Mn structure represented as a rod packing. (b) An individual Coxeter–Boerdijk tetrahelix rod is constructed from face-sharing tetrahedra. The three ridges running along the outside of the tetrahelix have been highlighted to show the triple-helix nature.

There are no easily identifiable clusters in β-Mn; one of the two atomic positions in the asymmetric unit has a distorted icosahedral environment and the other an irregular 14 coordination. An interesting alternative description of β-Mn is as a rod packing where the individual rods consist of Coxeter–Boerdijk helices (Figure 2.2) [25].

The fact that β-Mn is a cubic approximant to an axial quasicrystal is not really surprising, the essential feature of an approximant is the fact that it contains clusters similar to those in the quasicrystal itself, and in β-Mn these may well be the Coxeter–Boerdijk tetrahelices. Lifting the structure of β-Mn to six dimensional space produces a single-atom structure in which three 8_3 screw axes meet orthogonally [26].

2.5
Decagonal Approximants

Pentagonal quasicrystals are relatively rare and seem to occur only in systems that also yield decagonal quasicrystals. Their fundamental building units are clearly very similar and they share the same approximants. Decagonal quasicrystals conversely, are by far the most numerous of the axial quasicrystals, and similarly, their approximants are also legion. There is a modest number of high-quality X-ray studies of stable decagonal approximants and a very large number of reports from electron microscopy. For a few compounds electron diffraction has been used to solve and refine approximant structures. A representative group of approximants is found in the aluminum–cobalt system and in related

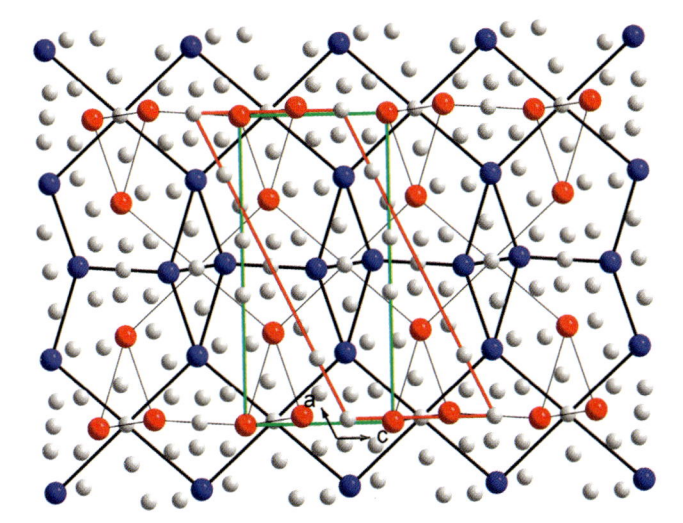

Figure 2.3 Structure of two layers of $Al_{13}(Co, Ni)_4$. The two layers are identical by symmetry, but shifted relative to each other. The transition metals of the bottom layer are shown in blue and joined by black lines to emphasize the pentagonal nature of the layer. Bonds to an Al position in the same layer completes the tiling of this ⟨2,2⟩ approximant. Transition metal positions in the upper layer are shown in red. The unit cell is outlined in red and the cell of the rational ⟨2,2⟩ approximant in green.

ternary systems. The high temperature phase Co_2Al_5 [27] (space group $P6_3/mmc$) is not a rational approximant, but coordination is largely icosahedral. For a large family of approximants, alternating layers exhibit the rational approximants of the decagonal system. Starting from the transition metal positions and sometimes including part of the aluminum positions highlights the similarity between the actual structure and the ideal rational approximant. The structure of $Al_{13}(Co,Ni)_4$ [28] crystallizes in space group $C2/m$ with the cell parameters $a = 17.0377$, $b = 4.0993$, $c = 7.4910$ Å, $\beta = 115.94°$ (Figure 2.3).

The structure of m-Co_4Al_{11} [29] crystallizes in the space group Cm with the cell parameters $a = 15.183$, $b = 8.122$, $c = 12.340$ Å; $\beta = 107.90°$ (Figure 2.4).

The orthorhombic phase o-Co_4Al_{13} [30] crystallizes in space group $Pmn2_1$ with the cell parameters $a = 8.1580$, $b = 12.3420$, $c = 14.4520$ Å (Figure 2.5).

Larger approximants include the τ^2-inflated o'-$Al_{13}Co_4$ [31] crystallizing in space group $Pnma$ with the cell parameters $a = 28.900$, $b = 8.1380$, $c = 12.3460$ Å (Figure 2.6).

Larger approximants have been reported from electron microscopy and electron diffraction studies in the system Al–Co–Ni. In the electron diffraction study by Grushko et al. [32], no less than five different phases are reported. For some the repeat along the pseudodecagonal direction is unknown, or badly defined because of diffuse layers, but the sizes within the decagonal plane are reported. A more complete list is given in the thesis of Döblinger [33] where nine phases are reported: PD_1 $a = 37.7$ $b = 39.7$ Å, PD_2 $a = 23.2$ $b = 32.0$ Å PD_3, PD_{3c} $a = 37.7$

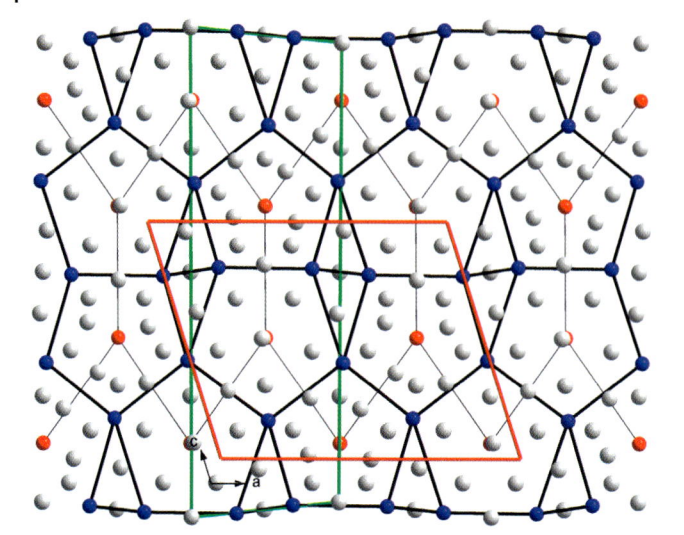

Figure 2.4 Structure of two layers of m-Co_4Al_{11}. Transition metal atoms in the flat layer are shown in blue and joined by black lines. Transition metal positions in the puckered layer are shown in red. The unit cell is outlined in red and the cell of the rational ⟨2,3⟩ approximant in green.

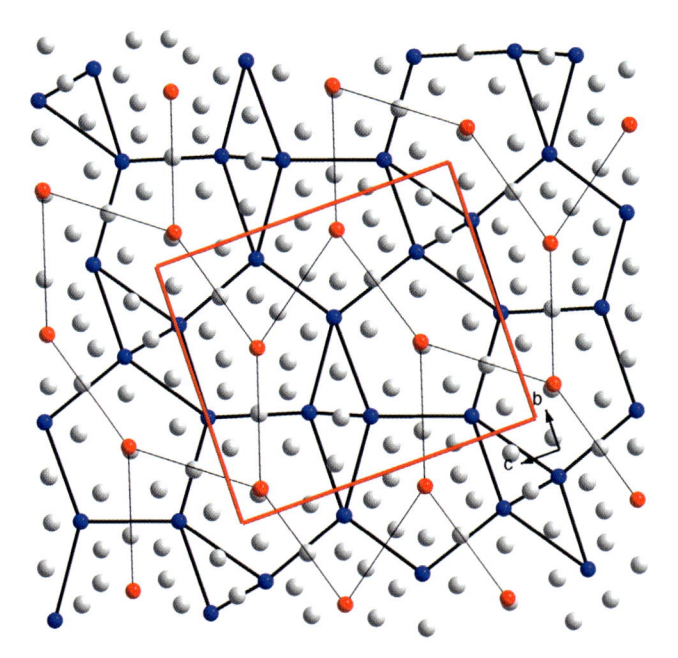

Figure 2.5 Structure of two layers of o-$Al_{13}Co_4$. Transition metal atoms in the flat layer are shown in blue and joined by black lines. Transition metal positions in the puckered layer are shown in red. The unit cell is outlined in red and it identical to the cell of the rational ⟨3,2⟩ approximant.

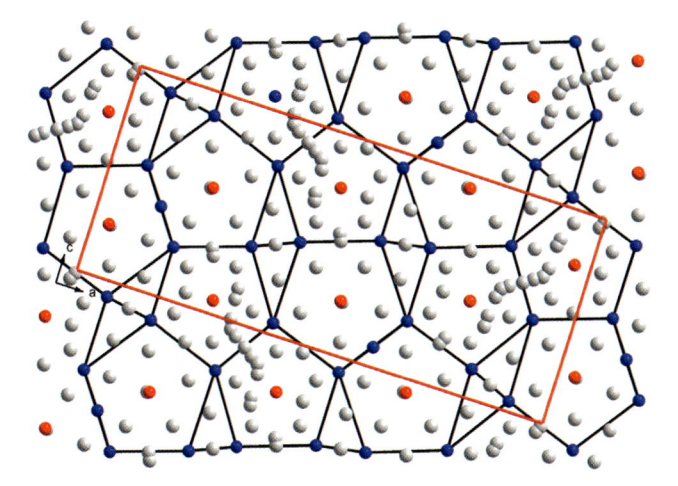

Figure 2.6 Structure of two layers of o′-Al$_{13}$Co$_4$. Transition metal atoms in the flat layer are shown in blue and joined by black lines. Transition metal positions in the puckered layer are shown in red. The unit cell is outlined in red.

$b = 51.8$ Å, PD$_4$ $a = 101.3$ $b = 32.0$ Å, PD$_5$ $a = 61.0$ $b = 32.0$ Å, PD$_6$ $a = 98.7$ $b = 32.0$ Å, PD$_7$ $a = 159.7$ $b = 51.8$ Å, PD$_8$ $a = 23.2$ $b = 19.8$ Å. The structure of the PD$_{3c}$ has been studied by high-resolution electron microscopy [34] and solved from electron diffraction data [35] while PD$_4$ [36] and PD$_8$ [37] have been solved from single-crystal X-ray data. Atomic coordinates are available for PD$_2$ from a solution from electron diffraction data (Figure 2.7) [38]. The pseudodecagonal arrangement becomes increasingly clear as the size of the approximant cell grows ever larger.

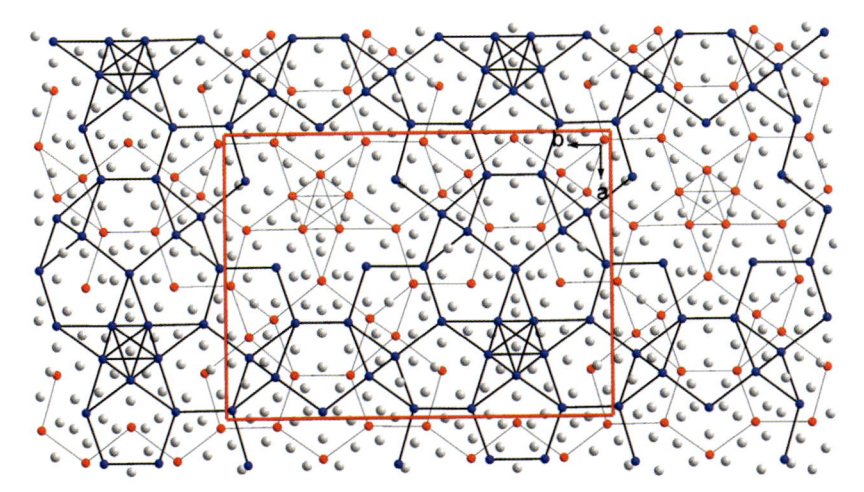

Figure 2.7 Structure of two layers of PD$_2$ Al–Co–Ni. Transition metal atoms in the lower layer are shown in blue and joined by black lines. Transition metal positions in upper layer are shown in red. The unit cell is outlined in red.

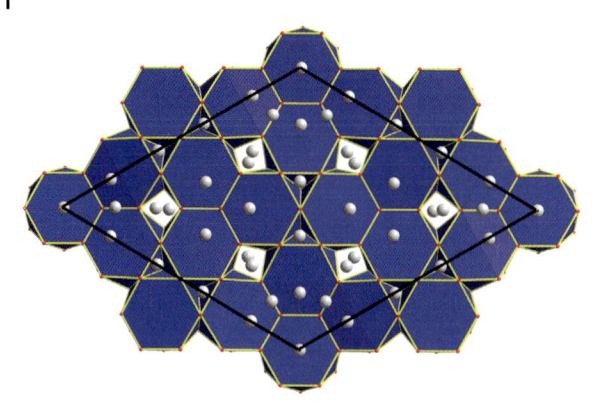

Figure 2.8 The tantalum-centered tantalum hexagonal antiprisms (blue) share faces to form a puckered surface in the hexagonal approximant to the Ta–Te dodecagonal quasicrystal. The tellurium atoms (gray) are interspersed between such layers.

2.6
Dodecagonal Approximants

Dodecagonal quasicrystals have been found to form in two widely different kinds of metallic systems. As classical intermetallics, dodecagonal quasicrystals are found in binary transition metal systems and a relatively large approximant has been identified in the system Mn–Si–V [39]. A very different system yielding dodecagonal quasicrystals is Ta/V–Te where several well-ordered approximants have also been identified. In these approximant phases, tantalum-centered hexagonal antiprisms of tantalum share triagonal faces to form puckered layers interspersed between layers of tellurium (Figures 2.8–2.10) [40,41].

2.7
Icosahedral Approximants

The original discovery of quasicrystals concerned icosahedral samples and they also furnish the most spectacular examples of the principle of quasicrystallinity.

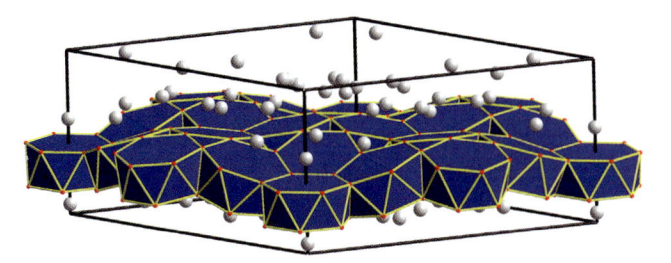

Figure 2.9 The warpedness of the layers of face-sharing hexagonal antiprisms is best appreciated from a side view.

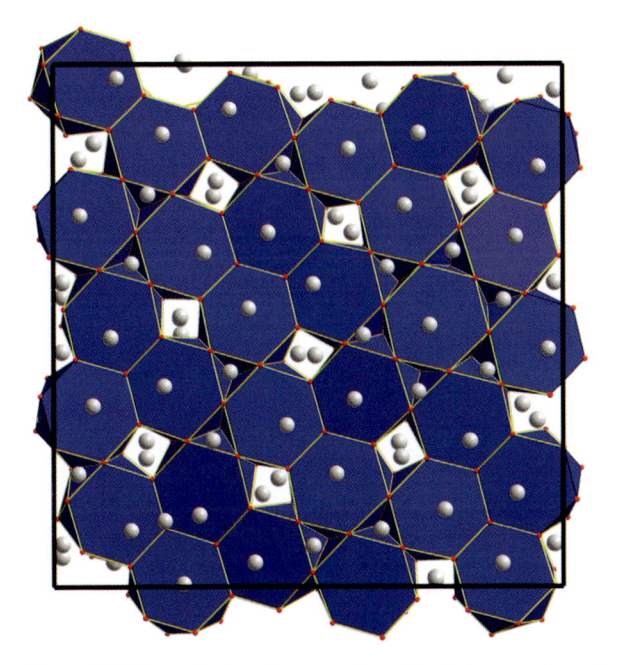

Figure 2.10 The orthorhombic approximant to dodecagonal quasicrystals contains even larger dodecagonal sections than the hexagonal approximant.

Approximants to icosahedral phases were known long before the advent of quasicrystals but they only appeared center stage as a result of that discovery. There are in fact several different kinds of icosahedral quasicrystals and one way of distinguishing between them is to classify them according to the most frequently occurring clusters in the structure. This also creates a direct relation between the quasicrystals and their approximants, although quasicrystal structures contain more than one type of cluster. Based on the prevalence of clusters, we can thus classify the quasicrystals as Bergman type, Mackay type, or Tsai type, and these names are directly related to cluster types that define approximants. The Bergman-type clusters are the basis for the 1/1 approximant first described by Bergman *et al.* [42] in the structure of $Mg_{32}(Zn,Al)_{49}$, but it has also been reported for a large number of systems: Li–Cu–Al [43], Li–Mg–Zn–Al [44], Li–Zn–Al [45], Mg–Ag–Al [46], and Mg–Cu–Al [47]. Additionally, the structure type $Li_{13}Zn_{6.5}Al_{20.5}$ [48] represents a different coloring of the same cluster structure. This cluster consists of a sequence of concentric shells, the outmost being a truncated icosahedron (soccer ball) followed by a rhombic triacontahedron, an icosahedron, and finally a single central atom (Figure 2.10). Higher order approximants have been found in the systems Al–Zn–Mg [49,50] (2/1), Zn–Mg–Er [51] (2/1), and Zn–Mg–Ga–Al [52] (3/2,2/1,2/1). In the 1/1 approximant, symmetry equivalent clusters sit at the vertices and the center of the I-centered cubic cell (Figure 2.11). Figure 2.12 shows the connection between clusters

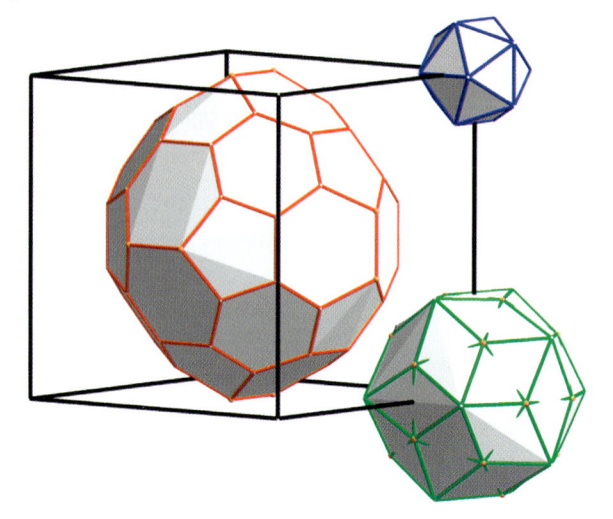

Figure 2.11 The consecutive shells of the Bergman cluster in the icosahedral 1/1 approximant. Blue edges denote the innermost icosahedral shell, green edges denote the rhombic triacontahedron, and red edges denote the truncated icosahedron. Note how the degree of deformation increases with increasing shell size as the shells are shoehorned into the translational repeat of the unit cell.

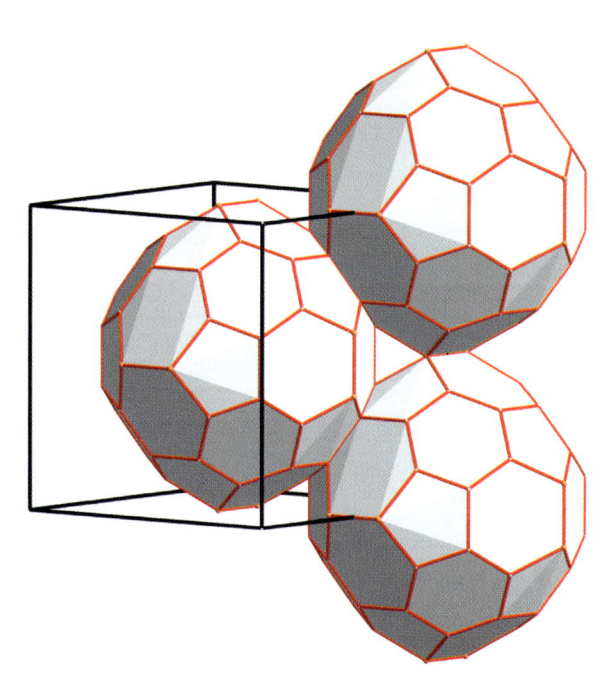

Figure 2.12 The truncated icosahedra share edges along the ⟨100⟩ directions and hexagonal faces along the ⟨111⟩ directions. Note that the polygons adjacent to the shared face are in an *anti*-configuration, hexagons facing pentagons.

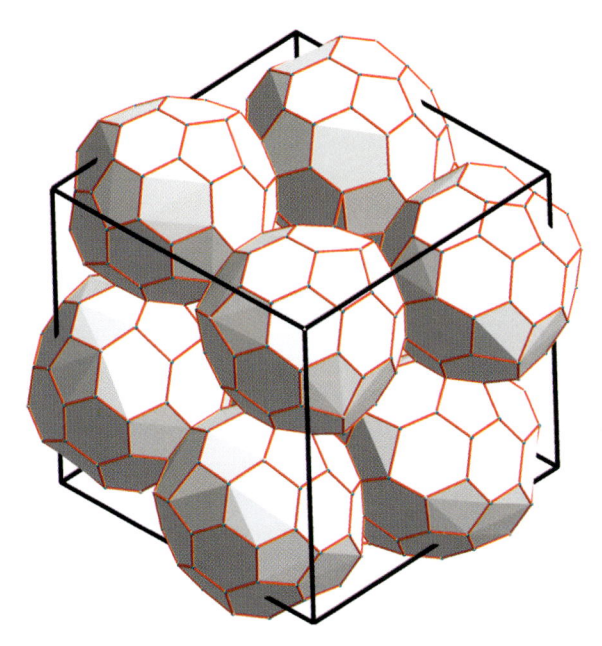

Figure 2.13 The unit cell of the 2/1 approximant shown with the connection between truncated icosahedra.

in the approximant. Along the cubic ⟨111⟩ direction truncated icosahedra share hexagonal faces, while along the cell edges, they share edges.

In the 2/1 approximant all connections between clusters are across hexagonal faces. The eight (Figure 2.13) clusters in the unit cell sit at the vertices of an oblate rhombohedron and the connection along the threefold direction of the short space diagonal is also across a hexagonal face.

In this larger approximant, it is also obvious that larger clusters may be identified, clusters that are too distorted in the 1/1 approximant to be properly seen. In Figure 2.14, the next shell consists of triacontahedra, and these polyhedra interpenetrate to share oblate rhombohedra.

For quasicrystals based on the Mackay cluster, a smaller approximant, 1/0 has been identified in the Al–Cu–Ru system [53] along with the 1/1 approximants in the same system [54], as well as in the systems Al–Cu–Fe–Si [55], Al–Mn–Fe–Si [56], Al–Mn–Si [57], Al–Fe–Si [58], Al–Re–Si [59], and Al–Pd–Mn–Si [60]. 2/1 approximants have been found in Al–Rh–Si [61] and Al–Pd–Mn–Si [62]. The structure of the 1/1 approximant is rather different from that of the Bergman cluster-type system since the unit cell is often primitive, allowing for different clusters at the cell vertices and the cell center. The reported disorder is quite pronounced, both in terms of positions and in terms of occupancy. The quaternary Al–Cu–Fe–Si and Al–Mn–Fe–Si approximants are I-centered while most other Mackay-type approximants are primitive. In Figure 2.15 the consecutive shells of the centered approximant are shown. It consists of a first coordination

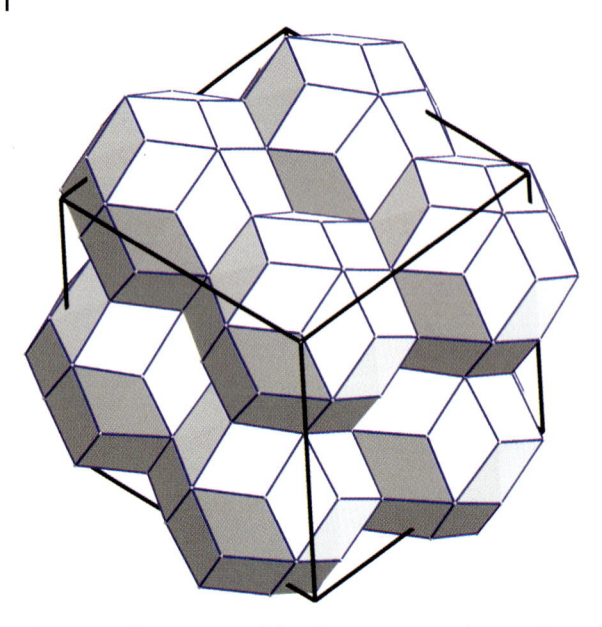

Figure 2.14 The structure of the 2/1 approximant shown as interpenetrating rhombic triacontahedra of the fourth shell.

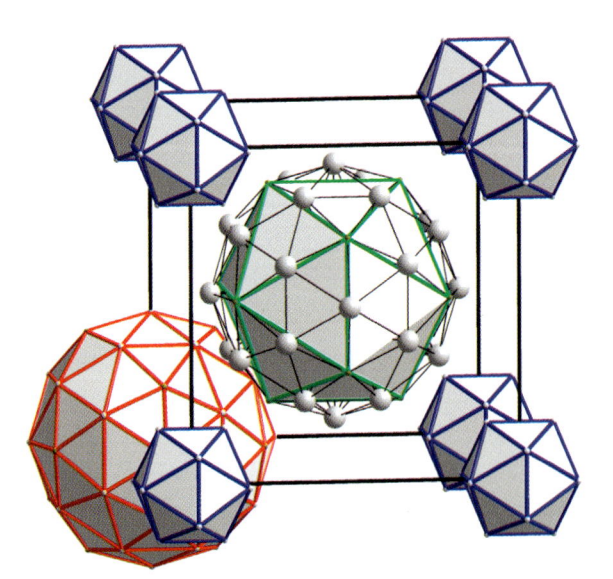

Figure 2.15 The first shell of the centered 1/1 Mackay-type approximant is an empty icosahedron (blue edges) followed by an icosahedron (green edges) and an icosidodecahedron (gray spheres outside the edges of the green-edged icosahedron). Together the last two shells form a pentakis icosidodecahedron with 42 vertices and 80 faces.

shell that is an empty icosahedron, followed by icosahedral and icosidodecahedral shells that are very similar in radius. Outside of these cluster shells, the structure is rather disordered.

For the noncentered 1/1 approximants, several different type structures exist. In the primitive structure of Al–Cu–Ru, the cluster centered at the origin is much more ordered and the shells of icosahedral symmetry extend further out. The first shells are identical with those from I-centered Al–Cu–Fe–Si, an icosahedron is followed by a pentakis icosidodecahedron, but here two further shells, an icosahedron and a rhombicosidodecahedron, complement the picture. Those last shells share atoms with neighboring clusters, the icosahedra share edges and the rhombicosidodecahedra share square faces. At the position ½ ½ ½, the picture is very different. Here, a central Ru atom is surrounded by a partially occupied dodecahedron, in turn surrounded by an icosidodecahedron (Figure 2.16)

Moving on to the 2/1 approximant, there are again two different sites – at the origin and at the cell center. Around the origin, a large, multishelled icosahedral cluster sits. It consists of an inner empty icosahedron surrounded by a dodecahedron. The following shells are an icosahedron, a truncated icosahedron, a dodecahedron, and finally yet another icosahedron. Around the cell center, a smaller cluster with a center built around a close-packed core forms an icosahedron and around this a dodecahedron.

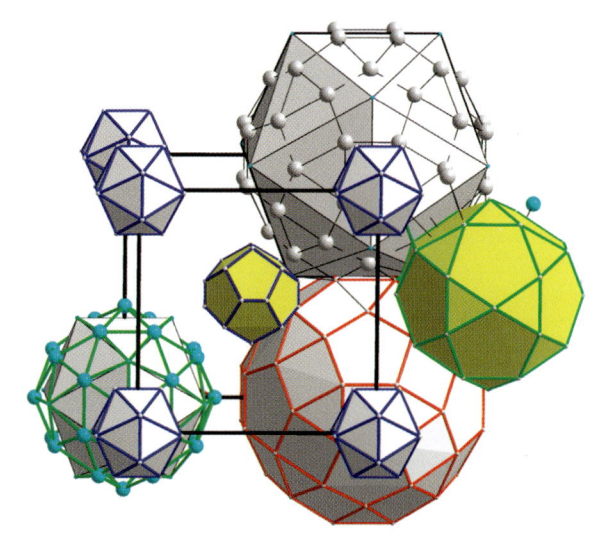

Figure 2.16 The clusters in the primitive Mackay 1/1 approximant. The shells in the cluster around the origin are shown with white faces. They are in order of size: an icosahedron (blue edges), pentakis icosidodecahedron (green edges, faces of the icosahedron shown), rhombicosidodecahedron (red edges), and finally a large icosahedron (gray edges) For the final icosahedral shell, the positions of the atoms from the interpenetrating rhombicosidodecahedral shell are shown. The two shells of the cluster surrounding the cell center are shown with yellow faces. They are a dodecahedron (blue edges) and an icosidodecahedron (green edges).

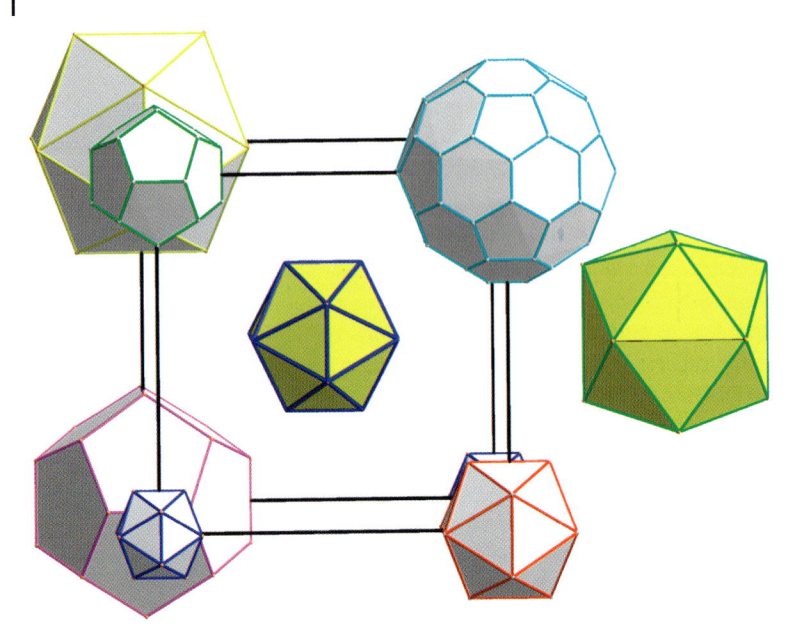

Figure 2.17 In the 2/1 Mackay approximant, the shells of the cluster at the origin are in order of increasing size: an icosahedron (blue edges), dodecahedron (green edges), icosahedron (red edges), truncated icosahedron (cyan edges), a large dodecahedron (purple edges), and finally a large icosahedron (yellow edges). The two easily identifiable shells of the cluster surrounding the cell center are shown with yellow faces. They are an icosahedron (blue edges) and a large icosahedron (green edges). These cluster have a center with nonicosahedral symmetry.

Naturally, the remaining atoms lend themselves to interpretation in terms of other more or less well-ordered shells. The important point is that in this 2/1 approximant there is no emergence of new clusters, but rather a growth of the clusters already present in the 1/1 approximant (Figure 2.17). From the differences between the clusters in the 1/1 and the 2/1 approximants, it is clear that Mackay- and Bergman-type clusters are interchangeable to some extent and as has been pointed out by several authors [50,63–65], the clusters do not actually translate directly to a particular type of quasicrystal.

The final group of approximants to be treated in this chapter are those related to the Tsai-type quasicrystals. These approximants form a particularly rich zoo of structures as the central part of the Tsai-type cluster, a disordered tetrahedron strongly influences its environment and may indeed order under certain conditions. There is a large group of 1/1 approximants known. Cadmium has been shown to form the 1/1 approximant with all the rare earths except Pm [66–69] as well as with Ca, Sr [70], Y, and some of the actinides (Np [71], Pu [72]). There are also systematic studies of ternary systems, where Cd has been replaced by Au/Ga [73], Au/In [74,75], Au/Sn [74,76], Ag/Ga [77], and Ag/In [74,78]. Zn only forms the 1/1 approximant with the smaller Sc [79] and Yb [80] while Be forms the same structure in the compound $Be_{17}Ru_3$ [81] and $Be_{17}Os_3$ [82]. There is a single high-quality study of a binary 2/1 approximant in this system $Cd_{76}Ca_{13}$ [83] and a number of studies where Cd is again replaced by a binary mixture to study site preference [73,74,84,85]. In addition to these rational approximants (or superstructures thereof), there is a set of hexagonal approximants that form between Cd and Ce [86], Eu [87], Gd [88], La [89], Nd [87], Sm [87], Sr [90], and Y [87] and between Zn and all the lanthanides save Pm and Eu [91–94]. Additionally, this phase forms in the systems Zn–Sc [95], Zn–Pu [96], Hg–La [97], and Hg–Sr [98] as well as a few ternary systems.

The 1/1 approximants are all closely related, but there are some subtle structural differences between them. The cluster center is a main group metal tetrahedron that is disordered at ambient pressure and temperature, but that may order at high pressures and/or low temperatures [99–103]. Outside of this is found a dodecahedron, an icosahedron, an icosidodecahedron, a slightly defect truncated icosahedron, and a pentakis icosidodecahedron (Figure 2.18).

The connectivity between the clusters is simple. If we consider the truncated icosahedra, these share hexagonal faces. Note that this is a different arrangement from the Bergman 1/1 approximant, where truncated icosahedra also share hexagonal faces, but where the surrounding polyhedral are in an *anti*-configuration. In the Tsai-type approximant, the faces adjacent to the shared hexagon are in *syn*-configuration. (Figure 2.19) The corresponding linkages for the 2/1 approximant are shown in Figure 2.20.

In the 2/1 approximant (Figure 2.20), the linkages between neighboring polyhedra are all the same. If truncated icosahedra are used in the representation, these polyhedra interpenetrate. For the early lanthanides, the 1/1

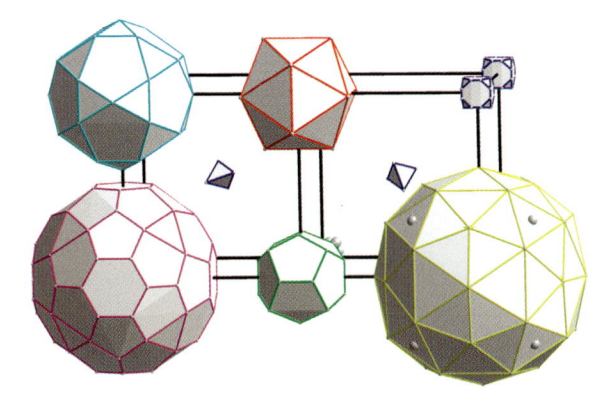

Figure 2.18 The successive shells of the Tsai-type approximants are shown. They are in order of size: first a disordered tetrahedron (blue edges). Because of the disorder the tetrahedra appear as truncated cubes (top right), but it is possible to select 4 atoms out of the 24 to define tetrahedra (center of the two unit cells). The next shell is a dodecahedron (green edges) followed by an icosahedron (red edges), an icosidodecahedron (cyan edges), a truncated icosahedron (purple edges), and a pentakis icosidodecahedron (yellow edges).

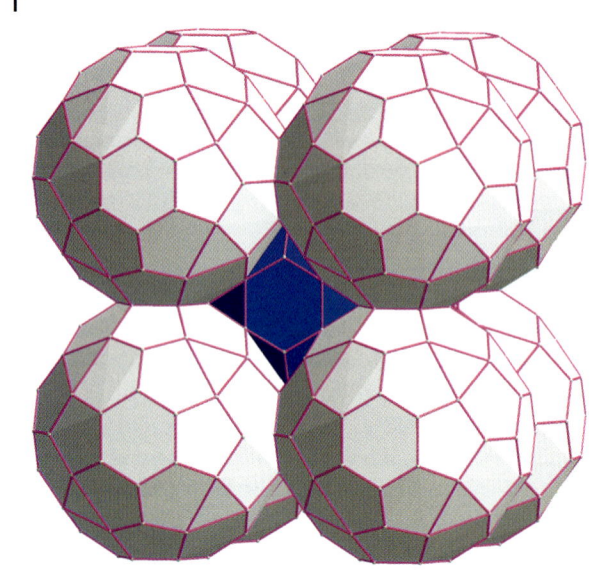

Figure 2.19 The arrangement of truncated icosahedra in the 1/1 Tsai-type approximant. The structure is I-centered and the blue polyhedron nesting in the center is identical to those at the vertices. While the polyhedra share hexagonal faces along the <100> directions, they interpenetrate along the ⟨111⟩ directions.

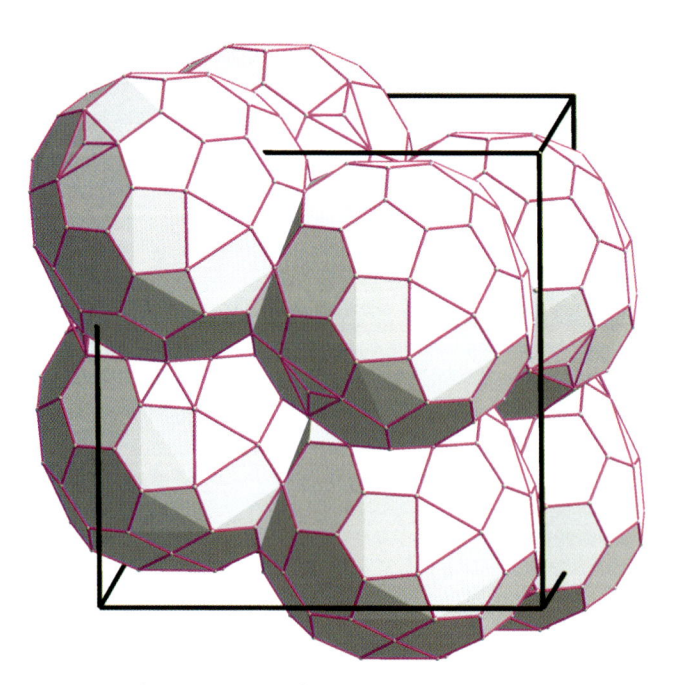

Figure 2.20 The arrangement of truncated icosahedra in the 2/1 Tsai-type approximant.

approximants with Cd contain extra Cd in an interstitial position that couples to the rest of the structure. For several compounds, this leads to severe disorder for the surrounding sites and for Cd–Ce and for Cd–Eu it leads to superstructure ordering. Interestingly, the kind of interstitial ordering leads to superstructures also in the hexagonal approximants; the compounds $Zn_{\sim 58}RE_{13}$, RE = Ce–Dy all exhibit subtly different ordering, $Zn_{\sim 58}RE_{13}$, RE = Er, Tm, and Lu show unusual occupational waves and $Cd_{58}Ce_{13}$ and $Zn_{58+\delta}Ho_{13}$ show complex modulated behavior.

2.8 Conclusion

The structures of the approximants of known intermetallic quasicrystals reveal a remarkably rich tapestry of arrangements, but there are some common features that underlie the formation of all approximants and quasicrystals alike. The first is the presence of highly preserved clusters or fused cluster fragments. The ubiquitous presence of these clusters is strongly indicative of their role in the early stages of precipitation from the metallic melt. For pentagonal, decagonal, and icosahedral approximants, the stability of clusters with local icosahedral symmetry (or interestingly, broken local icosahedral symmetry for the Tsai-type approximants) seems to be a prerequisite. Octagonal and dodecagonal phases are clearly based on other local arrangements but the pattern is clear; the existence of local clusters with symmetries incompatible with translational periodicity leads to large, complex arrangements that allow retention of that local symmetry. Such arrangements may be periodic approximants or aperiodic quasicrystals.

References

1 Dehlinger, U. (1927) *Z. Kristallogr.*, **65** (5/6), 615.
2 de Wolff, P.M.D. and van Aalst, W. (1972) *Acta Crystallogr. A*, **28** (S), S111.
3 Janner, A. (1972) *Acta Crystallogr. A*, **28** (S), S111.
4 Report of the Executive Committee, (1992) *Acta Crystallogr.*, **A48**, 928.
5 Boström, M. and Lidin, S. (2004) *J. Alloys Compd.*, **376**, 49–57.
6 Shechtman, D., Blech, I., Gratias, D., and Cahn, J.W. (1984) *Phys. Rev. Lett.*, **53**, 1951.
7 Shechtman, D. and Blech, I. (1985) *Metall. Trans.*, **A16**, 1005.
8 Bendersky, L. (1987) *J. Microsc.*, **146**, 303–312.
9 Feng, Y.C., Lu, G., Ye, H.Q., Kuo, K.H., Withers, R., and van Tendeloo, G. (1990) *J. Phys. Condens. Matter.*, **2** (49), 9749.
10 Xiao, C., Fujita, N., Miyasaka, K., Sakamoto, Y., and Terasaki, O. (2012) *Nature*, **487**, 349–353.
11 Zeng, X., Ungar, G., Liu, Y., Persec, V., Dulcey, A.E., and Hobbs, J.K. (2004) *Nature*, **428**, 157–160.
12 Hayashida, K., Dotera, T., Takano, A., and Matsushita, Y. (2007) *Phys. Rev. Lett.*, **98**, 195502.
13 Fischer, S., Exner, A., Zelske, K., Perlich, J., Deloudi, S., Steurer, W., Lindner, P., and Förster, S. (2010) *Proc. Natl. Acad. Sci.*, **108**, 1810–1814.

14 Wang, N., Chen, H., and Kuo, K.H. (1987) *Phys. Rev. Lett.*, **59**, 1010–1013.
15 Wang, N., Fung, K.K., and Kuo, K.H. (1988) *Appl. Phys. Lett.*, **52**, 2120–2121.
16 Cao, W., Ye, H.Q., and Kuo, K.H. (1988) *Phys. Status Solidi*, **A107**, 511–519.
17 Wang, Z. and Kuo, K.H. (1988) *Acta Crystallogr.*, **A44**, 857–863.
18 Villars, P. and Cenzual, K. (2016) Pearson's crystal data: crystal structure database for inorganic compounds, ASM International®, Materials Park, OH.
19 Walters, F.M., Jr., and Wells, C. (1935) *Trans. Am. Soc. Met.*, **23**, 727–750.
20 Dinega, D.P. and Bawendi, M.G. (1999) *Angew. Chem., Int. Ed.*, **38**, 1788–1791.
21 Vavilova, V.V., Galkin, L.N., and Glazov, M.V. (1991) *Inorg. Mater.*, **27**, 1801–1805.
22 Westin, L. and Edshammar, L.-E. (1973) *Chem. Scr.*, **3**, 15–22.
23 Pöttgen, R., Hlukhyy, V.H., Baranov, A., and Grin, Y. (2008) *Inorg. Chem.*, **47**, 6051–6055.
24 Elenius, M., Zetterling, H.M., Dzugutov, M., Fredrickson, D., and Lidin, S. (2009) *Phys. Rev. B*, **79**, 144201.
25 Nyman, H., Carrol, C.E., and Hyde, B.G. (1991) *Z. Kristallogr.*, **196**, 39–46.
26 Lidin, S. and Fredrickson, D. (2012) *Symmetry*, **4** (3), 537–544.
27 Burkhardt, U., Ellner, M., Grin, Yu., and Baumgartner, B. (1998) *Powder Diffr.*, **13**, 159–162.
28 Zhang, B., Gramlich, V., and Steurer, W. (1995) *Z. Kristallogr.*, **210**, 498–503.
29 Hudd, R.C. and Taylor, W.H. (1962) *Acta Crystallogr.*, **15**, 441–442.
30 Grin, J., Burkhardt, U., Ellner, M., and Peters, K. (1994) *J. Alloys Compd.*, **206**, 243–247.
31 Fleischer, F., Weber, T., Jung, D.Y., and Steurer, W. (2010) *J. Alloys Compd.*, **500**, 153–160.
32 Grushko, B., Holland, M.D., Wittman, R., and Wilde, G. (1998) *J. Alloys Compd.*, **280**, 215–230.
33 Döblinger, M. (2002) Elektronenmikroskopische Untersuchung Zum Umvandlungsverhalten dekagonaler Quasikristalle in der Systemen Al-Ni-Co und Al-Ni-Fe, Ph.D. thesis, Universität Karlsruhe, Logos Verlag, Berlin.
34 Yasuhara, A., Yubota, K., and Hiraga, K. (2014) *Philos. Mag. Lett.*, **94**, 539–547.
35 Hovmöller, S., Zou, L.H., Zou, X., and Grushko, B. (2012) *Philos. Trans. R. Soc. Lond. A*, **370**, 2949–2959.
36 Oleynikov, P., Demchenko, I., Christensen, J., Hovmöller, S., Yokosawa, T., Döblinger, M., Grushko, B., and Zou, X. (2006) *Philos. Mag.*, **86**, 457–462.
37 Sugiyama, K., Nishimura, S., and Hiraga, K. (2002) *J. Alloys Compd.*, **342**, 65–71.
38 Singh, D., Yun, Y., Wan, W., Grushko, B., Zou, X., and Hovmoller, S. (2014) *J. Appl. Crystallogr.*, **47**, 215–221.
39 Ishimasa, T. and Ishii, Y. (2011) *Philos. Mag.*, **91**, 2419–2420.
40 Conrad, M., Krumeich, F., Reich, C., and Harbrecht, B. (2000) *Mater. Sci. Eng. A*, **294**, 37–40.
41 Conrad, M. and Harbrecht, B. (2000) *Chem. Eur. J.*, **8** (14), 3094–3102.
42 Bergman, G., Waugh, J.L.T., and Pauling, L. (1952) *Nature*, **169**, 1057–1058.
43 Leblanc, M., LeBail, A., and Audier, M. (1991) *Physica B*, **173**, 329–355.
44 Lee, C.S. and Miller, G.J. (2001) *Inorg. Chem.*, **40**, 338–345.
45 Cherkashin, E.E., Kripyakevych, P.I., and Oleksiv, G.I. (1964) *Sov. Phys. Crystallogr.*, **8**, 681–685.
46 Auld, J.H. and Williams, B.E. (1966) *Acta Crystallogr.*, **21**, 830–831.
47 Phragmén, G. (1950) *J. Inst. Met.*, **77**, 489–552.
48 Tillard-Charbonell, M.M. and Belin, C. (1991) *J. Solid State Chem.*, **90**, 270–278.
49 Sugiyama, K., Sun, W., and Hiraga, K. (2002) *J. Alloys Compd.*, **342**, 139–142.
50 Lin, Q.S. and Corbett, J.D. (2006) *Proc. Natl. Acad. Sci. USA*, **103**, 13589–13594.
51 Kounis, A., Miehe, G., Saitoh, K., Fuess, H., Sterzel, R., and Assmus, W. (2001) *Philos. Mag. Lett.*, **81**, 395–403.
52 Kreiner, G. (2002) *J. Alloys Compd.*, **338**, 261–273.
53 Kirihara, K., Kimura, K., Ino, H., and Dmitrienko, V.E. (1998) *Proceedings of the 6th International Conference on Quasicrystals* (eds S. Takeuchi and T. Fujiwara), World Scientific, Singapore, pp. 243–246.

54 Shield, J.E., Chumbley, L.S., Mccallum, R.W., and Goldman, A.I. (1993) *J. Mater. Res.*, **8**, 44–48.
55 Takeuchi, T. and Mizutani, U. (2002) *J. Alloys Compd.*, **342**, 416–421.
56 Cooper, M.A. (1967) *Acta Crystallogr.*, **23**, 1106–1107.
57 Sugiyama, K., Kaji, N., and Hiraga, K. (1998) *Acta Crystallogr. C*, **54**, 445–447.
58 Phragmén, G. (1950) *J. Inst. Met.*, **77**, 489–552.
59 Onogi, T., Takeuchi, T., Sato, H., and Mizutani, U. (2002) *J. Alloys Compd.*, **342**, 397–401.
60 Sugiyama, K., Kaji, N., Hiraga, K., and Ishimasa, T. (1998) *Z. Kristallogr.*, **213**, 168–173.
61 Sugiyama, K., Sun, W., and Hiraga, K. (2004) *J. Non Cryst. Solids*, **334**, 156–160.
62 Sugiyama, K., Kaji, N., Hiraga, K., and Ishimasa, T. (1998) *Z. Kristallogr.*, **213**, 90–95.
63 Steurer, W. and Deloudi, S. (2009) *Crystallography of Quasicrystals*, Springer Series in Material Science, **126**, Springer, Berlin, 293.
64 Gratias, D., Puyraimond, F., Quiquandon, M., and Katz, A. (2001) *Phys. Rev. B*, **63**, doi: 10.1103/PhysRevB.63.024202.
65 Loreto, L., Farinato, R., Catallo, S., Janot, C., Gerbasi, G., and Angelis, G.De. (2003) *Phys. B*, **328**, 193–203.
66 Armbruster, M. and Lidin, S. (2000) *J. Alloys Compd.*, **307**, 141–148.
67 Gomez, C.P. and Lidin, S. (2003) *Phys. Rev. B*, **68**, doi: 10.1103/PhysRevB.68.024203.
68 Piao, S.Y., Gomez, C.P., and Lidin, S. (2006) *Z. Naturforsch.*, **61**, 644–649.
69 Gomez, C.P. and Lidin, S. (2004) *Chem. Eur. J.*, **10**, 3279–3285.
70 Köster, W. and Meixner, J. (1965) *Z. Metallkd.*, **56**, 695–703.
71 Krumpelt, M., Johnson, I., and Heiberger, J.J. (1969) *J. Less Common Met.*, **18**, 35–40.
72 Johnson, I., Chasanov, M.G., and Yonco, R.M. (1965) *Trans. Soc. Met. AIME*, **233**, 1408–1414.
73 Lin, Q.S. and Corbett, J.D. (2008) *Inorg. Chem.*, **47**, 7651–7659.
74 Lin, Q.S. and Corbett, J.D. (2007) *J. Am. Chem. Soc.*, **129**, 6789–6797.
75 Morita, Y. and Tsai, A.P. (2008) *Jpn. J. Appl. Phys.*, **47**, 7975–7979.
76 Kenzari, S., Demange, V., Boulet, P., Weerd, M.C.De., Ledieu, J., Dubois, J.M., and Fournee, V. (2008) *J. Phys. Condens. Matter*, **20**, 095218.
77 Kashimoto, S., Maezawa, R., Kaneko, Y., and Ishimasa, T. (2003) *Meet. Abstr. Physical Soc. Jpn.*, **58**, 716.
78 Sysa, L.V., Kalychak, Y.M., Galadzhun, Y.V., Zaremba, V.I., Akselrud, L.G., and Skolozdra, R.V. (1998) *J. Alloys Compd.*, **266**, 17–21.
79 Andrusyak, R.I., Kotur, B.Y., and Zavodnik, V.E. (1989) *Kristallografiya*, **34**, 996–998.
80 Bruzzone, G., Fornasini, M.L., and Merlo, F. (1970) *J. Less Common Met.*, **22**, 253–264.
81 Sands, D.E., Johnson, Q.C., Krikorian, O.H., and Kromholtz, K.L. (1962) *Acta Crystallogr.*, **15**, 1191–1195.
82 Matyushenko, N.N., Verkhorobin, L.F., Serykh, V.P., and Pugachev, N.S. (1982) *Russ. Metall.*, **6**, 146.
83 Gomez, C.P. and Lidin, S. (2001) *Angew. Chem., Int. Ed.*, **40**, 4037–4039.
84 Deng, B.B. and Kuo, K.H. (2004) *J. Alloys Compd.*, **366**, L1–L5.
85 Li, M.R., Hovmoller, S., Sun, J.L., Zou, X.D., and Kuo, K.H. (2008) *J. Alloys Compd.*, **465**, 132–138.
86 Piao, S.Y., Palatinus, L., and Lidin, S. (2008) *Inorg. Chem.*, **47**, 1079–1086.
87 Bruzzone, G., Fornasini, M.L., and Merlo, F. (1973) *J. Less Common Met.*, **30**, 361–375.
88 Bruzzone, G., Fornasini, M.L., and Merlo, F. (1971) *J. Less Common Met.*, **25**, 295–301.
89 Bruzzone, G. and Merlo, F. (1973) *J. Less Common Met.*, **30**, 303–305.
90 Bruzzone, G. (1972) *Gaz. Chim. Ital.*, **102**, 234–242.
91 Piao, S.Y., Gomez, C.P., and Lidin, S. (2006) *Z. Kristallogr.*, **221**, 391–401.
92 Gomez, C.P. and Lidin, S. (2002) *Solid State Sci.*, **4**, 901–906.
93 Piao, S.Y. and Lidin, S. (2007) *Inorg. Chem.*, **46**, 6452–6463.
94 Piao, S.Y. and Lidin, S. (2007) *Philos. Mag.*, **87**, 2693–2699.
95 Palenzona, A. and Manfretti, P. (1997) *J. Alloys Compd.*, **247**, 195–197.

96 Larsson, A.C. and Cromer, D.T. (1967) *Acta Crystallogr.*, **23**, 70–77.
97 Bruzzone, G. and Merlo, F. (1976) *J. Less Common Met.*, **44**, 259–265.
98 Bruzzone, G. and Merlo, F. (1974) *J. Less Common Met.*, **35**, 153–157.
99 Tamura, R., Nishimoto, K., Takeuchi, S., Edagawa, K., Isobe, M., and Ueda, Y. (2005) *Phys. Rev. B*, **71**, 092203.
100 Yamada, T., Tamura, R., Muro, Y., Motoya, K., Isobe, M., and Ueda, Y. (2010) *Phys. Rev. B*, **82**, 134121.
101 Yamada, T., Euchner, H., Gòmez, C.P., Takakura, H., Tamura, R., and de Boissieu, M. (2013) *J. Phys. Condens. Matter*, **25**, 205405.
102 Ishimasa, T., Kasano, Y., Tachibana, A., Kashimoto, S., and Osaka, K. (2007) *Philos. Mag.*, **87**, 2887–2897.
103 Yamada, T., Garbarino, G., Takakura, H., Gòmez, C.P., Tamura, R., and de Boissieu, M. (2014) *Z. Kristallogr.*, **229** (3), 230–235.

3
Medium-Range Order in Oxide Glasses

Hellmut Eckert[1,2]

[1]University of São Paulo, Institute of Physics in São Carlos, Av. Trabalhador Sãocarlense 400, 13566-590 São Paulo, Brazil
[2]WWU Münster, Institut für Physikalische Chemie, Corrensstraße 28-30, 48149 Münster, Germany

3.1
Introduction

Compared to crystalline solids, glasses present a formidable challenge to structure elucidation. Owing to the lack of translational symmetry in the glassy state, the concept of a "unit cell" is irrelevant, and structural models need to be cast in terms of interatomic distance statistics, the so-called pair correlation functions, which must be specified for each possible pair of atomic species occurring in the material. The most radical reference state is that of a random close packing of spheres, which is appropriate for the structural description of some *metallic glasses*. Ideally, such materials are characterized by no type of ordering at all. Abundant experimental evidence shows, however, that this extreme model is not applicable for the more common *covalent network glasses*, which are the subject of the present contribution. These glasses are formed by polyvalent oxides, chalcogenides, and halides of the Group 13–15 elements (e.g., B_2O_3, SiO_2, GeO_2, P_2O_5, $GeS(e)_2$, BeF_2, etc.), termed *network formers*, and may include a large variety of metal oxides, chalcogenides, and halides (termed *network modifiers*), which break up the network and create anionic sites (typically nonbridging oxygen (or sulfur) atoms), whose negative charges compensate the charges of the metal cations. In these glass systems, the clear tendency toward forming directional chemical bonds leads to the construction of a random assembly of well-defined coordination polyhedra. This is the famous *continuous random network model* of Zachariasen [1], which despite numerous challenges during the past decades, has withstood the test of time. Obtaining a comprehensive picture of glass structure then involves a detailed description of how the assembly of well-defined coordination polyhedra leads to loss of correlation as a function of distance. It is therefore convenient to discuss this subject on different length-scale

Handbook of Solid State Chemistry, First Edition. Edited by Richard Dronskowski, Shinichi Kikkawa, and Andreas Stein.
© 2017 Wiley-VCH Verlag GmbH & Co. KGaA. Published 2017 by Wiley-VCH Verlag GmbH & Co. KGaA.

domains: (1) *short-range order* involving only the first atomic coordination spheres (distance region 0.15–0.3 nm), (2) *intermediate-range order* comprising distance correlation in the second coordination sphere (0.3–0.5 nm), (3) *medium-range order* (0.5–1 nm) involving the formation of larger structural units (including clusters, chains, and rings), (4) *mesostructure* (1–200 nm) and, finally, (5) *microstructure* (>200 nm). This division may strike the reader as somewhat arbitrary, but it is closely related to the informational content of the experimental techniques available to probe glass structure. Despite the fact that the power of diffraction techniques is limited by the absence of translational symmetry, important structural concepts have emerged from information about second coordination spheres as extracted from X-ray and neutron diffraction data. Other relevant experimental techniques include vibrational spectroscopic methods (infrared absorption and Raman scattering), which have proven their utility in identifying certain ring structures in glasses, even though their quantification may be difficult. Further powerful techniques bearing on second coordination spheres include solid-state nuclear magnetic resonance (NMR) spectroscopy, X-ray photoelectron spectroscopy (XPS), extended X-ray absorption fine structure (EXAFS), and molecular dynamics simulations. All of these methods respond most sensitively to changes and details concerning the first coordination spheres, whereas extracting information on the second coordination spheres requires a substantially increased effort. On the other hand, the usual transmission and scanning electron microscopies cover the regimes of meso- and microstructure. Finally, the region within 0.5 to a few nanometers distance must be considered "unknown territory" as there is no ideally suited physical examination technique for deriving specific structural information in this length scale domain. Therefore, structural concepts describing medium-range order in glasses usually emerge from extrapolation, via computational modeling and/or energy minimization, respecting constraints posed by the information on first and second coordination spheres, meso- and microstructure, as well as knowledge deduced from macroscopic physicochemical observables. Within this length scale domain we cannot hope to "prove," but only to corroborate a postulated glass structure and help to eliminate potential alternative scenarios.

The present contribution is organized as follows: Chapter 2 introduces the chief concepts of structural ordering phenomena in glasses and current approaches toward their quantitative parametrization. Chapter 3 reviews the principal experimental methods by means of which medium-range order phenomena can be characterized. These include (1) X-ray and neutron diffraction, (2) vibrational spectroscopy, and (3) nuclear magnetic resonance (NMR) spectroscopy. While medium-range order concepts have traditionally emerged from modeling diffraction data, the use of solid state NMR for probing structural correlations beyond the first coordination sphere has been much less common. Therefore, the presentation in Chapter 3 places emphasis upon a description of the NMR methodology, which has been subject to substantial evolution during the past decades. In Chapter 4, we will review the state of the literature regarding the archetypical oxide glass systems SiO_2, GeO_2, and B_2O_3, and of some binary and ternary glass systems comprising them. Chapter 5 will be devoted to simple binary oxide glass systems containing network modifier species as well as ternary systems containing more than one network-former species. Even though much of the spectroscopic information reviewed in Chapters 4 and 5 pertains to quantitative analyses of the second coordination spheres in glasses, the latter often reflect ordering phenomena at longer distance scales, such as inhomogeneous ion distributions and the formation of larger structural entities. The state of knowledge concerning these more extended distance ranges will be discussed in terms of the ring statistics and size distributions in strongly polymerized networks, as deduced chiefly from molecular dynamics simulations. Finally, Chapter 6 will give a critical assessment of the current accomplishments and the challenges remaining, setting the stage for further experimental effort in this area. We hope this present account of the current state of the art in this field might serve as a useful starting point for subsequent reviews in the future.

3.2
General Concepts of Structural Order in Glasses

As mentioned in Section 3.1, Zachariasen's continuous random network model is based on covalently bonded well-defined structural units. Figure 3.1 summarizes such units as encountered in the most common glass systems based on the

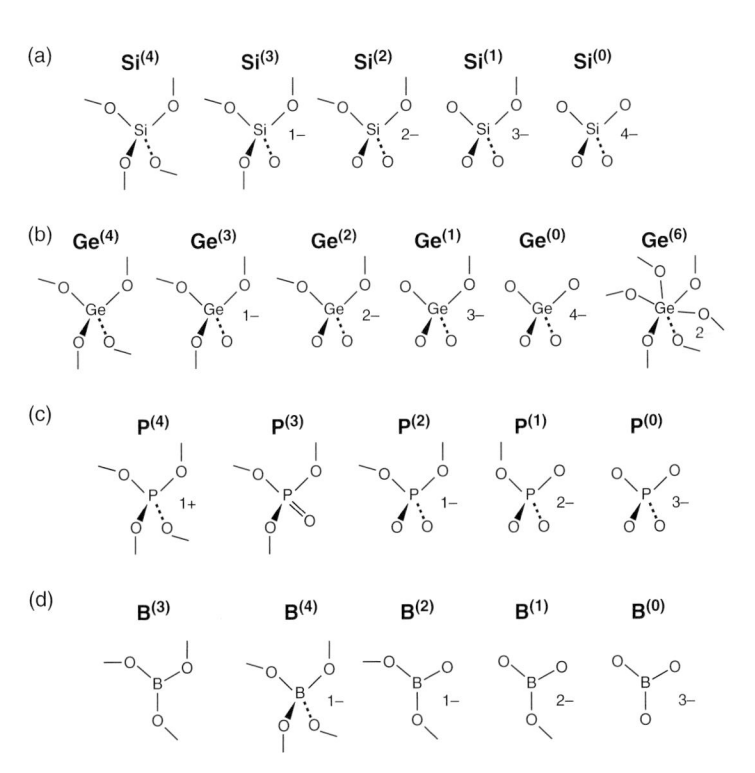

Figure 3.1 Possible short-range order units occurring in (a) silicate, (b) germanate, (c) phosphate, and (d) borate glasses.

network formers SiO_2, B_2O_3, P_2O_5, and GeO_2. Analogous structural units can be drawn up for sulfide, oxysulfide, and halide glasses. Within the first task of structural analysis we need to identify and quantify these local coordination polyhedra and to characterize their average bond lengths and bond angles as well as their distributions. From X-ray and neutron diffraction experiments it is now well-known that the typical bond lengths of these units are similar to those encountered in the structures of corresponding crystalline model compounds and that their distribution functions are rather narrow.

However, a glass is more than the sum of its individual coordination polyhedra, and this is the topic of this chapter. Concerning the second coordination sphere, we are interested in information such as (1) distance correlations between network formers, (2) distance correlations between network formers and network modifiers, and (3) distance correlations between the network modifiers (see Figure 3.2). Any distance distribution that is nonrandom in this domain can be classified as a manifestation of intermediate-range order. In covalent network glasses, the existence of intermediate-range order is a natural consequence of the distance and angle-dependent interaction potentials defining chemical bonds. For example, the pair correlation function defining next-nearest neighbor correlations cannot be random, as these atoms are linked by chemical bonds, whose three-body interaction potential $V_{abc}(\theta)$ clearly depends on the a, b, c bond angle θ (see Figure 3.3). If $V(\theta)$ has a sharp minimum, such as is the case for the tetrahedral O—Si—O bond angles in silica glasses, then there is a rather

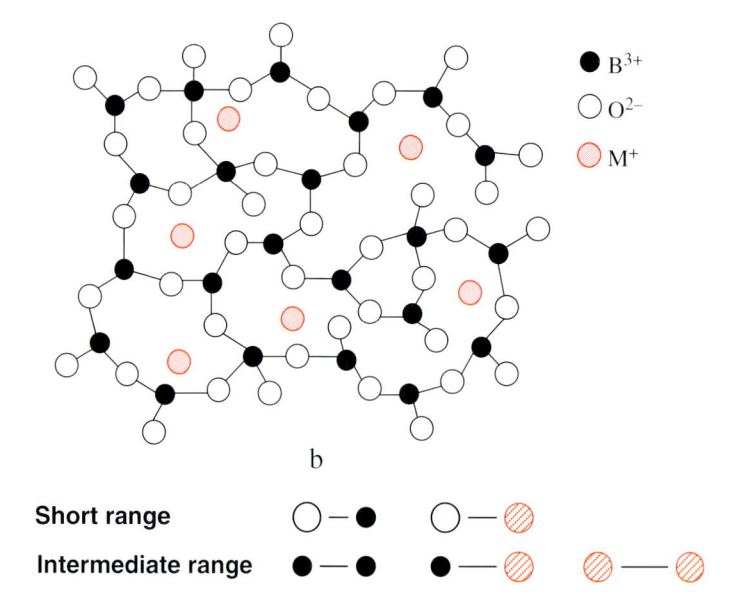

Figure 3.2 Sketch of the structural organization in covalent oxide glasses, depicting short- and intermediate-range distance correlations. Black, open, and red circles denote network formers, oxygen, and network modifier species, respectively.

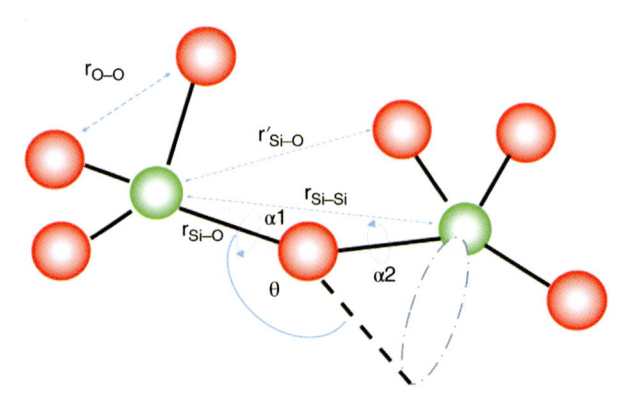

Figure 3.3 Next-nearest neighbor correlations in tetrahedral network glasses (such as SiO_2), T–O–T angle θ, defining second-nearest neighbor correlations and the two torsional angles α_1 and α_2 defining third- and fourth-nearest neighbor correlations.

narrow distribution of second-nearest neighbor distances. This is indeed the case for the O—O next nearest neighbor correlations in most oxide glasses. On the other hand, $V(\theta)$ functions with broad and shallow minima result in a wider distribution of θ, and hence a wider distribution of next-nearest neighbor distances. This is the case for the Si–Si pair correlation function in silicate glasses. Therefore, characterizing such bond angle distributions is an important part of describing intermediate range order in glasses. As discussed below, both X-ray and neutron diffraction as well as nuclear magnetic resonance experiments can provide important data on this topic. Based on these arguments, corroborated by experimental results, the atomic arrangement in the second coordination spheres is neither expected nor found to be random [2]. A related aspect of network-former connectivity concerns the formation of preferred linkages in mixed network-former glasses, owing to chemical preferences. As discussed in this chapter, such bonding preferences can be quantified by advanced dipolar solid-state NMR experiments.

As pointed out in the literature, the connectivity of individual structural units also involves two torsional angles, defining the spatial relations between atoms that are within the third and fourth coordination spheres of each other [3] (see Figure 3.3). While it is essentially impossible to probe such "dihedral angles" from an experimental point of view, this information can be extracted from structural models generated by molecular dynamics simulations [4]. Other ordering phenomena beyond the second coordination sphere involve the formation of larger structural motifs such as clusters, chains, or rings, which again are a consequence of the directionality of chemical bonding. This distance regime is called "medium-range order," which can be identified in chemically very diverse glass systems [5]. In highly polymerized glasses, a common way of characterizing such effects is specifying the rings formed within the structure (see Figure 3.4) and the distribution of their sizes [4,6]. The discussion has mostly focused on silicate glasses, where each network-former species is a four-coordinated $Si^{(n)}$

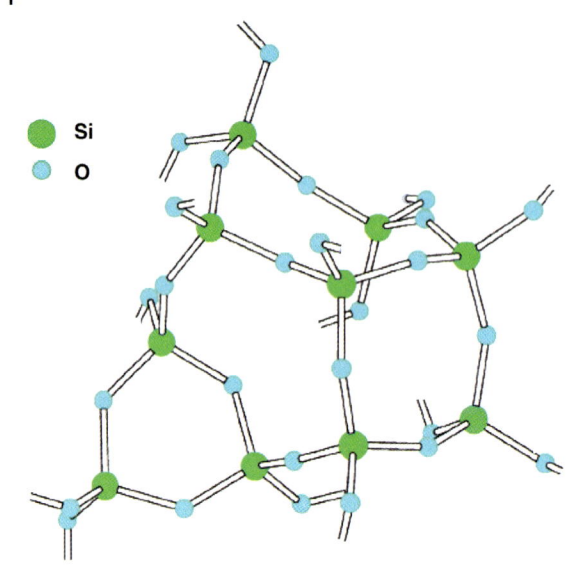

Figure 3.4 Ring structure motifs in tetrahedral network glasses illustrated at the example of SiO_2.

species containing 4-n nonbridging oxygen atoms. Each such $Si^{(n)}$ species is considered as a corner and the connection of two vertices is an edge. A ring is primitive if the number of edges connecting any two corners forms the shortest path. The smallest possible primitive ring is formed by two tetrahedra if they share a common edge (ring size 2). If all the tetrahedra are corner-sharing, the smallest ring size is three. Highly modified glass structures also contain $Si^{(1)}$ (terminal tetrahedra or dimers) and $Si^{(0)}$ (isolated tetrahedra) units that cannot take part in rings. From an experimental viewpoint, average ring sizes or ring size distributions have been deduced from X-ray or neutron diffraction data, using Reverse Monte Carlo modeling. In more recent years, this information has also become available from molecular dynamics simulation studies [6]. Further, in glasses containing network modifiers, the distribution of these modifiers is not necessarily random, and clustered arrangements have been postulated in some systems on the basis of such simulations. This is the *modified random network model* proposed by Greaves for alkali silicate glasses, leading to the postulate of ion-conduction channels [7]. It is important to realize that such larger-scale segregation effects will generally also manifest themselves at the level of the second coordination sphere, resulting in nonrandom network former – network modifier correlations as well as network modifier – network modifier correlations [8].

Nanoscale segregation effects of the above-mentioned kind are the consequence of phase separation tendencies, which are usually kinetically suppressed at and below the liquidus temperature, resulting in glasses that appear more homogeneous than expected on the basis of equilibrium thermodynamics. For such systems, the thermal history of the glass is an important control parameter, by means of which the size distribution of nanosegregated domains, and hence the macroscopic physical properties of the glass can be manipulated. As phase separation phenomena frequently lead to subsequent crystallization, an understanding of medium-range order also provides an important bridge toward the elucidation and elaboration of functional *glass ceramics* [9,10].

3.3 Experimental Characterization of Medium-Range Order

Below, we will summarize the principal features of the most important experimental techniques yielding insights into intermediate-range and medium-range order in glasses. All of them have in common that their dominant informational content relates to short-range order. Nevertheless, using more sophisticated experimental approaches and data analysis methods, these techniques can provide important insights into distance correlations beyond the first coordination sphere. Only brief summaries will be given on diffraction experiments and vibrational spectroscopy, as the application of these methods to the analysis of medium-range order phenomena in glasses is already well established and many reviews are available. Advanced solid state NMR spectroscopy, will be covered in somewhat more detail, as this method offers an enormous experimental flexibility, making its application toward studies of medium-range order in glasses particularly promising.

3.3.1 X-Ray and Neutron Diffraction Techniques

X-ray and neutron diffraction techniques measure the total pair distribution function, which represents the probability of finding an atom at a distance r away from a central atom. The theoretical background, experimental details of the data acquisition and subsequent corrections and further processing, have been excellently described in Refs [11,12], and the reader is referred to these articles for further information. The raw data consist of the measured scattering intensity $S(Q)$ as a function of momentum transfer Q, which is given by the expression

$$Q = 4\pi/\lambda \times \sin\beta, \tag{3.1}$$

where λ is the wavelength of the electromagnetic radiation and β is ½ the scattering angle. Fourier transformation yields the total pair distribution function

$$T(r) = 4\pi r g(r), \tag{3.2}$$

which is a superposition of all possible pair distribution functions $t(r)$ involving the various atomic species, averaged over the entire volume investigated. These are then extracted with the help of structure simulation routines, such as Reverse Monte Carlo modeling or molecular dynamics simulations. In the older diffraction literature and in papers focusing on MD simulations, the distribution

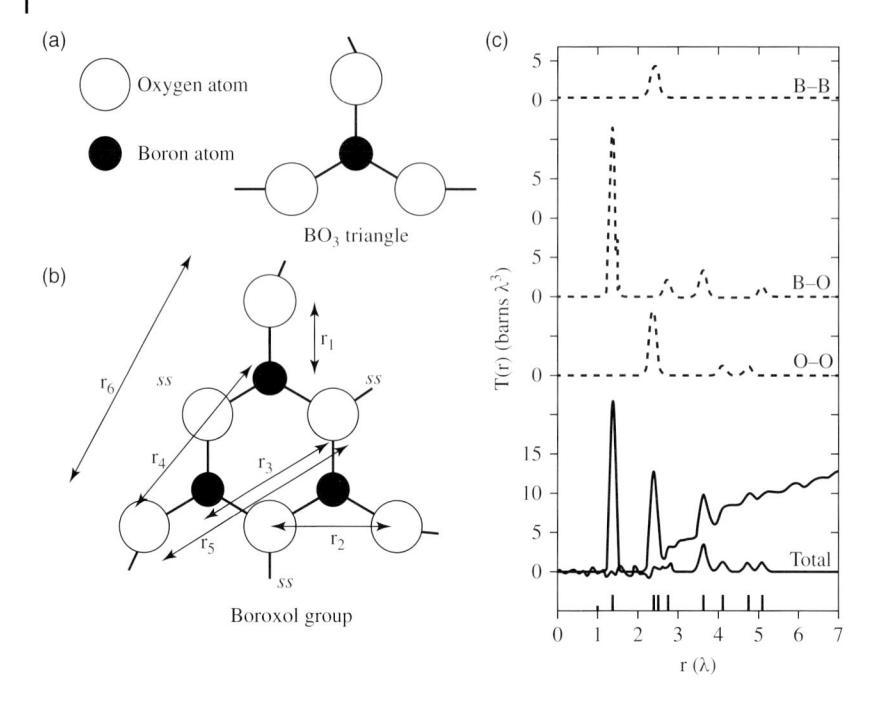

Figure 3.5 (a,b) Relevant local structural units and interatomic distances involving the boroxol groups in glassy B_2O_3. (c) Total distribution function obtained by neutron diffraction and partial distribution functions extracted from an analysis of the data obtained on glassy B_2O_3. (Reproduced with permission from Ref. [13].)

function $g(r)$ is frequently published instead of $T(r)$. Figure 3.5 shows an instructive application to B_2O_3 glass, where the abundant six-atom ("boroxol") rings with their rather well-defined B$\cdots$B, B$\cdots$O, and O$\cdots$O distances in the second, third, and fourth coordination spheres produce corresponding maxima in the experimental $T(r)$ data.

X-ray and neutron diffraction complement each other by measuring the electron and nuclear distribution functions, respectively. Thus, frequently, more detailed information is available by interpreting both experiments jointly. In neutron diffraction experiments, it is further possible to extract partial pair correlation functions belonging to a specific atom by conducting experiments with isotopically labeled samples. As the different nuclear isotopes of a given element normally feature widely different neutron scattering lengths, any difference observed between the $S(Q)$ functions of the different isotopologues investigated arises from pair correlation functions involving that element for which the isotopic replacement was done.

Considerable attention has been drawn to an ubiquitous phenomenon commonly observed in the $S(Q)$ data from neutron and X-ray diffraction experiments in many different types of glasses, which has been (mis)-labeled as first sharp diffraction peak (FSDP) [2,5]. Such peaks, which occur at scattering vectors smaller than Q_P, the position of the principal peak of the diffraction pattern determined by the nearest neighbor distance in real space, show rather anomalous behavior as a function of pressure, temperature, and composition. Unlike other diffraction peaks, the intensity of this feature increases with increasing temperature and decreases with increasing pressure and densification. Molecular dynamics simulations reveal such prepeaks are a commonplace occurrence in network glasses, and in fact an expected feature of glasses organized according to Zachariasen's continuous random network model, as long as directional forces are included [14]. As such sharp intense features at small wave vectors translate into very broad features in the corresponding Fourier transforms, their manifestations in the $T(r)$ plots are difficult to discern. Nevertheless, it is clear that these prepeaks correspond to real-space structural correlations, reflecting periodic longer-scale density fluctuations or periodic voids on the nanometer scale [15,16]. Structural interpretations in terms of system-specific features have thus far remained elusive. The present review will not discuss the interpretations of FSDP phenomena, as this subject has been extensively covered in the literature [2,5,14–16].

3.3.2
Vibrational Spectroscopy

Vibrational spectroscopy has enjoyed long-standing use for the structural analysis of many oxide glasses. From an experimental point of view, data are obtained either in an absorbance (or reflectance) mode, using polychromatic excitation in the infrared region, followed by Fourier transformation (FTIR spectroscopy) or in a scattering mode, exploiting the inelastic scattering of photons with atomic or molecular vibrations in the sample (Raman spectroscopy). As FTIR absorption and inelastic scattering intensities are governed by different selection rules, the spectra are generally complementary, and a full picture is obtained only by means of a joint interpretation of both experiments. Excellent texts detailing the fundamentals of both techniques have recently appeared [17,18]. The principal information drawn from these spectra concerns features of the first coordination spheres. Thus, the information about bonds involving the Si—O, B—O and P—O linkages in the glassy network is quite secure. The corresponding stretching modes are quite sensitive to the specific connectivities in which these local structural units are involved, and this particular property makes vibrational spectroscopy also a useful probe of the second coordination sphere. In addition, vibrational spectroscopy has also been very useful for probing larger-scale units such as ring structures. For example, the three- and four-coordinate units in crystalline borate compounds are usually found within distinct ring-membered units, which give characteristic signatures in IR and Raman spectra [19,20]. As the spectra of the glasses show similar features, much effort has been devoted to identifying such analogous units in borate glasses. Figure 3.6 gives an early example for a series of sodium borate glasses [20]. The spectrum of the glass containing 5 mol% Na_2O is completely dominated by the signal at 808 cm^{-1},

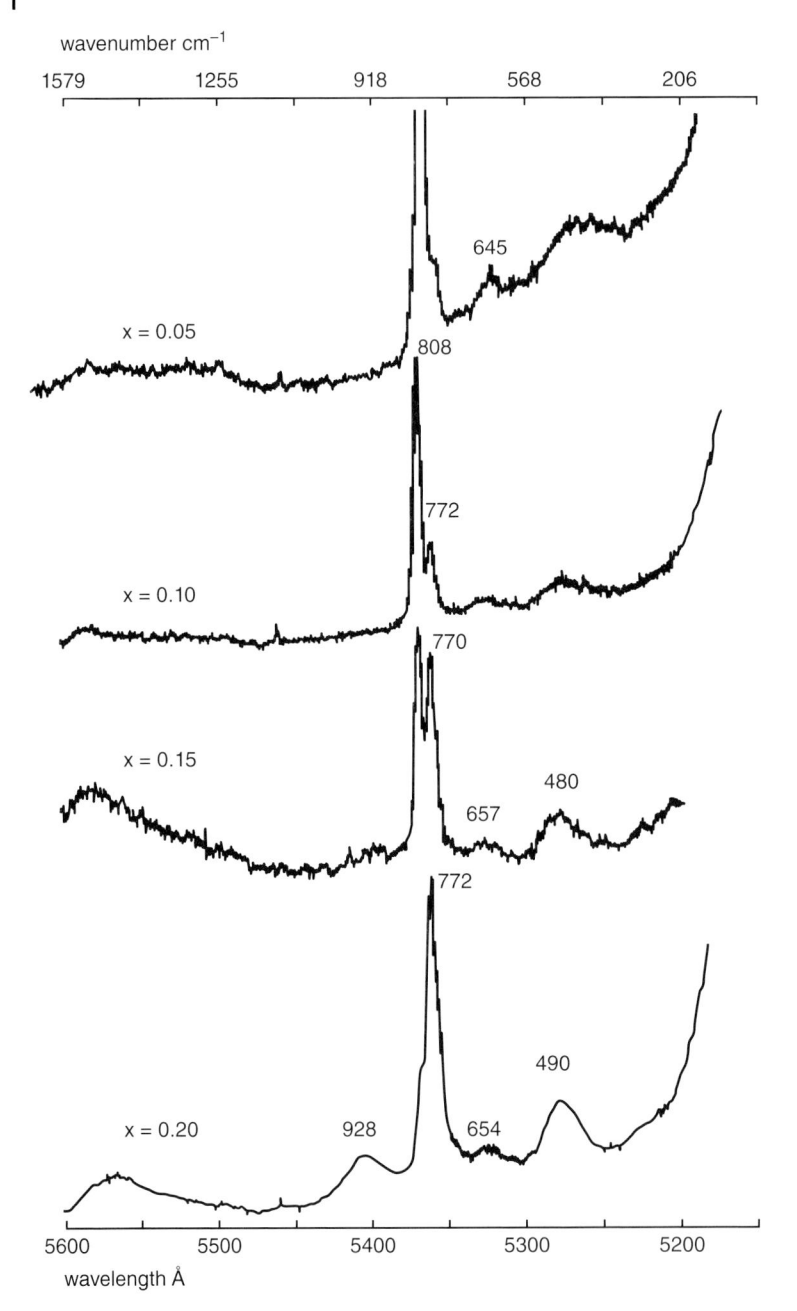

Figure 3.6 Raman spectra of $(Na_2O)_x(B_2O_3)_{1-x}$ glasses ($0.05 \leq x \leq 0.20$). The feature at 808 cm^{-1} reflects the symmetric breathing mode of the boroxol group, containing $B^{(3)}$ units, while the feature near 770 cm^{-1} growing with increasing modifier content, reflects the response of another type of ring structure containing $B^{(4)}$ units. (Reproduced with permission from Ref. [20].)

which is assigned to the symmetric vibrations of the six-atom boroxol group containing three $B^{(3)}$ units. At higher Na_2O contents, the boroxol peak diminishes and is replaced by a peak near 770 cm^{-1}, which is assigned to another six-atom ring unit containing two $B^{(3)}$ and one $B^{(4)}$ species. As discussed in Ref. [20] and summarized later in Section 5.1.2, there are a number of different possibilities of ring formation involving $B^{(3)}$ and $B^{(4)}$ units, which can be distinguished by their Raman spectra.

An unfortunate drawback of vibrational spectroscopic methods is the potentially large difference of intrinsic oscillator strengths associated with specific absorption or scattering peaks, which depend on the changes of dipole moments (or polarizabilities) as a function of the vibrational coordinate. For this reason, FTIR absorption and Raman spectra are generally only good for qualitative conclusions. Furthermore, as the data contain the spectroscopic response of the entire atomic/molecular inventory, the interpretation of spectra becomes increasingly difficult with increasing compositional complexity of the sample under study.

Finally, most glasses exhibit intense Raman scattering bands in the very low frequency range ($1/\lambda < 100$ cm^{-1}), which are attributed to cooperative vibrational motion of larger structural entities. In the literature, this phenomenon has been labeled "boson peak" and, similar to the FSDP in diffraction, divergent structural hypotheses have been proposed on various systems. This literature will not be covered in the present review, as comprehensive discussions have appeared elsewhere [21].

3.3.3
Solid-State NMR

Solid-state NMR is an element-selective, inherently quantitative method ideally suited for the structural characterization of glasses. The method relies on the orientational quantization of nuclear magnetic moments, which gives rise to quantized energy levels in applied magnetic fields. By applying radio waves fulfilling the Bohr condition

$$\omega = \gamma B_o \tag{3.3}$$

transitions between these quantized energy levels can be stimulated, producing the spectroscopic signal. As a matter of fact, the actual resonance frequency is not given by Eq. (3.3) alone, but is rather influenced by a four distinct and different *internal interaction mechanisms*, whose parameters reflect the details of the local structural environment. The internal interactions include (1) *magnetic screening effects* produced by the electronic environment of the nuclei, (2) *nuclear electric quadrupolar interactions* between the quadrupole moment of spin > ½ nuclei and electrostatic field gradients generated by the local distribution of charges surrounding the nuclei, (3) *magnetic dipole–dipole interactions* of the nucleus under observation with nuclear magnetic moments in their neighborhood, and (4) *internuclear spin–spin coupling* via the polarization of electron density within covalent bonds linking the two nuclei. For a more detailed presentation of the theoretical foundations of solid state NMR and the specific manifestations of these interactions upon the spectroscopic lineshapes we refer to some introductory texts [21–24]. All of the above-mentioned internal interactions influencing the Zeeman energy levels are anisotropic, and their simultaneous presence generally results in indistinct line-broadening effects in powdered or glassy samples, which leads to poor spectroscopic resolution. The great power of solid-state NMR for structural analysis derives from the fact that these interactions can be studied and related to structural information in a selective manner, using special *selective averaging experiments*. Applications of this concept to structural studies of glasses have been covered by numerous reviews in the literature [25–30].

3.3.3.1 Magic-Angle Spinning NMR

The most common selective averaging experiment is magic angle spinning (MAS-NMR) involving data acquisition under mechanical rotation of the sample about an axis inclined by 54.7° (the so-called "magic angle") relative to the magnetic field direction [31]. This operation eliminates the inhomogeneous spectral broadening caused by the anisotropic interactions influencing NMR spectra in the solid state, and hence, produces high-resolution spectra, from which each distinct structural unit specified in Figure 3.1 can be identified and characterized by an *isotropic chemical shift*. The latter is defined via

$$\delta_{iso} = (\omega_{sample} - \omega_{ref})/\omega_{ref}, \quad (3.4)$$

as the normalized difference between the resonance frequency under consideration and that of a suitably chosen reference standard. Isotropic chemical shifts are widely used for characterizing different local coordination environments in glasses. The isotropic chemical shifts are further influenced by atoms in the second coordination sphere, which makes MAS-NMR a suitable probe for studying next-nearest neighbor environments as well.

Nuclei with spin quantum number $> \frac{1}{2}$ feature an asymmetric charge distribution, which can be mathematically described by an *electric quadrupole moment* eQ. This quadrupole moment interacts with local *electric field gradients* (EFGs) at the nuclear site, described by a traceless diagonal tensor with components V_{xx}, V_{yy}, and V_{zz}, with V_{zz} being the largest component by definition. The quadrupolar interaction influences the Zeeman energy levels and leads to additional line broadening effects and shifts in resonance frequencies in the MAS-NMR spectra. By means of detailed lineshape simulations, we can extract the chief observables characterizing this interaction [32]. They are the nuclear electric quadrupolar coupling constant eQV_{zz}/h (in H_z) and the asymmetry parameter $\eta_Q = (V_{xx} - V_{yy})/V_{zz}$, which characterizes the overall magnitude of the electric field gradient and its deviation from cylindrical symmetry. Examples of such NMR lineshape fits are seen in Figure 3.7, for the MAS-NMR spectrum of glassy B_2O_3, which is comprised of two distinct lineshape components, corresponding to the $B^{(3)}$ units within boroxol rings, $B^{(3)}$-I and of $B^{(3)}$ units outside of boroxol rings, $B^{(3)}$-II.

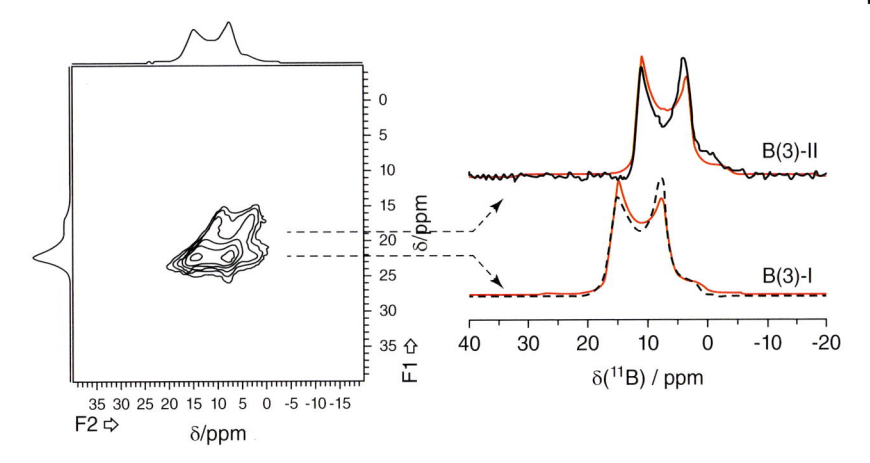

Figure 3.7 ^{11}B Triple-quantum (TQ-) MAS-NMR spectrum of glassy B_2O_3, recorded at 11.7 T. The top spectrum (projection along the frequency axis F2) corresponds to the anisotropically broadened standard MAS-NMR spectrum, whereas the vertical spectrum (projection along the frequency axis F1 corresponds to the isotropic high-resolution MAS-NMR spectrum. The individual contributions comprising the MAS-NMR lineshape can be obtained from the "slices" along F2 located at the peak maxima observed in the F1 dimension. (From Ref. [34].)

To resolve the superimposed MAS-NMR spectrum (shown at the top of the figure) into these two distinct components, a two-dimensional experiment was done, called TQMAS (triple-quantum MAS-NMR), which is specifically designed for eliminating anisotropic broadening due to second-order quadrupolar effects in the indirect dimension [33]. The isotropic spectrum is seen on the vertical axis of the 2D plot, and allows for a facile deconvolution into two individual components [33]. As will be seen further below, many conclusions concerning medium-range order in glasses are drawn from such TQMAS spectra, when quadrupolar nuclei are involved, specifically in NMR studies of the quadrupolar ^{17}O nuclei in glasses.

Both isotropic chemical shifts and electric field gradients can be calculated from structural input by first principles using density functional theory (DFT) [35,36]. In this fashion, structural hypotheses can be tested against experimental data. While their informational content relates primarily to short-range order, important information on the second coordination sphere is indirectly available, for example, through a relation between ^{17}O NMR electric field gradients and Si—O—Si bond angles (see Section 3.4).

3.3.3.2 Dipolar NMR Methods

The nuclear magnetic resonance frequencies are further influenced by magnetic dipole–dipole interactions, as the spins experience the magnetic moments of their neighbors. The strength of the interaction depends in a rather straightforward way on the distance r_{ij} between the nuclei i and j involved. Depending on whether the surrounding nuclei are of the same type as the ones, whose NMR signal is being observed or of different types, one differentiates between *homo- and heteronuclear* interactions. In systems characterized by multiple-spin interactions, a convenient quantity characterizing the average strength of these interactions is the average mean square of the local field, the so-called *second moment*, which can be measured selectively both under static and magic-angle spinning conditions by using special *dipolar recoupling* techniques. These second moments relate to the internuclear distance distributions, in a rather straightforward manner, by the van Vleck formulae [37]:

$$M_{2\text{homo}} = \frac{3}{5}\left(\frac{\mu_o}{4\pi}\right)^2 S(S+1)\gamma_S^4\hbar^2 \sum_{i\neq j} r_{ij}^{-6}, \quad (3.5)$$

$$M_{2\text{hetero}} = \frac{4}{15}\left(\frac{\mu_o}{4\pi}\right)^2 I(I+1)\gamma_I^2\gamma_S^2\hbar^2 \sum_S r_{is}^{-6}, \quad (3.6)$$

for homo- and heteronuclear interactions, respectively. In these equations, the quantities γ_S and γ_I denote the gyromagnetic ratios of the observed nuclei and those of the surrounding heteronuclei, and I and S are their spin quantum numbers, while all the other symbols have their usual meanings. The most important experimental technique for measuring such second moments is the *rotational echo double resonance* (REDOR) experiment [38], which has been widely used to measure heteronuclear dipolar coupling strengths under the high-resolution conditions of MAS-NMR. Because of its general importance, a brief description is given here. As illustrated in Figure 3.8, MAS causes the dipolar coupling constant to oscillate during the rotation period T_r and is averaged out over a complete rotor cycle. However, if we invert the sign of the dipolar Hamiltonian by

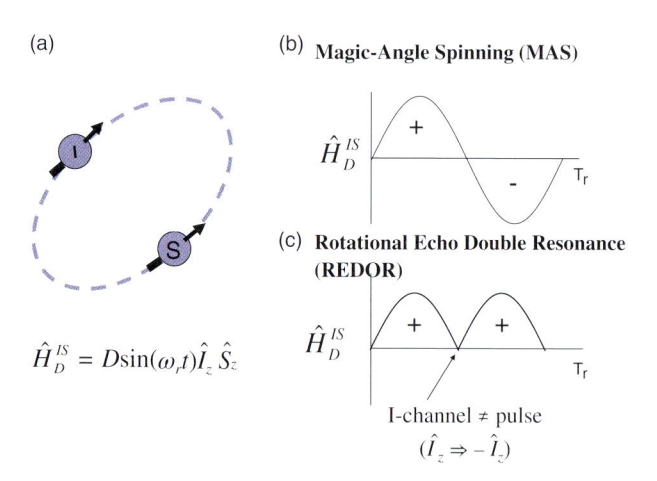

Figure 3.8 Principle of the REDOR experiment. Under MAS conditions the heteronuclear dipolar Hamiltonian H_D^{IS} oscillates sinusoidally with rotor orientation, leading to cancellation upon completion of the rotor cycle (b). Sign inversion created by a π-pulse applied to the nonobserved I spins (turning $\hat{I}_z$ into $-\hat{I}_z$) interferes with this cancellation (c), resulting in the dipolar recoupling effect.

applying a π-pulse to the nonobserved *I*-spins, this average is nonzero and the interaction is recoupled.

In the most common REDOR pulse sequence, the recoupling is accomplished by 180° pulse trains that are synchronized with the MAS rotation period [39]. The inversion pulse trains are applied to the I-spins, while the S-spin signal is detected by a rotor-synchronized Hahn spin echo sequence. One measures the normalized difference signal $\Delta S/S_o = (S_o - S)/S_o$ in the absence (intensity S_o) and the presence (intensity S) of the recoupling pulses. A REDOR curve is then generated by plotting $\Delta S/S_o$ as a function of dipolar evolution time NT_r, that is, the duration of one rotor period multiplied by the number of rotor cycles (Figure 3.9).

For isolated two-spin–1/2 pairs, this REDOR curve has universal character and can be directly used to extract the internuclear distance [38]. In contrast, for larger spin-clusters the REDOR curve depends on the detailed shape and distance geometry of the spin system [40]. In glasses, one generally expects a distribution of spin geometries and magnetic dipole coupling strengths. As previously shown, these complications can be avoided by limiting the REDOR data analysis to the initial curvature, where $\Delta S/S_o < 0.2$ [41–44]. In this limit of short dipolar

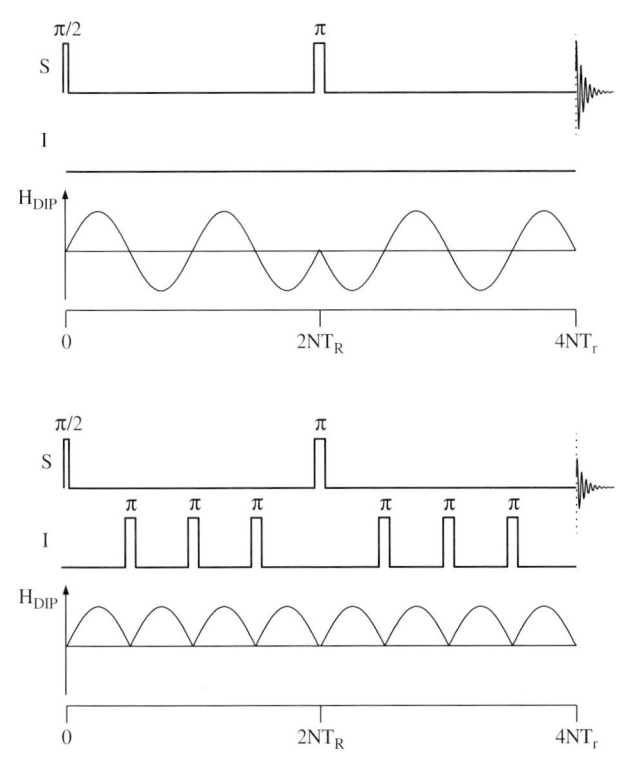

Figure 3.9 Timing of the REDOR pulse sequence and the corresponding evolution of the dipolar Hamiltonian.

evolution times, the REDOR curve is found to be geometry-independent and can be approximated by

$$\frac{\Delta S}{S_0} = \frac{1}{2I+1}\left(\sum_{m=-I}^{I}(2m)^2\right)\frac{1}{\pi^2(I+1)I}(NT_r)^2 M_2, \quad (3.7)$$

where the summation extends over the $2I+1$ quantum states with corresponding orientational quantum numbers m. The average heteronuclear van-Vleck second moment M_2 can then be extracted from a simple parabolic fit of the experimental data.

Using this method, a separate M_2-value can be specified for each separate signal observed in the MAS-NMR spectrum (site-resolved dipolar spectroscopy). A very useful application of the M_2 values obtained from such REDOR data is estimating the average number of heteroatomic connectivities within a mixed glass former system. This application assumes that the next-nearest neighbor distances r_{XY} within such a network are well defined and comparable to those found in crystalline model compounds with analogous network structures and connectivities.

To this end, the van Vleck equation can be recast in terms of the expression

$$\langle m_{X\text{-O-}Y}\rangle = M_{2XY}/\langle r_{XY}^6\rangle. \quad (3.8)$$

Important examples are the quantifications of B–O–P linkages, $\langle m_{B-O-P}\rangle$, in borophosphate glasses from ^{11}B{^{31}P} REDOR and of Al–O–P connectivities $\langle m_{Al-O-P}\rangle$ in aluminophosphate glasses from ^{27}Al{^{31}P} REDOR (see Section 3.5.2). For quantifying dipole–dipole interactions with quadrupolar nuclei, one preferentially utilizes a variant of the REDOR method, the so-called *rotational echo adiabatic passage double resonance* (REAPDOR) technique [45]. This method is able to recouple dipolar interactions of the observed nuclei with neighboring nuclei in all of the Zeeman states, thereby enhancing the overall dephasing effect.

If MAS affords no site resolution, the nonspinning variant, called *spin echo double resonance* (SEDOR), may be preferred as a simpler experimental alternative [46]. Analogous experiments are available for measuring homonuclear magnetic dipole–dipole couplings on spinning [47–51] and nonspinning samples [52–56]. As discussed below, such dipolar experiments have contributed a lot of quantitative information about connectivities in mixed network-former glasses and about the spatial distribution of network modifiers in ion conducting glasses. Figure 3.10 shows an illustrative example for a silver borophosphate glass [57]. As the signals of the B$^{(3)}$ and B$^{(4)}$ units are spectroscopically well-resolved, the strength of the ^{11}B/^{31}P dipole–dipole interactions and hence the average extent of B–O–P connectivity can be assessed separately for these two species. In Figure 3.10, the strong signal depletion observed for the signal of the B$^{(4)}$ species at relatively short NT_r values indicates the presence of several B–O–P linkages, whereas the curve observed for the B$^{(3)}$ species suggests fewer such linkages.

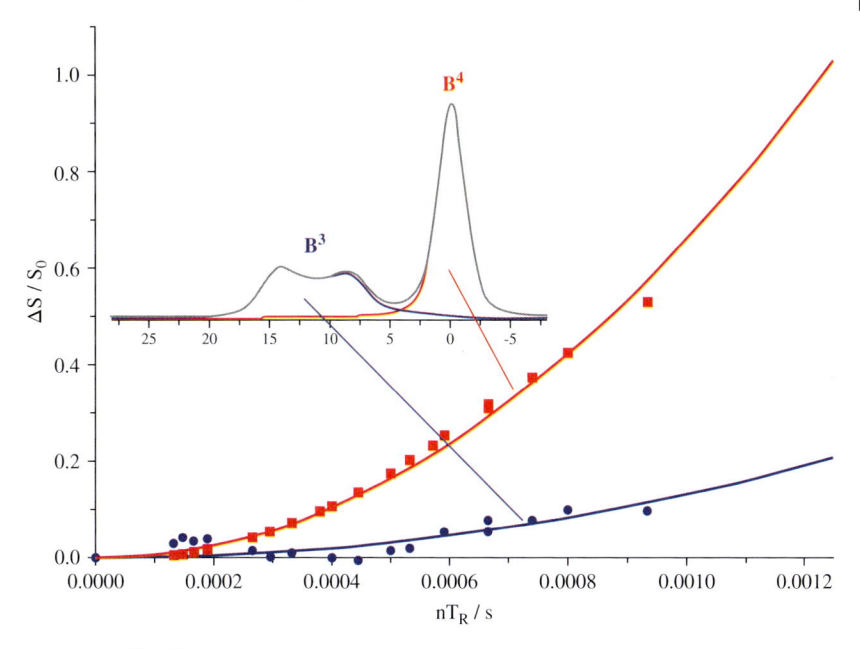

Figure 3.10 ^{11}B{^{31}P} REDOR experiment on a borophosphate glass, suggesting different extents of B–O–P linking for the three- and the four-coordinate boron species. (From Ref. [57].)

3.4
Medium Range Order in Covalent Network Glasses

3.4.1
Glassy SiO$_2$, GeO$_2$, and the Binary System SiO$_2$–GeO$_2$

3.4.1.1 T—O—T Bond Angle Distributions

Glassy silica is the archetype of the vitreous state. The structure is based on tetrahedrally coordinated fully polymerized Si$^{(4)}$ units that are linked via bridging oxygen atoms. As the potential $V(\theta)$ relating to the Si—O—Si bond angle θ is rather shallow, there are strong local variations in that particular bond angle, resulting in a distribution of Si$\cdots$Si next nearest neighbor distances $r_{\text{Si–Si}}$. Thus, the bond angle distribution function is an important criterion for assessing the extent of medium-range order in glassy silica. Experimental estimates have been deduced from X-ray and neutron diffraction data, solid-state NMR experiments, and molecular dynamics simulations. The first foray into the subject was the X-ray diffraction analysis of Mozzi and Warren, producing an asymmetric distribution with a maximum at 144° and a full width at half height (*FWHM*) of 37° [58]. These original data have been reanalyzed and reinterpreted several times [4]. The diversity of the proposed experimental distribution functions obtaind via different methods has been emphasized in Ref. [3]; in particular, these authors have pointed out that the choice of the potential used in the MD simulations can lead to large differences in distribution width. A more complete summary of all the results obtained up to 2008 was given in Ref. [59].

The information obtained from X-ray or neutron diffraction data arises from the determination of average r_{Si-Si} values that are available from the total scattering function $T(r)$. In principle, these values can be combined with Si—O distances to yield average Si—O—Si bond angles, $\langle \theta_{Si-O-Si} \rangle$. As pointed out by Wright such an analysis must take into account that the Si—O bond lengths r_{Si-O} and $\theta_{Si-O-Si}$ are likely to be correlated [4]. Furthermore, neutron and X-ray diffraction data can be modeled jointly to yield estimates of the Si—O—Si distribution functions as well. In this manner, Neuefeind and Liss were able to improve the definition of the Si⋯Si correlation in the $T(r)$ data, obtaining a more accurate distribution function of Si—O—Si bond angle [60]. Their results have been cast in terms of a Gaussian distribution with $\langle \theta \rangle = 148.3°$ and $\sigma_\theta = 7.5°$ (FWHM = 17.7°), which is much narrower than that of the initial work. Similar results (146° and FWHM 17°) have been obtained from Reverse Monte Carlo Modeling (RMC) studies of high-energy X-ray and neutron diffraction measurements [61] and by partial structure factor ND analysis based on isotope substitution [62]. A compilation of distribution functions using different approaches is given in Figure 3.11.

Early on NMR has been proposed and applied as an alternative method for extracting distribution functions of θ [65,67–74]. In the present review, we will focus on the most recent results only, which build, however, on continuous progress made by numerous research groups over the span of nearly 30 years. The width of the ^{29}Si MAS-NMR spectrum of glassy SiO$_2$ is given by a

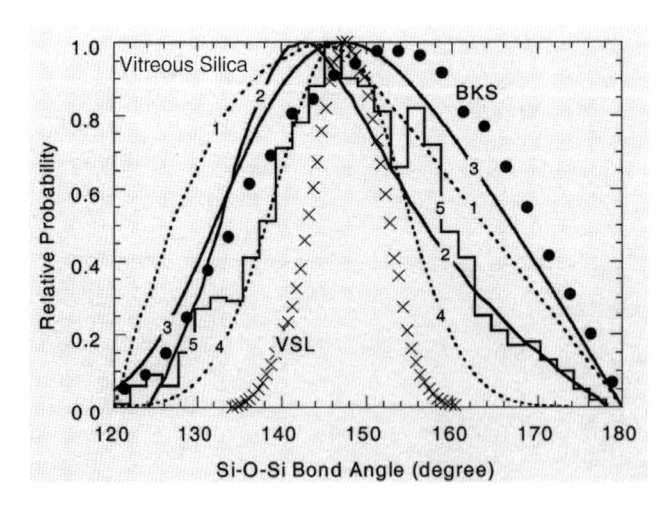

Figure 3.11 Compilation by Yuan and Cormack of different Si—O—Si bond angle distribution functions proposed in the literature [3]. BKS: MD simulation based on the two-body potential of van Beest et al. [63], VSL: MD simulation based on the three-body potential of Vessal et al. [64]. The curves 1, 2, 3, 4, and 5 originate from Refs [58,60,65,66,67], respectively. (Reproduced with permission from Ref. [3].)

distribution of isotropic chemical shifts, which can be mapped into a distribution of Si—O—Si bond angles, using theory-based relations between both parameters [65,67–70]. Assuming the contributions $F_i(\theta)$ of the four Si—O—Si bond angles involving a central Si atom to be independent, the chemical shift is given by

$$\delta_{iso}(^{29}Si) = 1/4 \Sigma F_i(\theta). \tag{3.9}$$

Using the semi-empirical relation

$$F(\theta) = -93.12 + 866 \cos\theta - 22.7 \cos 2\theta, \tag{3.10}$$

which represents the best fit of *ab initio* calculations obtained for the model compound cristobalite, the analysis of the experimental spectrum of glassy SiO$_2$ produces $\langle\theta\rangle = 151.0°$ and $\sigma_\theta = 11.4°$ (FWHM = 27°) [70]. It has been pointed out, however, that there is a correlation between ^{29}Si spin-lattice relaxation times T_1 and $\delta_{iso}(^{29}Si)$ in ^{29}Si-enriched glassy silica: the ^{29}Si nuclei contributing to the low-frequency components near −115 to −120 ppm have significantly shorter T_1 values (6–7 h) than those contributing to the high-frequency components near −100 ppm (~10 h) [59]; in addition, the results may be affected by shorter relaxation times for ^{29}Si nuclei near defect or impurity sites. These circumstances may compromise the accuracy of ^{29}Si NMR-derived bond angle distribution functions unless they were obtained on fully relaxed specimens [59]. A full analysis of the potential errors made in the bond angle distribution functions that could be caused by these effects has not yet been published.

As a seemingly more direct access to the Si—O—Si bond angle distribution, solid state ^{17}O NMR can be used. The chief observable here is the electric field gradient tensor at the ^{17}O nucleus. Using two-dimensional dynamic spinning NMR on isotopically labeled samples, it is then possible to extract the distribution functions of C_Q and η_Q from such ^{17}O NMR spectra [71]. Restricted Hartree–Fock calculations on (OH)$_3$Si—O—Si(OH)$_3$ clusters showed that the ^{17}O quadrupolar coupling constant $C_Q(^{17}O)$ is strongly dependent on both θ and r_{Si-O}, whereas the asymmetry parameter $\eta_Q(^{17}O)$ only depends on θ. Using the relations [72]

$$\begin{aligned} C_q(d_{TO},\Omega) &= a'\left(\frac{1}{2} + \frac{\cos\Omega}{\cos\Omega - 1}\right)^\alpha + m_d(d_{TO} - d^0_{TO}) \\ n_q(\Omega) &= b\left(\frac{1}{2} - \frac{\cos\Omega}{\cos\Omega - 1}\right)^\beta \end{aligned} \tag{3.11}$$

(with $\Omega = \theta_{Si-O-Si}$ and $d_{TO} = r_{Si-O}$, $d^0_{TO} = 165.4$ pm, and a, b, α, and β adjustable fit parameters) Clark *et al.* were able to map the quadrupolar coupling parameters into two-dimensional histograms, correlating $\theta_{Si-O-Si}$ with r_{Si-O} and r_{Si-Si} and also r_{Si-O} with r_{Si-Si} [73]. Based on these correlations, their analysis yields $\langle\theta\rangle = 146.6°$ and $\sigma_\theta = 3.8°$ (FWHM = 9°), suggesting a substantially narrower distribution than obtained from the diffraction analyses. A strong correlation between $\theta_{Si-O-Si}$ and r_{Si-O} was found, which was, however, opposite to experimental data on crystalline silica polymorphs [73]. Using a more sophisticated model-building approach of the SiO_2 network in conjunction with advanced DFT calculations of the NMR parameters, Charpentier et al. found a much weaker correlation between $C_Q(^{17}O)$ and $\eta_Q(^{17}O)$ and hence a much weaker correlation between $\theta_{Si-O-Si}$ and r_{Si-O} [74]. Their resulting distribution is characterized by $\langle\theta\rangle = 147.1°$ and $\sigma_\theta = 11.2°$ (FWHM = 26°). A convincing advantage of the modeling approach described in Ref. [74] is the fact that the ^{17}O NMR analysis is also found to be consistent with the analysis of the ^{29}Si MAS NMR lineshape based on the same modeling approach, which yielded $\langle\theta\rangle = 148.4°$ and $\sigma_\theta = 10.8°$ (FWHM = 25°). The agreement with the more recent diffraction analyses is quite good, even though the latter seem to indicate somewhat narrower distributions than the NMR results.

In a similar vein, these diffraction and NMR analyses have been extended to glassy GeO_2, even though the modeling approaches have been somewhat less sophisticated to date. Based on this work, there is general consensus that $\langle\theta_{Ge-O-Ge}\rangle$ is significantly smaller than $\langle\theta_{Si-O-Si}\rangle$ [75–80]. Probable $\langle\theta\rangle$ values from both diffraction and ^{17}O NMR experiments range from 130° to 135°, but few authors have specified the distribution width. From their diffraction data, Neuefeind and Liss suggest a distribution width FWHM of 20° in glassy GeO_2 [60] while a ^{17}O solid-state NMR analysis based again on the correlation between C_Q, η_Q, and θ concluded $\langle\theta\rangle = 130°$, with a FWHM of only 9.5° [79].

3.4.1.2 Ring Size Distributions

Clearly, the bond angle distribution is related to (and indeed a consequence of) the network topology, which can be described in terms of a distribution of ring sizes. Figure 3.12 summarizes the results from various MD simulations

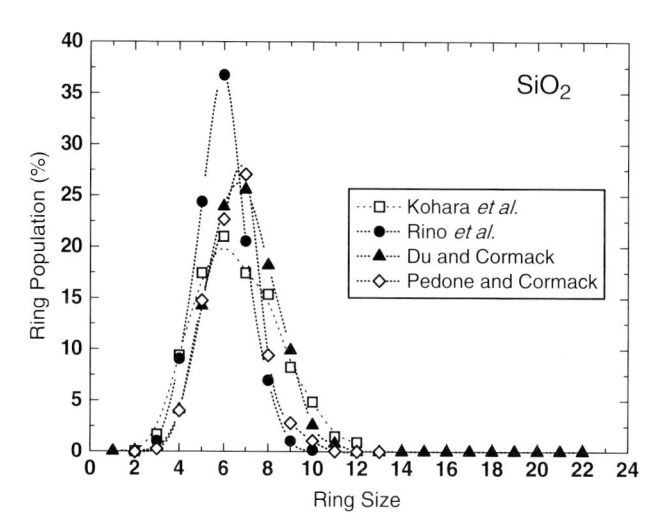

Figure 3.12 Ring size distributions in glassy silica as proposed by various authors. (Refs [60–63].)

conducted on glassy SiO$_2$, showing a maximum near six- or seven-membered rings [80–83]. Very few three- and four-membered rings are observed in this material, and their occurrence at all in glassy silica has been inferred only from the observation of some sharp "defect" features (commonly labeled the defect bands D1 and D2 at 491 and 606 cm$^{-1)}$ observed in the Raman spectra [78]. In glassy GeO$_2$, the main T—O—T band at 420 cm^{-1} has been attributed to the bridging oxygen stretching mode of six-membered rings [78]. The defect bands have been observed as well (at 347 and 520 cm^{-1}) albeit with significantly higher intensity than in glassy silica. Based on this difference, it has been suggested that GeO$_2$ glass contains a significantly larger fraction of three-membered rings than SiO$_2$ does [78]; and this suggestion is also in agreement with MD simulations [61].

As smaller rings naturally imply smaller T—O—T bond angles, the large difference in $\langle \theta \rangle$ between GeO$_2$ glass and SiO$_2$ glass has been attributed to the larger fraction of three-membered rings in the former material [61].

3.4.1.3 Network Connectivity in SiO$_2$–GeO$_2$ Glasses

Finally, medium-range structural features of binary SiO$_2$–GeO$_2$ glasses have been examined by diffraction [81,83], ^{17}O NMR [84], Raman spectroscopy [78,85], and MD simulations [61,86]. While the preponderance of evidence suggests that the Si$^{(4)}$ and Ge$^{(4)}$ units are randomly linked, certain features in the Raman spectrum have suggested to Henderson *et al.* that the mixing may be nonideal [78]. A particularly strong argument against this claim is the ^{17}O NMR spectrum [84], from which the ratio of the Si—O—Si, Si—O—Ge, and Ge—O—Ge oxygen sites has been quantified as being very close to 1 : 2 : 1, which is just the result expected for random mixing. Clearly, the issue of network mixing in binary silica–germania glasses requires further experimentation and analysis.

3.4.2
Boron Oxide, B$_2$O$_3$–SiO$_2$, and B$_2$O$_3$–GeO$_2$ Glasses

Boron oxide is another archetypical glassy material that has been of particular interest for fundamental study because of its particularly well-developed medium-range structure. The state of knowledge on this material was reviewed some time ago by Hannon *et al.* [13] and the most recent update has been given by Wright [87]. All the available experimental evidence from neutron diffraction, NMR, and Raman spectroscopy indicates that the structure is dominated by six-atom "boroxol" rings (ring size three). As already pointed out by Krogh-Moe [88] and reviewed in Section 3.3, the boroxol group can be identified by an extremely intense polarized Raman scattering peak near 808 cm^{-1}, which arises from the symmetric breathing mode of this symmetric ring structure and also is nicely consistent with the $T(r)$ function obtained from diffraction experiments [88] (see Figure 3.5). Responding to various occasional challenges of the boroxol model, the 1994 study by Hannon *et al.* confirmed all its essential features on the basis of high-quality diffraction data [13]. Static low-field [89] as well as high-field high-resolution [90] ^{11}B solid state NMR spectra show two distinct resonances in an area ratio of roughly 3:1, which have been assigned to boroxol-ring and nonring $B^{(3)}$ units, respectively. The assignment of the dominant resonance to the boroxol groups was confirmed by *ab initio* ^{11}B chemical shift calculations [91] and, independently, by an ingenious magic-angle flipping experiment designed by Zwanziger [92]. In this two-dimensional experiment, the spinning angle is flipped for an incremented mixing time, allowing a recoupling of ^{11}B – ^{11}B magnetic dipole–dipole interactions between neighboring nuclei. This mechanism leads to magnetization transfer between nonisochronous ^{11}B nuclei, and hence a decrease in the signal amplitude observed in the final detection step. The frequency dispersion of the ^{11}B resonance under MAS conditions is dominated by the anisotropy of second-order quadrupolar perturbations, and the actual resonance frequency of a given spin packet depends on the orientation of the electric field gradient (EFG) tensor relative to the rotor frame. As the $B^{(3)}$ groups of a boroxol ring are co-planar and the tensor is approximately axially symmetric, each of the three ^{11}B spin packets of the boroxol group (defined by a certain orientation in space) will have the same resonance frequency, and this will be true for all the possible molecular orientations. As a consequence, no magnetization transfer is observed between the ^{11}B nuclei within a boroxol group. In contrast, the EFG orientation of the nonring $B^{(3)}$ group is usually uncorrelated to that of the boroxol group, and therefore, these $B^{(3)}$ neighbors have usually different resonance frequencies from the boroxol species at most orientations, thus allowing for magnetization transfer. Zwanziger's quantitative analysis of this angle-flipping experiment confirmed that the dominant ^{11}B NMR signal in B_2O_3 indeed arises from the boroxol rings. Going even a step further, the interlinking between boroxol rings and nonring species was analyzed by high-resolution ^{17}O double rotation NMR on an isotopically enriched sample [93]. In principle, four distinct signals can be expected, corresponding to oxygen atoms within boroxol rings, oxygen atoms linking two adjacent boroxol rings, oxygen atoms linking a boroxol ring with a nonring boron species, and oxygen atoms linking two nonring boron species. From the detailed analysis of the observed intensities at 11.7 T, it was concluded that boroxol rings and nonring species may tend to form separate domains. A confirmation and further refinement of these results would be desirable however, at the higher magnetic field strengths nowadays available.

Finally, high-resolution ^{11}B NMR has been used to derive bond angle distribution functions of θ_{B-O-B} (ring) and θ_{B-O-B} (nonring) [94]. Within the boroxol rings the average angle is found to be very close to the ideal value of 120°, and the distribution is rather narrow (FWHM = 7.5°). For the nonring B atoms an average angle of 135° with a somewhat wider distribution (FWHM = 17°) was found. The latter presents an average of four contributions, depending on how many rings (three, two, one, or zero) the nonring B atom is connected to.

The binary system B_2O_3–SiO_2 has been subject to considerable attention as it is known to contain a metastable region of immiscibility and phase separation tendency. Nevertheless, high-resolution ^{17}O MAS-NMR studies of binary glasses have provided clear-cut peak assignments for bridging oxygen atoms within Si—O—Si linkages, Si—O—B linkages, and B—O—B linkages [95]. The concentration of the Si—O—B linkages is, however, lower than that expected for statistical mixing [96]. The data have been modeled in terms of the hypothetical reaction scheme Si—O—Si + B—O—B ⇒ 2Si—O—B yielding good agreement with a positive mixing energy of $2W = 8$ kJ/mol. The numerical value of this energy was independently predicted by *ab initio* cluster calculations. According to the NMR data obtained, the phase separation tendency appears to be greater in glasses with high B_2O_3 contents. The conclusions from ^{17}O NMR are nicely compatible with results from ^{11}B and ^{29}Si NMR [97]: ^{29}Si{^{11}B} REAPDOR results indicate that the strength of the ^{29}Si – ^{11}B magnetic dipole–dipole coupling attains a limiting value in glasses with B/Si near unity, corresponding to an average of about two Si—O—B linkages per Si$^{(4)}$ species, and it does not increase upon further increase of the boron concentration. In glasses with B/Si > unity, high-resolution ^{11}B MAS-NMR spectra show substantial amounts of boron atoms within boroxol units (in agreement with old Raman data [98]), confirming the increased phase separation tendency in this compositional region. This effect can be understood on the basis of the lower equilibration temperatures accessible to these glasses because of their lower glass transition temperatures, where the positive mixing enthalpy term outweighs the effect of the mixing entropy (Figure 3.13).

Older Raman studies on binary B_2O_3–GeO_2 [99] and ternary B_2O_3–SiO_2–GeO_2 glasses [100] have led to conceptually similar conclusions, but the corresponding solid state high-resolution ^{11}B and ^{17}O NMR or diffraction work needs yet to be done.

3.4.3
P_2O_5, P_2O_5-GeO_2, and Ternary P_2O_5-Based Network Glasses

Synchrotron X-ray diffraction studies on glassy P_2O_5 reveal a clear P···P correlation at 294.5 pm, from which a mean P—O—P bond angle of 137° can be deduced. From the width of this peak the FWHM of the distribution function is estimated at 21° [101]. Both NMR [102] and diffraction data [103] have also been taken on the binary P_2O_5–GeO_2 glass system, however, the discussion in both of these studies has focused on the potential change in Ge coordination number. To date the homo- versus hetero-atomic bond distribution remains unknown in these glasses. This issue could be addressed by measuring ^{31}P – ^{31}P magnetic dipole–dipole coupling strengths, a work to be conducted in the future. Such dipolar NMR methods have been successfully applied to characterize the connectivity distribution in melt-prepared ternary glasses with the composition 70 SiO_2–30 [$(Al_2O_3)_x(P_2O_5)_{1-x}$] [104]. The connectivity distribution in this system shows an extremely strong tendency toward being nonstatistical. The average number of P atoms linked to aluminum, $n_P(Al)$, was estimated by comparing the $M_2(^{27}Al\{^{31}P\})$ values to the value measured (4.7×10^6 rad^2/s^2) for

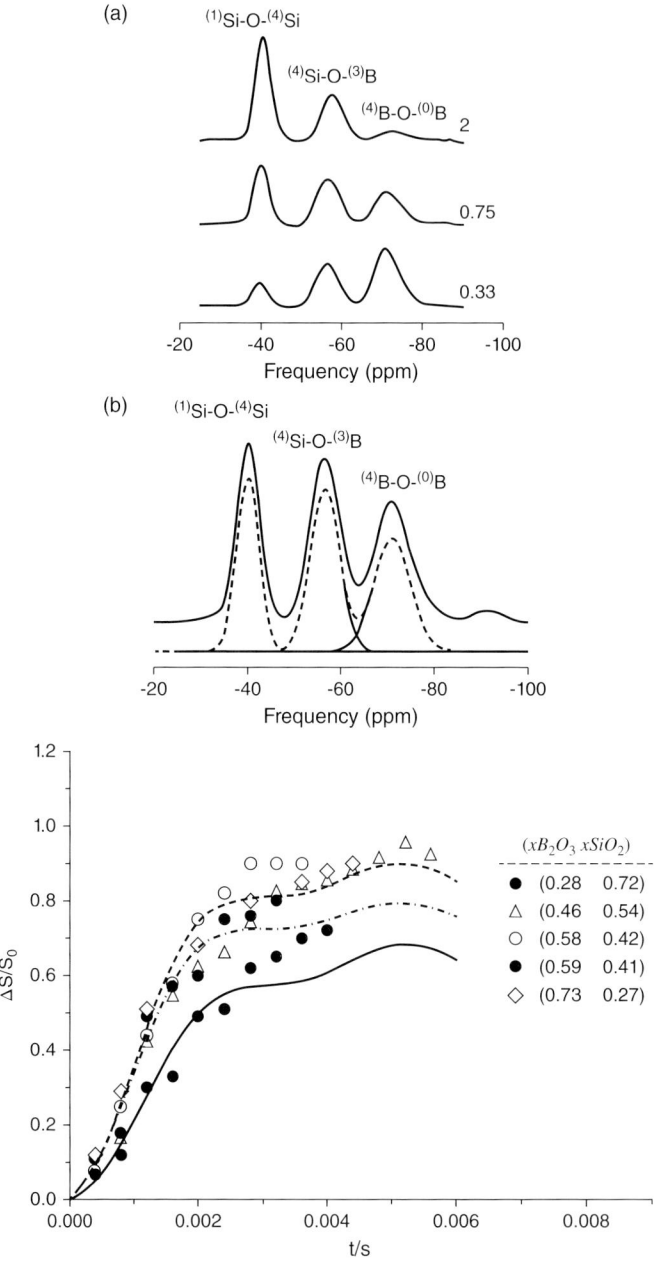

Figure 3.13 Solid-state NMR results on B_2O_3–SiO_2 glasses as a function of glass composition: (a) isotropic projection of ^{17}O TQMAS-NMR data and corresponding peak assignments for three glasses with different Si/B ratios listed to the right of each trace [96]. For Si/B = 0.75 fits of the projections to three invidvidual Gaussians for quantification purposes, are shown. (b) $^{29}Si\{^{11}B\}$ REDOR results and comparison with simulations: solid curve: $^{29}Si - ^{11}B$ two-spin system, dashed curve: $^{29}Si - (^{11}B)_2$ three-spin system. Dotted curve: statistical distribution of connectivities expected at a B/Si ratio of unity [97]. (Reproduced with permissions from Refs [96,97].)

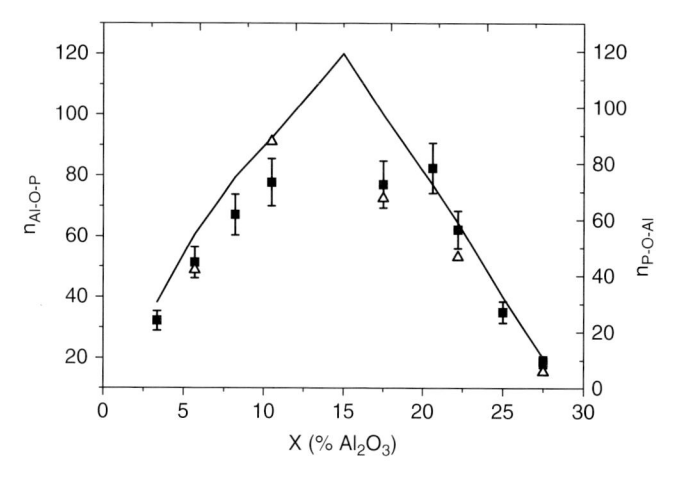

Figure 3.14 Total number of P–O–Al linkages n_{P-O-Al} as extracted from $^{27}Al\{^{31}P\}$ REDOR (squares) and $^{31}P\{^{27}Al\}$ REAPDOR analysis (open triangles). Solid curves illustrate the prediction from the extreme aluminum phosphate cluster scenario.

sol–gel prepared AlPO$_4$ glass, in which each aluminum atom is tetrahedrally surrounded by four P atoms. In a similar vein, the average number of Al atoms linked to phosphorus, $n_{Al}(P)$, can be extracted from $M_2(^{31}P\{^{27}Al\})$ values. Figure 3.14 summarizes the total numbers of P—O—Al linkages, n_{P-O-Al}, obtained either by multiplying $n_P(Al)$ with the aluminum content $2x$, or by multiplying $n_{Al}(P)$ with the phosphorus content $2\times(30-x)$. Note that these values, obtained via two independent experiments, are in excellent agreement with each other, thereby validating the above analysis. Figure 3.14 compares these values to those predicted from a clustering scenario, which maximizes the number of Al–O–P linkages possible based on the glass composition. It is evident that the numbers n_{P-O-Al} are always consistently lower than those predicted by the cluster scenario, suggesting that the glass structure is somewhat more disordered than that predicted from an AlPO$_4$ segregation scenario.

In the above study, no NMR data were available for glass with Al/P = 1:1 ($x = 15$) as no glass could be prepared by melt quenching. Alternatively, homogeneous glasses along the composition line $(AlPO_4)_x-(SiO_2)_{1-x}$ can be prepared by a sol–gel technique, using H$_3$PO$_4$ and aluminum lactate as precursor species [105]. The NMR results obtained on these materials were found to be essentially identical to those measured for pure glassy AlPO$_4$ [106], clearly indicating the formation of separated AlPO$_4$ and SiO$_2$ domains [105]. Most recently, this work was extended to the analogous GaPO$_4$–SiO$_2$ sol–gel system [107]. While the results again indicate a great deal of heterogeneity as well, they also reveal a quantifiable amount of Si—O—P (and by inference Ga—O—Si) linkages, suggesting a significantly larger extent of network mixing than in the AlPO$_4$–SiO$_2$ case. A similarly strong tendency for chemical ordering is observed in sol–gel prepared glasses in the system Al$_2$O$_3$–B$_2$O$_3$–P$_2$O$_5$ [108]. Both Al and B strive to maximize the number of linkages to phosphorus, in the attempt to establish an idealized tetrahedral ("silica-like") structure of composition $(B_{1-x},Al_x)PO_4$. Homoatomic B—O—B, Al—O—Al, and P—O—P linkages are formed only at compositions that deviate from this stoichiometry and B—O—Al linkages tend to be avoided.

3.5
Medium Range Order in Modified Oxide Glasses

Modified oxide glasses are formed by melting covalent oxide glass formers such as SiO_2, B_2O_3, or P_2O_5 with network modifier oxides such as M_2O (M = Li^+, Na^+, K^+, Rb^+, Cs^+, or Ag^+) or MO (M = Mg^{2+}, Ca^{2+}, Sr^{2+}, Ba^{2+}, Zn^{2+}, or Pb^{2+}). Upon incorporation of these latter compounds into the glasses the structure is converted into a polyanionic network compensating the charges of the cations. Monovalent ions often have high ionic mobilities in such systems, offering application prospects in the solid electrolyte area. Therefore, the local cation environments and spatial distributions are important issues, because they govern activation barriers and average jump distances controlling the mobility of the ions. In glasses containing more than one type of network former, new anionic units may be created by the interactions between the network-former species present, bearing upon the cation environments as well. We will review the current state of knowledge on how such network-former mixing effects influence the structure of the network and the medium-range ordering of the cations. Section 3.5.1 focuses on simple binary model glasses, $M_{(2)}O$–SiO_2, $M_{(2)}O$–B_2O_3, and $M_{(2)}O$–P_2O_5, where $M_{(2)}O$ is the oxide of a monovalent or divalent cation, and Section 3.5.2 addresses mixed network-former glasses.

3.5.1
Medium-Range Order Effects in Binary Model Glasses

Much of the fundamental studies have concentrated on the detailed investigation of model glasses in the systems $M_{(2)}O$–SiO_2, $M_{(2)}O$–B_2O_3, and $M_{(2)}O$–P_2O_5, M being a monovalent (most commonly Li^+, Na^+, Ag^+) or a divalent (most commonly Mg^{2+}, Ca^{2+}, Zn^{2+}) cation. Neutron and X-ray diffraction experiments have generally yielded reliable average nearest neighbor distances and coordination numbers, both for the network former and the network modifier ions, while information about the second coordination spheres, particularly those of the cations, has been sparse. In this regard, advanced solid-state NMR experimentation has produced significant new information about average network former – network modifier distances as well as network modifier – network modifier correlations.

3.5.1.1 Cation Next-Nearest Neighbor Environments and Spatial Distributions Studied by Neutron Diffraction

The most specific information regarding the cation–oxygen coordination numbers and M–O distances has come from X-ray and neutron diffraction

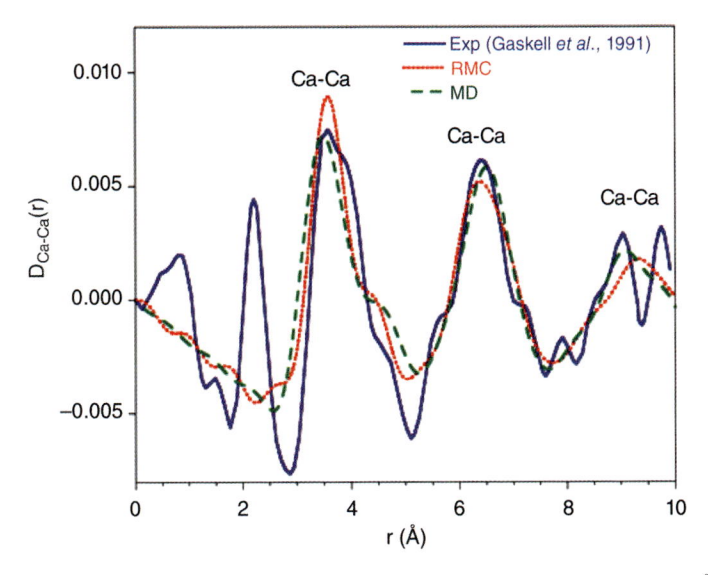

Figure 3.15 Comparison between the experimental partial structure factors D_{Ca-Ca} (equal to $g_{Ca-Ca}(r)$) and those calculated from molecular dynamics simulations and Reverse Monte Carlo fitting. The peak near 2 Å is considered an artifact. (Reproduced with permission from Ref. [111].)

experiments as well as EXAFS. Not unexpectedly, these environments are generally similar to those found in compositionally related crystalline model compounds. In contrast, it has been more difficult to gain information on the cation next-nearest neighbour environments such as cation-network former or cation–cation distance distributions by diffraction methods. Usually, neutron diffraction difference experiments with isotopically substituted samples are required. One of the first studies of this kind was conducted on $CaSiO_3$ glass [109]. As illustrated by Figure 3.15, the partial correlation function $T_{Ca-Ca}(r)$ shows strong Ca–Ca correlations extending up to 900 pm, hence suggesting a high degree of cation ordering. The experimental data are well-reproduced by both RMC and MD simulations [110,111] (see Figure 3.15). However, in contrast to what was originally proposed, the correlations do not match those observed in crystalline $CaSiO_3$, (wollastonite), suggesting that the ordering in the glass differs substantially from that in the crystal [111].

Special interest has been devoted to medium-range order effects in glasses containing mobile ions such as Li^+ and Na^+, as their local environments and spatial distributions bear on the mobility of these ions. For alkali silicate glasses, rather inhomogeneous distributions of the modifier ions have been proposed on the basis of EXAFS and molecular dynamics simulations [7,112]. In this model, which is being referred to in the literature (somewhat obliquely) as the modified random network (MRN) model the modifier ions form percolation channels with short jump distances and low activation energies. Obviously, such ion clustering implies that the average inter-cationic distances are much shorter than expected for a random cation arrangement. Neutron diffraction experiments done on lithium silicate glasses of composition $(Li_2O)_x(SiO_2)_{1-x}$ ($0 \leq x \leq 0.33$) have qualitatively confirmed this idea [113]. As pointed out by these authors, this compositional region of the Li_2O–SiO_2 phase diagram is characterized by metastable liquid immiscibility. More details about the lithium short- and medium-range order environments could be inferred from the partial radial distribution function of the lithium ions, $T_{Li}(r)$, extracted from neutron diffraction experiments with isotopic substitutions [114]: each Li is surrounded by 3.2 ± 0.2 oxygen atoms at an average distance of 197 pm, and 0.8 ± 0.5 oxygen atoms at a distance of 220 pm, thus suggesting a distorted four-coordinated environment. Regarding medium-range order, the partial correlation function suggests that each Li atom is surrounded by an average of 4 ± 1 lithium neighbors at an average Li–Li distance of 310 pm [114]. This value is considerably larger than the $g(r)$ maximum near 280 pm suggested by MD simulations [115,116]. Medium-range order effects concerning the sodium ions have also been analyzed from combined neutron/X-ray diffraction experiments on $(Na_2O)_{0.30}(SiO_2)_{0.70}$ glass, revealing a broad asymmetric interatomic Na–Na distribution in $g_{Na-Na}(r)$ extending from 260 to 350 pm [117]. MD simulations on a sodium silicate glass with variable compositions show a pronounced maximum in $g(r)$ near 300 pm for a glass containing 10 mol% Na_2O, with only a modest peak shift toward 280 pm for glass containing 30 mol% Na_2O [118]. As discussed below, this finding is consistent with solid-state NMR results and strongly supports the MRN model with cation clustering for sodium silicate glasses with low Na contents. Partial pair correlation functions $g_M(r)$ or $T_M(r)$ ($M = Li, Na$) have also been extracted from analogous diffraction experiments on lithium and sodium diborate glasses. In these systems, however, no M–M correlations below 400 pm have been observed, giving no evidence for cation clustering [119] and arguing against the applicability of the MRN model for borate glasses. More definitive evidence for the structural differences between alkali silicate and alkali borate glasses has come from solid-state NMR experiments, to be discussed in Section 3.5.1.2.

3.5.1.2 Cation Next-Nearest Environments and Spatial Distributions Studied by Dipolar NMR

The MRN hypothesis has also been tested by NMR spectroscopy, probing the average strengths of the dipole–dipole interactions associated with the constituent nuclei ^6Li, ^7Li, and ^{23}Na [27,29]. Sodium ion distributions have been measured by analyses of the $^{23}Na - ^{23}Na$ homonuclear dipolar couplings, while in the case of lithium-containing glasses, the measurement of heteronuclear $^6Li - ^7Li$ dipolar interactions proves to be more convenient for technical reasons [29,120]. Figure 3.16 compares the results obtained for silicate [120,121] and borate [120,122] glasses, containing both lithium [120] and sodium [121,122]. Note that the M_2 values and their compositional dependences are found to be fundamentally different. The large concentration-independent $M_2(^{23}Na - ^{23}Na)$ and $M_2(^6Li - ^7Li)$ values observed in both the lithium and sodium silicate

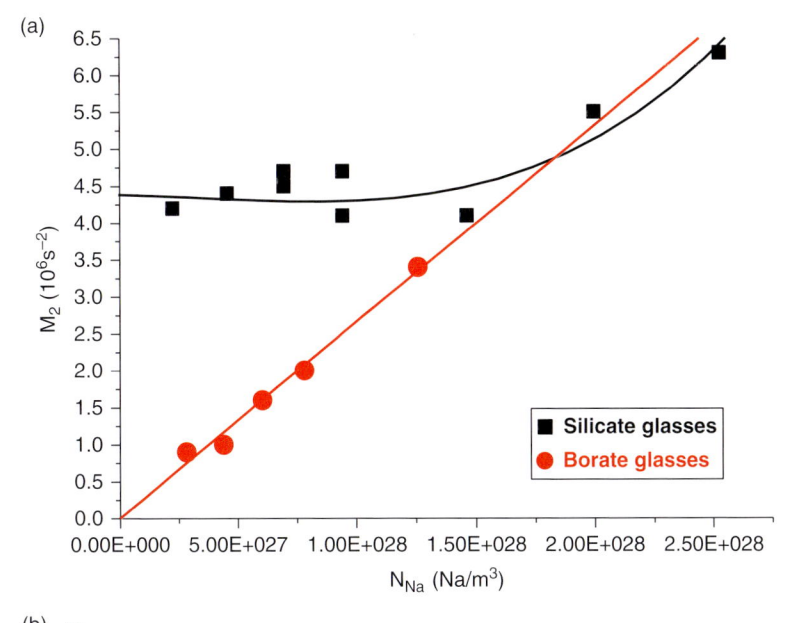

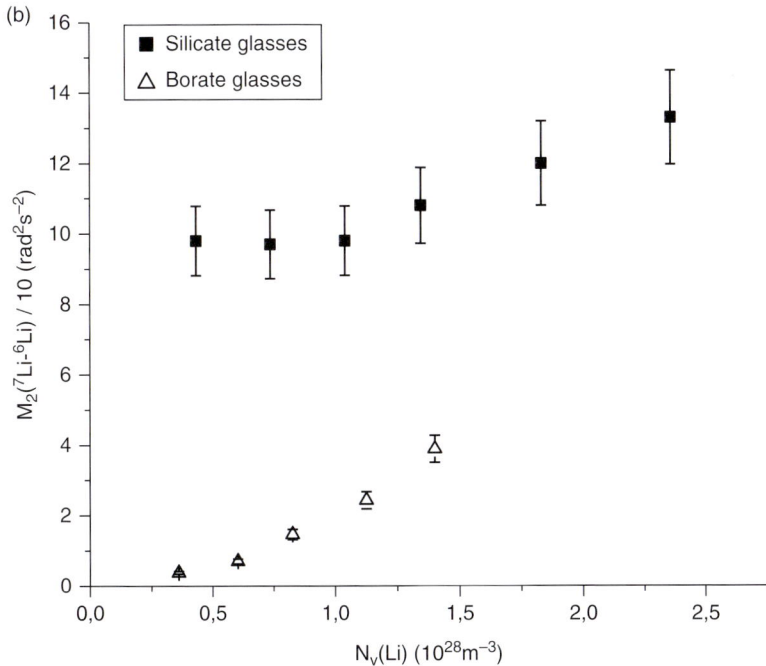

Figure 3.16 (a) Dipolar second moment values $M_2(^{23}\text{Na} - ^{23}\text{Na})$ measured via the ^{23}Na spin echo decay method, for sodium silicate and sodium borate glasses as a function of cation number density. (b) Dipolar second moment values $M_2(^6\text{Li} - ^7\text{Li})$ measured by $^7\text{Li}\{^6\text{Li}\}$ spin echo double resonance, for lithium silicate and lithium borate glasses as a function of cation number density [29,120].

glasses are a clear manifestation of cation clustering, consistent with the formation of percolation channels. In contrast, the M_2 values for lithium and sodium borate glasses are considerably smaller and strongly concentration dependent, which is consistent with a more random distribution.

On the basis of these fundamentally different ion distributions it is now possible to understand why at a given number density the ionic conductivities of alkali silicate are significantly higher than those of the corresponding alkali borate glasses, as reported in Ref. [123].

A complementary NMR approach related to the same issue of cation clustering can be used by studying the interaction between the network modifiers and the network-former species. If the cations are clustered, the principle of local charge compensation demands that the charge compensating anionic species (e.g., the $Si^{(3)}$ units in silicate glasses) would also be clustered [8]. As a consequence, the cations are expected to interact much more strongly with the anionic species than with the neutral units in this case. In the lithium silicate glass system, this idea was tested by measuring the $M_2(^{29}Si - {}^7Li)$ values separately for the neutral $Si^{(4)}$ and the anionic $Si^{(3)}$ sites, using a $^{29}Si\{^7Li\}$ REDOR experiment for a series of low-lithium silicate glasses [29,124]. The results (see Figure 3.17) confirm that there is indeed a big difference: the $Si^{(4)}$ species reside in spatial regions that are largely cation-depleted, consistent with the known tendency of lithium silicate glasses to phase-separate. By combining the $Si^{(3)}$–Li and $Si^{(4)}$–Li pair distribution functions $f_{SiLi}(r)$ extracted from complementary MD simulations with these NMR results new information is available about the structural organization of the lithium-rich nanophase. Based on the average $Si^{(3)}$–Li distances determined from the maxima in these curves (~318 pm), we

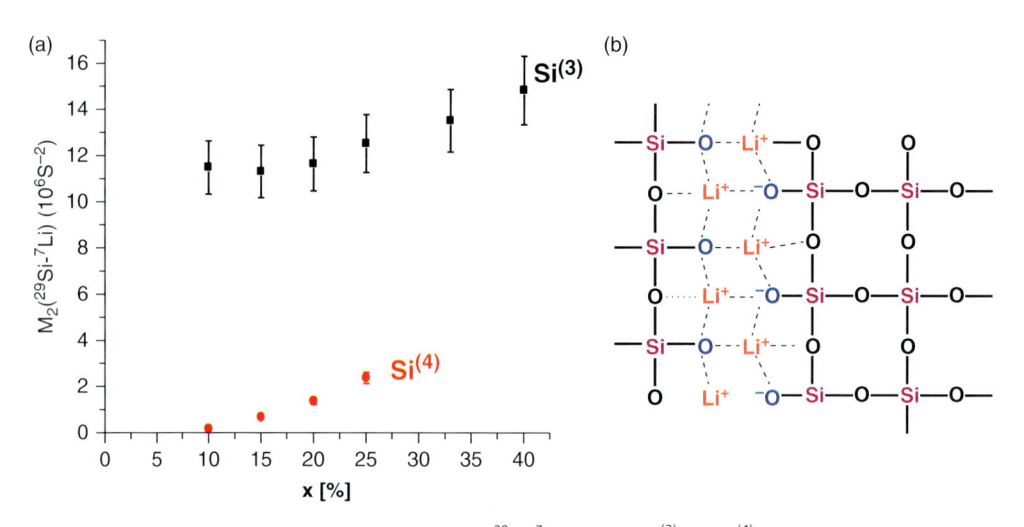

Figure 3.17 (a) Dependence of $M_2(^{29}Si\{^7Li\})$ for the $Si^{(3)}$ and $Si^{(4)}$ units on lithium oxide content in the lithium silicate glass system. (b) Schematic suggesting lithium clustering in lithium silicate glasses, bringing multiple Li-ions into proximity of multiple $Si^{(3)}$ units.

can estimate the number of Li nearest neighbors to a $Si^{(3)}$ silicon unit from the experimental M_2 values. Thus, taking $r_{SiLi} = 318$ pm as the relevant average $Si^{(3)}$–Li closest distance in the $x = 0.1$ glass, each closest lithium neighbor would – on average – produce a contribution of 3.3×10^6 rad^2/s^2 to $M_2(^{29}Si\{^7Li\})$. Based on the experimental M_2-values near $11–12 \times 10^6$ rad^2/s^2 and considering that these values also include minor contributions from more remote ^{7}Li spins, we can conclude that the $Si^{(3)}$ units are surrounded by three closest lithium ions. When combining this result with the M_2 values near 10×10^6 rad^2/s^2 from $^7Li\{^6Li\}$ SEDOR [120], one can estimate that there are four lithium ions in the second coordination sphere of a lithium ion at an average Li$\cdots$Li distance of 273 pm [29]. This distance is shorter than that (310 pm) inferred from the neutron diffraction study [114] but longer than the shortest Li–Li distance of 256 pm in crystalline lithium disilicate.

Analogous results have been obtained for sodium silicate glasses [121]. Again, the combined analysis of ^{23}Na spin echo decay spectroscopy and $^{29}Si\{^{23}Na\}$ REDOR experiments indicates that the sodium arrangement is highly nonstatistical, with large and constant dipolar second moments at low Na concentrations (≤ 20 mol%). In contrast, analogous $^{11}B\{^{23}Na\}$ REDOR results on sodium borate glasses reveal a strikingly different picture [122]. In this case, the neutral species are trigonal $B^{(3)}$ units and the anionic species are four-coordinate $B^{(4)}$ species. As already shown in Figure 3.13, the ^{11}B resonances of both structural units in borate glasses are generally well-resolved by MAS, thereby facilitating site-resolved $^{11}B\{^{23}Na\}$ REDOR measurements. With increasing sodium content, the M_2 values characterizing the ^{23}Na – ^{11}B dipole–dipole interactions for the $B^{(3)}$ groups increase linearly (Figure 3.18), consistent with a random distribution. With the exception of the glasses with the lowest Na contents, the neutral $B^{(3)}$ units and the anionic $B^{(4)}$ units interact equally strongly with the sodium ions, indicating that they are in close proximity to each other. This situation, which is completely different from that in alkali silicate glasses, argues strongly against cation clustering and the MRN model for alkali borate glasses. The $M_2(^{23}$Na – $^{11}B^{(4)})$ and $M_2(^{23}$Na – $^{11}B^{(3)})$ data diverge only at the lowest sodium contents, where there exists a finite probability that there are neutral $B^{(3)}$ units remote from sodium ions. In contrast, each of the anionic $B^{(4)}$ units requires proximity of a sodium ion for charge compensation. Therefore, the $M_2(^{23}$Na – $^{11}B^{(4)})$ data approach a constant "baseline" value at low Na contents. The second moment analysis suggests that at these low concentrations each $B^{(4)}$ has one Na$^+$ neighbor at an average Na–B distance r_{Na-B} near 316 ± 10 pm [122]. It is of interest to compare this result with Swenson's earlier neutron diffraction result of 340 ± 10 pm in sodium diborate glass [114].

Results obtained on sodium phosphate glasses are conceptually similar to the situation in sodium borate glasses. Figure 3.19 shows the dependence of the $M_2(^{31}P\{^{23}Na\})$ values on ion concentration in the ultraphosphate region. No significant clustering is observed, and at low cation concentrations, the M_2 values for the anionic $P^{(2)}$ sites are larger than those for the neutral $P^{(3)}$ sites as expected. In this limit, the $P^{(2)}$–Na distance of $r_{Na-P} = 330$ pm is estimated

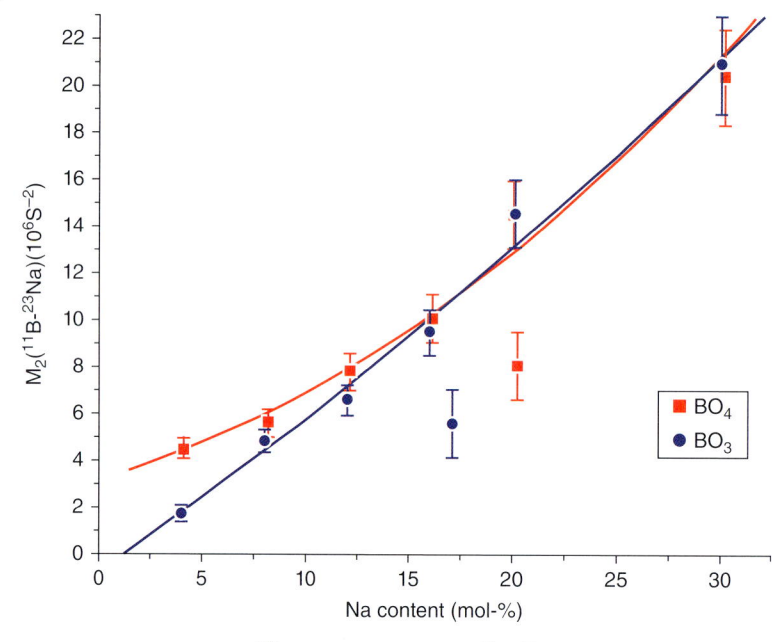

Figure 3.18 Site-selective $M_2^{\text{B-Na}}$ values extracted from $^{11}\text{B}\{^{23}\text{Na}\}$ REDOR results for trigonal and tetrahedral boron species in sodium borate glasses as a function of sodium oxide content [122].

assuming a dominant two-spin interaction [125]. This value agrees well with typical Na···P distances in crystalline sodium phosphates. Based on these considerations, we conclude that in alkali phosphate glasses, the cations tend to be isolated at low concentrations. Furthermore, an analysis of $M_2(^{23}\text{Na} - ^{23}\text{Na})$ values in this glass system is consistent with a statistical distribution of the sodium ions. The same conclusion was reached for the sodium ion distributions in sodium germanate and sodium tellurite glasses [29]. These results indicate that the clustering behavior found in the silicate glasses and described by the MRN model is not observed in other modified oxide glasses.

Medium-range order effects have also been studied in a number of mixed-alkali glasses. Measurements of both $M_2(^{23}\text{Na} - ^{23}\text{Na})$ and $M_2(^{23}\text{Na} - ^{6/7}\text{Li})$ values in mixed lithium sodium silicate, borate, and phosphate glasses consistently suggest that the relative Li^+-/Na^+-ion distributions are close to random [29]. In contrast, there is mounting evidence toward some tendency for like-cation segregation effects in mixed alkali borate and phosphate glasses containing Na^+ and larger cations such as K^+, Rb^+, and Cs^+ [126,127].

3.5.1.3 Ring-Size Distributions and "Super-Structural Units"

The medium-range structure of highly polymerized glasses is frequently discussed in terms of the distribution of the rings that can be identified in the network. The extent of medium-range order can then be assessed by comparing the ring size distribution determined in the glass with that present in the

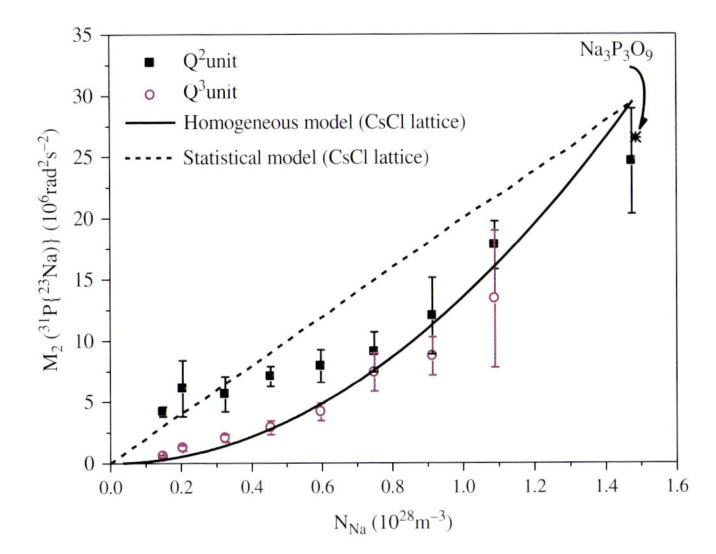

Figure 3.19 Dipolar second moments $M_2(^{31}\text{P}\{^{23}\text{Na}\})$ in sodium ultraphosphate glasses. Separate values are plotted for the $P^{(3)}$ and the $P^{(2)}$ units. Dashed line illustrates the compositional dependence expected for a statistical distribution, while the solid curve shows the predicted behavior for a homogeneous model. The latter model describes the relative distribution of $P^{(3)}$ units and Na^+ ions remarkably well. (From Ref. [125].)

isochemical crystalline compounds. Various authors have carried out studies using MD simulations and energy minimization methods for some alkali silicate glasses [115,116,118,128]. Figure 3.20 shows a typical result. With increasing extent of network depolymerization as the concentration of network modifier increases, the overall population of rings tends to decrease and both their average sizes, and the width of their size distribution tend to increase. In highly modified glasses such as sodium or calcium metasilicate glasses generally few rings are found. This is in agreement with the structures of their corresponding isochemical crystals, which are often based on infinite chain-type units.

As mentioned in Section 3.2, extensive ring formation has also been postulated to occur in modified borate glasses, mostly on the basis of Raman spectroscopic data and thermodynamic modeling studies [19,20,87,129,130]. Figure 3.21 shows the structural formulae of these units. They arise from the interaction between the neutral $B^{(3)}$ and the negatively charged $B^{(4)}$ units and are well-documented in the crystal chemistry of borate compounds. The most common ones are the pentaborate, the triborate, and the diborate units, which can also be interlinked by bridging oxygen atoms. The vibrational frequencies associated with these units differ somewhat between. Accordingly, changes in the Raman spectra as a function of glass composition have suggested that the ring speciations change as a function of modifier-to-B_2O_3 ratio. Considerable effort has been undertaken to quantify these units by modeling diffraction data [129] or by performing Raman scattering peak deconvolution exercises [130]. In general, the speciations derived

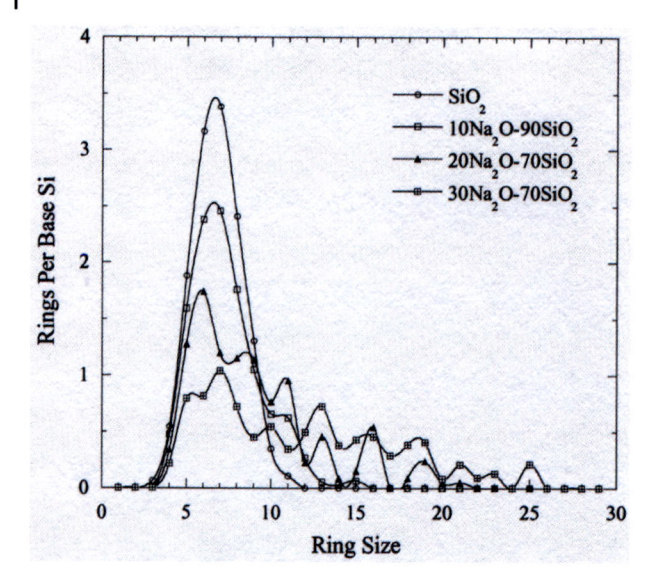

Figure 3.20 Ring size distributions in glassy SiO_2 (open circles), and binary sodium silicate glasses containing 10 (open squares), 20 (triangles), and 30 mol% Na_2O (crossed squares). (Reproduced with permission from Ref. [118].)

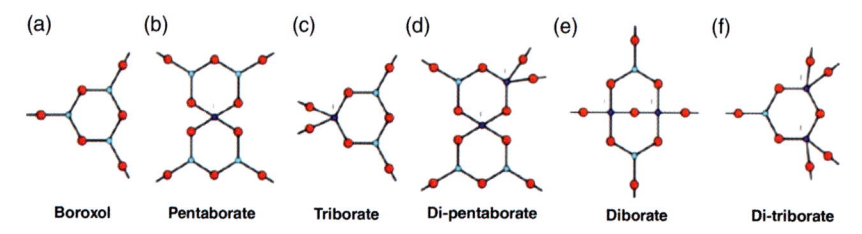

Figure 3.21 Ring structures postulated for modified borate glasses. (Reproduced with permission from Ref. [87].)

from such analyses are consistent with expectations based on the modifier-to-boron oxide ratio of the glass considered, meaning that the respective concentration of a particular structural unit reaches a maximum value at a glass composition that corresponds to its particular stoichiometry. Unfortunately, even though the occurrence of these "super-structural units" is highly plausible, neither diffraction nor solid-state NMR experiments can provide unambiguous structural signatures for them, let alone offer an approach for their quantification.

3.5.2
Connectivities and Cation Distributions in Glasses with Multiple Network Formers

The majority of technologically relevant glasses are based on more than one network-former species. The combination of several network formers usually offers the possibility of fine-tuning physical property combinations to special technological demands, and in certain cases the interaction between the various network former components results in dramatically altered physical properties when compared to the corresponding binary ion conducting glasses having the same ion concentrations. Such network former mixing (NFM) effects usually manifest themselves as strongly nonlinear changes of physical, chemical, and mechanical properties upon linear changes in composition. For optimizing glass formulations for specific applications, there is obviously a great need of developing and understanding the relation between glass composition and properties on a structural basis. In this connection, one needs to identify and quantify the different coordination polyhedra present (short-range order), as well as their respective linkages (*connectivities*) with each other (intermediate-range order). Another important structural aspect concerns the interactions (and the corresponding distance correlations) between the different network former species and the mobile cations. In general, these modifier species are not shared proportionally between both networks, but rather one of the two network-former species attracts the modifier preferentially. This process can lead to phase separation or at least nanosegregated domains, whose compositional details and structural features must be identified. Nuclear magnetic resonance is a powerful element-selective, inherently quantitative method to address these questions. Among the numerous mixed network-former systems investigated to date, the presentation will focus on recent results obtained on ionically conducting borophosphate glasses, and give some outlook on the various other systems that have been studied along similar lines.

3.5.2.1 Borophosphate Glasses and Related Systems

Strong medium-range order correlations have been identified in a range of alkali borophosphate glasses, both by O-1s XPS and $^{11}B\{^{31}P\}$ REDOR studies. Figure 3.22 shows results for glass system $(K_2O)_{0.33}((B_2O_3)_x(P_2O_5)_{1-x})_{0.67}$ [131]. The numbers of heteroatomic B–O–P linkages are found distinctly higher than those predicted from a random linkage model (dashed curves), suggesting preferred heteroatomic interactions in this glass system. Furthermore the $^{11}B\{^{31}P\}$ REDOR experiments indicate that the $B^{(4)}$ units interact much more strongly with the phosphate species than the $B^{(3)}$ units do. Figure 3.22 also shows the predicted connectivities based on an alternative structural scenario, in which the concentration of $B^{(4)}$–O–P linkages is maximized while $B^{(3)}$–O–P linkages are excluded based on the experimentally observed REDOR data. This model agrees quite well with the experimental borate and phosphate speciations [131]. Most recently, however, experiments sensitive to indirect heteronuclear $^{11}B - ^{31}P$ spin–spin interactions have proven that B^3–O–P linkages are also formed, albeit at lower concentrations [132]. Analogous data have been obtained in compositionally analogous germanophosphate [133] (Figure 3.23) and tellurophosphate glasses [134] (Figure 3.24). In those glasses, the connectivity distribution appears to be closer to random, with a small (but detectable) preference for heteroatomic bond connectivity. Finally, results obtained on $(NaPO_3)_{1-x}(Al_2O_3)_x$ [135] and

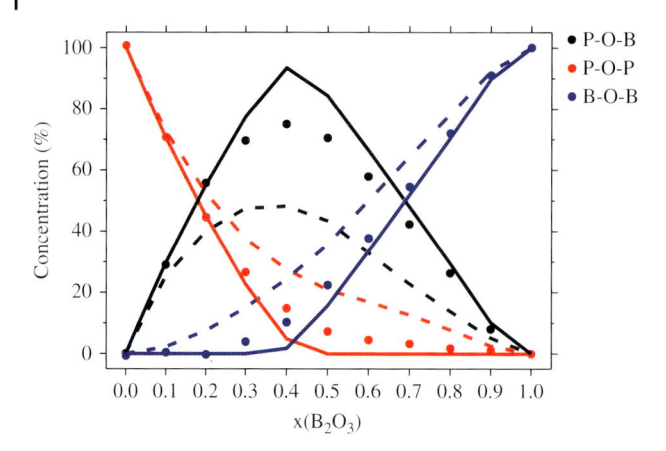

Figure 3.22 Connectivity distribution of the bridging oxygen atoms in $(K_2O)_{0.33}((B_2O_3)_x(P_2O_5)_{1-x})_{0.67}$ glasses derived from NMR data and comparison with different linkage scenarios. Dashed curves: random linkage scenario; solid curves: scenario corresponding to the maximum possible number of $B^{(4)}$–O–P linkages excluding the formation of $B^{(3)}$–O–P linkages.

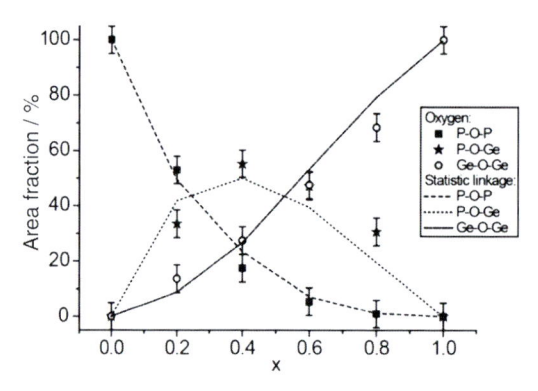

Figure 3.23 Connectivity distribution of the bridging oxygen atoms in $(Na_2O)_{0.33}((Ge_2O_4)_x(P_2O_5)_{1-x})_{0.67}$ glasses derived from NMR data and comparison with a statistical linkage scenario.

$(NaPO_3)_{1-x}(Ga_2O_3)_x$ [136] glasses indicate very strong Al—O—P and Ga—O—P bonding preferences. In these glass systems, the concentration of these linkages is always close to the maximum number that is possible at the specific compositions chosen, and the glass-forming range is limited to those compositions, for which a phosphorus-dominated second coordination sphere of the triel atom can be realized.

New insights on the second coordination sphere of alkali cations in borophosphate glasses have been recently obtained by advanced dipolar NMR methods [137]. As discussed above, the structure of these glasses features anionic $B^{(4)}$ and $P^{(2)}$ sites as well as neutral $B^{(3)}$ and $P^{(3)}$ sites. The interaction of these sites

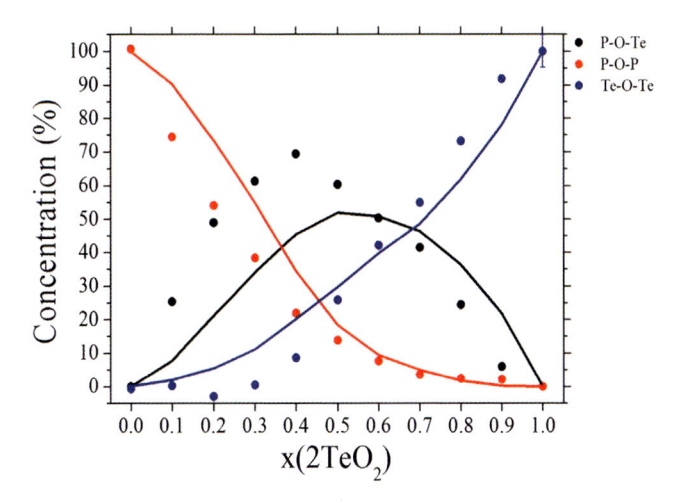

Figure 3.24 Connectivity distribution of the bridging oxygen atoms in $(Na_2O)_{0.33}((Te_2O_4)_x(P_2O_5)_{1-x})_{0.67}$ glasses derived from NMR data and comparison with a statistical linkage scenario.

with the network modifier ion Na^+ has been recently investigated using $^{11}B\{^{23}Na\}$ and $^{31}P\{^{23}Na\}$ REDOR spectroscopy, showing that the sodium ions interact significantly more strongly with the formally neutral $P^{(3)}$ units than with the formally anionic $B^{(4)}$ units [137]. Figure 3.25 shows that the distance sums Σr_{P-Na}^{-6} calculated from the measured $M_2(^{11}B\{^{23}Na\})$ and $M_2(^{31}P\{^{23}Na\})$ values in sodium borophosphate glasses are close to those measured in sodium ultraphosphate glasses whereas the corresponding Σr_{B-Na}^{-6} values are significantly smaller than those measured in sodium borate glasses with comparable sodium concentrations per network former. Thus, the network modifier–network former interactions in these sodium borophosphate glasses are much more similar to those in sodium phosphate than to those in sodium borate glasses. These findings can be explained on the basis of the preferred $P^{(3)}$–$B^{(4)}$ interactions in these systems as discussed earlier, leading to structural arrangements of the kind as shown in Figure 3.26. For such structures, bond valence considerations predict a partial displacement of the negative charge away from $B^{(4)}$ toward the nonbridging oxygen atoms of the $P^{(3)}$ sites. This charge dispersal mechanism most likely results in shallower Coulomb traps than in binary phosphate glasses, significantly enhancing the cationic mobility of the sodium ions. Indeed, the electrical conductivity of these glasses is significantly higher (by several orders of magnitude) compared to binary sodium phosphate glasses with comparable sodium contents.

3.5.2.2 Nanosegregation and Phase Separation Effects in Borosilicate Glasses:

Another system that has been attracting great interest is the alkali borosilicate glass system. At low-alkali compositions a well-documented phase separation

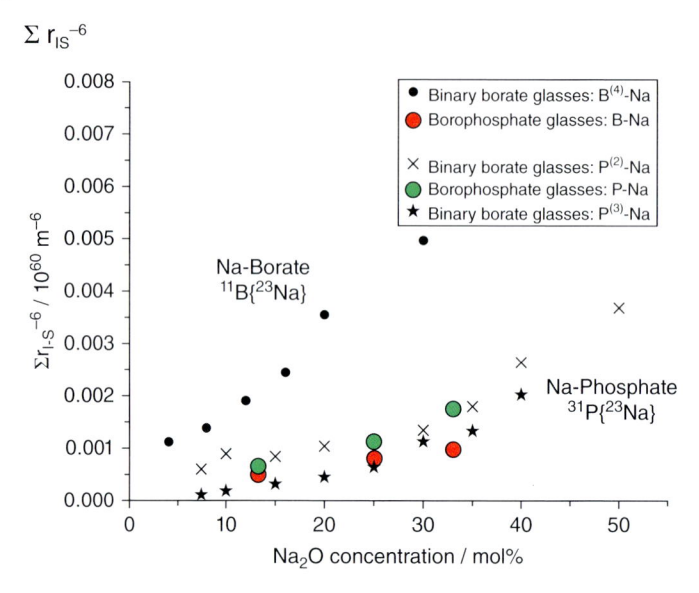

Figure 3.25 Values of Σr_{ij}^{-6} calculated from the experimental second moments of sodium borate, sodium phosphate, and sodium borophosphate glasses.

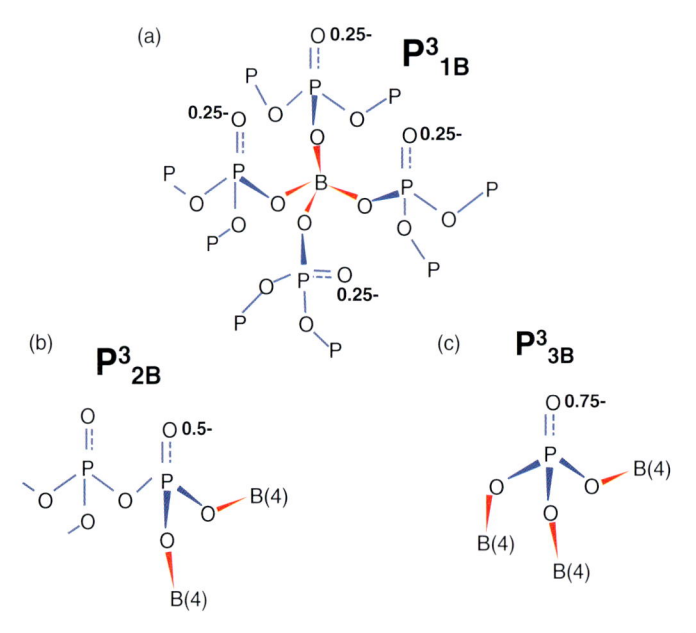

Figure 3.26 Postulated charge redistribution occurring in alkali borophosphate glasses following bond valence considerations. Blue and red arrows symbolize corresponding electron distributions within P—O and B—O single bonds associated with P—O—B linkages. (a) In $B^{(4)}_{4P}$ units linked to four $P^{(3)}_{1B}$ units, the negative formal charge on $B^{(4)}$ is delocalized on the nonbridging oxygen atoms of each $P^{(3)}$ unit, giving each a charge of 0.25. (b and c) Analogous considerations showing negative charges of 0.5 and 0.75 on the NBOs of $P^{(3)}_{2B}$ and $P^{(3)}_{3B}$ units, respectively.

effect is known to occur [138]: these glasses separate into an amorphous glassy alkali borate microphase, containing most of the modifier ions and an amorphous silica phase, which is essentially cation-free. The effect can be described in terms of the hypothetical reaction scheme,

$$Si^{(3)} + B(3) \rightarrow Si^{(4)} + B^{(4)} \qquad (3.12)$$

which transfers the anionic charge from the silicate to the borate species. Consistent with this finding, ^{11}B MAS-NMR studies of $(Na_2O)_{0.2}((Si_2O_4)_{1-x}-(B_2O_3)_x)$ glasses have shown that the entire alkali ion content is used to modify the boron species, leaving no modifier available for generating nonbridging oxygen atoms on silicon [139]. To further substantiate this model, ^{23}Na dipolar NMR experiments and temperature dependent ionic mobility studies need to be conducted in this composition region. In addition, the extent of phase separation should be studied by quantitative analyses of the B–O–B, B–O–Si, and Si–O–Si connectivities, both within and outside of the phase separation regime. Unfortunately, the interpretation of ^{29}Si chemical shifts is ambiguous, as both anionic $Si^{(3)}$ units and neutral $Si^{(4)}$ units having a Si–O–B$^{(4)}$ linkage have comparable chemical shifts [140]. Experiments sensitive to ^{11}B – ^{29}Si magnetic dipole–dipole couplings are hampered by the low natural abundance of the ^{29}Si isotope and long spin-lattice relaxation times. In principle, these restrictions can be overcome by isotopic enrichment. Furthermore, as shown by Stebbins and coworkers, ^{17}O NMR is an excellent probe for the spectroscopic distinction of such connectivities [141–144]. As far as such studies have been conducted on glasses in the phase separated region, they indeed show a lower than statistically expected concentration of Si—O—B linkages, which can be reduced even further by extended annealing [144]. Studies of this kind offer the prospect of deriving the exact compositions of the two coexisting nanophases as well as its dependence on thermal treatment.

3.6
Conclusions and Recommendations for Further Work

In summary, the present review has shown that very specific and quantitative information about the second coordination sphere and beyond has been obtained in numerous oxide glasses. These advances include a convergent picture of the bond angle distributions and ring size statistics of simple network glasses, knowledge about the structural arrangements and spatial distributions of network modifier cations in simple binary oxide glass systems, and the quantification of chemical ordering effects in glasses having more than one network former species. A review of these advances indicates, however, that ordering models that go beyond the second (or third) coordination sphere of a given network former or modifier species are very difficult to support unambiguously by either NMR or diffraction methods and more input from structural modeling is needed. For example, no good NMR or diffraction observables are available to sustain the well-developed ideas regarding the torsion angle distributions in glassy silica or the "superstructural units" in binary borate glasses. These ideas remain at the frontier of this field, awaiting new, more specific experimental approaches to be developed in the future. Certainly, the tools of molecular dynamics simulations and DFT-first principles calculations will be needed to formulate specific structural hypotheses for these and other ordering phenomena on the nanometer length scale.

Both X-ray/neutron diffraction and solid-state NMR spectroscopies have played a key role in making these advances possible, in part because there have been considerable advances in technique development during the past 25 years. It must further be emphasized that the incorporation of increasingly sophisticated molecular dynamics modeling techniques, aided by first-principles calculations, have been opening new opportunities for probing more deeply into medium-range ordering processes in disordered materials. To date, however, the benefits of these methodological advances have not yet resulted in the development of an integrated experimental/theoretical strategy combining all these methods, even though the idea had been proposed many years ago for glasses [145]. In the characterization of highly disordered crystalline compounds, such a development is currently indeed just happening, and has resulted in the establishment of a new research field, termed *NMR Crystallography*. NMR crystallographers solve crystal structures based on the Rietveld refinement of XRD powder patterns, which are constrained by input from NMR spectroscopic observables, and cross checked against DFT calculations. In principle, various elements of NMR crystallography are transferable from crystalline compounds to glasses and amorphous materials. One may anticipate that with a considerably enhanced role of molecular dynamics simulations and DFT-based modeling approaches, significant further advances can be made toward our knowledge of medium-range order in glasses.

Acknowledgments

The author acknowledges the contributions made by numerous doctoral students and postdoctoral fellows working on this subject. They include (in alphabetical order) Dr. Carla Cavalca Araujo (Ph.D., 2005), Dr. Frederik Behrends (Ph. D., 2013), PD Dr. Marko Bertmer (Ph.D., 1999, now University of Leipzig), Dr. Christian Brinkmann (Ph.D., 2003), PD Dr. Gunther Brunklaus (Ph.D., 2003, now WWU Münster), Dr. Alice Cattaneo, Prof. Dr. J. C. C. Chan (now National Taiwan University), Dr. Rashmi Deshpande (Ph.D., 2009), Dr. Marcos de Oliveira, Dr. Heinz Deters (Ph.D., 2011), Mr. Carsten Doerenkamp (M.Sc., 2013), Dr. Stefan Elbers (Ph.D., 2004), Dr. Jan Dirk Epping (Ph.D., 2004), Dr. Sandra Faske (Ph.D., 2007), Prof. Dr. Deanna Franke (Ph.D., 1993, now Moorpark College), Ms. Lena Funke (M.Sc., 2015), Dr. Becky Gee (Ph.D., 1997), Prof. Dr. Sophia Hayes (Ph.D., 1999, now Washington University), Dr. Christopher Hudalla (Ph.D., 1997), Dr. Michael Janssen (Ph.D., 1998), Dr. Martin Kalwei (Ph. D., 2002), Dr. Dirk Larink (Ph.D., 2011), Dr. David Lathrop (Ph.D., 1991), Dr.

Carri Lyda (Ph.D., 1994), Dr. Robert Maxwell (Ph.D., 1993, now Lawrence Livermore Laboratory), Dr. Christine Mönster (Ph.D., 2007), Dr. Daniel Mohr (2010), Dr. Paul Mutolo (Ph.D., 2000, now Cornell University), Dr. Stefan Puls (Ph.D., 2006), Dr. Devidas Raskar (Ph.D., 2008), Prof. Dr. Eva Ratai (Ph.D., 2000, now Harvard Medical School), Prof. Dr. Jinjun Ren (now Shanghai Institute of Optics and Fine Mechanics), Dr. Matthias Rinke (Ph.D., 2010), Dr. Carsten Rosenhahn (Ph.D., 2000), Dr. Silvia Santagneli (UNESP, Araraquara), Prof. Dr. Jörn Schmedt auf der Günne (Ph.D., 2000, now Universität Siegen), Dr. Cornelia Schröder (Ph.D., 2014), Prof. Dr. Robert Shibao (Ph.D., 1995, now El Camino College), Dr. Wenzel Strojek (Ph.D., 2006), Dr. Tobias Uesbeck (Ph.D., 2016), Dr. Julia Vannahme (Ph.D., 2007), Prof. Dr. Leo van Wüllen (now Universität Augsburg), Prof. Dr. Michael Vogel (now TU Darmstadt), Dr. Ulrike Voigt (Ph.D., 2004), Prof. Dr. Long Zhang (now Shanghai Institute of Optics and Fine Mechanics), and Dr. Lars Züchner (Ph.D., 1997). Fruitful collaborations with Professors Roland Böhmer (Universität Dortmund, Germany), Andrea de Camargo (University of Sao Paulo, Sao Carlos, Brazil), Dominik Eder (WWU Münster, Günther Heinz Frischat (University of Clausthal, *in memoriam*), Klaus Funke (WWU Münster), Andreas Heuer (WWU Münster), John H. Kennedy (UC Santa Barbara), Ladislav Koudelka (University of Pardubice), Steven W. Martin (Iowa State University, USA), Helmut Mehrer (WWU Münster), Annie Pradel and Michel Ribes (University of Montpellier, France), Gregory Tricot (University of Lille), and with Drs. Bruce Aitken and Randall Youngman (Corning Glassworks), Wolfram Höland (Ivoclar-Vivadent), Shingo Nakane (Nippon Electric Glass) are gratefully acknowledged. Some of this collaborative work has been reviewed here. I also acknowledge many useful discussions with Professor P. Maass (University of Osnabrück, Germany) on this subject during the past few years. A very special "Thank you" goes to Professor Werner Müller-Warmuth (Professor Emeritus, WWU Münster), who many years ago introduced the author to the subject of NMR spectroscopy of glasses. Over the years the work has been funded by the US National Science Foundation-Division of Materials Research, by the Deutsche Forschungsgemeinschaft (DFG), and by the Brazilian research support institutions CNPq and FAPESP. Our current research programme is funded by FAPESP via the CEPID programme, process number 07793-6.

References

1 Zachariasen, W.H. (1932) *J. Am. Chem. Soc.*, **54**, 3841.
2 Wright, A.C. (1990) *J. Non Cryst. Solids*, **123**, 129.
3 Yuan, X. and Cormack, A.N. (2003) *J. Non Cryst. Solids*, **319**, 31.
4 Wright, A.C. (1994) *J. Non Cryst. Solids*, **179**, 84.
5 Price, D.L. (1996) *Curr. Opin. Solid State Mater Sci.*, **1**, 572.
6 Yuan, X. and Cormack, A.N. (2002) *Comp. Mater. Sci*, **24**, 343.
7 Greaves, G.N. (1985) *J. Non Cryst. Solids*, **71**, 203.
8 Gurman, S.J. (1990) *J. Non Cryst. Solids*, **125**, 151.

9 Deubener, J. (2005) *J. Non Cryst. Solids*, **351**, 1500.
10 Zanotto, E.D., Tsuchida, J., Schneider, J.F., and Eckert, H. (2015) *Int. Mater. Rev.*, **60**, 380–394.
11 Hannon, A.C. (2015) *Modern Techniques of Glass Characterization*, (ed. M. Affatigato), John Wiley & Sons, Inc., Hoboken, p. 158.
12 Benmore, C.J. (2015) *Modern Techniques of Glass Characterization* (ed. M. Affatigato), John Wiley & Sons, Inc., Hoboken, p. 241.
13 Hannon, A.C., Grimley, D.J., Hulme, R.A., Wright, A.C., and Sinclair, R.N. (1994) *J. Non Cryst. Solids*, **177**, 299.
14 Vashishta, P., Kalia, R.K., Rino, J.P., and Ebbsjiö, I. (1990) *Phys. Rev B.*, **41**, 12197.
15 Elliott, S.R. (1991) *Phys. Rev. Lett.*, **57**, 711.
16 Elliott, S.R. (1995) *J. Non Cryst. Solids*, **182**, 40.
17 Kamitsos, E.I. (2015) *Modern Techniques of Glass Characterization* (ed. M. Affatigato), John Wiley & Sons, Inc., Hoboken, p. 32.
18 Almeida., R.M. and Santos, L.F. (2015) *Modern Techniques of Glass Characterization* (ed. M. Affatigato), John Wiley & Sons, Inc., Hoboken, p. 74.
19 Krogh Moe, J. (1962) *Phys. Chem. Glasses*, **3**, 101.
20 Konijendijk, W.L. and Stevels, J.M. (1975) *J. Non Cryst. Solids*, **18**, 307.
21 Parshin, D.A. (2004) *Phys. Status Solidi (c)*, **1**, 2860.
22 Slichter, C.P. (1978) *Principles of Magnetic Resonance*, Springer, Heidelberg.
23 Duer, M.J. (2004) *Introduction into Solid State NMR Spectroscopy*, Blackwell Publishing Ltd.
24 Schmidt-Rohr, K. and Spiess, H.W. (1996) *Solid State NMR and Polymers*, Academic Press, London.
25 Edén, M. (2012) *Ann. Rep. Prog. Chem. Sect. C*, **108**, 177.
26 Eckert, H. (1992) *Prog. NMR Spectrosc.*, **24**, 159.
27 Eckert, H., Elbers, S., Epping, J.D., Janssen, M., Kalwei, M., Strojek, W., and Voigt, U. (2005) *Top. Curr. Chem.*, **246**, 195.
28 Stebbins, J.F. and Xue, X. (2014) *Rev. Mineral. Geochem.*, **78**, 605.
29 Eckert, H. (2010) *Z. Phys. Chem*, **224**, 1591, and references therein.
30 Kroeker, S. (2015) *Modern Techniques of Glass Characterization* (ed. M. Affatigato), John Wiley & Sons, Inc., Hoboken, p. 315.
31 Andrew, E.R., Bradbury, A., and Eades, R.G. (1958) *Nature*, **182**, 1659.
32 Freude, D. and Haase, J. (1993) *NMR-Basic Principles and Progress*, vol. 29, Springer, Berlin, p. 1.
33 Medek, A., Harwood, J.S., and Frydman, L. (1995) *J. Am. Chem. Soc.*, **117**, 12779.
34 Deters, H. (2011) PhD Dissertation. WWU Münster.
35 Kaupp, M., Bühl, M., and Malkin, V.G. (2004) *Calculation of NMR and EPR parameters*, Wiley-VCH Verlag GmbH.
36 Cremer, D. and Grafenstein, J. (2007) *Phys. Chem. Chem. Phys.*, **9**, 2791.
37 van Vleck, J.H. (1948) *Phys. Rev.*, **74**, 1168.
38 Gullion, T. (1998) *Conc. Magn. Reson.*, **10**, 277.
39 Gullion, T. and Schaefer, J. (1989) *J. Magn. Reson.*, **81**, 196.
40 Naito, A., Nishimura, K., Tuzi, S., and Saito, H. (1994) *Chem. Phys. Lett.*, **229**, 506.
41 Bertmer, M. and Eckert, H. (1999) *Solid State Nucl. Magn. Reson.*, **15**, 139.
42 Chan, J.C.C. and Eckert, H. (2000) *J. Magn. Reson.*, **147**, 170.
43 Chan, J.C.C., Bertmer, M., and Eckert, H. (1999) *J. Am. Chem. Soc*, **121**, 5238.
44 Strojek, W., Kalwei, M., and Eckert, H. (2004) *J. Phys. Chem. B*, **108**, 7061.
45 Gullion, T. and Vega, A. (2005) *Prog. Nucl. Magn. Reson. Spectrosc.*, **47**, 123.
46 Kaplan, D.E. and Hahn, E. (1958) *J. Phys. Radium*, **19**, 821.
47 Schmedt auf der Günne, J. and Eckert, H. (1998) *Chem. Eur. J.*, **4**, 1762.
48 Saalwächter, K., Lange, F., Matyjaszewski, K., Huang, C.-F., and Graf, R. (2011) *J. Magn. Reson.*, **212**, 204, and references therein.
49 Schmedt auf der Günne, J. (2003) *J. Magn. Reson.*, **165**, 18.
50 Ren, J. and Eckert, H. (2012) *Angew. Chem., Int. Ed.*, **51**, 12888.
51 Ren, J. and Eckert, H. (2013) *J. Chem. Phys.*, **138**, 164201, and references therein.

52 Engelsberg, M. and Norberg, R.E. (1972) *Phys. Rev. B*, **5**, 3395.
53 Lathrop, D., Franke, D., Maxwell, R., Tepe, T., Flesher, R., Zhang, Z., and Eckert, H. (1992) *Solid State Nucl. Magn. Reson.*, **1**, 73, and references therein.
54 Lathrop, D. and Eckert, H. (1989) *J. Am. Chem. Soc.*, **111**, 3536.
55 Haase, J. and Oldfield, E. (1993) *J. Magn. Reson. A*, **101**, 30.
56 Gee, B. and Eckert, H. (1995) *Solid State Nucl. Magn. Reson.*, **5**, 113.
57 Elbers, S., Strojek, W., Koudelka, L., and Eckert, H. (2005) *Solid State Nucl. Magn. Reson.*, **27**, 65–76.
58 Mozzi, R.L. and Warren, B.E. (1969) *J. Appl. Crystallogr.*, **2**, 164.
59 Malfait, W.J., Halter, W.E., and Verel, R. (2008) *Chem. Geol.*, **256**, 269.
60 Neuefeind, J. and Liss, K.D. (1996) *Ber. Bunsenges. Phys. Chem.*, **100**, 1341.
61 Kohara, S. and Suzuya, K. (2005) *J. Phys. Cond. Matt*, **17**, S77.
62 Mei, Q., Benmore, J., Sen, S., Sharma, R., and Yarger, J.L. (2008) *Phys. Rev. B*, **78**, 144204.
63 van Beest, B.W.H., Kramer, G.J., and van Santen, R.A. (1990) *Phys. Rev. Lett.*, **64**, 1955.
64 Vessal, B., Amini, M., Fincham, D., and Catlow, C.R.A. (1989) *Philos. Mag. B*, **60**, 753.
65 Pettifer, R.F., Dupree, R., Farnan, I., and Sternberg, U. (1988) *J. Non Cryst. Solids*, **106**, 408.
66 Poulsen, H.F., Neuefeind, J., Neumann, H.B., Schneider, J.R., and Zeidler, M.D. (1995) *J. Non Cryst. Solids*, **188**, 63.
67 Gladden, L.F. (1992) *The Physics of Noncrystalline Solids* (eds L.D. Pye, W.C. La Course, and H.J. Stevens), Taylor and Francis, London, p. 91.
68 Gladden, L.F. (1990) *J. Non Cryst. Solids*, **119**, 318.
69 Dupree, E. and Pettifer, R.F. (1984) *Nature*, **308**, 523.
70 Mauri, F., Pasquarello, A., Pfrommer, B.G., Yoon, Y.G., and Louie, S.G. (2000) *Phys. Rev B*, **62**, R4786.
71 Farnan, I., Grandinetti, P.J., Baltisberger, J.H., Stebbins, J.F., Werner, U., Eastman, M.A., and Pines, A. (1992) *Nature*, **358**, 31.
72 Clark, T.M. and Grandinetti, P.J. (2003) *J. Phys. Condens. Matter*, **15**, S2387.
73 Clark, T.M., Grandinetti, P.J., Florian, P., and Stebbins, J.F. (2004) *Phys. Rev B*, **70**, 064202.
74 Charpentier, T., Kroll, P., and Mauri, F. (2009) *J. Phys. Chem. C*, **113**, 7917.
75 Tucker, M.G., Keen, D.A., Dove, M.T., and Trachenko, K. (2005) *J. Phys. Condens. Matter*, **17**, S67.
76 Micoulaut, M., Comier, L., and Henderson, G.S. (2006) *P. Phys. Condens. Matter*, **18**, R753.
77 Salmon, P.S., Barnes, A.C., Martin, R.A., and Cuello, G.J. (2007) *J. Phys. Condens. Matter*, **19**, 415110.
78 Henderson, G.S., Neuville, D.R., Chochain, B., and Cormier, L. (2009) *J. Non Cryst. Solids*, **355**, 468.
79 Hussin, R., Dupree, R., and Holland, D. (1999) *J. Non Cryst. Solids*, **246**, 159.
80 Kohara, S., Akola, J., Morita, H., Suzuya, K., Weber, J.K.R., Wilding, M.C., and Benmore, C.J. (2011) *Proc. Natl. Acad. Sci USA*, **108**, 14780.
81 Rino, J.P., Ebbsjö, I., Kalia, R.K., Nakano, A., and Vashishta, P. (1993) *Phys. Rev. B*, **47**, 3053.
82 Du, J. and Cormack, A.N. (2004) *J. Non Cryst. Solids*, **349**, 66.
83 Pedone, A., Malavasi, G., Cormack, A.N., Segre, U., and Menziani, M.C. (2007) *Chem. Mater.*, **19**, 3144.
84 Du, L.S., Peng, L., and Stebbins, J.F. (2007) *J. Non Cryst. Solids*, **353**, 2910.
85 Majerus, O., Cormier, L., Neuville, D.R., Galoisy, L., and Calas, G. (2008) *J. Non Cryst. Solids*, **354**, 2004.
86 Bernard, C., Chassedent, S., Monteil, A., Balu, N., Obriot, J., Duverger, C., Ferrari, M., Bouazaoui, M., Kinowski, C., and Turrell, S. (2001) *J. Non Cryst. Solids*, **284**, 68.
87 Wright, A.C. (2015) *Int. J. Appl. Glass Sci.*, **6**, 45.
88 Krogh-Moe, J. (1969) *J. Non Cryst. Solids*, **1**, 269.
89 Jellison J Jr, G.E., Panek, L.W., Bray, P.J., Rouse, G.B., Jr, and Risen, W.M. (1977) *J. Chem. Phys.*, **66**, 802.
90 Youngman, R.E. and Zwanziger, J.W. (1994) *J. Non Cryst. Solids*, **168**, 293.
91 Zwanziger, J.W. (2005) *Solid State Nucl. Magn. Reson.*, **27**, 5.

92 Joo, C., Werner-Zwanziger, U., and Zwanziger, J.W. (2000) *J. Non Cryst. Solids*, **261**, 282.

93 Youngman, R.E., Haubrich, S.T., Zwanziger, J.W., Jaenicke, M.T., and Chmelka, B.F. (1995) *Science*, **269**, 1416.

94 Hung, I., Howes, A.P., Parkinson, B.G., Anupold, T., Samoson, A., Brown, S.P., Harrison, P.F., Holland, D., and Dupree, R. (2009) *J. Solid State Chem.*, **182**, 2402.

95 Wang, S. and Stebbins, J.F. (1999) *J. Am. Ceram. Soc.*, **82**, 1519.

96 Lee, S.K., Musgrave, C.B., Zhao, P., and Stebbins, J.F. (2001) *J. Phys. Chem. B*, **105**, 12583.

97 van Wüllen, L. and Schwering, G. (2002) *Solid State Nucl. Magn. Reson.*, **21**, 134.

98 Bell, R.J., Carnevale, A., Kurkjian, C.R., and Peterson, G.E. (1980) *J. Non Cryst. Solids*, **35/36**, 1185.

99 Chakraborty, I.N. and Chondrate, R.A. Sr (1986) *J. Non Cryst. Solids*, **81**, 271.

100 Chakraborty, I.N. and Chondrate, R.A. Sr (1986) *J. Mater. Sci. Lett.*, **5**, 361.

101 Hoppe, U., Kranold, R., Barzz, A., Stachel, D., and Neuefeind, J. (2000) *Solid State Commun.*, **115**, 559.

102 Hoppe, U., Brow, R.K., Tischendorf, B.C., Jovari, P., and Hannon, A.C. (2006) *J. Phys. Condens. Matter*, **18**, 1847.

103 Zwanziger, J.W., Shaw, J.L., Werner-Zwanziger, U., and Aitken, B.G. (2006) *J. Phys. Chem. B*, **110**, 20123.

104 Aitken, B.G., Youngman, R.E., Deshpande, R.R., and Eckert, H. (2009) *J. Phys. Chem. C*, **113**, 3322.

105 Araujo, C., Zhang, L., and Eckert, H. (2006) *J. Mater. Chem.*, **16**, 1323.

106 Zhang, L., Bögershausen, A., and Eckert, H. (2005) *J. Am. Ceram. Soc.*, **88**, 897.

107 Ren, J., Doerenkamp, C., and Eckert, H. (2016) *J. Phys. Chem. C*, **120**, 1758–1769.

108 Zhang, L. and Eckert, H. (2005) *J. Mater. Chem.*, **15**, 1640.

109 Gaskell, P.H., Eckersley, M.C., Barnes, A.C., and Chieux, P. (1991) *Nature*, **350**, 675.

110 Mead, R.N. and Mountjoy, G. (2006) *J. Phys. Chem. B*, **110**, 14273.

111 Cormier, L. and Cuello, G.J. (2013) *Geochim. Cosmochim. Acta*, **122**, 498.

112 Huang, C. and Cormack, A.N. (1991) *J. Chem. Phys.*, **95**, 3634.

113 Ellison, A.J.G., Price, D.L., Dickinson J Jr., J.E., and Hannon, A.C. (1995) *J. Chem. Phys.*, **102**, 9647.

114 Zhao, J., Gaskell, P.H., Cluckie, M.M., and Soper, A.K. (1998) *J. Non Cryst. Solids*, **232–234**, 721.

115 Du, J. and Corrales, L.R. (2006) *J. Chem. Phys.*, **125**, 114702.

116 Du, J. and Corrales, L.R. (2006) *J. Non Cryst. Solids*, **352**, 3255.

117 Fabian, M., Jovari, P., Svab, E., Meszaros, G., Proffen, T., and Veress, E. (2007) *J. Phys. Cond. Matter*, **19**, 335209.

118 Du, J. and Cormack, A.N. (2004) *J. Non Cryst. Solids*, **349**, 66.

119 Swenson, J., Börjesson, L., and Howells, W.S. (1995) *Phys. Rev. B*, **52**, 9310.

120 Puls, S. and Eckert, H. (2007) *Phys. Chem. Chem. Phys.*, **9**, 3992.

121 Voigt, U. (2005) Doctoral dissertation. WWU Münster.

122 Epping, J.D., Strojek, W., and Eckert, H. (2005) *Phys. Chem. Chem. Phys.*, **7**, 2384.

123 Bunde, A., Funke, K., and Ingram, M.D. (1998) *Solid State Ionics*, **105**, 1.

124 Voigt, U., Eckert, H., Lammert, H., and Heuer, A. (2005) *Phys. Rev. B*, **72**, 064207.

125 Strojek, W. and Eckert, H. (2006) *Phys. Chem. Chem. Phys.*, **8**, 2276.

126 Epping, J.D., Eckert, H., Imre, A., and Mehrer, H. (2005) *J. Non Cryst. Solids*, **351**, 3521.

127 Schneider, J., Tsuchida, J., Deshpande, R., and Eckert, H. (2013) *Phys. Chem. Chem. Phys.*, **15**, 14328.

128 Du, J. and Corrales, L. (2005) *Phys. Rev. B*, **72**, 092201.

129 Dwivedi, B.P., Rahman, M.H., Kumar, Y., and Khanna, B.N. (1993) *J. Phys. Chem. Solids*, **54**, 621.

130 Yano, T., Kunimine, N., Shibata, S., and Yamane, M. (2003) *J. Non Cryst. Solids*, **321**, 137.

131 Larink, D., Eckert, H., Reichert, M., and Martin, S.W. (2012) *J. Phys. Chem. C*, **116**, 26162.

132 Tricot, G., Raguenet, B., Silly, G., Ribes, M., Pradel, A., and Eckert, H. (2015) *Chem. Commun.*, **51**, 9284.

133 Behrends, F. and Eckert, H. (2014) *J. Phys. Chem. C*, **118**, 10271.

134 Larink, D., Rinke, M.T., and Eckert, H. (2015) *J. Phys. Chem. C*, **119**, 17539.
135 Zhang, L. and Eckert, H. (2006) *J. Phys. Chem. B*, **110**, 8946.
136 Ren, J. and Eckert, H. (2012) *J. Phys. Chem. C*, **116**, 12747.
137 Funke, L. and Eckert, H. (2016) *J. Phys. Chem. C.*, **120**, 3196.
138 Doweidar, H., El-Shahawi, M.S., Reicha, F.M., Silim, H.A., and Ei-Elgaly, K. (1990) *J. Phys. D*, **23**, 1441.
139 Wu, X., Youngman, R., and Dieckmann, R. (2013) *J. Non Cryst. Solids*, **378**, 168.
140 Martens, R. and Müller-Warmuth, W. (2000) *J. Non Cryst. Solids*, **265**, 167.
141 Wang, S. and Stebbins, J.F. (1998) *J. Non Cryst. Solids*, **231**, 286.
142 Du, L.S. and Stebbins, J.F. (2003) *J. Phys. Chem. B*, **107**, 10063.
143 Du, L.S. and Stebbins, J.F. (2003) *J. Non Cryst. Solids*, **315**, 239.
144 Aguiar, P.M., Michaelis, V.K., Mc Kinley, C.M., and Kroeker, S. (2013) *J. Non Cryst. Solids*, **363**, 50.
145 Zwanziger, J.W. (1998) *Int. Rev. Phys. Chem.*, **17**, 65.

4
Suboxides and Other Low-Valent Species

Arndt Simon

MPI Stuttgart, Max-Planck-Institut für Festkörperforschung, Heisenbergstrasse 1, 70569 Stuttgart, Germany

4.1
Materials and Structure of Solids

4.1.1
Bulk Materials

4.1.1.1 Suboxides and Other Low-Valent Species

The periodic table is the main guide in chemistry. We see the behavior of the elements change stepwise along horizontal periods and, rather more subtle, down vertical groups with some diagonal shifts between them. The alkali metals lead the dance followed by the majority of metallic elements up to the Zintl border with comparatively few nonmetals on the right-hand side. A peculiar scenario arises when a small number of nonmetals get incorporated into a metal atom matrix creating quite unusual structures and compound compositions. Admittedly, there is a kind of Babylonian confusion of tongues in using the appropriate terms to describe such phases as "metal-rich," "electron-rich," "low-valent," or "subvalent." Is the molecule Li_6C a metal-rich species according to the large number of metal atoms attached to the central carbon atom or is it electron rich and low valent, because it carries a surplus of electrons according to $Li^+_6C^{4-}\cdot 2e^-$? Or, is it even hypervalent as repeatedly described, because six metal atoms are bonded to one carbon atom? The prefix "sub" is often applied to quite conventional bonding situations as, for example, in the case of C_3O_2, the suboxide of carbon that is actually a normal valence compound described by the Lewis notation O=C=C=C=O.

The following discussion will be focused on classes of compounds that bear the combined signatures of being metal rich and in addition having a surplus of electrons, and thus are characterized by metal–metal bonds that might be itinerant as well as localized. The necessary condition for covalent M—M bonding is a low oxidation state that leaves valence electrons with the metal atom. Such

Handbook of Solid State Chemistry, First Edition. Edited by Richard Dronskowski, Shinichi Kikkawa, and Andreas Stein.
© 2017 Wiley-VCH Verlag GmbH & Co. KGaA. Published 2017 by Wiley-VCH Verlag GmbH & Co. KGaA.

compounds need to be metal rich to allow close enough contacts between the metal atoms, and the extension of the valence electron orbitals in space must be large in order to provide good overlap. Therefore, it does not surprise that M—M bonding and cluster formation dominates in the lower region of the periodic table.

Starting from the alkali metals and their peculiar suboxides, the subnitrides of the alkaline earth metals will be addressed next. The structure and bonding principles found in groups 1 and 2 are amazingly similar as one might expect. When proceeding to group 3 certain features are kept and new features enter with the rare earth metals and lanthanoids that prepare the ground for an extended area of transition metal chemistry based on metal–metal bonding.

Alkali Metal Suboxides
When cesium comes in contact with water a violent explosion occurs, and in air the metal starts to burn. Better do not let a cesium-filled ampoule slip off your hand and break on the floor in a chemistry laboratory, which nowadays is frequently covered with polyvinyl chloride (PVC) instead of old-fashioned ceramic tiles. A grand firework is guaranteed. A specially designed vacuum line is needed to oxidize the metal under carefully controlled conditions. When the metal is exposed to oxygen at a sufficiently low pressure to avoid too much heat to be evolved, accompanied by some vaporization of the sample, it quickly becomes liquid and a black surface layer forms that dissolves to form a homogeneous melt. Pure cesium has a light golden color that sometimes has been attributed to contamination by oxygen; however, even at a concentration of less than 0.5 ppm oxygen, it preserves its native golden color. Upon oxidation, it gradually darkens to bronze, violet, and finally bluish black tinges accompanied by a significant volume contraction. Distilling off the excess of cesium leaves transparent orange crystals of Cs_2O. There has been early evidence for the formation of suboxides of the heavy alkali metal cesium in admirable investigations by Rengade [1]. It took more than half a century to verify and understand the nature of these compounds [2]. Thermal analysis reveals a phase diagram shown in Figure 4.1 that is

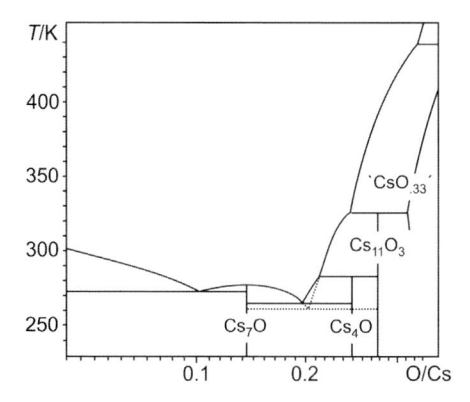

Figure 4.1 Phase diagram Cs/O.

characterized by the stoichiometric compounds Cs_7O, Cs_4O, and $Cs_{11}O_3$ [3]. The first two of them only exist below room temperature. The crystallization of Cs_4O is frequently missed in experimental studies due to its kinetically hindered formation. At further increased oxygen content a broad range of homogeneity with approximate composition "Cs_3O" exists.

Silver-colored rubidium joins cesium in forming suboxides. It also vigorously reacts with oxygen to form a brass-colored melt from which copper-red crystals Rb_9O_2 separate upon further oxidation. The phase analysis turns out to be particularly tricky as a few tenths of a degree below the Rb/Rb_9O_2 eutectic the more metal-rich compound Rb_6O forms in a kinetically delayed solid-state reaction. A sophisticated thermal analysis equipment had to be constructed that allows control of temperature within a few hundredth of a degree, and in order to get hold of Rb_6O in a reproducible way supercooling experiments have been performed down to 120 K. Quenching experiments result in amorphous and metastable phases in the crystallization process of this phase. It is amazing to see pronounced atom mobility with the heavy alkali metals at such low temperatures. Even at 180 K, the atom mobility is so large that an ordered metastable crystalline phase $Rb_{6.33}O$ forms from quenched amorphous precursors. The stability of the rubidium suboxides is quite low. Rb_9O_2 decomposes peritectically at 313 K, very close to the melting temperature of rubidium, 312 K, with the formation of the normal salt Rb_2O (see Figure 4.2). Experiments with potassium in combination with the heavier homologues rubidium and cesium give clear evidence for the existence of mixed suboxides; however, they could not yet be characterized as they decompose at temperatures around 210 K and crystalline K_2O separates. Suboxides of lithium and sodium are unknown.

The alkali metals are very much alike, rubidium and cesium in particular. However, when binary alloys of both metals are oxidized, a strange behavior is observed. In the quasibinary section shown in Figure 4.3, stoichiometric compounds exist that exhibit the compositions $Cs_{11}O_3$, $Cs_{11}O_3Rb$, $Cs_{11}O_3Rb_2$, and $Cs_{11}O_3Rb_7$. The Cs/O ratio does not deviate from 11 to 3 but only the amount of rubidium varies. It is amazing that rubidium and cesium exhibit such distinctly different functionalities in their crystal structures. To a certain extent the rubidium can also be substituted that results in a limited homogeneity range according to $Cs_{11}O_3Cs_xRb_y$. Only in the case $x, y = 0$, partial replacement of cesium by rubidium is possible and leads to phases $Cs_{11-z}Rb_zO_3$. The analogous behavior is found for $Rb_{9-z}Cs_zO_2$.

So far, thermal analysis data revealed the existence of numerous compounds of the alkali metals with strange compositions but with no detailed understanding. The formulas described in the text indeed anticipate knowledge that essentially originates from single-crystal structure investigations. Specially designed low-temperature devices were employed to grow crystals from liquid samples in sealed X-ray capillaries, using a miniaturized Bridgman technique. These investigations opened a unique area of cluster chemistry characterized by two types of entities: Rb_9O_2 and $Cs_{11}O_3$, respectively (see Figure 4.4).

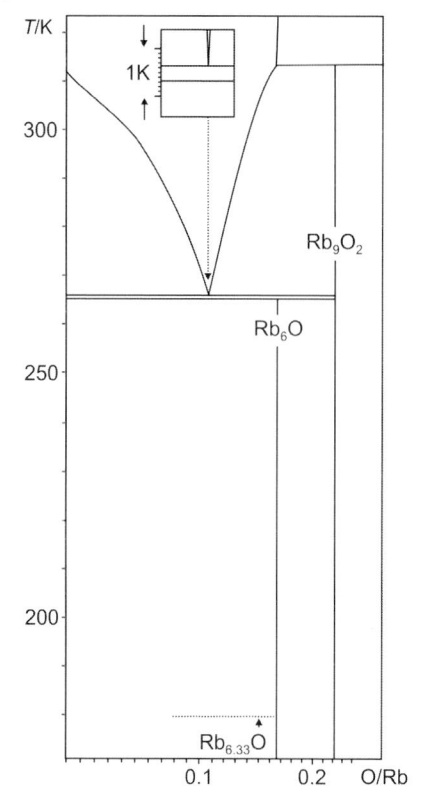

Figure 4.2 Phase diagram Rb/O.

The oxygen atoms are octahedrally coordinated by alkali metal atoms. Two octahedra join a common face to form the Rb_9O_2 cluster, and three such octahedra are condensed via faces resulting in the $Cs_{11}O_3$ cluster. Significant shifts of the oxygen atoms away from the centers of the octahedra toward the periphery

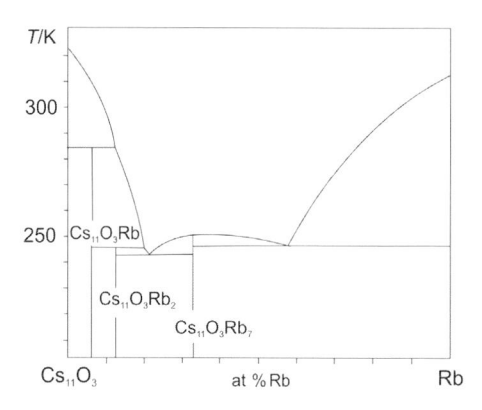

Figure 4.3 Phase diagram $Cs_{11}O_3$/Rb.

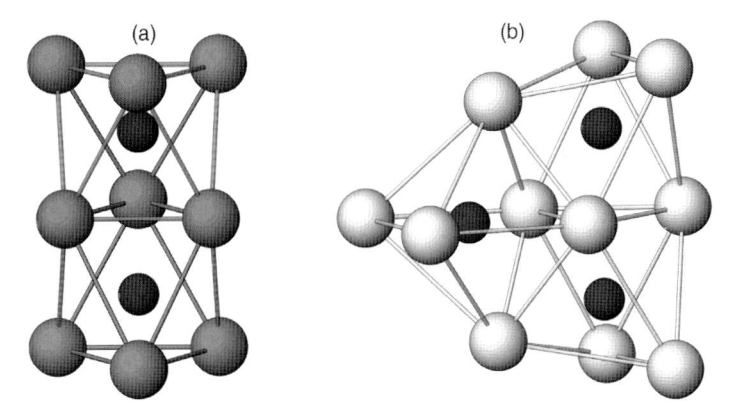

Figure 4.4 (a) Rb_9O_2 and (b) $Cs_{11}O_3$ clusters in alkali metal suboxides, small spheres O.

indicate repulsive interactions between the oxygen atoms according to an ionic description of bonding as $Rb^+{}_9O^{2-}{}_2 \cdot 5e^-$ and $Cs^+{}_{11}O^{2-}{}_3 \cdot 5e^-$, respectively. The interatomic distances d_{Rb-O} and d_{Cs-O} range from 264 to 285 and 269 to 298 pm, respectively, and therefore roughly correspond to the ionic radii of M^+ and O^{2-}. While the intercluster M–M distances cumulate around 530 and 550 pm, the intracluster distances are significantly shorter, ranging from 354 to 403 and 367 to 416 pm in Rb_9O_2 and $Cs_{11}O_3$, respectively. The details qualitatively agree with a bond model that implies metallic bonding within and between the clusters that are densely packed in the crystal structure and joined by itinerant metallic bonding. Detailed investigations mentioned later corroborate the model in terms of cluster metals that can form intermetallic compounds with additional rubidium or cesium to be inserted between the clusters. Figure 4.5 shows the known stable rubidium suboxides that have been characterized via crystal structure analysis.

The close-packing of Rb_9O_2 clusters is realized in the monoclinic crystal structure of the compound Rb_9O_2. The more metal-rich Rb_6O contains the identical type of cluster arranged in hexagonal sheets with rubidium atoms "intercalated" between them corresponding to $Rb_9O_2 \cdot Rb_3$. Those rubidium atoms in the intermediate layers exhibit an atomic volume increment very close to that of the elemental metal, hence form sheets of a two-dimensional metal. As one may expect from the earlier mentioned difficulties in getting hold of Rb_6O complications arise due to structural disorder. Pronounced rotational disorder between adjacent layers of the Rb_9O_2 clusters is observed. Moreover, hardly reproducible variations of thermal effects in the narrow range below the Rb/Rb_9O_2 eutectic indicate that additional layers or multilayers of rubidium may enter the cluster sheets and lead to phases with still higher metal content. It should be mentioned that the Rb_9O_2 cluster is obviously no stable entity in the liquid because in quenching experiments a metastable phase $Rb_{6.33}$ results that exhibits a complex structure based on Rb_6O octahedra and centered Rb_{13} icosahedra and thus bears no structural relationship with the other rubidium suboxides.

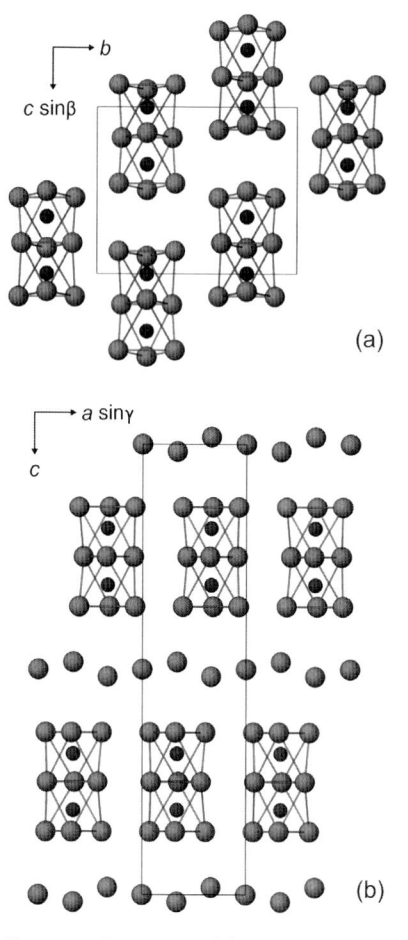

Figure 4.5 Projections of the crystal structures of (a) Rb_9O_2 and (b) $Rb_9O_2Rb_3$.

The structural principle characterizing the rubidium suboxides Rb_6O and Rb_9O_2 is the same as found for the suboxides of cesium except for the larger $Cs_{11}O_3$ cluster forming the essential constituent in all compounds (see Figure 4.6). Cs_7O might be selected to highlight the principle. In its hexagonal structure, rigid $Cs_{11}O_3$ clusters are stacked to form piles that are embedded in a soft matrix of purely metallic cesium to result in a compound composition $Cs_{11}O_3 \cdot Cs_{10}$. In the case of Cs_4O, the cesium atom matrix is reduced to just one cesium atom per cluster according to $Cs_{11}O_3 \cdot Cs$, and the single atom exhibits the same atomic volume as in elemental cesium. When this single cesium atom is replaced by rubidium in the compound $Cs_{11}O_3 \cdot Rb$ a different arrangement of the clusters and the single rubidium atom is observed. The question arises as to which factors determine the special stoichiometries of these suboxides. Obviously, an optimized packing of the clusters together with the additional alkali metal atoms is essential in organizing the different, yet closely related structures.

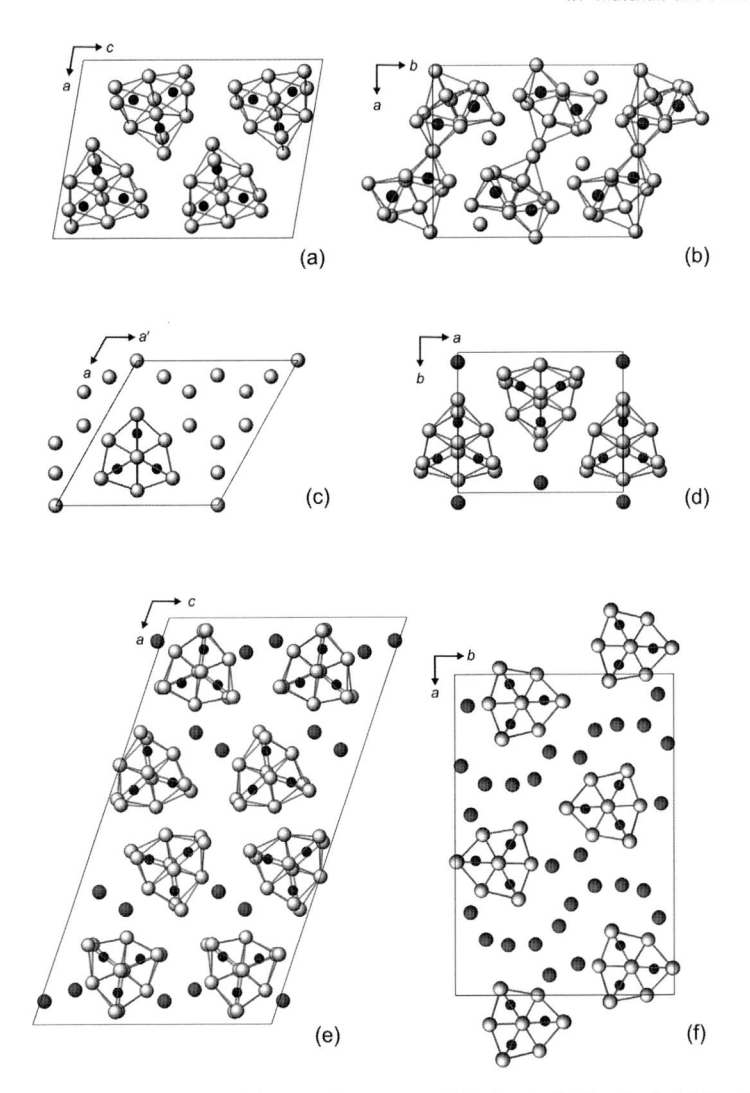

Figure 4.6 Projections of the crystal structures of (a) $Cs_{11}O_3$, (b) $Cs_{11}O_3 \cdot Cs$, (c) $Cs_{11}O_3 \cdot Cs_{10}$, (d) $Cs_{11}O_3 \cdot Rb$, (e) $Cs_{11}O_3 \cdot Rb_2$, and (f) $Cs_{11}O_3 \cdot Rb_7$.

Numerous investigations have been performed in order to understand the special situation in terms of such unique separation of ionic and metallic bonding in crystal space. As one approach, the vibrational modes of the clusters have been calculated for a simple ionic model on the basis of Born–Mayer potentials. In comparison with experimental Raman spectra measured for representative suboxides, their dynamic behavior closely conforms to the intuitive bond model. The calculations also reveal that the surplus of electrons establishes a kind of peripheral skin of electrons that is necessary to stabilize the cluster and to provide the bonding to the additional metal atoms. The term "low valent" mentioned in the introduction seems rather misleading because all atoms occur in their usual oxidation states +1 and −2 and as such are immersed in a free electron gas according to $Cs^+_{11}O^{2-}_3Cs^+_{10} \cdot 15e^-$. The characteristic colors of the suboxides are directly related to the number of free electrons. Measurement of the metallic reflectivity shows well-defined plasma edges that systematically shift with increasing oxygen content and decreasing electron concentration toward the infrared region in the spectrum. A rough estimate of the concentration of free electrons can be derived from these reflectivity data. Photoelectron spectroscopy allows a much more precise evaluation of the free carrier concentration.

Under inert conditions, liquid cesium and Cs_7O were introduced into the spectrometer and crystallized at low temperature while a compressed solid sample of $Cs_{11}O_3$ was used as a target. The spectra were measured at a basis pressure of 10^{-9} mbar. Repeated mechanical cleaning of the surface was necessary to reach reproducible and reliable results. Representative spectra of Cs, $Cs_{11}O_3 \cdot Cs_{10}$ and $Cs_{11}O_3$ are depicted in Figure 4.7, and they will now be discussed step-by-step. The features near the Fermi level are magnified by a factor 10 to show the density of states for the conduction band region. Its width qualitatively correlates with the number of conduction electrons. The prominent peak at 2.7 eV arises due to photoemission from the O2p band. The close position near the Fermi level signals the lowest known binding energy of the O^{2-} ion in oxides and indicates its essentially ionic character and only weak stabilization by the surrounding cations. Proceeding to the spectral features of elemental cesium, an Auger peak at 10 eV is observed near to the spin orbit split $Cs5p^{3/2}/5p^{1/2}$ core levels

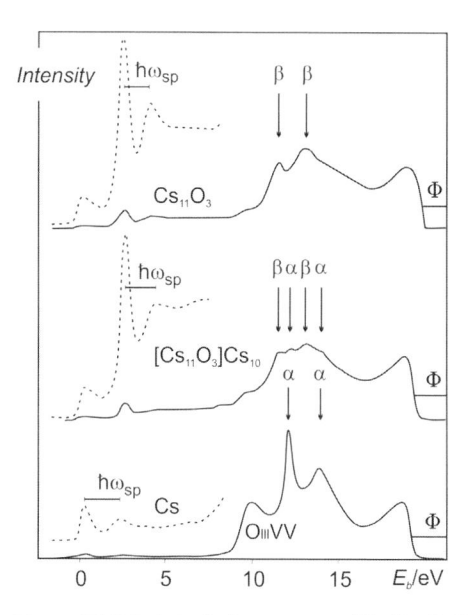

Figure 4.7 HeI photoelectron spectra of Cs, $Cs_{11}O_3 \cdot Cs_{10}$, and $Cs_{11}O_3$.

marked as α in the region between 10 and 15 eV. Photoemission terminates at a threshold slightly below 20 eV, and opens the gap toward the excitation energy of 21.2 eV indicated by the work function Φ. It is evident that Φ decreases from $Cs_{11}O_3 \cdot Cs_{10}$ to $Cs_{11}O_3$ upon oxidation. A remarkable feature is associated with the spin orbit split Cs5p levels marked as β that are shifted with respect to cesium to lower binding energies for $Cs_{11}O_3$. Even more interesting, a combined level structure $\alpha + \beta$ is seen in the spectrum of $Cs_{11}O_3 \cdot Cs_{10}$ that originates from the cluster on one hand and pure cesium on the other and leads to the illusion that the compound is actually a heterogeneous mixture of two phases. However, the continuous changes of the work function for the different measured samples and the features associated with the O2p peak corroborate the single phase character of $Cs_{11}O_3 \cdot Cs_{10}$. The humps aside the O2p peak are energy loss structures originating from emitted photoelectrons that escape from a depth of a few unit cells. Once the nature of these excitations could be verified as surface plasmons by applying different UV wavelengths, their measured energies provide a quantitative proof of the bond model for alkali metal suboxides. Based on the experimental values for the surface plasmon energies 2.0, 1.75, and 1.55, for Cs, $Cs_{11}O_3 \cdot Cs_{10}$, and $Cs_{11}O_3$, respectively, the numbers of free electrons per formula unit are calculated to be 1.0, 14.0, and 5.1 by referring to $Cs^+ \cdot e^-$ in excellent agreement with the bond model.

In conclusion, the description of phase relationships, crystal structures, and numerous physical investigations of the alkali metal suboxides are all consistent, however, certain details remain as puzzles that are still left to be solved or at least addressed. Unsolved problems in the chemistry of alkali metal suboxides still await further clarification, for example, the briefly mentioned observations made in the Rb/O system. Irreproducible thermal effects in the narrow region below the eutectic line seem to witness the existence of more metal-rich phases other than just Rb_6O. To mention another unsolved problem, the crystal structures shown in Figure 4.6 are well refined in contrast to the most oxidized suboxide described as Cs_3O in literature that actually shows a range of homogeneity $CsO_{0.31-0.36}$. The structure has been described as *anti*-TiI_3 type in the literature, and indeed the single-crystal refinement based only on Bragg reflection converges to an *R*-value as low as 4%. However, neglecting pronounced diffuse scattering effects result in an averaged structure that hardly discloses its essentials and its description might be entirely wrong. Beyond these unsolved details in the structural chemistry of alkali metal suboxides further open questions are related to quite fundamental aspects. Cesium metal exhibits the lowest work function, 2 eV, of all elements. Oxidation reduces the work function to 1.35 eV for $Cs_{11}O_3$ and even further down to approximately 1 eV for $CsO_{0.31-0.36}$. Hence, electrons are emitted from the suboxides by light in the near-infrared region. Early night vision devices made use of this fact on purely empirical reasons long before the actual importance of cesium suboxides in these devices was realized. What is the reason for the low work function? Is it the presence of O^{2-} ions just below the surface and their Coulomb repulsion that drives the electrons off [4]? Or do we rather meet the fundamental scenario of a void metal where the accumulation of the O^{2-} ions inside the clusters create repulsive Coulomb bubbles and the conduction electrons are confined to the space between the clusters? A quantum size effect like in highly doped semiconductor antiquantum dot arrays, although at much smaller length scales, could be the reason for the remarkable low work function [5].

Except for these still open questions the chemistry of alkali metal suboxides seemed to have reached a kind of final state. Quite recently, however, an unexpected discovery came as a surprise. Cesium suboxides easily dissolve in a high-temperature ceramic material like alumina and other well-crystallized oxides M_2O_3 (M = Al, Sc, Ga, In, Fe) at moderate temperatures as low as 550 K or even at 450 K in the case of Fe_2O_3. In rational syntheses, stoichiometric mixtures of these oxides were reacted together with Cs_2O and cesium to form the isotypic suboxometallates with compositions Cs_9MO_4 and characteristic colors ranging, for example, from brass via copper to violet bronze for Cs_9AlO_4, Cs_9GaO_4, and Cs_9InO_4, respectively [6]. A projection of the tetragonal structure is shown in Figure 4.8.

InO_4^{5-} anions are surrounded by cesium atoms to form columns of composition Cs_8InO_4, and the voids are occupied by single cesium atoms. The charge balance for the columns can be formulated as $Cs^+{}_8(InO_4)^{5-} \cdot 3e^-$. So far, the bonding is quite reminiscent of that in the cesium suboxides. Instead of O^{2-} ions in the suboxides, the structures of the suboxometallates contain InO_4^{5-} anions surrounded by columns of cesium atoms. Metallic bonding in the periphery of the columns as well as bonding to an additional cesium atom between the columns with no contact to the anions mimic the building principle realized in the alkali metal suboxides. To a certain extent the additional cesium atom can be

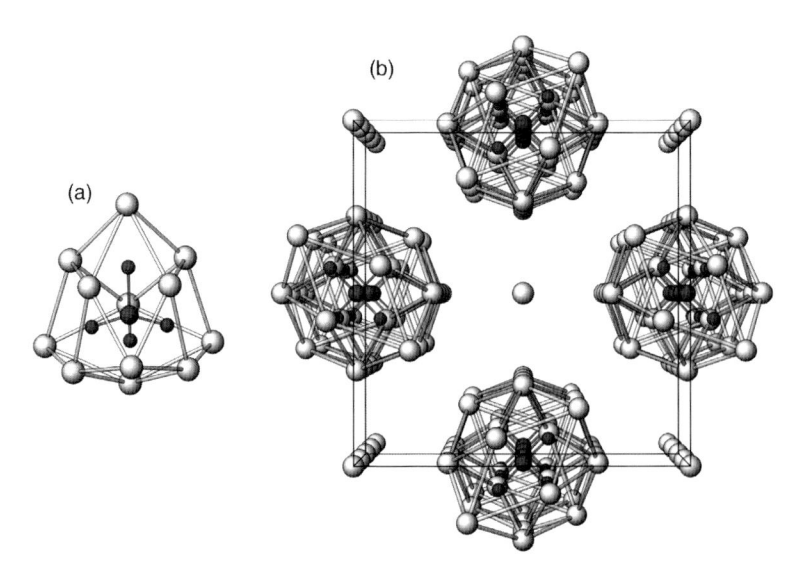

Figure 4.8 Unit cell of the suboxometalate Cs_9InO_4 and coordination of InO_4^{5-} ion by Cs atoms.

replaced by rubidium, whereas the columns contain only cesium, a fact that again reminds of the earlier described substitution experiments with alkali metal suboxides. The unusual bonding situation in these suboxometallates offers additional surprises: For Cs_9FeO_4, the oxidation state of iron is +3 as calculations of the lattice energy and Mössbauer studies clearly indicate, in spite of the highly reducing environment of a surplus of cesium in the compound. Yet, another interesting feature is concerned with the cesium atom positioned between the columns. In band structure calculations, a fully occupied 6 s band lies below the Fermi level that indicates that the electronic state of the additional cesium atom comes near to Cs^-.

The suboxometallates of the alkali metals opened an unexpected new field of exploration. It seems that this area of research will broaden as the recently discovered closely related phases $Cs_{10}MO_5$ testify [7].

Subnitrides of Alkaline Earth Metals

Corrosion by liquid sodium used as coolant in nuclear power stations urged intensive studies on relevant trace amounts of impurities, and in this context metal-rich alkaline earth nitride species were postulated [8]. In more recent investigations, the reaction of nitrogen with barium dissolved in liquid sodium led to numerous novel subnitrides that closely resemble the alkali metal suboxides with respect to the peculiar separation of metallic and ionic bonding in crystal space. Meanwhile, a plethora of such subnitrides could be prepared and structurally characterized [9]. The crystal structure of $Ba_6N \cdot Na_{10}$ may illustrate the obvious building principle (see Figure 4.9).

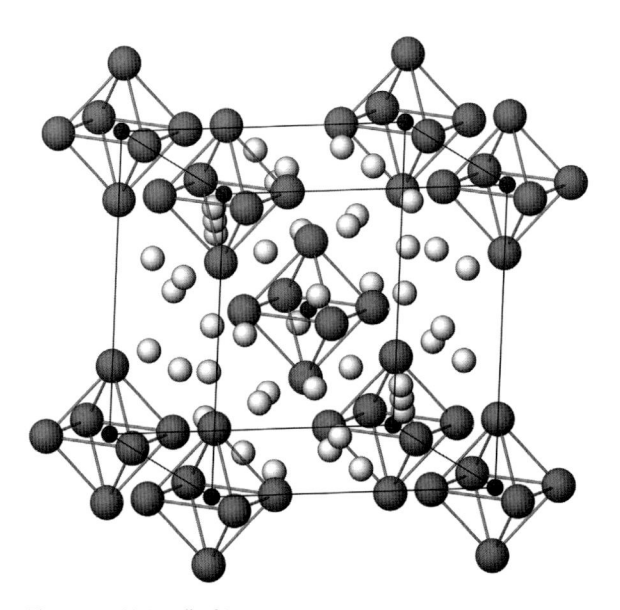

Figure 4.9 Unit cell of $Ba_6N \cdot Na_{10}$.

The crystal structure contains electron-rich Ba_6N clusters embedded in a purely metallic matrix of sodium. The building principle of ionic entities bound to additional metal atoms according to $Ba^{2+}{}_6N^{3-}Na^+{}_{10} \times 19e^-$ is just the same as with the alkali metal suboxides except for the difference concerning the discrete octahedral unit. A new structural feature arises when the amount of nitrogen is increased to a ratio $Ba/N = 3$. Instead of absorbing gaseous nitrogen it is more convenient to react the metals with stoichiometric amounts of the azides NaN_3 and/or $Ba(N_3)_2$, thus providing the appropriate amount of nitrogen for the reaction. By slowly cooling the solution of barium and nitrogen in liquid sodium, brittle black crystals of $Ba_3N \cdot Na$ separate (see Figure 4.10). The crystal structure exhibits a hexagonal rod packing of linear columns formed by Ba_6N units that are condensed via opposite faces according to $Ba_{6/2}N$, and these rods are surrounded by sodium atoms. At around 50 K above room temperature, another phase, $Ba_3N \cdot Na_5$, crystallizes whose structure contains the same kind of Ba_3N rods immersed in the matrix of pure sodium metal. The sodium can be distilled off under vacuum to leave polycrystalline Ba_3N with a parallel alignment of the rods alone. As a next step of condensation, the layered subnitride Ba_2N with the

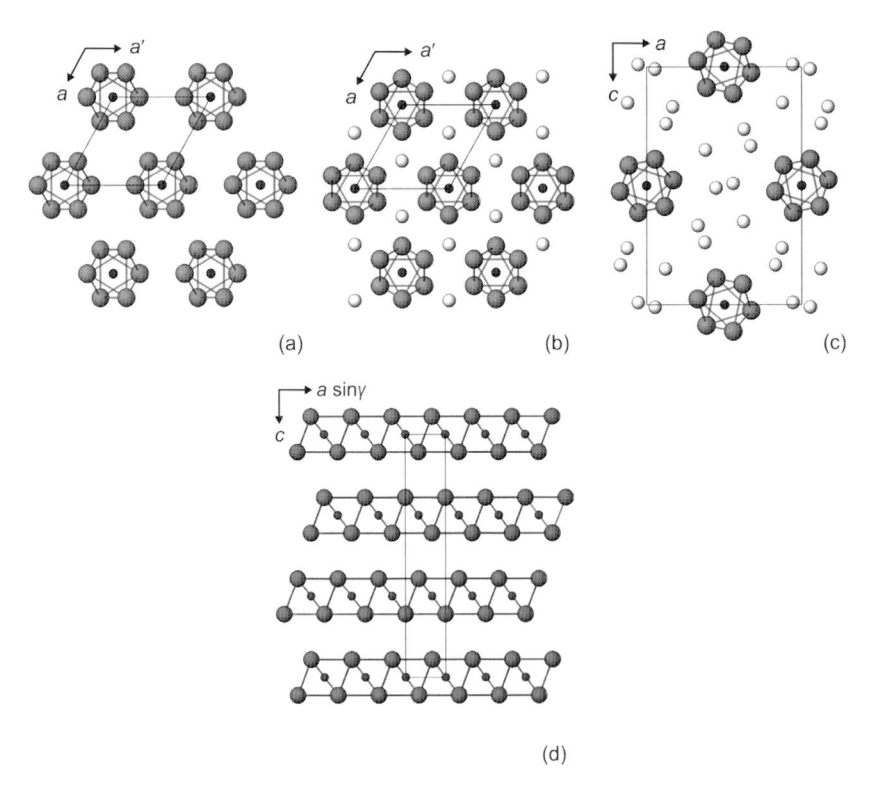

Figure 4.10 Projections of the rod structures of (a) Ba_3N, (b) $Ba_3N \cdot Na$, and (c) $Ba_3N \cdot Na_5$ and the layered structure of (d) Ba_2N.

same structure as the corresponding subnitrides of calcium and strontium can be prepared by reacting barium with nitrogen at 700 K or by decomposing stoichiometric amounts of $Ba(N_3)_2$ and barium in a sealed tantalum capsule.

Investigations of these novel subnitrides were performed in two directions. On one hand, electronic band structures were measured by photoelectron spectroscopy in great detail accompanied by theoretical band structure calculations to compare the results with the observations for the alkali metal suboxides [10]. On the other hand, it was tempting to extend the subnitride chemistry by applying the heavier alkali metals as solvents instead of sodium, but preliminary experiments failed because of the immiscibility with barium. As a surprise, the addition of potassium opened a new preparative route. It turned out that solid $Ba_3N \cdot Na_5$ when treated with potassium looses part of the sodium, and solid $Ba_3N \cdot Na$ remains besides liquid sodium–potassium alloy. Following this line, experiments with solutions of Ba/N/Na and some addition of potassium led to a few crystals that could be separated, and the structure analysis revealed a remarkable compound $Ba_{14}KN_6 \cdot Na_{14}$ composed of $Ba_{14}KN_6$ clusters in a matrix of pure sodium. However, the compound could not be prepared in convincing yield. It took some while until it was realized that the anticipated potassium atom in the compound actually represented calcium captured as trace contamination in barium. Once recognized, the compound could be prepared in a near quantitative yield by adding the appropriate amount of calcium or strontium that resides in the cluster center.

The characteristic cluster $Ba_{14}CaN_6$ shown in Figure 4.11 is formed from six M_6N octahedra condensed via faces. Formally, the central cube Ba_8CaN_6 can be viewed as one unit cell of an *anti*-type perovskite and as such represents an

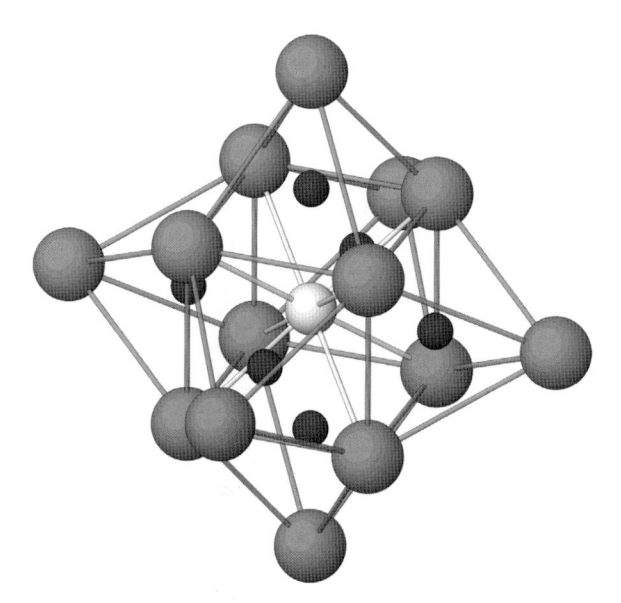

Figure 4.11 $Ba_{14}CaN_6$ cluster, increasing sizes for N, Ca, and Ba.

electroneutral piece of salt that is embedded in a metallic environment in the compound $Ba_{14}CaN_6 \cdot Na_{14}$. The description as a nanodispersion of a salt in a metal might be appropriate. One may speculate whether the Ba_8CaN_6 unit is stable in the melt as it is observed repeatedly in numerous closely related phases. Following the above-mentioned observation that potassium extracts sodium from $Ba_3N \cdot Na_5$ to leave $Ba_3N \cdot Na$, a series of compounds $Ba_{14}CaN_6 \cdot Na_x$ with varying sodium content could be prepared by adding appropriate amounts of potassium. In a systematic way the amount of potassium in the melt controls the compound compositions in such a way that an increase of potassium concentration leads to compounds that contain less sodium. Figure 4.12 shows a compilation of crystal structures that could be characterized.

So far, subnitrides of the alkaline earth metals in combination with sodium have been presented. The replacement of sodium by lithium allows another surprising extension of subnitride chemistry and provides compounds that are closely related to those described above as well as quite different types. As an example of the first kind, preparations with lithium instead of sodium result in $Ba_3N \cdot Li$ that is isotypic with $Ba_3N \cdot Na$, and there is evidence that lithium can partly substitute barium in the mixed system $Ba_3N \cdot (Li,Na)$. An example of the second kind is $LiBa_2N$. In the novel crystal structure, flattened $LiBa_5N$ octahedra

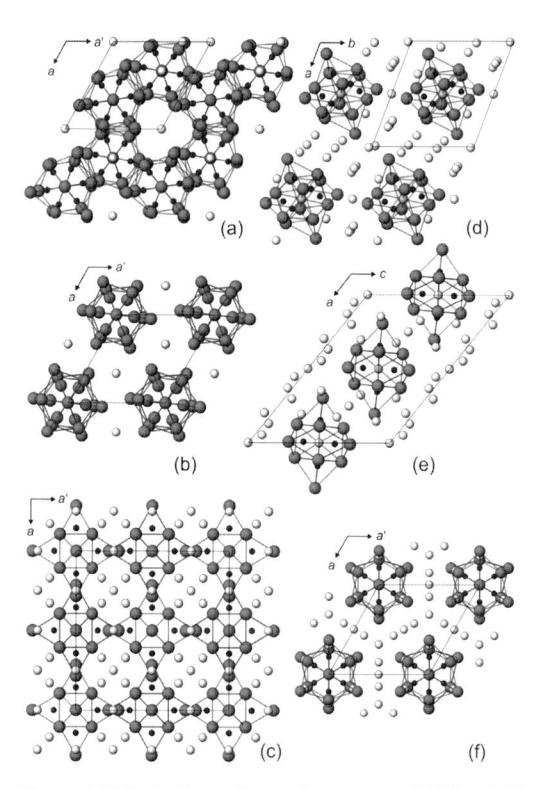

Figure 4.12 Projections of crystal structures of (a) $Ba_{14}CaN_6 \cdot Na_7$, (b) $Ba_{14}CaN_6 \cdot Na_8$, (c) $Ba_{14}CaN_6 \cdot Na_{14}$, (d) $Ba_{14}CaN_6 \cdot Na_{17}$, (e) $Ba_{14}CaN_6 \cdot Na_{21}$, and (f) $Ba_{14}CaN_6 \cdot Na_{22}$.

form *trans*-edge sharing chains that are condensed in an orthogonal mode with identical chains in the tetragonal structure (see Figure 4.13).

The flattened LiBa$_5$N octahedron is also the characteristic motif in the Ba$_{14}$LiN$_6$ cluster formed from six face-sharing octahedra as in the case of the analogous Ba$_{14}$(Ca,Sr)N$_6$ clusters. Lithium can now play a different game. It acts as cation in the center of the cluster and as a metal in the region between the clusters. Lithium and sodium atoms together occupy this region in two marginally different ordered cubic structures, Ba$_{14}$LiN$_6 \cdot$Li$_4$Na$_{11}$ and Ba$_{14}$LiN$_6 \cdot$Li$_5$Na$_{10}$, where the metal atoms in the special position 4b are simply exchanged.

With the latter examples the introduction of lithium into subnitride chemistry leads to great complexity that can be exaggerated even further as a few examples illustrate. When stoichiometric amounts of barium, lithium, and Ba(N$_3$)$_2$ are heated in a closed tantalum container at 600 K, large brass colored crystals form and they are characterized as Li$_{80}$Ba$_{39}$N$_9$ by structure analysis. The complicated structure can be described as composed of four different building units that are drawn as polyhedra representations for the sake of clarity. The prominent motif is a centered Li$_{13}$ icosahedron arranged in an fcc packing with interconnections between and additional lithium atoms attached to the icosahedra. The tetrahedral voids in the packing are alternatively occupied by Ba$_4$ tetrahedra, Ba$_6$N octahedra that have been identified earlier in Ba$_6$N$\cdot$Na$_{10}$, and Ba$_5$N$_6$Li$_{12}$ clusters where the lithium atoms are bonded to nitrogen. The subnitride structure of Li$_{80}$Ba$_{39}$N$_9$ with depicted fragments is shown in Figure 4.14.

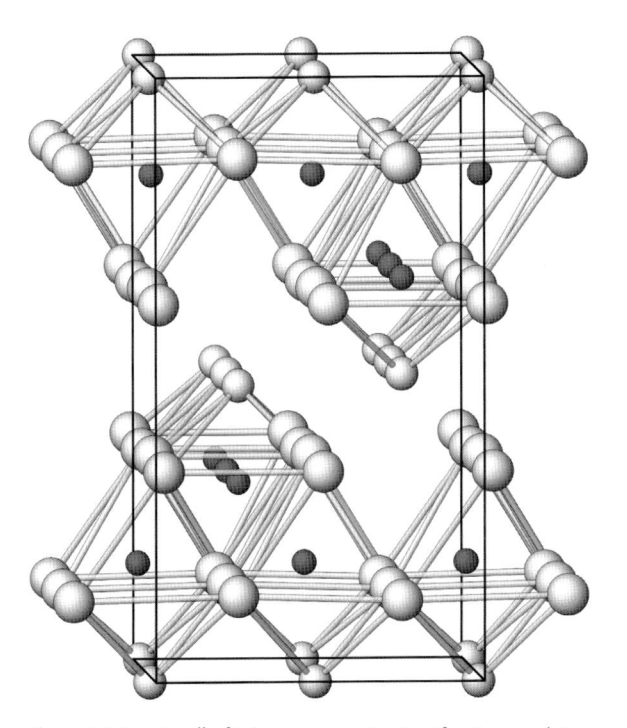

Figure 4.13 Unit cell of LiBa$_2$N, increasing sizes for N, Li, and Ba.

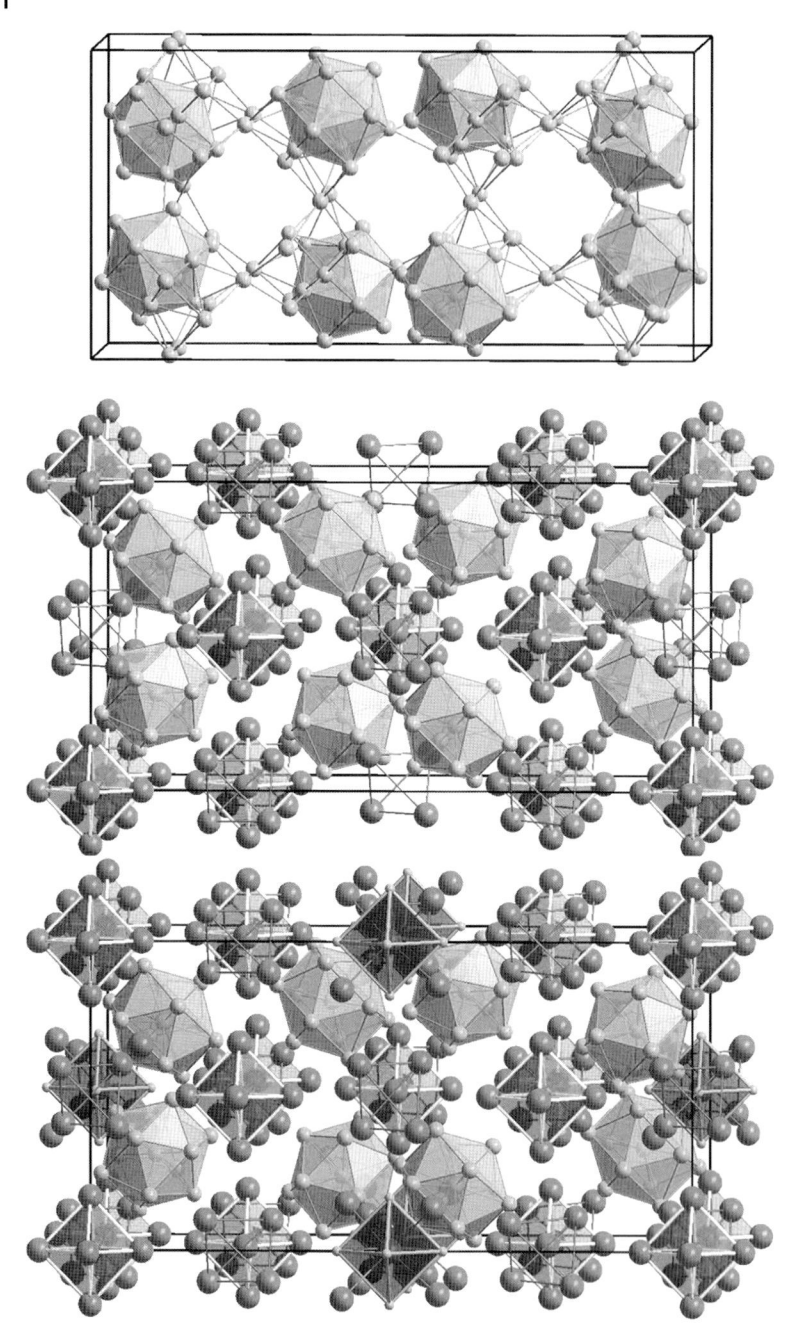

Figure 4.14 Schematic decomposition of the $Li_{80}Ba_{39}N_9$ structure into Li_{13}, Ba_4, Ba_6N, and $Ba_5N_6Li_{12}$ fragments.

The rather equal affinity of both lithium and barium atoms to bond with nitrogen shows up in a new cluster discovered in the quaternary subnitride $Li_8Ba_{12}N_6 \cdot Na_{15}$. In the structure, six Li_3Ba_3N octahedra are condensed via edges and vertices. The eight lithium atoms of the inner core are arranged in a constricted hexagonal bipyramid (see Figure 4.15). The arrangement reflects the large difference in the atomic sizes of lithium and barium that clearly favors site selectivity for the different kinds of metal atoms. The clusters are again embedded in a matrix of pure sodium with close bonding contacts to the barium atoms but no contacts with the lithium atoms.

In the introduction a general scenario was addressed, namely, that a few non-metal atoms may get incorporated in metals forming subcompounds of quite

Figure 4.15 Unit cell of $Li_8Ba_{12}N_6 \cdot Na_{15}$ with $Li_8Ba_{12}N_6$ clusters.

individual character. A special situation occurs when a pure well-defined intermetallic phase incorporates nonmetal atoms with hardly any recognizable change of the basic structure. Such a situation is met with the compound $Li_{13}Na_{29}Ba_{19}$ that crystallizes in a complicated new structure composed of characteristic fragments. One prominent building unit is a Li_{26} cluster composed of four interpenetrating centered icosahedra known from the typical atom arrangement M_{26} in γ-brass. These units are arranged in an fcc packing, and all of the octahedral and half of the tetrahedral voids are filled with Ba_5Na_{12} polyhedra and additional barium and sodium atoms to result in the composition $Li_{13}Na_{29}Ba_{19}$. For clarity, only the Li_{26} and Ba_5Na_{12} units are drawn in Figure 4.16.

Variations in the lattice constants prompted a more detailed investigation of the compound, particularly focusing on metric changes of the Li_{26} cluster that is composed of 56 close-packed tetrahedra. In a set of preparations, it turned out that this central tetrahedron could be occupied by an appropriate stoichiometric amount of nitrogen that exclusively enters this position in the subnitride $Li_{26}Na_{58}Ba_{38}N$.

These last examples, particularly involving lithium, give a flavor of the plethora of new subnitrides of the electropositive metals that have been discovered recently, and for sure there will be many more to come. Obviously, the tip of an iceberg has just been scratched at its top (Table 4.1).

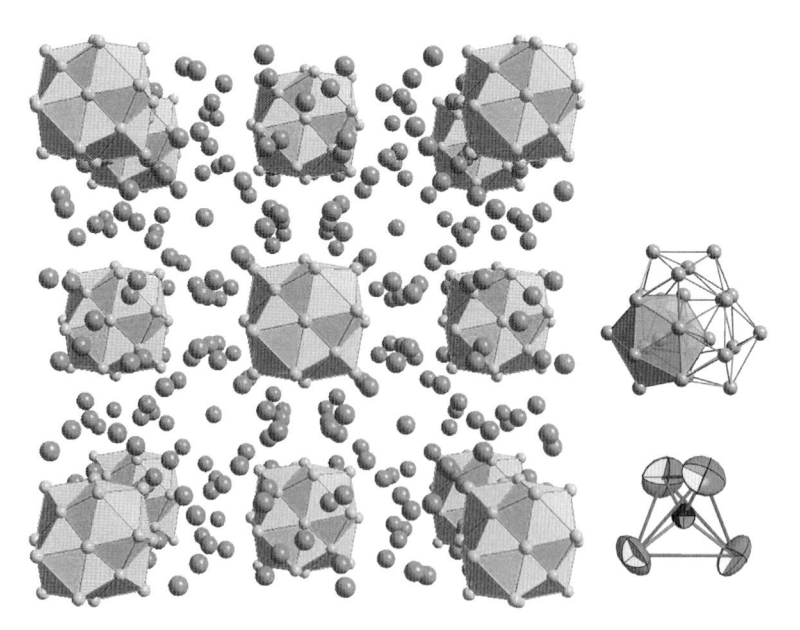

Figure 4.16 Packing of characteristic Li_{26} and Ba_5Na_{12} units in $Li_{13}Na_{29}Ba_{19}$ and central Li_4N unit.

Table 4.1 Compilation of alkali metal suboxides and alkaline earth subnitrides described in the text, references given for details.

Formula	Space group		Unit cell/pm	T/K	Reference
Rb_9O_2	$P2_1/m$	$z=2$	832.3, 1398.6, 1165.4, 104.39°	223	[11]
$Rb_9O_2 \cdot Rb_3$	$P6_3/m$	$z=2$	939.3, 3046.7	223	[12]
$RbO_{6.33}$	$Fm3c$	$z=24$	2315.2	121	[13]
$Cs_{11}O_3$	$P2_1/c$	$z=4$	1753.6, 919.2, 2401.7, 100.42°	223	[14]
$Cs_{11}O_3 \cdot Cs$	$Pna2$	$z=4$	1682.3, 2052.5, 1237.2	173	[15]
$Cs_{11}O_3 \cdot Cs_{10}$	$P\text{-}6m2$	$z=1$	1630.9, 915.4	223	[16]
$Cs_{11}O_3 \cdot Rb$	$Pmn2_1$	$z=2$	1646.4, 1368.3, 909.2	243	[17]
$Cs_{11}O_3 \cdot Rb_2$	$P2_1$	$z=8$	4231.1, 919.4, 2415.6, 108.94°	223	[3]
$Cs_{11}O_3 \cdot Rb_7$	$P212121$	$z=4$	3228.1, 2187.7, 902.5	183	[18]
$Cs_8(InO_4) \cdot Cs$	$I4/mcm$	$z=4$	1551.7, 1290.0	293	[6]
$Cs_{10}(Ga_4O \cdot O)$	$C2/c$	$z=8$	2138.7, 1246.1, 2406.4, 98.56°	293	[7]
$Ba_6N \cdot Na_{16}$	$Im\text{-}3m$	$z=2$	1252.7	293	[19]
$Ba_{14}CaN_6 \cdot Na_7$	$R\text{-}3c$	$z=6$	1136.0, 6306.1	293	[20]
$Ba_{14}CaN_6 \cdot Na_8$	$P6_3/m$	$z=2$	1141.9, 2154.3	293	[21]
$Ba_{14}CaN_6 \cdot Na_{14}$	$Fm\text{-}3m$	$z=8$	1789.5	293	[22]
$Ba_{14}CaN_6 \cdot Na_{17}$	$P\text{-}1$	$z=1$	1114.2, 1206.5, 1372.5 66.65°, 67.79°, 78.88°	293	[23]
$Ba_{14}CaN_6 \cdot Na_{21}$	$C2/m$	$z=2$	2150.0, 1266.4, 1629.5, 129.48°	293	[23]
$Ba_{14}CaN_6 \cdot Na_{22}$	$P6_3/mmc$	$z=1$	1266.6, 1263.5	293	[24]
Ba_2N	$R\text{-}3m$	$z=3$	402.9, 2242.5	293	[25]
Ba_3N	$P6_3/mcm$	$z=2$	764.2, 705.0	293	[26]
$Ba_3N \cdot Na$	$P6_3/mmc$	$z=2$	844.1, 698.2	293	[27]
$Ba_3N \cdot Na_5$	$Pnma$	$z=4$	1189.7, 705.6, 1780.1	293	[28]
$LiBa_2N$	$P4_2/nmc$	$z=8$	798.0, 1426.3	293	[29]
$Ba_{14}LiN_6 \cdot Na_{14}$	$Fm\text{-}3m$	$z=4$	1779.6	293	[30]
$Ba_{14}LiN_6 \cdot Li_4Na_{11}$	$F\text{-}43m$	$z=4$	1787.4	293	[30]
$Ba_{14}LiN_6 \cdot Li_5Na_{10}$	$F\text{-}43m$	$z=4$	1779.9	293	[30]
$Li_{80}Ba_{39}N_9$	$I\text{-}42m$	$z=2$	1607.6, 3226.7	293	[31]
$Li_8Ba_{12}N_6 \cdot Na_{15}$	$R\text{-}3m$	$z=3$	1234.1, 333.1	293	[32]
$Li_{26}N \cdot Na_{29}Ba_{19}$	$F\text{-}43m$	$z=4$	2752.8	293	[33]

4.2
Epilogue

Suboxides of the alkali metals and subnitrides of the alkaline earth metals represent a unique class of compounds based on naked metal clusters embedded in a metallic matrix, and itinerant electrons provide bonding between them. In the case of alkali metals, the clusters occur as discrete units; in the case of alkaline

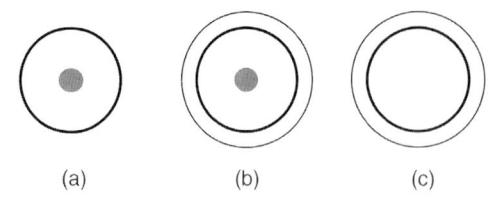

Figure 4.17 Sketches of different bonding situations with metal-rich compounds. (a) Metal–metal bonding with metals of groups 1 and 2 and encapsulated nonmetal atoms. (b) Clusters with interstitial atoms in an environment of nonmetal atoms. (c) Empty metal clusters in an environment of nonmetal atoms.

earth metals, discrete clusters as well as one- and two-dimensional condensed cluster systems also occur. It is essential that the clusters formed from the valence-poor metals in groups 1 and 2 need stabilization by interstitial nonmetal atoms. Figure 4.17a describes the bonding situation that is characterized by nonmetal atoms encapsulated by metal atoms forming a kind of metallic skin around.

Moving in the periodic table to the next group of metals, the rare earth elements (including the lanthanoids) come up with numerous reduced phases containing discrete metal clusters and condensed cluster systems. Certain structural principles carry on from groups 1 and 2, although in a modified manner. As the number of valence electrons has increased, clusters now get surrounded by halogen atoms and interstitial atoms stabilize the weakly metal–metal bonded clusters. Figure 4.17b shows this new scenario in rare earth metal chemistry. Besides other topologies, octahedra of metal atoms are quite common and they are surrounded by 12 halogen atoms in the M_6ZX_{12}-type cluster. Interstitials Z range from elements across the Zintl border down to heavy transition metals as well as molecular entities like C_2 units. A wealth of condensed cluster structures exist that are derived from the M_6ZX_{12} building unit. As a rare example, the amazing subchloride Gd_2Cl_3 composed of chains of edge-sharing Gd_6 octahedra surrounded by halogen atoms with no interstitials inside the octahedra as well as some isotypic phases is known.

Proceeding one step further to the right, the analogous structural chemistry of metal clusters encapsulated by halogen atoms and stabilized by interstitials is found with zirconium, for example, Zr_6CCl_{12} and numerous other compounds of discrete and condensed character. The scenario changes again for group 5 elements because now the metal atom core with strong metal–metal bonding lacks the interstitial and is encapsulated by halogen atoms as sketched in Figure 4.17c. A rare case of a M_6X_8-type cluster for a group 5 metal is found in the case of electron deficiency for the compound Nb_6I_{11} that reversibly incorporates hydrogen as interstitial into the cluster core Nb_6HI_8. Finally, the M_6X_8 unit of the group 6 metals becomes dominant and reaches the maximum number of electrons for metal–metal bonding in a very stable configuration followed by group 7 metals that due to the further increased number of valence electrons

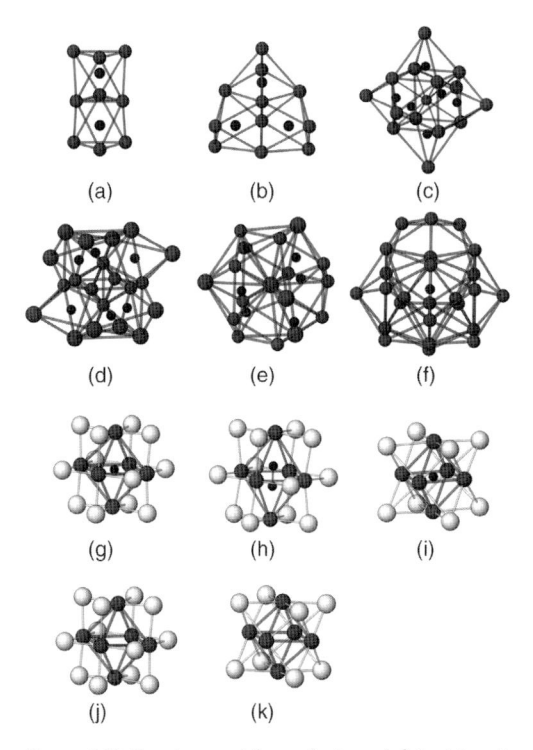

Figure 4.18 Development from electron deficient to saturated metal–metal-bonded clusters.

now allow the replacement of halogen versus chalcogen atoms. Within a few decades a tremendous number of compounds have come to light, and they cover discrete as well as condensed cluster systems. In Figure 4.18 the stepwise progression in our knowledge of cluster chemistry is briefly aligned in a chain of examples starting from the electron-poor alkali metal suboxides and ending with saturated metal–metal-bonded cluster compounds.

References

1 Rengade, E. (1909) Sur les sous – oxydes de cesium. *C.R. Acad. Sci.*, **148**, 1199–1202.
2 Simon, A. (1973) Suboxide der metalle rubidium und cäsium. *Z. Anorg. Allg. Chem.*, **395**, 301–319.
3 Simon, A. (1979) Structure and bonding with alkali metal suboxides. *Struct. Bond.*, **36**, 81–127.
4 Woratschek, B., Sesselmann, J., Küppers, J., Ertl, G., and Haberland, H. (1986) The interaction of cesium with oxygen. *J. Chem. Phys.*, **86**, 2411–2422.
5 Burt, M.G. and Heine, V. (1977) The theory of the work function of caesium suboxides and caesium films. *J. Phys. C*, **11**, 961–968.
6 Hoch, C., Bender, J., and Simon, A. (2009) Suboxides with complex anions: the suboxoindate Cs_9InO_4. *Angew. Chem., Int. Ed.*, **48**, 2415–2417.
7 Worsching, M. and Hoch, C. (2015) Alkali metal suboxometalates – structural chemistry between salts and metals. *Inorg. Chem.*, **54**, 7058–7064.

8 Addison, C.C. (1984) *The Chemistry of the Liquid Alkali Metals*, John Wiley & Sons, Ltd, Chichester.

9 Simon, A. (2004) Alkali and alkaline earth metal suboxides and subnitrides, in *Molecular Clusters of the Main Group Elements* (eds M. Dries and H. Nöth), Wiley-VCH Verlag GmbH, Weinheim, pp. 246–266.

10 Steinbrenner, U., Adler, P., Hölle, W., and Simon, A. (1998) Electronic structure and chemical bonding in alkaline earth metal subnitrides: photoemission studies and band structure calculations. *J. Phys. Chem. Solids*, **59**, 1527–1536.

11 Simon, A. (1977) Das "komplexe Metall" Rb_9O_2. *Z. Anorg. Allg. Chem.*, **431**, 5–16.

12 Simon, A. and Deiseroth, H.-J. (1976) Das Rubidiumsuboxid Rb_6O. *Rev. Chim. Miner.*, **13**, 98–112.

13 Deiseroth, H.-J. and Simon, A. (1978) Kristallisationsvorgänge bei metallreichen Rubidiumoxiden. *Z. Naturforsch.*, **33b**, 714–722.

14 Simon, A. (1977) Das komplexe Metall $Cs_{11}O_3$. *Z. Anorg. Allg. Chem.*, **428**, 187–198.

15 Simon, A., Deiseroth, H.-J., Westerbeck, E., and Hillenkötter, B. (1976) Untersuchungen zum Aufbau von Tetracäsiumoxid. *Z. Anorg. Allg. Chem.*, **423**, 203–211.

16 Simon, A. (1976) Das metallreichste Cäsiumoxid – Cs_7O. *Z. Anorg. Allg. Chem.*, **422**, 208–218.

17 Deiseroth, H.-J. and Simon, A. (1980) Struktur und Bildung von $RbCs_{11}O_3$. *Z. Anorg. Allg. Chem.*, **463**, 14–26.

18 Simon, A., Brämer, W., and Deiseroth, H.-J. (1978) Structural separation of rubidium and cesium in an alkali metal suboxide. *Inorg. Chem.*, **17**, 875–879.

19 Snyder, G.J. and Simon, A. (1994) Discrete M_6N octahedra in the subnitrides $Na_{16}Ba_6N$ and $Ag_{16}Ca_6N$: a reconsideration of the Ag_8Ca_3 structure type. *Angew. Chem., Int. Ed. Engl.*, **33**, 689–691.

20 Vajenine, G.V. and Simon, A. (2001) Preparation and crystal structure of $Na_7Ba_{14}CaN_6$: a new subnitride in the $Na_xBa_{14}CaN_6$ series. *Eur. J. Inorg. Chem.*, **5**, 1189–1193.

21 Vajenine, G.V., Steinbrenner, U., and Simon, A. (1999) A new subnitride in the series $Na_nBa_{14}CaN_6$ with $n = 8$. *C.R. Acad. Sci. IIC*, **2**, 583–589.

22 Steinbrenner, U. and Simon, A. (1996) $Na_{14}Ba_{14}CaN_6$ – a nanodispersion of a salt in a metal. *Angew. Chem., Int. Ed. Engl.*, **35**, 552–554.

23 Simon, A. and Steinbrenner, U. (1996) Subnitrides with the new cluster $Ba_{14}CaN_6$. *J. Chem. Soc., Faraday Trans.*, **92**, 2117–2123.

24 Steinbrenner, U. and Simon, A. (1997) Structural frustration in a rod packing – an analogy to the disordered triangular Ising net. *Z. Kristallogr.*, **212**, 428–438.

25 Reckeweg, O. and DiSalvo, F.J. (2005) Crystal structure of dibarium mononitride, Ba_2N, an alkaline earth metal subnitride. *Z. Kristallogr.*, **220**, 519–520.

26 Steinbrenner, U. and Simon, A. (1998) Ba_3N – a new binary nitride of an alkaline earth metal. *Z. Anorg. Allg. Chem.*, **624**, 228–232.

27 Rauch, P.E. and Simon, A. (1992) The new subnitride $NaBa_3N$ – an extension of alkali metal suboxide chemistry. *Angew. Chem., Int. Ed. Engl.*, **31**, 1519–1521.

28 Snyder, G.J. and Simon, A. (1995) The infinite chain nitride Na_5Ba_3N: a one-dimensional void metal. *J. Am. Chem. Soc.*, **117**, 1996–1999.

29 Smetana, V., Babizhetskyy, V., Vajenine, G.V., and Simon, A. (2006) Synthesis and structure of $LiBa_2N$ and identification of $LiBa_3N$. *J. Solid State Chem.*, **180**, 1889–1893.

30 Smetana, V., Babizhetskyy, V., and Simon, A. (2008) $Li_xNa_y Ba_{14}LiN_6$: new representatives of the subnitride family. *Z. Anorg. Allg. Chem.*, **634**, 629–632.

31 Smetana, V., Babizhetskyy, V., Vajenine, G.V., and Simon, A. (2006) $Li_{80}Ba_{39}N_9$: the first Li/Ba subnitride. *Inorg. Chem.*, **45**, 10786–10789.

32 Smetana, V., Babizhetskyy, V., Vajenine, G.V., and Simon, A. (2007) Synthesis and crystal structure of the new quaternary subnitride $Na_{15}Li_8Ba_{12}N_6$. *Z. Anorg. Allg. Chem.*, **633**, 2296–2299.

33 Smetana, V., Babizhetskyy, V., Vajenine, G.V., and Simon, A. (2008) Darstellung und Kristallstruktur der Phasen $Li_{26}Na_{58}Ba_{38}E_x$. *Z. Anorg. Allg. Chem.*, **634**, 849–852.

5
Introduction to the Crystal Chemistry of Transition Metal Oxides

J.E. Greedan

McMaster University, Department of Chemistry and Chemical Biology and the Brockhouse Institute for Materials Research, 1280 Main St W, Hamilton L8S 4L8, Canada

5.1
Introduction

This is of course a vast subject and only an outline of what is known can be described in the allotted space. We can begin by identifying the elements involved in transition metal oxides (TMO). These are the elements of the three d-group transition series and the two f-group series as illustrated in the periodic table in Figure 5.1, a total of 60 elements or 58% of the relatively stable chemical elements. The fourth d-group series is excluded. Generally, the transition elements are those with partially filled d or f shells that lead to characteristic properties such as magnetism due to the presence of unpaired electrons and an unusual number of accessible stable oxidation states per element. For example, in oxides manganese exhibits oxidation states from +2 to +7.

Also, in this chapter TMO are taken to include only those compounds which are solids and for which the structure is extended rather than discretely molecular. Also excluded are compounds in which the oxygen is associated with complex anions such as sulfates, nitrates, phosphates, and so on. That is, our only ligand is O^{2-}.

It is difficult to overstate the importance of this class of materials in both basic science and applications. Beginning with the discovery of the compass material lodestone millennia ago, Fe_3O_4 or magnetite, by the Greeks, Chinese, or the Olmec people of Central America, (according to Wikipedia), TMO has been a key source of magnetic materials. Take for example, the ubiquitous refrigerator door magnets that are based on the complex TMO $BaFe_{12}O_{19}$, which is also of interest for microwave absorption properties [1]. The "high temperature" superconducting copper oxides, such as "BSCCO 2112" or $Bi_2Sr_2CaCu_2O_8$ are beginning to find application in the generation of very high (>30 T) continuous magnetic fields [2]. Strong candidates for application as "spintronic" materials, where information is conveyed both by the charge and spin of the electron, are

---

Handbook of Solid State Chemistry, First Edition. Edited by Richard Dronskowski, Shinichi Kikkawa, and Andreas Stein.
© 2017 Wiley-VCH Verlag GmbH & Co. KGaA. Published 2017 by Wiley-VCH Verlag GmbH & Co. KGaA.

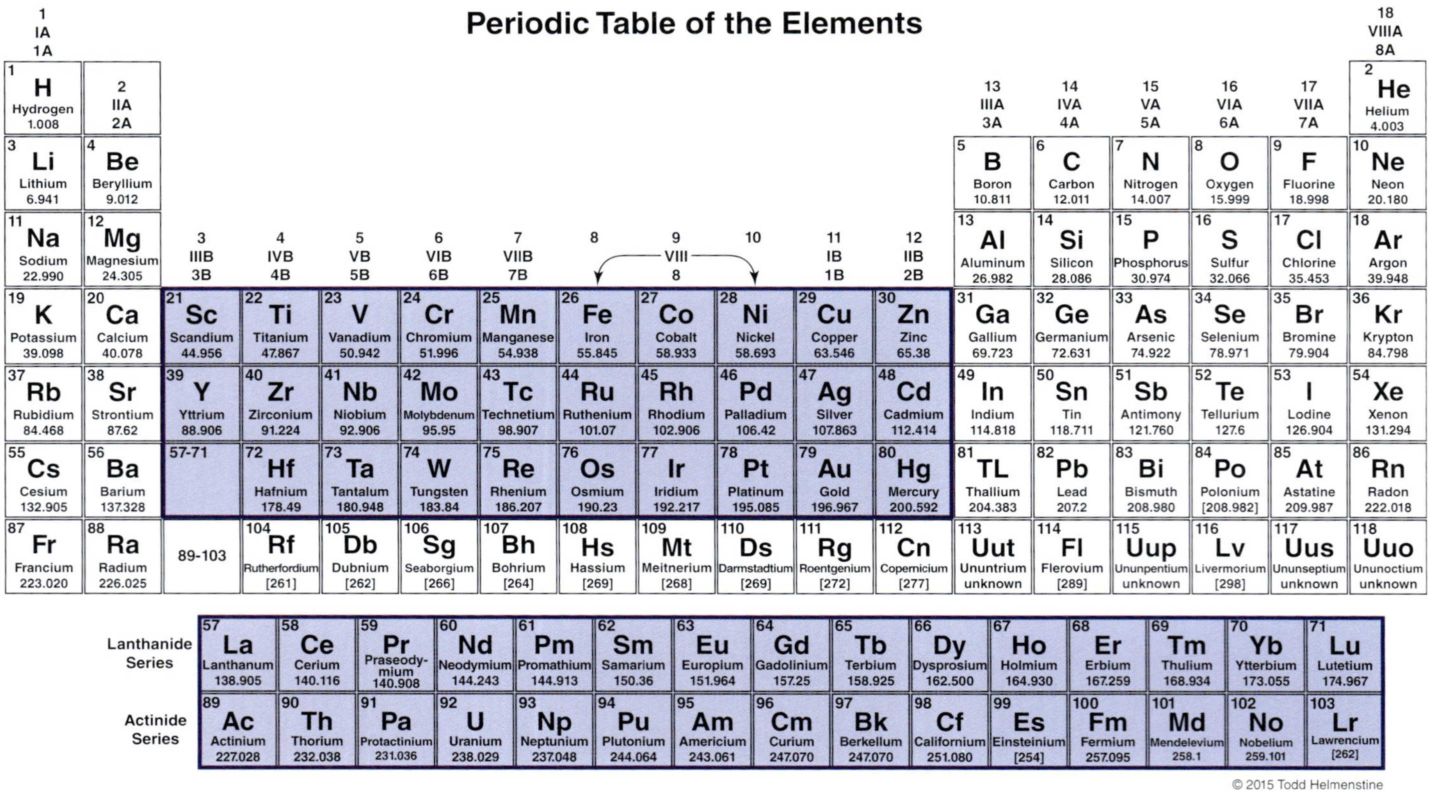

5 Introduction to the Crystal Chemistry of Transition Metal Oxides

Figure 5.1 The periodic table with the transition elements identified.

TMO such as Sr_2FeMoO_6 and related compounds [3]. Lithium battery cathodes are always TMO with materials such as $LiCoO_2$ and $LiMn_2O_4$ [4]. The chemical industry makes significant use of TMO as heterogeneous catalysts in a wide variety of processes, especially those involving redox. Although the mechanisms and in some cases the composition of the active form of the TMO catalysts are not known in detail, there are many examples such as the oxidation of methanol to formaldehyde over $Fe_2Mo_3O_{12}$ [5].

Moving to more basic research, the fundamental ideas of electron correlation embodied in the Mott–Hubbard model began with the puzzle that NiO was an insulator and not a metal as demanded by the band theory [6]. The Goodenough–Kanamori rules for the signs and magnitude of magnetic superexchange in insulators were based on observations of the magnetic orderings in TMO with the perovskite, spinel and rock salt structures [7]. The superconducting cuprates mentioned earlier have had a much greater impact on basic science leading to an unprecedented search for new materials and an intense investigation of the mechanism for such high critical temperatures leading to $>5 \times 10^4$ publications since 1986!

As already noted, there exist literally thousands of TMO, so, no attempt is made to be exhaustive. Examples will be selected based in part on the author's experience with these materials over the past decades. Many figures will be presented illustrating key crystal structures. Only limited crystallographic data are included in this chapter. Crystallographic Information Files (CIF) can be obtained for nearly all known structures from data bases such as the Inorganic Crystal Structure Database (ICSD) which can be inserted into visualization platforms. Readers are strongly encouraged to make use of this opportunity to view the structures.

Finally, the relationship between structural features and physical and chemical properties will be mentioned for specific TMO, but will not be described in a separate section. Also, methods of synthesis, covered in other chapters in this handbook, will not be described in detail.

5.2
Crystal Chemistry of TMO: Some Basics

5.2.1
Close Packing and Hole Filling

The structural chemistry of many binary TMO and indeed more complex oxides can be understood in terms of a simple model, namely, that of close packing of a three-dimensional lattice of oxide ions with various interstices occupied by transition metal ions. This is a well-known concept that often appears in first year university textbooks but is nonetheless very powerful and a brief review is warranted. Two fundamental close packing (CP) schemes exist, hexagonal (HCP) and cubic (CCP) along with many variations. These are illustrated in Figure 5.2.

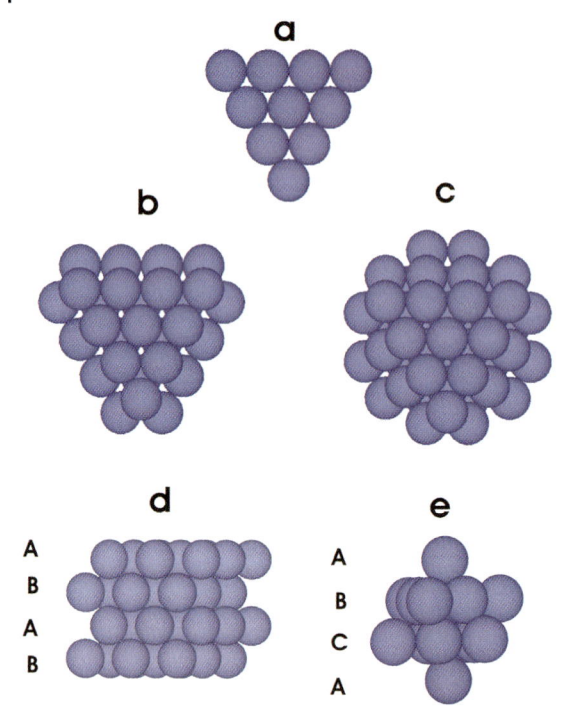

Figure 5.2 (a) One layer of close packed identical spheres. (b) hcp Stacking viewed along the stacking axis. (c) The ccp stacking viewed along the stacking axis. (d) hcp (ABAB) Stacking viewed normal to the stacking axis (e) ccp (ABC) Stacking viewed normal to the stacking axis.

In Figure 5.2a, part of a layer of identical close packed spheres is shown. Figure 5.2b and d show the HCP or ABAB stacking sequence viewed both along and normal, respectively, to the stacking axis and Figure 5.2c and e illustrate the CCP stacking both along and normal, respectively, to the stacking axis. Of course, the stacking of spheres cannot fill all space and interstices or holes exist in both schemes. There are two types of holes, one with fourfold coordination, the tetrahedral hole (T), and one with sixfold coordination, the octahedral hole (O). For both HCP and CCP, there are two T holes and one O hole per close packed (CP) sphere. An attempt to illustrate this situation is presented in Figure 5.3a–d. This is rather easy to do for the CCP lattice which is in fact the face-centered cubic (FCC) unit cell, Figure 5.3a, where it is clear that O holes are located at the body center and at the center of each edge. As each edge is shared by four cells, the total number of O holes/fcc unit cell is $1 + 12/4 = 4$. Given that there are four CP atoms/fcc cell there is one O hole/atom. From Figure 5.3c, the T holes are all located within the unit cell for a total of eight or two/CP atom. Note that the coordination polyhedra of both the O holes share edges while those of the T holes share corners and edges. In the HCP lattice, Figure 5.3b and

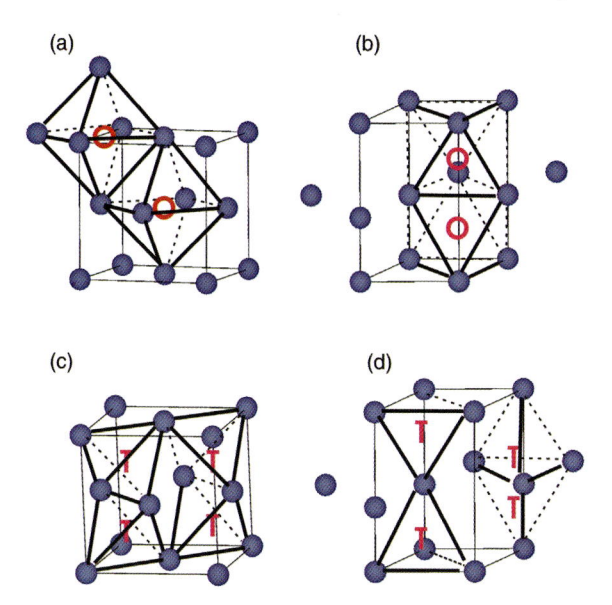

Figure 5.3 (a) Location of the O holes and (c) the T holes in the CCP lattice. Half of the T sites are shown. (b) Location of the O holes and (d) the T holes in the HCP lattice. For all cases slightly more than one unit cell is shown when appropriate.

d, there are two atoms per unit cell. The two O-sites, Figure 5.3b, are within the unit cell and there are two of these, giving one O hole/CP atom. There are two T-sites within the unit cell and two on each of four edges, Figure 5.3d, giving a total of $2 + 8/4 = 4$ T holes/unit cell and 2 T holes/CP atom. In the HCP scheme, the coordination polyhedra of the O holes share faces while those of the T holes share corners and faces.

For those with a crystallographic background, in Table 5.1 are given the space group and positions of the atoms and the O and T holes for both the CCP (5.1a) and HCP (5.1b) stacking sequences. A perusal of Table 5.1 shows that the site symmetries for O and T holes in the CCP lattice are ideal for the polyhedra involved but are of lower symmetry in the HCP lattice.

Table 5.1a Atomic positions and symmetry information for the CCP lattice. $Fm\text{-}3m$ (#225).

Sites	Wyckoff symbol	Site symmetry	Fractional coordinates
CP atoms	4a	$m\text{-}3m(O_h)$	$(0\ 0\ 0) + F^*$
O-holes	4b	$m\text{-}3m(O_h)$	$(1/2\ 1/2\ 1/2) + F^*$
T-holes	8c	$\text{-}43m(T_d)$	$(1/4\ 1/4\ 1/4)\ (1/4\ 1/4\ 3/4) + F^*$

$F^* = (0\ 0\ 0), (1/2\ 1/2\ 0), (1/2\ 0\ 1/2), (0\ 1/2\ 1/2)$.

Table 5.1b Atomic positions and symmetry information for the HCP lattice. $P6_3/mmc$ (#194).

Sites	Wyckoff symbol	Site symmetry	Fractional coordinates
CP atoms*	2c	$\bar{6}m2(D_{3h})$	(0 0 0) (2/3 1/3 1/2)
O-holes	2a	$\bar{6}m2(D_{3h})$	(1/3 2/3 1/4) (1/3 2/3 3/4)
T-holes	4f	$3m(C_{3v})$	(2/3 1/3 z) (2/3 1/3 − z) (0 0 1/2 − z) (0 0 1/2 + z) z ~ 1/8

*The origin has been shifted from the standard setting in the International Tables by (2/3 1/3 3/4).

5.2.2
Ionic Radii, Radius Ratios, and Site Preferences

Site occupation or hole filling in CP lattices is directed in part by consideration of the size of the transition metal ion and the dimensions of the hole involved, which of course depends on the size of the oxide ion. The concept of radius ratios (RR) is also a venerable one and it appears in first year textbooks as well. It is easily shown that the ideal radius for an ion to fit into an O-site or hole is 0.414 $r(O^{2-})$ and 0.225 $r(O^{2-})$ for a T hole. Note that in these considerations we treat the ions as hard spheres rather than fuzzy quantum mechanical ones and clearly the hard sphere approach has limitations. The problem of determining ionic radii has existed since the earliest days of crystallography. Of course, the only empirical data available are those of interatomic distances between two ions and the issue is how to partition these distances into self-consistent contributions from the ions involved in the pair. For oxides and fluorides, this problem was solved some time ago by Shannon and Prewitt, and later by Shannon alone [8,9]. A brief history of previous efforts is contained in those papers. Using methodology described in Refs [8,9], sets of self-consistent radii have been derived from voluminous interatomic distance data, >900 observations. The radii were found to reproduce interatomic distances subject to the following conditions: (1) both cation and anion radii change with coordination number (CN), (2) cation ion radii change with spin state (high-spin HS or low-spin LS), (3) covalency, (4) repulsive forces (when cations are in close proximity), and (5) polyhedral distortion. Examples of the table entries are given in Table 5.2 for O^{2-} and Fe^{3+}. Two self-consistent sets of radii are listed: CR and IR. The former is based on the assignment of an O^{2-} radius of 1.21 Å and the latter, 1.35 Å, that is, for CN = 2. Both sets are self-consistent and of course differ by 0.14 Å.

Note that these entries confirm expectations in that (1) radii increase with CN, (2) radii decrease for the same CN and spin state (SP) with an increase in oxidation state or charge, and (3) radii are smaller for LS ions than HS ions for the same CN.

Returning to the RR concept, while not all oxide lattices are CP, there is a clear expectation that ions with a larger RR will prefer O-sites and those with a smaller RR will be found in T-sites. These trends are illustrated in Figure 5.4 in

Table 5.2 Example of entries in the Shannon Tables of Effective Ionic Radii (Å) [9].

ION	EC	CN	SP	CR	IR
O^{2-}	2P6	II		1.21	1.35
		III		1.22	1.36
		IV		1.24	1.38
		VI		1.26	1.40
		VIII		1.28	1.42
Fe +3	3D5	IV	HS	0.63	0.49
		V		0.72	0.58
		VI	LS	0.69	0.55
		VI	HS	0.785	0.645
		VIII	HS	0.92	0.78
Fe +2	3D6	IV	HS	0.77	0.63
		IVSQ	HS	0.78	0.64
		VI	LS	0.75	0.61
		VI	HS	0.920	0.780
		VIII	HS	1.06	0.92

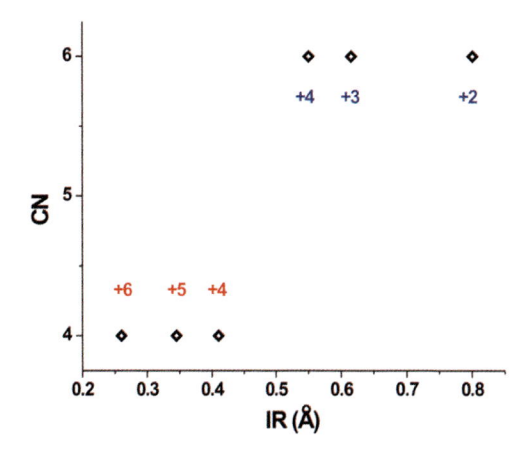

Figure 5.4 A comparison of the IR for chromium ions from +2 to +6 in oxides with the observed CN.

which the CN of a set of chromium oxides is compared to the IR of the chromium ion in five oxidation states, +2 to +6, which are stable in oxides. Note that Cr^{4+} can exist in either CN.

One should not take away the impression that CN = 4 and 6 in roughly octahedral and tetrahedral point symmetries are the only possibilities for TM ions. In fact, these are most common for the d-group TM elements that have appropriate IR for such sites. On the other hand, the f-group TM elements are generally too large to fit into such small sites and tend to exhibit CN of 8, 9 or even higher in oxides. These cases will be discussed separately in a later section.

5.2.3
Electronic Influences on Site Symmetry

As already mentioned, the actual site symmetry associated with O or T holes is not necessarily the full m-$3m$ (O_h) or $-43m$ (T_d), respectively, especially for the HCP case. Structural distortions, often due to size mismatch issues, are important in lowering the symmetry and these will be identified in later sections. However, there are notable single ion contributions that have electronic origins. Among these are Jahn–Teller (J–T) phenomena, both first- and second order, and the peculiar case of certain d^{10} ions.

The first-order J–T (FOJT) effect concerns ions with orbital degeneracy, as the original J–T theorem states that such systems are subject to a structural distortion that acts to lift the orbital degeneracy [10]. The effect is most prominent for ions for which the e_g orbitals are either singly or triply filled, such configurations as $t_{2g}^3 e_g^1$ (Cr^{2+}, Mn^{3+}, or Fe^{4+}), $t_{2g}^6 e_g^1$ (LS Co^{2+} or Ni^{3+}), or $t_{2g}^6 e_g^3$ (Cu^{2+}) in nominally O_h symmetry. The distortion is typically a large tetragonal elongation of the octahedron, which lowers the energy of the $3d_{z^2}$ orbital relative to the $3d_{x^2-y^2}$ orbital as shown in many inorganic textbooks. A first order J–T is possible for O_h ions with t_{2g} orbital degeneracy such as t_{2g}^1, t_{2g}^2, t_{2g}^5, or t_{2g}^4 but is less prominent and more difficult to separate from other influences.

The second order J–T effect (SOJT) is seen for nd^0 ions, particularly, those at the beginning of the 3d, 4d, and 5d series such as Sc^{3+}, Ti^{4+}, Zr^{4+}, Hf^{4+}, Nb^{5+}, Ta^{5+}, Mo^{6+}, and W^{6+}. Its origin has been discussed in some theoretical detail [10] but, basically, in the language of MO theory, it arises when a HOMO (valence band) lies just below a LUMO (empty conduction band) and the two can engage in a symmetry-allowed mixing. The nature of the distortion is not easily predicted but can occur along either the twofold, threefold, or fourfold axes of the octahedron [10].

There is one further effect which is of electronic origin and that is the octahedral site preference energy (OSPE). For certain ions, irrespective of IR, there exists a strong tendency to prefer O-sites over T-sites. Again, this is a standard topic in undergraduate courses in inorganic chemistry. The OSPE for HS ions of the 3d group are shown in Figure 5.5. Ions with configurations $3d^3$ and $3d^8$ have unusually large values and the effect is often seen in $3d^4$(Mn^{3+}) and $3d^9$(Cu^{2+}) as well. A zero value of OSPE occurs for the exactly half-filled shell configuration $3d^5$ (Mn^{2+}, Fe^{3+}).

Two final cases concern ions with d^{10} and d^8 configurations. Ions such as Cu^+, Ag^+, Au^+, and Hg^{2+} often prefer the very rare linear twofold coordination, while Pd^{2+} and Pt^{2+} are often found in CN = 4 square planar environments.

5.2.4
Representation of Crystal Structures of TMO

Crystal structures included in this chapter will be described in three ways: (1) close packing and hole filling (when appropriate), (2) a "ball and stick" model

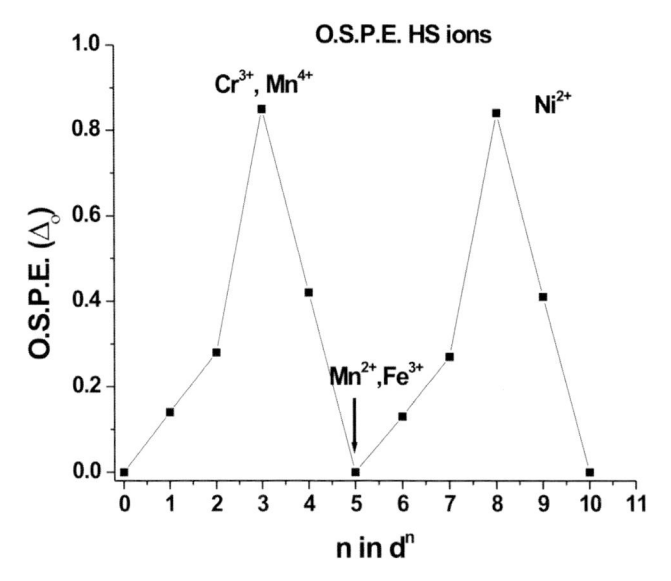

Figure 5.5 Octahedral site preference energies (OPSE) in units of Δ_0, the overall octahedral crystal field splitting energy, for TM ions in the high-spin (HS) state.

showing individual ions, and (3) the polyhedral condensation or linkage scheme. In the latter case, we treat the structures as arising from a grand polymerization of coordination polyhedra, usually octahedra or tetrahedra.

5.3 Most Important Structure Types for TMO

5.3.1 NaCl or Rock Salt

In this structure type, all of the O-sites in a ccp lattice of oxide ions are filled and the space group is $Fm\text{-}3m$ with the oxides ions in 4a and the TMO ions in 4b as shown previously in Table 5.1. From Figure 5.3a and c the coordination polyhedra share all 12 edges forming a three-dimensional array. Clearly, the TMO to ligand ratio is 1:1.

5.3.1.1 Binary Oxides: MO

The NaCl structure is found for most of the binary oxides of the divalent TM ions of the 3d group, except CrO (which does not exist), CuO, and ZnO (Table 5.3). It is also found for the divalent 4f oxide, EuO. YbO can be prepared under high-pressure stabilizing Yb^{2+}. LnO with Ln = La – Sm are also high-pressure phases but contain Ln^{3+} and are metallic [11]. For the 5f oxides, NpO, PuO, AmO, CmO, and BkO have been reported. While the divalent state is of limited

Table 5.3 Known binary TM oxides, MO, with the NaCl structure.

M	a_0 (Å)	IR- M^{2+} (Å)	Reference	M	a_0 (Å)	IR-M^{2+} (Å)	Reference
Ti[a]	4.177(1)	0.86	[13]	Ce[b]	5.089(5)		[11]
V[a]	4.063	0.79	[14]	Pr[b]	5.031(5)		[11]
Mn	4.446(1)	0.83	[15]	Nd[b]	4.994(5)		[11]
Fe[a]	4.3088	0.78	[12]	Sm[b]	4.943(5)		[16]
Co	4.263(1)	0.745	[15]	Np	5.01(1)	1.10	[17]
Ni	4.178(1)	0.69	[15]	Pu	4.960(3)		[18]
Eu	5.1439(5)	1.17	[16]	Am	5.045(3)	1.16	[19]
Yb[b]	4.877(5)	1.02	[11]	Cm	5.091(2)		[20]
La[b]	5.144(5)		[11]	Bk	5.002(1)		[20]

a) Highly nonstoichiometric oxides with a range of compositions.
b) Stable only at high pressure. Only YbO contains a divalent cation. The others involve $Ln^{3+}(e^-)O^{2-}$ and are metallic.

stability for the 4d and 5d series TM elements in oxides, NbO is well known but crystallizes in an unique structure described in *Pm-3m* (#221) not related to NaCl. Returning to the 3d group materials, it should be noted that the majority of these are nonstoichiometric and the 1 : 1 composition is either just one of a wide range (M = Ti, V) or does not exist at all (M = Fe), the limiting composition of the mineral wüstite is near $FeO_{0.97}$ [12]. Only MnO is reliably stoichiometric. Black forms of $Ni_{1-x}O$ exist with Ni deficient compositions in contrast to transparent, green stoichiometric NiO.

NbO has been mentioned as an outlier [21] but PdO (Group 10), CuO (Group 11), and the Group 12 oxides ZnO and HgO do not form the NaCl structure, although from IR considerations this would be expected. The IR for Zn^{2+} (0.74 Å), Cu^{2+} (0.73 Å), Cd^{2+}(0.95 Å), Hg^{2+}(1.02 Å), and Pd^{2+}(0.86 Å) are very close to those for Mg^{2+}(0.72 Å) and Ca^{2+}(1.00 Å) and MgO, CaO, and CdO are all rock salt. ZnO crystallizes in the zincite or würzite structure, which involves an hcp stacking of oxide ions with Zn^{2+} in 1/2 of the T-sites [22]. The structure of HgO is difficult to relate to a CP scheme and Hg^{2+} shows the rare linear CN = 2 [23]. PdO, which was first solved by Moore and Pauling [24], is tetragonal. The structures of NbO, PdO, and CuO are shown in Figure 5.6a–c. Those for ZnO and HgO are shown in Figure 5.6d and e. For two cases NbO (Figure 5.6a) and PdO (Figure 5.6b), CN = 4 square planar (SQP) coordination is clearly evident. While this is not unexpected for Pd^{2+} ($4d^8$) it is certainly odd for Nb^{2+} ($4d^3$). NbO is metallic and is superconducting below $T_c = 1.5$ K while PdO is an insulator. The CuO structure is quite interesting and is monoclinic [25]. The Cu—O coordination is highly anisotropic due to the first order JT effect. Four Cu—O planar bonds are at an average distance of 1.955 Å while two axial bonds are at 2.809 Å, 0.85 Å longer. If one ignores the long bonds, CN = 4 and the structure (part (i and ii) of Figure 5.6c) appears to be a monoclinically distorted version of PdO. Of course the distortion is due to the two long bonds, which are

5.3 Most Important Structure Types for TMO | 171

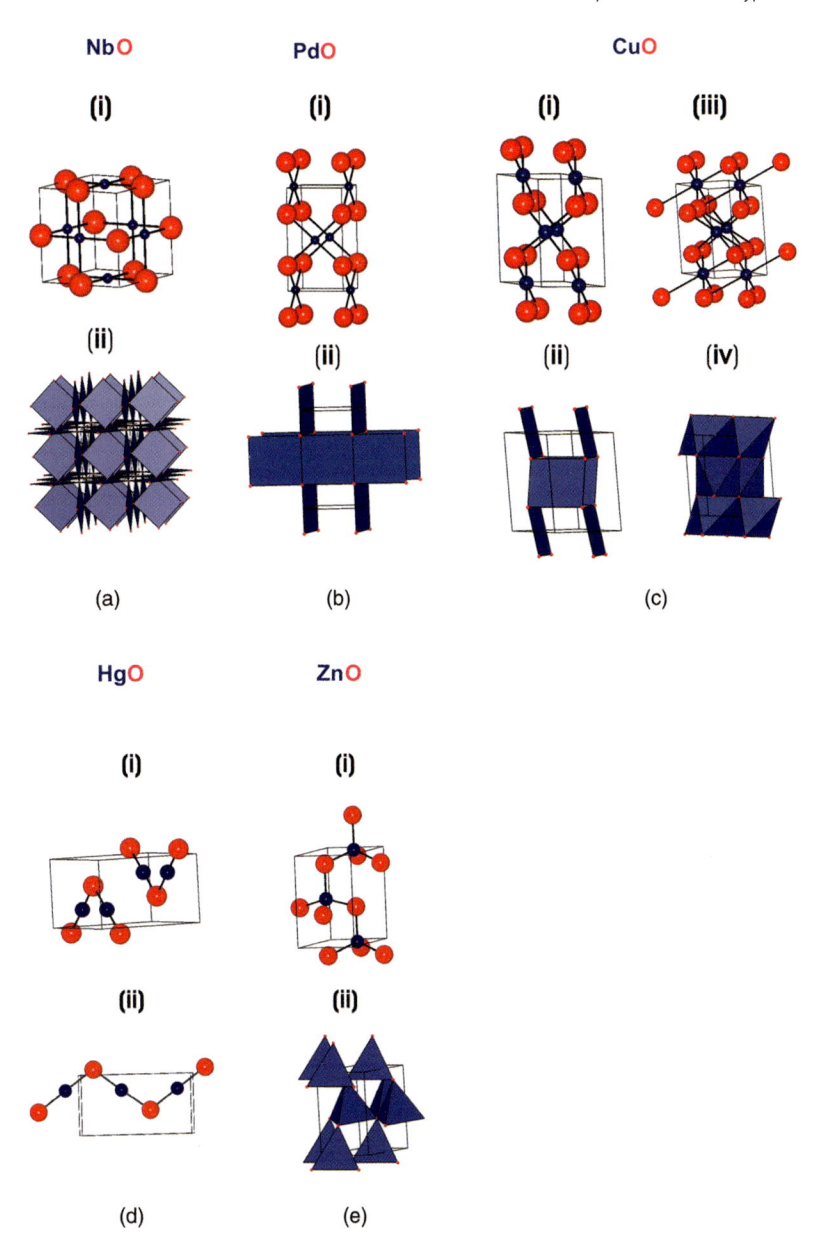

Figure 5.6 (a) (i) The unit cell of NbO showing the unexpected square planar (SQP) coordination of $Nb^{2+}(4d^3)$. (ii) The three-dimensional network of corner sharing square planes. (b) (i) The unit cell of PdO with the expected SQP coordination of $Pd^{2+}(4d^8)$ and (ii) the three-dimensional network of edge and corner sharing square planes. (c) Two views of the CuO structure. (i) CN = 4 ignoring the two long axial bonds. (ii) The resulting three-dimensional network of edge and corner sharing square planes, a distorted version of the PdO structure. (iii) CN = 6, including the two long axial bonds. (iv) The resulting three-dimensional network of edge sharing octahedra, a distorted rock salt structure. (d) (i) The unit cell of HgO and the unusual CN = 2 linear coordination. (ii) The zigzag O—Hg—O chains along the a-axis. (e) (i) The unit cell of ZnO showing the tetrahedral coordination. (ii) The three-dimensional array of corner sharing tetrahedra.

absent in PdO and it is also possible to view CuO as comprised of distorted Cu—O octahedra, CN=6, (part (iii) and (iv)) of Figure 5.6c) and from this perspective, the octahedra share all edges and now the structure can be regarded as a highly distorted rock salt.

5.3.1.2 Ternary and More Complex Oxides with the Ordered-Rock Salt Structure (ORS)

Mather *et al.* have pointed out that many ternary and more complex oxides can be described as derivatives of the rock salt structure, if cation site ordering is taken into account [26]. They show that ternary oxides of general composition $A_aB_bO_{a+b}$ satisfy Pauling's rule of electroneutrality for the cases where A is monovalent and B is either tri-, tetra-, or heptavalent, leading to compositions ABO_2, AB_2O_3, and AB_5O_6. For oxides, the Pauling rule means that the sum of the "electrovalences" at the O-site, z/n, where z = the charge on the cation and n is CN = 6 must be equal to 2, to balance the formal charge on the oxide ion. In the case of ABO_2, each O has three A and three B neighbors, so $\Sigma z/n = 3 \times 1/6 + 3 \times 3/6 = 2$. Similar arguments work for the other two cases. Even when the Pauling rule does not hold, such as A_3BO_4 or A_4BO_5 where B is pentavalent and hexavalent, respectively, a description in terms of a rock salt model can still be valid. In all of these cases the A and B cations must order in distinct crystallographic sites. This ordering is driven largely by differences in formal charge and to a lesser extent, IR differences. Repulsive electrostatic interactions between B cations of high formal charge play an important role. The stability of these ordered rock salt (ORS) phases has been discussed and the interested reader may refer to Ref. [27]. Each case will be discussed and some selected examples are also included.

ABO_2

There are three common structure types for this composition, γ-LiFeO$_2$, α-NaFeO$_2$, and ordered LiTiO$_2$ (Li$_2$Ti$_2$O$_4$). β-LiFeO$_2$ also exists but is unique to that phase and will not be discussed in detail here. For more information, see Ref. [26]. The structures of γ-LiFeO$_2$ and α-NaFeO$_2$ are shown in Figure 5.7a and b, respectively. The cation ordering scheme is emphasized. The unit cell for γ-LiFeO$_2$ is tetragonal ($I4_1/amd$) with dimensions a_{RS} and $c \sim 2a_{RS}$, that is, a doubled RS cell along one axis. The single RS cell is outlined in bold in Figure 5.7a. α-NaFeO$_2$, on the other hand, is best visualized as RS cells oriented along body diagonal (111) axes. Figure 5.7b shows that the stacking sequence along (111) is O—Na—O—Fe—O, that is, Na ions and Fe ions appear in alternating layers. The symmetry is rhombohedral, R-$3m$. Li$_2$Ti$_2$O$_4$ crystallizes in cubic symmetry, Fd-$3m$, with $a = 2a_{RS}$, that is, a cell doubled along all three axes. The layer stacking is shown in Figure 5.7c both normal and parallel to the (111) stacking direction (i and iii) and (ii and iv), respectively. The cation layers are seen to have mixed Li/Ti occupancy but within each layer a distinct Li/Ti ordering occurs.

Of these three ABO_2 structures, α-NaFeO$_2$ is the most common and is of great technological interest. LiVO$_2$, NaVO$_2$, LiCrO$_2$, NaCrO$_2$, KCrO$_2$, LiCoO$_2$, and

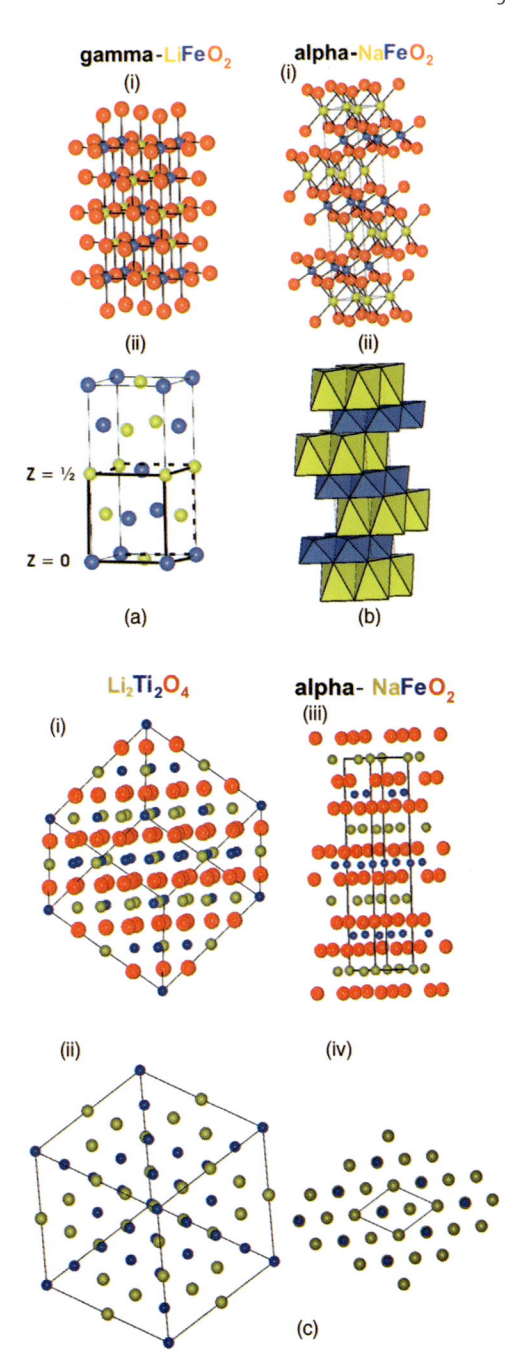

Figure 5.7 (a) (i) The unit cell of γ-LiFeO$_2$. (ii) The cation ordering between Li$^+$ (gold) and Fe^{3+} (blue). The single RS cell is outlined in bold. (b) (i) The unit cell of alpha-NaFeO$_2$ viewed normal to the stacking axis. The O—Na—O—Fe—O layering is clear. (ii) Polyhedral representation of the structure emphasizing the layered nature. (c) (i) The unit cell of Li$_2$Ti$_2$O$_4$ viewed normal to a (111) direction to facilitate comparison with α-NaFeO$_2$ (iii). (ii) The Li$^+$ (gold) and Ti^{3+} (blue) ordering pattern viewed parallel to a (111) direction to facilitate comparison with α-NaFeO$_2$ (iv).

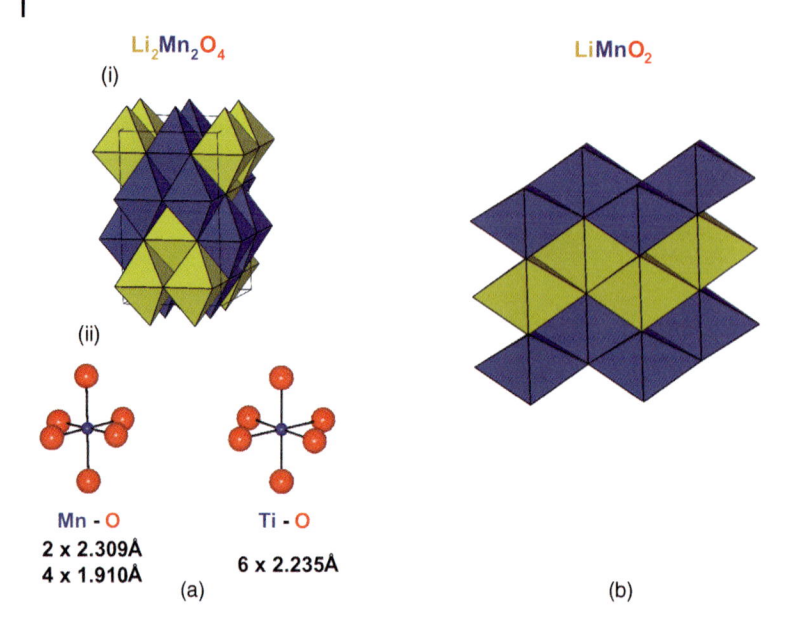

Figure 5.8 (a) (i) The tetragonal, $I4_1/amd$, unit cell of $Li_2Mn_2O_4$ showing the ordering of the Li^+ (gold) and Mn^{3+} (blue) ions in polyhedral representation. About 25% of the Li^+ ions are in tetrahedral sites (not shown). (ii) The local distortion of the octahedron about Mn^{3+}, a tetragonal elongation, compared with the more regular octahedron for Ti^{3+}. (b) The stacking of buckled layers of Li^+ (gold) and Mn^{3+} (blue) ions in orthorhombic, $Pmmn$, $LiMnO_2$.

$LiNiO_2$ all form with this structure. The relative ease of removal of the Li^+ ions either by chemical or electrochemical means is well-known and has led to the application of some of these phases, notably $LiCoO_2$, in lithium batteries as the cathode material. In most $LiMO_2$ phases, some degree of Li/M interlayer mixing occurs. $Li_2V_2O_4$, with the cubic $Li_2Ti_2O_4$ structure can be prepared by insertion of Li into LiV_2O_4 [28]. $LiMnO_2$ and $Li_2Mn_2O_4$ also exist but contain the Mn^{3+} ion and the first-order JT effect drives structural distortions as illustrated in Figure 5.8a and b. Orthorhombic (Pmmn) $LiMnO_2$ is the thermodynamically stable polymorph, meta-stable $Li_2Mn_2O_4$ is obtained by room temperature lithiation of $LiMn_2O_4$. The unit cell of $Li_2Mn_2O_4$ is tetragonal ($I4_1/amd$) and is related to that of cubic $Li_2Ti_2O_4$ but with dimensions $a \sim 2\, a_{RS}/\sqrt{2}$, and $c \sim 2a_{RS}$. The local JT induced distortion of the Mn—O octahedron, a pronounced tetragonal elongation, is illustrated in Figure 5.8a (ii) and drives the cooperative crystallographic distortion. $LiMnO_2$, $Pmmn$, Figure 5.8b, is layered similar to α-$NaFeO_2$, but the layers are now buckled and not flat along the stacking direction. The Mn—O octahedron is similarly distorted with distances 2×2.291 Å and 4×1.959 Å.

A_2BO_3

This class generally involves A^+ and B^{4+} ions. Three structure types exist, although there are relatively few examples. The most common is β-Li_2SnO_3

Li$_2$SnO$_3$

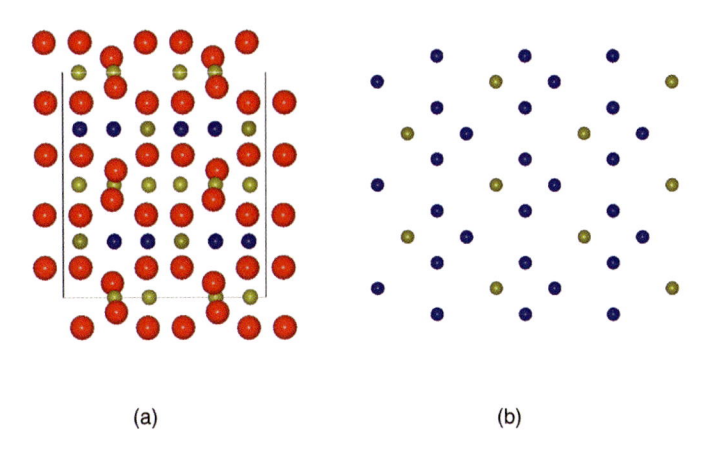

(a) (b)

Figure 5.9 (a) The stacking sequence of O—Li—O—Li/Sn—O layers in the β-Li$_2$SnO$_3$ structure. (b) Li-ion ordering(gold) and Sn (blue) ion ordering within one of the mixed layers.

which is found for Li$_2$TiO$_3$ [29]. β-Li$_2$ZrO$_3$, which is only partially ordered, and β-Na$_2$PtO$_3$ are also known but are unique to those compositions. Only β-Li$_2$SnO$_3$ will be described. It is related to α-NaFeO$_2$ but with layers of only Li alternating with layers of composition Li$_{1/3}$Sn$_{2/3}$ as might be expected, given the stoichiometry. The symmetry is monoclinic C2/c reduced from the R-3m symmetry of α-NaFeO$_2$ due to Li/Sn ordering within the mixed layer. The stacking sequence and the Li/Sn ordering in the mixed layer are shown in Figure 5.9.

Li$_2$ReO$_3$ has been reported but with a rhombohedral symmetry R3c and the structure is also related to α-NaFeO$_2$ [30]. Figure 5.10a shows the stacking sequence involving mixed Li/Re layers and in Figure 5.10b the intra layer Li/Re order is indicated for various z values.

A$_5$BO$_6$

For this class of ORS oxides, if A is monovalent, a heptavalent B ion is needed which reduces the possibilities rather severely. In fact, only Li$_5$BO$_6$ with B=Re and Os and Na$_5$OsO$_6$ have been reported [31,32]. However, allowing for mixed composition on the A-sites, others can be generated containing hexavalent ions such as Li$_4$MgReO$_6$ and pentavalent ions Li$_3$Ni$_2$TaO$_6$ [33,34]. Other examples of this latter type include TMO involving substitution of Mg and Co for Ni and Nb for Ta. As well, Li$_3$Mg$_2$RuO$_6$ and Li$_3$Mg$_2$OsO$_6$ both exist [35,36]. In all of these cases, the A-site ions occupy a number of octahedral sites, roughly statistically, while the B cation is on an unique site. There are two basic structure types illustrated in Figure 5.11 by Li$_4$MgReO$_6$ with C2/m symmetry and Li$_3$Mg$_2$RuO$_6$ described in Fddd. For the monoclinic structure, the layer stacking sequence

Li$_2$ReO$_3$

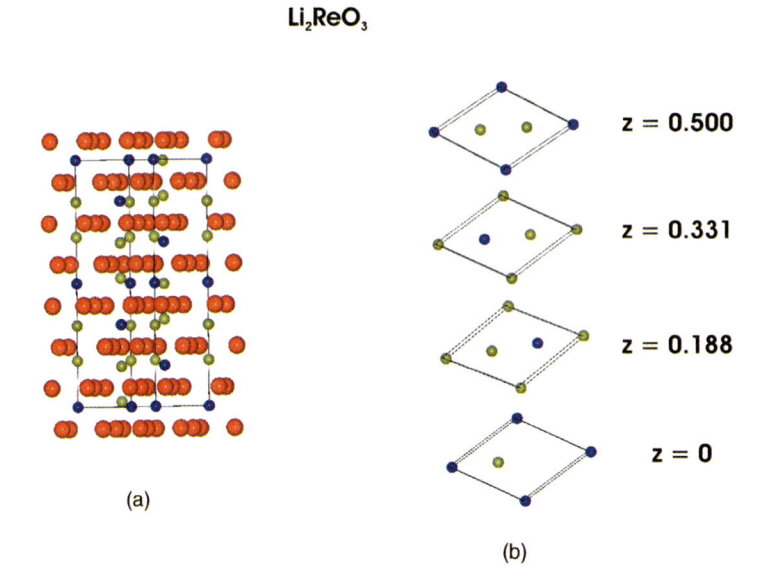

Figure 5.10 (a) The stacking sequence for Li$_2$ReO$_3$ showing mixed Li(gold)/Re (blue) layers. (b) Composition of the mixed layers at various z values.

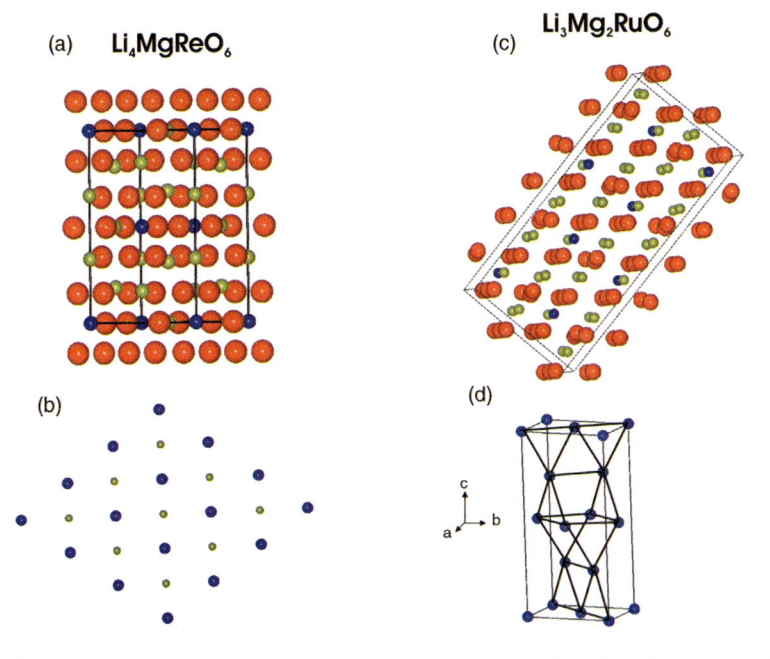

(a) **Li$_4$MgReO$_6$**

(c) **Li$_3$Mg$_2$RuO$_6$**

Figure 5.11 (a) The unit cell of Li$_4$MgReO$_6$ showing the stacking sequence. Re(blue), Li (Mg) (gold) O (red). (b) One Re/Li(Mg) layer revealing the cation site ordering. The Re^{6+} ions form a layer of edge-sharing triangles. (c) The unit cell of Li$_3$Mg$_2$RuO$_6$ oriented to display the stacking sequence. (d) The Ru sublattice, a three-dimensional array of edge-sharing triangular ribbons linked by corners.

involves O—Re—O—Li(Mg)—O—Re—O. Within the Re containing layers, the Re sublattice consists of a two-dimensional array of edge-sharing triangles. On the other hand, for the orthorhombic structure, the layers have mixed Re/Li(Mg) occupancy and the Ru sublattice is a three-dimensional array of edge-sharing triangular ribbons linked by corner-sharing. Other structures exist which can be regarded as of the ORS type and the reader is referred to Ref. [26] for further details.

5.3.2
Corundum or Al_2O_3

While the NaCl-related structures involve a c.c.p. stacking of oxide ions, there is of course a large class of TMO for which the stacking scheme is h.c.p. These are exemplified by the structure of the mineral corundum or Al_2O_3, also known as sapphire when in single crystal form. In this case, octahedral sites are also occupied and the fraction is 2/3 rather than 1/1 for NaCl. The symmetry is $R\text{-}3c$, rhombohedral, and the structure is shown in Figure 5.12. Note the tendency for occupation of adjacent O-sites along the c-axis, giving rise to dimeric face-sharing octahedral units. All forms of sharing are exhibited, face, edge, and corners.

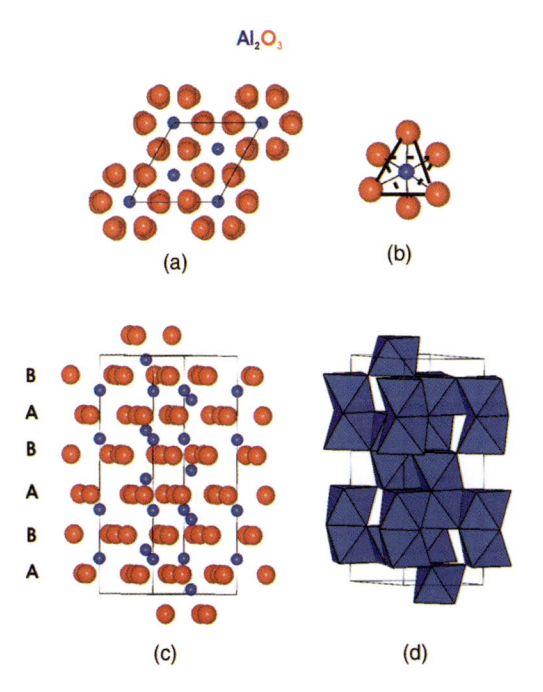

Figure 5.12 The corundum or Al_2O_3 structure. (a) A view parallel to the stacking axis. Al^{3+} (blue), O^{2-} (red). (b) One octahedral site viewed along a 3-axis. (c) View of the stacking pattern normal to the stacking axis, with the ABAB sequence. (d) Polyhedral representation illustrating face, edge, and corner sharing octahedra.

Table 5.4 Unit cell constants and cell volumes for binary TMO with the Al_2O_3 structure [a].

TMO	a (Å)	c (Å)	V (Å^3)	R (TM^{3+}) (Å)	Reference
Ti_2O_3	5.1570	13.610	313.46	0.67	[39]
V_2O_3	4.9525	14.0003	297.32	0.64	[40]
Cr_2O_3	4.9600	13.5982	289.72	0.615	[41]
Fe_2O_3	5.038	13.772	302.72	0.645	[42]
Rh_2O_3	5.127	13.853	315.36	0.665	[43]

a) Hexagonal setting of R-$3c$.

5.3.2.1 Binary Oxides

The Al_2O_3 structure is adopted mainly by TMO of this composition in the 3d series, except for Co, Ni, and Cu, in spite of some stability of the 3+ state for all of these elements. Among the 4d and 5d elements, the 3+ state is less stable and only Rh_2O_3 is well documented apart from Au_2O_3, which has a different structure, Au^{3+} is a 5d^8 ion, so square planar coordination is observed. TMO of the trivalent rare earths, Sc_2O_3 and Mn_2O_3 also form different structures to be mentioned later. While there are only a small number of binary TMO with the Al_2O_3 structure, they have been studied intensively.

Both Ti_2O_3 and V_2O_3 undergo subtle structural changes generating metal/insulator transitions as a function of temperature and with small levels of doping by, for example, Cr^{3+} ions. V_2O_3 is considered to be a textbook case of a Mott–Hubbard insulator [37].

Fe_2O_3 is of course the mineral hematite. Table 5.4 collects known data for the binary TMO with the Al_2O_3 structure. The cell volumes scale roughly with the CN = 6 IR for the trivalent ions. In Figure 5.13, the unusual structure of Au_2O_3 is

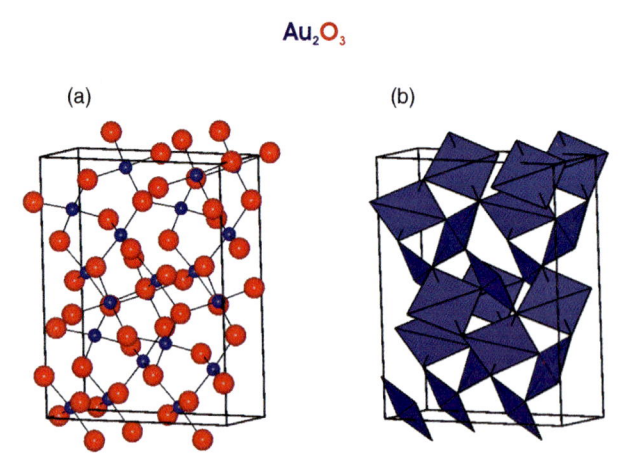

Figure 5.13 The structure of Au_2O_3 showing (a) square planar coordination for the Au^{3+}(5d^8) ion and (b) a topology of corner sharing [38].

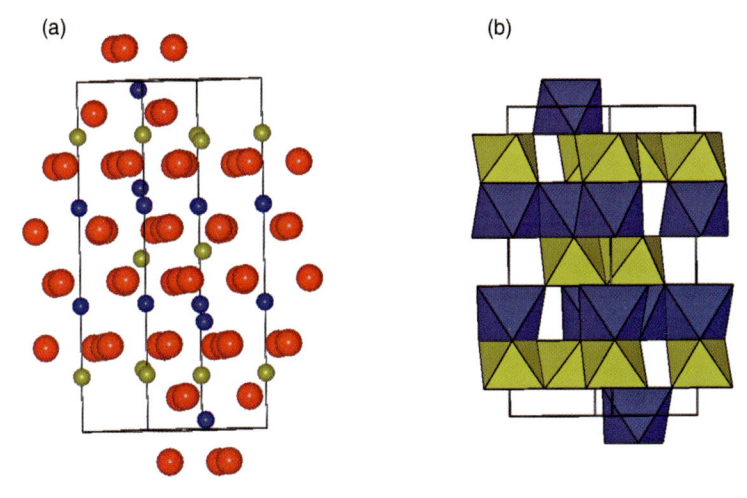

Figure 5.14 (a) The unit cell of FeTiO$_3$ showing the site ordering of Ti^{4+} (gold) and Fe^{2+} ions. The ABAB stacking of the oxide ions is evident. (b) A polyhedral view.

displayed with attention to the square planar coordination expected for a 5d^8 ion. The square planes share corners as in NbO but the topology is quite different. In PdO, the square planes share edges (Fig. 5.6b).

5.3.2.2 Ilmenite and Other Ordered Corundum-Like Structures

When two ions occupy the Al site, ordering can occur. This is illustrated by the ilmenite structure exemplified by FeTiO$_3$. While one might normally expect a solid solution between isostructural Ti$_2$O$_3$ and Fe$_2$O$_3$, in fact a redox occurs resulting in Fe^{2+} and Ti^{4+}, which order crystallographically in a corundum-like structure, Figure 5.14. The symmetry is reduced from $R\text{-}3c$ to $R\text{-}3$. Formation of the ilmenite structure requires stable tetravalent and divalent ions. Within the 3d series, stable tetravalent ions are found for Ti, V, and Mn. Table 5.5 lists the

Table 5.5 TMO with the ilmenite structure, $R\text{-}3$[a].

TMO	a (Å)	c (Å)	V (Å^3)	Reference
MgTiO$_3$	5.0547	13.8992	307.56	[45]
MnTiO$_3$	5.1386	14.2857	326.68	[46]
FeTiO$_3$	5.0884	14.0855	315.84	[47]
CoTiO$_3$	5.0650	13.9199	309.26	[48]
NiTiO$_3$	5.0321	13.7924	302.46	[46]
ZnTiO$_3$	5.0796	13.9328	311.34	[49]
CoMnO$_3$	4.933	13.7106	288.94	[50]
NiMnO$_3$	4.9054	13.5820	283.23	[50]

a) Hexagonal setting.

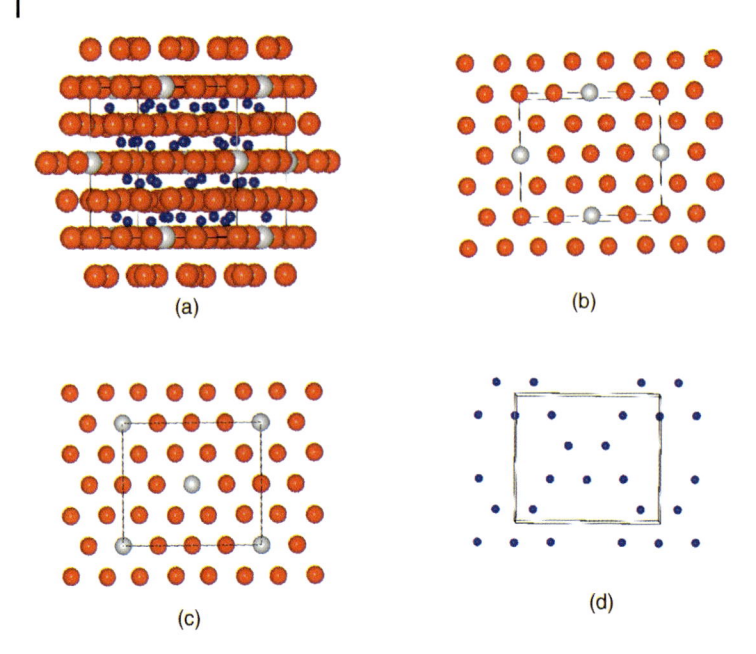

Figure 5.15 BaV$_{10}$O$_{15}$. (a) Stacking along the c-axis showing inclusion of Ba^{2+} ions (silver) into CP layers of O^{2-} (red). (b) The CP layer at $z=0$. The ions are not strictly coplanar. (c) The CP layer at $z=\frac{1}{2}$. (d) The pattern of octahedral site occupation by V^{2+}/V^{3+} ions (blue) in one layer near $z=0.86$. Note the avoidance of sites near the Ba^{2+} sites. The ions are not strictly co-planar.

known TMO with this structure. There are no vanadium examples. Zr and Hf from the 4d and 5d series have stable 4+ states but do not form compounds with the ilmenite structure. Interestingly, a partial solid solution exists between V and Ni, Ni$_x$V$_{2-x}$O$_3$, $x_{max}=0.66$, but it is likely that both ions are trivalent [44].

At this stage, it is worth considering CP schemes which involve admixtures of oxide ions and large cations, such as Ba^{2+}, Sr^{2+}, or K$^+$ that have IR similar to that of O^{2-} [9].

These ions can fit into a CP layer of oxide ions and there is always an ordering of cations and anions within the layer, as the cations will tend to avoid each other, presumably due to electrostatic repulsion. This has implications for the pattern of occupation of octahedral sites between the mixed CP layers, arising again from electrostatic issues. New structures are formed that are not seen in conventional CP schemes. This class of TMO can be illustrated by BaV$_{10}$O$_{15}$ that can be formulated as V$_{10}$O$_{15}$Ba and contains an admixture of V^{2+} and V^{3+} ions. Structural details are shown in Figure 5.15 [51]. The symmetry is orthorhombic described in *Cmca* at ambient temperature. CP layers of O/Ba are stacked along the c-axis, alternating with CP layers of O^{2-}, Figure 5.15a. Ba^{2+} ions are ordered within the CP layers as indicated, Figure 5.15b and c. The presence of ordered Ba^{2+} ions within the CP layers directs the ordering of the V ions between the layers, Figure 5.15d. The pattern consists of V_5 units consisting of edge-sharing triangles connected by V—V bonds and gives rise to frustrated magnetic interactions and unusual magnetic properties [52,53]. There is also a structural phase transition below 135 K to a structure described in *Pbca* driven by an apparent trimerization of V—V bonds [52,54]. $BaCr_{10}O_{15}$ also exists, providing a very rare example of Cr^{2+} in an oxide [53].

Several other similar phases have been reported including AV_6O_{11} with A = Na, K, Ba, Sr, and Pb [55]. $SrCo_6O_{11}$ is also known, as are $Sr(Ba)V_{13}O_{18}$ which also contain V^{2+}/V^{3+} layers that result in exotic occupation patterns for the TM ions [56]. Many of these compounds exhibit structural phase transitions with temperature and a rich variety of electrical transport and magnetic properties. The interested reader is directed to the cited papers.

5.3.3
Rutile and Related Structures

Rutile is one of the mineral forms of TiO_2. It is sometimes described in terms of a distorted HCP lattice with occupation of half of the O-sites, but this relationship is not easy to visualize.

5.3.3.1 Binary Oxides

The rutile structure generally forms for TM with IR appropriate for an O-site. Of course a stable tetravalent state is needed and this requirement is satisfied by several 3d, 4d, and 5d TM. The structure is tetragonal, $P4_2/mnm$, and is shown in Figure 5.16.

Table 5.6 lists the known binary TMO with the rutile structure. The d^1 ion TMO VO_2 and NbO_2 have a distorted rutile structure at ambient temperature but transform to the ideal form at higher temperatures, as will be discussed later. MoO_2 and ReO_2, containing d^2 and d^3 ions, are isostructural with the ambient temperature form of VO_2, $P2_1/c$, but do not transform to the ideal structure at higher temperatures. PtO_2 also crystallizes in a distorted orthorhombic form, *Pnnm*, of the rutile structure but the origin of the distortion is distinct from that of VO_2. CrO_2 can only be prepared under high O_2 pressure. ZrO_2, HfO_2, PrO_2, TbO_2, ThO_2, UO_2 and others crystallize in other structures with higher local coordination numbers due to larger IR to be discussed in a later section.

In the ideal rutile structure, the CN of O^{2-} is three with an IR of 1.36 Å. Thus, the ideal range of radius ratios for this structure type appears to be 0.39–0.50. The IR of Mo^{4+}(0.65 Å), Re^{4+}(0.63 Å), and Pt^{4+}(0.625) also lie within the range for a rutile structure.

VO_2 famously undergoes a first-order structural phase transition at ~333 K (60 °C) from $P2_1/c$ to $P4_2/mnm$ that generates an insulator to metal transition and has been studied intensively for many years [66]. ReO_2 and MoO_2 are isostructural, $P2_1/c$, [67], and NbO_2, [68] while crystallizing in a larger tetragonal cell, $I4_1/a$, shares similar features. The critical aspect of the structure of all three exceptions can be illustrated by ambient temperature VO_2,

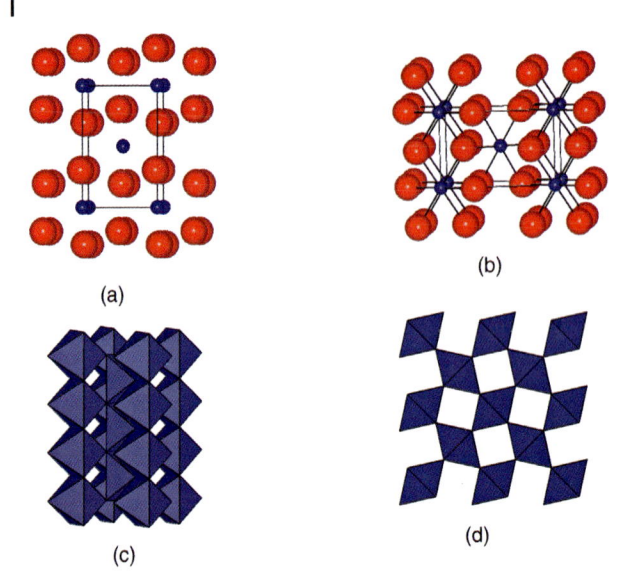

Figure 5.16 The rutile (TiO$_2$) structure. (a) Stacking view normal to the a-axis, showing the buckled CP layers and O-site occupation. (b) Conventional view of the unit cell with the shorter c-axis vertical. (c) Polyhedral connectivity: chains of edge sharing octahedra viewed normal to the c-axis. (d) View along c-axis showing corner sharing of octahedra.

Table 5.6 Binary TMO with the ideal rutile structure, $P4_2/mnm$.

TM	a (Å)	c (Å)	V (Å^3)	IR (TM^{4+}) (Å)	Reference
Ti	4.5922	2.9590	62.40	0.605	[57]
V[a]	4.5561	2.8598	59.36	0.58	[58]
Cr[b]	4.4190	2.9154	56.93	0.55	[59]
Mn	4.4041	2.8765	55.79	0.53	[60]
Nb[a]	4.8463	3.0315	71.20	0.68	[61]
Ru	4.4919	3.1066	62.68	0.62	[62]
Rh	4.4862	3.0884	62.16	0.60	[63]
Os	4.5003	3.1839	64.48	0.63	[64]
Ir	4.4990	3.1546	63.85	0.625	[65]

a) High temperature form.
b) Stable only under high O$_2$ pressure.

shown in Figure 5.17 and compared with undistorted rutile. The structures are nearly indistinguishable from rutile in terms of the polyhedral linkages, however, the V—V distances along the edge sharing chains alternate short–long–short as indicated.

The same patterns are seen for NbO$_2$, Nb—Nb distances of 2.743 Å and 3.256 Å, MoO$_2$, 2.513 and 3.109, and ReO$_2$, 2.528 Å and 3.638 Å. The most

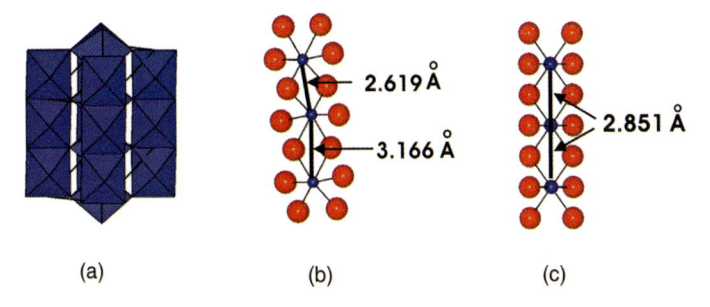

Figure 5.17 The ambient temperature VO$_2$ structure, P2$_1$/c. (a) Polyhedral view showing the same edge-sharing chains as in undistorted rutile. (b) Alternation of V—V bond distances along the chains. (c) Uniform V—V bond distances in the high temperature rutile form.

straightforward interpretation of this phenomenon is in terms of metal – metal bond formation involving the partially filled t$_{2g}$ orbitals – V^{4+} and Nb^{4+} are t$_{2g}^1$, Mo^{4+} is t$_{2g}^2$ and Re^{4+} is t$_{2g}^3$. As mentioned earlier, the distortion in PtO$_2$ is not the result of Pt—Pt bond formation, in fact the Pt—Pt distances are uniform at 3.138 Å along the chains. Pt^{4+} has configuration t$_{2g}^6$ and all of the potential bonding orbitals are filled [69].

5.3.3.2 Site-Ordered Rutiles: the Trirutile Structure and Columbite

TMO of composition AB$_2$O$_6$ can crystallize in the trirutile structure. Two valence combinations are most common, in one case one has A^{2+}B^{5+} (B = Ta, Nb, Sb) and in the other A^{6+}B^{3+} (A = W, Te, Re, Rh). In general, the structure is tetragonal with the same symmetry as rutile itself, P4$_2$/mnm, but with a tripled c-axis, hence the name, but there are exceptions in the cases of J–T ions such as Cu^{2+} and Cr^{2+} which induce monoclinic, P2$_1$/n, symmetry. In fact, CrTa$_2$O$_6$ is one of the rare examples of a stable Cr^{2+} state in oxides. The trirutile structure is shown in Figure 5.18 and some examples are collected in Table 5.7. Another structure is competitive with trirutile, columbite, also shown in Figure 5.18 with examples included in Table 5.8. The symmetry is lower, orthorhombic, *Pbcn*. While in trirutile the edge-sharing octahedral chains are linear and with mixed chain occupation A—B—B—A . . . those in columbite have a zigzag topology and the composition is all A or all B. Shown at the bottom of Figure 5.18 are the Ta—O and Nb—O coordination octahedra for CoTa$_2$O$_6$ (trirutile) and CoNb$_2$O$_6$ (columbite), respectively. The symmetry is high, 2 mm, at the Ta site but there is no symmetry at the Nb site. A view of the Nb—O octahedron along a pseudo threefold axis shows three long and three short bonds, a typical distortion seen in materials in which the second-order JT effect is active. As the IR of Ta^{5+} and Nb^{5+} are identical, 0.64 Å, an electronic factor must be involved. Note that for the A^{2+}B^{5+} series, nearly all tantalates are trirutile while the niobates are columbite. ZnTa$_2$O$_6$ crystallizes in the α-PbO$_2$ structure, which has the same symmetry as columbite, *Pbcn*, but a different unit cell and polyhedral connectivity [70].

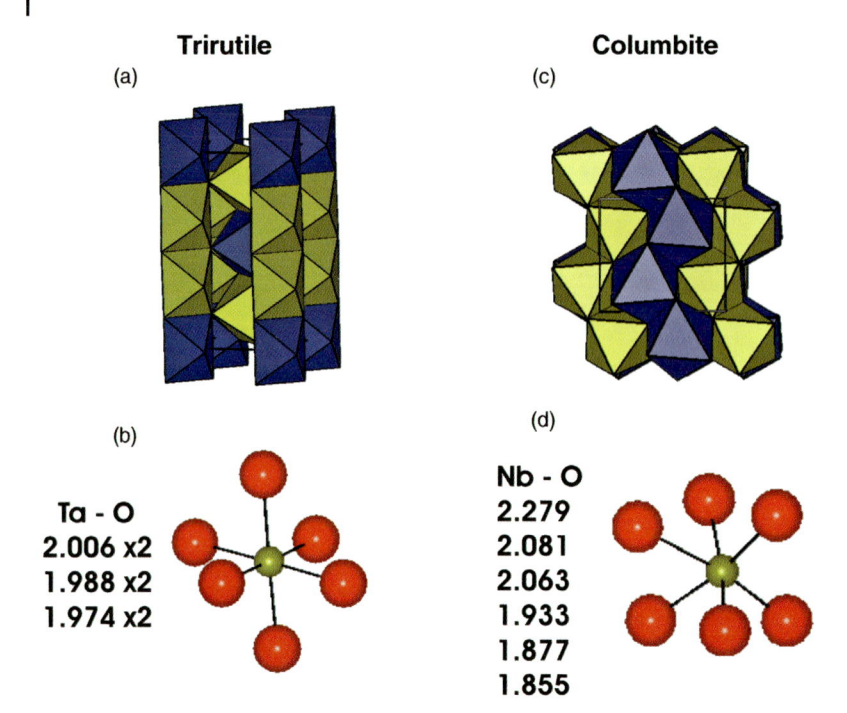

Figure 5.18 (a) The trirutile structure, AB_2O_6, showing the linear chains of edge-sharing octahedra parallel to the c-axis, with site ordering between the A (blue) and B (gold) ions. (b) The Ta—O octahedron in $CoTa_2O_6$, distances in Å. The site symmetry is 2 mm and the octahedron is quite regular. (c) The columbite structure of $CoNb_2O_6$ showing zigzag chains of edge sharing octahedra parallel to the c-axis. (d) The Nb—O octahedron in $CoNb_2O_6$, distances in Å, seen along a pseudo threefold axis. Nb is at a general position with no symmetry. The three long Nb—O bonds are pointing out of the plane of the page while the three shorter ones point inward. This is a typical result of a second order JT distortion.

5.3.3.3 Structures Compositionally Related to Rutile: Hollandites and Others

There exists a large family of TMO which are compositionally related to TiO_2 in that the TM/O ratio is ~2, the TM ions are octahedrally coordinated and edge sharing is involved. An example is the hollandite family. Often the TM is Ti or Mn but many others based on V and Ru are known. Other TM such as Fe or Cr can also be substituted on the B-sites. A quasi-general formula is $A_x(BO_2)_y$, where A is usually a large cation from Group 1 or 2. They are often called "tunnel oxides" due to a characteristic feature shown in Figure 5.19 for $Ba_2Mn_8O_{16}$ [90]. The Mn oxidation state in this material is +3.5, so there are 4 Mn^{4+} and 4 Mn^{3+}. The symmetry is $I2/m$ with $a = 10.006$ Å, $b = 2.866$ Å, $c = 9.746$ Å, and $\beta = 91.17°$. There are indeed two sets of fourfold sites for Mn, but the Mn—O distances do not suggest complete charge ordering between the two states. The distinguishing feature of all hollandites and related mineral types is clearly seen, a tunnel formed by 2×2 blocks of edge-sharing octahedra which share corners, in this case running parallel to the short b-axis. The large cation

Table 5.7 TMO with the trirutile, $P4_2/mnm$ structure.

TMO	a (Å)	c (Å)	Reference
$MgTa_2O_6$	4.7189	9.2003	[75]
VTa_2O_6	4.758	9.159	[76]
$CrTa_2O_6$ [a]	4.7381	9.297	[77]
	4.7413	$\beta = 91.05$	
$FeTa_2O_6$	4.749	9.192	[78]
$CoTa_2O_6$	4.7358	9.1708	[79]
$NiTa_2O_6$	4.7219	9.150	[80]
$CoSb_2O_6$	4.6535	9.2804	[79]
$NiSb_2O_6$	4.641	9.219	[81]
$CuSb_2O_6$ [a]	4.6349	9.2931	[82]
	4.6370	$\beta = 91.12$	
$ZnSb_2O_6$	4.6638	9.263	[83]
$TeCr_2O_6$	4.546	9.014	[84]
WCr_2O_6	4.582	8.870	[84]
$ReCr_2O_6$	4.542	8.9173	[85]
$TeFe_2O_6$	4.601	9.087	[84]
WV_2O_6	4.6213	8.8864	[86]
$TeMn_2O_6$	4.6079	9.1144	[87]
$RhMo_2O_6$	4.606	9.063	[88]
RhU_2O_6	4.744	9.360	[89]

a) $P2_1/n$.

Table 5.8 TMO with the columbite structure, $Pbcn$.

TMO	a (Å)	b (Å)	c (Å)	Reference
$MnTa_2O_6$	14.4219	5.5713	5.0816	[71]
$MnNb_2O_6$	14.4236	5.7609	5.0839	[71]
$FeNb_2O_6$	14.2448	5.7276	5.0421	[71]
$CoNb_2O_6$	14.1475	5.7120	5.0446	[71]
$NiNb_2O_6$	14.0217	5.6752	5.0150	[71]
$CuNb_2O_6$ [a]	14.1027	5.6093	5.1223	[72]
$MgNb_2O_6$	14.1875	5.7001	5.0331	[73]

a) High temperature form. Below 740 °C a $P2_1/c$, structure is stable [74].

composition is variable, for example, the structure of $Ba_{1.2}Mn_8O_{16}$ is essentially the same as for $Ba_2Mn_8O_{16}$, [91] so the Mn^{3+}/Mn^{4+} ratio can be tuned, and the tunnels can accommodate H_2O and other small molecules. The tunnel dimensions can also be engineered by varying composition on the A- and B-sites. The romanechite and todorokite structures involve 2×3 and 3×3 corner linked

Ba$_2$Mn$_8$O$_{16}$

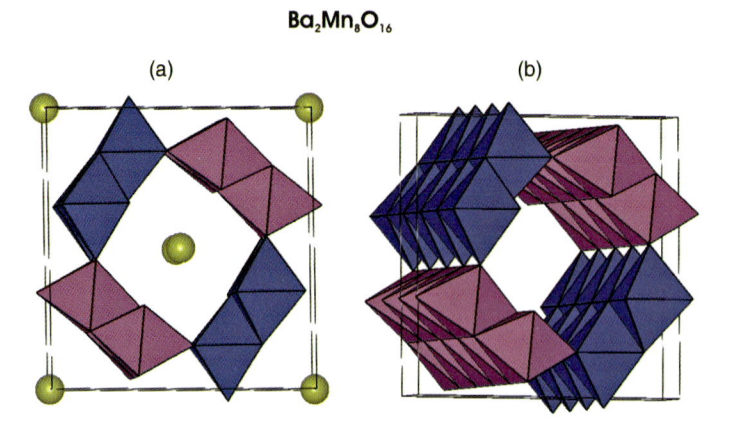

Figure 5.19 The hollandite structure exemplified by Ba$_2$Mn$_8$O$_{16}$. (a) The Ba ions (gold) occupy the tunnels created by the edge and corner sharing Mn—O octahedra (blue and magenta). (b) The tunnel parallel to the short b-axis.

edge-sharing groups [92]. The materials in this class show a rich variety of electric transport and magnetic properties and have been intensively studied.

5.3.3.4 Other Forms of TiO$_2$ and MnO$_2$

While rutile is the thermodynamically stable form of TiO$_2$, at least two other forms occur naturally, anatase ($I4_1/amd$) and brookite (*Pbca*). These structures are also based on edge and corner sharing octahedra, Figure 5.20, but the connectivity is different from rutile. One-dimensional channels exist as in rutile but the geometries are different. Anatase is considerable less dense than rutile. MnO$_2$ also shows an orthorhombic form, ramsdellite – *Pnma*, but again the

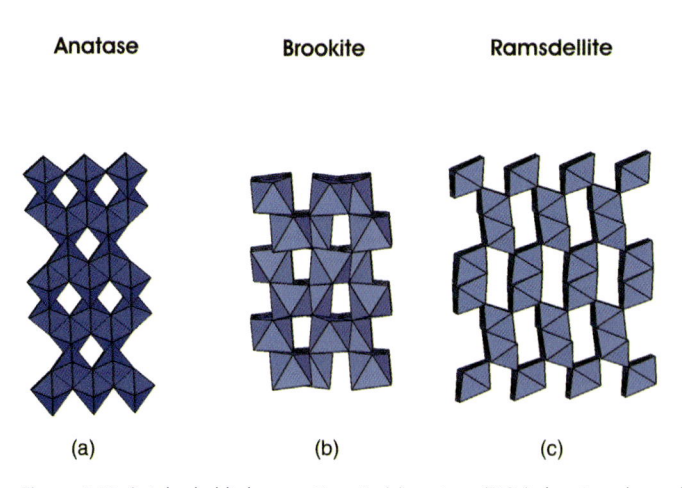

Anatase **Brookite** **Ramsdellite**

Figure 5.20 Octahedral linkage pattern in (a) anatase (TiO$_2$) showing channels along the a and b axes, (b) brookite (TiO$_2$) showing channels along the c-axis, (c) ramsdellite (MnO$_2$) with 2 × 1 channels along the b-axis.

connectivity is different. Here there are 2×1 channels parallel to the b-axis. Thus, ramsdellite can be thought of as a precursor to the hollandites and related structures.

A total of five polymorphs of MnO_2 have been reported, α-MnO_2 (hollandite structure), β-MnO_2 (pyrolusite, the rutile form), R-MnO_2 (ramsdellite), δ-MnO_2 (a layered structure), and λ-MnO_2 (a spinel related structure). Of these the α, δ, and λ forms require special synthetic techniques. For a summary of the structures and synthetic methods for the MnO_2 family, see Ref. [93].

5.3.4
Fluorite, CaF$_2$, and Related Structures

5.3.4.1 Binary Oxides

The fluorite structure can also be described in terms of a CP lattice with hole filling, but there is a twist. In this case the cations form a CCP (FCC) lattice with the anions occupying all of the tetrahedral sites as shown in Figure 5.21. While the oxide ion has CN = 4, the cationic site is cubic with CN = 8. It is thus not surprising that only the larger TM^{4+} ions form binary oxides with this structure. Examples are the tetravalent lanthanides CeO_2, PrO_2, and TbO_2 along with the stable tetravalent actinides ThO_2 to CfO_2. *Es* and *Fm* do not apparently form in the 4+ state. Some cell constants are listed in Table 5.9.

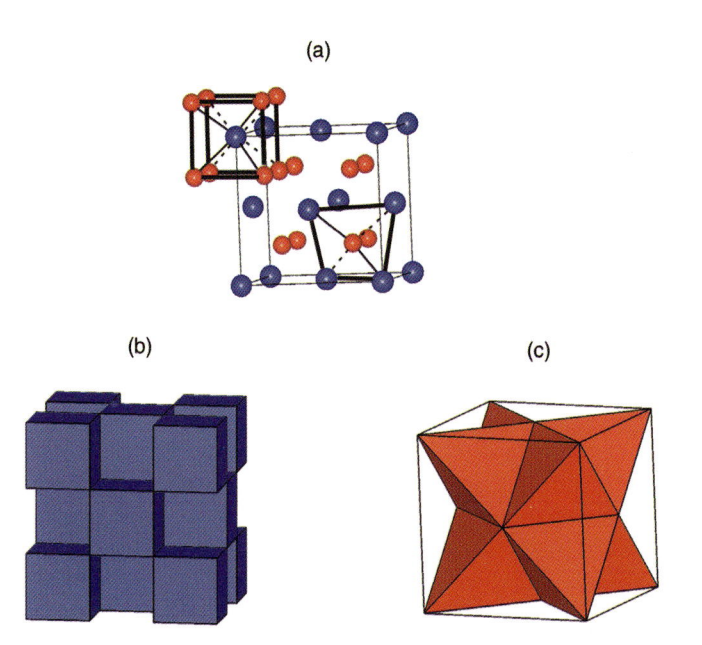

Figure 5.21 The CaF_2 fluorite structure. (a) The unit cell showing cubic CN = 8 about the cation (blue) and tetrahedral CN = 4 about the anion (red). (b) Polyhedral connectivity for the cation sites (edge-sharing cubes). (c) Polyhedral connectivity for the anion sites (edge-sharing tetrahedra). Note that this is the topology of the FCC lattice cation positions.

Table 5.9 Some unit cell constants for binary TMO oxides with the CaF_2 structure, $Fm\text{-}3m$.

TMO	a (Å)	Reference	TMO	a (Å)	Reference
CeO_2	5.411	[94]	NpO_2	5.436	[95]
PrO_2	5.394	[96]	PuO_2	5.397	[95]
TbO_2	5.213	[97]	AmO_2	5.388	[95]
ThO_2	5.5957	[98]	CmO_2	5.372	[99]
PaO_2	5.505	[100]	BkO_2	5.336	[101]
UO_2	5.468	[102]	CfO_2	5.310	[103]

Notable by their absence from Table 5.9 are ZrO_2 and HfO_2. Both Zr^{4+} and Hf^{4+} are slightly too small for CN = 8 and the binary oxides crystallize in the monoclinic ($P2_1/c$) baddeleyite structure, although polymorphism exits with ZrO_2. In the baddeleyite structure, the cation has CN = 7 as illustrated in Figure 5.22a [104]. This polyhedron has no symmetry and the seven Zr—O bond distances vary from 2.051 to 2.285 Å; however, it can be regarded as a cube with one missing corner. While this is the low temperature form, ZrO_2 transforms at ambient pressure to a tetragonal structure ($P4_2/nmc$) beginning at about 1000 K via a first-order transition with a broad hysteresis [105]. Here, Zr has CN = 8 in the form of a distorted cube. The local symmetry at the Zr site is -4m2 or D_{2d} as shown in Figure 5.22b and there are two Zr—O distances at 2.103 Å and 2.035 Å. HfO_2 does not undergo this phase transformation.

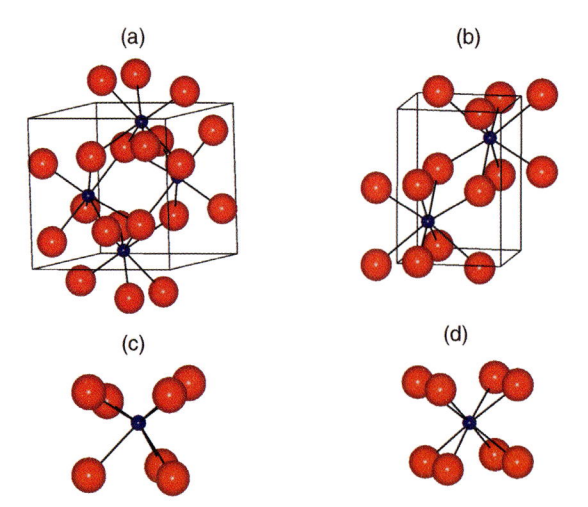

Figure 5.22 (a) The baddeleyite ($P2_1/c$) unit cell for ZrO_2 and HfO_2. (b) The tetragonal ($P4_2/nmc$) unit cell of the high temperature form of ZrO_2. (c) The CN = 7 polyhedron in baddeleyite. (d) The distorted CN = 8 cubic polyhedron.

Upon substitution of about 10% Ca^{2+} or Y^{3+} or indeed nearly any trivalent lanthanide for Zr^{4+}, one obtains the well-known calcia-(CSZ) or yttria-(YSZ) stabilized zirconia materials that have the cubic CaF_2 structure but with of course vacancies on the oxide ion sites, that is, $Ca_xZr_{1-x}O_{2-x}$ and $Y_xZr_{1-x}O_{2-x/2}$ [106]. These compounds find application as oxide ion conductors in various devices including high-temperature fuel cells [107].

5.3.4.2 Ordered Defect Fluorite Strutures; -Bixbyite, Pyrochlore, and Weberite

In CSZ and YSZ, the average structure as determined by conventional diffraction methods is cubic $Fm\text{-}3m$, implying that the vacancies on the oxide sites are disordered, although the local structure is likely more complex. There exists a large family of TMO with structures that can be derived from fluorite with ordered oxide ion vacancies.

Three possibilities are shown in Figure 5.23. These are the bixbyite, pyrochlore, and weberite structures and all involve high levels of substitution of trivalent and or pentavalent ions for tetravalent ions on the cation site of CaF_2. All three can be regarded as superlattices of fluorite, as the unit cell parameters are simple multiples of the fluorite cell constant a_F.

Bixbyites

The mineral from which this family takes its name has the composition $Mn_{2-x}Fe_xO_3$. Pure $Mn_{2-x}Fe_xO_3$ is evidently orthorhombic, but very small levels of Fe induce the ideal cubic lattice symmetry, which is now $Ia\text{-}3$. The unit cell is doubled over that of the parent fluorite cell. The general composition is A_3BO_6. Often the A and B cations are the same and thus, this reduces to A_2O_3. Bixbyite is the most stable structure for the larger TM^{3+} ions such as the lanthanides and

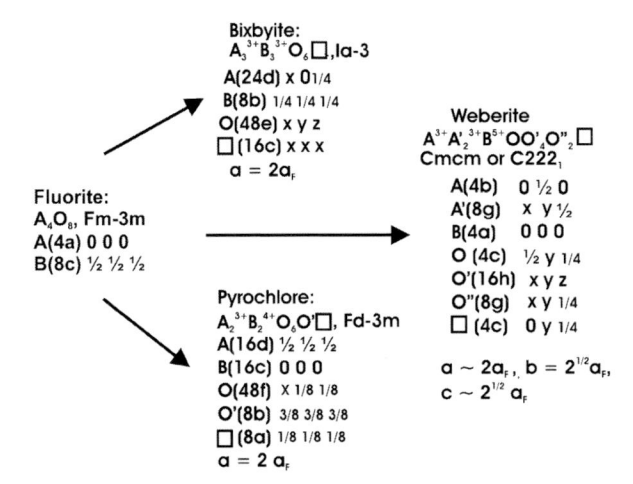

Figure 5.23 Three structural families, which result from the ordering of oxide ion vacancies, □, within the fluorite structure. The space groups are given along with atomic positions and Wyckoff symbols for both the occupied and vacant sites. (This figure is reproduced from the Journal of Materials A (2013) 1, 10487.)

Table 5.10 Some TMO with the bixbyite structure, Ia-3.

TMO	a (Å)	Reference	TMO	a (Å)	Reference
Sm_2O_3	10.920	[109]	Tm_2O_3	10.448	[110]
Eu_2O_3	10.859	[109]	Yb_2O_3	10.4345	[110]
Gd_2O_3	10.8170	[111]	Lu_2O_3	10.391	[109]
Tb_2O_3	10.7281	[112]	Sc_2O_3	9.846	[113]
Dy_2O_3	10.6706	[114]	Am_2O_3	11.03	[115]
Ho_2O_3	10.606	[114]	Cm_2O_3	11.006	[116]
Y_2O_3	10.604	[110]	Bk_2O_3	10.889	[101]
Er_2O_3	10.536	[117]	Cf_2O_3	10.839	[103]

some actinides. In Table 5.10 some bixbyite structure TMO are listed. The bixbyite unit cell contains 80 atoms. There are two TM sites in the ratio of 24/8 = 3 and the CN of each must be six, as dictated by the formula. The coordination polyhedra are essentially the same and are illustrated in Figure 5.24. It is common to describe these in terms of the original cubic coordination and the location of the two oxide vacancies per cube, although a description in terms of a highly distorted octahedron is also correct. In terms of the former, the two vacancies per cube are located on opposite ends of a body diagonal. These defect cubic units form an ordered pattern in the full bixbyite unit cell as has been illustrated elsewhere [108].

Notice that for the lanthanide sesquioxides, the bixbyite structure is listed only for those with smaller radii. The binary lanthanide oxides are polymorphic and two other structures are found for TM = La to Nd, the so-called A (P-3m1) and B (C2/m) types. In these structures the cation sublattice is better described as HCP rather than CCP as in fluorite. A discussion of these structures can be found in Ref. [118].

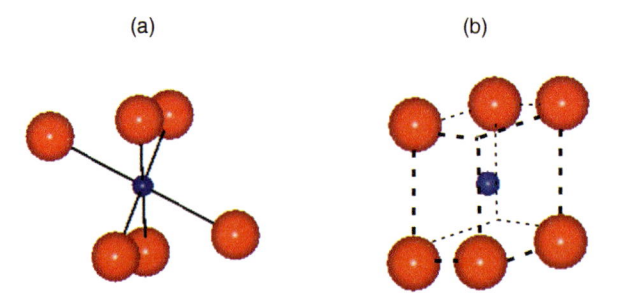

(a)

(b)

Figure 5.24 The cation coordination polyhedra for the 8a site in bixbyite. (a) The distorted octahedron. All bond distances are equal and the bond angles are roughly 80° and 100° rather than the ideal 90°. (b) The cube with two anion vacancies along one body diagonal of the cube. The polyhedron for the 24d site is essentially the same but more distorted.

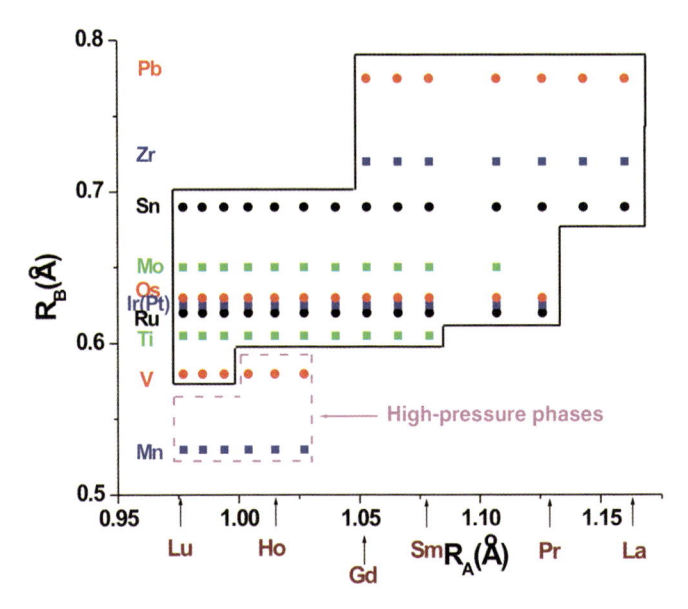

Figure 5.25 A structure field map for the stability of pyrochlore structure TMO with the A^{3+}/B^{4+} combination. The phases within the solid lines are stable at ambient temperature and pressure. High-pressure pyrochores have been reported recently for $B = Ge^{4+}$ and $A = Er^{3+}$, and other lanthanides [124].

Pyrochlores

This is certainly the largest family of defect fluorites with at least 100 examples. Here, there are two cation sites of different formal charge (3+ and 4+ or 2+ and 5+) and radii. The vast majority falls into the 3+/4+ group and the A^{3+} cation is usually a lanthanide and the B^{4+} cation a transition element. The crystal chemistry and physical properties have been described in several review articles and no attempt at listing all of the known examples can be made here [119–123]. Instead, a recent structure-field map for the $A^{3+}{}_2B^{4+}{}_2O_7$ family is shown in Figure 5.25. Note that in general the pyrochlore structure is not stable for a given B ion for all lanthanide A ions. The only exception is for $B = Sn^{4+}$.

Most pyrochlore TMO have the formula $A_2B_2O_7$ and the A^{3+} and B^{4+} or A^{2+} and B^{5+} ions order crystallographically as indicated in Figure 5.23. The unit cell, described in $Fd\text{-}3m$ symmetry, contains 88 atoms. The cation ordering is driven by the very different ionic radii of the A and B ions as suggested in Figure 5.25. The B ions seek CN = 6 octahedral coordination while the larger A ions have CN = 8 in a very unusual coordination geometry best described as a slightly distorted hexagonal bi-pyramid as shown in Figure 5.26. The axial A—O bond is one of the shorter known in lanthanide oxide chemistry. There is thus a strong axial component to the crystal field seen by the A-site cation. This has major implications for the magnetic properties of pyrochlore oxides containing lanthanides on the A-site.

Y - O2 = 2.215
Y - O1 = 2.452

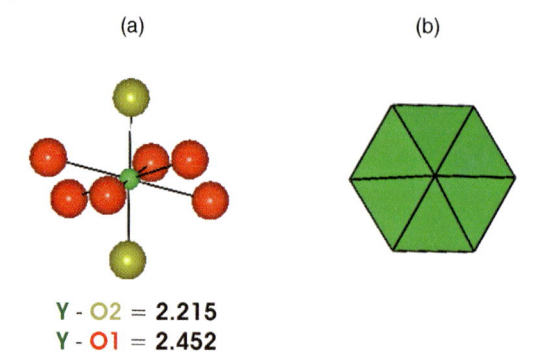

Figure 5.26 (a) The coordination polyhedron of the A-site in pyrochlore that can be described as a distorted hexagonal bipyramid. The O2 ions form a pattern similar to the "chair" configuration of cyclohexane and the A—O2 bonds are normal to the average plane of the A—O1 bonds. The values quoted (in Å) are for $Y_2Mo_2O_7$. (b) A view of the polyhedron parallel to the -3 axis.

As mentioned, the B-site CN = 6 and is a distorted octahedron compressed slightly along the -3 axis. Both the A (16d) and B (16c) sites have -3 (D_{3d}) symmetry. Each site is half of a CCP lattice that can be best appreciated by comparing a projection of the pyrochlore and fluorite unit cells as in Figure 5.27. The main difference between the two structures is the location of the oxide ions, which is driven by the differences in radii between the A- and B-site ions. Individually, the A- and B-site sublattices display a remarkable topology that consists

CaF_2

Pyrochlore

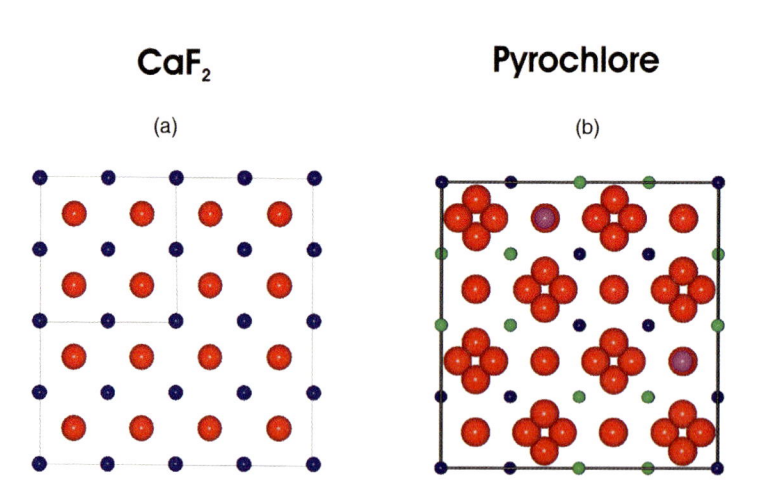

Figure 5.27 Projections along a (100) axis for the (a) CaF_2 (doubled cell) and (b) pyrochlore structures. Blue and green spheres are the cations and red spheres are the oxide ions. It is easily seen that in pyrochlore the A and B sites taken together form a FCC lattice. The main difference between the two structures is the location of the oxide ions, which is driven by the difference in radii between the A and B site ions in pyrochlore.

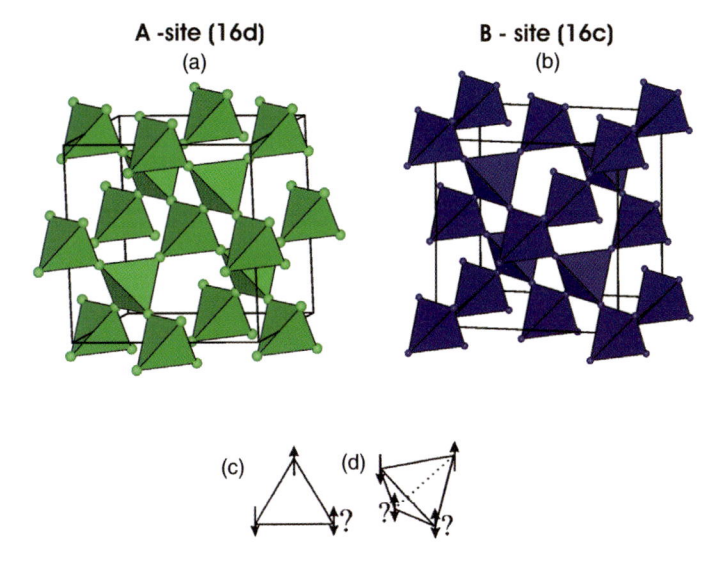

Figure 5.28 The topology of the (a) A (16d) and (b) B (16c) sublattices in the pyrochlore structure, each consisting of a three-dimensional array of corner sharing tetrahedral. At the bottom is illustrated the geometric magnetic frustration inherent in such a topology for spins on a triangle (c) or a tetrahedron (d).

of a three-dimensional array of corner sharing tetrahedra (Figure 5.28). This feature has enormous implications for the magnetic properties of pyrochlore oxides due the inherent geometric magnetic frustration associated with a tetrahedron of spins, each subject to antiferromagnetic exchange, as also illustrated in Figure 5.28. An extensive literature on the remarkable magnetic properties of pyrochlore oxides exists that continues to grow at a rapid pace [122].

A further important feature of the pyrochlore structure is displayed in Figure 5.29 in which the corner sharing connectivity of the $BO(1)_6$ octahedra is emphasized. Both very small triangular and large hexagonal tunnels are formed. The A ions reside in the geometric centers of the hexagonal tunnels and are coordinated linearly by the O(2) ions which form zigzag chains. The O(2)—A—O(2) angle is 109.7°. The B—O(1)—B angle is near 130° and does not vary greatly with the radius of the A ion.

In addition to lanthanide ions, the A^{3+} site can be occupied by Bi^{3+} or Tl^{3+}, such as $Bi_2Ru_2O_7$, $Bi_2Ti_2O_7$, and $Tl_2Ru_2O_7$ [125]. As well, the B^{4+} site can be occupied by B^{3+}/B^{5+} combinations such as $A_2Cr^{3+}Sb^{5+}O_7$ or $A_2Fe^{3+}Sb^{5+}O_7$ [126]. There is no evidence for crystallographic ordering on the B site in these materials.

There also exist two other pyrochlore families in addition to the $A^{3+}_2B^{4+}_2O_7$ group. A few $A^{2+}_2B^{5+}_2O_7$ pyrochlores as well as a small number of "defect" pyrochlores $A^{2+}_2B^{4+}_2O_6$ and $AB^{5.5+}_2O_6$ are known. For the $A^{2+}_2B^{5+}_2O_7$ pyrochlores, A^{2+} is most often Cd^{2+} and more rarely Pb^{2+}. The B^{5+} ion is usually a 4d or 5d

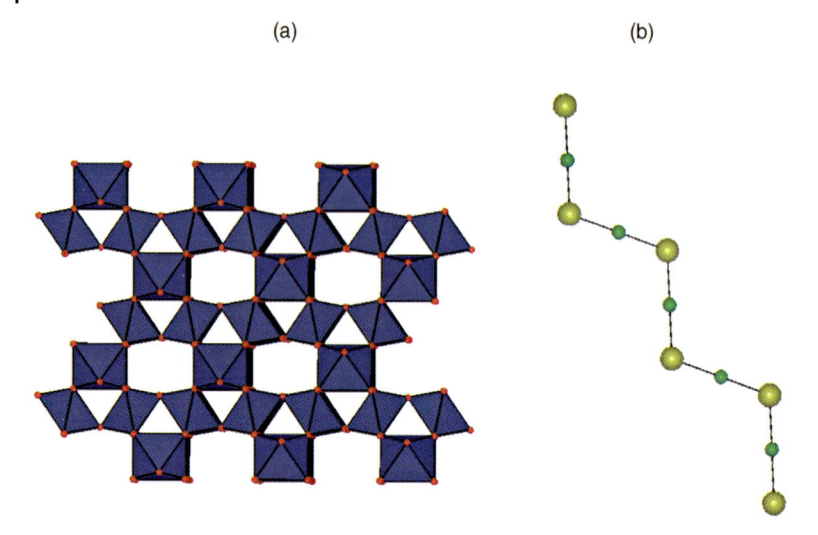

Figure 5.29 (a) The B—O(1) sublattice consists of corner sharing B—O(1) octahedra which form triangular and hexagonal tunnels parallel to (110) directions. (b) The A—O(2) sublattice consists of zigzag chains with the A ions residing in the hexagonal tunnels. The A ions are green and the O(2) ions are gold.

element such as Nb, Ru, Ta, Re, and Os [127]. Regarding the defect pyrochlores, it can be inferred from the discussion surrounding Figure 5.29 that the stability of the pyrochlore structure is due mainly to the rigid corner sharing $BO(1)_3$ network. Thus, both the O(2) and A-sites can be vacant with retention of the basic pyrochlore framework. Phases such as $Pb_2Ru_2O_{7-x}$ with $x=0.5$ as well as $Pb_2Re_2O_6$ and $Pb_2Ir_2O_6$ [128] are known. Especially for materials with A = Bi or Pb, small vacancy levels can appear on the A-site, for example, $Pb_{1.45}Nb_2O_{6.26}$ and $Bi_{1.92}Ir_2O_{6.82}$ [129].

For the cases where A^+ is present, the normal A-site (16d) is vacant and the 8b site is now occupied by the A^+ ion [130]. In these examples, A^+ is a large Group 1 ion such as K^+, Rb^+, or Cs^+ and the B-site, still 16c is Os that must exist in a mixed valent 5+/6+ state. This family of defect pyrochlores, exemplified by KOs_2O_6, has been of much interest recently as superconductivity has been observed with T_c near 10 K [130].

A final type of defect pyrochlore is represented by the so called "stuffed pyrochlores" [131]. These occur for certain lanthanide titanates where the difference in radii between the normal A-site ion and Ti^{4+} is small. In these cases, the A^{3+} (A = Dy, Ho, Yb) ion can substitute on the B-site, resulting in formulae such as $A_2(A_xTi_{2-x})O_{7-x/2}$. The limit is $x=2/3$, giving an end member of composition $A_{8/3}Ti_{4/3}O_{20/3}$ which reduces to A_2TiO_5.

This structure is still described in *Fd-3m* with the 16c, 16d, 48f, and 8b appropriately occupied.

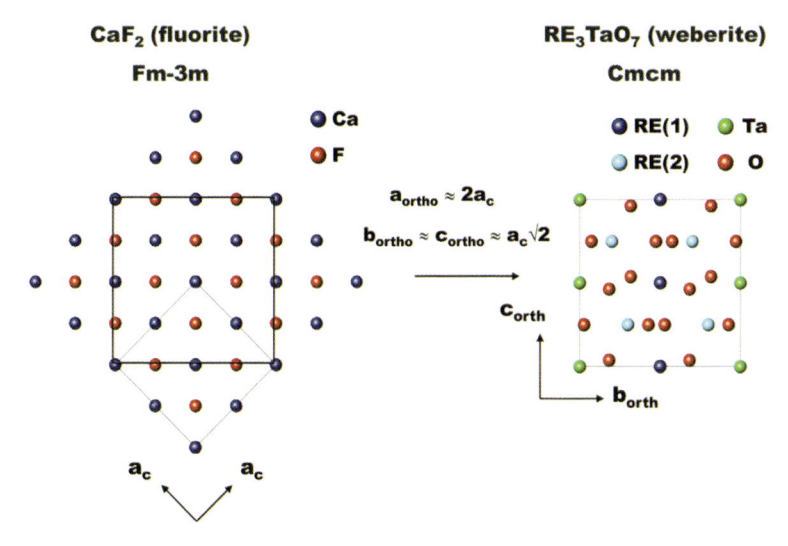

Figure 5.30 (100) Projections of the CaF$_2$ (left) and AA'$_2$BO$_7$ weberite (right) unit cells showing the structural relationship.

Weberites

This is also a large family of TMO. It should be noted that there are probably more weberites in fluoride chemistry than among oxides. The structural situation has been reviewed recently [132]. In this section, TMO with TM on both the A- and B-sites will be emphasized, although many examples exist with other combinations. For the most part, the compositions of interest will involve A^{3+} lanthanide ions and B^{5+} ions from the 4d and 5d series, such as Nb, Ta, Ru, and Os, and formulae A^{3+}A$_2'^{3+}$B^{5+}O$_7$. Often, A and A' are the same element. In general, these phases crystallize in one of two orthorhombic space groups, *Cmcm* or *C*222$_1$. Weberites are closely related compositionally to pyrochlores. A structure-field map involving both weberites and pyrochores published in Ref. [132] shows that the weberite structure is favored for larger A-site ions. As has been demonstrated, see Figure 5.27, the total cation sublattices for both bixbyite and pyrochlore are FCC, this is also nearly the case for weberite in spite of the lower crystal symmetry. Figure 5.30 shows projections along (100) for both fluorite and weberite cells. Typical atomic positions for both are given in Figure 5.23. The A (dark blue) and B (green) sublattices conform to the FCC pattern while the A' ions (light blue) are shifted from ideal positions. Of course the oxide ion (red) positions are strongly shifted as well. For this type of weberite, there are three cation coordination geometries as illustrated in Figure 5.31. For the A-site, this is a distorted cube (CN = 8), for the A'-site a distorted pentagonal bipyramid (CN = 7), and the B-site is octahedral (CN = 6). These octahedra share two trans corners forming zigzag chains along the c-axis. If B is the only magnetic ion as, for example, if A = A' = Y or La, and B = Ru or Os, such materials exhibit one-dimensional magnetic behavior [133].

cation sites in weberite AA'$_2$BO$_7$

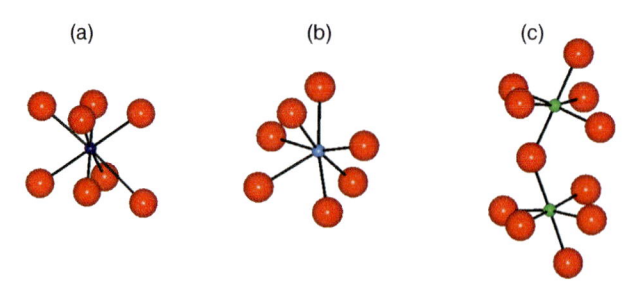

(a)   (b)   (c)

A: CN = 8   A': CN = 7   B: CN = 6
distorted cube   distorted pentagonal   zig-zag octahedral
               bipyramid   chains

Figure 5.31 The three-cation coordination polyhedra in AA'$_2$BO$_7$ weberites.

5.3.4.3 Other Defect Fluorite Structures

Bixbyites, pyrochlores, and weberites do not exhaust the possibilities for defect fluorites. Other structures are found especially in the systems Ce—O, Pr—O and Tb—O, where these lanthanides exhibit stable 3+ and 4+ valences. In fact under ambient conditions, the stable oxides for Pr and Tb are Pr_6O_{11} and Tb_4O_7, not PrO_2 or TbO_2. Such systems have been studied intensively and it is found that a large number of phases with fixed oxygen stoichiometries exist as a function of temperature and oxygen partial pressure. These have apparently nonstoichiometric formulae such as $MO_{1.818}$ or $MO_{1.833}$, implying the presence of high concentrations of oxide ion vacancies, yet are found to be perfectly ordered structures. These phases form sets of homologous series which for M = Pr are given as Pr_nO_{2n-2m} with $m=1$, 2, and 3. Bizarre formulae and large unit cells are found. $Pr_{40}O_{72}$, for example, is monoclinic, $P2_1/c$, with $a=6.73$ Å, $b=19.39$ Å, $c=12.79$ Å, and $\beta=100.2°$. The interested reader is directed to a recent review [118].

5.3.5
Perovskite and Related Structures

5.3.5.1 ABO$_3$

Almost certainly, the most intensively studied TMO are those with the perovskite structure. Entering "perovskite" into a database search engine will yield nearly 10^5 hits. This vast family takes its name from the mineral perovskite, nominally $CaTiO_3$, which was discovered by a Swiss mineralogist, Gustav Rose, in the Ural mountains and named for a Russian member of the Academy of Sciences, Perovski Lev Alekseevic. The general formula is ABO_3 with A being a large radius cation and B a smaller radius cation.

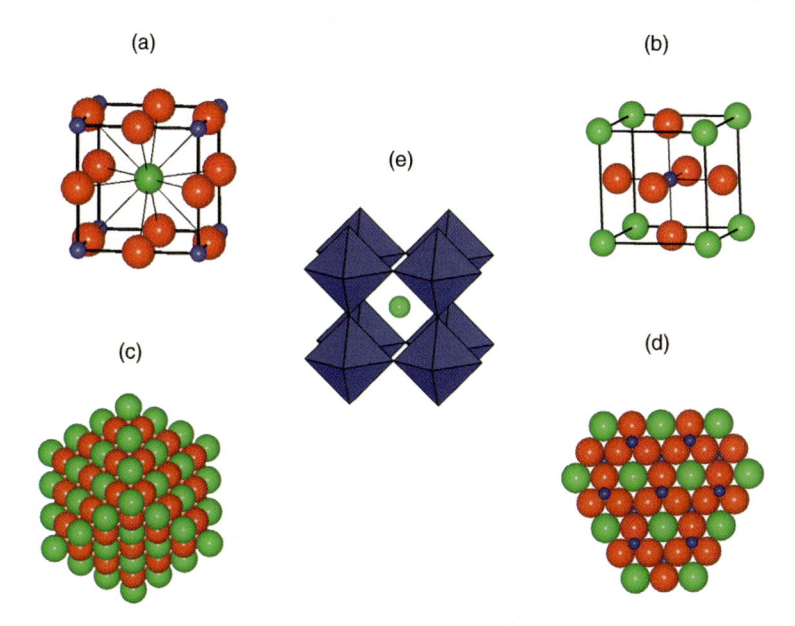

Figure 5.32 The cubic (*Pm-3m*) perovskite structure. (a) Setting 1, showing CN = 12 cubo-octahedral for the A site. (b) Setting 2, showing CN = 6 octahedral coordination for the B-site. (c) Layers of AO$_3$ (A – green, O – red) viewed parallel to (111) directions showing the CCP stacking. (d) Ordered occupation of 1/4 of the available octahedral sites by the B (blue) cation. (e) Corner sharing three-dimensional network of BO$_6$ octahedra.

The ideal or aristotypic structure is cubic *Pm-3m* and is illustrated in a number of ways in Figure 5.32. This structure can be described using two settings in *Pm-3m*. In setting 1 B is in 1a (000), A is in 1b (1/2 1/2 1/2), and O is in 3d (1/2 0 0) while for setting 2, B is in 1b, A is in 1a, and O is in 3c (1/2 1/2 0). The AO$_3$ sublattice forms a CCP stacking sequence and the B cations occupy 1/4 of the available octahedral sites in an ordered manner directed by the presence of the A cation in the CCP layers. Recall a similar situation with structures related to corundum where the mixed cation/oxide layers are HCP or mixed stacking sequences.

Conditions for the stability of the perovskite structure were stated in 1926 by the Swiss-born, Norwegian geochemist, Victor M. Goldschmidt in the form of what is now universally known as the Goldschmidt tolerance factor *t*:

$$(A\text{—}O)/\sqrt{2}(B\text{—}O) = t$$

where A—O and B—O are the relevant interionic distances [134]. These can also be expressed in terms of the radii involved, $(A\text{—}O) = r(A^{n+}) + r(O^{2-})$ and $(B\text{—}O) = r(B^{n+}) + r(O^{2-})$. It is observed that perovskite is stable over the range $\sim 0.9 < t < 1.0$, but the actual values vary with choice of radii used. As will be discussed in a later paragraph, cubic *Pm-3m* symmetry is not conserved over the entire range but the structures formed are clearly recognizable as perovskites.

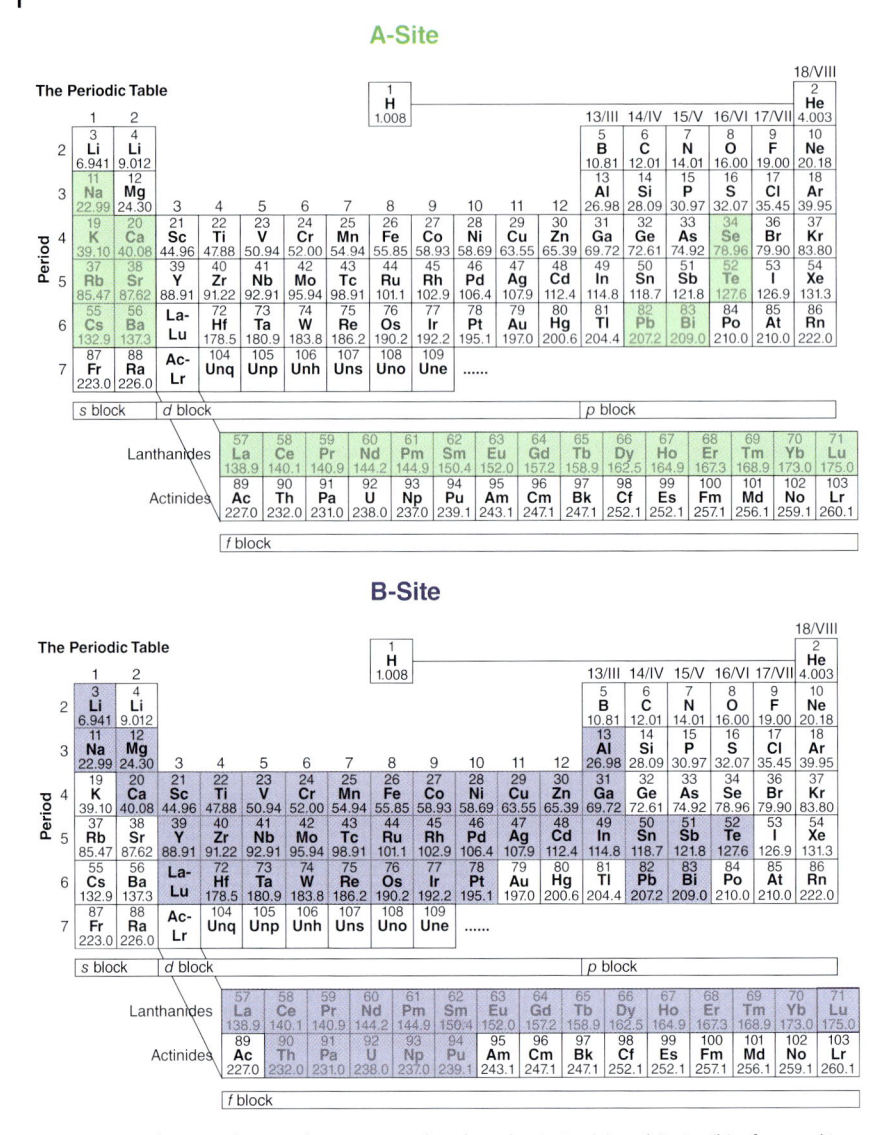

Figure 5.33 Elements that can be accommodated on the A-site (a) and B-site (b) of perovskite structure oxides under ambient pressure conditions.

Perhaps, this is somewhat surprising but the perovskite structure can accommodate ~80% of the stable elements in the periodic table as indicated in Figure 5.33. There is, thus, enormous chemical flexibility at one's disposal in the design of perovskite materials that accounts in part for the extraordinary interest in this class of TMO, as already noted. Additionally, perovskite oxides play a central role in many important areas of both basic and applications research. An attempt to illustrate this point is shown in Table 5.11.

Table 5.11 Basic and applied research topics in which TMO perovskites are featured.

Selected examples	
Topic	TMO perovskites
High T_c superconductivity	$La_{1.85}Ba_{0.15}CuO_4$, $YBa_2Cu_3O_7$
Colossal magnetoresistance, charge, and orbital ordering	$La_{1-x}Ca_xMnO_3$, Sr_2FeMoO_6
Ferroelectricity, relaxors	$BaTiO_3$, $PbTiO_3$ - $Pb(Zn_{1/3}Nb_{2/3})O_3$
Mott/Hubbard physics Metal/insulator transitions	$Nd_{1-x}TiO_3$, $LaNiO_3$
2D magnetism	La_2CuO_4
Geometric magnetic frustration	Ba_2YMoO_6
Fuel cell cathodes	$La_{1-x}Sr_xMnO_3$, $La_{0.6}Sr_{0.4}Co_{0.2}Fe_{0.8}O_{3-x}$

As already mentioned, perovksites exist in many symmetries other than the ideal *Pm-3m*. Glazer was the first to demonstrate that in fact 14 symmetries can result if one takes into account all of the possible tiltings of the octahedra of Figure 5.32e about each of the three crystallographic axes [135]. There are 23 different tilt possibilities that are described by what are now known as the Glazer symbols. Each has six components, three of which, a, b, and c, refer to the three crystallographic axes and each has a superscript that gives information about the relative magnitudes of the three possible tilt angles.

An example is presented in Figure 5.34 for the cases of $a^0a^0c^+$ and $a^0a^0c^-$ [136]. These describe tetragonal symmetries and the view is parallel to the unique c(z) axis, that is, in the (001) plane. Both have a^0a^0 that means that there are zero tilt angles with respect to these two axes, a necessary requirement to preserve tetragonal symmetry. There is, however, a tilt about c(z) which forces the other octahedra in (001) to rotate in the manner shown. Whether the system

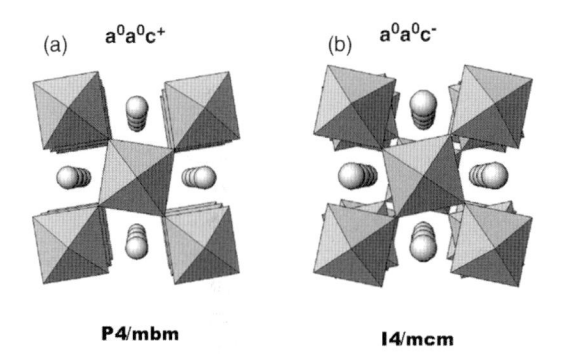

Figure 5.34 The (a) $a^0a^0c^+$ and (b) $a^0a^0c^-$ Glazer tilt systems. (Reproduced with the permission from Ref. [136]. Copyright 2001, International Union of Crystallography.)

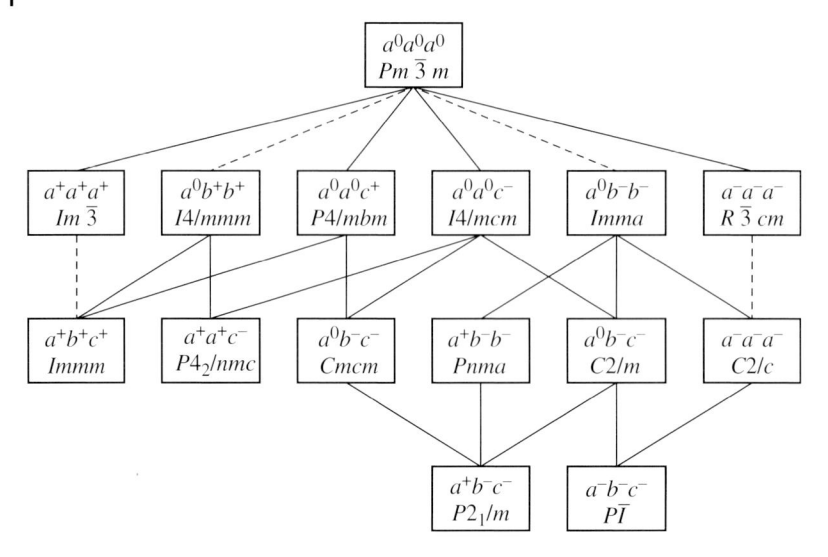

Figure 5.35 The space group "family tree" for perovskites with a single B-site ion based on Glazer's octahedral tilt systems. (Reproduced with the permission from Ref. [135]. Copyright 1972, International Union of Crystallography.) Note that a center of symmetry is preserved in all cases.

is c^+ or c^- is determined by the tilts of the octahedral in the adjacent layers. If these tilt by the same angle in the same sense, then it is c^+, if they tilt by the same angle but in the opposite sense, we have c^-. The full symmetry "family tree" for primitive perovskite tilt systems is shown in Figure 5.35. It should be emphasized that not all distortions from Pm-$3m$ symmetry are covered in Figure 5.35, as the resulting space groups are all centrosymmetric, that is, no allowance is made for distortions within the octahedra themselves. In $BaTiO_3$, the famous ferroelectric material, the Ti—O octahedron is distorted such that Ti is not at the center of the octahedron and the resulting space group is $P4mm$.

The distribution of tilt systems among the hundreds of known perovskites with a single B-site ion is displayed in Figure 5.36 [136]. By far, the most common is $a^+b^-b^-$ and permutations, corresponding to $Pnma$ (#62). In general, the tilts are driven by the degree to which the A cation fits into the large 12-fold site. As the relative size of the A cation decreases, as tracked by a decrease in the tolerance factor from unity, the reality of a smaller CN for A causes the tilting of the B-centered octahedra. For example, the CN for the A ion in $Pnma$ materials such as $GdFeO_3$ approaches 8 or 9, depending on the cutoff bond distance for coordination. This situation is illustrated in Figure 5.37 for isostructural $NdTiO_3$. A major consequence of octahedral tilting is the reduction in the B—O—B angle from 180° in Pm-$3m$ symmetry to as low as ~140°. Properties, such as magnetic super exchange and bandwidth, are strongly influenced. For example, in the series $LnTiO_3$, tuning of this angle from ~160° in $LaTiO_3$ to ~140° in $YTiO_3$, changes the sign of the magnetic exchange from negative (antiferromagnetic) to positive (ferromagnetic) [137].

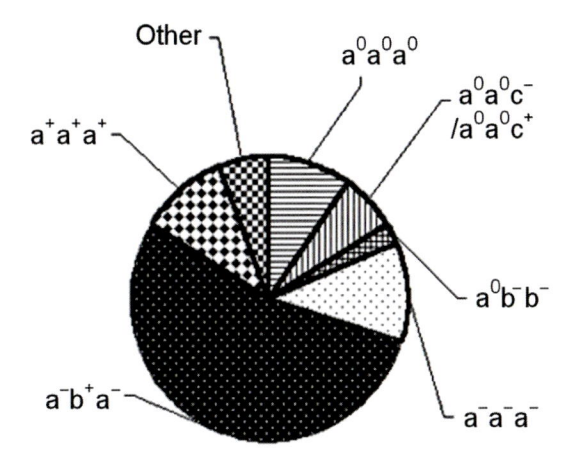

Figure 5.36 The distribution of tilt systems among known perovskites with a single B-site ion. (Reproduced with the permission from Ref. [136]. Copyright 2001, International Union of Crystallography). Note that $a^-b^+a^-$ and $a^+b^-b^-$ are equivalent upon permutation of the orthorhombic axes.

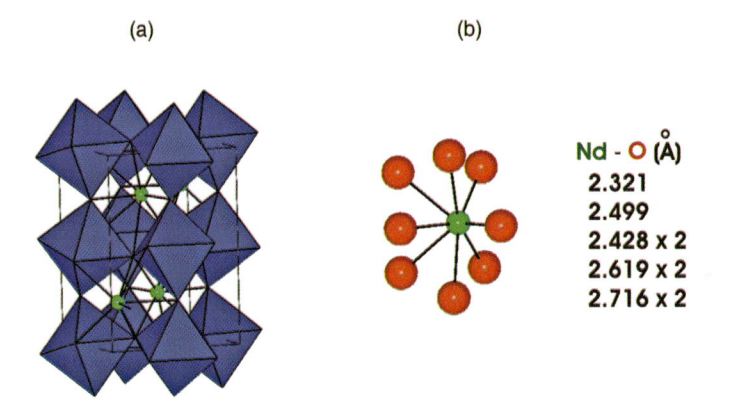

(a)   (b)

Nd - O (Å)
2.321
2.499
2.428 x 2
2.619 x 2
2.716 x 2

Figure 5.37 (a) The unit cell of NdTiO$_3$ (Pnma) $a^-b^+a^-$. (b) The Nd—O coordination polyhedron with CN = 8 taken up to a Nd—O limit of 2.90 Å [138].

Thus in general, one should expect a descent in symmetry according to Figure 5.35 as t decreases from unity, but no universal pathway is known and the details depend on the oxide in question. At this stage, some very useful software packages developed for the study and prediction of new perovskite phases based on the tilting phenomenon, developed by Woodward et al., should be mentioned. These are POTATO and SPuDS [136,139]. (Professor Woodward is a native of Idaho). The interested reader is strongly encouraged to become familiar with these.

5.3.5.2 Perovskite Related Structures

While there are literally hundreds of perovskites of the type illustrated in Figure 5.32, that is, with a single B-site cation, a single A cation and no vacancies

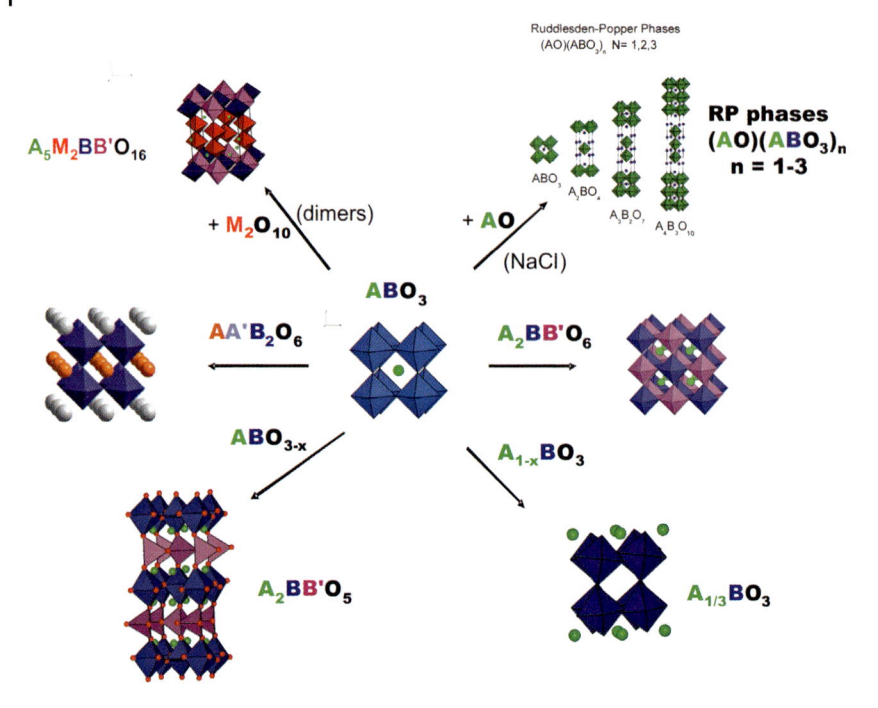

Figure 5.38 Some of the structural families that can be derived from ABO_3 perovskites by various means: site ordering, vacancy creation on the A, O, and B sites, and intergrowth with other structures.

on any site, the basic perovskite structure can be modified or engineered to produce many new families of TMO. An attempt to illustrate at least a few of these possibilities is shown in Figure 5.38.

This figure emphasizes three basic pathways to new structures: (1) site ordering on the A- and B-site, (2) vacancy creation and ordering on the A- and O-sites, and (3) intergrowth with other structure types.

Before discussing these in some detail, it is important to return, briefly, to the simplest perovskites, that is, those with a single A-site and B-site cation or, allowing for multiple occupation of each site by various elements, those which show no indication of site ordering. The possible valence combinations are A^+/B^{5+}, A^{2+}/B^{4+}, and A^{3+}/B^{3+}. Typical examples are $KTaO_3$, $SrTiO_3$, and $GdFeO_3$. There are few, if any, examples of A^{4+}/B^{2+} or A^{5+}/B^+ oxide perovskites. Examination of Figure 5.33 suggests hundreds of possible combinations, including those with aliovalent or nonaliovalent substitutions on each site such as $Sr_{1-x}Ba_xTiO_3$ or $Sr_{1-x}La_xTiO_3$ or even $La_{0.6}Sr_{0.4}Co_{0.2}Fe_{0.8}O_3$. However, when site ordering occurs on either the A- or B-sites, new structures and symmetries result.

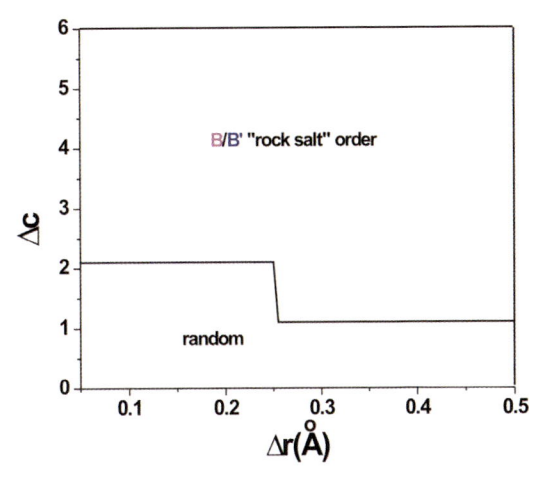

Figure 5.39 A phase diagram for $A_2BB'O_6$ double perovskites showing the existence range for the ordered "rock salt" structures.

B-Site-Ordered Double Perovskites

This is probably the largest family of Figure 5.38 and a recent review has appeared which indicates that more than 10^3 such materials exist [140]. These are oxides with composition $A_2BB'O_6$ or $AA'BB'O_6$, assuming no A-site ordering. When B and B' order as in Figure 5.38, each forms a FCC lattice and the perovskite cell is doubled along all axes, hence the name "double perovskites." The criteria for B-site ordering were stated in the 1990s in terms of the difference in B/B' cation formal charge Δc and radii Δr [141]. A phase diagram, modified by more recent results is shown in Figure 5.39. Generally, if $\Delta c = 1$, there is no or only partial ordering. For $\Delta c = 2$, full ordering occurs for $\Delta r > 0.26$ and if $\Delta c = 3$ or greater ~100% order should always be seen. As Figure 5.39 is a rough guide, it is always advisable to check each individual case by either refinement of site occupations from diffraction data or in favorable cases local probes such as MAS NMR [142]. Partial B/B'-site order is possible and often found [143].

As for the case of the "single" perovskites, several different symmetries can result from octahedral tilting as demonstrated in Figure 5.40 [144]. There are 12 possible space groups. The Glazer tilt notation is also given. In this case, the aristotype is $Fm\text{-}3m$ and as the tolerance factor decreases from unity lower symmetry space groups are found. Only two of these are common, $I4/m$ and $P2_1/c$, usually in the $P2_1/n$ setting. For example, Ba_2LiOsO_6 ($t = 1.04$) is $Fm\text{-}3m$, Sr_2MgReO_6 ($t = 0.98$) is $I4/m$ and Sr_2CaReO_6 ($t = 0.92$) is $P2_1/n$ [145]. The valence permutations for these double perovksites include, for A^{2+}: $B^+B'^{7+}$, $B^{2+}B'^{6+}$, and $B^{3+}B'^{5+}$, for A^{3+}: $B^+B'^{5+}$ and $B^{2+}B'^{4+}$ and for $A^{2+}A'^{3+}$: $B^+B'^{6+}$ and $B^{2+}B'^{5+}$. See Ref. [141] for more details, there are about 650 such materials known.

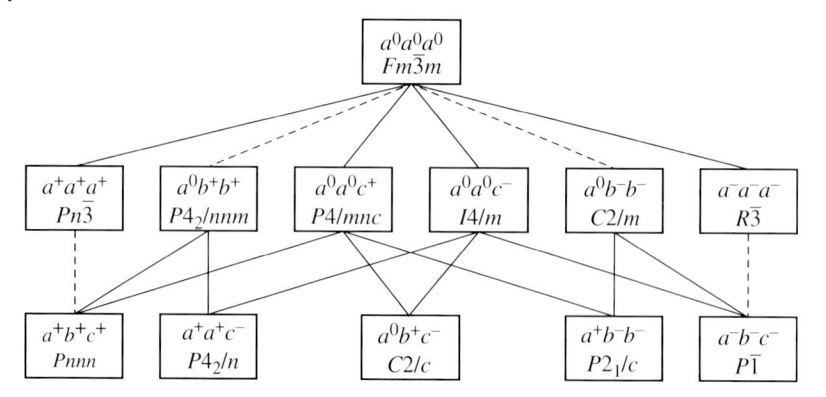

Figure 5.40 The space group "family tree" for double perovskites $A_2BB'O_6$. (Reproduced with the permission from Ref. [144]. Copyright 2003, International Union of Crystallography).

B-site-ordered double perovskites are of interest in at least two areas. First, when only B′ is a magnetic ion, the magnetic sublattice is FCC, which was shown earlier, Figure 5.21, to be an array of edge-sharing tetrahedra. As for the pyrochlore lattice in which the tetrahedra share corners, this leads to geometric magnetic frustration for antiferromagnetic exchange (Figure 5.28). Unprecedented magnetic properties can be found, for example, Ba_2YMoO_6, instead of showing simple antiferromagnetic order, has a magnetic ground state consisting of a gapped collective singlet [146]. If both B and B′ are magnetic as in Sr_2FeMoO_6, Sr_2FeReO_6, Ba_2FeReO_6 and many others, ferrimagnetic interactions are found with ordering temperatures well above ambient. These materials are usually metallic and show strong spin polarization and so-called half-metallic magnetism. This means that the carriers of the current have a specific spin that is of great interest for the developing science and technology of spintronics [3].

A-Site-Ordered Perovskites
A-site ordering is more difficult to achieve than B-site ordering and there are fewer examples. A review and analysis has recently appeared [147]. This paper also discusses B-site ordering, including two very rare types not described in the preceding section. By far, the most common A-site order results in a layered A/A′ motif as depicted in Figure 5.38. A/A′ rock salt ordering is particularly rare. In part, this is due to the fact that $\Delta c = 1$ in most cases. The chemical bonding requirements for A/A′ site ordering and the factors which favor layered ordering are discussed in detail in Ref. [147] and the reader is strongly encouraged to consult this paper for details. The largest class of materials exhibiting A/A′-site-layered ordering are those in which B/B′ rock salt ordering is also present, that is, doubly ordered double perovskites, $AA'BB'O_6$, the structure of which is exemplified by $NaLaMnWO_6$ ($P2_1$) in Figure 5.41. Common features are $\Delta c = 2$ for the A-site ions and the presence of W^{6+} ($5d^0$) on the B′ site which enables an SOJT induced distortion and a tilt system, $a^-a^-c^+$. Note the distortion of the W polyhedron, typical for a SOJT d^0 ion.

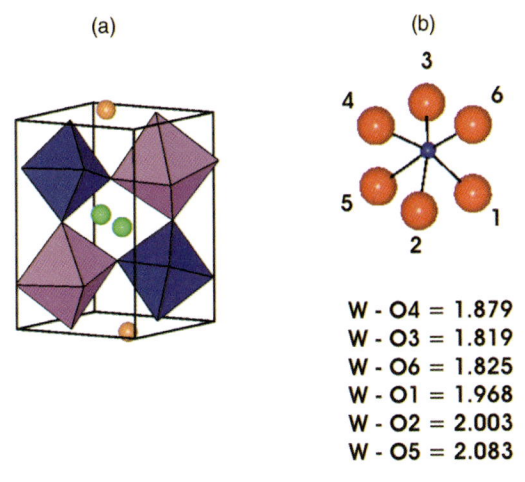

Figure 5.41 (a) The unit cell of the doubly ordered double perovskite NaLaMnWO$_6$ [148]. Na (orange), La (green), Mn (magenta), and W (blue). (b) The W—O octahedron showing bond distances (Å) typical of a SOJT induced distortion.

W - O4 = 1.879
W - O3 = 1.819
W - O6 = 1.825
W - O1 = 1.968
W - O2 = 2.003
W - O5 = 2.083

Even more complex A-site-ordered perovskites have been synthesized at high pressures, notably a class of materials with Cu^{2+} and a group 2 ion ordering on the A-site, such as CaCu$_3$B$_4$O$_{12}$, B = Ti,Sn and Ge [149]. The interested reader is directed to this paper and others by the same author.

A-Site Deficient Perovskites

Stable perovskite related structures can result upon removal of significant fractions of the A-site ion, A$_{1-x}$BO$_3$, with $x = 1$, 2/3 and 1/3, somewhat similar to the case of defect pyrochlores.

BO$_3$ In the most extreme case, one finds BO$_3$ with B = Re and W, for example. ReO$_3$ (*Pm-3m*) based on Re^{6+} (5d^1, t$_{2g}^1$) is a magenta red metal with one of the highest electrical conductivities known among oxide materials, due to broad conduction bands enabled by the presence of 5d orbitals. WO$_3$, 5d^0, on the other hand forms a perovskite like network of corner sharing octahedra but with triclinic symmetry (*P-1*), with 4 W^{6+} sites and 12 O sites [150]. Each W—O polyhedron shows the typical pattern of a SOJT distortion, three long and three short bonds [10]. Interestingly, MoO$_3$, 4d^0, has an entirely different structure unrelated to perovskite featuring layers of edge-sharing octahedra (*Pnma*) [151].

Insertion of Group 1 atoms into WO$_3$ results in the formation of the famous "tungsten bronzes," Na$_x$WO$_3$, in which electrons are now present in the 5d bands and the symmetry depends strongly on x. For example, $x = 0.1$ (*P4/nmn*), $x = 0.33$ (*P-42m*), $x = 0.48$ (*P4/mbm*), $x = 0.54$ (*Im-3*), and for $x = 0.77$ and above (*Pm-3m*) [152]. This series displays a remarkable variation in color and properties with increasing x and has been studied for many years [153].

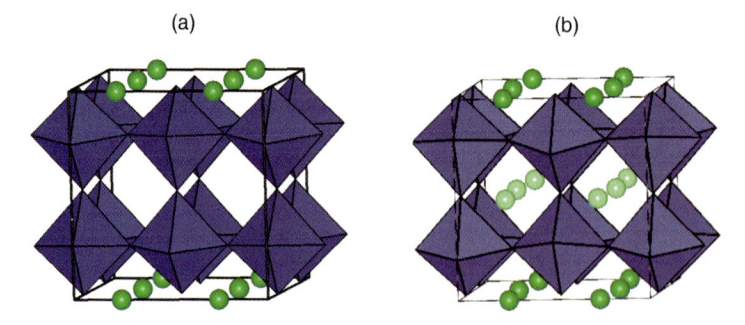

Figure 5.42 (a) The unit cell of $La_{0.33}NbO_3$, La (green), Nb (blue) showing a layer of La vacancies [154]. (b) The unit cell of $Nd_{0.67}TiO_3$, Nd (green and light green), (Ti blue) showing the partial occupation of adjacent layers by Nd (light green) ions phenomena [155]. While both end members are insulators, intermediate phases are metallic [156].

$A_{1/3}BO_3$ and $A_{2/3}BO_3$ These can be discussed together as the unit cell and symmetry are essentially the same, *Cmmm*, with $a \sim 2a_p$, $b \sim 2a_p$, and $c \sim 2a_p$, that is, a doubled cell relative to a primitive perovskite. The structures are shown in Figure 5.42. The only difference is that for $A_{1/3}BO_3$ the A-site is completely vacant in adjacent layers, while for $A_{2/3}BO_3$, there is partial A-site occupation of these layers. In nearly all cases, A is a lanthanide ion and B is an early transition element from the 3d, 4d, or 5d series. Examples are $La_{1/3}NbO_3$, $Ce_{1/3}NbO_3$, $La_{1/3}TaO_3$, $La_{2/3}TiO_3$, $Nd_{2/3}TiO_3$, and so on. These materials are all insulators as the B ion has a nd^0 configuration. Interestingly, there is evidence for only a weak SOJT distortion at the B-site for B = Nb and essentially none for B = Ti. Solid solutions exist for certain titanates between $Ln_{2/3}TiO_3$ and the fully occupied $LnTiO_3$ (*Pnma*) phase which have been studied to probe the details of metal insulator transitions and Mott Hubbard.

B-Site-Deficient Perovskites

There are almost no examples of perovskites with significant levels of B-site vacancies. Two cases are represented by the B-site-ordered double perovskites $Ba_2Y_{2/3}WO_6$ and $Ba_2Y_{2/3}ReO_6$. Both retain *Fm-3m* symmetry [157].

Oxygen Defect Perovskites

While most perovskites containing an element with variable valence on the B-site can exhibit low levels of O-site defects, that is, ABO_{3-x}, and these can have profound effects on physical properties, for example, $SrTiO_{3-x}$ can change from an insulator to a metal with increasing x, these vacancies rarely order in a long range sense to form new structures. For $SrTiO_{2.72}$ ($x = 0.28$), the symmetry changes to *P4/mmm* but the distortion is very slight [158].

Brownmillerites The most important structure type in which oxide vacancies order is that of brownmillerite, $A_2B_2O_5$, in which the vacancy concentration is

5.3 Most Important Structure Types for TMO

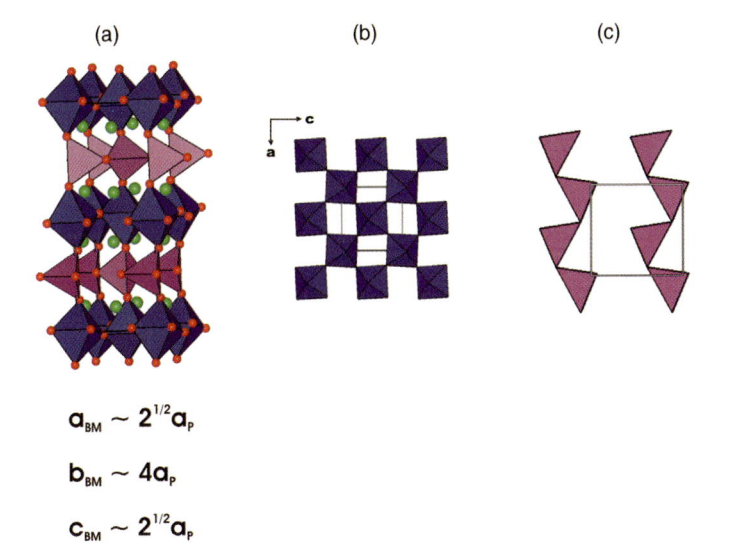

(a)   (b)   (c)

$a_{BM} \sim 2^{1/2} a_P$

$b_{BM} \sim 4 a_P$

$c_{BM} \sim 2^{1/2} a_P$

Figure 5.43 (a) The vacancy ordered brownmillerite structure (*I2mb*). Note the large supercell formed by the vacancy ordering. (b) A view of the perovskite corner sharing octahedral layer along *b*. (c) A view of the corner sharing tetrahedral chain layer along *b*. If the setting *b*>*c*>*a* is chosen, the chain direction is along *a*.

16.67%. A typical orthorhombic brownmillerite unit cell is shown in Figure 5.43. The location of the oxide vacancies creates adjacent layers of corner sharing octahedra and chains of corner sharing tetrahedra. There is a large supercell with the indicated dimensions relative to a primitive perovskite cell. The specific symmetry of any brownmillerite is determined by the relative orientations of the tetrahedral chains and four orthorhombic (*Pcmb*, *Pnma*, *I2mb*, and *Imma*) and one monoclinic (*C2/c*) space groups have been reported. The chains can be regarded as either right (R)- or left (L)-handed and their relative orientations both within one layer and relative to the adjacent layer determine the overall symmetry, as indicated in Figure 5.44 for three orthorhombic space groups: *I2bm*, *Pnma*, and *Pcmb*. In *Imma*, there is disorder within the chains and no long-range orientation can be assigned. In fact, both *Pcmb* and *C2/c* symmetry are quite rare and most brownmillerites show the other three space groups. A crucial feature of the tetrahedral site chains is that the individual tetrahedra are highly distorted, the cation is not in a center of symmetry and, thus, there is a dipole moment. These add vectorially to result in a net chain dipole moment oriented parallel to the chain direction, the magnitude of which depends on the B—O—B angle. The dipole orientation of L- and R chains is exactly opposite assuming the same B—O—B angle [159,160]. Thus, one has a polar space group when the chains have the same orientation within and between the layers, *I2mb*, and a nonpolar symmetry when both R- and L- chains are present as for *Pnma*.

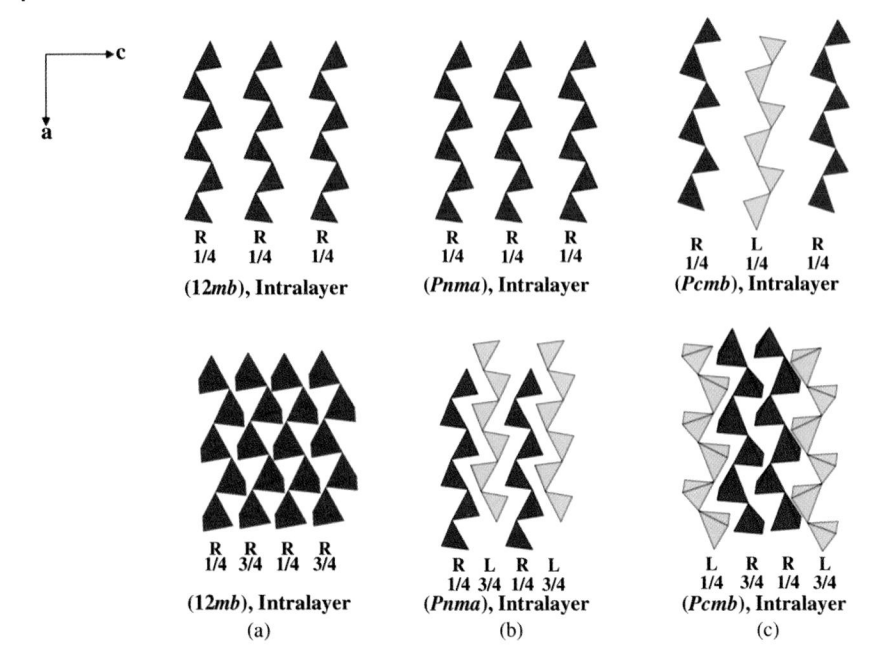

R R R	R R R	R L R
1/4 1/4 1/4	1/4 1/4 1/4	1/4 1/4 1/4
(12mb), Intralayer	(Pnma), Intralayer	(Pcmb), Intralayer

R R R R	R L R L	L R R L
1/4 3/4 1/4 3/4	1/4 3/4 1/4 3/4	1/4 3/4 1/4 3/4
(12mb), Intralayer	(Pnma), Intralayer	(Pcmb), Intralayer
(a)	(b)	(c)

Figure 5.44 The dependence of space group symmetry of Brownmillerites on the intra- and interlayer tetrahedral chain handedness, L or R. The view is normal to the long b-axis. (Reprinted with permission from: F. Ramezanipour et al, Chem. Mater. (2010) 22, 6008. Copyright 2010, American Chemical Society.)

A structure field map for brownmillerite symmetries has been published, as shown in Figure 5.45 [160]. The axes are the tetrahedral interlayer distance and 180° − the B—O—B angle, which is proportional to the chain dipole moment. For large interlayer separations and small dipole moments, the *Imma* structure, with chain disorder, prevails, while more ordered structures occur as the separation decreases and the dipole moment increases. A brownmillerite periodic table, Figure 5.46, shows the relatively small number of elements, which can form this structure, much reduced from the parent perovskite.

As the brownmillerite structure presents both an octahedral and a tetrahedral site, site preferences come into play. Consistent with Figure 5.5, Cr^{3+}, Mn^{3+}, and Ni^{2+} show a strong octahedral site preference, as does Sc^{3+}, while Fe^3, Mn^{2+}, and Co^{3+} (high spin) can be found in either site. Al^{3+}, Ga^{3+}, and In^{3+} prefer the tetrahedral site [161].

A few brownmillerites can form ordered structures with lower vacancy levels. $Sr_2Fe_2O_5$ can be oxidized to $Sr_2Fe_2O_{5.5}$ ($Sr_4Fe_4O_{11}$) and $Sr_2Fe_2O_{5.75}$ ($Sr_8Fe_8O_{23}$) [162]. The vacancy ordering patterns are different and the interested reader is invited to consult these references.

Finally, a few TMO with a composition consistent with brownmillerite show no apparent vacancy or cation ordering and the *Pm-3m* space group. Examples are Sr_2FeMnO_5 and $Sr_2Fe_{1.5}Cr_{0.5}O_5$ [163]. Such TMO, when examined with

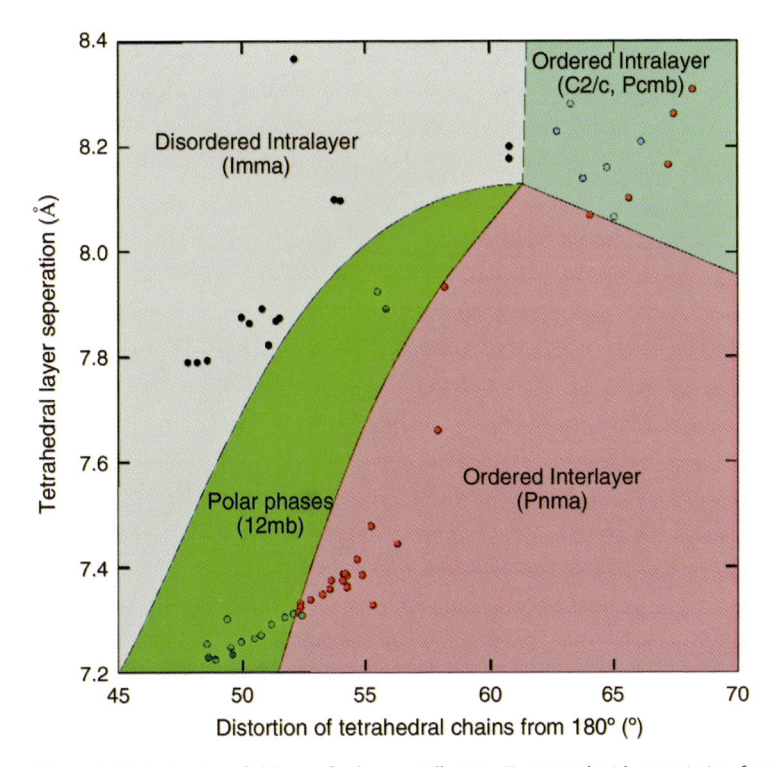

Figure 5.45 A structure field map for brownmillerites. (Reprinted with permission from Ref. [160], Copyright 2009, American Chemical Society.)

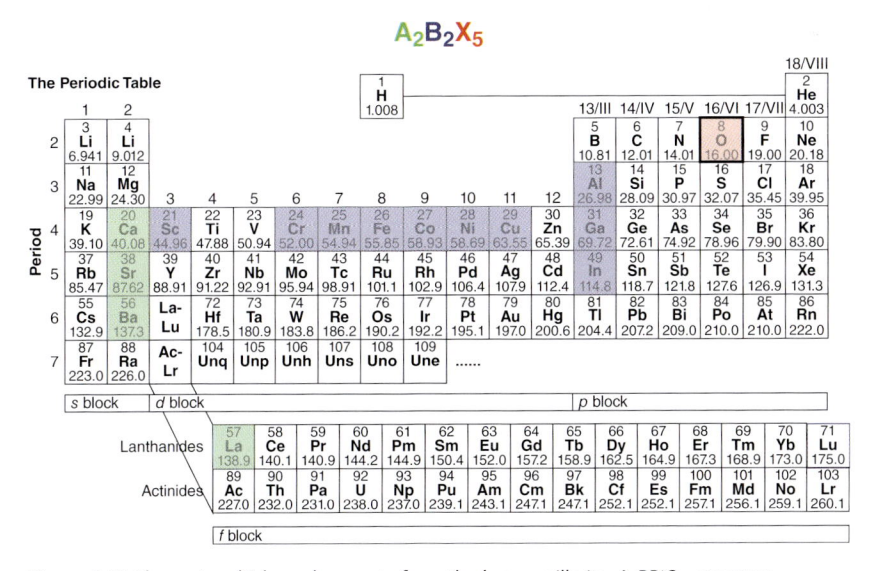

Figure 5.46 Elements, which are known to form the brownmillerite, $A_2BB'O_5$, structure.

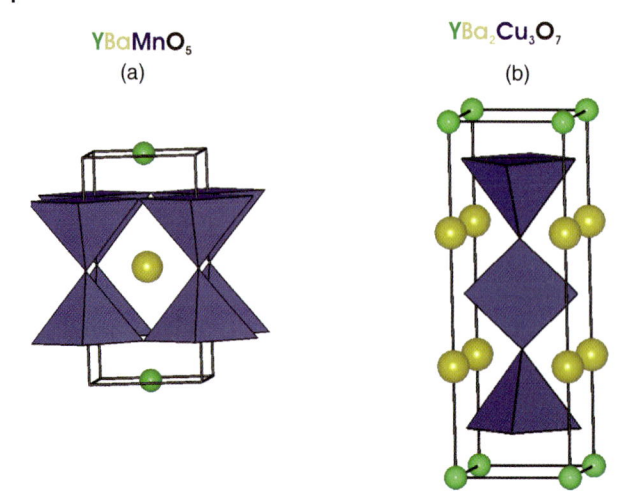

Figure 5.47 The structures of YBa$_2$MnO$_5$ (a) and YBa$_2$Cu$_3$O$_7$ (b).

local structure probes such as neutron pair distribution function (NPDF) techniques, [164] are found to have short-range order ($r \sim 5$–10 Å) which can be similar to that of a brownmillerite [165].

Ordered Perovskite Structures Involving Combinations of A- and B-Site and Oxygen Vacancy Ordering

It is of course possible to combine the various site-ordering mechanisms just described to create new perovskite related structures. Among these is the relatively large family with composition LnBaB$_2$O$_{5+x}$ where $x=0$ and 0.5 and Ln^{3+} is a lanthanide ion and B = Fe or Mn. For $x=0$, one has B$^{+2.5}$ or equal concentrations of B^{2+} and B^{3+}. There is simultaneous layered A-site order between Ba^{2+} and Ln^{3+} and oxide vacancy order. An even more complex case is represented by the so-called triple perovskites, YBa$_2$FeO$_8$ [166]. By far, the most famous example ($\sim 3 \times 10^4$ scifinder hits) is YBa$_2$Cu$_3$O$_7$, the first superconductor with $T_c >$ the b.p. of N$_2$ ($T_c = 93$ K) in which A-site, B-site, and oxide vacancy order are all present [167]. This material contains, formally, 2 Cu^{2+}, which occupy the square pyramidal sites, and one Cu^{3+}, which is found in the square planar site, that is, B-site order. It is even possible to create quadruple perovskites such as YLaBa$_2$Cu$_2$Ti$_2$O$_{11}$ [168]. Some of these are illustrated in Figure 5.47.

One interest in these phases, apart form superconductivity, is as oxygen storage materials. Considerable concentrations of oxygen can be intercalated reversibly into the oxide vacancy layers, involving redox processes associated with the B = Mn and Fe sites [169].

Intergrowths

The final mechanism to be discussed for generation of new structures from a primitive perovskite is that of intergrowth with other structure types. This

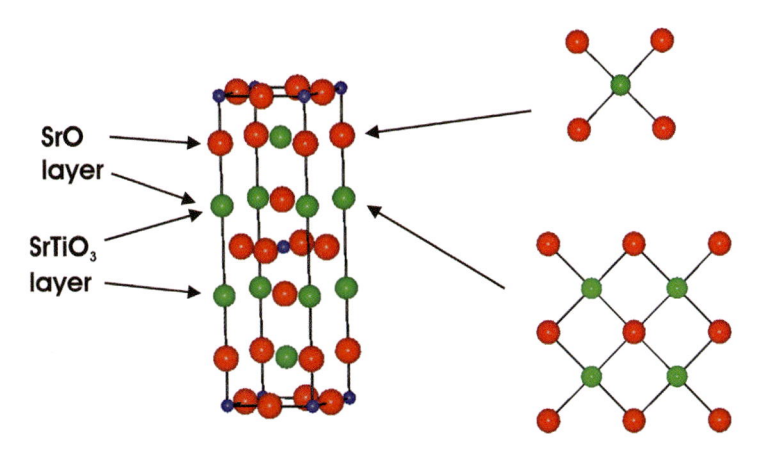

Figure 5.48 The structure of the $n = 1$ RP phase, Sr_2TiO_4, as an intergrowth of one perovskite and ¼ of a NaCl structure layer. The structural and dimensional similarity at the interface of the two blocks is also illustrated with views parallel to the c-axis.

mechanism is not exclusive to perovskites but the most famous example in TMO chemistry results from this approach when perovskite intergrows with NaCl(AO) structure oxide blocks to form the Ruddlesden–Popper phases (RP) [170]. This is illustrated for Sr_2TiO_4 in Figure 5.48. The composition of the RP phases is written as $(AO)(ABO_3)_n$, where $n = 1–3$ or 4, so Sr_2TiO_4 is an $n = 1$ RP. The key requirement is a structural and dimensional similarity between the two structures. Note that ¼ of a full NaCl (A_4O_4) cell is involved.

It is generally straightforward to prepare $n = 1 (A_2BO_4)$, $n = 2 (A_3B_2O_7)$, and $n = 3 (A_4B_3O_{10})$ in bulk form but phases for $n > 3$ are rare. Usually, A is from Group 2 or 3 and B is tetravalent or mixed valent if the A-site involves both 2+ and 3+ ions. These materials have not been reviewed recently, but one from the late 1990s is still useful and describes a very large number of RP materials [171]. The symmetry can be tetragonal ($I4/mmm$) or orthorhombic. RP phases have quite interesting properties. Of course, one the earliest cuprate superconductors, $Sr_{1.8}La_{0.2}CuO_4$ is a prime example [172]. The perovskite $SrRuO_3$ is a ferromagnetic metal but Sr_2RuO_4 is an exotic superconductor [173].

Another example of intergrowth involving a perovskite base, shown in Figure 5.38, is that of the "pillared" perovskites, $A_5M_2BB'O_{16}$, where $A = Ln^{3+}$, $M = Mo^{5+}$, Re^{5+} or Os^{5+}, and B and B' can be from the 3d, 4d, or 5d series [174]. Here single perovskite layers, $BB'O_6^{5-}$, are pillared by dimeric $M_2O_{10}^{10-}$ units. The $M_2O_{10}^{10-}$ dimers involve M–M multiple bonds, double bonds for $M = Mo^{5+}(4d^1)$ and $Re^{5+}(5d^2)$, and triple bonds for $Os^{5+}(5d^3)$ and are diamagnetic. The separation between perovskite layers is ~10 Å, leading to interesting low dimensional magnetic properties.

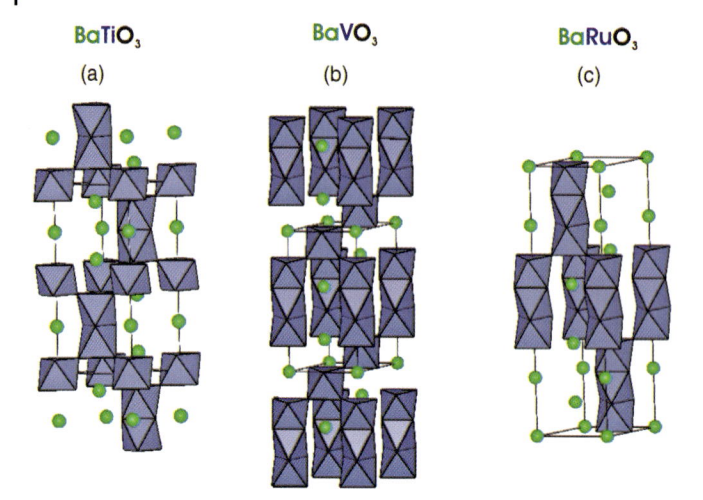

Figure 5.49 Three hexagonal perovskites: (a) BaTiO$_3$, (b) BaVO$_3$, and (c) BaRuO$_3$, showing face-sharing octahedra and various linking schemes.

Hexagonal Perovskites

Before leaving the perovskite family of TMO, the so-called hexagonal perovskites should be mentioned. While these are not structurally related to perovskite, the compositions are very similar, ABO$_3$, and occur when A = Ba and tolerance factor a >1. There is considerable structural variety as illustrated by BaTiO$_3$ (t = 1.07), BaVO$_3$ (t = 1.08), and BaRuO$_3$ (t = 1.06), Figure 5.49. (Note that BaTiO$_3$ exists in both a tetragonal and hexagonal form.) In general, mixed face and corner sharing octahedra are involved, rather than strictly corner sharing as in true perovskites. BaTiO$_3$ has dimeric face shared units and single octahedra, BaVO$_3$ has trimeric units linked by corner sharing octahedra, while in BaRuO$_3$, only trimeric corner sharing units appear [175]. As the nomenclature indicates, hexagonal (P6$_3$/mmc – BaTiO$_3$), trigonal (P-3m1 – BaVO$_3$), or rhombohedral (R-3m – BaRuO$_3$) symmetry is found. A large number of polytypes of hexagonal perovskites exist which are too numerous to mention here.

5.3.6
Spinels

This will be the final TMO structural family discussed in this chapter. Spinels could have been included under the NaCl heading, as the O – sublattice is CCP, but the tetrahedral sites are also occupied in this case. Again, the family is vast with many detailed studies of structure and properties. The mineral spinel has a composition MgAl$_2$O$_4$ and is found in nature. Highly colored forms, often regarded as gemstones, exist and, indeed, one resides in the British Crown Jewels as the "Black Prince's Ruby" in George V's Imperial State Crown.

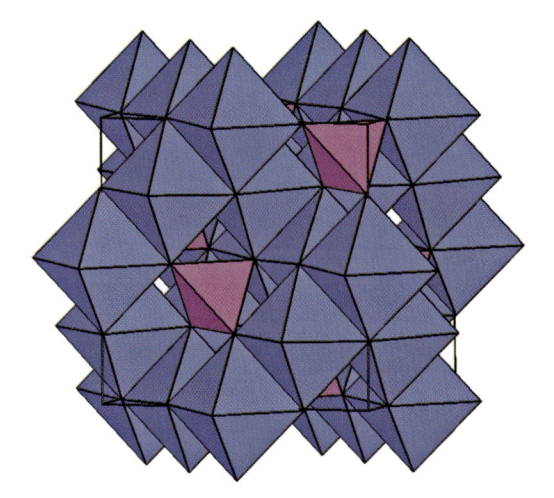

Figure 5.50 The structure of spinel AB_2O_4 showing the edge-sharing octahedra of the B-sites (blue) and the A-site tetrahedra sharing corners with the octahedra (magenta).

One half of the O sites and one quarter of the T sites of the CCP lattice are occupied in an ordered manner, Figure 5.50, with symmetry *Fd-3m*, the same as cubic pyrochlore. Writing the general formula as AB_2O_4, A is in 8a, (1/8 1/8 1/8), B in 16d (1/2 1/2 1/2), and O in 32e, (x x x) with x ~0.24, using the setting of *Fd-3m* with the origin at the center of symmetry. It has been noted previously that the B sublattice, 16d, is half of a FCC lattice with the topology of corner sharing tetrahedra.

The allowed valence combinations are A^{2+}/B^{3+}, A^{4+}/B^{2+}, and A^{6+}/B^{+} for integral valences, while many more possibilities occur for nonintegral cases such as $A^{+}/B^{3.5+}$ and nonintegral site occupations. By far, the most common is A^{2+}/B^{3} and the most famous of these is the mineral magnetite, Fe_3O_4 ($Fe^{2+}Fe_2^{3+}O_4$) or lodestone as the ancients called it.

O/T site preference is a major issue in spinel oxides. There are two limiting cases called "normal," where A^{2+} resides in the T site and the B^{3+} ions are on the O site and "inverted" where one B^{3+} occupies the T site and A^{2+} and B^{3+} occupy, randomly, the O site. Real materials may deviate from these limits with partial site occupations and each situation should be examined in detail. Using Figure 5.5 as a guide, fearless predictions can be made for certain cases. For $B = Cr^{3+}$, the normal case is always found even for $NiCr_2O_4$ [176]. This is also true for $B = Rh^{3+}$, where the low spin ion has configuration $4d^6(t_{2g}^6)$ and the largest OSPE of any ion, although this case is not addressed in Figure 5.5. If $B^{3+} = Fe^{3+}$, the inverted pattern is found for nearly any A^{2+} and this is the situation for magnetite. For the interesting situation of $MnFe_2O_4$, a statistical occupation of both sites is reported [177]. Interestingly, $ZnFe_2O_4$ presents a counter example, as it is normal [178]. Zn^{2+} spinels of the 2+/3+ type are usually normal, indicating a strong T site preference for Zn^{2+}. Distortions from cubic *Fd-3m* symmetry are observed when $A = Cu^{2+}$ and $B = Mn^{3+}$, both strong FOJT ions, usually described in the sub group *I4$_1$/amd*. Mn_3O_4 haussmanite, ($Mn^{2+}Mn_2^{3+}O_4$) adopts this distorted structure [179].

Considering other valence combinations such as $A^{4+}B_2^{2+}O_4$ spinels, the inverted form is normally found, except where $A = Ge^{4+}$ and the strong T site preference of Ge^{4+} results in a normal distribution [180].

Historically, the magnetic properties of spinel TMO drew much attention, which continues to the present time. When both A and B are magnetic TM, two magnetic sublattices exist which interact or couple antiferromagnetically to result in a ferrimagnetic material with a net magnetic moment. For example, in magnetite, $Fe^{2+}Fe_2^{3+}O_4$, the T-site Fe^{3+} ions have a magnetic moment of ~$5\mu_B$ (Bohr Magnetons) while on the O-site there is a total moment of $4\mu_B$ from Fe^{2+} plus $5\mu_B$ from the other Fe^{3+} ion. The T-site and O-site sublattice moments couple antiparallel for a net moment of ~$4\mu_B$ per formula unit. For most other Fe^{3+} spinels, historically known as "ferrites" similar magnetic properties occur with ordering temperatures considerably above ambient that attracts application in various devices. As well, Fe_3O_4 undergoes a number of interesting phase transitions upon cooling, notably one called the Verwey transition near 125 K, studies of which consititute a considerable literature [181]. More recently, spinels have been investigated in studies of geometric magnetic frustration, as the B-sublattice is potentially frustrated (corner sharing tetrahedra), in particular the normal spinel $ZnCr_2O_4$ [182]. $LiMn_2O_4$ is a candidate for the cathode in Li-batteries [183]. This material also undergoes interesting structural phase transitions, driven by charge ordering between Mn^{3+} and Mn^{4+} ions [184] and Li can be both added and removed using mild techniques at ambient temperature to produce $Li_2Mn_2O_4$ and λ-MnO_2 in which the basic spinel lattice is preserved [185]. Finally, certain spinels are metallic and show superconductivity, $LiTi_2O_4$, [186] or so-called "heavy Fermion" behavior, LiV_2O_4 [187].

5.4
Conclusion

It is impossible to provide more than a bare introduction to a subject as vast as TMO structural crystal chemistry in such a brief chapter as this. For example, notable topics which have been omitted include the "Magneli" phases of the rutile structure oxides, garnet structure oxides, and many more. There has been a resurgence of interest in garnet structure materials, especially as Li-ion conductors [188]. This chapter should be used as an introductory passageway to the ever-expanding literature on Transition Metal Oxides, a remarkable and complex class of materials which will remain of intense interest to many investigators in many disciplines for the foreseeable future.

References

1. (a) Sixtus, K.J., Kronenberg, K.J., and Tenzer, R.K. (1956) *J. Appl. Phys.*, **27**, 1051. (b) Peng, Z., Hwanng, J.-Y., and Matthew, A. (2013) *IEEE Trans. Magn.*, **49** (3 Pt2), 1163.
2. Shen, T., Li, P., Jiang, J., Cooley, L., Tompkins, J., McRae, D., and Walsh, R. (2015) *Supercond. Sci. Technol.*, **28**, 1.
3. Rubi, D., Frontera, C., Roig, A., Nogues, J., Munoz, J.S., and Fontcuberta, J. (2006) *Mat. Sci. Eng.*, **B126**, 139.
4. Shokoohi, F.K., Tarascon, J.M., Wilkens, B.J., Guyomard, D., and Chang, C.C. (1992) *J. Electrochem. Soc.*, **139**, 1845.
5. Machiels, C.J. and Sleight, A.W. (1982) *J. Catal.*, **76**, 238.
6. Mott, N.F. (1968) *Rev. Mod. Phys.*, **40**, 677.
7. (a) Goodenough, J.B. (1963) Magnetism and the chemical bond. *Interscience.* (b) Kanamori, J. (1959) *J. Phys. Chem. Solids*, **10**, 87.
8. Shannon, R.D. and Prewitt, C.T. (1969) *Acta. Cryst. B*, **25**, 925.
9. Shannon, R.D. (1976) *Acta. Cryst. A*, **32**, 751.
10. Halasyamani, P.S. and Poeppelmeier, K.R. (1998) *Chem. Matter.*, **10**, 2753.
11. Leger, J.M., Yacoubi, N., and Loriers, J. (1981) *J. Solid State Chem.*, **36**, 261.
12. Willis, B.T.M. and Rooksby, H.P. (1953) *Acta. Cryst.*, **6**, 827.
13. Christensen, A.N. (1990) *Acta. Chem. Scand.*, **44**, 851.
14. loehman, R.E., Rao, C.N.R., and Honig, J.M. (1969) *J. Phys. Chem.*, **73**, 1781.
15. Sasaki, S., Fujino, K., and Takeuchi, Y. (1995) *Proc. Jpn Acad.*, **55**, 43.
16. Eick, H.A. (1956) *J. Am. Chem. Soc.*, **78**, 5147.
17. Zachariasen, H. (1949) *Acta. Cryst.*, **2**, 388.
18. Chikalla, T.D., McNeilly, C.E., and Skavdahl, R.E. (1964) *J. Nucl. Mater.*, **12**, 131.
19. Akimoto, Y. (1967) *J. Inorg. Nucl. Chem.*, **29**, 2650.
20. Seleznev, A.G., Radchenko, V.M., Shusakov, V.D., Ryabin, M.A., Droznik, R.R., Lebedeva, L.S., and Vasil'ev, V.Ya. (1989) *Radiokhimiya*, **31**, 20.
21. Bowman, A.L., Wallace, T.C., Yarnell, J.L., and wenzel, R.G. (1966) *Acta. Cryst.*, **21**, 843.
22. Abrahams, S.C. and Bernstein, J.L. (1969) *Acta. Cryst. B*, **25**, 1233.
23. Aurivillius, K. (1964) *Acta. Chem. Scand.*, **18**, 1305.
24. (a) Waser., J., Levy, H.A. and Peterson, S.W. (1953) *Acta. Cryst.*, **6**, 661. (b) Moore, W.J. and Pauling, L. (1941) *J. Am. Chem. Soc.*, **63**, 1392.
25. Asbrink, S. and Norrby, L.J. (1970) *Acta. Cryst. B*, **26**, 8.
26. Mather, G.C., Dussarrat, C., Etourneau, J., and West, A.R. (2000) *J. Mater. Chem.*, **10**, 2219.
27. (a) Brunel, M., de Bergevin, F. and Gondrand, M. (1972) *J. Phys. Chem. Solids*, **33**, 1927. (b) Hauck, J. (1980) *Acta. Cryst. A*, **36**, 228.
28. De Picciotto, L.A. and Thackeray, M.M. (1985) *Mat. Res. Bull.*, **20**, 1409.
29. Hodeau, J.L., Marezio, M., Santoro, A., and Roth, R.S. (1982) *J. Solid State Chem.*, **45**, 170.
30. Cava, R.J., Santoro, A., Murphy, D.W., Zahurak, S.M., and Roth, R.S. (1981) *Solid State Ion.*, **5**, 323.
31. Betz, T. and Hoppe, R. (1984) *Z. Anorg. Allgem. Chem.*, **512**, 19.
32. Betz, T. and Hoppe, R. (1985) *Z. Anorg. Allgem. Chem.*, **524**, 17.
33. Bieringer, M., Greedan, J.E., and Luke, G.M. (2000) *Phys. Rev. B*, **62**, 6521.
34. Fletcher, J.G., Mather, G.C., West, A.R., Castellanos, M., and Gutierrez, M.P. (1994) *J. Mater. Chem.*, **4**, 1305.
35. Derakhshan, S., Greedan, J.E., Katsumata, T., and Cranswick, L.M.D. (2008) *Chem. Matter.*, **20**, 5714.
36. Nguyen, P.-H., Ramezanipour, F., Greedan, J.E., Cranswick, L.M.D., and Derakhshan, S. (2012) *Inorg. Chem.*, **51**, 11493.
37. (a) McWhan, B.D. and Remeika, J.P. (1970) *Phys. Rev. B*, **2**, 3734. (b) McWhan, B.D., Menth, A., Remeika, J.P., Brinkman, W.F., and Rice, T.M. (1973) *Phys. Rev. B*, **7**, 1920.

38 Jones, P.G., Rumpel, H., Schwarzmann, E., and Sheldrick, G.M. (1979) *Acta. Cryst. B*, **35**, 1435.
39 Rice, C.E. and Robinson, W.R. (1977) *Acta. Cryst.*, **B33**, 1342.
40 Vincent, M.G., Yvon, K., and Ashkenasi, J. (1980) *Acta. Cryst. A*, **36**, 808.
41 Sawada, H. (1994) *Mat. Res. Bull.*, **29**, 239.
42 Blake, R.L., Hessevick, R.E., Zoltai, T., and Finger, L.W. (1966) *Am. Miner.*, **51**, 123.
43 Coey, J.M.D. (1970) *Acta. Cryst. B*, **26**, 1876.
44 Rozier, P., Ratuszna, A., and Galy, J. (2002) *Z. Anorg. Allgem. Chem.*, **628**, 1236.
45 Weschler, B.A. and von Dreele, R.B. (1989) *Acta. Cryst. B*, **45**, 524.
46 Liferovich, R.P. and Mitchell, R.H. (2005) *Phys. Chem. Miner.*, **32**, 442.
47 Wechsler, B.A. and Prewitt, C.T. (1984) *Am. Miner.*, **69**, 176.
48 Newnham, R.E., Fang, J.H., and Santoro, R.P. (1964) *Acta. Cryst.*, **17**, 240.
49 Kennedy, B.J., Zhou, Q., and Avdeev, M. (2011) *J. Solid State Chem.*, **184**, 2987.
50 Cloud, W.H. (1958) *Phys. Rev.*, **111**, 1046.
51 de Beaulieu, D.C. and Mueller-Buschbaum, H. (1980) *Z. Naturfor. Teil B*, **35**, 669.
52 (a) Bridges, C.A. and Greedan, J.E. (2004) *J. Solid State Chem.*, **177**, 1098. (b) Bridges, C.A., Hansen, T., Wills, A.S., Luke, G.M., and Greedan, J.E. (2006) *Phys. Rev. B*, **74**, 024426.
53 Liu, G. and Greedan, J.E. (1996) *J. Solid State Chem.*, **122**, 416.
54 Kajita, T., Kanzaki, T., Suzuki, T., Kim, J.E., Kato, K., Takata, M., and Katsufuji, T. (2010) *Phys. Rev. B*, **81**, 060405.
55 Friese, K. and Kanke, Y. (2006) *J. Solid State Chem.*, **179**, 3277, and references therein.
56 Iwasaki, K., Takizawa, H., Yamane, H., Kubota, S., Takahashi., J., Uheda, J., and Endo, T. (2003) *Mat. Res. Bull.*, **38**, 141, and references therein.
57 Baur, W.H. (1956) *Acta. Cryst.*, **9**, 515.
58 Ghedira, M., Vincent, H., Marezio, M., and Launay, J.C. (1977) *J. Solid State Chem.*, **22**, 423.
59 Cloud, W.H., Schreiber, D.S., and Babcock, K.R. (1962) *J. Appl. Phys.*, **33**, 1193.
60 Bauer, W.H. (1976) *Acta. Cryst. B*, **32**, 2200.
61 Bolzan, A.A., Fong, C., kennedy, B.J., and Howard, C.J. (1994) *J. Solid State Chem.*, **113**, 9.
62 Bowman, C.E. (1970) *Acta. Chem. Scand.*, **24**, 116.
63 Shannon, R.D. (1968) *Solid State Commun.*, **6**, 139.
64 Bowman, C.E. (1970) *Acta. Chem. Scand.*, **24**, 123.
65 Bolzan, A.A., Fong, C., kennedy, B.J., and Howard, C.J. (1997) *Acta. Cryst.*, **B53**, 373.
66 Morin, F. (1959) *Phys. Rev. Lett.*, **3**, 34, The literature on VO_2 comprises more than 10^3 papers.
67 (a) Tribalat, S., Jungfleisch, M.-L. and Delafosse, D. (1964) *C. R. Acad. Sci. Ser.*, **259**, 2109. (b) Cox, D.E., Cava, R.J., McWhan, D.B., and Murphy, D.W. (1982) *J. Phys. Chem. Solids*, **43**, 657.
68 Cheetham, A.K. and Rao, C.N.R. (1976) *Acta. Cryst.*, **B32**, 1579.
69 Siegel, S., Hoekstra, H.R., and Tani, B.S. (1969) *J. Inorg. Nucl. Chem.*, **31**, 3803.
70 Warburg, M. and Mueller-Buschbaum, H. (1984) *Z. Anorg. Allgem. Chem.*, **508**, 55.
71 Weitzel, H. (1976) *Z. Kristall.*, **144**, 238.
72 Kratzheller, B. and Gruehn, R. (1992) *J. Alloys Compd.*, **183**, 75.
73 Pagola, S., Carbonio, R.E., Fernandez Diaz, R.E., and Alonso, J.A. (1998) *J. Solid State Chem.*, **137**, 359.
74 Drew, M.G.B., Hobson, R.J., and Padayatchy, V.T. (1995) *J. Mater. Chem.*, **5**, 1779.
75 Halle, G. and Mueller-Buschbaum, H. (1988) *J. Less Common Met.*, **142**, 263.
76 Bernier, J.C. and Poix, P. (1967) *C. R. Acad. Sci. Ser. C*, **265**, 1164.
77 Massard, P., Bernier, J.C., and Michel, A. (1972) *J. Solid State Chem.*, **4**, 269.
78 Eicher, S.M., Greedan, J.E., and Lushington, K.J. (1986) *J. Solid State Chem.*, **62**, 220.
79 Reimers, J.N., Greedan, J.E., Stager, C.V., and Kremer, R. (1989) *J. Solid State Chem.*, **83**, 20.

80 Wichmann, R. and Mueller-Buschbaum, H. (1986) *Z. Anorg. Allgem. Chem.*, **536**, 15.
81 Ramos, E.M., Viega, M.L., Fernandez, F., Saez-Puche, R., and Pico, C. (1991) *J. Solid State Chem.*, **91**, 113.
82 Nakua, A.M., Yun, H., Reimers, J.N., Greedan, J.E., and Stager, C.V. (1991) *J. Solid State Chem.*, **91**, 105.
83 Ercit, T.S., Foord, E.E., and Fitzpatrick, J.J. (2002) *Can. Mineral.*, **40**, 1207.
84 Kunnmann, W., La Placa, S.J., Corliss, L.M., Hastings, J.M., and Banks, E. (1968) *J. Phys. Chem. Solids*, **29**, 1359.
85 Mikhailova, D., Ehrenberg, H., Trots, D., Brey, G., Oswald, S., and Fuess, H. (2009) *J. Solid State Chem.*, **182**, 1506.
86 Hodeau, J.L., Gondrand, M., Labeau, M., and Jobert, J.C. (1978) *Acta. Cryst. B*, **34**, 3543.
87 Fruchart., D., Montmory, M.C., Bertaut, E.F., and Bernier, J.C. (1980) *J. Phys. (Paris)*, **41**, 141.
88 Badaud, J.P., Fournier, J.P., and Omaly, J. (1977) *C. R. Acad. Sci. Ser. C*, **284**, 921.
89 Omaly, J. and Badaud, J.P. (1972) *C. R. Acad. Sci. Ser. C*, **275**, 371.
90 Miura, H. (1986) *Mineral. J. Jpn.*, **13**, 119.
91 Ishiwata, S., Bos, J.W.G., Huang, Q., and Cava, R.J. (2006) *J. Phys. Condens. Matter*, **18**, 3745.
92 Hwang, G.C., Post, J.E., and Lee, Y. (2015) *Phys. Chem. Miner.*, **42**, 405.
93 Robinson, D.M., Go, Y.B., Muiu, M., Gardner, G., Zhang, Z., Mastrogiovanni, D., Garfunkel, E., Li, J., Greenblatt, M., and Dismukes, G.C. (2013) *J. Am. Chem. Soc.*, **135**, 3404.
94 Kummerle, E.A. and Heger, G. (1999) *J. Solid State Chem.*, **147**, 485.
95 Zachariesen, W.H. (1949) *Acta. Cryst*, **2**, 388.
96 Brauer, G. and Gradinger, H. (1954) *Z. Anorg. Allgem. Chem.*, **276**, 226.
97 Baenziger, N.C., Eick, H.A., Shuldt, H.S., and Eyring, L. (1961) *J. Am. Chem. Soc.*, **83**, 2219.
98 Leigh, H.D. and McCartney, E.R. (1974) *J. Am. Cer. Soc.*, **57**, 192.
99 Asprey, L.B., Ellinger, F.H., Fried, S., and Zachariesen, W.H. (1955) *J. Am. Chem. Soc.*, **77**, 1707.
100 Elson, R., Fried, S., Sellers, P.A., and Zachariesen, W.H. (1950) *J. Am. Chem. Soc.*, **72**, 5791.
101 Peterson, J.R. and Cunningham, B.B. (1967) *Inorg. Nucl. Chem. Lett.*, **3**, 327.
102 Barrett, S.A., Jacobson, A.J., Tofield, B.C., and Fender, B.E.F. (1982) *Acta. Cryst. B*, **38**, 2775.
103 Baybarz, R.D., Haire, R.G., and Fahey, J.A. (1972) *J. Inorg. Nucl. Chem.*, **34**, 557.
104 Smith, D.K., Jr. and Newkirk, H.W. (1965) *Acta. Cryst*, **18**, 983.
105 Martin, U., Boysen, H., and Frey, F. (1993) *Acta. Cryst. B*, **48**, 403.
106 (a) Horiuchi, H., Schultz, A.J., Leung, P.C., and Williams, J.M. (1984) *Acta. Cryst. B*, **40**, 367. (b) Rabenau, A. (1956) *Z. Anorg. Allgem. Chem.*, **288**, 221.
107 Birke, P. and Weppner, W. (2011) Solid electrolytes, in *Handbook of Battery Materials*, 2nd edn, vol. **2**, Wiley-VCH Verlag GmbH, p. 657.
108 Galasso, F.S. (1970) *Structure and Properties of Inorganic Solids*, Pergamon Press, pp. 99–100.
109 Saiki, A., Ishizawa, N., Mizutani, N., and Kato, M. (1985) *J. Ceram. Assoc. Jpn*, **93**, 649.
110 Ishibashi, H., Shimomoto, K., and Nakahigashi, K. (1994) *J. Phys. Chem. Solids*, **55**, 809.
111 Kennedy, B.J. and Avdeev, M. (2011) *Aust. J. Chem.*, **64**, 119.
112 Gasgnier, M., Schiffmacher, G., Caro, P., and Eyring, L. (1986) *J. Less Common Met.*, **116**, 31.
113 Knop, O. and Hartley, J.M. (1968) *Can. J. Chem.*, **46**, 1446.
114 Maslen, E.N., Strel'tsov, V.A., and Ishikawa, N. (1996) *Acta. Cryst. B*, **52**, 414.
115 Templeton, D.H. and Dauben, C.H. (1953) *J. Am. Chem. Soc.*, **75**, 4560.
116 Noe, M., Fuger, J., and Duyckaerts, G. (1970) *J. Inorg. Nucl. Lett.*, **6**, 111.
117 Malinovskii, Yu.A. and Bondareva, O.S. (1991) *Kristallografiya*, **36**, 1558.
118 Schweda, E. and Kang, Z.C. (2004) Structural features of rare earth oxides, in *Binary Rare Earth Oxides* (eds G. Adachi, N. Imanaka, and Z.C. Kang), Kluwer Acad. Publ., p. 57.

119 Subramanian, M.A., Aravamudan, G., and Subba Rao, G.V. (1983) *Prog. Solid State Chem.*, **15**, 55.
120 Kennedy, B.J. (1998) *Physica B Condens. Matter.*, **214–243**, 303.
121 Greedan, J.E. (2006) *J. Alloys Compd.*, **408–412**, 408.
122 Gardner, J.S., Gingras, M.J.P., and Greedan, J.E. (2010) *Rev. Mod. Phys.*, **82**, 53.
123 Seshadri, R. (2006) *Solid State Sci.*, **8**, 259.
124 Li, X., Li, W., Matsubayashi, K., Sato, Y., Jin, C., Uwatoko, Y., Kawae, T., Hallas, A.M., Wiebe, C.R., and Arevalo-Lopez, A.M. (2014) *Phys. Rev. B*, **89**, 064409.
125 (a) Kanno, R., Takeda, Y., Yamamoto, T., Kamamoto, Y., and Yamamoto, O. (1993) *J. Solid State Chem.*, **102**, 106. (b) Hector, A.L. and Wiggin, S.B. (2004) *J. Solid State Chem.*, **177**, 139. (c) Takeda, T., Nagata, M., Kobayashi, H., Kanno, R., Kawamoto, Y., Takano, M., Kamiyama, T., Izumi, F., and Sleight, A.W. (1998) *J. Solid State Chem.*, **140**, 182. (d) Bongers, P.F. and van Muers, E.R. (1967) *J. Appl. Phys.*, **38**, 944.
126 (a) Brisse, F., Stewart, D.J., Seidl, V., and Knop, O. (1972) *Can. J. Chem.*, **50**, 3648. (b) Wang, R. and Sleight, A.W. (1988) *Mater. Res. Bull.*, **33**, 1005. (c) Donohue, P.C., Longo, J.M., Rothstein, R.D., and Katz, L. (1965) *Inorg. Chem.*, **4**, 1152. (d) Reading, J. and Weller, M.T. (2001) *J. Mater. Chem.*, **11**, 2373.
127 (a) Beyerlein, R.A., Horowitz, H.S., Longo, J.M., Leonowicz, M.E., Jorgensen, J.D., and Rotella, F.J. (1984) *J. Solid State Chem.*, **51**, 253. (b) Reading, J., Knee, C.S. and Weller, M.T. (2002) *J. Mater. Chem.*, **12**, 2376. (c) Bernotat-Wulf, H. and Hoffmann, W. (1982) *Z. Kristall.*, **158**, 101.
128 Kennedy, B.J. (1996) *J. Solid State Chem.*, **123**, 14.
129 Galati, R., Hughes, R.W., Knee, C.S., Henry, P.F., and Weller, M.T. (2007) *J. Mater. Chem.*, **17**, 160.
130 Hiroi, Z., Yamaura, J.-I., and Hattori, K. (2012) *J. Phys. Soc. Jpn*, **81**, 011012.
131 Lau, G.C., McQueen, T.M., Huang, Q., Zandbergen, H.W., and Cava, R.J. (2008) *J. Solid State Chem.*, **181**, 45.

132 Cai, L. and Nino, J.C. (2009) *Acta. Cryst. B*, **65**, 269.
133 Lam, R., Wiss, F., and Greedan, J.E. (2002) *J. Solid State Chem.*, **167**, 182.
134 Goldschmidt, V.M. (1926) *Naturwissenschaften*, **21**, 477.
135 Glazer, A.M. (1972) *Acta. Cryst. B*, **28**, 3384.
136 Lufaso, M.W. and Woodward, P.M. (2001) *Acta. Cryst. B*, **57**, 725.
137 Greedan, J.E. (1985) *J. Less. Common Met.*, **111**, 335.
138 Sefat, A.S., Greedan, J.E., and Cranswick, L. (2006) *Phys. Rev. B*, **74**, 104418.
139 Woodward, P.M. (1997) *Acta. Cryst. B*, **53**, 32.
140 Vasala, S. and Karppinen, M. (2015) *Prog. Solid State Chem.*, **43**, 1.
141 Anderson, M.T., Greenwood, K.B., Taylor, G.A., and Poeppelmeier, K.R. (1993) *Prog. Solid State Chem.*, **22**, 197.
142 Battle, P.D., grey, C.P., Hereviue, M., Martin, C., Moore, C.A., and Paik, Y. (2003) *J. Solid State Chem.*, **175**, 20.
143 (a) Barnes, P.W., Lufaso, M.E. and Woodward, P.M. (2006) *Acta. Cryst. B*, **62**, 384. (b) Woodward, P.M., Hoffman, R.-D. and Sleight, A.W. (1994) *J. Mater. Res.*, **9**, 2118.
144 Howard, C.J., Kennedy, B.J., and Woodward, P.M. (2003) *Acta. Cryst.*, **B59**, 463.
145 (a) Stitzer, K.E., Smiath, M.D. and zur Loye, H.-C. (2002) *Solid State Sci.*, **4**, 311. (b) Wiebe, C.R., Greedan, J.E., Luke, G.M., and Gardner, J.S. (2002) *Phys. Rev. B*, **65**, 144413. (c) Wiebe, C.R., Greedan, J.E., Kyriakou, P.P., Luke, G.M., Gardner, J.S., Fuyaka, A., Gat-Malureanu, I.M., Russo, P.L., Savici, A.T., and Uemura, Y.J. (2003) *Phys. Rev. B*, **68**, 134410.
146 (a) Aharen, T., Greedan, J.E., Imai, T., Bridges, C.A., Aczel, A.A., Rodriguez, J., MacDougall, G., Luke, G.M., Michaelis, V.K., Kroeker, S., Wiebe, C.R., Zhou, H., and Cranswick, L.M.D. (2010) *Phys. Rev B*, **81**, 224409. (b) de Vries, M.A., Mclaughlin, A.C., and Bos, J.-W.G. (2010) *Phys. Rev. Lett.*, **104**, 177202.
147 King, G. and Woodward, P.M. (2010) *J. Mater. Chem.*, **20**, 5785.

148 King, G., Thimmaiah, S., Dwivedi, A., and Woodward, P.M. (2007) *Chem. Mater.*, **19**, 6451.
149 Shimikawa, Y. (2008) *Inorg. Chem.*, **47**, 8562.
150 Woodward, P.M., Sleight, A.W., and Vogt, T. (1995) *J. Phys. Chem. Solids*, **56**, 1305.
151 Negishi, H., Negishi, S., Kuroiwa, Y., Sato, N., and Aoyagi, S. (2004) *Phys. Rev. B*, **69**, 064111.
152 (a) Trantafyllou, S.T., Chistidis, P.C., and Lioutas, C.B. (1997) *J. Solid State Chem.*, **133**, 479. (b) Takusagawa, F. and Jacobson, R.A. (1976) *J. Solid State Chem.*, **18**, 163. (c) Wiseman, P.J. and Dickens, P. (1976) *J. Solid State Chem.*, **17**, 91. (d) Darlington, C.N.W., Hriljac, J.A., and Knights, K.S. (2003) *Acta. Cryst. B*, **59**, 584.
153 Guo, J.D. and Whittingham, M.S. (1993) *Intl. J. Mod. Phys. B*, **7**, 4145.
154 Kennedy, B.J., Howard, C.J., Kubota, Y., and Kato, K. (2004) *J. Solid State Chem.*, **177**, 4552.
155 Sefat, A.S., Amow, G., Wu, M., Botton, G.A., and Greedan, J.E. (2005) *J. Solid State Chem.*, **178**, 1008.
156 Sefat, A.S., Greedan, J.E., Luke, G.M., and Niewczas, M. (2006) *Phys. Rev. B*, **74**, 104419.
157 (a) Treiber, U. and Kemmler-Sack, S. (1981) *Z. Anorg. Allgem. Chem.*, **478**, 22. (b) Marjerrison, C.A., Thompson, C.M., Sala, G., Maharaj, D.D., Cai, E.Y., Hallas, A.M., Munsie, T.S.J., Granroth, G.E., Flacau, R., Greedan, J.E., Gaulin, B.D., and Luke, G.M. (2016) *Inorg. Chem.*, **55**, 10701.
158 Gong, W., Yu, H., Ning, Y.B., Greedan, J.E., Datars, W.R., and Stager, C.V. (1991) *J. Solid State Chem.*, **90**, 320.
159 (a) Abakumov, A.M., Kalyuzhnaya, A.S., Rozova, M.G., Antipov, E.V., Hadermann, J., and van Tendeloo, G. (2005) *Solid State Sci.*, 7, 801. (b) Lambert, S., Leligny, H., Grebille, D., Pelloquin, D., and Raveau, B. (2002) *Chem. Mater.*, **14**, 1818.
160 Parsons, T.G., D'Hondt, H., Hadermann, J., and Hayward, M.A. (2009) *Chem. Mater.*, **21**, 5527.
161 Grosvenor, A.P. and Greedan, J.E. (2009) *J. Phys. Chem. C*, **113**, 11366, and references therein.
162 Hodges, J.P., Short, S., Jorgensen, J.D., Xiong, X., Dabrowski, B., Mini, S.M., and Kimball, C.W. (2000) *J. Solid State Chem.*, **151**, 190.
163 (a) Ramezanipour, F., Greedan, J.E., Sewenie, J., Proffen, Th., Ryan, D.H., Grosvenor, A.P., and Donaberger, R.L. (2011) *Inorg. Chem.*, **50**, 7779. (b) Gibb, T.C. and Matsuo, M. (1990) *J. Solid State Chem.*, **86**, 164. (c) Ramezanipour, F., Greedan, J.E., Sewenie, J., Donaberger, R.L., Turner, S., and Botton, G.A. (2012) *Inorg. Chem.*, **51**, 2638.
164 Egami, T. and Billinge, S.J.L. (2003) *Underneath the Bragg Peaks: Structural Analysis of Complex Materials*, Plenum, Oxford.
165 King, G., Ramezanipour, F., Llobet, A., and Greedan, J.E. (2013) *J. Solid State Chem.*, **198**, 407.
166 Karen, P., Kjekshus, A., Huang, Q., Karen, V.L., Lynn, J.W., Rosov, N., Sora, I.N., and Santoro, A. (2003) *J. Solid State Chem.*, **174**, 87.
167 (a) Beno, M.A., Soderholm, L., Capone, D.W. II, Hinks, D.G., Jorgensen, J.D., Grace, J.D., Schuller, I.K., Segre, C.U., and Zhang, K. (1987) *Appl. Phys. Lett.*, **51**, 57. (b) Greedan, J.E., O'Reilly, A.H., and Stager, C.V. (1987) *Phys. Rev. B*, **35**, 8770.
168 Greenwood, K.B., Sarjeant, G.M., Poeppelmeier, K.R., Salvador, P.A., Mason, T.O., Dabrowski, B., Rogacki, K., and Chen, Z. (1995) *Chem. Mater.*, 7, 1355.
169 (a) Klimkowicz, A., Swierczek, K., Takasaki, A., Molenda, J., and Dabrowski, B. (2015) *Mater. Res. Bull.*, **65**, 116. (b) Jeamjumnunja, K., Gong, W., Makarenko, T., and Jacobson, A.J. (2015) *J. Solid State Chem.*, **230**, 397.
170 (a) Ruddlesden, S.N. and Popper, P. (1958) *Acta. Cryst.*, **11**, 54. (b) (1957) *Acta. Cryst*, **10**, 538.
171 Sharma, I.B. and Singh, D. (1998) *Bull. Mater. Sci.*, **5**, 463.
172 Cava, R.J., van Dover, R.B., Batlogg, B., and Rietman, E.A. (1987) *Phys. Rev. Lett.*, **58**, 408.

173 (a) Kanbayasi, A. (1976) *J. Phys. Soc. Jpn*, **41**, 1876. (b) Maeno, Y., Nishizaki, S., Yoshida, K., Ikeda, S., and Fujita, T. (1996) *J. Low Temp. Phys.*, **105**, 1577. The literature on these two oxides exceeds 2000 papers.

174 (a) Ledesert, M., Labbe, Ph., McCarroll, W.H., Leligny, H., and Raveau, B. (1993) *J. Solid State Chem.*, **105**, 165. (b) Chi, L., Green, A.E.C., Hammond, R., Wiebe, C.R., and Greedan, J.E. (2003) *J. Solid State Chem.*, **170**, 165.

175 (a) Akimoto, J., Gotoh, Y., and Oosawa, Y. (1994) *Acta. Cryst. C*, **50**, 160. (b) Liu, G. and Greedan, J.E. (1994) *J. Solid State Chem.*, **110**, 274. (c) Donohue, P.C., Katz, L., and Ward, R. (1965) *Inorg. Chem.*, **4**, 303.

176 Armbruster, T., Lager, G.A., Ihringer, J., Rotella, F.J., and Jorgensen, J.D. (1983) *Z. Kristall.*, **162**, 162.

177 Denecke, M.A., Gunssner, W., Buxbaum, G., and Kuske, P. (1992) *Mater. Res. Bull.*, **27**, 514.

178 Schaefer, W., Kockelmann, W., Kirfel, A., Potzel, W., Burghart, F.J., Kalvius, G.M., Martin, A., Kaczmarek, W.A., and Campbell, S.J. (2000) *Mat. Sci. Forum*, **321**, 802.

179 Baron, V., Gutzmer, J., Rundloef, H., and Tellgren, R. (1998) *Am. Miner.*, **83**, 786.

180 Furuhashi, H., Inagaki, M., and Naka, S. (1973) *J. Inorg. Nucl. Chem.*, **35**, 3009.

181 (a) Verwey, E.J.W. (1939) *Nature*, **144**, 327. (b) Kolodziej, T., Kozlowski, A., Przemyslaw, P., tabis, W., Kakol, Z., Zajac, M., Tarnawski, Z., and Honig, J.M. (2012) *Phys. Rev. B*, **85**, 104301. (c) Senn, M.S., Wright, J.P., Cumby, James, and Attfield, J.P. (2015) *Phys. Rev. B*, **92**, 024104, The literature on the Verwey transition in Fe_3O_4 comprises ~500 papers.

182 (a) Matsuda, M. (2006) *J. Phys. Conf. Ser.*, **51**, 483. (b) Ji, S., Lee, S.-H., Broholm, C., Koo, T.Y., Ratcliff, W., Cheong, S.-W., and Zschack., P. (2009) *Phys. Rev. Lett.*, **103**, 037201.

183 Dou, S. (2015) *Ionics*, **21**, 3001.

184 Rousse, G., Masquelier, C., Rodriguez-Carvajal, J., Elkaim, E., Lauriat, J.-P., and Martinez, J.L. (1999) *Chem. Mater.*, **11**, 3629.

185 (a) Hunter, J.C. (1981) *J. Solid State Chem.*, **39**, 142. (b) Wills, A.S., Raju, N.P., Morin, C., and Greedan, J.E. (1999) *Chem. Mater.*, **11**, 1936.

186 Johnston, D.C., Prakash, H., Zachariasen, W.H., and Viswanathan, R. (1973) *Mater. Res. Bull.*, **8**, 777.

187 Kondo, S., Johnston, D.C., and Miller, L.L. (1999) *Phys. Rev. B*, **59**, 2609.

188 Cussen, E.J. (2010) *J. Mater. Chem.*, **20**, 5167.

6
Perovskite Structure Compounds

Yuichi Shimakawa

Kyoto University, Institute for Chemical Research, Uji, Kyoto 611-0011, Japan

Perovskite is a mineral of calcium titanium oxide with the chemical formula $CaTiO_3$. The mineral was discovered in Russia's Ural Mountains, and was named after the Russian mineralogist Lev Perovski. "Perovskites" here represent the class of compounds that have the same type of crystal structure as $CaTiO_3$. The general chemical formula of the perovskites is ABX_3, in which A is a relatively large cation (such as alkaline metal, alkaline earth metal, or lanthanide ion), B is generally transition metal ion, and X is an oxide or halide ion. The vast majority of perovskites are oxides and fluorides [1], but chlorides, bromides, hydrides, and sulfides with the perovskite structure are also known.

Compounds with the perovskite structure have been studied extensively because they show wide variety of physical and chemical properties, and many of the properties are widely used in technological applications. $BaTiO_3$, for example, is used to make capacitors in electronic devices and $LaCoO_3$ is used industrially as a catalyst for carbon monoxide oxidation. Novel physical properties such as superconductivity, magnetoresistance, and multiferroism found in some perovskite structure oxides shed light on new areas of fundamental condensed matter physics and facilitate researches of exploring new compounds. Perovskites and perovskite-related compounds arguably represent the most important, fundamental, and intriguing family of complex materials.

In this chapter, a large variety of the perovskite structure compounds are reviewed and structural features of the fundamental compounds are highlighted. Because the diverse varieties of physical and chemical properties that the perovskite materials show are described in the following chapters, they will not be described in detail in this chapter. Some synthesis techniques for preparing perovskites are also focused on in different chapters.

Handbook of Solid State Chemistry, First Edition. Edited by Richard Dronskowski, Shinichi Kikkawa, and Andreas Stein.
© 2017 Wiley-VCH Verlag GmbH & Co. KGaA. Published 2017 by Wiley-VCH Verlag GmbH & Co. KGaA.

6.1
Chemical Variety of Perovskites

Most of the perovskites can basically be considered as ionic crystals. There are some perovskites in which the A—X and B—X bonds have covalent characters, and some show metallic behaviors that can be well-described with the electronic band structures. Nevertheless, the "ionic crystal" concept is very useful for the perovskites.

The ABX_3 simple perovskites adopt a few variations of the A and B cation combination in the charge neutral condition. As listed in Table 6.1, typical cation combinations for oxides ($X = O^{2-}$) are $A^+B^{5+}O^{2-}{}_3$, $A^{2+}B^{4+}O^{2-}{}_3$, and $A^{3+}B^{3+}O^{2-}{}_3$. A rare case is $B^{6+}O^{2-}{}_3$ with complete vacancy at the A-site (e.g., ReO_3). Less variety of the charge combinations is known for simple perovskite fluorides and chlorides, and $A^+B^{2+}X^-{}_3$ ($X = F^-$ and Cl^-) combinations are seen in $KMnF_3$, $TlMnCl_3$, and $NaMgF_3$.

In addition to typical valence states of B transition metal cations in oxides, unusually high valence states can also be stabilized in a few perovskites synthesized in strongly oxidizing atmospheres. For instance, although Fe ions in oxides usually have a +2 or +3 valence state, unusually high-valence Fe^{4+} is stabilized in the perovskites $SrFeO_3$ [2] and $CaFeO_3$ [3,4] that are synthesized at high pressures and temperatures. Since the energy levels of the 3d-orbitals in Fe^{4+} are rather low because of the reduced screening effect of the core electrons, the low-lying Fe 3d orbitals strongly hybridize with O 2p orbitals, and as a result, oxygen p-holes (ligand holes, L) are produced in the electronic structure of the perovskites. Thus, a realistic electronic structure of Fe^{4+} with d^6 electron configuration is considered as d^5L, and such an unusual electronic state is accepted in the perovskite structure oxide [5]. The instability of the unusually high valence state is relieved at low temperatures by a characteristic charge behavior (e.g., charge disproportionation) [3].

Table 6.1 Typical charge combinations of simple perovskite oxides.

Charge combinations	Compounds
$B^{6+}O^{2-}{}_3$	ReO_3 [a]
$A^+B^{5+}O^{2-}{}_3$	$KTaO_3$,[a] $NaNbO_3$,[a] $KNbO_3$ [a]
$A^{2+}B^{4+}O^{2-}{}_3$	$SrTiO_3$,[a] $BaTiO_3$,[b] $CaTiO_3$ [c]
$A^{3+}B^{3+}O^{2-}{}_3$	$NdGaO_3$,[c] $LaAlO_3$,[d] $BiFeO_3$ [d]

a) Crystal structures of the compounds at room temperature are cubic with the space group $Pm\text{-}\bar{3}m$ ($a = b = c$). See Section 6.2.
b) The crystal structures at room temperature are tetragonal with the space group $P4mm$ ($a = b \ne c$).
c) The crystal structures at room temperature are orthorhombic with the space group $Pnma$ ($a \ne b \ne c$).
d) The crystal structures at room temperature are rhombohedral with the space groups of $R\text{-}\bar{3}c$ ($LaAlO_3$) and $R3c$ ($BiFeO_3$).

Not only typical but also mixed valence states of the B transition metal ions are often stabilized. When the A cation is partially substituted by another cation with a different valence state, a mixed valence state is given for the B-site cation. For instance, aliovalent substitution at the A-site for $A^{3+}B^{3+}O^{2-}{}_3$ gives $A^{3+}{}_{1-x}A^{2+}{}_x B^{(3+x)+}O^{2-}{}_3$, producing a mixed valence state for $B^{(3+x)+}$ [$(1-x)B^{3+} + xB^{4+}$]. Different from the typical valence states, a mixed valence state of the B cation often induces a conduction behavior in the system. This chemical doping effect is important in controlling the compound's physical properties. While LaMnO$_3$ with Mn^{3+} is a Mott insulator because of the strong on-site Coulomb interaction at the B-site ions, Sr^{2+} substitution make the La$_{1-x}$Sr$_x$MnO$_3$ perovskite conducting [6,7]. The double-exchange interaction between Mn^{3+} and Mn^{4+} mediated by the itinerant holes results in unusual magnetic and transport properties, including a colossal magnetoresistance effect [8,9]. Similar chemical doping to Cu^{2+} oxides produces exotic electronic structures in the strongly electron correlation systems, leading to high-T_c superconductivity in the perovskite-related compounds [10–12].

An interesting class of perovskites, one in which the A-site is occupied by the methylammonium cation (CH$_3$NH$_3$)$^+$, recently attracts much attention because of its use in solar cells. Solar cells that use the methylammonium lead halides as the light absorber material are called "perovskite solar cells" [13,14]. The most commonly studied compounds are CH$_3$NH$_3$PbX$_3$ (X = I, Br, and/or Cl) with optical bandgaps between 1.5 and 2.3 eV. Solar cell efficiencies of devices using these compounds have increased to more than 20% in 2015. The perovskite solar cells are thus becoming commercially attractive. Note also that CsPbCl$_3$ and CsPbBr$_3$ show rather high ionic conductivity of X$^-$ anions [15].

6.2 Crystal Structure of Perovskite

As shown in Figure 6.1,[1] the ideal simple ABX$_3$ perovskite structure is cubic with space group $Pm\text{-}3m$. The B cations have six coordinations by X anions, forming BX$_6$ octahedra. The corner-sharing BX$_6$ octahedra make a three-dimensional network structure with the A-site cavity that is formed by 12 coordinated X anions. Distorted variations of the structure, with lower space group symmetry, are often seen. These variations could be due to distortion of the octahedra, cation displacements within the octahedra, and/or tilting of the octahedra. The octahedral distortion is the result of relieving the electronic instabilities of the B cation, and the so-called Jahn–Teller distortion is an example. The B-site Ti displacement in the TiO$_6$ octahedra in BaTiO$_3$ is also related to the electronic instability [17,18], and the resultant ferroelectric property is now widely utilized in the electroceramics industry. The most frequently occurring distortion is the tilting of the linked BX$_6$

---

1) The crystal structure figures in this chapter are drawn with the VESTA software, which is a three-dimensional visualization program for structural models [16].

(a)   (b)

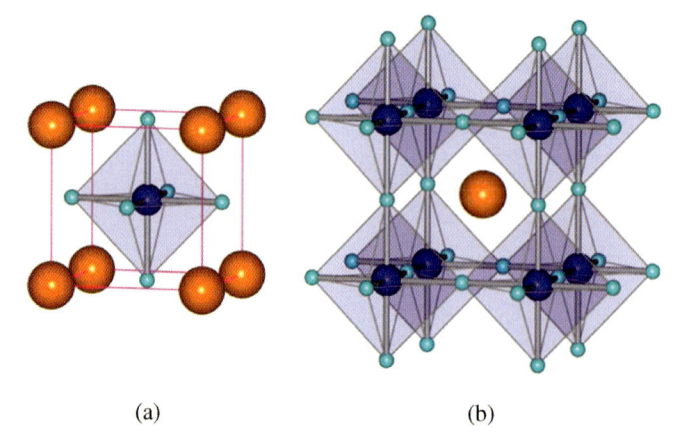

Figure 6.1 Ideal cubic crystal structure of ABX_3 perovskite. The B cations (blue spheres) are coordinated by 6 X anions (small light blue spheres), forming BX_6 octahedra. The A cation (a brown sphere) is located in the corner-sharing BX_6 octahedra cavity. Views of (a) the B cation at the center of the unit cell and (b) the A cation at the center of the unit cell.

octahedra. Because the BX_6 octahedra remains rigid with only slight changes in the B—O—B bond angle, this type of distortion requires less lattice energy. Such distortions are driven by a mismatch in the size of the constituent ions, and were first described by Goldschmidt with a "tolerance factor":

$$t = (r_A + r_X)/\sqrt{2}(r_B + r_X),$$

where r_A, r_B, and r_X are the radii of the A, B, and X ions [19]. When the ABX_3 perovskite structure is considered as a close-packed array of hard ionic spheres, the ideal size combination ($t = 1$) of the ions requires the A—X bond distance to be $\sqrt{2}$ of the B—X bond distance (Figure 6.2). A tolerance factor less than 1 implies that the A cation is smaller than the cavity made by the BX_6 octahedra network, and $t > 1$ implies that the A cation is larger than the cavity. The perovskite structure is often stabilized for ion combinations when $0.75 \leq t < 1$, as shown in Figure 6.3 for $A^{3+}B^{3+}O^{2-}_3$ [20]. The more the t deviates from the unity, the greater the distortion from the ideal cubic structure.

A convenient notation to describe the tilts and rotations was developed by Glazer [21,22]. The notation specifies the magnitude and phase of the octahedral rotations about the three orthogonal Cartesian axes of the cubic perovskite. A superscript is used to denote the phase of the octahedral tilting in neighboring layers. A positive superscript means the neighboring octahedra tilt in the same direction (are in phase) and a negative superscript means the neighboring octahedra tilt in opposite directions (are out of phase). Rotation of one octahedron causes the four adjacent octahedra in the same layer to rotate by the same angle in the opposite direction. Lattice connectivity is such that rotations of the octahedra in the layers above and below are not geometrically constrained to the initial rotation and can occur in phase with it or out of phase with it.

(a) ABO₃

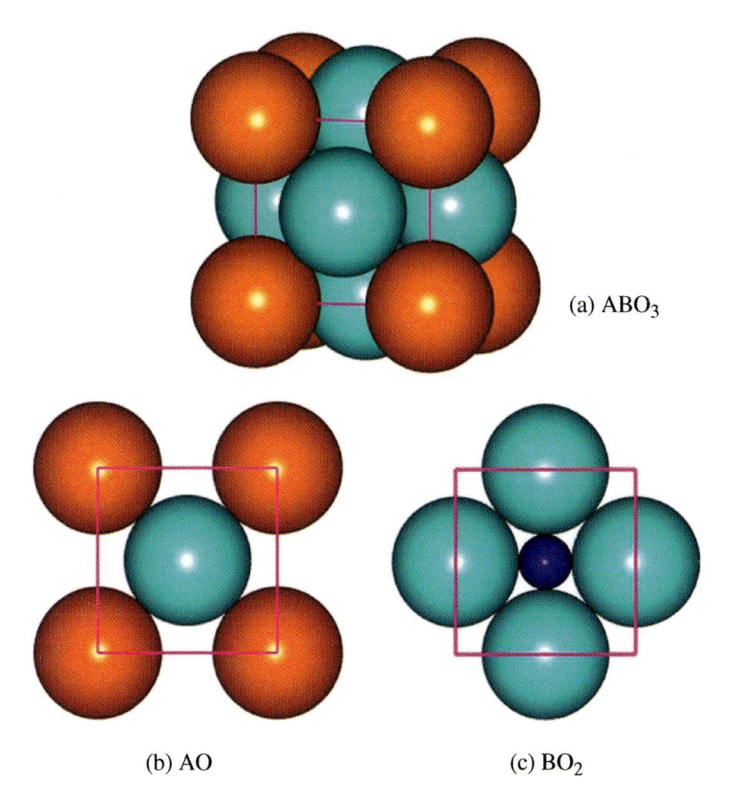

(b) AO

(c) BO₂

Figure 6.2 (a) Schematic drawing of simple perovskite structure of ABX$_3$ with each ionic sphere size proportional to the ionic radius. Views of (b) (0 0 1) plane with AO and (c) (0 0 1/2) plane with BO$_2$.

An example of such distortion is seen when comparing the perovskite crystal structures of SrTiO$_3$ and CaTiO$_3$ (Figure 6.4). The ionic radii of 1.44 Å for Sr^{2+} with 12-fold oxygen coordination, 0.605 Å for Ti^{4+} with sixfold oxygen coordination, and 1.40 Å for O^{2-} [23] give a tolerance factor of 0.996 for SrTiO$_3$. The crystal structure of SrTiO$_3$ is cubic with the $Pm\text{-}3m$ ($a^0a^0a^0$) space group symmetry, and the structure parameters are listed in Table 6.2. The observed cubic lattice constant 3.905 Å is quite close to $2(r_{Ti}+r_O)=4.010$ Å and $\sqrt{2}\times(r_{Sr}+r_O)=4.016$ Å. For CaTiO$_3$, on the other hand, the ionic radius of 1.34 Å for Ca^{2+} with 12-fold oxygen coordination gives a t of 0.966, indicating greater mismatch of the close-packed array in the ionic crystal. The crystal structure of CaTiO$_3$ is orthorhombic with the $Pnma$ ($a^-b^+a^-$) space group symmetry. The lattice constants are $a=5.4423$ Å, $b=7.6401$ Å, and $c=5.3796$ Å.

The distortion caused by the mismatch in the size of A-site cations to a B-site cation often refers to "chemical pressure." The A-site cation plays a role in not modifying the size of the structure but also tuning the physical properties. A good example is seen in the insulator–metal transition in the ANiO$_3$ perovskites

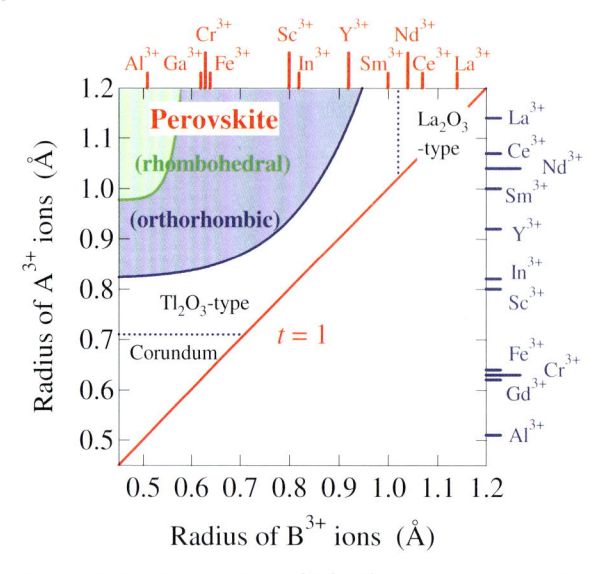

Figure 6.3 Classification of the $A^{3+}B^{3+}O^{2-}{}_3$ compounds according to the constituent radii. The red line indicates the compounds with the tolerance factor $t = 1$. (The figure is reproduced with the data in the reference [20].)

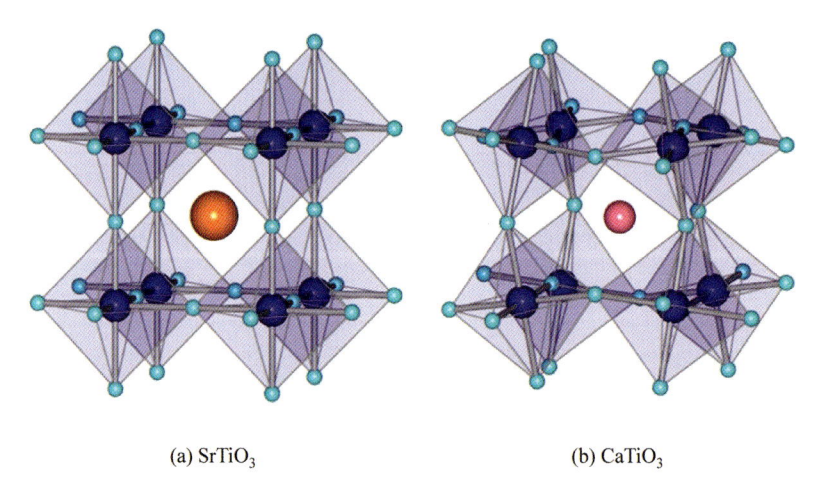

(a) SrTiO$_3$

(b) CaTiO$_3$

Figure 6.4 Crystal structures of (a) cubic SrTiO$_3$ and (b) orthorhombic CaTiO$_3$.

with lanthanide ions at the A-site [24–26]. For the small A-site ions like Eu and Sm, as shown in Figure 6.5, the insulator–metal transition occurs at a higher temperature than the antiferromagnetic ordering. By increasing the size of the lanthanide ions, only an insulator–metal transition is observed. These significant changes in the magnetic and electronic properties are caused by the change of the charge-transfer energy gap induced by an increase in the electronic bandwidth depending on the size of the A-site lanthanide ions.

6.2 Crystal Structure of Perovskite

Table 6.2 Structure parameters of SrTiO₃ and CaTiO₃.

Atom	Site	x	y	z
SrTiO₃: Space group; Pm-3m, a = 3.905 Å				
Sr^{2+}	1b	1/2	1/2	1/2
Ti^{4+}	1a	0	0	0
O^{2-}	3d	1/2	0	0
CaTiO₃: Space group; Pnma, a = 5.4419 Å, b = 7.6400 Å, c = 5.3785 Å				
Ca^{2+}	4c	0.5330	1/4	0.5072
Ti^{4+}	4a	1/2	0	0
$O^{2-}(1)$	4c	0.4893	1/4	0.0722
$O^{2-}(2)$	8d	0.2842	0.0346	0.7174

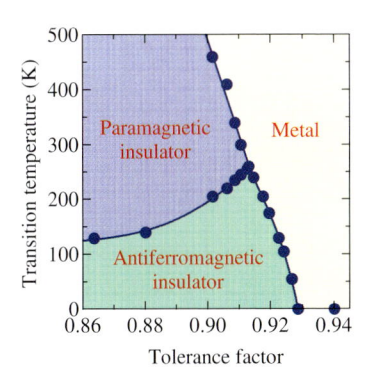

Figure 6.5 Phase diagram for ANiO₃, with lanthanide ions at the A-site, as a function of the tolerance factor. (The figure is reproduced with the data in the reference [24].)

From the ideal $Pm\text{-}3m$ cubic perovskite structure, 23 tilt systems are derived by the tilting of rigid octahedra, and 15 possible distorted structures are lead based on a group theoretical analysis [27–29]. The schematic diagram Figure 6.6 shows the group–subgroup relationships among the 15 space groups. A solid line indicates a transition that is allowed to be continuous and a dashed line indicates one that is required by Landau theory to be first order.

As mentioned above, most of the perovskites can basically be considered as ionic crystals. Therefore, the bond valence analysis from the crystal structure data is quite useful. The bond valence is related to the number of electrons contributing to the making of chemical bonds, and is defined as

$$S_{ij} = \exp[(R_{ij} - R_0)/B],$$

where R_{ij} is the observed bond length, R_0 is a tabulated parameter, and B is an empirical constant, typically 0.37 Å [30,31]. The sum of the bond valences of the

6 Perovskite Structure Compounds

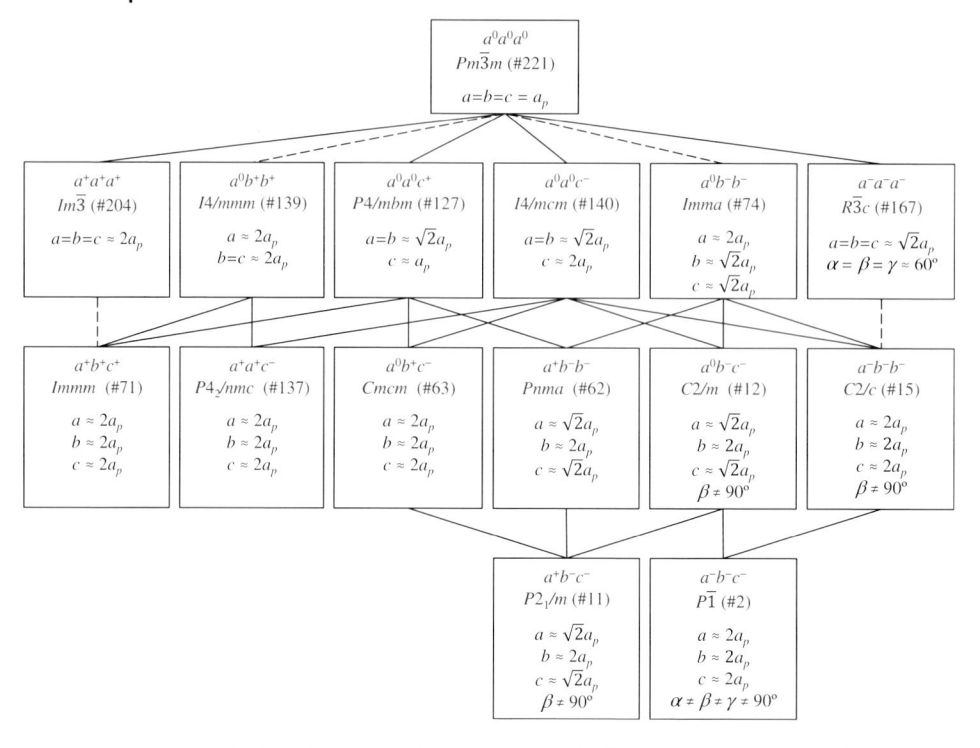

```
                          a⁰a⁰a⁰
                         Pm3̄m (#221)
                         a=b=c = aₚ
```

$a^+a^+a^+$	$a^0b^+b^+$	$a^0a^0c^+$	$a^0a^0c^-$	$a^0b^-b^-$	$a^-a^-a^-$
$Im\bar{3}$ (#204)	$I4/mmm$ (#139)	$P4/mbm$ (#127)	$I4/mcm$ (#140)	$Imma$ (#74)	$R\bar{3}c$ (#167)
$a=b=c \approx 2a_p$	$a \approx 2a_p$ $b=c \approx 2a_p$	$a=b \approx \sqrt{2}a_p$ $c \approx a_p$	$a=b \approx \sqrt{2}a_p$ $c \approx 2a_p$	$a \approx 2a_p$ $b \approx \sqrt{2}a_p$ $c \approx \sqrt{2}a_p$	$a=b=c \approx \sqrt{2}a_p$ $\alpha = \beta = \gamma \approx 60°$

$a^+b^+c^+$	$a^+a^+c^-$	$a^0b^+c^-$	$a^+b^-b^-$	$a^0b^-c^-$	$a^-b^-b^-$
$Immm$ (#71)	$P4_2/nmc$ (#137)	$Cmcm$ (#63)	$Pnma$ (#62)	$C2/m$ (#12)	$C2/c$ (#15)
$a \approx 2a_p$ $b \approx 2a_p$ $c \approx 2a_p$	$a \approx 2a_p$ $b \approx 2a_p$ $c \approx 2a_p$	$a \approx 2a_p$ $b \approx 2a_p$ $c \approx 2a_p$	$a \approx \sqrt{2}a_p$ $b \approx 2a_p$ $c \approx \sqrt{2}a_p$	$a \approx \sqrt{2}a_p$ $b \approx 2a_p$ $c \approx \sqrt{2}a_p$ $\beta \neq 90°$	$a \approx 2a_p$ $b \approx 2a_p$ $c \approx 2a_p$ $\beta \neq 90°$

$a^+b^-c^-$	$a^-b^-c^-$
$P2_1/m$ (#11)	$P\bar{1}$ (#2)
$a \approx \sqrt{2}a_p$ $b \approx 2a_p$ $c \approx \sqrt{2}a_p$ $\beta \neq 90°$	$a \approx 2a_p$ $b \approx 2a_p$ $c \approx 2a_p$ $\alpha \neq \beta \neq \gamma \neq 90°$

Figure 6.6 Schematic diagram showing the group–subgroup relationships of space groups for possible octahedral tilt systems in simple perovskites. A dashed line joining a group with its subgroup indicates that the corresponding phase transition is required by Landau theory to be first order.

nearest neighbor 6 B—X bonds gives the valence state of B, and the sum of the bond valences of the 12 A—X bonds gives the valence state of the A cation. Based on this bond valence sum analysis, a structure prediction software SPuDS (Structure Prediction Diagnostic Software) was developed, and it predicts the crystal structures of perovskites, including those distorted by tilting of the octahedral [32]. The calculation algorithm of the program is to optimize the crystal structure in order to minimize the difference between the calculated bond valence sum and the formal valence of the A and B cations. To simplify the optimization process, SPuDS restricts the octahedra to remain rigid with six equivalent B—X bonds and 90° X–B–X angles. Because most distorted perovskites show very little distortion of the BX_6 octahedra, this seems to be a reasonable restriction. The size of the octahedron and the magnitude of the octahedral tilting distortion are then considered for possible tilt systems in ABX_3. The stability of the compound is judged by the quantity of the difference between the calculated bond valence sum and the formal valence of the A and B cations that is referred to a global instability index obtained from the discrepancy factor [33,34]. It is empirically known that compounds with a global instability index less than

0.1 valence unit can highly be synthesized in an unstrained structure, and that compounds with a global instability index as large as 0.2 valence unit can be synthesized in a structure with lattice-induced strains. Many of the perovskite structure compounds predicted have actually been obtained.

6.3 Ion Order in Perovskites

Perovskites show considerable compositional flexibility. Not only can they be oxides, fluorides, chlorides, bromides, or sulfides, but partial chemical substitution can also be made to the anion site producing compounds like oxynitrides and oxyfluorides. The oxynitrides and the oxyfluorides are focused in other chapters and are not described here. Cation order in oxides is representatively highlighted here because there are huge varieties of compounds that give good examples to describe the extended structure family. The observed pronounced properties are strongly related to the cation order.

Both A- and B-sites in the perovskite structure oxides can accommodate chemical substitutions. Although A and B cation arrangements in ABO_3 have the same simple cubic topology in the perovskite structure, they typically adopt different patterns of chemical order. The B cation order occurs more readily than the A cation order. In the partially cation-substituted perovskites like $AA'B_2O_6$ and $A_2BB'O_6$, the ordered arrangements readily occur if the difference between the valence states and/or ionic sizes of the two cations at the site is significant.

6.3.1 B-Site Cation Order

The ways of order of the B cations in the double perovskites, in which half of the B cation is substituted by another element, resulting in the chemical formula of $A_2BB'O_6$, are first described. Three simple patterns of the order of the B cations are shown in Figure 6.7. In addition to random (disordered) arrangement of the cations, rock salt, layered, and columnar orders are known. Simple perovskites

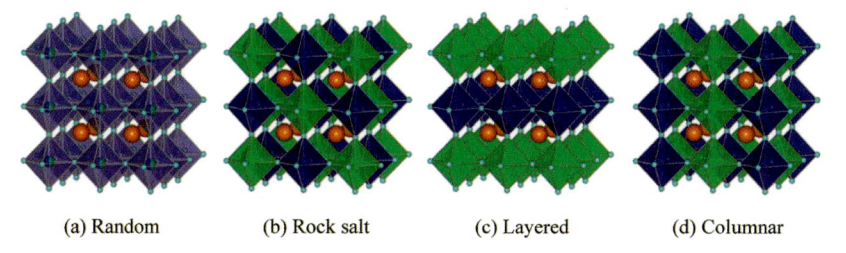

Figure 6.7 B-site cation order patterns in the $A_2BB'O_6$ double perovskites: (a) random, (b) rock salt, (c) layered, and (d) columnar.

Table 6.3 B-site cation orders and the charge differences of double perovskite oxides [35–37].

Charge difference	Charge combinations	Compounds
(1) B-site cation rock salt order		
(1–1) 0	$A^{2+}_2B^{4+}B'^{4+}O_6$	Ba_2PrPtO_6
(1–2) +1	$A^{2+}A''^{3+}B^{3+}B'^{4+}O_6$	$CaLaMnCoO_6$
(1–3) +2	$A^{2+}_2B^{3+}B'^{5+}O_6$	Sr_2FeMoO_6, Ca_2AlNbO_6
	$A^{3+}_2B^{2+}B'^{4+}O_6$	La_2CuMnO_6, Bi_2NiMnO_6
(1–4) +3	$A^{2+}A''^{3+}B^{2+}B'^{5+}O_6$	$CaLaMgMoO_6, SrLaCuTaO_6$
(1–5) +4	$A^{2+}_2B^{2+}B'^{6+}O_6$	Sr_2CuWO_6, Ba_2MgWO_6
	$A^{3+}_2B^{+}B'^{5+}O_6$	La_2LiVO_6, La_2LiIrO_6
(1–6) +5	$A^{2+}A''^{3+}B^{+}B'^{6+}O_6$	$SrLaLiWO_6, BaLaLiMoO_6$
(1–7) +6	$A^{2+}_2B^{+}B'^{7+}O_6$	Ca_2LiReO_6, Sr_2NaOsO_6
(2) B-site cation layered order		
(2–1) 0	$A^{2+}_2B^{4+}B'^{4+}O_6$	Ca_2FeMnO_6
(2–2) +2	$A^{3+}_2B^{2+}B'^{4+}O_6$	La_2CuSnO_6, La_2CuZrO_6
(3) B-site cation columnar order		
(3–1) +1	$A^{2+}A''^{3+}B^{3+}B'^{4+}O_6$	$SrNdMn^{3+}Mn^{4+}O_6$

with the charge combinations $A^{2+}B^{4+}O_3$ and $A^{3+}B^{3+}O_3$ adopt such B cation orders. From the $A^{2+}B^{4+}O_3$ perovskites, $A^{2+}_2B^{4+}B'^{4+}O_6$, $A^{2+}_2B^{3+}B'^{5+}O_6$, $A^{2+}_2B^{2+}B'^{6+}O_6$, and $A^{2+}_2B^{+}B'^{7+}O_6$ are known. From the $A^{3+}B^{3+}O_3$ perovskites, $A^{3+}_2B^{3+}B'^{3+}O_6$, $A^{3+}_2B^{2+}B'^{4+}O_6$, and $A^{3+}_2B^{+}B'^{5+}O_6$ are possible charge combinations. In those ordered double perovskites, the charge differences of +2, +4, and +6 are the main driving force of the structure order. The rock salt-type order is the most common, while the other two arrangement patterns are rare. Because each B cation has six nearest neighbor B' cations, the separation of the two cations minimizes the electrostatic lattice energy. When the B and B' cations have the same charges (e.g., $A^{2+}_2B^{4+}B'^{4+}O_6$), the difference in the cation size plays an important role in stabilizing the ordered structure. Interestingly, with the random arrangement of the A-site cations with different charges ($AA''BB'O_6$), an even number of charge difference can also lead to the B cation order in the double perovskites. The examples are $A^{2+}A''^{3+}B^{3+}B'^{4+}O_6$, $A^{2+}A''^{3+}B^{2+}B'^{5+}O_6$, and $A^{2+}A''^{3+}B^{+}B'^{6+}O_6$. Representative compounds with the B cation ordered arrangements in those typical charge combinations are listed in Table 6.3. Actually, large numbers of the B-cation order double perovskites have been obtained, and there are good review articles on them [35–37].

B cation rock salt order, which is the most frequently stabilized, usually requires the unit cell to be doubled, giving the face-centered cubic symmetry (Fm-$\bar{3}m$). The octahedral tilt is also induced, and hence lower symmetry structures are stabilized. From the group theoretical analysis, 12 different structures are identified (Figure 6.8) [38]. Layered and columnar B cation orders are less

6.3 Ion Order in Perovskites

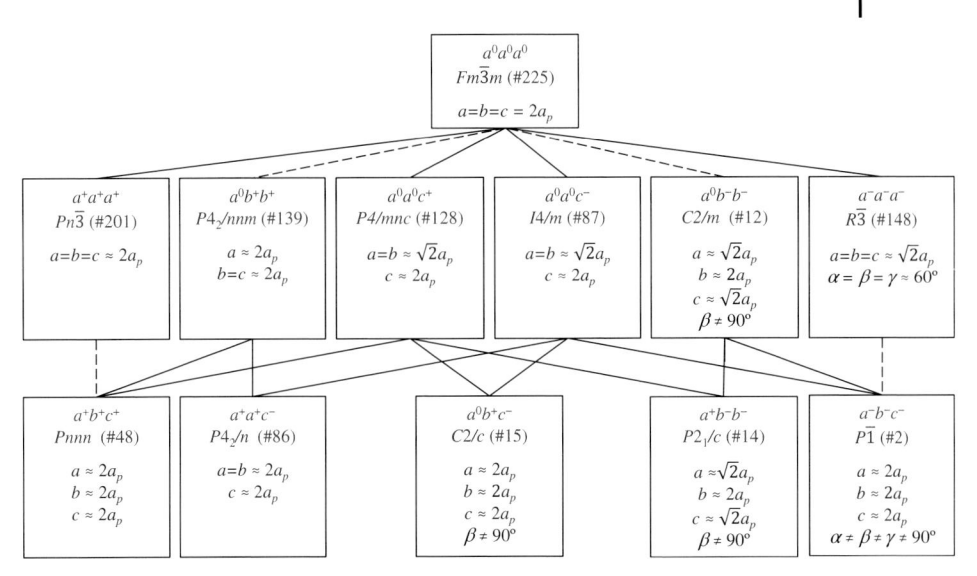

```
                          a⁰a⁰a⁰
                         Fm3̄m (#225)
                         a=b=c = 2aₚ
```

$a^+a^+a^+$	$a^0b^+b^+$	$a^0a^0c^+$	$a^0a^0c^-$	$a^0b^-b^-$	$a^-a^-a^-$
$Pn\bar{3}$ (#201)	$P4_2/nnm$ (#139)	$P4/mnc$ (#128)	$I4/m$ (#87)	$C2/m$ (#12)	$R\bar{3}$ (#148)
$a=b=c \approx 2a_p$	$a \approx 2a_p$ $b=c \approx 2a_p$	$a=b \approx \sqrt{2}a_p$ $c \approx 2a_p$	$a=b \approx \sqrt{2}a_p$ $c \approx 2a_p$	$a \approx \sqrt{2}a_p$ $b \approx 2a_p$ $c \approx \sqrt{2}a_p$ $\beta \neq 90°$	$a=b=c \approx \sqrt{2}a_p$ $\alpha = \beta = \gamma \approx 60°$

$a^+b^+c^+$	$a^+a^+c^-$	$a^0b^+c^-$	$a^+b^-b^-$	$a^-b^-c^-$
$Pnnn$ (#48)	$P4_2/n$ (#86)	$C2/c$ (#15)	$P2_1/c$ (#14)	$P\bar{1}$ (#2)
$a \approx 2a_p$ $b \approx 2a_p$ $c \approx 2a_p$	$a=b \approx 2a_p$ $c \approx 2a_p$	$a \approx 2a_p$ $b \approx 2a_p$ $c \approx 2a_p$ $\beta \neq 90°$	$a \approx \sqrt{2}a_p$ $b \approx 2a_p$ $c \approx \sqrt{2}a_p$ $\beta \neq 90°$	$a \approx 2a_p$ $b \approx 2a_p$ $c \approx 2a_p$ $\alpha \neq \beta \neq \gamma \neq 90°$

Figure 6.8 Schematic diagram showing the group–subgroup relationships of space groups for possible octahedral tilt systems in ordered double perovskites. A dashed line joining a group with its subgroup indicates that the corresponding phase transition is required by Landau theory to be first order.

common, and only a few compounds with such orders are known. Ca_2FeMnO_6 has a layered arrangement of Fe^{4+} and Mn^{4+} in the perovskite structure [39], but it can neither exist in nature nor be obtained by a normal synthesis method. Low-temperature topotactic oxidation with ozone changes the oxygen-deficient perovskite (brownmillerite structure) $Ca_2Fe^{3+}Mn^{3+}O_5$ to $Ca_2Fe^{4+}Mn^{4+}O_6$ without disturbing the B cation arrangement. On the other hand, the Jahn–Teller distortion of the Cu^{2+} cation plays an important role in stabilizing the unusual layered structure of La_2CuSnO_6 [40].

The B cation order in the perovskites often modifies their chemical and physical properties significantly, and the magnetism in the double perovskites $A_2BB'O_6$ is of particular interest. A simple perovskite with the magnetic cations at the B-site usually shows antiferromagnetism due to superexchange interaction between the B cations via oxygen (B–O–B interaction). On the other hand, in a double perovskite with the B cations in the rock salt order, the magnetic interaction is dominated by the linear B–O–B' superexchange interaction and according to the Kanamori–Goodenough rule can be ferromagnetic [41–43]. La_2CuMnO_6 and La_2NiMnO_6 are the examples and they show ferromagnetism [44]. When the Bi cations with the $6s^2$ electron lone pair are included at the A-site, as in Bi_2NiMnO_6, multiferroic properties, in which both ferroelectricity and ferromagnetism coexist, are found [45]. The rock salt-ordered double perovskite Sr_2FeMoO_6 with a half-metallic electronic structure, which produces fully spin-polarized conduction electrons, attracts considerable attention owing to not only its intriguing fundamental physics but also its potential utility in

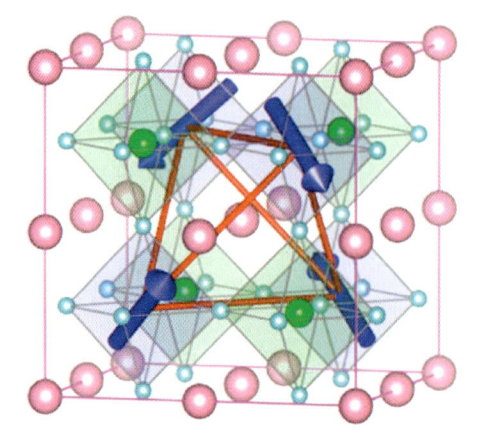

Figure 6.9 Spin glass magnetic structure of the double perovskite $A_2BB'O_6$. Both magnetic B and nonmagnetic B' cations are ordered in a rock salt-type manner forming a tetrahedral spin sublattice.

technological applications of large magnetoresistance [46]. Another interesting example concerns spin frustration. The B (or B') cation arrangement in the double perovskite $A_2BB'O_6$ with the B/B' rock salt ordering forms a tetrahedral spin sublattice. When the magnetic interaction of B cations is antiferromagnetic, this spin arrangement results in geometrical spin frustration as shown in Figure 6.9. In the pseudocubic double perovskites, Sr_2FeSbO_6 and Ca_2FeSbO_6 with the magnetic Fe^{3+} and nonmagnetic Sb^{5+} cations, for example, frustration of the Fe^{3+} spins leads to spin glass-like behaviors at low temperatures [47,48].

Other types of B cation order such as 1:2 and 1:3 orders are known, although relatively few examples have been reported compared with the many examples with the 1:1 order described above. $Ba_3MgTa_2O_9$ and $Ba_3ZnNb_2O_9$ have the B cation orders in layered manner that run perpendicular to the $\langle 111 \rangle$ direction of the simple perovskite structure. The B cation layered sequence of B–B'–B'–B is essentially the closest analogues to rock salt order (Figure 6.10). The microstructure of the compound is composed of nanosized domains with all possible orientation variations of the 1:2 order, and the resultant high volume of disordered and elastically strained domain boundaries can affect the dielectric properties adversely. The observed very low dielectric losses in the compounds are suited for dielectric resonators in microwave communication devices. There is a good review of the unusual B cation order perovskites [49].

6.3.2
A-Site Cation Order

A-site cation order in the double perovskites is much less common than B-site cation order [36]. One of the main reasons for this observation is related to possible differences in the valence states of the cations. While the B- and B'-site

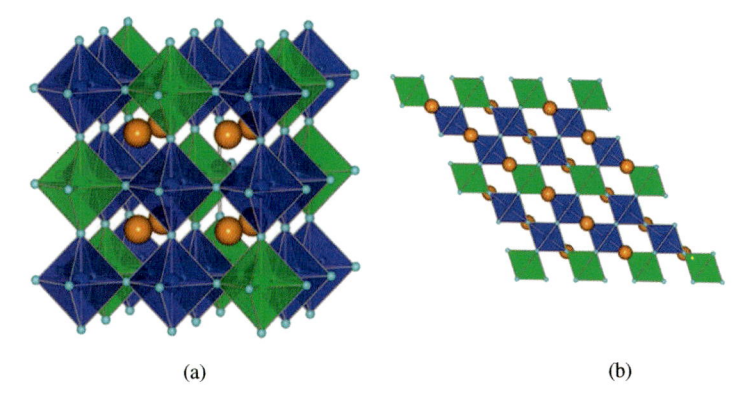

Figure 6.10 (a) Crystal structure of the B-site cation 1:2 order perovskite $A_3BB'_2O_9$. (b) The B cations order in a layered manner that runs perpendicular to the ⟨111⟩ direction of the simple perovskite structure.

cation valence state difference can be as large as +6 (e.g., as seen in Ca_2LiReO_6), the A- and A′-site cation valence state difference is limited to +1 or +2. The exceptional order of the isovalent A-site cations concerns $PbCaTi_2O_6$ that is reported to have short-range rock salt order of Pb^{2+} and Ca^{2+} at the A-site [50]. A-site cation columnar order is found in $CaFeTi_2O_6$, and the large size difference between Ca^{2+} and Fe^{2+} drives the order by the $a^+a^+c^-$ octahedral tilt [51]. The layered order pattern is relatively common in the A-site cation arrangement, but simple compounds with the $AA'B_2O_6$ chemical formula are extremely rare. The A-site cation layered order is often found in combination with B cation rock salt order ($NaLaMgWO_6$ and $NaLaScNbO_6$ [52]) or anion vacancy order ($YBaMn_2O_5$ [53]). Low-temperature topotactic oxidation of the A-site cation and oxygen vacancy order compound, $LaBaMn_2O_5$, produces a metastable phase of the simple A-site cation layered perovskite, $LaBaMn_2O_6$, where the A-site cation layered order is kept intact [54,55].

A significant in-phase $a^+a^+a^+$ tilt of the BO_6 octahedra from the ABO_3 perovskite gives 1:3 ordered arrangement of the A-site cations, producing the chemical formula of $AA'_3B_4O_{12}$. As shown in the crystal structure in Figure 6.11, the structure has the $Im\text{-}3$ cubic space group symmetry, and 4 of the originally 12 near neighbor A—O bonds become short, forming square-coordinated units that align perpendicular to each other. This special alignment of the $A'O_4$ square units and the significant tilting of the corner-sharing BO_6 octahedra make the $2a \times 2a \times 2a$ unit cell. The A′-site in $AA'_3B_4O_{12}$, unlike that in other perovskite structure oxides, can accommodate transition metal cations. A large number of the compounds are reported to crystallize with this structure. While some of the compounds, such as $CaCu_3Ti_4O_{12}$ [56], can be prepared at ambient conditions, most need to be synthesized under high pressure [57].

Cu and Mn ions in various valence states are typical transition metal cations accommodated at the A′-site. Many of the known compounds have

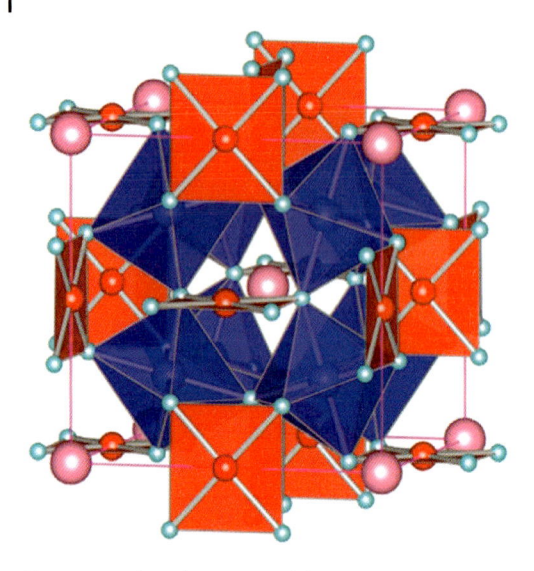

Figure 6.11 Crystal structure of the A-site cation-ordered quadruple perovskite AA′$_3$B$_4$O$_{12}$. The A cation (large pink sphere) and A′ cation (middle-size red sphere) are ordered at the A-site at a ratio of 1 : 3, and the B-site cations form BO$_6$ octahedra. The A′-site cations form square-coordinated units that align perpendicular to each other.

the charge combinations A^{2+}A′$^{2+}_3$B′$^{4+}_4$O$_{12}$ (e.g., Ca^{2+}Cu$^{2+}_3$Ti$^{4+}_4$O$_{12}$ and Ca^{2+}Cu$^{2+}_3$Mn$^{4+}_4$O$_{12}$ [58]) and A^{3+}A′$^{3+}_3$B′$^{3+}_4$O$_{12}$ (e.g., YMn$^{3+}_3$Al$^{3+}_4$O$_{12}$ [59]). Mixed valence states are also often introduced into the B-site cations as seen in compounds with the charge combination like A^{3+}A′$^{2+}_3$B′$^{3.75+}_4$O$_{12}$ (e.g., La^{3+}Cu$^{2+}_3$Mn$^{3.75+}_4$O$_{12}$ [60] and Bi^{3+}Cu$^{2+}_3$Mn$^{3.75+}_4$O$_{12}$ [61]). Unusual valence states of cations can also be stabilized at the square-planar coordinated A′-site: the unusually high-valence Cu^{3+} seen in intersite charge-transferred Ca^{2+}Cu$^{3+}_3$Fe$^{3+}_4$O$_{12}$ [62], for instance, and the unusually low valence of Mn$^{1.75+}$ in La^{3+}Mn$^{1.67+}_3$V$^{4+}_4$O$_{12}$ [63]. Unusually, high valence states of the cations are also stabilized at the B-site, such as Fe^{4+} in CaCu$_3$Fe$_4$O$_{12}$ [64] and Fe$^{3.75+}$ in LaCu$_3$Fe$_4$O$_{12}$ [62]. With decreasing temperature, the instabilities of such high valence states of Fe are relieved by distinct charge transitions: charge disproportionation (4Fe^{4+} → 2Fe^{3+} + 2Fe^{5+}) in CaCu$_3$Fe$_4$O$_{12}$ and intersite charge transfer (3Cu^{2+} + 4Fe$^{3.75+}$ → 3Cu^{3+} + 4Fe^{3+}) in LaCu$_3$Fe$_4$O$_{12}$. Although these two kinds of charge transitions look completely different from each other in simple ionic models, they can both be explained by the localization of ligand holes. The ligand holes in the charge-disproportionated CaCu$_3$Fe$_4$O$_{12}$ are localized at the Fe-O sites alternately (4d^{5}L → 2d^5 + 2d^{5}L^2), and the ligand holes in the charge-transferred LaCu$_3$Fe$_4$O$_{12}$ are localized at the Cu-O sites (3d^9 + 4d^{5}L$^{0.75}$ → 3d^{9}L + 4d^5) as shown in Figure 6.12 [65–67].

Because the transition metal ions are included at both the A′- and B-sites in AA′$_3$B$_4$O$_{12}$, competitive and/or cooperative interactions of A′–A′ and A′–B in

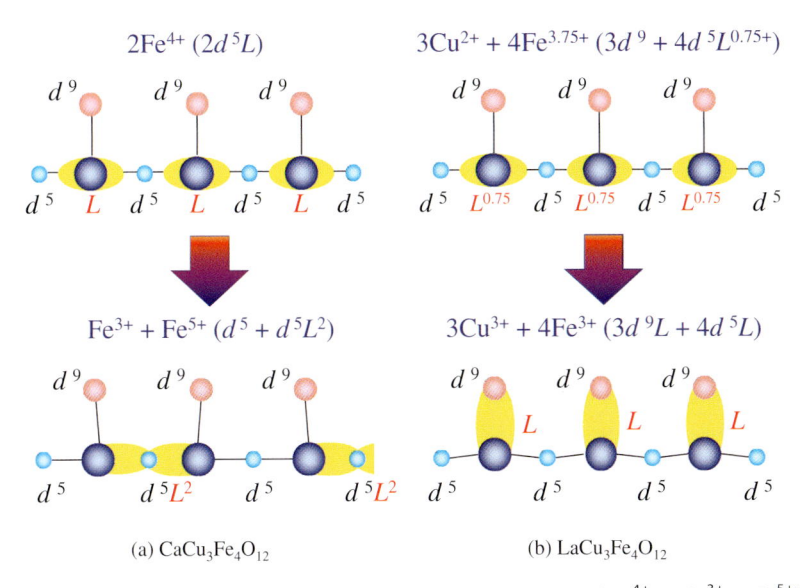

Figure 6.12 Schematic drawings of (a) charge disproportionation ($4Fe^{4+} \rightarrow 2Fe^{3+} + 2Fe^{5+}$) in $CaCu_3Fe_4O_{12}$ and (b) intersite charge transfer ($3Cu^{2+} + 4Fe^{3.75+} \rightarrow 3Cu^{3+} + 4Fe^{3+}$) in $LaCu_3Fe_4O_{12}$, based on the ligand hole models.

addition to B–B usually seen in ABO_3 simple perovskites give rise to diverse and intriguing physical properties. An interesting observation in $AA'_3B_4O_{12}$ is that, unusual A-site cation magnetism can be seen when the B cation is nonmagnetic, whereas the magnetic properties of many of the perovskites originate from the interaction of the B-site transition metal cations. The $A'(Cu^{2+})$–$A'(Cu^{2+})$ magnetic interaction can be either ferromagnetic (as in $CaCu_3Ge_4O_{12}$ [68]) or antiferromagnetic (as in $CaCu_3Ti_4O_{12}$ [69]). The special arrangement of the square CuO_4 units that align perpendicular to each other induces direct-exchange or weak-superexchange ferromagnetic interaction between the nearest-neighbor A'-site Cu^{2+} spins [70]. In $CaCu_3Ti_4O_{12}$, strong hybridization of Ti-3d, Cu-3d, and O-2p orbitals mediates antiferromagnetic superexchange interaction through the Cu–O–Ti–O–Cu path that overcomes the ferromagnetic interaction, giving the G-type antiferromagnetic spin structure [71–73]. Another recent interesting finding with this structure compound concerns multiferroism in $CaMn_3Mn_4O_{12}$ [74,75]. The $Im\text{-}\bar{3}$ cubic cell is distorted to the rhombohedral one at low temperatures and a giant improper ferroelectric polarization is induced by an incommensurate helical magnetic structure. The multiferroicity can be explained by the ferroaxial coupling of the magnetic chirality to the macroscopic structural rotations associated with the $R\text{-}\bar{3}$ paramagnetic phase. There are review papers on compounds with this structure that have intriguing physical properties [36,57,76].

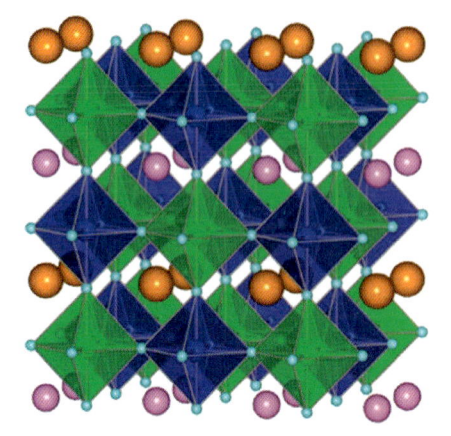

Figure 6.13 Crystal structure of the A- and B-site cation-ordered perovskite AA′BB′O$_6$. The A-site cations are ordered in a layered manner and the B-site cations in a rock salt manner.

6.3.3
A- and B-Site Cation Order

Both A-site cations and B-site cations in double perovskites can be simultaneously ordered. The layered order at the A site and the rock salt order at the B site are found in AA′BB′O$_6$, and the A cation layered order appears to be linked to the B cation order (Figure 6.13). The NaLaMgWO$_6$ and NaLaScNbO$_6$ [52] described above are the examples and the crystal structure has $P2_1$ space group symmetry with $a^-a^-c^+$ octahedral tilt.

The body-centered 1 : 3 A-site cation order and the rock salt B-site cation order lead to AA′$_3$B$_2$B′$_2$O$_6$ with the $Pn\text{-}\bar{3}$ cubic space group symmetry. CaCu$_3$B$_2$Sb$_2$O$_{12}$ (B = Ga, Cr, Fe) [77–79] have this type of A- and B-site cation orders. Considering that the B-site Fe spin sublattice in the double perovskite Ca$_2$FeSbO$_6$ adopts a tetrahedral arrangement and that the Fe^{3+}–Fe^{3+} antiferromagnetic interaction results in geometrical spin frustration, in CaCu$_3$Fe$_2$Sb$_2$O$_{12}$ the antiferromagnetic spin frustration is relieved by introducing the Cu^{2+} spins at the A′ site. When the magnetic Re^{5+} is included at the B′-site, the ferrimagnetic spin structure Ca^{2+}Cu^{2+}(↑)$_3$Fe^{3+}(↑)Re^{5+}(↓)$_2$O$_{12}$ is stabilized below the very high magnetic transition temperature of 560 K (Figure 6.14). The compound has a half-metallic electronic structure in which fully spin-polarized conduction electrons are produced [80].

6.3.4
X Anion Vacancy Order

The perovskite structure often accepts anion vacancy. In oxides, anion-deficient perovskites ABO$_{3-\delta}$ are of technological interest as oxide-ion conductors in which conduction is caused by the oxide-ion hopping through the vacancies.

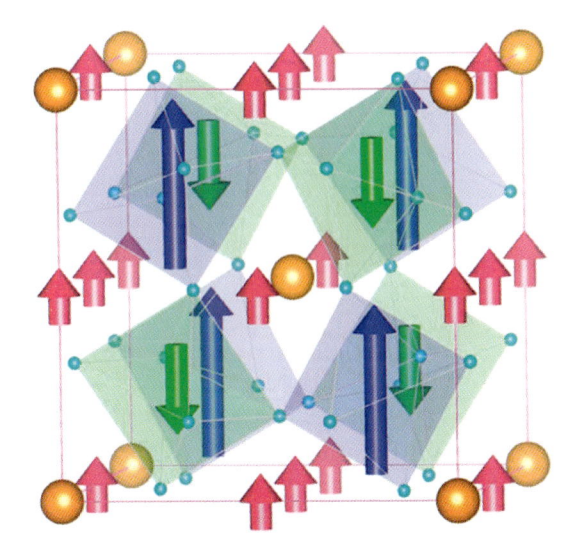

Figure 6.14 Ferrimagnetic spin structure of the A- and B-site cation-ordered perovskite $CaCu_3Fe_2Re_2O_{12}$. Cu^{2+} ($S = 1/2$) spins at the A'-site couple ferromagnetically with Fe^{3+} ($S = 5/2$) spins at the B-site and antiferromagnetically with Re^{5+} ($S = 1$) spins at the B'-site, leading to the ferrimagnetism.

Although at high temperature the oxygen vacancies are distributed randomly, strong electrostatic forces between the vacancies usually create anion-ordered phases at temperatures typically below 700–800 °C [81,82].

The anion vacancy reduces the anion coordination of the cations. In $ABaB_2O_5$, the oxygen vacancies, which lie exclusively within the A layers, are ordered in combination with the A-site cation layered order. In this compound, the B-site cation coordination environment is reduced to square pyramidal and the A-site cation coordination number is reduced from 12 to 8 (Figure 6.15). The A^{3+} cation is typical for $ABaB_2O_5$ and thus the mixed valence state is induced for the B-site cation. This 2.5+ half integer oxidation state of the B cation is particularly favorable for charge ordering. When B = Mn, as in $YBaMn_2O_5$, the charge order of Mn^{2+} and Mn^{3+} occur in a rock salt-type manner [53]. The charge order of $Fe^{2.5+}$ in $YBaFe_2O_5$ [83] and $Co^{2.5+}$ in $YBaCo_2O_5$ [84] gives rather columnar patterns at low temperature. $ABaB_2O_5$ compounds can also incorporate additional oxygen into the A-site cation layers. The amount of additional oxygen depends on the size of the A-site rare earth cation. In $NdBaFe_2O_{5.5}$, $SmBaFe_2O_{5.5}$, and $LaBaMn_2O_{5.5}$, the oxygen vacancies order so that half of the B-site cations are five-coordinated (square pyramids) and the other half are six-coordinated [85,86]. More complicated oxygen vacancy-ordered patterns in the perovskite structure are seen in $Sr_nFe_nO_{3n-d}$ ($n = 2, 4, 8$, and ∞) [87].

Complete vacancy in the A-site layer gives ABO_2 with square-planar coordination of the B-site cation. The so-called infinite-layer structure compound $ACuO_2$ shown in Figure 6.16 is an example [88], and the stacking of the square-planar

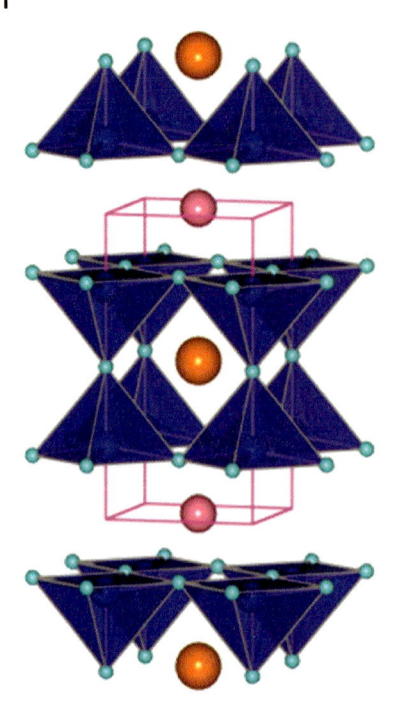

Figure 6.15 Crystal structure of the oxygen vacancy-ordered perovskite $ABaB_2O_5$. The B-site cation coordination is square pyramidal and the A-site cation coordination number decreases from twelve to eight.

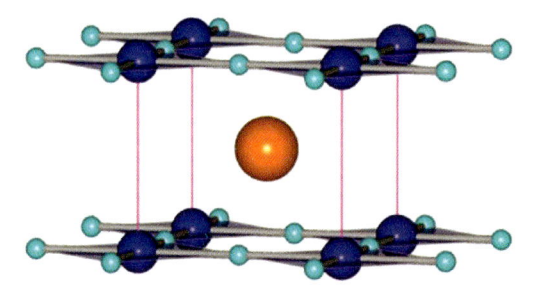

Figure 6.16 Crystal structure of ABO_2 with square-planar coordination of the B-site cation.

CuO_2 units in it is essential for high-T_c superconductivity. Low-temperature topotactic reaction of a few perovskite structure oxides with strong reducing agents like CaH_2 and NaH_2 can also produce $SrFeO_2$ [89] and $LaNiO_2$ [90] with square-planar coordinations of the B-site cations.

In $A_2B_2O_5$, if the B-site cation is more stable in fourfold oxygen coordination than in fivefold coordination, then the oxygen vacancies order so as to leave the B-site cations in sixfold and fourfold oxygen coordination, as seen in $Ca_2Fe_2O_5$ [91] (Figure 6.17). The structure is called brownmillerite type – after

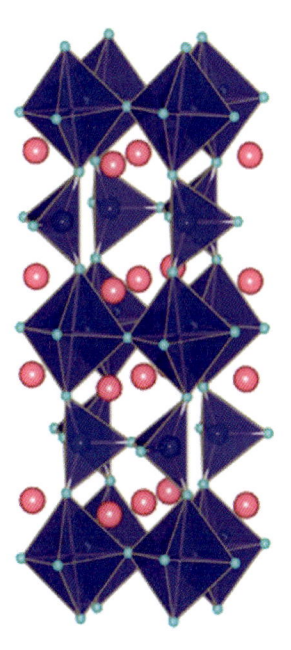

Figure 6.17 Crystal structure of brownmillerite-type $A_2B_2O_5$ with alternating layers of BO_6 octahedra and BO_4 tetrahedra.

the mineral brownmillerite, $Ca_2(Al,Fe)_2O_5$ – and consists of alternating layers of BO_6 octahedra and BO_4 tetrahedra. The crystal structure is *Imma* orthorhombic, but two possible ordered arrangements of the BO_4 tetrahedral chains with *I2mb* and *Pnma* symmetries are reported [92]. On the other hand, very unusual oxygen vacancy arrangement is achieved in $La_2Ni_2O_5$ that consists of alternate units of NiO_6 octahedra and NiO_4 square planes with oxygen vacancies ordering along the $\langle 110 \rangle$ direction, as shown in Figure 6.18 [93].

Oxygen vacancy order with the A cation-layered order in so-called triple perovskites gives a mother compound of high-T_c superconducting copper oxides[2]: $YBa_2Cu_3O_7$. As shown in Figure 6.19, full oxygen vacancy in the Y layers makes the CuO_2 pyramidal coordination planes, which are essential for high-T_c superconductivity, and partial oxygen vacancy in the CuO_2 layers makes the CuO linear chains sandwiched by BaO layers. The structure can also be considered as the stacking of $-CuO-BaO-CuO_2-Y-CuO_2-BaO-CuO-$. Chemical doping by inducing a small amount of oxygen vacancy in the CuO chains ($YBa_2Cu_3O_{7-\delta}$) makes the compound superconducting at temperatures as high as 90 K. A similar ordered triple perovskite but with the FeO_6 octahedral layer sandwiched by the BaO layers is found in $YBa_2Fe_3O_{8+\delta}$ [94]. Ordered quadruple

---

2) It is hard to identify an appropriate reference to cite for the high-T_c superconducting copper oxides because many papers on the same compound were published at almost the same time. Therefore, none are cited here.

Figure 6.18 Crystal structure of $La_2Ni_2O_5$ consisting of alternate units of NiO_6 octahedra and NiO_4 square planes with oxygen vacancies ordering along the $\langle 110 \rangle$ direction.

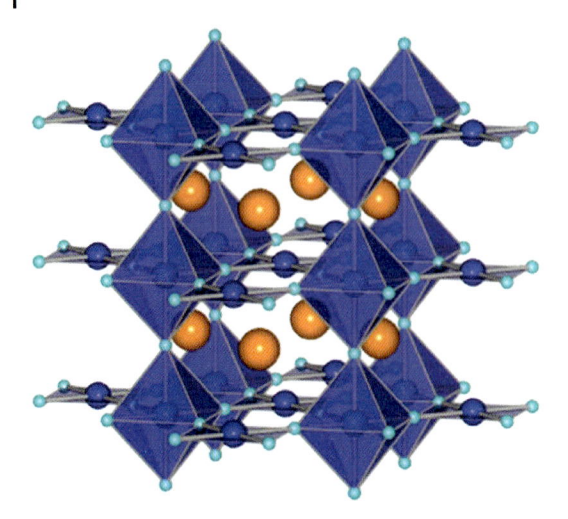

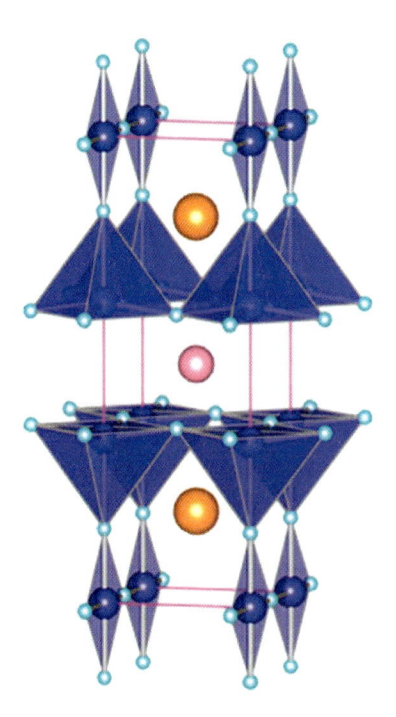

Figure 6.19 Crystal structure of high-T_c superconducting copper oxides $YBa_2Cu_3O_7$.

perovskites with oxygen vacancy order and A-cation layered order are reported in Ln'Ln''Ba$_2$Cu$_2$Ti$_2$O$_{11}$ (Ln'Ln'' = LaY, LaHo, LaEr, and NdDy), where the B cations are arranged into bilayers consisting of double TiO$_6$ octahedral layers alternating with the double CuO$_5$ square-pyramid layers [95]. The structures are also described as the layer sequence of Ln''–CuO$_2$–BaO–TiO$_2$–Ln'O–TiO$_2$–Ln'O–CuO$_2$–Ln.''–

6.4 Perovskite-Related Compounds

6.4.1 Hexagonal Perovskite Compounds

When the tolerance factor $t>1$ in ABO$_3$, the structure mismatch is often relieved by introducing hexagonal stacking, giving a series of the layered-structure compounds called "hexagonal perovskites." While the simple perovskite structure is described by a cubic stacking of close-packed AX$_3$ layers with the B cations at the octahedral sites between the layers, some of the cubic layers are shifted by (1/3, 2/3, 0), introducing hexagonal layers. The B cation octahedra shares common octahedral faces in isolated columns and the large A cations are accommodated between columns of face-shared octahedra, as shown in Figure 6.20. Because the shared octahedral face increases the B–B cation electrostatic repulsive energy, the hexagonal layer stackings are introduced in stages. The stage structures are described with the numbers of cubic to hexagonal stacking layers, and the series typically goes from 3C (cubic perovskite) to 6H to 4H to 9R to 2H, where a letter designates symmetry as cubic (C), hexagonal (H), and rhombohedral (R), as the fraction of hexagonal stackings increases. A two-to-one cubic to hexagonal stacking (cchcch) has the repeat distances of 6 AX$_3$ layers and is referred to as 6H. A one-to-one stacking (hchc) and a one-to-two cubic to hexagonal stacking (chhchhchh) are referred to as 4H and 9R, respectively [96].

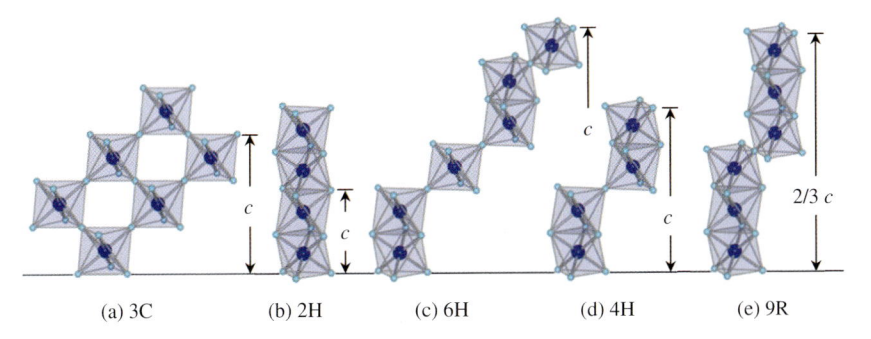

Figure 6.20 Projections of octahedra network of the hexagonal perovskite structures. (a) Cubic 3C, (b) hexagonal 2H, (c) hexagonal 6H, (d) hexagonal 4H, and (e) rhombohedral 9R perovskites.

BaNiO$_3$ ($t=1.13$) crystallizes the 2H hexagonal perovskite structure [97]. It is interesting to note that the unusually high-valence Ni^{4+} is stabilized in this hexagonal perovskite. Several hexagonal perovskites with other stacking sequences are also reported, for example, 8H-type Ba$_4$LiNb$_3$O$_{12}$ [98,99].

High pressure usually stabilizes cubic stacking relative to hexagonal stacking in this class of compounds. BaRuO$_3$ with the 9R structure, which is stabilized at the ambient condition, transforms at 15 GPa to 4H and at 30 GPa to 6H [100]. The volume of the unit cell decreases by the change from 9R to 4H to 6H, and the density of the compounds increases accordingly. The pressures required for these transformations were systematically lowered by substitution, at the A-site, of Ba by Sr, which has a smaller ionic radius. And the transformation from 6H to the 3C cubic perovskite is reported to occur in Ba$_{1-x}$Sr$_x$RuO$_3$. The 2H hexagonal perovskite BaMnO$_3$ transforms at 30 GPa to the 9R structure and at 90 GPa to the 4H structure, while the 4H SrMnO$_3$ transforms at 50 GPa to the 6H structure [101].

6.4.2
Post-Perovskite Compounds

Post-perovskite is a high-pressure phase of the simple perovskite MgSiO$_3$. The crystal structure of the post-perovskite is *Cmcm* orthorhombic and Mg consists of SiO$_6$-octahedra and eightfold oxygen coordinated Mg. The edge-sharing SiO$_6$-octahedral sheets stack along the *b*-axis with interlayer Mg ions (Figure 6.21) [102,103]. This structure is the same as that of CaIrO$_3$ [104]. The

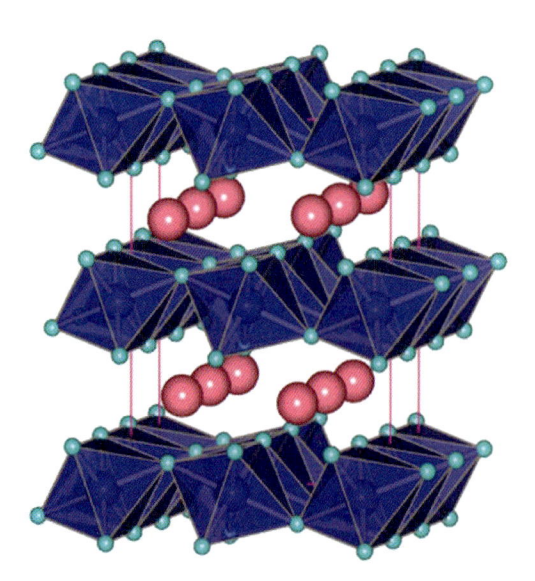

Figure 6.21 Crystal structure of the post-perovskite MgSiO$_3$ with SiO$_6$-octahedral sheets stacked along the *b*-axis.

prefix "post-" refers to the fact that the phase is stable above 120 GPa at 2500 K after the simple perovskite $MgSiO_3$ with increasing pressure. This also implies that the density of the post-perovskite is higher than that of the simple perovskite. The post-perovskite phase is especially important in geoscience research because the temperature and pressure conditions under which it is stable correspond to a depth of about 2600 km in the earth, where the "D" seismic discontinuity occurs.

After the discovery of the post-perovskite phase of $MgSiO_3$ in 2004, the research on its analogues was accelerated. Many were discovered and their perovskite to post-perovskite transitions were studied. Examples include $MgGeO_3$ [105], $CaPtO_3$ [106], and $CaRuO_3$ [107].

6.4.3
Antiperovskite Structure

The antiperovskite crystal structure is similar to the normal perovskite structure ABX_3 but X and B interchange their roles. That is, the anion is at the B-site instead of the X-site and becomes the center of an octahedron formed by six cations. For the structure to be stable, the tolerance factor t is also between 0.7 and 1.0. Some of the antiperovskite compounds show interesting and useful physical properties. Li_3OBr and Li_3OCl show superionic conductivity of lithium ions [108], and $CuNNi_3$ and $ZnNNi_3$ show superconductivity. Ge-doped antiperovskite manganese nitrides Mn_3AN (A = Cu, Zn, Ga) were reported to show a large magnetovolume effect [109] and also show giant negative thermal expansion [110].

6.4.4
Perovskite-Related Layered Structure Compounds

Introducing additional layers into the perovskite structure, which can be described as alternately stacking AX and BX_2 layers, gives layered-perovskite structures. Depending on the introduced layers, a few series of layered structures are known.

A series of compounds with the general formula $A'A_{(n-1)}B_nO_{(3n+1)}$ are known as Dion–Jacobson phases [111,112]. An alkali metal A' ion layer is inserted every n perovskite structure units. $KLaNb_2O_7$ and $CsLaNb_2O_7$ are the $n = 2$ compounds. Interestingly, in $KLaNb_2O_7$ the perovskite structure units shift in (1/2, 0) while in $CsLaNb_2O_7$ no phase shift is introduced (Figure 6.22) [113,114].

Another series of compounds with the general formula of $A'_2A_{(n-1)}B_nX_{(3n+1)}$ are known as Ruddlesden–Popper phases, named after S.N. Ruddlesden and P. Popper, who first synthesized and described the structure [115,116]. The A cations are located in the perovskite layer and have a twelvefold coordination to the anions. The A' cations have a ninefold coordination and are located at the perovskites boundary with an intermediate block layer. The structure can also be seen as an intergrowth on one of the perovskite-type and rock salt-type units.

(a) KLaNb$_2$O$_7$ (b) CsLaNb$_2$O$_7$

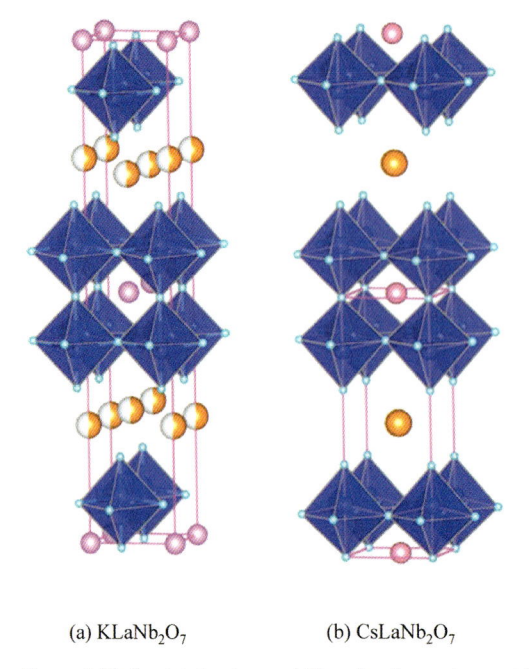

Figure 6.22 Crystal structures of Dion–Jacobson phases: (a) KLaNb$_2$O$_7$ and (b) CsLaNb$_2$O$_7$. The perovskite structure units shift in (1/2, 0) in KLaNb$_2$O$_7$, while no phase shift is introduced in CsLaNb$_2$O$_7$. A half of the K-site in KLaNb$_2$O$_7$ is vacant.

Sr$_2$TiO$_4$, Sr$_3$Ru$_2$O$_7$, and Sr$_4$Ru$_3$O$_{10}$ are the examples (A = A′) of n = 1, 2, and 3 members, respectively (Figure 6.23). The n = 1 end member structure is often called a K$_2$NiF$_4$-type crystal structure. The perovskite structure units shift in (1/2, 1/2). La$_2$CuO$_4$ is the mother compound of high-T_c copper oxide superconductors, and substituting Ba^{2+} or Sr^{2+} chemical doping for La^{3+} gives rise to superconductivity with a transition temperature higher than 40 K [10].

When the intruding layer is composed of [Bi$_2$O$_2$]$^{2+}$, so-called bismuth Aurivillius phases with the general formula (Bi$_2$O$_2$)A$_{(n-1)}$B$_n$X$_{(3n+1)}$ are stabilized [117–119]. The crystal structures consist of intergrowths of the perovskite structure units and the Bi$_2$O$_2$ units with edge-sharing BiO$_2$ square pyramids. Introducing the Bi$_2$O$_2$ layer units causes the perovskite structure units to shift in (1/2, 1/2). The Aurivillius phases represent a large family of oxides and oxyfluorides with B cations of Ti, V, Nb, Ta, and Mo. Ta in Bi$_2$SrTa$_2$O$_9$ (n = 2) [120] and Ti in Bi$_4$Ti$_3$O$_{12}$ (n = 3) [121] are statistically dispersed on the B-site in the BO$_6$ octahedra, and the noncentrosymmetric crystal structures with space groups of $A2_1am$ for Bi$_2$SrTa$_2$O$_9$ (Figure 6.24) and $B1a1$ for Bi$_4$Ti$_3$O$_{12}$ make the compounds displacive-type ferroelectrics useful for nonvolatile memory devices [122,123]. The B = Cu compounds are layered-structure high-T_c superconductors with transition temperatures higher than the temperature of liquid nitrogen, and the T_cs become higher than 100 K in Bi$_2$Sr$_2$CaCu$_2$O$_7$ (n = 2) and

(a) A$_2$BO$_4$
n = 1

(b) A$_3$B$_2$O$_7$
n = 2

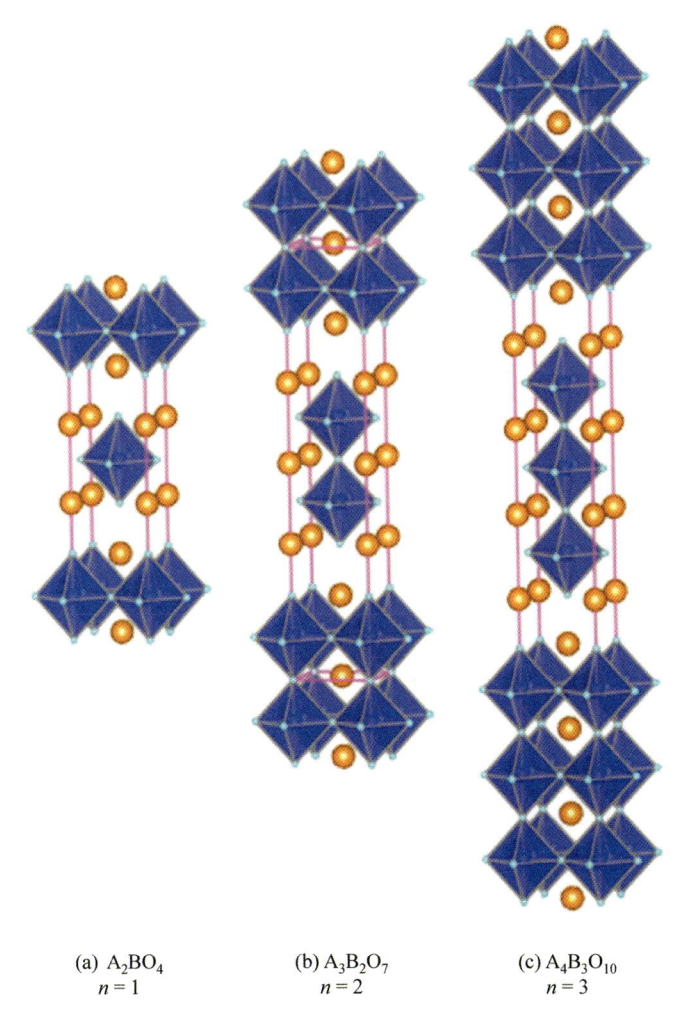

(c) A$_4$B$_3$O$_{10}$
n = 3

Figure 6.23 Crystal structures of Ruddlesden-Popper phases with the general formula A$'_2$A$_{(n-1)}$B$_n$X$_{(3n+1)}$. (a) A$_2$BO$_4$ ($n = 1$) (B) A$_3$B$_2$O$_7$ ($n = 2$), and (c) A$_4$B$_3$O$_{12}$ ($n = 3$).

Bi$_2$Sr$_2$Ca$_2$Cu$_3$O$_{10}$ ($n = 3$), where the A-site cations are Sr and Ca and where oxygen vacancies are introduced for making the CuO$_5$ pyramidal layers. Similar structure compounds with TlO and HgO layers instead of Bi$_2$O$_2$ layers were also discovered and many of them show high-T_c superconductivity. The ideal compositions of the series of compounds can be written in (Tl,Hg)$_m$Ba$_2$Ca$_{(n-1)}$Cu$_n$O$_{(2+m+2n)}$ and abbreviated as m:2:$n-1$:n like Tl-2212 and Hg-2223. The layer stackings of the compounds are $-(TlO)_m-(BaO)-(CuO_2)-$Ca$-(CuO_2)-\cdots-$Ca$-(CuO_2)-(BaO)-(TlO)_m$ as shown in Figure 6.25. The CuO$_2$ planes in CuO$_4$ squares and CuO$_5$ pyramids play essential roles in giving rise to high-T_c superconductivity. Single-phase samples with $n \geq 4$ were difficult to synthesize because the compounds tend to include intergrowth structures.

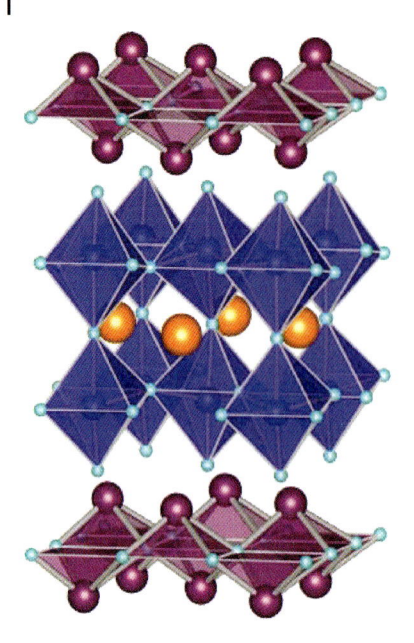

Figure 6.24 Crystal structure of ferroelectric Aurivillius phases of (a) $Bi_2SrTa_2O_9$ and (b) $Bi_4Ti_3O_{12}$.

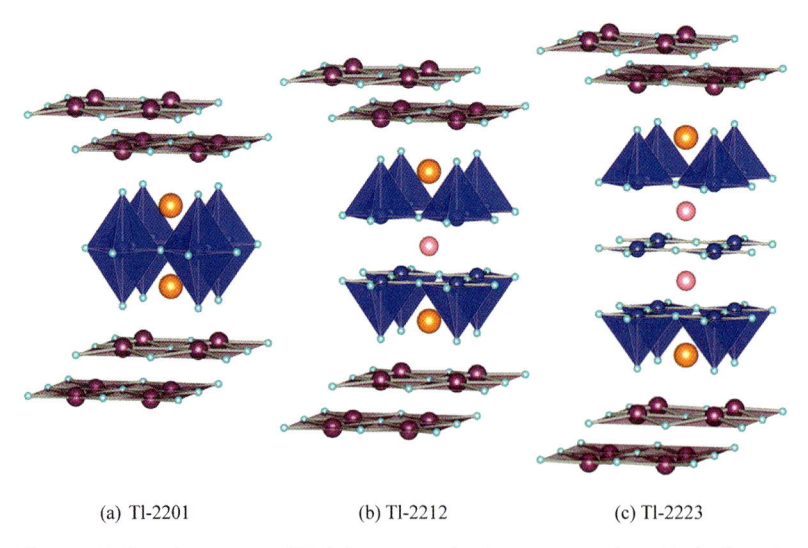

(a) Tl-2201 (b) Tl-2212 (c) Tl-2223

Figure 6.25 Crystal structures of high-T_c superconducting copper oxides with the formula $(Tl,Hg)_m Ba_2 Ca_{(n-1)} Cu_n O_{(2+m+2n)}$. (a) $Tl_2Ba_2CuO_6$ (Tl-2201: $m=2; n=1$) (B) $Tl_2Ba_2CaCu_2O_8$ (Tl-2212: $m=2; n=2$), and (c) $Tl_2Ba_2Ca_2Cu_3O_{10}$ (Tl-2223: $m=2; n=3$).

References

1 Goodenough, J.B. and Longo, J.M. (1970) *Landolt-Börnstein*, **4**, Springer, Berlin, 126.
2 MacChesney, J.B., Sherwood, R.C., and Potter, J.F. (1965) *J. Chem. Phys.*, **43**, 1907.
3 Takano, M., Nakanishi, N., Takeda, Y., Naka, S., and Takada, T. (1977) *Mater. Res. Bull.*, **12**, 923.
4 Takeda, Y., Naka, S., Takano, M., Shinjo, T., Takada, T., and Shimada, M. (1978) *Mater. Res. Bull.*, **13**, 61.
5 Bocquet, A.E., Fujimori, A., Mizokawa, T., Saitoh, T., Namatame, H., Suga, S., Kimizuka, N., Takeda, Y., and Takano, M. (1992) *Phys. Rev. B*, **45**, 1561.
6 Zener, C. (1951) *Phys. Rev.*, **82**, 403.
7 P. G., deGennes. (1960) *Phys. Rev.*, **118**, 141.
8 Tokura, Y., Urushibara, A., Moritomo, Y., Arima, T., Asamitsu, A., Kido, G., and Furukawa, N. (1994) *J. Phys. Soc. Jpn.*, **63**, 3931.
9 Ramirez, A.P. (1997) *J. Phys. Condens. Matter*, **9**, 8171.
10 Bednorz, J.G. and Müller, K.A. (1986) *Z. Physik B Condens. Matter*, **64**, 189.
11 Imada, M., Fujimori, A., and Tokura, Y. (1998) *Rev. Mod. Phys.*, **70**, 1039.
12 Lee, P.A., Nagaosa, N., and Wen, X.-G. (2006) *Rev. Mod. Phys.*, **78**, 17.
13 Kojima, A., Teshima, K., Shirai, Y., and Miyasaka, T. (2009) *J. Am. Chem. Soc.*, **131**, 6050.
14 Green, M.A., Ho-Baillie, A., and Snaith, H.J. (2014) *Nat. Photonics*, **8**, 506.
15 Mizusaki, J., Arai, K., and Fueki, K. (1983) *Solid State Ion.*, **11**, 203.
16 Momma, K. and Izumi, F. (2011) *J. Appl. Crystallogr.*, **44**, 1272.
17 Shirane, G., Danner, H., and Pepinsky, R. (1957) *Phys. Rev.*, **105**, 856.
18 Slater, J.C. (1950) *Phys. Rev.*, **78**, 748.
19 Goldschmidt, V.M. (1926) *Naturwissenschaften*, **14**, 477.
20 Roth, R.S. (1957) *J. Res. Natl. Bur. Stand.*, **58**, 2736.
21 Glazer, A. (1972) *Acta Crystallogr. B*, **28**, 3384.
22 Glazer, A. (1975) *Acta Crystallogr. A*, **31**, 756.
23 Shannon, R. (1976) *Acta Crystallogr. A*, **32**, 751.
24 Torrance, J.B., Lacorre, P., Nazzal, A.I., Ansaldo, E.J., and Niedermayer, Ch. (1992) *Phys. Rev. B*, **45**, 8209.
25 Demazeau, G., Marbeuf, A., Pouchard, M., and Hagenmuller, P. (1971) *J. Solid State Chem.*, **3**, 582.
26 Lacorre, P., Torrance, J.B., Pannetier, J., Nazzal, A.I., Wang, P.W., and Huang, T.C. (1991) *J. Solid State Chem.*, **91**, 225.
27 Woodward, P. (1997) *Acta Crystallogr. B*, **53** (1), 32.
28 Woodward, P. (1997) *Acta Crystallogr. B*, **53**, 44.
29 Howard, C.J. and Stokes, H.T. (1998) *Acta Crystallogr. B*, **54**, 782.
30 Brown, I.D. (1978) *Chem. Soc. Rev.*, **7**, 359.
31 Brown, I.D. and Altermatt, D. (1985) *Acta Crystallogr. B*, **41**, 244.
32 Lufaso, M.W. and Woodward, P.M. (2001) *Acta Crystallogr. B*, **57**, 725.
33 Rao, G.H., Barner, K., and Brown, I.D. (1998) *J. Phys. Condens. Matter*, **10**, L757.
34 Salinas-Sanchez, A., Garcia-Muñoz, J.L., Rodriguez-Carvajal, J., Saez-Puche, R., and Martinez, J.L. (1992) *J. Solid State Chem.*, **100**, 201.
35 Anderson, M.T., Greenwood, K.B., Taylor, G.A., and Poeppelmeier, K.R. (1993) *Prog. Solid State Chem.*, **22**, 197.
36 King, G. and Woodward, P.M. (2010) *J. Mater. Chem.*, **20**, 5785.
37 Vasala, S. and Karppinen, M. (2015) *Prog. Solid State Chem.*, **43**, 1.
38 Howard, C.J., Kennedy, B.J., and Woodward, P.M. (2003) *Acta Crystallogr. B*, **59**, 463.
39 Hosaka, Y., Ichikawa, N., Saito, T., Haruta, M., Kimoto, K., Kurata, H., and Shimakawa, Y. (2015) *Bull. Chem. Soc. Jpn.*, **88**, 657.
40 Anderson, M.T. and Poeppelmeier, K.R. (1991) *Chem. Mater.*, **3**, 476.
41 Goodenough, J.B. (1955) *Phys. Rev.*, **100**, 564.
42 Kanamori, J. (1959) *J. Phys. Chem. Solids*, **10**, 87.

43 Goodenough, J.B. (1963) *Magnetism and the Chemical Bond*, John Wiley & Sons, Inc., New York.
44 Blasse, G. (1965) *J. Phys. Chem. Solids*, **26**, 1969.
45 Azuma, M., Takata, K., Saito, T., Ishiwata, S., Shimakawa, Y., and Takano, M. (2005) *J. Am. Chem. Soc.*, **127**, 8889.
46 Kobayashi, K.I., Kimura, T., Sawada, H., Terakura, K., and Tokura, Y. (1998) *Nature*, **395**, 677.
47 Battle, P.D., Gibb, T.C., Herod, A.J., and Hodges, J.P. (1995) *J. Mater. Chem.*, **5** (1), 75.
48 Battle, P.D., Gibb, T.C., Herod, A.J., Kim, S.-H., and Munns, P.H. (1995) *J. Mater. Chem.*, **5**, 865.
49 Davies, P.K., Wu, H., Borisevich, A.Y., Molodetsky, I.E., and Farber, L. (2008) *Annu. Rev. Mater. Res.*, **38**, 369.
50 King, G., Goo, E., Yamamoto, T., and Okazaki, K. (1988) *J. Am. Ceram. Soc.*, **71**, 454.
51 Leinenweber, K. and Parise, J. (1995) *J. Solid State Chem.*, **114**, 277.
52 Knapp, M.C. and Woodward, P.M. (2006) *J. Solid State Chem.*, **179**, 1076.
53 Chapman, J.P., Attfield, J.P., Molgg, M., Friend, C.M., and Beales, T.P. (1996) *Angew. Chem., Int. Ed. Engl.*, **35**, 2482.
54 Millange, F., Caignaert, V., Domengès, B., Raveau, B., and Suard, E. (1998) *Chem. Mater.*, **10**, 1974.
55 Nakajima, T., Kageyama, H., Yoshizawa, H., and Ueda, Y. (2002) *J. Phys. Soc. Jpn.*, **71**, 2843.
56 Subramanian, M.A., Li, D., Duan, N., Reisner, B.A., and Sleight, A.W. (2000) *J. Solid State Chem.*, **151**, 323.
57 Shimakawa, Y. (2008) *Inorg. Chem.*, **47**, 8562.
58 Chenavas, J., Joubert, J.C., Marezio, M., and Bochu, B. (1975) *J. Solid State Chem.*, **14**, 25.
59 Tohyama, T., Saito, T., Mizumaki, M., Agui, A., and Shimakawa, Y. (2010) *Inorg. Chem.*, **49**, 2492.
60 Alonso, J.A., Sánchez-Benítez, J., De Andrés, A., Martínez-Lope, M.J., Casais, M.T., and Martínez, J.L. (2003) *Appl. Phys. Lett.*, **83**, 2623.
61 Takata, K., Yamada, I., Azuma, M., Takano, M., and Shimakawa, Y. (2007) *Phys. Rev. B*, **76**, 024429.
62 Long, Y.W., Hayashi, N., Saito, T., Azuma, M., Muranaka, S., and Shimakawa, Y. (2009) *Nature*, **458**, 60.
63 Tohyama, T., Senn, M.S., Saito, T., Chen, W.-T., Tang, C.C., Attfield, J.P., and Shimakawa, Y. (2013) *Chem. Mater.*, **25**, 178.
64 Yamada, I., Takata, K., Hayashi, N., Shinohara, S., Azuma, M., Mori, S., Muranaka, S., Shimakawa, Y., and Takano, M. (2008) *Angew. Chem., Int. Ed.*, **47**, 7032.
65 Shimakawa, Y. and Takano, M. (2009) *Z. Anorg. Allg. Chem.*, **635**, 1882.
66 Chen, W.T., Saito, T., Hayashi, N., Takano, M., and Shimakawa, Y. (2012) *Sci. Rep.*, **2**, 449.
67 Shimakawa, Y. (2015) *J. Phys. D Appl. Phys.*, **48**, 504006.
68 Shiraki, H., Saito, T., Yamada, T., Tsujimoto, M., Azuma, M., Kurata, H., Isoda, S., Takano, M., and Shimakawa, Y. (2007) *Phys. Rev. B*, **76**, 140403(R).
69 Kim, Y.J., Wakimoto, S., Shapiro, S.M., Gehring, P.M., and Ramirez, A.P. (2002) *Solid State Commun.*, **121**, 625.
70 Toyoda, M., Yamauchi, K., and Oguchi, T. (2013) *Phys. Rev. B*, **87**, 224430.
71 Shimakawa, Y., Shiraki, H., and Saito, T. (2008) *J. Phys. Soc. Jpn.*, **77**, 113702.
72 Shimakawa, Y. and Saito, T. (2012) *Phys. Status Solidi B*, **249**, 423.
73 Shimakawa, Y. and Mizumaki, M. (2014) *J. Phys. Condens. Matter*, **26**, 473203.
74 Zhang, G., Dong, S., Yan, Z., Guo, Y., Zhang, Q., Yunoki, S., Dagotto, E., and Liu, J.M. (2011) *Phys. Rev. B*, **84**, 174413.
75 Johnson, R.D., Chapon, L.C., Khalyavin, D.D., Manuel, P., Radaelli, P.G., and Martin, C. (2012) *Phys. Rev. Lett.*, **108**, 067201.
76 Vasil'ev, A.N. and Volkova, O.S. (2007) *J. Low Temp. Phys.*, **33**, 895.
77 Byeon, S.-H., Lee, S.-S., Parise, J.B., Woodward, P.M., and Hur, N.H. (2005) *Chem. Mater.*, **17**, 3552.
78 Byeon, S.-H., Lufaso, M.W., Parise, J.B., Woodward, P.M., and Hansen, T. (2003) *Chem. Mater.*, **15**, 3798.

79 Chen, W.T., Mizumaki, M., Saito, T., and Shimakawa, Y. (2013) *Dalton Trans.*, **42**, 10116.

80 Chen, W.T., Mizumaki, M., Seki, H., Senn, M.S., Saito, T., Kan, D., Attfield, J.P., and Shimakawa, Y. (2014) *Nat. Commun.*, **5**, 3909.

81 Goodenough, J.B. (2004) *Rep. Prog. Phys.*, **67**, 1915.

82 Nowick, A.S. and Du, Y. (1995) *Solid State Ion.*, **77**, 137.

83 Woodward, P.M. and Karen, P. (2003) *Inorg. Chem.*, **42**, 1121.

84 Vogt, T., Woodward, P.M., Karen, P., Hunter, B.A., Henning, P., and Moodenbaugh, A.R. (2000) *Phys. Rev. Lett.*, **84**, 2969.

85 Karen, P. and Woodward, P.M. (1999) *J. Mater. Chem.*, **9**, 789.

86 Caignaert, V., Millange, F., Domengès, B., Raveau, B., and Suard, E. (1999) *Chem. Mater.*, **11**, 930.

87 Hodges, J.P., Short, S., Jorgensen, J.D., Xiong, X., Dabrowski, B., Mini, S.M., and Kimball, C.W. (2000) *J. Solid State Chem.*, **151**, 190.

88 Takano, M., Takeda, Y., Okada, H., Miyamoto, M., and Kusaka, T. (1989) *Physica C Supercond.*, **159**, 375.

89 Tsujimoto, Y., Tassel, C., Hayashi, N., Watanabe, T., Kageyama, H., Yoshimura, K., Takano, M., Ceretti, M., Ritter, C., and Paulus, W. (2007) *Nature*, **450**, 1062.

90 Hayward, M.A., Green, M.A., Rosseinsky, M.J., and Sloan, J. (1999) *J. Am. Chem. Soc.*, **121**, 8843.

91 Bertaut, E.F., Blum, P., and Sagnieres, A. (1959) *Acta Crystallogr.*, **12**, 149.

92 Lambert, S., Leligny, H., Grebille, D., Pelloquin, D., and Raveau, B. (2002) *Chem. Mater.*, **14**, 1818.

93 Moriga, T., Usaka, O., Imamura, T., Nakabayashi, I., Matsubara, I., Kinouchi, T., Kikkawa, S., and Kanamaru, F. (1994) *Bull. Chem. Soc. Jpn.*, **67**, 687.

94 Karen, P., Kjekshus, A., Huang, Q., Karen, V.L., Lynn, J.W., Rosov, N., Sora, I.N., and Santoro, A. (2003) *J. Solid State Chem.*, **174**, 87.

95 Greenwood, K.B., Sarjeant, G.M., Poeppelmeier, K.R., Salvador, P.A., Mason, T.O., Dabrowski, B., Rogacki, K., and Chen, Z. (1995) *Chem. Mater.*, **7**, 1355.

96 Katz, L. and Ward, R. (1964) *Inorg. Chem.*, **3**, 205.

97 Krischner, H., Torkar, K., and Kolbesen, B.O. (1971) *J. Solid State Chem.*, **3**, 349.

98 Negas, T., Roth, R.S., Parker, H.S., and Brower, W.S. (1973) *J. Solid State Chem.*, **8**, 1.

99 Jendrek, E.F., Potoff, A.D., and Katz, L. (1974) *J. Solid State Chem.*, **9**, 375.

100 Longo, J.M. and Kafalas, J.A. (1968) *Mater. Res. Bull.*, **3**, 687.

101 Syono, Y., Akimoto, S., and Kohn, K. (1969) *J. Phys. Soc. Jpn.*, **26**, 993.

102 Murakami, M., Hirose, K., Kawamura, K., Sata, N., and Ohishi, Y. (2004) *Supramol. Sci.*, **304**, 855.

103 Oganov, A.R. and Ono, S. (2004) *Nature*, **430**, 445.

104 Rodi, F. and Babel, D. (1965) *Z. Anorg. Allg. Chem.*, **336**, 17.

105 Hirose, K., Kawamura, K., Ohishi, Y., Tateno, S., and Sata, N. (2005) *Am. Mineral.*, **90**, 262.

106 Inaguma, Y., Hasumi, K., Yoshida, M., Ohba, T., and Katsumata, T. (2008) *Inorg. Chem.*, **47**, 1868.

107 Kojitani, H., Shirako, Y., and Akaogi, M. (2007) *Phys. Earth Planet. Inter.*, **165**, 127.

108 Zhao, Y. and Daemen, L.L. (2012) *J. Am. Chem. Soc.*, **134**, 15042.

109 Fruchart, D. and Bertaut, E.F. (1978) *J. Phys. Soc. Jpn.*, **44**, 781.

110 Takenaka, K. and Takagi, H. (2005) *Appl. Phys. Lett.*, **87**, 261902.

111 Dion, M., Ganne, M., and Tournoux, M. (1981) *Mater. Res. Bull.*, **16**, 1429.

112 Jacobson, A.J., Johnson, J.W., and Lewandowski, J.T. (1985) *Inorg. Chem.*, **24**, 3727.

113 Sato, M., Abo, J., Jin, T., and Ohta, M. (1992) *Solid State Ion.*, **51**, 85.

114 Gopalakrishnan, J., Bhat, V., and Raveau, B. (1987) *Mater. Res. Bull.*, **22**, 413.

115 Ruddlesden, S.N. and Popper, P. (1957) *Acta Crystallogr.*, **10**, 538.

116 Ruddlesden, S.N. and Popper, P. (1958) *Acta Crystallogr.*, **11**, 54.

117 Aurivillius, B. (1949) *Arkiv. Kemi.*, **1**, 463.

118 Aurivillius, B. (1950) *Arkiv. Kemi.*, **2**, 499.

119 Aurivillius, B. (1950) *Arkiv. Kemi.*, **2**, 519.

120 Shimakawa, Y., Kubo, Y., Nakagawa, Y., Kamiyama, T., Asano, H., and Izumi, F. (1999) *Appl. Phys. Lett.*, **74**, 1904.

121 Shimakawa, Y., Kubo, Y., Tauchi, Y., Asano, H., Kamiyama, T., Izumi, F., and Hiroi, Z. (2001) *Appl. Phys. Lett.*, **79**, 2791.

122 C. A., P.A., Cuchiaro, J.D., McMillan, L.D., Scott, M.C., and Scott, J.F. (1995) *Nature*, **374**, 627.

123 Park, B.H., Kang, B.S., Bu, S.D., Noh, T.W., Lee, J., and Jo, W. (1999) *Nature*, **401**, 682.

7
Nitrides of Non-Main Group Elements

P. Höhn[1] and R. Niewa[2]

[1]Max-Planck-Institut für Chemische Physik fester Stoffe, Chemische Metallkunde, Nöthnitzer Str. 40, 01187 Dresden, Germany
[2]Universität Stuttgart, Institut für Anorganische Chemie, Pfaffenwaldring 55, 70569 Stuttgart, Germany

7.1
Introduction

In the last two decades, there has been a renewed interest in the chemistry of nitrides and nitridometalates [1–6]. Both binary and higher nitrides M_xN_y have already featured prominently as refractory materials, corrosion- and mechanical wear-resistant coatings, hard materials, and hard magnets; thin films are used as diffusion barriers in integrated circuits. Recently, they have attracted interest as promising electrocatalysts [7], possibly able to replace noble metals of high price and limited supply.

Whereas research on binary transition metal nitrides is mostly driven by technical and economic interests, the investigation on nitridometalates primarily focuses on exploration with respect to the development of new synthetic strategies and the design of new materials. Within this field of interest, especially the nitride chemistry of rare earth metals is still comparably undeveloped [8,9].

Here we report on binary transition and rare earth metal nitrides and compounds thereof. However, mixed anion nitrides (any compound with interactions M–X with $X \neq N$; for example, nitride carbides [10], nitride oxides, and oxonitridometalates and combinations, nitride chlorides, nitride fluorides, and imides) as well as nitridocompounds of main group elements (nitridoborates [11,12], nitridosilicates [13,14], nitridogermanates, nitridophosphates and related nitrides as, for example, Cr_3GeN [15] or V_3PN [16]) are not covered. The same applies to compounds containing N in the form of diazenides, pernitrides, or azides (comprising $[N_2]^{4-}$, $[N_2]^{2-}$, $[N_3]^{-}$ ions) as sole nitrogen-containing species. Therefore, the pernitrides MN_2 (M = Ti [17], Os [18,19], Ir [19,20], Pd [21,22], and Pt [20,23]) as well as ternary azidometalates (e.g., $K[Au(N_3)_4]$ [24]) are not considered, the same holds for actinides other than U and Th.

Handbook of Solid State Chemistry, First Edition. Edited by Richard Dronskowski, Shinichi Kikkawa, and Andreas Stein.
© 2017 Wiley-VCH Verlag GmbH & Co. KGaA. Published 2017 by Wiley-VCH Verlag GmbH & Co. KGaA.

Chemical bonding in binary nitrides varies from primarily salt-like (e.g., LaN) via covalent (e.g., Ta_3N_5) to metallic (e.g., Fe_4N), whereas nitridometalates are best described as containing covalent complex anions $[M_xN_y]^{z-}$ with alkali, alkaline earth, or rare earth metal cations providing electroneutrality.

7.2
Preparative Aspects

Due to space constraints, a complete discussion of the preparative methods is outside the scope of this contribution; so only general techniques with typical examples are presented. Additionally, this list does not claim to be complete.

A plethora of methods have been reported for the synthesis of binary transition metal nitrides, most of them involving high-temperature reactions of different starting materials with reactive gases for extended periods of time. In general, the following four methods are chiefly considered for the preparation of bulk material of binary transition metal and rare earth nitrides:

1) The easiest – but not necessarily most successful – method for the preparation of bulk samples of pure nitrides is the direct reaction of metal and nitrogen at elevated temperatures. This works well with lithium [25], alkaline earth [26,27], and rare earth metals; however, only few transition metal nitrides may be effectively prepared by this conventional method, for example, VN at 1200 °C [28,29]. Under the label "self-propagating high-temperature synthesis" or "combustion synthesis," the formation of TiN and TaN among others was reported. Altogether, this mode of preparation works well for nitrides of groups 3–6. Another method to prepare multinary nitrides from metals and molecular nitrogen gas employs the electrical explosion of coated wires and subsequent nitriding of the products in N_2 atmosphere (e.g., (Ti, Cr)N) [30].

2) The reaction of pure metal with ammonia was investigated in detail both under ambient pressure and under ammonothermal conditions at elevated pressures [31] in autoclaves made of highly heat-resistant alloys [32] and is applicable for many transition metals.

3) Many nitrides are easily prepared by ammonolysis of solid precursors (oxides [33], sulfides [34], and halides [33]) at elevated temperatures.

4) Solid-state metathesis reactions have been reported using a variety of starting materials: metal chlorides (M_xCl_y) and azides (NaN_3 [35]), amides ($NaNH_2$ [36]), or nitrides (Ca_3N_2 [37]) as nitrogen sources.

The preparation of high melting or unreactive elements is facilitated by reactions in arc furnaces. Due to the high reactivity and volatility of the alkali and alkaline earth metals, these techniques are predominantly restricted to syntheses of binary refractory nitrides of the early transition metals (e.g., η-Ti_3N_{2-x} [38]).

Reactive molecular nitrogen at high pressures in stainless steel autoclaves was used for the investigation of binary nitrides and alkaline earth metal diazenides (e.g., SrN_2 [39]) as well as nitride metalides (e.g., Sr_2AuN [40]).

Other methods include the urea route using halides ($M_xCl_y + (NH_2)_2CO$ [41]) or oxides ($M_xM'_yO_z + (NH_2)_2CO$ [42]) for both binary and multinary nitrides. Carbothermal reduction of oxides under nitrogen proceeds at high temperatures above 1500 °C [43]. Under extreme conditions, the formation of binary nitrides has been studied by laser heating the constituting elements in diamond anvil cells [44]; nitrogen also acts as the pressure-transmitting medium in these experiments.

The preparation of thin films generally proceeds via chemical vapor deposition (CVD) of metal chlorides and ammonia at 900 °C. High temperatures and formation of HCl as a by-product render this process unsuitable for microelectronic applications. Utilizing $M(N(CH_3)_2)_x$ instead of MCl_x leads to far lower reaction temperatures between 200 and 450 °C and prevents release of corrosive agents; however, organic contaminants are observed [45].

The gas-phase interstitial modification technique has been developed to investigate the formation of hard magnets based on $RE_2M_{17}N_x$ and similar phases. Within a limited temperature range and by carefully controlling the nitridation conditions using N_2 or NH_3, it is possible to produce the metastable interstitial compounds from RE_2M_{17} without degradation to the stable binary rare earth metal nitrides and M [46]. Both hydrogen treatment prior to the nitridation reaction and high-pressure nitriding [47] minimize formation of impurity phases and improve product quality. In this respect, the HDDR process (hydrogenation–decomposition–desorption–recombination) has been studied in detail [48–50].

Chemical transport reactions using a small amount of a transport agent (e.g., halides) seem to be rather unsuitable for nitride synthesis due to the lack of suitable volatile species. Although successful in the case of Ta_3N_5 [51], such experiments often resulted in nonvolatile nitride halides (e.g., Sr_2NCl [52]) only.

The preparation of nitridometalates provides different challenges compared to binary transition metal nitrides. Most of the nitridometalates as well as many of the starting materials are extremely sensitive to air and moisture; so, all manipulations have to be performed under dry argon, and great care has to be taken in preparation and handling procedures. A further challenge is the high reactivity of the alkali and alkaline earth metals, providing difficulties in the selection of suitable crucible materials. The preparation of multinary phases containing rare earth metals is still another issue: In most cases, their binary nitrides are so stable that only rarely formation of ternary compounds or higher phases is observed.

Nitrogen easily reacts to form nitrides and nitridometalates in the presence of Li or alkaline earth metals, making the traditional solid (liquid)/gas route feasible. Nitridometalates containing lithium or alkaline earth metals are synthesized from the binary (sub/diazenide-)nitrides (Li_3N, AE_3N_2 (AE = Be, Mg, Ca), AE_4N_3 (AE = Sr, Ba), AE_2N (AE = Ca, Sr, Ba), or mixtures thereof) together with the respective transition metal powders at temperatures between 500 and 1200 °C. In general, reaction temperatures need to be increased within the sequence of Li → Ba → Sr → Ca [1].

Generally, by employing excess nitrogen in these reactions, the highest (possible) oxidation state for M in nitridometalates is achieved. However, within a given ternary system, higher reaction temperatures seem to result in nitridometalates with lower AE/M ratios (e.g., $Sr_3[FeN_3]$ [53] at 750 °C → $Sr_8[FeN_2][FeN_3]_2$ [54] at 850 °C → $Sr_{10}[FeN_2][Fe_2N_4]_2$ [55] at 950 °C). Other factors may also influence the reaction paths, such as the concentration of the reactants or nitrogen pressure, which may stabilize phases otherwise having a tendency to decompose.

In typical experiments, mixtures of the finely powdered reactants are pressed into pellets, filled into open crucibles, and heated to the reaction temperature by use of a simple tube reactor under nitrogen atmosphere. Typical annealing times are 10–168 h; the reaction products are either quenched (for metastable phases, for example, $Li_4[FeN_2]$ [56]) or slowly cooled (e.g., $Li_6[WN_4]$ [57,58]) to ambient temperature. Since diffusion processes are the rate-limiting factors in these solid (liquid)/gas reactions, high temperatures, long reaction times, and repeated treatments are often necessary to achieve complete reaction of the starting materials. However, due to the small energy of formation of nitridometalates, many of these phases already decompose close to the reaction temperatures.

The choice of crucible materials always provides a challenge during nitride synthesis. Li_3N and the binary alkaline earth nitrides are highly corrosive toward nearly all materials. Refractory materials, such as Al_2O_3, ZrO_2, BN, or C, are not feasible as crucible materials in nitridometalate synthesis due to the easy formation of oxides, such as $AEAl_2O_4$, cyanamides, and carbodiimides $AECN_2$, or related compounds (e.g., $Ca_{4-x}Sr_xN_2[CN_2]$ [59]), even at rather low temperatures.

Even small amounts of nonmetallic impurities in the starting mixtures and the crucible materials (e.g., C in the form of Fe_3C in iron crucibles) often result in the formation of stable phases other than nitrides and nitridometalates. In the case of carbon, detailed studies on the effect of this undesirable foreign component in nitridometalate synthesis have been performed, resulting in a plethora of different phases. Sources of carbon impurities range from carbide admixtures in crucible materials to carbonate contaminations on educt surfaces; for on-purpose preparation attempts, a variety of carbon sources, including graphite, carbides, cyanides, and melamine, were employed. The essential factor affecting the reaction paths is given by the amount of nitrogen in the system, independent of the preparation route. Under inert or reducing conditions, in the presence of excess alkaline earth metal, highly reduced cyanide ions $[CN^{1.67-}]$ are preferably formed (e.g., $AE_3[M(CN)_3]$, AE = Sr, Ba; M = Co [60], Rh [61], Ir [62]); in the presence of excess nitrogen under oxidizing conditions, symmetric carbodiimide or asymmetric cyanamide $[CN_2]^{2-}$ ions prevail (e.g., $(Sr_6N)[MN_2][CN_2]_2$, M = Fe [63], Co [64], Ni [65]; $Sr_6[MN_2]_2[CN]_2$, M = Co [66], Ni [65]). "Intermediate" compounds such as acetonitrilides, for example, $Ba_5[TaN_4][C_2N]$ [67], or cyanonitrido-metalates, for example, $Sr_2[NNi(CN)]$ [68] and $Ba_2[NNi(CN)]$ [69], may be prepared employing narrowly defined amounts of nitrogen under strictly controlled conditions; quite likely, disordered $Sr_2[NiN_2]$ [70], prepared from a NaN_3 flux, is a misinterpretation of the latter compound. Altogether, the presence of carbon is not always obvious, so misinterpretations are common consequences, for example, $Ba_9[(Ta, Nb)N_4]_2[CN_2]O$ [71], was first described as nitridometalate-azide-nitride $Ba_9[(Ta, Nb)N_4]_2[N_3]N$ [72,73].

Using crucibles of other transition metals generally leads to impurities in the form of the respective nitridometalates. Therefore, as a rule, it is always attempted to use crucibles directly machined from the respective transition metal, or, if this is not possible as in the case of, for example, cobalt, it is advantageous to use at least crucibles that are covered with the respective metal foil.

The preparation of phase-pure nitridometalates is still a challenge. Even small amounts of impurities, especially in case of the ferromagnetic elements Fe, Co, and Ni, massively interfere with the intrinsic magnetic properties of nitridometalates. Traditional purification methods such as the removal of Ni metal impurities in the form of $[Ni(CO)_4]$ (the so-called Mond process [74]) are not advisable, since nitridometalates easily decompose on contact with CO by forming cyanides and other phases not yet identified [75].

Due to these difficulties, other methods of preparation were developed to overcome at least some of the most pressing problems depending on the field of application:

- Ammonolysis reactions at ambient pressure proceed at somewhat lower temperatures compared to the standard method (e.g., $Na_3[MoN_3]$ [76], $Ca_3[CrN_3]$ [77]), but do not provide any other advantages.
- The amides of the heavier alkali metals have been used in autoclaves (400 °C $< T < 800$ °C, $p < 2$ kbar) in order to prepare ternary nitridometalates from metal powders. The amides decompose at these temperatures to form N_2, H_2, and alkali metals, which in turn then act as a reactive flux [3]. Hydrogen leaves the autoclaves via a Ni membrane; excess alkali metal is removed with liquid ammonia after the reaction. Until now, only nitridometalates with the transition metals in their highest oxidation states were obtained under these strongly reducing conditions (e.g., $Rb[TaN_2]$ [78], $Na_3[MoN_3]$ [76]). In some cases, nitridometalates containing hydride, amide, or imide species were also observed (e.g., $Li_{14}[CrN_4]_2NH$ [57]).
- An improved reaction route is based on sodium azide instead of the amides, thereby providing both nitrogen for the reaction and the metal as a flux and mineralizer without the risk of hydrogen impurities. In general, this method

leads to nitridometalates with the transition metal in preferably lower oxidation states (e.g., $Ba_2[Ni_3N_2]$ [75,79], $Sr_6[Mn_{20}N_{20}]N$ [80,81]); also some controversially discussed phases (e.g., $Sr_{39}[CoN_2]_{12}N_7$ [82]) have been prepared using this method. The crystallization process of alkaline earth metal nitridometalates within sodium flux has been studied in detail: Whereas nitrogen is virtually insoluble in pure Na (10^{-10} wt%) [83], small amounts of alkaline earth metals significantly increase the solubility to about 1 mol% [84] due to the formation of complexes like AE_6N [85] or similar species with the solubility increasing with the size of the cation Ca → Sr → Ba. The solubility of M in sodium flux may be similarly enhanced by nitrogen. After the reactions, the flux can be removed from the nitridometalates by extraction with liquid ammonia or via sublimation at temperatures as low as 300 °C [80,81], which may prevent decomposition effects in case of heat-sensitive materials.

- In order to prepare reduced compounds independent of the reaction temperatures, reactions under argon in open systems (inert conditions) as well as reactions in sealed ampoules (the ampoule material act as a nitrogen sink) have been employed with success, for example, $Ba_2[Ni_3N_2]$ [75,79].
- The growth of single crystals of decent size is a challenge for all nitridometalates. Hexagonal single crystals of about 0.3 mm in diameter of some nitridoferrates (e.g., $Ba_3[FeN_3]$ [86]) have been grown using mixtures of the alkaline earth metal, the respective nitride, and the transition metal by first heating and annealing the reaction mixtures under argon, thereby forming a highly reactive melt. After the onset of the reaction and a grace period of generally 12 h, the atmosphere is changed to nitrogen to complete the reaction. This method seems to be suitable to obtain larger single crystals, but is useless to prepare single-phase samples.
- Li_3N melts congruently at 813 °C and readily reacts with most transition metal powders at lower temperatures. Excess Li_3N as a flux medium and growth techniques such as Czochralski and Bridgman methods are feasible in order to grow thermodynamically stable lithium nitridometalate phases (e.g., $Li_3[FeN_2]$) [87].
- Solid-state metathesis reactions of Li_3N have been successfully used as a reactive flux to grow single crystals of alkaline earth metal nitridometalates (e.g., $Ca_2[VN_3]$ [88]), thereby significantly lowering the reaction temperatures by several hundred degrees.
- The use of alkaline earth and transition metals together with Li_3N as a nitrogen source in Li metal fluxes [89–91] highly improves single-crystal size and quality. The samples are reacted in tantalum ampoules at temperatures around 750 °C, slowly cooled down to 300 °C with 1 °C/h, and finally quenched during centrifugation of the melt. Maximum size of crystals is limited by container size; however, most often rather highly reduced phases are being formed (e.g., $Li_2[(Fe_{1-x}Li_x)N]$ [92], $Li_2Sr[(Fe_{1-x}Li_x)N]_2$ [90]). The extent of transition metal incorporation and oxidation is directly correlated with nitrogen content of the melt.

- Other metals being used as flux for single-crystal growth of nitrides and nitridometalates include sodium [93] (e.g., $Sr_6Cu_3N_5$ [94]) and gallium (e.g., $La_3[Cr_2N_6]$ [95]). Metal fluxes have also been shown to enable exchange of the cationic components of nitridometalates (e.g., $Na_3[WN_3] + K \rightarrow Na_2K[WN_3] + Na$ [96]).
- Salt melts of alkali or alkaline earth metal halides have been successfully employed as fluxes for single-crystal growth of a limited number of nitridometalates. In the case of nitridomolybdates (e.g., $LiBa_5[Mo_2N_7]Cl_2$ [97]) and nitridotungstates (e.g., $Ba_4[WN_4]Cl_2$ [98]), the halide flux method results in stable reaction products due to incorporation of flux material. However, the nitridometalates of $M = Mn - Zn$ especially seem to be instable in basic molten salt fluxes, as the complex anions decompose and only transition metal halides are formed.
- The suitability as electrode materials has been investigated for a variety of Li nitridometalates. Starting from well-characterized thermodynamically stable compounds and by adding mixed binding (e.g., polytetrafluoroethylene) and/or conducting agents (e.g., acetylene black), oxidation of M was observed by electrochemical routes, thereby opening new regions in the respective chemical systems, which are not accessible by conventional methods (e.g., $Li_{7-x}[MnN_4]$ from $Li_7[MnN_4]$ [99,100] and $Li_{3-x}[FeN_2]$ from $Li_3[FeN_2]$ [101]).
- Solid-state metathesis reactions have been proven useful in the synthesis of RE nitridoborates (e.g., $3\ RECl_3 + 3\ Li_3[BN_2] \rightarrow RE_3[B_3N_6] + 9\ LiCl$ [102]) and related compounds (e.g., $3\ CeCl_3 + 2\ NiCl_2 + 10\ Li + BN + Li_3[BN_2] \rightarrow Ce_3Ni_2(BN)_2N + 13\ LiCl$ [103]).

Despite all these efforts, the understanding of the factors that determine multinary nitride phase formation is rather poor.

7.3 Structural Chemistry of Nitrides

The discussion of binary and multinary nitrides is based on the environment of the transition/rare earth metals in the compounds. In general, four distinctively different groups of compounds are observed, which are arranged according to their distinctive features:

1) All binary nitrides are discussed in Section 7.3.1, although a wide range of compositions, physical properties, and bonding behavior (metallic to ionic to covalent) are observed. The discussion is arranged by groups in the periodic table of elements.
2) Section 7.3.2 summarizes multinary transition metal nitrides with low N content. N is located within a matrix of transition metals, bonding behavior is generally metallic.
3) Multinary nitrides and nitridometalates contain isolated transition metal atoms or clusters thereof, which are fully coordinated by N. Electroneutrality

is achieved by electropositive cations, for example, alkali, alkaline earth, or rare earth metals. Generally, covalent, metallic, and ionic bonding are observed in these compounds. Respective phases are listed in Section 7.3.3.
4) In metallide nitrides, no interaction of M with N is evident; in fact, M is located in a metallic matrix, and N is exclusively coordinated by alkaline earth metals. Although these phases are outside the scope of this treatise, they are summarized in Section 7.3.4.

All paragraphs discussing ternary compounds start with a graphical abstract in the form of a periodic table of elements highlighting the involved elements. These periodic tables are intended to offer a fast first glance on the content and provide a second criterion next to crystal structural considerations to rationalize the large number of compounds.

7.3.1
Binary Nitrides of Transition and Rare Earth Metals

In binary transition metal nitrides (Tables 7.1 and 7.2), some simple trends can be observed: The highest thermodynamic stability is reached in groups 4 and 5. Only for elements of period 5 and 6 of these groups, oxidation states of +4 and +5 are reached with nitrogen, namely, in Zr_3N_4, Hf_3N_4, and Ta_3N_5. Particularly for tantalum, the largest number of different compositions and modifications thereof is currently known. Decreasing thermodynamic (and thermal) stability with increasing group number is expressed by an increasing molar ratio $n(M)/n(N)$ until group 10 is reached. For Mn and Fe this results in a larger number of distinct phases with maximum oxidation state of the transition metal of +3. The noble metals typically do not form stable nitrides at all; however, application of high pressure may lead to formation of compounds with covalently bonded dinitrogen anions.

A trend of dominating ionic bonding from group 3 compounds via metallic phases to eventually covalently bonded compounds in group 9 is obvious. Particularly for Cu and Hg in the monovalent state, the transformation to covalent compounds may be seen from linear coordination of copper in Cu_3N and Hg in $[Hg_2N]^+$ ions, respectively. The following paragraph tries to give an overview of the binary transition metal nitrides, including rare earth metal nitrides and those of U and Th.

Most crystal structures of binary nitrides are based on simple prototypes. From a crystal structural viewpoint, mutual sixfold coordination of M and N in octahedra and trigonal prisms are the dominating motifs in binary 1 : 1 transition metal nitrides. Typically, close-packed layers with various stacking variants are realized, most frequently in a rock salt (Figure 7.1a) or NiAs-type (Figure 7.1b) variant. For molar ratios $n(M)/n(N) \geq 2$, interstitial phases form with N in a metal *fcc*, *hcp*, or *bcc* packing, or ordered variants thereof. To this group belong most compounds of Fe, Co, and Ni with their γ'-Fe_4N, ε-Fe_3N, and ζ-Fe_2N-type

7.3 Structural Chemistry of Nitrides

Table 7.1 Enthalpies of formation of some selected binary nitrides (kJ/mol).

Li$_3$N	Be$_3$N$_2$											BN	
−199	−280											−132	
	Mg$_3$N$_2$											AlN	Si$_3$N$_4$
	−241											−240	−188
	Ca$_3$N$_2$	ScN	TiN	VN	CrN	Mn$_5$N$_2$	Fe$_2$N	Co$_3$N	Ni$_3$N	Cu$_3$N	Zn$_3$N$_2$	GaN	
	−216	−338	−336	−173	−124	−101	−13	+8	−0.8	+75	−11	−110	
		YN	ZrN	Nb$_2$N	Mo$_2$N					Ag$_3$N	Cd$_3$N$_2$	InN	
		−299	−344	−256	−70					+255	+81	−17	
		LaN	HfN	TaN									
		−302	−369	−251									

CeN	PrN	NdN	PmN	SmN	EuN	GdN	TbN	DyN	HoN	ErN	TmN	YbN	LuN
−326	−327	−327	−333	−330	−234	−331	−332	−332	−356	−334	−334	−293	−339

phases as well as ordered variants of the solution phases of N in the different metal modifications.

7.3.1.1 Group 3, Sc, Y, La, and Rare Earth Metals, U, Th

In the binary systems of group 3 elements – Sc, Y, and La – as well as for all lanthanides Ce–Lu, nitrogen compounds with the equimolar composition R_EN crystallizing in the rock salt structure (Figure 7.1a) are well established. Careful studies indicate potential homogeneity ranges R_EN_{1-x} of $x < 0.05$ at lower and $x < 0.10$ at higher temperatures close to the melting point (where $1 - x$ denotes the total nonmetal content including oxygen impurities) [104]. However, many reports discuss much broader homogeneity ranges. Apparently, finely dispersed metal inclusions may in fact simulate wider homogeneity ranges than would follow from, for example, X-ray diffraction data.

Particularly, ScN has attracted a number of studies addressing its hardness, semiconducting properties, and the possible application as additive for improved materials properties of ceramics. Largely different data on the homogeneity range of ScN$_x$ were reported (e.g., for bulk samples: $0.87 \leq x \leq 1.00$ [105], $0.71 \leq x \leq 1.00$ [106], $0.74 \leq x \leq 0.87$ [107]; for thin layers: no homogeneity range [108]) in line with strongly varying unit cell parameters (440 pm $\leq a \leq 451$ pm) [105,109,110]. An often reported high electrical conductivity and the very different colors observed ranging from red to gray and black may also be provoked by this large and variable degree of vacancies on the nitrogen site. Samples obtained from thermal decomposition of the precursor [ScCl$_2$[N(SiMe$_3$)$_2$](THF)$_2$] are red, but not well crystallized [109]. Single crystals grown from molten Sc in nitrogen gas are transparent reddish-brown in color at thinner edges indicating semiconducting behavior connected with a small concentration of defects x in ScN$_{1-x}$ [111].

Table 7.2 Compositions of binary transition metal nitrides.

ScN	δ'-Ti$_2$N	V$_{16}$N$_{1.5}$	β-Cr$_2$N	ε-Mn$_4$N	α'-Fe$_8$N	Co$_4$N	Ni$_4$N	Cu$_4$N	Zn$_3$N$_2$
	ε-Ti$_2$N	β-V$_2$N	γ-CrN	$\zeta(\zeta')$-Mn$_2$N	α''-Fe$_{16}$N$_2$	Co$_3$N	Ni$_3$N	Cu$_3$N	
	η-Ti$_3$N$_{2+x}$	δ-VN		ε-Mn$_2$N	γ-Fe$_4$N	Co$_2$N			
	ζ-Ti$_4$N$_{3+x}$			η-Mn$_3$N$_2$	ε-Fe$_3$N	CoN			
	δ-TiN			θ-Mn$_6$N$_{5+x}$	ζ-Fe$_2$N				
					α'-FeN				
					α''-FeN				
YN	ZrN	β-Nb$_2$N	β-Mo$_2$N					Ag$_3$N	Cd$_3$N$_2$
	c-Zr$_3$N$_4$	γ-Nb$_4$N$_3$	γ-Mo$_2$N						
		δ-NbN	δ-MoN						
		ε-NbN	δ'-MoN						
		η-NbN	δ''-MoN						
		Nb$_{0.84}$N							
		Nb$_5$N$_6$							
		Nb$_4$N$_5$							
LaN	$\eta(\varepsilon)$-Hf$_3$N$_2$	β-Ta$_2$N	α-W$_2$N	Re$_3$N					
La$_2$N$_3$	ζ-Hf$_4$N$_3$	δ-TaN	β-W$_2$N	Re$_2$N					
LaN$_2$	δ-HfN	η-TaN	WN						
	c-Hf$_3$N$_4$	ε-TaN	δ-WN						
		Ta$_5$N$_6$	h-W$_2$N$_3$						
		Ta$_4$N$_5$	r-W$_2$N$_3$						
		Ta$_2$N$_3$	c-W$_3$N$_4$						
		η-Ta$_2$N$_3$	(WN$_2$)						
		Ta$_3$N$_5$							
	ThN		UN						
	(Th$_3$N$_4$)		α-U$_2$N$_{3+x}$						
			β-U$_2$N$_{3+x}$						
			UN$_2$						

All rare earth mononitrides, with exception of CeN, contain the rare earth metal in the +3 valence state, and represent semimetals or semiconductors [112]. CeN exhibits a significantly smaller unit cell volume than expected from trivalent ionic radius, golden metallic luster, and metallic characteristic in the electrical resistivity [112,113]. Electronic structure calculations support the presence of the remaining valence electron in a delocalized d-centered band rather than a localized f^1 state [114–116], also consistent with the absence of a localized magnetic moment. Thus, CeN is rather similar to the rock salt-type nitrides of group 4 elements.

Single crystals of the binary nitrides REN of the more volatile rare earth metals R = Eu–Dy were successfully grown by sublimation [117–119], while EuN single crystals were also obtained ammonothermally from europium with potassium serving as mineralizer [120]. Single crystals of EuN and YbN could be grown in

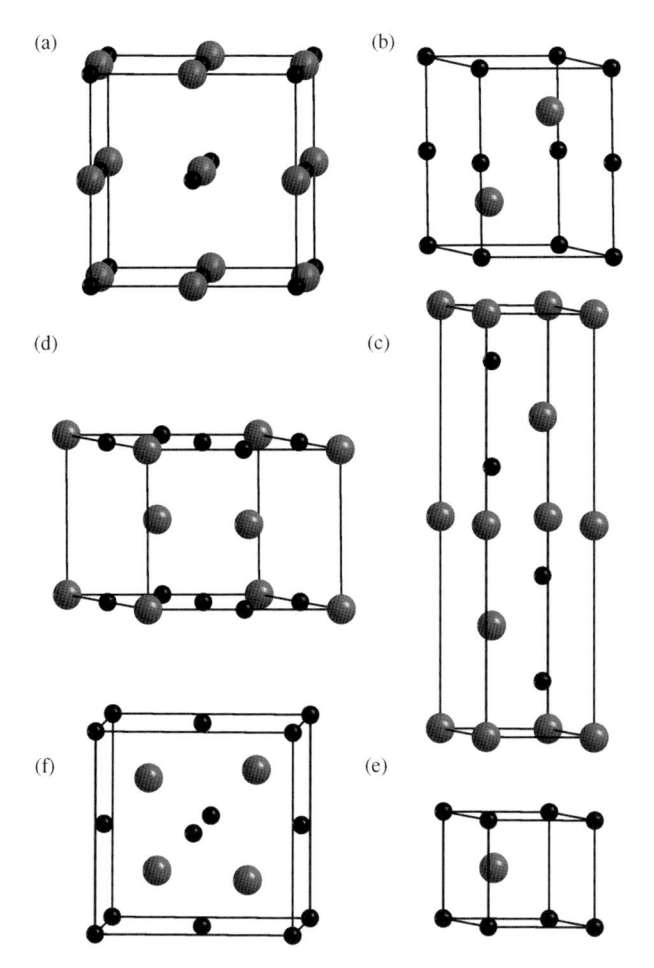

Figure 7.1 Simple prototypes dominate the crystal structures of binary 1:1 nitrides. (a) NaCl. (b) NiAs. (c) ε-NbN. (d) CoSn. (e) WC. (f) Sphalerite α-ZnS. Small black spheres: metal atoms; larger gray spheres: N.

sealed tantalum ampoules from the respective metals and Li_3N as a nitrogen source in lithium melts [90].

For LaN, a reversible pressure-induced tetragonally distorted CsCl-type high-pressure modification above 22 GPa was obtained [121]. Electronic structure calculations indicate a further tetragonal phase at slightly lower pressures [122]. Earlier, it was observed that LaN may exhibit a simple tetragonal distortion from cubic rock salt structure when synthesized at elevated temperature of 1273 K *in vacuo* [123].

Compounds with higher nitrogen content up to compositions of R_EN_2 (R_E = Y, Ce–Nd, Tb–Ho, Tm, Lu) were originally prepared at nitrogen pressures in the megapascal range [124]. They exhibited unit cell parameters and structure types nearly identical with the data of the respective binary oxides of the composition RE_2O_3 (A-type La_2O_3 structure for RE = Ce–Nd (Figure 7.2a), bixbyite-type Mn_2O_3 structure type for RE = Tb–Lu). Recently, Hasegawa et al. succeeded in preparing two novel lanthanum nitrides with $La_2N_{3+\delta}$ composition using laser-heated diamond-anvil cells applying N_2 simultaneously as reaction agent and pressure medium [125]. One of these materials is discussed to crystallize in the La_2O_3 (A) type, the second similar to the bixbyite type.

With thorium, the cubic rock salt-type ThN and rhombohedral anti-Al_4C_3-type Th_3N_4 (Figure 7.2b) with Th in sixfold and in $3+1+3$ coordination, respectively, are established [126–131]. The existence of Th_2N_3 with Th in an average oxidation state of +4.5 being isostructural to $Th_2N_2(NH)$ and Th_2N_2O is in question [131–135].

Rock salt-type UN is the uranium nitride with lowest nitrogen content and it apparently hardly tolerates any deviation from the equimolar composition [136]. UN represents a collinear antiferromagnet with $T_N = 53$ K. Above 29 GPa, UN transforms to a high-pressure phase with a rhombohedrally distorted rock salt-type structure [137]. α-U_2N_{3+x} crystallizes in the bixbyite type with a homogeneity range up to $x = 0.48$ [138–140]. With higher nitrogen content, the cubic fluorite structure type of UN_2 (Figure 7.3a) is realized. Interestingly, both α-U_2N_{3+x} and UN_2 show decreasing cubic unit cell parameters with increasing nitrogen contents. On heating to 1100 °C under a reduced nitrogen pressure, α-U_2N_{3+x} transforms to β-U_2N_{3+x} with the A-type La_2O_3 structure [141].

7.3.1.2 Group 4, Ti, Zr, Hf

One of the rare nitride minerals is osbornite [142,143], a golden metallic rock salt-type δ-TiN_x with a broad range of compositions in the approximate range of $0.4 < x < 1$ [144] and random distribution of vacancies [145]. Despite the fact that most reports on the binary phase diagram Ti–N indicate the nitrogen-rich border of solid phases close to 50 at.% N independent of temperature, a phase with the composition $Ti_{0.76}N = Ti_{3.04}N_4$ was found as a defect variant of δ-TiN [146]. ε-Ti_2N with rutile-type structure (Figure 7.3b) apparently has a narrow homogeneity range [147], although conflicting data exist in literature on this issue [148]. The same applies for δ'-Ti_2N with anatase structure (Figure 7.3c) and probably a somewhat enlarged nitrogen content compared to ε-Ti_2N [148–150]. Rapid cooling of samples Ti–N with about 30 at.% N has led to two rhombohedral phases η-Ti_3N_{2-x} (Figure 7.2e) and ζ-Ti_4N_{3-x} (Figure 7.2c) being isotypes of $\eta(\varepsilon)$-Hf_3N_2 and ζ-Hf_4N_3 with close-packed metal atom stackings of the sequences ... *ABABCBCAC* ... and ... *ABABCACABCBC* ... , respectively, and ordered but partial occupation of 50% of the octahedral voids [38,151].

Golden yellow metallic rock salt-type ZrN [152] and reddish-brown orthorhombic Zr_3N_4 [153,154], with Zr in octahedral, capped octahedral, and capped trigonal prismatic coordination, are known at ambient pressure. An earlier

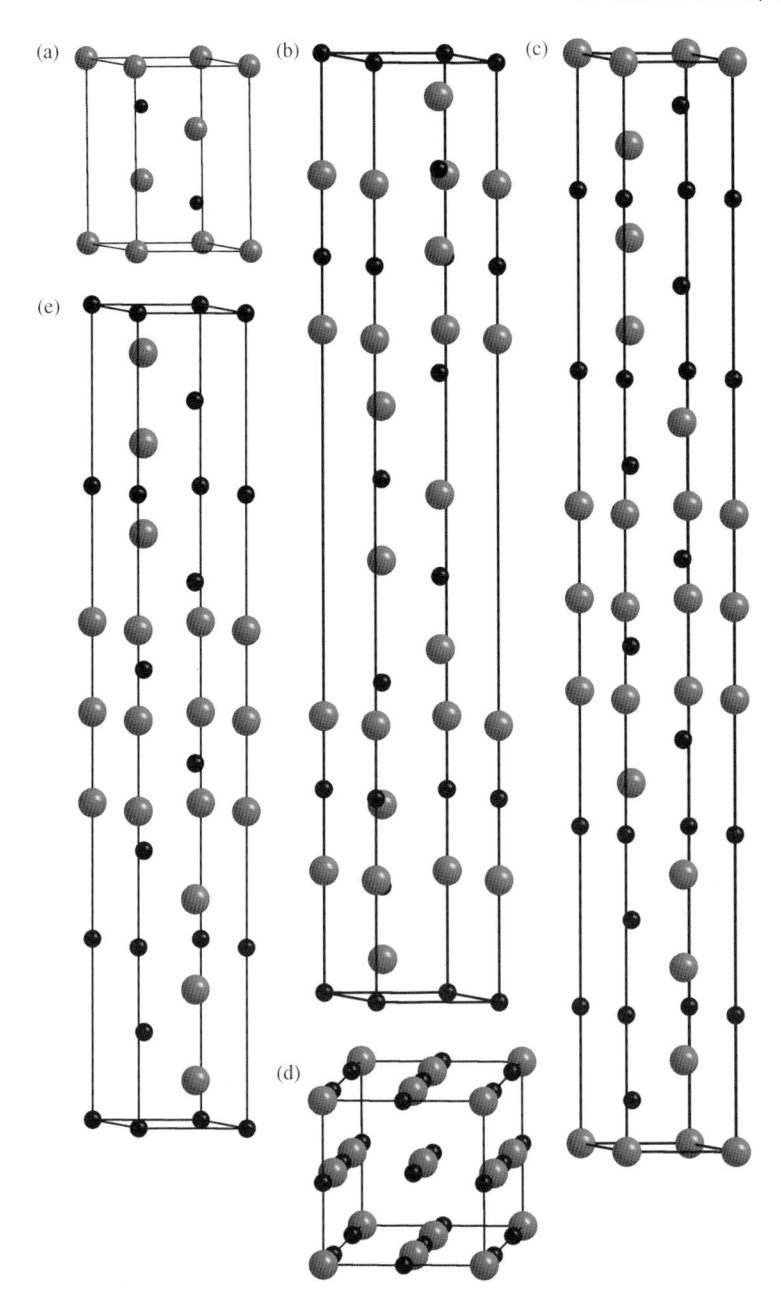

Figure 7.2 Idealized crystal structures of selected binary nitrides with compositions A_2X_3, A_3X_4, A_4X_3, and A_3X_2. (a) A-type La_2O_3. (b) Anti-Al_4C_3. (c) ζ-Ti_4N_3. (d) Nb_4N_3. (e) η-Ti_3N_2. Small black spheres: metal atoms; larger gray spheres: N.

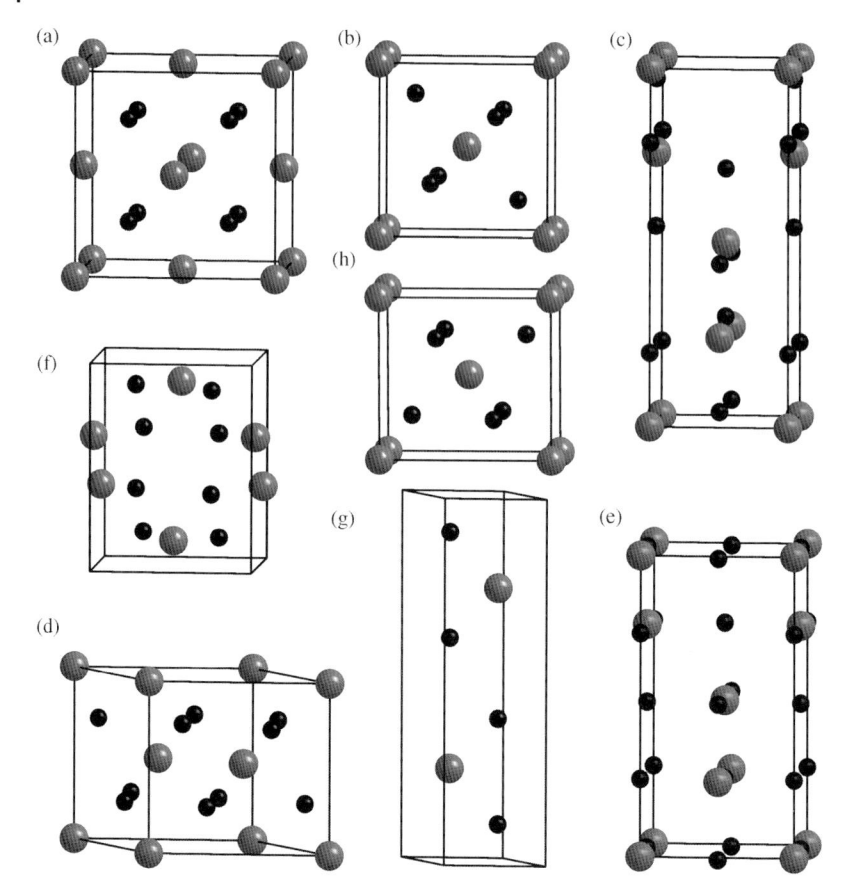

Figure 7.3 Crystal structures of binary nitrides with composition A_2X. (a) Anti-CaF$_2$. (b) Rutile anti-TiO$_2$. (c) Anatase anti-TiO$_2$. (d) V$_2$N. (e) Mo$_2$N. (f) Anti-PbO$_2$/Fe$_2$N. (g) Re$_2$N. (h) Fe$_2$C. Small black spheres: metal atoms; larger gray spheres: N.

reported blue-phase ZrN$_x$ probably represents a mixture of these two compounds [155]. Above 18 GPa, a cubic modification c-Zr$_3$N$_4$ forms upon laser heating. This quite incompressible material crystallizes in the Th$_3$P$_4$ structure type with eightfold coordination of Zr by N, which itself resides in an octahedral surrounding. For binary hafnium nitrides, quite close correspondence to zirconium and titanium nitrides was derived: δ-HfN$_{1-x}$, η(ε)-Hf$_3$N$_2$, and ζ-Hf$_4$N$_3$ as well as the high-pressure phase c-Hf$_3$N$_4$ were characterized to be isotypes of the respective phases of Ti and Zr already discussed [156–158].

7.3.1.3 Group 5, V, Nb, Ta

In the binary system V–N, the thermodynamically stable phases β-V$_2$N and δ-VN occur in addition to the disordered border solution phase of N in *bcc* vanadium (probably up to well above 14 at.% N close to the melting point) [159]. Rock salt-type δ-VN can exhibit a quite significant nitrogen deficiency down to about VN$_{0.68}$ (close to 2000 °C) depending on temperature. Below 520 °C the vacancies may order [160]. Below 205 K, VN suffers a minor tetragonal distortion [161]. β-V$_2$N (Figure 7.3d) similarly can be described as *hcp* of vanadium with occupation of octahedral voids by N [159,162,163], isostructural to β-Nb$_2$N [164] and β-Ta$_2$N [165]. Both stable nitrides of vanadium reach full nitrogen content only after application of an elevated nitrogen pressure at synthesis temperatures [166,167]. A number of further phases (V$_{16}$N [168–172], V$_{16}$N$_{1.5}$ [173], V$_9$N [159], V$_8$N [174,175], V$_9$N [159], and V$_4$N [175]) were described to precipitate from the supersaturated solution phase below 550 °C, and represent order variants of N within vacancies in *bcc* vanadium. Only V$_{16}$N$_{1.5}$ recently proved to represent a stable phase missing in the equilibrium phase diagram [176].

On reaction of the elements and further heat treatment, β-Nb$_2$N, γ-Nb$_4$N$_{3+x}$, and δ-NbN can be obtained, whereby δ-NbN in nitrogen atmosphere decomposes into the two aforementioned nitrides at elevated temperatures [164]. β-Nb$_2$N (Figure 7.3d) crystallizes in an *hcp* motif of Nb with occupation of octahedral holes by N. γ-Nb$_4$N$_{3+x}$ (Figure 7.2d) represents a tetragonally distorted vacancy-ordered variant [164,177,178] of the rock salt-type structure of δ-NbN. Both rock salt-based phases exhibit a slight N deficiency compared to the equimolar composition [164]. With larger nitrogen deficiency, NbN$_{0.95}$ was observed to realize the NiAs-type structure with nitrogen in octahedral holes; thus, Nb localized in trigonal prismatic coordination [179]. Hexagonal ε-NbN (Figure 7.1c) has a stacking sequence of ... *AABB* ... with N alternating in octahedral and trigonal prismatic voids [179]. η-NbN apparently can exist with some nitrogen excess, that is, with vacancies in the Nb substructure of a NiAs-type and Nb residing in trigonal prismatic voids (Nb$_{0.95}$N) [180]. From ammonolysis of NbO$_2$, a phase with composition Nb$_{0.84}$N, stacking of ... *ABABCBCACA* ... , and N in octahedral holes, thus Nb in octahedral and trigonal prismatic coordination, was prepared [181].

Tetragonal Nb$_4$N$_5$ and hexagonal Nb$_5$N$_6$ were obtained from reaction of metal films with ammonia [182]. Ammonolysis of sodium or potassium oxoniobates results in the formation of bulk Nb$_4$N$_5$ under evaporation of the alkali metal [183]. Nb$_4$N$_5$ crystallizes in an ordered defect rock salt variant [183]. Both compounds are isotypes of the respective tantalum nitrides.

β-Ta$_2$N (Figure 7.3d) is an isotype of β-Nb$_2$N and tolerates a significant nitrogen deficiency up to some 20% [164,165,184–186]. In hexagonal ε-TaN (CoSn structure type) (Figure 7.1d), the nitrogen atoms are displaced from the ideal octahedrally coordinated positions resulting in a square pyramidal environment [184,185,187]. ε-TaN transforms into υ-TaN at elevated pressures above 20 kbar with applied temperatures above 800 °C or under mechanical stress [188–190]. υ-TaN crystallizes hexagonal in the WC-type structure (Figure 7.1e) with a simple hexagonal stacking of the metal layers and nitrogen atoms in the resulting trigonal prisms. Above 3000 K and 16 GPa, a further new phase was indicated, although it could not be further elucidated so far [190]. Rock salt-type δ-TaN apparently is a high-temperature modification [191–193].

Ta_2N_3 forms from $TaCl_5$ and nitrogen gas using the plasma-enhanced chemical vapor deposition technique and crystallizes in the bixbyite type [194]. Apparently, this compound is identical to earlier reported phases, obtained via reactive sputtering, which were proposed to realize a structure based on the fluorite type [195,196]. A second modification, η-Ta_2N_3, was obtained under high-pressure high-temperature conditions from Ta_3N_5. It crystallizes in the U_2S_3 structure type (variant of the stibnite structure) with Ta in 7 + 1 coordination [197,198].

Orange-red Ta_3N_5 is the nitrogen-rich tantalum nitride (Figure 7.4) with a pseudobrookite (Fe_2TiO_5)-type structure and nitrogen in both three- and four-coordinate surroundings [51,199–201]. It is currently under evaluation as a promising semiconductor photocatalyst for solar water splitting [202] and represents the rare case of a binary transition metal nitride that can be obtained from its respective binary oxide via ammonolysis [200]. Single crystals can be grown by chemical vapor transport employing NH_4Cl as transporting agent [51]. Upon heating, Ta_3N_5 decomposes to subsequently yield hexagonal Ta_5N_6 and tetragonal Ta_4N_5 and finally produce ε-TaN and β-Ta_2N [186,203]. Two dense polymorphs of Ta_3N_5 with U_3Se_5 and U_3Te_5 structure types have been prepared by a high-pressure–high-temperature route employing laser heating at 22 GPa [204].

Ta_5N_6 and Ta_4N_5 were also obtained from reaction of metal films with ammonia [205]; Ta_5N_6 additionally forms from Ta powder and nitrogen at 300 bar and 1200 °C in an autoclave [206]. Vacancy-ordered rock salt-type Ta_4N_5 is an

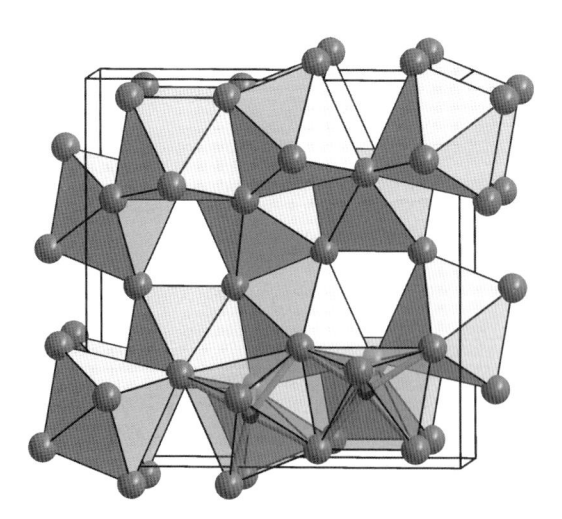

Figure 7.4 The crystal structure of Ta_3N_5 showing the connectivity of polyhedra surrounding Ta. Small black spheres: Ta; larger gray spheres: N.

isotype of Nb$_4$N$_5$ [186], while Ta$_5$N$_6$ as isotype of Nb$_5$N$_6$ is based on a stacking of tantalum in the sequence ... *ABAC* ... [186].

7.3.1.4 Group 6, Cr, Mo, W

The formation and precipitation of chromium nitrides play a fundamental role in the steel-hardening process of chromium-containing steels [207–213]. Additionally, γ-CrN [214] recently gained interest for application in proton exchange membranes [215,216], as substrate material for oxidation catalysts [217], and anodes for lithium ion batteries [218,219]. Chromium with nitrogen forms a *bcc* solid solution [220–222]. Next to this solid solution, only two thermodynamic stable binary nitrides are known, namely, β-Cr$_2$N and γ-CrN [223]. In β-Cr$_2$N, up to 50% of the octahedral holes in an *hcp* motif of Cr are filled with nitrogen [223]. An energetic preference of nitrogen ordering rather than a statistical distribution is favored in thermodynamic models [224,225], although no experimental indication of any ordering could be observed until now [226]. The N-poor phase boundary shows a significant variation with temperature (Cr$_2$N$_{0.83}$ at 950 °C and Cr$_2$N$_{0.645}$ at 1350 °C), well in contrast to the N-rich phase boundary [227]. Carlsbergite, γ-CrN, crystallizes in the rock salt-type structure [223,228]. For bulk samples of γ-CrN, only very small deviations from the ideal composition occur [229]; for thin films, vacancies in the chromium substructure were observed [230,231]. On cooling below approximately 15 °C, γ-CrN suffers a phase transformation from paramagnetic to an antiferromagnetic state with orthorhombic distortion [232,233]. The formation of CrN via ammonolysis of binary halides and sulfides was recently revisited and revealed a complex reduction–reoxidation sequence via intermediates depending on the chemical nature of the starting material [234].

β-Mo$_2$N (Figure 7.3e) crystallizes in a *fcc* motif of Mo with ordered occupation of 50% of the octahedral holes by N, leading to a T-shaped threefold coordination of Mo by N. In the cubic high-temperature phase γ-Mo$_2$N, the nitrogen atoms are randomly distributed over octahedral voids, thus realizing a cubic rock salt motif. Both phases tolerate some homogeneity range [235–238]. γ-Mo$_2$N is a semimetal and becomes superconducting below 5.5 K [239].

Three hexagonal modifications of δ-MoN were reported, all synthesized by ammonolysis of MoCl$_5$, nitridation of Mo using ammonia, or reaction of nitrogen plasma with MoCl$_5$ at various temperatures [236,240–242]. Two of these polymorphs can be understood as WC-type (Figure 7.1e) and NiAs-type (Figure 7.1b) structures with Mo in trigonal prismatic and octahedral coordination by N, respectively, and the third as slightly distorted superstructure of the NiAs-type [235,238,241–246]. The latter modification can be obtained from the WC-type form by annealing at elevated temperatures [246]. δ-MoN also becomes superconducting below about 12–15 K, depending on composition and degree of disorder [240,246,247]. Finally, hexagonal Mo$_5$N$_6$ may be described as filled MoS$_2$-type structure, thus 60% of Mo in trigonal prismatic coordination arranged within layers and the remaining 40% in octahedral holes in-between [34,244]. However, the exact distribution of Mo in the octahedral voids still needs to be determined.

The binary system W–N apparently is not well studied. Early investigations indicated β-W_2N (Figure 7.3e) isotypic to β-Mo_2N [236], a cubic α-W_2N (Figure 7.3a) [248], a WC-type δ-WN (Figure 7.1e) [245], and a rock salt-type WN [249], while further reports on W_3N_2, W_2N_3, and WN_2 [250,251] partly were just recently picked up. In metathesis reactions from sodium oxotungstate(VI) with boron nitride followed by thermal decomposition in high-temperature high-pressure experiments, two modifications of W_2N_3, namely h-W_2N_3 and r-W_2N_3, c-W_3N_4, and δ-WN were obtained [252]. The WC-type structure for δ-WN was confirmed. c-W_3N_4 is reported to realize a cubic ordered defect variant of the rock salt structure. h-W_2N_3 is discussed to crystallize in an unusual layered structure with W and one-third of N in trigonal prismatic coordination of the other kind, and two-third of N terminating the layers forming a van der Waals gap. r-W_2N_3 similarly contains N in trigonal prismatic and W both in trigonal prismatic and octahedral coordination. However, all these structures were solely deduced from powder X-ray diffraction patterns [252]. The two forms of W_2N_3 might be identical to W_5N_6 earlier synthesized from ammonolysis of WS_2 [253] or $W_{4.6}N_4$ prepared in thin films [254,255].

7.3.1.5 Group 7, Mn, Re

ε-Mn_4N crystallizes in the γ′-Fe_4N-type perovskite structure (Figure 7.5d) with a noncollinear ferrimagnetic spin structure and a Curie temperature of 738–748 K [256–262]. It exists with a rather narrow homogeneity range attributed to Frenkel-type disorder in the nitrogen substructure [263].

ζ(ζ′)-Mn_2N (also called Mn_5N_2) forms with an extremely broad range of compositions from about 30 to 36 at.% N at 573 K [264,265], at higher temperatures (1473 K), and even down to about 11 at.% N [266,267]. The crystal structure is

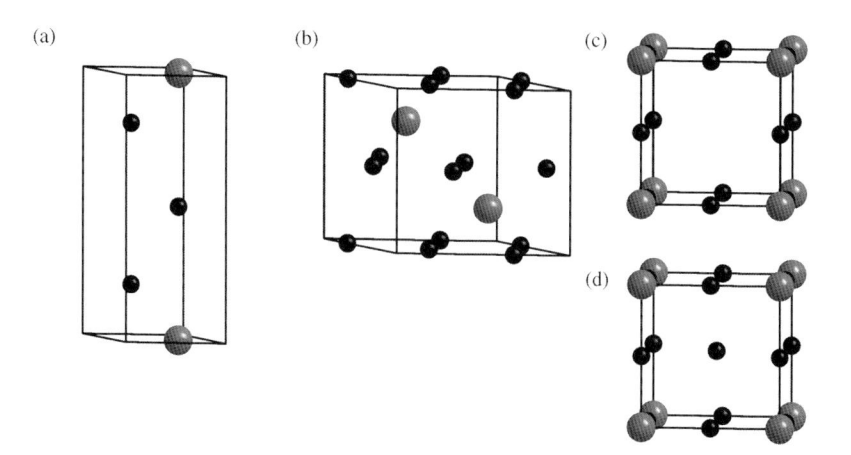

Figure 7.5 Crystal structures of binary nitrides with compositions A_3N and A_4N. (a) Re_3N. (b) ε-Fe_3N. (c) Cu_3N. (d) γ′-Fe_4N. Small black spheres: metal atoms; larger gray spheres: N.

based on an *hcp* arrangement of Mn. For two different samples with the compositions $Mn_2N_{0.98}$ [268] and $Mn_2N_{0.86}$ [269], different ordering variants of the occupation of octahedral holes with nitrogen were detected. $Mn_2N_{0.98}$ was described with an α-PbO_2 structure (Figure 7.3f), and thus with manganese threefold coordinated by nitrogen and nitrogen in turn in octahedral coordination by Mn. In $Mn_2N_{0.86}$, the octahedral holes are only partly occupied leading to a defective NiAs-type arrangement. The occurrence of different ordering variants leads to the suspicion that the α-PbO_2-type phase exists only in a narrow region of compositions, while the remaining part of the homogeneity range more likely belongs to an ε-type phase region [270]. In fact, an "ε-type" phase was confirmed for the composition $Mn_3N_{1.17}$ ($= Mn_2N_{0.78}$) [271]. At room temperature, it has a partly ordered structure, while the nitrogen atoms randomize over the octahedral holes upon increasing temperature. For the composition $Mn_2N_{0.98}$ (ζ-Mn_2N), a noncollinear antiferromagnetic spin structure with a Néel temperature of 310 K [268] and for $Mn_2N_{0.86}$ (ε-type) a collinear antiferromagnetic spin structure were reported [269,270]. However, an even more complicated order behavior of nitrogen was indicated, since four different superstructures were detected depending on temperature and nitrogen content for samples close to ζ-Mn_2N in composition [272].

η-Mn_3N_2 with a homogeneity range from $Mn_3N_{1.84}$ to $Mn_3N_{2.08}$ [264,265] crystallizes in an ordered defect variant of the rock salt-type structure, leading to a tetragonal distortion. It contains one fivefold coordinated Mn site and one Mn site linearly coordinated by two nitrogen atoms [273]. η-Mn_3N_2 at room temperature orders antiferromagnetic along [001] and ferromagnetic within (001) [273–276].

θ-Mn_6N_5 has a rock salt-type structure with random vacancies on the nitrogen sites [274,275,277]. Antiferromagnetic order below the Néel temperature at about 670 K leads to a tetragonal distortion due to magnetostriction, while the compound is cubic above T_N [275,276,278].

The formation of rhenium nitrides ReN_x with $0.06 < x < 1.35$ has been accomplished using precursors at high temperatures [279–281] and in thin films [282–284]. However, due to small crystallite sizes, the chemical identity and composition of these samples remained elusive. Recently, the highly incompressible materials Re_3N (Figure 7.5a) and Re_2N (Figure 7.3g) were synthesized during laser heating of Re foil under nitrogen pressures above 10 GPa. Both compounds can be understood as stacking variants of hexagonal close-packed layers of Re with N introduced layerwise into trigonal prismatic coordination (. . . ABB . . . and . . . AABB . . . stacking sequences of Re, respectively) [285–287].

7.3.1.6 Group 8, Fe, Ru, Os
Iron nitrides are well established for various technical applications. Especially, γ'-Fe_4N (Figure 7.5d) and ε-Fe_3N_{1+x} (Figure 7.5b) are of great importance in the industrial surface hardening of steel work pieces, improving hardness and tribological properties as well as corrosion and wear resistance [288,289]. Furthermore, ferromagnetic iron nitrides gained interest in magnetic data recording due to their tunable magnetic properties [290,291].

Important early attempts to determine the Fe–N phase diagram were carried out during the initial investigations concerning the ammonia synthesis in the Haber–Bosch process [214,292,293]. Three thermodynamic stable binary iron nitride phases are known: γ'-Fe$_4$N (roaldite [294]) with only a narrow compositional variation, ε-Fe$_3$N with a broad homogeneity range, and ζ-Fe$_2$N with an again narrow range of compositions [295]. Formation and decomposition sequences of these stable nitrides in inert as well as ammonia gas dominated by kinetics were studied via *in situ* diffraction [296–298]. Furthermore, four metastable phases are reported in literature: bct α'-Fe$_8$N, α''-Fe$_{16}$N$_2$ as an ordered variant thereof [299], and FeN in rock salt [300] and in sphalerite structure types [301].

Cubic γ'-Fe$_4$N (Figure 7.5d) represents a perovskite, Fe(Fe$_3$N), with three Fe per formula unit coordinated by N and one Fe exclusively surrounded by 12 Fe of the other kind in a cuboctahedron. Polycrystalline γ'-Fe$_4$N transforms into single-crystalline ε-Fe$_3$N$_{0.75}$ under high-pressure high-temperature conditions (8.5 GPa and 1373 K) and reforms upon heating to 525 K without application of external pressure [302,303].

In the ε-phase (Figure 7.5b), iron atoms are arranged in an *hcp* motif and nitrogen occupies octahedral voids in a partially ordered arrangement [304–306]. Through varying grades of filling of these crystallographically distinct octahedral voids, a very large homogeneity range from ε-Fe$_3$N$_{0.65}$ up to ε-Fe$_3$N$_{1.47}$ depending on temperature develops [307,308]. Diffraction data of microcrystalline samples were usually treated in space group $P6_322$ [224]; however, according to electronic band structure calculations, a symmetry reduction to space group $P312$ decreases the repulsion forces between nitrogen atoms [309,310].

The ζ-phase (Figure 7.3f) also exhibits the *hcp* motif of iron atoms, but the nitrogen atoms are differently ordered within the octahedral holes leading to an orthorhombic α-PbO$_2$-type structure [295,311–313]. Upon annealing at temperatures between 300 and 450 °C, ζ-Fe$_2$N transforms into ε-Fe$_3$N$_{1+x}$ [310]. Furthermore, at 15 GPa and 1600 K, microcrystalline ζ-Fe$_2$N was transformed into single-crystalline ε-Fe$_3$N$_{1.47}$ [308].

Especially, the metastable phase α''-Fe$_{16}$N$_2$ [299,314] caused great attention in the scientific community since largely enhanced saturation magnetization and magnetic moment on thin films were discovered, which even exceed the theoretically predicted maximum values [315]. The magnetic properties of α''-Fe$_{16}$N$_2$ were a topic of debate ever since, partly due to the fact that it is difficult to prepare pure samples. Very recently, quality of obtained bulk samples greatly improved [316,317]. For bulk samples, the enlarged magnetic moment was not confirmed in agreement with the majority of electronic structure calculations [318,319]. The reversible order–disorder transition to α'-Fe$_8$N could be traced.

For Ru and Os, no binary nitrides have been reported. The binary phases RuN$_2$ and OsN$_2$ represent pernitrides containing anionic dinitride species [320].

7.3.1.7 Group 9, Co, Rh, Ir

Elemental Co only dissolves minor amounts of nitrogen, but forms at least four stable binary nitrides. Co_4N crystallizes in a perovskite structure (γ'-Fe_4N type) (Figure 7.5d) and Co_3N realizes the ideal ordered structure of ε-Fe_3N (Figure 7.5b) with only small deviations from ideal composition [321], with nitrogen located in exclusively vertex-sharing octahedral voids of an *hcp* of Co. Co_4N and Co_3N are ferromagnetically ordered up to their decomposition temperatures above 650 K [322]. With Co_2N, the resemblance of iron nitrides is continued. Co_2N crystallizes similar to ζ-Fe_2N (α-PbO_2 type) (Figure 7.3f), with nitrogen again located in octahedral voids of an *hcp* of Co, although with a different order motif leading to edge-connected coordination polyhedra centered by N [321]. CoN in a rock salt-type structure is thermally stable only up to 160 °C [323], thus typically accessible only via mild decomposition of precursor compounds. In thin layers, the additional occurrence of a second modification with sphalerite structure was observed [324–326]. Both forms strongly differ in their magnetic properties since only the rock salt modification suffers an antiferromagnetic order below 37 °C [326]. The early reports on Co_3N_2 [327,328] are not confirmed so far [321].

For Rh and Ir, no binary nitrides have been confirmed. The binary phases RhN_2 and IrN_2 again represent pernitrides [320].

7.3.1.8 Group 10, Ni, Pd, Pt

Ni forms two binary nitrides, namely Ni_4N and Ni_3N. Ni_4N (Figure 7.5d) is an isotype of γ-Fe_4N and Co_4N with perovskite-type structure and similarly orders ferromagnetically [329,330]; however, the magnetic moment decreases within this row of perovskite compounds due to increasing number of 3d electrons and thus covalent localization. According to electronic structure calculations, Ni_4N is metastable [330]. Ni_3N (Figure 7.5b) like Co_3N crystallizes in the ideal structure of ε-Fe_3N [331]. Additionally, the formation of Ni_3N_2 was claimed [328,332].

Pd_2N [333] apparently is obtained in impure form as a by-product in synthesis attempts for a π-phase solid solution $Ni_{2-x}Pd_xMo_3N$ [334] and crystallizes in the Fe_2C-type structure (Figure 7.3h) [335].

The synthesis of PtN (Figure 7.1f) with sphalerite-type structure by high-temperature high-pressure techniques was asserted [23], but later shown to rather represent PtN_2, a pernitride-containing covalently-bonded dinitrogen species realizing a pyrite (FeS_2)-type structure [20,336–339].

7.3.1.9 Group 11, Cu, Ag, Au

Metastable semiconducting Cu_3N (Figure 7.5c) represents the exclusive example of a transition metal nitride with cubic ReO_3 structure [340–343]. Due to its properties as a thermolabile but corrosion-resistive semiconductor, it has been discussed for various applications, for example, as oxidation-resistant coating, as optical recording medium, for random access memory chips, conductive ink, as well as for maskless printing of electronic devices [344–348]. Microcrystalline samples are obtained from ammonolysis of CuF_2, nanoparticles from copper(II) acetate and ammonia in alcohols, and crystals can be grown ammonothermally [344,349]. Above 5 GPa, a reversible pressure-induced phase transition via transient steps leads to a tetragonal defect rock salt type with metallic properties [350–352]. The perovskite Cu_4N was obtained as a thin film by plasma deposition [353].

Black explosive so-called "Berthollet'sches Knallsilber" or "fulminating silver" was obtained from aqueous solutions of AgCl, KOH, and NH_4OH or action of aqueous ammonia on Ag_2O [354,355]. It exhibits a cubic unit cell and is believed to represent Ag_3N [356,357]. A binary nitride of gold seems to be unknown.

Resulting only from theoretical *ab initio* calculations, the recently reported $CuN[N_2]$ [358] and $AgN[N_2]$ [358] as well as $Au[N_2]$ [359] are outside the scope of this work.

7.3.1.10 Group 12, Zn, Cd, Hg

Binary Zn_3N_2 in the bixbyite structure [360,361] exhibits semiconducting properties [343] with probably a direct bandgap of around 1 eV [362,363]. Black Cd_3N_2 was obtained via thermolysis of $Cd(NH_2)_2$ [332,361] or $Cd(N_3)_2$ [364]. It also crystallizes in the bixbyite structure type and apparently is thermolabile. Chocolate-brown Hg_3N_2 was claimed to be produced in a reaction of HgI_2 with KNH_2 in liquid ammonia or via electric discharge in a $Hg-N_2$ gas mixture, which is however explosive [365]. Heavy shaking of metallic Hg in the presence of electric discharged N_2 at room temperature supposedly leads to small amounts (2% of the starting material) of Hg_3N [366]. Explosive yellow "Hg_2N_4" represents the azide of Millon's base $[Hg_2N]N_3$ with the typical linear complex ion $[Hg_2N]^+$ (Figure 7.6). It may be prepared from mercury azide and aqueous ammonia [367,368].

7.3.2
Multinary Derivatives of Metal-Rich Compounds

For metal-rich ternary nitrides of transition or rare earth metals, few structure types dominate, which can be classified according to the following categories: Layered structures with compositions $MM'N_2$ frequently contain both metals in sixfold coordination and will be discussed in the nitridometalate section in Section 7.3.3.7. Perovskites based on the γ'-Fe_4N structure type constitute a quite large class and spread from essentially fixed compositions to phases with substitutional disorder on one or both metal sites depending on the combination of metallic constituents (Section 7.3.2.1). ε-Fe_3N-type phases are a comparably new field with small number of members so far. The actually known phases in this class can simply be designated as $Fe_{3-x}M_xN$ with typically $x \leq 1$ and are summarized in Section 7.3.2.2. Compounds $M_2M'_3N$ crystallizing in a filled β-Mn variant (Section 7.3.2.3) are structurally closely related to η-carbide derivatives $M_3M'_3N$ (Section 7.3.2.4). The so-called H-phases currently gain a renewed interest within the calls of so-called MAX phases due to their superior mechanical properties paired with chemical stability (Section 7.3.2.5). Particularly,

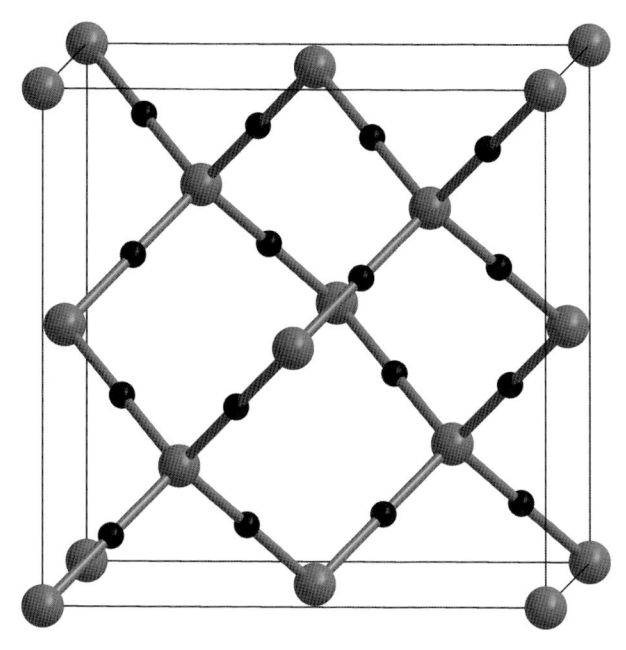

Figure 7.6 The complex anion $[Hg_2N]^+$ in mercury-containing nitride phases. Small black spheres: Hg; larger gray spheres: N.

Cr-based Z-phases precipitate from creep-resistant steels and probably limit their performance (Section 7.3.2.6). The large class of $RE_2M_{17}N_y$ and related phases can be structurally derived from the hard magnetic prototype SmCo$_5$ with CaCu$_5$-type structure. Predominately, these groups of compounds will be discussed in Section 7.3.2.7. Finally, a plethora of substituted rock salt-type phases are collected in Section 7.3.2.8.

7.3.2.1 γ'-Fe$_4$N Derivatives

H												B	C	N	O	F	He
Li	Be											B	C	N	O	F	Ne
Na	Mg											Al	Si	P	S	Cl	Ar
K	Ca	Sc	Ti	V	Cr	Mn	Fe	Co	Ni	Cu	Zn	Ga	Ge	As	Se	Br	Kr
Rb	Sr	Y	Zr	Nb	Mo	Tc	Ru	Rh	Pd	Ag	Cd	In	Sn	Sb	Te	I	Xe
Cs	Ba	La	Hf	Ta	W	Re	Os	Ir	Pt	Au	Hg	Tl	Pb	Bi	Po	At	Rn
Fr	Ra	Ac															

		Ce	Pr	Nd	Pm	Sm	Eu	Gd	Tb	Dy	Ho	Er	Tm	Yb	Lu
		Th	Pa	U											

As already discussed, the cubic γ'-Fe$_4$N structure type (Figure 7.7a) represents a perovskite structure, with three Fe atoms per formula unit coordinated by N

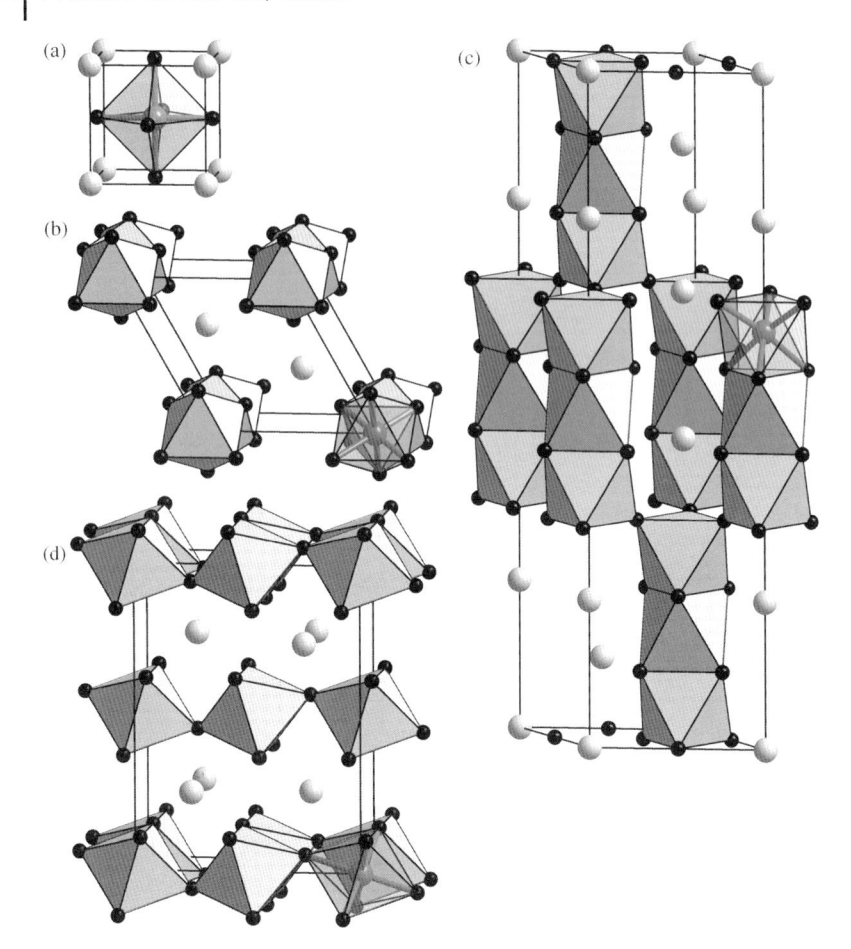

Figure 7.7 Crystal structure types of MM'_3N phases. (a) Anti-CaTiO$_3$. (b) Anti-2H-BaNiO$_3$. (c) Anti-4H-BaRuO$_3$. (d) Post-perovskite AM_3N. Small black and larger light gray spheres: metal atoms; octahedra surround N.

and 1 Fe atom exclusively surrounded by 12 Fe of the other kind within a cuboctahedron. Both iron sites may be substituted by a variety of metal atoms: Due to the fundamental difference in surrounding and thus electronic situation, the majority position in this respect is typically occupied by a more nitrophilic metal; the minority site may be even adopted by metals that themselves do not form stable binary nitrides. As a general formula, $M_{4-x}M'_xN$ naturally develops. For metal combinations with similar interaction toward nitrogen, the whole range of $0 \leq x \leq 4$ can be accessible, particularly for Co and Ni, which themselves form binary nitrides with γ'-Fe$_4$N structure type [261,369–377]. In combination with a less nitrophilic metal, typically $x \leq 1$ and as a substitution limit $(M_3N)M'$ may be obtained, even though slightly larger $x \leq 1.2$ were reported for, for example, Fe$_{4-x}$Sn$_x$N [291,378,379]. Although alkaline earth metal atoms may take over the function of M [380], only one such combination with a transition metal at the M' position, namely, $(Ca_3N)Au$, has been reported [381]. The latter metallic compound may be described as a subnitride auride [382]. Analogous rare earth metal nitrides are known with M' constituting from a group 13 or rarely a group 14 element with integral composition $(RE_3N)M'$ and some possibility to form with minor nitrogen deficiency $(RE_3N_{1-x})M'$ [383].

As prototype compound for this group, γ'-Fe_4N is particularly susceptible for substitution of iron by a large variety of different metals forming solid solutions $Fe_{4-x}M'_xN$. The substitution can occur on both sites and in the full range of $0 \leq x \leq 4$ for metals that form γ'-M_4N phases themselves, for example, for Co [369,370] and Ni (although with some site preference) [261,371–377]. Other metals such as Ru [384,385], Os [384], Ir [384,386], Mn [387,388], Cu [389,390], Ag [391], and Zn [392–394] apparently substitute only to a small degree with $x < 1$ or in the range of $0 \leq x \leq 1.2$ for $M = Sn$ [291,378,379,394,395]. At least Mn and Cu are again able to substitute both sites, while Ru, Os, and Ir appear to be restricted to the site not coordinated to N. Transition metals with minor nitrophilicity, for example, Pd [371,387,389,396–398], Pt [261], Au [391], Rh [386,399,400], Ge [394,395], Ga [394,395,401–403], In [392,394,404], or Mg [394,395], lead to apparently ordered perovskite-type nitrides with $x \leq 1$, that is, a maximum substitution up to a composition of $Fe_3M'N$ depending on the specific element M'. All these latter metals are not coordinated by N in the perovskite-type structures and, except for Ga, they themselves do not form a stable binary nitride; even InN is supposed to be metastable [405,406]. Quite similar to the situation in binary γ'-Fe_4N, in these ternary phases the metals at this site act as electron donors for the other metal sites [374]. Fe_3AlN, however, was found to realize a rock salt structure with Fe/Al disorder and concomitant to only 25% occupation of the octahedral holes by N [403].

Particularly, these iron-based nitrides are studied in order to improve the magnetic properties of the prototype binary nitride, for example, for application for high-density storage materials due to high coercive fields [369,384,388]. On substitution with group 13 metals, $Fe_{3.2}In_{0.8}N$ remains ferromagnetically ordered [404]. $Fe_{3.1}Ga_{0.9}N$, however, is paramagnetic with magnetic order below about 8 K [402]. With increasing Ga content, the magnetic behavior changes, either from ferromagnetism of γ'-Fe_4N to antiferromagnetism [401,403] or just due to weakening of ferromagnetic Fe–Fe interactions [402]. Due to the intrinsic disorder, Fe_3AlN is a soft ferromagnet up to the thermal decomposition above 540 °C [403]. For a hypothetically ordered structure, Fe_3AlN was predicted not to exhibit any magnetic moments [407]. Fe_3GeN_x suffers a distortion to tetragonal via rotation of the $Fe_{6/2}N$ octahedra [408].

Most phases with a second transition metal order ferromagnetically. This is true, for example, for Fe_3CoN according to electronic structure calculations [409]. The more cobalt-rich disordered phase $FeCo_3N$ orders ferromagnetically according to XMCD measurements [370], while the (hypothetical) ordered compound with this composition was predicted to represent a half-metal with a bandgap of 0.2 eV at the Fermi level and thus present potential for spintronic applications [410]. Fe_3NiN shows coexistence of localized and itinerant magnetism [411] with dominating ferromagnetic interactions [412]. Fe_3MnN is a ferrimagnet according to band structure calculations [387]. Fe_3RhN was shown to represent a semihard itinerant ferromagnet [399] and Fe_3PdN a ferromagnet according to band structure calculations [387,397]. Small particles Fe_3MN (e.g., with $M=$ Ru, Os, Ir) show superparamagnetic behavior [384]. Several compounds Fe_3MN (e.g., with $M=$ Sn, Mn, Ni, Pd, Pt) were suggested to exhibit invar-like behavior [371,396,413,414].

Similar to γ'-Fe_4N, the Mn atoms in Mn_4N are known to be readily substituted by a variety of metallic elements, such as Cr, Fe, Rh, Ir, Ni, Pd, Pt, Cu, Ag, Zn, Hg, Ga, In, Ge, Sn, As, and Sb [260,394,395,408,415–429], leading to cubic or distorted perovskite arrangements. For some combinations of elements, the occupation of the nitrogen site can be substantially lower than unity, as seen in, for example, $Mn_3RhN_{0.1}$ or $Mn_3PtN_{0.07}$, often connected with the possibility of a hexagonal structure [430–432]. Basically, all phases suffer magnetic transitions with typically strong magnetostriction leading to crystallographic distortions and various magnetic structures including spin glass-like behavior [433].

Mn_3AlN_x exhibits a Curie temperature of 818 K [434,435], while Mn_3InN shows only weak ferromagnetism below 366 K [436]. Curie temperature and magnetic moment of Mn_4N decrease upon substitution by In. A compensation point is reached at a substitution level of about 0.25 [437]. Mn_3GaN in contrast orders antiferromagnetically at 298 K [427,436,438]. At lower temperatures, a magnetic rearrangement was observed [422–425]. With smaller substitution level by Ga, ferrimagnetism develops at lower temperatures [439]. Somewhat deviating behavior of different samples was speculated to be due to slightly varying compositions [426]. Mn_3AsN and Mn_3SbN undergo a first-order transition from paramagnetic to ferromagnetic concomitant to a distortion to tetragonal metric at 360 K due to large magnetostriction [418,419,440]. Upon substitution by Sn, the magnetic moment as well as the Curie temperature of Mn_4N drops with a compensation point at a substitution level of about 0.38 [437].

With M being a 3d transition metal and smaller substitution levels $x<1$ in $Mn_{4-x}M_xN$, some M show different site preferences depending on the interaction with nitrogen compared to Mn [415,420]. Mn_3NiN, Mn_3CoN, Mn_3RhN, and Mn_3PdN order antiferromagnetically below about 262 K [436,441,442], 252 K [436], 226 K [436], and 316 K [436], respectively. Similarly, Mn_3ZnN exhibits antiferromagnetic order below 191 K, with a magnetic rearrangement at 127.5 K [422,424,436,443]. Mn_3AgN only shows weak ferromagnetism below 366 K [436] and Mn_3CuN orders ferromagnetically below about 143 K, concomitant to a distortion to tetragonal metric due to magnetostriction [444–446]. Apparently, due to small nitrogen content, $Mn_3PtN_{0.25}$ exists as cubic perovskite and with hexagonal 2 H perovskite structure ($BaNiO_3$ type) (Figure 7.7b), both probably exhibiting ferromagnetism [420,431,432,447]. $Mn_3RhN_{0.2}$ similarly realizes the 2 H structure [432].

Recently, Mn-based perovskite nitrides gained considerable interest due to tunable isotropic thermal expansion of near zero to largely negative in several compositions, namely, of Mn_3Cu_{1-x}(Ga, Si, Ge, Sn, Ag)$_x$N [448–464] and

Mn$_3$(Zn, Sn)N [462,465]. Further elements for doping of Mn$_3$CuN at the Cu site were studied [436]. The properties of these phases were recently surveyed [466]. A related ferromagnetic shape memory effect of Mn$_3$CuN$_x$ based on a martensitic phase transition was described [467,468].

Similar to binary rhenium nitrides, a perovskite Re$_3$ZnN$_x$ with 4H-BaRuO$_3$-type structure (Figure 7.7c) and N in face-sharing octahedra formed exclusively by Re was obtained under high-temperature high-pressure conditions [469].

Since the discovery of type II superconductivity in Ni$_3$CuN [470,471] and Ni$_3$ZnN [472] with T_c around 3 K, Ni-based perovskites Ni$_3$MN recently gained substantial attention. In this respect, Ni$_3$ZnN is potentially an unconventional superconductor according to the temperature dependence of muon spin relaxation rate [473]. Upon partial substitution with Co phase separation and formation of ferromagnetic, Ni$_{0.6}$Co$_{2.4}$ZnN$_y$ ($y < 1$) occurs [474]. Ni$_3$CdN was reported to exhibit paramagnetism [475] or very soft and weak ferromagnetism [476]. It forms a complete solid solution with ferromagnetic Co$_3$CdN [476]. Fully ordered Ni$_3$InN shows spin glass-like behavior [477]. Solid solutions Ni$_3$Cd$_{1-x}$In$_x$N and Ni$_3$Cd$_{1-x}$Cu$_x$N exhibit very soft and weak ferromagnetism [478]. In the course of these studies, several element combinations of partly hypothetical perovskites were studied using various theoretical methods, for example, Ni$_3$MN with $M =$ Mg, Al, Ga, In, Sn, Sb, Pd, Pt, Cu, Ag, Zn, Cd [393,479–493].

The knowledge on cobalt-based perovskite nitrides appears to be more limited. Apparently, compounds with Zn and Mg form [393] in addition to the ferromagnet Co$_3$CdN, which is able to form a complete solid solution with Ni$_3$CdN [475]. Metallic Co$_3$InN is completely ordered and shows spin glass-like behavior [477].

Cubic Cu$_3$N with ReO$_3$-type structure can be intercalated with Pd during ammonothermal synthesis to form metastable perovskite-type Cu$_3$Pd$_x$N with $x = 0.020$ and 0.989 [494]. Colloidal Cu$_3$PdN nanoparticles obtained from organic solvents were successfully used as catalysts for oxygen reduction [495]. Cu$_3$PdN represents a semimetal indicating a semiconducting-to-semimetallic transition with increasing x in Cu$_3$Pd$_x$N [496]. Thin films with composition Cu$_3$Pd$_{0.238}$N thus exhibit a vanishing temperature coefficient of the electrical resistivity due to a delicate balance between the opposite changes of the number of carriers and the carrier mobility with temperature [497]. A similar situation was observed for thin films Cu$_3$M$_x$N with $M =$ Cu, Ag, Au [498], while for nanocrystalline Cu$_3$Zn$_x$N films only small x were obtained [499]. Still the Zn intercalation provokes a rapid drop of electrical resistivity. A narrow-gap semiconductor Cu$_3$Au$_{0.5}$N was additionally reported [500]. Again, these compounds provoked a number of theoretical studies on compounds resulting from intercalation of various metals into Cu$_3$N [501–504]. Intercalation of Li into Cu$_3$N is a special case, since Li partially replaces Cu, which enters the site not coordinated to N in the developing perovskite structure. Phases Cu$_3$Li$_x$N in the whole range of $0 < x \leq 1$ were obtained [505].

For Cr-based ternary perovskite nitrides, little information can be found in literature. Cr$_3$GaN crystallizes as cubic perovskite, while the Cr$_{6/2}$N octahedra in

Cr$_3$GeN and Cr$_3$AsN rotate to result in a tetragonal distorted variant (stuffed U$_3$Si-type structure) [15,408,417]. Upon substitution of some Cr by Mn in Cr$_{3-x}$Mn$_x$GaN ($0 < x < 1.5$), some antiferromagnetic contributions in the paramagnetic parent compound with $x = 0$ develop [506]. Electronic structures of Cr$_3$GaN and Cr$_3$RhN were studied with focus on magnetism and possible superconductivity [507–509]. For Cr$_3$GeN, an antiferromagnetic-to-paramagnetic transition was observed around 418 K [510]. Cr$_3$PtN forms as protective layer upon nitridation of Cr$_3$Pt [420,511,512].

A phase with composition V$_3$AuN$_x$ was briefly mentioned [513].

Although no binary perovskite nitride of Ti is known, apparently phases with Ti-excess Ti$_3$(M$_{1-x}$Ti$_x$)N$_y$ form readily with a large range of further metals, apparently typically tolerating some nitrogen deficiency ($y < 1$): Ti$_{3.1}$Zn$_{0.9}$N$_{0.5}$, Ti$_{3.2}$Zn$_{0.8}$N$_{0.8}$, Ti$_3$(Ti, Cd)N, Ti$_{3.2}$Hg$_{0.8}$N [514], Ti$_3$AlN [515,516], Ti$_3$InN, and Ti$_3$TlN [517]. Particularly, Ti$_3$AlN as a closely related compound to a series of so-called MAX phases was often a subject of studies using various theoretical methods, most recently due to its outstanding hardness and abrasive properties [518–522].

Sc$_3$AlN [523] and Sc$_3$InN [8] directly connect to a series of rare earth metal perovskites (RE$_3$N)M with RE = La – Lu and M = Al, Ga, In, Sn, Pb [8,10,383,524,525], whereas, for example, in the ternary system Y–Al–N no ternary phase formation was observed [526]. Both Sc-based compounds attracted a number of experimental and theoretical works focusing on electronic structures and mechanical properties [527–532]. Perovskite nitrides with Al were reported with RE = La – Nd, Sm, Gd, Tb [10,383,525,533,534]. Compounds (RE$_3$N)Ga similarly form at least with RE = Ce, Pr, Nd, Sm, Gd, Tb [10,383,533]. For M = In, all nitrides with RE = Sc, La–Nd, Sm, Gd–Tm, Lu are known. Magnetic properties are dominated by the exclusive presence of RE^{3+} magnetic ions, realizing complex magnetic structures and transitions with increasing external magnetic field [8,383,535,536]. Superconductivity is suppressed below 2 K in paramagnetic (La$_3$N)In, while the isotypic oxide (La$_3$O)In exhibits a transition to superconductivity at 10 K [8,524,533,537]. (Eu$_3$N)In was obtained only in multiphase samples and according to the unit cell dimension, it contains Eu^{2+} [383]. For (RE$_3$N)Sn, only the compounds with larger rare earth metal ions were reported (RE = La, Ce, Pr, Nd, Sm, Eu), so far, for Pb exclusively (Nd$_3$N)Pb [10,538]. The europium nitride of Sn again contains Eu^{2+} according to the unit cell parameter derived from a multiphase sample [383]. However, in the quaternary phase (Ca$_2$EuN$_x$)Sn with $x = 0.75$ and disorder of Ca and Eu, a more complex mixed-valent situation with the presence of Eu^{2+} and Eu^{3+} develops [539].

7.3.2.2 ε-Fe$_3$N Derivatives

Nanosized powders ε-Fe$_{3-x}$Co$_x$N, ε-Fe$_{3-x}$Ni$_x$N ($0 \leq x \leq 0.8$), and ε-Fe$_{3-x}$Cr$_x$N ($0 \leq x \leq 0.2$) have been synthesized via ammonolysis of precursors with small particle sizes (Figure 7.5b). The cobalt and nickel phases have been shown to contain the transition metal atoms in random distribution and to behave superparamagnetic using Mößbauer spectroscopy [412,540–546]. The magnetic properties of these nanoparticles are apparently largely dominated by the small particle size and surface states of the magnetic spins.

By means of high-pressure high-temperature treatments, the metastable bulk nitrides ε-Fe$_2$CoN and ε-Fe$_2$NiN recently have been prepared and subsequently characterized [547]. Due to the rising numbers of electrons, the saturation magnetization as well as the Curie temperature of ε-Fe$_3$N increasingly drops upon substitution with Co and Ni. Electronic structure calculations confirm a decreasing stability with increasing substitution level [548]. Through the same preparation technique, Fe$_2$IrN$_{0.24}$ was obtained [549]. The reduced nitrogen content in the latter phase is a direct consequence of the limited affinity of the noble metal toward nitrogen, expressed in the fact that no simple binary nitride is formed by Ir.

7.3.2.3 β-Mn Derivatives

A large group of metallic nitrides with the general composition of $M_2M'_3$N crystallize in the cubic Al$_2$Mo$_3$C structure type, which may be described as a filled β-Mn variant [550]. In β-Mn, two distinct sites for Mn exist with different multiplicities of 8 and 12 in the unit cell and icosahedral 12- and 14-fold coordination, respectively, and unlike electronic states, which are prone to occupation with different metallic elements [551]. In the nitrides, the majority site is typically occupied by a group five or six transition metal (V, Nb, Ta, Cr, Mo, W), the minority site by a late transition metal (Fe, Co, Ni, Pt, Pd, Au, Zn), or a main group 13 element (Al, Ga) leading to a variety of combinations [513,550,552–565]. Nitrogen occupies distorted octahedral voids formed exclusively by the majority element, which in turn is twofold nonlinear coordinated by N. The structure can alternatively be described by two interpenetrating frameworks: one composed from mutually triangular coordinated M atoms, and the other from vertex-sharing $M'_{6/2}$N polyhedra [553].

Like their isostructural carbides, such intermetallic nitrides can precipitate from austenitic steels and thus may play an important role in defining the nitrogen content [566]. All show metallic temperature dependence of the electric resistivity and paramagnetism at room temperature [333,553–558]. Any magnetic order strongly depends on the combination of metallic constituents, with iron compounds typically showing ferromagnetism [555,556] and, for example, Ni_2Mo_3N antiferromagnetism [553] at low temperatures. The Mo-based compounds hold some promises for ultrahard materials applications according to their low compressibilities [334]. Interestingly, Ni_2Mo_3N has attracted some interest as catalyst for water-splitting and hydrogenation of aromatic compounds [567,568].

7.3.2.4 η-Carbide Derivatives

H																	He
Li	Be											B	C	N	O	F	Ne
Na	Mg											Al	Si	P	S	Cl	Ar
K	Ca	Sc	Ti	V	Cr	Mn	Fe	Co	Ni	Cu	Zn	Ga	Ge	As	Se	Br	Kr
Rb	Sr	Y	Zr	Nb	Mo		Ru	Rh	Pd	Ag	Cd	In	Sn	Sb	Te	I	Xe
Cs	Ba	La	Hf	Ta	W	Re	Os	Ir	Pt	Au	Hg	Tl	Pb	Bi			

Ce	Pr	Nd	Pm	Sm	Eu	Gd	Tb	Dy	Ho	Er	Tm	Yb	Lu
Th		U											

The cubic η-carbide Fe_3W_3C structure type is closely related to the Al_2Mo_3C-type of filled β-Mn nitrides with general compositions $M_2M'_3N$ discussed above. The prototype nitride of the η-carbide type probably studied in most detail is represented by Fe_3Mo_3N. The crystal structure consists of two interpenetrating frameworks, one constituted from vertex-sharing octahedra $Mo_{6/2}N$, the other from highly condensed Fe tetrahedra. Two symmetry-independent Fe are both situated in 12-fold icosahedral coordination equally by 6 Fe and 6 Mo, the Mo atoms in twofold nonlinear coordination by N (although with a somewhat smaller angle than in the filled β-Mn-type nitrides) and additionally in 14-fold surrounding by 6 Fe and 8 Mo. In the nitrides, the latter metal site is typically occupied by a more nitrophilic early transition metal (Ti, Zr, Hf, V, Nb, Ta, Mo, W), the other metal sites by a less nitrophilic later transition metal (V, Cr, Mn, Re, Fe, Ru, Os, Co, Rh, Ir, Ni, Pt, Pd, Zn) [516,558,564,569–575]. Deviations of the composition to $M_4M'_2N$ and $M_2M'_4N$ exist, implying some disorder of both metals on specific sites, including the introduction of, for example, Fe in the coordination sphere of N in Fe_4W_2N [576]. However, an excess of the early transition metal apparently can stabilize the phases from a thermodynamic point of view [577]. Introduction of some main group metal on the second framework in Ni_2GaMo_3N also succeeded [571,578]. Upon heating in H_2/Ar gas mixtures, Co_3Mo_3N transforms into a so-called η-12-type Co_6Mo_6N with virtually identical metal atom arrangement, although with N content reduced to 50% and now located on a distinctly different site in octahedral coordination by Mo. These isolated octahedra are embedded in the cobalt atom framework [579]. This process can be reversed via heat treatment in a N_2/H_2 gas mixture. An analogue η-12-type Fe_6W_6N was obtained from the metals under elevated nitrogen pressures [580].

The metallic η-carbide-type compounds show paramagnetic behavior, becoming particularly complex if involving metals from the iron group, due to geometric frustration in the tetrahedral arrangement [558,571,575,578,580–585]. According to their low compressibilities, Fe$_3$Mo$_3$N and Co$_3$Mo$_3$N may represent materials with high hardness [334]. Again, the η-carbide-type nitrides may have some relevance as precipitates from steels, since, for example, formation of Fe$_3$Nb$_3$N was observed upon hydrogen treatment of an Fe–Nb alloy [586].

7.3.2.5 MAX Phases $M_{n+1}AN_n$ and Related Compounds M_3AN

H												B	C	N	O	F	He
Li	Be											Al	Si	P	S	Cl	Ne
Na	Mg											Ga	Ge	As	Se	Br	Ar
K	Ca	Sc	Ti	V	Cr	Mn	Fe	Co	Ni	Cu	Zn	In	Sn	Sb	Te	I	Kr
Rb	Sr	Y	Zr	Nb	Mo	Tc	Ru	Rh	Pd	Ag	Cd	Tl	Pb	Bi	Po	At	Xe
Cs	Ba	La	Hf	Ta	W	Re	Os	Ir	Pt	Au	Hg						Rn
Fr	Ra	Ac															

Ce	Pr	Nd	Pm	Sm	Eu	Gd	Tb	Dy	Ho	Er	Tm	Yb	Lu
Th	Pa	U											

So-called MAX phases are nitrides and carbides (X = N, C) with the general formula $M_{n+1}AX_n$ and n = 1, 2, or 3 currently gaining large interest due to their excellent mechanical properties and machinability coupled with high thermal and electrical conductivity, small thermal expansion, and high chemical stability [587–589]. Earlier, such hexagonal compounds with n = 1 were frequently called Hägg phases (H-phase, Cr$_2$AlC structure type). For nitrides, apparently only compounds with n = 1 and 3 and M = Ti, Zr, Hf, V, Cr and A = Al, Ga, In, Tl, Sn are known so far. For completeness, Figure 7.8 depicts sections of the hexagonal crystal structures of all MAX phases with n = 1, 2, 3. These consist of two-dimensional slabs of close-packed alternating layers of M and A. The difference between the three structures lies in the number of M layers separating the A layers: In the compounds M_2AX (n = 1), there are two (Figure 7.8a); in M_3AX_2 (n = 2) there are three (Figure 7.8b), and in M_4AX_3 (n = 3) there are four (Figure 7.8c). The stacking of the metal atoms perpendicular to the layer planes is represented by ... ABcBAc ... , ... ABCbCBAb ... , and ... ABCAcACBAb ... with capital letters denoting layers of M and small letters denoting layers of A for n = 1, 2, and 3, respectively (a second polymorph with simple AB-stacking of M and A layers was observed in carbides with n = 3). In this way, the A atoms reside in trigonal prismatic coordination by M. X occupies octahedra exclusively formed by M leading to single, double, and triple layers of edge- and vertex-sharing octahedra, thus layered sections of the rock salt structure.

In case of nitrides, most attention was directed to Ti$_2$AlN [590] and Ti$_4$AlN$_3$ [591–595]. The latter compound apparently was earlier designated with the composition Ti$_3$Al$_2$N$_2$, although there were indications for a deviating composition before [515]. Also, the existence of Ti$_3$AlN$_2$ is in question [596–598]. One advantage of these nitrides is the formation of protective titanium nitride layers upon high-temperature decomposition [599], which might be similarly given for further early transition metal MAX nitrides [600]. Electronic structure calculations were performed with emphasis on mechanical, electrical, and

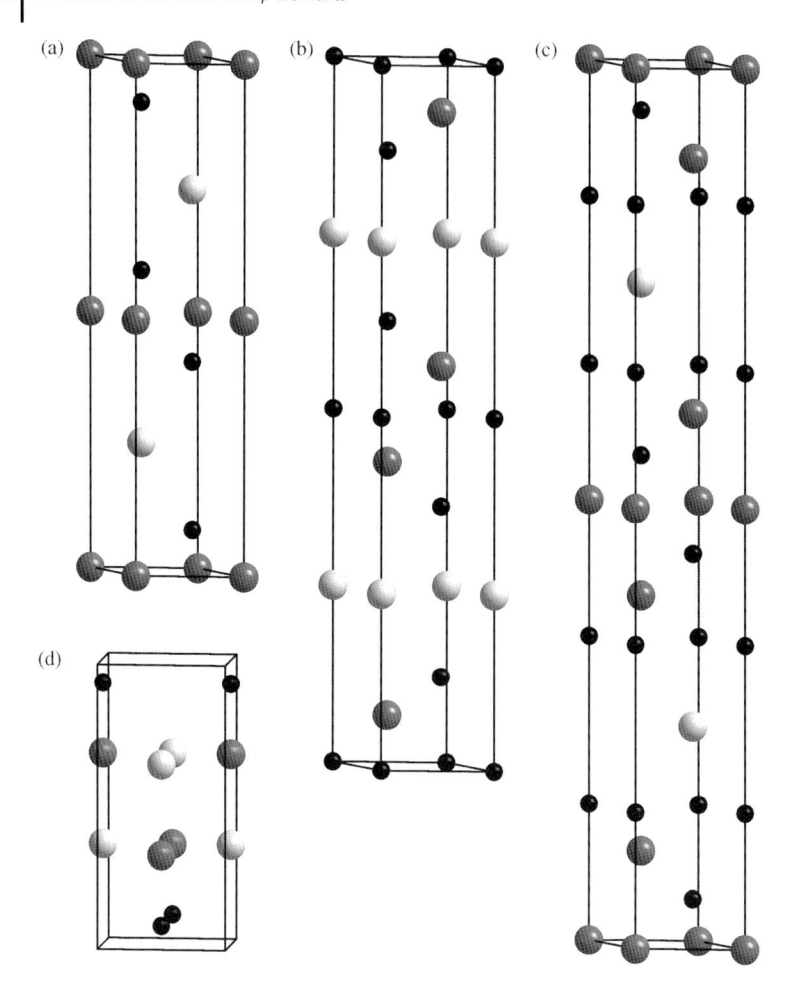

Figure 7.8 Crystal structures of MAX-phases $M_{n+1}AN_n$ with (a) $n=1$; (b) $n=2$; (c) $n=3$, and (d) of Z-phases MM'N. Small black spheres: M; larger light gray spheres: A; larger dark gray spheres: N.

bonding properties, resulting in metallic to semimetallic properties [601–605]. Ti$_2$InN apparently becomes superconducting below 7.3 K [606], while other combinations with group 4 metals are little studied [564,607–609].

Magnetic MAX phases such as Cr$_2$GaN may have some impact for applications via two-dimensional geometric frustration within the layer planes [514,610]. However, Cr$_2$GaN tends to decomposition under growth of Ga whiskers even at room temperature conditions [611].

Orthorhombic compounds with filled Re$_3$B structure (the so-called post-perovskite type) (Figure 7.7d) and composition M_3AN feature layers build of edge-sharing octahedral strands joined to layers via further vertices $M_{6/2}$N and thus are closely related to MAX phases in structure and properties [612]. Compositions comprise V$_3$PN, V$_3$AsN, Cr$_3$PN [613], V$_3$GeN, V$_3$GaN [614], and

Zr$_3$AlN and Hf$_3$AlN [615–618]. V$_3$PN and V$_3$AsN become superconducting below 5.6 and 2.6 K, respectively [619–622].

7.3.2.6 Z-Phases *MM'*N

H																	He
Li	Be											B	C	**N**	O	F	Ne
Na	Mg											Al	Si	P	S	Cl	Ar
K	Ca	Sc	Ti	**V**	**Cr**	Mn	Fe	Co	Ni	Cu	Zn	Ga	Ge	As	Se	Br	Kr
Rb	Sr	Y	Zr	**Nb**	**Mo**	Tc	Ru	Rh	Pd	Ag	Cd	In	Sn	Sb	Te	I	Xe
Cs	Ba	La	Hf	**Ta**	**W**	Re	Os	Ir	Pt	Au	Hg	Tl	Pb	Bi	Po	At	Rn
Fr	Ra	Ac															

	Ce	Pr	Nd	Pm	Sm	Eu	Gd	Tb	Dy	Ho	Er	Tm	Yb	Lu
	Th	Pa	U											

Tetragonal phases of the general composition *MM'*N (Figure 7.8d) have been realized to be of importance, since they precipitate from martensitic creep-resistant steels upon thermal strain, associated with a dramatic reduction of the creep strength and thus leading to failure of structural parts [623]. In the crystal structure, quadratic double layers of both metals are stacked alternately in [001], forming a distorted motif of a *bcc* structure. N is located within the double layers of the more nitrophilic metal *M*, thus coordinated octahedrally by five *M* and one *M'*. Compositions comprise like (V$_x$Nb$_{1-x}$)CrN [213,623–629], (V$_x$Ta$_{1-x}$)CrN [623,625,630], Nb(Mo$_{1-x}$W$_x$)N ($0 \leq x \leq 0.5$) [631,632], and TaMoN [633,634]. CaGaN also fits into this scheme [635].

7.3.2.7 *RE*$_2$*M*$_{17}$N$_y$ and Related Phases

H																	He
Li	Be											B	C	**N**	O	F	Ne
Na	Mg											Al	Si	P	S	Cl	Ar
K	Ca	Sc	**Ti**	V	Cr	Mn	**Fe**	**Co**	**Ni**	Cu	Zn	Ga	Ge	As	Se	Br	Kr
Rb	Sr	**Y**	Zr	Nb	Mo	Tc	Ru	Rh	Pd	Ag	Cd	In	Sn	Sb	Te	I	Xe
Cs	Ba	La	Hf	Ta	W	Re	Os	Ir	Pt	Au	Hg	Tl	Pb	Bi	Po	At	Rn
Fr	Ra	Ac															

	Ce	Pr	Nd	Pm	Sm	Eu	Gd	Tb	Dy	Ho	Er	Tm	Yb	Lu
	Th	Pa	U											

The same reasoning applies to phases $RE M_{12}N_x$, $RE_2 M_{17}N_x$, and $RE_3 M_{29}N_x$ (*M* = Fe, transition metal) [636–638]; currently under worldwide intensive scrutiny due to their superior hard magnetic properties with higher Curie temperatures at lower costs compared to SmCo$_5$, Sm$_2$Co$_{17}$, and Nd$_2$Fe$_{14}$B-type compounds. A general method to modify these rather new magnetic materials is to interstitially introduce small nonmetallic atoms such as N into the host compounds; the first observed dramatic increase in magnetic properties in such interstitial nitride compounds was observed in Fe$_{16}$N$_2$ [290,639].

In the case of nitrides, four distinct groups of compounds have been established, all of them based on the CaCu$_5$-type structure with a fraction of *RE* replaced by dumbbells M_2, namely, $RE M_{12}N_x$ (tetragonal ThMn$_{12}$ type) [640],

$RE_2M_{17}N_x$ (rhombohedral Th$_2$Zn-type, R-3m, RE until Gd) [641], $RE_2M_{17}N_x$ (hexagonal Th$_2$Ni$_{17}$-type, P6$_3$/mmc, RE later than Tb) [642], and monoclinic $RE_3M_{29}N_x$. Formally, the latter one may be regarded as a linear combination of $REM_{12}N_x$ and $RE_2M_{17}N_x$.

The binary phase $REFe_{12-x}M_x$ (RE = Y, Ce–Nd, Sm, Gd–Tm, Lu) can only be stabilized with additional metals M being the main group elements Al [643], Ga [644], and Si [645] as well as the transition metals Ti, V, Cr, Mn, Nb. Mo, W, and Re in ratios Fe:M between 11: 1 and 4: 8. Nitrogen in $REFe_{12-x}M_xN_y$ (RE = Y, Ce–Nd, Sm, Gd–Tm, Lu) is observed only for M = Ti [646–652], V [653–655], Co [652,656], and Mo [653,654,657–661]; full occupation corresponds to $REFe_{12-x}M_xN$ and lattice expansion of about 3%.

With RE = Ce, Gd, Tb, and Y, $RE_2M_{17}N_x$ compounds may exist in both rhombohedral and hexagonal types [641,642]. Although first investigations on $RE_2Fe_{17}N_x$ nearly exclusively centered on Fe-containing compounds, the substitution of Fe by other elements has been studied subsequently in detail. Complete substitution of Fe by Co is possible in many $RE_2Fe_{17}N_x$ phases, for example, $RE_2Co_{17}N_2$ (RE = Dy, Er, Ho, Tb, Tm) [662–665], whereas only a small fraction of Fe may be substituted by main group elements Al, Ga, and Si or transition metals Ti, V, Nb, Mo, and Ni [664,666–674]; RE substitution was also studied, for example, $Sm_{1-x}RE_xFe_{17}N_{2+y}$ (RE = Y, Ce, Nd, Tb, Dy, Er, Lu) [675–681].

Interstitial nitrogen, in general $2 < x < 3$, leads to an increase of lattice parameters without a change in symmetry; the cell volume increases by about 6–8% compared to the intermetallic compound. In case of $Sm_2Fe_{17}N_x$ [682], the nitrogen content follows a solid solution pattern and is dependent on the amount of nitrogen available in the system; however, in the case of Y_2Fe_{17} [683] and Nd_2Fe_{17} [684], it has been shown that only the fully nitrided phase coexists with the nitrogen free phase without any indication of any intermediate solid solution. In this context, fully nitrided does not necessarily imply that all possible interstitial sites are fully occupied by nitrogen, but that a certain value is immediately reached without any further significant change [683,684].

In all these phases, nitrogen is preferentially located on the so-called octahedral site [657,685]. Depending on compound and thermal history, occupation of the octahedral site ranges from 0.4 to 0.96 in $REM_{12}N_x$ [651,657,659,686–689], 0.6 to full occupation in rhombohedral $RE_2M_{17}N_x$ [685], and about 0.56–0.83 in hexagonal $RE_2M_{17}N_x$ [685]. Only few authors indicate mostly minor participation of other sites [656].

RE_3(Fe, M)$_{29}$ also does not exist as a binary phase [690]; stabilization is achieved by M = Ti, V, Cr, Mn, and Mo; up to 4 N are located in octahedral voids [691].

Owing to their commercial importance, the magnetic properties of the $RE_2M_{17}N_x$ and the $REM_{12}N_x$ phases have been investigated in great detail [637,692]. These materials are also of interest for energy conversion and power generation [693]. Due to the relative simplicity of the $REM_{12}N_x$ crystal structure, this class of compounds also provides good model systems for fundamental studies concerning exchange interaction, magnetocrystalline anisotropy, thermal expansion [660], or electronic state as well as computational methods in $RE_xFe_yN_z$ intermetallics [692].

7.3.2.8 Related Systems Based on Close-Packed Arrangements of M

The rock salt-type group 3 and rare earth metal nitrides exhibit complete mutual solubility in solid state [694–699]. Similar solid solutions of these nitrides form with UN and ThN [700–704], typically showing only small and tentatively negative deviations from linear extrapolation of the cubic unit cell parameter (Vegard's law). Similarly, ThN and UN are completely miscible [705,706]. Only the system ThN–CeN appears to present an exception indicating a positive deviation, probably connected to the +4 valence state of Ce in metallic CeN [114,115]. ScN forms solid solutions with CrN and VN [698] and additionally a defect rock salt variant $Zr_3Sc_4N_8$ (=$Zr_{0.375}Sc_{0.5}N$), to be understood as solid solution with Zr_3N_4 [707]. UN shows complete miscibility with ZrN and HfN, while it dissolves only few percent TiN at temperatures as high as 2400 °C [708]. Similar mutual rock salt-type solid solutions are indicated in most intertransition metal systems, where binary nitrides are known with rock salt-type (defect) structures of Ti, Zr, Hf, V, Nb, Ta, Cr, Mn [192,698,709–722]. On formation, $(Cr_{0.5}W_{0.5})N$ was found to nucleate with a disordered rock salt-type crystal structure, from which an ordered layered phase $CrWN_2$ with W in trigonal prismatic coordination grows (see Section 7.3.3.7) [723]. A rock salt-type phase $Mo_{1-x}Nb_xN$ in the full range of compositions was also obtained in thin films via reactive sputtering [724]. A further inter-transition metal rock salt-type solid solution is known for $Co_{1-x}Ni_xN$, where a substitution of 33% Co by Ni was accomplished [725]. Interestingly, even AlN and GaN with wurtzite-type structures as stable form for their binary nitrides realize large miscibility ranges with ScN, TiN, ZrN, and NbN in rock salt-type phases at least in thin films [715,726,727]. Similar is true for Fe_3AlN with a disordered *fcc* lattice of the metal atoms and 25% occupation of octahedral holes [403].

Substitution variants of binary TaN with (approximate) compositions Ta_3MnN_4 and $Ta_2MN_{2.5}$ (M = Fe, Co, Ni) were characterized to exhibit close-packed metal atom arrangements, although only for Ta_3MnN_4 a complete crystal structure in the β-$RbScO_2$ type (= $Ta(Mn_{0.5}Ta_{0.5})N_2$) was proposed (Figure 7.39f) [728]. Similarly, a partially ordered metal atom arrangement was derived for Mo $(Mo_{0.5}Ta_{1.5})N_4$, but with unknown positions of N [729]. Furthermore, ternary phases with WC structure and metal atom disorder in $Ti_{0.7}M_{0.3}N$ and $Mo_{0.8}M_{0.2}N$ with M = Ni, Co were once described [730]. Next to the solution phases of common binary nitrides [192,722,731], the ternary system Nb–Ta–N contains one phase $(Nb, Ta)_8N_9$, characterized by a close-packed arrangement of N in the ...ABACBC... sequence and statistical distribution of Nb and Ta over 8 of 9 possible octahedral holes [732].

For FeNiN, a tetragonally distorted cubic closed packing of Fe and Ni with ordered 50% occupation of octahedral voids by N was discussed. The N-centered octahedra thus form layers via edge sharing, which are connected in the third direction via vertex sharing [733].

The hexagonal Mn_5Si_3 structure type is known to be extremely susceptible for uptake of a huge number of very different elements to form ternary compounds $M_5M'_3X_{1-x}$ ($x \leq 1$) [734]. The interstitial X in this respect occupies voids in channels within the host structure in octahedral coordination by M. While studies for X = transition metal exist and for X = C studies are numerous, information on nitrides is comparably rare. In the nitrides, M typically selects from the group 4 or 5 transition metals or the rare earth metals, and M' from the main group metals or metametals, leading, for example, to the compositions $Ti_5Si_3N_x$, $Zr_5Al_3N_x$, Zr_5Sn_3N, Hf_5Al_3N, V_5P_3N, $Nb_5Ga_3N_x$, $Nb_5Ge_3N_x$, $Ta_5Ga_3N_x$, $Ta_5Ge_3N_x$, or La_5Ge_3N [16,615,616,735–739]. A closely related chemistry develops for the filled tetragonal Cr_5B_3 type, for example, in La_5Pb_3N with N located in tetrahedral voids exclusively formed from La within the host structure [740].

7.3.3
Multinary Nitrides and Nitridometalates

Nitridometalates of transition metals represent a class of compounds containing complex anions $[M_xN_y]^{n-}$, charge balanced by cationic species, generally elements from groups 1 and 2 of the periodic table, and in a limited number of instances also rare earth metal elements. From the chemical viewpoint, a complete transfer of electrons to the complex anions is implied, and the interactions between cationic and anionic components are clearly predominantly ionic.

The discussion of the nitridometalates in general focuses on the coordination geometry of the transition metal M. The subsequent sections are generally divided by increasing coordination number of M by nitrogen. Sections are arranged by rising dimensionality of the complex anion. Whereas most phases contain one type of complex anions only, there is a growing number of examples showing a variety of both different coordination geometries and oxidation states of M. These phases are therefore presented in two or more of the respective sections; this conception enables an adequate representation of the specific characteristics of the complex anions. However, a detailed discussion of specific synthesis conditions, the physical properties, or other important details are listed at the first appearance of a particular phase.

Some systems (e.g., Li–Fe–N [87,741–743]) have already been investigated in detail. They exhibit a quite complex structural chemistry concerning composition, coordination, and oxidation state of the transition metal. Under the scope of this presentation, the different phases are described in the respective chapters, but no in-depth discussion of these systems is intended.

7.3.3.1 CN(M) = 2

H																	He
Li	Be											B	C	**N**	O	F	Ne
Na	Mg											Al	Si	P	S	Cl	Ar
K	Ca	Sc	Ti	V	Cr	**Mn**	**Fe**	**Co**	**Ni**	**Cu**	**Zn**	Ga	Ge	As	Se	Br	Kr
Rb	Sr	Y	Zr	Nb	Mo		Ru	Rh	Pd	**Ag**	Cd	In	Sn	Sb	Te	I	Xe
Cs	Ba	La	Hf	Ta	W	Re	Os	Ir	Pt	Au	Hg	Tl	Pb	Bi			

Ce	Pr	Nd	Pm	Sm	Eu	Gd	Tb	Dy	Ho	Er	Tm	Yb	Lu
Th		U											

Linear coordination of transition metal atoms M by nitrogen in nitridometalates is generally limited to the late 3d metals Mn–Zn and one single phase containing Ag. The structural variety of the complex anions containing low-valent M ranges from isolated dumbbells via chain fragments to chains and 2D networks.

Li$_3$N Superstructures: From Dumbbell Anions to Chains

α-Li$_3$N is the exclusive example of a nitride that forms from the elements already at ambient conditions. [25,744] Its crystal structure (Figure 7.9a) consists of eightfold coordinated N by Li in the form of a hexagonal pyramid and two different types of Li: twofold Li(1) and threefold coordinated Li(2), respectively. Li(2)$_2$[Li(1)N] is a well-known lithium ion conductor [744–746]. The linearly coordinated Li may be substituted by late 3d transition metals, forming a variety of ternary phases, whereas the occupation of the trigonal-planar coordinated positions with Li varies with the oxidation state of M. Strong covalent bonding dominates between M and the nitrogen atoms within infinite chains, causing weakening of the predominantly ionic bonding within the planes perpendicular to the chains [747]. Generally, the degree of substitution correlates linearly with the decrease of the Li/M-N distance.

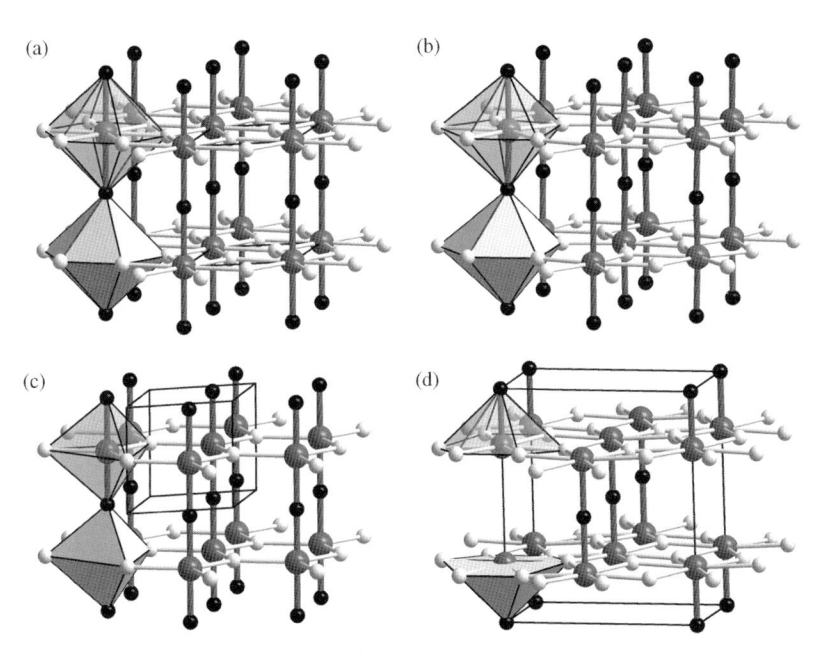

Figure 7.9 Li$_3$N superstructures. (a) Crystal structure of Li$_3$N and the substitution series Li$_2$[Li$_{1-x}$$M^I_x$N] ($M$ = Mn, Fe, Co, Ni, Cu). (b) Li$_5$[(Li$_{1-x}$$M^{I,II}_x$)N]$_3$ (M = Mn, Ni). (c) Idealized Li[Ni^{II}N]. (d) Li$_4$[Fe^{II}N$_2$]. Small black spheres: transition metal atoms; small white spheres: Li; larger dark gray spheres: N.

The first investigations in systems $Li_2[Li_{1-x}M_xN]$ (Figure 7.9a) were undertaken with $M = Co$ [748], Ni [748–750], and Cu [748,750,751]; more recently, the systems with $M = Fe$ [742] and Mn [752,753] followed. Above the maximum x-parameter of the substitution series, segregation of M or other ternary phases occurs [754]. Successful preparation attempts generally start with Li_3N and the respective transition metals. The Fe substitution series was investigated by use of a low-temperature solid-state sintering route [755]. Crystal growth experiments in lithium melts employing an improved high-temperature centrifugation-aided filtration (HTCAF) [756] technique resulted in single-crystalline specimens of up to 12 mm in length and 150 mm^3 in volume [90,92]. Applying this technique, a compound precipitates from a liquid phase, which is subsequently removed by filtration at elevated temperatures; achievable crystal size is often limited by container dimensions only.

Due to optimization of reaction conditions, the amount of substitution of Li by M could be greatly increased [750,757] compared to initial observations [748], but in long-term crystal growth experiments, only much lower values of x for the thermodynamically stable products were achieved [92]. Full exchange of Li by M is not observed in any system $Li_2[Li_{1-x}M_xN]$.

Electrical resistivity measurements indicate a gradual insulator metal transition around $x \approx 0.8$ for the solid solution series $Li_2[(Li_{1-x}Ni_x)N]$ [758]. XANES data of the substitution series $Li_2[(Li_{1-x}M_x)N]$ ($M = Fe$, Co, Ni) were used to determine the amount of Li vacancies within the "Li_2N" layers, which may result from Li/M substitution within the infinite chains in case that the oxidation state of the transition metal is $>+1$. The main K-absorption edges in XAS spectra are only slightly shifted from the reference metals (Mn, Fe) or coincide with that of the metal (Ni) indicating that the substituted phases crystallizing in the $Li_2[LiN]$-type structure predominantly contain M^I species [750,759]. In case of the Ni phase, only a minor part of the transition metal may exist as Ni^{II}.

Investigations of magnetic susceptibility on several members of the $Li_2[(Li_{1-x}M_x)N]$ series are often hampered by minor amounts of impurities in the form of the respective transition elements. In general, the data are consistent with the presence of M^I species [750,752,758,760]. The DC magnetization of $Li_2[(Li_{0.5}Co_{0.5})N]$ [761] revealed the characteristics of a spin glass with a maximum in the zero-field cooling (ZFC) curve shifting to lower temperatures with increasing applied field. In the Co- and Ni-series [750,758], the effective moments μ_{eff}/Co, Ni atom decrease almost linearly approaching zero for $x \to 0$. For low x values, the magnetic moments μ_{eff}/Co, Ni atom strongly exceed the spin-only values expected for Co^I and Ni^I in $3d^8$ and $3d^9$ configurations, respectively. The interplay of correlation and orbital effects is assumed to be responsible for these strong variations of the effective magnetic moments in the Co and Ni series [747,750,758].

In the series $Li_2[(Li_{1-x}Fe_x)N]$ ($x = 0.16$ and 0.21), large effective moments and ferromagnetic saturation moments are observed [92,760]. In addition, spectacular large hyperfine magnetic fields of nearly 700 kOe were determined from Mößbauer spectra for Fe atoms exclusively surrounded by lithium ions within the hexagonal plane of the crystal structure, which are common in samples with small x values but are very rare in samples with $x \to 1$. The observed effects could be explained by electronic structure calculations [762] where orbital effects and strong interatomic correlations play an important role in enhancing the magnetic moments beyond the spin-only value. Investigations on single crystals of $Li_2[(Li_{1-x}Fe_x)N]$ [92] with $x \ll 1$ show an extreme, uniaxial magnetic anisotropy field of 220 T and a coercivity field of 11 T at 2 K outperforming all known hard magnets and single-molecular magnets and indicating nanoscale magnetic centers.

The overall transport properties of the substitution series depend on the nature of the substituting transition metal (Fe, Co, Ni), the amount of substitution, and the resulting conduction behavior [763]. Although the exact relationship between Li vacancy concentrations and Li ion mobility is not definitely known until now, the vacancy levels can obviously be controlled by reaction temperatures and the duration of temperature treatment [764]. Li vacancies within the "Li_2N" layers can be produced in ternary phases by Li/M substitution within the infinite chains, if the oxidation state of the transition metal is $>+1$. Three different variants for this Li/M substitution are reported with an increasing amount of Li vacancies within the "Li_2N" layer:

$Li_{2-y}\square_y[(Li_{1-x}M_x)N]$ (M = Co [757], Ni [765], Cu [766]) crystallize as isotypes of $Li_2[LiN]$ and are characterized by both a homogeneity range within the fully occupied $[(Li_{1-x}M_x)N]$ chain and a random distribution of Li vacancies within the "Li_2N" layers (Figure 7.9a). The chemical compositions of some members of these series were reported to be $Li_{1.52}\square_{0.48}[Li_{0.47}Co_{0.53}N]$ [757], $Li_{1.55}\square_{0.45}[Li_{0.4}Co_{0.6}N]$ [757], $Li_{1.14}\square_{0.86}[Li_{0.21}Ni_{0.79}N]$ [765], $Li_{1.98}\square_{0.02}[Li_{0.578}Cu_{0.423}N]$ [766], and $Li_{1.98}\square_{0.02}[Li_{0.596}Cu_{0.404}N]$ [766]. The extent of the vacancies does not directly correspond to the degree of substitution of the d metal within the chains, thus the average oxidation states of M correspond to values between +1 and +2. Whether these oxidation states of M should be addressed as mixed valent with distinct M^I and M^{II} species, or as intermediate valent remains speculative considering the data available until now.

The crystal structures of $Li_5[(Li_{1-x}M_x)N]_3$ (M = Mn [752], Ni [767,768]) represent ordered defect variants (superstructures) of the $Li_2[LiN]$ type (Figure 7.9b). Their chemical formulas can be derived from $Li_{2-y}\square_y[(Li_{1-x}M_x^{(1+y)+})N]$ with $y = 0.33$. The ordered distribution of Li vacancies within the "Li_2N" planes of the $Li_2[LiN]$ crystal structure leads to a three times larger unit cell and the coordination polyhedron around nitrogen is changed from hexagonal bipyramidal to pentagonal bipyramidal. For M = Mn [752], the degree of substitution within the infinite chains ($Li_{0.41}Mn_{0.59}$) leads to an oxidation state of about $Mn^{+1.6}$; for M = Ni, the values are ($Li_{0.23}Ni_{0.77}$) and $Ni^{+1.4}$, respectively [767]. The bond lengths $d(Li_{0.41}Mn_{0.59})$-N = 190.9(1) pm and $d(Li_{0.23}Ni_{0.77})$-N = 179.0(5) pm are significantly reduced compared to the respective distance in pure $Li_2[LiN]$ (Li(1)-N: 193.8(1) pm) [25]. However, compared to the substituted phases, $Li_2[Li_{1-x}M_xN]$, $d(Li_{0.41}Mn_{0.59})$-N is shorter than the respective value in the distance in

$Li_2[Li_{1-x}Mn_xN]$ (~192 pm), but $d(Li_{0.23}Ni_{0.77})$-N is significantly longer than the extrapolated value for $Li_2[Li_{1-x}Ni_xN]$ (~176 pm).

$Li[Ni^{II}N]$ (=$Li_{2-1}\square_1[(Li_{1-1}M_1N)]$) (Figure 7.9c) may be considered as the end member of the substitution series $Li_{2-y}\square_y[(Li_{1-x}M_xN]$ with $x = y = 1$ and ordered vacancies in the "Li_2N" layer resulting in a fivefold trigonal bipyramidal coordination of N (3 Li + 2 Ni) [768,769].

In striking contrast to the Li_3N substitution variants, $Li_5[(Li_{1-x}M_x)N]_3$ ($M =$ Mn [752], Ni [767,768]) and $Li_{2-y}\square_y[(Li_{1-x}M_x)N]$ ($M =$ Co [757], Ni [765], Cu [766]), in which Li deintercalation is observed within the Li_2N layer, $Li_4[Fe^{II}N_2]$ [56] forms an ordered and orthorhombically distorted variant of the hexagonal $Li_2[Li_{1-x}M_xN]$ substitution series, which may be described as $Li_4[MN\square N]$ by complete substitution of 50% of all Li atoms within the linear chain $[Li_{1-x}M_xN]$ by Fe in an ordered fashion and vacating the remaining positions (Figure 7.9d). Formally, charge balance is achieved by complete Li deintercalation from the linear coordinated Li position within the chain. The Li positions within the Li_2N layer remain fully occupied.

Both the substitution series $Li_2[(Li_{1-x}M_x)]N$ and the corresponding oxidized phases have been investigated concerning their suitability as materials for energy applications experimentally and theoretically [769,770]. The versatility of the systems with regard to constituting elements and composition through modification by substitution provides a wide variety in defect concentration and the number of charge carriers, both ionic and electronic. The implementation of these phases as anode materials of high reversible capacity has already been tested [771]. For $Li_2[(Li_{1-x}Fe_x)N]$, a maximum reversible capacity of about 700 mAh/g was obtained for $x = 0.2$ in the range of 0.05 – 1.3 V (versus Li/Li^+). Fe K-edge XANES as well as Mößbauer spectra indicates the presence of Fe^{+1}, Fe^{+2}, and Fe^{+3} species during charge and discharge cycling modes. Reversible intercalation/deintercalation of Li ions in the $1 - 0.02$ V potential range is possible in both the Co and Ni solid solution series combined with the corresponding redox reactions [772–774]. The structure of the host lattice is stable without significant changes in the lattice parameters over several intercalation/deintercalation cycles.

In contrast, in $Li_2[(Li_{1-x}Cu_x)]N$ [751], no redox process forming Cu^{II} from Cu^I is observed, but breakdown of the Li_3N structure and formation of Cu_3N and other still unknown phases.

Dumbbell Anions $[MN_2]^{n-}$, Chain Fragments $[M_xN_{x+1}]^{n-}$, Linear Chains $[MN]^{n-}$, and Linear Substituted Chains $[(Li_{1-x}M_x)N]^{n-}$ in AE Containing Phases

Isolated dumbbell anions are observed in a variety of compounds; all phases containing AE and M share the structural motif of two octahedra NAE_5M connected by a common apex.

The crystal structures of the isotypic and transparent phases $Ca_2[ZnN_2]$ [775], $Sr_2[ZnN_2]$ [776], and $Ba_2[ZnN_2]$ (Figure 7.10a) [776] crystallize in the $Na_2[HgO_2]$-type structure [777], which may be derived from the rock salt-type structure $NaCl = (AE, M)N$ in terms of a threefold superstructure containing

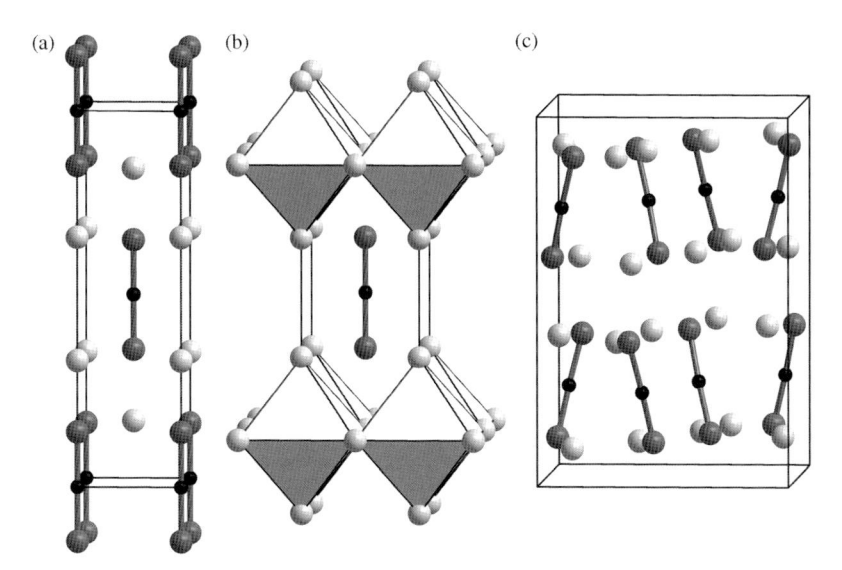

Figure 7.10 The crystal structures of (a) $AE_2[ZnN_2]$; (b) $AE_3[ZnN_2]O$, octahedra $OAE_2AE_{4/4}$ emphasized; (c) $AE_5[MN_2]_2$ (M = Co, Cu). Small black spheres: transition metal atoms; larger light gray spheres: AE; larger dark gray spheres: N.

double layers AEN (layers AB) running parallel to the (001) plane with M atoms and voids on the anion position (layers CD) interconnecting these layers in the sequence ... $ABCBAD$ .... The nitridometalate oxide $Ba_3[ZnN_2]O$ [778] – although outside the scope of this publication – may also be described on the basis of a rock salt-type structure (Figure 7.10b) by addition of a layer BaO (E) in the sequence ... $ACAE$ .... The description of $Sr_2[NiN_2]$ [70], which has also been published on the basis of a $Na_2[HgO_2]$-type [777] structure and employing a twinned and disordered model, has been shown to be erroneous [68]. Instead, it has to be regarded as a phase containing additional carbon with the composition $Sr_2[NNiCN]$ [68]. An isotypic phase $Ba_2[NNiCN]$ also exists [69].

In alkaline earth metal-rich systems (AE:M>1: 1) with Co and Cu, the isotypic phases $Ca_5[CoN_2]_2$ [64,779] (a phase mixture containing both $Ca_5[CoN_2]_2$ and CaO was first falsely described as $Ca_3[CoN_3]$ containing trigonal-planar nitridocobaltate ions [780], a phase that could not be presented until now), $Ca_4Ba[CoN_2]_2$ [64,781], $Sr_5[CoN_2]_2$ [782], and $CaBa_4[CuN_2]_2$ [783] are known. The crystal structures of all these compounds $AE_5[MN_2]_2$ (Figure 7.10c) can again be derived from the rock salt-type structure NaCl = (AE, M)N in terms of a 5 × 3 superstructure containing double layers $AE_5(N_4\square)$ (layers AB) running parallel to the (001) plane with the M species ($M_4\square$)($\square_5$)) (layer C) interconnecting these layers in the sequence ... $ABCBAC$ .... Magnetic susceptibility data at high temperatures for $Ca_5[CoN_2]_2$ and $BaCa_4[CoN_2]_2$ follow a Curie–Weiss law. Effective magnetic moments of 3.22 μ_B per cobalt atom for $Ca_5[CoN_2]_2$ and 3.10 μ_B per cobalt atom for $BaCa_4[CoN_2]_2$ indicate an $S=1$ state (spin-only value: $2.83\,\mu_B$) and are consistent with the presence of a $3d^8$ configuration that is in agreement with the assignment of a Co^I oxidation state by XAS data [64]. At low temperatures and high fields, a rounded maximum in $\chi(T)$ is visible followed by a sharp decrease indicating low-dimensional antiferromagnetic interactions at high temperatures and long-range antiferromagnetic ordering below 50 K [64]. Electric conductivity data for $Ca_5[CoN_2]_2$ and $BaCa_4[CoN_2]_2$ show semiconducting behavior with a small bandgap (down to 0.03 eV) [64]. From topology-based examinations [784], a double bond character was derived for the strongly covalent M-N bonds in the dumbbells $[CoN_2]^{5-}$.

Another controversial phase is $Sr_{39}Co_{12}N_{31}$ ($\equiv Sr_{39}[CoN_2]_{12}N_7$) [82], which crystallizes in a complex cubic structure containing dumbbell anions $[CoN_2]$ with mixed-valent $Co^{I/II}$ and both Sr and N on partially occupied or disordered sites. Recent research suggests [785] that complex anions [NCoCN] are present in this phase similar to $Sr_2[NNiCN]$ [68].

In alkaline earth metal nitridocuprates, dumbbells and chain fragments are simultaneously observed in $Sr_6[CuN_2][Cu_2N_3]$ [94] (Figure 7.11a) and $Ba_{16}[CuN]_8[Cu_2N_3][Cu_3N_4]$ [783] (Figure 7.11b), the latter phase also supporting helical chains at the same time. The nitridocuprate indides $Sr_8[CuN_2]_2[CuN]In_4$ [786] and $Ba_8[CuN_2]_2[CuN]In_4$ [787] also contain two different linear nitridometalate anions, dumbbells $[CuN_2]$ and linear chains [CuN]. Isolated dumbbells $[CuN_2]$ are also present in the crystal structure of $Ba_{14}N_3[CuN_2]_2In_4$ [788], a nitridocuprate nitride indide.

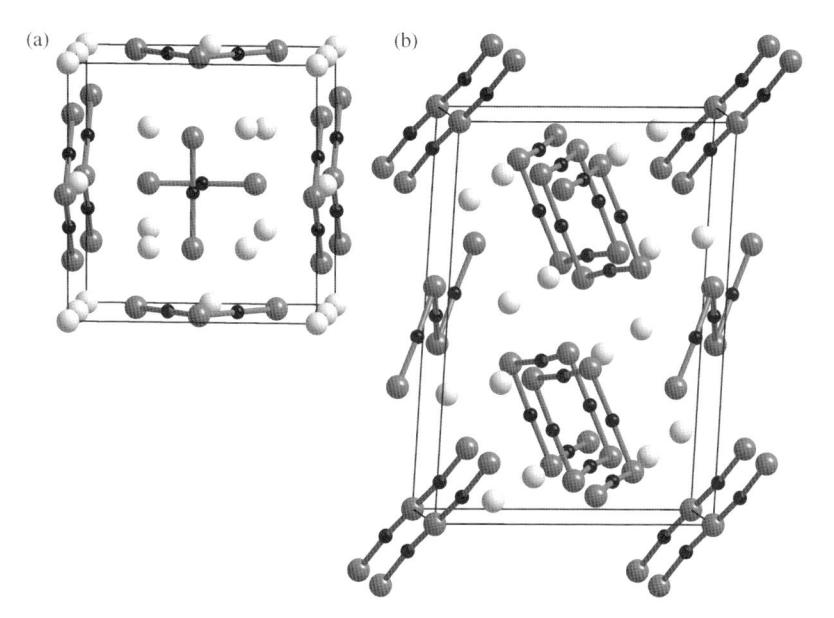

Figure 7.11 Sections of the crystal structures of (a) $Sr_6[CuN_2][Cu_2N_3]$ and (b) $Ba_{16}[CuN]_8[Cu_2N_3][Cu_3N_4]$. Small black spheres: Cu; larger light gray spheres: AE; larger dark gray spheres: N.

Until now, no in-depth investigations have been performed concerning the formation of dumbbells, chain fragments, and chains, considering phases containing several different motifs at the same time; however, a straightforward explanation based solely on composition and ion size may fall short.

In the Delafossite-type nitrides $CuNbN_2$ [789], $CuTaN_2$ [790], and $AgTaN_2$ [791], the group 11 metal is linearly coordinated by N, whereas the group 5 metal is octahedrally coordinated. $AgTaN_2$ is the only nitride known constituting direct Ag-N contacts. N is tetrahedrally coordinated by one group 11 and three group 5 atoms (Figure 7.39g). For further information, refer to Section 7.3.3.7.

Linear chains [MN] and [$Li_{1-x}M_x$] may be regarded as 1D fragments of the Li_2[LiN]-type structure. The simple 1:1:1 phases Ca[NiN] [780,792,793] and Ca[CuN] [794,795] as well as the substituted phase $Ca_{1-x}Sr_x$[NiN] ($1 < x < 0.50$) [780,793] and the solid solution series Ca[$Ni_{1-x}Li_x$N] ($1 < x < 0.58$) [796] crystallize in the YCoC-type structure [797] (Figure 7.12a). From topology-based examinations, a double bond character was derived for the infinite linear chains $^1_\infty$[$NiN_{2/2}$] in Ca[NiN] [784]. The linear arrangement of polymeric anions in the crystal structure of Ca[NiN] [780,792,793] is assumed to be stabilized by weak through-space couplings between the $^1_\infty$[$NiN_{2/2}$] chains [798]; valence charge density calculations of Ca[NiN] [799] are consistent with an almost ionic character of calcium and with weak Ni–Ni bonding between neighboring chains along [001].

Sr[LiN] [800] is another isotype of the YCoC-type structure, corresponding phases Ca[LiN] and Ba[LiN] in this structure type are not known. Upon substitution of Li by M = Fe, Co, Cu, the AE[LiN]-type structure is stabilized and partial ordering of Li and M occurs forming a twofold tetragonal superstructure with formal composition AE_2[Li($M_{1-x}Li_x$)N_2] (Figure 7.12b) indicating a preferred order [-Li-N-M-N-] within the chains. Whereas for most phases the extent of the substitution series has not been investigated in detail, generally $x = 0$ is not reached, as observed in Ca_2[Li($Fe_{1-x}Li_x$)N_2] ($x \leq 0.18$) [801], Ca_2[Li($Co_{1-x}Li_x$)N_2] ($x \leq 0.15$) [90], Ca_2[Li($Cu_{1-x}Li_x$)N_2] ($x \leq 0.18$) [795] (here an alternative substitution model according to Ca_2[($Li_{1-y}Cu_y$)($Cu_{1-x}Li_x$)N_2] is also discussed [794]), and Sr_2[Li($Fe_{1-x}Li_x$)N_2] ($x \leq 0.03$) [90]. Sr_2[($Li_{1-y}Co_y$)($Co_{1-x}Li_x$)N_2] is insofar special that at lower x values Sr_2[Li($Co_{1-x}Li_x$)N_2] is formed [90], at $x = 1$ order within the chains is achieved (Sr_2[LiNCoN] [802]), and at even higher amounts of cobalt the Co position is fully occupied, whereas the Li position may be substituted by Co up to $y \approx 0.15$ [90]. The fully ordered Sr_2[LiNCoN] [802] is isotypic to α-Li_3[BN_2] (= Li_2[LiNBN] [803,804]). Investigations of the magnetic properties of Sr_2[Li($Fe_{1-x}Li_x$)N_2] and Sr_2[($Li_{1-y}Co_y$)($Co_{1-x}Li_x$)N_2] show antiferromagnetic ordering ($T_N = 11$ K) for the Fe-containing phase and ferromagnetic ($T_C = 45$ K) ordering for Co, respectively [90]. For both compounds, the values $\mu_{eff} = 4.68\,\mu_B$/Fe (Fe^I d^7, spin-only 3.87 μ_B) and $\mu_{eff} = 3.2\,\mu_B$/Co (Co^I d^8, spin-only 2.83 μ_B) strongly exceed the spin-only values, so pronounced spin–orbit coupling has to be assumed [90], similar to the case in Li_2[$Li_{1-x}Fe_x$N] (see Section "7.3.3.1 Li_3N Superstructures: From Dumbbell Anions to Chains").

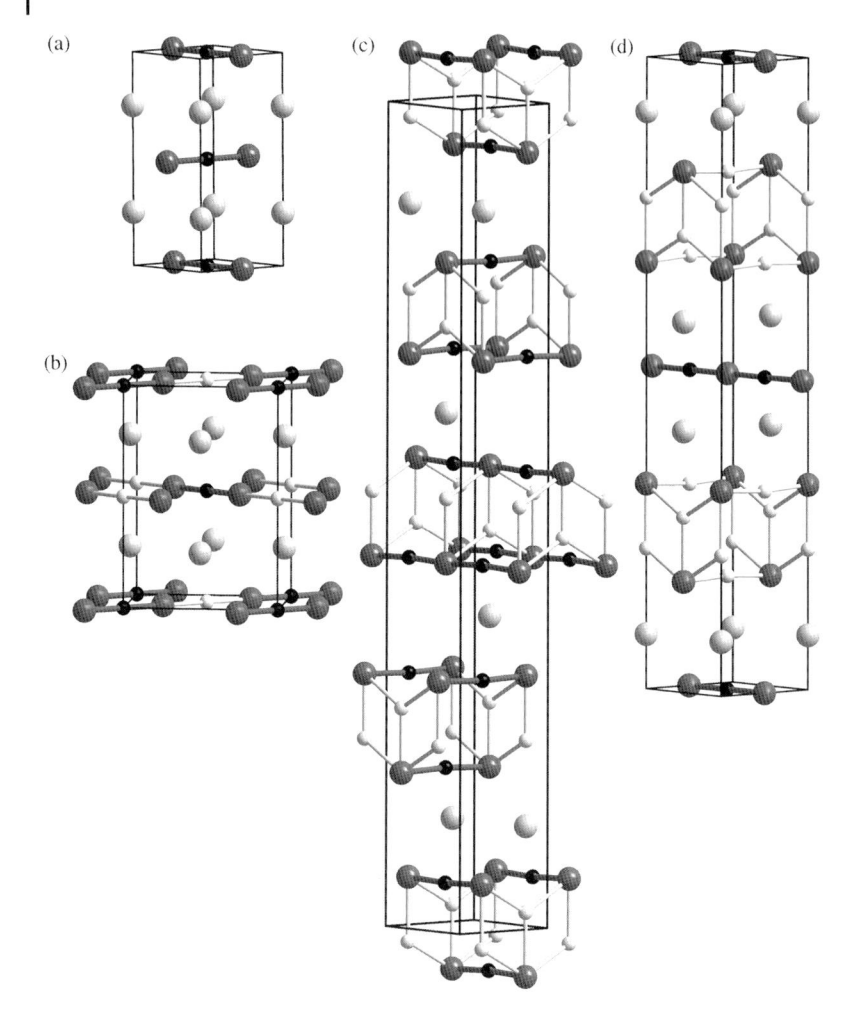

Figure 7.12 Linear chains [Li/MN] in lithium alkaline earth metal nitridometalates. (a) AE[Li$_{1-x}$M$_x$N]. (b) AE_2[Li(M$_{1-x}$Li$_x$)N$_2$]. (c) AE(Li[Li$_{1-x}$M$_x$N])$_2$. (d) Sr$_2$(Li$_4$N$_2$[Li$_{1-x}$Ni$_x$N]). Small black spheres: transition metal atoms disordered with Li; larger light gray spheres: AE; small white spheres: Li; larger dark gray spheres: N.

Although the solid solution series Sr$_2$[Li(M$_{1-x}$Li$_x$)N$_2$] have been reported for M = Fe, Co, Cu, no similar substitution of Li by Ni [805] or large amounts of Cu [795] has been observed, instead the substitution series Sr[Ni$_{1-x}$Li$_x$N] [749] and Sr[Cu$_{1-x}$Li$_x$N] [795] crystallizing in the Ba[NiN]-type [805] structure (see Section "7.3.3.1 Infinite Zigzag Chain Anions [MN]$^{n-}$ and Substituted Versions [(M$_{1-x}$Li$_x$)N]$^{n-}$") are formed.

In the Li-rich region of the quaternary system Li-Sr-M-N, the aristotype Li$_4$SrN$_2$ [806] (=AE(Li[Li$_{1-x}$M$_x$N])$_2$) may be regarded as a stacking variant of structural elements characteristic for Li$_2$[LiN] [25] and Sr[LiN] [800] in ratio

1:1. Two-dimensional fragments of the Li$_2$[LiN]-type structure are connected by fragments of the Sr[LiN] structure with the [LiN] chains representing the interface (Figure 7.12c). Within the linear chains, substitution has been observed for M = Fe, Co, Ni, and Cu; however, its extent is generally still unclear: Sr(Li[Li$_{1-x}$Fe$_x$N])$_2$, x = 0.46 [807], Sr(Li[Li$_{1-x}$Co$_x$N])$_2$, x = 0.50 [90], Sr(Li[Li$_{1-x}$Ni$_x$N])$_2$, x = 0.1 [806], and Sr(Li[Li$_{1-x}$Cu$_x$N])$_2$, x = 0.39 [795]. The isotypic ternary phases Li$_4$CaN$_2$ and Li$_4$BaN$_2$ are still unknown, but substituted phases have been reported as Ca(Li[Li$_{1-x}$Fe$_x$N])$_2$, x = 0.30 [807] and Ba(Li[Li$_{1-x}$Cu$_x$N])$_2$, x = 0.31 [795].

Sr$_2$(Li$_4$N$_2$[Li$_{1-x}$Ni$_x$N]) [749,808] is an example of another stacking variant of 2D fragments of the Li$_2$[LiN]-type structure connected by double layers of fragments of the Sr[LiN] structure with the [LiN] chains representing the interface (Figure 7.12d), the ratio Li$_2$[LiN]:Sr[LiN] corresponds to 1:2. Only in the linear chains within the Sr[LiN] double layer, substitution of Li by Ni (x = 0.2) has been observed; the un-substituted pure ternary nitride Li$_5$Sr$_2$N$_3$ is still elusive.

Besides Li$_5$[(Li$_{1-x}$Mn$_x$)N]$_3$ [752], Ca(Li[Mn$_{1-x}$Li$_x$N])$_2$ (x = 0.06) [809] is the only further example of a nitridomangante containing linear chains. The pure manganese compound Ca(Li[MnN])$_2$ (Figure 7.13a), formally Ca[MnN]·Li$_2$[MnN], is still elusive, and the extent of the substitution series has not been investigated either. The tetragonal crystal structure exhibits a close structural relation to that of Li$_2$[LiN], the square-planar coordination of Ca by N is similarly observed in Ca$_5$[CoN$_2$]$_2$ [64,779]. The partial structure [Li$_2$N$_2^{4-}$] is both isoelectronic (16 e$^-$ system) and with the same topology of the nets (4.8^2) compared to the layers [B$_2$C$_2^{2-}$] in the crystal structures of the ternary phases M[B$_2$C$_2$] [810,811].

Predominant structural features of Li$_3$Sr$_3$[NiN]$_4$ (Figure 7.13b) are nearly linear chains [NiN] connected by square-planar or trigonal-planar coordinated Sr and linear coordinated Li species [749,812]. No substitution of Li by Ni is observed. Ni$^{+0.75}$ is formally in an oxidation state lower than +1, distances Ni–N (177.3(1) pm and 177.8(1) pm) are slightly shorter compared to nitridonickelates containing Ni^I (Ca[NiN]: 179.0 pm [780,792,793]; Sr[NiN]: 182.0 pm [749,793]).

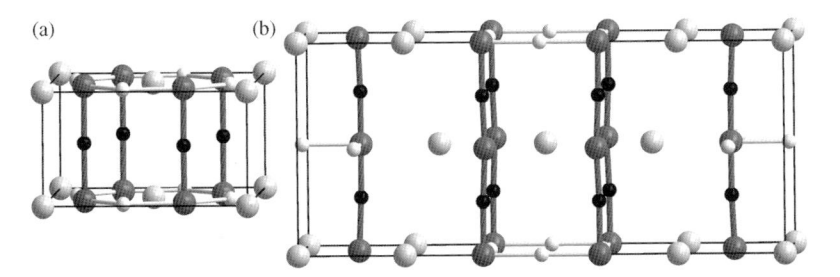

Figure 7.13 Sections of the crystal structures of (a) Ca(Li[Mn$_{1-x}$Li$_x$N])$_2$ and (b) Li$_3$Sr$_3$[NiN]$_4$. Small black spheres: transition metal atoms, for Mn disordered with Li; larger light gray spheres: AE; small white spheres: Li; larger dark gray spheres: N.

Infinite Zigzag Chain Anions $[MN]^{n-}$ and Substituted Versions $[(M_{1-x}Li_x)N]^{n-}$

In phases of general composition $A_E[MN]$ with A_E = Sr, Ba and M = Co, Ni, Cu, four different types of planar zigzag anionic nitridometalate chains $[MN^{2-}]$ are observed (Figure 7.14), which may be classified by the respective number of M atoms between kinks before repetition is reached as [2121], [22], [221221], and [33], respectively. Altogether, the crystal structures are complex, but rather closely related to each other.

Depending on the elements involved, different chains are formed and different stacking variants are realized. The simple [2121] chain is observed in Ba[CuN] [783], the symmetric [22] chain in $Sr_{0.53}Ba_{0.47}$[CuN] [786]. Both Ba[CoN] [813] and Ba[Ni$_{1-x}$Cu$_x$N] [749] constitute complex chains [221221], whereas Sr[NiN] [749,793], Sr[CuN] [94], the substituted phases Sr[Ni$_{1-x}$Li$_x$N] (x = 0.48) [749], and Sr[Cu$_{1-x}$Li$_x$N] (x = 0.66) [794], as well as Ba[NiN] [805] contain [33] chains.

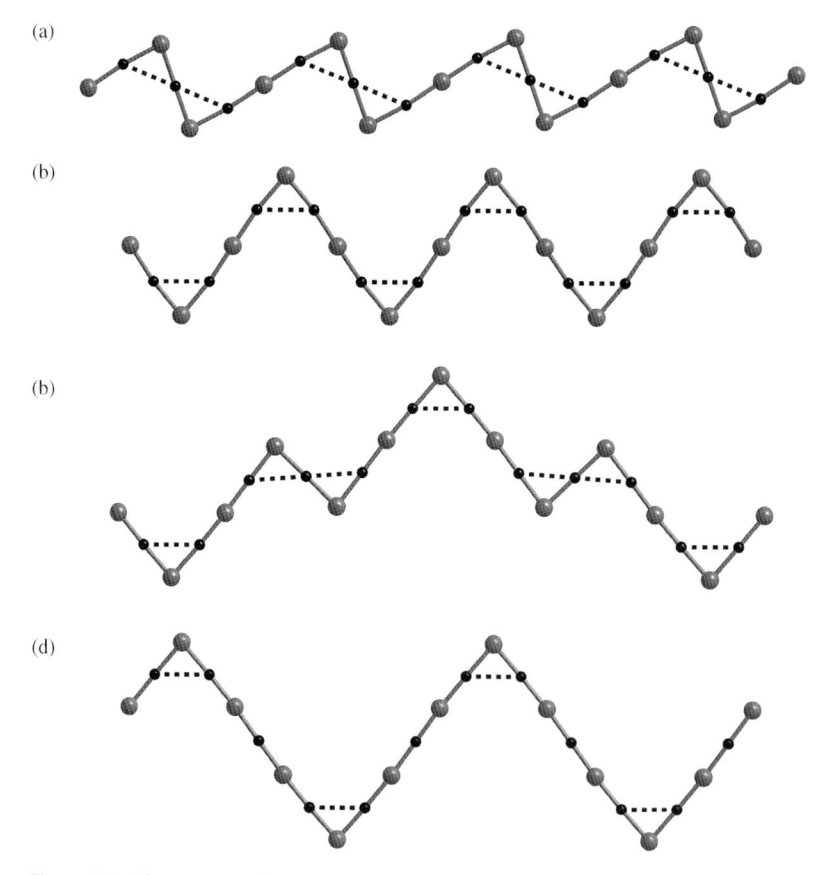

Figure 7.14 Planar zigzag chains in nitridometalates, classified as (a) [2121], (b) [22], (c) [221221], and (d) [33]. Small black spheres: transition metal atoms; larger dark gray spheres: N.

Until now, no clear understanding of the formation mechanism is evident, both atomic size and electronic effects may play a role. In all phases, heteroatomic distances M-N at the kink positions are elongated in comparison to the values for the linear part, but on average in the same range compared to other ternary nitridometalates with M with the same coordination number. In contrast, the M–M distances at the kink positions are short in comparison with the pure metal. The electronic structure of Ba[NiN] has been investigated in detail by use of *extended Hückel* and *ab initio* methods [814] with respect to the bonding topologies [784]. The nature of the M—N bonds has to be classified as strongly covalent. A double bond character was derived for the linear parts of the zigzag chains in Ba[NiN]. At the kink positions, the nitrogen atoms are assumed to bear a lone pair together with one double and one single bond to the transition metals. The Ni···Ni contacts of the kink positions in the crystal structure of Ba[NiN] correspond to a net bonding interaction that is mainly σ in character.

Besides planar zigzag chains, also helical chains [MN] are observed in nitridometalates. The crystal structure of $Ba_8[NiN]_6N$ [815] appears quite complex: Besides octahedra NBa_6 formed by the isolated nitride ions, it contains chains $^1_\infty[NiN_{2/2}]$ in a double-spiral arrangement (Figure 7.15a). Considering Ba^{2+} and N^{3-}, the oxidation state of Ni is +5/6, slightly reduced compared with other nitridonickelates. Whereas distances Ni–N are in the same range as in other nitridonickelates, the homoatomic Ni···Ni contacts at the kink positions (256.8(3) pm) are significantly longer compared to Ba[NiN] (242.6(1) pm) [805]. In $Ba_{16}[CuN]_8[Cu_2N_3][Cu_3N_4]$ [783], helical chains (Figure 7.15b) are observed; as expected for a Cu^I phase, average distances Cu-N are similar in all nitridocuprate anions.

Layered Anions ($M^{<+1}$)
Orthorhombic $Ba_2[Ni_3N_2]$ [75,79] ($\equiv Ba_2[Ni(1)^1_2Ni(2)^0N_2]$) provides the first example of a nitridonickelate with a corrugated two-dimensional complex anion

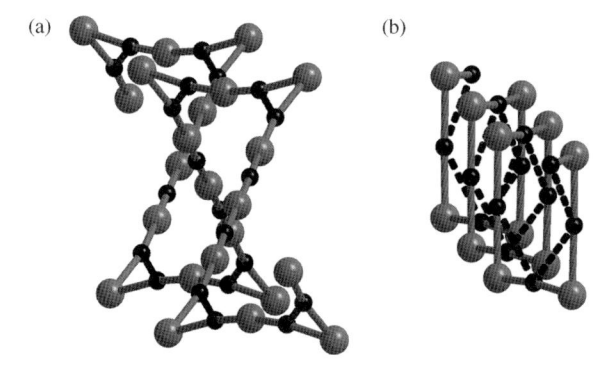

Figure 7.15 (a) Spiral chains $^1_\infty[NiN]$ in the crystal structure of $Ba_8[NiN]_6N$. (b) Helical chains $^1_\infty[CuN]$ in the crystal structure of $Ba_{16}[CuN]_8[Cu_2N_3][Cu_3N_4]$. Small black spheres: transition metal atoms; larger dark gray spheres: N.

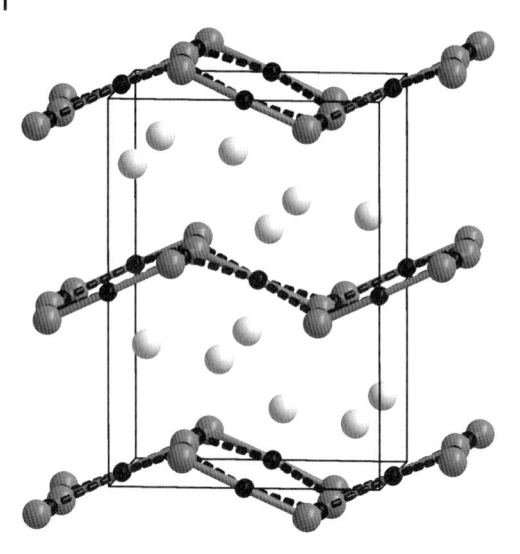

Figure 7.16 Section of the crystal structure of $Ba_2[Ni_3N_2]$. Small black spheres: Ni; larger light gray spheres: Ba; larger dark gray spheres: N.

(Figure 7.16) consisting of near-linear [Ni-N] chains connected by [NiN$_2$] dumbbells described with a static split position. The average oxidation state of Ni in this compound is +0.67, the lowest average value observed in nitridonickelates so far; however, the assignment of definite oxidation states remains controversial. $Ba_2[Ni_3N_2]$ is a bad metal with a large resistivity of ≈ 2.7 mΩ cm at 300 K and displays a structural and magnetic phase transition at $T \approx 90$ K, which is interpreted as a long-range antiferromagnetic ordering of Ni^I species accompanied by a symmetry-retaining structural distortion. XAS investigations correspond well with low-valent Ni states.

Upon substitution of Ni by Li, only the dumbbell position is substituted, forming $Ba_2[Ni_2(Ni_{1-x}Li_x)N_2]$ [816]. The distance (Ni, Li)-N increases with increasing Li content in accordance with comparable substituted phases; however, the extent of the split position Ni, Li as evidenced in the ratio of the anisotropic displacement parameters decreases with increasing Li content. Recently, the isotypic phase $Sr_2[Ni_2(Ni_{1-x}Li_x)N_2]$ was described [90].

7.3.3.2 CN(M) = 2 + 3

H																	He
Li	Be											B	C	**N**	O	F	Ne
Na	Mg											Al	Si	P	S	Cl	Ar
K	Ca	Sc	Ti	V	Cr	**Mn**	**Fe**	Co	Ni	Cu	Zn	Ga	Ge	As	Se	Br	Kr
Rb	**Sr**	Y	Zr	Nb	Mo		Ru	Rh	Pd	Ag	Cd	In	Sn	Sb	Te	I	Xe
Cs	**Ba**	La	Hf	Ta	W	Re	Os	Ir	Pt	Au	Hg	Tl	Pb	Bi			

Ce	Pr	Nd	Pm	Sm	Eu	Gd	Tb	Dy	Ho	Er	Tm	Yb	Lu
Th		U											

In the system Sr–Fe–N, three different phases are observed. All of them are prepared under oxidizing conditions in excess nitrogen. In two of these phases, Fe is observed in different coordination spheres: Triclinic $Sr_{10}[FeN_2][Fe_2N_4]_2$ [55] (Figure 7.17a) is obtained above 1323 K and contains dumbbell anions $[Fe^{II}N_2]^{4-}$ and planar $[Fe^{II}{}_2N_4]^{8-}$ anions consisting of two trigonal-planar units sharing a common edge. Between 1223 and 1253 K, upon fast cooling mixed-valent monoclinic $Sr_8[Fe^{III}N_3]_2[Fe^{II}N_2]$ [54] (Figure 7.17b) forms. It features trigonal-planar $[Fe^{III}N_3]^{6-}$ and linear $[Fe^{II}N_2]^{4-}$ anions, therefore containing a transition metal with different coordination spheres and in different oxidation states in an ordered fashion. $Sr_3[FeN_3]$ is obtained in pure form below 1000 K (see Section 7.3.3.3). The coordination spheres of the $[FeN_2]^{4-}$ anions in both $Sr_{10}[FeN_2][Fe_2N_4]_2$ and $Sr_8[Fe^{III}N_3]_2[Fe^{II}N_2]$ show the same features as other nitridometalate dumbbell anions, the peculiarities of the $[FeN_3]^{6-}$ and $[Fe_2N_4]^{8-}$ anions are discussed in comparison with similar units in Section 7.3.3.3.

$Sr_8[Fe^{III}N_3]_2[Fe^{II}N_2]$ is a semiconductor (ρ (300 K) $\approx 1 \times 10^{-1}$ Ωm) and revealed Curie–Weiss behavior at high temperatures with an effective magnetic moment $\mu_{eff} = 7.02\,\mu_B$/f.u. ($\Theta_p = -55$ K) [785]. This value is compatible with either two high-spin (HS) $[Fe^{III}N_3]^{6-}$ units and one low-spin (LS) $[Fe^{II}N_2]^{4-}$ complex, two LS $[Fe^{III}N_3]^{6-}$ and one HS $[Fe^{II}N_2]^{4-}$, or both $[Fe^{III}N_3]^{6-}$ and $[Fe^{II}N_2]^{4-}$ showing intermediate spin (IS). Pronounced spin–orbit coupling cannot be ruled out. Below 56 K, antiferromagnetic ordering is observed.

Upon substitution of Fe by Mn in $Sr_8[MN_3]_2[MN_2]$, the isotypic phases $Sr_8[MnN_3]_2[FeN_2]$ [817] and $Sr_8[MnN_3]_2[MnN_2]$ [80] are observed. The assignment of Mn and Fe in $Sr_8[MnN_3]_2[FeN_2]$ was based on comparison with other nitridomanganates and nitridoferrates, but substitution of Fe by Mn on both positions cannot be ruled out [785]. Whereas $Sr_8[Fe^{III}N_3]_2[Fe^{II}N_2]$ and $Sr_8[Mn^{III}N_3]_2[Fe^{II}N_2]$ may be prepared under oxidizing conditions in a N_2 stream, great care has to be taken concerning the amount of N in the system in the case of $Sr_8[Mn^{III}N_3]_2[Mn^{II}N_2]$. Upon nitrogen excess, mixed-valent $Sr_8[Mn^{III,IV}N_3]_3$ ($P2_1/c$) [818] is formed (see Section 7.3.3.3), the equilibrium depends on the N_2 partial pressure [80].

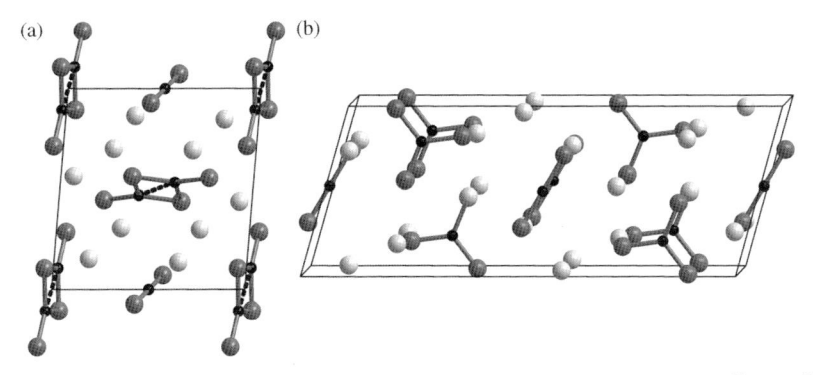

Figure 7.17 Sections of the crystal structures of (a) $Sr_{10}[FeN_2][Fe_2N_4]_2$ and (b) $Sr_8[Fe^{III}N_3]_2[Fe^{II}N_2]$. Small black spheres: Fe; larger light gray spheres: Sr; larger dark gray spheres: N.

$Sr_8[MnN_3]_2[FeN_2]$ is a semiconductor (ρ (300 K) $\approx 1\times10^{-2}$ Ωm) and revealed Curie–Weiss behavior at high temperatures with an effective magnetic moment $\mu_{eff} = 7.16\,\mu_B$/f.u. ($\Theta_p = -72$ K) [785]. Below 20 K, antiferromagnetic ordering is observed. $Sr_8[MnN_3]_2[MnN_2]$ displays semiconducting behavior in the whole temperature range (ρ (300 K) $\approx 4\times10^{-2}$ Ωm) [80]; the $\rho(T)$ dependence corresponds to a small bandgap of 0.22(1) eV in agreement with the gray color of the sample. Curie–Weiss behavior of the magnetic susceptibility is observed down to 10 K with $\Theta = -62(3)$ K and $\mu_{eff} = 5.1(1)\,\mu_B$/f.u.; an AFM transition occurs at 10 K. An in-depth analysis of the magnetic behavior of $Sr_8[MnN_3]_2[MnN_2]$ based on EPR data ($g = 2.29$) led to two IS $[Mn^{III}N_3]^{6-}$ and one LS $[Mn^{II}N_2]^{4-}$ anions taking the considerable orbital contribution to the magnetic moment into account [80].

7.3.3.3 CN(M) = 3

Nitridometalates with M in a threefold coordination by N show, despite the rather large number of elements involved, a relatively small structural variety. The oxidation state +III leads most often to ideal or distorted isolated trigonal planar units $[MN_3]^{6-}$.

Isolated Planar Anions $[MN_3]^{n-}$

The alkaline earth metal nitridometalates with M^{III} containing trigonal-planar coordinated complex anions crystallize in one of three structure types: (a) $(Ca_3N)_2[MN_3]$ (M = Mn [819], Fe [820]) (Figure 7.18); (b) $Ca_3[MN_3]$ (M = V [821,822], Cr [77], Mn [823]) (Figure 7.19a); and (c) $AE_3[MN_3]$ (AE = Sr, Ba, M = Cr [824], Mn [825], Fe [53,86]) (Figure 7.19b). For $(Ba_{1-x}Sr_x)_3[FeN_3]$ [826], the extent of the solid solution series was also investigated. In all these phases, large voids are present directly above and below the complex $[MN_3]^{6-}$ anions. Whereas in $AE_3[MN_3]$ compounds for AE = Sr, Ba, the complex anions $[MN_3]^{6-}$ show ideal D_{3h} symmetry, for AE = Ca, symmetry reduction results in a distortion of the still planar $[MN_3]^{6-}$ units to C_{2v}. A structure map of the $AE_3[MN_3]$ compounds [824] shows the clear dependence on the bond distances and, consequently, on cation size. However, for $Ca_3[CrN_3]$, this symmetry reduction is also discussed on the basis of a Jahn–Teller distortion induced by the low-spin Cr^{3+} ion [77] and validated by *ab initio* extended Hückel calculations [827].

A stacking variant of the $AE_3[MN_3]$-type structure with additional octahedra Ba_6O between every other $Ba_3[MN_3]$ layer is realized in $Ba_6[Re^{IV}N_3]_2O$ [828] and $Ba_6[Os^{IV}N_3]_2O$ [829].

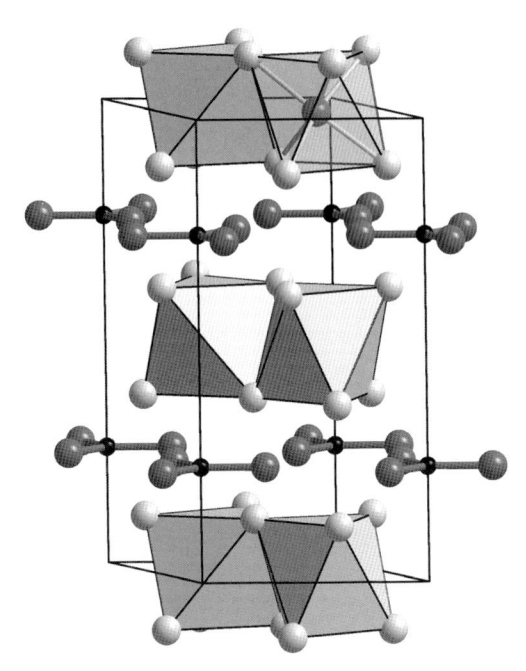

Figure 7.18 Isolated trigonal-planar nitridometalate anions $[MN_3]^{6-}$ in $(Ca_3N)_2[MN_3]$. Small black spheres: transition metal atoms; larger light gray spheres: Ca; larger dark gray spheres: N.

The physical properties of only a fraction of these phases have been investigated:

$(Ca_3N)_2[FeN_3]$ is a semiconductor (ρ (300 K) $\approx 1 \times 10^0$ Ωm) and shows Curie–Weiss behavior above 50 K [785]; the effective magnetic moment of $\mu_{eff} = 4.03$ μ_B/f.u. corresponds to three unpaired electrons indicating an intermediate spin state for Fe in the $[Fe^{III}N_3]^{6-}$ anion. The nature of low-dimensional AFM ordering in weak magnetic fields below 50 K and of ferromagnetic ordering below $\Theta_p = 28$ K is still unclear.

$Ca_3[VN_3]$ [821] appears to be essentially intrinsically diamagnetic and insulating, whereas for insulating $Ca_3[CrN_3]$ AFM ordering at 240 K has been reported [77].

In $Ca_3[MnN_3]$, magnetic susceptibility increases almost linearly down to 85 K, where three-dimensional AFM ordering occurs [785]. Due to a certain frustration, Curie–Weiss behavior is not achieved up to 400 K, so no reliable information on the magnetic moment can be obtained. The effective magnetic moment starts saturating at around 2.5 μ_B/f.u., not far from the expected value of 2.83 μ_B/f.u. for $[Mn^{III}N_3]^{6-}$ units with $S = 1$.

$Ba_3[FeN_3]$ is a bad metal (ρ (300 K) $\approx 1 \times 10^{-4}$ Ωm) with nearly temperature-independent resistivity [785]; magnetic susceptibility data imply non-Curie–Weiss behavior and still defy interpretation.

Upon nitrogen excess, $Sr_8[Mn^{III}N_3]_2[Mn^{II}N_2]$ (Figure 7.19c) is oxidized to mixed-valent $Sr_8[Mn^{III,IV}N_3]_3$ [818] with doubling the unit cell (Figure 7.19d); the equilibrium depends on N_2 partial pressure [80]. Isotypic $Ba_8[MnN_3]_3$ is formed from $Ba_3[MnN_3]$ upon extended reaction times (due to decomposition of $Ba_3[MnN_3]$ and reaction of Ba with the crucible material) or elevated nitrogen pressure [81]. Although the oxidation state of Mn is changed and Mn is located on a low-symmetry position, the complex anions retain trigonal planar symmetry.

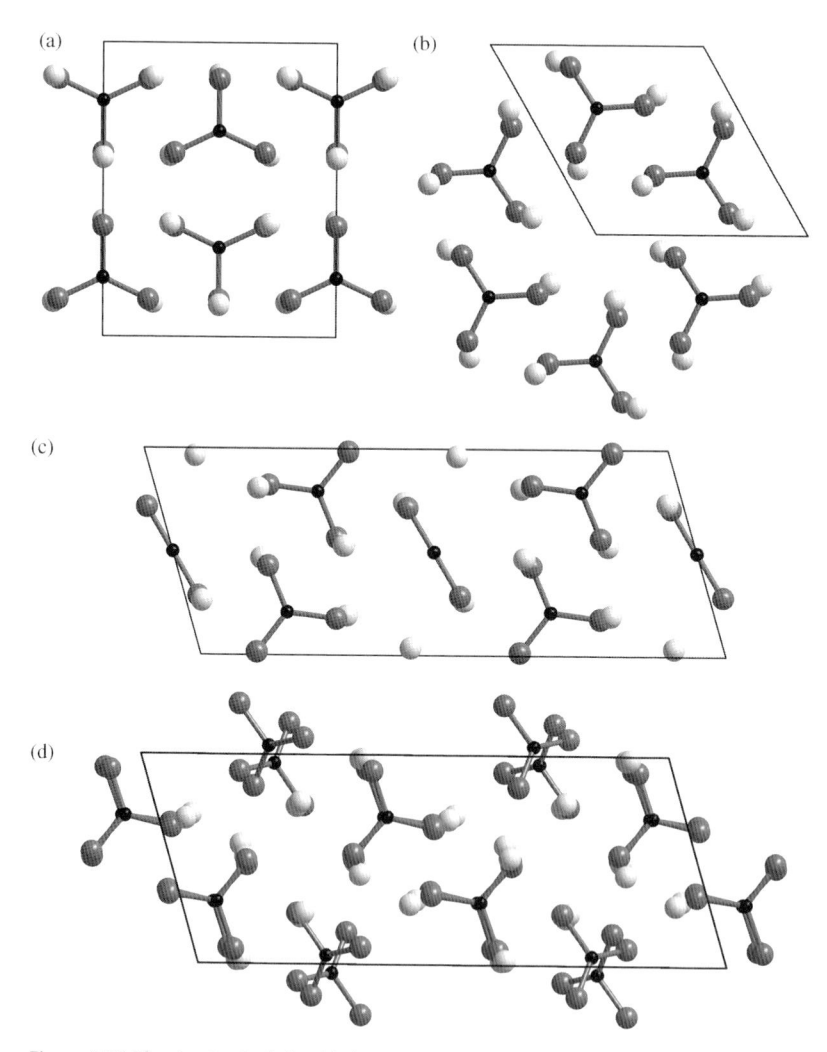

Figure 7.19 The structural relationship between (a) $Ca_3[MnN_3]$, (b) $Sr_3[MnN_3]$, (c) $Sr_8[MnN_2][MnN_3]_2$, and (d) $Sr_8[MnN_3]_3$. Small black spheres: Mn; larger light gray spheres: A_E; larger dark gray spheres: N.

Oligomeric Anions and 1D Chain Anions

Condensation of $[MN_3]$ units may either proceed via common corners forming $[M_2N_5]$ units (these units have not been observed in nitridometalates) or via common edges forming $[M_2N_4]$ units. The only examples of edge-sharing anions $[M_2N_4]^{n-}$ known until now are the nitridoferrates(II) $Ca_2[FeN_2]$ [55] (= $Ca_4[Fe_2N_4]$) (Figure 7.20a) and $Sr_2[FeN_2]$ (= $Sr_{10}[FeN_2][Fe_2N_4]_2$ (see Section 7.3.3.2) [55], which contain planar dimers $[Fe^{II}_2N_4]^{8-}$. The local surrounding of Fe^{II} in these ions is considerably distorted to a more T-like arrangement compared to the $[Fe^{III}N_3]^{6-}$ anions, although planarity is retained. Additional corner-sharing of the $[Fe^{II}_2N_4]^{8-}$ units leads to 1D chains in $LiSr_2[Fe_2N_3]$ [830] and

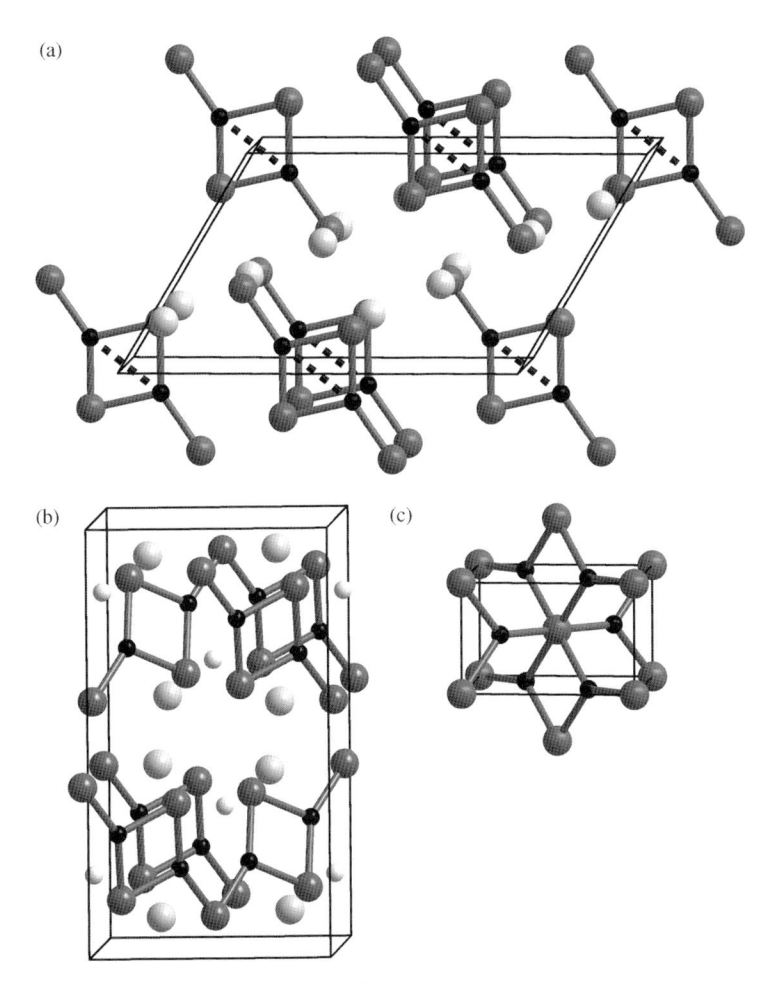

Figure 7.20 Complex anions $[Fe_2N_4]^{8-}$, 1D $[Fe_2N_3]^{5-}$, and 3D $[Mn_2N]$ in the crystal structures of (a) $Ca_2[FeN_2]$, (b) $LiAE_2[Fe_2N_3]$ (AE = Sr, Ba), and (c) $Li_x[Mn_{2-x}N]$. Small black spheres: transition metal atoms, for (c) disordered with Li; larger light gray spheres: AE; small white spheres: Li; larger dark gray spheres: N.

LiBa$_2$[Fe$_2$N$_3$] [830]. In these phases, the [Fe$_2$N$_4$] units are still planar, but a kink occurs at the connecting terminal N (Figure 7.20b). Unfortunately, no information on magnetism and related physical properties of these phases has been reported as yet.

3D–Anions

Li$_x$[Mn$_{2-x}$N] [831] has an anti-rutile structure with Li and Mn disordered on the Ti position in a roughly 1:2 ratio, thereby forming a 3D structure of edge- and corner-sharing [MnN$_3$] units (Figure 7.20c). Li$_x$[Mn$_{2-x}$N] is considered a degenerate semiconductor or a poor metal with a room temperature resistivity $\rho \approx 8.5 \times 10^{-5}$ Ωm; magnetic susceptibility measurements suggested localized moments on the Mn ions and showed an antiferromagnetic ordering at 115 K.

Isolated Nonplanar Anions [MN$_3$]$^{n-}$

Most [MN$_3$] units are planar, either in D_{3h} or C_{2v} symmetry. In contrast, nonplanar anions are rare. Trigonal Li$_{24}$[Mn^{III}N$_3$]$_3$N$_2$ (Figure 7.21a) [752], which is obtained by reaction of Li$_7$[MnN$_4$] (see Section "7.3.3.5 Li$_2$O-Type Defect and Order Variants") with Li at 1100 K < T < 1115 K, is still another example for the rich structural chemistry of lithium nitridomanganates, ranging from Mn^V to Mn^I and coordination numbers 4 to 2. It contains slightly but significantly nonplanar trigonal [MnN$_3$]$^{6-}$ units with C_{3v} symmetry. The isolated nitrogen species are located in distorted cubes built up by eight lithium ions. According to DFT calculations, due to restrictions in the Li-N substructure, Mn is located about 28 pm out of the plane defined by the three coordinating nitrogen atoms. Measurements of the magnetic susceptibility reveal a d^4 ($S=1$) spin-state for the manganese.

Na$_4$[Re^{V}N$_3$] contains pyramidal anions [ReN$_3$]$^{5-}$ (Figure 7.21b) [832], which correspond well with a nonbonding electron pair at Re. The noncentrosymmetric space group Cc was validated via measurements of the second harmonic generation effect employing an Nd:YAG laser.

Extended Nonplanar Anions

The anion [Mn$_2^{IV}$N$_6$]$^{10-}$ with a nonbridged Mn–Mn bond may be considered as a dimer of the pyramidal anion [Re^{V}N$_3$]$^{4-}$ upon reducing the metal by 1 e$^-$ each and forming a metal–metal bond to keep the full electron shell. Such anions were described in the crystal structures of Li$_6$Ca$_2$[Mn$_2$N$_6$] [833], Li$_6$Sr$_2$[Mn$_2$N$_6$] [834], and the solid solution series Li$_6$(Ca$_{1-x}$Sr$_x$)$_2$[Mn$_2$N$_6$] [834] (Figure 7.22a). Magnetization measurements of the unsubstituted phases show the typical behavior of antiferromagnetically coupled Mn–Mn entities (dimers) with high coupling constants of $J = -739$ and -478 cm^{-1}, respectively. Analysis of the calculations using different theoretical methods shows that the magnetic coupling between the local π orbitals is not caused by a direct interaction but by the spin-polarized σ bond [835].

A similar situation is observed in Ca$_6$[Cr$_2$N$_6$]H (Figure 7.22b) [836], the complex anion contains Cr^{III} and Cr^{IV}.

7.3.3.4 CN(M) = 3 + 4

H																	He
Li	Be											B	C	N	O	F	Ne
Na	Mg											Al	Si	P	S	Cl	Ar
K	Ca	Sc	Ti	V	Cr	Mn	Fe	Co	Ni	Cu	Zn	Ga	Ge	As	Se	Br	Kr
Rb	Sr	Y	Zr	Nb	Mo	Tc	Ru	Rh	Pd	Ag	Cd	In	Sn	Sb	Te	I	Xe
Cs	Ba	La	Hf	Ta	W	Re	Os	Ir	Pt	Au	Hg	Tl	Pb	Bi	Po	At	Rn
Fr	Ra	Ac															

	Ce	Pr	Nd	Pm	Sm	Eu	Gd	Tb	Dy	Ho	Er	Tm	Yb	Lu
	Th	Pa	U											

Only a few phases are known that contain M in both trigonal and tetrahedral coordination by N. All these phases are nitridomanganates with average Mn oxidation states between +2 and +3 and N-bridged Mn–Mn contacts in 2D and 3D complex anions.

Manganese-rich $Ca_{12}[Mn_{19}N_{23}]$ [81] (Figure 7.23a) and $Ca_{133}[Mn_{216}N_{260}]$ [81] (Figure 7.23b) contain 2D anions $[Mn_{19}N_{23}]^{24-}$ and $[Mn_{216}N_{260}]^{266-}$, respectively, separated by Ca along [001]. In $Ca_{12}[Mn_{19}N_{23}]$, four of seven symmetry-independent Mn positions are tetrahedrally coordinated, the others show trigonal coordination by N. Between all Mn atoms, close distances interpreted as bonds are observed except for one Mn (highlighted ∗ in Figure 7.23a), which is coordinated by N in the form of a trigonal pyramid and exclusively features long distances to all other Mn. The average oxidation state calculates to $Mn^{2.37+}$, both

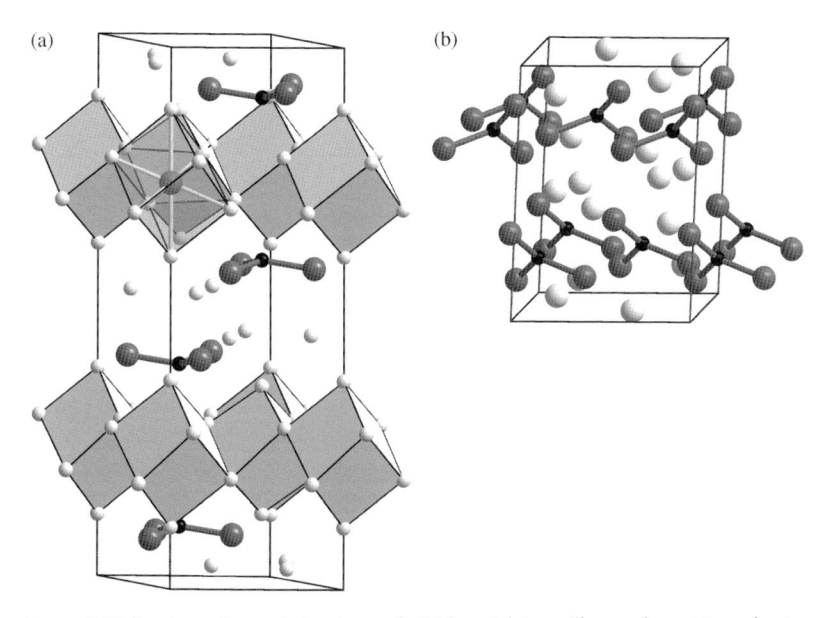

Figure 7.21 Sections of crystal structures of nitridometalates with nonplanar trigonal anions [MN_3]. (a) $Li_{24}[Mn^{III}N_3]_3N_2$. (b) $Na_4[Re^VN_3]$. Small black spheres: transition metal atoms; small white spheres: Li; larger light gray spheres: Na; larger dark gray spheres: N.

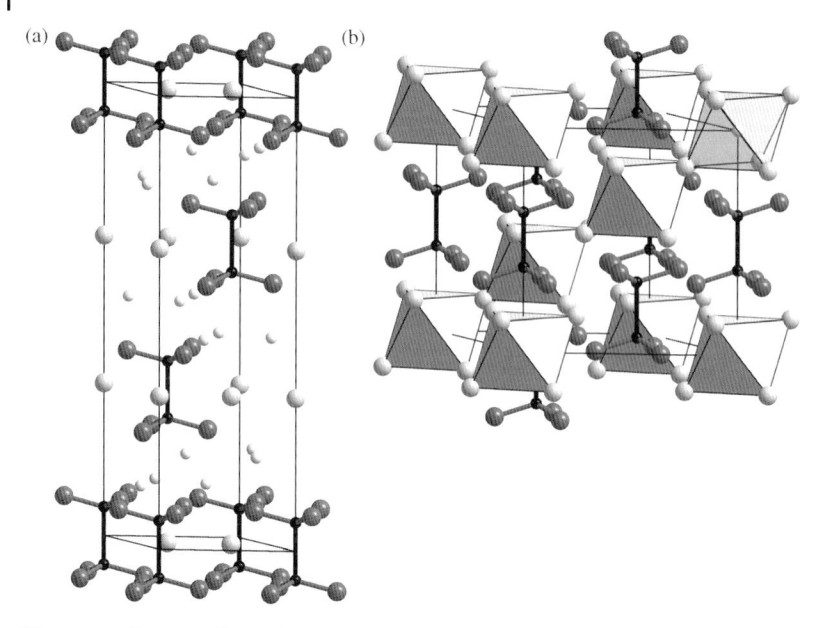

Figure 7.22 Sections of crystal structures of (a) $Li_6Ca_2[Mn_2N_6]$; (b) $Ca_6[Cr_2N_6]H$, HCa_6 octahedra emphasized. Small black spheres: transition metal atoms; larger light gray spheres: Ca; small white spheres: Li in (a) or H in (b); larger dark gray spheres: N.

magnetic susceptibility and EPR measurements show $\mu_{eff} = 2.7(1)$ μ_B corresponding with only one 1 of 19 Mn atoms per formula unit carrying a localized magnetic moment. The same holds true for $Ca_{133}[Mn_{216}N_{260}]$; 1 out of 36 Mn is trigonal nonplanar coordinated with long distances to the next-nearest Mn neighbors (highlighted * in Figure 7.23b), the remaining Mn are trigonally or tetrahedrally coordinated by N and feature short distances to other Mn species. Magnetic susceptibility data correspond well with every 1 of 36 Mn atoms holding a localized magnetic moment, the average oxidation state calculates to be $Mn^{2.38+}$. In contrast, this particular structural feature of one trigonal nonplanar coordinated Mn with long distances to the next-nearest Mn neighbors is not observed in $Sr_{25}[Mn_{42}N_{50}]$ [81] (Figure 7.23c) with an average oxidation state $Mn^{2.38+}$, and magnetic susceptibility data show no localized magnetic moments as well. However, the structure contains rather large channels (highlighted + in Figure 7.23b and c) along [001] similar to $Sr_3[MnN_3]$ (Figure 7.19b), a characteristic also observed in $Ca_{133}[Mn_{216}N_{260}]$.

The phases $(AE_6N_{1-\delta})Mn_{20}N_{20+x}$ (AE = Sr, Ba) [81] show a close structural relation to the rare earth metal compounds $La_6Cr_{20-x}N_{22}$ [837–839], $Ce_6Cr_{20-x}N_{22}$ [838,839], and $Pr_6Cr_{20-x}N_{22}$ [838,839] (see Section "7.3.3.5 Alkaline Earth and Rare Earth Metal Nitridometalates"). In all these phases, the metal substructure corresponds to the Cu_2AlMn-type structure – a Heusler-phase – with AE_6 octahedra and Mn_8 cubes occupying the Al and Mn positions, and two Mn_6 octahedra the Cu positions, respectively (Figure 7.24a). In the strontium

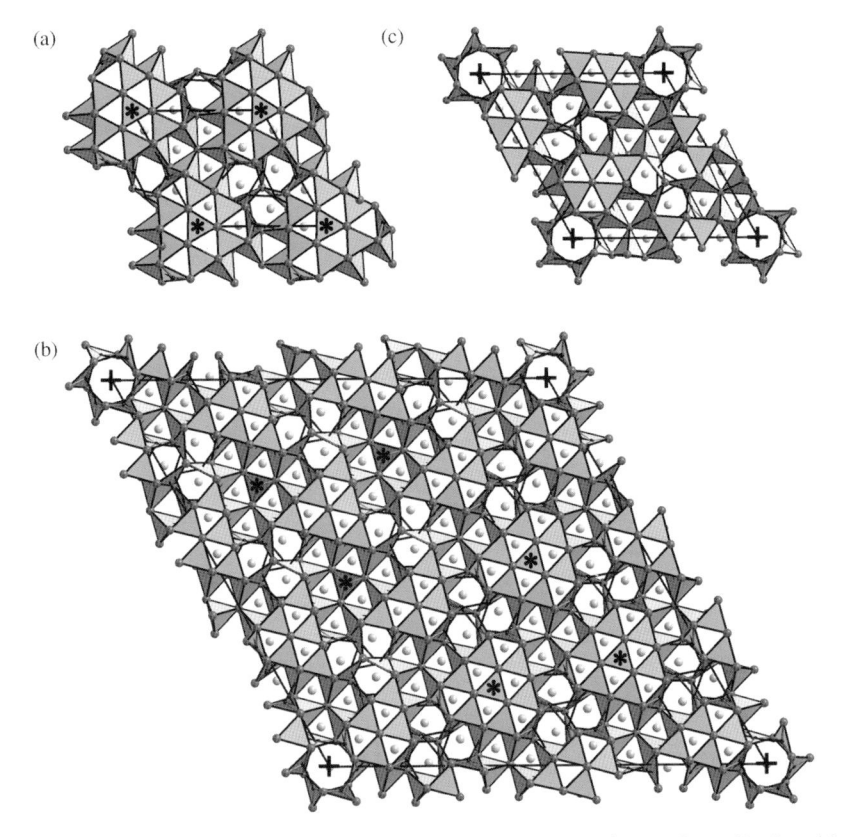

Figure 7.23 Sections of the crystal structures of (a) $Ca_{12}[Mn_{19}N_{23}]$, (b) $Ca_{133}[Mn_{216}N_{260}]$, and (c) $Sr_{25}[Mn_{42}N_{50}]$; centers of similar building blocks (∗, +) are emphasized. Small black spheres: Mn; larger light gray spheres: *AE*; larger dark gray spheres: N.

phase $Sr_6Mn_{20}N_{21}$ ($x = \delta = 0$), the position within the Sr_6 octahedron is fully occupied by N, whereas the center of the Mn cube is empty, so all the Mn species in the Mn_6 octahedra are tetrahedrally coordinated by N, whereas the N coordination of the Mn species in the Mn_8 cubes is trigonal-planar. In the barium phase, $Ba_6Mn_{20}N_{21.5-\delta}$, the center of the Mn cube is half occupied by N ($x = 0.5$). This results in a split position of the Mn cube (Figure 7.24b), with the *x*-values for Mn refining to similar values as N: In an empty cube, Mn is trigonal-planar coordinated by N, whereas in a N-filled cube Mn moves toward the center of the cube resulting in a tetrahedral coordination. Depending on nitrogen pressure during synthesis, the center of the barium octahedra may be only partially occupied ($0 < \delta < 0.9$) by N, resulting in a wide range of differing physical properties. For $x = 0.9$, a broad symmetric maximum in the magnetic susceptibility around 240 K suggests short-range ordering, long-range AFM ordering occurs at 80 K, and a possible ferro-, ferri-, or canted AFM ordering is observed at 25 K. An alternative description of the crystal structure is based on Mn_{10} supertetrahedra (Figure 7.24c), in which all face centers are capped by one and all

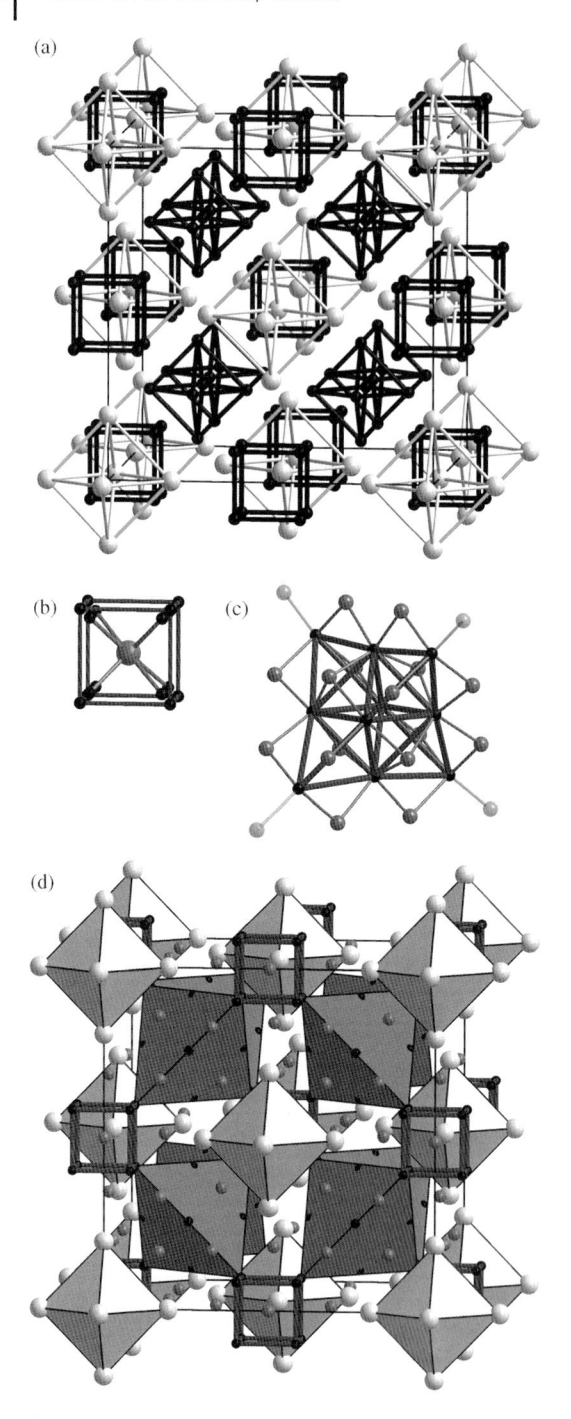

Figure 7.24 (a) Metal substructure of Cu$_2$AlMn-type (fcc A_{E6} (light gray)/Mn$_8$ (black), Mn$_6$ (black) filling all tetrahedral voids) in phases $A_{E6}M_{20}N_{21+x}$. (b) The split position of the Mn cube as an effect of the partial occupation of the center by N in Ba$_6$Mn$_{20}$N$_{21.5}$. (c) Mn$_{10}$N$_{10}$ cluster in A_{E6}Mn$_{20}$N$_{21.5}$. (d) The crystal structure of A_{E6}Mn$_{20}$N$_{21.5}$. Small black spheres: Mn; larger light gray spheres: A_E; larger dark gray spheres: N.

edges by two N species forming $Mn_{10}N_{10}$ units. These units are arranged in a cubic primitive array with alternating large cuboctahedral and small cubic voids (Figure 7.24d). $AE_6N_{1-\delta}$ octahedra are located in the centers of the cuboctahedral voids; in the Sr phase, the cubic void is empty, whereas in the Ba phase, it is half occupied by N.

7.3.3.5 CN(M) = 4

In nitridometalate chemistry, fourfold coordination by nitrogen is generally connected with the "early" transition metals (groups 3–7, although further examples with iron and zinc are known) and high oxidation states, meaning small ionic radii of M. The majority of these compounds contain M in tetrahedral coordination; however, a few examples with square-planar coordination have also been observed and are treated in a separate section.

Tetrahedral Coordination

H																	He
Li	**Be**											B	C	**N**	O	F	Ne
Na	**Mg**											Al	Si	P	S	Cl	Ar
K	**Ca**	**Sc**	**Ti**	**V**	**Cr**	**Mn**	**Fe**	**Co**	Ni	Cu	**Zn**	Ga	Ge	As	Se	Br	Kr
Rb	**Sr**	**Y**	**Zr**	**Nb**	**Mo**	Tc	**Ru**	**Rh**	Pd	Ag	Cd	In	Sn	Sb	Te	I	Xe
Cs	**Ba**	**La**	**Hf**	**Ta**	**W**	**Re**	**Os**	Ir	Pt	Au	Hg	Tl	Pb	Bi	Po	At	Rn
Fr	Ra	Ac															

	Ce	**Pr**	**Nd**	**Pm**	**Sm**	**Eu**	**Gd**	**Tb**	**Dy**	**Ho**	**Er**	**Tm**	**Yb**	**Lu**
	Th	Pa	**U**											

Nitridometalates with M in tetrahedral coordination show a wide variety of anionic substructures ranging from isolated tetrahedra, oligomeric units of edge- or vertex-sharing tetrahedra, infinite chains, and two- and three-dimensional frameworks of edge- and/or vertex-sharing tetrahedra. Despite sometimes widely different compositions and dimensionalities, several of these phases may be described as defect or order variants of simple structure types, that is, the Li_2O (anti-CaF_2) defect/order variants $Li_{2-x-y}M_x\square_yN$ or the Na_3As/Li_3Bi superstructures $AE_3[MN_4]$, which are discussed in separate sections.

Li_2O-Type Defect and Order Variants All phases containing Li and M in exclusively tetrahedral coordination crystallize in ordered or defect variants $Li_{2-x-y}M_x\square_yN$ of the Li_2O-type structure [840]. Generally, the molar ratio Li:M within a given system shrinks with raising preparation temperatures, that is, $Li_{15}[Cr^{VI}N_4]_2N$ [57] (730 K) to $Li_6[Cr^{VI}N_4]$ [57] (850 K) [841]. Isolated tetrahedra $[MN_4]^{n-}$ are present in the phases $Li_8[Ti^{IV}N_4]$ [383], the three different polymorphs of α-, β-, γ-$Li_7[V^VN_4]$ [842–845], $Li_7[Nb^VN_4]$ [844,846], $Li_7[Ta^VN_4]$ [847], $Li_{15}[Cr^{VI}N_4]_2N$ [57], $Li_6[Cr^{VI}N_4]$ [57], $Li_6[Mo^{VI}N_4]$ [57], $Li_6[W^{VI}N_4]$ [57,58], $Li_7[Mn^VN_4]$ [752,845,848], and $Li_5[Re^{VII}N_4]$ [849]. $Li_5Mg[VN_4]$, also containing an alkaline earth metal cation, fits into this description as well [850]. Chains of vertex- and edge-sharing tetrahedra are observed in $Li_4[TaN_3]$ [851] and $Li_3[FeN_2]$ [87,101], respectively.

Two nitridometalates are known with an ordered 3D anion: $Li_3[ScN_2]$ [852] is an isotype to $Li_3[AlN_2]$ [853,854] and $Li_3[GaN_2]$ [853,854] and crystallizes cubic in a $2\times2\times2$ superstructure of Li_2O. Whereas the Sc arrangement is topologically equivalent to the Si arrangement in γ-Si [855], no similar SiO_2 modification for $[ScN_2]^{3-}$ is known. The $[ZnN^-]$ substructure in $Li[ZnN]$ [856,857] corresponds to a Sphalerite-type arrangement, the remaining tetrahedral voids are occupied by Li.

In all of the above phases except $Li_7[MnN_4]$ and $Li_3[FeN_2]$, the transition metal M features the highest possible oxidation state, the samples often contain transparent yellow to orange crystals and reveal diamagnetic behavior. In $Li_7[MnN_4]$, according to magnetic susceptibility data, Mn ($3d^2$) is in a $S=1$ state ($\mu_{eff} = 2.79\,\mu_B$/f.u.); below 8 K weak ferromagnetic interactions are observed corresponding to $\Theta = 10.3$ K. For $Li_3[FeN_2]$, magnetic susceptibility data indicate a $3d^5$ low-spin behavior and antiferromagnetic order below $T_N = 10$ K [87].

$Li_7[Mn^VN_4]$ decomposes at elevated temperatures to $Li_{24}[Mn^{III}N_3]_3N_2$ (1100 K $< T < 1115$ K), $Li_5[(Li_{1-x}Mn_x)N]_3$ (1123 K), and $Li_2[(Li_{1-x}Mn_x^I)N]$ (1173 K) [752]. Decomposition of $Li_3[FeN_2]$ proceeds via $Li_4[FeN_2]$ and $Li_2[(Li_{1-x}Fe_x)N]$ to metallic Fe [841]. All other phases decompose to binary nitrides upon losing Li at high temperatures, whereas a tendency to form intermediate phases containing Li with lower oxidation states of M has not been observed.

A systematic description of these Li_2O-type superstructures was proposed [383] on the basis of the order patterns of M and voids □ in the crystal structures, but also disordered structures may be fitted within this system.

Within the constraints of the Li_2O structure, full occupation of all positions is observed in the following phases containing isolated tetrahedra: α-$Li_7[V^VN_4]$ [843] (Figure 7.25a), β-$Li_7[V^VN_4]$ [842,843], and isotypes $Li_7[Nb^VN_4]$ [844,846] and $Li_7[Ta^VN_4]$ [847] (Figure 7.25b), as well as γ-$Li_7[V^VN_4]$ [843–845] and isotypic $Li_7[Mn^VN_4]$ [752,845,848] (Figure 7.25c). A completely filled Li_2O structure with isolated $[MN_4]$ tetrahedra and further isolated anions is realized in the isotypic phases $Li_{16}[NbN_4]_2O$ [858,859] and $Li_{16}[TaN_4]_2O$ [860] (Figure 7.25d). Full occupation of all positions is also realized in $Li_3[FeN_2]$ [87,101] (Figure 7.25e), containing 1D chains of edge-sharing tetrahedra as well as in $Li_3[ScN_2]$ [852] (Figure 7.25f) and $Li[ZnN]$ [856,857] (Figure 7.25g) with 3D networks of corner-sharing tetrahedra of different complexity, respectively.

Ordering of both isolated tetrahedra $[MN_4]$ and voids is observed in the isotypic phases $Li_6\square[Cr^{VI}N_4]$ [57], $Li_6\square[Mo^{VI}N_4]$ [57], and $Li_6\square[W^{VI}N_4]$ (Figure 7.26a) [57,58], as well as in $Li_5\square_2[Re^{VII}N_4]$ (Figure 7.26b) [849] and $Li_{14}\square_2[CrN_4]_2O$ (Figure 7.26c) [57,861]. In $Li_4\square[TaN_3]$ (Figure 7.26d) [851], chains of vertex-sharing tetrahedra $[TaN_2N_{2/2}]$ and voids are observed. In $Li_{15}\square[Cr^{VI}N_4]_2N$ (Figure 7.27a) [57] and $Li_5Mg\square[V^VN_4]$ (Figure 7.27b) [850] – representing the endmember of the substitution series $Li_{7-2x}Mg_x\square_x[VN_4]$ – the largest parts of the structures are perfectly ordered, but in each phase one of the sites is only half occupied resulting in disordered Li/□ and Mg/□ substructures, respectively.

7.3 Structural Chemistry of Nitrides | **311**

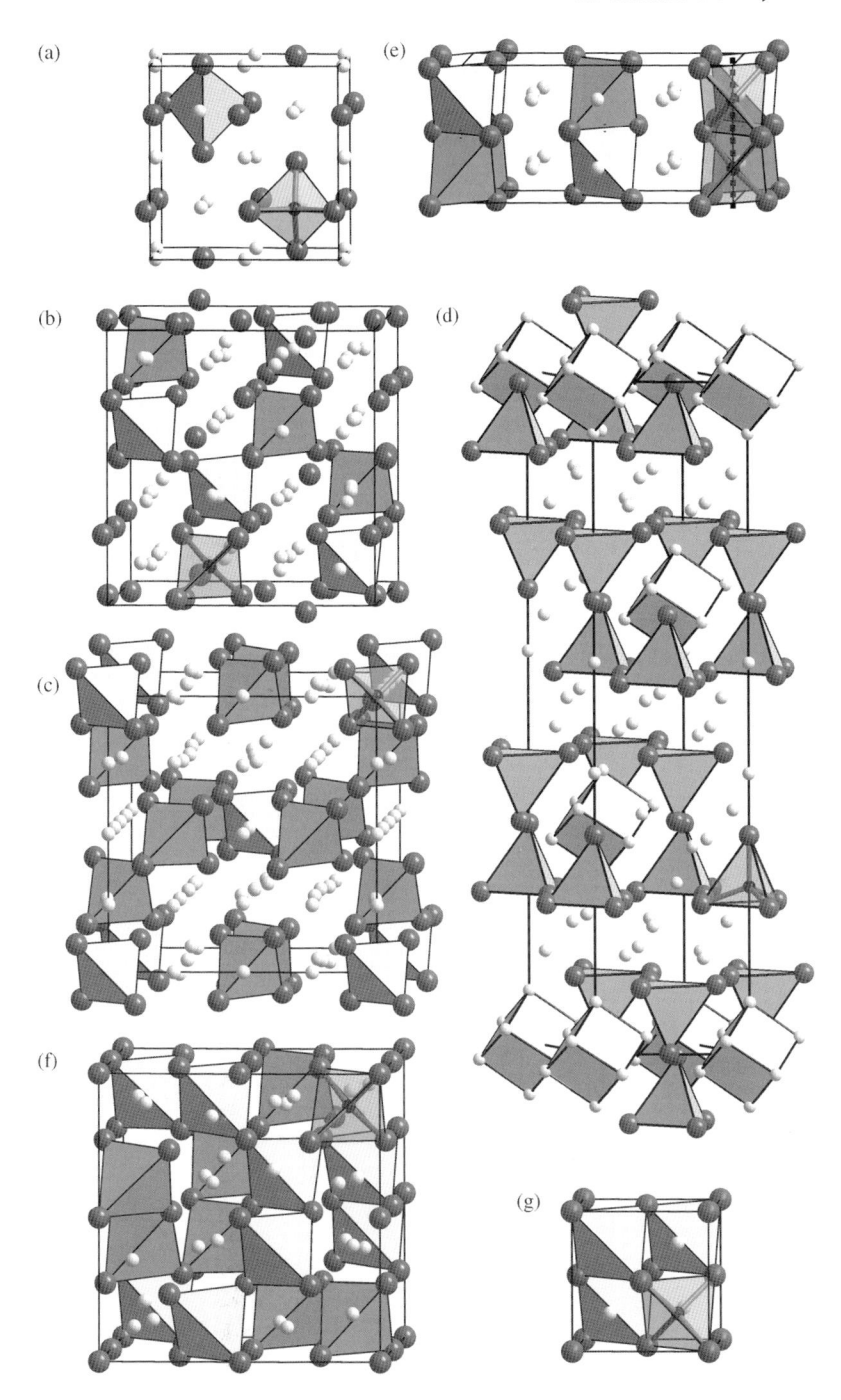

Figure 7.25 Order patterns of [MN_4] (gray tetrahedra) and isolated anions (white cubes) in the completely ordered and filled crystal structures of (a) α-Li_7[V^VN_4]; (b) β-Li_7[V^VN_4]; (c) γ-Li_7[V^VN_4]; (d) Li_{16}[TaN_4]$_2$O, cubes OLi_8 emphasized; (e) Li_3[FeN_2]; (f) Li_3[ScN_2]; and (g) Li[ZnN]. For corresponding isotypes, see text. Small black spheres: transition metal atoms; small white spheres: Li; larger dark gray spheres: N.

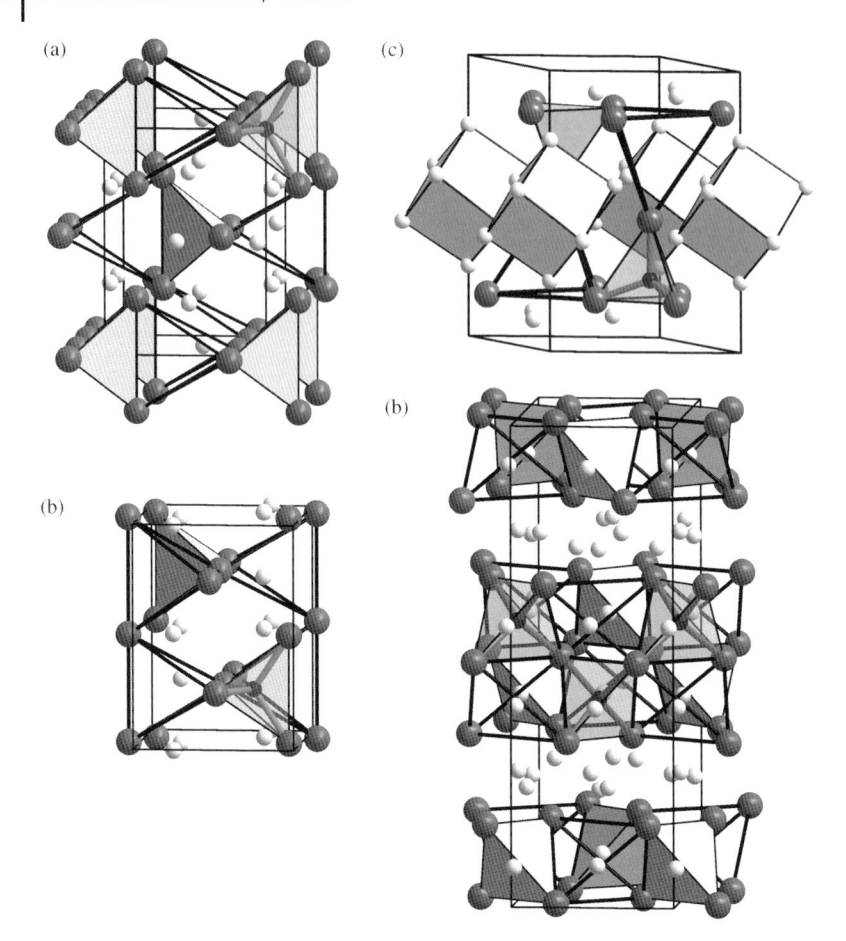

Figure 7.26 Order patterns of [MN_4] (gray tetrahedra), tetrahedral voids (empty black tetrahedra), and isolated anions (white cubes) in the crystal structures of (a) $Li_6[Cr^{VI}N_4]$, (b) $Li_5[Re^{VII}N_4]$, (c) $Li_{14}[CrN_4]_2O$, and (d) $Li_4[TaN_3]$. For corresponding isotypes, see text. Small black spheres: transition metal atoms; small white spheres: Li; larger dark gray spheres: N; cubes are centered by O.

Besides these phases featuring exclusively ordered M sites, phases with M/Li substitution sites are also known. Here a definite assignment within the order patterns is difficult. Due to M/Li substitution, M site occupation is systematically overestimated and therefore connection patterns are more complex: Charge-balanced $Li_5[TiN_3]$ (Figure 7.27c) [862], isotypic to $Li_5[SiN_3]$ [863] and $Li_5[GeN_3]$ [864], has been investigated in detail [383]. According to neutron diffraction and ^{7}Li NMR spectroscopy, one position is occupied by 2/3 Ti and 1/3 Li, resulting in a 3D pattern of edge- and corner-sharing tetrahedra.

Upon heating of $Li_7[VN_4]$, at 1370 K charge-balanced $Li_{18}[V_3N_{11}]$ and at about 1670 K $Li_4[VN_3]$ are obtained. Their crystal structures are closely related to $Li_{15}[Cr^{VI}N_4]_2N$ and $Li_{14}\square_2[CrN_4]_2O$, respectively; however, both show

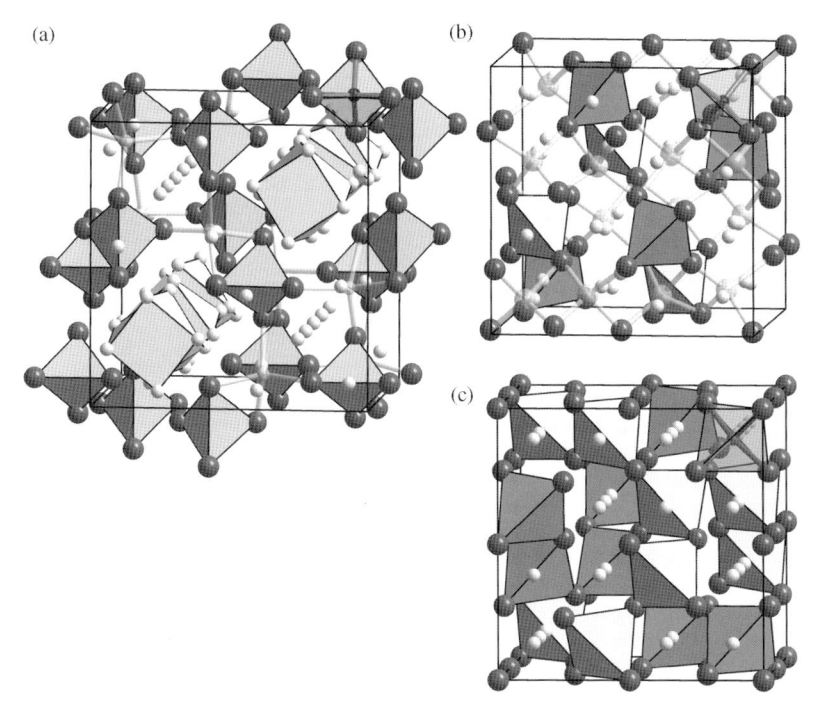

Figure 7.27 Order patterns of [MN_4] (gray tetrahedra), isolated anions (white cuboids), and partial occupation sites (half transparent atoms with half transparent bonds) in the crystal structures of (a) $Li_{15}[Cr^{VI}N_4]_2N$ and (b) $Li_5Mg[VN_4]$; and mixed occupation sites [M/LiN_4] (gray tetrahedra) in (c) $Li_5[TiN_3]$. Small black spheres: transition metal atoms; larger light gray spheres: Mg; small white spheres: Li; larger dark gray spheres: N.

pronounced V/Li disorder as well as partial occupancy on several sites well in accordance with a charge-balanced model [383]. Other samples Li-V-N quenched from 1370 K showed only the X-ray diffraction pattern of the Li_2O parent structure [844], so these highly disordered phases may represent examples for decreasing order with increasing temperatures upon crystallization. Similar observations were also reported on the completely disordered (Li_2O) nitridometalate oxides $Li_{11}[NbN_4]O_2$ [865], $Li_{10}[CrN_4]O_2$ [861], and $Li_{7.9}MnN_{5-y}O_y$ [848,866].

Lithium extraction in an electrochemical cell provides access to higher oxidation states otherwise not achieved: For $Li_{7-x}[MnN_4]$, a composition of $Li_{5.45}[Mn^{+6.55}N_4]$ was reported [100]; for $Li_{3-x}[FeN_2]$, less-pronounced results were achieved [101].

Other Phases Containing Li, M, and N In contrast to the already discussed phases crystallizing in Li_2O superstructures, most other nitridometalates with M in tetrahedral coordination and containing Li crystallize in unique structures and need to be discussed on the basis of the connectivity of the [MN_4] tetrahedra.

The quaternary nitrides $Li_3Sr_2[NbN_4]$ [867] and $Li_3Sr_2[TaN_4]$ [867] are isotypic to orthorhombic $Li_3Na_2[FeO_4]$ [868] and contain isolated tetrahedra

$[M^VN_4]^{7-}$. Together with the also tetrahedrally coordinated Li atoms, a 3D arrangement of edge- and corner-sharing tetrahedra is formed with Sr atoms occupying channels formed by N atoms (Figure 7.28a). Magnetic susceptibility data for $Li_3Sr_2[NbN_4]$ between 4 and 300 K exhibit temperature-independent paramagnetism [867].

The isotypic structures $Li_3Ba_2[NbN_4]$ [869] and $Li_3Ba_2[TaN_4]$ [869,870] crystallize in a proprietary structure type. The barium compounds also feature isolated tetrahedra $[M^VN_4]^{7-}$; however, in contrast to $Li_3Sr_2[MN_4]$, not all Li are also tetrahedrally coordinated, thereby forming columns of edge-sharing tetrahedra. These columns are connected by (distorted) trigonally coordinated Li resulting in a 3D network with channels occupied by barium (Figure 7.28b).

Both $Li_2Sr_5[MoN_4]_2$ [826] (Figure 7.28c) and $Li_2Sr_5[WN_4]_2$ [826] (Figure 7.28d) crystallize in related proprietary structure types. They feature isolated tetrahedra $[M^{VI}N_4]^{6-}$, which are connected by trigonally and tetrahedrally

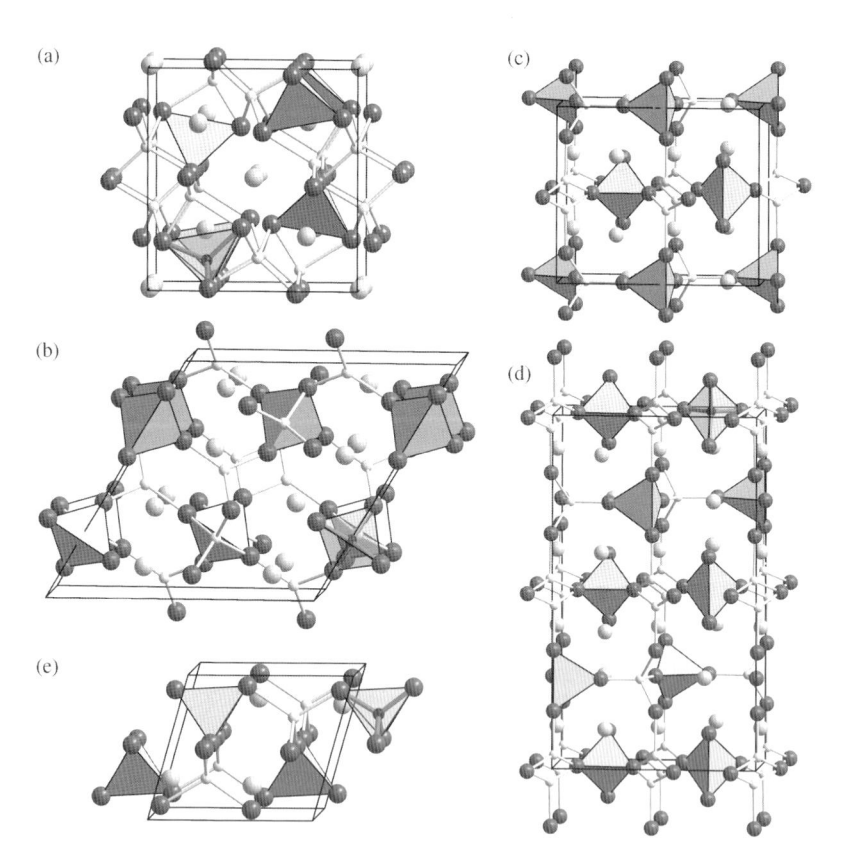

Figure 7.28 Sections of the crystal structures of quaternary nitridometalates containing Li. (a) $Li_3Sr_2[NbN_4]$. (b) $Li_3Ba_2[NbN_4]$. (c) $Li_2Sr_5[MoN_4]_2$. (d) $Li_2Sr_5[WN_4]_2$. (e) $LiSr_2[ReN_4]$. For corresponding isotypes, see text. Small black spheres: transition metal atoms; larger light gray spheres: A_E, small white spheres: Li, larger dark gray spheres: N.

7.3 Structural Chemistry of Nitrides

coordinated Li to form a 3D network with voids occupied by strontium. Whereas Li$_2$Sr$_5$[WN$_4$]$_2$ crystallizes centrosymmetric in *Pbcm*, Li$_2$Sr$_5$[MoN$_4$]$_2$ crystallizes acentric in *Pmc*2$_1$ resulting in a halved unit cell.

In the crystal structures of the isotypic quaternary nitridorhenates(VII) LiSr$_2$[ReN$_4$] [871] and LiBa$_2$[ReN$_4$] [871], the alkaline earth metals together with the nitride ions are arranged in the motif of the InNi$_2$-type structure. N builds up layers of edge-sharing trigonal prisms centered by the alkaline earth metals. One half of the tetrahedral voids between these layers is alternately occupied by rhenium, thereby forming [Re^{VII}N$_4$]$^{5-}$ tetrahedra; lithium takes the positions of the remaining tetrahedral sites (Figure 7.28e).

Higher ratios M/AE lead to condensation of the [MN$_4$] tetrahedra. In LiBa$_4$[Mo$_2$N$_7$] [872] and LiBa$_4$[W$_2$N$_7$] [872], [M_2^{VI}N$_7$]$^{9-}$ units are connected via tetrahedrally coordinated Li to layers; the Ba atoms are incorporated in the structure to complete the generally octahedral coordination spheres of nitrogen (Figure 7.29a).

The quaternary hexanitridodichromate(V) Li$_4$Sr$_2$[Cr$_2$N$_6$] [873] contains isolated complex anions [Cr$_2$N$_6$]$^{8-}$ consisting of two tetrahedra [CrN$_4$] sharing a common edge. Strontium is located in a distorted octahedral and lithium in a distorted tetrahedral coordination by nitrogen (Figure 7.29b). The diamagnetic behavior of Li$_4$Sr$_2$[Cr$_2$N$_6$] corresponds well with a strong AFM coupling of the Cr^V species in the complex anions [Cr$_2$N$_6$]$^{8-}$ corresponding to bonding interactions Cr–Cr as indicated by the analysis of the electron localization function (ELF) [873].

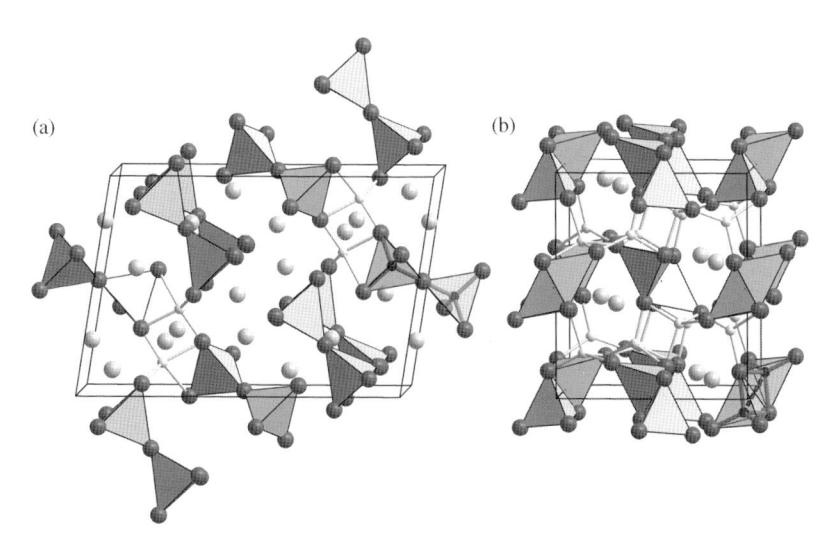

Figure 7.29 Sections of the crystal structures of quaternary nitridometalates with extended complex anions containing Li. (a) LiBa$_4$[M_2N$_7$]. (b) Li$_4$Sr$_2$[Cr$_2$N$_6$]. For corresponding isotypes, see text. Larger light gray spheres: A_E; small white spheres: Li; larger dark gray spheres: N; tetrahedra are centered by transition metal atoms.

Nitridometalates Containing Heavier Alkali Metals The heavier alkali metals are too large to fit into the tetrahedral voids of the Li_2O superstructure, so no structures with isolated [MN_4] tetrahedra are realized. Instead, nitridometalate tetrahedra linked via common vertices are observed in several group 6 compounds. Phases containing infinite chains with quite different conformations are featured in the isotypic phases $Na_3[MoN_3]$ [76] and $Na_3[WN_3]$ [76,874] (Figure 7.30a), Na_2K [WN_3] (Figure 7.30b) [96], isotypic $Na_5Rb[(WN_3)_2]$ [875] and Na_5Cs [$(WN_3)_2$] [875] (Figure 7.30c), and $Na_{11}Rb[(WN_3)_4]$ (Figure 7.30d) [96].

Going the lower atomic ratios A:M, in $K_{14}[W_6N_{16}NH]$ [876] [$WN_2N_{2/2}$] tetrahedra build six-membered rings in approximate boat confirmation that are connected via the NH function to form 1D bands (Figure 7.31a). Further condensation of these bands via another tetrahedron to 2D layers is achieved in $Na_2K_{13}[W_7N_{19}]$ (Figure 7.31b) [877]. In the crystal structures of both $NaCs_5[W_4N_{10}]$ [878] and the isotypic phases $Rb_{9+x}[W_6N_{15}]$ [879], $Cs_{9+x}[Mo_6N_{15}]$ [879], and $Cs_{9+x}[W_6N_{15}]$ [879], all nitridometalate tetrahedra are

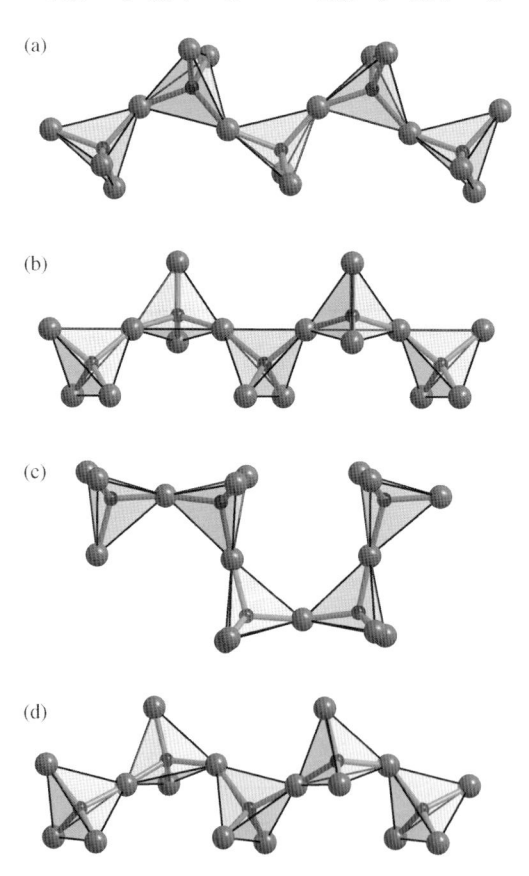

Figure 7.30 Infinite chains of vertex-sharing tetrahedra in the crystal structures of (a) $Na_3[MoN_3]$, (b) $Na_2K[WN_3]$, (c) $Na_5Rb[(WN_3)_2]$, and (d) $Na_{11}Rb[(WN_3)_4]$. For corresponding isotypes, see text. Small black spheres: transition metal atoms; larger dark gray spheres: N.

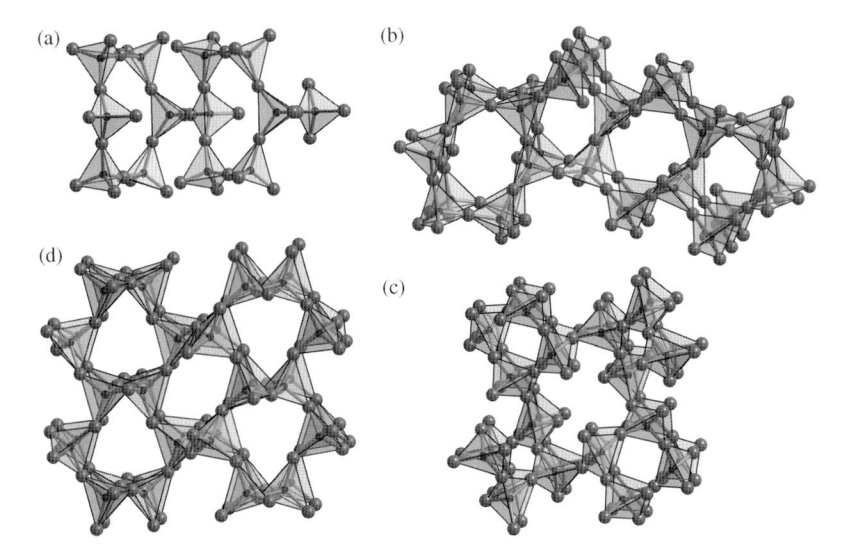

Figure 7.31 Dimensionalities of anionic tetrahedral networks in nitridometalates and related compounds. (a) $K_{14}[W_6N_{16}NH]$. (b) $Na_2K_{13}[W_7N_{19}]$. (c) $NaCs_5[W_4N_{10}]$. (d) $Rb[TaN_2]$. For corresponding isotypes, see text. Small black spheres: transition metal atoms; larger dark gray spheres: N.

linked at three vertices to three other tetrahedra. The resulting network of six rings (Figure 7.31c) corresponds to the anionic part of the synthetic silicate $K_2Ce[Si_6O_{15}]$ [880]; the nature of the partial occupation of the cationic structure in $A_{9+x}[W_6N_{15}]$ is still unclear. The crystal structures of $K[TaN_2]$ [78], $Rb[TaN_2]$ [78], $Cs[NbN_2]$ [881], and $Cs[TaN_2]$ [78] correspond to a filled ß-cristobalite type (Figure 7.31d); however, the phases containing K and Rb feature an orthorhombic distortion and have not been investigated in detail until now. While all these compounds containing Nb, Ta, Mo, and W in their highest oxidation states +5 or +6, respectively, are bright red, orange, or yellow, the reduced Mo and W compounds feature dark black, red, or brownish colors.

Alkaline Earth and Rare Earth Metal Nitridometalates In alkaline earth metal containing compounds, generally the same trends compared to alkali metal phases prevail. For nitridometalates of transition metals, numerous examples of anionic substructures containing isolated tetrahedra, oligomeric units of edge- or vertex-sharing tetrahedra, and infinite chains have been reported; in contrast, in the respective nitridometalates of main group metals, 2D layered units and 3D frameworks are also observed.

The lower dimensionality of the complex anions in systems containing highly charged transition metal as well as alkaline earth metal cations might be attributed to a lower stability of these phases. In most nitridometalates featuring low molar ratios $A_E:M \leq 1$, the transition metal is in five- or sixfold coordination by N (see Sections 7.3.3.6 and 7.3.3.7). In case the ratio $M:N > 4$, the additional N atoms are located within NA_{E_6} octahedra.

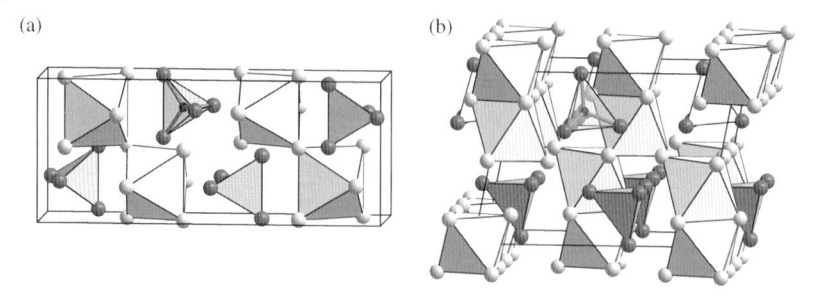

Figure 7.32 The crystal structures of (a) Ca$_5$[VN$_4$]N and (b) Ca$_5$[NbN$_4$]N contain [MN$_4$] tetrahedra and NCa$_6$ octahedra. For corresponding isotypes, see text. Small black spheres: transition metal atoms; larger light gray spheres: Ca; larger dark gray spheres: N.

Isolated tetrahedra [MN$_4$] besides octahedra NAE_6 are observed in the isotypic phases Ca$_5$[VN$_4$]N [882] and Sr$_5$[NbN$_4$]N [883] (Figure 7.32a), as well as the isotypic phases Ca$_5$[NbN$_4$]N [884], Ba$_5$[VN$_4$]N [885], Ba$_5$[NbN$_4$]N [886], and Ba$_5$[CrN$_4$]N [887] (Figure 7.32b), the latter of black color and containing a not fully oxidized Cr^V.

Exclusively isolated tetrahedra [MN$_4$]$^{x-}$ (Figure 7.33a) are observed in a plethora of phases, among them being Ca$_4$[TiN$_4$] [884], which crystallizes isotypic to Na$_4$[TiO$_4$] [888], and Ba$_7$[V$_{1-x}$Ta$_x$N$_4$]$_2$ [885], which is isotypic to Cs$_7$[W$_2$N$_3$O$_5$] [889].

Taking only the metal positions into account, most phases of AE_3[MN$_4$] composition crystallize in one of three different superstructures of the Na$_3$As type with M building up a *hcp* network and AE occupying all trigonal-planar (the common face of the octahedral voids) and tetrahedral voids. The structures differ only in the arrangement of the N atoms, which generally maintain coordination spheres in the form of – sometimes highly distorted – octahedra. Isotypic Sr$_3$[CrN$_4$] [890], Ca$_2$Sr[WN$_4$] [891], (Ca, Ba)$_3$[MoN$_4$] [892], Ca, Ba)$_3$[WN$_4$] [892], LT-(β)-Sr$_3$[MoN$_4$] [893,894], (Sr, Ba)$_3$[MoN$_4$] [895], (Sr, Ba)$_3$[WN$_4$] [895], α-Ba$_3$[MoN$_4$] [896–898], and α-Ba$_3$[WN$_4$] [896,899]

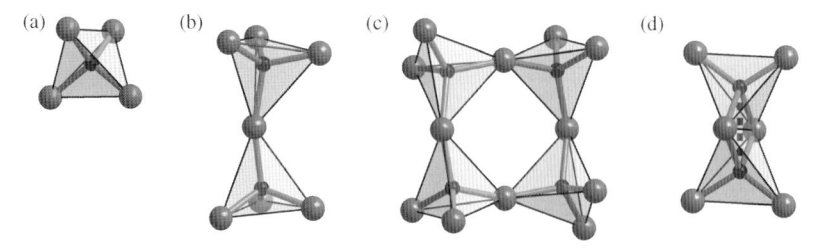

Figure 7.33 Tetrahedral anions in ternary nitridometalates (a) [MN$_4$]$^{x-}$ (M = Ti, V, Nb, Ta, Cr, Mo, W); (b) [Nb$_2$N$_7$]$^{x-}$; (c) [Ti$_4$N$_{12}$]$^{x-}$; (d) [M_2N$_6$]$^{x-}$ (M = Cr, Mo, Mn). Small black spheres: transition metal atoms; larger dark gray spheres: N.

crystallize in the Na$_3$[NO$_4$]-type structure [900]. The isotypic high-temperature phases β-Ba$_3$[MoN$_4$] [896–898] and β-Ba$_3$[WN$_4$] [896,899] as well as LT-α-Sr$_3$[MoN$_4$] [893,894] represent unique structure types. In contrast, the heavy atom arrangement in CaBa$_2$[WN$_4$] [901] corresponds to the Li$_3$Bi type with W building up a *fcc* network in which half of the Ba occupy the octahedral voids and Ca and the other half of Ba occupying the tetrahedral voids. Despite the different metal coordination, the coordination spheres of both Ba atoms by N are strikingly similar.

It might be worthwhile to note that some nitridometalate oxides also crystallize in superstructures derived from simple A_xB_y types, for example, (Li$_2$Ca$_4$O)$_3$[MN$_4$]$_4$ (M = Mo [902], W [902], Re [903]), in which the polyhedra centers are arranged according to the Th$_3$P$_4$-type structure.

Condensation of [MN$_4$] tetrahedra via common vertices to oligomeric units is quite rare: In Ba$_{16}$[NbN$_4$]$_3$[Nb$_2$N$_7$] [904], both single [NbN$_4$]$^{7-}$ and double [Nb$_2$N$_7$]$^{11-}$ (Figure 7.33b) units are present, whereas in Ba$_{10}$[Ti$_4$N$_{12}$] [905] tetranuclear cyclic units [Ti$_4$N$_{12}$]$^{20-}$ (Figure 7.33c) are observed.

Isolated double tetrahedra [M_2N$_6$]$^{8-}$ (Figure 7.33d) with a common edge as observed in Li$_4$Sr$_2$[Cr$_2$N$_6$] [873] (see Section "7.3.3.5 Other Phases Containing Li, M, and N") are associated with M^V and a strong Cr–Cr bonding interaction. In accordance with the Cr^V state, the phases Ca$_4$[Cr$_2$N$_6$] (Figure 7.34a) [906] and Sr$_4$[Cr$_2$N$_6$] [906] show a brilliant green color. In contrast, orange Sr$_4$[Mn$_2$N$_6$] (Figure 7.34b) [81] crystallizes in a twofold unit cell compared to Sr$_4$[Cr$_2$N$_6$], but does not show any sign of metal–metal bonding. Sr$_{10}$[Mo$_2$N$_6$][MoN$_4$]$_2$ [894] is the sole example of a phase containing both Mo^V and Mo^{VI} in tetrahedral coordination by N. It may be regarded as "2 Sr$_3$[Mo^{VI}N$_4$]·Sr$_4$[Mo$_2^V$N$_6$]", the double tetrahedra [Mo$_2$N$_6$]$^{8-}$ with a common edge is again associated with M^V and a strong bonding interaction.

Chains of vertex-sharing tetrahedra are observed in the isotypic phases Ca$_2$[VN$_3$] [88], Sr$_2$[VN$_3$] [907], Sr$_2$[NbN$_3$] [908], Sr$_2$[TaN$_3$] [909,910],

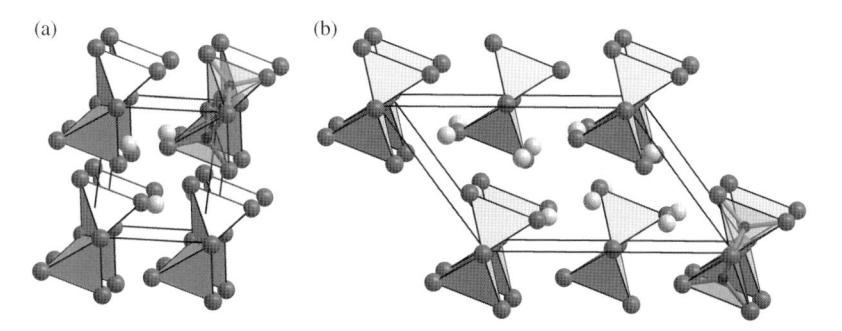

Figure 7.34 Edge-sharing double tetrahedra in the crystal structures of (a) Ca$_4$[Cr$_2$N$_6$] and (b) Sr$_4$[Mn$_2$N$_6$]. For corresponding isotypes, see text. Small black spheres: transition metal atoms; larger light gray spheres: AE; larger dark gray spheres: N.

Ba$_2$[NbN$_3$] [911], and Ba$_2$[TaN$_3$] [909] (Figure 7.35a); a different arrangement of the chains is realized in Ba$_2$[VN$_3$] (Figure 7.35b) [907]. All these fully oxidized phases form yellow to orange transparent crystals.

Ba$_4$[Mn$_3$N$_6$] (Figure 7.35c) [912], a mixed-valent nitridomanganate with a one-dimensional anionic framework, contains infinite corrugated [MnN$_{4/2}$]$^{8/3-}$ chains of edge-sharing tetrahedra with an ordered arrangement of Mn(II) and Mn(IV) atoms featuring semiconducting behavior and a spin-chain behavior with an AFM ordering at 68 K.

Whereas no nitrodometalates with alkaline earth metals and two-dimensional complex anions with M in exclusively tetrahedral coordination are known, several different examples containing three-dimensional frameworks have been reported. Due to similar coordination spheres and potential substitution of the M site, Mg is considered as part of the anionic partial structure.

In the crystal structure of MgEu$_4$[TaN$_4$]N [913], isolated tetrahedra [TaN$_4$]$^{5-}$ are observed. Taking Mg into account, zigzag chains of vertex-sharing tetrahedra [MgN$_{4/2}$] connect the [TaN$_2$N$_{2/2}$] units via two common vertices to a 3D network (Figure 7.36a); according to this description, in Eu$_4$[TaMgN$_5$] no isolated

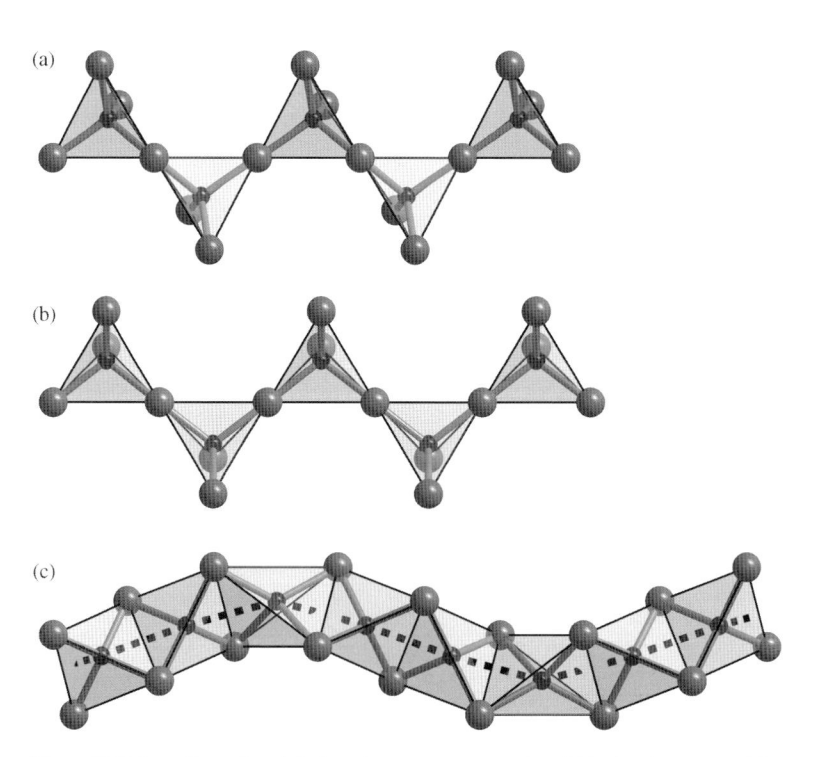

Figure 7.35 One-dimensional chains of vertex- and edge-sharing tetrahedra in the crystal structures of (a) Ca$_2$[VN$_3$], (b) Ba$_2$[VN$_3$], and (c) Ba$_4$[Mn$_3$N$_6$]. For corresponding isotypes, see text. Small black spheres: transition metal atoms; larger light gray spheres: AE; larger dark gray spheres: N.

nitride species are present. The N-coordination of the europium atoms is rather complex.

Isotypic Ba[Mg$_{3.33}$Nb$_{0.67}$N$_4$] [913], Ba[Mg$_{3.33}$Ta$_{0.67}$N$_4$] [913], and Sr[Mn$_4$N$_4$] [81] (Figure 7.36b) crystallize in the UCr$_4$C$_4$-type structure, which consists of a 3D network of chains of edge-sharing tetrahedra along [001] connected to other chains via common vertices. The framework contains small and large tetragonal channel-like voids. Whereas the small channels are empty, the large voids are occupied by Sr in eightfold coordination by N.

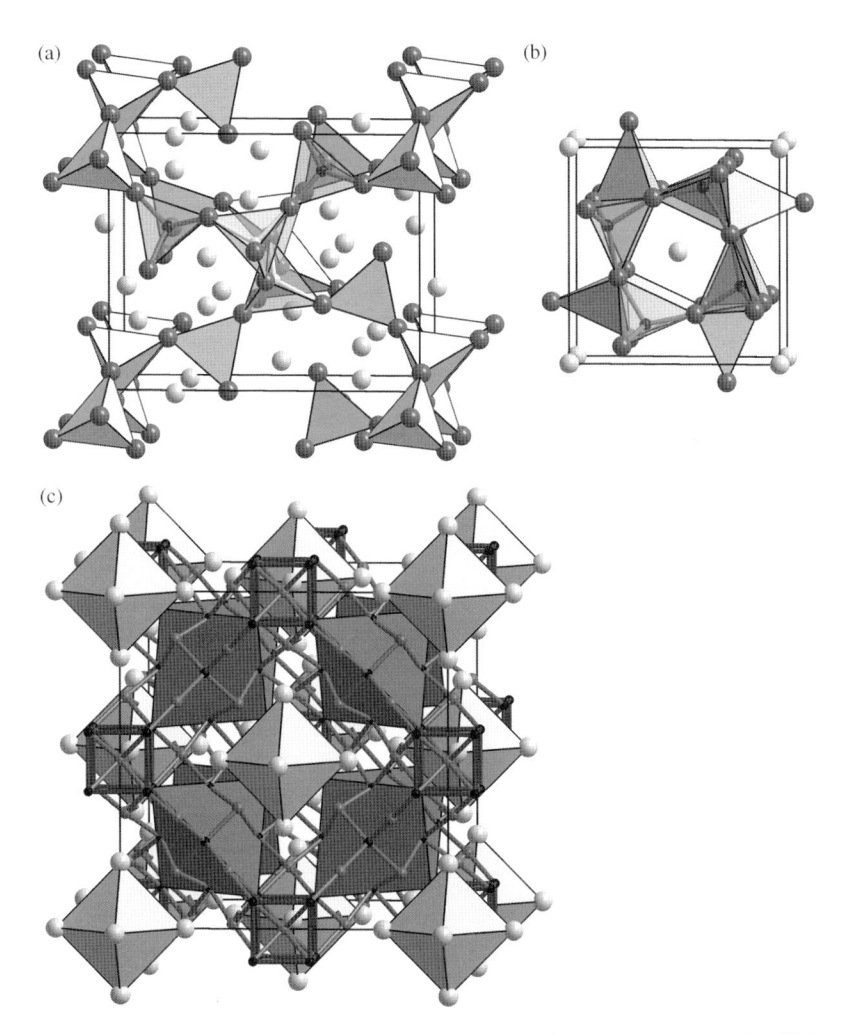

Figure 7.36 Three-dimensional anionic networks of vertex- and edge-sharing tetrahedra in the crystal structures of (a) Eu$_4$[TaMgN$_5$]; (b) Sr[Mn$_4$N$_4$], and (c) La$_6$[Cr$_{20-x}$N$_{22}$]. For corresponding isotypes, see text. Small black spheres: transition metal atoms; larger light gray spheres: A_E or R_E; larger dark gray spheres: N.

In the rare earth metal compounds $La_6Cr_{20-x}N_{22}$ [837–839], $Ce_6Cr_{20-x}N_{22}$ [838,839], and $Pr_6Cr_{20-x}N_{22}$ [838,839], the metal substructure corresponds to the Cu_2AlMn-type structure – a Heusler-phase – with RE_6 octahedra and Cr_8 cubes occupying the Al and Mn positions, and two Cr_6 octahedra the Cu positions, respectively. All Cr atoms are tetrahedrally coordinated by N. However, this description fails to hit the mark (see Section 7.3.3.4), since the Cr atoms of a Cr_6 octahedron and the neighboring corners of the four surrounding Cr_8 cubes form a Cr_{10} supertetrahedral cluster that is connected to N only on its outer faces. In contrast to the alkaline earth metal nitridomanganates, $Sr_6[Mn_{20}N_{21}]$ and $Ba_6Mn_{20}N_{21.5-\delta}$ discussed in Section 7.3.3.4, in $La_6Cr_{20-x}N_{22}$ the center of the Cr cube is fully occupied by N, while the positions forming the Cr_8 cube are only occupied to about 80%; otherwise, the structure follows the description given above (Figures 7.24 and 7.36c).

Square-Planar Coordination

Square-planar coordination of M by N is observed exclusively in the isotypic phases $Ce_2[CrN_3]$ [839], $U_2[CrN_3]$ [914], $Th_2[CrN_3]$ [914], $Ce_2[MnN_3]$ [915], $U_2[MnN_3]$ [914], and $Th_2[MnN_3]$ [914], which crystallize in the Sr_2CuO_3-type structure [916]. Earlier reported phases with the tentative compositions "$Ce_3Cr_2N_5$" [838] and "$UCrN_{2\pm x}$" [917] also turned out to be likely of the formula type $RE_2[CrN_3]$.

While there is little known about the compounds containing 5f elements (R = U, Th), the dark black cerium nitridometalates $Ce_2[MN_3]$ of Cr and Mn were investigated in greater detail [114,915,918–920]. Predominant features of these phases (Figure 7.37) are one-dimensional planar chains of vertex-sharing nearly square units $^1_\infty[MN_2N_{2/2}]$ arranged in the motif of a hexagonal rod packing and bridged by the cationic species.

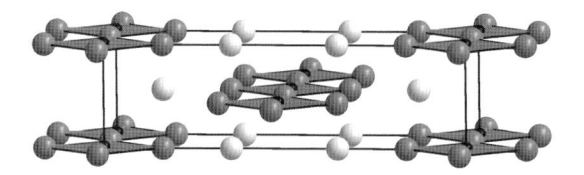

Figure 7.37 The crystal structure of $Ce_2[MnN_3]$ and its isotypes is comprised of hexagonal close-packed planar chains [$MnN_2N_{2/2}$] bridged by Ce atoms. Small black spheres: Mn; larger light gray spheres: Ce; larger dark gray spheres: N.

Terminal M–N contacts are longer than contacts within the chains, corresponding to single-bond and substantial double-bond character, respectively, well in accordance with *extended Hückel* calculations. Electric resistivity measurements suggest metallic behavior ($\rho \approx 1 \times 10^{-3}\,\Omega$ cm). The absence of a significant magnetic moment ($\mu_{\text{eff}} = 0.53\,\mu_B$) together with a clearly Ce^{4+} (4f^0) state deduced from XAS data indicates a low-valence situation for both Cr and Mn (3d^6) in full agreement with electronic structure calculations and the fact that the nitridomanganate develops a magnetic moment upon fluorination to Ce$_2$MnN$_3$F$_{2-\delta}$ [921].

Whereas Ce$_2$[CrN$_3$] decomposes at elevated temperatures to Ce$_3$Cr$_{10-x}$N$_{11}$ [838,839], no similar behavior was observed for Ce$_2$[MnN$_3$] [383].

7.3.3.6 CN(M) = 5

H																	He
Li	Be											B	C	**N**	O	F	Ne
Na	Mg											Al	Si	P	S	Cl	Ar
K	Ca	Sc	**Ti**	**V**	**Cr**	Mn	Fe	Co	Ni	Cu	Zn	Ga	Ge	As	Se	Br	Kr
Rb	**Sr**	Y	**Zr**	**Nb**	Mo	Tc	Ru	Rh	Pd	Ag	Cd	In	Sn	Sb	Te	I	Xe
Cs	**Ba**	La	**Hf**	**Ta**	W	Re	Os	Ir	Pt	Au	Hg	Tl	Pb	Bi	Po	At	Rn
Fr	Ra	Ac															

Ce	Pr	Nd	Pm	Sm	Eu	Gd	Tb	Dy	Ho	Er	Tm	Yb	Lu
Th	Pa	U											

Nitridometalates with transition metal species in fivefold coordination are rather rare; with alkaline earth metals only the isotypes Sr[TiN$_2$] [922,923], Ba[ZrN$_2$] [911], Ba[HfN$_2$] [924], and the solid solution Ba[(Zr$_{1-x}$Hf$_x$)N$_2$] [924] are known. These phases crystallize in the layered K[CoO$_2$]-type structure, which itself has rather few members, with the transition metal coordinated in a square-pyramidal fashion (Figure 7.38a). These polyhedra are edge-connected at the bases to form $^2_\infty[MN_{4/4}N^{2-}]$ layers with the apical N ligands in a checkerboard manner alternately located on the opposite sides of the layers. All isotypes were reported to exhibit temperature-independent paramagnetism above $T = 20$ K, even though M is formally in a d^0 state, indicating either Pauli or van Vleck paramagnetism. Since Ba[ZrN$_2$] originally was reported to exhibit a dark red color [911], metallic properties, that is, Pauli paramagnetism, can be excluded for this compound. Additionally, the Zr and Hf samples were reported to show superconductivity below $T_c = 8$ K [925]. Because neither the Zr nor the Hf containing phases were prepared as pure samples, one might speculate that these properties are due to impurities of binary transition metal nitrides. A high electric conductivity observed for Sr[TiN$_2$] as well as an only occasionally observed drop in the magnetic susceptibility [925] also seems likely to be caused by impurities, while thin films show semiconducting behavior [926]. Similarly, a high density of states at E_F appears to be an artifact from the electronic structure calculation [923], since several later studies indicated all nitrides of this section to represent semiconductors with indirect bandgaps of about 1 eV [927,928], but with excellent thermoelectric properties [929,930]. Thermoelectric transport properties of Sr[TiN$_2$] were recently investigated [931–933].

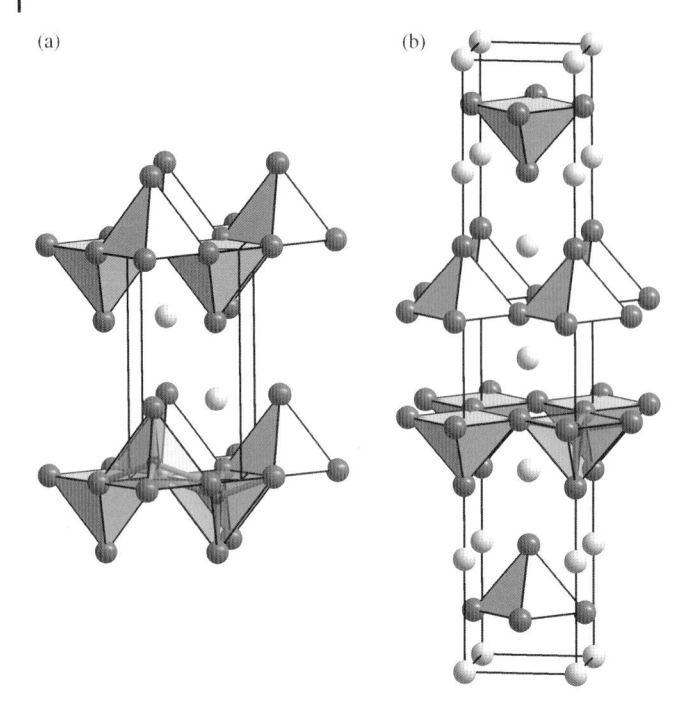

Figure 7.38 Sections of the crystal structures of (a) Sr[TiN$_2$] and (b) La$_3$[Cr$_2$N$_6$] as prototypes for fivefold coordination of M in nitridometalates. Small black spheres: transition metal atoms; larger light gray spheres: Sr or La; larger dark gray spheres: N.

The metallic compounds La$_3$[Cr$_2$N$_6$] [95], La$_3$[V$_2$N$_6$] [934], R_{E3}[Nb$_2$N$_6$], and R_{E3}[Ta$_2$N$_6$] with R_E = La, Ce, Pr [935] crystallize in the anion defect variant of the Ruddlesden–Popper phase Sr$_3$Ti$_2$O$_7$ also realized by the cuprate La$_{2-x}$Sr$_x$CaCu$_2$O$_6$ [936]. Layers of quadratic–pyramidally coordinated transition metal sandwich the rare earth metal ions in square prismatic and single-capped square antiprismatic coordination by nitrogen (Figure 7.38b). According to the classic counting approach, mixed valency is evident for all crystallographically equivalent independent $M^{IV,V}$.

7.3.3.7 CN(M) = 6

7.3 Structural Chemistry of Nitrides

With alkaline earth metals, disordered rock salt structures were found for, for example, $(Ca_{0.65}Th_{0.35})(N_{0.78}O_{0.18})$ [937], $Ca_xTi_{1-x}N$ [775], Ca_2NbN_3 [383], Ca_3UN_4 [938], and $CaCeN_2$ [939]. The latter three compounds apparently show no significant homogeneity ranges. However, the ordered rock salt variant LT-$SrCeN_2$ [939] with α-$NaFeO_2$-type structure (Figure 7.39a) at increased temperatures above 1000 K suffers a slow and continuous monotrope order–disorder transition to a high-temperature form, connected with some loss of Sr due to evaporation, leading to the formation of a rock salt-type solid solution $Sr_{1-x}Ce_{1+x}N_2$ ($x \approx 0.10$) [939]. In the disordered cubic rock salt-type phase

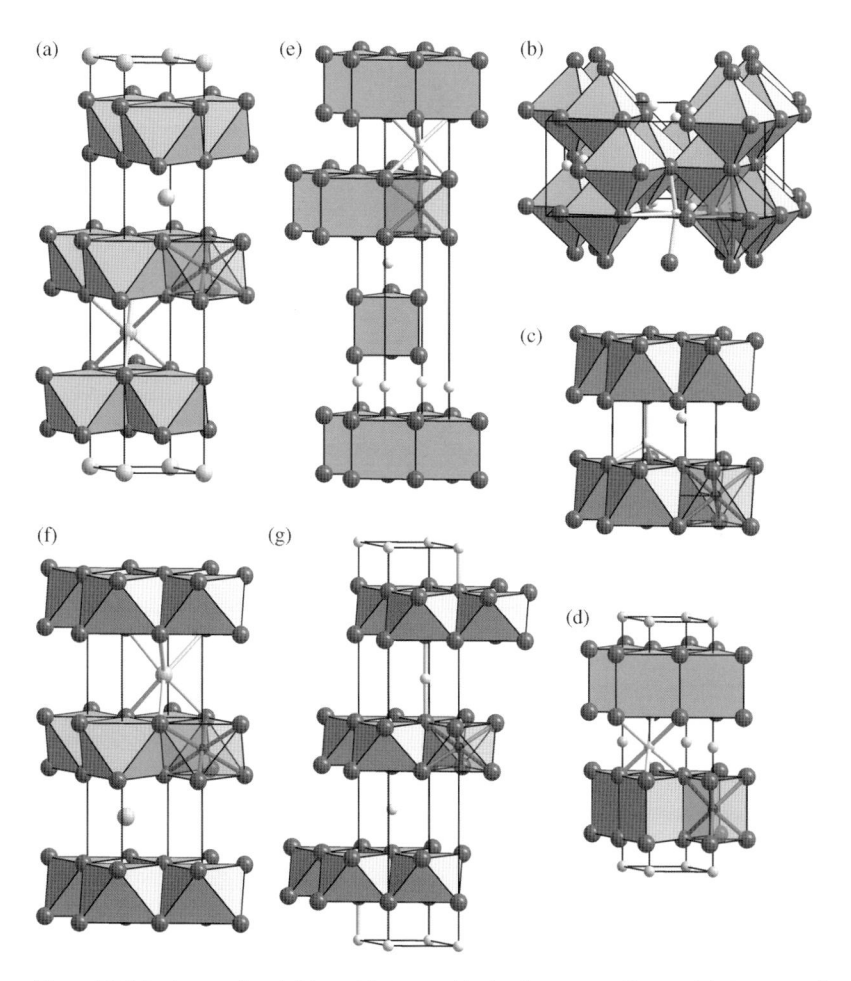

Figure 7.39 Prototypes for nitridometalates with M in sixfold coordination: (a) LT-$SrCeN_2$, (b) $Li_2Ta_3N_5$, (c) Li_2ZrN_2, (d) $FeMoN_2$, (e) $LiMoN_2$, (f) $MnMoN_2$, and (g) $CuTaN_2$; for non-nitride prototypes and isotypes, see text. Small black spheres: transition metal atoms; small light gray spheres: Li; larger light gray spheres: Sr, Fe, Mn, or Cu in (a), (d), (f), and (g), respectively; larger dark gray spheres: N.

Ca$_x$La$_{1-x}$N$_{1-x/3}$ ($0 < x < 0.7$) in contrast, the charge balance is attained by defects in the anionic substructure [940].

Li$_{2-x}$Ta$_{2+x}$N$_4$ ($0.2 \leq x \leq 1$) and Mg$_{2.6-x}$Ta$_{1.3+x}$N$_4$ ($0.3 \leq x \leq 1$) with disordered rock salt-type crystal structures realize rather large homogeneity ranges [941]. With $x = 0$, the compositions LiTaN$_2$ and approximately Mg$_2$TaN$_3$ with Ta in the oxidation state +5 would result, but as is indicated by the formula, both phases exhibit homogeneity ranges toward Ta excess, reducing the oxidation state of Ta. Li$_2$Ta$_3$N$_5$ [942] with a filled Ta$_3$N$_5$-type [51] crystal structure containing Ta and Li in sixfold, distorted octahedral coordination can be regarded as an ordered variant of the rock salt structure, thus as a low-temperature phase compared to the orthorhombic Li$_{2-x}$Ta$_{2+x}$N$_4$ ($x = 0.4$ Li$_2$Ta$_3$N$_5$) [941]. The unit cell of Li$_2$Ta$_3$N$_5$ (Figure 7.39b) is distorted to monoclinic symmetry compared to the orthorhombic crystal structure of Ta$_3$N$_5$, although electron diffraction gives indications for a monoclinic symmetry of Ta$_3$N$_5$ [201]. The reduced oxidation state of Ta (+4.3) is evidenced in short distances d (Ta–Ta), indicating directed Ta–Ta bonding interactions in the diamagnetic compound. For comparison, Na$_x$Ta$_3$N$_5$ ($0 \leq x \leq 1.4$) [943] crystallizes in a partially filled, but undistorted orthorhombic structure of Ta$_3$N$_5$. Na$_x$Ta$_3$N$_5$ was prepared from Na and Ta$_3$N$_5$ in an intercalation reaction, which consequently might conduct topotactically. Compared to Ta$_3$N$_5$, the phase with $x = 1$ (NaTa$_3$N$_5$) has only slightly different unit cell parameters, but appears to be a metallic electrical conductor, indicating that Ta–Ta directed bonding does not play a major role in this phase. Again, a disordered rock salt-type phase Na$_2$Ta$_3$N$_5$ can be obtained at higher temperatures [943].

Alkali and alkaline earth metal nitrides of the transition metals with the general formula AMN$_2$ or AEMN$_2$, respectively, frequently crystallize in the layered α-NaFeO$_2$ structure type, representing a rhombohedrally distorted order variant of a rock salt-type structure (Figure 7.39a), in which layers of A or AE and M alternate along [001]. This group comprises LT-SrCeN$_2$ [939], SrZrN$_2$ [944], SrHfN$_2$ [944], NaNbN$_2$ [945,946], NaTaN$_2$ [78], CaNbN$_2$ [947], CaTaN$_2$ [947–949], and LiUN$_2$ [950,951]. CaTaN$_2$ and CaNbN$_2$ contain the transition metal formally in a d^1 state [947–949]. These compounds show both metal-like electrical resistivities and superconducting transitions near $T_c = 9$ K; however, the superconductivity might have its origin in binary nitride impurities [947]. Higher synthesis temperatures apparently favor some Ca deficiency modifying the properties of Ca$_{0.74}$NbN$_2$ [947] and Ca$_{0.74}$TaN$_2$ [948,949]. The semiconducting LiUN$_2$ exhibits weak paramagnetism with an antiferromagnetic order below $T = 120$ K and localized magnetic moments at U below $\mu_{\text{eff}} = 1$ μ$_B$ [950,951].

Transition metals and rare earth elements in the oxidation state +4 (Zr [952–954], Hf [955], Ce [955,956], and U [950–952]) in ternary lithium nitrides of the formula type Li$_2M$N$_2$ frequently crystallize in an ordered (anti) La$_2$O$_3$-type structure (CaAl$_2$Si$_2$ structure type) (Figure 7.39c) with Li and M located at the tetrahedrally and octahedrally coordinated sites of oxygen, respectively. N occupies the La site in A-type La$_2$O$_3$ and forms an *hcp* arrangement.

7.3 Structural Chemistry of Nitrides

This structure can alternatively be derived from the CdI_2 type: Li is intercalated in all tetrahedral voids between $^2_\infty[MN^{2-}_{6/3}]$ layers of edge-sharing octahedra, which are stacked according to an *hcp* arrangement of N. A superstructure variant is realized in the crystal structure of Li_2ThN_2 [952,955].

Several ternary nitrides can be derived from the MoS_2-type structure, with one metal species in trigonal prismatic and the other in octahedral coordination, comprising three main structural groups based on the stacking of nitride ions:

In the crystal structures of $FeMoN_2$ [957,958] and $FeWN_2$ [516,957,959], the stacking of N is double primitive (... *AABB* ...) with tungsten located in trigonal prisms and iron in octahedra, resulting in face-sharing alternating polyhedra along the stacking direction ($MoNiP_2$ structure type) (Figure 7.39d). These compounds tolerate both Fe deficiency and substitution of some Fe by W or Mo in $Fe_{1-x}WN_2$, $Fe_{1-x}W_{1+x}N_2$, and $Fe_{1-x}Mo_{1+x}N_2$ with magnetic properties strongly dependent on the Fe content [958,960–962]. Additionally, $CoMoN_2$ and $FeMoN_2$, for example, were obtained as nanoparticles [963,964].

In $LiMoN_2$ [965], $LiWN_2$ [966], and $CrWN_2$ [723,967], N is stacked according to (... *AABBCC* ...), resulting in alternating layers of Mo and W in trigonal prismatic coordination and octahedrally coordinated Li and Cr atoms, respectively, however, arranged in a way that face-sharing pairs of octahedra and trigonal prisms appear ($AgCrSe_2$-type structure) (Figure 7.39e). For $LiMoN_2$, electronic structure calculations show the stability of this structure being due to strongly covalent bonding within layers, essentially described as $^1_\infty[MoN_2^-]$, and additional directed bonding between N atoms in opposing layers, although the compound exhibits three-dimensional metallic properties [968].

The largest group constitutes from $Li_{1-x}W_{1+x}N_2$ ($x = 0.16$) [969]. $LiNb_3N_4$ [183] and $Li_{1-x}Ta_{3+x}N_4$ ($x \ll 0.1$) [941], $Mg_{1-x}Ta_{2+x}N_3$ ($x \ll 0.1$) [941], $MgMoN_2$ [970], $ScNbN_2$ [383], $ScTaN_2$ [383,971] (probably identical with earlier reported $ScNbN_{1-x}$ [972] and $ScTaN_{1-x}$ [972,973]), $MnMoN_2$ [957,974], β-$MnWN_2$, [957,974] and $BaCeN_2$ [975]. The nitride ions are stacked according to (... *AABB* ...) with occupation of octahedra and trigonal prisms preventing any face-sharing of polyhedra. The simplest variant with β-$RbScO_2$-type structure (Figure 7.39f) is realized for $ScNbN_2$, $ScTaN_2$, $MnMoN_2$, β-$MnWN_2$, and $BaCeN_2$ with Sc, Mn, and Ce located in octahedra. While the trigonal prismatic coordination of Nb, Ta, Mo, and W is well known from the binary nitride systems, for Ba it is probably due to the enlarged ionic size.

It may be speculated that the formation of layers of not fully oxidized transition metal species in trigonal prismatic coordination is due to metal–metal bonding within these layers. In $LiNb_3N_4$ [183], $Li_{1-x}Ta_{3+x}N_4$ [941], and $Mg_{1-x}Ta_{2+x}N_3$ [941], the octahedral voids are occupied by Li or Mg and *M* in a disordered manner ($LiM_3N_4 \equiv (Li_{0.5}M_{0.5})MN_2$, M = Nb, Ta; $MgTa_2N_3 \equiv (Mg_{2/3}Ta_{1/3})TaN_2$). Homogeneity ranges as observed in the Ta and W phases and corresponding antisite defects in $LiMoN_2$ [965] and $ScTaN_2$ [383,971] can be regarded as related phenomena.

$ScNbN_2$ and $ScTaN_2$ are diamagnetic small bandgap semiconductors or semimetals. The Li phases were typically reported to exhibit a moderate, nearly temperature-independent electrical resistivity and thus interpreted as metals. All further compounds composed of two different transition metals typically show poor electrical conductivity ($\delta > 1.0\,\Omega\,cm$), with only a minor temperature dependence [383,971].

Until now, the Delafossite-type nitrides $CuNbN_2$ [789], $CuTaN_2$ [790], and $AgTaN_2$ (Figure 7.39g) [791] were exclusively obtained via ionic exchange from the respective sodium compounds $NaNbN_2$ and $NaTaN_2$ with the related α-$NaFeO_2$ structure type. During structure transformation, the rigid $^1_\infty[NbN_2^-]$ and $^1_\infty[TaN_2^-]$ layers need to shift relative to each from an original ... ABC ... stacking of N in the ordered rock salt variant, to eventually provide linear coordination for monovalent Cu and Ag in a ... $AABBCC$ ... stacking of N. The octahedral surrounding of Nb and Ta remains essentially unchanged. These compounds are metastable against decomposition into the elements and binary nitrides. They represent semiconductors with dark red to brown colors and indirect bandgaps around 1 eV, which may make them interesting for application as catalysts in water splitting [789,791,976,977].

$ThTaN_3$ is the only known perovskite structure in the nitride field; Ta is octahedrally coordinated by N, whereas Th occupies the large cuboctahedral void [978]. In a DFT study on $ThTaN_3$, the potential existence of corresponding post-perovskite nitrides is discussed [979].

7.3.3.8 Further Unexplored Systems

The ternary compound $La_2U_2N_5$ was indicated to form with a tetragonal unit cell closely related to the CsCl type, however, the exact crystal structure is unknown [980]. The crystal structure of hexagonal $BeThN_2$ is also still unknown [952].

Selected binary nitrides also are investigated concerning lithium intercalation and lithium storage behavior and application as anodes in lithium batteries, however, structural details for Li_xCrN [981] and Li_xCoN [325,982] are still unclear. The potential existence of $LiZrN_2$ is discussed based on theoretical calculations [954].

7.3.4
Metallide Nitrides

The last group of phases discussed is not necessarily an integral part of this work. Although these phases exclusively contain – similar to the nitridometalates – the constituting elements M and N as well as A or A_E, they miss any M–N interactions.

In the perovskite Ca$_3$AuN [381], isolated gold atoms are located in a network of corner-sharing Ca$_{6/2}$N octahedra. The metallic character of the phase was discussed according to a formula Ca$_3{}^{2+}$Au$^-$N^{3-}·2e$^-$ [381], agreeing with the concept of the chemical bond in the hexagonal perovskite NaBa$_3$N [983].

Cubic Ca$_{19}$Ag$_8$N$_7$ (Figure 7.40a) [984] contains tetrahedral clusters Ag$_4$ in a 3D matrix built up by octahedra NCa$_{6/2}$ and superoctahedra CaN$_6$Ca$_{12}$Ca$_{6/2}$ and is isotypic to Ca$_{19}$Ga$_8$N$_7$ [985].

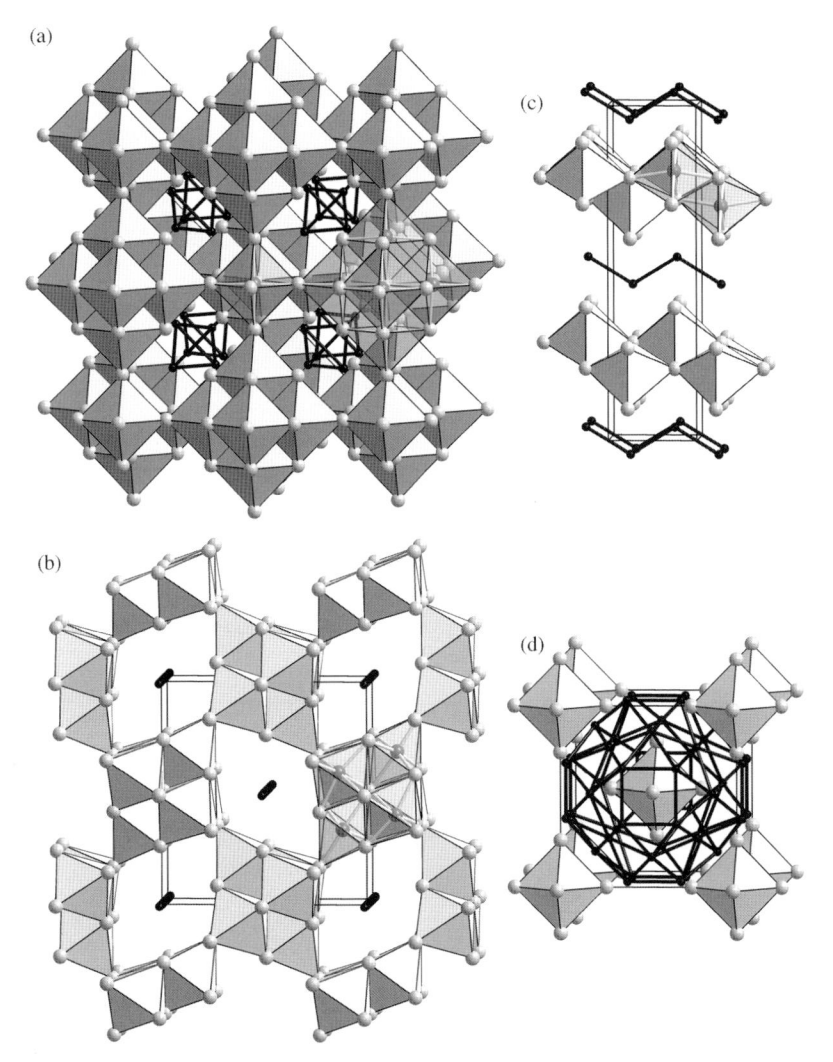

Figure 7.40 Selected crystal structures of nitride–metalide phases. (a) Ca$_{19}$Ag$_8$N$_7$. (b) Ca$_7$Ag$_{2.72}$N$_4$. (c) Ca$_2$AuN. (d) Ca$_6$Ag$_{16}$N. Small black spheres: coin metal atoms; larger light gray spheres: Ca; larger dark gray spheres: N.

$Ca_7Ag_{2.72}N_4$ [986] is best described as incommensurate linear Ag chains with equidistant atoms in channels of the 3D Ca_7N_4 host fragments built up by corner and edge-sharing NCa_6 octahedra (Figure 7.40b); due to the metallic character of the resistivity, a Peierls distortion can be ruled out.

Ca_2AuN [987] and Sr_2AuN [40] are built up by zigzag Au chains located between corrugated layers of edge-sharing octahedra $NCa_{6/3}$ and $NSr_{6/3}$, respectively (Figure 7.40c).

In $Ca_6Ag_{16}N$ [85], isolated octahedra Ca_6N are located in a 3D matrix of Ag (Figure 7.40d). First described as Ag_8Ca_3 [988], this phase is isotypic to $Na_{16}Ba_6N$ [85].

In the system Ba–Au–N, an orthorhombic phase $Ba_3Au_2N_x$ has been prepared by high-pressure reactions, but no further information is known [40].

All these compounds may be regarded as intermediates between intermetallic phases A_xM_y or AE_xM_y and AE_xN_y nitrides. Up to now, only a limited number of such phases have been reported; in all of them, N is generally located in isolated AE_6N octahedra or networks thereof. Considering AE^{2+} and N^{3-} and even M^- as s^2d^{10} (Ag^-, Au^-), there are still sufficient excess electrons in the metallic compounds (except Ca_2NAu [987], Sr_2NAu [40]).

7.4
Conclusion and Future Trends

With this contribution, an extensive survey on binary and multinary nitrides and nitridometalates as well as on nitride–metalides of the transition metals is provided. The investigation of structures and properties of these phases is still a fast-growing area of solid-state science, although great progress has been made in recent years. However, in contrast to the vast numbers of oxo-, thio-, and higher pnictide compounds, there is still only limited knowledge about the related nitrides, which may be attributed to the two following reasons:

- Due to the high bonding energy of molecular nitrogen (941 kJ/mol) and the unfavorable electron affinity (N –> N^{3-} + 2300 kJ/mol), nitrides are inherently thermodynamically less stable than comparable oxides.
- The synthesis of many if not most nitrides and nitridometalates is experimentally challenging. Many starting materials and reaction products are sensitive toward air and moisture, the temperature ranges of phase formation are quite small, and especially in low-valency systems, the N_2 partial pressure in the reaction chamber plays an important part in phase formation.

Altogether, binary nitrides may be classified according to the character of the prevalent bonding type into the following:

Ionic nitrides: Besides the alkali metal nitrides Li_3N and Na_3N, the alkaline earth metal nitrides Be_3N_2, Mg_3N_2, and Ca_3N_2, the highly oxidized transition metal nitrides ScN, YN, and LaN, and the other binary rare earth nitrides $RE^{III}N$, $Zr_3^{IV}N_4$, $Ta_3^{V}N_5$, and Zn_3N_2, as well as the uranium nitrides UN, U_2N_3, and UN_2

and thorium nitride $Th^{IV}_3N_4$ are members of this class. Although all these compounds show salt-like character, only the alkali and alkaline earth metal nitrides readily react with water in an acid–base reaction to produce ammonia and the corresponding hydroxide.

Interstitial nitrides: The largest groups of binary nitrides (as well as most multinary metal-rich nitrides) are the interstitial nitrides formed exclusively with transition metals. Close relations apply to the interstitial carbides, with nitrogen atoms occupying the interstices, or holes, in the lattice of more-or-less close-packed metal atoms. The most prominent representatives of this class are MN, M_2N, and M_4N, although their compositions may vary widely. These compounds are chemically inert, have high melting points, are extremely hard, and are usually opaque materials that have metallic luster and high conductivities. Refractory applications include tools (TiN, CrN), protective coatings (Fe_4N), or turbine blades (TiN); TaN is used in semiconductor industry as a diffusion barrier.

Covalent nitrides: Cu_3N and the main-group nitrides are characterized by covalent bonding and possess a wide range of properties depending on the constituting element to which nitrogen is bonded and the stoichiometry. Whereas BN, AlN, and Si_3N_4 are refractory materials characterized by a high degree of thermal stability and exceptional chemical stability and used as highly inert and heat-resistant ceramics or crucibles, GaN is of growing importance in semiconductor and LED industries. Within the same elements, different covalent nitrides exist such as C_3N_4, theoretically harder than diamond, and cyanogen, $(CN)_2$, a colorless and poisonous gas chemically similar to a halogen.

Nitridometalates of d metals M represent an interesting class of solid-state phases, which feature isolated complex anions or anionic frameworks $[M_xN_y]^{n-}$ of different dimensionality with M exhibiting coordination numbers by nitrogen typically between two and four. Whereas the bonding within these complex anions and frameworks is essentially covalent, the structures are stabilized by predominantly ionic bonding through counterions such as alkali (A), alkaline earth (A_E), or rare earth (R_E) metal cations.

An overall summary of the distribution of oxidation states of transition metals in nitridometalates is presented in Figure 7.41. Hereby, the assignment of oxidation states to transition metals in nitridometalates is based on classical counting rules (A: 1+; A_E: 2+; R_E: 3+ (in $Ce_2[CrN_3]$ [839] and $Ce_2[MnN_3]$ [915]: Ce: 4+); nitride ion: 3−).

For nitridometalates, the pattern emerging many years ago [826] still holds: for groups 3–6, M in nitridometalates exhibit intermediate to high, most often the highest possible, oxidation states. Group 7 shows a very wide variety of oxidation states from +1 in nitridomanganates to +7 in nitridorhenates, the highest oxidation state observed in nitridometalates so far. Nitridometalates of groups 8–12 are insofar special that only compounds with M as a 3d element are known. For iron, +3 is the highest oxidation state observed so far; for Co, Ni, and Cu only low oxidation states $\leq$ +1 are realized. Apart from some exceptional examples in

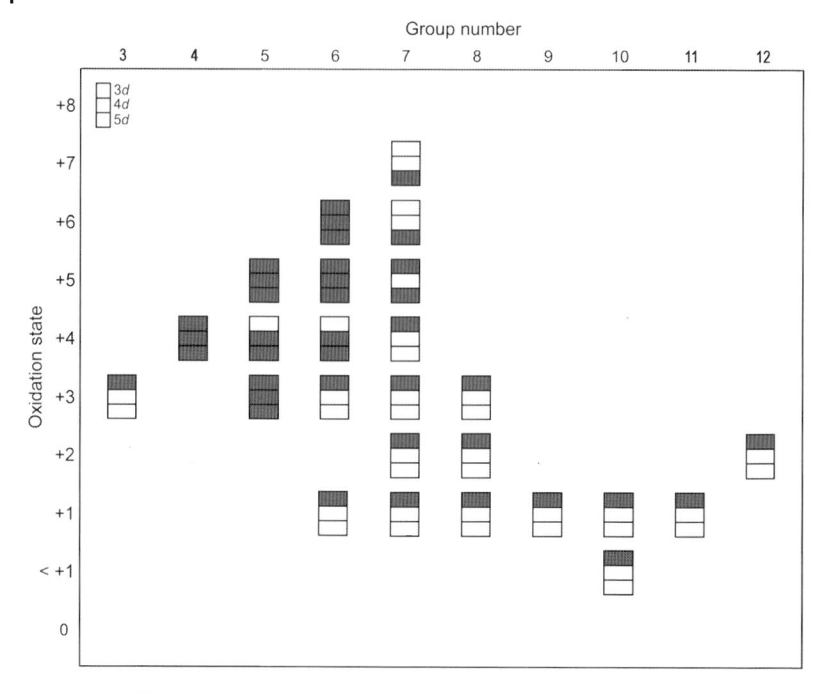

Figure 7.41 Observed oxidation states (gray) of M in nitridometalates of transition metals.

oxides ($K_3[FeO_2]$ [989], $Rb_3[CoO_2]$ [990], $La[NiO_2]$ [991]/$Nd[NiO_2]$ [992]) containing the transition metal in the low oxidation state +1, M oxidation states $\leq$ +1 are preferentially realized in nitrido- and carbometalates [993].

This tendency toward stabilization of low oxidation states of M in nitridometalates is exemplified in a series of complex nickelates. All contain complex anions and monoatomic nonmetal ligands, and the maximum oxidation states of nickel significantly decrease from fluoro (+4, $Rb_2[NiF_6]$ [994]) via oxo- (+4, $Ba[NiO_3]$ [995]) and nitrido- (+1, $Ba[NiN]$ [805]) to finally carbometalates (0, $Th_2[CNiC]$ [996]). In parallel, the ligand sequence $F^- \rightarrow O^{2-} \rightarrow N^{3-} \rightarrow C^{4-}$ represents decreasing electronegativity and increasing polarizability of the monoatomic ligands (L) together with increasing covalency of the M–L bonds.

It needs to be emphasized that charge distributions resulting from oxidation state assignments are clearly overestimated. In order to get a more realistic impression on the charge distribution in nitridometalates, the QTAIM (Quantum Theory of Atoms in Molecules) method [997] was used to determine the volumes of the basins (volumes of atoms or ions). By integration of the electron densities within these volumes, the effective charges are obtained defining the neutral or ionic character of the QTAIM species. By comparing the calculated charges of the nitrido ligands with the oxidation states of the transition metals deduced from their chemical formulas presuming electroneutrality, it is evident that especially the charge of the transition metals is significantly reduced [998].

The charges range between 2.2+ and 0 for the transition metals in spite of the big differences in the oxidation states (+1 to +7), and between 1.4– and 1.9– for the nitrido ligands.

For the shape of the nitridometalate anions, another correlation also becomes evident: For those compounds with high or even the highest possible oxidation states of the transition elements, tetrahedral complexes are formed, whereas trigonal-planar or T-shaped complexes are observed for nitridometalate anions with intermediate oxidation states of the transition metals. Low-valency nitridometalates (oxidation states $M \leq +1$) exclusively contain the transition elements in twofold coordination by the nitrido ligands, thereby forming dumbbells or chains.

Altogether, the chemistry of nitrides and nitridometalates of transition metals provides great potential for interesting chemical, structural, and physical properties as well as innovative applications. Taken recent developments into account, there is a good chance for substantial progress and exciting discoveries in the years to come.

Abbreviations

A Alkali metal
A_E Alkaline earth metal
R_E Rare earth metal
M Transition metal
X Nonmetal

References

1 Kniep, R. and Höhn, P. (2013) in *Comprehensive Inorganic Chemistry II. Volume 2: Transition Elements, Lanthanides and Actinides* (eds J. Reedijk and K. Poeppelmeier), 2nd edn, Elsevier, pp. 137–160.
2 Kniep, R. (1997) *Pure Appl. Chem.*, **69**, 185–191.
3 Niewa, R. and Jacobs, H. (1996) *Chem. Rev.*, **96**, 2053–2062.
4 Niewa, R. and DiSalvo, F.J. (1998) *Chem. Mater.*, **10**, 2733–2752.
5 Niewa, R. (2002) *Z. Kristallogr.*, **217**, 8–23.
6 Brese, N.E. and O'Keeffe, M. (1992) *Struct. Bond.*, **79**, 307–378.
7 Ham, D.J. and Lee, J.S. (2009) *Energies*, **2**, 873–899.
8 Kirchner, M., Schnelle, W., Wagner, F.R., and Niewa, R. (2003) *Solid State Sci.*, **5**, 1247–1257.
9 Marchand, R. (1998) *Handbook on the Physics and Chemistry of Rare Earths*, vol. **25**, Elsevier, pp. 51–99.
10 Haschke, H., Nowotny, H., and Benesovs, F. (1967) *Monatsh. Chem.*, **98**, 2157–2163.
11 Rogl, P. and Klesnar, H. (1992) *J. Solid State Chem.*, **98**, 99–104.
12 Höhn, P., Prots, Y., Kokal, I., and Somer, M. (2009) *Z. Kristallogr.*, **224**, 379–380.
13 Schnick, W. and Huppertz, H. (1997) *Chem. Eur. J.*, **3**, 679–683.
14 Woike, M. and Jeitschko, W. (1995) *Inorg. Chem.*, **34**, 5105–5108.

15 Boller, H. (1969) *Monatsh. Chem.*, **100**, 1471–1476.
16 Boller, H. and Nowotny, H. (1968) *Monatsh. Chem.*, **99**, 672–675.
17 Bhadram, V.S., Kim, D.Y., and Strobel, T.A. (2016) *Chem. Mater.*, **28**, 1616–1620.
18 Montoya, J.A., Hernandez, A.D., Sanloup, C., Gregoryanz, E., and Scandolo, S. (2007) *Appl. Phys. Lett.*, **90**, 011909.
19 Young, A.F., Sanloup, C., Gregoryanz, E., Scandolo, S., Hemley, R.J., and Mao, H.K. (2006) *Phys. Rev. Lett*, **96**, 155501.
20 Crowhurst, J.C., Goncharov, A.F., Sadigh, B., Evans, C.L., Morrall, P.G., Ferreira, J.L., and Nelson, A.J. (2006) *Science*, **311**, 1275–1278.
21 Crowhurst, J.C., Goncharov, A.F., Sadigh, B., Zaug, J., Aberg, D., Meng, Y., and Prakapenka, V.B. (2008) *J. Mater. Res.*, **23**, 1–5.
22 Aberg, D., Erhart, P., Crowhurst, J., Zaug, J.M., Goncharov, A.F., and Sadigh, B. (2010) *Phys. Rev. B*, **82**, 104116.
23 Gregoryanz, E., Sanloup, C., Somayazulu, M., Badro, J., Fiquet, G., Mao, H.K., and Hemley, R.J. (2004) *Nat. Mater.*, **3**, 294–297.
24 Ayfon, S., Höhn, P., Armbrüster, M., Baranov, A., Wagner, F.R., Somer, M., and Kniep, R. (2006) *Z. Anorg. Allg. Chem.*, **632**, 1671–1680.
25 Rabenau, A. and Schulz, H. (1976) *J. Less Common Met.*, **50**, 155–159.
26 Höhn, P., Hoffmann, S., Hunger, J., Leoni, S., Nitsche, F., Schnelle, W., and Kniep, R. (2009) *Chem. Eur. J.*, **15**, 3419–3425.
27 Brese, N.E. and O'Keeffe, M. (1990) *J. Solid State Chem.*, **87**, 134–140.
28 Toth, L.E. (1971) *Transition Metal Carbides and Nitrides*, Academic Press, pp. 1–279.
29 Oyama, S.T. (ed.) (1996) *The Chemistry of Transition Metal Carbides and Nitrides*, Chapman & Hall, London.
30 Lee, S., Kim, W., Suh, C.y., Cho, S.w., Ryu, T., Park, J.S., and Shon, I.J. (2011) *Mater. Trans.*, **52**, 261–264.
31 Richter, M.T. and Niewa, R. (2014) *Inorganics*, **2**, 29–78.
32 Jacobs, H. and Schmidt, D. (1982) *Current Topics in Materials Science*, vol. **8**, North Holland, pp. 379–425.

33 Chen, X.Z., Dye, J.L., Eick, H.A., Elder, S.H., and Tsai, K.L. (1997) *Chem. Mater.*, **9**, 1172–1176.
34 Marchand, R., Tessier, F., and DiSalvo, F.J. (1999) *J. Mater. Chem.*, **9**, 297–304.
35 Feng, X., Bai, Y.J., Lu, B., Wang, C.G., Qi, Y.X., Liu, Y.X., Geng, G.L., and Li, L. (2004) *Inorg. Chem.*, **43**, 3558–3560.
36 Chen, L.Y., Gu, Y.L., Shi, L., Yang, Z.H., Ma, J.H., and Qian, Y.T. (2004) *Solid State Commun.*, **132**, 343–346.
37 Janes, R.A., Aldissi, M., and Kaner, R.B. (2003) *Chem. Mater.*, **15**, 4431–4435.
38 Lengauer, W. (1986) *J. Less Common Met.*, **125**, 127–134.
39 Prots', Yu., Auffermann, G., Tovar, M., and Kniep, R. (2002) *Angew. Chem., Int. Ed.*, **41**, 2288–2290.
40 Prots', Yu., Auffermann, G., and Kniep, R. (2002) *Z. Anorg. Allg. Chem.*, **628**, 2205.
41 Gomathi, A. and Rao, C.N.R. (2006) *Mater. Res. Bull.*, **41**, 941–947.
42 Gomathi, A. (2007) *Mater. Res. Bull.*, **42**, 870–874.
43 Tripathy, P.K., Sehra, J.C., and Kulkarni, A.V. (2001) *J. Mater. Chem.*, **11**, 691–695.
44 Friedrich, A., Winkler, B., Juarez-Arellano, E.A., and Bayarjargal, L. (2011) *Materials*, **4**, 1648–1692.
45 Fix, R., Gordon, R.G., and Hoffman, D.M. (1993) *Chem. Mater.*, **5**, 614–619.
46 Liu, J.P., Bakker, K., de Boer, F.R., Jacobs, T.H., de Mooij, D.B., and Buschow, K.H.J. (1991) *J. Less Common Met.*, **170**, 109–119.
47 Fukuno, A., Ishizaka, C., and Yoneyama, T. (1991) *J. Appl. Phys.*, **70**, 6021–6023.
48 Zinkevich, M., Mattern, N., Handstein, A., and Gutfleisch, O. (2002) *J. Alloys Compd.*, **339**, 118–139.
49 Xiang-Zhong, W., Donnelly, K., Coey, J.M.D., Chevalier, B., Etourneau, J., and Berlureau, T. (1988) *J. Mater. Sci.*, **23**, 329–331.
50 Coey, J.M.D. and Sun, H. (1990) *J. Magn. Magn. Mater.*, **87**, L251–L254.
51 Strähle, J. (1973) *Z. Anorg. Allg. Chem.*, **402**, 47–57.
52 Reckeweg, O. and DiSalvo, F.J. (2002) *Solid State Sci.*, **4**, 575–584.
53 Bendyna, J.K., Höhn, P., and Kniep, R. (2008) *Z. Kristallogr.*, **223**, 109–110.

54 Bendyna, J.K., Höhn, P., and Kniep, R. (2008) *Z. Kristallogr.*, **223**, 181–182.

55 Höhn, P. and Kniep, R. (1992) *Z. Naturforsch. B*, **47**, 477–481.

56 Gudat, A., Kniep, R., and Rabenau, A. (1991) *Angew. Chem., Int. Ed.*, **30**, 199–200.

57 Gudat, A., Haag, S., Kniep, R., and Rabenau, A. (1990) *Z. Naturforsch. B*, **45**, 111–120.

58 Yuan, W.X., Hu, J.W., Song, Y.T., Wang, W.J., and Xu, Y.P. (2005) *Powder Diffr.*, **20**, 18–21.

59 Höhn, P., Niewa, R., and Kniep, R. (2000) *Z. Kristallogr.*, **215**, 323–324.

60 Höhn, P., Jach, F., Karabiyik, B., Prots', Yu., Agrestini, S., Wagner, F.R., Ruck, M., Tjeng, L.H., and Kniep, R. (2011) *Angew. Chem., Int. Ed.*, **50**, 9361–9364.

61 Jach, F., Höhn, P., Prots, Y., and Ruck, M. (2015) *Z. Anorg. Allg. Chem.*, **641**, 998–1001.

62 Jach, F., Höhn, P., Senyshyn, A., Ruck, M., and Kniep, R. (2012) *Z. Anorg. Allg. Chem.*, **638**, 1959–1961.

63 Bendyna, J.K., Höhn, P., and Kniep, R. (2009) *Z. Kristallogr.*, **224**, 5–6.

64 Bendyna, J.K., Höhn, P., Ormeci, A., Schnelle, W., and Kniep, R. (2009) *J. Alloys Compd.*, **480**, 138–140.

65 Höhn, P., Bendyna, J.K., Nitsche, F., Schnelle, W., and Kniep, R. (2008) Book of Abstracts, SCTE-16, Dresden, p. 296.

66 Bendyna, J.K., Höhn, P., Prots', Yu., and Kniep, R. (2010) *Z. Anorg. Allg. Chem.*, **636**, 1297–1300.

67 Jach, F., Groh, M., Höhn, P., and Ruck, M. (2014) *Z. Anorg. Allg. Chem.*, **640**, 2359.

68 Höhn, P., Armbrüster, M., Auffermann, G., Burkhardt, U., Haarmann, F., Mehta, A., and Kniep, R. (2006) *Z. Anorg. Allg. Chem.*, **632**, 2129.

69 Höhn, P. and Kniep, R. (2010) *Z. Anorg. Allg. Chem.*, **636**, 2104.

70 Kowach, G.R., Brese, N.E., Bolle, U.M., Warren, C.J., and DiSalvo, F.J. (2000) *J. Solid State Chem.*, **154**, 542–550.

71 Reckeweg, O. and DiSalvo, F.J. (2003) *Z. Naturforsch. B*, **58**, 201–204.

72 Clarke, S.J. and DiSalvo, F.J. (1997) *Z. Kristallogr.*, **212**, 309–310.

73 Höhn, P., Kniep, R., and Maier, J. (1993) *Angew. Chem., Int. Ed.*, **32**, 1350–1352.

74 Mond, L., Langer, C., and Quincke, F. (1890) *J. Chem. Soc. Trans.*, **57**, 749–753.

75 Mehta, A. (2005) Nitridonickelates: preparation, structure and properties. Dissertation, TU Dresden, pp. 1–184.

76 Ostermann, D., Zachwieja, U., and Jacobs, H. (1992) *J. Alloys Compd.*, **190**, 137–140.

77 Vennos, D.A., Badding, M.E., and DiSalvo, F.J. (1990) *Inorg. Chem.*, **29**, 4059–4062.

78 Jacobs, H. and von Pinkowski, E. (1989) *J. Less Common Met.*, **146**, 147–160.

79 Mehta, A., Höhn, P., Schnelle, W., Petzold, V., Rosner, H., Burkhardt, U., and Kniep, R. (2006) *Chem. Eur. J.*, **12**, 1667–1676.

80 Ovchinnikov, A., Schnelle, W., Bobnar, M., Borrmann, H., Sichelschmidt, J., Grin, Y., and Höhn, P. (2015) Book of Abstracts, ECSSC-15, Vienna, European Conference on Solid State Chemistry, Vienna, Austria, p. 288.

81 Ovchinnikov, A., Bobnar, M., Borrmann, H., Prots, Y., Schnelle, W., Sichelschmidt, J., Ormeci, A., Grin, Y., and Höhn, P. (2016) Book of Abstracts, SCTE-20, Zaragoza, International Conference on Solid Compounds of Transition Elements, Zaragoza, Spain, p. 309.

82 Kowach, G.R., Lin, H.Y., and DiSalvo, F.J. (1998) *J. Solid State Chem.*, **141**, 1–9.

83 Borgstedt, H.U. and Guminski, C. (2000) *Monatsh. Chem.*, **131**, 917–930.

84 Addison, C.C., Pulham, R.J., and Trevillion, E.A. (1975) *J. Chem. Soc., Dalton Trans.*, **1975**, 2082–2085.

85 Snyder, G.J. and Simon, A. (1994) *Angew. Chem., Int. Ed.*, **33**, 689–691.

86 Höhn, P., Kniep, R., and Rabenau, A. (1991) *Z. Kristallogr.*, **196**, 153–158.

87 Gudat, A., Kniep, R., Rabenau, A., Bronger, W., and Ruschewitz, U. (1990) *J. Less Common Met.*, **161**, 31–36.

88 Zherebtsov, D.A., Aksel'rud, L.G., and Niewa, R. (2002) *Z. Kristallogr.*, **217**, 469.

89 Jesche, A. and Canfield, P.C. (2014) *Philos. Mag.*, **94**, 2372–2402.

90 Höhn, P., Schnelle, W., and Zechel, K. (2014) Book of Abstracts, SCTE-19, Genoa, SCTE, Genoa, Italy, p. 71.

91 Höhn, P., Borrmann, H., Burkhardt, U., Ovchinnikov, A., Schnelle, W., and Zechel, K. (2015) Book of Abstracts, ECSSC-15, Vienna, European Conference on Solid State Chemistry, Vienna, Austria, p. 189.

92 Jesche, A., McCallum, R.W., Thimmaiah, S., Jacobs, J.L., Taufour, V., Kreyssig, A., Houk, R.S., Bud'ko, S.L., and Canfield, P.C. (2014) *Nat. Commun.*, **5**, 3333.

93 Yamane, H. (2009) *J. Ceram. Soc. Jpn.*, **117**, 1021–1027.

94 DiSalvo, F.J., Trail, S.S., Yamane, H., and Brese, N.E. (1997) *J. Alloys Compd.*, **255**, 122–129.

95 Chevire, F., Ranjan, C., and DiSalvo, F.J. (2009) *Solid State Commun.*, **149**, 273–276.

96 Niewa, R. and Jacobs, H. (1996) *J. Alloys Compd.*, **233**, 61–68.

97 Gudat, A., Kniep, R., and Maier, J. (1992) *Z. Naturforsch. B*, **47**, 1363–1366.

98 Wachsmann, C. and Jacobs, H. (1996) *Z. Anorg. Allg. Chem.*, **622**, 885–888.

99 Suzuki, S. and Shodai, T. (1999) *Solid State Ionics*, **116**, 1–9.

100 Nishijima, M., Tadokoro, N., Takeda, Y., Imanishi, N., and Yamamoto, O. (1994) *J. Electrochem. Soc.*, **141**, 2966–2971.

101 Nishijima, M., Takeda, Y., Imanishi, N., and Yamamoto, O. (1994) *J. Solid State Chem.*, **113**, 205–210.

102 Meyer, H.J. (2010) *Dalton Trans.*, **39**, 5973–5982.

103 Glaser, J., Mori, T., and Meyer, H.J. (2008) *Z. Anorg. Allg. Chem.*, **634**, 1067–1070.

104 Brown, R.C. and Clark, N.J. (1974) *J. Inorg. Nucl. Chem.*, **36**, 2507–2514.

105 Lengauer, W. (1988) *J. Solid State Chem.*, **76**, 412–415.

106 Lyutaya, M.D., Goncharuk, A.B., and Timofeeva, I.I. (1975) *J. Appl. Chem. USSR*, **48**, 757–760.

107 Aivazov, M.I., Rezchikova, T.V., and Gurov, S.V. (1977) *Inorg. Mater.*, **13**, 999–1001.

108 Porte, L. (1985) *J. Phys. C*, **18**, 6701.

109 Karl, M., Seybert, G., Massa, W., and Dehnicke, K. (1999) *Z. Anorg. Allg. Chem.*, **625**, 375–376.

110 Hajek, B., Brozek, V., and Duvignea, P.H. (1973) *J. Less Common Met.*, **33**, 385–386.

111 Niewa, R., Zherebtsov, D.A., Kirchner, M., Schmidt, M., and Schnelle, W. (2004) *Chem. Mater.*, **16**, 5445–5451.

112 Didchenko, R. and Gortsema, F.P. (1963) *J. Phys. Chem. Solids*, **24**, 863–870.

113 O'Dell, K.D. and Hensley, E.B. (1972) *J. Phys. Chem. Solids*, **33**, 443–449.

114 Landrum, G.A., Dronskowski, R., Niewa, R., and DiSalvo, F.J. (1999) *Chem. Eur. J.*, **5**, 515–522.

115 Olcese, G.L. (1979) *J. Phys. F*, **9**, 569–578.

116 Aerts, C.M., Strange, P., Horne, M., Temmerman, W.M., Szotek, Z., and Svane, A. (2004) *Phys. Rev. B Condens. Matter*, **69**, 045115.

117 Busch, G., Kaldis, E., Schaufelberger-Teker, E., and Wachter, P. (1969) *Les Éléments des Terres Rares*, vol. **1**, CNRS, pp. 359–374.

118 Kordis, J., Kaldis, E., Bischof, R., Seyse, R.J., and Gingeric, K.A. (1972) *J. Cryst. Growth*, **17**, 53–60.

119 Busch, G., Junod, P., and Vogt, O. (1967) Book of Abstracts, Colloques Internationaux du CNRS-157, Paris, p. 337.

120 Jacobs, H. and Fink, U. (1978) *Z. Anorg. Allg. Chem.*, **438**, 151–159.

121 Schneider, S.B., Baumann, D., Salamat, A., and Schnick, W. (2012) *J. Appl. Phys.*, **111**, 093503.

122 Mukherjee, D., Sahoo, B., Joshi, K., and Gupta, S.C. (2013) *High Press. Res.*, **33**, 563–571.

123 Kikkawa, S., Ohmura, T., Takahashi, M., Kanamaru, F., and Ohtaka, O. (1997) *J. Eur. Ceram. Soc.*, **17**, 1831–1835.

124 Kieffer, R., Ettmayer, P., and Pajakoff, S. (1972) *Monatsh. Chem.*, **103**, 1285–1298.

125 Hasegawa, M., Niwa, K., and Yagi, T. (2007) *Solid State Commun.*, **141**, 267–272.

126 Chiotti, P. (1952) *J. Am. Ceram. Soc.*, **35**, 123–130.

127 Olson, W.M. and Mulford, R.N.R. (1965) *J. Phys. Chem.*, **69**, 1223–1226.

128 Uno, M., Katsura, M., and Miyake, M. (1986) *J. Less Common Met.*, **121**, 615–619.

129 Uno, M., Katsura, M., and Miyake, M. (1987) *Inorg. Chim. Acta*, **140**, 123–126.
130 Uno, M., Katsura, M., and Miyake, M. (1987) *J. Less Common Met.*, **135**, 25–38.
131 Benz, R., Hoffman, C.G., and Rupert, G.N. (1967) *J. Am. Chem. Soc.*, **89**, 191–197.
132 Silva, G., Yeamans, C.B., Weck, P.F., Hunn, J.D., Cerefice, G.S., Sattelberger, A.P., and Czerwinski, K.R. (2012) *Inorg. Chem.*, **51**, 3332–3340.
133 Zachariasen, W.H. (1949) *Acta Crystallogr.*, **2**, 388–390.
134 Benz, R. and Zacharia, W.H. (1966) *Acta Crystallogr.*, **21**, 838–840.
135 Blunck, H. and Juza, R. (1974) *Z. Anorg. Allg. Chem.*, **410**, 9–20.
136 Rundle, R.E., Baenziger, N.C., Wilson, A.S., and McDonald, R.A. (1948) *J. Am. Chem. Soc.*, **70**, 99–105.
137 Olsen, J.S., Gerward, L., and Benedict, U. (1985) *J. Appl. Crystallogr.*, **18**, 37–41.
138 Masaki, N., Tagawa, H., and Tsuji, T. (1972) *J. Nucl. Mater.*, **45**, 230–234.
139 Tagawa, H. and Masaki, N. (1974) *J. Inorg. Nucl. Chem.*, **36**, 1099–1103.
140 Masaki, N. and Tagawa, H. (1975) *J. Nucl. Mater.*, **57**, 187–192.
141 Masaki, N. and Tagawa, H. (1975) *J. Nucl. Mater.*, **58**, 241–243.
142 Bannister, F.A. (1941) *Mineral. Mag.*, **26**, 36–44.
143 Tatarintsev, V.I., Sandomirskaya, S.M., and Tsymbal, S.M. (1987) *Dokl. Akad. Nauk SSSR*, **296**, 1458–1461.
144 Ehrlich, P. (1949) *Z. Anorg. Chem.*, **259**, 1–41.
145 Christensen, A.N. (1978) *Acta Chem. Scand.*, **32**, 89–90.
146 Christensen, A.N. (1975) *Acta Chem. Scand.*, **29**, 563–564.
147 Holmberg, B. (1962) *Acta Chem. Scand.*, **16**, 1255–1261.
148 Khidirov, I. (2010) *Int. Sci. J. Altern. Energy Ecol.*, **3**, 10–13.
149 Christensen, A.N., Alamo, A., and Landesman, J.P. (1985) *Acta Crystallogr. C*, **41**, 1009–1011.
150 Gusev, A.I. and Rempel, A.A. (1997) *Phys. Status Solidi A*, **163**, 273–304.
151 Lengauer, W. and Ettmayer, P. (1988) *Mater. Sci. Eng. A*, **105**, 257–263.
152 Bittner, H., Goretzki, H., Benesovsky, F., and Nowotny, H. (1963) *Monatsh. Chem.*, **94**, 518–526.
153 Lerch, M., Füglein, E., and Wrba, J. (1996) *Z. Anorg. Allg. Chem.*, **622**, 367–372.
154 Juza, R., Rabenau, A., and Nitschke, I. (1964) *Z. Anorg. Allg. Chem.*, **332**, 1–4.
155 Juza, R., Gabel, A., Rabenau, H., and Klose, W. (1964) *Z. Anorg. Allg. Chem.*, **329**, 136–145.
156 Zerr, A., Miehe, G., and Riedel, R. (2003) *Nat. Mater.*, **2**, 185–189.
157 Rudy, E. (1970) *Metall. Trans.*, **1**, 1249–1252.
158 Lengauer, W., Rafaja, D., Zehetner, G., and Ettmayer, P. (1996) *Acta Mater.*, **44**, 3331–3338.
159 Carlson, O.N., Smith, J.F., and Nafziger, R.H. (1986) *Metall. Trans. A*, **17**, 1647–1656.
160 Onozuka, T. (1978) *J. Appl. Crystallogr.*, **11**, 132–136.
161 Kubel, F., Lengauer, W., Yvon, K., Knorr, K., and Junod, A. (1988) *Phys. Rev. B*, **38**, 12908–12912.
162 Hahn, H. (1949) *Z. Anorg. Chem.*, **258**, 58–68.
163 Christensen, A.N. and Lebech, B. (1979) *Acta Crystallogr. B*, **35**, 2677–2678.
164 Christensen, A.N. (1976) *Acta Chem. Scand.*, **30**, 219–224.
165 Conroy, L.E. and Nørlund Christensen, A. (1977) *J. Solid State Chem.*, **20**, 205–207.
166 Kieda, N., Mizutani, N., and Kato, M. (1988) *J. Less Common Met.*, **144**, 293–299.
167 Lengauer, W. and Ettmayer, P. (1985) *J. Less Common Met.*, **109**, 351–359.
168 Potter, D. and Altstett, C. (1971) *Acta Metall.*, **19**, 881–886.
169 Potter, D. and Altstett, C. (1972) *Mater. Sci. Eng.*, **9**, 43–46.
170 Epstein, H., Goldstein, B., and Potter, D. (1973) *Scr. Metall.*, **7**, 717–719.
171 Potter, D.I., Epstein, H.D., and Goldstein, B.M. (1974) *Metall. Trans.*, **5**, 2075–2082.
172 Khayenko, B.V. and Frenkel, O.A. (1977) *Fiz. Met. Metalloved.*, **44**, 105–115.
173 Khaenko, B.V. (1980) *Dopov. Akad. Nauk Ukr. RSR Ser. A*, 85–89.

174 Cambini, M. (1974) *Mater. Res. Bull.*, **9**, 1469–1480.

175 Nouet, G., Vicens, J., and Delavignette, P. (1980) *Phys. Status Solidi A*, **62**, 449–457.

176 Widenmeyer, M., Wessel, C., Dronskowski, R., and Niewa, R. (2015) *Z. Anorg. Allg. Chem.*, **641**, 2610–2616.

177 Brauer, G. and Esselborn, R. (1961) *Z. Anorg. Allg. Chem.*, **309**, 151–170.

178 Oya, G. and Onodera, Y. (1974) *J. Appl. Phys.*, **45**, 1389–1397.

179 Schönberg, N. (1954) *Acta Chem. Scand.*, **8**, 208–212.

180 Joguet, M., Lengauer, W., Bohn, M., and Bauer, J. (1998) *J. Alloys Compd.*, **269**, 233–237.

181 Frenzel, N., Irran, E., Lerch, M., and Buchsteiner, A. (2011) *Z. Naturforsch. B*, **66**, 1–6.

182 Terao, N. (1971) *J. Less Common Met.*, **23**, 159–169.

183 Tessier, F., Assabaa, R., and Marchand, R. (1997) *J. Alloys Compd.*, **262–263**, 512–515.

184 Schönberg, N. (1954) *Acta Chem. Scand.*, **8**, 199–203.

185 Brauer, G. and Zapp, K.H. (1954) *Z. Anorg. Allg. Chem.*, **277**, 129–139.

186 Terao, N. (1971) *Jpn. J. Appl. Phys.*, **10**, 248–259.

187 Christensen, A.N. and Lebech, B. (1978) *Acta Crystallogr. B*, **34**, 261–263.

188 Brauer, G., Skokan, A., Neuhaus, A., and Mohr, E. (1972) *Monatsh. Chem.*, **103**, 794.

189 Mashimo, T. and Tashiro, S. (1994) *J. Mater. Sci. Lett.*, **13**, 174–176.

190 Friedrich, A., Morgenroth, W., Bayarjargal, L., Juarez-Arellano, E., Winkler, B., and Konopkova, Z. (2013) *High Press. Res.*, **33**, 633–641.

191 Gerstenberg, D. and Calbick, C.J. (1964) *J. Appl. Phys.*, **35**, 402–407.

192 Gatterer, J., Dufek, G., Ettmayer, P., and Kieffer, R. (1975) *Monatsh. Chem.*, **106**, 1137–1147.

193 Mashimo, T., Tashiro, S., Toya, T., Nishida, M., Yamazaki, H., Yamaya, S., Ohishi, K., and Syono, Y. (1993) *J. Mater. Sci.*, **28**, 3439–3443.

194 Ganin, A.Y., Kienle, L., and Vajenine, G.V. (2004) *Eur. J. Inorg. Chem.*, 3233–3239.

195 Buvinger, E.A. (1965) *Appl. Phys. Lett.*, **7**, 14–15.

196 Coyne, H.J. and Tauber, R.N. (1968) *J. Appl. Phys.*, **39**, 5585.

197 Zerr, A., Miehe, G., Li, J., Dzivenko, D.A., Bulatov, V.K., Hoefer, H., Bolfan-Casanova, N., Fialin, M., Brey, G., Watanabe, T., and Yoshimura, M. (2009) *Adv. Funct. Mater.*, **19**, 2282–2288.

198 Friedrich, A., Winkler, B., Bayarjargal, L., Juarez Arellano, E.A., Morgenroth, W., Biehler, J., Schroeder, F., Yan, J., and Clark, S.M. (2010) *J. Alloys Compd.*, **502**, 5–12.

199 Brese, N.E., O'Keeffe, M., Rauch, P., and DiSalvo, F.J. (1991) *Acta Crystallogr. C*, **47**, 2291–2294.

200 Brauer, G. and Weidlein, J.R. (1965) *Angew. Chem., Int. Ed.*, **4**, 241–242.

201 Terao, N. (1977) *C. R. Acad. Sci. B*, **285**, 17–20.

202 Ma, S.S.K., Hisatomi, T., Maeda, K., Moriya, Y., and Domen, K. (2012) *J. Am. Chem. Soc.*, **134**, 19993–19996.

203 Gilles, J.C. (1968) *C. R. Acad. Sci. C*, **266**, 546–547.

204 Salamat, A., Woodhead, K., Shah, S.I., Hector, A.L., and McMillan, P.F. (2014) *Chem. Commun.*, **50**, 10041–10044.

205 Fontbonn, A. and Gilles, J.C. (1969) *Rev. Int. Hautes Temp. Refract.*, **6**, 181–192.

206 Vendl, A. (1978) *Monatsh. Chem.*, **109**, 1009–1012.

207 Kikuchi, M., Kajihara, M., and Choi, S.K. (1991) *Mater. Sci. Eng. A*, **146**, 131–150.

208 Lee, T.H., Kim, S.J., and Jung, Y.C. (2000) *Metall. Mater. Trans. A*, **31**, 1713–1723.

209 Lee, T.H., Oh, C.S., Lee, C.G., Kim, S.J., and Takaki, S. (2004) *Scr. Mater.*, **50**, 1325–1328.

210 Lee, T.H., Oh, C.S., Han, H.N., Lee, C.G., Kim, S.J., and Takaki, S. (2005) *Acta Crystallogr. B*, **61**, 137–144.

211 Lee, T.H., Kim, S.J., and Takaki, S. (2006) *Acta Crystallogr. B*, **62**, 190–196.

212 Lee, T.H., Kim, S.J., Shin, E., and Takaki, S. (2006) *Acta Crystallogr. B*, **62**, 979–986.

213 Lv, Z., Shi, Z., Gao, Y., Wang, Z., Sun, S., and Fu, W. (2014) *J. Alloys Compd.*, **583**, 79–84.

214 Baur, E. and Voerman, G.L. (1905) *Z. Phys. Chem.*, **52**, 467–478.

215 Nguyen, D.N., Jo, D.S., Kim, J.G., and Yoon, D.H. (2011) *Thin Solid Films*, **519**, 6787–6791.

216 Paulauskas, I., Brady, M., Meyer, H., Buchanan, R., and Walker, L. (2006) *Corros. Sci.*, **48**, 3157–3171.

217 Yang, M., Guarecuco, R., and DiSalvo, F.J. (2013) *Chem. Mater.*, **25**, 1783–1787.

218 Sun, Q. and Fu, Z.W. (2007) *Electrochem. Solid State Lett.*, **10**, A189–A193.

219 Sun, Q. and Fu, Z.W. (2008) *Electrochem. Solid State Lett.*, **11**, A233–A237.

220 Seybolt, A.U. and Oriani, R.A. (1956) *Trans. Am. Inst. Min. Metall. Eng.*, **206**, 556–562.

221 Sun, G., Meissner, E., Berwian, R., and Mueller, G. (2008) *Thermochim. Acta*, **474**, 36–40.

222 Mills, T. (1971) *J. Less Common Met.*, **23**, 317–324.

223 Blix, R. (1929) *Z. Phys. Chem. B*, **3**, 229–239.

224 Leineweber, A., Jacobs, H., Hüning, F., Lueken, H., and Kockelmann, W. (2001) *J. Alloys Compd.*, **316**, 21–38.

225 Shohoji, N., Katsura, M., and Sano, T. (1974) *J. Less Common Met.*, **38**, 59–70.

226 Kim, S.J., Marquart, T., and Franzen, H.F. (1990) *J. Less Common Met.*, **158**, L9–L10.

227 Mills, T. (1972) *J. Less Common Met.*, **26**, 223–234.

228 Buchwald, V.F. and Scott, E.R.D. (1971) *Nat. Phys. Sci.*, **233**, 113–114.

229 Widenmeyer, M. (2014) Synthese und Charakterisierung (meta)stabiler 3*d*-Übergangsmetallnitride MN_x (M=V–Cu) und deren Zwischenstufen mittels in-situ Neutronenbeugung. Dissertation, Universität Stuttgart.

230 Tsuchiya, Y., Kosuge, K., Ikeda, Y., Shigematsu, T., Yamaguchi, S., and Nakayama, N. (1996) *Mater. Trans.*, **37**, 121–129.

231 Prieto, P., Fernandez, A., Soriano, L., Yubero, F., Elizalde, E., Gonzalezelipe, A.R., and Sanz, J.M. (1995) *Phys. Rev. B*, **51**, 17984–17987.

232 Corliss, L.M., Elliott, N., and Hastings, J.M. (1960) *Phys. Rev.*, **117**, 929–935.

233 Browne, J.D., Liddell, P.R., Street, R., and Mills, T. (1970) *Phys. Status Solidi A*, **1**, 715–723.

234 Widenmeyer, M., Meissner, E., Senyshyn, A., and Niewa, R. (2014) *Z. Anorg. Allg. Chem.*, **640**, 2801–2808.

235 Ettmayer, P. (1970) *Monatsh. Chem.*, **101**, 127–140.

236 Hägg, G. (1930) *Z. Phys. Chem. B*, **7**, 339–362.

237 Evans, D.A. and Jack, K.H. (1957) *Acta Crystallogr.*, **10**, 833–834.

238 Kawashima, T., Takayama-Muromachi, E., and McMillan, P.F. (2007) *Physica C*, **460**, 651–652.

239 Bull, C.L., Kawashima, T., McMillan, P.F., Machon, D., Shebanova, O., Dalsenberger, D., Soignard, E., Takayama-Muromachi, E., and Chapon, L.C. (2006) *J. Solid State Chem.*, **179**, 1762–1767.

240 Lengauer, W. (1988) *J. Cryst. Growth*, **87**, 295–298.

241 Ganin, A.Y., Kienle, L., and Vajenine, G.V. (2005) *Electrochem. Soc. Proc.*, **9**, 449–456.

242 Ganin, A.Y., Kienle, L., and Vajenine, G.V. (2006) *J. Solid State Chem.*, **179**, 2339–2348.

243 Troickaja, N.V. and Pinsker, Z.G. (1961) *Kristallografiya*, **6**, 43–48.

244 Troickaja, N.V. and Pinsker, Z.G. (1963) *Kristallografiya*, **8**, 548–555.

245 Schönberg, N. (1954) *Acta Chem. Scand.*, **8**, 204–207.

246 Bezinge, A., Yvon, K., Muller, J., Lengauer, W., and Ettmayer, P. (1987) *Solid State Commun.*, **63**, 141–145.

247 Bull, C.L., McMillan, P.F., Soignard, E., and Leinenweber, K. (2004) *J. Solid State Chem.*, **177**, 1488–1492.

248 Kiessling, R. and Liu, Y.H. (1951) *Trans. Am. Inst. Min. Metall. Eng.*, **191**, 639–642.

249 Khitrova, V.I. (1959) *Sov. Phys. Crystallogr.*, **4**, 513–520.

250 Smithells, C.J., Rooksby, H.P., and Pitkin, W.R. (1926) *J. Inst. Met.*, **36**, 107–120.

251 Langmuir, I. (1913) *J. Am. Chem. Soc.*, **35**, 931–945.

252 Wang, S., Yu, X., Lin, Z., Zhang, R., He, D., Qin, J., Zhu, J., Han, J., Wang, L., Mao, H.K., Zhang, J., and Zhao, Y. (2012) *Chem. Mater.*, **24**, 3023–3028.

253 Tessier, F. and Marchand, R. (1997) *J. Alloys Compd.*, **262–263**, 410–415.

254 Khitrova, V.I. and Pinsker, Z.G. (1958) *Kristallografiya*, **3**, 545–553.
255 Khitrova, V.I. and Pinsker, Z.G. (1961) *Kristallografiya*, **6**, 882–891.
256 Juza, R., Puff, H., and Wagenknecht, F. (1957) *Z. Elektrochem.*, **61**, 804–809.
257 Juza, R. (1966) *Adv. Inorg. Chem. Radiochem.*, **9**, 81–131.
258 Takei, W.J., Shirane, G., and Heikes, R.R. (1962) *Phys. Rev.*, **125**, 1893–1897.
259 Guillaud, C. and Wyart, T. (1948) *Rev. Metall.*, **45**, 271–276.
260 Juza, R. and Puff, H. (1957) *Z. Elektrochem.*, **61**, 810–819.
261 Wiener, G.W. and Berger, J.A. (1955) *Trans. Am. Inst. Min. Metall. Eng.*, **203**, 360–368.
262 Takei, W.J., Shirane, G., and Frazer, B.C. (1960) *Phys. Rev.*, **119**, 122–126.
263 Iyer, S.K. and Grabke, H.J. (1973) *Arch. Eisenhüttenwes*, **44**, 720.
264 Lihl, F., Ettmayer, P., and Kutzelnigg, A. (1962) *Z. Metallkd.*, **53**, 715–719.
265 Hägg, G. (1929) *Z. Phys. Chem. B*, **4**, 346–370.
266 Kudielka, H. and Grabke, H.J. (1975) *Z. Metallkd.*, **66**, 469–471.
267 Jarl, M. (1979) *Metall. Trans. A*, **10**, 511–512.
268 Mekata, M., Haruna, J., and Takaki, H. (1968) *J. Phys. Soc. Jpn.*, **25**, 234–238.
269 Eddine, M.N., Bertaut, E.F., and Maunaye, M. (1977) *Acta Crystallogr. B*, **33**, 2696–2698.
270 Leineweber, A. (1999) Ordnungsverhalten von Stickstoff sowie Magnetismus in binären Nitriden einiger 3d-Metalle: Mn/N, Fe/N und Ni/N. Dissertation, Universität Dortmund, pp. 1–221.
271 Leineweber, A., Jacobs, H., Kockelmann, W., and Hull, S. (2000) *Physica B*, **276**, 266–267.
272 Leineweber, A., Jacobs, H., and Kockelmann, W. (2004) *J. Alloys Compd.*, **368**, 229–247.
273 Jacobs, H. and Stüve, C. (1984) *J. Less Common Met.*, **96**, 323–329.
274 Kreiner, G. and Jacobs, H. (1992) *J. Alloys Compd.*, **183**, 345–362.
275 Leineweber, A., Niewa, R., Jacobs, H., and Kockelmann, W. (2000) *J. Mater. Chem.*, **10**, 2827–2834.
276 Tabuchi, M., Takahashi, M., and Kanamaru, F. (1994) *J. Alloys Compd.*, **210**, 143–148.
277 Suzuki, K., Kaneko, T., Yoshida, H., Obi, Y., Fujimori, H., and Morita, H. (2000) *J. Alloys Compd.*, **306**, 66–71.
278 Otsuka, N., Hanawa, Y., and Nagakura, S. (1977) *Phys. Status Solidi A*, **43**, K127–K129.
279 Hahn, H. and Konrad, A. (1951) *Z. Anorg. Chem.*, **264**, 174–180.
280 Clark, P., Dhandapani, B., and Oyama, S.T. (1999) *Appl. Catal. A*, **184**, L175–L180.
281 Kojima, R. and Aika, K. (2001) *Appl. Catal. A*, **209**, 317–325.
282 Haq, A.U. and Meyer, O. (1983) *J. Low Temp. Phys.*, **50**, 123–133.
283 Soto, G., Rosas, A., Farias, M., De la Cruz, W., and Diaz, J. (2007) *Mater. Char.*, **58**, 519–526.
284 Fuchigami, M., Inumaru, K., and Yamanaka, S. (2009) *J. Alloys Compd.*, **486**, 621–627.
285 Friedrich, A., Winkler, B., Bayarjargal, L., Morgenroth, W., Juarez-Arellano, E.A., Milman, V., Refson, K., Kunz, M., and Chen, K. (2010) *Phys. Rev. Lett*, **105**, 088504.
286 Friedrich, A., Winkler, B., Refson, K., and Milman, V. (2010) *Phys. Rev. B*, **82**, 224106.
287 Friedrich, A., Winkler, B., Refson, K., and Milman, V. (2012) *Phys. Rev. B*, **86**, 014114.
288 Fry, A. (1923) *Stahl Eisen*, **43**, 1271–1279.
289 Prenosil, B., Holub, J., and Koutnik, M. (1973) *Härterei. Tech. Mitt.*, **28**, 157–164.
290 Kim, T.K. and Takahash, M. (1972) *Appl. Phys. Lett.*, **20**, 492–494.
291 Andriamandroso, D., Fefilatiev, L., Demazeau, G., Fournes, L., and Pouchard, M. (1984) *Mater. Res. Bull.*, **19**, 1187–1194.
292 Noyes, A.A. and Smith, L.B. (1921) *J. Am. Chem. Soc.*, **43**, 475–481.
293 Lehrer, E. (1930) *Z. Elektrochem.*, **36**, 383–392.
294 Buchwald, V.F. and Nielsen, H.P. (1981) *Lunar Planet. Sci.*, **12**, 112–114.

295 Wriedt, H.A., Gokcen, N.A., and Nafziger, R.H. (1987) *Bull. Alloy Phase Diagrams*, **8**, 355–377.
296 Widenmeyer, M., Niewa, R., Hansen, T.C., and Kohlmann, H. (2013) *Z. Anorg. Allg. Chem.*, **639**, 285–295.
297 Widenmeyer, M., Hansen, T.C., Meissner, E., and Niewa, R. (2014) *Z. Anorg. Allg. Chem.*, **640**, 1265–1274.
298 Widenmeyer, M., Shlyk, L., Senyshyn, A., Moenig, R., and Niewa, R. (2015) *Z. Anorg. Allg. Chem.*, **641**, 348–354.
299 Jack, K.H. (1951) *Proc. R. Soc. Lond. A*, **208**, 200–215.
300 Suzuki, K., Morita, H., Kaneko, T., Yoshida, H., and Fujimori, H. (1993) *J. Alloys Compd.*, **201**, 11–16.
301 Suzuki, K., Yamaguchi, Y., Kaneko, T., Yoshida, H., Obi, Y., Fujimori, H., and Morita, H. (2001) *J. Phys. Soc. Jpn.*, **70**, 1084–1089.
302 Niewa, R., Rau, D., Wosylus, A., Meier, K., Wessel, M., Hanfland, M., Dronskowski, R., and Schwarz, U. (2009) *J. Alloys Compd.*, **480**, 76–80.
303 Guo, K., Rau, D., von Appen, J., Prots, Y., Schnelle, W., Dronskowski, R., Niewa, R., and Schwarz, U. (2013) *High Pressure Res.*, **33**, 684–696.
304 Hägg, G. (1928) *Nature*, **121**, 826–827.
305 Osawa, A. and Iwaizumi, S. (1928) *Z. Kristallogr.*, **69**, 26–34.
306 Hendricks, S.B. and Kosting, P.R. (1930) *Z. Kristallogr.*, **74**, 511–533.
307 Burdese, A. (1957) *Metall. Ital.*, **49**, 195–199.
308 Schwarz, U., Wosylus, A., Wessel, M., Dronskowski, R., Hanfland, M., Rau, D., and Niewa, R. (2009) *Eur. J. Inorg. Chem.*, 1634–1639.
309 Niewa, R., Rau, D., Wosylus, A., Meier, K., Hanfland, M., Wessel, M., Dronskowski, R., Dzivenko, D.A., Riedel, R., and Schwarz, U. (2009) *Chem. Mater.*, **21**, 392–398.
310 Jack, K.H. (1952) *Acta Crystallogr.*, **5**, 404–411.
311 Jack, K.H. (1948) *Proc. R. Soc. Lond. A*, **195**, 34–40.
312 Jack, K.H. (1948) *Proc. R. Soc. Lond. A*, **195**, 41–55.
313 Rechenbach, D. and Jacobs, H. (1996) *J. Alloys Compd.*, **235**, 15–22.
314 Tessier, F., Navrotsky, A., Niewa, R., Leineweber, A., Jacobs, H., Kikkawa, S., Takahashi, M., Kanamaru, F., and DiSalvo, F.J. (2000) *Solid State Sci.*, **2**, 457–462.
315 Wang, J.P., Ji, N., Liu, X., Xu, Y., Sanchez-Hanke, C., Wu, Y., de Groot, F., Allard, L.F., and Lara-Curzio, E. (2012) *IEEE Trans. Magn.*, **48**, 1710–1717.
316 Ogawa, T., Ogata, Y., Gallage, R., Kobayashi, N., Hayashi, N., Kusano, Y., Yamamoto, S., Kohara, K., Doi, M., Takano, M., and Takahashi, M. (2013) *Appl. Phys. Express*, **6**, 073007.
317 Widenmeyer, M., Hansen, T.C., and Niewa, R. (2013) *Z. Anorg. Allg. Chem.*, **639**, 2851–2859.
318 Sims, H., Butler, W., Richter, M., Koepernik, K., Sasioglu, E., Friedrich, C., and Bluegel, S. (2012) *Phys. Rev. B*, **86**, 174422.
319 Shi, Y., Du, Y., and Chen, G. (2013) *Scr. Mater.*, **68**, 976–979.
320 Yu, R., Zhan, Q., and De Jonghe, L.C. (2007) *Angew. Chem., Int. Ed.*, **46**, 1136–1140.
321 Juza, R. and Sachsze, W. (1945) *Z. Anorg. Chem.*, **253**, 95–108.
322 Lourenco, M., Carvalho, M., Fonseca, P., Gasche, T., Evans, G., Godinho, M., and Cruz, M. (2014) *J. Alloys Compd.*, **612**, 176–182.
323 Schmitzdumont, O. and Kron, N. (1955) *Angew. Chem., Int. Ed.*, **67**, 231–232.
324 Suzuki, K., Kaneko, T., Yoshida, H., Morita, H., and Fujimori, H. (1995) *J. Alloys Compd.*, **224**, 232–236.
325 Das, B., Reddy, M., Malar, V.P., Osipowicz, T., Rao, G., and Chowdari, V.B. (2009) *Solid State Ionics*, **180**, 1061–1068.
326 Liu, X., Lu, H., He, M., Jin, K., Yang, G., Ni, H., and Zhao, K. (2014) *J. Alloys Compd.*, **582**, 75–78.
327 Bergstrom, F.W. (1924) *J. Am. Chem. Soc.*, **46**, 2631–2636.
328 Vournasos, A.C. (1919) *C. R. Séances Acad. Sci.*, **168**, 889–891.
329 Nagakura, S., Otsuka, N., and Hirotsu, Y. (1973) *J. Phys. Soc. Jpn.*, **35**, 1492–1495.
330 Fang, C.M., Koster, R.S., Li, W.F., and van Huis, M.A. (2014) *RSC Adv.*, **4**, 7885–7899.

331 Fratczak, E., Prieto, J., and Moneta, M. (2014) *J. Alloys Compd.*, **586**, 375–379.
332 Bohart, G.S. (1915) *J. Phys. Chem.*, **19**, 537–563.
333 Prior, T.J. and Battle, P.D. (2003) *J. Solid State Chem.*, **172**, 138–147.
334 Errandonea, D., Ferrer-Roca, C., Martinez-Garcia, D., Segura, A., Gomis, O., Munoz, A., Rodriguez-Hernandez, P., Lopez-Solano, J., Alconchel, S., and Sapina, F. (2010) *Phys. Rev. B*, **82**, 174105.
335 Hirotsu, Y. and Nagakura, S. (1972) *Acta Metall.*, **20**, 645–655.
336 Yu, R. and Zhang, X.F. (2005) *Appl. Phys. Lett.*, **86**, 121913.
337 Yu, R., Zhan, Q., and Zhang, X.F. (2006) *Appl. Phys. Lett.*, **88**, 051913.
338 Fan, C.Z., Sun, L.L., Wang, Y.X., Wei, Z.J., Liu, R.P., Zeng, S.Y., and Wang, W.K. (2005) *Chin. Phys. Lett.*, **22**, 2637–2638.
339 von Appen, J., Lumey, M.W., and Dronskowski, R. (2006) *Angew. Chem., Int. Ed.*, **45**, 4365–4368.
340 Juza, R. and Hahn, H. (1938) *Z. Anorg. Allg. Chem.*, **239**, 282–287.
341 Juza, R. and Hahn, H. (1939) *Z. Anorg. Allg. Chem.*, **241**, 172–178.
342 Juza, R. (1941) *Z. Anorg. Allg. Chem.*, **248**, 118–120.
343 Juza, R. and Rabenau, A. (1956) *Z. Anorg. Allg. Chem.*, **285**, 212–220.
344 Nakamura, T., Hayashi, H., Hanaoka, T.A., and Ebina, T. (2014) *Inorg. Chem.*, **53**, 710–715.
345 Zhu, W., Zhang, X., Fu, X., Zhou, Y., Luo, S., and Wu, X. (2012) *Phys. Status Solidi A*, **209**, 1996–2001.
346 Asano, M., Umeda, K., and Tasaki, A. (1990) *Jpn. J. Appl. Phys. Pt 1*, **29**, 1985–1986.
347 Wu, H. and Chen, W. (2011) *J. Am. Chem. Soc.*, **133**, 15236–15239.
348 Ji, A., Lu, N., Gao, L., Zhang, W., Liao, L., and Cao, Z.X. (2013) *J. Appl. Phys.*, **113**, 043705.
349 Zachwieja, U. and Jacobs, H. (1990) *J. Less Common Met.*, **161**, 175–184.
350 Zhao, J.G., Yang, L.X., Yu, Y., You, S.J., Liu, J., and Jin, C.Q. (2006) *Phys. Status Solidi B*, **243**, 573–578.
351 Yang, L.X., Zhao, J.G., Yu, Y., Li, F.Y., Yu, R.C., and Jin, C.Q. (2006) *Chin. Phys. Lett.*, **23**, 426–427.
352 Wosylus, A., Schwarz, U., Akselrud, L., Tucker, M.G., Hanfland, M., Rabia, K., Kuntscher, C., von Appen, J., Dronskowski, R., Rau, D., and Niewa, R. (2009) *Z. Anorg. Allg. Chem.*, **635**, 1959–1968.
353 Blucher, J., Bang, K., and Giessen, B.C. (1989) *Mater. Sci. Eng. A*, **117**, L1–L3.
354 Hahn, H. (1948) *Angew. Chem.*, **60**, 254.
355 Raschig, F. (1886) *Ann. Chem.*, **233**, 93–101.
356 Hahn, H. and Gilbert, E. (1949) *Z. Anorg. Chem.*, **258**, 77–93.
357 Shanley, E.S. and Ennis, J.L. (1991) *Ind. Eng. Chem. Res.*, **30**, 2503–2506.
358 Zhuravlev, Y., Lisitsyn, V., and Morozova, Y. (2012) *Phys. Status Solidi B*, **249**, 2096–2107.
359 Chen, W., Tse, J.S., and Jiang, J.Z. (2010) *Solid State Commun.*, **150**, 181–186.
360 Juza, R., Neuber, A., and Hahn, H. (1938) *Z. Anorg. Allg. Chem.*, **239**, 273–281.
361 Juza, R. and Hahn, H. (1940) *Z. Anorg. Allg. Chem.*, **244**, 125–132.
362 Paniconi, G., Stoeva, Z., Smith, R.I., Dippo, P.C., Gallagher, B.L., and Gregory, D.H. (2008) *J. Solid State Chem.*, **181**, 158–165.
363 Suda, T. and Kakishita, K. (2006) *J. Appl. Phys.*, **99**, 076101.
364 Karau, F. and Schnick, W. (2007) *Z. Anorg. Allg. Chem.*, **633**, 223–226.
365 Franklin, E.C. (1905) *J. Am. Chem. Soc.*, **27**, 820–851.
366 Tiede, E. and Knoblauch, H.G. (1935) *Ber.Dtsch. Chem. Ges. A/B*, **68**, 1149–1154.
367 Strecker, W. and Schwinn, E. (1939) *J. Prakt. Chem.*, **152**, 205–218.
368 Lund, H., Oeckler, O., Schröder, T., Schulz, A., and Villinger, A. (2013) *Angew. Chem., Int. Ed.*, **52**, 10900–10904.
369 Matar, S., Fournes, L., Cherubinjeannette, S., and Demazeau, G. (1993) *Eur. J. Solid State Inorg. Chem*, **30**, 871–881.
370 Ito, K., Sanai, T., Yasutomi, Y., Zhu, S., Toko, K., Takeda, Y., Saitoh, Y., Kimura, A., and Suemasu, T. (2014) *J. Appl. Phys.*, **115**, C712.

371 Mohn, P., Schwarz, K., Matar, S., and Demazeau, G. (1992) *Phys. Rev. B Condens. Matter*, **45**, 4000–4007.
372 Li, F.S., Yang, J.B., Xue, D.S., and Zhou, R.J. (1995) *Appl. Phys. Lett.*, **66**, 2343–2345.
373 Rochegude, P. and Foct, J. (1983) *Ann. Chim. Matériaux*, **8**, 533–540.
374 Shirane, G., Takei, W.J., and Ruby, S.L. (1962) *Phys. Rev.*, **126**, 49–52.
375 Li, F.S., Yang, J.B., Xue, D.S., and Zhou, R.J. (1995) *J. Magn. Magn. Mater.*, **151**, 221–224.
376 Li, F.S., Yang, J.B., Xue, D.S., and Zhou, R.J. (1995) *J. Mater. Sci.*, **30**, 4857–4860.
377 Yang, J.B., Xue, D.S., Zhou, R.J., and Li, F.S. (1996) *Phys. Status Solidi A*, **153**, 307–312.
378 Zhao, Z.J., Xue, D.S., Chen, Z.Y., and Li, F.S. (1999) *Phys. Status Solidi A*, **174**, 249–253.
379 Zhao, Z.J., Xue, D.S., and Li, F.S. (2001) *J. Magn. Magn. Mater.*, **232**, 155–160.
380 Niewa, R. (2013) *Z. Anorg. Allg. Chem.*, **639**, 1699–1715.
381 Jäger, J., Stahl, D., Schmidt, P.C., and Kniep, R. (1993) *Angew. Chem., Int. Ed.*, **32**, 709–710.
382 Boca, R. and Kniep, R. (1993) *Solid State Commun.*, **88**, 391–394.
383 Niewa, R. (2005) Habilitation, TU Dresden.
384 Andriamandroso, D., Matar, S., Demazeau, G., and Fournes, L. (1993) *IEEE Trans. Magn.*, **29**, 2–6.
385 dos Santos, A. and Kuhnen, V.C. (2009) *J. Solid State Chem.*, **182**, 3183–3187.
386 von Appen, J. and Dronskowski, R. (2005) *Angew. Chem., Int. Ed.*, **44**, 1205–1210.
387 Kuhnen, C.A. and dosSantos, A.V. (1994) *J. Magn. Magn. Mater.*, **130**, 353–362.
388 Siberchicot, B., Matar, S.F., Fournes, L., Demazeau, G., and Hagenmuller, P. (1990) *J. Solid State Chem.*, **84**, 10–15.
389 Stadelmaier, H.H. and Fraker, A.C. (1960) *Trans. Am. Inst. Min. Metall. Eng.*, **218**, 571–572.
390 de Figueiredo, R.S., Foct, J., dos Santos, A.V., and Kuhnen, C.A. (2001) *J. Alloys Compd.*, **315**, 42–50.
391 de Figueiredo, R.S., Kuhnen, C.A., and dos Santos, A.V. (1997) *J. Magn. Magn. Mater.*, **173**, 141–154.
392 Kuhnen, C.A., de Figueiredo, R.S., and dos Santos, A.V. (2000) *J. Magn. Magn. Mater.*, **219**, 58–68.
393 Stadelmaier, H.H. and Yun, T.S. (1961) *Z. Metallkd.*, **52**, 477–480.
394 Stadelmaier, H.H. (1961) *Z. Metallkd.*, **52**, 758–762.
395 Stadelmaier, H.H. and Fraker, A.C. (1962) *Z. Metallkd.*, **53**, 48–51.
396 Music, D., Burghaus, J., Takahashi, T., Dronskowski, R., and Schneider, J. (2010) *Eur. Phys. J. B*, **77**, 401–406.
397 Takahashi, T., Music, D., and Schneider, J.M. (2012) *J. Vac. Sci. Technol. A*, **30**, 030602.
398 dos Santos, A.V., Padilha, G., and Moncalves, M. (2012) *Solid State Sci.*, **14**, 269–275.
399 Houben, A., Muller, P., von Appen, J., Lueken, H., Niewa, R., and Dronskowski, R. (2005) *Angew. Chem., Int. Ed.*, **44**, 7212–7215.
400 Houben, A., Sepelak, V., Becker, K.D., and Dronskowski, R. (2009) *Chem. Mater.*, **21**, 784–788.
401 Burghaus, J., Wessel, M., Houben, A., and Dronskowski, R. (2010) *Inorg. Chem.*, **49**, 10148–10155.
402 Burghaus, J., Sougrati, M.T., Moechel, A., Houben, A., Hermann, R.P., and Dronskowski, R. (2011) *J. Solid State Chem.*, **184**, 2315–2321.
403 Houben, A., Burghaus, J., and Dronskowski, R. (2009) *Chem. Mater.*, **21**, 4332–4338.
404 Burghaus, J., Kleemann, J., and Dronskowski, R. (2011) *Z. Anorg. Allg. Chem.*, **637**, 935–939.
405 Ranade, M.R., Tessier, F., Navrotsky, A., and Marchand, R. (2001) *J. Mater. Res.*, **16**, 2824–2831.
406 Onderka, B., Unland, J., and Schmid-Fetzer, R. (2002) *J. Mater. Res.*, **17**, 3065–3083.
407 Kuhnen, C.A. and dos Santos, A.V. (2004) *J. Alloys Compd.*, **384**, 80–87.
408 Boller, H. (1968) *Monatsh. Chem.*, **99**, 2444–2449.
409 dos Santos, A.V. and Kuhnen, C.A. (2001) *J. Alloys Compd.*, **321**, 60–66.
410 Takahashi, Y., Imai, Y., and Kumagai, T. (2011) *J. Magn. Magn. Mater.*, **323**, 2941–2944.

411 Matar, S., Mohn, P., Demazeau, G., and Schwarz, K. (1991) *J. Magn. Magn. Mater.*, **101**, 251–252.
412 Gajbhiye, N. and Bhattacharyya, S. (2007) *Indian J. Pure Appl. Phys.*, **45**, 834–838.
413 Kuhnen, C.A. and dos Santos, A.V. (2000) *J. Alloys Compd.*, **297**, 68–72.
414 Matar, S., Mohn, P., and Kubler, J. (1992) *J. Magn. Magn. Mater.*, **104**, 1927–1928.
415 Juza, R., Deneke, K., and Puff, H. (1959) *Z. Elektrochem.*, **63**, 551–557.
416 Burdese, A., Firrao, D., Rolando, P., and Rosso, M. (1984) *Chim. Ind. (Milan)*, **66**, 456–460.
417 Samson, C., Bouchaud, J.P., and Fruchart, R. (1964) *C. R. Hebd. Séances Acad. Sci.*, **259**, 392–393.
418 Barberon, M., Madar, R., Fruchart, E., Lorthioir, G., and Fruchart, R. (1970) *Mater. Res. Bull.*, **5**, 1–8.
419 Barberon, M., Fruchart, E., Fruchart, R., Lorthioir, G., Madar, R., and Nardin, M. (1972) *Mater. Res. Bull.*, **7**, 109–118.
420 Nardin, M., Lorthioir, G., Barberon, M., Madar, R., Fruchart, E., and Fruchart, R. (1972) *C. R. Séances Acad. Sci. C*, **274**, 2168–2171.
421 Fruchart, D., l'Héritier, P., and Fruchart, R. (1980) *Mater. Res. Bull.*, **15**, 415–420.
422 García, J., Bartolomé, J., González, D., Navarro, R., and Fruchart, D. (1983) *J. Chem. Thermodyn.*, **15**, 1041–1057.
423 García, J., Bartolomé, J., González, D., Navarro, R., and Fruchart, D. (1983) *J. Chem. Thermodyn.*, **15**, 1169–1180.
424 García, J., Navarro, R., Bartolomé, J., Burriel, R., González, D., and Fruchart, D. (1980) *J. Magn. Magn. Mater.*, **15–18**, 1155–1156.
425 García, J., Rojo, J.A., Navarro, R., Bartolomé, J., and González, D. (1983) *J. Magn. Magn. Mater.*, **31–34**, 1401–1403.
426 García, J., González, D., Bartolomé, J., Navarro, R., and Rojo, J.A. (1985) *J. Magn. Magn. Mater.*, **51**, 365–374.
427 Navarro, R., Rojo, J.A., García, J., González, D., Bartolomé, J., and l'Héritier, P. (1986) *J. Magn. Magn. Mater.*, **59**, 221–234.
428 Ettmayer, P. (1967) *Monatsh. Chem.*, **98**, 1881–1883.
429 Fruchart, R., Bouchaud, J.P., Fruchart, E., Lorthioi, G., Madar, R., and Rouault, A. (1967) *Mater. Res. Bull.*, **2**, 1009–1020.
430 Barberon, M., Lorthioi, G., and Fruchart, R. (1969) *Bull. Soc. Chim. Fr.*, **7**, 2329–2331.
431 Krén, E., Zsoldos, É., Barberon, M., and Fruchart, R. (1971) *Solid State Commun.*, **9**, 27–31.
432 Krén, E., Kadar, G., Barberon, M.M., and Fruchart, R. (1971) *Int. J. Magn.*, **1**, 341–344.
433 Fruchart, D. and Bertaut, E.F. (1978) *J. Phys. Soc. Jpn.*, **44**, 781–791.
434 Lin, J., Wang, B., Tong, P., Lu, W., Zhang, L., Zhu, X., Yang, Z., Song, W., Dai, J., and Sun, Y. (2011) *Appl. Phys. Lett.*, **98**, 092507.
435 Shao, D., Lu, W., Lin, J., Tong, P., Jian, H., and Sun, Y. (2013) *J. Appl. Phys.*, **113**, 023905.
436 Takenaka, K., Ichigo, M., Hamada, T., Ozawa, A., Shibayama, T., Inagaki, T., and Asano, K. (2014) *Sci. Technol. Adv. Mater.*, **15**, 015009.
437 Mekata, M. (1962) *J. Phys. Soc. Jpn.*, **17**, 796–803.
438 Bertaut, E.F., Fruchart, D., Bouchaud, J.P., and Fruchart, R. (1968) *Solid State Commun.*, **6**, 251–256.
439 Aoki, M., Yamane, H., Shimada, M., and Kajiwara, T. (2004) *J. Alloys Compd.*, **364**, 280–282.
440 Shimizu, T., Shibayama, T., Asano, K., and Takenaka, K. (2012) *J. Appl. Phys.*, **111**, 07A903.
441 Fruchart, D., Bertaut, E.F., Madar, R., Lorthioir, G., and Fruchart, R. (1971) *Solid State Commun.*, **9**, 1793–1797.
442 Wu, M., Wang, C., Sun, Y., Chu, L., Yan, J., Chen, D., Huang, Q., and Lynn, J.W. (2013) *J. Appl. Phys.*, **114**, 123902.
443 Madar, R., Gilles, L., Rouault, A., Bouchaud, J.P., Fruchart, E., Lorthioir, G., and Fruchart, R. (1967) *C. R. Séances Acad. Sci. C*, **264**, 308–311.
444 Asano, K., Koyama, K., and Takenaka, K. (2008) *Appl. Phys. Lett.*, **92**, 161909.
445 Yang, C., Tong, P., Lin, J., Lin, S., Cui, D., Wang, B., Song, W., Lu, W., and Sun, Y. (2014) *J. Appl. Phys.*, **116**, 033902.

446 Madar, R., Barberon, M., Lorthioi, G., Fruchart, E., and Fruchart, R. (1968) *C. R. Séances Acad. Sci. C*, **267**, 1404–1406.

447 Goumri-Said, S., Kanoun, M., and Calvayrac, F. (2009) *J. Magn. Magn. Mater.*, **321**, 1012–1014.

448 Nakamura, Y., Takenaka, K., Kishimoto, A., and Takagi, H. (2009) *J. Am. Ceram. Soc.*, **92**, 2999–3003.

449 Huang, R.J., Li, L.F., Cai, F.S., Xu, X.D., and Qian, L.H. (2008) *Appl. Phys. Lett.*, **93**, 081902.

450 Qu, B., He, H., and Pan, B. (2014) *Appl. Phys. A*, **114**, 785–791.

451 Huang, R., Wu, Z., Chu, X., Yang, H., Chen, Z., and Li, L. (2010) *Solid State Sci.*, **12**, 1977–1980.

452 Huang, R., Wu, Z., Yang, H., Chen, Z., Chu, X., and Li, L. (2010) *Cryogenics*, **50**, 750–753.

453 Chen, Z., Huang, R., Chu, X., Wu, Z., Liu, Z., Zhou, Y., and Li, L. (2012) *Cryogenics*, **52**, 629–631.

454 Tan, J., Huang, R., Li, W., Huang, C., Han, Y., and Li, L. (2014) *Cryogenics*, **63**, 122–124.

455 Sun, Z. and Song, X. (2014) *J. Mater. Sci. Technol.*, **30**, 903–909.

456 Iikubo, S., Kodama, K., Takenaka, K., Takagi, H., Takigawa, M., and Shamoto, S. (2008) *Phys. Rev. Lett*, **101**, 205901.

457 Kodama, K., Iikubo, S., Takenaka, K., Takigawa, M., Takagi, H., and Shamoto, S. (2010) *Phys. Rev. B*, **81**, 224419.

458 Iikubo, S., Kodama, K., Takenaka, K., Takagi, H., and Shamoto, S. (2008) *Phys. Rev. B*, **77**, 020409.

459 Takenaka, K. and Takagi, H. (2005) *Appl. Phys. Lett.*, **87**, 261902.

460 Takenaka, K. and Takagi, H. (2006) *Mater. Trans.*, **47**, 471–474.

461 Takenaka, K., Asano, K., Misawa, M., and Takagi, H. (2008) *Appl. Phys. Lett.*, **92**, 011927.

462 Takenaka, K. and Takagi, H. (2009) *Appl. Phys. Lett.*, **94**, 131904.

463 Song, X., Sun, Z., Huang, Q., Rettenmayr, M., Liu, X., Seyring, M., Li, G., Rao, G., and Yin, F. (2011) *Adv. Mater.*, **23**, 4690–4694.

464 Dai, Y., Song, X., Huang, R., Li, L., and Sun, Z. (2015) *Mater. Lett.*, **139**, 409–413.

465 Sun, Y., Wang, C., Wen, Y., Chu, L., Pan, H., and Niez, M. (2010) *J. Am. Ceram. Soc.*, **93**, 2178–2181.

466 Takenaka, K. (2012) *Sci. Technol. Adv. Mater.*, **13**, 013001.

467 Takenaka, K., Shibayama, T., Asano, K., and Koyama, K. (2010) *J. Phys. Soc. Jpn.*, **79**, 073706.

468 Takenaka, K., Hamada, T., Shibayama, T., and Asano, K. (2013) *J. Alloys Compd.*, **577**, S291–S295.

469 Serghiou, G., Guillaume, C.L., Thomson, A., Morniroli, J.P., and Frost, D.J. (2009) *J. Am. Chem. Soc.*, **131**, 15170–15175.

470 He, B., Dong, C., Yang, L., Chen, X., Ge, L., Mu, L., and Shi, Y. (2013) *Supercond. Sci. Technol.*, **26**, 125015.

471 Hui, Z., Tang, X., Shao, D., Lei, H., Yang, J., Song, W., Luo, H., Zhu, X., and Sun, Y. (2014) *Chem. Commun.*, **50**, 12734–12737.

472 Uehara, M., Uehara, A., Kozawa, K., and Kimishima, Y. (2009) *J. Phys. Soc. Jpn.*, **78**, 033702.

473 Ohishi, K., Ito, T.U., Higemoto, W., Yamazaki, T., Uehara, A., Kozawa, K., Kimishima, Y., and Uehara, M. (2010) *Physica C*, **470**, S705–S706.

474 Yamazaki, T., Uehara, A., Kozawa, K., Kimisima, Y., and Uehara, M. (2012) *Adv. Cond. Mat. Phys.*, **2012**, 902812.

475 Uehara, M., Uehara, A., Kozawa, K., Yamazaki, T., and Kimishima, Y. (2010) *Physica C*, **470**, S688–S690.

476 He, B., Dong, C., Yang, L., Ge, L., and Chen, H. (2011) *J. Solid State Chem.*, **184**, 1939–1945.

477 Cao, W., He, B., Liao, C., Yang, L., Zeng, L., and Dong, C. (2009) *J. Solid State Chem.*, **182**, 3353–3357.

478 He, B., Dong, C., Yang, L., Ge, L., Mu, L., and Chen, X. (2012) *Chin. Phys. B*, **21**, 047401.

479 Shein, I., Bannikov, V., and Ivanovskii, V.A. (2010) *Phys. Status Solidi B*, **247**, 72–76.

480 Okoye, C. (2010) *Physica B*, **405**, 1562–1570.

481 Bannikov, V., Shein, V.I., and Ivanovskii, A. (2010) *Physica B*, **405**, 4615–4619.

482 Bannikov, V., Shein, V.I., and Ivanovskii, A. (2010) *Comput. Mater. Sci.*, **49**, 457–461.

483 Tütüncü, H.M. and Srivastava, G. (2013) *Philos. Mag.*, **93**, 4469–4487.
484 Tütüncü, H.M. and Srivastava, G. (2014) *Physica C*, **507**, 10–16.
485 Tütüncü, H.M. and Srivastava, G. (2006) *J. Phys. Condens. Matter*, **18**, 11089–11101.
486 Li, C., Chen, W., Wang, F., Li, S., Sun, Q., Wang, S., and Jia, Y. (2009) *J. Appl. Phys.*, **105**, 123921.
487 Shim, J.H., Kwon, S.K., and Min, B.I. (2001) *Phys. Rev. B Condens. Matter*, **6418**, 180510.
488 Hou, Z. (2010) *Solid State Commun.*, **150**, 1874–1879.
489 Helal, M. and Islam, A. (2011) *Physica B*, **406**, 4564–4568.
490 Xu, Y., Gao, F., Hao, X., and Li, Z. (2010) *Comput. Mater. Sci.*, **50**, 737–741.
491 Liu, L., Wu, X., Wang, R., Gan, L., and Wei, Q. (2014) *J. Supercond. Nov. Magn.*, **27**, 2607–2615.
492 Liu, L., Wu, X., Wang, R., Gan, L., and Wei, Q. (2014) *J. Supercond. Nov. Magn.*, **27**, 1851–1859.
493 Zhai, H., Li, X., Du, J., and Ji, G. (2012) *Chin. Phys. B*, **21**, 057102.
494 Jacobs, H. and Zachwieja, U. (1991) *J. Less Common Met.*, **170**, 185–190.
495 Vaughn, D.D., Araujo, J., Meduri, P., Callejas, J.F., Hickner, M.A., and Schaak, R.E. (2014) *Chem. Mater.*, **26**, 6226–6232.
496 Hahn, U. and Weber, W. (1996) *Phys. Rev. B Condens. Matter*, **53**, 12684–12693.
497 Ji, A., Li, C., and Cao, Z. (2006) *Appl. Phys. Lett.*, **89**, 252120.
498 Lu, N., Ji, A., and Cao, Z. (2013) *Sci. Rep.*, **3**, 3090.
499 Gao, L., Ji, A., Zhang, W., and Cao, Z.X. (2011) *J. Cryst. Growth*, **321**, 157–161.
500 Soto, G., Ponce, I., Moreno, M., Yubero, F., and de la Cruz, W. (2014) *J. Alloys Compd.*, **594**, 48–51.
501 Cui, X., Soon, A., Phillips, A., Zheng, R., Liu, Z., Delley, B., Ringer, S., and Stampfl, C. (2012) *J. Magn. Magn. Mater.*, **324**, 3138–3143.
502 Moreno-Armenta, M., Lopez Perez, W., and Takeuchi, N. (2007) *Solid State Sci.*, **9**, 166–172.
503 Moreno-Armenta, M.G., Soto, G., and Takeuchi, N. (2011) *J. Alloys Compd.*, **509**, 1471–1476.
504 Arbey, R.M., Guadalupe Moreno-Armenta, M., and Takeuchi, N. (2013) *J. Alloys Compd.*, **576**, 285–290.
505 Gulo, F., Simon, A., Köhler, J., and Kremer, R.K. (2004) *Angew. Chem., Int. Ed.*, **43**, 2032–2034.
506 Lin, S., Tong, P., Wang, B., Huang, Y., Shao, D., Lu, W., and Sun, Y.P. (2014) *J. Solid State Chem.*, **209**, 127–134.
507 Wiendlocha, B., Tobola, J., Kaprzyk, S., and Fruchart, D. (2007) *J. Alloys Compd.*, **442**, 289–291.
508 Tütüncü, H.M. and Srivastava, G. (2012) *J. Appl. Phys.*, **112**, 093914.
509 Tütüncü, H.M. and Srivastava, G. (2013) *J. Appl. Phys.*, **114**, 053905.
510 Lin, S., Tong, P., Wang, B., Huang, Y., Song, W., and Sun, Y. (2014) *J. Alloys Compd.*, **584**, 308–314.
511 Brady, M.P., Hoelzer, D.T., Payzant, E.A., Tortorelli, P.F., Horton, J.A., Anderson, I.M., Walker, L.R., and Wrobel, S.K. (2001) *J. Mater. Res.*, **16**, 2784–2787.
512 Brady, M.P., Wrobel, S.K., Lograsso, T.A., Payzant, E.A., Hoelzer, D.T., Horton, J.A., and Walker, L.R. (2004) *Chem. Mater.*, **16**, 1984–1990.
513 Rieger, W., Nowotny, H., and Benesovs, F. (1965) *Monatsh. Chem.*, **96**, 232–241.
514 Beckmann, O., Boller, H., Nowotny, H., and Benesovs, F. (1969) *Monatsh. Chem.*, **100**, 1465–1470.
515 Schuster, J.C. and Bauer, J. (1984) *J. Solid State Chem.*, **53**, 260–265.
516 Weil, K.S. and Kumta, P.N. (1996) *Mater. Sci. Eng. B*, **38**, 109–117.
517 Jeitschko, W., Nowotny, H., and Benesovsky, F. (1964) *Monatsh. Chem.*, **95**, 436–438.
518 Vogtenhuber-Pawelczak, D. and Herzig, P. (1992) *J. Solid State Chem.*, **99**, 85–94.
519 Ivanovskii, A.L., Medvedeva, N.I., and Novikov, D.L. (1997) *Phys. Solid State*, **39**, 929–931.
520 Kanchana, V. (2009) *Europhys. Lett.*, **87**, 26006.
521 Cherrad, D., Selmani, L., Maouche, D., and Maamache, M. (2011) *J. Alloys Compd.*, **509**, 4357–4362.

522 Wang, J., Chen, Z., Li, C., Li, F., and Nie, C. (2014) *J. Solid State Chem.*, **216**, 1–8.
523 Höglund, C., Birch, J., Beckers, M., Alling, B., Czigany, Z., Muecklich, A., and Hultman, L. (2008) *Eur. J. Inorg. Chem.*, 1193–1195.
524 Zhao, J.T., Dong, Z.C., Vaughey, J.T., Ostenson, J.E., and Corbett, J.D. (1995) *J. Alloys Compd.*, **230**, 1–12.
525 Schuster, J.C., Bauer, J., and Nowotny, H. (1985) *Rev. Chim. Minér.*, **22**, 546–554.
526 Schuster, J.C. and Bauer, J. (1985) *J. Less Common Met.*, **109**, 345–350.
527 Magnuson, M., Mattesini, M., Hoeglund, C., Abrikosov, I.A., Birch, J., and Hultman, L. (2008) *Phys. Rev. B*, **78**, 235102.
528 Hossain, M., Ali, M., Parvin, F., and Islam, A. (2013) *Comput. Mater. Sci.*, **73**, 1–8.
529 Kanchana, V. and Ram, S. (2012) *Intermetallics*, **23**, 39–48.
530 Mikhaylushkin, A., Höglund, C., Birch, J., Czigany, Z., Hultman, L., Simak, S., Alling, I.B., Tasnadi, F., and Abrikosov, I. (2009) *Phys. Rev. B*, **79**, 134107.
531 Bagci, S., Yalcin, B.G., Tütüncü, H.M., and Srivastava, G. (2010) *Phys. Rev. B*, **81**, 054523.
532 Mattesini, M., Magnuson, M., Tasnadi, F., Höglund, C., Abrikosov, I.A., and Hultman, L. (2009) *Phys. Rev. B*, **79**, 125122.
533 Kacimi, S., Mekam, D., Djermouni, M., Azzouz, M., Hallouche, A., and Zaoui, A. (2013) *Mater. Sci. Semicond. Process.*, **16**, 1971–1976.
534 Schuster, J.C. (1985) *J. Less Common Met.*, **105**, 327–332.
535 Gäbler, F., Schnelle, W., Senyshyn, A., and Niewa, R. (2008) *Solid State Sci.*, **10**, 1910–1915.
536 Bettahar, N., Nasri, D., Benalia, S., Merabet, M., Abidri, B., Benkhettou, N., Khenata, R., Rached, D., and Rabah, M. (2013) *Int. J. Thermophys.*, **34**, 434–449.
537 Ram, S. and Kanchana, V. (2014) *Solid State Commun.*, **181**, 54–59.
538 Kirchner, M., Schnelle, W., and Niewa, R. (2006) *Z. Naturforsch. B*, **61**, 813–819.
539 Gäbler, F., Bräunling, D., Schnelle, W., Schellenberg, I., Pöttgen, R., and Niewa, R. (2011) *Z. Anorg. Allg. Chem.*, **637**, 977–982.
540 Gajbhiye, N.S., Ningthoujam, R.S., and Weissmuller, J. (2004) *Hyperfine Interact.*, **156**, 51–56.
541 Gajbhiye, N., Ningthoujam, R., and Bhattacharyya, S. (2005) *Hyperfine Interact.*, **164**, 17–26.
542 Gajbhiye, N. and Bhattacharyya, S. (2005) *Hyperfine Interact.*, **165**, 147–151.
543 Gajbhiye, N. and Bhattacharyya, S. (2006) *Prog. Cryst. Growth Charact.*, **52**, 132–141.
544 Gajbhiye, N. and Bhattacharyya, S. (2008) *Mater. Chem. Phys.*, **108**, 201–207.
545 Gajbhiye, N., Bhattacharyya, S., and Sharma, S. (2008) *Pramana*, **70**, 367–373.
546 Ningthoujam, R. and Gajbhiye, N. (2010) *Mater. Res. Bull.*, **45**, 499–504.
547 Guo, K., Rau, D., Toffoletti, L., Mueller, C., Burkhardt, U., Schnelle, W., Niewa, R., and Schwarz, U. (2012) *Chem. Mater.*, **24**, 4600–4606.
548 Zhang, C., Yan, M., You, Y., Che, H., Zhang, F., Bai, B., Chen, L., Long, Z., and Li, R. (2014) *J. Alloys Compd.*, **615**, 854–862.
549 Guo, K., Rau, D., Schnelle, W., Burkhardt, U., Niewa, R., and Schwarz, U. (2014) *Z. Anorg. Allg. Chem.*, **640**, 814–818.
550 Jeitschko, W., Nowotny, H., and Benesovsky, F. (1964) *Monatsh. Chem.*, **95**, 1212–1218.
551 Sliwko, V., Mohn, P., and Schwarz, K. (1994) *J. Phys. Condens. Matter*, **6**, 6557–6564.
552 Evans, D.A. and Jack, K.H. (1957) *Acta Crystallogr.*, **10**, 769–770.
553 Weil, K.S., Kumta, P.N., and Grins, J. (1999) *J. Solid State Chem.*, **146**, 22–35.
554 Herle, P.S., Hegde, M.S., Sooryanarayana, K., Guru Row, T.N., and Subbanna, G.N. (1998) *Inorg. Chem.*, **37**, 4128–4130.
555 Oldham, S.E., Battle, P.D., Blundell, S.J., Brooks, M.L., Pratt, F.L., and Prior, T.J. (2005) *J. Mater. Chem.*, **15**, 3402–3408.
556 Prior, T.J., Oldham, S.E., Couper, V.J., and Battle, P.D. (2005) *Chem. Mater.*, **17**, 1867–1873.
557 Prior, T.J., Nguyen-Manh, D., Couper, V.J., and Battle, P.D. (2004) *J. Phys. Condens. Matter*, **16**, 2273–2281.

558 Alconchel, S., Sapina, F., Beltran, D., and Beltran, A. (1998) *J. Mater. Chem.*, **8**, 1901–1909.

559 El-Himri, A., Sapiña, F., Ibañez, R., and Beltrán, A. (2001) *J. Mater. Chem.*, **11**, 2311–2314.

560 El-Himri, A., Marrero-Lopez, D., and Nunez, P. (2004) *J. Solid State Chem.*, **177**, 3219–3223.

561 Wang, H., Wu, Z., Kong, J., Wang, Z., and Zhang, M. (2012) *J. Solid State Chem.*, **194**, 238–244.

562 Herle, P.S., Hegde, M.S., Sooryanarayana, K., Row, T.N.G., and Subbanna, G.N. (1998) *J. Mater. Chem.*, **8**, 1435–1440.

563 Alconchel, S., Sapiña, F., Beltrán, D., and Beltrán, A. (1999) *J. Mater. Chem.*, **9**, 749–755.

564 Jeitschko, W., Nowotny, H., and Benesovsky, F. (1964) *Monatsh. Chem.*, **95**, 156–157.

565 Jeitschko, W., Holleck, H., Nowotny, H., and Benesovsky, F. (1965) *Monatsh. Chem.*, **95**, 1004–1006.

566 Ono, N., Kajihara, M., and Kikuchi, M. (1992) *Metall. Trans. A*, **23**, 1389–1393.

567 Chen, W.F., Sasaki, K., Ma, C., Frenkel, A.I., Marinkovic, N., Muckerman, J.T., Zhu, Y., and Adzic, R.R. (2012) *Angew. Chem., Int. Ed.*, **51**, 6131–6135.

568 Li, W., Wang, Z.Q., Zhang, M.H., and Tao, K.Y. (2005) *Catal. Commun.*, **6**, 656–660.

569 Holleck, H. and Thümmler, F. (1967) *Monatsh. Chem.*, **98**, 133–134.

570 Kotyk, M. and Stadelma, H.H. (1970) *Metall. Trans.*, **1**, 899–903.

571 Prior, T.J. and Battle, P.D. (2004) *J. Mater. Chem.*, **14**, 3001–3007.

572 Bem, D.S., Gibson, C.P., and zur Loye, H.C. (1993) *Chem. Mater.*, **5**, 397–399.

573 Weil, K.S. and Kumta, P.N. (1997) *Acta Crystallogr. C*, **53**, 1745–1748.

574 Weil, K.S. and Kumta, P.N. (1998) *J. Alloys Compd.*, **265**, 96–103.

575 Jackson, S.K., Layland, R.C., and zur Loye, H.C. (1999) *J. Alloys Compd.*, **291**, 94–101.

576 Weil, K.S. and Kumta, P.N. (1997) *J. Solid State Chem.*, **134**, 302–311.

577 Chen, Y., Shen, J., and Chen, N.X. (2009) *Solid State Commun.*, **149**, 121–125.

578 Battle, P.D., Sviridov, L.A., Woolley, R.J., Grandjean, F., Long, G.J., Catlow, C., Sokol, A.A., Walsh, A., and Woodley, S.M. (2012) *J. Mater. Chem.*, **22**, 15606–15613.

579 Hunter, S.M., Mckay, D., Smith, R.J., Hargreaves, J.S., and Gregory, D.H. (2010) *Chem. Mater.*, **22**, 2898–2907.

580 Waki, T., Terazawa, S., Tabata, Y., Murase, Y., Kato, M., Hirota, K., Ikeda, S., Kobayashi, H., Sato, K., Kindo, K., and Nakamura, H. (2011) *J. Alloys Compd.*, **509**, 9451–9455.

581 Waki, T., Terazawa, S., Tabata, Y., Oba, F., Michioka, C., Yoshimura, K., Ikeda, S., Kobayashi, H., Ohoyama, K., and Nakamura, H. (2010) *J. Phys. Soc. Jpn.*, **79**, 043701.

582 Waki, T., Terazawa, S., Yamazaki, T., Tabata, Y., Sato, K., Kondo, A., Kindo, K., Yokoyama, M., Takahashi, Y., and Nakamura, H. (2011) *Europhys. Lett.*, **94**, 37004.

583 Waki, T., Terazawa, S., Tabata, Y., Sato, K., Kondo, A., Kindo, K., and Nakamura, H. (2014) *Phys. Rev. B*, **90**, 014416.

584 Panda, R.N. and Gajbhiye, N.S. (1997) *J. Alloys Compd.*, **256**, 102–107.

585 Panda, R., Balaji, G., Pandey, P., and Gajbhiye, N. (2008) *Hyperfine Interact.*, **184**, 245–250.

586 Lü, F.X. and Jack, K.H. (1985) *J. Less Common Met.*, **114**, 123–127.

587 Barsoum, M.W. (2000) *Prog. Solid State Chem.*, **28**, 201–281.

588 Barsoum, M.W. and Radovic, M. (2011) *Annu. Rev. Mater. Res.*, **41**, 195–227.

589 Eklund, P., Beckers, M., Jansson, U., Hogberg, H., and Hultman, L. (2010) *Thin Solid Films*, **518**, 1851–1878.

590 Jeitschko, W., Nowotny, H., and Benesovsky, F. (1963) *Monatsh. Chem.*, **94**, 1198–1200.

591 Barsoum, M.W., Rawn, C.J., El-Raghy, T., Procopio, A.T., Porter, W.D., Wang, H., and Hubbard, C.R. (2000) *J. Appl. Phys.*, **87**, 8407–8414.

592 Procopio, A.T., Barsoum, M.W., and El-Raghy, T. (2000) *Metall. Mater. Trans. A*, **31**, 333–337.

593 Rawn, C.J., Barsoum, M.W., El-Raghy, T., Procipio, A., Hoffmann, C.M., and

Hubbard, C.R. (2000) *Mater. Res. Bull.*, **35**, 1785–1796.

594 Barsoum, M.W., Farber, L., Levin, I., Procopio, A., El-Raghy, T., and Berner, A. (1999) *J. Am. Ceram. Soc.*, **82**, 2545–2547.

595 Procopio, A.T., El-Raghy, T., and Barsoum, M.W. (2000) *Metall. Mater. Trans. A*, **31**, 373–378.

596 Lee, H.D. and Petuskey, W.T. (1997) *J. Am. Ceram. Soc.*, **80**, 604–608.

597 Lee, H.D. and Petuskey, W.T. (1998) *J. Am. Ceram. Soc.*, **81**, 787–788.

598 Barsoum, M.W. and Schuster, J.C. (1998) *J. Am. Ceram. Soc.*, **81**, 785–786.

599 Low, I., Pang, W., Kennedy, S., and Smith, R.I. (2011) *J. Eur. Ceram. Soc.*, **31**, 159–166.

600 Li, C. and Wang, Z. (2011) *Phys. Status Solidi B*, **248**, 1639–1644.

601 Holm, B., Ahuja, R., Li, S., and Johansson, B. (2002) *J. Appl. Phys.*, **91**, 9874–9877.

602 Sun, Z. and Zhou, Y. (2002) *J. Phys. Soc. Jpn.*, **71**, 1313–1317.

603 Hug, G., Jaouen, M., and Barsoum, M.W. (2005) *Phys. Rev. B*, **71**, 024105.

604 Hug, G. and Fries, E. (2002) *Phys. Rev. B*, **65**, 113104.

605 Khazaei, M., Arai, M., Sasaki, T., Estili, M., and Sakka, Y. (2014) *J. Phys. Condens. Matter*, **26**, 505503.

606 Bortolozo, A., Serrano, G., Serquis, A., Rodrigues, D., dos Santos, C., Fisk, Z., and Machado, A. (2010) *Solid State Commun.*, **150**, 1364–1366.

607 Jeitschko, W., Nowotny, H., and Benesovsky, F. (1964) *Monatsh. Chem.*, **95**, 178–179.

608 Jeitschko, W., Nowotny, H., and Benesovsky, F. (1964) *Monatsh. Chem.*, **95**, 431–435.

609 Daoudi, B., Yakoubi, A., Beldi, L., and Bouhafs, B. (2007) *Acta Mater.*, **55**, 4161–4165.

610 Liu, Z., Waki, T., Tabata, Y., Yuge, K., Nakamura, H., and Watanabe, I. (2013) *Phys. Rev. B*, **88**, 134401.

611 Barsoum, M.W. and Farber, L. (1999) *Science*, **284**, 937–939.

612 Wang, J., Wang, J., Li, F., and Zhou, Y. (2009) *J. Mater. Res.*, **24**, 3523–3532.

613 Boller, H. and Nowotny, H. (1968) *Monatsh. Chem.*, **99**, 721–725.

614 Boller, H. (1971) *Monatsh. Chem.*, **102**, 431–437.

615 Schuster, J.C., Bauer, J., and Debuigne, J. (1983) *J. Nucl. Mater.*, **116**, 131–135.

616 Schuster, J.C. and Bauer, J. (1984) *J. Nucl. Mater.*, **120**, 133–136.

617 Schuster, J.C. (1986) *Z. Kristallogr.*, **175**, 211–215.

618 Li, F., Hu, C., Wang, J., Liu, B., Wang, J., and Zhou, Y. (2009) *J. Am. Ceram. Soc.*, **92**, 476–480.

619 Wang, B. and Ohgushi, K. (2013) *Sci. Rep.*, **3**, 3381.

620 Wang, B., Cheng, J., Matsubayashi, K., Uwatoko, Y., and Ohgushi, K. (2014) *Phys. Rev. B*, **90**, 139901.

621 Wang, B., Cheng, J., Matsubayashi, K., Uwatoko, Y., and Ohgushi, K. (2014) *Phys. Rev. B*, **89**, 144510.

622 Wang, B. and Ohgushi, K. (2015) *J. Phys. Soc. Jpn.*, **84**, 044707.

623 Danielsen, H.K. and Hald, J. (2006) *Energy Mater.*, **1**, 49–57.

624 Jack, D.H. and Jack, K.H. (1972) *J. Iron Steel Inst.(London)*, **210**, 790–792.

625 Ettmayer, P. (1971) *Monatsh. Chem.*, **102**, 858–863.

626 Strang, A. and Vodarek, V. (1996) *Mater. Sci. Technol.*, **12**, 552–556.

627 Danielsen, H.K., Hald, J., Grumsen, F.B., and Somers, M.A. (2006) *Metall. Mater. Trans. A*, **37**, 2633–2640.

628 Karlsson, L., Henjered, A., Andren, H.O., and Norden, H. (1985) *Mater. Sci. Technol.*, **1**, 337–343.

629 Hughes, H. (1967) *J. Iron Steel Inst. (London)*, **205**, 775–778.

630 Danielsen, H.K. and Hald, J. (2009) *Scr. Mater.*, **60**, 811–813.

631 Vendl, A. (1979) *Monatsh. Chem.*, **110**, 103–108.

632 Vendl, A. (1979) *Monatsh. Chem.*, **110**, 1099–1107.

633 Vendl, A. (1978) *Monatsh. Chem.*, **109**, 1001–1004.

634 Vendl, A. (1979) *Monatsh. Chem.*, **110**, 879–885.

635 Verdier, P., L'Haridon, P., Maunaye, M., and Marchand, R. (1974) *Acta Crystallogr. B*, **30**, 226–228.

636 Li, H.S. and Coey, J.M.D. (1991) *Handbook of Magnetic Materials*, vol. **6**, Elsevier, pp. 1–83.

637 Fujii, H. and Sun, H. (1995) *Handbook of Magnetic Materials*, vol. **9**, Elsevier, pp. 303–404.

638 Coey, J.M.D. and Smith, P.A.I. (1999) *J. Magn. Magn. Mater.*, **200**, 405–424.

639 Toda, K., Honda, M., Orihara, Y., Sato, M., and Kanamaru, F. (2000) *Key Eng. Mater.*, **181–182**, 213–216.

640 Florio, J.V., Rundle, R.E., and Snow, A.I. (1952) *Acta Crystallogr.*, **5**, 449–457.

641 Iandelli, A. and Palenzon, A. (1967) *J. Less Common Met.*, **12**, 333.

642 Florio, J.V., Baenziger, N.C., and Rundle, R.E. (1956) *Acta Crystallogr.*, **9**, 367–372.

643 Felner, I. (1980) *J. Less Common Met.*, **72**, 241–249.

644 Gryniv, I.A., Godovanec, O.I., Lapunova, R.V., Grin', Yu., and Yarmolyuk, Ya.P. (1983) *Dopov. Akad. Nauk Ukr. RSR Ser. A*, **45**, 75–78.

645 de Mooij, D.B. and Buschow, K.H.J. (1988) *J. Less Common Met.*, **136**, 207–215.

646 Akayama, M., Fujii, H., Yamamoto, K., and Tatami, K. (1994) *J. Magn. Magn. Mater.*, **130**, 99–107.

647 Coey, J.M.D., Otani, Y., Sun, H., and Hurley, D.P.F. (1991) *J. Magn. Soc. Jpn.*, **15**, 769–772.

648 Hurley, D.P.F., Coey, M., Osakaba, T., Kuroiwa, Y., and Kohgi, M. (1994) *J. Phys. Soc. Jpn.*, **63**, 3048–3052.

649 Yang, Y., Zhang, X., Kong, L., Pan, Q., and Ge, S. (1991) *Appl. Phys. Lett.*, **58**, 2042–2044.

650 Yang, Y.C., Zhang, X.D., Ge, S.L., Pan, Q., Kong, L.S., Li, H., Yang, J.L., Zhang, B.S., Ding, Y.F., and Ye, C.T. (1991) *J. Appl. Phys.*, **70**, 6001–6005.

651 Yang, Y.C., Zhang, X.D., Kong, L.S., Pan, Q., Ge, S.L., Yang, J.L., Ding, Y.F., Zhang, B.S., Ye, C.T., and Jin, L. (1991) *Solid State Commun.*, **78**, 313–316.

652 Kim, Y.B., Kim, H.T., Lee, K.W., Kim, C.S., and Kim, T.K. (1992) *IEEE Trans. Magn.*, **28**, 2566–2568.

653 Anagnostou, M., Christides, C., Pissas, M., and Niarchos, D. (1991) *J. Appl. Phys.*, **70**, 6012–6014.

654 Chen, X., Liao, L.X., Altounian, Z., Ryan, D.H., and Stromolsen, J.O. (1992) *J. Magn. Magn. Mater.*, **111**, 130–134.

655 Wang, Y.Z., Hadjipanayis, G.C., Kim, A., Sellmyer, D.J., and Yelon, W.B. (1992) *J. Magn. Magn. Mater.*, **104**, 1132–1134.

656 Fujii, H., Miyazaki, Y., Tatami, K., Sun, H., Morii, Y., Akayama, M., and Funahashi, S. (1995) *J. Magn. Magn. Mater.*, **140**, 1089–1090.

657 Psycharis, V., Anagnostou, M., Christides, C., and Niarchos, D. (1991) *J. Appl. Phys.*, **70**, 6122–6124.

658 Anagnostou, M., Christides, C., and Niarchos, D. (1991) *Solid State Commun.*, **78**, 681–684.

659 Sun, H., Morii, Y., Fujii, H., Akayama, M., and Funahashi, S. (1993) *Phys. Rev. B*, **48**, 13333–13339.

660 Du, H., Shan, Y., Liu, S., Wang, C., Han, J., Yang, Y., and Yue, M. (2007) *J. Appl. Phys.*, **101**, 09D514.

661 Gou, C., Zhang, B.S., Cheng, Z.X., Cheng, Y.F., Du, H., and Sun, K. (1999) *Yuanz. Kexue Jishu*, **33**, 515–520.

662 Liu, J.P., Brabers, J.H.V.J., Winkelman, A.J.M., Menovsky, A.A., de Boer, F.R., and Buschow, K.H.J. (1993) *J. Alloys Compd.*, **200**, L3–L6.

663 Katter, M., Wecker, J., Kuhrt, C., Schultz, L., and Grossinger, R. (1992) *J. Magn. Magn. Mater.*, **114**, 35–44.

664 Hu, B.P., Rao, X.L., Xu, J.M., Liu, G.C., Cao, F., Dong, X.L., Li, H., Yin, L., and Zhao, Z.R. (1992) *J. Magn. Magn. Mater.*, **114**, 138–144.

665 Mulder, F.M., Thiel, R.C., Coehoorn, R., Jacobs, T.H., and Buschow, K.H.J. (1992) *J. Magn. Magn. Mater.*, **117**, 413–418.

666 Li, X.W., Tang, N., Lu, Z.H., Zhao, T.Y., Lin, W.G., Zhao, R.W., and Yang, F.M. (1993) *J. Appl. Phys.*, **73**, 5890–5892.

667 Yang, F.M., Li, X.W., Tang, N., Wang, J.L., Lu, Z.H., Zhao, T.Y., Li, Q.A., Liu, J.P., and Deboer, F.R. (1995) *J. Alloys Compd.*, **221**, 248–253.

668 Middleton, D.P. and Buschow, K.H.J. (1994) *IEEE Trans. Magn.*, **30**, 699–701.

669 Middleton, D.P. and Buschow, K.H.J. (1994) *J. Alloys Compd.*, **203**, 217–220.

670 Plugaru, N., Valeanu, M., and Burzo, E. (1994) *IEEE Trans. Magn.*, **30**, 663–665.

671 Valeanu, M. and Plugaru, N. (1994) *Mater. Lett.*, **18**, 331–335.

672 Valeanu, M., Plugaru, N., and Burzo, E. (1994) *Solid State Commun.*, **89**, 519–522.

673 Rogalski, M.S., Morariu, M., Valeanu, M., and Plugaru, N. (1995) *Solid State Commun.*, **96**, 573–577.

674 Tang, N., Liu, Y.L., Yu, M.J., Lu, Y., Tegus, O., Li, Q.A., Ji, S.Q., and Yang, F.M. (1992) *J. Magn. Magn. Mater.*, **104**, 1086–1087.

675 Huang, M.Q., Zheng, Y., Miller, K., Elbicki, J.M., Sankar, S.G., and Wallace, W.E. (1991) *J. Appl. Phys.*, **70**, 6024–6026.

676 Huang, M.Q., Zheng, Y., Miller, K., Elbicki, J.M., Sankar, S.G., Wallace, W.E., and Obermyer, R. (1991) *J. Magn. Magn. Mater.*, **102**, 91–95.

677 Katter, M., Wecker, J., Kuhrt, C., Schultz, L., Kou, X.C., and Grossinger, R. (1992) *J. Magn. Magn. Mater.*, **111**, 293–300.

678 Liang, J.Z., Yu, M.J., Tang, N., Liu, Y.L., Liu, J.F., Ji, S.Q., and Chen, G.M. (1991) *J. Magn. Magn. Mater.*, **102**, 217–220.

679 Lu, Y., Tegus, O., Li, Q.A., Tang, N., Yu, M.J., Zhao, R.W., Kuang, J.P., Yang, F.M., Zhou, G.F., Li, X., and Deboer, F.R. (1992) *Physica B*, **177**, 243–246.

680 Tegus, O., Lu, Y., Tang, N., Wu, J.X., Yu, M., Li, Q.A., Zhao, R.W., Jian, Y., and Yang, F. (1992) *IEEE Trans. Magn.*, **28**, 2581–2583.

681 Yu, M.J., Tang, N., Liu, Y.L., Tegus, O., Lu, Y., Kuang, J.P., Yang, F.M., Li, X., Zhou, G.F., and Deboer, F.R. (1992) *Physica B*, **177**, 238–242.

682 Coey, M.D., Skomski, R., and Wirth, S. (1992) *IEEE Trans. Magn.*, **28**, 2332–2337.

683 Coey, J.M.D., Lawler, J.F., Sun, H., and Allan, J.E.M. (1991) *J. Appl. Phys.*, **69**, 3007–3010.

684 Isnard, O., Soubeyroux, J.L., Miraglia, S., Fruchart, D., Garcia, L.M., and Bartolome, J. (1992) *Physica B*, **180**, 624–626.

685 Ibberson, R.M., Moze, O., Jacobs, T.H., and Buschow, K.H.J. (1991) *J. Phys. Condens. Matter*, **3**, 1219–1226.

686 Yelon, W.B. and Hadjipanayis, G.C. (1992) *IEEE Trans. Magn.*, **28**, 2316–2321.

687 Tomey, E., Isnard, O., Fagan, A., Desmoulins, C., Miraglia, S., Soubeyroux, J.L., and Fruchart, D. (1993) *J. Alloys Compd.*, **191**, 233–238.

688 Wang, Y.Z., Hadjipanayis, G.C., Tang, Z.X., Yelon, W.B., Papaefthymiou, V., Moukarika, A., and Sellmyer, D.J. (1993) *J. Magn. Magn. Mater.*, **119**, 41–48.

689 Morii, Y., Sun, H., Fujii, H., Miyazaki, Y., Akayama, M., and Funahashi, S. (1995) *Physica B*, **213**, 291–293.

690 Psycharis, V., Kalogirou, O., Niarchos, D., and Gjoka, M. (1998) *Mater. Sci. Forum*, **278-2**, 526–531.

691 Hu, Z., Yelon, W.B., Kalogirou, O., and Psycharis, V. (1996) *J. Appl. Phys.*, **80**, 2955–2959.

692 Drebov, N., Martinez-Limia, A., Kunz, L., Gola, A., Shigematsu, T., Eckl, T., Gumbsch, P., and Elsässer, C. (2013) *New J. Phys.*, **15**, 125023.

693 Lewis, L.H. and Jimenez-Villacorta, F. (2013) *Metall. Mater. Trans. A*, **44**, 2–20.

694 Talbi, K., Cherchab, Y., and Sekkal, N. (2012) *Eur. Phys. J. Appl. Phys.*, **58**, 30103.

695 Vendl, A. (1979) *J. Nucl. Mater.*, **79**, 246–248.

696 Holleck, H., Smailos, E., and Thuemmler, F. (1969) *J. Nucl. Mater.*, **32**, 281–289.

697 Ihara, S., Tanaka, K., Suzuki, M., and Akimoto, Y. (1971) *J. Nucl. Mater.*, **39**, 203–208.

698 Aivazov, M.I. and Rezchikova, T.V. (1977) *Zh. Neorg. Khim.*, **22**, 458–463.

699 Ettmayer, P., Waldhart, J., Vendl, A., and Banik, G. (1980) *J. Nucl. Mater.*, **91**, 293–296.

700 Kouhsen, C., Naoumidis, A., and Nickel, H. (1976) *J. Nucl. Mater.*, **61**, 88–98.

701 Holleck, H., Smailos, E., and Thummler, F. (1968) *J. Nucl. Mater.*, **28**, 105–109.

702 Holleck, H. and Smailos, E. (1980) *J. Nucl. Mater.*, **91**, 237–239.

703 Ettmayer, P., Waldhart, J., Vendl, A., and Banik, G. (1980) *Monatsh. Chem.*, **111**, 945–948.

704 Ettmayer, P., Waldhart, J., and Vendl, A. (1979) *Monatsh. Chem.*, **110**, 1109–1112.

705 Venard, J.T., Spruiell, J.E., and Cavin, O.B. (1967) *J. Nucl. Mater.*, **24**, 245–246.

706 Venard, J.T. and Spruiell, J.E. (1968) *J. Nucl. Mater.*, **27**, 257–263.

707 Meta, N.M., Schweda, E., Boysen, H., Haug, A., Chasse, T., and Hoelzel, M. (2007) *Z. Anorg. Allg. Chem.*, **633**, 790–794.
708 Holleck, H., Smailos, E., and Thummler, F. (1968) *Monatsh. Chem.*, **99**, 985–989.
709 Nowotny, H., Benesovsky, F., and Rudy, E. (1960) *Monatsh. Chem.*, **91**, 348–356.
710 Kieffer, R., Ettmayer, P., and Petter, F. (1971) *Monatsh. Chem.*, **102**, 1182–1196.
711 Duwez, P. and Odell, F. (1950) *J. Electrochem. Soc.*, **97**, 299–304.
712 Yen, C.M., Toth, L.E., Shy, Y.M., Anderson, D.E., and Rosner, L.G. (1967) *J. Appl. Phys.*, **38**, 2268–2271.
713 Aivazov, M.I., Rezchikova, T.V., Agababyan, E.V., and Domashnev, I.A. (1976) *Inorg. Mater.*, **12**, 368–370.
714 Aivazov, M.I., Rezchikova, T.V., and Degtyareva, V.F. (1976) *J. Struct. Chem.*, **17**, 267–270.
715 Yamamoto, T., Kikkawa, S., Takahashi, M., Miyamoto, Y., and Kanamaru, F. (1991) *Physica C*, **185**, 2719–2720.
716 Straumanis, M.E. and Faunce, C.A. (1967) *Z. Anorg. Allg. Chem.*, **353**, 329–335.
717 Anderson, D.E., Toth, L.E., Rosner, L.G., and Yen, C.M. (1965) *Appl. Phys. Lett.*, **7**, 90–92.
718 Toth, L.E., Yen, C.M., Rosner, L.G., and Anderson, D.E. (1966) *J. Phys. Chem. Solids*, **27**, 1815–1819.
719 Eddine, M.N., Bertaut, E.F., Roubin, M., and Paris, J. (1977) *Acta Crystallogr. B*, **33**, 3010–3013.
720 Aivazov, M.I., Rezchikova, T.V., and Degtyareva, V.F. (1975) *Inorg. Mater.*, **11**, 201–203.
721 Ettmayer, P., Schebesta, W., Vendl, A., and Kieffer, R. (1978) *Monatsh. Chem.*, **109**, 929–941.
722 Brauer, G. and Mohr-Rosenbaum, E. (1974) *Z. Anorg. Allg. Chem.*, **405**, 225–229.
723 Weil, K.S. (2004) *J. Solid State Chem.*, **177**, 1976–1986.
724 Lai, G.C., Takahashi, M., Nobugai, K., and Kanamaru, F. (1989) *J. Am. Ceram. Soc.*, **72**, 2310–2313.
725 Das, B., Reddy, M.V., and Chowdari, B.V. (2013) *Nanoscale*, **5**, 1961–1966.
726 Inamura, S., Nobugai, K., and Kanamaru, F. (1987) *J. Solid State Chem.*, **68**, 124–127.
727 Höglund, C., Birch, J., Alling, B., Bareno, J., Czigany, Z., Persson, P.O., Wingqvist, G., Zukauskaite, A., and Hultman, L. (2010) *J. Appl. Phys.*, **107**, 123515.
728 Schönberg, N. (1954) *Acta Chem. Scand.*, **8**, 213–220.
729 Ettmayer, P. and Vendl, A. (1980) *Monatsh. Chem.*, **111**, 547–550.
730 Schönberg, N. (1954) *Acta Metall.*, **2**, 427–432.
731 Brauer, G. and Kiliani, W. (1979) *Z. Anorg. Allg. Chem.*, **452**, 17–26.
732 Ettmayer, P. and Vendl, A. (1980) *J. Less Common Met.*, **72**, 209–217.
733 Arnott, R.J. and Wold, A. (1960) *J. Phys. Chem. Solids*, **15**, 152–156.
734 Nowotny, H. (1963) *Electronic Structure and Alloy Chemistry of the Transition Elements*, John Wiley & Sons, Inc., New York, pp. 179–235.
735 Jeitschko, W., Nowotny, H., and Benesovsky, F. (1964) *Monatsh. Chem.*, **95**, 1242–1246.
736 Kwon, Y.U. and Corbett, J.D. (1992) *Chem. Mater.*, **4**, 1348–1355.
737 Guloy, A.M. and Corbett, J.D. (1993) *Inorg. Chem.*, **32**, 3532–3540.
738 Corbett, J.D., Garcia, E., Guloy, A.M., Hurng, W.M., Kwon, Y.U., and Leon-Escamilla, E.A. (1998) *Chem. Mater.*, **10**, 2824–2836.
739 Thom, A.J., Young, V.G., and Akinc, M. (2000) *J. Alloys Compd.*, **296**, 59–66.
740 Guloy, A.M. and Corbett, J.D. (1992) *Z. Anorg. Allg. Chem.*, **616**, 61–66.
741 Frankenburger, W., Andissow, L., and Ludwigs, F.D. (1928) *Z. Elektrochem.*, **34**, 632–637.
742 Klatyk, J. and Kniep, R. (1999) *Z. Kristallogr.*, **214**, 447–448.
743 Gudat, A., Kniep, R., and Rabenau, A. (1991) *Angew. Chem.*, **103**, 217–218.
744 Rabenau, A. (1982) *Solid State Ionics*, **6**, 277–293.
745 von Alpen, U. (1979) *J. Solid State Chem.*, **29**, 379–392.
746 Zucker, U.H. and Schulz, H. (1982) *Acta Crystallogr. A*, **38**, 568–576.

747 Novák, P. and Wagner, F.R. (2004) *J. Magn. Magn. Mater.*, **272–276** (Suppl.), e269–e270.
748 Sachsze, W. and Juza, R. (1949) *Z. Anorg. Allg. Chem.*, **260**, 278–290.
749 Gudat, A., Kniep, R., and Rabenau, A. (1990) *Thermochim. Acta*, **160**, 49–56.
750 Niewa, R., Huang, Z.L., Schnelle, W., Hu, Z., and Kniep, R. (2003) *Z. Anorg. Allg. Chem.*, **629**, 1778–1786.
751 Bach, S., Pereira-Ramos, J.P., Ducros, J.B., and Willmann, P. (2009) *Solid State Ionics*, **180**, 231–235.
752 Niewa, R., Wagner, F.R., Schnelle, W., Hochrein, O., and Kniep, R. (2001) *Inorg. Chem.*, **40**, 5215–5222.
753 Klatyk, J. and Kniep, R. (1999) *Z. Kristallogr.*, **214**, 445–446.
754 Martem'yanov, N.A., Tamm, V.Kh., Obrosov, V.P., and Martem'yanova, Z.S. (1995) *Inorg. Mater.*, **31**, 65–68.
755 Yamada, A., Matsumoto, S., and Nakamura, Y. (2011) *J. Mater. Chem.*, **21**, 10021–10025.
756 Boström, M. and Hovmöller, S. (2001) *J. Alloys Compd.*, **314**, 154–159.
757 Gordon, A.G., Smith, R.I., Wilson, C., Stoeva, Z., Gregory, D.H., and Duncan, H. (2004) *Chem. Commun.*, **2004**, 2812–2813.
758 Schnelle, W., Niewa, R., and Wagner, F.R. (2004) *J. Magn. Magn. Mater.*, **272**, 828–829.
759 Niewa, R., Hu, Z.W., and Kniep, R. (2003) *Eur. J. Inorg. Chem.*, **2003**, 1632–1634.
760 Klatyk, J., Schnelle, W., Wagner, F.R., Niewa, R., Novák, P., Kniep, R., Waldeck, M., Ksenofontov, V., and Gütlich, P. (2002) *Phys. Rev. Lett*, **88**, 207202.
761 Liu, D.-L., Du, F., Wei, Y.-J., Wang, C.-Z., Huang, Z.-F., Meng, X., Chen, G., Chen, Y., and Feng, S.-H. (2009) *Mater. Lett.*, **63**, 133–135.
762 Novák, P. and Wagner, F.R. (2002) *Phys. Rev. B*, **66**, 184434.
763 Stoeva, Z., Gomez, R., Gregory, D.H., Hix, G.B., and Titman, J.J. (2004) *Dalton Trans.*, **2004**, 3093–3097.
764 Stoeva, Z., Smith, R.I., and Gregory, D.H. (2005) *Chem. Mater.*, **18**, 313–320.
765 Gregory, D.H., O'Meara, P.M., Gordon, A.G., Hodges, J.P., Short, S., and Jorgensen, J.D. (2002) *Chem. Mater.*, **14**, 2063–2070.
766 Gordon, A.G., Gregory, D.H., Blake, A.J., Weston, D.P., and Jones, M.O. (2001) *Int. J. Inorg. Mater.*, **3**, 973–981.
767 Klatyk, J., Höhn, P., and Kniep, R. (1998) *Z. Kristallogr.*, **213**, 31.
768 Barker, M.G., Blake, A.J., Edwards, P.P., Gregory, D.H., Hamor, T.A., Siddons, D.J., and Smith, S.E. (1999) *Chem. Commun.*, **1999**, 1187–1188.
769 Hu, C.H., Yang, Y., and Zhu, Z.Z. (2009) *Solid State Sci.*, **11**, 1898–1902.
770 Hu, C.H., Yang, Y., and Zhu, Z.Z. (2010) *Solid State Commun.*, **150**, 669–674.
771 Gregory, D.H. (2008) *Chem. Rec.*, **8**, 229–239.
772 Ducros, J.B., Bach, S., Pereira-Ramos, J.P., and Willmann, P. (2007) *Electrochem. Commun.*, **9**, 2496–2500.
773 Wu, S., Dong, Z., Wu, P., and Boey, F. (2011) *J. Mater. Chem.*, **21**, 165–170.
774 Ducros, J.B., Bach, S., Pereira-Ramos, J.P., and Willmann, P. (2007) *Electrochim. Acta*, **52**, 7035–7041.
775 Chern, M.Y. and DiSalvo, F.J. (1990) *J. Solid State Chem.*, **88**, 528–533.
776 Yamane, H. and DiSalvo, F.J. (1995) *J. Solid State Chem.*, **119**, 375–379.
777 Hoppe, R. and Rohrborn, H.J. (1964) *Z. Anorg. Allg. Chem.*, **329**, 110–122.
778 Yamane, H. and DiSalvo, F.J. (1996) *J. Alloys Compd.*, **234**, 203–206.
779 Bendyna, J.K., Höhn, P., and Kniep, R. (2007) *Z. Kristallogr.*, **222**, 165–166.
780 Yamamoto, T., Kikkawa, S., and Kanamaru, F. (1993) *Solid State Ionics*, **63-5**, 148–153.
781 Bendyna, J.K., Höhn, P., Prots', Yu., and Kniep, R. (2007) *Z. Kristallogr.*, **222**, 167–168.
782 Ovchinnikov, A., Höhn, P., Borrmann, H., Kazancioglu, M., and Kniep, R. (2015) *Z. Kristallogr.*, **230**, 3–4.
783 Niewa, R. and DiSalvo, F.J. (1998) *J. Alloys Compd.*, **279**, 153–160.
784 King, R.B. (1995) *Can. J. Chem.*, **73**, 963–971.
785 Bendyna, J.K. (2009) New developments in nitridometalates, cyanamides and cyanides: chemical and physical properties. Dissertation, TU Dresden, pp. 1–227.

786 Yamane, H., Sasaki, S., Kubota, S., Shimada, M., and Kajiwara, T. (2003) *J. Solid State Chem.*, **170**, 265–272.

787 Yamane, H., Sasaki, S., Kubota, S., Inoue, R., Shimada, M., and Kajiwara, T. (2002) *J. Solid State Chem.*, **163**, 449–454.

788 Yamane, H., Sasaki, S., Kubota, S., Kajiwara, T., and Shimada, M. (2002) *Acta Crystallogr. C*, **58**, i50–i52.

789 Zakutayev, A., Allen, A.J., Zhang, X., Vidal, J., Cui, Z., Lany, S., Yang, M., DiSalvo, F.J., and Ginley, D.S. (2014) *Chem. Mater.*, **26**, 4970–4977.

790 Zachwieja, U. and Jacobs, H. (1991) *Eur. J. Solid State Inorg. Chem.*, **28**, 1055–1062.

791 Miura, A., Lowe, M., Leonard, B.M., Subban, C.V., Masubuchi, Y., Kikkawa, S., Dronskowski, R., Hennig, R.G., Abruna, H.D., and DiSalvo, F.J. (2011) *J. Solid State Chem.*, **184**, 7–11.

792 Chern, M.Y. and DiSalvo, F.J. (1990) *J. Solid State Chem.*, **88**, 459–464.

793 Yamamoto, T., Kikkawa, S., and Kanamaru, F. (1995) *J. Solid State Chem.*, **115**, 353–359.

794 Jäger, J. and Kniep, R. (1992) *Z. Naturforsch. B*, **47**, 1290–1296.

795 Jäger, J. (1995) Nitride und Nitridoverbindungen in Systemen Li–(Ca, Sr, Ba)–(Co, Ag, Au)–N. Dissertation, TH Darmstadt, pp. 1–186.

796 Gudat, A., Kniep, R., and Maier, J. (1992) *J. Alloys Compd.*, **186**, 339–345.

797 Gerss, M.H. and Jeitschko, W. (1986) *Z. Naturforsch. B*, **41**, 946–950.

798 Massidda, S., Pickett, W.E., and Posternak, M. (1991) *Phys. Rev. B*, **44**, 1258–1265.

799 Hannebauer, B., Schmidt, P.C., Kniep, R., Jansen, N., Walcher, D., Gütlich, P., Gottschall, R., Schollhorn, R., and Methfessel, M. (1996) *Z. Naturforsch. A*, **51**, 515–526.

800 Cordier, G., Gudat, A., Kniep, R., and Rabenau, A. (1989) *Angew. Chem., Int. Ed.*, **28**, 201–202.

801 Klatyk, J. and Kniep, R. (1999) *Z. Kristallogr.*, **214**, 451–452.

802 Höhn, P. and Kniep, R. (1992) *Z. Naturforsch. B*, **47**, 434–436.

803 Yamane, H., Kikkawa, S., and Koizumi, M. (1987) *J. Solid State Chem.*, **71**, 1–11.

804 Cenzual, K., Gelato, L.M., Penzo, M., and Parthé, E. (1991) *Acta Crystallogr. B*, **47**, 433–439.

805 Gudat, A., Haag, S., Kniep, R., and Rabenau, A. (1990) *J. Less Common Met.*, **159**, 29–31.

806 Cordier, G., Gudat, A., Kniep, R., and Rabenau, A. (1989) *Angew. Chem., Int. Ed.*, **28**, 1702–1703.

807 Klatyk, J. and Kniep, R. (1999) *Z. Kristallogr.*, **214**, 449–450.

808 Gudat, A. and Kniep, R. (1992) *J. Alloys Compd.*, **179**, 333–338.

809 Klatyk, J., Niewa, R., and Kniep, R. (2000) *Z. Naturforsch. B*, **55**, 988–991.

810 van Duijn, J., Suzuki, K., and Attfield, J.P. (2000) *Angew. Chem., Int. Ed.*, **39**, 365–366.

811 Albert, B. and Schmitt, K. (1999) *Inorg. Chem.*, **38**, 6159–6163.

812 Gudat, A., Kniep, R., and Rabenau, A. (1991) *Z. Anorg. Allg. Chem.*, **597**, 61–67.

813 Tennstedt, A. and Kniep, R. (1994) *Z. Anorg. Allg. Chem.*, **620**, 1781–1785.

814 Green, M.T. and Hughbanks, T. (1993) *Inorg. Chem.*, **32**, 5611–5615.

815 Gudat, A., Milius, W., Haag, S., Kniep, R., and Rabenau, A. (1991) *J. Less Common Met.*, **168**, 305–312.

816 Gudat, A., Kniep, R., and Rabenau, A. (1992) *Z. Anorg. Allg. Chem.*, **607**, 8–12.

817 Bendyna, J.K., Höhn, P., and Kniep, R. (2008) *Z. Kristallogr.*, **223**, 183–184.

818 Bendyna, J.K., Höhn, P., and Kniep, R. (2008) *Z. Kristallogr.*, **223**, 185–186.

819 Gregory, D.H., Barker, M.G., Edwards, P.P., and Siddons, D.J. (1995) *Inorg. Chem.*, **34**, 5195–5198.

820 Cordier, G., Höhn, P., Kniep, R., and Rabenau, A. (1990) *Z. Anorg. Allg. Chem.*, **591**, 58–66.

821 Vennos, D.A. and DiSalvo, F.J. (1992) *J. Solid State Chem.*, **100**, 401.

822 Vennos, D.A. and DiSalvo, F.J. (1992) *J. Solid State Chem.*, **98**, 318–322.

823 Tennstedt, A., Röhr, C., and Kniep, R. (1993) *Z. Naturforsch. B*, **48**, 1831–1834.

824 Barker, M.G., Begley, M.J., Edwards, P.P., Gregory, D.H., and Smith, S.E. (1996) *J. Chem. Soc., Dalton Trans.*, **1996**, 1–5.

825 Tennstedt, A., Röhr, C., and Kniep, R. (1993) *Z. Naturforsch. B*, **48**, 794–796.

826 Höhn, P. (1993) Ternäre und quaternäre Nitridometallate: Verbindungen in den Systemen Lithium–Erdalkalimetall–Übergangsmetall–Stickstoff (Übergangsmetall=Ta, Mo, W, Fe, Co). Dissertation, TH Darmstadt, pp. 1–284.

827 Yee, K.A. and Hughbanks, T. (1992) *Inorg. Chem.*, **31**, 1921–1925.

828 Schmidt, C.L., Dinnebier, R., and Jansen, M. (2010) *Z. Anorg. Allg. Chem.*, **636**, 2529–2531.

829 Schmidt, C.L., Wedig, U., Dinnebier, R., and Jansen, M. (2008) *Chem. Asian. J.*, **2008**, 3 1983.

830 Höhn, P., Haag, S., Milius, W., and Kniep, R. (1991) *Angew. Chem., Int. Ed.*, **30**, 831–832.

831 Niewa, R., DiSalvo, F.J., Yang, D.K., Zax, D.B., Luo, H., and Yelon, W.B. (1998) *J. Alloys Compd.*, **266**, 32–38.

832 Chaushli, A., Wickleder, C., and Jacobs, H. (2000) *Z. Anorg. Allg. Chem.*, **626**, 892–896.

833 Hochrein, O., Grin', Y., and Kniep, R. (1998) *Angew. Chem., Int. Ed.*, **37**, 1582–1585.

834 Hochrein, O., Höhn, P., and Kniep, R. (2003) *Z. Anorg. Allg. Chem.*, **629**, 923–927.

835 Bolvin, H. and Wagner, F.R. (2012) *Inorg. Chem.*, **51**, 7112–7118.

836 Bailey, M.S., Obrovac, M.N., Baillet, E., Reynolds, T.K., Zax, D.B., and DiSalvo, F.J. (2003) *Inorg. Chem.*, **42**, 5572–5578.

837 Marchand, R. and Lemarchand, V. (1981) *J. Less Common Met.*, **80**, 157–163.

838 Pollmeier, P. (1989) Ternäre Pnictide der Lanthanoide, Actinoide und des Zirkoniums mit Chrom und Mangan. Dissertation, Westfälischen Wilhelms-Universität Münster, pp. 1–138.

839 Broll, S. and Jeitschko, W. (1995) *Z. Naturforsch. B*, **50**, 905–912.

840 Juza, R., Langer, K., and von Benda, K. (1968) *Angew. Chem.*, **80**, 373–384.

841 Gudat, A. (1990) Ternäre und quaternäre Nitride und Nitridometallate in Systemen Lithium–Erdalkalimetall–Übergangsmetall–Stickstoff. Dissertation, Heinrich-Heine-Universität Düsseldorf, pp. 1–187.

842 Niewa, R. and Kniep, R. (2001) *Z. Kristallogr.*, **216**, 5–6.

843 Niewa, R., Zherebtsov, D., and Hu, Z. (2003) *Inorg. Chem.*, **42**, 2538–2544.

844 Juza, R., Gieren, W., and Haug, J. (1959) *Z. Anorg. Allg. Chem.*, **300**, 61–71.

845 Juza, R., Anschütz, E., and Puff, H. (1959) *Angew. Chem.*, **71**, 161.

846 Vennos, D.A. and DiSalvo, F.J. (1992) *Acta Crystallogr. C*, **48**, 610–612.

847 Wachsmann, C. and Jacobs, H. (1992) *J. Alloys Compd.*, **190**, 113–116.

848 Cabana, J., Dupre, N., Rousse, G., Grey, C.P., and Palacin, M.R. (2005) *Solid State Ionics*, **176**, 2205–2218.

849 Chaushli, A., Jacobs, H., Weisser, U., and Strähle, J. (2000) *Z. Anorg. Allg. Chem.*, **626**, 1909–1914.

850 Niewa, R. and Zherebtsov, D.A. (2004) *Z. Anorg. Allg. Chem.*, **630**, 229–233.

851 Niewa, R., Zherebtsov, D.A., Borrmann, H., and Kniep, R. (2002) *Z. Anorg. Allg. Chem.*, **628**, 2505–2508.

852 Niewa, R., Zherebtsov, D.A., and Leoni, S. (2003) *Chem. Eur. J.*, **9**, 4255–4259.

853 Juza, R. (1948) *Angew. Chem.*, **60**, 74.

854 Juza, R. and Hund, F. (1948) *Z. Anorg. Chem.*, **257**, 13–25.

855 Wentorf, R.H. and Kasper, J.S. (1963) *Science*, **139**, 338–339.

856 Juza, R. and Hund, F. (1946) *Naturwissenschaften*, **33**, 121–122.

857 Juza, R. and Hund, F. (1948) *Z. Anorg. Chem.*, **257**, 1–12.

858 Chen, X.Z. and Eick, H.A. (1996) *J. Solid State Chem.*, **127**, 19–24.

859 Wachsmann, C. and Jacobs, H. (1996) *Z. Kristallogr.*, **211**, 477.

860 Wachsmann, C., Brokamp, T., and Jacobs, H. (1992) *J. Alloys Compd.*, **185**, 109–119.

861 Cabana, J., Ling, C.D., Oro-Sole, J., Gautier, D., Tobias, G., Adams, S., Canadell, E., and Palacin, M.R. (2004) *Inorg. Chem.*, **43**, 7050–7060.

862 Juza, R., Weber, H.H., and Meyer-Simon, E. (1953) *Z. Anorg. Allg. Chem.*, **273**, 48–64.

863 Lang, J. and Charlot, J.P. (1970) *Rev. Chim. Miner.*, **7**, 121–131.

864 David, J., Charlot, J.P., and Lang, J. (1974) *Rev. Chim. Miner.*, **11**, 405–413.

865 Cabana, J., Casas-Cabanas, M., Santner, H.J., Fuertes, A., and Palacin, M.R. (2010) *J. Solid State Chem.*, **183**, 1609–1614.

866 Cabana, J., Rousse, G., Fuertes, A., and Palacín, M.R. (2003) *J. Mater. Chem.*, **13**, 2402–2404.

867 Chen, X.Z. and Eick, H.A. (1997) *J. Solid State Chem.*, **130**, 1–8.

868 Luge, R. and Hoppe, R. (1985) *Z. Anorg. Allg. Chem.*, **520**, 39–50.

869 Chen, X.Z. and Eick, H.A. (1994) *J. Solid State Chem.*, **113**, 362–366.

870 Chen, X.Z., Ward, D.L., and Eick, H.A. (1994) *J. Alloys Compd.*, **206**, 129–132.

871 Hochrein, O., Borrmann, H., and Kniep, R. (2001) *Z. Anorg. Allg. Chem.*, **627**, 37–42.

872 Höhn, P., Kniep, R., and Maier, J. (1994) *Z. Naturforsch. B*, **49**, 5–8.

873 Hochrein, O., Kohout, M., Schnelle, W., and Kniep, R. (2002) *Z. Anorg. Allg. Chem.*, **628**, 2738–2743.

874 Jacobs, H. and Niewa, R. (1994) *Eur. J. Solid State Inorg. Chem*, **31**, 105–113.

875 Niewa, R. and Jacobs, H. (1996) *J. Alloys Compd.*, **234**, 171–177.

876 Ostermann, D. and Jacobs, H. (1994) *J. Alloys Compd.*, **206**, 15–19.

877 Niewa, R. and Jacobs, H. (1996) *J. Alloys Compd.*, **236**, 13–18.

878 Niewa, R. and Jacobs, H. (1996) *Z. Anorg. Allg. Chem.*, **622**, 881–884.

879 Stegen, H. and Jacobs, H. (2000) *Z. Anorg. Allg. Chem.*, **626**, 639–644.

880 Strelkova, E.E., Karpov, O.G., Litvin, B.N., Pobedimskaya, E.A., and Belov, N.V. (1977) *Kristallografiya*, **22**, 174.

881 Jacobs, H. and Hellmann, B. (1993) *J. Alloys Compd.*, **191**, 277–278.

882 Ovchinnikov, A. and Höhn, P. (2016) *Z. Kristallogr.*, **231**, 797–798.

883 Höhn, P. and Kniep, R. (2002) *Z. Anorg. Allg. Chem.*, **628**, 463–467.

884 Hunting, J.L., Szymanski, M.M., Johnson, P.E., Kellar, C.B., and DiSalvo, F.J. (2007) *J. Solid State Chem.*, **180**, 31–40.

885 Ovchinnikov, A. and Höhn, P. (2016) Inorganics, submitted.

886 Reckeweg, O., Lind, C., Simon, A., and DiSalvo, F. (2004) *J. Alloys Compd.*, **384**, 98–105.

887 Tennstedt, A., Kniep, R., Hüber, M., and Haase, W. (1995) *Z. Anorg. Allg. Chem.*, **621**, 511–515.

888 Kissel, J. and Hoppe, R. (1990) *Z. Anorg. Allg. Chem.*, **582**, 103–110.

889 Stegen, H. and Jacobs, H. (2000) *Z. Anorg. Allg. Chem.*, **626**, 536–539.

890 Niewa, R., Zherebtsov, D.A., and Höhn, P. (2003) *Z. Kristallogr.*, **218**, 163.

891 Berger, U., Schultz-Coulon, V., and Schnick, W. (1995) *Z. Naturforsch. B*, **50**, 213–216.

892 Baker, C.F., Barker, M.G., Blake, A.J., Wilson, C., and Gregory, D.H. (2003) *Dalton Trans.*, **2003**, 1065–1069.

893 Höhn, P. and Kniep, R. (2000) *Z. Kristallogr.*, **215**, 325–326.

894 Bailey, M.S., McGuire, M.A., and DiSalvo, F.J. (2004) *Z. Anorg. Allg. Chem.*, **630**, 2177–2185.

895 Baker, C.F., Barker, M.G., Blake, A.J., Wilson, C., and Gregory, D.H. (2002) *J. Chem. Soc., Dalton Trans.*, **2002**, 3961–3966.

896 Gudat, A., Höhn, P., Kniep, R., and Rabenau, A. (1991) *Z. Naturforsch. B*, **46**, 566–572.

897 Höhn, P. and Kniep, R. (2000) *Z. Kristallogr.*, **215**, 327–328.

898 Francesconi, M.G., Barker, M.G., Cooke, P.A., and Blake, A.J. (2000) *J. Chem. Soc., Dalton Trans.*, **2000**, 1709–1713.

899 Höhn, P. and Kniep, R. (2000) *Z. Kristallogr.*, **215**, 329–330.

900 Jansen, M. (1982) *Z. Anorg. Allg. Chem.*, **491**, 175–183.

901 Höhn, P. and Kniep, R. (2000) *Z. Kristallogr.*, **215**, 331–332.

902 Wachsmann, C., Höhn, P., Kniep, R., and Jacobs, H. (1997) *J. Alloys Compd.*, **248**, 1–6.

903 Hochrein, O. and Kniep, R. (2001) *Z. Anorg. Allg. Chem.*, **627**, 301–303.

904 Höhn, P. and Kniep, R. (2001) *Z. Naturforsch. B*, **56**, 604–610.

905 Seeger, O. and Strähle, J. (1995) *Z. Anorg. Allg. Chem.*, **621**, 761–764.

906 Bailey, M.S. and DiSalvo, F.J. (2003) *Dalton Trans.*, **2003**, 2621–2625.

907 Gregory, D.H., Barker, M.G., Edwards, P.P., and Siddons, D.J. (1995) *Inorg. Chem.*, **34**, 3912–3916.

908 Chen, X.Z., Eick, H.A., and Lasocha, W. (1998) *J. Solid State Chem.*, **138**, 297–301.

909 Helmlinger, F.K.J., Höhn, P., and Kniep, R. (1993) *Z. Naturforsch. B*, **48**, 1015–1018.

910 Bowman, A. and Gregory, D.H. (2003) *J. Alloys Compd.*, **348**, 80–87.
911 Seeger, O., Hofmann, M., Strähle, J., Laval, J.P., and Frit, B. (1994) *Z. Anorg. Allg. Chem.*, **620**, 2008–2013.
912 Ovchinnikov, A., Kniep, R., Grin, Y., and Höhn, P. (2014) Book of Abstracts, SCTE-19, Genoa, SCTE, Genoa, Italy, p. 124.
913 Reckeweg, O., Molstad, J.C., and DiSalvo, F.J. (2001) *J. Alloys Compd.*, **315**, 134–142.
914 Benz, R. and Zachariasen, W.H. (1970) *J. Nucl. Mater.*, **37**, 109–113.
915 Niewa, R., Vajenine, G.V., DiSalvo, F.J., Luo, H.H., and Yelon, W.B. (1998) *Z. Naturforsch. B*, **53**, 63–74.
916 Teske, C. and Müller-Buschbaum, H. (1969) *Z. Anorg. Allg. Chem.*, **371**, 325–332.
917 Spear, K.E. and Leitnaker, J.M. (1971) *High Temp. Sci.*, **3**, 26–28.
918 Niewa, R., Hu, Z., Grazioli, C., Rößler, U., Golden, M.S., Knupfer, M., Fink, J., Giefers, H., Wortmann, G., de Groot, F.M.F., and DiSalvo, F.J. (2002) *J. Alloys Compd.*, **346**, 129–133.
919 Tessier, F., Ranade, M.R., Navrotsky, A., Niewa, R., DiSalvo, F.J., Leineweber, A., and Jacobs, H. (2001) *Z. Anorg. Allg. Chem.*, **627**, 194–200.
920 Bräunling, D. and Niewa, R. (2011) *Z. Anorg. Allg. Chem.*, **637**, 1853–1857.
921 Headspith, D.A., Sullivan, E., Greaves, C., and Francesconi, M.G. (2009) *Dalton Trans.*, 9273–9279.
922 Gregory, D.H., Barker, M.G., Edwards, P.P., and Siddons, D.J. (1998) *Inorg. Chem.*, **37**, 3775–3778.
923 Farault, G., Gautier, W., Baker, C.F., Bowman, A., and Gregory, D.H. (2003) *Chem. Mater.*, **15**, 3922–3929.
924 Gregory, D.H., Barker, M.G., Edwards, P.P., Slaski, M., and Siddons, D.J. (1998) *J. Solid State Chem.*, **137**, 62–70.
925 Gregory, D.H., O'Meara, P.M., Gordon, A.G., Siddons, D.J., Blake, A.J., Barker, M.G., Hamor, T.A., and Edwards, P.P. (2001) *J. Alloys Compd.*, **317–318**, 237–244.
926 Luo, H., Wang, H., Bi, Z., Feldmann, D.M., Wang, Y., Burrell, A.K., McCleskey, T., Bauer, E., Hawley, M.E., and Jia, Q. (2008) *J. Am. Chem. Soc.*, **130**, 15224–15225.
927 Tian, H., Liu, Z.T., Liu, Q.J., Zhang, N.C., and Liu, F.S. (2014) *Comput. Mater. Sci.*, **93**, 249–254.
928 Orisakwe, E., Fontaine, B., Gregory, D.H., Gautier, R., and Halet, J.F. (2014) *RSC Adv.*, **4**, 31981–31987.
929 Al Orabi, R.A.R., Orisakwe, E., Wee, D., Fontaine, B., Gautier, R., Halet, J.F., and Fornari, M. (2015) *J. Mater. Chem. A*, **3**, 9945–9954.
930 Ohkubo, I. and Mori, T. (2014) *Chem. Mater.*, **26**, 2532–2536.
931 Ohkubo, I. and Mori, T. (2014) *Inorg. Chem.*, **53**, 8979–8984.
932 Ohkubo, I. and Mori, T. (2015) *Chem. Mater.*, **27**, 7265–7275.
933 Ohkubo, I. and Mori, T. (2015) *Eur. J. Inorg. Chem.*, 3715–3722.
934 Gál, Z.A., Cario, L., and DiSalvo, F.J. (2003) *Solid State Sci.*, **5**, 1033–1036.
935 Cario, L., Gál, Z.A., Braun, T.P., DiSalvo, F.J., Blaschkowski, B., and Meyer, H.J. (2001) *J. Solid State Chem.*, **162**, 90–95.
936 Cava, R.J., Batlogg, B., Vandover, R.B., Krajewski, J.J., Waszczak, J.V., Fleming, R.M., Peck, W.F., Rupp, L.W., Marsh, P., James, A.C.W.P., and Schneemeyer, L.F. (1990) *Nature*, **345**, 602–604.
937 Brese, N.E. and DiSalvo, F.J. (1995) *J. Solid State Chem.*, **120**, 372–377.
938 Heckers, U., Jacobs, H., and Kockelmann, W. (2003) *Z. Anorg. Allg. Chem.*, **629**, 2431–2432.
939 Prots', Y., Niewa, R., Schnelle, W., and Kniep, R. (2002) *Z. Anorg. Allg. Chem.*, **628**, 1590–1596.
940 Clarke, S.J. and DiSalvo, F.J. (1997) *J. Solid State Chem.*, **129**, 144–146.
941 Brokamp, T. and Jacobs, H. (1992) *J. Alloys Compd.*, **183**, 325–344.
942 Brokamp, T. and Jacobs, H. (1991) *J. Alloys Compd.*, **176**, 47–60.
943 Clarke, S.J. and DiSalvo, F.J. (1997) *J. Solid State Chem.*, **132**, 394–398.
944 Gregory, D.H., Barker, M.G., Edwards, P.P., and Siddons, D.J. (1996) *Inorg. Chem.*, **35**, 7608–7613.
945 Rauch, P.E. and DiSalvo, F.J. (1992) *J. Solid State Chem.*, **100**, 160–165.
946 Jacobs, H. and Hellmann, B. (1993) *J. Alloys Compd.*, **191**, 51–52.

947 Wachsmann, C. (1995) Synthese und Charakterisierung von Verbindungen in den Systemen Lithium–Erdalkalimetall–Übergangsmetall– Stickstoff (Übergangsmetall=Nb, Ta, Mo, W). Dissertation, Universität Dortmund, pp. 1–223.

948 Brokamp, T. (1991) Darstellung und Charakterisierung einiger ternärer Tantalnitride mit Lithium, Magnesium oder Calcium. Dissertation, Universität Dortmund, pp. 1–168.

949 Balbarin, V., Van Dover, R.B., and DiSalvo, F.J. (1996) *J. Phys. Chem. Solids*, **57**, 1919–1927.

950 Jacobs, H., Heckers, U., Zachwieja, U., and Kockelmann, W. (2003) *Z. Anorg. Allg. Chem.*, **629**, 2240–2243.

951 Heckers, U. (2001) Beiträge zur Darstellung binärer und ternärer Nitride der Elemente Ga, Ru, Rh, Pd, Au, U. Dissertation, Universität Dortmund, pp. 1–108.

952 Palisaar, A.P. and Juza, R. (1971) *Z. Anorg. Allg. Chem.*, **384**, 1–11.

953 Niewa, R., Jacobs, H., and Mayer, H.M. (1995) *Z. Kristallogr.*, **210**, 513–515.

954 Matar, S.F., Pöttgen, R., Al Alam, A.F., and Ouaini, N. (2012) *J. Solid State Chem.*, **190**, 191–195.

955 Barker, M.G. and Alexande, I.C. (1974) *J. Chem. Soc., Dalton Trans.*, 2166–2170.

956 Halot, D. and Flahaut, J. (1971) *C. R. Séances Acad. Sci. Sér. C*, **272**, 465–467.

957 Bem, D.S., Lampe-Önnerud, C.M., Olsen, H.P., and zur Loye, H.C. (1996) *Inorg. Chem.*, **35**, 581–585.

958 Bem, D.S., Olsen, H.P., and zur Loye, H.C. (1995) *Chem. Mater.*, **7**, 1824–1828.

959 Bem, D.S. and zur Loye, H.C. (1993) *J. Solid State Chem.*, **104**, 467–469.

960 Houmes, J.D., Deo, S., and zur Loye, H.C. (1997) *J. Solid State Chem.*, **131**, 374–378.

961 Miura, A., Wen, X.D., Abe, H., Yau, G., and DiSalvo, F.J. (2010) *J. Solid State Chem.*, **183**, 327–331.

962 Miura, A., Takei, T., Kumada, N., Magome, E., Moriyoshi, C., and Kuroiwa, Y. (2014) *J. Alloys Compd.*, **593**, 154–157.

963 Bhattacharyya, S., Kurian, S., Shivaprasad, S., and Gajbhiye, N. (2010) *J. Nanopart. Res.*, **12**, 1107–1116.

964 Panda, R.N. and Gajbhiye, N.S. (1998) *J. Cryst. Growth*, **191**, 92–96.

965 Elder, S.H., Doerrer, L.H., DiSalvo, F.J., Parise, J.B., Guyomard, D., and Tarascon, J.M. (1992) *Chem. Mater.*, **4**, 928–937.

966 Herle, S.P., Hegde, M.S., Vasanthacharya, N.Y., Gopalakrishnan, J., and Subbanna, G.N. (1994) *J. Solid State Chem.*, **112**, 208–210.

967 Weil, K.S. and Kumta, P.N. (1997) *J. Solid State Chem.*, **128**, 185–190.

968 Singh, D.J. (1992) *Phys. Rev. B Condens. Matter*, **46**, 9332–9335.

969 Kaskel, S., Hohlwein, D., and Strähle, J. (1998) *J. Solid State Chem.*, **138**, 154–159.

970 Wang, L., Tang, K., Zhu, Y., Li, Q., Zhu, B., Wang, L., Si, L., and Qian, Y. (2012) *J. Mater. Chem.*, **22**, 14559–14564.

971 Niewa, R., Zherebtsov, D.A., Schnelle, W., and Wagner, F.R. (2004) *Inorg. Chem.*, **43**, 6188–6194.

972 Lengauer, W. (1989) *J. Solid State Chem.*, **82**, 186–191.

973 Lengauer, W. and Ettmayer, P. (1988) *J. Less Common Met.*, **141**, 157–162.

974 Houmes, J.D., Bem, D.S., and zur Loye, H.-C. (1994) *MRS Proc.*, **327**, 153–164.

975 Seeger, O. and Strähle, J. (1994) *Z. Naturforsch. B*, **49**, 1169–1174.

976 Miura, A., Wessel, M., and Dronskowski, R. (2011) *J. Ceram. Soc. Jpn.*, **119**, 663–666.

977 Yang, M., Zakutayev, A., Vidal, J., Zhang, X., Ginley, D.S., and DiSalvo, F.J. (2013) *Energy Environ. Sci.*, **6**, 2994–2999.

978 Brese, N.E. and DiSalvo, F.J. (1995) *J. Solid State Chem.*, **120**, 378–380.

979 Matar, S.F. and Demazeau, G. (2010) *J. Solid State Chem.*, **183**, 994–999.

980 Waldhart, J. and Ettmayer, P. (1979) *Monatsh. Chem.*, **110**, 21–26.

981 Das, B., Reddy, M.V., Rao, G.V., and Chowdari, B.V. (2012) *RSC Adv.*, **2**, 9022–9028.

982 Das, B., Reddy, M.V., Rao, G.V., and Chowdari, B.V. (2012) *J. Mater. Chem.*, **22**, 17505–17510.

983 Rauch, P.E. and Simon, A. (1992) *Angew. Chem., Int. Ed.*, **31**, 1519–1521.

984 Reckeweg, O., Braun, T.P., DiSalvo, F.J., and Meyer, H.J. (2000) *Z. Anorg. Allg. Chem.*, **626**, 62–67.

985 Cordier, G. and Rönninger, S. (1988) *Z. Kristallogr.*, **182**, 60–61.
986 Höhn, P., Auffermann, G., Ramlau, R., Rosner, H., Schnelle, W., and Kniep, R. (2006) *Angew. Chem.*, **45**, 6681–6685.
987 Henry, P.F. and Weller, M.T. (1998) *Angew. Chem., Int. Ed.*, **37**, 2855–2857.
988 Calvert, L.D. and Rand, C. (1964) *Acta Crystallogr.*, **17**, 1175–1176.
989 Bernhardt, F. and Hoppe, R. (1993) *Z. Anorg. Allg. Chem.*, **619**, 969–975.
990 Sofin, M. and Jansen, M. (2001) *Z. Anorg. Allg. Chem.*, **627**, 2115–2117.
991 Crespin, M., Isnard, O., Dubois, F., Choisnet, J., and Odier, P. (2005) *J. Solid State Chem.*, **178**, 1326–1334.
992 Hayward, M.A. and Rosseinsky, M.J. (2003) *Solid State Sci.*, **5**, 839–850.
993 Dashjav, E., Kreiner, G., Schnelle, W., Wagner, F.R., Kniep, R., and Jeitschko, W. (2007) *J. Solid State Chem.*, **180**, 636–653.
994 Bode, H. and Voss, E. (1956) *Z. Anorg. Allg. Chem.*, **286**, 136–141.
995 Takeda, Y., Kanamaru, F., Shimada, M., and Koizumi, M. (1976) *Acta Crystallogr. B*, **32**, 2464–2466.
996 Moss, M.A. and Jeitschko, W. (1991) *Z. Anorg. Allg. Chem.*, **603**, 57–67.
997 Bader, R.F.W. (1994) *Atoms in Molecules: A Quantum Theory*, Clarendon Press, pp. 1–456.
998 Bronger, W., Baranov, A., Wagner, F.R., and Kniep, R. (2007) *Z. Anorg. Allg. Chem.*, **633**, 2553–2557.

8
Fluorite-Type Transition Metal Oxynitrides[1]

Franck Tessier

Université de Rennes 1, Institut des Sciences Chimiques de Rennes, UMR CNRS 6226, 263 avenue du Général Leclerc, 35042 Rennes cedex, France

8.1
Introduction

(Oxy)nitride materials present attractive properties in relation to the role played by nitrogen. A comparison between oxynitrides and oxides highlights the characteristics of the nitrogen N^{3-}/oxygen O^{2-} substitution: an increase in the anionic formal charge, a stronger covalent character, a higher cross-linking density in glasses, a reducing character due to the N^{3-}/N^0 redox couple, and modified acid–base properties. The increase in the covalent character gives rise to interesting optical properties, for which some applications are developed in this chapter.

Although nitride-type compounds are less numerous than oxides, their chemistry received a broader interest in the last decade due to the interesting properties brought by nitrogen as a dopant in oxides (TiO_2, ZnO, etc.) or in larger quantities in nitrides and oxynitrides for emerging applications related to energy and environment. Several review papers of reference cover different aspects of the (oxy)nitrides chemistry: about their crystal chemistry [1–6], optical properties [7,8], perovskite systems [9–11], vitreous compositions [12], SiAlON system [13,14], PON compounds [15], or more generally about their physical and chemical properties [16,17].

This chapter focuses on a less reviewed (oxy)nitride family based on transition metals with structures related to (or derived from) the fluorite type. The scope of this chapter is to present an overview – as much exhaustive as possible – of the existing oxynitride compositions and their interest illustrated with some application examples.

---

1) Dedicated to Dr. Roger Marchand on the occasion of his 75th birthday.

Handbook of Solid State Chemistry, First Edition. Edited by Richard Dronskowski, Shinichi Kikkawa, and Andreas Stein.
© 2017 Wiley-VCH Verlag GmbH & Co. KGaA. Published 2017 by Wiley-VCH Verlag GmbH & Co. KGaA.

Fluorite is an important crystal structure type in solid-state chemistry for which oxynitride compounds exist and continue to be discovered by studying new systems or developing novel synthetic approaches. The great flexibility of this structure is illustrated in detail with the case of two solid solutions R–W–O–N and R–Ta–O–N (R = rare earth). The presence of defects in the anionic lattice and distorted environments around the cations give access to several structures and open a large range of properties.

8.2
Fluorite-Type Structures

8.2.1
From Diamond to Fluorite Structures

The fluorite structure results from a filiation deriving from the diamond arrangement where carbon atoms form a cubic face-centered lattice. Four extra carbon atoms complete the unit cell by occupying tetrahedral sites at the octants center in a diagonal opposition, as indicated in Figure 8.1. Then, ordering of the diamond structure with two different atoms leads to the ZnS blende structure (or sphalerite). And finally, the fluorite structure may be considered as a filled blende structure where four extra anions complete the remaining empty tetrahedral sites (Figure 8.1).

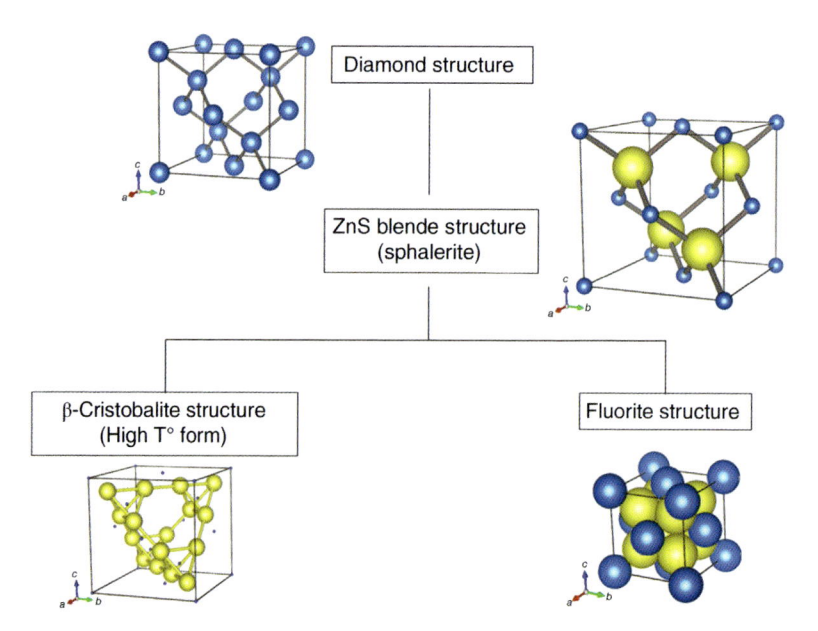

Figure 8.1 Structural filiation to the fluorite-type.

8.2 Fluorite-Type Structures

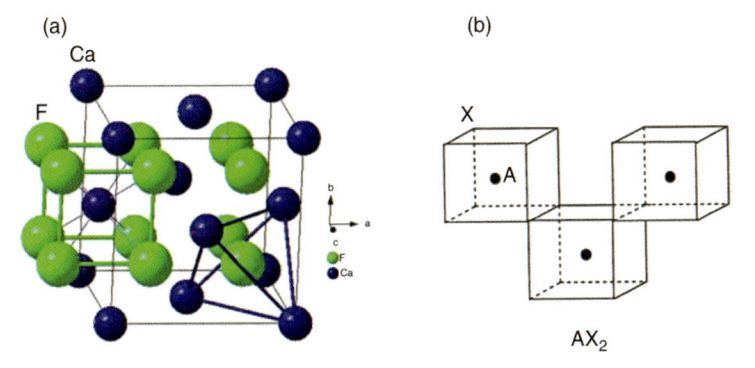

Figure 8.2 The fluorite structure.

The fluorite (also named fluorspar) is a mineral (ideal formula CaF_2) mainly composed of calcium fluoride. However, the presence of impurities, Y, Ce, Si, Al, Fe, Mg, Eu, Sm, O, Cl, and so on, as traces offers a large panel of color variations. As illustrated in Figure 8.2, CaF_2 forms a quite basic arrangement that has been described in several reference crystal chemistry books [18,19]. Cations are placed in cubic sites and anions occupy tetrahedral ones. Accordingly, fluorites can be represented by a stacking of cubes alternating empty and full ones. Calcium and fluorine atoms occupy the positions 4a and 8c in the space group $Fm\text{-}3m$, respectively. The cubic parameter reaches values close to 5.463 Å. This structure is generally adopted by AB_2-type phases with the radii condition $r_A/r_B > 0.73$. Oxides or fluorides with large cations Ce, Th, or U, crystallize in the fluorite structure at ambient conditions, but only at high temperatures for the smaller ones Zr or Hf. For even smaller elements – Sn or Ti – the radii ratio r_A/r_B in AB_2 phases belongs to the range 0.73–0.41 and the resulting structure is often of the rutile type due to the lower coordination number of the A atoms (6 instead of 8).

Fluorite-type materials offer a rich chemistry by the possibility to substitute cations and anions, for which the charge and the size of the elements affect the properties, as well as the relation between energetics and defect chemistry [20]. Many applications such as solid electrolyte, catalysts, and sensors related to their high anionic conduction properties, nuclear materials, or thermal coatings are based on modified (or not) zirconia, hafnia, ceria, thoria, or urania systems.

8.2.1.1 Antifluorite Structure

Several alkali metal oxides and sulfides present an antifluorite structure A_2B where the positions occupied by the cations and anions are interchanged [18]. Anions form a fcc lattice, while cations occupy all the tetrahedral sites. An interesting family of lithium transition metal nitrides results from the exhaustive research work of Robert Juza including $LiZnN$, Li_5TiN_3, $Li_7Mn(V,Nb)N_4$, and Li_9CrN_5 [2,21]. Note also the preparation of the antifluorite compositions Li_5ReN_4 and $Li_{11}NbN_4O_2$ [22].

Transition metal-based materials, where the antifluorite structure is kept from Ti to Fe, are particularly studied as alternative anode materials in Li-ion batteries. The transition metal is held at the highest oxidation state and Li atoms are in tetrahedral sites. This structure provides easy movement for Li along the cationic vacancies for intercalation and desintercalation. Interesting and promising electrochemical performances were obtained on nitrides Li_3FeN_2 [23], $Li_{2.7}Fe_{0.3}N$ [24], Li_7MnN_4 [25,26], and oxynitride Li–M–O–N systems (M = Ti, V, Cr, Mn) [27–32]. A series of compounds were prepared in the Li_3N–Mg_3N_2 system from the binaries [33]. At 673 K, LiMgN undergoes a structural transition from orthorhombic (Pnma) at ambient temperature to a simple cubic antifluorite structure (Fm-3m). Nitrogen-deficient compositions $Li_{1.11}Mg_{0.89}N_{0.96}$ as well as $Li_{1.09}Mg_{0.91}N_{0.97}$ keep this antifluorite structure. Note also the reaction of Li_3N with AlN, BN, and Si_3N_4 that leads to Li-ion conductors (Li_3AlN_2, Li_3BN_2, $LiSi_2N_3$, Li_2SiN_3, $Li_{18}Si_3N_{10}$, $Li_{21}Si_3N_{11}$, and Li_8SiN_4) with structures related to the antifluorite type [34].

8.2.2
Defect Fluorites

Many oxides and (oxy)fluorides crystallize in structures related to the fluorite type by removing a fraction of the anions [18]. Cationic deficiencies are quite rare in this structure type.

Defective oxynitride fluorites are easily prepared when nitriding an oxide without cationic substitutions. Vacancies are thus introduced to keep electroneutrality and the oxide structure following the charge balance equation: $3\,O^{-II} \rightarrow 2\,N^{-III} + \square$ ($\square$ = vacancy). The following rare earth (R) tungstate example, developed in Section 8.4.1, illustrates quite well the formation of defect fluorite oxynitride solid solutions starting from R_2WO_6: $R_2WO_6 \rightarrow R_2W(O,N,\square)_6$.

This section reviews and illustrates some of the defect structures like bixbyite, scheelite, and pyrochlore mostly encountered for transition metal oxynitrides (Figure 8.3).

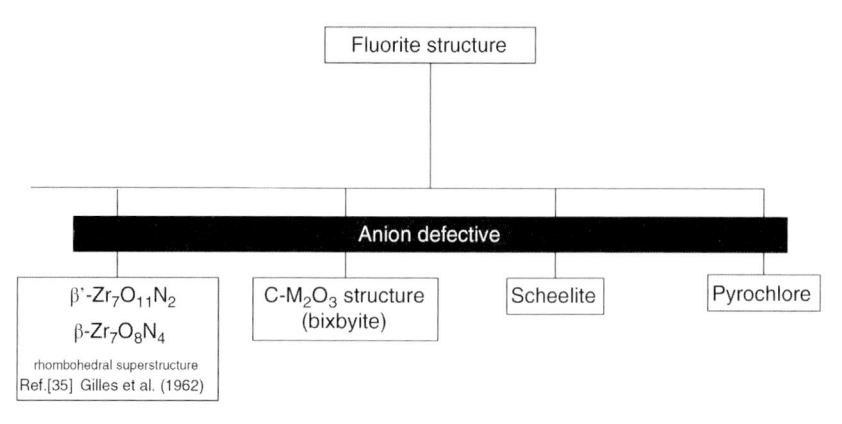

Figure 8.3 Defect fluorite-type structures.

8.2.2.1 Zirconium-Based Oxynitrides

The nitridation of the fluorite ZrO_2 at temperatures $T > 1400\,°C$ under nitrogen or $T > 900\,°C$ under ammonia leads to $ZrO_{2-x}N_{2x/3}\square_{x/3}$ oxynitrides ($\square$ = anionic vacancy) with structures deriving from the fluorite structure. After the first work of Gilles and Collongues [35], Lerch revisited the crystal chemistry of the prepared phases in the ZrO_2-Zr_3N_4 system: γ-Zr_2ON_2, β- $Zr_7O_8N_4$, and β'- $Zr_7O_{11}N_2$ [36,37].

Zr_2ON_2 crystallizes in the bixbyite type (C-M_2O_3) in a cubic unit cell (Ia-3, $a = 10.13940(7)$ Å) [38]. The combination of neutron and X-ray diffraction refinement concludes to a statistical distribution of oxygen and nitrogen on only one 16f site without any ordering. However, it was difficult to affirm if a possible anionic ordering may occur in the orthorhombic Ibca space group [39]. β-$Zr_7O_8N_4$ and β'-$Zr_7O_{11}N_2$ present rhombohedral superstructures of the fluorite structure. A β''-$Zr_7O_{9.2}N_{3.2}$ phase was also prepared by Lerch et al. by carbothermal reaction under ammonia at $1900\,°C$ [36,37,40], the structure of which is a nonregular stacking of units of β-phase and ZrO_2.

The introduction of nitrogen within ZrO_2 and Zr–M–O–N (M = Ca, Mg, Y, etc.) systems allows controlling the formation of vacancies in the anionic network [41]. The presence of vacancies induces the possibility to stabilize metastable cubic and tetragonal phases. It strongly enhances the anion mobility at high temperatures where anions and vacancies are disordered. These materials are studied as solid electrolyte for fuel cells or gas sensors.

The nitridation of $ZrTiO_4$ (α-PbO_2 structure type) under NH_3 conducts, respectively, two oxynitride phases: $ZrTiO_{1.92}N_{1.23}$, obtained at $735\,°C$ (three cycles of 20 h), and $ZrTiO_{1.06}N_{1.90}$, obtained at higher temperatures ($860\,°C$, 20 h) [42]. The first compound shows a fluorite structure ($a = 4.9505(4)$ Å) containing a large concentration of vacancies, and for the nitrogen-rich phase obtained at higher temperature an ordering between anions and vacancies changes the structure into the bixbyite type (described later).

Another impact of the nitrogen insertion was reported on the optical properties of Zr–M–O–N systems (M = Eu, Ce) [43–45]. Nitrogen has a direct sensitizing effect for the Eu(III) luminescence and this property depends on nitrogen/europium concentration ratio. No luminescence was observed in Zr–Ce(IV)–O–N phases for excitations in the 220–500 nm range. It appears that the oxidation state of the dopant remains unchanged during nitridation. However, the reaction under NH_3 modifies clearly the color of the resulting compounds due to N^{3-} charge transfer transitions or color center formation. Zr–Ce–O powders display a large panel of colors depending on the nitridation temperature and nitrogen content, with coloration intensity and reflectance higher than for other dopants (Eu, Er, etc.).

A higher degree of complexity in defect fluorites is brought by the compound $Zr_4O_5N_2$ characterized by a layered intergrowth of fluorite and bixbyite structures with anionic vacancies ordering [46].

8.2.2.2 Bixbyite and Antibixbyite Structures

When only six of the eight vertices of the anionic cubes are occupied, two sites remain empty along the diagonal of the cube ([111] direction) and of a face

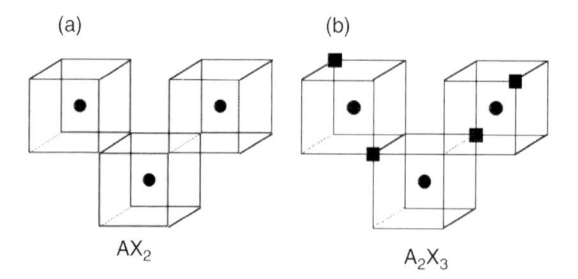

Figure 8.4 AX$_2$ fluorite and A$_2$X$_3$ bixbyite structures.

(Figure 8.4). From this arrangement results an A$_4$X$_6$ stoichiometry corresponding to the bixbyite structure (C-M$_2$O$_3$ type) found in the mineral (Mn,εFe)$_2$O$_3$. This structure is described as a slightly distorted close-packed array of A atoms with X atoms occupying 3/4 of the tetrahedral holes.

Zr$_2$ON$_2$ as previously mentioned and the isostructural phase Hf$_2$ON$_2$ present the bixbyite type [38]. The nitridation of the defect fluorite Zr$_3$Er$_4$O$_{12}$ leads to a bixbyite-type oxynitride Zr$_{0.43}$Er$_{0.57}$O$_{1.07}$N$_{0.43}$ [47]. This phase may belong to a solid solution but unfortunately no end terms were given within a temperature range. Bixbyite-type phases, among other structure types (anatase and anosovite), were reported in the system Sc–Ta–O–N. A Sc$_x$Ta$_{1-x}$(O,N)$_y$ solid solution was clearly identified with $0.33 \leq x \leq 1$ and $1.7 \leq y \leq 1.9$ [48]. As no order–disorder transition was observed until 900 °C, the ordering between vacancies and anions is thus detrimental to fast anion conduction at high temperatures. The bixbyite type was also evidenced in the solid solution V$_2$O$_{3.08}$N$_{0.02}$–V$_2$O$_{3.07}$N$_{0.13}$ resulting from the nitridation of the sulfide precursor V$_{5.45}$S$_8$ at temperatures lower than 550 °C [49]. With the same transition metal, a multi-anion phase V$_2$O$_{2.70}$N$_{0.15}$F$_{0.15}$ was referenced as a bixbyite [50].

Among nitrides, some binary (α-Be$_3$N$_2$, α-Ca$_3$N$_2$, α-Mg$_3$N$_2$) [51], Zn$_3$N$_2$ [52], and ternary (Li$_3$Al(Sc)N$_2$) [53] nitrides adopt the antibixbyite structure. In this antistructure each A atom is tetrahedrally coordinated by X atom, while each X atom is coordinated by six A atoms. The difference in colors between α-Be$_3$N$_2$ (white), α-Ca$_3$N$_2$ (yellow), and α-Mg$_3$N$_2$ (red-brown) was explained by a significant anion–cation covalent interaction decreasing from Be to Mg and Ca to interpret the corresponding decrease in the bandgaps, respectively [51]. In the Li–Mg–N system, the phase Li$_{0.24}$Mg$_{2.76}$N$_{1.92}$ is isostructural with Mg$_3$N$_2$ and, therefore, adoptsa cubic antibixbyite structure [33].

8.2.2.3 Scheelite Structure

This structure is another derivative of the fluorite, where two fluorite cells are stacked on each other to build the scheelite cell. The scheelite representative is the mineral CaWO$_4$ built with isolated tetrahedra [WO$_4$]$^{2-}$ linked together by eight coordinated Ca^{2+}. The polyhedra arrangement is represented in Figure 8.5 (space group $I4_1/a$, $a = 5.242$ Å, $c = 11.372$ Å – JCPDS file 41–1431). Ca and W occupy the position of Ca in CaF$_2$ in an alternate way on each vertical row;

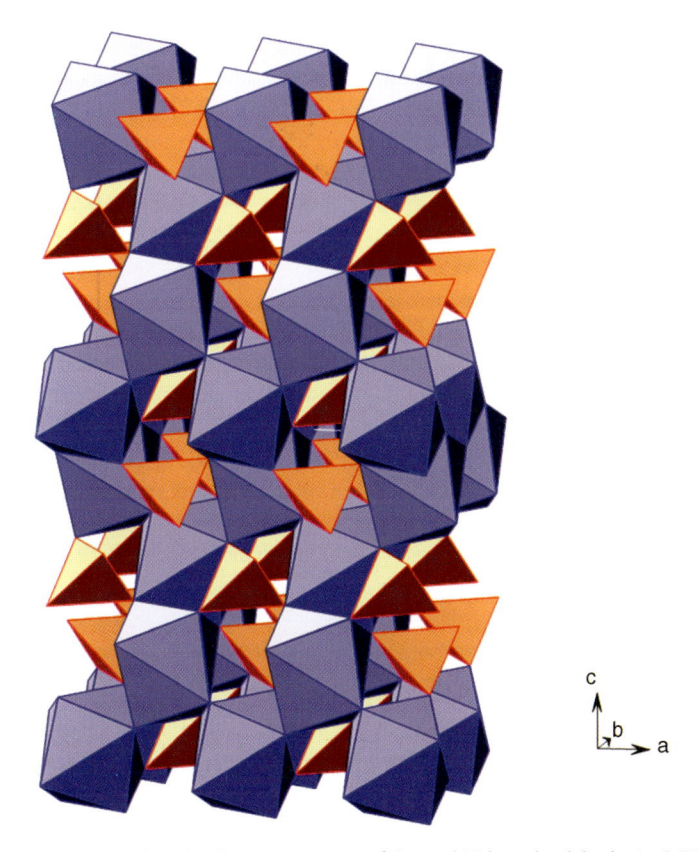

Figure 8.5 The scheelite arrangement of Ca- and W-based polyhedra in CaWO$_4$.

O atoms replace F and move toward tungsten to form tetrahedra [19]. Two coordinating elements – osmium and tungsten – are known to give the scheelite structure type to oxynitrides [12].

KOsO$_3$N ($I4_1/a$, $a = 5.646(2)$ Å, $c = 13.027(4)$ Å) [54] where Os^{VIII} is stabilized at its highest oxidation state is probably the only one (or one of the rare case) oxynitride prepared in aqueous solution. This trend is more a specificity of the osmium chemistry than a generality. Yellow monocrystals are obtained by reacting a solution containing OsO$_4$ and KOH with NH$_4$OH (Figure 8.6). Replacing osmium by rubidium leads to the metallate RbOsO$_3$N having the CsReO$_4$ structure closely related to the scheelite. Structural studies did not reveal oxygen–nitrogen ordering in the [OsO$_3$N] tetrahedron.

Scheelites with W^{VI} were evidenced in the RWO$_3$N series with the rare earth elements R = Nd, Sm, Gd, and Dy [1,55,56]. This series derives from CaWO$_4$ with the following cross-substitution: $R^{3+} + N^{3-} = Ca^{2+} + O^{2-}$. RWO$_3$N compounds are prepared from a solid–gas (NH$_3$) reaction of the corresponding tungstates R$_2$W$_2$O$_9$ at 700–750 °C. They present an isolating behavior due to the VI oxidation state stabilized in tetrahedral sites. However, the reducing feature

Figure 8.6 Yellow KOsO$_3$N monocrystals.

of ammonia and the nitrogen environment of W^{VI} – unknown in binaries WN and W$_2$N – were not favorable to maintain tungsten at its highest oxidation state. Only brown colored SmWO$_3$N and GdWO$_3$N were isolated as single phases. NdWO$_3$N formation is subject to a competition with a nitrogen-rich NdW(O,N)$_3$ perovskite phase as soon as the nitridation begins at 700 °C. The perovskite phase being more stable at high temperature presents a black color due to the mixed-valent tungsten.

It was possible to stabilize NdWO$_3$N ($I4_1/a$, $a = 5.2821(3)$ Å, $c = 11.5893(8)$ Å) and PrWO$_3$N ($I4_1/a$, $a = 5.299(3)$ Å, $c = 11.6314(9)$ Å) as single phases after thermal ammonolysis of reactive precursors prepared using the citrate complexation/calcination route [57]. This method used the complexing properties of citric acid that after calcination produces homogeneous and reactive fine precursor powders. After reaction at 600 °C, the resulting oxynitrides display a light gray color, different from the brown color observed for Sm and Gd. Neutron diffraction performed on the neodynium phase did not prove the existence of any anionic ordering. By analogy to the well-known luminescent properties of CaWO$_4$ (emission at 420 nm), tests were performed on oxynitride scheelites. The origin of the luminescence mechanism rely on charge transfers within [WO$_4$]$^{2-}$ tetrahedra at 250 nm. However, no luminescence property was measured from partially nitrided [WO$_3$N]$^{3-}$ environments in NdWO$_3$N and PrWO$_3$N [58].

Cheviré et al. have extended the scheelite type to the solid solution Nd$_{1-x}$A$_x$WO$_{3+x}$N$_{1-x}$ (A = Ca,Sr) but in a narrow range of temperatures close to 550 °C due to the presence of a perovskite phase at 600 °C [57].

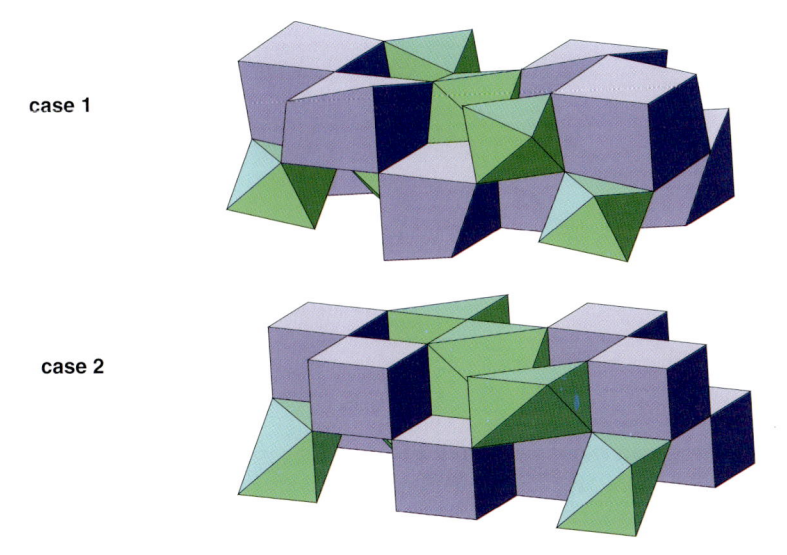

Figure 8.7 Representation of the pyrochlore structure with a regular octahedra around B atoms ($x = 0.3125$) for case 1 and with a regular cube around A atoms ($x = 0.375$) in case 2.

8.2.2.4 Pyrochlore Structure

The pyrochlore unit cell has a cubic parameter twice that of a fluorite cell and contains eight $A_2B_2X_6X'$ entities in the space group Fd-$3m$. A full cube with anions at the vertices alternates with a defect cube containing two vacancies along the main diagonal [18,19]. The A cation retains the eightfold coordination and the B cation, smaller than A, the sixfold coordination. The position (48f) of the six X anions determines the regularity of the crystallographic sites. There is only one variable parameter in this structure, the x parameter of the 48f position (x, 1/8, 1/8). Two limit cases are possible if $x = 5/16$ (0.3125): a distorted cube around A atoms and a regular octahedron around B ones; when $x = 3/8$ (0.375) a regular cube around A atoms and a distorted octahedron around B ones, as displayed in Figure 8.7.

Oxynitrides with the pyrochlore structure were prepared by reaction of rare earth tantalates $RTaO_4$ (R = Nd-Yb, Y) [59]. We have revisited this series of oxynitrides considered as pyrochlores. While the pyrochlore structure is kept for the larger rare earth (Nd, Sm) due to the presence of low-intensity peaks in the XRD profile, a fine study led for the smaller rare earth rather involves the indexation of the XRD profile in a defect fluorite-type unit cell [60]. These results are developed in Section 8.4.2.

A molybdenum-based pyrochlore phase – $Sm_2Mo_2O_{3.83}N_{3.17}$ – was prepared by reacting $Sm_2MO_2O_7$ in ammonia at 625 °C [61]. A semiconducting behavior and a temperature-dependent magnetic susceptibility following a Curie–Weiss law are reported for this phase. The reaction of $R_2Ti_2O_7$ (R = Y, Dy, Sm) prepared from R_2O_3 and TiO_2 under dry NH_3 at $T < 1000$ °C gave oxynitrides of the composition $R_2Ti_2O_{5.5}N$ [62].

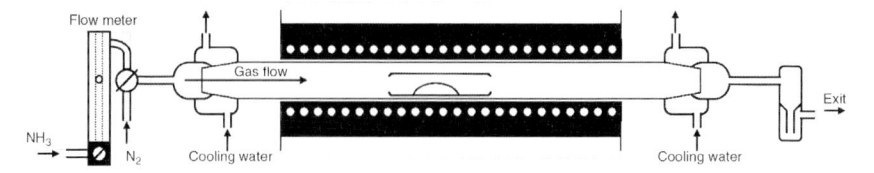

Figure 8.8 Nitridation tubular furnace.

8.3
Synthesis

By experience, it is well known that the syntheses of nitrides and oxynitrides do not follow well established rules [56]. The main problem is that there exists no ammonolysis systematics: Each system represents a particular case and is likely to react differently as a function of several intrinsic (reacting components) or extrinsic (temperature, reaction time, furnace atmosphere, etc.) parameters. Solid/gas reactions are the most usual routes to prepare oxynitrides. Nitridation reactions are generally carried out in an alumina boat containing the oxide precursor powder placed inside an alumina (or silica) tube through which ammonia gas flows generally at a rate up to 30–40 l/h (Figure 8.8). The temperature may be raised in the range 500–1400 °C, depending on the precursor and system used, with a heating rate of about 1–10 °C/min. Ammonia acts as both a reducing and a nitriding (oxidizing) agent. The ammonia flow rate depends on the reaction temperature: the higher the temperature, the higher the flow rate, in order to minimize the dissociation of NH_3 into dinitrogen and dihydrogen before reaching the product. According to the reaction temperature and the nature of the used oxide precursor, a reducing character may predominate or not. This dual behavior – nitriding and reducing – is essential for the ammonolysis reaction. Indeed, while nitrogen is incorporated, hydrogen combines with oxygen atoms from the oxide and eliminates them as water vapor:

$$\text{oxide} + NH_3(g) \rightarrow \text{oxynitride} + H_2O \uparrow (g)$$

Typically after reaction, the furnace is switched off and the powder is allowed to cool to room temperature under a nitrogen atmosphere.

The success of a solid/gas ammonolysis depends highly on the nature of the precursors. The preparation of ammonolysis precursors by classic ceramic routes has shown limits regarding their homogeneity and reactivity. Pure single phases are often difficult to prepare: as an example, five to six–6 heating steps at 1300 °C during 20 h each are necessary to obtain pure tungstates R_6WO_{12} from binary oxides. The difficulty increases with multinary cationic stoichiometries, due to the numerous compositions potentially accessible. The specific surface areas of the powders are generally small, around a few m^2/g, inducing low reactivity. High temperatures may also cause loss of one of the components, such as WO_3 that sublimes at 800 °C. So, to obtain R_6WO_{12} powders, a first heating step between 700 and 800 °C is necessary to combine WO_3 and R_2O_3, so that intermediate by-products need to be dissociated to reach the desired stoichiometry R/W = 6.

Numerous *chimie douce*-type processes have been developed to prepare oxide powders in order to improve their quality (purity, chemical homogeneity, etc.) and reactivity. For example, the "Pechini" and the "amorphous citrate" routes have often been used to synthesize several oxynitride compounds mentioned in this chapter. This process utilizes citric acid as a complexing agent and is not, strictly speaking, a classic sol–gel process, in the usual sense that the gel is not formed by a metal–oxygen–metal network, but rather from a calcination of metal–organic complexes, thus producing ultrafine reactive powders with a good chemical homogeneity [63]. The citrate route can be generalized to synthesize a large number of compositions that cannot be obtained via traditional methods. The use of citric acid presents several advantages such as the formation of very stable solutions of more or less complex stoichiometries. The cation stoichiometry being the same in the solution and in the powdered residue after calcination, all designed powder compositions may be easily and rapidly synthesized by this aqueous process. Thus, complexation–calcination-type methods, and particularly the citrate route, constitute a suitable answer to the above-mentioned problematics. The experimental route to the tungstates R_6WO_{12} is described in the flowchart drawn in Figure 8.9. Similar experimental method was also proposed for the synthesis of rare earth-doped zirconium oxides [44]. This experimental approach is at the origin of novel phases described in Section 3.

8.4
Defect-Fluorite Solid Solutions

This part of the chapter focuses on two interesting defect solid solutions based on rare earth, tungsten and tantalum oxynitrides. Anionic vacancies are represented by a square □. The resulting materials were studied for their optical properties in different areas, including pigments, UV absorbers, and visible light-driven photocatalysis.

8.4.1
R–W–O–N System

Among the numerous stoichiometries in the R–W–O system (R = rare earth) listed in Figure 8.10, the compositions highlighted in a box are those available for a large number of rare earth elements. The most frequent compositions are $R_2W_3O_{12}$, $R_2W_2O_9$, R_2WO_6, $R_{14}W_4O_{33}$, $R_{10}W_2O_{21}$, and R_6WO_{12}.

Considering the ratio R/W = 0.66 ($R_2W_3O_{12}$), Nillson *et al.* have shown that the nitridation of $Gd_2W_3O_{12}$ leads to a defect fluorite $(Gd_{3-x}W_{1+x}(O,N)_y)$ and a nitride compound isostructural with β-W_2N [64].

The reaction of $R_2W_2O_9$ under NH_3 produces different phases depending on the rare earth radius. For larger radii (R = La, Ce, Pr, and Nd), black

8 Fluorite-Type Transition Metal Oxynitrides

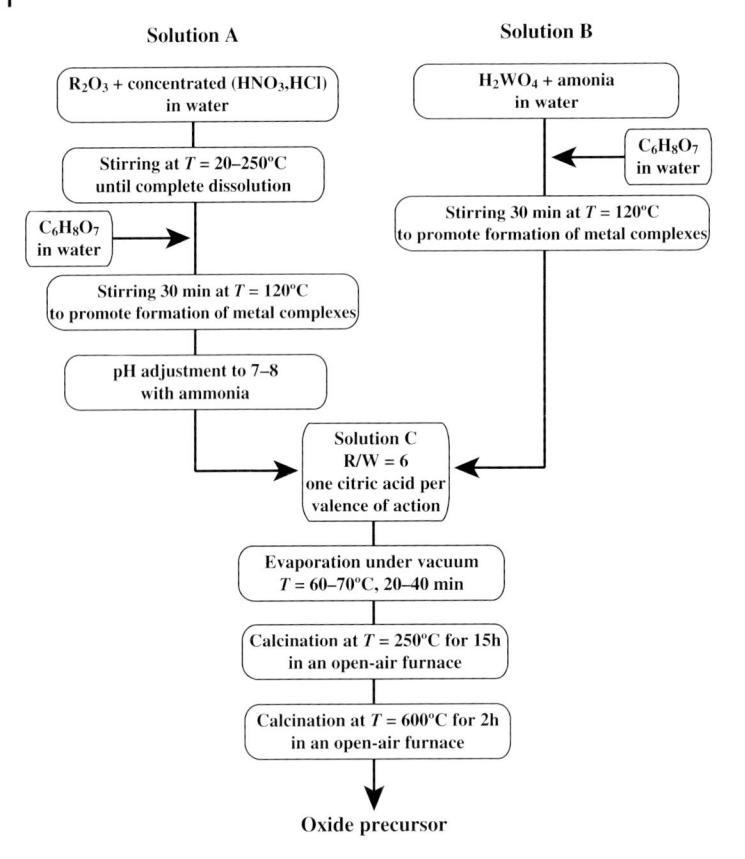

Figure 8.9 Preparation of R_6WO_{12} tungstates by the amorphous citrate method.

$RW_{1-x}^{VI}W_x^VO_xN_{3-x}$ perovskites are obtained $(0.6 < x < 0.8)$ where tungsten is partially reduced [65]. For smaller rare earth elements, we observe the formation of RWO_3N scheelites with only W^{VI} described in Section 8.2.2.3.

$R_{10}W_2O_{10}$ oxides are extremely difficult to prepare as single phase, so their nitridation was not undertaken.

For all rare earth elements R, an oxide composition exists with the R_2WO_6 stoichiometry. The nitridation of such phases at 800 °C during 15 h gives

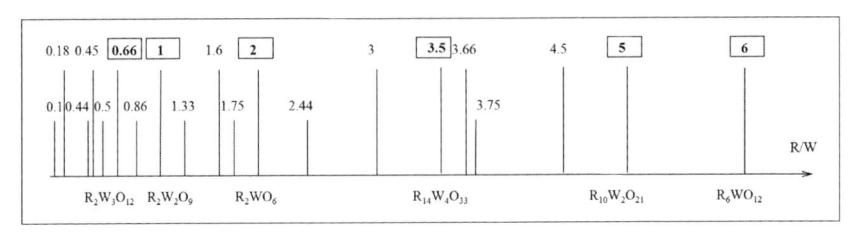

Figure 8.10 Existing stoichiometries in the rare earth–tungsten–oxygen system.

8.4 Defect-Fluorite Solid Solutions

Table 8.1 Composition and unit cell parameter of nitrided R_2WO_6 precursors [66].

Element (R)	A_3X_5 formulation	a (Å)	A_4X_8 formulation
Nd	$Nd_2WO_{3.03}N_{1.97}$	5.383(1)	$Nd_{2.67}W_{1.33}O_{4.04}N_{2.63}\square_{1.33}$
Sm	$Sm_2WO_{2.89}N_{2.07}$	5.326(3)	$Sm_{2.67}W_{1.33}O_{3.86}N_{2.76}\square_{1.38}$
Gd	$Gd_2WO_{3.00}N_{2.00}$	5.289(2)	$Gd_{2.67}W_{1.33}O_{4.00}N_{2.66}\square_{1.34}$
Dy	$Dy_2WO_{2.84}N_{2.10}$	5.240(2)	$Dy_{2.67}W_{1.33}O_{3.78}N_{2.80}\square_{1.42}$
Ho	$Ho_2WO_{2.83}N_{2.11}$	5.221(3)	$Ho_{2.67}W_{1.33}O_{3.77}N_{2.81}\square_{1.42}$
Er	$Er_2WO_{2.81}N_{2.12}$	5.201(3)	$Er_{2.67}W_{1.33}O_{3.75}N_{2.83}\square_{1.42}$
Tm	$Tm_2WO_{2.93}N_{2.05}$	5.178(3)	$Tm_{2.67}W_{1.33}O_{3.90}N_{2.73}\square_{1.37}$
Yb	$Yb_2WO_{2.85}N_{2.09}$	5.164(2)	$Yb_{2.67}W_{1.33}O_{3.80}N_{2.78}\square_{1.42}$
Y	$Y_2WO_{3.31}N_{1.78}$	5.215(3)	$Y_{2.67}W_{1.33}O_{4.41}N_{2.37}\square_{1.22}$

oxynitride compounds with a cubic symmetry for R = Nd–Yb, Y (Table 8.1). Corresponding powders are brown with colorimetric coordinates $L^*a^*b^* = 25, 8, 5$ for R = Yb. They present an insulating behavior attesting that tungsten keeps the VI oxydation state after treatment under ammonia flow [66,67].

These oxynitrides are stable under ambient conditions and present a defect fluorite structure formulated $R_{2.67}W_{1.33}(O,N,\square)_8$ according to the A_4X_8 unit cell content (A = Σcations, X = Σanions). Whatever the rare earth is, anions and cations once gathered involve a same formulation $A_4X_{6.6}$ intermediate between the fluorite (A_4X_8) and bixbyite (A_4X_6) stoichiometries. No solid solution domain was evidenced with the ratio R/W = 2. The XRD pattern of such phases can be indexed in a small cubic unit cell of the fluorite type. Although it is quite logical to place rare earth elements in a cubic or seven-coordinated site, we expect for W^{VI} a lower coordination (4 or 6). R and W occupy the same 4a position of the $Fm\text{-}3m$ space group, in spite of different radii, and O and N occupy the same 8c position. Neutron diffraction analyses confirm the small fluorite unit cell, but a high value of the anion isotropic displacement parameters indicate more than likely a more complex solution [67].

Considering now the rare earth-rich oxides $R_{14}W_4O_{33}$ (R/W = 3.5) and R_6WO_{12} (R/W = 6), there is a stoichiometry analogy between these ternary oxide compositions and mixed valent rare earth binary oxides (R = Ce, Pr, Tb). Thus, $R_6O_{11}\square_1$ and $R_{14}W_4O_{33}$ have the same Σanions/Σcations ratio = 1.83; similarly $R_7O_{12}\square_2$ and R_6WO_{12} with the Σanions/Σcations ratio = 1.71. The structure of R_6WO_{12} (R = Y, Ho) was refined in a three-dimensional rhombohedral structure close to that of the binary R_7O_{12} deriving from the ideal fluorite structure [68]. Nilsson et al. also reported about the nitridation of phases of the Gd–W–O system with Gd/W = 0.1–3.5 [64].

Diot et al. studied the thermal nitridation of the tungstates $R_{14}W_4O_{33}$ and R_6WO_{12} from 700 to 1000 °C. Novel oxynitride compositions were formed in the corresponding defect fluorite-type solid solutions $R_{14}W_4O_{33-3x}N_{2x}\square_{3+x}$ (with R = Nd, Sm, Ho, Y) and $R_6WO_{12-3x}N_{2x}\square_{2+x}$ (with R = La, Sm, Ho, Y),

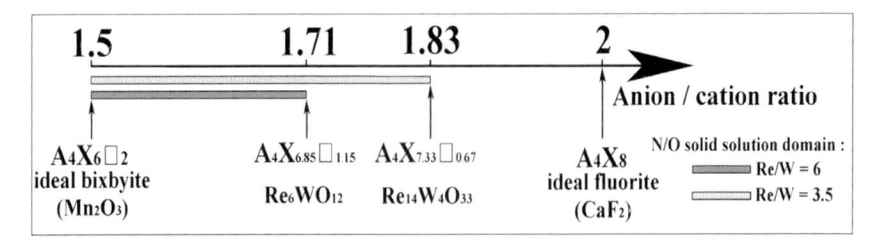

```
  1.5            1.71    1.83         2
  ├──────────────────────────┤        ──▶
                                      Anion / cation ratio
  ↑              ↑       ↑            ↑
  A₄X₆□₂         A₄X₆.₈₅□₁.₁₅  A₄X₇.₃₃□₀.₆₇   A₄X₈         N/O solid solution domain :
  ideal bixbyite                      ideal fluorite  ▬▬▬ Re/W = 6
  (Mn₂O₃)        Re₆WO₁₂   Re₁₄W₄O₃₃  (CaF₂)      ▭▭▭ Re/W = 3.5
```

Figure 8.11 Formation of solid solution domains during the thermal ammonolysis reaction of $R_{14}W_4O_{33}$ (R/W = 3.5) and R_6WO_{12} (R/W = 6).

respectively [67,69]. The inserted nitrogen content increases with temperature and reaction time from $A_4X_{7.33}\square_{0.67}$ (R/W = 3.5) and $A_4X_{6.85}\square_{1.15}$ (R/W = 6) to a limit formulation of the bixbyite stoichiometry ($A_4X_6\square_2$), $R_{3.11}W_{0.89}O_{3.4}N_{2.6}\square_2$ and $R_{3.43}W_{0.57}O_{4.3}N_{1.7x}\square_2$, respectively (Figure 8.11).

The study of these compositions was motivated by the development of new colored pigments in the field of nitride-type compounds [67]. The introduction of nitrogen within an oxide network gives rise to additional energy levels just over the 2p orbitals of oxygen, and involves a decrease in the bandgap in the visible part of the solar spectrum. The resulting greater covalent behavior is at the origin of interesting properties in the visible range. A direct application is the preparation of colored samples for pigments and photocatalytic applications [8].

As a function of the nitrogen enrichment, the R/W ratio, and the rare earth element these compositions offer a range of colors from pale yellow to brown or khaki. The diffuse reflectance spectra for the $Y_6WO_{12-3x}N_{2x}$ samples presented in Figure 8.12 show a progressive shift of the absorption edge with increasing nitrogen content toward higher wavelengths. Therefore, there is a possibility to tune the color to a defined shade by only adjusting the nitrogen content.

Among this colored materials, the following compositions $Y_6WO_{9.1}N_{1.9}\square_{3.0}$, $Sm_6WO_{8.3}N_{2.5}\square_{3.2}$, and $Sm_{14}W_4O_{23.4}N_{6.4}\square_{6.2}$ with high nitrogen contents are particularly attractive by their spectral characteristics (Table 8.2) and can compete with $BiVO_4$, a yellow industrial pigment, as nontoxic potential challengers.

New fluorite-type solid solution domains were evidenced in the $Y_6(W,Mo)(O,N)_{12}$ system [70]. The O/N substitution brings the possibility to tune the optical absorption properties of Y_6WO_{12} in a different context related to the preparation of novel inorganic UV absorbers. In Figure 8.12, we observe a continuous shift of the absorption edge from 340 to 525 nm and in parallel a decrease of the diffuse reflectance maximum intensity with nitrogen enrichment. By adjusting the amount of nitrogen, it is possible to find a composition with an absorption edge cutting exactly at 400 nm, limit between UV and visible ranges. Nevertheless, due to a not enough steep slope the spectral selectivity of these oxynitride tungstates appears still low to make these materials suitable for UV absorption applications.

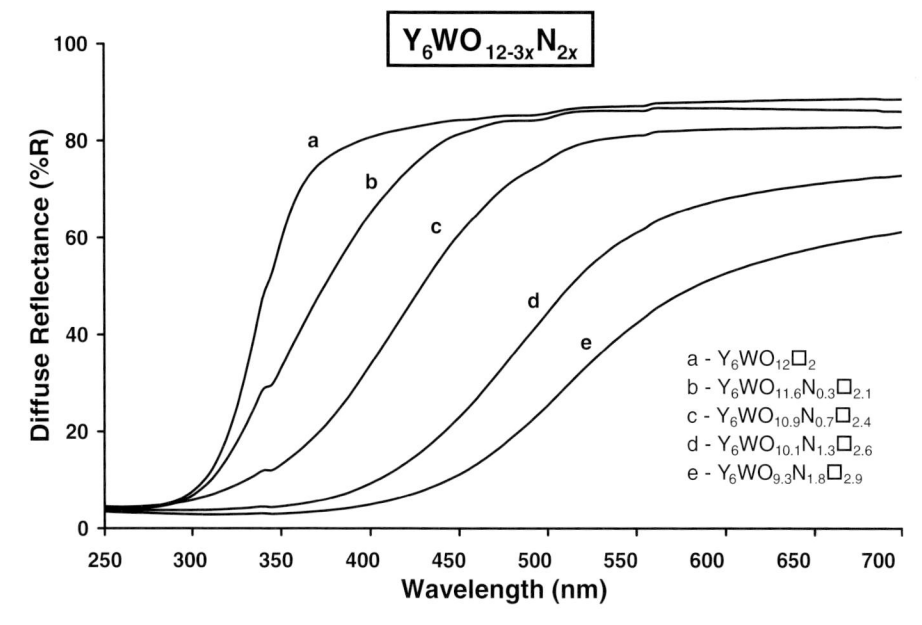

Figure 8.12 Diffuse reflectance spectra of $Y_6WO_{12-3x}N_{2x}$ oxynitride powders.

Curves:
- a - $Y_6WO_{12}\square_2$
- b - $Y_6WO_{11.6}N_{0.3}\square_{2.1}$
- c - $Y_6WO_{10.9}N_{0.7}\square_{2.4}$
- d - $Y_6WO_{10.1}N_{1.3}\square_{2.6}$
- e - $Y_6WO_{9.3}N_{1.8}\square_{2.9}$

8.4.2
R–Ta–O–N System

Rare earth-containing tantalum oxynitride phases were studied to stabilize tetragonal or cubic fluorite-type phases of TaON. With Y_2O_3, as a dopant, a minimum of 15% yttrium allows forming a fluorite phase ($Y_{0.15}Ta_{0.85}O_{0.62}N_{1.15}$) [71]. Although macroscopically these phases crystallize in a cubic fluorite-type defective structure, DFT calculations showed at the microscopic level triclinically distorted unit cells due to the migration of anions toward vacant sites to optimize the charge distribution [72].

A systematic study was performed in the R_xTa–O–N system with different rare earth elements and diverse R/Ta ratio corresponding to $x = 1$, 2, and 3.

Table 8.2 CIE $L^*a^*b^*$ coordinates of three oxynitride powders in comparison with the industrial pigment $BiVO_4$.

Compound	L^*	a^*	b^*
$Sm_6WO_{8.3}N_{2.5}\square_{3.2}$	75	4	60
$Y_6WO_{9.1}N_{1.9}\square_{3.0}$	74	4	56
$Sm_{14}W_4O_{23.4}N_{6.4}\square_{6.2}$	81	−2	64
$BiVO_4$	76	−4	81

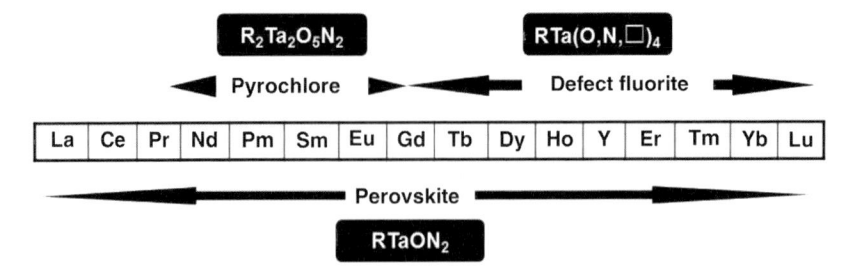

```
┌─────────────────┐                    ┌─────────────────┐
│  R₂Ta₂O₅N₂      │                    │  RTa(O,N,□)₄    │
└─────────────────┘                    └─────────────────┘
      ◄──── Pyrochlore ────►       ◄──── Defect fluorite ────►
┌──┬──┬──┬──┬──┬──┬──┬──┬──┬──┬──┬──┬──┬──┐
│La│Ce│Pr│Nd│Pm│Sm│Eu│Gd│Tb│Dy│Ho│Y │Er│Tm│Yb│Lu│
└──┴──┴──┴──┴──┴──┴──┴──┴──┴──┴──┴──┴──┴──┘
      ◄──────────────── Perovskite ────────────────►
                  ┌─────────────┐
                  │   RTaON₂    │
                  └─────────────┘
```

Figure 8.13 Domains of existence of RTa–O–N phases versus rare earth elements.

The nitridation of $RTaO_4$ oxides leads to oxynitrides for which the structure type is depending on the rare earth radius. Thus, $LaTaO_4$ reacts with ammonia at 900/950 °C to form only a distorted $LaTiO_2N$ perovskite. But with decreasing R radii in $RTaO_4$ fergusonites, a $R_2Ta_2O_5N_2$ pyrochlore phase (R = Nd, Sm, Gd, Dy, Ho, Er, Yb, Y) with lower nitrogen content coexists with the perovskite phase [73]. The stability of the pyrochlore phase increases with smaller radii so that single phases are isolated from R = Er. However, if the pyrochlore phase is confirmed for the larger rare earth elements Nd and Sm, a finer study has shown, using X-ray and neutron diffractions, a different result for the small rare earth-containing precursors (R = Ho, Er, Yb, Y) [60]. Indeed, corresponding XRD patterns can be indexed in a defect-type fluorite unit cell ($RTa(O,N,□)_4$) with halved parameter (Figure 8.13). Pyrochlore and defect-fluorite unit cells differ only by an additional low-intensity peak (*hkl*: 311) at $2\theta = 28.8°$ (Cu Kα). The presence of this weak peak implies to double the fluorite cubic unit cell parameter.

The citrate route does not produce the expected fergusonite $RTaO_4$ precursors, usually obtained via the ceramic route, but metastable scheelite polymorphs corresponding to the high temperature form [60]. Their nitridation as low as 900 °C leads easily to single fluorite-type phases. Nitrogen insertion is tuned with increasing nitridation reaction temperature and time so that solid solution domains are delimited in composition and yellow hues (Table 8.3).

Table 8.3 Characterization of $RTa(O,N,□)_4$ oxynitride solid solutions.

R	Ammonolysis conditions	N wt %	Experimental formulation	Unit cell parameter (Å)	Color
Ho	15 h, 900 °C	2.08	$HoTaO_{3.10}N_{0.60}□_{0.30}$	5.170(2)	Yellow
	6 × 15 h, 950 °C	3.24	$HoTaO_{2.61}N_{0.93}□_{0.46}$	5.1572(6)	Dark yellow
Y	15 h, 900 °C	3.56	$YTaO_{2.75}N_{0.52}□_{0.43}$	5.1675(5)	Yellow
	6 × 15 h, 950 °C	4.66	$YTaO_{2.39}N_{1.08}□_{0.53}$	5.1555(2)	Brown
Er	15 h, 900 °C	2.97	$ErTaO_{2.83}N_{0.78}□_{0.39}$	5.156(2)	Dark yellow
	6 × 15 h, 950 °C	3.19	$ErTaO_{2.62}N_{0.92}□_{0.46}$	5.1507(7)	Dark yellow

Pyrochlore and fluorite structures are quite close. The cationic radii ratio is the geometric parameter that influences the ordering of cations. However, it is surprising, as observed in the R–W–O–N system, to index the XRD patterns in a small fluorite-type unit cell that implies necessarily a single cationic position for atoms R and Ta with different radii. The Rietveld refinements, both from X-rays and neutron diffractions, tend toward the same conclusions: a cubic fluorite-type structure, no superstructure peaks, and high atomic displacements parameters. The wavy background (neutrons) indicates a more complex structure where anions and vacancies may be ordered. These defect fluorite unit cells are totally disordered and reveal a more complex atoms arrangement more likely in agreement with the conclusions of the DFT analysis previously mentioned [72]. The local structure of the cations – R and Ta – was studied by Exafs and photoluminescence with no clear conclusions regarding the structure adopted by smaller rare earth elements. The defect-fluorite structure appears to be only an apparent structure [74].

Novel fluorite-type phases were prepared with the stoichiometry "$R_2TaO_{5.5}$" (R = Nd–Yb, Y) [75]. These reactive precursors are only obtained from the citrate route and lead to an oxynitride defect-fluorite series: $R_{2.67}Ta_{1.33}(O,N,\square)_8$. Nd- and Gd-based oxides produce a mixture of R_2O_3 and $RTaON_2$ (perovskite). Starting from Gd, colored single-phase fluorites are isolated. Structural analogies were made with the R_2W–O–N and R_2MoO_5 systems in the determination of limit compositions: $A_4X_{6.6}$ or A_3X_5 (R_2TaO_4N).

Oxynitride solid solutions with the fluorite type were evidenced in the $R_3Ta(O,N,\square)_8$ and $R_3Nb(O,N,\square)_8$ systems (R = Eu, Gd, Ho, Y, Yb) within a limited range of stability compared to the RTa and R_2Ta ratios [76]. Within a same set of elements, for example, Gd- and Ta-based oxynitrides, a blueshift characterizes the position of the absorption edge with increasing Gd/Ta ratio, from GdTa to Gd_2Ta and Gd_3Ta, respectively. The niobium/tantalum substitution is at the origin of a modification of the absorption properties toward higher wavelengths. Corresponding R_3Ta phases present whitish colors, and R_3Nb ones are yellowish.

In the context of the study of oxynitrides with pyrochlore and defect-fluorite structures, attempts were made to stabilize a cationic ordering within the R–Ta–O–N (R = Y, Nd, Sm, Gd, Yb) system by substituting the Zr^{4+}/W^{6+} couple to Ta^{5+} [77]. The pyrochlore composition $R_2Zr_{1.5}W_{0.5}O_6N$ was targeted but without any experimental success. However, by modifying the Zr/W ratio, novel fluorite-type phases were prepared in a panel of different colors.

Among all these colored defect-fluorites phases, photocatalytic tests were performed under visible light to achieve the overall water splitting reaction. This application emerged a decade ago as an original way to produce hydrogen from water [8,78]. Although some of the powders manifest bright colors, none of them produce significant results for the photoreduction and photooxidation of water. Many parameters influence the photoactivity– in the case of these defect-fluorites compositions: the poor crystallization state of the samples, the presence of defects and vacancies in the anionic network may act as recombination centers for holes and electrons, and limit the photocatalytic activity.

8.5 Conclusion

This chapter explored the growing interest of transition metal oxynitrides for the fluorite structure. In the two last decades, novel compositions were evidenced in structure types deriving from the fluorite unit cell. Specific features were pointed out regarding the great flexibility of fluorites for substitution both in cationic and anionic networks. The diversity of resulting compositions makes them promising for applications currently studied in areas related to energy and environment.

Vacancies play an important role in defect fluorites for the targeted properties (conduction, interaction with light, etc.). From a crystallographic viewpoint, the insertion of nitrogen engenders modifications in the anionic arrangements and leads to some ambiguities in the structural resolution. Deeper investigations deserve to be performed to solve the problematics of the *apparent* unit cell. This fluorite-type family opens a large field for novel nitrided compositions, fine structural studies, and emerging applications.

Acknowledgments

Oxynitride fluorites were the subject of several studies in our Laboratory in Rennes during the Ph.D. works of Dr Philippe Antoine (defense: 1987), Dr Laurent Le Gendre (1997), Dr Nadège Diot (1999), Dr Olivier Larcher (2001), Dr Emmanuelle Orhan (2002), Dr François Cheviré (2004), Dr Pascal Maillard (2006), and Dr Erwan Ray (2011) under the supervision of Dr Roger Marchand and Dr Franck Tessier (for the last three theses). Financial supports (Ph.D. grants) from the French Ministry of Higher Education and Research and from the Région Bretagne are acknowledged.

References

1 Marchand, R. (1998) Ternary and higher order nitride materials, in *Handbook on the Physics and Chemistry of Rare Earths*, vol. 25 (eds K.A. Gschneider, Jr., and L. Eyring), Elsevier Science B.V., Amsterdam, pp. 51–99.

2 Brese, N.E. and O'Keeffe, M. (1992) Crystal chemistry of inorganic nitrides. *Struct. Bond.*, **79**, 307–378.

3 Gregory, D.H. (2001) Nitride chemistry of the s-block elements. *Coord. Chem. Rev.*, **215**, 301–345.

4 Gregory, D.H. (1999) Structural families in nitride chemistry. *J. Chem. Soc., Dalton Trans.*, 259–270, doi: 10.1039/A807732K.

5 Niewa, R. and DiSalvo, F.J. (1998) Recent developments in nitride chemistry. *Chem. Mater.*, **10**, 2733–2752.

6 Marchand, R., Laurent, Y., Guyader, J., L'Hardion, P., and Verdier, P. (1991) Nitrides and oxynitrides: preparation, crystal chemistry and properties. *J. Eur. Ceram. Soc.*, **8**, 197–213.

7 Xie, R.-J. and Hintzen, H.T. (2013) Optical properties of (oxy)nitride materials: a review. *J. Am. Ceram. Soc.*, **96** (3), 665–687.

8 Tessier, F., Maillard, P., Cheviré, F., Domen, K., and Kikkawa, S. (2009) Optical properties of oxynitride powders. *J. Ceram. Soc. Japan*, **117** (1), 1–5.

9 Fuertes, A. (2012) Chemistry and applications of oxynitride perovskites. *J. Mater. Chem.*, **22**, 3293–3299.

10 Ebbinghaus, S.G., Abicht, H.-P., Dronskowski, R., Müller, T., Reller, A., and Weidenkaff, A. (2009) Perovskite-related oxynitrides: recent developments in synthesis, characterization and investigations of physical properties. *Prog. Solid State Chem.*, **37**, 173–205.

11 Masubuchi, Y., Sun, S.-K., and Kikkawa, S. (2015) Processing of dielectric oxynitride perovskites for powders, ceramics, compacts and thin films. *Dalton Trans.*, **44** (23), 10570–10581.

12 (a) Hampshire, S. (2008) Oxynitride glasses. *J. Eur. Ceram. Soc.*, **28**, 1475–1483. (b) Das, T. (2000) Oxynitride glasses: an overview. *Bull. Mater. Sci.*, **23**, 499–507. (c) Becher, P.F., Hampshire, S., Pomeroy, M.J., Hoffmann, M.J., Lance, M.J., and Satet, R.L. (2011) An overview of the structure and properties of silicon-based oxynitride glasses. *Int. J. Appl. Glass Sci.*, **2**, 63–83.

13 Jack, K.H. (1976) Review Sialons and related nitrogen ceramics. *J. Mater. Sci*, **11**, 1135–1158.

14 Cao, G.Z. and Metselaar, R. (1991) α′-SiAlON ceramics: a review. *Chem. Mater.*, **3**, 242–252.

15 Marchand, R., Schnick, W., and Stock, N. (2000) Molecular, complex ionic, and solid state PON compounds. *Adv. Inorg. Chem.*, **50**, 193–233.

16 Fuertes, A. (2010) Synthesis and properties of functional oxynitrides: from photocatalysts to CMR materials. *Dalton Trans.*, **39**, 5942–5948.

17 Oyama, S.T. (ed.) (1996) *The Chemistry of Transition Metal Carbides and Nitrides*, Blackie Academic & Professional.

18 Wells, A.F. (1975) *Structural Inorganic Chemistry*, Clarendon Press.

19 Galasso, F.S. (1970) *Structure and Properties of Inorganic Solids*, Pergamon Press.

20 Navrotsky, A. (2010) Thermodynamics of solid electrolytes and related oxide ceramics based on the fluorite structure. *J. Mater. Chem.*, **20**, 10577–10587.

21 Juza, R., Langer, K., and Von Benda, K. (1968) Ternary nitrides, phosphides, and arsenides of lithium. *Angew. Chem., Int. Ed.*, **7**, 360–370.

22 (a) Chaushli, A., Jacobs, H., Weisser, U., and Strähle, J. (2000) Li_5ReN_4, a lithium nitridorhenate(VII) with anti fluorite superstructure. *Z. Anorg. Allg. Chem.*, **626**, 1909–1914 (b) Cabana, J., Casas-Cabanas, M., Santner, H.J., Fuertes, A., and Palacin, M.R. (2010) Exploring order–disorder structural transitions in the Li-Nb-O-N system: the new antifluorite oxynitride $Li_{11}NbN_4O_2$. *J. Solid State Chem.*, **183**, 1609–1614.

23 Nishijima, M., Takeda, Y., Imanishi, N., and Yamamoto, O. (1994) Li deintercalation and structural change in the lithium transition metal nitride Li_3FeN_2. *J. Solid State Chem.*, **113**, 205–210.

24 Rowsell, J.L.C., Pralong, V., and Nazar, L.F. (2001) Layered lithium iron nitride: a promising anode material for Li-ion batteries. *J. Am. Chem. Soc.*, **123**, 8598–8599.

25 Panabiere, E., Emery, N., Bach, S., Pereira-Ramos, J.-P., and Willmann, P. (2013) Ball-milled Li_7MnN_4: an attractive negative electrode material for lithium-ion batteries. *Electrochim. Acta*, **97**, 393–397.

26 Emery, N., Panabiere, E., Crosnier, O., Bach, S., Brousse, T., Willmann, P., and Pereira-Ramos, J.-P. (2014) In operando X-ray diffraction study of Li_7MnN_4 upon electrochemical Li extraction-insertion: a reversible three-phase mechanism. *J. Power Sources*, **247**, 402–405.

27 Cabana, J., Rousse, G., Fuertes, A., and Palacin, M.R. (2003) The first lithium manganese oxynitride $Li_{7.9}MnN_{5-y}O_y$: preparation and use as electrode material in lithium batteries. *J. Mater. Chem.*, **13**, 2402–2404.

28 Cabana, J., Mercier, C., Gautier, D., and Palacin, M.R. (2005) Synthesis and electrochemical study of antifluorite-type phases in the Li-M-N-O (M = Ti, V) systems. *Z. Anorg. Allg. Chem.*, **631**, 2136–2141.

29 Barker, M.G. and Frankham, S.A. (1982) The effects of carbon and nitrogen on the corrosion resistance of type 316 stainless steel to liquid lithium. *J. Nucl. Mater.*, **107**, 218–221.

30 Barker, M.G., Hubberstey, P., Dadd, A.T., and Frankham, S.A. (1983) The interaction of chromium with nitrogen dissolved with nitrogen in liquid lithium. *J. Nucl. Mater.*, **114**, 143–149.

31 Gudat, A., Haag, S., Kniep, R., and Rabenau, A. (1990) Ternäre Nitride des Lithiums mit den Elementen Cr, Mo und W. *Z. Naturforsch.*, **45b**, 111–120.

32 Niewa, R., Zherebtsov, D., and Hu, Z. (2003) Polymorphism of heptalithium nitridovanadate(V) $Li_7[VN_4]$. *Inorg. Chem.*, **42**, 2538–2544.

33 Bailey, A.S., Hubberstey, P., Hughes, R.W., Ritter, C., and Gregory, D.H. (2010) Tunable defect structure in the Li-Mg-N ternary phase system: a powder neutron diffraction study. *Chem. Mater.*, **22**, 3174–3182.

34 (a) Yamane, H., Kikkawa, S. and Koizumi, M. (1985) Lithium aluminum nitride, Li_3AlN_2 as a lithium solid electrolyte. *Solid State Ion.*, **15**, 51–54; (b) Yamane, H., Kikkawa, S., Horiuchi, H., and Koizumi, M. (1986) Structure of a new polymorph of lithium boron nitride Li_3BN_2. *J. Solid State Chem.*, **65**, 6–12; (c) Yamane, H., Kikkawa, S. and Koizumi, M. (1987) High and low temperature phases of lithium boron nitride, Li_3BN_2: preparation, phase relation, crystal structure and ionic conductivity. *J. Solid State Chem.*, **71**, 1–11; (d) Yamane, H., Kikkawa, S. and Koizumi, M. (1987) Preparation and electrochemical properties of double metal nitrides containing lithium. *J. Power Source*, **20**, 311–315; (e) Yamane, H., Kikkawa, S. and Koizumi, M. (1987) Preparation of lithium silicon nitrides and their lithium ionic conductivity. *Solid State Ion.*, **25**, 183–191.

35 Gilles, J.-C. and Collongues, R. (1962) Sur la structure et les propriétés des phases formées au cours de l'action de l'ammoniac sur la zircone. *C. R. Acad. Sci.*, **254**, 1084–1086.

36 Lerch, M. (1996) Nitridation of zirconia. *J. Am. Ceram. Soc.*, **79**, 2641–2644.

37 Lerch, M. (1998) Phase relationships in the ZrO_2-Zr_3N_4 system. *J. Mater. Sci. Lett.*, **17**, 441–443.

38 Clarke, S.J., Michie, C.W., and Rosseinsky, M.J. (1999) Structure of Zr_2ON_2 by neutron powder diffraction: the absence of nitride-oxide ordering. *J. Solid State Chem.*, **146**, 399–405.

39 Füglein, E., Hock, R., and Lerch, M. (1997) Crystal structure and high temperature behavior of Zr_2ON_2. *Z. Anorg. Allg. Chem.*, **623**, 304–308.

40 Lerch, M., Krumeich, F., and Hock, R. (1997) Diffusion controlled formation of b type phases in the system ZrO_2-Zr_3N_4. *Solid State Ion.*, **95**, 87–93.

41 (a) Lerch, M., Wrba, J. and Lerch, J. (1996) Vacancy ordering in the ZrO_2 rich part of the systems Ca-Zr-O-N, Mg-Zr-O-N, and Y-Zr-O-N. *J. Solid State Chem.*, **125**, 153–158; (b) Lerch, M., Janek, J., Becker, K.D., Berendts, S., Boysen, H., Bredow, T., Dronskowski, R., Ebbinghaus, S.G., Kilo, M., Lumey, M.W., Martin, M., Reimann, C., Schweda, E., Valov, I., and Wiemhöfer, H.D. (2009) Oxide nitrides: from oxides to solids with mobile nitrogen ions. *Prog. Solid State Chem.*, **37**, 81–131.

42 Clarke, S.J., Michie, C.W., and Rosseinsky, M.J. (2000) New zirconium titanium oxynitrides obtained by ammonolysis of zirconium titanate. *Chem. Mater.*, **12**, 863–865.

43 Gutzov, S. and Lerch, M. (2001) Preparation and optical properties of Zr-Ce-O-N materials. *J. Eur. Ceram. Soc.*, **21**, 595–601.

44 Perez-Estebanez, M., Pastrana-Fabregas, R., Isasi-Marin, J., and Saez-Puche, R. (2006) Inorganic pigments based on fluorite-type oxynitrides. *J. Mater. Res.*, **21**, 1427–1433.

45 Gutzov, S., Kohls, M., and Lerch, M. (2000) The luminescence of Zr-Eu-O-N materials. *J. Phys. Chem. Solids*, **61**, 1301–1309.

46 Michie, C.W., Claridge, J.B., Clarke, S.J., and Rosseinsky, M.J. (2003) $Zr_4O_5N_2$: intergrowth of fluorite and bixbyite anion layers formed by coupled site-selective anion and vacancy ordering. *Chem. Mater.*, **15**, 1547–1553.

47 Meyer, S., Martinez Meta, N.J., Schweda, E., and Eibl, O. (2009) Synthesis and crystal structure of a bixbyite-type solid solution $Zr_{0.43}Er_{0.57}O_{1.07}N_{0.43}$. *Z. Anorg. Allg. Chem.*, **635**, 2470–2473.

48 Stork, A., Schilling, H., Wessel, C., Wolff, H., Börger, A., Baehtz, C., Becker, K.-D., Dronskowski, R., and Lerch, M. (2010) Bixbyite- and antase-type phases in the system Sc-Ta-O- N. *J. Solid State Chem.*, **183**, 2051–2058.

49 Nakhal, S., Hermes, W., Ressler, T., Pöttgen, R., and Lerch, M. (2009) Synthesis, crystal structure and magnetic properties of bixbyite-type vanadium oxide nitrides. *Z. Naturforsch.*, **64b**, 281–286.

50 Tanguy, B., Pezat, M., Wold, A., and Portier, J. (1976) Sur une nouvelle phase dérivée de V_2O_3 à structure bixbyite. *C. R. Acad. Paris C*, **282**, 291–293.

51 Orhan, E., Jobic, S., Brec, R., Marchand, R., and Saillard, J.-Y. (2002) Binary nitrides α-M_3N_2 (M = Be, Mg, Ca): a theoretical study. *J. Mater. Chem.*, **12**, 2475–2479.

52 Paniconi, G., Stoeva, Z., Smith, R.I., Dippo, P.C., Gallagher, B.L., and Gregory, D.H. (2008) Synthesis, stoichiometry and thermal stability of Zn_3N_2 powders prepared by ammonolysis reactions. *J. Solid State Chem.*, **181**, 158–165.

53 Vegas, A., Martin, R.L., and bevan, D.J.M. (2009) Compounds with a 'stuffed' anti-bixbyite-type structure, analysed in terms of Zintl-klemm and coordination-defect concepts. *Acta Crystallogr.*, **B65**, 11–21.

54 Laurent, Y., Pastuszak, R., L'Haridon, P., and Marchand, R. (1982) Etude du Nitrurotrioxoosmate(VllI) de Potassium: Une Structure de Type Scheelite. *Acta Crystallogr.*, **B38**, 914–916.

55 Antoine, P., Marchand, R., and Laurent, Y. (1987) Oxynitrures à structure scheelite dans les systèmes lanthanide- tungstène-oxygène-azote. *Rev. Int. Hautes Tempér. Réfract.*, **24**, 43–46.

56 Tessier, F. and Marchand, R. (2003) Ternary and higher order rare-earth nitride materials: synthesis and characterization of ionic-covalent oxynitride powders. *J. Solid State Chem.*, **171**, 143–151.

57 Cheviré, F., Tessier, F., and Marchand, R. (2004) New scheelite-type oxynitrides in systems RWO_3N-AWO_4 (R = rare-earth element; A = Ca, Sr) from precursors obtained by the citrate route. *Mater. Res. Bull.*, **39**, 1091–1101.

58 Cheviré, F. (2004) Nouvelle génération d'absorbeurs UV inorganiques: Etude de solutions solides oxydes et oxynitrures. Application à la protection du bois,, Université de Rennes 1, France.

59 Pors, F., Marchand, R., and Laurent, Y. (1993) Une nouvelle famille de pyrochlores: les oxynitrures $Ln_2Ta_2O_5N_2$: préparation et étude cristallochimique. *J. Solid State Chem.*, **107**, 39–42.

60 Maillard, P., Tessier, F., Orhan, E., Cheviré, F., and Marchand, R. (2005) Thermal ammonolysis study of the rare-earth tantalates $RTaO_4$. *Chem. Mater.*, **17**, 152–156.

61 Veith, G.M., Greenblatt, M., Croft, M., and Goodenough, J.B. (2001) Synthesis and characterization of the oxynitride pyrochlore: $Sm_2Mo_2O_{3.83}N_{3.17}$. *Mater. Res. Bull.*, **36**, 1521–1530.

62 Dolgikh, V.A. and Lavut, E.A. (1991) Synthesis of new oxynitrides with the pyrochlore structure. *Russ. J. Inorg. Chem.*, **36**, 1389–1392.

63 Douy, A. and Odier, P. (1989) The polyacrylamide gel: a novel route to ceramic and glassy oxide powders. *Mater. Res. Bull.*, **24**, 1119–1126.

64 Nilsson, M., Grins, J., Kall, P.O., and Svensson, G. (1996) Synthesis, structural characterisation and magnetic properties of $Gd_{14}W_4O_{33-x}N_y$ ($0 \leqslant x \leqslant 17 \pm 2$, $0 \leqslant y \leqslant 9 \pm 2$), a new fluorite- related oxynitride. *J. Alloys Compd.*, **240**, 60–69.

65 Antoine, P., Marchand, R., Laurent, Y., Michel, C., and Raveau, B. (1988) On the electrical properties of the perovskites $LnWO_xN_{3-x}$. *Mater. Res. Bull.*, **23**, 953–957.

66 Marchand, R., Antoine, P., and Laurent, Y. (1993) Préparation et caractérisation d'oxynitrures $Ln_2W(O,N)_{6-x}$ à structure fluorine déficitaire. *J. Solid State Chem.*, **107**, 34–38.

67 Diot, N., Larcher, O., Marchand, R., Kempf, J.Y., and Macaudière, P. (2001) Rare-earth and tungsten oxynitrides with a defect fluorite-type structure as new pigments. *J. Alloys Compd.*, **323–324**, 45–48.

68 Diot, N., Bénard-Rocherullé, P., and Marchand, R. (2000) X-ray powder diffraction data and Rietveld refinement

for Ln_6WO_{12} (Ln = Y, Ho) *Powder Diffr.*, **15** (4), 220–226.

69 Diot, N. (2001) Nouveaux oxynitrures de type perovskite ou fluorine dans les systèmes M-Ta(Nb)-O-N (M: alcalinoterreux) et Ln-W-O-N (Ln: élément de terre rare) en vue d'une application comme pigments colorés, Université de Rennes1, France.

70 Cheviré, F., Clabau, F., Larcher, O., Orhan, E., Tessier, F., and Marchand, R. (2009) Tunability of the optical properties in the $Y_6(W,Mo)(O,N)_{12}$ system. *Solid State Sci.*, **11**, 533–536.

71 Schilling, H., Wolff, H., Dronskowski, R., and Lerch, M. (2006) Fluorite-type solid solutions in the system Y-Ta-O-N: a nitrogen-rich analogue to yttria-stabilized zirconia (YSZ). *Z. Naturforsch.*, **61b**, 660–664.

72 Wolff, H., Schilling, H., Lerch, M., and Dronskowski, R. (2006) A density-functional and molecular- dynamics study on the physical properties of yttrium-doped tantalum oxynitride. *J. Solid State Chem.*, **179**, 166–171.

73 Pors, F., Marchand, R., and Laurent, Y. (1993) Une nouvelle famille de pyrochlore: les oxynitrures $Ln_2Ta_2O_5N_2$: préparation et étude cristallochimique. *J. Solid State Chem.*, **107**, 39–42.

74 Kikkawa, S., Takeda, T., Yoshiasa, A., Maillard, P., and Tessier, F. (2008) Crystal structure and optical properties of oxynitride rare-earth tantalates RTa-(O,N) (R = Nd, Gd, Y) *Mater. Res. Bull.*, **43**, 811–818.

75 Maillard, P., Merdrignac-Conanec, O., and Tessier, F. (2008) Characterization of fluorite-type oxynitride phases in the R_2Ta-O-N system (R = rare-earth element) *Mater. Res. Bull.*, **43**, 30–37.

76 Ray, E., Tessier, F., and Cheviré, F. (2011) Preparation and optical characteritics of novel oxynitride phases in the R_3(Ta/Nb)-O-N system (R = La, Eu, Gd, Ho, Y, Yb) *Solid State Sci.*, **13**, 1031–1035.

77 Tessier, F., Maillard, P., Orhan, E., and Cheviré, F. (2010) Study of the R-(Zr,W)-(O,N) (R = Y, Nd, Sm, Gd, Yb) oxynitride sytem. *Mater. Res. Bull.*, **45**, 97–102.

78 Moriya, Y., Takata, T., and Domen, K. (2013) Recent progress in the development of (oxy)nitride photocatalysts for water splitting under visible-light-irradiation. *Coord. Chem. Rev.*, **257**, 1957–1969.

9
Mechanochemical Synthesis, Vacancy-Ordered Structures and Low-Dimensional Properties of Transition Metal Chalcogenides

Yutaka Ueda[1] and Tsukio Ohtani[2]

[1]*Toyota Physical and Chemical Research Institute, 41-1 Yokomichi, Nagakute, Aichi 480–1192, Japan*
[2]*Okayama University of Science, Laboratory for Solid State Chemistry, Ridai-cho 1-1, Kita-ku, Okayama 700-0005, Japan*

9.1
Introduction

The chalcogens are the chemical elements in the periodic table group 16, consisting of O, S, Se, Te, and Po. More commonly, the term "chalcogen" is restrictively used for the three elements of S, Se, and Te. Their compounds are referred to as "chalcogenides."

The term "chalcogens" originates from the Greek word "chalcos" (meaning "ore formers"). There are a lot of chalcogenide ore minerals: pyrite (FeS_2), chalcopyrite ($CuFeS_2$), chalcocite (Cu_2S), sphalerite (zinc blend: ZnS), galena (PbS), stibnite (Sb_2S_3), cinnabar (HgS), molybdenite (MoS_2), and so on [1]. The hydrothermal minerals in the deep sea mainly consist of the metal chalcogenides.

Metal chalcogenides are ordinarily prepared by heating the elemental mixtures of metals (M) and chalcogens (X). Transition metal chalcogenides have a large variety of structures: M_2X (antifluorite-type), MX (NaCl-type, NiAs-type, zinc blend-type, wurtzite-type, etc.), and MX_2 (CdI_2-type, $CdCl_2$-type, pyrite-type, marcasite-type) [2]. Besides these chalcogenides, transition metals form various chalcogenides of the formulae: M_3X_4, M_2X_3, MX_3, and so on. M_3X_4 compounds have the unique structures such as Cr_3S_4-type, Mo_3Se_4-type (Chevrel phase), Nb_3Te_4-type, and so on. Generally, sulfide, selenide, and telluride of a transition metal having the same composition tend to possess an identical structure. Various preparation methods of metal chalcogenides are described in the literature [3].

Compared to metal oxides possessing a strong ionic character, the bonding nature of metal chalcogenides is highly covalent. As the atomic number of chalcogen increases, the chalcogenides tend to become more metallic. In transition metal chalcogenides, the covalency reduces the formal charge of transition metals, and tends to form the metal–metal bonds. This nature favors the formation

Handbook of Solid State Chemistry, First Edition. Edited by Richard Dronskowski, Shinichi Kikkawa, and Andreas Stein.
© 2017 Wiley-VCH Verlag GmbH & Co. KGaA. Published 2017 by Wiley-VCH Verlag GmbH & Co. KGaA.

of berthollide compounds having the wide nonstoichiometric compositions. This nature also favors the formation of mixed-valence compounds. These characteristics are typically exemplified in the system MX (NiAs-type)-MX_2 (CdI_2-type) (M = early 3d transition metals), which show various mixed-valence phases with nonstoichiometric compositions. In addition, d electrons participating in covalent bonding have the itinerant nature in most cases, exhibiting a large variety of physical and chemical properties.

The metal chalcogenides have a lot of low-dimensional structures possibly due to the directionality of the covalent bonds. The two-dimensional chalcogenides of MoS_2 and WS_2 are used as solid lubricants. The layered structure presents various chemical and physical properties such as intercalation, charge density wave instability, superconductivity, and so on. In addition, there exist many quasi-one-dimensional chalcogenides exhibiting a large variety of physical and chemical properties. For instance, the quasi-one-dimensional selenide $NbSe_3$ shows the CDW instability as well as the superconductivity. The more detailed structures and properties of transition metal chalcogenides are described in Refs [4–6].

In this chapter, we shall focus on several interesting topics of metal chalcogenides: mechanochemical synthesis of Cu, Ag, Zn, and Cd chalcogenides, phase diagram and order–disorder transition of 3d transition metal chalcogenides with NiAs- or CdI_2-related structure, charge density wave transition and superconductivity in quasi-one-dimensional chalcogenides, and so on.

9.2
Synthesis of Metal Chalcogenides by Mechanical Alloying

As mentioned in Section 9.1, stable phases of metal chalcogenides can be prepared by heating the elemental mixtures. The metastable phases, however, require the special techniques to be prepared. Mechanical alloying (MA) method is one of the versatile techniques for preparation of unique materials such as metastable materials [7], nonequilibrium solid solution [8], amorphized intermetallics [9,10], and so on. MA is a high-energy ball milling process involving repeated welding, fracturing, and rewelding of powder particles. This method was originally developed for preparation of nickel-based superalloys [11]. In these processes particles are strongly stressed by colliding with each other and contacting with vials and balls. The mechanical energy supplied by the milling is transferred to the particles, resulting in introduction of strain into the powders. Such a highly activated state promotes to start the reactions at ordinary temperature.

The mechanochemical reactions were probably first found in ancient Greece; the reaction of $HgS + Cu \rightarrow Hg + CuS$ took place when HgS was rubbed with vinegar using a copper pestle and a mortar [12]. It is of interest that the metal chalcogenides were concerned in the first observation of mechanochemistry. Now, many metal chalcogenides have been prepared by the MA technique [13–17]. The transition metal sulfides (ZnS, NiS, FeS) are easily formed on high-energy grinding [13]. MA study on Ni–S system revealed that the high-temperature phase of NiS (stable above 577 K) is formed by milling, indicating that the relatively high temperatures are generated during grinding metal–sulfur mixtures [14].

Sonochemical synthesis using ultrasound is known as a unique technique of MA [18]. This method was applied for the preparation of Cu and Ag chalcogenides [19,20]. In addition, Cu_3Se_2 was prepared only by strongly contacting α-Cu_2Se and α-CuSe pellets at room temperature [21]. This method can be regarded as the ultimately simple MA method, which requires only the static contact between particles.

9.2.1
Synthesis of Copper Chalcogenides by Ball Milling

Copper has a cultural significance, as it is the first metal used by man. Copper chalcogenides are important sources of copper metal [1]. A large number of copper chalcogenides including metastable phases were prepared by MA from elemental mixtures using a planetary ball mill (Fritsch P-7). Detailed experimental conditions are described in literatures [15,16].

A large variety of copper chalcogenides were obtained by MA.

1) *Stable phases*: CuS (covellite), β-$Cu_{2-x}Se$, Cu_3Se_2, CuSe, $Cu_{2-x}Te$, and Cu_3Te_2
2) *Metastable phases*: hexagonal $Cu_{2-x}S$ ($Cu_{1.71}S$ – $Cu_{1.89}S$: transformed into djurleite by aging at room temperature)
3) *High-pressure phases*: tetragonal $Cu_{2-x}S$ ($Cu_{1.92}S$–$Cu_{1.99}S$: transformed into djurleite by aging at room temperature) and $CuSe_2$ (pyrite-type structure)
4) *Ternary chalcogenides*: $CuInX_2$ (X = S, Se, Te) and $AgCuX$ (X = S, Se)
5) *Decomposition of γ-AgCuS through the reaction*: γ-AgCuS + Cu → Ag + Cu_2S

The main results are described as follows.

β-$Cu_{2-x}Se$
Figure 9.1 gives X-ray diffraction (XRD) pattern for $Cu_{1.75}Se$ obtained by 60 min of milling of the elemental mixture. The pattern shows good crystallinity. The obtained sample was identified to be β-$Cu_{2-x}Se$ phase (space group: $Fm\overline{3}m$) [22] by XRD measurements. Figure 9.2 shows temperature dependences of the electrical resistivity for the β-$Cu_{2-x}Se$. The sample of β-$Cu_{1.75}Se$ showed a sudden drop in the resistivity at about 8 K, indicative of superconductivity. This compound shows no superconductivity, when prepared by the normal method. The anomaly at 8 K, therefore, would be owing to a surface superconductivity since the surface suffers strong stress induced by the ball milling. This behavior was not always observed.

High-Pressure Phase of Tetragonal $Cu_{2-x}S$
Figure 9.3 shows XRD pattern of $Cu_{1.96}S$ obtained by milling the elemental mixture with desired ratio for 60 min at room temperature [15]. The pattern shows

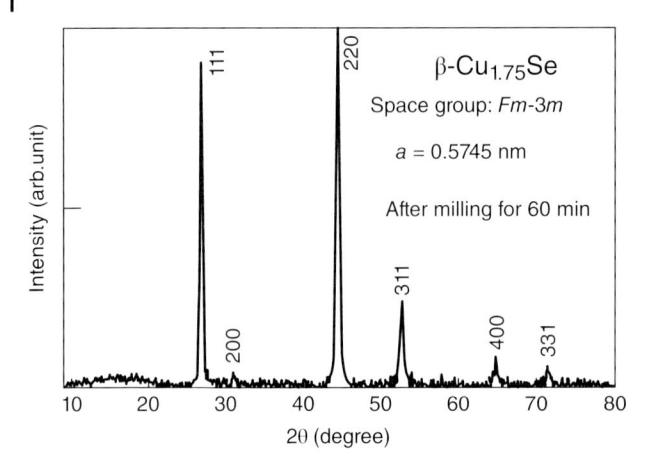

Figure 9.1 XRD pattern of the β-$Cu_{1.75}$Se obtained by milling the elemental mixture for 60 min.

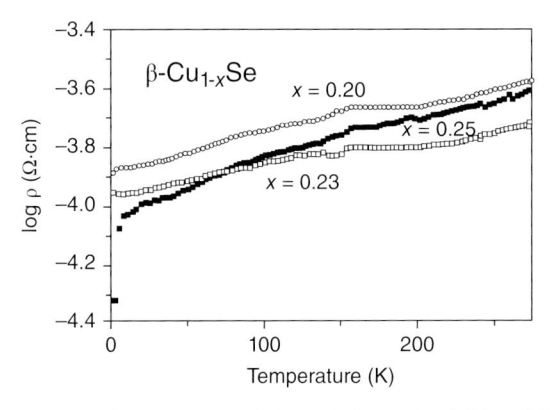

Figure 9.2 Temperature variations of electrical resistivity of the β-Cu_{2-x}Se prepared by MA.

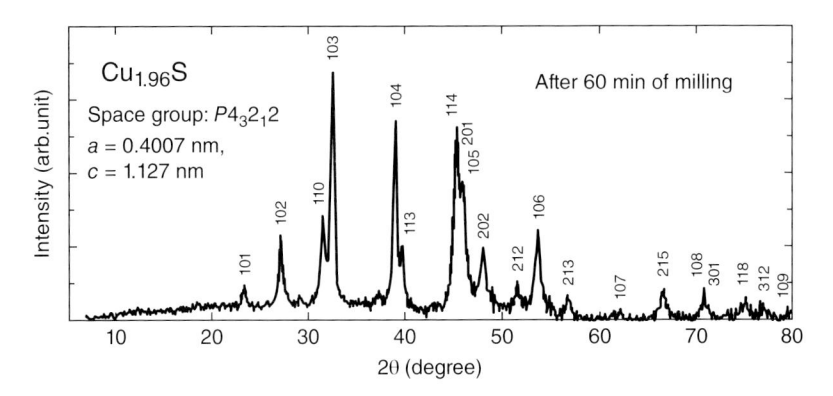

Figure 9.3 XRD pattern of the tetragonal $Cu_{1.96}$S (high-pressure phase) obtained by milling the elemental mixture for 60 min at room temperature.

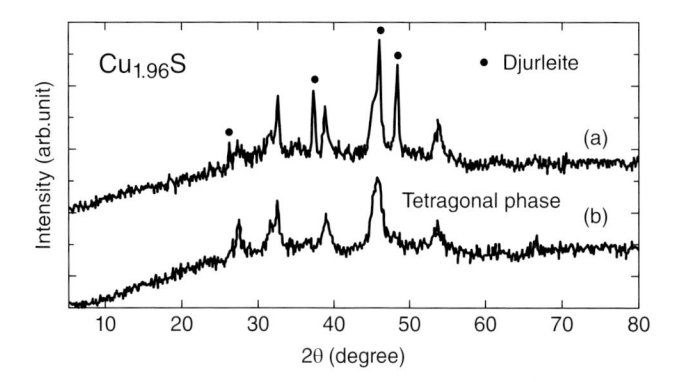

Figure 9.4 XRD patterns of $Cu_{1.96}S$ after aging of the metastable (high pressure) tetragonal phase at room temperature for 6 months (a), and after 30 min of milling of the above aged sample (b).

the tetragonal $Cu_{2-x}S$ phase (space group: $P4_32_12$) [23], which is stable at high pressures [24]. It is most likely that this phase was formed under the high pressure induced by the ball milling. The composition of obtained $Cu_{2-x}S$ was observed to extend from $Cu_{1.92}S$ to $Cu_{1.99}S$. Figure 9.4a shows XRD pattern of $Cu_{1.96}S$ after aging the metastable high-pressure phase for 6 months at room temperature. The stable phase of djurleite [25] appeared as indicated by closed circles. The high-pressure phase was entirely transformed to the stable djurleite after 10 months of aging at room temperature. The aged sample reverted to the high-pressure phase after 30 min of milling as shown in Figure 9.4b.

High-Pressure Phase of Pyrite-Type $CuSe_2$

$CuSe_2$ has two polymorphs of a stable marcasite type and a metastable pyrite type. The pyrite phase of $CuSe_2$ was obtained at high pressure of $6–6.5 \times 10^3$ MPa [26]. Figure 9.5 shows XRD pattern of $CuSe_2$ obtained by milling the elemental mixture for 120 min. The pattern shows two-phase mixture

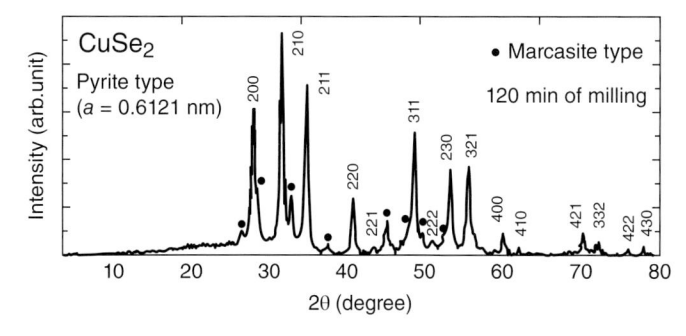

Figure 9.5 XRD pattern of $CuSe_2$ obtained by milling the elemental mixture for 120 min.

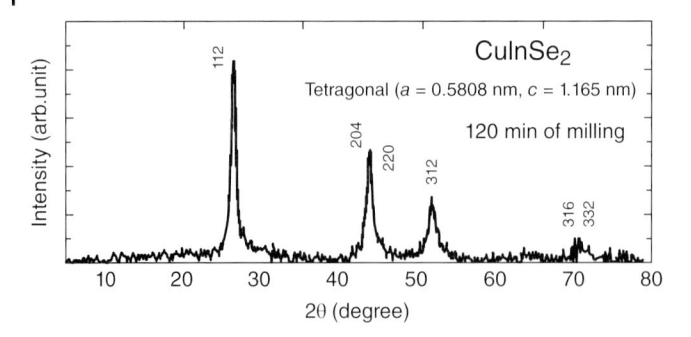

Figure 9.6 XRD pattern of the CuInSe$_2$ obtained by milling the elemental mixture of Cu, In, and Se for 120 min.

consisting of the majority of pyrite phase and the minority of marcasite phase. The proportion of pyrite phase in the obtained samples did not increase on further milling, and did not decrease by aging at room temperature. The pyrite phase shows the superconductivity at 2.30–2.43 K [26]. The present samples showed a small drop of the resistivity at about 3 K [15].

CuInX$_2$ (X = S, Se, Te)

Ternary selenides of CuInS$_2$, CuInSe$_2$, and CuInTe$_2$ were obtained by milling the elemental mixture for 120–180 min [16]. Figure 9.6 shows XRD patterns of CuInSe$_2$ obtained by ball-milling for 120 min at room temperature. CuInSe$_2$ is now used for high-performance solar cell. The present results present an easy way to obtain CuInSe$_2$.

Decomposition of γ-AgCuS through the Reaction: γ-AgCuS + Cu → Ag + Cu$_2$S[1)]

AgCuS was obtained by the following method. At first, CuS was prepared by milling the elemental mixture of Cu and S for 30 min. Then the obtained CuS and Ag were milled together for 30 min. Obtained sample was identified to be γ-AgCuS (stromeyerite) [27] by XRD measurements. This phase was not able to be prepared by milling the elemental mixture of Ag, Cu, and S. Figure 9.7a shows the XRD pattern of the mixture of γ-AgCuS and Cu with equimolar ratio. Figure 9.7b shows XRD pattern of the sample obtained by milling the mixture (γ-AgCuS + Cu) for 180 min. The elemental Ag was segregated from the γ-AgCuS and the Cu$_{2-x}$S phase newly appeared. These results indicate that the reaction of γ-AgCuS + Cu → Ag + Cu$_2$S took place by the milling. The same reaction was observed when the mixture of γ-AgCuS and Cu was pressed into the pellet and then was aged at room temperature. These reactions would be owing to the fact that Cu has a strong affinity to S.

1) Ohtani, T., Makita, Y., and Tamoto, T. unpublished.

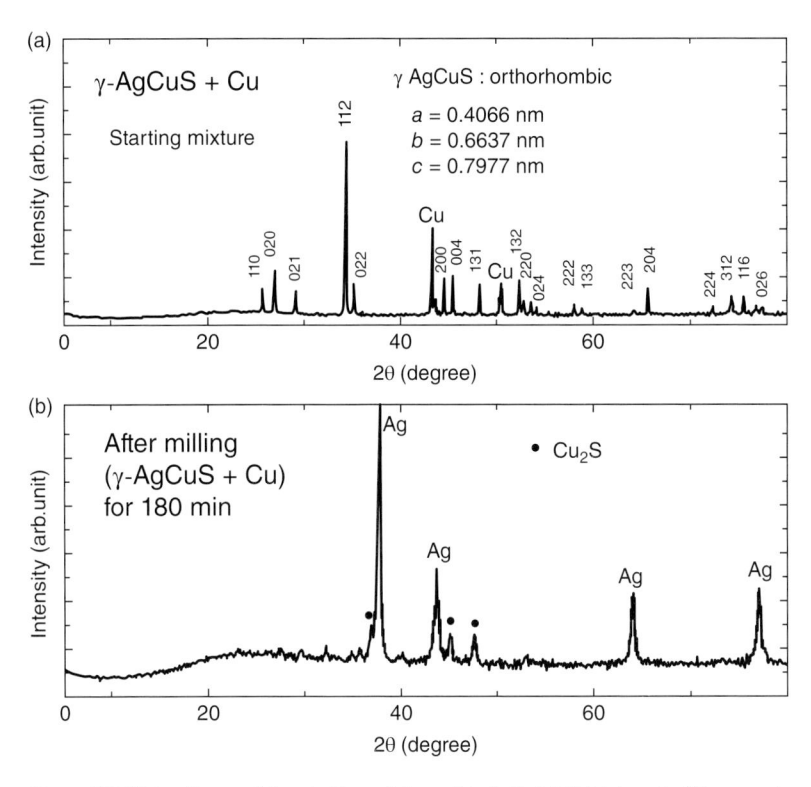

Figure 9.7 XRD patterns of the starting mixture of (γ-AgCuS + Cu) (a), and of the sample obtained by milling the mixture of (γ-AgCuS + Cu) for 180 min (b).

9.2.2
Sonochemical Synthesis of Cu and Ag Chalcogenides

Sonochemistry is the application of ultrasound to chemical reactions and processes [18,28–31]. Sonochemical reactions arise from acoustic cavitation: the formation, growth, and implosive collapse of bubbles in a liquid. The collapse of bubbles produces intense local heating (~5000 K), high pressures (~50 MPa), and lifetime of a few microseconds. Shock waves induced by the cavitation in liquid–solid slurries produce high-velocity interparticle collisions, leading to the reactions between particles. Figure 9.8 gives scanning electron microscopy (SEM) images of Cu particles obtained by irradiating Cu powders in methanol with ultrasound at 28 kHz for 1 h. The neck formation of Cu particles would be caused by the localized melting at the impact points.

The sonochemical method was applied to the preparation of Cu and Ag chalcogenides from elemental mixtures. $Cu_{1.96}S$, Cu_3Se_2, α-Cu_2Se, and Ag_2Se were successfully prepared by this method [19,20]. Experimental procedures are as follows. The powdered mixture of the elements with desired ratio (0.5 g in weight) was immersed in 0.3 cm^3 of methanol contained in a silica glass tube

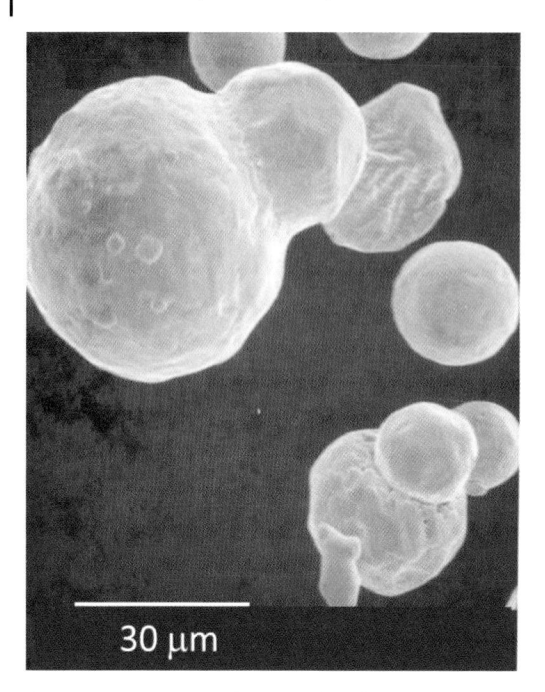

Figure 9.8 SEM image of Cu particles obtained by irradiating Cu–methanol slurry with ultrasound at 28 kHz for 1 h.

with inner diameter of 0.8 mm and a wall thickness of 1.0 mm. The tube was partly submerged in 1.5 dm^3 of water in a common ultrasonic cleaner (Shimazu, SUS-103; 100 W). The methanol slurry was irradiated with ultrasound at 28 kHz for several hours at room temperature.

Figure 9.9 exhibits XRD patterns for the products obtained by irradiating the mixture of (3Cu + 2Se) in methanol with ultrasound at 28 kHz for 0.5 h (a), 4 h (b), and 8 h (c). The starting mixture reacted gradually with the irradiation time, and entirely reacted into Cu_3Se_2 after 8 h of irradiation. Contrary to the ball milling process, the particles are not fractured during the sonochemical reactions. This enables us to regard the reactions as the simple collision processes.

Figure 9.10a shows SEM image of starting mixture of (3Cu + 2Se), where spherical particles are Cu and irregular particles are Se. Figure 9.10b gives SEM image of the final products of Cu_3Se_2 obtained by irradiating the elemental mixture with ultrasound for 8 h. After the reaction, the starting particles of Cu and Se are not able to be distinguished to each other. Figure 9.11 shows SEM images of Cu particles obtained by irradiating with ultrasound for 1 min (a) and for 1 h (b). After 1 min of irradiation, the surface of Cu particle was covered with many small granules, which seem to be produced by the collisions with Se particles. EPMA (electron probe microanalysis) measurements revealed that the surface has the composition of Cu : Se = 11 : 1.0. After 1 h of irradiation, the surface of Cu particle was constructed with many pentagonal and hexagonal domains. The composition of this surface was observed to be Cu : Se = 7.6 : 1.0.

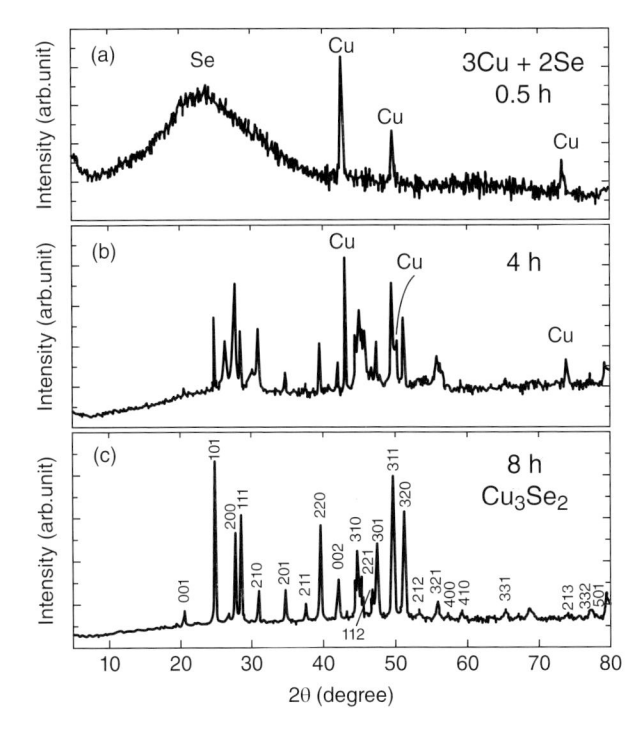

Figure 9.9 XRD patterns for the products obtained by irradiating the mixture of (3Cu + 2Se) in methanol with ultrasound at 28 kHz for 0.5 h (a), 4 h (b), and 8 h (c).

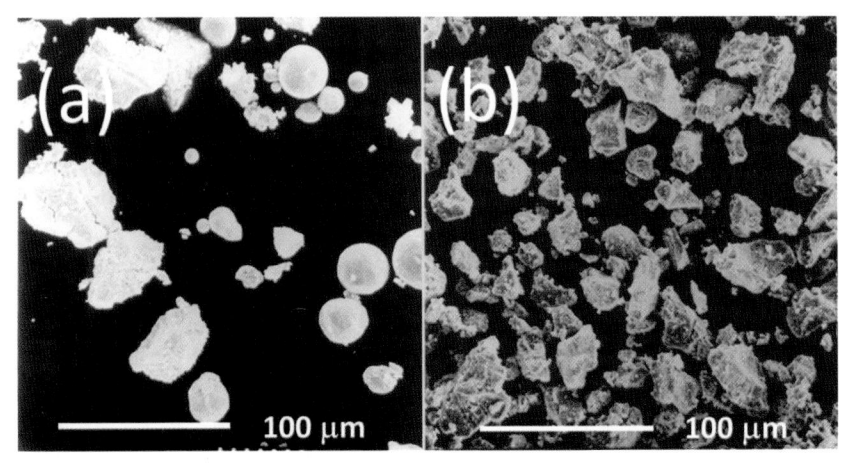

Figure 9.10 SEM images of starting mixture of (3Cu + 2Se) (a), and the Cu_3Se_2 obtained by irradiating the elemental mixture in methanol with ultrasound at 28 kHz for 8 h (b).

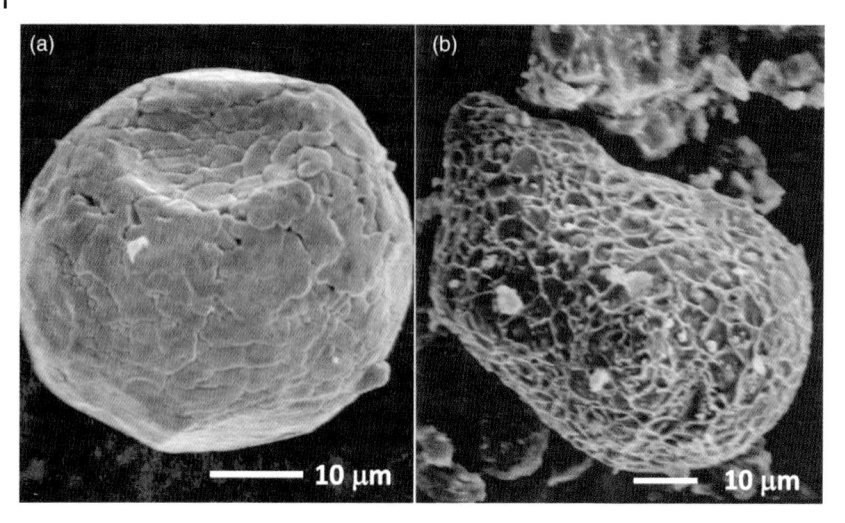

Figure 9.11 SEM images of Cu particles obtained by irradiating the mixture of (3Cu + 2Se) in methanol with ultrasound at 28 kHz for 1 min (a), and for 1 h (b).

In order to observe the inside of reacted particles, each of Cu and Se particles obtained by sonochemical reaction was crashed by a pestle in agate mortar. Figure 9.12a gives SEM image of crashed Cu particle obtained by irradiating the elemental mixture with ultrasound for 3 h. The particle consists of the inner spherical portion and the peeled outer portion. EPMA measurements revealed that the inner portion was pure copper and the outer portion had the composition of Cu : Se = 1.8 : 1.0. The outer reacted portion spread toward the inside on further ultrasonic irradiation, keeping the clear boundary between the reacted and the unreacted portion. Finally, the particle was uniformly transformed to the Cu_3Se_2. Figure 9.12b shows SEM image of cross section of Se particle obtained by irradiating the elemental mixture with ultrasound for 3 h. There is also a clear boundary between the inner portion with smooth surface and the outer portion consisting of multiple grains. EPMA measurements showed that the inner portion was pure selenium and the outer portion had the homogeneous composition of Cu : Se = 4 : 5. On further irradiation, the boundary moves toward the inside of the particle as shown in Figure 9.12c, where only a small amount of pure Se remains around the center, which is just like a seed of fruit. The surrounding reacted region of this particle was almost homogeneous, possessing near the stoichiometric composition of Cu_3Se_2. It is noted that there is no concentration gradient in the outer reacted portion for both Cu and Se particles. Such clear boundaries between the reacted and the unreacted parts in the particles during the reaction have been observed in many solid-state reactions [32].

Ag_2Se was obtained by irradiating the elemental mixture of (2Ag + Se) in methanol with ultrasound at 28 kHz for 240 min at room temperature [20]. XRD pattern of the obtained sample is shown in Figure 9.13. The pattern is well

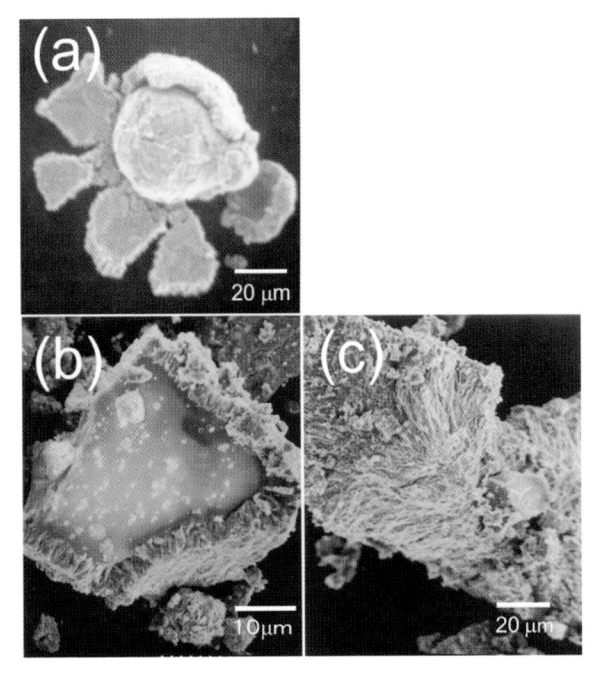

Figure 9.12 SEM images of crashed Cu and Se particles: (a) the Cu particle obtained by irradiating the mixture of (3Cu + 2Se) in methanol with ultrasound for 3 h, (b) the Se particle obtained by 3 h of irradiation, and (c) the Se particle obtained by 6 h of irradiation.

consistent with the published data for β-Ag$_2$Se (space group: $P2_12_12_1$) [33]. This compound has the potential applications for high-sensitivity magnetic sensors, due to the large linear magnetoresistance (LMR) [34]. Figure 9.14a shows SEM image of the cross section of crashed Se particle after irradiating the mixture of (2Ag + Se) with ultrasound for 1 min. There is a clear boundary in the particle between the inner portion with smooth surface and the outer portion with many

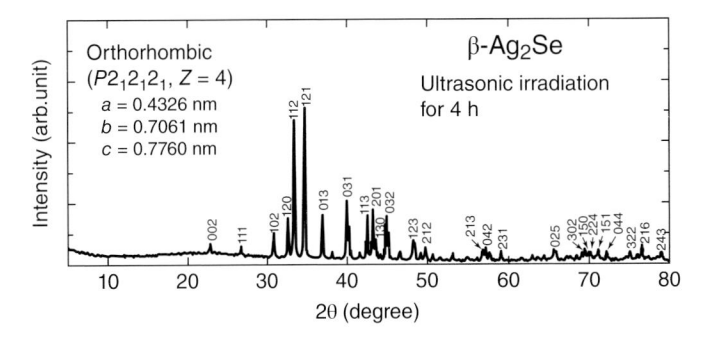

Figure 9.13 XRD pattern of β-Ag$_2$Se obtained by irradiating the elemental mixture of (2Ag + Se) in methanol with ultrasound for 240 min at room temperature.

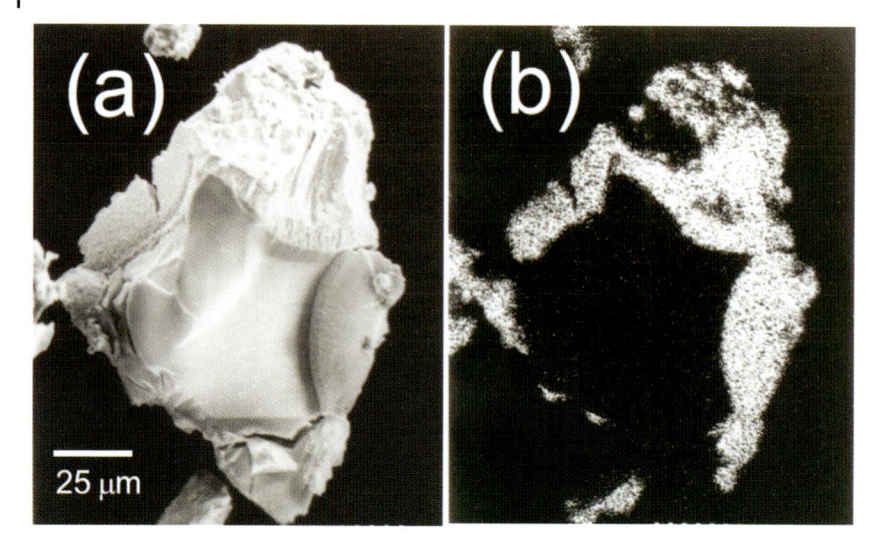

Figure 9.14 SEM image of the cross section of crushed Se particle after irradiating the mixture of (2Ag + Se) with ultrasound for 1 min (a), and the distribution map of Ag (b). The inner portion is pure Se.

granules, as similarly observed in the (3Cu + 2Se) system. Figure 9.14b shows the Ag-distribution map of the particle corresponding to that shown in Figure 9.14a. The inner part was found to be pure Se and the outer portion was observed to have the final composition of Ag_2Se. This means that the reacted portion always has the composition of Ag_2Se. This observation differs from that of the (3Cu + 2 Se) system, where the composition of outer portion gradually approaches to the final composition of Cu_3Se_2 with increasing the irradiation time. This discrepancy would be owing to that the β-Ag_2Se is only an existing phase in the Ag–Se system at ordinary temperature.

9.2.3
Solid-State Reaction: Formation of Cu_3Se_2 by Making α-CuSe and α-Cu_2Se Pellets Contact

α-CuSe and α-Cu_2Se pellets reacted entirely forming Cu_3Se_2 simply by being brought into contact with each other at room temperature [21]. Figure 9.15 shows the partial phase diagram of Cu–Se system, including Cu_2Se, Cu_3Se_2, and CuSe phases [35]. The stoichiometric Cu_2Se has α- and β-phases below and above about 400 K, respectively. The α–β transition temperature is sharply lowered with increasing selenium content. Both α- and β-phases basically have an antifluorite structure [25]. Statistically disordered Cu atoms in the β-$Cu_{1.75}$Se showed an atomic ordering at about 250 K, forming a superstructure of $2a \times 2a \times 2a$ [36,37]. The Cu_3Se_2 phase is stable only at the stoichiometric composition below 408 K. This phase has a tetragonal unit cell (space group: $P42_1m$), primarily composed of a tetrahedral network of Cu atoms [25]. The α-CuSe phase with

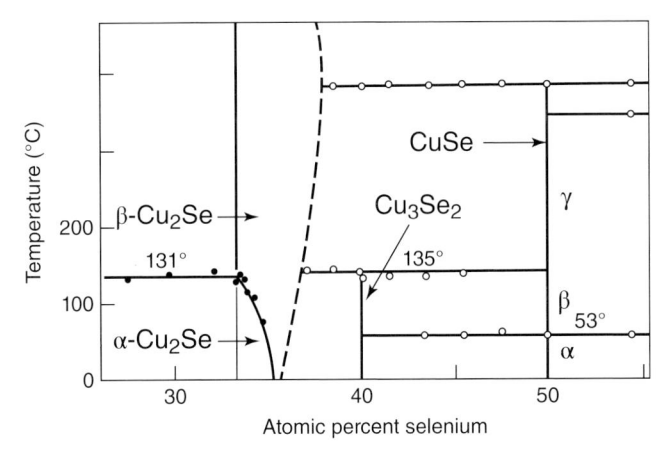

Figure 9.15 Partial phase diagram of the Cu–Se system [35].

a covellite structure is also a line phase stable at a stoichiometric composition below about 325 K. It should be noted that there is no obvious similarity among the crystal structures of α-Cu_2Se, Cu_3S_2, and α-$CuSe$.

Experiments were carried out in following ways.

1) The compounds of α-Cu_2Se and α-$CuSe$ were prepared by heating the elemental mixtures sealed in silica tubes.
2) Equimolar ratios of powdered samples of α-Cu_2Se and α-$CuSe$ (1.3×10^{-3} mol) were separately pressed into rectangular pellets (15 mm × 5 mm × 0.6 mm) under the pressure of 4×10^3 kg/cm^2.
3) The α-Cu_2Se and α-$CuSe$ pellets were brought into contact through their rectangular faces (15 mm × 5 mm) under a pressure of 10.0 kg/cm^2, and allowed to react for the periods as long as 30 days at room temperature.
4) The chemical composition of the lateral face (15 mm × 0.6 mm) of each pellet was analyzed by EPMA at certain intervals of the reaction. The pellets were subsequently ground into powder, and were identified by XRD method.

Figure 9.16 shows XRD patterns of α-$CuSe$, α-Cu_2Se, and Cu_3Se_2. Here the Cu_3Se_2 was obtained by tightly making α-$CuSe$ and α-Cu_2Se pellets contact for 10 days under pressure of 10.0 kg/cm^2 at room temperature. These results apparently indicate that Cu_3Se_2 is formed by the solid-state reaction between α-$CuSe$ and α-Cu_2Se.

Figure 9.17 shows XRD patterns of the samples obtained after being allowed to react for 60 min at room temperature: (a) the α-$CuSe$ pellet, (b) the α-Cu_2Se pellet. The pattern of α-$CuSe$ pellet shows two phase mixture of the α-$CuSe$ and Cu_3Se_2 phases, indicating the increase of Se content. The pellet of α-Cu_2Se changed to the β-Cu_2Se phase, indicating the decrease of Se content. According to the phase diagram of Cu–Se system (Figure 9.15), these results show that the compositions of pellets are both approaching to Cu_3Se_2.

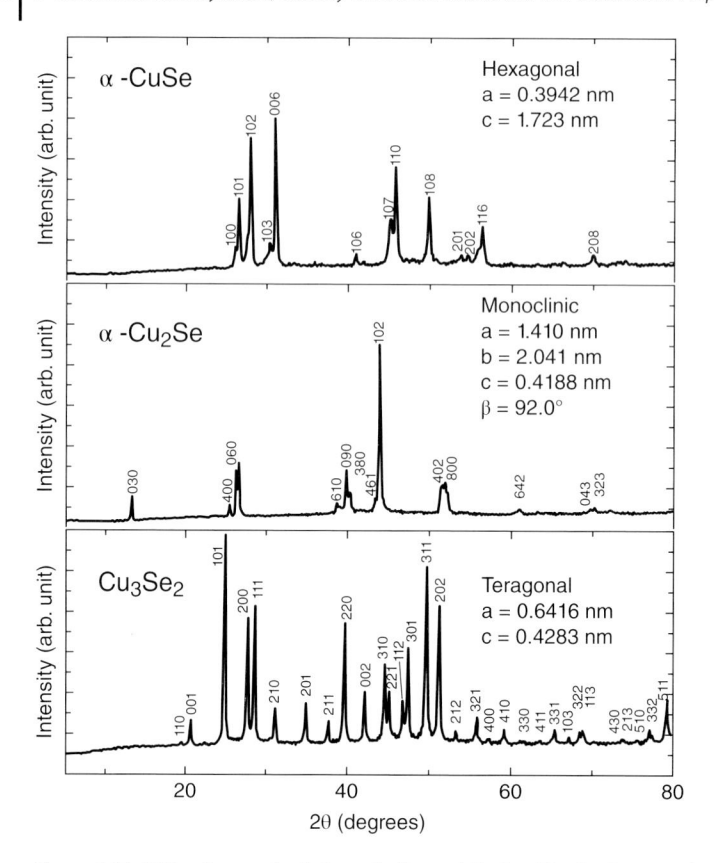

Figure 9.16 XRD patterns of α-CuSe, α-Cu$_2$Se, and Cu$_3$Se$_2$. The Cu$_3$Se$_2$ was obtained by contacting α-CuSe and α-Cu$_2$Se pellets with each other at room temperature under pressure of 10.0 kg/cm^2 for 10 days.

Figure 9.18 gives the distribution map of Se observed by EPMA for lateral face of the α-CuSe (upper) and the α-Cu$_2$Se (lower) pellets after the reaction progressed for 10 min (a), 60 min (b), and 240 min (c). Two pellets were tightly contacted during the reaction under pressure of 10.0 kg/cm^2. With increasing the reaction time, the Se concentration of the two pellets becomes closer to each other, which is parallel to the XRD measurements.

Figure 9.19 gives the chemical compositions of the lateral faces (15 mm × 0.6 mm) of α-CuSe and α-Cu$_2$Se pellets after allowed to react for 10, 60, and 240 min. The abscissa shows the distance from the end of the α-Cu$_2$Se pellet and the ordinate chemical composition x in Cu$_x$Se$_2$. The surface was scanned in a direction normal to the plane of contact between the α-CuSe and α-Cu$_2$Se pellets. Each of five scans was made in increment of 0.125 mm. With increasing the reaction time, the compositions of the two pellets became closer to each other. After allowed to react for 240 min, both pellets reacted to form the final compound of Cu$_3$Se$_2$. It is noteworthy that the lateral faces scarcely show the

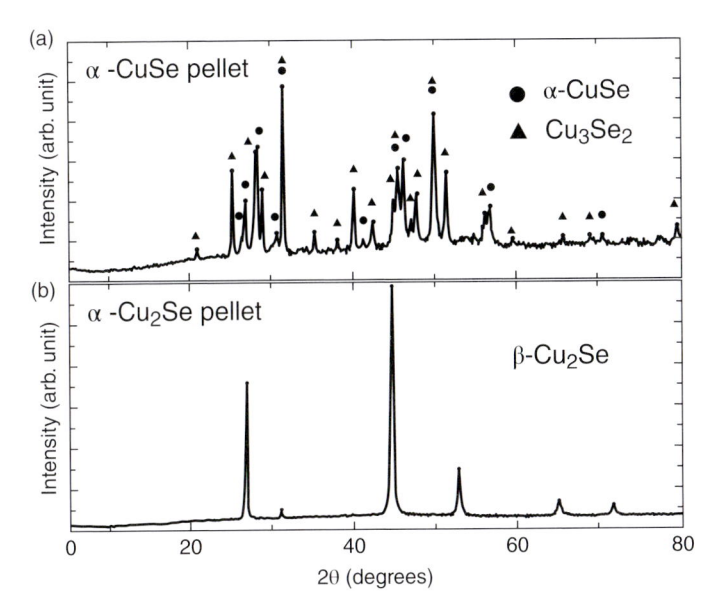

Figure 9.17 XRD patterns of the samples obtained by contacting α-CuSe and α-Cu$_2$Se with each other for 60 min at room temperature under pressure of 10.0 kg/cm^2: (a) the α-CuSe pellet, (b) the α-Cu$_2$Se pellet.

compositional gradient during the reaction, as also seen in the distribution map of Se (Figure 9.18). These results suggest the high mobility of Cu or/and Se ions in both pellets.

In order to see which type of ions dominantly migrated, the weight change was measured for each pellet. After the reaction for 20 days, the α-Cu$_2$Se pellet with 0.200 g was changed to the Cu$_3$Se$_2$ pellet with 0.173 g. The weight loss is consistent with the value calculated by assuming that only Cu ions migrate from the α-Cu$_2$Se pellet to the α-CuSe pellet. It is well known that the Cu ions are in the monovalent state in almost all copper chalcogenides [38]. The present reactions,

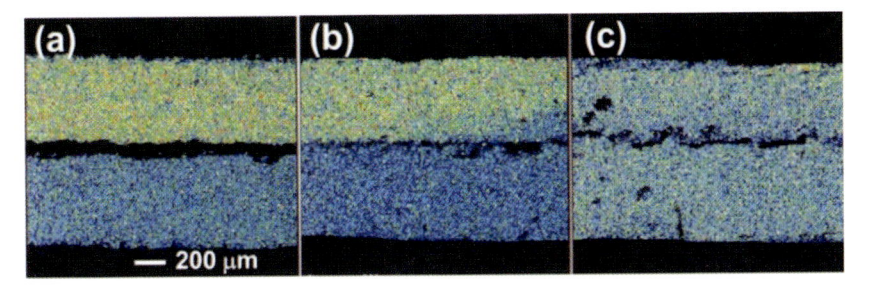

Figure 9.18 The distribution maps of Se observed by EPMA for lateral face of the α-CuSe (upper) and the α-Cu$_2$Se (lower) pellets after the reaction progressed for 10 min (a), 60 min (b) and 240 min (c).

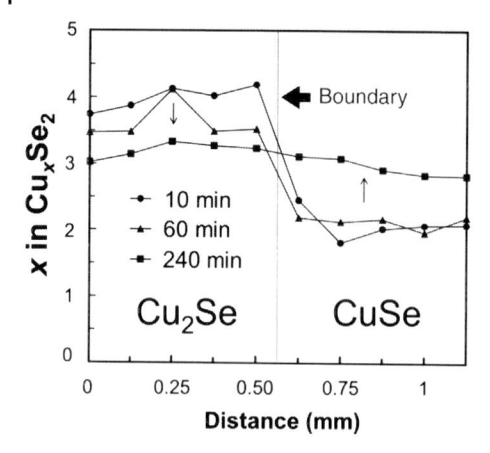

Figure 9.19 Chemical composition of the lateral faces (15 mm × 0.6 mm) of α-CuSe and α-Cu$_2$Se pellets allowed to react for 10, 60, and 240 min. The abscissa shows the distance from the end of the α-Cu$_2$Se pellet.

thus, would be represented as α-Cu$_2$Se − 1/2Cu$^+$ = Cu$_3$Se$_2$ in the α-Cu$_2$Se pellet, and α-CuSe + 1/2Cu$^+$ = 1/2(Cu$_3$Se$_2$) in the α-CuSe pellet, as shown in Figure 9.20.

Similar solid-state reactions were observed in the systems of (α-Cu$_2$Se and Cu$_3$Se$_2$) and (α-Cu$_2$Se and CuS). On the other hand, solid-state reactions were not observed in the systems of (Cu$_3$Se$_2$ and α-CuSe), (Cu$_2$S and CuS), and (α-CuSe and Cu$_2$S). These results show that the solid-state reactions take place only in the systems including the α-Cu$_2$Se phase. It was reported that both the α- and β-Cu$_2$Se phases show the high-ionic conductivity for Cu ions [39]. It is most likely, thus, that the high ionic conductivity of Cu ions in α-Cu$_2$Se and β-Cu$_2$Se is primarily responsible for the present solid-state reactions at room temperature.

The above model suggests that this system would work as a solid-state battery. Figure 9.21 represents the time dependence of electromotive force (EMF) of the α-Cu$_2$Se | β-CuSe system. To obtain the higher values of EMF, the

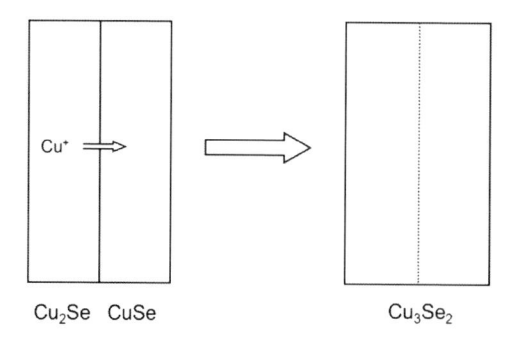

Figure 9.20 Schematic figure showing the migration of Cu$^+$ ions when α-Cu$_2$Se and α-CuSe pellets are in contact with each other at room temperature.

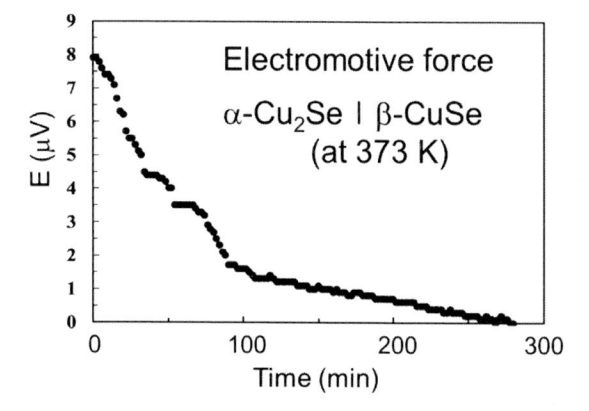

Figure 9.21 Electromotive force of the α-Cu$_2$Se | β-CuSe system as a function of time at 373 K.

measurements were carried out at 373 K. The α-CuSe phase changes to the β-CuSe phase at this temperature. As described above, the reaction takes place when the α-Cu$_2$Se is included in the system. Therefore, the solid-state reaction also occurs in the present system, because the α-Cu$_2$Se phase remains stable at this temperature. The value of EMF was observed to be + 8.0 µV immediately after the contact of two pellets. The β-CuSe side was observed to be the cathode, which is consistent with the migration of Cu$^+$ ions from the α-Cu$_2$Se pellet to the β-CuSe pellet. The values of EMF decreased with time; the relaxation time was 86 min. The EMF values were reversible in subsequent charge–discharge cycles. This reaction was observed to be sensitive to the conditions of compounds and the setup conditions; sometimes no reaction was observed, and sometimes an abrupt reaction took place with emitting a spark.

9.2.4
Mechanically Induced Self-Propagating Reactions (MSR) of CdSe and ZnSe: Premilling Effects

Mechanically induced self-propagating reactions (MSR) are observed in powder mixture whose reaction is highly exothermic, where the reaction starts suddenly at a critical time, which is termed as the ignition time [40]. No reaction is observed in the induction period, and after the ignition most of the reactants are consumed within seconds. MSR resembles to self-propagating high-temperature synthesis (SHS) [41]. Many metal chalcogenides have been prepared by MSRs [42]. The absence of a reaction in the induction period suggests that the mechanical energy supplied by ball milling is employed only for activation of the elemental powders. Therefore, MSR would occur for much shorter milling times, if the elemental powders were activated by milling separately beforehand. This idea was applied to MSR of CdSe and ZnSe according to the following procedures; each elemental powder was separately premilled for a certain time, and ball milling was then carried out for the mixture of the premilled elemental

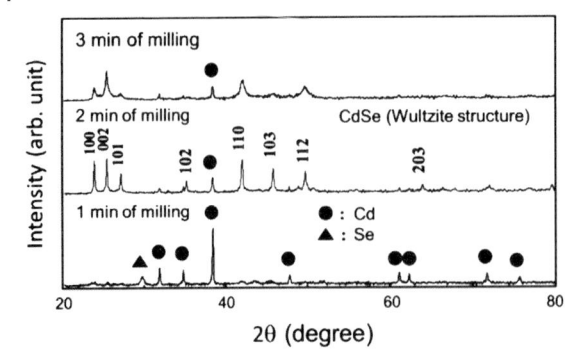

Figure 9.22 XRD patterns of Cd–Se obtained by milling the elemental mixture (Cd: Se = 1 : 1) for 1, 2, and 3 min.

powders [43]. Experiments were carried out using tungsten carbide vial and balls. For more detailed conditions, see Ref. [43].

Figure 9.22 shows XRD patterns of Cd–Se obtained by milling the elemental mixture (Cd : Se = 1 : 1) for 1, 2, and 3 min. The mixture showed no reaction up to 1 min of milling, but it then suddenly reacted to form CdSe with the wurtzite structure after 2 min of milling, suggesting that the product was formed via MSR. For the SHS processes, the adiabatic temperature, defined as the final temperature, is usually used as a measure of self-heating. For the case of MSR processes, the quantity $-\Delta H_{298}/C_{298}$ (ΔH_{298}: reaction enthalpy, C_{298}: heat capacity) is often used as a simpler substitute for the adiabatic temperature. MSR occurs for the condition: $-\Delta H_{298}/C_{298} > 2000$ K for all materials [40]. Using available data of ΔH_{298} and C_{298}, the value of $-\Delta H_{298}/C_{298}$ for CdSe (wurtzite type) was estimated to be 2919 K, satisfying the above condition.

Shown in Figure 9.23 are the XRD patterns for CdSe obtained by milling a mixture of the elements that were each premilled separately for 15 min.

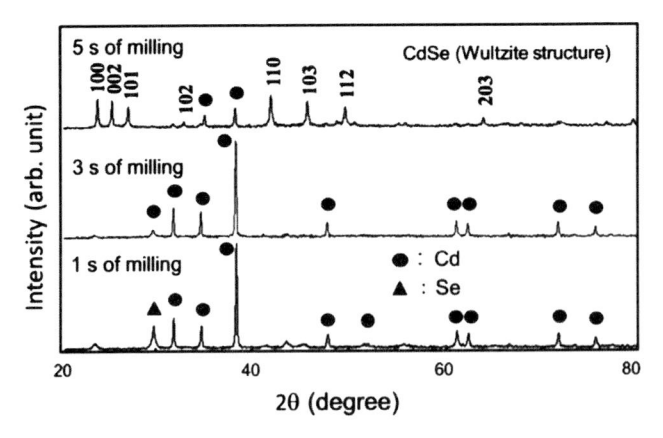

Figure 9.23 XRD patterns for Cd–Se obtained by milling the elemental mixture. Each elemental powder was separately premilled for 15 min.

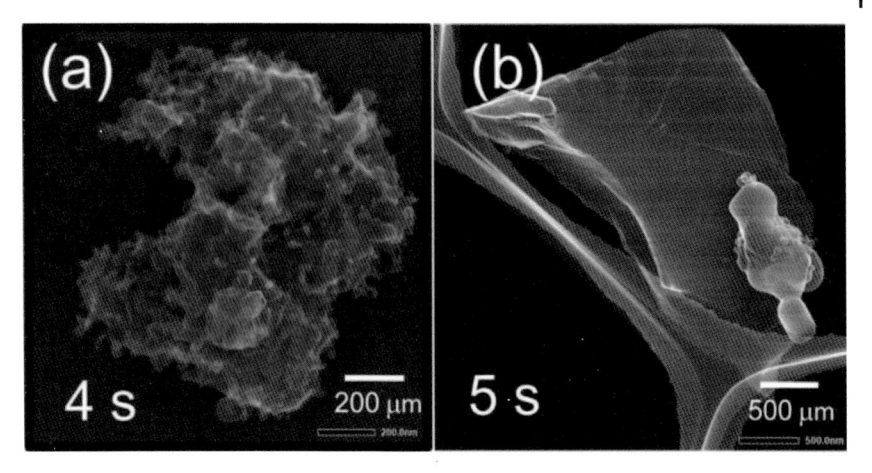

Figure 9.24 SEM images of Cd–Se obtained by milling the elemental mixture of Cd and Se for 4 s (a) and for 5 s (b). Prior to the milling, each elemental powder was separately premilled for 15 min.

Wurtzite-type CdSe formed following 3–5 s of milling, which is much shorter than for MSR without the premilling process. This suggests that the elemental powders are transformed into a highly activated state by the premilling process and, as a result, the final compound forms rapidly. Figure 9.24 gives SEM images of Cd–Se obtained by milling the elemental mixture of Cd and Se, which were separately premilled for 15 min, for 4 s (a) and for 5 s (b). The particle obtained by milling for 4 s seems to be an aggregate of much smaller particles. EPMA measurements revealed that this particle was the mixture of elemental Cd and Se. The particle obtained by milling for 5 s appears to be a broken piece of an ingot or a single crystal with smooth fracture surface, suggesting the melting of the sample. EPMA measurements showed that the particle was CdSe. These observations clearly show that the elemental mixture reacts to form the CdSe by milling for 5 s, when the elements are separately premilled. These results are consistent with the XRD observations. It is supposed, thus, that in the induction period, the mechanical energy supplied by the ball-milling process is employed primarily for activation of the particles.

ZnSe with wurtzite structure was synthesized by milling the elemental mixture for 8 min via MSR. Figure 9.25 shows optical micrographs of ZnSe samples obtained by milling the elemental mixture for 4 s (a) and 5 s (b). Prior to the milling, each elemental powder was separately premilled for 15 min. One can see the remarkable color change between 4 s and 5 s of milling, indicating the sudden formation of ZnSe at 5 s of milling. This was confirmed by the XRD measurements.

During the premilling processes, each particle is subjected to the strong stresses and is activated by many factors such as induced strain, formation of lattice defects and dislocation, partial amorphization, and so on. Such highly

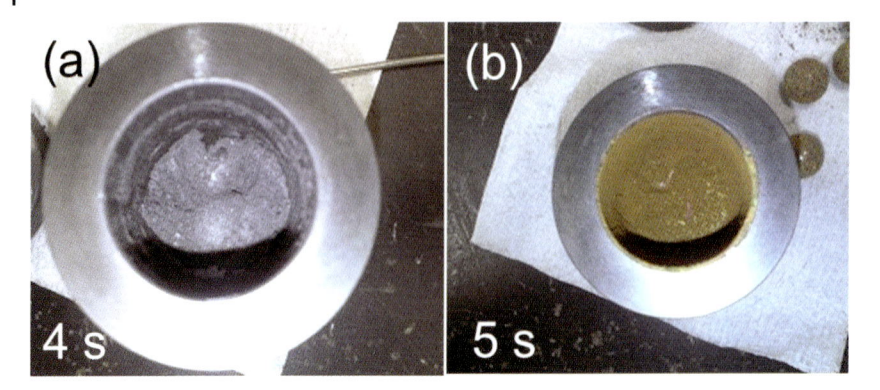

Figure 9.25 Optical micrographs of ZnSe in the vial after: (a) 4 s of milling, (b) 5 s of milling. The elemental powders of Zn and Se were separately premilled for 15 min.

activated state can reduce the height of energy barrier required to start the reaction. In the usual MA processes the reactions are caused by mutual collisions between different elemental particles and the starting mixtures are gradually transformed into the final compounds. On the contrary, the present results show that the most important factor for promoting MSR is not the mutual collisions between different elemental particles but the activation of each particle by the ball milling. These premilling effects are anticipated to be observed in many other MSR systems.

9.3
Vacancy-Ordered Structure, Order–Disorder Transition and Site Preference in the 3d Transition Metal Chalcogenides

9.3.1
Vacancy-Ordered Structures in Binary MX–MX$_2$ System

Most 3d transition metal chalcogenides MX (X = S, Se, Te) take the NiAs-type structure, while some of MX_2 capture the CdI$_2$-type structure. There is a close relation between the two structures, as shown in Figure 9.26. In both MX and MX_2, X ions form a hexagonal closed packed structure and M ions occupy the octahedral holes. In MX, M ions occupy all octahedral holes of the X lattices and the structure has an alternative stacking of M-layer and X-layer which are both two-dimensional hexagonal lattices, as $MXMXM$. Whereas the M-layers in MX_2 are vacant every other M-layer, resulting in the stacking of $MX\square XM$, where $\square$ is the vacant layer (van der Waals gap).

In the intermediate composition $M_{1+y}X_2$ ($0 < y < 1$) (or MX_x ($1 < x < 2$)), the vacant layers are partially occupied by metal atoms forming the layer stacking of $M^F X M^V X M^F$, where M^F denotes the metal layer fully occupied and M^V the metal layer partially occupied. Moreover, metal-vacancies are ordered in characteristic

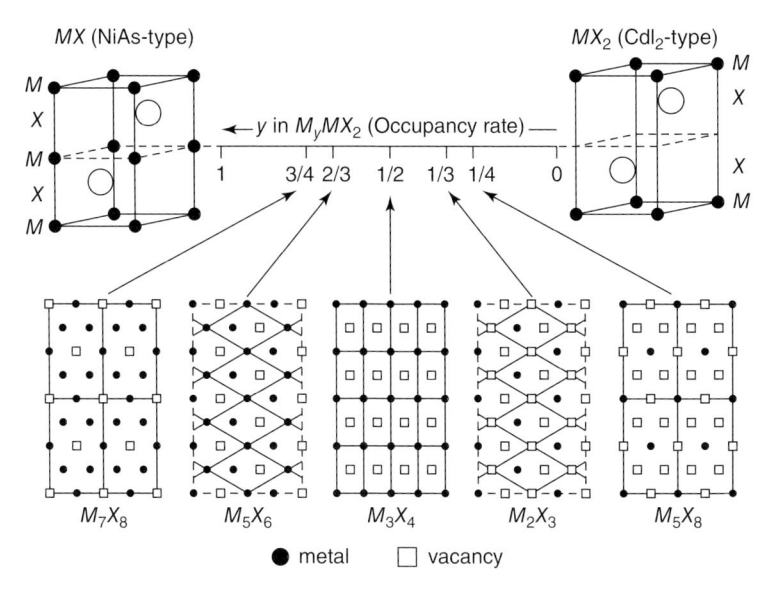

Figure 9.26 Structural relation between 3d-transition metal chalcogenides MX (NiAs-type) and MX_2 (CdI_2-type), and various vacancy-ordered structures of $M_{1+y}X_2$ ($0 < y < 1$).

manners at specific compositions both within M^V-layers and along the direction perpendicular to the layers. A number of vacancy-ordered phases with the characteristic structures, M_5X_8, M_2X_3, M_3X_4, M_5X_6, M_7X_8, and so on, appear in the composition range $0 < y < 1$, particularly in the selenide system $M_{1+y}Se_2$ most. Their intra-M^V-layer structures (vacancy-ordered manners) and the corresponding compositions are shown in Figure 9.26. In M_3X_4, just a half of M sites in the M^V-layer are occupied, giving the composition of $M_{1+1/2}X_2 = M_3X_4$. Centering on M_3X_4, M_5X_8 and M_7X_8 (M_2X_3 and M_5X_6) have a just opposite relation with each other in the ratio of occupied/vacant sites and hence in the vacancy-ordered manner of the M^V-layers, namely, M_7X_8 (M_5X_6) is equivalent to M_5X_8 (M_2X_3) if the vacant and occupied sites are exchanged to each other.

These vacancy-ordered manners can be naively understood in the light of repulsive interaction between metal atoms ($0 < y < 0.5$) or between vacancies ($0.5 < y < 1$). In M_5X_8 with $y = 1/4$ (M_7X_8 with $y = 3/4$), the first- and second-nearest sites centered at an M atom (vacancy) in the hexagonal lattice are vacant (occupied) and the third nearest sites are occupied (vacant). In M_2X_3 with $y = 1/3$ (M_5X_6 with $y = 2/3$), the second nearest sites are occupied (vacant). On the other hand, the vacancy-ordered manner in M_3X_4 with $y = 1/2$ is not explained from such a simple scenario based on the ideal hexagonal lattice, but it can be realized if the hexagonal lattice is distorted in a way that the first nearest six sites in the ideal hexagonal lattice become inequivalent as the first nearest four (or two) sites and the second nearest two (or four) sites. The fact that the M_3X_4-type structure is most popular in 3d transition metal chalcogenides indicates that the hexagonal lattices are actually distorted.

These vacancy-ordered structures allow partial occupancy of M atoms at the vacant sites and/or of vacancies at the occupied sites to a certain extent, namely, each phase has a certain nonstoichiometric composition range. In broad sense, therefore, each phase is defined as M_mX_n-type when the occupancy rate (p_o) of the originally occupied sites in the M^V-layers is larger than that (p_v) of the originally vacant sites. Similarly, CdI$_2$-type is defined as the phase in which the occupancy rate (p^F) of the M^F-layers is larger than that (p^V) of the M^V-layers without vacancy-ordering, and NiAs-type is defined as the phase with $p^F = p^V$. These vacancy-ordered states become disordered as temperature is raised, and at $p_o = p_v$ the vacancy-ordered phases are transformed to CdI$_2$-type (order–disorder transition) and further to NiAs-type at $p^F = p^V$.

9.3.2
Phase Diagram and Order–Disorder Transition in Binary MSe–MSe$_2$ System

The phase diagram of TiSe$_x$ is shown in Figure 9.27 as a typical example [44]. There are three vacancy-ordered phases, M_5X_8, M_2X_3, and M_3X_4, each of which has its own nonstoichiometric composition region and is separated by the two-phase region. On heating, each phase shows a transition to CdI$_2$-type phase by intra-M^V-layer disordering of metal vacancies, and in the high-temperature region the CdI$_2$-type phase applies to the wide composition range. In the composition region close to TiSe, the NiAs-type phase becomes stable rather than the CdI$_2$-type phase, although it is rather hard to distinguish the both phases by X-ray diffraction.

The phase diagram of VSe$_x$ is shown in Figure 9.28 as another example [45]. The V–Se system has two vacancy-ordered phases, M_5X_8 and M_3X_4. There is

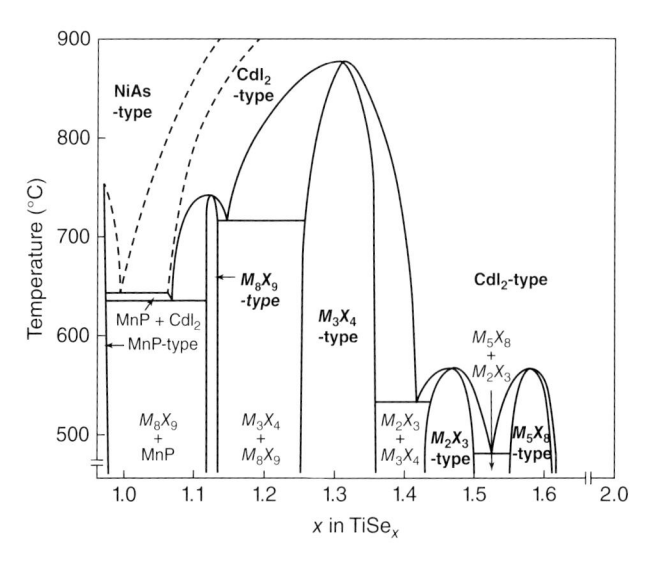

Figure 9.27 Phase diagram of TiSe$_x$.

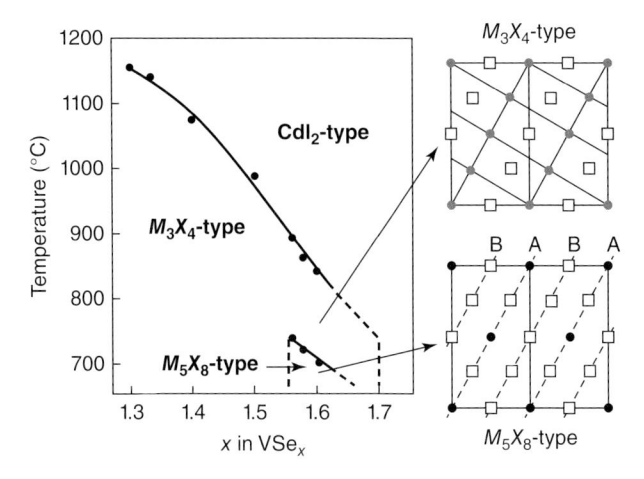

Figure 9.28 Phase diagram of VSe$_x$ and the relation between M_5X_8-type and M_3X_4-type vacancy-ordered structures.

no two-phase region between the two phases and the M_3X_4-type phase has a considerably wide nonstoichiometric region. On heating, the M_5X_8-type phase is successively transformed to M_3X_4-type and further to CdI$_2$-type phases. The transition from M_5X_8-type to M_3X_4-type can be understood by the disorder of metal vacancies along the one-dimensional chain, as shown in Figure 9.28 [45,46]. The M^V-layer of the M_5X_8-type structure consists of two chains, A and B, where the A-chain has an alternate arrangement of occupied (A1) and vacant (A2) sites, while the B-chain consists of only vacant sites. The perfectly ordered M_5X_8-type structure with the stoichiometric composition has $p_A = 1/2$ ($p_{A1} = 1$ and $p_{A2} = 0$) and $p_B = 0$, where p_A (p_{A1} and p_{A2}) and p_B are occupancy rate of the A-chain (A1 and A2 sites) and B-chain, respectively. When p_{A1} becomes equal to p_{A2} by the disorder of metal vacancies within the A-chains on heating, the structure of the M^V-layer has an alternate arrangement of the two chains with different occupancy rates ($p_A = 1/2$ and $p_B = 0$), which corresponds to a just arrangement of the M_3X_4-type phase in a broad sense. On further heating, the M_3X_4-type phase is transformed to CdI$_2$-type phase at $p_A = p_B$ by the disorder of metal vacancies between the A- and B-chains. Such a mechanism of the phase change from M_5X_8-type to M_3X_4-type is applicable to the phase change on the composition. With decreasing x, the A2 sites and the B-chains of the M_5X_8-type phase are occupied and the M_5X_8-type phase is transformed to the M_3X_4-type phase at $p_{A1} = p_{A2}$ ($p_A > p_B$). This indicates that the hexagonal lattice is considerably distorted in VSe$_x$ compared to TiSe$_x$.

The Cr–Se system has two vacancy-ordered phases, M_3X_4 and M_7X_8. The phase diagram of CrSe$_x$ is shown in Figure 9.29, where the M_7X_8-type phase is successively transformed to M_3X_4-type and further to CdI$_2$-type (or NiAs-type) phases on heating [47]. This transition can be understood by the same mechanism as for the transition, M_5X_8-type → M_3X_4-type → CdI$_2$-type in VSe$_x$, because

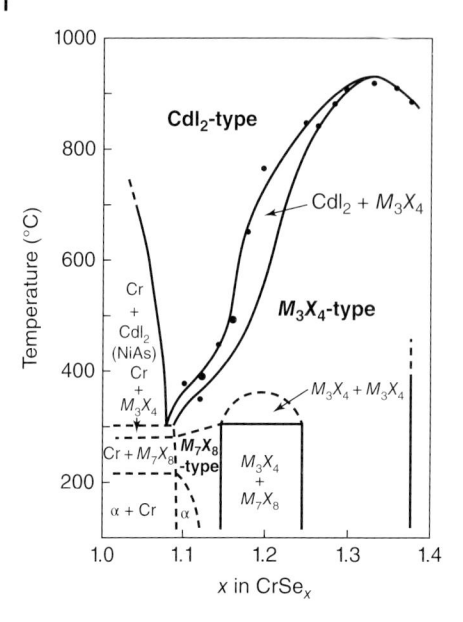

Figure 9.29 Phase diagram of CrSe$_x$.

the M_7X_8-type structure is transformed to the M_5X_8-type structure by changing the occupied and vacant sites one another. The Cr–Se system features the phase separation into two M_3X_4-type phases with different compositions, resulting from spinodal decomposition on cooling. One of the two M_3X_4-type phases is transformed to M_7X_8-type at lower temperatures, resulting in the phase separation into the M_3X_4-type and M_7X_8-type phases. Such a phase separation phenomenon indicates that the vacancy-ordered structures cannot be simply explained by the repulsive interaction between vacancies in $M_{1+y}X_2$ (0.5 < y < 1).

Figure 9.30 shows the phase diagram of FeSe$_x$, which is quite similar to the phase diagram of CrSe$_x$ [47]. The Fe–Se system has two vacancy ordered phases, M_3X_4-type and M_7X_8-type, and shows a successive transition; M_7X_8-type → M_3X_4-type → CdI$_2$-type and the phase separation into two M_3X_4-type phases (M_3X_4-type and M_7X_8-type phases at lower temperatures). FeSe is a mixture of M_7X_8-type phase and FeSe$_{1-\delta}$ phase with the α-PbO-type structure. Very recently, FeSe$_{1-\delta}$ has attracted attention as iron-based superconductor [48].

The Co–Se and Ni–Se systems have only M_3X_4-type phase, while neither Mn–Se system nor Cu–Se system has vacancy-ordered phase.

9.3.3
Phase Diagram and Site Preference in Pseudobinary M_3X_4–M'_3X_4 System

The M_3X_4-type structure (Cr$_3$S$_4$-structure; monoclinic $I/2m$) is most popular among these vacancy-ordered structures. Its structure is shown in Figure 9.31 and the compounds with M_3X_4-type structure are summarized in Table 9.1. The

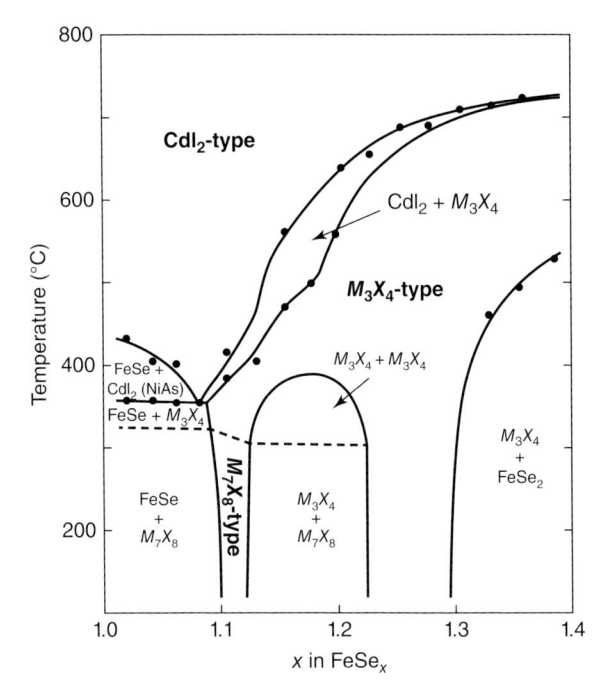

9.3 Vacancy-Ordered Structure, Order–Disorder Transition and Site Preference | 407

Figure 9.30 Phase diagram of FeSe$_x$.

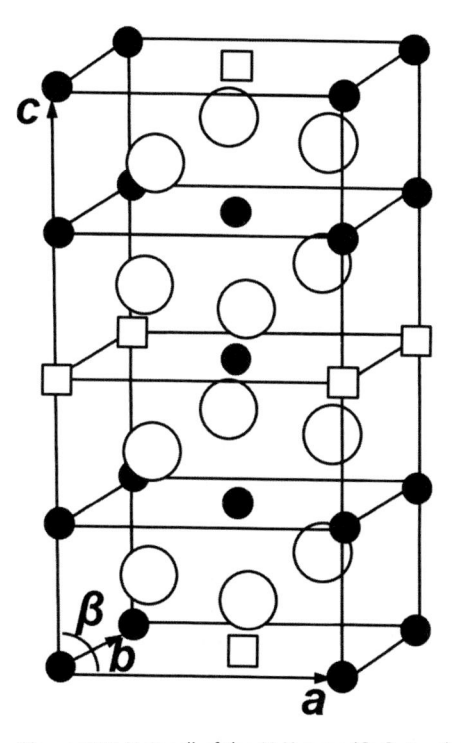

Figure 9.31 Unit cell of the M_3X_4-type (Cr$_3$S$_4$-type) structure (I/2m).

Table 9.1 M_3X_4-type 3d-transition metal chalcogenides and their physical properties.

X	M						
	Ti	V	Cr	Mn	Fe	Co	Ni
S	—	M(AF)	M(AF)	—	—	—	—
Se	M(PP)	M(AF)	SC(AF)	—	SC(Fr)	M(PP)	M(PP)
Te	M(PP)	M(P)	M(F)	—	—	—	—

M: metallic, SC: semiconductive, P: paramagnetic, PP: Pauli paramagnetic, AF: antiferromagnetic, F: ferromagnetic, and Fr: ferrimagnetic.

M_3X_4 is expressed as $(M)[M_2]X_4$, where parenthesis and bracket denote the M^V- and M^F-layers, respectively. In pseudobinary compound, $M'M_2X_4$, it is an interesting issue that which of M and M'' prefers which of the M^V-layer and M^F-layer, namely, $(M')[M_2]X_4$ or $(M)[M'M]X_4$, on the analogy of normal-type or inverse-type in the spinel compounds. Since $M'M_2X_4$ is an intermediate compound in M'_yMX_2 system or pseudobinary M_3X_4–M'_3X_4 system, as shown in Figure 9.32, the pseudobinary M_3X_4–M'_3X_4 system is expected to give us any information on the site preference. Selenide systems, $M'M_2Se_4$ are suitable in consideration of the site preference of M'/M, because all of M_3Se_4 except for $M =$ Mn and Cu take the M_3X_4-type structure.

Among the early 3d transition metals, each pseudobinary M_3X_4–M'_3X_4 system (M, $M' =$ Ti, V, and Cr) forms a solid solution over a whole composition range. The lattice parameters and the phase diagram of $(Cr_xTi_{1-x})_3Se_4$ are shown in Figure 9.33a and b, respectively [49]. The lattice parameter versus composition curve exhibits a bending around $x = 1/3$ in all lattice parameters. The order–disorder transition temperature from M_3X_4-type to CdI_2-type also shows a minimum around $x = 1/3$. These imply that Cr/Ti preferably occupies the M^V-layer/M^F-layer. Neutron diffraction study of $CrTi_2Se_4$ and $TiCr_2Se_4$ directly confirmed normal-type $(Cr)[Ti_2]Se_4$ and inverse-type $(Cr)[CrTi]Se_4$ with disorder to some extent [49–51]. Such a site-preference was observed in a sulfide and telluride as $(Cr)[CrTi]S_4$ [51,52] and $(Cr)[CrTi]Te_4$ [53]. On the other hand, both lattice parameters and transition temperature of $(V_xTi_{1-x})_3Se_4$ have no clear inflection in their composition dependences; nevertheless, neutron diffraction of VTi_2Se_4 reveals $(V_{0.60}Ti_{0.40})[V_{0.40}Ti_{1.60}]Se_4$ with a considerable disorder, indicating the

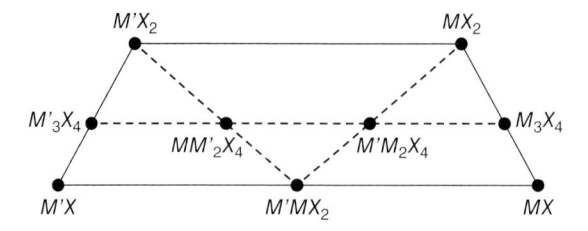

Figure 9.32 A part of phase diagram of ternary M'-M-X system.

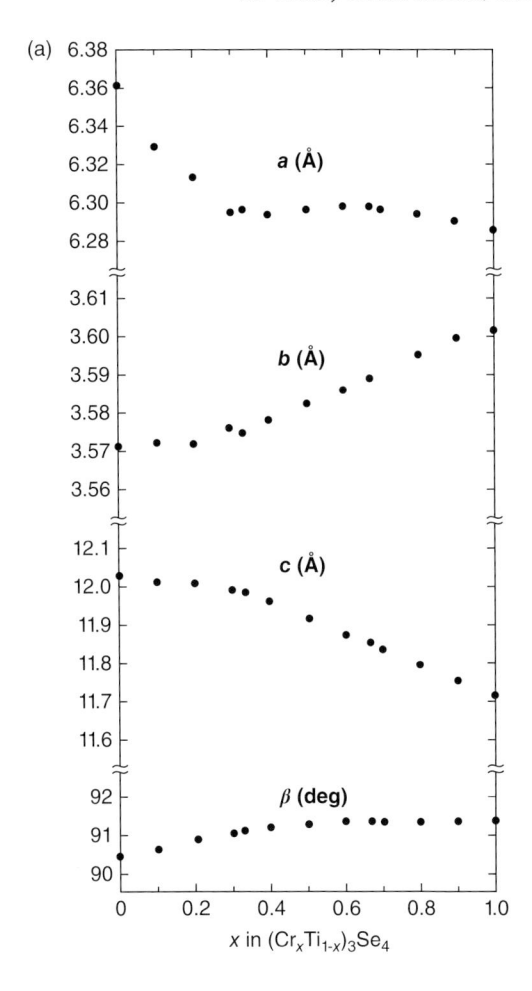

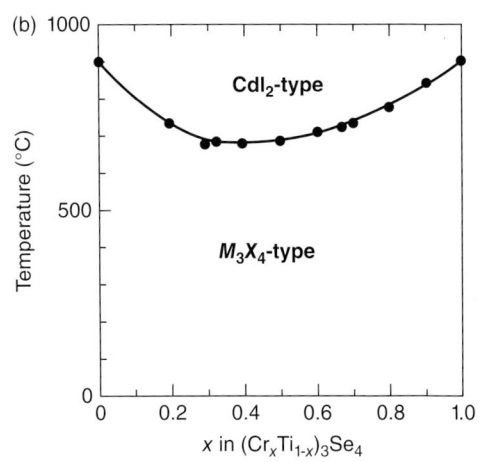

Figure 9.33 (a) Lattice parameters of $(Cr_xTi_{1-x})_3Se_4$. (b) Phase diagram of $(Cr_xTi_{1-x})_3Se_4$.

preference of V/Ti for the M^V-layer/M^F-layer [54,55]. A similar site-preference was observed in Ti–V–S system [56]. The transition temperature from M_3X_4-type to CdI_2-type in $(Cr_xV_{1-x})_3Se_4$ smoothly changes in a concave manner as a function of composition, but the lattice parameter curves show inflections around $x = 2/3$, suggesting that V/Cr tends to occupy the M^V-layer/M^F-layer [54]. Neutron diffraction of VCr_2Se_4 reveals $(V_{0.55}Cr_{0.45})[V_{0.45}Cr_{1.55}]Se_4$, which indicates the preference of V/Cr for the M^V-layer/M^F-layer [55]. In conclusion, the site preference of M^F-layer is Ti » Cr > V among the early 3d transition metals.

The pseudobinary systems, M_3Se_4–M'_3Se_4 of the late 3d transition metals (M, M' = Fe, Co, and Ni) form solid solution over a whole composition range and show a transition from M_3X_4-type to CdI_2-type phases at higher temperatures. The transition temperature in each M_3Se_4–M'_3Se_4 system smoothly changes as a function of composition in a convex manner [54]. The phase diagram of $(Co_xFe_{1-x})_3Se_4$ is shown in Figure 9.34 as a typical example. The composition dependence of lattice parameters does not show any inflection in $(Ni_xFe_{1-x})_3Se_4$, while only the a-axis of $(Co_xFe_{1-x})_3Se_4$ shows an inflection around $x = 2/3$ and all lattice parameters of $(Ni_xCo_{1-x})_3Se_4$ have inflections around $x = 1/3$ [54]. These results suggest that the site preference of the M^F-layer is roughly Co > Fe ≈ Ni among the late 3d transition metals.

In the M_3Se_4–M'_3Se_4 systems between the early (M) and late (M') 3d transition metals, there are two types; solid solution system (V_3Se_4-Fe_3Se_4, V_3Se_4-Co_3Se_4, and Cr_3Se_4-Fe_3Se_4) and phase separation system (Ti_3Se_4-Fe_3Se_4, Ti_3Se_4-Co_3Se_4, Ti_3Se_4-Ni_3Se_4, V_3Se_4-Ni_3Se_4, Cr_3Se_4-Co_3Se_4, and Cr_3Se_4-Ni_3Se_4) [54]. The phase separation systems are further categorized into two types. One shows the phase separation into two M_3X_4-types and the solid solution in CdI_2-type at higher temperatures; V_3Se_4-Ni_3Se_4, Cr_3Se_4-Co_3Se_4, and Cr_3Se_4-Ni_3Se_4. In the case of Ti_3Se_4-Fe_3Se_4 system, the two separated M_3X_4-type phases

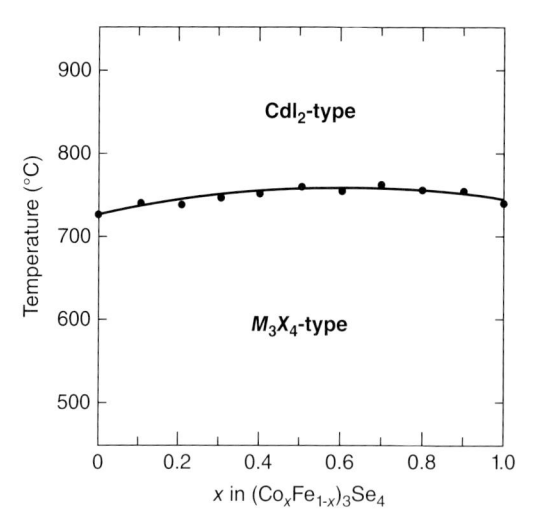

Figure 9.34 Phase diagram of $(Co_xFe_{1-x})_3Se_4$.

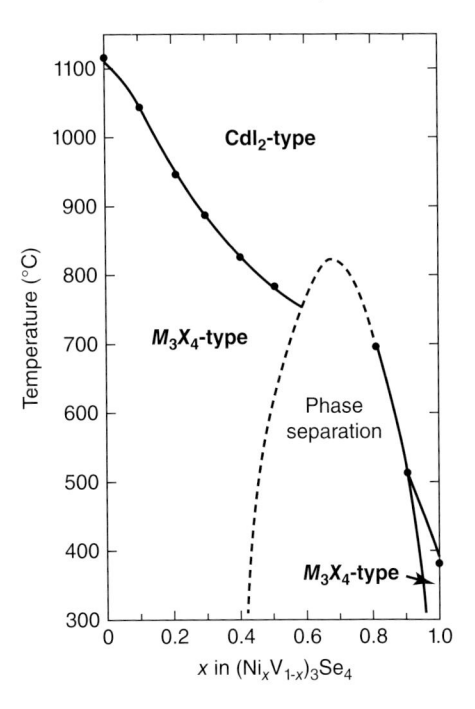

Figure 9.35 Phase diagram of $(Ni_xV_{1-x})_3Se_4$.

form solid solution prior to the transition to CdI_2-type phase on heating [57]. The other shows the phase separation that penetrates into the CdI_2-type phase from the M_3X_4-type phase; Ti_3Se_4–Co_3Se_4 and Ti_3Se_4–Ni_3Se_4.

Among the phase separation systems, Ti_3Se_4–Fe_3Se_4, Ti_3Se_4–Co_3Se_4, Ti_3Se_4–Ni_3Se_4, and V_3Se_4–Ni_3Se_4 have the phase separation region weighted toward Fe, Co, or Ni side, as typically seen in $(Ni_xV_{1-x})_3Se_4$ shown in Figure 9.35. This suggests the presence of the normal-type compounds, $(Fe)[Ti_2]Se_4$, $(Co)[Ti_2]Se_4$, $(Ni)[Ti_2]Se_4$, and $(Ni)[V_2]Se_4$, namely, Ti prefers the M^F-layer compared with Fe, Co, or Ni, and V does with Ni. On the other hand, the remaining phase separation systems, Cr_3Se_4–Co_3Se_4 and Cr_3Se_4–Ni_3Se_4 have the phase separation region widely spread centered on $x = 0.5$ in $(M'_xM_{1-x})_3Se_4$, and neither $M'M_2Se_4$ nor MM'_2Se_4 with the M_3X_4-type structure exists [54]. Hence, it is difficult to evaluate the site preference of Cr in Cr_3Se_4–Co_3Se_4 and Cr_3Se_4–Ni_3Se_4 systems.

Among the solid solution systems, the composition dependence of the M_3X_4–CdI_2 transition temperature shows no inflection in $(Fe_xV_{1-x})_3Se_4$, while it shows a broad inflection around $x = 1/3$ in both $(Co_xV_{1-x})_3Se_4$ and $(Fe_xCr_{1-x})_3Se_4$ [54]. The composition dependences of the lattice parameters show similar behaviors as no inflection in $(Fe_xV_{1-x})_3Se_4$ and an inflection around $x = 1/3$ in both $(Co_xV_{1-x})_3Se_4$ and $(Fe_xCr_{1-x})_3Se_4$. These results indicate that Ti, V, and Cr prefer the M^F-layer more than Fe, Co, and Ni. Actually, the normal type $(Fe)[Cr_2]Se_4$ [58], $(Ni)[Cr_2]Se_4$ [59], $(Fe)[V_2]S_4$ [60], and inverse-type $(Fe)[FeV]S_4$ [60] are reported. It can be concluded that Ti prefers the M^F-layer most among 3d transition metals.

9.3.4
The Site Preference and Magnetic Properties

The site preference is reflected to the magnetic properties. Ti_3Se_4, Co_3Se_4, and Ni_3Se_4 are Pauli paramagnets, indicating that 3d electrons are perfectly delocalized. Cr_3Se_4 is an antiferromagnet and Fe_3Se_4 is a ferrimagnet, where both Cr and Fe ions behave as magnetic ions with localized spin moments. V_3Se_4 is an itinerant antiferromagnet with a reduced spin moment. Hence, the combination between the localized and delocalized systems is a good system to evaluate the site preference, because the magnetic properties would be simply diluted by nonmagnetic metals.

The spin structure of Cr_3Se_4 with the Néel temperature $T_N = 82$ K consists of the ferromagnetic (101) layers coupled antiferromagnetically to each other, as shown in Figure 9.36 [61]. The metal layer stacking parallel to the (101) plane is

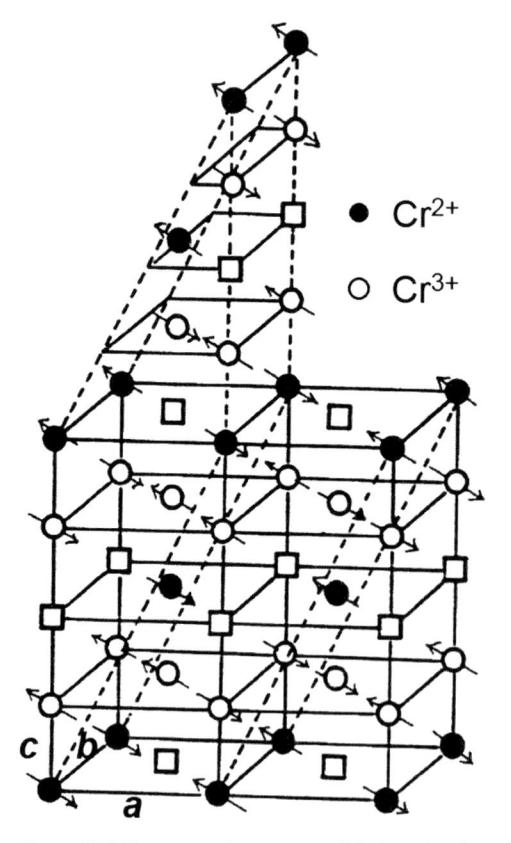

Figure 9.36 The magnetic structure of Cr_3Se_4. The dotted line denotes the unit cell constructed from the metal layer stacking parallel to the (101) plane.

expressed as $\square M_fM_vM_f\square M_fM_vM_f\square$, where M_f denotes the metal layer formed by a half of metals in the M^F-layer, and M_v and $\square$ represent the metal and vacant layers associated with the half-occupied M^V-layer, respectively. Therefore, the spin structure is expressed as $\square M_f^- M_v^+ M_f^- \square M_f^+ M_v^- M_f^+ \square$, where + and − specify spin directions of the interlayer antiferromagnetic coupling. The magnetic susceptibility versus temperature (χ-T) curves of $(Cr_xTi_{1-x})_3Se_4$ are shown in Figure 9.37. The T_N characterized by a kink in the χ-T curve is rapidly suppressed with decreasing x as $T_N = 82$ K for $x = 1.0$ and 38 K for $x = 0.9$. For $0.4 \leq x \leq 0.8$, the χ-T curves show broad maxima around 10 K suggesting glassy magnetic state, and below $x = 0.3$ they no longer show any kink suggesting no magnetic order. The χ-T curves in the paramagnetic state well obey Curie–Weiss (CW) law; $\chi = C/(T-\theta) + \chi_0$ (χ_0: temperature-independent term). The curie constant (C) and Weiss temperature (θ) obtained from CW fitting are shown in Figure 9.38a. The C decreases with increasing Ti-concentration toward to zero at $x = 0$. This implies that Ti is nonmagnetic metal and only Cr ions carry the CW behavior (spin part). Figure 9.38b shows the composition dependence of effective Bohr magnetons (P_{eff}) per Cr ion. The P_{eff} is constant at $4\mu_B$ up to $x \sim 2/3$ and then smoothly increases up to $4.6\mu_B$ at $x = 1.0$. On the other hand, the θ, which is a measure of magnetic interaction, continuously increases with decreasing x, changing its sign from negative to positive around $x = 0.8$, and reaches a maximum around $x = 1/3$. These results are well explained by the site preference of nonmagnetic Ti for the M^F-layer. The interlayer antiferromagnetic

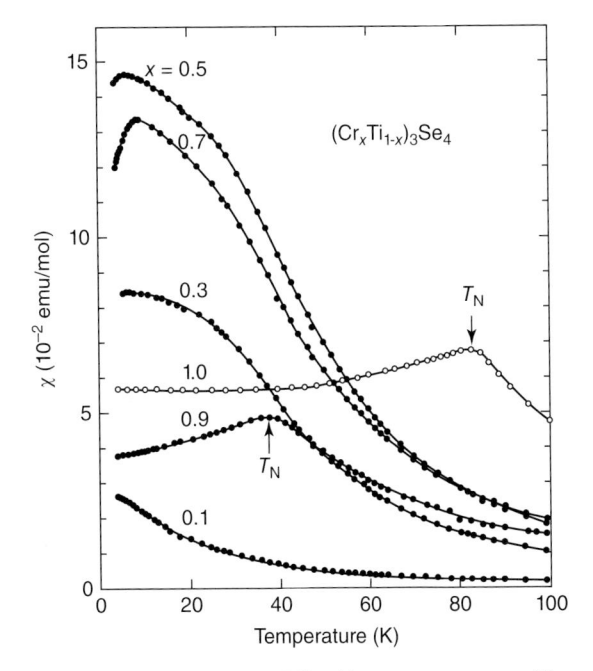

Figure 9.37 Magnetic susceptibility (χ) versus temperature (T) curves of $(Cr_xTi_{1-x})_3Se_4$.

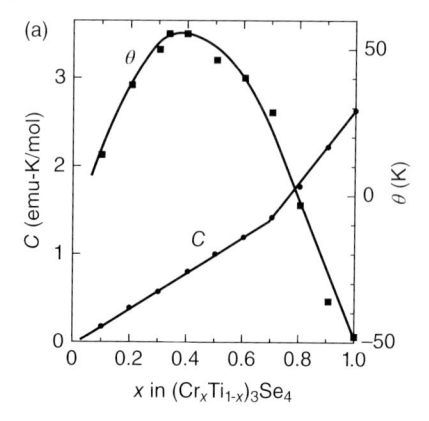

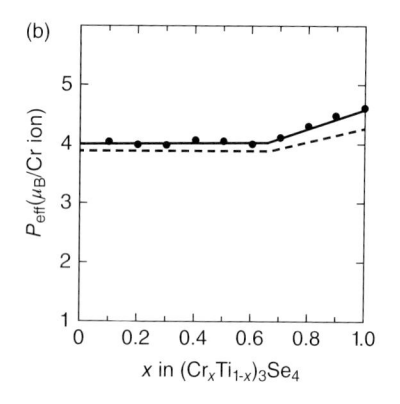

Figure 9.38 (a) Composition dependence of Curie constant (C) and Weis temperature (θ) in $(Cr_xTi_{1-x})_3Se_4$. (b) Composition dependence of effective Bohr magneton (P_{eff}) per Cr ion in $(Cr_xTi_{1-x})_3Se_4$.

interaction is weakened by the preferable occupation of nonmagnetic Ti for the M^F-layer, leading to the rapid suppression of the antiferromagnetic order and the rapid increase of θ changing the sign from negative (antiferromagnetic) to positive (ferromagnetic). Ultimately, in the magnetic structure of $\square Ti_fCr_v^+Ti_f\square Ti_fCr_v^-Ti_f\square$ at $x=1/3$, the interlayer antiferromagnetic interaction ($Cr_v^+ - Cr_v^-$) becomes far weaker and the intralayer ferromagnetic interaction becomes most predominant (positive and maximum θ). On further decreasing x, the intralayer ferromagnetic interaction is weakened, resulting in the decrease of θ. The Cr ions are expected to take $Cr^{3+}(d^3)$ state up to $x=2/3$ because $Cr^{3+}(d^3)$ state is most stable, but beyond $x=2/3$ they are forced to partly take $Cr^{2+}(d^4)$ state toward the final state of $Cr^{2+}Cr_2^{3+}Se_4$ at $x=1.0$, as shown with the dotted line in Figure 9.38b.

Fe_3Se_4 is a ferrimagnet with the Curie temperature $T_C=320$ K [62,63]. Its magnetic structure consists of ferromagnetic (001) layers coupled antiferromagnetically, as $M^{F+}M^{V-}M^{F+}M^{V-}$, with the spin direction parallel to the b-axis

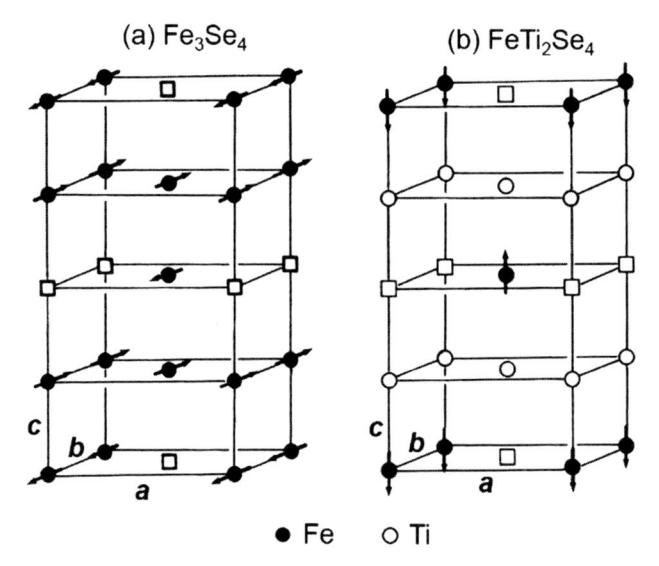

9.3 Vacancy-Ordered Structure, Order–Disorder Transition and Site Preference

Figure 9.39 The magnetic structures of (a) Fe$_3$Se$_4$ and (b) FeTi$_2$Se$_4$.

(Figure 9.39a) [62,63]. Since the number of Fe ions in the M^F-layer is twice as large as that in the M^V-layer, the ferrimagnetism with the net moments is realized. The χ-T curves of (Fe$_x$Ti$_{1-x}$)$_3$Se$_4$ are shown in Figure 9.40 [57]. With the substitution of Ti for Fe in Fe$_3$Se$_4$, both T_C and the saturated magnetization decrease and finally the magnetism becomes antiferromagnetic below $x = 0.3$. Figure 9.41 shows the composition dependence of average magnetic moment per Fe ion (μ_{net}) estimated from the saturated magnetization, assuming that Ti atoms have no magnetic moment. The solid lines, (a), (b), and (c) in Figure 9.41 show the compositional variations of μ_{net} calculated for the following substitution models;

a) Ti atoms substitute for Fe atoms in the M^F-layers preferentially,
b) Ti atoms substitute for Fe atoms in the M^V-layers preferentially, and
c) Ti atoms substitute for Fe atoms randomly,

using the equation as below;

$$\mu_{net} = (\mu_F N_F - \mu_V N_V)/(N_F + N_V)$$

where $\mu_F = 2.27\,\mu_B$ and $\mu_V = 2.89\,\mu_B$ are magnetic moments of Fe ions in the M^F- and M^V-layers of Fe$_3$Se$_4$, and N_F and N_V are the number of Fe ions in the M^F- and M^V-layers, respectively [57]. The observed μ_{net} in $0.8 \leq x \leq 1.0$ well agrees with the calculated one in the model (a). This indicates the site preference of Ti for the M^F-layers, agreeing with the conclusion from the phase diagram. Such a site preference of Ti for the M^F-layers leads to a normal type (Fe)[Ti$_2$]Se$_4$ at $x = 1/3$. FeTi$_2$Se$_4$ is an antiferromagnet with $T_N = 134$ K and its spin structure is expressed as (Fe)$^+$[Ti$_2$](Fe)$^-$[Ti$_2$], with its spin direction parallel to the c axis

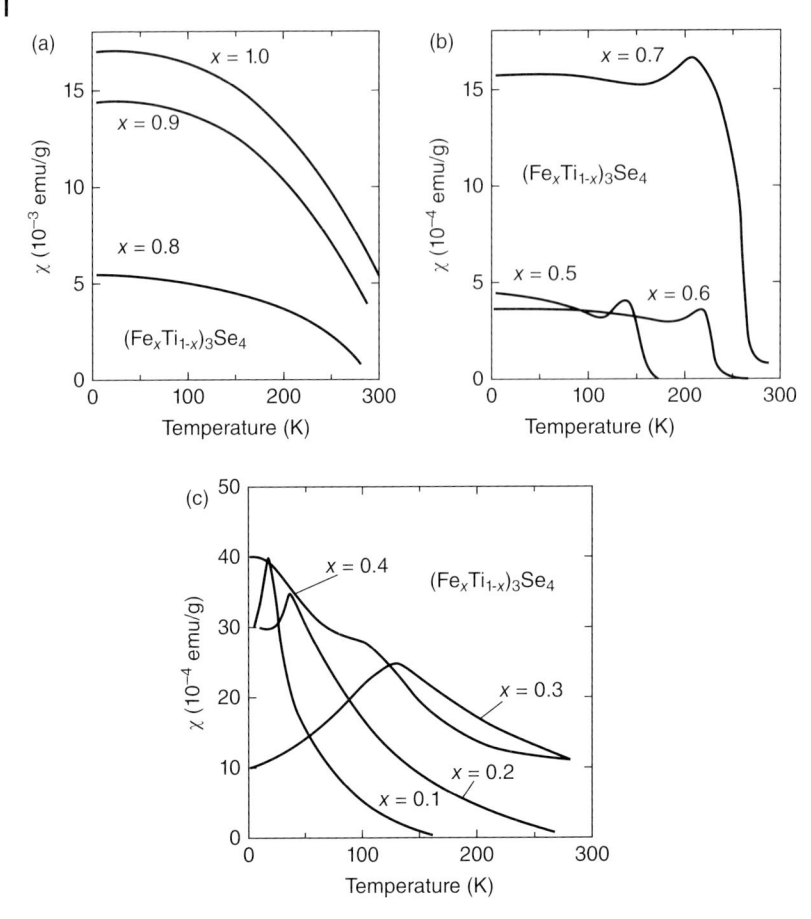

Figure 9.40 Magnetic susceptibility (χ) versus temperature (T) curves of $(Fe_xTi_{1-x})_3Se_4$. (a) $x = 1.0$, 0.9, and 0.8, (b) $x = 0.7$, 0.6, and 0.5, (c) $x = 0.4$, 0.3, 0.2, and 0.1.

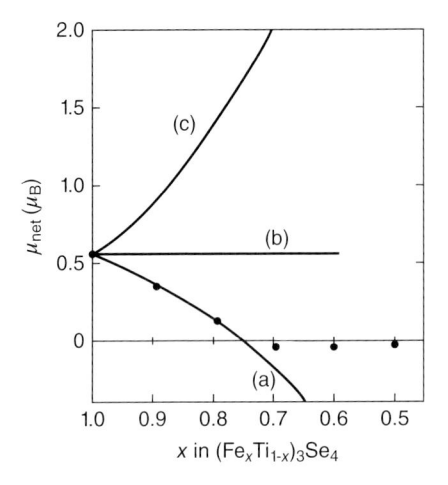

Figure 9.41 Composition dependences of average magnetic moment per Fe ion (μ_{net}) observed (circles) and calculated (lines) based on the models of (a), (b), and (c) (see the text).

(Figure 9.39b) [64]. This indicates that the magnetic coupling between Fe ions of the different M^V-layers becomes antiferromagnetic when it is mediated by Ti metals of the-M^F layers. The small magnetization and complex magnetic behaviors in $1/3 < x < 0.8$ come from the coexistence of ferromagnetic and antiferromagnetic couplings between the M^V-layers mediated by the M^F-layers as $(Fe)^+[Fe_x]^-(Fe)^+$ and $(Fe)^+[Ti_{2-x}](Fe)^-$.

9.4
Quasi-One-Dimensional Chalcogenides

9.4.1
Charge Density Wave (CDW) Instability in Quasi-One-Dimensional Compound NbSe$_3$

Peierls showed that a one-dimensional system with a half-filled band can lower its ground-state energy by a dimerization of the period, leading to the formation of energy gaps at the Fermi wave vectors $\pm k_F$ [65]. Such a system undergoes a metal-to-insulator transition, which is known as the Peierls transition. This is realized when the energy loss due to the distortion of the lattice is overcome by a gain in electronic energy owing to the opening of gaps in the Fermi surface. The doubling will lead to a periodic modulation of the electronic charge density, called charge density wave (CDW) [66,67]. The CDW is accompanied by a periodic distortion of the atomic lattice in one-dimensional or two-dimensional structures. The CDW appears in the commensurate or the incommensurate states, according to that the ratio of wave periods of CDW and lattice distortion is rational or irrational, respectively.

NbSe$_3$ is a typical chalcogenide exhibiting CDW instability [68,69]. NbSe$_3$ has a monoclinic unit cell (space group $P2_1/m$). Figure 9.42 shows the schematic crystal structure of NbSe$_3$ projected along the [010] direction, where the large

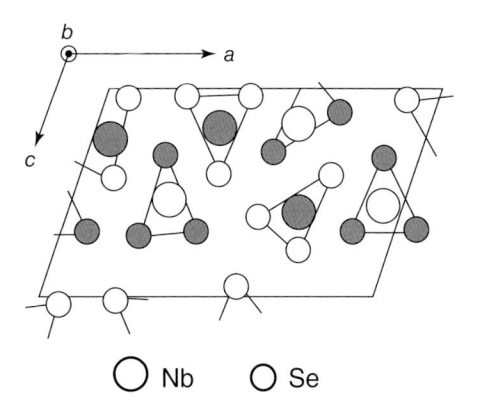

Figure 9.42 Schematic crystal structure of NbSe$_3$ projected along [010] direction of monoclinic unit cell. The gray circles represent the atoms situated at the height of $1/2b$.

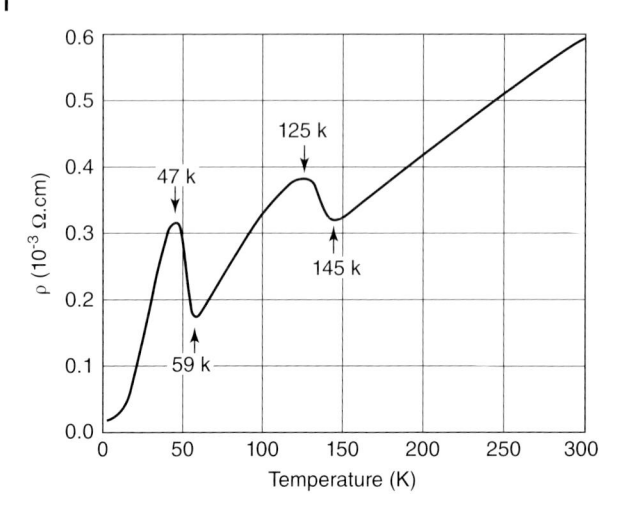

Figure 9.43 Temperature variations of electrical resistivity of NbSe$_3$.

circles and small circles represent Nb and Se atoms, respectively. The hatched circles represent the atoms situated at the height of 1/2b. Irregular NbSe$_6$ trigonal prisms are stacked to form NbSe$_3$ chains running parallel to the b-axis. There are three kinds of Nb–Nb chains in the unit cell. Since the intrachain distance of Nb atoms is shorter than the interchain distance, NbSe$_3$ can be considered to be a quasi-one-dimensional compound.

Temperature variation of electrical resistivity ρ of NbSe$_3$ is shown in Figure 9.43. Two large anomalies are observed at 145 and 59 K (onset temperature). Both anomalies were found to be due to the incommensurate CDWs associated with the different two types of Nb–Nb chains. The CDW transition at 59 K was progressively suppressed by the applied pressure and was smeared out at about 6 kbar [70]. When the CDW was suppressed, a superconducting transition appeared at 3–4 K [70]. This is owing to that the CDW and the superconductivity have the competitive nature with each other. It is most interesting that the NbSe$_3$ shows a nonohmic conductivity in the CDW state, indicating that the CDW is sliding due to the depinning [71].

9.4.2
CDW and Superconductivity in Quasi-One-Dimensional Compound In$_x$Nb$_3$Te$_4$

The niobium chalcogenides of Nb$_3$X$_4$ (X = S, Se, Te) have hexagonal unit cells with the Nb$_3$Te$_4$ type structure (space group: $P6_3$). In$_x$Nb$_3$X$_4$ has also the Nb$_3$Te$_4$ type structure, where x ranges from 0 to 1 [72]. The schematic crystal structure of InNb$_3$X$_4$ is shown in Figure 9.44. The structure is built up by NbX$_6$ octahedra linked together via shared faces in the a–b plane and via shared edges along the c-axis. The structure is characterized by zigzag Nb–Nb chains and the large hexagonal tunnels, both being running parallel to the c-axis. The tunnels are able to accommodate the ternary elements such as In, Cd, and so on. There

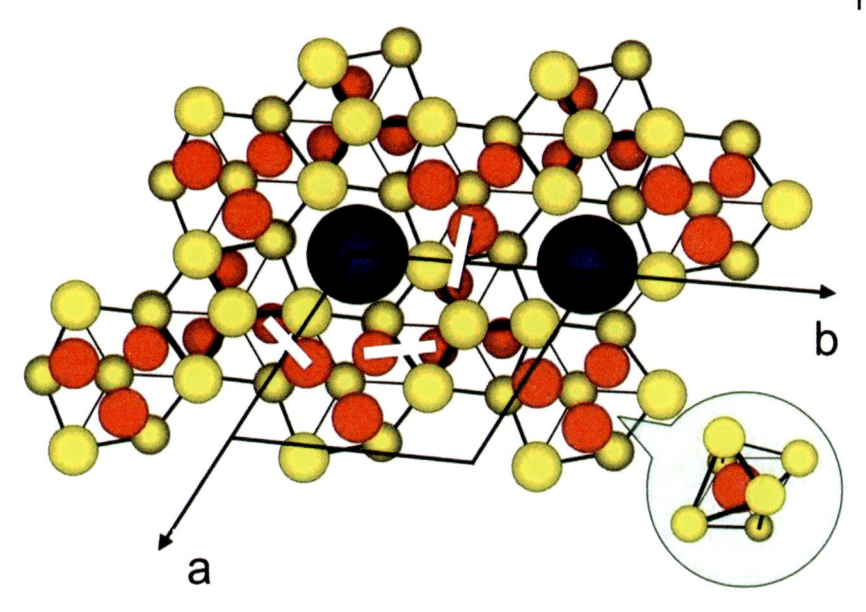

Figure 9.44 Schematic crystal structure of InNb$_3$X$_4$ (X: S, Se, Te) (Nb: red circles, X: yellow circles, and In: dark blue circles). The bold white lines show the different Nb–Nb chains running along the c-axis.

are three kinds of Nb–Nb chains, which are depicted by white bold lines in Figure 9.44. In the structure of InNb$_3$Te$_4$, the intrachain distance of Nb atoms (0.297 nm) is shorter than the interchain distance (0.385 nm) of Nb atoms, bringing about the one-dimensionality to this structure. Figure 9.45 gives a transmission electron microscope (TEM) image of Nb$_3$Se$_4$, where the incident beam is perpendicular to the a–b plane. One can see flower like patterns, where the tunnels are surrounded by six Se atoms (white spots).

Figure 9.45 TEM image of Nb$_3$Se$_4$ (a–b plane). The white spots are Se atoms.

Nb_3X_4 shows the superconducting transition at $T_c = 4.0$, 2.0, and 1.8 K for $X =$ S, Se, and Te, respectively [73]. Nb_3Te_4 shows phase transitions at 38 and 80 K [74]. The transition at 80 K is associated with CDW formation [75]. XRD measurements revealed that the superstructure of Nb_3Te_4 is characterized by commensurate wave vector $\boldsymbol{q} = \pm (1/3\boldsymbol{a}^* + 1/3\boldsymbol{b}^*) + 3/7\boldsymbol{c}^*$ [76]. It is of interest that the Nb_3Te_4 exhibits the coexistence of superconductivity and CDW under normal conditions in spite of conflicting nature of these phenomena.

The energy-band structure of Nb_3X_4 ($X =$ S, Se, Te) was calculated by using the numerical-based-set LCAO method [77,78]. The resulting Fermi surfaces of Nb_3X_4 are similar with each other. They consist of warped or undulating plane like sheets, indicative of the good one-dimensionality. Figure 9.46a shows the outermost Fermi surface of Nb_3Te_4 [77]. Figure 9.46b shows the Fermi surfaces of Nb_3Se_4 viewed from a direction perpendicular to the k_c vector, where three kinds of Fermi surfaces correspond to the three kinds of Nb–Nb chains [78].

Samples of $In_xNb_3Te_4$ were prepared by heating the elemental mixtures sealed in silica tubes at 1373 K. The In^+ ions were observed to be able to be accommodated in the tunnels up to $x = 1$ in $In_xNb_3Te_4$; the lattice parameters of a, b, and c monotonously increased with x. Figure 9.47 shows temperature variations of the electrical resistivity ρ of $In_xNb_3Te_4$ [79]. There are two kinds of phase transitions. The higher transition temperature (T_1) is almost constant at about 90 K (maximum point in $\rho - T$ curves) up to $x = 0.55$, and then shows the increase to about 140 K for $x \geq 0.60$. The lower transition temperature (T_2) is observed at about 25 K for $x \leq 0.60$, and at about 50 K for $x > 0.60$. The superconducting temperature T_c was observed at 4–5 K for all samples.

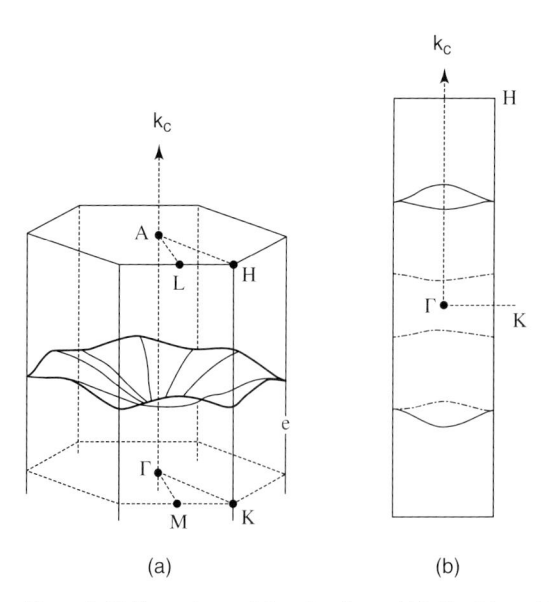

Figure 9.46 The outermost Fermi surface of Nb_3Te_4 (a), and the Fermi surfaces of Nb_3Se_4 viewed from a direction perpendicular to the k_c vector (b) [77,78].

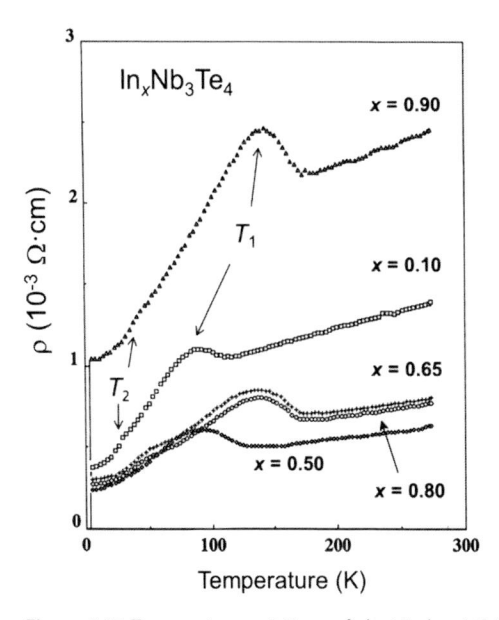

Figure 9.47 Temperature variations of electrical resistivity of $In_xNb_3Te_4$.

Figure 9.48 shows electron diffraction (ED) patterns of $In_{0.90}Nb_3Te_4$ taken at room temperature (a) and at about 90 K (b). The direction of incident electron beam is $[1\bar{1}00]$. The superlattice reflections appeared at about 90 K below T_1. The reflections are characterized by the wave vector $q = \pm(1/3a^* + 1/3b^*) + 0.4534c^*$. These values are essentially consistent with the earlier results for Nb_3Te_4 [76], except for slight increase in the c^* component. Boswell and Bennet

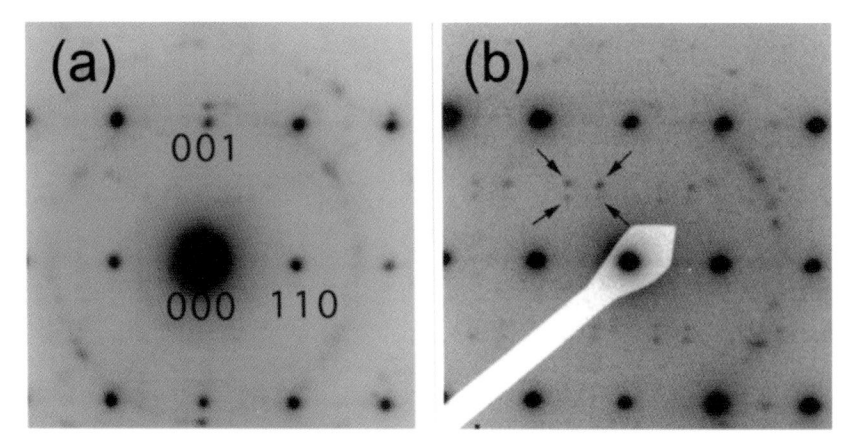

Figure 9.48 Electron diffraction patterns of $In_{0.9}Nb_3Te_4$ taken at room temperature (a), and at about 90 K (b). The direction of the incident electron beam is $[1\bar{1}00]$.

Table 9.2 CDW vectors q observed in $In_xNb_3Te_4$.

In content x	q
0	$\pm(1/3a^* + 1/3b^*) + 3/7c^*$ [76]
0.40	$\pm(1/3a^* + 1/3b^*) + 0.433c^*$
0.60	$\pm(1/3a^* + 1/3b^*) + 0.440c^*$
0.75	$\pm(1/3a^* + 1/3b^*) + 4/9c^*$
0.90	$\pm(1/3a^* + 1/3b^*) + 0.453c^*$
1.0	$\pm(1/3a^* + 1/3b^*) + 0.462c^*$

reported that the wave vectors in $In_xNb_3Te_4$ are represented as $q = \pm(1/3a^* + 1/3b^*) + zc^*$, where z ranges from 0.461 to 0.469 [80]. However, they did not observe the x dependence of the z value. Table 9.2 gives CDW q vectors observed at 90 K in ED measurements for the samples with various x in $In_xNb_3Te_4$ [81]. The CDW of $In_xNb_3Te_4$ would be incommensurate, except for Nb_3Te_4. Figure 9.49 shows the c^* component in the q vectors as a function of x. The c^* component continuously increases with increasing x.

In the one-dimensional metal, the Fermi surface has a flat surface perpendicular to the one-dimensional axis as shown in Figure 9.46. The CDW can be induced, if the Fermi surfaces get superimposed on each other on translation by the vector q $(= 2k_F)$, which is called as the nesting. Sekine *et al.* pointed out that the observed value of $3/7c^*$ of the q vector for Nb_3Te_4 is about two times of calculated values of $0.25c^*$ and $0.17c^*$ of the outer two Fermi surfaces of Nb_3Te_4 [76]. This is the strong evidence for that the transitions in $In_xNb_3Te_4$ are caused by the CDW instability. In a rigid band model, the Fermi energy of $In_xNb_3Te_4$ would increase with increasing x, because the electrons are donated

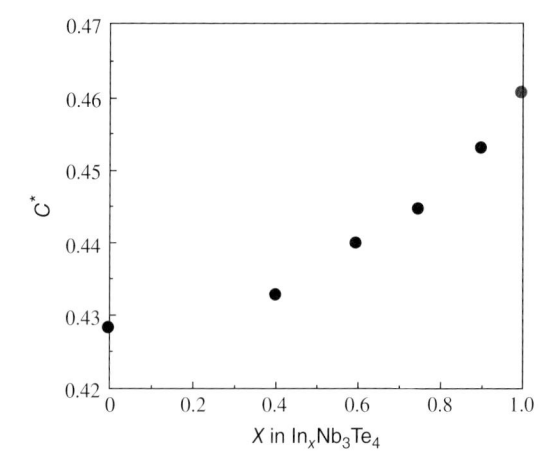

Figure 9.49 The c^* component in the CDW q vectors as a function of x in $In_xNb_3Te_4$.

from In into the conduction band (In ions are in the monovalent state). The present observation that the c^* component of q vector increases with increasing x is well consistent with this model. In this context, the CDW of this compound is possibly of the incommensurate type. To author's knowledge, this is the first case that the CDW q vector is continuously varied with increasing the Fermi energy while keeping the host structure.

The characteristic feature of this compound is the coexistence of the CDW and the superconductivity, which are competitive with each other. As mentioned above, CDW q vectors of this compound are associated with two Nb–Nb chains. The remaining Nb–Nb chain may be responsible for the superconductivity, suggesting that the present compound is a possible candidate for one-dimensional superconductor.

9.4.3
Successive Phase Transitions in Quasi-One-Dimensional Sulfides ACu$_7$S$_4$ (A = Tl, K, Rb)

ACu$_7$S$_4$ (A = Tl, K, Rb) has an NH$_4$Cu$_7$S$_4$ type structure (space group $I\bar{4}$) [82]. Figure 9.50 shows the schematic crystal structure of TlCu$_7$S$_4$ viewed along the tetragonal c-axis. There are two kinds of Cu sites, labeled as Cu1 and Cu2, which belong to the different 8 g sites. The Cu1 atoms are coordinated triangularly to form (Cu$_4$S$_4$) columns running parallel to the c-axis. The Cu2 atoms are nearly three-coordinated and form double "CuS$_3$" tetrahedral chains. The occupancy factor of the Cu2 atoms is 3/4. The (Cu$_4$S$_4$) columns are joined by the tetrahedral chains of "CuS$_3$" to form tunnels containing Tl ions.

Temperature variations of electrical resistivity ($\rho - T$ curve) of TlCu$_7$S$_4$ from 2 to 420 K are shown in Figure 9.51 [83]. This compound shows relatively good conduction in spite of the full bands of d^{10} state of Cu$^+$, suggesting the presence

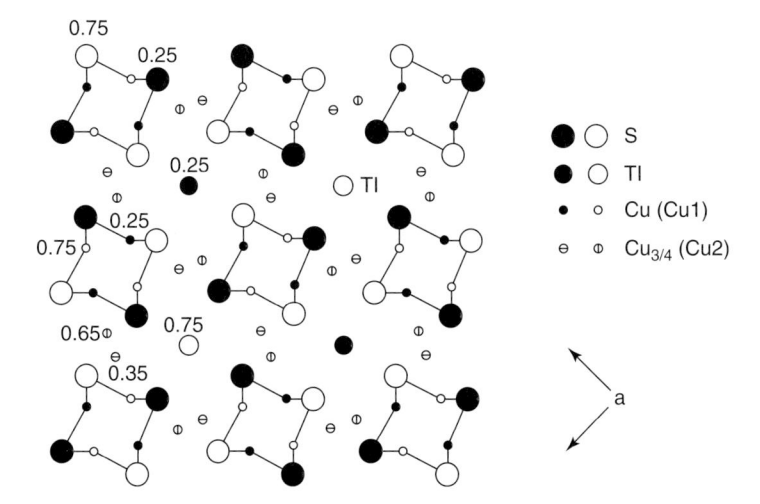

Figure 9.50 Schematic crystal structure of TlCu$_7$S$_4$ viewed along the tetragonal c-axis.

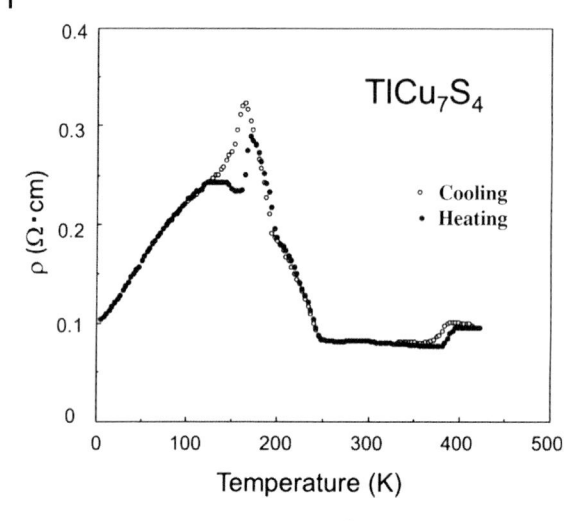

Figure 9.51 Temperature variations of electrical resistivity of TlCu$_7$S$_4$.

of Cu vacancies. It is noticed that many anomalies suggesting phase transitions are observed at 395, 245, 220, 190, 160–170, and 60 K. No superconductivity is observed down to 2 K. The $\rho - T$ curve shows the semiconductive-to-metallic transition around 170 K. Similar anomalies are observed in the $\rho - T$ curves of KCu$_7$S$_4$ and RbCu$_7$S$_4$, as shown in Figure 9.52, where the data of TlCu$_7$S$_4$ are

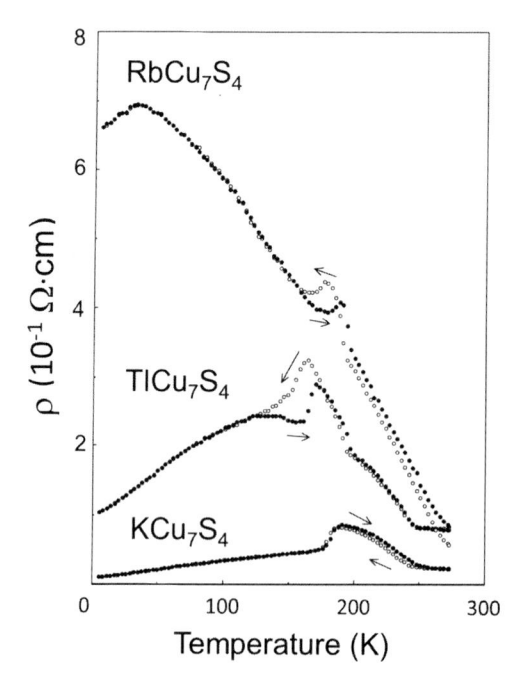

Figure 9.52 Temperature variations of electrical resistivity of ACu$_7$S$_4$ (A = Tl, K, Rb).

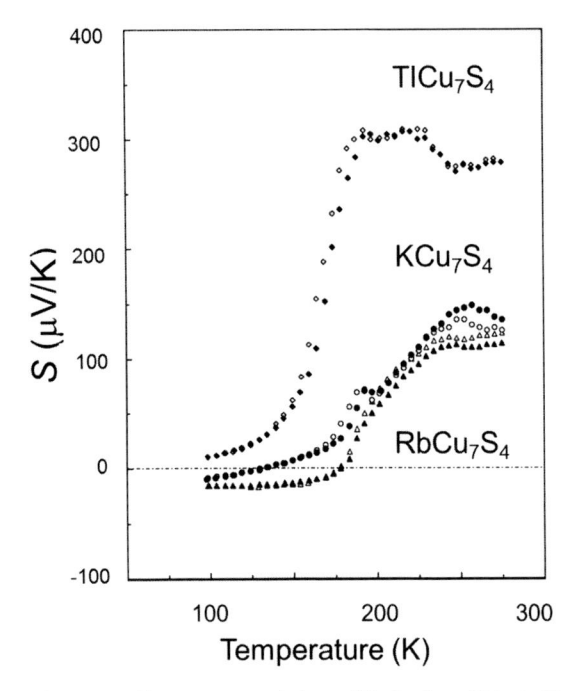

Figure 9.53 Temperature variations of Seebeck coefficients (S) of ACu_7S_4 (A = Tl, K, Rb).

shown again for comparison. The transition at 250 K is commonly observed in all compounds. From these measurements, it was revealed that KCu_7S_4 shows phase transitions at 445 K, 250 K, 220 K, 190 K, and 60 K, and $RbCu_7S_4$ at 402 K, 260 K, 220 K, 190 K, 160–170 K, 60 K, and 30 K. The ρ measurements of $(Tl_{1-x}K_x)Cu_7S_4$ showed that the 160–170 K transition of $TlCu_7S_4$ gradually approaches to the 190 K transition of KCu_7S_4.

Figure 9.53 shows temperature variations of Seebeck coefficients (S) of ACu_7S_4 (A = Tl, K, Rb) [83]. The conduction nature apparently changes from semiconductive to metallic around 170–190 K with cooling for all compounds, which is consistent with the resistivity measurements. The dominant carriers are holes above about 200 K. $TlCu_7S_4$ and KCu_7S_4 show the higher power factor (S^2/ρ) of about 1×10^{-4} W/(K^{2}m) at 250 K.

Magnetic susceptibility (χ) measurements for ACu_7S_4 (A = Tl, K, Rb) also show anomalies at the temperatures corresponding to the anomalies in the $\rho - T$ curves, as shown in Figure 9.54 [83]. A small kink at about 60 K in each compound can be seen more clearly than in other measurements. In the low temperature range below about 150 K, the $1/(\chi - \chi_0) - T$ curves obey Curie–Weiss law for all compounds. By assuming magnetic ion with $S = 1/2$, the concentration of magnetic ion can be estimated from the Curie constant to correspond to 0.09, 0.3, and 0.3% of the total number of Cu atoms, respectively, for A = Tl, K, and Rb. Such small values suggest that the observed CW behaviors come from accidentally mixed magnetic impurities or Cu^{2+} ions. The temperature-independent

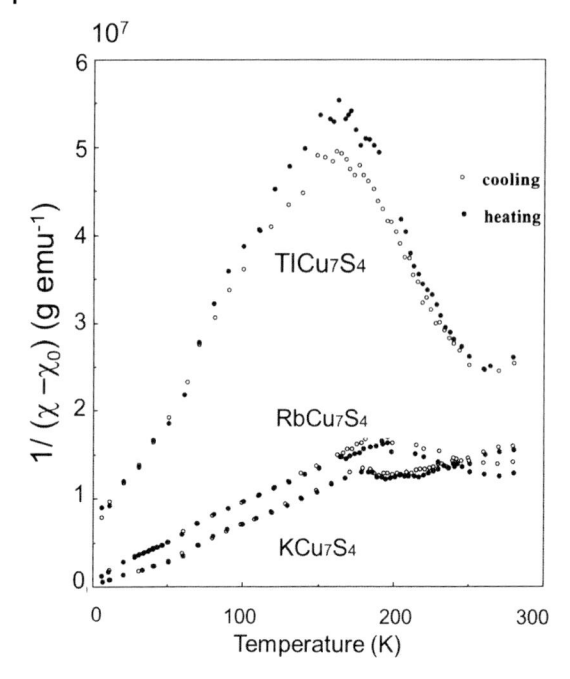

Figure 9.54 Inverse magnetic susceptibility of ACu_7S_4 (A = Tl, K, Rb) as a function of temperature.

term χ_0 is negative value in each compound, which indicates that the compounds ACu_7S_4 (A = Tl, K, Rb) are essentially band insulators and their good conduction originates from conduction carriers induced by the Cu vacancies.

Figure 9.55 represents the schematic drawing of electron diffraction (ED) patterns of KCu_7S_4 observed at room temperature (a), at about 210 K (b), and at about 100 K (c). The direction of the incident electron beam is [2$\bar{2}$0]. Only main reflection spots are observed at room temperature. The superlattice reflections observed at about 210 K are characterized by a wave vector of $q = 1/2c^*$, whose appearance would be associated with the transition at 250 K. The additional satellite spots appear at about 100 K, which is probably associated with the transition at 190 K. This pattern is characterized by the commensurate wave vector of $q = \pm(1/3a^* + 1/3b^*) + 2/3c^*$. $TlCu_7S_4$ showed similar ED patterns, though the satellite spots were weak. The DSC (differential scanning calorimetry) measurements for KCu_7S_4 showed the endothermic peak at 186 K on heating, indicating the first-order nature of the 190 K transition. The energy band calculation using the extended Hückel tight-binding method did not predict the CDW formation in these compounds [84]. The transition at 250 K, thus, would be associated with an ordering of Cu^+ ions (Cu2 site), which is caused by a pairing distortion of tetrahedral chains along the c-axis [84]. This was confirmed in X-ray diffraction analysis [85]. On the transition at 190 K, the latent heat was observed to increase with increasing conduction carriers (or decreasing the Cu content), suggesting

(a)

KCu$_7$S$_4$

[diffraction pattern with labels 110, 002, 000]

(b)

[diffraction pattern with labels 110, 002, 001, 000, q]

$q = 1/2c^*$

(c)

[diffraction pattern with labels 110, 002, 001, 000, q]

$q = \pm (1/3a^* + 1/3b^*) + 2/3c^*$

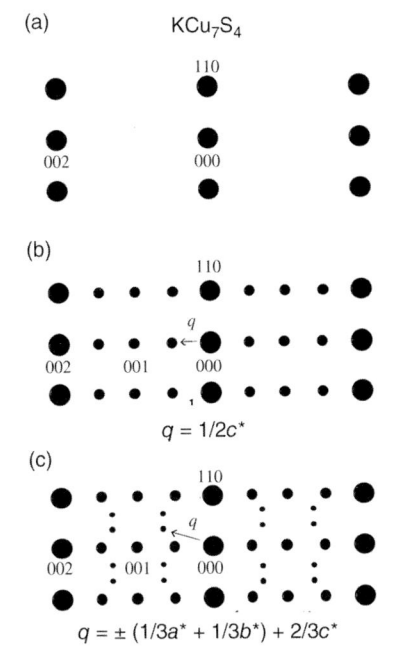

Figure 9.55 Schematic drawings of electron diffraction patterns of KCu$_7$S$_4$ observed at room temperature (a), at about 210 K (b), and at about 100 K (c). The direction of the incident electron beams is [$2\bar{2}0$].

that the conduction carriers are concerned with this transition [83]. As another possibility, this transition would be caused by a new type of ordering of Cu atoms. The origins of other phase transitions have not been clear yet.

Appendix: Enantioselective Crystallization of Amorphous Se in the Presence of Chiral Organic Fluids

Elemental selenium has a lot of polymorphs: amorphous Se (*a*-Se), hexagonal Se (*h*-Se), α-monoclinic Se (α-*m*-Se), and so on. The *h*-Se is most stable phase, which can be easily obtained by slowly cooling molten Se [86]. The α-*m*-Se is obtained as metastable phase by evaporation from CS$_2$ solution [87]. Because of the high photoconductivity, the elemental Se has been widely used in many applications such as photocells, xerography, electrical rectifiers, and so on. Se is also used for coloring glasses red or reddish yellow. The recent important use of Se is a high-sensitive camera tube using HARP (High-gain Avalanche Rushing Amorphous Photoconductor) imaging devices [88].

Figure 9.56 gives schematic crystal structures of *h*-Se (a) and α-*m*-Se (b). The *h*-Se crystal consists of parallel infinite chains of Se atoms, where Se atoms are arranged in screw-like spirals along the crystalline *c*-axis. The chains can be right-hand screws or left-hand screws, bringing about the chirality to the crystal.

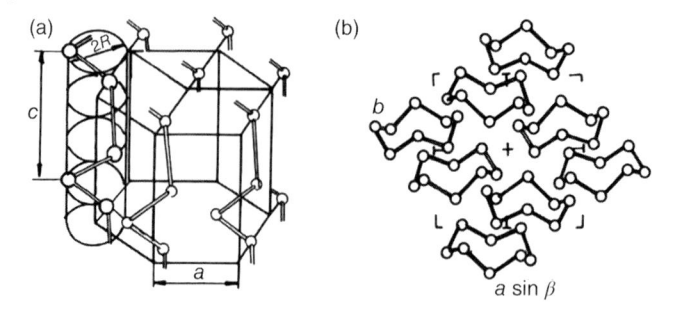

Figure 9.56 Schematic crystal structures of hexagonal selenium (h-Se) (a), and α-monoclinic selenium (α-m-Se) (b).

The α-m-Se has a molecular crystal, being composed of building blocks of Se_8 molecules arranged in a crown shape.

When a-Se was exposed to various organic fluids, the a-Se was found to be crystallized in h-Se or α-m-Se [89,90]. Saunders found similar selective crystal growth in various organic liquids more than 100 years ago [91]. Experimental procedures are follows; the a-Se powder was exposed to organic fluids (liquids or vapors) contained in a glass capsule, and allowed to stand at room temperature for a certain period. The kind of grown phase depends on the dielectric constant ε of the organic fluid [83]. The h-Se crystals were grown in the liquids with ε greater than about 4, such as acetone, methanol, aniline, and so on. The α-m-Se crystals formed in the liquids with ε less than about 3, such as benzene, hexane, toluene, and so on. The reactions take place even under the vapor of organic solid, suggesting that organic fluids do not act as solvent, but the organic molecules adsorbed on the surface of a-Se play a catalytic role for the crystal growth.

The crystallization of a-Se was investigated by using enantiomeric secondary alcohols [92]. The enantiomeric configuration of secondary alcohols is represented in Figure 9.57. For a pair of enantiomeric secondary alcohol having the carbon number less than 8, the h-Se crystals were obtained regardless of the enantiomer. The enantioselective crystallization was found for 2-octanol.

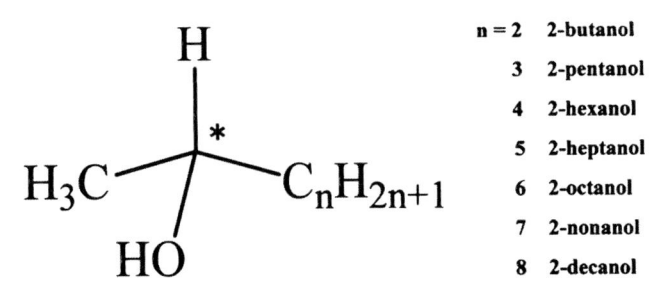

n	alcohol
2	2-butanol
3	2-pentanol
4	2-hexanol
5	2-heptanol
6	2-octanol
7	2-nonanol
8	2-decanol

Figure 9.57 Enantiomeric configuration of secondary alcohols.

Appendix: Enantioselective Crystallization of Amorphous Se in the Presence of Chiral Organic Fluids | 429

Figure 9.58 SEM images of h-Se crystals after exposing a-Se to (R)-(−)-2-octanol for 2 years (a), and α-m-Se crystals obtained by exposing a-Se to (S)-(+)-2-octanol for 2 years (b).

Figure 9.58a and b exhibit SEM images of Se crystals obtained by exposing a-Se to (R)-(−)-2-octanol for 2 years, and Se crystals after exposing a-Se to (S)-(+)-2-octanol for 2 years, respectively. Fibrous h-Se crystals were obtained for (R)-(−)-2-octanol, and α-m-Se crystals with polyhedral morphology were obtained for (S)-(+)-2-octanol. These enantioselective crystallizations were confirmed by XRD patterns as shown in Figure 9.59. Similar behaviors were observed when the samples of different makers were used. The enantioselective

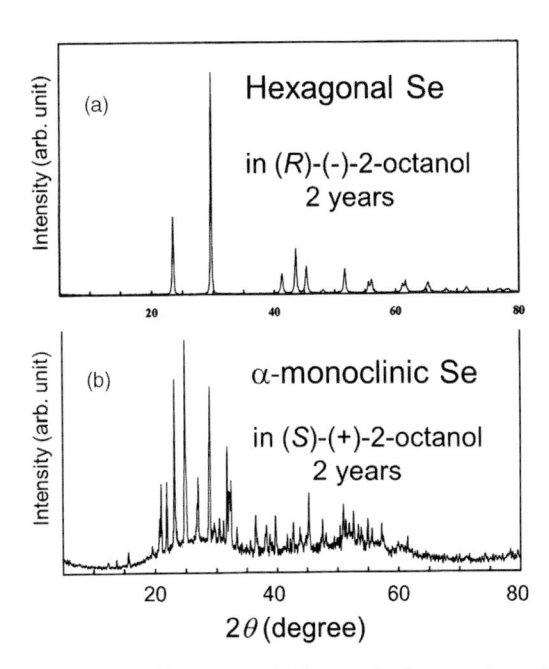

Figure 9.59 XRD patterns of h-Se crystals after exposing a-Se to (R)-(−)-2-octanol for 2 years (a), and α-m-Se crystals obtained by exposing a-Se to (S)-(+)-2-octanol for 2 years (b).

crystal growth of Se was also observed in the vapor of chiral proline. A pair of enantiomer possesses the same physical and chemical properties with each other, except for the optical rotation. However, the difference of optical rotation is not responsible for the present phenomena, because the reactions occur even in the dark. Unfortunately, the origin of these phenomena is not clear yet.

Gardner proposed the "Ozma problem," asking the way to communicate the meaning of "left or right" by a language to the aliens living in faraway space [93]. This problem was solved by the observations of nonpreservation of the parity in the β-decay of Co^{60} atoms. The present observation of enantioselective crystallization of a-Se may be the answer to this problem from the field of chemistry.

References

1 Greenwood, N.N. and Earnshaw, A. (1984) *Chemistry of the Elements*, Pergamon Press, Oxford.

2 Wells, A.F. (1984) *Structural Inorganic Chemistry*, 5th edn, Oxford University Press, Oxford.

3 Brauer, G. (ed.) (1963) *Handbook of Preparative Inorganic Chemistry*, vols. 1 and 2, Academic Press, New York (translation from German).

4 Rao, C.N.R. and Pisharody, P.R. (1975) Transition metal sulfides. *Prog. Solid State Chem.*, **10**, 207–270.

5 Jellinek, F. (1988) Transition metal chalcogenides; relationship between chemical composition, crystal structure and physical properties. *React. Solids*, **5**, 323–339.

6 Lévy F F. (ed.) (1979) *Intercalated Layered Materials*, Physics and Chemistry of Materials with Layered Structures, vol. 6, Springer, Netherlands, pp. 1–307.

7 Fecht, H.J., Han, G., Fu, Z., and Johnson, W.L. (1990) Metastable phase formation in the Zr–Al binary-system induced by mechanical alloying. *J. Appl. Phys.*, **67**, 1744–1748.

8 Aning, A.O., Wan, Z., and Courtney, T.H. (1993) Tungsten solution kinetics and amorphization of nickel in mechanically alloyed Ni–W alloys. *Acta Metall. Mater.*, **41**, 165–174.

9 Shultz, L. (1988) Formation of amorphous metals by mechanical alloying. *Mater. Sci. Eng.*, **97**, 15–23.

10 Schwarz, R.B. (1988) Formation of amorphous-alloys by solid-state reactions. *Mater. Sci. Eng.*, **97**, 71–78.

11 Benjamin, J.S. (1970) Dispersion strengthened superalloys by mechanical alloying. *Metall. Trans.*, **1**, 2943.

12 Baláž, P. (2010) *Mechanochemistry in Nanoscience and Minerals Engineering*, Springer-Verlag, Berlin.

13 Kosmac, T. and Courtney, T.H. (1992) Milling and mechanical alloying of inorganic nonmetallics. *J. Mater. Res.*, **7**, 1519–1525.

14 Kosmac, T., Maurice, D., and Courtney, T.H. (1993) Synthesis of nickel sulfides by mechanical alloying. *J. Am. Ceram. Soc.*, **76**, 2345–2352.

15 Ohtani, T., Motoki, M., Koh, K., and Ohshima, K. (1995) Synthesis of binary copper chalcogenides by mechanical alloying. *Mater. Res. Bull.*, **30**, 1495–1504.

16 Ohtani, T., Maruyama, K., and Ohshima, K. (1997) Synthesis of copper, silver, and samarium chalcogenides by mechanical alloying. *Mater. Res. Bull.*, **32**, 343–350.

17 Ohtani, T., Ikeda, K., Hayashi, Y., and Fukui, Y. (2007) Mechanochemical preparation of palladium chalcogenides. *Mater. Res. Bull.*, **42**, 1930–1934.

18 Suslick, K.S. (1990) Sonochemistry. *Science*, **247**, 1439–1445.

19 Ohtani, T., Nonaka, T., and Araki, M. (1998) Sonochemical synthesis of copper and silver chalcogenides. *J. Solid State Chem.*, **138**, 131–134.

20 Ohtani, T., Araki, M., and Shohno, M. (2004) Formation of Cu_3Se_2 and Ag_2Se by sonochemical solid-state reactions at room temperature. *Solid State Ion.*, **172**, 197–203.

21 Ohtani, T. and Shohno, M. (2004) Room temperature formation of Cu_3Se_2 by solid-state reaction between α-Cu_2Se and α-CuSe. *J. Solid State Chem.*, **177**, 3886–3890.

22 Heyding, R.D. and Murray, R.M. (1976) Crystal-structures of $Cu_{1.8}Se$, Cu_3Se_2, α-CuSe and γ-CuSe, $CuSe_2$. *Can. J. Chem.*, **54**, 841–848.

23 Janosi, A. (1964) La structure du sulfure cuivreux quadratique. *Acta Cryst.*, **17**, 311.

24 Skinner, B.J. (1970) Stability of tetragonal polymorph of Cu_2S. *Econ. Geol.*, **65**, 724.

25 Djurle, S. (1958) An X-ray study on the system Cu–S. *Acta Chem. Scand.*, **12**, 1415–1426.

26 Bither, T.A., Prewitt, C.T., Gillson, J.L., Bierstedt, P.E., Flippen, R.B., and Young, H.S. (1966) New transition metal dichalcogenides formed at high pressure. *Solid State Commun.*, **4**, 533.

27 Frueh, A.J., Jr (1955) The crystal structure of stromeyerite, AgCuS: a possible defect structure. *Zeits. Krist.*, **106**, 299.

28 Doktycz, S.J. and Suslick, K.S. (1990) Interparticle collisions driven by ultrasound. *Science*, **247**, 1067–1069.

29 Suslick, K.S. (1994) *Encyclopedia of Inorganic Chemistry*, vol. 7 (ed. R.B. King), John Wiley & Sons, Inc, New York, pp. 3890.

30 Suslick, K.S. and Doktycz, S.J. (1989) The sonochemistry of Zn powder. *J. Am. Chem. Soc.*, **111**, 2342–2344.

31 Suslick, K.S., Choe, S.B., Cichowias, A.A., and Grinstaff, M.W. (1991) Sonochemical synthesis of amorphous iron. *Nature*, **353**, 414–416.

32 Yanagida, H., Koumoto, K., and Miyayama, M. (1996) *The Chemistry of Ceramics*, John Willey & Sons, Ltd, Chichester.

33 Wiegers, G.A. (1971) Crystal-structure of low-temperature form of silver selenide. *Am. Mineral.*, **56**, 1882.

34 Xu, R., Husmann, A., Rosenbaum, T.F., Enderby, J.E., and Littlewood, P.B. (1997) Large magnetoresistance in non-magnetic silver chalcogenides. *Nature*, **390**, 57–60.

35 Glazov, V.M., Pashinkin, A.S., and Fedorov, V.A. (2000), Phase equilibria in the Cu–Se system. *Inorg. Mater.*, **36**, 641.

36 Ohtani, T., Tachibana, Y., Ogura, J., Miyake, T., Okada, Y., and Yokota, Y. (1998) Physical properties and phase transitions of β $Cu_{2-x}Se$ ($0.20 \leq x \leq 0.25$). *J. Alloys Compd.*, **279**, 136–141.

37 Okada, Y., Ohtani, T., Yokota, Y., Tachibana, Y., and Morishige, K. (2000) Crystal structure of the low-temperature phase of β $Cu_{1.75}Se$ analyzed by electron diffraction. *J. Electron Microsc.*, **49**, 25–29.

38 Folmer, J.C.W. and Jellinek, F. (1980) The valence of copper in sulfides and selenides – an X-ray photoelectron-spectroscopy study. *J. Less Common Met.*, **76**, 153–162.

39 Takahashi, T., Yamamoto, O., Matsuyama, F., and Noda, Y. (1976) Ionic-conductivity and coulometric titration of copper selenide. *J. Solid State Chem.*, **16**, 35–39.

40 Takacs, L. (2002) Self-sustaining reactions induced by ball milling. *Prog. Mater. Sci.*, **47**, 355–414, and references therein.

41 Merzhanov, A.G. (1994) Solid flames-discoveries, concepts, and horizons of cognition. *Combust. Sci. Technol.*, **98**, 307–336.

42 Tschakarov, Chr.G., Gospodinov, G.G., and Bontschev, Z. (1982) The mechanism of the mechanochemical synthesis of inorganic-compounds. *J. Solid State Chem.*, **41**, 244–252.

43 Ohtani, T., Kusano, Y., Ishimaru, K., Morimoto, T., Togano, A., and Yoshioka, T. (2015) Pre-milling effects on self-propagating reactions in mechanochemical synthesis of CdSe and ZnSe. *Chem. Lett.*, **44**, 1234–1236.

44 Hirota, T., Ueda, Y., and Kosuge, K. (1988) Phase diagram of the $TiSe_x$ System ($0.95 \leq x \leq 2.00$). *Mater. Res. Bull.*, **23**, 1641–1650.

45 Oka, Y., Kosuge, K., and Kachi, S. (1978) Order–disorder transition of metal vacancies in vanadium–sulfur system. I. An experimental-study. *J. Solid State Chem.*, **23**, 11–18.

46 Oka, Y., Kosuge, K., and Kachi, S. (1978) Order–disorder transition of metal vacancies in vanadium–sulfur system. II. A

statistical thermodynamic treatment. *J. Solid State Chem.*, **24**, 41–55.

47 Katsuyama, S., Ueda, Y., and Kosuge, K. (1990) Phase diagram and order–disorder transition of vacancies in the Cr–Se and Fe–Se systems. *Mater. Res. Bull.*, **25**, 913–922.

48 Hsu, F.-C., Luo, J.-Y., Yeh, K.-W., Chen, T.-K., Huang, T.-W., Wu, P.-M., Lee, Y.-C., Huang, Y.-L., Chu, Y.-Y., Yan, D.-C., and Wu, M.-K. (2008) Superconductivity in the PbO-type structure α-FeSe. *Proc. Natl. Acad. Sci. USA*, **105**, 14262–14264.

49 Ueda, Y., Kosuge, K., Urabayashi, M., Hayashi, A., and Kachi, S. (1985) Phase diagram and metal distribution of the $(Cr_xTi_{1-3})_3Se_4$ system ($0 \leq x \leq 1$) with the Cr_3S_4-type structure. *J. Solid State Chem.*, **56**, 263–267.

50 Hayashi, A., Ueda, Y., Kosuge, K., Murata, H., Asano, H., Watanabe, N., and Izumi, F. (1987) Cation distribution in $(M',M)_3Se_4$. I. $(Cr,Ti)_3Se_4$. *J. Solid State Chem.*, **67**, 346–353.

51 Lamberta, B., Berodias, G., and Chevreton, M. (1968) Neutron diffraction study of $TiCr_2Se_4$ and $TiCr_2S_4$. *Bull. Soc. Fr. Mineral. Crystallogr.*, **91**, 88.

52 Chevreton, M. and Sapet, A. (1965) Structure de V_3S_4 et de quelques sulfures ternaires isotypes. *C. R. Acad. Sci.*, **261**, 928.

53 Andron, B., Berodias, G., Chevreton, M., and Mollard, P. (1966) Etude par diffraction neutronique du compose ferromagnetique $TiCr_2Te_4$. *C. R. Acad. Sci.*, **263**, 621.

54 Hayashi, A., Imada, K., Inoue, K., Ueda, Y., and Kosuge, K. (1986) Phase diagram of $(M'_xM_{1-x})_3Se_4$ ($0 \leq x \leq 1$) (M, M' = 3d transition metals). *Bull. Inst. Chem. Res. Kyoto Univ.*, **64**, 186–206.

55 Hayashi, A., Ueda, Y., Kosuge, K., Murata, H., Asano, H., Watanabe, N., and Izumi, F. (1987) Cation distribution in $(M',M)_3Se_4$. II. $(V,Ti)_3Se_4$ and $(Cr,V)_3Se_4$. *J. Solid State Chem.*, **71**, 237–243.

56 Nozaki, H. (1981) Formation of local magnetic-moments in $(V,Ti)_3S_4$. *Mater. Res. Bull.*, **16**, 861–868.

57 Hayashi, A., Kishi, T., Ueda, Y., and Kosuge, K. (1989) Phase diagram and magnetic properties of the $(Fe_xTi_{1-x})_3Se_4$ system. *Mater. Res. Bull.*, **24**, 701–710.

58 Chevreton, M. and Andron, B. (1967) Etude par diffraction neutronique de $FeCr_2Se_4$ antiferromagnetique. *C. R. Acad. Sci.*, **B264**, 316.

59 Andron, B. and Bertaut, E.F. (1966) Etude par diffraction neutronique de Cr_2NiS_4. *J. Phys.*, **27**, 619.

60 Nozaki, H., Wada, H., and Yamaura, H. (1982) Site distribution on Fe and V in the system, $Fe_xV_{3-x}S_4$. *Solid State Commun.*, **44**, 63–65.

61 Bertaut, E.F., Roult, G., Aleonard, R., Pauthenet, R., Chevreton, M., and Jansen, R. (1964) Structures magnetiques de Cr_3X_4 (X = S, Se, Te). *J. Phys.*, **25**, 582.

62 Andresen, A.F. (1968) A Neutron diffraction investigation of Fe_3Se_4. *Acta Chem. Scand.*, **22**, 827.

63 Andresen, A.F. and van Laar, B. (1970) Magnetic structure of Fe_3Se_4. *Acta Chem. Scand.*, **24**, 2435.

64 Muranaka, S. and Takada, T. (1975) Magnetic-susceptibility and torque measurements of FeV_2S_4, FeV_2Se_4 and $FeTi_2Se_4$. *J. Solid State Chem.*, **14**, 291–298.

65 Peierls, R.E. (1955) *Quantum Theory of Solids*, Oxford University Press, Oxford.

66 Gor'kov, L.P. and Grüner, G. (1989) *Charge Density Waves in Solids*, North-Holland, Amsterdam.

67 Grüner, G. (1994) *Density Waves in Solids*, Addison-Wesley Publishing Company, New York.

68 Chaussy, J., Haen, P., Lasjaunias, J.L., Monceau, P., Waysand, G., Waintal, A., Meerschaut, A., Molinié, P., and Rouxel, J. (1976) Phase-transitions in $NbSe_3$. *Solid State Commun.*, **20**, 759–763.

69 Meerschaut, A. and Rouxel, J. (1986) *Crystal Chemistry and Properties of Materials with Quasi-One-Dimensional Structures* (ed. J. Rouxel), Reidel Publishing Company, Dordrecht, p. 205.

70 Briggs, A., Monceau, P., Nunez-Regueiro, M., Peyrard, J., Ribault, M., and Richard, J. (1980) Charge-density wave formation, superconductivity and Fermi-surface determination in $NbSe_3$ – a pressure study. *J. Phys. C Solid State Phys.*, **13**, 2117–2130.

71 Monceau, P., Ong, N.P., Portis, A.M., Meerschaut, A., and Rouxel, J. (1976) Electric-field breakdown of charge-density wave-induced anomalies in $NbSe_3$. *Phys. Rev. Lett.*, **37**, 602–606.

72 Selte, K. and Kjekshus, A. (1964) Crystal structures of $Nb_3Se_4 + Nb_3Te_4$. *Acta Crystallogr.*, **17**, 1568.

73 Amberger, E., Polborn, K., Grimm, P., Dietrick, M., and Obst, B. (1978) Superconductivity in Nb cluster compounds $Nb_3(X,Y)_4$ with X, Y = S, Se, Te. *Solid State Commun.*, **26**, 943–947.

74 Ishihara, Y. and Nakada, I. (1982) Electrical transport-properties of a quasi-one-dimensional Nb_3Te_4 single-crystal. *Solid State Commun.*, **45**, 129–132.

75 Suzuki, K., Ichihara, M., Nakada, I., and Ishihara, Y. (1986) Stripe and charge-density-wave domain-structure of Nb_3Te_4. *Solid State Commun.*, **59**, 291–293.

76 Sekine, T., Kiuchi, Y., Matsuura, E., Uchinokura, K., and Yoshizaki, R. (1987) Charge-density-wave phase-transition in the quasi-one-dimensional conductor Nb_3Te_4. *Phys. Rev. B*, **36**, 3153–3160.

77 Oshiyama, A. (1982) Energy-bands and Fermi surfaces of quasi-one-dimensional transition-metal chalcogenides Nb_3X_4. *Solid State Commun.*, **43**, 607–612.

78 Oshiyama, A. (1983) Electronic-structure of quasi-one-dimensional transition-metal chalcogenide Nb_3X_4. *J. Phys. Soc. Jpn.*, **52**, 587–596.

79 Ohtani, T., Sano, Y., and Yokota, Y. (1993) Electrical properties and phase transitions in superconducting quasi-one-dimensional chalcogenides $In_xNb_3X_4$ (X = S, Se, and Te). *J. Solid State Chem.*, **103**, 504–513.

80 Boswell, F.W. and Bennet, J.C. (1996) Density waves in Nb_3Te_4: effects of indium and thallium intercalation. *Mater. Res. Bull.*, **31**, 1083–1092.

81 Ohtani, T., Yokota, Y., and Sawada, H. (1999) Charge density waves in quasi-one-dimensional compound $In_xNb_3Te_4$ ($0 \leq x \leq 1.0$): electron diffraction observations. *Jpn. J. Appl. Phys.*, **38**, L142–L144.

82 von Gattow, G. (1957) Die kristallstruktur von $NH_4Cu_7S_4$. *Acta. Cryst.*, **10**, 549–553.

83 Ohtani, T., Ogura, J., Yoshihara, H., and Yokota, Y. (1995) Physical properties and successive phase transitions in quasi-one-dimensional sulfides ACu_7S_4 (A = Tl, K, Rb). *J. Solid State Chem.*, **115**, 379–389.

84 Whangbo, M.-H. and Canadell, E. (1992) Nonelectronic origin of superlattice modulations and resistivity anomalies in ternary copper sulfides. *Solid State Commun.*, **81**, 895–899.

85 Mackay, H., Li, R., Hwu, S.-J., Kuo, Y.-K., Skove, M.J., Yokota, Y., and Ohtani, T. (1998) On the electrochemically grown quasi-one-dimensional $KCu_{7-x}S_4$ series ($0 \leq x \leq 0.34$): nonstoichiometry, superlattice, and unusual phase transitions. *Chem. Mater.*, **10**, 3172–3183.

86 Cherin, P. and Unger, P. (1967) Crystal structure of trigonal selenium. *Inorg. Chem.*, **6**, 1589.

87 Cherin, P. and Unger, P. (1972) Refinement of crystal-structure of α-monoclinic Se. *Acta Crystallogr.*, **B28**, 313.

88 Tanioka, K., Yamazaki, J., Shidara, K., Taketoshi, K., Kawamura, T., Ishioka, S., and Takasaki, Y. (1987) An avalanche-mode amorphous selenium photoconductive layer for use as a camera tube target. *IEEE Electron Devices Lett.*, **EDL-8, 9**, 392–394.

89 Ohtani, T., Takayama, N., Ikeda, K., and Araki, M. (2004) Unusual crystallization behavior of selenium in the presence of organic molecules at room temperature. *Chem. Lett.*, **33**, 100–101.

90 Ikeda, K. and Ohtani, T. (2007) Peculiar crystallization of amorphous selenium in the presence of organic liquids: effects of addition of titanium oxides. *J. Alloys Comp.*, **434–435**, 275–278.

91 Saunders, A.P. (1900) The allotropic forms of selenium. *J. Phys. Chem.*, **4**, 423–513.

92 Ohtani, T., Ikeda, K., Takayama, N., Nishida, Y., and Iyama, Y. (2015) Enantioselective crystallization of amorphous Se in the presence of chiral organic molecules. *Bull. Okayama Univ. Sci.*, **50**, 21–29.

93 Gardner, M. (1964) *The Ambidextrous Universe*, Penguin Books Ltd.

10
Metal Borides: Versatile Structures and Properties

Barbara Albert and Kathrin Hofmann

Technische Universität Darmstadt, Eduard-Zintl-Institute of Inorganic and Physical Chemistry, Alarich-Weiss-Str. 12, 64287 Darmstadt, Germany

10.1
Introduction

Boron is a non-metallic, light element with three valence electrons. Its ability to form multi-centered bonds to overcome the electron deficiency is huge, and it manifests itself in an enormous variety of complicated structures, often characterized by boron atom polyhedra. Boron allotropes, such as α- and β-rhombohedral boron (these are the two modifications most stable under ambient conditions), contain icosahedra of boron atoms (Figure 10.1) that are three-dimensionally interconnected (Figure 10.2).

Many metals react with boron and form borides. Metal borides (M_xB_z, $M_xM'_yB_z$, etc.; M, M' = metals) are binary or multinary compounds in which boron is the most electronegative partner. Thousands of borides are known, their compositions ranging from $M_{\sim 7}B$ to $MB_{\sim 80}$: M_5B, M_3B, M_5B_2, M_7B_3, M_2B, M_5B_3, M_3B_2, $M_{11}B_8$, MB, $M_{10}B_{11}$, M_3B_4, M_2B_3, M_3B_5, MB_2, M_2B_5, MB_3, MB_4, MB_6, MB_{10}, MB_{12}, MB_{15}, MB_{18}, MB_{66}.

Borides represent an important class of inorganic solids. Often they exhibit interesting properties and can be used either as structural and functional materials or as catalysts. Table 10.1 gives a glimpse on binaries of the main group elements that were confirmed to exist [1]; many more have been described. Even more borides are formed by transition metals (Table 10.2) and rare earth elements (lanthanoides in Table 10.3) [2,3]. Among the borides of the actinoides, it is ThB_4, ThB_6, and UB_{12} that have been investigated most. These three compounds will be discussed in Chapter 10.4, but otherwise borides of radioactive elements are omitted in this book. Several of the formulae in Tables 10.1 and 10.2 are marked by an asterisk, indicating that they crystallize with structures similar to that of the so-called β-rhombohedral boron (Figure 10.2), where metal atoms are located on interstitial positions, and both boron and metal atom positions may be partially occupied only. This leads to some uncertainty in terms of

Handbook of Solid State Chemistry, First Edition. Edited by Richard Dronskowski, Shinichi Kikkawa, and Andreas Stein.
© 2017 Wiley-VCH Verlag GmbH & Co. KGaA. Published 2017 by Wiley-VCH Verlag GmbH & Co. KGaA.

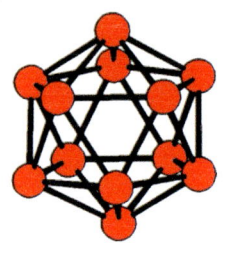

Figure 10.1 Boron atom icosahedron, the most important structural entity in elemental boron and boron-rich metal borides (boron atoms: red).

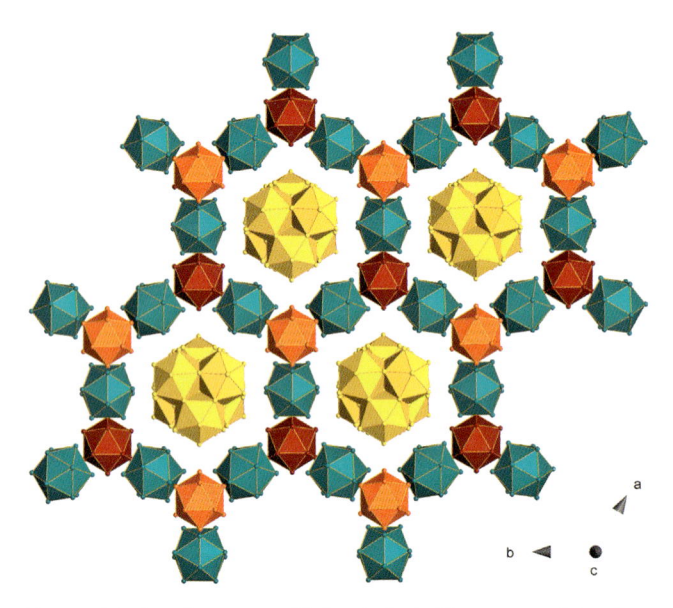

Figure 10.2 Crystal structure of β-rhombohedral boron, one of the allotropes of the element. (Figure from Ref. [1].)

Table 10.1 Binary borides of main group metals.

LiB_{1-x}	BeB_3	
Li_2B_6	Be_3B_{50}	
Li_3B_{14}		
LiB_{10}^*		
Na_3B_{20}	MgB_2	AlB_2
Na_2B_{29}	MgB_7	AlB_{12}
NaB_{28}^*	Mg_5B_{44}	AlB_{31}^*
	MgB_{12}	
	$MgB_{17.9}^*$	
KB_6	CaB_6	
	SrB_6	
	BaB_6	

* indicates a β-rhombohedral boron-type structure.

Table 10.2 Binary borides of d-group elements.

ScB_2	TiB	V_3B_2	Cr_2B	Mn_2B	$Fe_{23}B_6$	$Co_{23}B_6$	$Ni_{23}B_6$	CuB_{28}^*	ZnB_{25}^*
ScB_{12}	Ti_3B_4	VB	Cr_5B_3	MnB	Fe_3B	Co_3B	Ni_3B		
ScB_{15}	TiB_2	V_5B_6	CrB	Mn_3B_4	Fe_2B	Co_2B	Ni_2B		
ScB_{78}^*		V_3B_4	Cr_3B_4	MnB_2	FeB	CoB	NiB		
		V_2B_3	Cr_2B_3	MnB_4	FeB_4	Co_5B_{16}	Ni_4B_3		
		VB_2	CrB_2	MnB_{23}^*	FeB_{49}^*		Ni_7B_3		
			CrB_4				$NiB_{48.5}^*$		
YB_2 YB_4	ZrB	Nb_3B_2	Mo_2B		Ru_7B_3	Rh_7B_3	Pd_6B	—	—
YB_6	ZrB_2	NbB	Mo_3B_2		$Ru_{11}B_8$	Rh_5B_4	Pd_5B		
YB_{12}	ZrB_{12}	Nb_5B_6	MoB		RuB	RhB	Pd_3B		
YB_{66}		Nb_3B_4	Mo_2B_4		Ru_2B_3		Pd_5B_2		
		Nb_2B_3	MoB_2		RuB_2		Pd_2B		
		NbB_2	$Mo_{1-x}B_3$						
LaB_4	HfB	Ta_2B	W_2B	Re_3B	OsB	$IrB_{0.9}$	Pt_4B	—	—
LaB_6	HfB_2	Ta_3B_2	WB	Re_7B_3	Os_2B_3	IrB	Pt_2B		
		TaB	WB_2	ReB_2	OsB_2	$IrB_{1.1}$	PtB		
		Ta_5B_6	W_2B_4			$IrB_{1.35}$	$PtB_{0.7}$		
		Ta_3B_4	$W_{1-x}B_3$						
		TaB_2	WB_4						

* indicates a β-rhombohedral boron-type structure.

an accurate M:B ratio, and the compositions of these compounds were approximated only.

While almost all of the metals of the periodic system do form borides, no binaries are known for the systems Rb–B, Cs–B, Hg–B, Ag–B, Au–B, Ga–B, In–B, Tl–B, Sn–B, Pb–B, and Bi–B. Furthermore, these combinations do not show any (or almost no) solid solubility in each other.

Surely, it is the combination of certain physical properties that make metal borides in general interesting [3]. Common to many borides is their highly refractory character, and others have been described repeatedly: (super)hardness, (ultra)incompressibility, and thermal stability [4]. These properties lead to existing or potential applicability for the hardening of tools or thermal or

Table 10.3 Binary borides of the lanthanides.

CeB_4	PrB_4	NdB_4	—	SmB_2	EuB_6	GdB_2
CeB_6	PrB_6	NdB_6		SmB_4		GdB_4
		Nd_2B_5		SmB_6		GdB_6
				Sm_2B_5		Gd_2B_5
				SmB_{66}		GdB_{66}
TbB_2	DyB_2	HoB_2	ErB_2	TmB_2	YbB_2	LuB_2
TbB_4	DyB_4	HoB_4	ErB_4	TmB_4	YbB_4	LuB_4
TbB_6	DyB_6	HoB_6	ErB_{12}	TmB_{12}	YbB_6	LuB_{12}
TbB_{12}	DyB_{12}	HoB_{12}	ErB_{66}	TmB_{66}	YbB_{12}	LuB_{66}
TbB_{66}	DyB_{66}	HoB_{66}			YbB_{66}	

environmental barrier coatings of turbines. Furthermore, certain representatives of the metal borides have exceptional special properties. MgB_2 is a superconductor below its world record T_C of 39 K [5a]. TiB_2 is a refractory (ceramic) material with five times higher electrical conductivity than the metal, titanium. Neodymium iron boride, $Nd_2Fe_{14}B$, is the best hard magnet known so far [6]. Manganese borides may well have potential for magnetocaloric refrigeration [7]. Low thermal conductivities and high Seebeck coefficients of several metal borides point to potential applications in high-temperature thermoelectric generators [8]. Chemical properties of the metal borides, on the other hand, seem to be quite unexplored. Their catalytic properties, for example, still live in a niche [9]. When talking about boride characteristics, one might also mention important economical parameters that determine possible uses. Boron itself is easily available from boric acid and borates, minerals that are not rare and not extremely expensive. It is also a light element and not described to be toxic, contrary to certain boron–oxygen compounds that have been classified by the European Union as very toxic.

The huge variety of boride structures has been described many times and can be considered textbook knowledge [10,11]. It is reasonable to distinguish between, on the one hand, the more metal-rich borides in which boron atoms are either isolated or they form dumbbells or chains (Figure 10.3) and, on the other side, the boron-rich materials that display very exotic framework structures, often consisting of beautiful deltahedra (e.g., icosahedron in Figure 10.1). In between there are the layered structures with two-dimensional arrays of boron atoms. The degree of dimensionality of the arrangement of the boron atoms correlates closely with the (non)metallic character of the borides. Thus, the structural diversity is reflected by an extreme variability of electrical properties between metallic and ceramic. The huge influence of the boron atom structure on the electronic structure of the metal borides is quite unique and an interesting topic for research. It has its origin in the special electron-deficient character of the boron atom that often leads to a very special bonding situation in boron

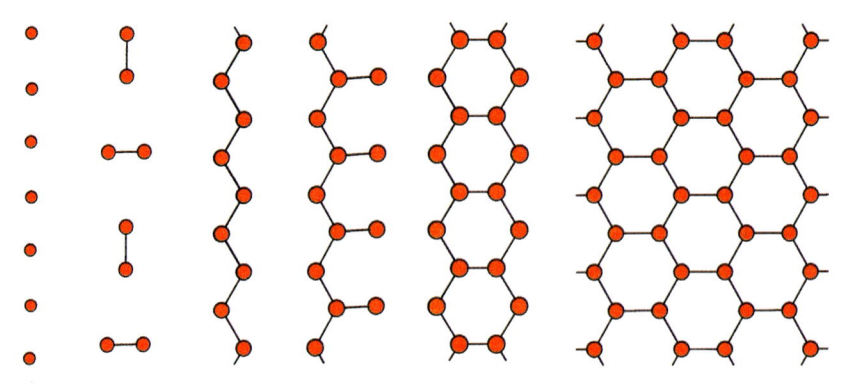

Figure 10.3 Boron atom arrangement in metallic borides: from isolated boron atoms to dumbbells, chains, and layers.

compounds. At the same time, the dominance of the boron atom framework results also in what can be seen as a drawback in the development of new materials: Simple modifications of existing borides by well-established methods like doping and substitution are not always possible or do not necessarily lead to the desired changes in properties. New borides' existence or characteristics are mostly unforeseeable, and boride materials cannot often be designed.

10.2 Synthesis of Borides

10.2.1 Powder Preparation

In boride synthesis, due to the refractory character and nonsolubility of both starting materials and products it is usually the classical solid-state high-temperature reactions ($T > 1000\,°C$) that are chosen to prepare and sinter metal boride powders, sometimes combined with the application of high pressure. Several combinations of starting materials are commonly used; most often the direct combination of the elements themselves ($z\,M + y\,B \rightarrow M_zB_y$), but also reactions between oxides, carbides, halides and hydrogen, carbon, and so on [1–3,12].

Metal oxides can be reduced by boron, carbon, or boron carbide. Boron oxide can be reduced by metals. Metal halides and boron halides react with hydrogen (e.g., $MCl_4 + 2\,BCl_3 + 5\,H_2 \rightarrow MB_2 + 10\,HCl$). Boron halides can also be thermally decomposed. Certain metal halides are also found to react with metal borides, forming a metathesis product.

Attention has to be paid to the choice of suitable crucible materials, since they may be found to react with the starting materials, instead of remaining inert.

The following are the examples of boride syntheses [10]:

- $Cr + n\,B \rightarrow CrB_n$ (1150 °C)
- $Sc_2O_3 + 7\,B \rightarrow 2\,ScB_2 + 3\,BO$ (1800 °C)
- $2\,TiCl_4 + 4\,BCl_3 + 10\,H_2 \rightarrow 2\,TiB_2 + 20\,HCl$ (1300 °C)
- $n\,BX_3 + (x+1)\,M \rightarrow MB_n + x\,MX_{3n/x}$, for example, $BCl_3 + W \rightarrow WB + Cl_2 + HCl$ (H_2/1200 °C)
- $V_2O_5 + B_2O_3 + 8\,C \rightarrow 2\,VB + 8\,CO$ (1500 °C)
- $Eu_2O_3 + 3\,B_{4+x}C_{1-x} \rightarrow 2\,EuB_6 + 3\,CO$ (1600 °C)
- $7\,Ti + B_2O_3 + 3\,B_{4+x}C_{1-x} \rightarrow 7\,TiB_2 + 3\,CO$ (2000 °C)
- $3\,CaO + 9\,B_2O_3 + 20\,Al \rightarrow 3\,CaB_6 + 10\,Al_2O_3$

10.2.2 Single-Crystal Growth

Small single crystals can be obtained by direct reaction of the elements through high temperature annealing over long periods of time. The incongruent melting of many borides is challenging, since it may prevent crystallization from melts and result in the formation of multiphase segregation. Thus, sintering of the solids is usually preferred to melting the starting materials. But, solid-state reactions require very high temperatures and extensive reaction times to produce large single crystals. The floating zone method has been employed successfully to synthesize boride crystals of several millimeter lengths [13]. Alternatively, the application of high pressure was proven to yield interesting results [14]. An important alternative is re-crystallization and synthesis in flux, using metals such as copper, indium, or aluminum [15].

Occasionally, chemical vapor transport reactions with iodine as transport agent were used to grow boride crystals [16].

10.2.3
Nanomaterials

There are two major routes that have been developed to synthesize nanoscale borides [17]: First, alkali metal tetrahydridoborates, such as $NaBH_4$, were used in solvents (water, organic solvents) to precipitate borides by reduction of metal salts. Powders produced like this are mostly amorphous, and were originally used as catalysts without further characterization [9,18]. Today, it is well established that nanoscale borides like VB_2, TiB_2, CrB_2, MnB, FeB, Ni_3B, and Co_2B are formed at temperatures as low as room temperature and can be crystallized by annealing at intermediate temperatures [7b,18]. Second, the synthesis in salt melts has been established and shown to be very successful for the synthesis of nanoscale metal borides [17,19].

10.3
Metal-Rich Borides: Interstitials or More?

10.3.1
General

The lattices of several of the elemental metals can take up small amounts of boron. Alloys with boron contents up to 10% are obtained at elevated temperatures and may be retained at room temperature by quenching. This proved successful, for example, in the system Ta–B, where body-centered cubic tantalum and boron form solid solutions with slightly increased lattice parameters *a* compared to the pure element.

According to Hägg [20], interstitial compounds can be formed at ratios of radii r_B/r_M between 0.59 and 0.41, and this requirement is fulfilled by the borides like Re_3B (also nitrides, carbides, etc.). A general phenomenon of the chemistry of boron-containing metal-rich solids is the variability of the boron atom arrangement – it manifests itself with special evidence in certain binary systems of transition metals [21]. An example is the Ta–B system [22] with Ta(B), an alloy

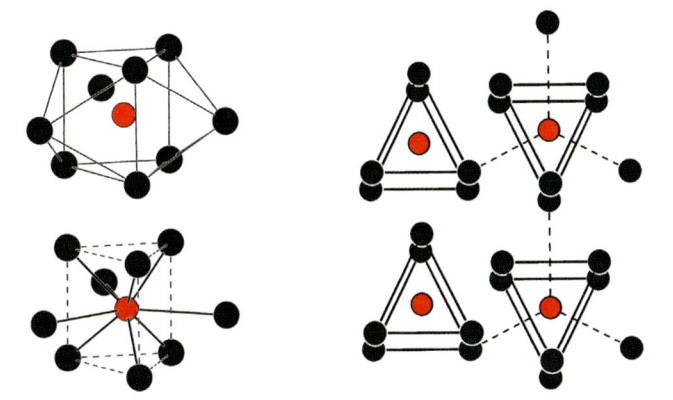

Figure 10.4 Coordination sphere of isolated boron atoms (red) in M$_3$B, and structural arrangement of the trigonal prisms (boron atoms: red, metal atoms: gray).

called α-phase in 1949, Ta$_2$B containing isolated B atoms, TaB with zigzag chains and isostructural to CrB, Ta$_3$B$_4$ with double chains, and TaB$_2$ that contains graphene-like sheets. Ta$_5$B$_6$ (and also Nb$_5$B$_6$) contains double chains, and its structure type is named after V$_5$B$_6$.

10.3.2
M$_3$B to MB$_2$: From Isolated Boron Atoms to Chains and Sheets

The most common structural motif for the coordination sphere of isolated boron atoms in metal borides (e.g., Ni$_3$B [18b]) is that of a three-capped trigonal prism of metal atoms around the boron atom, coordination number $6+3$ as shown in Figure 10.4. Typical distances between isolated boron atoms are around 210 pm.

If one of the rectangular faces of the prism is capped by a boron atom, B–B dumbbells result. If two are capped by boron atoms instead of metal atoms, a coordination number of the boron atoms of $6+1+2$ results, and zigzag chains of boron atoms are formed. Typical distances in B—B bonds like this are around 190 pm or smaller. An example is FeB, that crystallizes in the CrB structure type [23]. If all of the rectangular faces are capped with boron atoms, planar sheets of boron atoms are formed that are typical for AlB$_2$-type borides. This structure type will be discussed further (Figure 10.5).

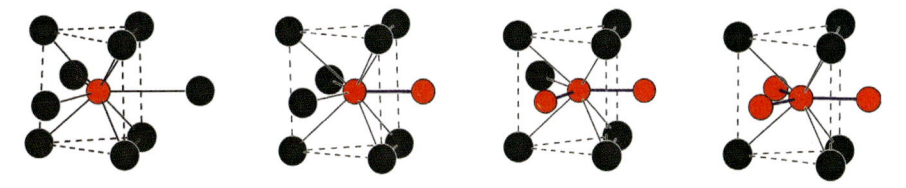

Figure 10.5 Coordination sphere of boron atoms (red) in metal-rich borides.

Occasionally, a square antiprism of metal atoms around an isolated boron atom is found. This results in an unusual coordination number of eight, as seen for Fe_2B [23].

10.3.3
τ-Borides

Borides of the general composition $M_{23}B_6$ are named τ-borides, and more than 80 different compounds have been described, usually being stabilized by a third component and thus ternary. They crystallize in the face-centered cubic crystal system with a $Cr_{23}C_6$-type structure and lattice parameters around 1050 pm. An example of a true binary τ-boride is $Co_{23}B_6$, (Figure 10.6) of which single crystals were obtained from indium melts [24].

10.3.4
Diborides and Related Compounds

Diborides, MB_2, are a striking example for a class of substances with a few existing but many potential applications. Several of them have outstanding properties, the most prominent being magnesium diboride, MgB_2, a metallic compound that becomes superconducting below a T_C of 39 K, which was a completely unexpected finding reported in 2001 [5a]. Wires and tapes of several kilometers in length have since been produced that withstand considerable critical currents, and prototypes of solenoids and magnetic/medical resonance imaging instruments/imagers were built to demonstrate that this boride may well enter the

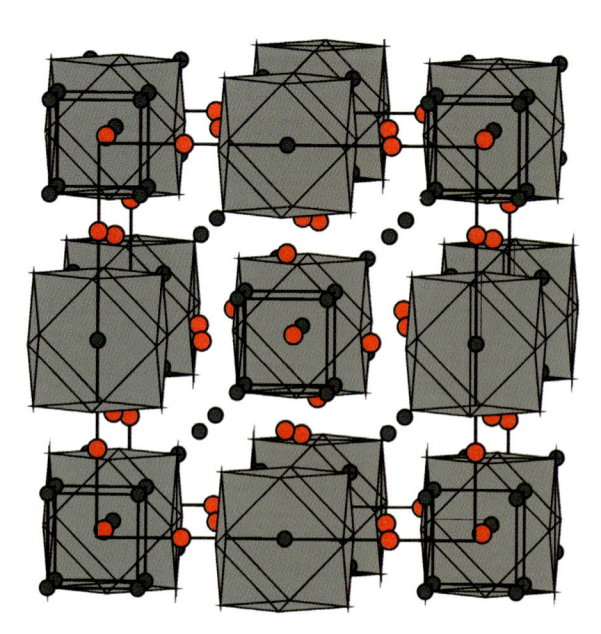

Figure 10.6 Crystal structure of $Co_{23}B_6$, a τ-boride.

market as a useful material. A recent world record current for a superconductor was determined to be 20 kA at 24 K for MgB_2 [5b]. Another diboride, TiB_2, is already widely needed. Produced on an industrial scale, mostly in the United States and Germany, the major part of it ends in vaporization boats for the coating industry, to produce, for example, metalized bags for potato chips. The basis of this application is of course the unique profile of properties that titanium diboride presents: A huge melting point combined with high electrical conductivity and exceptional chemical inertness.

Historically, probably MgB_2 was the first diboride synthesized, since it was Henry Moissan who obtained a mixture of elemental boron and magnesium borides when he reduced boron oxide magnesiothermally. It is interesting to note that the magnesium compound remains the only diboride known until today for all of the alkali and alkaline earth metals, despite some reports on LiB_2 and BeB_2 that could not be confirmed.

Diboride structures exhibit a varying stacking of different boron atom layers. These may be arranged as planar, boat-like, or chair-like condensed six-rings. Structural motifs emphasizing the coordination of the metal atoms are shown in Figure 10.7.

The dominant structure type of diborides is named after AlB_2, a compound that was structurally characterized in 1935 and crystallizes in the hexagonal crystal system (Figure 10.8) [25a]. There is one formula unit per unit cell, and the lattice parameters are around 300 pm, c being about 10% longer than a. It can be seen that its mean feature are planar sheets of condensed six-rings of boron atoms, very similar to the arrangement of the carbon atoms in graphite. The stacking of the layers resembles that in boron nitride.

Thermodynamically stable borides with this structure type are found for many metals (Mg, Sc, Y, Ti, Zr, V, Nb, Ta, Cr, Mn, Zr, Al, Mo, Hf, W, Sm, Tm, Er). Remarkable properties are their conductivities, hardnesses, stability ranges, and magnetism (e.g., MnB_2: ferromagnetic, CrB_2: antiferromagnetic). In some cases, the AlB_2-type phases represent the high-temperature modification (WB_2), in others they are metastable and were obtained by ball milling (OsB_2) [25b]. ZrB_2, HfB_2, NbB_2, and TaB_2 exhibit melting points >3000 °C, which is high compared even to the very high melting points of the corresponding metals (Zr: 1857 °C, Hf: 2222 °C, Nb: 2468 °C, Ta: 2980 °C). TbB_2 is ferromagnetic below 151 K and exhibits relatively strong magnetocrystalline anisotropy.

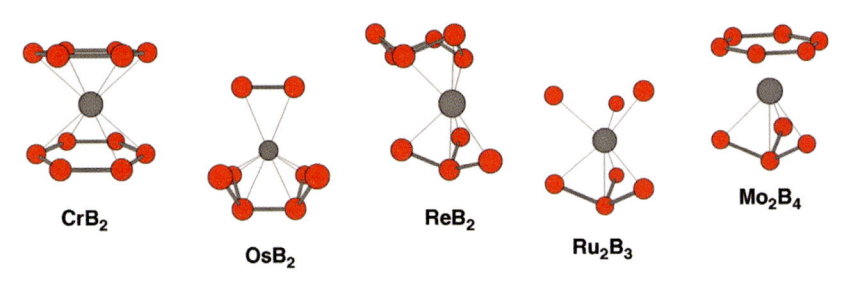

Figure 10.7 Metal atom coordination in diborides and related compounds. (CrB_2, OsB_2, ReB_2, Ru_2B_3, Mo_2B_4)

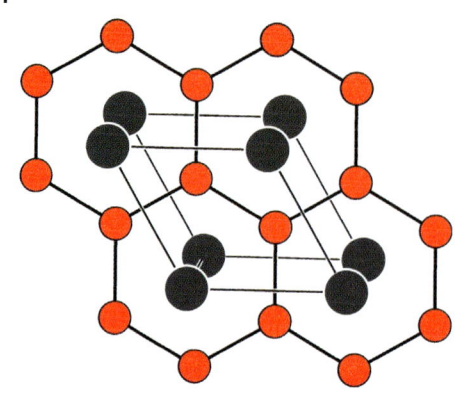

Figure 10.8 AlB$_2$-structure type, space group, $P6/mmm$ (No. 191).

Formally, the bonding situation of MgB$_2$ can be discussed in a Zintl-like manner as a transfer of two electrons per metal atom to the boron atom layers, resulting formally in $^2[\text{B}^-]_\infty$ sheets that are stacked directly above each other with only weak interactions perpendicular to the layers. However, the transfer of electrons certainly is not complete and the bonding between M and B is not predominately ionic. Despite its formal oxidation state of II, the Mg atom shows a temperature dependence of the electrical conductivity that is typical for a metal, and X-ray absorption spectroscopic measurements indicated an oxidation state of the Mg atoms smaller than II. Recently, VB$_2$ was advertised as electrode component that is extraordinarily redox-active, for example, for energy storage in redox flow technologies [26].

If the sheets of boron atoms are buckled instead of planar, with sheets of chair-like, condensed six-rings, the resulting structure type is named after ReB$_2$. It occurs for Re and Tc, and it is also hexagonal and shown in Figure 10.9(a).

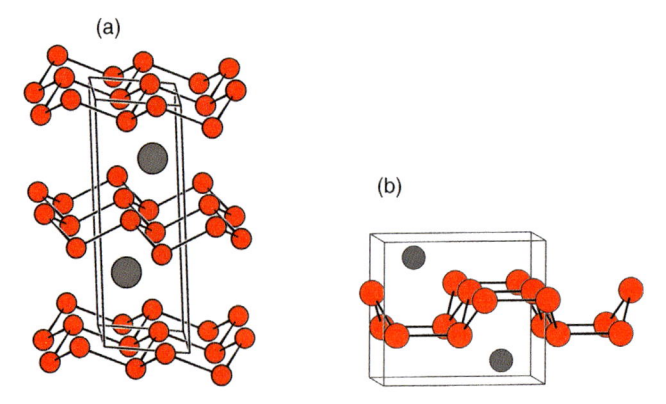

Figure 10.9 ReB$_2$-structure type, space group, $P6_3/mmc$ (No. 194) (a) and OsB$_2$-structure type, space group $Pmmn$ (No. 59) (b).

The volume of the unit cell is approximately twice that of the AlB_2-type compounds due to the doubled c-axis.

It should be mentioned that ReB_2-type crystallizing compounds have been discussed as superhard and ultraincompressible [4b,27], which is ascribed to a certain bonding situation with strong metal–boron interactions that have been characterized being predominately covalent.

The third simple structure type seen for metal diborides is that of OsB_2 and RuB_2, exhibiting two-dimensionally linked six-rings in boat-like conformation (Figure 10.9, b). The crystal system is orthorhombic, and the size of the unit cell is again approximately twice that of the AlB_2-type compounds ($a \sim 300$ pm, $b \sim 400$ pm, $c \sim 450$ pm). Accurate structure data for OsB_2 was obtained by neutron diffraction [28]. Neutron diffraction has been shown to be the appropriate method to characterize heavy metal borides, but requires the synthesis of isotopic pure (^{11}B) samples, since the ^{10}B isotope absorbs neutrons. The mechanical properties of OsB_2 were described as ultraincompressible but not superhard, again with special metal–boron interactions between the osmium atom and boron atoms [29].

Several variations of the three structure types already discussed do occur in metal borides with boron to metal ratios close to two. Thus, a molybdenum diboride modification (Mo_2B_4) formally described as "Mo_2B_5" was found to exhibit a structure with chair-like and planar B_6 rings (Figure 10.10a) [30].

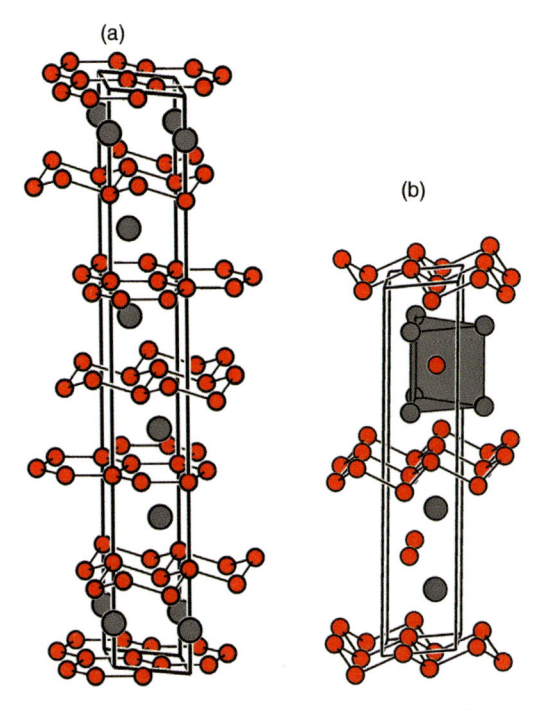

Figure 10.10 Mo_2B_4, space group, $R\text{-}3m$ (No. 166), formerly described as Mo_2B_5 (a), and Ru_2B_3 and Os_2B_3, space group, $P6_3/mmc$ (No. 194) (b).

Neutron diffraction data had to be collected to allow for an explicit assignment of a correct formula. The scattering power of the light element atoms is insufficient to determine the location and occupancy of the boron atom sites unambiguously next to tungsten or osmium atoms, if X-ray diffraction data is used. The W analogue is related.

Another variation of the stacking of planar and buckled boron atom layers is found for Ru_2B_3 and Os_2B_3 (Figure 10.10b) [31].

10.4
Boron-Rich Solids

10.4.1
General

The structures of boron and many boron-rich compounds are characterized by subunits of triangular-defined polyhedra, the so-called deltahedra (from the Greek letter Δ), for example, octahedra or icosahedra. In the case of closed polyhedra, one talks of *closo*-clusters, a nomenclature which makes reference to the Wade classification used for molecules. These polyhedra – mostly octahedra, icosahedra, or cube-octahedra – are interconnected either directly or through additional B–B chains.

Often, but not always it is these polyhedra that are the most striking structural entities in boron-rich solids. It should be emphasized that the B–B distances within these polyhedra are usually longer than the ones in between the polyhedra, thus indicating a weaker intrapolyhedral bonding situation compared to the interpolyhedral interaction [32]. Figure 10.11 shows polyhedra that occur in binary borides (B_6: octahedron, B_7: pentagonal bipyramid, B_8: dodecahedron, B_{12}: icosahedron).

10.4.2
Octahedra and Bipyramids in Borides: MB_4, MB_6, and Na_3B_{20}

The aforementioned ThB_4 (UB_4) structure type contains B_6 octahedra and B–B linkages between them, resulting in heptagone ring as an additional motif

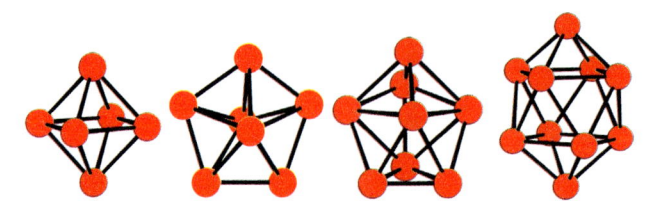

Figure 10.11 Polyhedra in boron-rich borides.

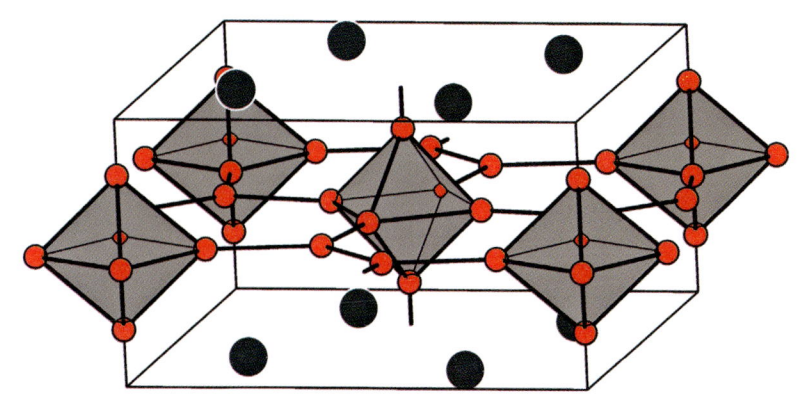

Figure 10.12 Structure type of UB$_4$, space group *P4/mbm* (No. 127).

(Figure 10.12) [33]. It exists with M = Ce, Gd, U, Th, and La, and a carbon-stabilized variant has been found for Ca.

ErB$_4$, TbB$_4$, and HoB$_4$ have strong anisotropic magnetic properties. Magnetic structures of MB$_4$ variants have been studied by neutron diffraction, and ErB$_4$ and DyB$_4$ were found to order antiferromagnetically with Néel temperatures of 16 and 21 K, respectively [34]. They crystallize in the UB$_4$ structure type as explained already, with the orthorhombic Shubnikov space group *Pb'am*.

The structure type of MB$_6$ is very well known since 1932 [35a]. Hexaborides crystallize (example EuB$_6$ in Figure 10.13a) in the cubic crystal system (space group *Pm-3m*, No. 221). Within this structure, octahedra are three-dimensionally interconnected as shown in Figure 10.13(b). Such borides have been

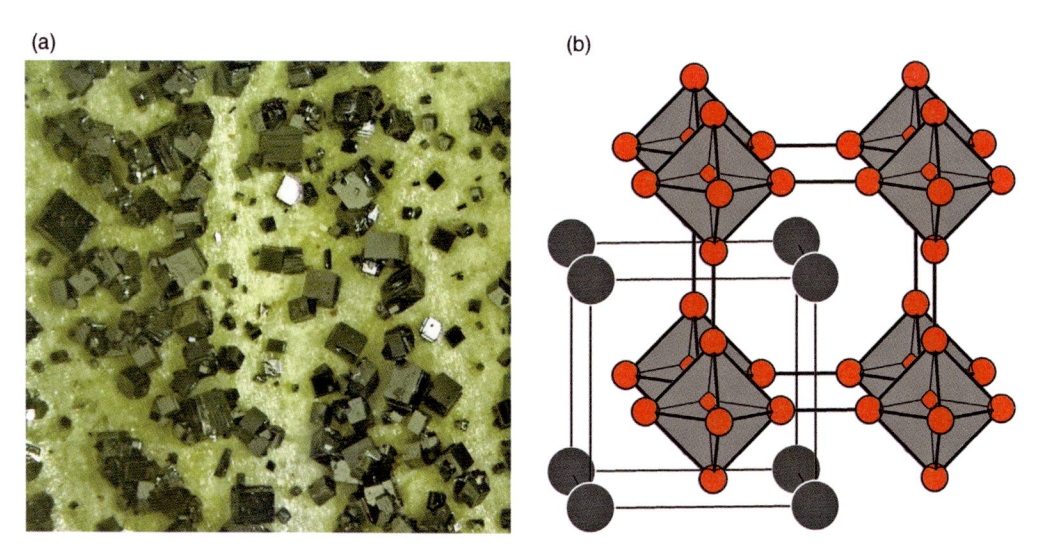

Figure 10.13 Crystals of EuB$_6$ (a) and structure type of CaB$_6$ (b).

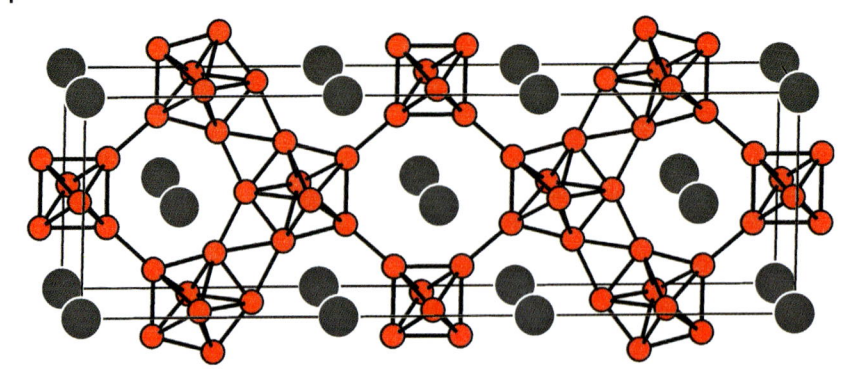

Figure 10.14 Structure of Na_3B_{20}, space group *Cmmm* (No. 65).

obtained with K, Ca, Sr, Ba, Th, and many lanthanides, and Li_2B_6 represents a slightly distorted variant [35b].

The stability, hardness, and good Seebeck coefficients of hexaborides make them interesting materials for thermoelectric applications [8b]. Further applications are those of lanthanum hexaboride that is a thermoionic emitter with a very low work function that can be used as an electrode in transmission electron microscopes, and especially for electron beam lithography. The properties of SmB_6 were described in 1959 by Samsonov *et al.* [35c] Recently, this compound was discussed to be a topological insulator [35d].

Na_3B_{20} is the only boride that contains pentagonal bipyramids and octahedra (Figure 10.14) [1,36,37]. Originally, it was believed to crystallize with MB_6-type structure, but X-ray and neutron diffraction proved the presence of extra boron atoms.

10.4.3
Dodecaborides and Borides with Icosahedra

Cube-octahedral entities are found in boron-rich borides such as YB_{12} and ZrB_{12} (UB_{12} structure type, Figure 10.15, $a \sim 740$ pm); [38] and further boron-rich borides show icosahedral entities, for example, Na_2B_{29} (Figure 10.16) [39]. Another binary boride containing B_{12} units is WB_4, which was also found to be very hard and incompressible [4b].

The UB_{12} structure was first proposed by Bertaut and Blum [38]. The cube-octahedral entity of boron atoms forms instead of the icosahedron when M is a large atomic number element.

As already mentioned, there are numerous examples of the incorporation of metal atoms in β-B, a process in which the β-B structure is largely preserved. This is caused by the electron deficit in β-B that has to be compensated in only partially occupied sites between the linked polyhedra in pure boron by interstitial boron atoms. Structures with 15–16 boron atom positions have been described. In most cases these phases – as far as they have been discovered and

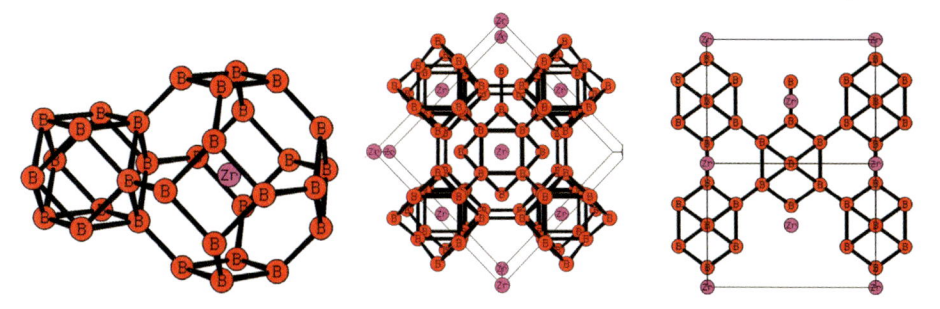

Figure 10.15 Structure of UB$_{12}$ with boron atom cube-octahedra, space group *Fm-3m* (No. 223).

described – are transition metal borides (Cr, Cu, Fe, Mn, Ni, Sc, V, Zn), but β-rhombohedral boron-type borides of main group metals have been found, too (Li, Na, Mg, Al).

10.4.4
Carbon-Like Frameworks of Boron Atoms in Borides

For boron-rich borides, the most unusual structure type is that of CrB$_4$, MnB$_4$, and FeB$_4$. The manganese and chromium variants were synthesized and

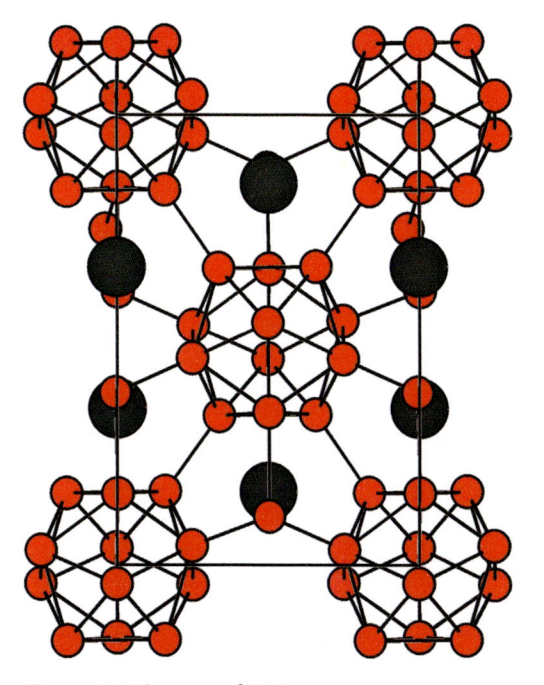

Figure 10.16 Structure of Na$_2$B$_{29}$, space group, *Im* (No. 8).

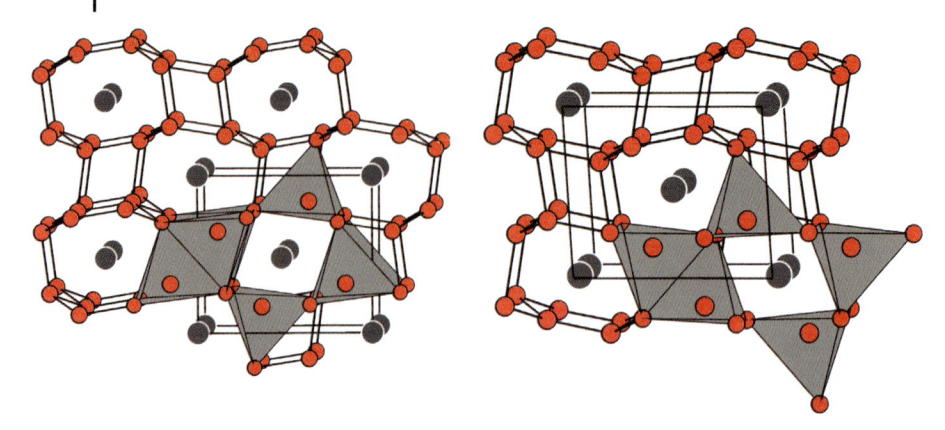

Figure 10.17 Structure of CrB$_4$ (a), space group, *Pnnm* (No. 58) and MnB$_4$ (b), space group, *P*2$_1$/*c* (No. 14). (Reproduced with permission from Ref. [13]. Copyright 1998, Elsevier.)

structurally described as early as 1968 [40], and since then theoretically investigated many times due to their interesting electronic situation and possible (super)hardness [41]. Their syntheses, correct structures (Figure 10.17), and properties were described recently [16,42–46]. The boron atom framework can be considered to be an electron-deficient variant of the hypothetical tetragonal diamond, and the boron atoms are indeed coordinated almost tetrahedrally by four more boron atoms, reminding of sp^3-hybridized carbon atoms. In addition to that, MnB$_4$ exhibits close Mn–Mn contacts indicating a metal–metal multibond caused by a Peierls-type distortion.

The iron variant, FeB$_4$, was first predicted to be stable and superconducting, and then synthesized at high pressures and found to become superconducting at low temperatures [46].

10.5
Higher Borides

A vast number of higher, multinary borides have been described [2,47,48]. Interesting examples – double perovskite-like Ti$_2$Rh$_6$B and Ti$_7$Rh$_4$Ir$_2$B$_8$ with a very unusual planar, isolated B$_6$-ring were published by Fokwa *et al.* [48,49].

10.6
Conclusion

The synthesis and crystallization of pure borides is challenging, but their bonding situations and crystal structures are fascinating and highly versatile. The current state of knowledge shows very interesting chemical and physical properties of borides. But many questions of boride chemistry are still open, and the future research will certainly prove borides a class of compounds interesting not only for basic science but also for high-end applications.

References

1 Albert, B. and Hillebrecht, H. (2009) *Angew. Chem., Int. Ed.*, **48**, 8640–8668.
2 Fokwa, B.P.T. (2014) *Encyclopedia of Inorganic and Bioinorganic Chemistry*, John Wiley & Sons, Inc., 1–14.
3 Matkovich, V.I. (1977) *Boron and Refractory Borides*, Springer, Berlin.
4 (a) Aronsson, B., Lundström, T. and Rundqvist, S. (1965) *Borides, Silicides and Phosphides*, Methuen & Co Ltd, London; (b) Yeung, M.T., Mohammadi, R. and Kaner, R.B. (2016) *Annu. Rev. Mater. Res.*, **46**, 465–485.
5 (a) Nagamatsu, J., Nakagawa, N., Muranaka, T., Zentani, Y., and Akimitsu, J. (2001) *Nature*, **410**, 63–64; (b) Del Rosso, A. (2014 *CERN document server*, April 14, 2014.
6 Herbst, J.F., Lee, R.W., and Pinkerton, F.E. (1986) *Annu. Rev. Mater. Sci.*, **16**, 467–458.
7 (a) Fries, M., Gercsi, Z., Ener, S., Skokov, K.P., and Gutfleisch, O. (2016) *Acta Mater.*, **113**, 213e220; (b) Klemenz, S., Fries, M., Dürrschnabel, M., Skokov, K., Kleebe, H.-J., Gutfleisch, O., and Albert, B., *submitted*.
8 (a) Mori, T. and Nishimura, T. (2006) *J. Solid State Sci.*, **179**, 2908–2915; (b) Gürsoy, M., Takeda, M., and Albert, B. (2015) *J. Solid State Chem.*, **221**, 191–195; (c) Gürsoy, M., Hempel, S., Reitz, A., Hofmann, K., and Albert, B. (2014) *Z. Anorg. Allg.Chem.*, **640**, 2714–2716.
9 Schlesinger, H.I., Brown, H.C., Finholt, A.E., Gilbreath, J.R., Hockstra, H.R., and Hyde, E.K. (1953) *J. Am. Chem. Soc.*, **75**, 215–219.
10 Greenwood, N.N. and Earnshaw, A. (1988) *Chemie der Elemente*, 1st edn, Wiley-VCH Verlag GmbH, Weinheim.
11 Janiak, C., Meyer, H.-J., Gudat, D., and Alsfasser, R. (2012) *Riedel Moderne Anorganische Chemie*, 4th edn, De Gruyter, Berlin.
12 Schwarzkopf, P. and Kieffer, R. (1953) *Refractory Hard Metals*, Macmillan, New York.
13 Otani, S., Kosrukova, M.M., and Mitsuhashi, T. (1998) *J. Cryst. Growth*, **186**, 582–586.
14 (a) Bykova, E., Tsirlin, A.A., Gou, H., Dubrovinsky, L., and Dubrovinskaia, N. (2014) *J. Alloys Comp.*, **608**, 69–72; (b) Gou, H., Dubrovinskaia, N., Bykova, E., Tsirilin, A., Kasinathan, D., Schnelle, W., Richter, A., Merlini, M., Hanfland, M., Abakumov, A.M., Batuk, D., Van Tendeloo, G., Nakajima, Y., Kolmogorov, A., and Dobrovinsky, L. (2013) *Phys. Rev. Lett.*, **111** 157002.
15 (a) Higashi, I., Takahashi, Y. and Atoda, T. (1977) *J. Cryst. Growth*, **33**, 207–211; (b) Sinel'nikova, V.S., Gurin, V.N., Pilyankevich, A.N., Strashinskaya, L.V., and Korsukova, M.M. (1976) *J. Less-Common Met.*, **46**, 265–272; (c) Vojteer, N., Schroeder, M., Röhr, C., and Hillebrecht, H. (2008) *Chem. Eur. J.*, **14**, 7331–7342; (d) Amsler, B., Fisk, Z., Sarrao, J.L., Molnar, S.von., Meisel, M.W., and Sharifi, F. (1998) *Phys. Rev. B* **57** 8747–8750.
16 (a) Knappschneider, A., Litterscheid, C., Kurzman, J., Seshadri, R., and Albert, B. (2011) *Inorg. Chem.*, **50**, 10540–10542; (b) Knappschneider, A., Litterscheid, C., George, N.C., Brgoch, J., Wagner, N., Beck, J., Kurzman, J.A., Seshadri, R., and Albert, B. (2014) *Angew. Chem., Int. Ed.*, **126**, 1710–1714.
17 Carenco, S., Portehault, D., Boissière, C., Mézailles, N., and Sanchez, C. (2013) *Chem. Rev.*, **113** (10), 7981–8065.
18 (a) Glavee, G.N., Klabunde, K.J., Sorensen, C.M., and Hadjipanayis, G.C. (1993) *Inorg. Chem.*, **32**, 474–477; (b) Kapfenberger, C., Hofmann, K. and Albert, B. (2003) *Solid*

State Sci., **5**, 925–930; (c) Hofmann, K., Kalyon, N., Kapfenberger, C., Lamontagne, L., Zarrini, S., Berger, R., Seshadri, R., and Albert, B. (2015) *Inorg. Chem.*, **54**, 10873–10877; (d) Rades, S., Kornowski, A., Weller, H., and Albert, B. (2011) *ChemPhysChem*, **12**, 1756–60; (e) Rades, S., Kraemer, S., Seshadri, R., and Albert, B. (2014) *Chem. Mater.*, **26** 1549–1552.

19 (a) Portehault, D., Devi, S., Beaunier, P., Gervais, C., Giordano, C., Sanchez, C., and Antonietti, M. (2011) *Angew. Chem., Int. Ed.*, **50**, 3262–3265; (b) Terlan, B., Levin, A.A., Börnert, F., Simon, F., Oschatz, M., Schmidt, M., Cardoso-Gil, R., Lorenz, T., Baburin, I.A., Joswig, J.-O., and Eychmüller, A. (2015) *Chem. Mater.*, **27**, 5106–5115; (c) Terlan, B., Levin, A.A., Börrnert, F., Zeisner, J., Kataev, V., Schmidt, M., and Eychmüller, A. (2016) *Eur. Inorg. Chem.*, 3460–3468.

20 Hägg, G. (1931) *Z. Phys. Chem.*, **B12**, 33.

21 Greenwood, N.N., Parish, R.V., and Thornton, P. (1966) *Chem. Soc. Q. Rev.*, **20**, 441–464.

22 (a) Kiessling, R. (1949) *Acta Chem. Scand.*, **3**, 603–615; (b) Bolmgren, H., Lundström, T., Tergenius, L.-E., Okada, S., and Higashi, I. (1990) *J. Less-Comm. Met.*, **161**, 341–345.

23 Kapfenberger, C., Albert, B., Pöttgen, R., and Huppertz, H. (2006) *Z. für Kristallogr.*, **221**, 477–481.

24 Kotzott, D., Ade, M., and Hillebrecht, H. (2009) *J. Solid State Chem.*, **182**, 538–546.

25 (a) Hofmann, W. and Jäniche, W. (1935) *Naturwiss*, **23**, 851; (b) Xie, Z., Graule, M., Orlovskaya, N., Payzant, E.A., Cullen, D.A., and Blair, R.G. (2014) *J. Solid State Chem.*, **215**, 16–21.

26 (a) Licht, S., Wu, H., Yu, X., and Wang, Y. (2008) *Chem. Commun.*, 3257–3259; (b) Stuart, J., Hohenadel, A., Li, X., Xiao, H., Parkey, J., Rhodes, C.P., and Licht, S. (2015) *J. Electrochem. Soc.*, **162**, A192–A197.

27 Cumberland, R.W., Weinberger, M.B., Gilman, J.J., Clark, S.M., Tolbert, S.H., and Kaner, R.B. (2005) *J. Am. Chem. Soc.*, **127**, 7264–7265.

28 Frotscher, M., Hölzl, M., and Albert, B. (2010) *Z. Anorg. Allg. Chem.*, **636**, 1783.

29 (a) Chung, H.-Y., Weinberger, M.B., Levine, J.B., Kavner, A., Yang, J.-M., Tolbert, S.H., and Kaner, R.B. (2007) *Science*, **316**, 436–439; (b) Kaner, R.B., Gilman, J.J. and Tolbert, S.H. (2005) *Science*, **308**, 1268–1269.

30 (a) Froscher, M., Klein, W., Bauer, J., Fang, C., Halet, J.-F., Senyshyn, A., Baehtz, C., and Albert, B. (2007) *Z. Anorg. Allg. Chem.*, **633**, 2626; (b) Kayhan, M., Hildebrand, E., Frotscher, M., Senyshyn, A., Hofmann, K., Alff, L., and Albert, B. (2012) *Solid State Sci.*, **14**, 1656.

31 Frotscher, M., Senyshyn, A., and Albert, B. (2012) *Z. Anorg. Allg. Chem.*, **638**, 2978.

32 Albert, B. (2001) *Eur. J. Inorg. Chem.*, 1679.

33 Zalkin, A. and Templeton, D.H. (1950) *J. Chem. Phys.*, **18**, 391–392.

34 Gianduzzo, J.C., Georges, R., Chevalier, B., Etourneau, J., Hagenmuller, P., Will, G., and Schäfer, W. (1981) *J. Less-Comm. Met.*, **82**, 29–35.

35 (a) von Stackelberg, M. and Neumann, F. (1932) *Z. Phys. Chem.*, **19**, 314–320; (b) von Schnering, H.G., Mair, G., Woerle, M., Nesper, R., and Anorg, Z. (1999) *Allg. Chem.*, **625**, 1207–1211; (c) Samsonov, G.V., Zhuravlev, N.N., Paderno, Yu.B., and Melik-Adamyan, V.R. (1959) *Kristallografiya*, **4**, 538–541 (english translation: *Sov. Phys. Crystallogr.* 1959, 4, 507–509); (d) Kim, D.J., Xia, J. and Fisk, Z. (2014) *Nat. Mater.*, **13**, 466–470.

36 Albert, B. (1998) *Angew. Chem., Int. Ed.*, **37**, 1117.

37 Albert, B. and Hofmann, K. (1999) *Z. Anorg. Allg. Chem.*, **625**, 709.

38 Bertaut, F. and Blum, P. (1949) *C. R. Acad. Sci. Paris*, **229**, 666–667.

39 Albert, B., Hofmann, K., Fild, C., Eckert, H., Schleifer, M., and Gruehn, R. (2000) *Chem. Eur. J.*, **6**, 2531–2536.

40 Andersson, S. and Lundstroem, T. (1968) *Acta Chem. Scand.*, **22**, 3103–3110.

41 Burdett, J.K. and Canadell, E. (1988) *Inorg. Chem.*, **27**, 4437–4444.

42 Knappschneider, A., Litterscheid, C., Dzivenko, D., Kurzman, J., Seshadri, R., Wagner, N., Beck, J., Riedel, R., and Albert, B. (2013) *Inorg. Chem.*, **52**, 540–542.

43 Niu, H., Wang, J., Chen, X.-Q., Li, D., Li, Y., Lazar, P., Podloucky, R., and Kolmogorov, A.N. (2012) *Phys. Rev. B*, **85**, 144116.

44 Knappschneider, A., Litterscheid, C., Brgoch, J., George, N.C., Henke, S., Cheetham, A.K., Hu, J.G., Seshadri, R., and Albert, B. (2015) *Chem. Eur. J.*, **21**, 8177–8181.

45 Gou, H., Tsirlin, A.A., Bykova, E., Abakumov, A.M., Tendeloo, G.Van., Richter, A., Ovsyannikov, S.V., Kurnosov, A.V., Trots, D.M., Konôpková, Z., Liermann, H.-P., Dubrovinsky, L., and Dubrovinskaia, N. (2014) *Phys. Rev. B*, **89**, 064108.

46 Gou, H., Dubrovinskaia, N., Bykova, E., Tsirilin, A., Kasinathan, D., Schnelle, W., Richter, A., Merlini, M., Hanfland, M., Abakumov, A.M., Batuk, D., Van Tendeloo, G., Nakajima, Y., Kolmogorov, A., and Dobrovinsky, L. (2013) *Phys. Rev. Lett.*, **111** 157002.

47 Mori, T. (2008) *Handbook on the Physics and Chemistry of Rare-Earths*, vol. **38** (eds K.A. Gschneider Jr, J.-C. Bunzli, and V. Pecharsky), North-Holland, Amsterdam.

48 Fokwa, B.P.T. (2010) *Eur J. Inorg. Chem.*, 3075.

49 Fokwa, B.P.T. and Hermus, M. (2012) *Angew. Chem., Int. Ed.*, **51**, 1702–1705.

11
Metal Pnictides: Structures and Thermoelectric Properties

Abdeljalil Assoud and Holger Kleinke

University of Waterloo, Institute for Nanotechnology, Department of Chemistry, 200 University Avenue West, Waterloo, ON N2L 3G1, Canada

11.1
Introduction

The elements of group 15 of the periodic table, namely, nitrogen, phosphorus, arsenic, antimony, and bismuth, are called pnictogens (Pn) since the 1950s. This is derived from the Ancient Greek word "to choke" because of the suffocating character of nitrogen gas. Pnictides are compounds containing a pnictogen as the anion, that is, nitrides, phosphides, arsenides, antimonides, and bismuthides. This includes the so-called III–V semiconductors, which are composed of the elements of groups 13 (III) and 15 (V), such as GaAs. In 2008, a whole book was devoted to III–V semiconductors, focusing on nitrides [1]. Several reviews exist on pnictides, which are all focused on different aspects of this large field of high complexity and versatility. An earlier review from the year 2000 was centered around the rich structural chemistry of metal-rich antimonides [2], and one from the year 2002 described the various extended Sb atom motifs in metal-poor antimonides [3]. Cameron *et al.* presented a review on pnictides with emphasis on hydrogen storage, anodes for lithium batteries, superconductivity, and thermoelectrics in 2011 [4]. The journal *Coordination Chemistry Reviews* published a special issue about "Chemistry and applications of metal nitrides" in 2013 that includes reviews on photocatalysis [5], heterogeneous catalysis [6], and lithium nitrides for Li-ion conduction as well as hydrogen storage [7]. Also in 2013, Johrendt *et al.* [8] reviewed transition metal pnictides with a focus on their crystal chemistry. In 2014, Greiwe and Nilges focused on polyantimonides as the active material in anodes for lithium batteries [9].

Evidently, pnictides are a very large family with over 5000 pnictides listed in Pearson's Handbook [10]. They are currently used in many different applications, as shown in the above-mentioned reviews, so we can only discuss selected materials here. This chapter thus presents an overview of the huge variety of pnictides that excel as thermoelectric (TE) materials.

Handbook of Solid State Chemistry, First Edition. Edited by Richard Dronskowski, Shinichi Kikkawa, and Andreas Stein.
© 2017 Wiley-VCH Verlag GmbH & Co. KGaA. Published 2017 by Wiley-VCH Verlag GmbH & Co. KGaA.

Thermoelectrics can convert a temperature gradient into electricity (Seebeck effect) as well as create a temperature gradient from electricity (Peltier effect) [11,12]. The former effect is used in spacecrafts such as Voyager since decades [13], and is currently under investigation to create electricity from the waste heat in automotives to ultimately enhance gas mileage by reducing the load on the alternator [14–16].

The efficiency of the TE materials is assessed by their figure of merit, $zT = T\alpha^2 \sigma \kappa^{-1}$. Therein, α is the Seebeck coefficient, σ is the electrical conductivity, κ is the thermal conductivity, and T is the average temperature. The best TE materials are heavily doped semiconductors comprising heavy elements such as Bi_2Te_3 for room temperature applications and PbTe for high temperature applications. A thermoelectric generator comprises several modules, in which n- and p-doped semiconducting legs are alternating, so that the electrons of the n-type legs move in the same direction as the holes of the p-type legs. The ZT of the device is a function of the zT values of all materials involved, that is, both on the n- and on the p-side. Equation (11.1) shows how the power generation conversion efficiency η depends on the device's average figure of merit ZT (with T_H = hot temperature, T_C = cold temperature) [11,12]:

$$\eta = \frac{T_H - T_C}{T_H} \frac{\sqrt{1 + \overline{ZT}} - 1}{\sqrt{1 + \overline{ZT}} + \frac{T_C}{T_H}}. \tag{11.1}$$

State-of-the-art materials may exhibit zT values in excess of 1. Notable, rare exceptions with $zT_{max} > 2$ include thin films made of superlattices of Sb_2Te_3/Bi_2Te_3, reported to achieve $zT_{max} = 2.4$ at room temperature [17], $AgPb_{18}SbTe_{20}$ with its nanodomains reaching $zT_{max} = 2.2$ at 800 K [18], and a composite of Na-doped PbTe with SrTe nanocrystals with $zT_{max} = 2.2$ at 915 K [19]. Assuming an average $ZT = 0.5$ for the whole temperature range, and $T_H = 800$ K and $T_C = 400$ K as attainable in the car exhaust, an efficiency of $\eta = 6.5\%$ could theoretically be reached. Doubling ZT would result in a significantly higher, though not doubled, $\eta = 10.8\%$, while doubling T_H – which may be impractical – would have a slightly larger impact, yielding $\eta = 11.4\%$. The materials under investigation and in current use are dominated by antimonides, the topic of this chapter, and tellurides, the most prominent examples being Bi_2Te_3 and PbTe. Compared to antimonides, nitrides and phosphides are more ionic, detrimental to the mobility of the charge carriers, and also are composed of lighter elements, disadvantageous for the required low thermal conductivity of advanced thermoelectrics. While the arsenides are intermediate, research into their properties is scarce because of the higher toxicity of arsenic. Bismuthides on the other hand tend to be metallic, and thus have a too small Seebeck coefficient to be considered as thermoelectrics. In this chapter, we therefore focus on antimonides, specifically, on bulk materials, while noting that they may very well experience further improvements via nanostructuring [12,20–24].

11.2
Binary Antimonides

11.2.1
Main Group Antimonides

The alkali metal antimonides, although numerous and mostly semiconducting, were rarely considered as thermoelectric materials, most likely because of their moisture and air sensitivity and chemical reactivity. On the other hand, the materials A_3Sb (A = alkali metal) such as Cs_3Sb, Na_2KSb, and Na_3Sb, are being used as photocathodes (in vacuum) because of their high photoemissivity since decades [25]. Among the alkaline earth metal antimonides, Mg_3Sb_2 stands out with good thermoelectric properties, once properly doped. Below 1200 K, α-Mg_3Sb_2 adopts the La_2O_3 type, and its crystal structure consists of alternating layers of Mg^{2+} cations (Mg1) and $Mg_2Sb_2^{2-}$ layers (formed by Mg2) [26], reminiscent of $CaAl_2Si_2$ [27]. An alternative description is based on layers of $Mg1Sb_6$ octahedra and $Mg2Sb_4$ tetrahedra alternating along the c-axis (Figure 11.1).

Earlier characterization attempts of Mg_3Sb_2 were mitigated by difficulties to minimize MgO formation in the grain boundaries [28]. Through mechanical milling, followed by spark plasma sintering of the series $Mg_3Sb_{2-x}Bi_x$ (with $0 \leq x \leq 0.4$), oxygen contamination was minimized and much better thermoelectric properties obtained. This family is bestowed with a low thermal conductivity $\kappa < 1.4$ W m^{-1}K^{-1} above 300 K for all values of x investigated. This is somewhat surprising, giving the relatively simple crystal structure and the large amount of the light element Mg, as well as considering earlier

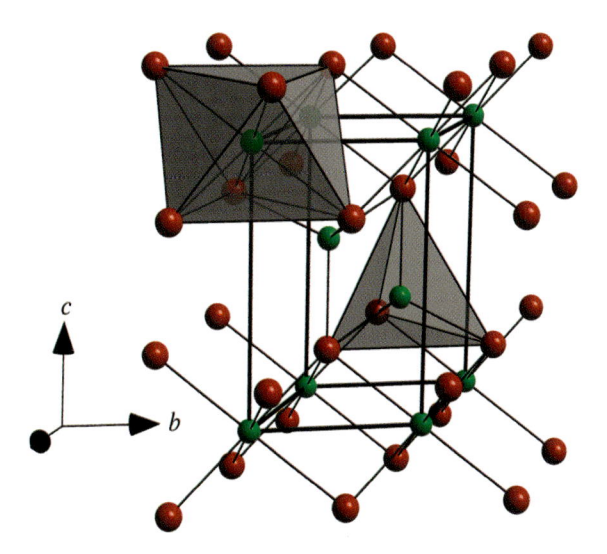

Figure 11.1 Crystal structure of α-Mg_3Sb_2. Green: Mg; red: Sb. Emphasized is one $Mg1Sb_6$ octahedron and one $Mg2Sb_4$ tetrahedron.

findings of significantly higher thermal conductivity, namely, $\kappa > 4\,\mathrm{W\,m^{-1}K^{-1}}$ around room temperature [28].

The thermal conductivity κ is composed of the sum of the lattice contribution κ_L and the electronic contribution κ_e: $\kappa = \kappa_L + \kappa_e$. κ_e is related to the electrical conductivity σ via the Wiedemann–Franz law, $\kappa_e = L_0 \sigma T$, with $L_0 =$ Lorenz number. It is constructive to compare the lattice thermal conductivity values in order to relate the structure to the thermal conductivity, as the carrier concentration and thus the electrical conductivity differ from sample to sample. Specifically, in the $Mg_3Sb_{2-x}Bi_x$ series, the carrier concentration increases with increasing Bi amount, although Sb and Bi are isovalent. As a consequence, the sample with the largest Bi content, that is, $x = 0.4$, exhibits the highest electrical conductivity, but the lowest thermal conductivity. Therefore, its lattice thermal conductivity is also the lowest in this series, likely a consequence of increasing mass fluctuation as well as increasing total mass. However, the $x = 0.4$ sample exhibits the smallest Seebeck coefficient of the series, and thus not the largest figure of merit, zT, the maximum was found for $Mg_3Sb_{1.8}Bi_{0.2}$ with $zT_{max} = 0.6$ at 750 K [29]. Doping with Pb caused a larger enhancement of the electrical conductivity without a significant decrease of the Seebeck coefficient, resulting in an enhanced $zT_{max} = 0.84$ at 773 K in case of $Mg_3Sb_{1.8}Pb_{0.2}$ [30].

The group 13 antimonides are dominated by the III–V semiconductors like GaSb and InSb. InSb is used in infrared detectors since decades [31], but its thermoelectric properties are not well explored. While it is a narrow-gap semiconductor (gap<0.2 eV), it adopts the very simple zinc blende structure type, and thus exhibits a relatively high thermal conductivity of $\kappa > 15\,\mathrm{W\,m^{-1}K^{-1}}$, and its zT was reported to reach only 0.23 at 723 K after optimizing via codoping with Zn and Si [32].

The only thermodynamically stable group 14 antimonide is SnSb with a large homogeneity range, adopting an incommensurate structure [33]. This material's thermoelectric properties have not been investigated yet to our knowledge, but it is presumably metallic and thus a poor thermoelectric. For group 15, the Bi–Sb alloy – though strictly speaking not an antimonide – deserves special mention here. While bismuth and antimony are semimetals, their alloys are semiconducting with gaps <0.03 eV for compositions between 6 and 20% Sb due to small energy shifts of the bands around the Fermi level. With such a small bandgap, these alloys are best suited for applications below room temperature [34]. Both bismuth and antimony crystallize in the structure-type of gray arsenic, where layers of three bonded Pn atoms are connected to a three-dimensional structure via weaker bonds. A first report from 1962 showed good thermoelectric properties below room temperature with the current flow perpendicular to these layers; for example, the (undoped) $Bi_{95}Sb_5$ alloy maintained a $zT \approx 0.52$ between 100 and 300 K [35]. During the next 30 years, no significant improvements were made [36]. Te doping was shown to enhance zT to 0.6 at 100 K that increased up to $zT = 1.35$ in a magnetic field of 0.5 T at 165 K [37]. Finally, the main groups 16 and 17 do not form any antimonides because Sb is the cation in these cases as in Sb_2Se_3 or $SbCl_3$.

11.2.2
Early Transition Metal Antimonides

To our knowledge, the antimonides (and bismuthides) of the groups 3–5 (and of the lanthanides) are all metallic, because the oxidation potential of Sb is not high enough to completely vacate the d-orbitals [38]. This is even true for LaSb adopting the NaCl type, although it is sometimes written as La^{3+}Sb^{3-}, because the La d-states slightly overlap with the Sb p-states [39]. TiSb$_2$ adopts the CuAl$_2$ structure with chains of square TiSb$_8$ antiprisms forming Sb$_2$ dumbbells with an Sb–Sb distance of 2.85 Å, as shown in Figure 11.2a [40]. The next shortest Sb–Sb distance is 3.44 Å that may correspond to van der Waals interactions. Sb–Sb distances <2.9 Å point toward regular two-electron – two-center (2e–2c) (single) bonds, as found in, for example, KSb. Here, each Sb atom participates in two bonds of 2.83 and 2.85 Å with neighboring Sb atoms, resulting in an infinite zig-zag Sb atom chain [41]. In accord with the Zintl concept [42], one can assign six valence electrons to each Sb atom, thus forming Sb$^-$ chains in K$^+$Sb$^-$, isovalent to elemental tellurium that contains similar (helical) Te atom chains. Using the same counting rules, the Sb$_2$ dumbbell is isoelectronic with Te$_2^{2-}$ and I$_2$, and thus an Sb$_2^{4-}$ unit. This would require Ti^{4+} and thus empty d-orbitals, but Ti participates in Ti—Ti bonds across the Sb$_4$ squares of the TiSb$_8$ chains, and the material is metallic [43]. Metallic character was also observed in the most Sb-rich antimonides of the heavier homologues, ZrSb$_2$ and HfSb$_2$ [44–46], as well as the unique Hf$_5$Sb$_9$ [47,48].

Among the group 5 antimonides, the most Sb-rich ones are VSb$_2$ [43], NbSb$_2$ [49], and TaSb$_2$ [50], again all with metallic character. The former is isostructural with TiSb$_2$, and the two latter ones are crystallizing in the OsGe$_2$ structure type. Two crystallographically different Sb atoms occur in NbSb$_2$, one with the shortest Sb–Sb distances of 3.30 Å, the second with a short contact of

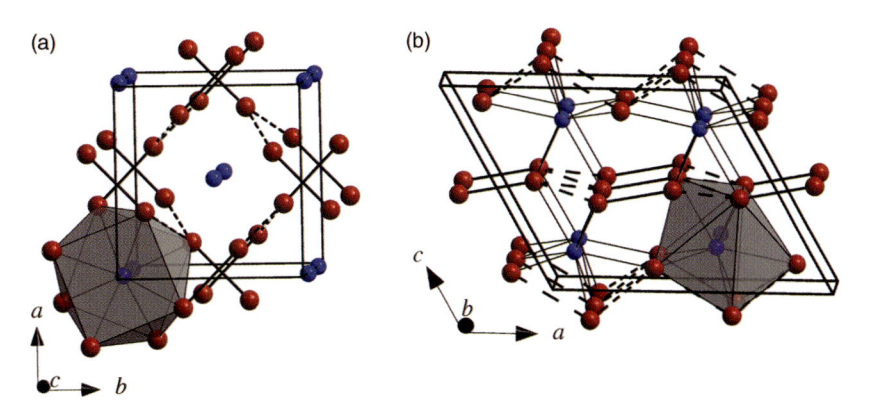

Figure 11.2 Crystal structures of (a) TiSb$_2$ and (b) NbSb$_2$. Blue: Ti/Nb; red: Sb. Emphasized are one TiSb$_8$ square antiprism and one NbSb$_8$ bicapped trigonal prism. Dashed lines denote the longer Sb–Sb distances of 3.0–3.4 Å.

2.77 Å and two longer ones of 3.04 Å (Figure 11.2b). These longish Sb—Sb bonds on the order of 3.0–3.2 Å are indicative of hypervalent 4e–3c (half) bonds [51], the simplest example being in the linear Sb_7^{3-} found in $Ca_{14}AlSb_{11}$ [52], isoelectronic with Se_3^{4-} [53] and I_3^- [54]. Treating two of these half bonds as one single bond with respect to the electron count [55], the two Sb atoms of $NbSb_2$ are best described as Sb^{3-} and Sb^-, respectively, leaving Nb in the $4d^1$ configuration ($Nb^{4+}Sb^-Sb^{3-}$). This is supported by the Nb—Nb bond between neighboring bicapped $NbSb_8$ trigonal prisms, as well as a deep local minimum in the density of states at the Fermi level in $NbSb_2$. We have succeeded in creating a gap by replacing the Nb pair with one atom of group 4 and one atom of group 6, realized with $Hf_{0.5}Mo_{0.5}Sb_2 \equiv HfMoSb_4$ [56] and $TiMoSb_4$ [38]. With negligible bandgap sizes, these materials exhibit too small Seebeck coefficient values to be useful as thermoelectrics.

Moving on to group 6, we find the first semiconducting transition metal antimonide in $CrSb_2$ [57], whereas CrSb adopting the zinc blend type is a halfmetal [58]. $CrSb_2$ crystallizes in the marcasite type, comprising Sb pairs (Sb—Sb distance 2.85 Å) and two longer contacts of 3.27 Å per Sb atom. Its thermoelectric properties are rather poor, in part due to a high thermal conductivity $\kappa > 6\,W\,m^{-1}K^{-1}$, and can only be slightly improved with Te doping to reach $zT = 0.014$ at 310 K [59].

While tungsten is not known to form any binary antimonide, Mo_3Sb_7 (Ir_3Ge_7 type) is the only binary antimonide of molybdenum. Mo_3Sb_7 is metallic [60], but exhibits a bandgap above the Fermi level in contrast to the antimonides already discussed. Adding two valence electrons per formula unit via doping with tellurium results in semiconducting properties realized in $Mo_3Sb_5Te_2$ with $3 \times 6 + 5 \times 5 + 2 \times 6 = 55$ valence electrons per formula unit [61], which can be understood based on the Zintl concept [62].

Like Ti in $TiSb_2$, the Mo atoms are coordinated by eight Sb atoms in form of a square antiprism (Figure 11.3). Two such antiprisms share a common square with long Sb1—Sb1 distance of 3.4 Å, resulting in an Mo—Mo bond of 3 Å across the $Sb1_4$ square. On the outsides of the $MoSb_8$ antiprism pair, the $Sb2_4$ square is connected to the corresponding $Sb2_4$ square of the next antiprism pair via Sb2—Sb2 contact of 3.1 Å. This formation results in an empty cube of Sb2 atoms whose faces are all capped by Mo atoms. In addition, neighboring cubes are interconnected via shorter Sb2—Sb2 bonds of 2.9 Å. The Te atoms of $Mo_3Sb_{7-x}Te_x$ prefer the sites without Sb—Sb bonds, namely, the Sb1 sites [63] like As in the isostructural $Re_3Ge_xAs_{7-x}$ [64].

Thus far, the best thermoelectric properties were found for p-type $Mo_3Sb_{7-x}Te_x$ when $x = 1.6$, culminating in $zT = 0.76$ at 1050 K [65]. n-type variants have not been prepared to date. Adding nickel atoms, presumed to go into the cubic holes, yielded a slightly higher $zT = 0.93$ in case of the nominal composition $Ni_{0.06}Mo_3Sb_{5.4}Te_{1.6}$ [66]. Instead of a substitution on the Sb site, one can also substitute on the transition metal site to add valence electrons. This was demonstrated with Fe [67] and Ru [68], as well as simultaneously on the Mo and Sb sites with Fe and Te [69] or Ru and Te [70].

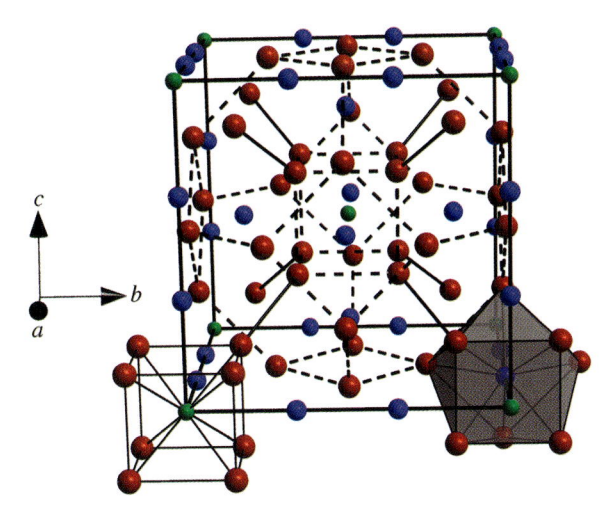

Figure 11.3 Crystal structure of Mo_3Sb_7. Green: cubic holes; blue: Mo; red: Sb. Emphasized is one $MoSb_8$ square antiprism. Dashed lines denote the longer Sb–Sb distances of 3.4 Å.

At least in case of nickel, there is a competition between substituting on the Mo site and adding into the voids of the $Sb2_8$ cubes [71]. Single crystal structure data proved that the Ni atoms occupy both the Mo site and the cubic void in case of the nominal composition of "$Mo_{2.8}Ni_{0.2}Sb_7$" [72]. The latter is disadvantageous for the thermoelectric properties because the substitution leads to more filled bands, thus a smaller carrier concentration is desired because of the metallic character of as-cast Mo_3Sb_7. On the other hand, adding Ni atoms into the cubes not only adds electrons but also five d-bands per added atom to the band structure. For Ni comes with ten valence electrons and five d-orbitals, filling the hole with Ni atoms is thus formally neutral with respect to the carrier concentration. This was supported by density functional theory calculations: The Mo-substituted model "$Mo_{2.5}Ni_{0.5}Sb_7$" is an intrinsic semiconductor, while distributing the same amount of Ni atoms equally over the Mo sites and the holes (model "$Ni_{0.25}Mo_{2.75}Ni_{0.25}Sb_7$") results in a p-doped semiconductor (Figure 11.4) [72].

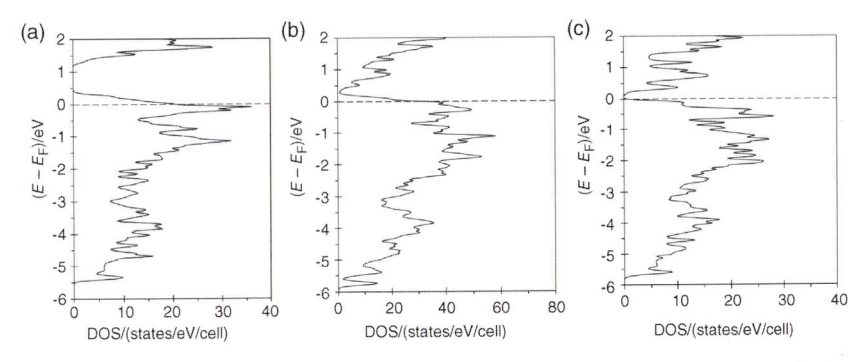

Figure 11.4 Density of states of (a) Mo_3Sb_7, (b) "$Ni_{0.25}Mo_{2.75}Ni_{0.25}Sb_7$," and (c) "$Mo_{2.5}Ni_{0.5}Sb_7$.".

The total thermal conductivity of these materials is relatively high for a thermoelectric material, with $\kappa > 4\,\mathrm{W\,m^{-1}K^{-1}}$ around room temperature in most cases. In these Mo_3Sb_7 variants, a high electrical conductivity of $\sigma > 1500\,\Omega^{-1}\,\mathrm{cm}^{-1}$ is found at 300 K, depending on the carrier concentration, which results in a high electronic contribution to the thermal conductivity. Still, κ_L is high as well and often the dominating part in these materials. Recent efforts into using carbon-containing nanocomposites have successfully lowered the lattice thermal conductivity and raised the figure of merit by 14% in case of C_{60} additions [73] to 25% in case of carbon nanotubes additions [74], depending on the concentration.

11.2.3
Late Transition Metal Antimonides

There are no Sb-rich antimonides of group 7. MnSb adopts the NiAs type, and is a metallic ferromagnet [75]. The rhenium arsenide, Re_3As_7, could be of interest, as it is isostructural with Mo_3Sb_7, and with its $3\times 7 + 7\times 5 = 56$ valence electrons per formula unit, it is an n-type semiconductor. Reducing the electron concentration via Ge/As substitution was proven to be beneficial; the zT curve of n-type $Re_3Ge_{0.6}As_{6.4}$ runs above that of $Mo_3Sb_{5.4}Te_{1.6}$ with a comparable slope as far as measured, namely, up to 700 K. Here, its zT is 0.30 compared to 0.27 for $Mo_3Sb_{5.4}Te_{1.6}$ [64].

Several diantimonides of groups 8–10 are narrow-gap semiconductors, adopting either the marcasite or the pyrite type. In both structures, the transition metal is coordinated by 6 pnictogen atoms in form of a distorted octahedron, and all pnictogen atoms form dumbbells with one homonuclear pnictogen–pnictogen bond. Thus, both may be viewed as distorted variants of the NaCl type, where the dumbbells replace the Cl atoms. The difference between the two distorted types lies in the orientation of the dumbbells, which are parallel to the space diagonal in case of the arsenopyrite, thus maintaining the cubic symmetry, but are parallel to the c-axis in case of the orthorhombic marcasite. The aristotype FeS_2 can crystallize in both of these structure types. An early overview was given in 1965, arguing that these and other isostructural compounds are semiconducting with 20 or less valence electrons, that is, all pnictides of these families with the exception of the bismuthide. Specifically mentioned were $MnTe_2$, $FeAs_2$, $FeSb_2$, FeS_2, $RuSe_2$, $RuTe_2$, $OsSb_2$, $OsTe_2$, $CoAsS$, $CoSe_2$, $CoTe_2$, $RhSb_2$, $NiSbS$, PtP_2, $PtAs_2$, $PtSb_2$, $\alpha\text{-}PtBi_2$, and $AuSb_2$ [76].

$FeSb_2$, adopting the marcasite structure like $CrSb_2$, is a strongly correlated semiconductor with a colossal Seebeck coefficient of $-45000\,\mu\mathrm{V\,K^{-1}}$ at 10 K measured on a single crystal. Its maximum power factor, $P.F. = \alpha^2\sigma$, occurs at 12 K, and is over 50 times larger than that of classic thermoelectrics. This may be of interest for cryogenic applications. However, its figure of merit, $zT = 0.005$ at 12 K, remains too low because of the high thermal conductivity [77]. Subsequently, it was revealed that the performance can be enhanced via Sb deficiencies, with $FeSb_{1.8}$ reaching $zT = 0.01$ at 30 K [78]. The properties of $RuSb_2$ [79]

were optimized via Se and Te doping, but overall no good thermoelectric performance was achieved because of the low carrier concentration [80].

$CoSb_2$ is a polymorph, the low temperature form arsenopyrite (FeAsS, a distorted variant of marcasite) turns into marcasite at 644 K. A study of its thermoelectric properties along with $IrSb_2$ and the corresponding solid solution yielded the best performance at 10 at.% $CoSb_2$, but only culminating in $zT_{max}=0.2$ at 900 K [81].

A special case is the skutterudite $CoSb_3$, the aristotype being $CoAs_3$ named after a town in Norway. $CoSb_3$ is a distorted variant of ReO_3, where the $CoSb_6$ octahedra are tilted in such a way to enable the formation of empty Sb_{12} icosahedra and planar Sb_4 rectangles, the latter with Sb—Sb bonds of 2.90 and 2.98 Å. Considering these bonds as single bonds, the Sb atoms are comprised of six valence electrons, that is, Sb^-, in line with the experimentally observed semiconducting properties of $Co^{3+}(Sb^-)_3$. With a thermal conductivity >6 W m^{-1}K^{-1} at room temperature, the thermoelectric properties of $CoSb_3$ are far from being special, for example, reported to reach 0.05 at 723 K [82] or 0.10 at 673 K after spark plasma sintering (SPS) [83]. However, the large icosahedral voids of $CoSb_3$ can be filled with a variety of atoms, leading to a huge family of filled skutterudites, the first antimonide being $LaFe_4Sb_{12}$ [84]. These materials are bestowed with outstanding thermoelectric properties [85], and are therefore under investigation for use in automobiles for energy recovery from the exhaust heat since over a decade [86].

In the filled skutterudites, the icosahedra are large even for the La atoms, as can be seen from the long La–Sb distances of 3.41 Å in $LaFe_4Sb_{12}$. This causes increased thermal vibration of the filler atoms, as can be measured in form of the anisotropic displacement parameters, ADPs. These are –three to six times larger than that of the Fe, Co, and Sb atoms in $LaFe_4Sb_{12}$ [84] and $LaFe_3CoSb_{12}$ [87]. This increased vibration has been called rattling effect, postulated to cause the significant decrease in the lattice thermal conductivity in the filled skutterudites. Although these effects are observable, evidence for significant La–Sb and Ce–Sb coupling was proven via inelastic neutron scattering experiments on $LaFe_4Sb_{12}$ and $CeFe_4Sb_{12}$ that mitigates against considering the lanthanide atoms, Ln, as rattlers [88].

Electronic structure calculations confirm the electron count discussed above: Filled skutterudites are semiconducting when the guest atom is fully oxidized, for example, Ln^{3+}, and six valence electrons fill the "t_{2g}" states of the transition metal atoms, as for example in $LaFe_3CoSb_{12} \equiv La^{3+}(Fe^{2+})_3Co^{3+}(Sb^-)_{12}$ with a bandgap of 0.5 eV [62,89–92]. Typical room temperature thermoelectric properties of the filled skutterudites are around $\alpha = \pm 100$ μV K^{-1}, σ values between 600 and 3000 Ω^{-1} cm^{-1}, and κ values between 3 and 5 W m^{-1}K^{-1}. Both n- and p-doped filled skutterudites are known to exceed a figure of merit of unity at elevated temperatures, for example, the n-types $Yb_{0.19}Co_4Sb_{12}$ with $zT_{max}=1.14$ at 640 K [93], $Ba_{0.24}Co_4Sb_{12}$ with $zT_{max}=1.10$ at 850 K [94], and the triple-filled $Ba_{0.08}La_{0.05}Yb_{0.04}Co_4Sb_{12}$ with $zT=1.66$ at 850 K [95], and the p-type $LaFe_3CoSb_{12}$ was postulated to reach $zT_{max}=1.4$ at 1000 K [96].

Similarly, no promising binary antimonides of groups 10 and 11 are known to exist [97]. The last binary transition metal antimonide of interest is β-Zn$_4$Sb$_3$ [98]. Its crystal structure contains ZnSb$_4$ tetrahedra and linear chains of Sb2 atoms with alternating distances of 2.82 and 3.38 Å (Figure 11.5). Several Zn positions with very small occupancy factors between 5 and 6% are a special feature of this material, while in turn the major Zn site, Zn1, exhibits about a 90% occupancy yielding a refined formula of Zn$_{3.83}$Sb$_3$ [99]. A later study resulted in a refined formula of Zn$_{3.55}$Sb$_3$, and an electron microprobe result of Zn$_{3.90}$Sb$_3$ [100].

For electron counting purposes, let us consider the crystallographic formula without the small Zn positions, which is (Zn1)$_6$(Sb1)$_3$(Sb2)$_2$. Treating again the isolated Sb1 atoms as Sb^{3-} and the Sb2 atoms as part of the Sb$_2^{4-}$ pairs, we find that "Zn$_6$Sb$_5$" ≡ Zn$_{3.6}$Sb$_3$ is electron deficient with one hole, h, per formula unit: (Zn^{2+})$_6$(Sb1^{3-})$_3$(Sb2)$_2^{4-}h^+$. This is in agreement with the p-type properties of β-Zn$_4$Sb$_3$ [62,101]. Compared to the filled skutterudites, β-Zn$_4$Sb$_3$ has a comparable Seebeck coefficient and much lower electrical and thermal conductivities, that is, $\kappa \approx 1$ W m^{-1}K^{-1} at room temperature. The several Zn sites of low occupancy, the complex crystal structure [99,100], as well as the intrinsically occurring nanodomains [102] contribute to this unusually low thermal conductivity. Overall, very competitive zT values were determined, on the order of 1.3 around 660 K [98,103]. Beyond 660 K, however, β-Zn$_4$Sb$_3$ transforms into γ-Zn$_4$Sb$_3$ that is detrimental to its thermoelectric performance [104].

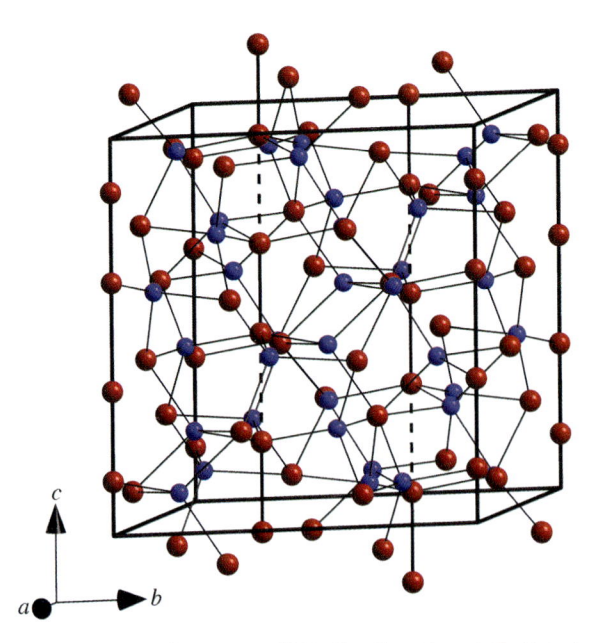

Figure 11.5 Crystal structure of β-Zn$_4$Sb$_3$. Blue: Zn; red: Sb. Only the major Zn site is shown. Dashed lines denote the longer Sb–Sb distances of 3.4 Å.

11.3
Ternary Antimonides

11.3.1
Ternary Antimonides without Sb—Sb Bonds

Ternary and higher antimonides are far too numerous to be discussed all here. Therefore, we will only present those with promising thermoelectric performance, namely, with a figure of merit $zT_{max}>0.5$. The simplest ternary representatives are the equiatomic ones, that is, MM'Sb with M and M' being different metal atoms. This includes the half-Heusler phases, which adopt the MgAgAs type. Therein, the anions are cubic closest packed, and the two different metal atoms fill all octahedral voids as well as half of the tetrahedral holes (Figure 11.6, with Ti atoms at the origin). The antimonides are semiconducting or semimetallic with 18 valence electrons per formula unit, as in LnPdSb with Ln = Ho, Er, and Dy [105] and MCoSb with M = Ti, Zr, and Hf, which may be written as M^{4+}Co$^-$Sb^{3-} [106,107]. Considering the simple high-symmetry crystal structures, their high thermal conductivity on the order of $10\,\text{W}\,\text{m}^{-1}\text{K}^{-1}$ is not surprising.

Nevertheless, reasonably high zT values can be achieved by these materials after alloying and doping because of the high power factor. Examples include n-type Ti$_{0.5}$Zr$_{0.25}$Hf$_{0.25}$Co$_{0.95}$Ni$_{0.05}$Sb with $zT_{max}=0.51$ at 813 K [108] and Ti$_{0.6}$Hf$_{0.4}$Co$_{0.87}$Ni$_{0.13}$Sb with $zT_{max}=0.70$ at 900 K [109]. An intrinsic phase separation was used to further enhance the properties of MCoSb, culminating in

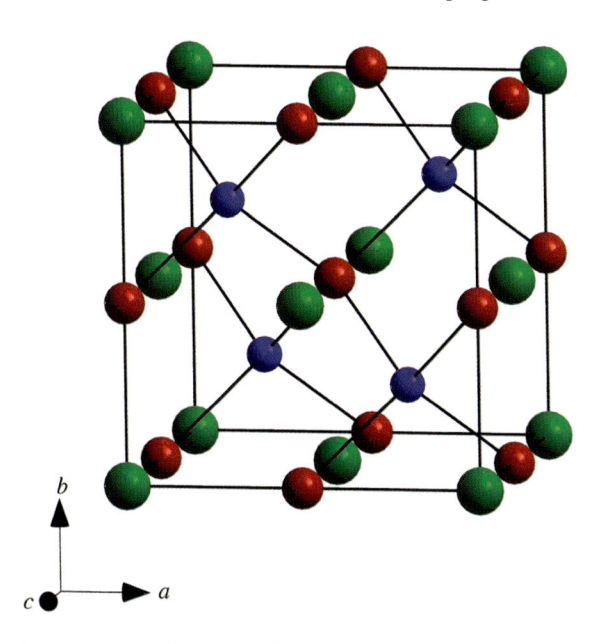

Figure 11.6 Crystal structure of TiCoSb. Green: Ti; blue: Co; and red: Sb.

$zT_{max} = 0.9$ at 973 K for p-type $Ti_{0.5}Hf_{0.5}CoSb_{0.8}Sn_{0.2}$ [110]. It should be noted that other half-Heusler materials, in particular stannides based on TiNiSn [111], exhibit comparable or even better thermoelectric performance. Here again, nanostructuring proved to be useful in further enhancing the thermoelectric performance by tackling the high thermal conductivity [23,112,113].

CaAgSb is another interesting equiatomic antimonide that crystallizes in the TiNiSi type (Figure 11.7a). The metal atoms Ca and Ag are in their common oxidation states, leading to $Ca^{2+}Ag^+Sb^{3-}$, with the Ag atoms being tetrahedrally coordinated by four Sb atoms at distances between 2.84 and 2.99 Å. Replacing part of the divalent Ca with trivalent lanthanide atoms results in an electron excess, which in part is then self-compensated by defects on the Ag site. Interestingly, $Ca_{1-x}Ln_xAg_{1-y}Sb$ adopts the layered LiGaGe type with a 3 + 1 coordination of the Ag atoms, with Ag–Sb distances of 3×2.77 and 1×3.35 Å (Figure 11.7b). A common feature of these two structures is the three-dimensional network of corner-sharing $AgSb_4$ "tetrahedra."

The transformation to the LiGaGe type results in a bandgap, and thus improved thermoelectric properties, up to $zT_{max} = 0.7$ at 1079 K for $Ca_{0.84}Ce_{0.16}Ag_{0.87}Sb$. Here, the thermal conductivity is below 1 W m^{-1}K^{-1}, that is, much lower than in case of the half-Heusler materials, likely a consequence of the significantly more complex structure [114].

The next group of antimonides is the so-called 1-2-2 family, based on the $CaAl_2Si_2$ type, as mentioned above during the Mg_3Sb_2 discussion. $CaAl_2Si_2$ can be understood based on alternating layers of (octahedrally coordinated) Ca^{2+} atoms and $Al_2Si_2^{2-}$ sheets, the latter being two-dimensional layers of corner-sharing $AlSi_4$ tetrahedra. The antimonides are formally charge-balanced when all metal atoms are divalent, like in $Yb^{2+}(Zn^{2+})_2(Sb^{3-})_2$, but often defects on the cation sites cause large carrier concentration. Its solid solution with Ca, that is, $Ca_xYb_{1-x}Zn_2Sb_2$, was investigated back in 2005, which showed that the defects

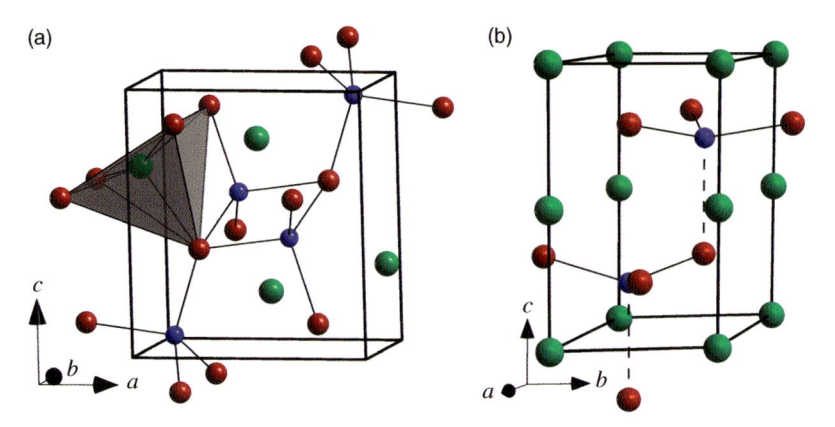

Figure 11.7 Crystal structures of (a) CaAgSb in the TiNiSi type, emphasizing one $CaSb_5$ square pyramid, and (b) $Ca_{1-x}Ln_xAg_{1-y}Sb$ in the LiGaGe type. Green: Ca/Ln; blue: Ag; and red: Sb. Dashed lines denote the longer Ag–Sb distances of 3.4 Å.

and the carrier concentration decrease with increasing x. At $x=0$, the best thermoelectric properties are observed with $zT_{max}=0.56$ at 773 K, while the lowest thermal conductivity of this series with $\kappa=1.3$ W m^{-1}K^{-1} at room temperature occurs in the $x=0.5$ case, a consequence of the high disorder of the two different cations on the symmetry equivalent site between the $Zn_2Sb_2^{2-}$ sheets. Subsequent work on $YbZn_2Sb_2$ led to further improvements, for example, via introducing vacancies on the Yb site ($zT_{max}=0.87$ at 775 K) [115], and introducing Cd on the Zn site (zT_{max} of 1.2–1.3 at 700 K) [116,117]. Similarly, the isostructural Cd variant, $YbCd_2Sb_2$, may attain $zT_{max}=1.1$ at 650 K when alloyed with Mn on the Cd site [118].

$BaGa_2Sb_2$, another narrow-gap semiconductor is a special case where the divalent Ga atoms form a Ga—Ga single bond because of the s^1 configuration of the Ga^{2+} atoms. Its crystal structure is composed of dimeric [Sb$_3$Ga–GaSb$_3$] units (in analogy to ethane) that are interconnected via common corners to a complex three-dimensional network (Figure 11.8). The latter gives rise to a low

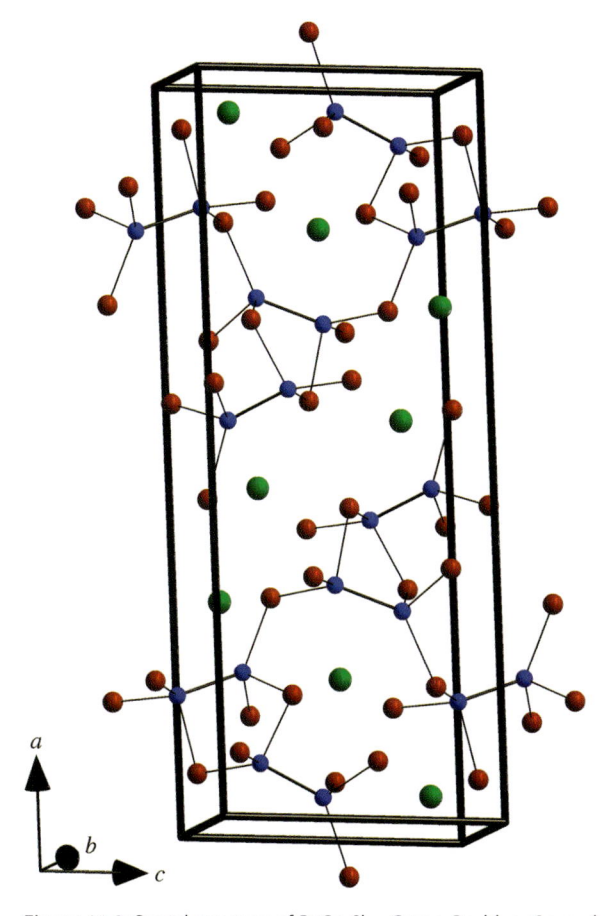

Figure 11.8 Crystal structure of $BaGa_2Sb_2$. Green: Ba; blue: Ga; and red: Sb.

thermal conductivity of $\kappa = 1.3\,\mathrm{W\,m^{-1}K^{-1}}$. Doping with Zn on the Ga site affords promising thermoelectric performance, culminating in $zT_{max} = 0.6$ at 800 K [119].

Moreover, the so-called 3-1-3 family, represented by Ca_3AlSb_3 and Sr_3GaSb_3, deserves attention in this context as well. Both structures comprise $CaSb_6/SrSb_6$ octahedra as well as 1D chains of corner-sharing $AlSb_4/GaSb_4$ tetrahedra, which are linear in case of Ca_3AlSb_3 but not in case of Sr_3GaSb_3 (Figure 11.9). Both materials are bestowed with a low thermal conductivity. In case of Ca_3AlSb_3, Zn-doping on the Al site leads to $zT_{max} = 0.5$ at 900 K [120], and Na-doping on the Ca site leads to $zT_{max} = 0.79$ at 1040 K [121]. Better performance is achieved in Zn-doped Sr_3GaSb_3, culminating in $zT_{max} = 0.97$ at 1000 K [122].

Of particular interest is the 9-4-9 family, adopting the (filled) $Ca_9Mn_4Bi_9$ type that encompasses a large variety of compounds, including $Ca_9Zn_{4+x}Sb_9$, $Yb_9Zn_{4+x}Sb_9$ [123], $Sr_9Cd_{4+x}Sb_9$, $A_9Zn_{4+x}Bi_9$ and $A_9Cd_{4+x}Bi_9$ with A = Ca, Sr, Eu, Yb [124], and $Yb_9Mn_{4+x}Sb_9$ and $Yb_9Mn_{4+x}Bi_9$ [125], all with $0<x<0.5$. Another unique structure type is adopted by $Ca_9Mn_{4+x}Sb_9$, whose properties are not known as yet [126]. The structure of $Yb_9Mn_{4+x}Sb_9$ is shown in Figure 11.10. Comparable to some cases discussed above, the $MnSb_4$ tetrahedra form linear chains via corner-sharing (here along the c-axis) that are interconnected via common corners to tetrameric units. An additional Mn site, Mn3, shows various occupancies up to 25%, and is threefold coordinated by Sb atoms. As no Sb—Sb bonds are present, an assignment of the most common oxidation states results in charge balance when $x = 0.5$ (occupancy of Mn3: 25%): $(Yb^{2+})_9(Mn^{2+})_{4.5}(Sb^{3-})_9$. Varying x then presents a convenient method to change the balance and thus vary the carrier concentration. This complex structure along with the Mn disorder was reported to result in "glass-like lattice thermal conductivity," that is, $\kappa_L < 0.4\,\mathrm{W\,m^{-1}K^{-1}}$, and a high figure of merit of $zT_{max} = 0.7$ at 950 K in case of $Yb_9Mn_{4.2}Sb_9$ [127].

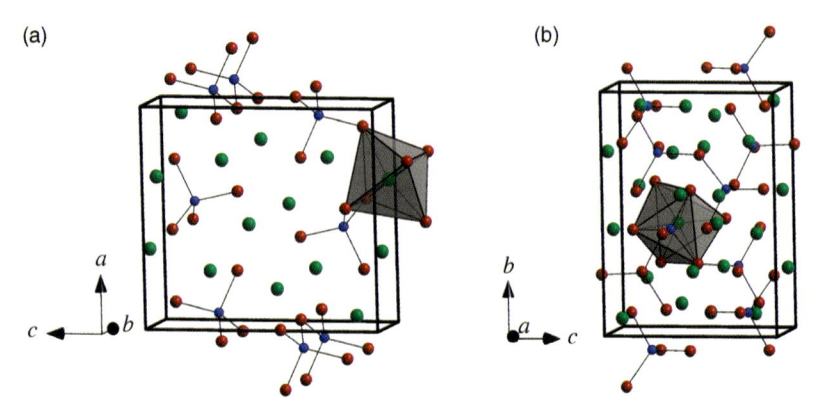

Figure 11.9 Crystal structures of (a) Ca_3AlSb_3 and (b) Sr_3GaSb_3. Green: Ca/Sr; blue: Al/Ga; and red: Sb. Emphasized are one $CaSb_6$ and one $SrSb_6$ octahedron.

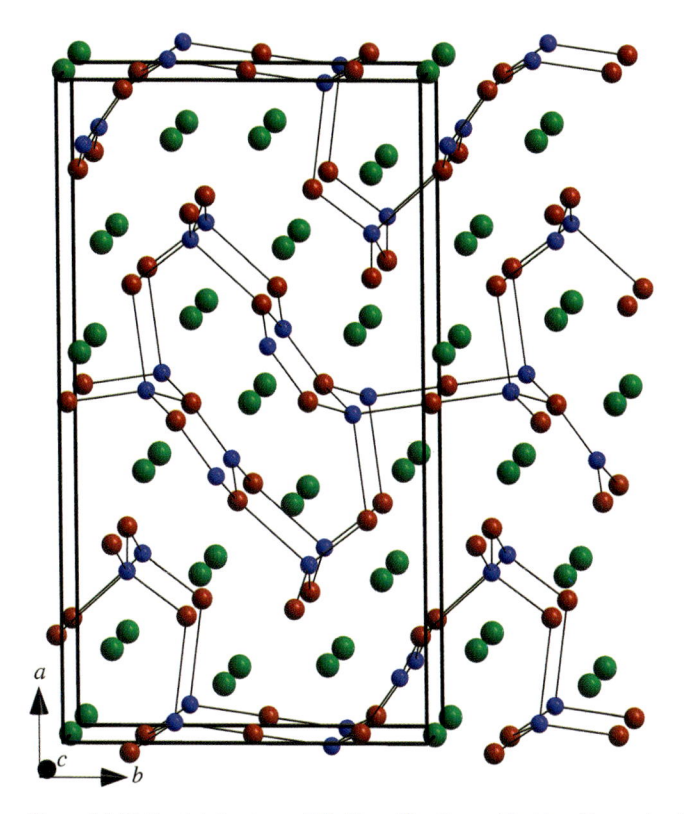

Figure 11.10 Crystal structure of $Yb_9Mn_{4+x}Sb_9$. Green: Yb; blue: Mn; and red: Sb. The threefold coordinated Mn3 site is the deficient site, occupied to different degrees depending on the reaction conditions.

11.3.2
Ternary Antimonides with Sb—Sb Bonds

The 5-2-6 antimonides may occur in two different crystal structures, namely, the $Ca_5Ga_2Sb_6$ and the $Ba_5Al_2Bi_6$ types. Both structures are comprised of linear chains running along the c-axis of corner-sharing $GaSb_4$ and $AlBi_4$ tetrahedra, respectively, which are connected to form pairs of these chains via Pn—Pn bonds. Of the six Pn atoms per formula unit, two form a $[Pn-Pn]^{4-}$ pair resulting in the charge-balanced formula of, for example, $(Ca^{2+})_5(Ga^{3+})_2Sb_2^{4-}(Sb^{3-})_4$. The structures differ in the relative orientations of these chains to each other, as shown in Figure 11.11 for example (a) $Ca_5Al_2Sb_6$ ($Ca_5Ga_2Sb_6$ type) and (b) $Yb_5In_2Sb_6$ ($Ba_5Al_2Bi_6$ type). In the latter case, the double chains have the same orientation along the a-axis, while the next chains along the b-axis have a different orientation (related by a glide plane). In case of $Ca_5Al_2Sb_6$, the chains change orientation along both the a- and the b-axes.

Like several previously discussed ternary antimonides, these materials exhibit low thermal conductivity as well, typically on the order of $1\,W\,m^{-1}K^{-1}$. While

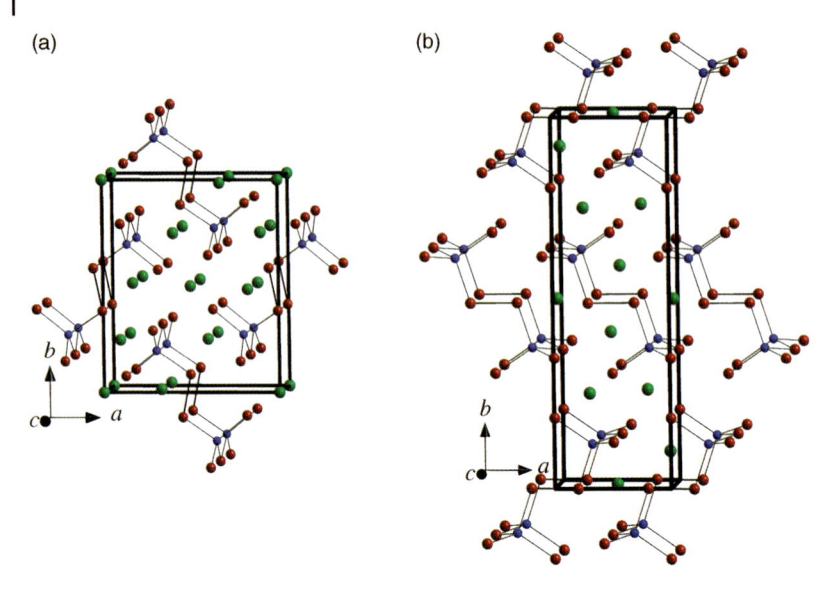

Figure 11.11 Crystal structures of (a) $Ca_5Al_2Sb_6$ and (b) $Yb_5In_2Sb_6$. Green: Ca/Yb; blue: Al/In; and red: Sb.

the undoped materials such as $Sr_5Al_2Sb_6$, $Sr_5In_2Sb_6$, and $Yb_5Mn_2Sb_6$ had low zT values [128–130], promising thermoelectric performance was demonstrated after p-type doping, including 3% Na-doping on the Ca site in $Ca_5Al_2Sb_6$ (zT_{max} = 0.62 at 1000 K) [131] and 5% Zn-doping on the In site in $Ca_5In_2Sb_6$ (zT_{max} = 0.72 at 940 K) [132].

Last, and possibly most exciting, is the 14-1-11 family, the aristotype being $Ca_{14}AlSb_{11}$. Its highly complex unit cell is comprised of isolated $AlSb_4$ tetrahedra and linear Sb_3 units. This Sb_3 unit is reminiscent of the linear I_3^- units of, for example, CsI_3 [54] and $[Ph_4As]I_3$ [133], and linear Se_3^{4-} of $Ba_2Ag_4Se_5$ [53], and thus most likely contains 22 valence electrons as well. With one such Sb_3 unit per formula unit, $Ca_{14}AlSb_{11}$ can then be understood as $(Ca^{2+})_{14}Al^{3+}Sb_3^{7-}(Sb^{3-})_8$, supported by band structure calculations on $Ca_{14}AlSb_{11}$ [62] and isostructural and isoelectronic $Ca_{14}AlAs_{11}$ [134]. The central atom of the Pn_3^{7-} unit (Pn = P, As, Sb, Bi) often exhibits enlarged thermal displacement parameters, ADPs, as observed in $Ba_{14}InP_{11}$ [135], $Ca_{14}GaAs_{11}$ [136], $Ca_{14}AlSb_{11}$ [52], $Sr_{14}ZnSb_{11}$ [137], and $Ba_{14}MnSb_{11}$ [138], likely caused by the tendency toward the formation of one regular Pn—Pn bond instead of two intermediate ones. Such enlarged displacement parameters are known to cause enhanced phonon scattering, thus lower thermal conductivity.

The isostructural $Yb_{14}MnSb_{11}$ is currently the best high-temperature p-type material above 1000 K, exhibiting a zT_{max} = 1.1 at 1275 K [139], and is therefore under investigation by NASA for use in future spacecrafts [140]. The high figure of merit of these materials is, again, a consequence of the glass-like thermal conductivity combined with outstanding high temperature stability. While one might propose isovalency with $Ca_{14}AlSb_{11}$, experiments showed the presence of Mn^{2+} along with a hole in the Sb 5p band [141], resulting in high p-type carrier concentration of the undoped material. These findings are also in line with the Mn^{2+} found in $Yb_9Mn_{4+x}Sb_9$ and $Yb_5Mn_2Sb_6$. Subsequently, various variants were investigated as well, in part leading to further enhanced thermoelectric performance. Examples include $Yb_{13.6}La_{0.4}MnSb_{11}$ with $zT_{max} = 1.15$ at 1150 K [142], $Yb_{14}Mn_{0.4}Al_{0.6}Sb_{11}$ with $zT_{max} = 1.32$ at 1275 K [143], $Yb_{13.82}Pr_{0.18}Mn_{1.01}Sb_{10.99}$ with $zT_{max} = 1.2$ at 1275 K [144], and the Mn-free $Yb_{14}MgSb_{11}$ with $zT_{max} = 1.02$ at 1075 K [145].

11.4 Conclusions

This chapter provides an overview of the crystal structure of various antimonides of interest for the thermoelectric energy conversion. When discussing the thermoelectric performance, it should be noted that these materials are far from being optimized with respect to, for example, grain size and consolidation conditions. On the other hand, many of the zT values were obtained using the Dulong–Petit approximation for the specific heat to determine the thermal conductivity that tends to overestimate zT at elevated temperatures. Therefore, the precise numbers should be treated with caution.

When comparing the zT_{max} values, it is important to realize that they often come from different temperatures. Depending on the temperature dependence of zT as well as the temperature range of the actual application, materials with lower zT_{max} values may be preferred. To demonstrate this, several of the best antimonides – excluding the far too numerous filled skutterudites – are compared in Figure 11.12: the material with the highest zT_{max}, namely, $Yb_{14}Mn_{0.4}Al_{0.6}Sb_{11}$, is outperformed by several others between 300 and 700 K.

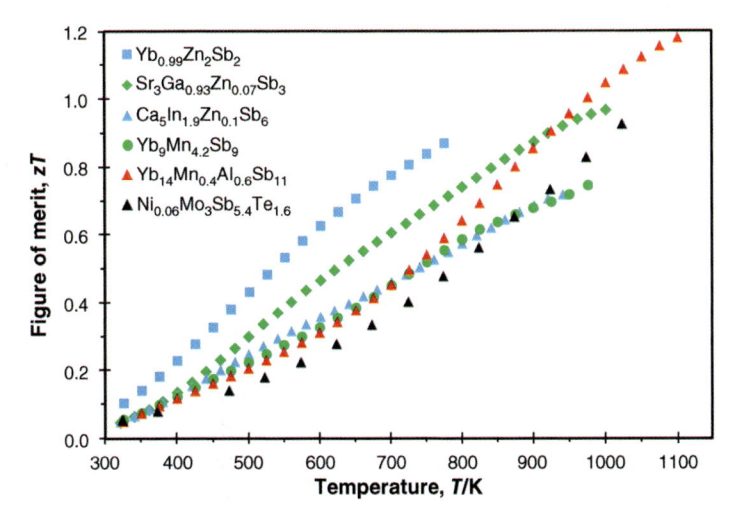

Figure 11.12 Temperature-dependent figure of merit of selected antimonides. (Data are courtesy of Dr A. Zevalkink, Michigan State University, Dr G.J. Snyder, Northwestern University, and Dr S.M. Kauzlarich, UC Davis (references are provided in the main text).)

To summarize, the best antimonides known to date are doped variants of the filled skutterudites, Mo_3Sb_7, half-Heusler phases, and several ternary complex antimonides that consist of metal-centered Sb_4 tetrahedra. The latter occur in many variants, from isolated (zero dimensional) to puckered or linear chains (1D) to pairs or tetramers of chains to layers (2D) and a truly 3D network of these tetrahedra in the TiNiSi and LiGaGe types.

Considering that the vast majority of these data were published within the last ten years, research into this area continues to attract more researchers. In the end, antimonides may well be able to replace the tellurides as the leading materials in thermoelectrics that would be very beneficial because of the lower price, higher availability, and lower toxicity of antimony compared to tellurium.

References

1 Morkoç, H. (2008) *Handbook of Nitride Semiconductors and Devices*, vol. **1**, Wiley-VCH Verlag GmbH, Weinheim, Germany.
2 Kleinke, H. (2000) *Chem. Soc. Rev.*, **29**, 411–418.
3 Mills, A.M., Lam, R., Ferguson, M.J., Deakin, L., and Mar, A. (2002) *Coord. Chem. Rev.*, **233–234**, 207–222.
4 Cameron, J.M., Hughes, R.W., Zhao, Y., and Gregory, D.H. (2011) *Chem. Soc. Rev.*, **40**, 4099–4118.
5 Moriya, Y., Takata, T., and Domen, K. (2013) *Coord. Chem. Rev.*, **257**, 1957–1969.
6 Hargreaves, J.S.J. (2013) *Coord. Chem. Rev.*, **257**, 2015–2031.
7 Tapia-Ruiz, N., Segalés, M., and Gregory, D.H. (2013) *Coord. Chem. Rev.*, **257**, 1978–2014.
8 Johrendt, D., Hieke, C., and Stürzer, T. (2013) *Comprehensive Inorganic Chemistry*, 2nd edn, Reference Module in Chemistry, Molecular Sciences and Chemical Engineering, **2**, 111–135.
9 Greiwe, M. and Nilges, T. (2014) *Prog. Solid State Chem.*, **42**, 191–201.
10 Villars, P. (1997) *Pearson's Handbook: Crystallographic Data for Intermetallic Phases*, American Society for Metals, Materials Park, OH.
11 Rowe, D.M. (1995) *CRC Handbook of Thermoelectrics*, CRC Press, Boca Raton, FL.
12 Rowe, D.M. (2006) *Thermoelectrics Handbook: Macro to Nano*, CRC Press, Taylor & Francis Group, Boca Raton, FL.
13 Furlong, R.R. and Wahlquist, E.J. (1999) *Nucl. News*, **42**, 26–35.
14 Yang, J. and Caillat, T. (2006) *MRS Bull.*, **31**, 224–229.
15 Yang, J. and Stabler, F.R. (2009) *J. Electron. Mater.*, **38**, 1245–1251.
16 Bell, L.E. (2008) *Science*, **321**, 1457–1461.
17 Venkatasubramanian, R., Slivola, E., Colpitts, T., and O'Quinn, B. (2001) *Nature*, **413**, 597–602.
18 Hsu, K.F., Loo, S., Guo, F., Chen, W., Dyck, J.S., Uher, C., Hogan, T., Polychroniadis, E.K., and Kanatzidis, M.G. (2004) *Science*, **303**, 818–821.
19 Biswas, K., He, J., Blum, I.D., Wu, C.-I., Hogan, T.P., Seidman, D.N., Dravid, V.P., and Kanatzidis, M.G. (2012) *Nature*, **489**, 414–418.
20 Baxter, J., Bian, Z.X., Chen, G., Danielson, D., Dresselhaus, M.S., Fedorov, A.G., Fisher, T.S., Jones, C.W., Maginn, E., Kortshagen, U., Manthiram, A., Nozik, A., Rolison, D.R., Sands, T., Shi, L., Sholl, D., and Wu, Y.Y. (2009) *Energy Environ. Sci.*, **2** 559–588.
21 Kanatzidis, M.G. (2010) *Chem. Mater.*, **22**, 648–659.
22 Liu, W., Yan, X., Chen, G., and Ren, Z. (2012) *Nano Energy*, **1**, 42–56.
23 Xie, W., Weidenkaff, A., Tang, X., Zhang, Q., Poon, J., and Tritt, T.M. (2012) *Nanomaterials*, **2**, 379–412.

24 Koumoto, K. and Mori, T. (2013) *Thermoelectric Nanomaterials: Materials Design and Applications*, vol. **182**, Springer, Heidelberg, Germany.

25 Sommer, A.H. (1968) *Photoemissive Materials*, John Wiley & Sons, New York, NY.

26 Martinez-Ripoll, M., Haase, A., and Brauer, G. (1974) *Acta Crystallogr. B*, **30**, 2006–2009.

27 Zheng, C., Hoffmann, R., Nesper, R., and Schnering, H.G.v. (1986) *J. Am. Chem. Soc.*, **108**, 1876–1884.

28 Condron, C.L., Kauzlarich, S.M., Gascoin, F., and Snyder, G.J. (2006) *J. Solid State Chem.*, **179**, 2252–2257.

29 Bhardwaj, A., Rajput, A., Shukla, A.K., Pulikkotil, J.J., Srivastava, A.K., Dhar, A., Gupta, G., Auluck, S., Misra, D.K., and Budhani, R.C. (2013) *RSC Adv.*, **3**, 8504–8516.

30 Bhardwaj, A. and Misra, D.K. (2014) *RSC Adv.*, **4**, 34552–34560.

31 Avery, D.G., Goodwin, D.W., and Rennie, A.E. (1957) *J. Sci. Instrum.*, **34**, 394–395.

32 Kim, D., Kurosaki, K., Ohishi, Y., Muta, H., and Yamanaka, S. (2013) *J. Electron. Mater.*, **42**, 2388–2392.

33 Norén, L., Withers, R.L., Schmid, S., Brink, F.J., and Ting, V. (2006) *J. Solid State Chem.*, **179**, 404–412.

34 Sofo, J.O. and Mahan, G.D. (1994) *Phys. Rev. B*, **49**, 4565–4570.

35 Smith, G.E. and Wolfe, R. (1962) *J. Appl. Phys.*, **33**, 841–846.

36 Lenoir, B., Dauscher, A., Devaux, X., Martin-Lopez, R., Ravich, Y.I., Scherrer, H., and Scherrer, S. (1996) Proceedings of the International Conference on Thermoelectrics, pp. 1–13.

37 Zemskov, V.S., Belaya, A.D., Beluy, U.S., and Kozhemyakin, G.N. (2000) *J. Cryst. Growth*, **212**, 161–166.

38 Derakhshan, S., Assoud, A., Kleinke, K.M., and Kleinke, H. (2007) *Inorg. Chem.*, **46**, 1459–1463.

39 Kasuya, T., Haga, Y., Kwon, Y.S., and Suzuki, T. (1993) *Physica B*, **186–188**, 9–15.

40 Armbrüster, M., Gil, R.C., Burkhardt, U., and Grin, Y. (2004) *Z. Kristallogr.*, **219**, 209–210.

41 Hönle, W. and von Schnering, H.-G. (1981) *Z. Kristallogr.*, **155**, 307–314.

42 Nesper, R. (1990) *Prog. Solid State Chem.*, **20**, 1–45.

43 Donaldson, J.D., Kjekshus, A., Nicholson, D.G., and Rakke, T. (1975) *J. Less Common Met.*, **41**, 255–263.

44 Kjekshus, A. (1972) *Acta Chem. Scand.*, **26**, 1633–1639.

45 Papoian, G. and Hoffmann, R. (2001) *J. Am. Chem. Soc.*, **123**, 6600–6608.

46 Soheilnia, N., Assoud, A., and Kleinke, H. (2003) *Inorg. Chem.*, **42**, 7319–7325.

47 Assoud, A., Kleinke, K.M., Soheilnia, N., and Kleinke, H. (2004) *Angew. Chem., Int. Ed.*, **43**, 5260–5262.

48 Xu, J., Kleinke, K.M., and Kleinke, H. (2008) *Z. Anorg. Allg. Chem.*, **634**, 2367–2372.

49 Rehr, A. and Kauzlarich, S.M. (1994) *Acta Crystallogr. C*, **50**, 1177–1178.

50 Hulliger, F. (1964) *Nature*, **204**, 775–775.

51 Kleinke, H. (1999) *J. Mater. Chem.*, **9**, 2703–2708.

52 Cordier, G., Schäfer, H., and Stelter, M. (1984) *Z. Anorg. Allg. Chem.*, **519**, 183–188.

53 Assoud, A., Xu, J., and Kleinke, H. (2007) *Inorg. Chem.*, **46**, 9906–9911.

54 Tasman, H.A. and Boswijk, K.H. (1955) *Acta Crystallogr.*, **8**, 59–60.

55 Papoian, G.A. and Hoffmann, R. (2000) *Angew. Chem., Int. Ed.*, **39**, 2408–2448.

56 Derakhshan, S., Kleinke, K.M., Dashjav, E., and Kleinke, H. (2004) *Chem. Commun.*, 2428–2429.

57 Harada, T., Kanomata, T., Takahashi, Y., Nashima, O., Yoshida, H., and Kaneko, T. (2004) *J. Alloys Compd.*, **383**, 200–204.

58 Chen, L. (2011) *Int. J. Mod. Phys. B*, **25**, 653–664.

59 Li, H.J., Qin, X.Y., and Li, D. (2008) *Mater. Sci. Eng. B*, **149**, 53–57.

60 Candolfi, C., Lenoir, B., Dauscher, A., Guilmeau, E., Hejtmánek, J., Tobola, J., Wiendlocha, B., and Kaprzyk, S. (2009) *Phys. Rev. B*, **79**, 035114.

61 Dashjav, E., Szczepenowska, A., and Kleinke, H. (2002) *J. Mater. Chem.*, **12**, 345–349.

62 Xu, J. and Kleinke, H. (2008) *J. Comput. Chem.*, **29**, 2134–2143.

63 Candolfi, C., Lenoir, B., Dauscher, A., Tobola, J., Clarke, S.J., and Smith, R.I. (2008) *Chem. Mater.*, **20**, 6556–6561.

64 Soheilnia, N., Xu, H., Zhang, H., Tritt, T.M., Swainson, I., and Kleinke, H. (2007) *Chem. Mater.*, **19**, 4063–4068.

65 Gascoin, F., Rasmussen, J., and Snyder, G.J. (2007) *J. Alloys Compd.*, **427**, 324–329.

66 Xu, H., Kleinke, K.M., Holgate, T., Zhang, H., Su, Z., Tritt, T.M., and Kleinke, H. (2009) *J. Appl. Phys.*, **105**, 053703/1-5.

67 Candolfi, C., Lenoir, B., Dauscher, A., Malaman, B., Guilmeau, E., Hejtmánek, J., and Tobola, J. (2010) *Appl. Phys. Lett.*, **96**, 262103/1-3.

68 Candolfi, C., Lenoir, B., Leszczynski, J., Dauscher, A., Tobola, J., Clarke, S.J., and Smith, R.I. (2009) *Inorg. Chem.*, **48**, 5216–5223.

69 Guo, Q. and Kleinke, H. (2014) *J. Solid State Chem.*, **215**, 253–259.

70 Candolfi, C., Lenoir, B., Dauscher, A., Guilmeau, E., and Hejtmánek, J. (2010) *J. Appl. Phys.*, **107**, 093709/1-2.

71 Deng, X., Lü, M., and Meng, J. (2013) *J. Alloys Compd.*, **577**, 183–188.

72 Guo, Q., Assoud, A., and Kleinke, H. (2016) *J. Solid State Chem.* **236**, 123–129.

73 Nandihalli, N., Lahwal, A., Thompson, D., Holgate, T.C., Tritt, T.M., Dassylva-Raymond, V., Kiss, L.I., Sellier, E., Gorsse, S., and Kleinke, H. (2013) *J. Solid State Chem.*, **203**, 25–30.

74 Nandihalli, N., Gorsse, S., and Kleinke, H. (2015) *J. Solid State Chem.*, **226**, 164–169.

75 Continenza, A., Picozzi, S., Geng, W.T., and Freeman, A.J. (2001) *Phys Rev. B*, **64**, 085204/1-7.

76 Johnston, W.D., Miller, R.C., and Damon, D.H. (1965) *J. Less Common Met.*, **8**, 272–287.

77 Bentien, A., Johnsen, S., Madsen, G.K.H., Iversen, B.B., and Steglich, F. (2007) *Europhys. Lett.*, **80**, 17008/1-5.

78 Sanchela, A.V., Thakur, A.D., and Tomy, C.V. (2015) *J. Materiomics*, **1**, 205–212.

79 Caillat, T., Borshchevsky, A., and Fleurial, J.-P. (1993) *AIP Conf. Proc.*, **271**, 771–776.

80 Caillat, T. (1998) *J. Phys. Chem. Solids*, **59**, 61–67.

81 Caillat, T. (1996) *J. Phys. Chem. Solids*, **57**, 1351–1358.

82 Kawaharada, Y., Kurosaki, K., Uno, M., and Yamanaka, S. (2001) *J. Alloys Compd.*, **315**, 193–197.

83 Zhang, J.X., Lu, Q.M., Liu, K.G., Zhang, L., and Zhou, M.L. (2004) *Mater. Lett.*, **58**, 1981–1984.

84 Braun, D.J. and Jeitschko, W. (1980) *J. Less Common Met.*, **72**, 147–156.

85 Nolas, G.S., Morelli, D.T., and Tritt, T.M. (1999) *Annu. Rev. Mat. Sci.*, **29**, 89–116.

86 Yang, J. and Caillat, T. (2006) *Mater. Res. Bull.*, **31**, 224–229.

87 Chakoumakos, B.C., Sales, B.C., Mandrus, D., and Keppens, V. (1999) *Acta Crystallogr. B*, **55**, 341–347.

88 Koza, M.M., Johnson, M.R., Viennois, R., Mutka, H., Girard, L., and Ravot, D. (2008) *Nat. Mater.*, **7**, 805–810.

89 Fornari, M. and Singh, D.J. (1999) *Appl. Phys. Lett.*, **74**, 3666–3668.

90 Fornari, M. and Singh, D.J. (1999) *Phys. Rev. B*, **59**, 9722–9724.

91 Anno, H., Ashida, K., Matsubara, K., Nolas, G.S., Akai, K., Matsuura, M., and Nagao, J. (2002) *Mater. Res. Soc. Symp. Proc.*, **691**, 49–54.

92 Uher, C. (2001) *Semicond. Semimet.*, **69**, 139–253.

93 Nolas, G.S., Kaeser, M., Littleton, R.T.I., and Tritt, T.M. (2000) *Appl. Phys. Lett.*, **77**, 1855–1857.

94 Chen, L.D., Kawahara, T., Tang, X.F., Goto, T., Hirai, T., Dyck, J.S., Chen, W., and Uher, C. (2001) *J. Appl. Phys.*, **90**, 1864–1868.

95 Shi, X., Yang, J., Salvador, J.R., Chi, M., Cho, J.Y., Wang, H., Bai, S., Yang, J., Zhang, W., and Chen, L. (2011) *J. Am. Chem. Soc.*, **133**, 7837–7846.

96 Sales, B.C., Mandrus, D., and Williams, R.K. (1996) *Science*, **272**, 1325–1328.

97 Dargys, A. and Kundrotas, J. (1983) *J. Phys. Chem. Solids*, **44**, 261–267.

98 Caillat, T., Fleurial, J.-P., and Borshchevsky, A. (1997) *J. Phys. Chem. Solids*, **58**, 1119–1125.

99 Snyder, G.J., Christensen, M., Nishibori, E., Caillat, T., and Iversen, B.B. (2004) *Nat. Mater.*, **3**, 458–463.

100 Mozharivskyj, Y., Janssen, Y., Harringa, J.L., Kracher, A., Tsokol, A.O., and

Miller, G.J. (2006) *Chem. Mater.*, **18**, 822–831.

101 Kim, S.-G., Mazin, I.I., and Singh, D.J. (1998) *Phys. Rev. B*, **57**, 6199–6203.

102 Kim, H.J., Božin, E.S., Haile, S.M., Snyder, G.J., and Billinge, S.J.L. (2007) *Phys. Rev. B*, **75**, 134103/1.

103 Chitroub, M., Besse, F., and Scherrer, H. (2008) *J. Alloys Compd.*, **460**, 90–93.

104 Ueno, K., Yamamoto, A., Noguchi, T., Inoue, T., Sodeoka, S., and Obara, H. (2005) *J. Alloys Compd.*, **388**, 118–121.

105 Mastronardi, K., Young, D., Wang, C.C., Khalifah, P., Cava, R.J., and Ramirez, A.P. (1999) *Appl. Phys. Lett.*, **74**, 1415–1417.

106 Xia, Y., Bhattacharya, S., Ponnambalam, V., Pope, A.L., Poon, S.J., and Tritt, T.M. (2000) *J. Appl. Phys.*, **88**, 1952–1955.

107 Xia, Y., Ponnambalam, V., Bhattacharya, S., Pope, A.L., Poon, S.J., and Tritt, T.M. (2001) *J. Phys. Condens. Matter*, **13**, 77–89.

108 Xie, W.J., Jin, Q., and Tang, X.F. (2008) *J. Appl. Phys.*, **103**, 043711/1-5.

109 Qiu, P., Huang, X., Chen, X., and Chen, L. n. (2009) *J. Appl. Phys.*, **106**, 103703/1-6.

110 Rausch, E., Balke, B., Quardi, S., and Felser, C. (2014) *Phys. Chem. Chem. Phys.*, **16**, 25258–25262.

111 Bhattacharya, S., Pope, A.L., Littleton, R.T.I., Tritt, T.M., Ponnambalam, V., Xia, Y., and Poon, S.J. (2000) *Appl. Phys. Lett.*, **77**, 2476–2478.

112 Xie, H., Wang, H., Pei, Y., Fu, C., Liu, X., Snyder, G.J., Zhao, X., and Zhu, T. (2013) *Adv. Funct. Mater.*, **23**, 5123–5130.

113 Yan, X., Joshi, G., Liu, W., Lan, Y., Wang, H., Lee, S., Simonson, J.W., Poon, S.J., Tritt, T.M., Chen, G., and Ren, Z.F. (2011) *Nano Lett.*, **11**, 556–560.

114 Wang, J., Liu, X.-C., Xia, S.-Q., and Tao, X.-T. (2013) *J. Am. Chem. Soc.*, **135**, 11840–11848.

115 Zevalkink, A., Zeier, W.G., Cheng, E., Snyder, J., Fleurial, J.-P., and Bux, S. (2014) *Chem. Mater.*, **26**, 5710–5717.

116 Wang, X.-J., Tang, M.-B., Chen, H.-H., Yang, X.-X., Zhao, J.-T., Burkhardt, U., and Grin, Y. (2009) *Appl. Phys. Lett.*, **94**, 092106/1-3.

117 Guo, K., Cao, Q., and Zhao, J. (2013) *J. Rare Earths*, **31**, 1029–1038.

118 Guo, K., Cao, Q.-G., Feng, X.-J., Tang, M.-B., Chen, H.-H., Guo, X., Chen, L., Grin, Y., and Zhao, J.-T. (2011) *Eur. J. Inorg. Chem*, **2011**, 4043–4048.

119 Aydemir, U., Zevalkink, A., Ormeci, A., Gibbs, Z.M., Bux, S., and Snyder, G.J. (2015) *Chem. Mater.*, **27**, 1622–1630.

120 Zeier, W.G., Zevalkink, A., Schechtel, E., Tremel, W., and Snyder, G.J. (2012) *J. Mater. Chem.*, **22**, 9826–9830.

121 Zevalkink, A., Toberer, E.S., Zeier, W.G., Flage-Larsen, E., and Snyder, G.J. (2011) *Energy Environ. Sci.*, **4**, 510–518.

122 Zevalkink, A., Zeier, W.G., Pomrehn, G., Schechtel, E., Tremel, W., and Snyder, G.J. (2012) *Energy Environ. Sci.*, **5**, 9121–9128.

123 Bobev, S., Thompson, J.D., Sarrao, J.L., Olmstead, M.M., Hope, H., and Kauzlarich, S.M. (2004) *Inorg. Chem.*, **43**, 5044–5052.

124 Xia, S.-Q. and Bobev, S. (2007) *J. Am. Chem. Soc.*, **129**, 10011–10018.

125 Xia, S.-Q. and Bobev, S. (2010) *Chem. Mater.*, **22**, 840–850.

126 Liu, X.-C., Wu, Z., Xia, S.-Q., Tao, X.-T., and Bobev, S. (2015) *Inorg. Chem.*, **54**, 947–955.

127 Ohno, S., Zevalkink, A., Takagiwa, Y., Bux, S.K., and Snyder, G.J. (2014) *J. Mater. Chem. A*, **2**, 7478–7483.

128 Zevalkink, A., Takagiwa, Y., Kitahara, K., Kimura, K., and Snyder, G.J. (2014) *Dalton Trans.*, **43**, 4720–4725.

129 Zevalkink, A., Chanakian, S., Aydemir, U., Ormeci, A., Pomrehn, G., Bux, S., Fleurial, J.-P., and Snyder, G.J. (2015) *J. Phys.*, **27**, 015801/1-7.

130 Aydemir, U., Zevalkink, A., Ormeci, A., Wang, H., Ohno, S., Bux, S., and Snyder, G.J. (2015) *Dalton Trans.*, **44**, 6767–6774.

131 Toberer, E.S., Zevalkink, A., Crisosto, N., and Snyder, G.J. (2010) *Adv. Funct. Mater.*, **20**, 4375–4380.

132 Zevalkink, A., Swallow, J., and Snyder, G.J. (2013) *Dalton Trans.*, **42**, 9713–9719.

133 Mooney Slater, R.C.L. (1959) *Acta Crystallogr.*, **12**, 187–196.

134 Gallup, R.F., Fong, C.Y., and Kauzlarich, S.M. (1992) *Inorg. Chem.*, **31**, 115–118.

135 Carrillo-Cabrera, W., Somer, M., Peters, K., and von Schnering, H.G. (1996) *Chem. Ber.*, **129**, 1015–1023.

136 Kauzlarich, S.M., Thomas, M.M., Odink, D.A., and Olmstead, M.M. (1991) *J. Am. Chem. Soc.*, **113**, 7205–7208.

137 Young, D.M., Torardi, C.C., Olmstead, M.M., and Kauzlarich, S.M. (1995) *Chem. Mater.*, **7**, 93–101.

138 Rehr, A., Kuromoto, T.Y., Kauzlarich, S.M., Castillo, J.Del., and Webb, D.J. (1994) *Chem. Mater.*, **6**, 93–99.

139 Brown, S.R., Kauzlarich, S.M., Gascoin, F., and Snyder, G.J. (2006) *Chem. Mater.*, **18**, 1873–1877.

140 Snyder, G.J. and Toberer, E.S. (2008) *Nat. Mater.*, **7**, 105–114.

141 Hermann, R.P., Grandjean, F., Kafle, D., Brown, D.E., Johnson, C.E., Kauzlarich, S.M., and Long, G.J. (2007) *Inorg. Chem.*, **46**, 10736–10740.

142 Toberer, E.S., Brown, S.R., Ikeda, T., Kauzlarich, S.M., and Snyder, G.J. (2008) *Appl. Phys. Lett.*, **93**, 062110/1-3.

143 Toberer, E.S., Cox, C.A., Brown, S.R., Ikeda, T., May, A.F., Kauzlarich, S.M., and Snyder, G.J. (2008) *Adv. Funct. Mater.*, **18**, 2795–2800.

144 Hu, Y., Bux, S.K., Grebenkemper, J.H., and Kauzlarich, S.M. (2015) *J. Mater. Chem. C*, **3**, 10566–10573.

145 Hu, Y., Wang, J., Kawamura, A., Kovnir, K., and Kauzlarich, S.M. (2015) *Chem. Mater.*, **27**, 343–351.

12
Metal Hydrides

Yaoqing Zhang,[1] Maarten C. Verbraeken,[2] Cédric Tassel,[3] and Hiroshi Kageyama[3]

[1]*Tokyo Institute of Technology, Materials Research Centre for Element Strategy, 4259 Nagatsuta, Midori-ku, Yokohama 226-8503, Japan*
[2]*University of St Andrews, School of Chemistry, North Haugh, St Andrews KY16 9ST, UK*
[3]*Kyoto University, Department of Energy and Hydrocarbon Chemistry, Nishikyo-ku, Kyoto 615-8581, Japan*

12.1
Introduction

Hydrogen is the first element in the periodic table and the third most abundant on Earth. It is the simplest species, made up of just one proton and one electron (deuterium and tritium have one and two additional neutrons within their nuclei, respectively). Hydrogen sits in the periodic table rather awkwardly. It is usually placed on top of the group 1 alkali metals, because of their mutual one valence electron. From a chemical and physical point of view, this position is not satisfactory however, especially because hydrogen is not a metal. Less frequently but maybe more reasonably, hydrogen is placed ahead of the group 17/VII halogens, since these only need one more electron to fill up their valence shells.

Chemical compounds in which hydrogen is present as partly or fully negatively charged are commonly referred to as hydrides. Despite an abundance of elements in the periodic table with which hydrogen will bond in this way, the hydride species seems poorly researched, especially when compared to its positively charged counterpart, the proton, H^+. Hydrides can exhibit covalent, metallic, and ionic bonding, depending on the electronegativity of the element with which hydrogen bonds. For instance, with elements from group 13, such as Al and B, a covalent bond is formed with hydrogen carrying a δ^- charge, whereas most transition and rare earth metals will form metallic bonds with hydrogen. These materials are named chemical and metallic hydrides, respectively. With the most electropositive elements, the alkali and alkaline earth metals, hydrogen forms ionic bonds and these materials are known as saline hydrides. Hydrides may also serve as anions within an oxide framework to form a particularly interesting materials subgroup called oxyhydrides.

Handbook of Solid State Chemistry, First Edition. Edited by Richard Dronskowski, Shinichi Kikkawa, and Andreas Stein.
© 2017 Wiley-VCH Verlag GmbH & Co. KGaA. Published 2017 by Wiley-VCH Verlag GmbH & Co. KGaA.

Compared to the proton, the negatively charged hydride species and its compounds have received little attention. Metal hydrides have been researched for their use as electrode material in nickel metal hydride batteries and more recently various saline, metallic, and chemical hydrides have received interest for their potential as hydrogen storage materials. Hydrogen storage is considered a key challenge for mobile transport based on hydrogen fuel, and hydrides have the potential to fulfill certain requirements pertaining to this challenge. Very few researchers have looked into the hydride ion's potential in electrochemical devices. Despite its strongly reducing character and therefore (electro)chemical potential, many believe this chemical reactivity makes it unwieldy as such. It is likely to be very mobile however and an early study of conductivity in calcium hydride corroborated this. This chapter will discuss the various classes of hydride compounds, with a special focus on saline and metallic hydrides as well as oxyhydrides. Topics will include thermodynamic stability, crystal chemistry, synthesis, and physical properties. Recent progress in understanding hydride ion mobility in alkaline earth hydrides will be highlighted.

12.2
Hydrogen Chemistry

12.2.1
Hydrogen Species

Elemental hydrogen exists predominantly as hydrogen gas, H_2. The covalent bond between the two nuclei is very strong, with an enthalpy of 436 kJ/mol and a bond length of 0.74 Å. The interaction between different hydrogen molecules, on the other hand, is very weak, because of the few electrons in the hydrogen atoms. Atomic hydrogen ($H^0 \bullet$) is a very unstable species. It combines readily with any chemical compound in order to complete or deplete its 1s valence shell. Accordingly, it will form molecular orbitals resulting in metallic or covalent bonds or, in the extreme case of complete charge separation, ionic bonds (as H^-). Isolated protons (H^+) are, strictly speaking, very unstable and therefore always found in combination with a Lewis base. The electronegativity of hydrogen is 2.2, which is comparable to the electronegativities of B, C, and Si, and so the E—H bonds in compounds with these elements are not expected to be very polar. On the other hand, with the strongly electropositive elements from groups 1 and 2, it forms ionic compounds in which the hydrogen acquires electron density and is best described as a negatively charged ion, the hydride ion. When combined with electronegative elements on the right-hand side of the periodic table, the E—H bond is best regarded as covalent and polar, with the H atom carrying a small positive (e.g., C—H, O—H, N—H, and F—H) or small negative charge (e.g., B—H and Al—H). As briefly discussed earlier, metallic bonding is observed with the majority of transition and rare earth metals.

The radius of atomic hydrogen (H^0•) is 0.37 Å, making it the smallest found for any atom of the chemical elements. The free hydride ion, H$^-$, on the other hand, has a much higher ionic radius of 2.08 Å, which is comparable to the ionic radii of large halides, such as bromide (1.95 Å) and iodide (2.16 Å) anions [1]. The hydride ion is extremely polarizable however and its ionic radius in alkali and alkaline earth hydrides can be as small as 1.25 Å, which would make it more similar to the ionic radius of the fluoride ion F$^-$ (1.31 Å in fourfold coordination) [2–5]. The electron affinity of the hydrogen atom is only −0.80 eV. This makes that the electron density of H$^-$ is loosely bound to the nucleus and therefore the hydride anion is a strong electron-donating species, a strong Lewis base and a potent reducing agent. In fact, with −2.25 V (versus the standard hydrogen potential), the electrochemical potential for the H$^-$/H$_2$ couple is similar to that of the Mg/Mg^{2+} couple. The proton, H$^+$, obviously has an ionic radius close to 0.00 Å, which gives it an extremely high charge/radius ratio. This in turn makes the proton a very strong Lewis acid. In the gas phase, it therefore readily combines with other molecules or atoms. In condensed phases, it is always found in combination with a Lewis base, from which it spontaneously extracts some electron density. The strong Lewis acidity is also emphasized by the large ionization energy of the hydrogen atom, that is, 13.6 eV. It becomes evident that despite hydrogen's seemingly simple makeup, its chemical properties are versatile, ranging from being a very strong Lewis acid to a strong Lewis base. And although the hydrogen atom has only one single electron, it still can bind to more than one atom simultaneously [6].

12.2.2
Hydride Compounds

Hydrogen forms stable compounds with the major part of the elements in the periodic table, but in this chapter we are concerned with hydride compounds and in particular those containing alkali, alkaline earth, and transition and rare earth metals. The saline hydrides, that is, AH and AeH_2 (with A = Li, Na, K, Rb, and Cs; Ae = Mg, Ca, Sr, and Ba) are proper ionic materials, in which hydrogen is present as hydride anions, H$^-$. Because of the strong Lewis base character of the hydride ion, combination with any Lewis acid will result in decomposition. Exposure to protonated compounds in particular will lead to violent reactions with the release of H$_2$, meaning that all of these compounds are pyrophoric. They are however stable in dry inert atmospheres. The metallic hydrides, formed with most transition and rare earth metals, exhibit hydrogen dissolved as single atoms. In these materials, hydrogen donates some of its electron density to lower the energy of the metals' d and f bands, while retaining most of its hydridic character. Their reactivity toward water and other protonated compounds however is much reduced compared to saline hydrides.

Aluminum and boron hydrides (AlH$_3$ and BH$_3$, respectively) are considered chemical hydrides, due to the covalent nature of their bonding. Although not the focus of this chapter, the closely related and more ionic compounds, the complex hydrides with general formula $AX\text{H}_4$ (A = Li, Na, K, Mg, Ca; X = B, Al) have been the focus of much research relating to their hydrogen storage capabilities. The strong ionic bonding in the "pure" saline hydrides, that is, those compounds comprising alkali and alkaline earth cations only, generally results in thermodynamically too stable compounds to be of interest for most hydrogen storage applications.

12.3
Saline Hydrides

Saline hydrides show many similarities with their halide analogues, especially concerning crystal and electronic structures and, perhaps to a lesser extent, physical attributes such as brittleness, hardness, and optical properties [7–10]. The degree of ionicity for saline hydrides compared to their halide analogues is a much debated issue, due to the much reduced electronegativity of the hydride anion. However, even for LiH, which is expected to have the highest covalent character of the alkali hydrides, the degree of ionicity is reported to be between 0.8 and 1 [7,11,12]. More recently, it has been suggested that the bonding in MgH_2, which was traditionally believed to be intermediate between covalent and ionic, is in fact predominantly ionic in nature [13]. The electronic states of these hydride materials are relevant as will become clear when discussing their electric properties. Any metallic or covalent bonding could give rise to mobile electronic defects, whereas in purely ionic crystals, transport properties can be assumed to be due to mobile ions only (or ion vacancies). Several studies have found that bond valence sum values in alkaline earth hydrides, in which especially the hydride ions show nontrivial values [14–16], leading to suggestions of limited degrees of covalency in such materials. We will see in the structural section, however, that there may be different reasons for this discrepancy and therefore it is suggested in this chapter that alkali and alkaline earth hydrides are proper ionic crystals.

12.3.1
Thermodynamics of Hydride Compounds

The formation of hydrides occurs via simple absorption of either hydrogen gas or hydrogen atoms from an electrolyte. The latter is typical for metal hydride batteries, using a standard hydroxide solution to transfer the hydrogen atoms from one electrode to the other. Absorption/desorption of hydrogen in the metal is typically shown as composition/hydrogen pressure isotherms. These show some generalities, and typical isotherms are shown in Figure 12.1. At low pressures, hydrogen simply dissolves into the metal lattice, or α-phase, or in Kröger–Vink notation:

$$\frac{1}{2}H_2 + \square_i^x \rightarrow H_i^x. \tag{12.1}$$

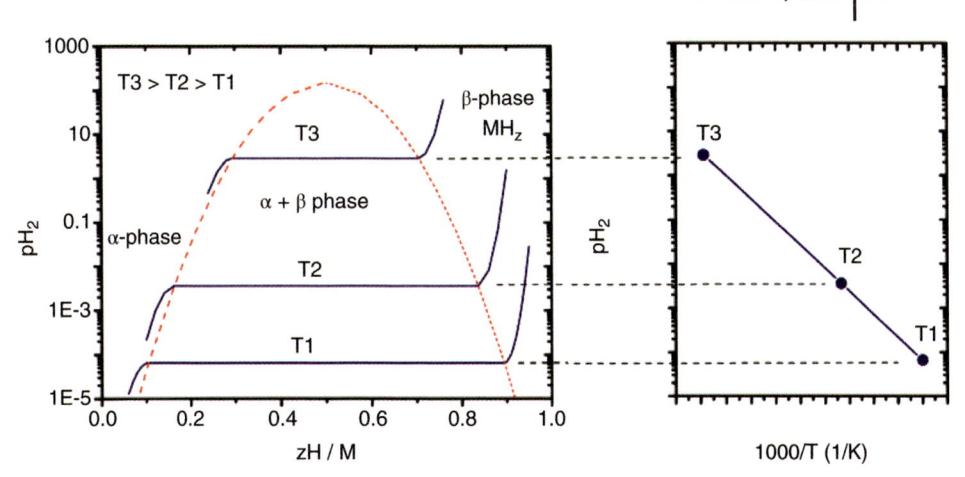

Figure 12.1 General representation of a metal–metal hydride phase diagram. Plateau pressures for formation/decomposition move up with increasing temperature.

During this stage, hydrogen fills vacant octahedral or tetrahedral holes (*Entityismissing$_i^x$*) and is present as an interstitial species. With increasing pH$_2$, the concentration of hydrogen in the metal increases and nucleation of the actual hydride phase (β-phase) starts. According to Gibb's phase rule, where f is the degree of freedom, C is the number of components, and P is the number of phases,

$$f = C - P + 2, \tag{12.2}$$

now due to the coexistence of α- and β-phases, a plateau in the hydrogen partial pressure is observed. When equilibrium has been reached at a specific temperature and hydrogen partial pressure, only the β-phase will remain, with often substoichiometric hydrogen content. Increasing the hydrogen content further to obtain full stoichiometry is now accompanied with high hydrogen pressures; these stoichiometric phases are therefore often thermodynamically unstable under realistic conditions. The plateau pressure increases with increasing temperature and the combination of these two variables determines the thermodynamic stability of the hydride phase. Many hydrides however are kinetically stable in the absence of hydrogen and only start decomposing when the material's internal hydrogen pressure starts to become significant. This kinetic barrier plays an important role in hydrogen storage materials.

So-called higher hydrides, for instance, GdH$_3$ or Th$_4$H$_{15}$, have pressure-composition isotherms with two plateaus. The lower plateau resembles the formation of the lower hydride (MH_2), whereas the higher hydride starts to form on the higher plateau, coexisting with the lower hydride.

The logarithm of the dissociation pressure generally shows a linear relationship with the reciprocal temperature, as shown in Figure 12.1. The reason for this behavior can be found in the system's thermodynamics. Consider the next dissociation reaction for the hydride or β-phase:

$$MH_x \rightarrow M + \frac{x}{2}H_2. \tag{12.3}$$

For the free enthalpy of hydride formation, the following equation holds:

$$\Delta G_f = \Delta G_f^0 + RT \ln Q_p, \quad \text{with} \quad Q_p = \frac{a_{MH_x}}{a_M \cdot pH_2^{x/2}}. \tag{12.4}$$

In equilibrium, that is, with $\Delta G = 0$, this becomes

$$-\Delta G_f^0 = RT \ln K_p. \tag{12.5}$$

Using $\Delta G_f^0 = \Delta H_f^0 - T\Delta S_f^0$, and doing some basic differentiation, this becomes

$$-\Delta H_f^0 + T\Delta S_f^0 = RT \ln K_p, \tag{12.6}$$

$$\frac{\Delta H_f^0}{RT^2} = \frac{\partial \ln K_p}{\partial T}. \tag{12.7}$$

The last equation is better known as the Van 't Hoff equation. When the metal and metallic hydrides are assumed to be pure solids, their activities become unity. Therefore, with $K_p = pH_2^{-(x/2)}$, the Van 't Hoff equation can also be written as

$$\ln pH_2 = \frac{2}{xR}\left(\frac{\Delta H_f^0}{T} - \Delta S_f^0\right). \tag{12.8}$$

From this last equation, it can clearly be seen that the logarithm of the dissociation pressure is linearly dependent on the reciprocal temperature. Unfortunately, the significant nonstoichiometry of hydrides yields nonunity values for the activities of the α- and β-phases. This means that the slope in the Van 't Hoff plot can only be used as an approximation to calculate the enthalpy of formation. Still it is found that all hydrides obey an equation of a similar form experimentally:

$$\log pH_2 = -\frac{A}{T} + B. \tag{12.9}$$

Figure 12.2 shows the Van 't Hoff plots for a selection of saline hydrides. Table 12.1 gives an overview of enthalpies and entropies of formation for the various saline hydrides, along with their reported decomposition temperatures. It can easily be seen that even the least stable saline hydride, that is, MgH_2, still has a relatively high decomposition temperature at relevant partial hydrogen pressures for onboard hydrogen storage. Numerous studies have attempted to destabilize MgH_2, chiefly by chemical substitution strategies, to lower its decomposition temperature and capitalize on its promising gravimetric hydrogen density of 7.7 wt%, but challenges remain, not least the poor kinetics for hydrogen absorption/desorption.

Excellent overviews of some of the early thermodynamic work on various saline hydrides are provided in Refs [10,28]. This chapter is not attempting to rewrite those earlier findings, but merely recapture the basics and add some recent developments in the field.

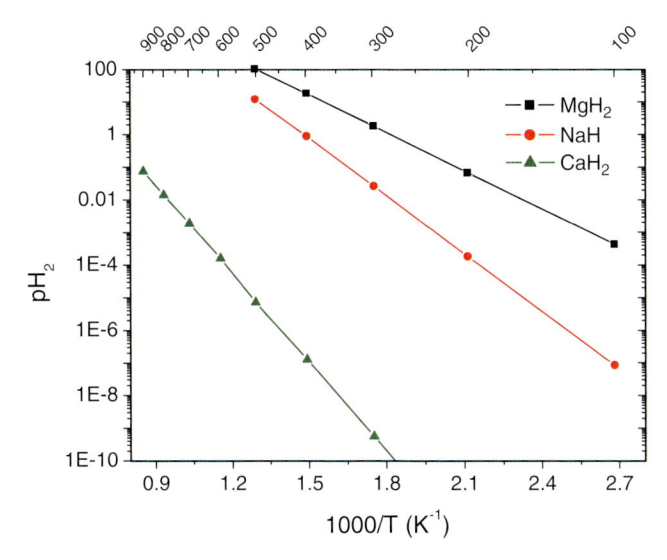

Figure 12.2 Typical Van't Hoff plot for a selection of saline hydrides, showing large variety in thermodynamic stability across this range of materials.

12.3.2
Crystal Structures

To obtain the correct crystal structure information on hydride phases, neutron diffraction data are generally required in combination with X-ray diffraction. Due to hydrogen's low electron density, it is a weak scatterer in X-rays and thus very little information can be obtained on the hydride positions using this

Table 12.1 Formation enthalpies and entropies for saline hydrides, determined/extrapolated at 298 K, unless stated otherwise.

	ΔH_f^0	ΔS_f^0	T_{decomp} (°C at $pH_2 = 1$ bar)	References
LiH	−90.6	−68.6	985	[17]
NaH	−56.4	−83.2	428	[18]
KH	−57.8	−83.4	425	[19]
RbH	−54.4	−85.4	363	[20]
CsH	−56.5	−85.3	389	[21]
MgH_2	−74.3[a]	−135[a]	277	[22–24]
CaH_2	−174.3	−127.2	1061	[25]
SrH_2	−182.4	−191.4[b]	1054	[26]
BaH_2	−191.2	−148.4[c]	1015	[27]

a) Determined at 565–645 K.
b) Determined at 973–1173 K.
c) Determined at 1008–1223 K.

technique. Hydrogen, on the other hand, has a sizable coherent and negative scattering length for neutrons (−3.74 fm), which contrasts with oxygen with a positive scattering length (+5.8 fm). Unfortunately, a much larger incoherent scattering length (25.3 fm) of hydrogen causes poor signal-to-noise ratios during diffraction experiments [29]. Although this effect can be partially reduced by optimizing diffractometer geometry, counting times, and finding suitable neutron wavelengths [30], the hitherto preferred solution is to use isotopically substituted samples, where hydrogen has been replaced by deuterium. Deuterium has a much smaller incoherent cross section compared to hydrogen (2.05 versus 80.3 barns, respectively) coupled with a larger (and positive) scattering length of 6.67 fm. This means that most available structural information on hydrides has in fact been obtained on deuterides while assuming that the structures are identical. Caution obviously needs to be exercised then when comparing the two, especially considering that the hydrogen isotopes have the highest mass ratio of all isotope combinations. From X-ray diffraction data, it can, for instance, be seen that lattice parameters for hydrides are generally larger than their deuteride phases, due to the larger zero-point motion of hydrogen. It was also found that atomic positions may shift significantly between the two isotopic variations [31]. No effect on symmetry has ever been observed however.

All alkali metal hydrides have the cubic rocksalt (NaCl type, $Fm3m$ space group) structure, completely analogous to their fluoride counterparts. The rocksalt structure can be described as face centered closed packed (usually for the larger anions), with the other ions occupying the octahedral holes. A graphical representation of this structure is shown in Figure 12.3, and Table 12.2 shows a list of lattice parameters.

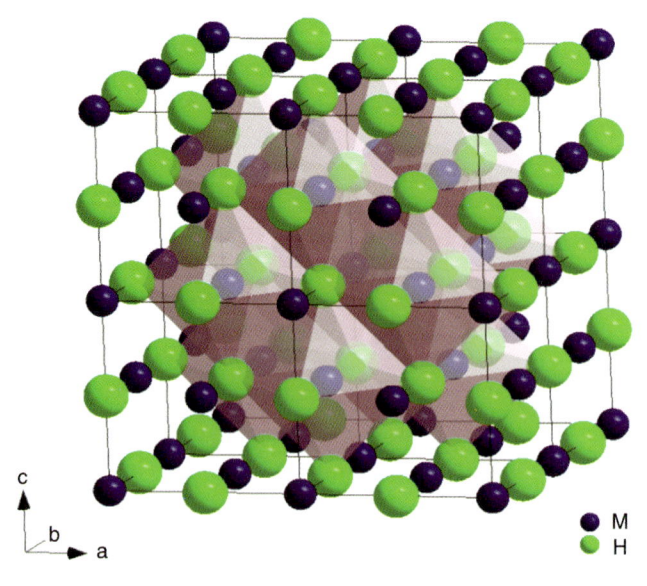

Figure 12.3 Rocksalt structure of alkali hydrides.

Table 12.2 Cubic lattice parameters for alkali hydride.

Alkali hydride	Cubic lattice parameter (Å)	Reference
LiH	4.083	[32]
NaH	4.872	[33]
KH	5.698	[33]
RbH	6.037	[34]
CsH	6.388	[35]

As shown in Table 12.3, the heavy alkaline earth (Ca, Sr, and Ba) hydrides are all isostructural, crystallizing in the orthorhombic cotunnite ($PbCl_2$ type) structure (space group *Pnma*) [36]. This is in contrast with their fluoride analogues, which all exhibit the cubic fluorite structure. However, a high-pressure variant of BaF_2 shares the cotunnite structure. The cotunnite structure of the alkaline earth

Table 12.3 Key structural information for heavy alkaline earth hydrides and deuterides obtained at room temperature.

	CaH_2	CaD_2	SrH_2	SrD_2	BaH_2	BaD_2
Lattice parameters						
a (Å)	5.936	5.9475	6.3799(2)	6.3673	6.792	6.789
b (Å)	3.600	3.5933	3.8708(1)	3.8629	4.168	4.171
c (Å)	6.838	6.8019	7.3234(2)	7.3089	7.858	7.852
V (Å^3)	146.1	145.36	180.8	179.77	222.5	222.3
Fractional coordinates						
Ca/Sr/Ba (4c)						
x	0.2424	0.2397	0.2601(5)	0.2399	0.2397	0.240
y	1/4	1/4	1/4	1/4	1/4	1/4
z	0.1086	0.1093	0.1109(3)	0.1107	0.1114	0.113
H1/D1 (4c)						
x	0.3543	0.3551	—	0.3542	0.35	0.345
y	1/4	1/4	—	1/4	1/4	1/4
z	0.4264	0.4268	—	0.4266	0.414	0.427
H2/D2 (4c)						
x	0.9727	0.9743	—	0.9742	0.96	0.971
y	1/4	1/4	—	1/4	1/4	1/4
z	0.6775	0.6759	—	0.6780	0.67	0.687
Reference	[15]	[37]	a)	[14]	[38]	[4]

a) Unpublished refined powder X-ray diffraction data obtained at University of St Andrews (H positions were not refined).

hydrides, shown in Figure 12.4a, is based on a distorted hexagonal array of alkaline earth ions as can easily be seen by comparing the cotunnite structure with the hexagonal variant in Figure 12.4b. The AeH$_2$ cotunnite structure involves two H types: H1 that is located at the corner-shared vertices between different AeH$_x$ polyhedra and H2 that can be viewed as isolated or only close coordinated to one Ae atom. The ideal hexagonal structure (Figure 12.4c) can be seen as based upon trigonally coordinated layers containing Ae and H1 alternating with layers of H2. The cotunnite structure (Figure 12.4a) is derived from the hexagonal variant by buckling of the AeH1 layers, resulting in a square pyramidal Ae–H coordination involving three H1 atoms derived from the trigonal Ae–H1 layer

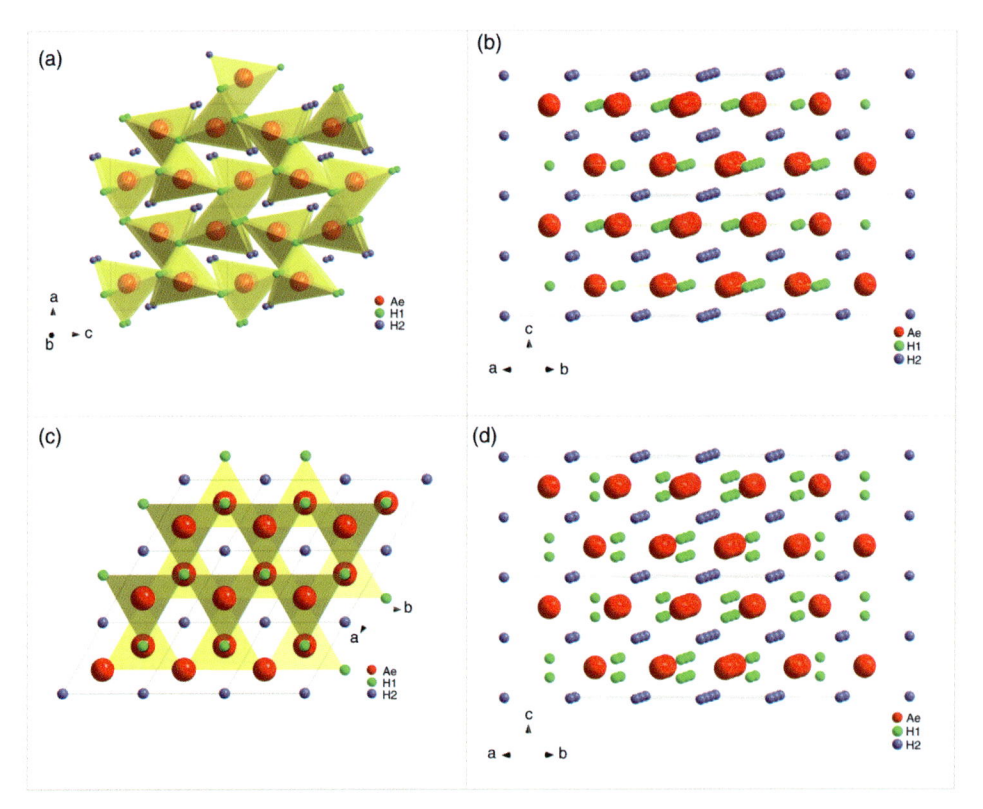

Figure 12.4 Representations of the alkaline earth hydride crystal structures (a and b). The cotunnite structure consists of inverted layers of corner-sharing AeH$_5$-distorted square pyramids (a). Corner sharing is through H1, whereas H2 is only coordinated to one Ae. On going to hexagonal symmetry, a regular layered structure emerges, with alternating Ae-H1 and H2 layers (b). Looking down the c-axis of the ideal hexagonal, unit cell shows the trigonal coordination of H1 to Ae (c). The ideal hexagonal structure does not provide the best fit; instead, the H1 sites are split and H1 occupies a partially occupied disordered site resulting in three Ae-H1 bonds of 2.63 Å and one Ae-H1 bond of 2.84 Å (d). (Reproduced with permission from Ref. [16]. Copyright 2015, Nature Publishing Group.)

with the addition of a H1 from an adjacent Ae–H1 layer and a H2 from a H2 layer. While all H1 atoms can be viewed as corner shared, the H2 is only in close coordination with one Ae atom. The trigonal arrangement of the Ae–H1 layers in the hexagonal variant is made clear in Figure 12.4c.

A structural phase change was in fact observed for BaD_2 at high temperatures [16], giving rise to a hexagonal variant. This phase change exhibits a certain degree of hysteresis and therefore both low- and high-temperature phases coexist between 500 and 610 °C. The high-temperature hexagonal phase has the Ni_2In-type structure, space group $P6_3/mmc$. Unusually, H1/D1 was observed to occupy a site slightly off the high symmetry 2d site, shown in Figure 12.4b, causing it to be split with partial occupation (see Figure 12.4d). This high-temperature phase change has also been observed to take place in SrH_2 and CaH_2 at 855 and 780 °C, respectively, using thermal studies. Thus far, however, the phase change has not been observed directly using structural methods. It is therefore unknown whether the calcium and strontium analogues transform to a similar hexagonal structure, or whether the difference in cation sizes results in different high-temperature symmetries, such as fluorite. Experimental difficulties due to chemical reactivity at high temperatures have precluded such studies until now.

A feature, which is well known to exist for metallic hydrides, is flexible anion stoichiometry. Due to reasons explained in the thermodynamics section, these materials are almost always substoichiometric in hydrogen. Due to the ionic nature of alkali and alkaline earth hydrides, this substoichiometry is often totally ignored or deemed absent in this particular class of materials. Although this seems justified for the alkali hydrides [39,40], neutron diffraction studies, in combination with thermal analyses on CaD_2, SrD_2, and BaD_2, however, have consistently shown that this substoichiometry is indeed significantly present in the alkaline earth hydrides as well. The neutron data are very sensitive to crystal lattice occupancies and, with the exception of Sr, the neutron scattering length of deuterium is significantly different from the cations, thus allowing such analyses. Correct formulas at room temperature would be $CaD_{1.91}$, $SrD_{1.90}$, and $BaD_{1.84}$, with the substoichiometry increasing with increasing temperature. This substoichiometry is also in line with the phase diagrams of the alkaline earth hydrides, as obtained by thermal analysis [41–43]. These substoichiometries obviously suggest the existence of a large number of deuteride vacancies, $V_D^{\bullet\bullet}$, raising questions about the neutralizing negatively charged defects. As it is hard to fathom reduced cation species or mobile electronic defects within such ionic materials, it is suggested instead that the materials contain no deuteride vacancies (other than intrinsic Schottky and Frenkel defects), but rather trapped electrons, occupying the deuteride sites. These defects are also known as F-centers, e_D^x, and their formation can be described as follows:

$$H_H^x \rightarrow \frac{1}{2}H_2 + e_H^x. \tag{12.10}$$

F-centers are common in alkali halide crystals, albeit in smaller concentrations. The hydride substoichiometry may also provide an explanation for the unusual bond valence sums (BVS) found for the hydride ions in some studies [14–16], as the ionic model used to generate BVS generally assumes stoichiometry. Deviations should therefore be expected in hydride materials.

The local coordination and behavior of saline hydrides has been studied through various techniques, including solid-state NMR, infrared spectroscopy, inelastic neutron scattering, and computational methods. Straightforward spectra are obtained for the alkali hydrides, due to their simplicity and high-symmetry structure [44,45]. Studies on the alkaline earth hydrides, on the other hand, corroborate the presence of two distinct H sites and more recently even managed to resolve a degree of anisotropy within these sites [37,46–48]. The inelastic neutron spectra for CaH_2, SrH_2, and BaH_2 show great similarities, including strong absorptions in the 65–100 and 100–140 meV bands, caused by vibration and rotation of the H2 and H1 sites, respectively. Solid-state NMR studies on these materials are still few, but they have been used to assess chemical shifts and proton/deuteron relaxation times [49–51]. The latter results were shown to be strongly thermally activated and could be related to diffusion of hydride ions in the lattice.

Magnesium hydride, MgH_2, is often considered as an intermediate between covalent and ionic hydrides, but several studies have suggested its bonds are in fact predominantly ionic, similar to LiH [13,52–54], with Mg showing a fully ionized state. Therefore, some reference should be made to its structure and properties. Unlike the heavy alkaline earth hydrides, MgH_2 crystallizes in a tetragonal, rutile (TiO_2 type) structure, in which chains of MgH_6 octahedra are both face and corner sharing [55,56]. This structure is shown in Figure 12.5. The

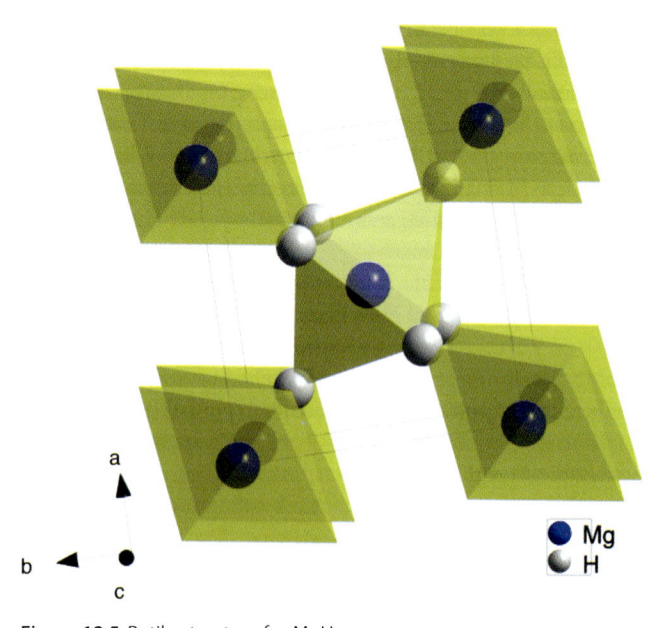

Figure 12.5 Rutile structure for MgH_2.

hydrogen atoms, as determined by neutron diffraction, are such that the MgH_6 octahedra are slightly distorted, resulting in two Mg—H bonds of 1.955 Å and four with distance 1.954 Å. Many polymorphs of MgH_2 exist at high pressures, including a cotunnite structure, as will be discussed in the following section.

12.3.3
High-Pressure Behavior

There has long been an interest in finding metallic behavior and indeed room-temperature superconductivity in solid hydrogen after it was first proposed in 1935 [57]. This metallic behavior would originate at high pressures, when hydrogen is so compressed and densified as to causing enough of a "squeeze" on its electrons to close the large bandgap of >14 eV as experienced at ambient pressures. If such a metallic phase would exist and could be stabilized to ambient pressures, it would mean a breakthrough in many fields, including magnetic levitation, electricity transport, and hydrogen storage. In reality, however, it has proven hard to turn hydrogen into a metal, even at pressures more than 10 times higher than initially calculated [58–60]. The idea of densifying hydrogen until it becomes superconductive has also sparked research into finding new high-pressure phases in the alkali and alkaline earth hydrides, as it is believed that the cations in these materials may aid in effecting the "squeeze" of hydrogen's electronic structure. Alternatively, additional cation or hydridic orbitals are introduced into the electronic structure, which can act as donor/acceptor impurities, effectively lowering the pressure at which metallization of hydrogen would occur (see Figure 12.6). Metallic properties have in fact been observed in a number of these materials so far.

High-pressure diffraction experiments have shown that rocksalt AH (B1 structure, with A = Na, K, Rb, and Cs) adopt the primitive cubic CsCl (B2) structure between 29 and 1 GPa, analogous to their halide counterparts [61,62]. The same transition has been predicted to happen in LiH at pressures over 300 GPa [63,64], although a recent study has found that alternative structures form well before the formation of the B2 phase (see Figure 12.7) [65]. Another transition to the orthorhombic CrB structure has been reported for RbH and CsH at 85 and 18 GPa, respectively. It was also predicted that the B1 → B2 transition in LiH gives rise to a metal–insulator transition, but for the heavier alkali hydrides, bandgaps have not been observed to reduce to less than 1.9 eV, even in the CrB structure [66].

MgH_2 was shown to have several phase transitions at high pressures [56,67]. The high-pressure phases include γ-MgH_2 (coexisting with rutile α-MgH_2 up to 9 GPa with orthorhombic *Pbcn* structure), δ-MgH_2 (orthorhombic phase between 9 and 17 GPa, space group *Pbc2$_1$*), and above 17 GPa cotunnite (*Pnma*), which is stable up to 57 GPa [67]. A pseudocubic β-MgH_2 phase was also found experimentally and computationally at both high pressures and high temperatures [52,54,68]. All MgH_2 polymorphs were shown to be ionic, insulating materials. The heavy alkaline earth hydrides (CaH_2, SrH_2, and BaH_2) also transform at high pressures. These materials adopt a hexagonal Ni_2In structure at 15, 10, and

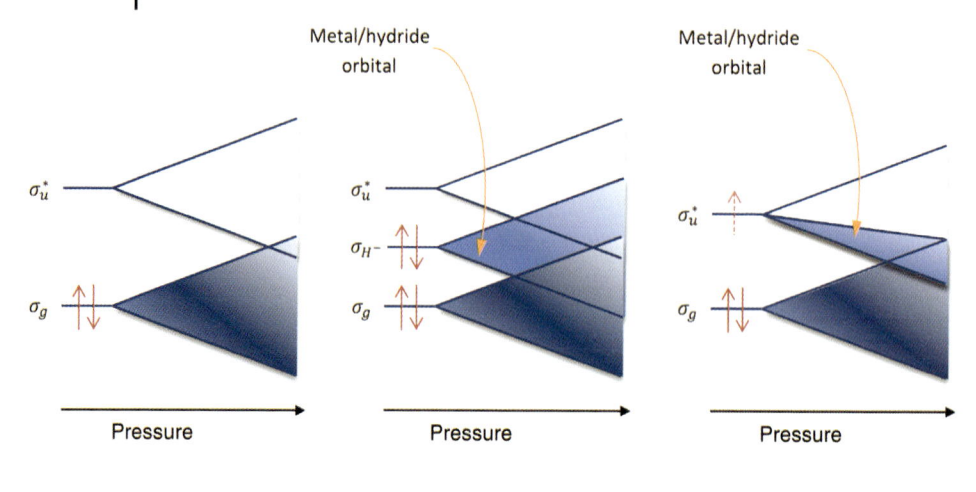

Figure 12.6 Diagram showing concept of and possible scenarios of metallization in hydrogen (compounds). Valence (filled σ_g) and conduction (empty antibonding σ_u^*) band broadening under pressure causes eventual overlap and degeneracy in solid hydrogen (a). In presence of metals, such as Li, an impurity band of H$^-$ is introduced between σ_g and σ_u^* lowering the pressure required for band overlap (b). Alternatively, Li may donate electrons to the empty band, effecting the lowering of the required pressure for metallization (c).

2.5 GPa, respectively [69–72]. Although this Ni$_2$In structure was found to be insulating, another transition to simple hexagonal (AlB$_2$ type) at 50 GPa for BaH$_2$ (predicted to occur at 115 GPa for SrH$_2$) was shown to give rise to an electronic transition and concomitant covalent bonding (see Figure 12.8). Electronic

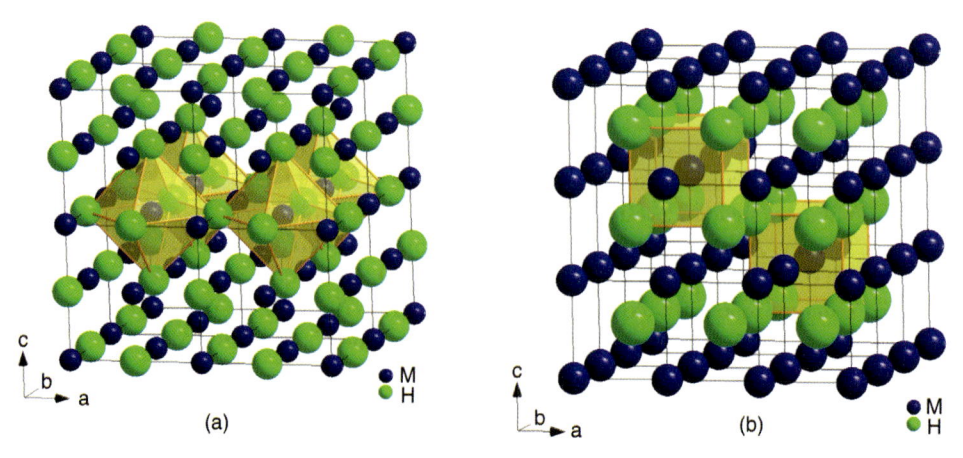

Figure 12.7 High-pressure phase transition in alkali hydrides from NaCl structure type (a) to CsCl structure type (b). Hydride coordination around cations increases from sixfold in part (a) to eightfold in part (b).

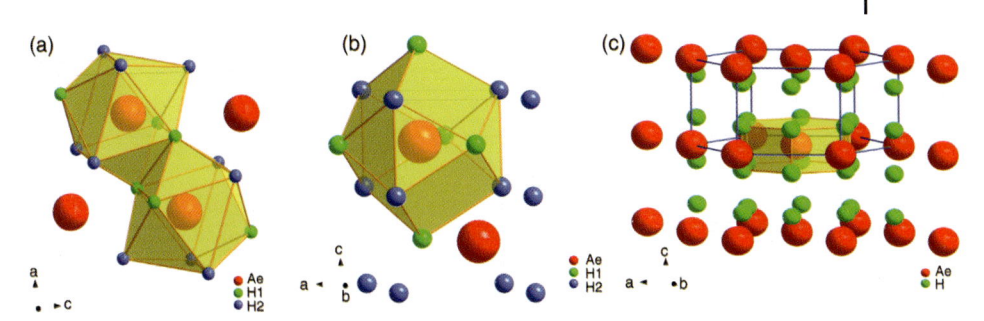

Figure 12.8 High-pressure phase transitions in heavy alkaline earth hydrides, going from cotunnite to Ni$_2$In to AlB$_2$. Hydride coordination around Ae increases from 9-fold to 11-fold in Ni$_2$In and 12-fold in AlB$_2$.

structure calculations suggest possible superconductivity, but with a very low critical temperature T_c of a few millikelvin [73]. Interestingly, the Ni$_2$In structure is also a predicted polymorph for MgH$_2$, but has so far not been observed experimentally. This does suggest more similarities exist between MgH$_2$ and the heavy alkaline earth hydrides than has traditionally been thought.

Due to the difficulty in finding superconducting hydrides with nominal stoichiometry, research is now also focusing on so-called polyhydrides. These are formed by pressurizing alkali and alkaline earth metals (or their hydrides) either alone or in a pressurized/liquid hydrogen medium, resulting in phases that are effectively solid hydrogen doped with the metals [74]. In these polyhydrides, hydrogen starts forming chain-like structures consisting of H$^-$, H$_2$, H$_3^-$, and even H$_4$ subunits and structures as hydrogen rich as LiH$_6$, BaH$_{10}$, and CaH$_{12}$ have been predicted [63,65,75]. Calculations have suggested that these compounds may have high superconducting critical temperatures T_c, with CaH$_6$ having one of the highest predicted T_c so far (220–235 K at 150 GPa) [76]. Polyhydrides for Li and Na have actually been found experimentally, whereas most other hydrogen rich compounds are still based on theoretical work. LiH$_6$ was found to be optically transparent at 215 GPa and therefore insulating, despite predictions that it should be metallic. This method of finding superconducting phases is in fact not exclusive to alkali and alkaline earth metals, but has been applied to other metals or covalent compounds as well, such as FeH$_n$ and H$_n$S, the latter ($n=3$) yielding a superconducting T_c of 203 K, which is higher than those observed for cuprates [77].

12.3.4
Hydride Ion Mobility in Saline Hydrides

Considering the ionic nature of the saline hydrides and the reducing potential of the hydride ion, it is surprising that very few studies have attempted to describe the electrical properties of this class of materials. LiH is probably the best-studied alkali hydride in terms of its electrical behavior. It exhibits appreciable electrical conductivity at elevated temperatures, up to 1×10^{-3} S cm^{-1} at 600 °C; but with an activation energy of 1.7 eV, it is believed the charge carriers are a mixture of lithium and hydrogen ions [7,78]. The heavier alkali hydrides are believed to behave much like their halide counterparts, meaning good ionic conductivity in the melt. The heavy alkaline earth hydrides (CaH$_2$, SrH$_2$, and BaH$_2$) were initially studied for their electrical properties at room temperature at elevated pressures, presumably in search for superconducting phases [79]. The conductivity of CaH$_2$ at ambient pressure was studied by Gorelov and Pal'Guev, who reported particularly large values of 10 S/cm above 780 °C, where the material undergoes a phase transition [80]. The conductivity was assumed to arise from mobile hydride ions, but no further evidence for this was provided. Verbraeken *et al.* studied all of these materials in more detail and found an increase in bulk conductivity on going from CaH$_2$ to SrH$_2$ to BaH$_2$, that is, 5×10^{-3}, 2×10^{-2}, and 0.2 S/cm at 630 °C, respectively [14,16]. The materials were found to exhibit Arrhenius-type behavior as observed in many typical ionic conductors, and their conductivity therefore shows the following relationship with temperature, where σ denotes the conductivity, σ_0 a pre-exponential factor, E_{act} the activation energy, and k_B the Boltzmann constant:

$$\sigma T = \sigma_0 e^{-E_{act}/k_B T}. \tag{12.11}$$

The high conductivity in BaH$_2$ was shown to be related to the structural phase transition as described in the crystallographic section while electromotive force (emf) experiments suggest that the main charge carriers are indeed hydride ions. Considering the structural and chemical similarity of these materials, it is believed the observed conductivities for all materials are due to mobile hydride ions. The bulk conductivity data for these materials have been reproduced in Figure 12.9, showing the jump by an order of magnitude in BaH$_2$, as a result of going to the high-symmetry phase. Unpublished results on CaH$_2$ have shown a similar jump in conductivity around 800 °C (albeit by less than an order of magnitude), which would coincide with the suggested phase transition for this material. The conductivity at this temperature was on the order of 0.1 S/cm, that is, two orders of magnitude lower than reported [81]. The sudden increase in conductivity on going to the high-symmetry phase is believed to result from the creation of a large concentration of vacant H$^-$ sites on the split H1 crystallographic site, as discussed previously. As the activation energy for conductivity remained unchanged across the phase change, it is likely that an increase in charge carrier concentration gives rise to the observed jump. The conductivity can be expressed as a function of defect concentration n, with effective charge e and mobility μ:

$$\sigma = \sum n_i e_i \mu_i, \tag{12.12}$$

where i is the ith type of defect contributing toward electrical transport.

The exact mechanism for conduction in the orthorhombic phase is still a matter of debate and could arise from both interstitial sites and H$^-$ vacancies as

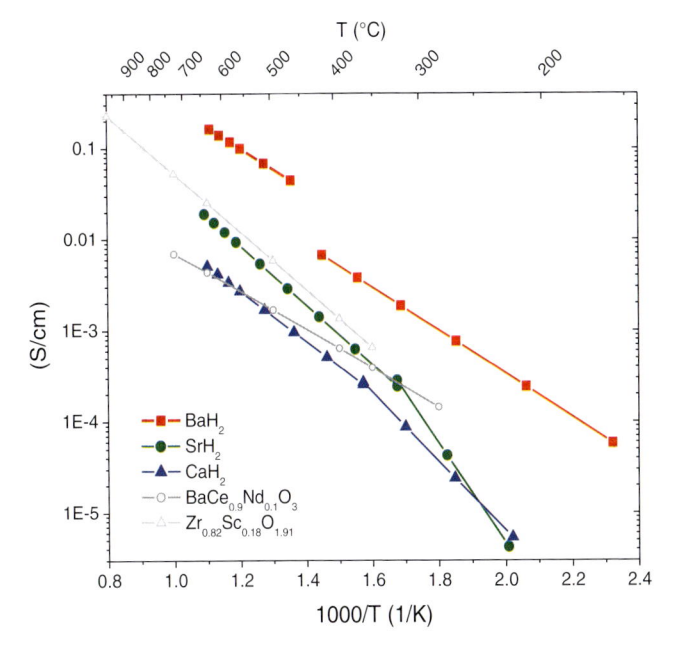

Figure 12.9 Bulk conductivities for alkaline earth hydrides and two typical oxide ion and proton conductors for comparison.

observed for fluoride conduction in alkaline earth fluorides [82,83]. The activation energy for bulk hydride ion conductivity at elevated temperatures in the cotunnite phase is similar for all the heavy alkaline earth hydrides, that is, 0.5–0.7 eV, suggesting the same mechanism(s) for all materials, as expected. A study on hydrogen diffusion in CaH_2 by solid-state NMR gives a similar value for the activation energy [49]. Calcium and strontium hydride show an increase in activation energy below 300–400 °C, which was tentatively assigned to an order–disorder transition within the hydride sublattice, but may equally arise from a transition between conduction mechanisms when changing temperature. Barium hydride shows nonlinear behavior below 200 °C, which could also be an indicator for this change in mechanism. Doping with aliovalent cations would be the obvious choice to create extrinsic defects and observe their effect on conductivity, but such studies have been hampered by poor solubility of such impurities [14,84]. NaH could be dissolved into CaH_2 and SrH_2 in small quantities and resulted in an increase in conductivity for the latter at low temperatures, suggesting vacancies do play a role in the hydride conduction [14]. Introducing trivalent cations, for instance, through LaH_3, has so far not resulted in anything concrete.

The activation energy for bulk hydride ion conductivity is relatively low compared to traditional oxide ion conductors [85,86]. This may normally offer the potential to extend the highly conductive regime to lower temperatures, especially if suitable dopants could be found to stabilize the high-symmetry phase at such temperatures. However, as discussed, it has proven difficult to find aliovalent dopants with sufficient solubility in the host lattices. It may simply be that the cotunnite structure is too rigid to allow significant substitution. Nonetheless, the results still show that this class of materials would make a suitable high-temperature electrolyte for electrochemical devices utilizing the potent hydride ion.

Due to the reactive nature of hydrides and their seemingly unaccommodating lattices, some researchers have attempted to create hydride ion conductivity in mixed halide – hydride compounds. It was found, for instance, that below 200 °C, $CaF_{2-x}H_x$ ($0.22 < x < 0.94$) has similar conductivity values with similar activation energies (0.47–0.68 eV) to pure CaH_2 [87]. A ^1H and ^{19}F NMR study of $CaF_{1.06}H_{0.94}$ suggested the ionic conductivity should be due to mobile hydride ions [88]. Similarly, introducing the hydride ion into $CaCl_2$ seems to result in a poorly conducting material, even at elevated temperatures [89]. Recently, pure hydride ion conduction was reported for the first time in an oxide-based material with a K_2NiF_4-type structure (as described later) [90]. Such materials would really bridge the gap between exciting new electrochemistry and phase stability/flexibility.

12.4
Ternary Ionic Hydrides

New ternary phases consisting of a mixture of alkali and alkaline earth metal hydrides have been synthesized over the last five decades. Apart from their structural properties, very little is generally known, but recent computational studies suggest that these ternary compounds largely retain the ionic nature of their constituent parts [91–93]. Such studies have also attempted to describe these materials' physical properties (electronic, optical, mechanical/elastic [92–96]) in more detail, but few experimental studies seem to have been performed to this end so far. Many of these ternary hydride phases form perovskite-related structures, which could be relevant in terms of physical properties. Oxide perovskite materials are known for having some of the best ferroelectric, dielectric, magnetic, electronic/ionic transport, and optical properties. So far the following hydride perovskites have been synthesized and characterized: $NaMgH_3$, $KMgH_3$, $RbMgH_3$, $RbCaH_3$, $CsCaH_3$, $SrLiH_3$, and $BaLiH_3$. The latter two have the inverse perovskite structure. Table 12.4 summarizes the structural properties of these materials, and Figure 12.10 shows the various (distorted) perovskite crystal structures. It should be stressed that not all of these materials have been characterized using neutron diffraction techniques. Due to hydrogen's weak scattering in X-ray diffraction, the atomic positions of hydrogen in $KMgH_3$ and $SrLiH_3$ are still unknown and presumed to be in high-symmetry places. Other ternary or quaternary hydrides exist as well, but have lower symmetry complicated

Table 12.4 Structural information for perovskite hydrides at room temperature.

	Technique	Space group	Lattice parameters			Distortion	Reference
			a	b	c		
NaMgH$_3$	XRD	Pnma	5.460	7.698	5.409	Octahedral tilting	[100]
NaMgD$_3$	NPD	Pnma	5.449	7.647	5.397	Octahedral tilting	[101]
KMgH$_3$	XRD	Pm3m	4.025	—	—	n/a	[102]
KMgD$_3$	—	—	—	—	—	—	
RbMgH$_3$	XRD	P6$_3$/mmc				Hexagonal perovskite	
RbMgD$_3$	NPD	P6$_3$/mmc	5.903	—	14.316	Combination of cubic and hexagonal close-packed MgD$_6$ octahedra	[103]
			5.910	—	14.333		[104]
RbCaH$_3$	XRD	Pm3m	4.547	—	—	n/a	[105]
RbCaD$_3$	NPD	Pm3m	4.532	—	—	n/a	[104]
CsCaH$_3$	XRD	Pm3m	4.617	—	—	n/a	[103]
CsCaD$_3$	NPD	Pm3m	4.609	—	—	n/a	[103]
SrLiH$_3$	XRD	Pm3m	3.833	—	—	n/a	[106]
SrLiD$_3$	—	—	—	—	—	—	
BaLiH$_3$	XRD	Pm3m	4.023	—	—	n/a	[107]
BaLiD$_3$	NPD	Pm3m	—	—	—	n/a	[107]

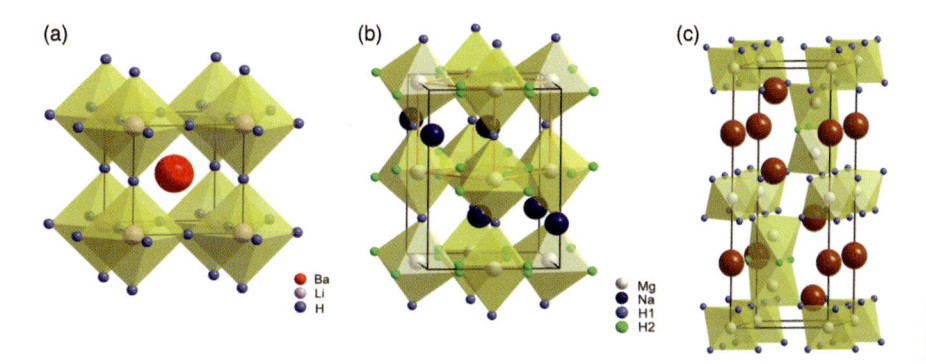

Figure 12.10 Graphical representation of ideal perovskite structure (a) and distorted variants (b and c). NaMgH$_3$ exhibits distorted, tilted octahedra, resulting in an orthorhombic structure (b). RbMgH$_3$ exhibits hexagonal stacking of Rb and MgH$_6$ layers (c).

structures. Due to the low stability of the alkali hydrides, many of these ternary hydrides can only be synthesized under significant hydrogen pressure (50–200 bar; hydrostatic pressures up 35 kbar are required for forming phases such as $Cs_4Mg_3H_{10}$), although more recently reactive ball milling is also being used as a viable synthesis route, allowing much reduced hydrogen pressures [97,98]. A comprehensive overview of alternative structure types to perovskite found in magnesium-based ternary hydrides can be found elsewhere [99].

Synthesizing ternary hydride phases is an effective method in controlling the thermodynamic stability of the resulting materials. This method has been used effectively to create new hydrogen storage materials, in which the temperatures and pressures for hydrogenation/dehydrogenation are generally lower than may be expected from the constituent hydride parts, making them better suited toward typical application such as low-temperature fuel cells in electric vehicles. Prime examples of such phases are alanates, borohydrides, and the amide/imide system [108–111]. So far however, the most promising hydride perovskite in terms of its gravimetric potential to store 6 wt% hydrogen, $NaMgH_3$, has shown an increase in thermodynamic stability compared to MgH_2 and may therefore be unsuitable as a storage material.

Various authors have calculated the electronic structures of some ternary hydrides, and in particular $MMgH_3$ (M = Na, K, Rb) [91–93]. They have found predominant ionic bonding, with total ionization for the alkali metal, but with some covalency existing within the Mg—H bonds. This is analogous to the small degree of covalent bonding observed in MgH_2 and would also be in line with some nonionic bonding observed in oxide perovskite materials. Especially the tightly coordinated MO_6 octahedra are known to exhibit reduction/oxidation, polarization, magnetic moments, and so on, giving the perovskites their unique properties. From electronic structure calculations, the $MMgH_3$ materials have also been suggested to be insulators with bandgaps of 2.5–3.5 eV, and due to the nature of the computational methods, it is assumed that these estimates are in fact too low [92]. $BaLiH_3$ and $SrLiH_3$ were similarly calculated to have significant indirect bandgaps of 2.3–3.8 and 1.9–3.6 eV, respectively; the large spread in values arises from different calculation techniques [95,96,112]. Despite these large values, Greedan reports $SrLiH_3$ to be a good electric conductor with relatively low activation energy, that is, 0.4–0.7 eV [113]. This may perhaps suggest that the experimentally observed electric properties would be due to migration of ions, as opposed to electrons. The calculated optical properties, such as absorption and refraction, seem to be in line with experimental findings.

12.5
Metallic Hydrides

Many of the transition metals and rare earth elements react with hydrogen gas to produce metallic hydrides. These hydrides are referred to as metallic hydrides because they generally exhibit metallic conductivities as well as other metallic properties, although some of the rare earth hydrides become semiconductors at high hydrogen contents, such as CeH_3, which is more like a saline hydride and sharply different from metallic CeH_{2-x}. Unlike metals, however, most metallic hydrides lack the characteristic metallic property of ductility and are quite brittle. They usually have a smaller density than that of their parent metal [6]. The presence of hydrogen has a significant impact on the band structure of the metal [114]. As a result, the conductivity of the hydride is also different from that of its pure metal. Most metallic hydrides are flexible in composition. For example, titanium hydrides allow stoichiometries from TiH to TiH_2. Since hydrogen atoms are small enough, they are often regarded as occupying interstitial sites within the metal framework despite the fact that this occupation rarely occurs without an expansion or an occasional distortion of the lattice. In many ways, metallic hydrides are similar to their carbide or nitride analogues.

The metallic hydrides show large domains of existence [115]. Metallic hydrides are formed by all elements of groups 3, 4, and 5 and almost all f-block elements. As can be seen in Figure 12.11, chromium is the only element of group 6 that forms hydride materials. There is a gap region in the periodic table covered by groups 7, 8, and 9 whose elements do not form binary hydride phases. Moreover, if copper and zinc hydrides are not considered as the official members of the current hydride category, groups 11 and 12 may represent the other gap region. Palladium hydride is probably the most well-known and studied metallic

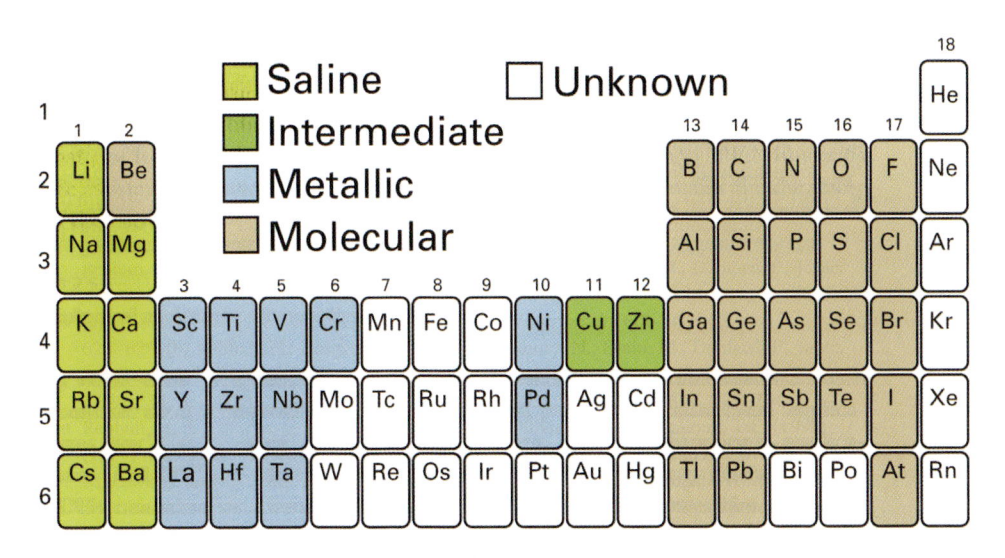

Figure 12.11 Graphical representation of binary hydrogen compounds in the periodic table in terms of the materials type of metal hydrides (From Ref. [116], with permission from Oxford University Press).

hydride material, partly because both palladium and nickel metals in group 10 are important hydrogenation catalysts that involve the formation of hydrides during the catalysis process. Actually, palladium takes up large volumes of hydrogen at room temperature and pressure. In contrast, very high pressures are required for nickel to form hydrides. For rare earth metals, promethium is the only element that does not form binary hydrides.

12.5.1
Crystal Structures and Compositions

The structures of many metallic hydride materials have been determined through combined neutron and X-ray diffraction techniques. Table 12.5

Table 12.5 Structural information for metallic hydrides at room temperature, based on data from Ref. [10].

Metal	Metal hydride	Space group	Structure	References
La	LaH_2	Fm-$3m$	Cubic	[117]
	LaH_3	Fm-$3m$	Cubic	[118]
Ce	CeH_2	Fm-$3m$	Cubic	[117]
	CeH_3	Fm-$3m$	Cubic	[118]
Pr	PrH_2	Fm-$3m$	Cubic	[117]
	PrH_3	Fm-$3m$	Cubic	[118]
Nd	NdH_2	Fm-$3m$	Cubic	[119]
	NdH_3	P-$3c1$	Trigonal	[119]
Sm	SmH_2	Fm-$3m$	Cubic	[120]
	SmH_3	P-$3c1$	Trigonal	[10]
Eu	EuH_2	$Pnma$	Orthorhombic	[121,122]
Gd	GdH_2	Fm-$3m$	Cubic	[123]
	GdH_3	$P6_3/mmc$; P-$3c1$	Hexagonal/trigonal	[124]
Tb	TbH_2	Fm-$3m$	Cubic	[125]
	TbH_3	$P6_3/mmc$	Hexagonal	[126]
Dy	DyH_2	Fm-$3m$	Cubic	[127]
	DyH_3	P-$3c1$	Trigonal	[126]
Ho	HoH_2	Fm-$3m$	Cubic	[120]
	HoH_3	P-$3c1$	Trigonal	[126]
Er	ErH_2	Fm-$3m$	Cubic	[120]
	ErH_3	$P6_3/mmc$; P-$3c1$	Hexagonal/trigonal	[126]
Tm	TmH_2	Fm-$3m$	Cubic	[120]
	TmH_3	P-$3c1$	Trigonal	[126]
Yb	YbH_2	$Pnma$	Orthorhombic	[127]
	YbH_{2+x}	Fm-$3m$	Cubic	[128]
Lu	LuH_2	Fm-$3m$	Cubic	[120]
	LuH_3	$P6_3/mmc$	Hexagonal	[126]

Table 12.5 (Continued)

Metal	Metal hydride	Space group	Structure	References
Sc	ScH$_2$	Fm-3m	Cubic	[122]
	ScH$_3$	P6$_3$/mmc	Hexagonal	[129]
Y	YH$_2$	Fm-3m	Cubic	[120]
	YH$_3$	P6$_3$/mmc	Hexagonal	[126]
Ti	TiH$_2$	Fm-3m; I4/mmm	Cubic or tetragonal	[130]
Zr	ZrH$_2$	Fm-3m; I4/mmm	Cubic or tetragonal	[130]
Hf	HfH$_2$	Fm-3m; I4/mmm	Cubic or tetragonal	[130]
V	VH	Fm-3m	Cubic	[130]
	VH$_2$	Fm-3m	Cubic	[130]
Nb	NbH	Im-3 m; Pnnn	Tetragonal or orthorhombic	[130]
	NbH$_2$	Fm-3m	Cubic	[130]
Ta	TaH	I4/mmm	Tetragonal	[130]
	TaH$_2$	Fm-3m	Cubic	[130]
Cr	CrH$_2$	Fm-3m	Cubic	[130]
	CrH	P63/mmc	Hexagonal	[130]
Ni	NiH	Fm-3m	Cubic	[130]
Pd	PdH	Fm-3m	Cubic	[130]

summarizes the crystal structure and composition of known metallic hydrides for d- and f-block elements. Most transition metal hydrides with formula MH$_2$, such as ScH$_2$, YH$_2$, TiH$_2$, and HfH$_2$, crystallize in the cubic fluorite structure, in which the hydride ions occupy the tetrahedral positions. Moreover, this cubic structure survives a wide composition range. For example, titanium hydrides are in the range from TiH to TiH$_2$. In contrast, zirconium and hafnium hydride phases tolerate a limited composition range to retain the cubic structure. Beyond this composition limit (for ZrH$_x$ and HfH$_x$, $x = 1.61$ and 1.8, respectively), the structure undergoes a tetragonal distortion. Yttrium and scandium are transition metals, but their hydride behavior is very similar to that observed for the hydrides of the rare earth metals.

Most rare earths form cubic fluorite-type dihydrides and trihydrides that are usually hexagonal. It is interesting to note that europium and ytterbium form saline dihydrides that are isostructural with CaH$_2$ [131]. This is because of Hund's rule that stabilizes the 4f shell in these metals and the availability of the two 6s electrons for bonding. Although ytteribium dihydride appears to be ionic, Yb also forms complex higher hydrides under higher hydrogen pressures. The trihydrides of the light rare earth metals (La, Ce, Nd, and Pr) are cubic as well, although for LnH$_{2+x}$, the hydride ions are distorted from their ideal tetrahedral/octahedral positions and form an ordered superlattice, which is better described as body-centered tetragonal [117,118,124]. NdH$_3$ and the trihydrides of the heavy rare earth metals, like GdH$_3$ and SmH$_3$, usually have hexagonal/trigonal structures [119,120,132]. This hexagonal structure starts to form at different values of x in LnH$_{2+x}$ for different rare earth metals, but can only be obtained as a single phase at high hydrogen contents, that is, $x=1$. So often a mixture of the cubic and hexagonal phase is obtained when synthesizing the heavy rare earth trihydrides.

The actinide hydrides show a variety of different crystal structures, ranging from the fluorite structure for actinium and plutonium dihydride to a tetragonal structure for thorium dihydride. The higher actinide hydrides, like Th$_4$H$_{15}$ and UH$_3$, have complex cubic structures. A range of compositions with various crystal structures is possible, analogous to the rare earth metal hydrides.

12.5.2
Bonding, Stoichiometry, and Properties

The nature of the bonding in metallic hydrides remains a subject of debate. There are different models to account for various hydride cases in which hydrogen is either present as protons, hydride ions, or bound covalently [10]. PdH$_{0.6}$ is a protonic model example, in which the d-band of palladium seems to be filled with electrons from hydrogen. Moreover, upon electrolysis, hydrogen evolution is observed at the cathode side. This is, however, not the case for real metallic hydrides, but true for solid solutions of hydrogen in metal. There are also indications for the hydride anion model. The heat of formation of metallic hydrides and their brittleness are both similar to those of the saline hydrides, suggesting a certain degree of ionic bonding nature. Furthermore, the interatomic distances between the metal atoms and hydrogen agree well with calculations using the ionic radii of both metal cations and hydride ions. The high electronic conductivity in this ionic model can be understood by the fact that metallic hydride compositions often depart from the preferred composition (i.e., TiH$_2$ and ZrH$_2$ rather than TiH$_4$ and ZrH$_4$). This may be an indicator to a possibility that after bonding with hydrogen, available d-electrons remain enough for the metallic bonding. So in spite of bonding with hydrogen, sufficient d-orbital overlap is present to maintain the metallic conduction. In the covalent model, the hydrogen electrons are delocalized. As the hydrogen lies much lower in energy than the electronic bands of the metal, hybridization occurs between the metal and hydrogen orbitals. As a result, the Fermi level is shifted downward, which may lead to interesting phase transitions (e.g., metallic– semiconducting, magnetic–nonmagnetic, order–disorder) [109]. This last covalent model seems to agree most with the observed nonstoichiometry of metallic hydrides. For example, in ZrH$_x$, x can be values between 1.30 and 2.00. Within this range, the compound is clearly a metallic hydride and not just a solid solution of hydrogen in the zirconium metal. Physical properties change as a function of the hydrogen content.

Obviously, the electronic structure largely relies on the amount of hydrogen stored inside the metallic lattice.

The study of metal hydrides has been closely related to the energy storage. When the d-block metal hydrides are heated, H_2 can be released and this property means that they are convenient storage vessels for dihydrogen. This is a well-studied area and readers are recommended to refer to an excellent reviewing article [109]. Metal hydrides are of interest to the community of superconductivity. This is because hydrogen is believed to be a candidate material for room-temperature superconductivity at high pressures. Yet the pressure required is out of range for the current experimental facilities. As a result, pressurized hydrogen-containing materials become alternative targets. As mentioned earlier, hydrogen sulfide has been recently found to superconduct at an unprecedented high transition temperature. For metallic hydrides, PdH_x ($x \geq 0.8$) was reported by Skoskiewice to be a superconductor, although Pd metal itself is not superconducting. When the H content is increased, $T_c = 8.8$ K was realized for PdH. The H-induced T_c enhancement in Pd is related to a rather unusual combination of a soft H sublattice and the Fermi energy electrons spending considerable time around the H sites. This favorable combination is not present in various other systems and consequently H is not expected to raise T_c in these systems [133]. Hydride substitution in 1111-type iron arsenide compounds yielded superconducting $LnFeAsO_{1-x}H_x$. The superconductivity is a result of the electron doping to the FeAs layer. Of particular interest is the higher hydride substitution level than that of fluoride, which leads to unprecedented shape and width of the superconducting region [134,135]. As in Figure 12.12, the original $LaFeAsO_{1-x}H_x$ with $x = 0$ exhibits a structural transition at 155 K, followed by an antiferromagnetic state (AF1) at $T_N = 137$ K. With increasing x, two superconductivity domes emerge. The first one with $0.05 \leq x \leq 0.20$ (SC1) shows $T_{c,max} = 26$ K and the other one with $0.20 \leq x \leq 0.42$ (SC2) holds a maximum T_c of 36 K. Beyond the composition, another antiferromagnetic phase (AF2) is developed.

12.6
Oxyhydrides: The Hydride Anion in Oxide Structures

Hydrogen in oxides in general takes a protonic state H^+, such as in OH^- groups [136]. Much less encountered are oxyhydrides that contain hydride and oxide anions. Coexistence of H^- and O^{2-} in the same framework is rather counterintuitive since oxides often imply oxidized species while hydrides are typically related with their reducing ability. A number of differences are found between oxygen and hydrogen: anionic charge, ionic radii, electronegativity, atomic polarizability, and bond energy. One of the unique features of hydride anion is the fact that it consists only of 1s-orbital (with $(1s)^2$ electronic configurations), which is the smallest closed shell and spherical. This situation is contrasting

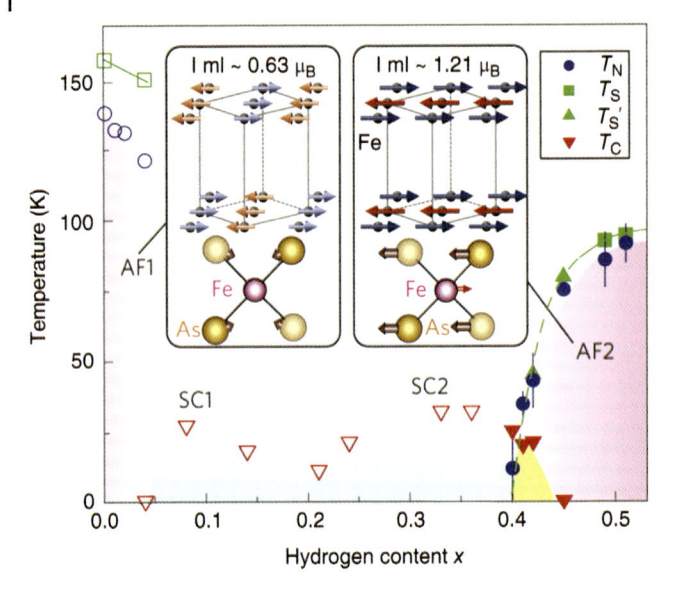

Figure 12.12 Superconducting phase diagram for LaFeAsO$_{1-x}$H$_x$. ○: antiferromagnetic transition temperature; □: structural transition temperature; △: the c-axis upturn temperature observed in X-ray measurement; ▽: superconducting transition temperature. The filled and open marks are used to distinguish data collected at different time. (With permission from Ref. [135].)

with oxygen and other anionic elements (N, S, F, Cl, Br, etc.) that belong to p-block in the periodic table, in which the outer shell p-orbitals play a significant role in chemical bonding. Furthermore, H$^-$ adopts a flexible ionic radius (known to vary from 127 to 152 pm) depending on its environment [10]. One can therefore expect that these differences give rise to exotic chemical and physical properties that could never be realized in simple oxides.

The research on oxyhydrides has blossomed over the last several decades. Oxyhydride compounds in early days have been exclusively limited to those containing p-block and f-block elements, which are hard to reduce and thus synthesized by high-temperature solid-state reactions under reducing atmosphere. Recent synthetic developments, namely, low-temperature topochemical reaction and high-pressure reaction, have allowed expansion of the oxyhydride family to those with transition metal. Transition metal oxyhydrides offer various possibilities toward exotic chemical and physical properties, including catalysts for hydrogenation reaction, (proton-assisted) hydride conductivity, extensive carrier doping, and dimensional control by O^{2-}/H$^-$ order. Here, we will review history and recent advances of oxyhydrides to show how these relatively new compounds have contributed or will contribute to materials sciences, in particular, putting special emphasis on novel hydride-derived functions.

12.6.1
p- and f-Block Oxyhydrides

12.6.1.1 Rare Earth Oxyhydride

The earliest report on oxyhydrides is found in the work of Brice and coworkers who prepared LaHO, CeHO, and PrHO [137,138]. These rare earth oxyhydrides are synthesized via high-temperature solid-state reaction, using a mixture of binary hydride and oxide (e.g., $LaH_{2.5}$ and La_2O_3), typically at 900 °C under hydrogen atmosphere. Recently, Norby and coworkers obtained NdHO with additional reaction pathways, "$Nd_2O_3 + CaH_2 \rightarrow 3NdHO + CaO$" and "$2Nd + Nd(OH)_3 \rightarrow 3NdHO$" [139]. Calculations of Gibbs free energy for reaction revealed that the latter (metal/hydroxide) route is more favorable and predicts an extension to heavier rare earth oxyhydrides [140]. The REHO compounds (RE = rare earth) crystallize in a structure that is derived from the fluorite (CaF_2) type with a complete ordering of oxide and hydride anions, as found in LaOF [141]. It possesses three distinct REH_4O_4 cubes that share edges with each other, as shown in Figure 12.13. Unlike the oxyfluoride, REHO does not have rhombohedral and tetragonal polymorphs.

12.6.1.2 Antiperovskite

Huang and Corbett presented Ba_3AlO_4H, which can be prepared at 1100 °C under H_2 atmosphere using BaH_2, BaO, and Al_2O_3 [142]. As shown in

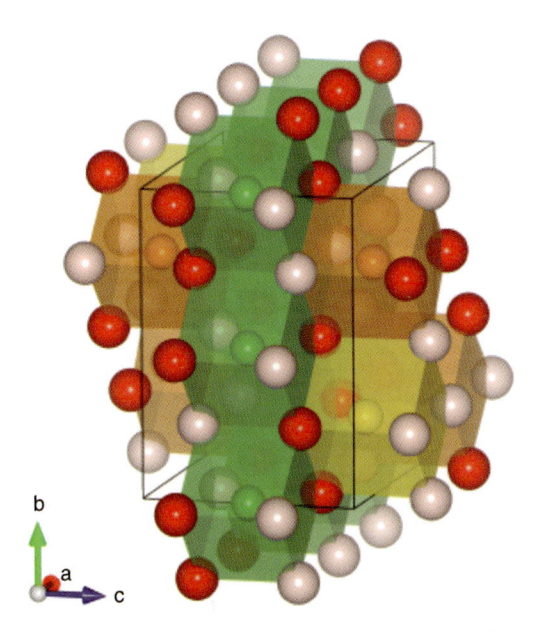

Figure 12.13 Crystal structure of LaHO. The red and pink spheres represent the oxide and hydride anions, respectively. The yellow balls represent La^{3+} with different coordination geometries of LaH_4O_4.

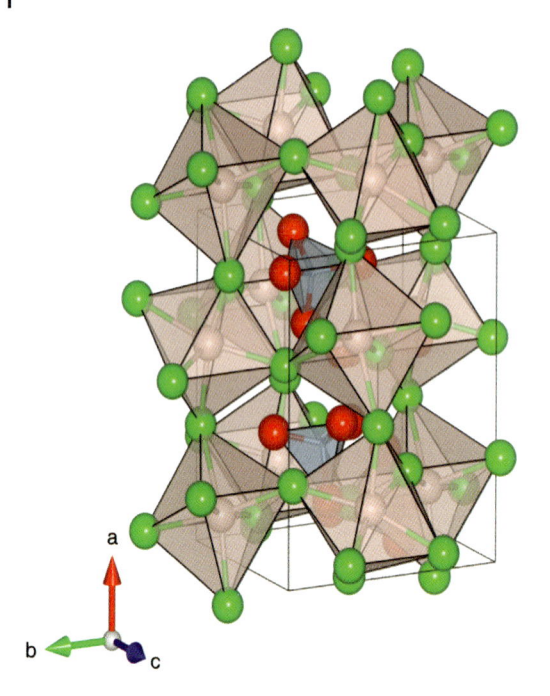

Figure 12.14 Crystal structure of Ba_3AlO_4H with the antiperovskite-type structure. The red, blue, pink, and green balls represent the oxide, aluminum, hydride, and barium ions, respectively.

Figure 12.14, this material adopts the orthorhombic *Pnma* space group with a framework that can be related to an antiperovskite BXA_3, where Ba^{2+} lies in the classical anionic site, the $(AlO_4)^{5-}$ polyanionic unit lies in the classical A site, and the classical B site is occupied by H^-. The hydride anion is surrounded by six barium cations forming an HBa_6 octahedral cage with Ba–H lengths of 2.79–2.99 Å. Although there is only one oxyhydride of this type, several oxyfluorides forming the antiperovskite structure, K_3SO_4F [143], Na_3MoO_4F [144], Na_3WO_4F [145], Sr_3AlO_4F [146], and Sr_3GaO_4F [146], have been reported. Similar to Ba_3AlO_4H with the HBa_6 octahedron, the hydride ion in $Rb_2Ba_6Sb_5HO$ is surrounded by four rubidium (3.26 Å) and two barium cations (2.36 Å) bearing a *trans*-HBa_4Rb_2 octahedron (Figure 12.15) [147]. It consists of alternate stacking of an antiperovskite layer of corner-shared $H(Ba_4Rb_2)$ octahedra and a layer of OBa_4 tetrahedra and Sb atoms. $Ba_8Sb_4OH_2$ and $Ba_8Bi_4OH_2$ have a similar crystal structure with layers of antiperovskite Ba_4Sb_2O sandwiched by anti-$KCoO_2$ block $SbHBa_2$ with HBa_5 pyramids (Figure 12.15) [147].

12.6.1.3 Suboxide Hydrides

Suboxide metal hydrides $A_{21}M_2O_5H_{12}$ ($Ba_{21}Ge_2O_5H_{24}$, $Ba_{21}Si_2O_5H_{24}$, $Ba_{21}Ga_2O_5H_{24}$, and $Ba_{21}Tl_2O_5H_{24}$) were reported by Huang and Corbett [148]. Röhr

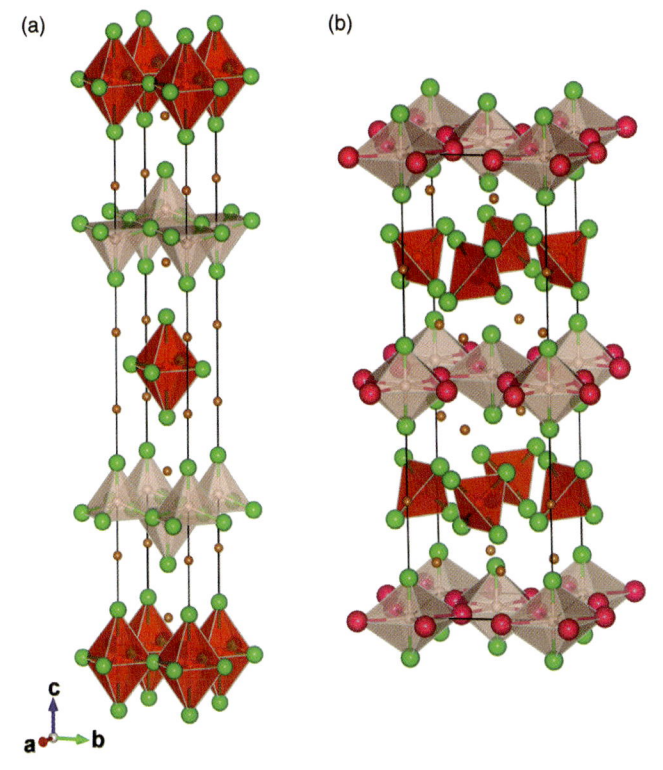

Figure 12.15 Crystal structures of Ba$_8$Sb$_4$OH$_2$ and Rb$_2$Ba$_6$Sb$_5$OH. The red, gold, pink, and green balls represent the oxide, antimony, hydride, and barium ions, respectively.

and coworkers expanded the family to compounds with Sr at the Ba site, as well as most of groups 12–15 elements at the M site [149]. They also revealed that the composition is better described as $A_{21}M_2O_5H_{12+x}$. It is suggested that compounds with groups 12 and 13 elements contain 12H$^-$ per formula units, while those with groups 14 and 15 elements contain 14H$^-$ and 16H$^-$. In this framework, the hydride anions mostly occupy an octahedral site surrounded by Sr or Ba, together with two tetrahedral sites.

12.6.1.4 Mayenite

Mayenite with a $(A_{24}Al_{28}O_{64})^{4+} \cdot 4X^-$ composition ($A = Ca^{2+}$, Sr^{2+} and $X^- = H^-$, F$^-$, OH$^-$, Cl$^-$) is a cage structure formed by alkali earth cations and AlO$_4$ polyanions (Figure 12.16). The unit cell can accommodate four anionic X^- charges distributed randomly within 12 cages. This structure has attracted much attention owing to its capability to contain electrons as anions in the cage $(A_{24}Al_{28}O_{64})^{4+} \cdot 4e^-$, with properties ranging from ammonia catalysis [150] to superconductivity [151]. The hydride version ($X^- = H^-$) of this material can be prepared topochemically from the reaction of the oxide with TiH$_2$ or CaH$_2$ at

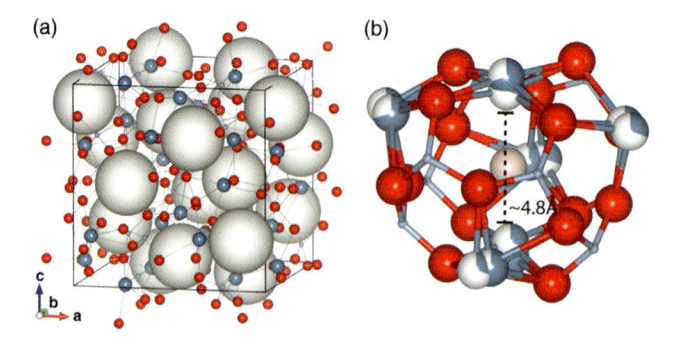

Figure 12.16 Crystal structure of mayenite with free space shown by large gray spheres (a). An example of hydride anion in a Ca–Al–O cage (b). The red, pink, blue, and light blue balls represent the oxide, hydride, calcium, and aluminum ions, respectively.

high temperatures [152,153]. In $(Ca_{24}Al_{28}O_{64})^{4+} \cdot 4H^-$ and $(Sr_{24}Al_{28}O_{64})^{4+} \cdot 4H^-$, the hydride anions sit between two cationic poles in four cages randomly distributed with bond distances of $d_{Ca-H} = 2.40$ Å and $d_{Sr-H} = 2.60$ Å [152]. The hydride content can be tuned by reaction conditions. Application of UV light onto insulating films of $(Ca_{24}Al_{28}O_{64})^{4+} \cdot xH^-$ (H^-: 2.5×10^{20} cm^{-3}) leads to color change accompanied by an electronic conductivity of ~0.3 S/cm at room temperature [153,154]. The reaction mechanism proposed involves the conversion of the H^- anion to "$H^0 + e^-$" with the resulting H^0 partially recombined with e^- (70%) to recover a H^- state. The remaining H^0 (30%) reacts with O^{2-} to yield OH^- and the second extra electron contributes to the conductivity. This reaction is reversible at ~600 K, indicating that this material could be used, for instance, to write or erase data using UV light [154].

Detecting hydride anion in oxides is not straightforward since its chemical shift of nuclear magnetic resonance (NMR) ranges from positive to negative. Using NMR studies on mayenite and pure hydrides (Figure 12.17), combined with theoretical calculations, Hayashi *et al.* demonstrate that the chemical shift of the hydride anion increases linearly with respect to its distance to the closest cationic center [152]. The increase of chemical shift with the M–H distance can be ascribed to the reduced electron density around the nucleus of hydrogen. On the contrary, the chemical shift of hydrogen in OH group decreases linearly with increasing distance to the next-nearest oxygen (O—H···O). This study not only allows one to understand the impact of the chemical bond on the state of the hydride but can also be applied toward differentiating NMR signals of a sample containing both H^+ and H^-.

12.6.2
Low-Temperature Synthesis of Transition Metal Oxyhydrides

The presence of the hydride anion in transition metal oxide has long been an issue of solid-state ionics. Based on electrochemical characterizations, it was

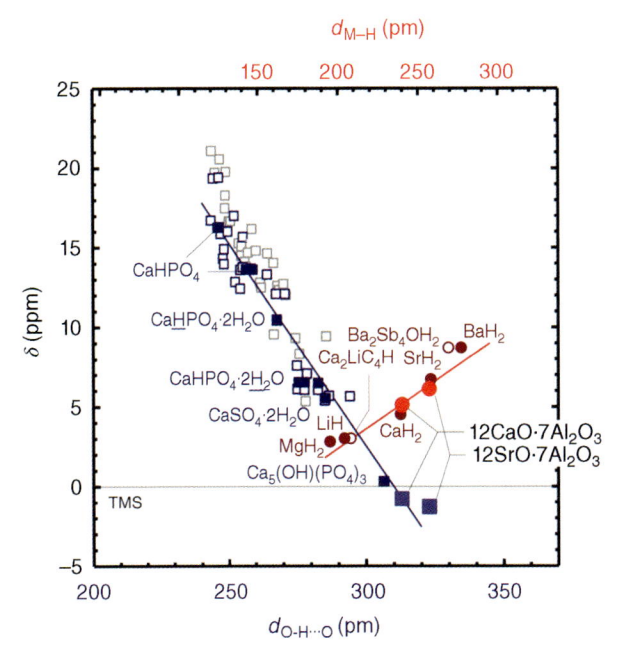

Figure 12.17 NMR chemical shifts of several hydride and hydroxides plotted against the distance between hydride and metal ligand as well as against the distance between hydroxide and neighboring oxide. (With permission from [152].)

speculated that the charge transport involving hydrogen-containing ions in SrTiO$_3$-based oxides and some other oxide materials appears to get a contribution from a negatively charged species under reducing conditions [155–158]. In 2001, Poulsen stated from a thermodynamic perspective that H$^-$ is only stable in combination with certain transition metal ions in their lowest positive oxidation state [159]. One year later, the first transition metal oxyhydride LaSrCoO$_3$H$_{0.7}$ with a very low oxidation state of Co$^{1.7+}$ was indeed stabilized by Hayward and coworkers who employed a low-temperature topochemical reaction using CaH$_2$ [160]. This hydride reduction technique using CaH$_2$, NaH, and LiH had been extensively used to synthesize highly oxygen-deficient phases such as LaNiO$_2$ and SrFeO$_2$, respectively, from LaNiO$_3$ and SrFeO$_3$ [161,162], but the reaction for LaSrCoO$_3$H$_{0.7}$ indicates that CaH$_2$ serves also as a hydrogen source. Subsequent discovery of oxyhydride BaTiO$_{2.4}$H$_{0.6}$ with a moderate titanium oxidation state of +3.4 [163] and a series of vanadium oxyhydrides Sr$_{n+1}$V$_n$O$_{2n+1}$H$_n$ ($n = 1, 2$, and ∞) [164] ensures that transition metal oxyhydrides will be ubiquitously found, opening a broad range of applications in various fields.

12.6.2.1 Cobalt Oxyhydrides

A powder specimen of LaSrCoO$_3$H$_{0.7}$ was obtained by reacting the oxide precursor LaSrCoO$_4$ possessing a Ruddlesden–Popper-type layered perovskite

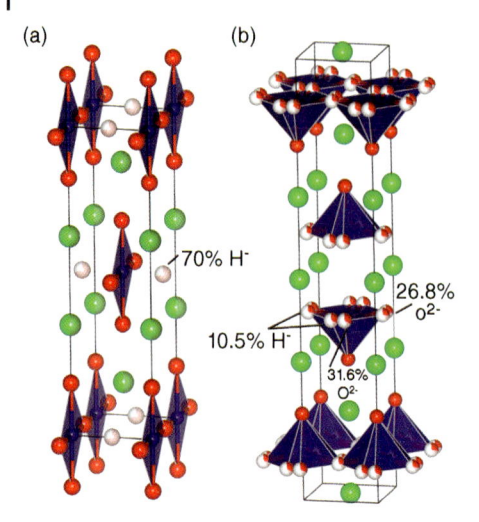

Figure 12.18 The crystal structures of LaSrCoO$_3$H$_{0.7}$ (a) and Sr$_3$Co$_2$O$_{4.33}$H$_{0.84}$ (b). The blue, red, pink, and green balls represent the cobalt, oxide, hydride, and strontium ions, respectively.

structure with CaH$_2$ at 450 °C in an evacuated Pyrex tube [160,165]. Contrary to the tetragonal precursor, LaSrCoO$_3$H$_{0.7}$ is orthorhombic (Figure 12.18) owing to H$^-$/O^{2-} order within the equatorial plane, leading to a *trans*-CoO$_4$H$_2$ octahedral coordination with two shorter Co—H bonds and four longer Co—O bonds. LaSrCoO$_3$H$_{0.7}$ exhibits an antiferromagnetic order at 380 K. The high Néel temperature is attributed to the strong σ-type interactions through Co e_g-H 1s-Co e_g path, along with Co—O—Co path, as supported by spin wave calculations [163,166]. Sr$_3$Co$_2$O$_{4.33}$H$_{0.84}$ with a double-layer Ruddlesden–Popper perovskite structure (Figure 12.18) is obtained by the topochemical reduction of Sr$_3$Co$_2$O$_{5.80}$ using CaH$_2$ [167]. This double-layer analogue is a mixture of CoO$_3$H tetrahedra (68%), CoO$_4$H pyramids (16%), and CoO$_4$ tetrahedra (16%). The anionic disorder appears to be responsible for the lack of magnetic order down to 5 K.

Another important feature of LaSrCoO$_3$H$_{0.7}$□$_{0.3}$ (□ = vacancy) is hydride mobility at mild temperatures as revealed by quasi-inelastic neutron scattering (QENS) [168]. QENS is a powerful tool to detect motion of hydrogen owing to a large incoherent scattering cross section for H. As shown in Figure 12.19, the quasi-elastic scattering arising from hydride hopping via vacancy (30%) starts to evolve at around 685 K. The site-to-site hopping distance extracted from the fits is within error equal to the length of the *a* lattice parameter. Surprisingly, an upper estimate for the conductivity, 3–5 S/cm, is significantly higher than that observed in proton conducting materials. Tuning anion (H$^-$/O^{2-}) order–disorder is shown by means of epitaxial strain in thin films [169]. The CaH$_2$ reduction is conducted for LaSrCoO$_4$ films epitaxially deposited on LaSrAlO$_4$ substrates with (100) and (001) orientation. It turned out that the oxyhydride

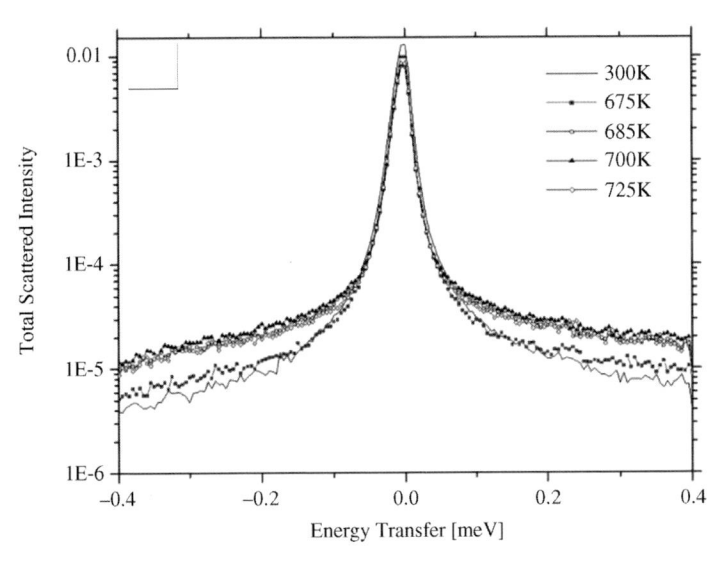

Figure 12.19 QENS data for LaSrCoO$_3$H$_{0.7}$. (With permission from Ref. [34], Advanced Materials.)

film on the (001) substrate is orthorhombic, similar to that of the powder sample, while that on the (100) substrate stayed tetragonal due to the strain from the substrate, implying anion disorder within the plane.

12.6.2.2 Titanium Oxyhydrides

The low-temperature topochemical technique also works for titanium perovskite BaTiO$_3$; it yields BaTiO$_{3-x}$H$_x$ with x as large as 0.6 (i.e., 20% substitution) [163]. As opposed to the Co case, upon hydride reaction the structure converts from tetragonal to cubic with a complete anion disorder. The symmetry change is explained by electron donation to Ti 3d conduction band, resulting in the loss of the second-order Jahn–Teller (SOJT) effect, which is the origin of the ferroelectricity in BaTiO$_3$. This chemistry was expanded to CaTiO$_3$ and SrTiO$_3$ and their solid solutions [170].

An epitaxial thin film of SrTiO$_{2.7}$H$_{0.3}$ deposited on a (LaAlO$_3$)$_{0.3}$(SrAl$_{0.5}$Ta$_{0.5}$O$_3$)$_{0.7}$ substrate was used to investigate transport properties [171]. As a result of electron donation, the oxyhydride thin film shows a metallic conductivity down to 2 K. In-depth studies on ATiO$_{3-x}$H$_x$ films (A = Sr, Ba) reveal that the metallic phase of SrTiO$_{3-x}$H$_x$ is stable over a wide range of H$^-$ concentration, while for BaTiO$_{3-x}$H$_x$, a semiconducting state is observed for $x < 0.3$, beyond which metallic behavior appears. Given a similar contrasting behavior in the Nb-substituted ATi$_{1-x}$Nb$_x$O$_3$ (A = Sr, Ba) [172], BaTiO$_{3-x}$H$_x$ in a low doped region likely exhibits Ti off-centering to give polarons and in-gap states [173]. EuTiO$_{3-x}$H$_x$ ($x < 0.3$) is a ferromagnetic metal with $T_c \sim 12$ K, which is higher than those of metal-substituted EuTiO$_3$ [174]. The ferromagnetism was interpreted in terms of the RKKY (Ruderman–Kittel–Kasuya–Yosida) mechanism with ferromagnetic interactions between Eu^{2+} 4f-electrons mediated by Ti 3d conduction electrons. These examples suggest that the aliovalent anion substitution by hydride (O^{2-}/H^- exchange) is an alternative and highly efficient approach for electron doping in insulating titanium (and possibly other transition metal) oxides.

Hydride anion in the oxide framework provides novel functions that are not realized before in simple oxides. It is found that heating at 400 °C in flowing D_2 gas converts $BaTiO_{2.4}H_{0.6}$ to $BaTiO_{2.4}D_{0.6}$ as monitored by quadrupole mass spectroscopy (Figure 12.20) and neutron diffraction [163]. This H^-/D^- exchange in $BaTiO_{2.4}H_{0.6}$ demonstrates the ability of the hydride anion to react with gaseous hydrogen species at moderate temperatures. In other words, the hydride anion is quite labile (versus oxide anion), offering a potential of oxyhydrides as a precursor for multistep topochemical reactions for anion exchange. Such hydride exchange chemistry is indeed possible. For instance, the low-temperature (350–500 °C) reaction of $BaTiO_{2.4}H_{0.6}$ under flowing ammonia allows a H^-/N^{3-} exchange, yielding the fully oxidized

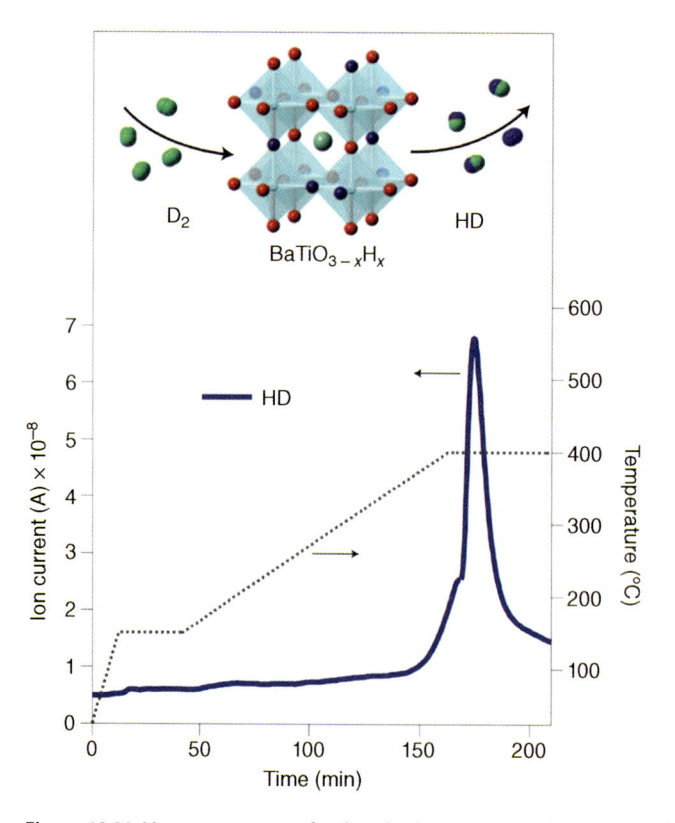

Figure 12.20 Mass spectrometry for desorbed gaseous species upon heating $BaTiO_{3-x}H_x$ under D_2 gas. The solid line indicates the ion current signal corresponding to HD, suggesting the H^-/D^- exchange reaction. (Reproduced with permission from Ref. [163]. Copyright 2012, Nature Publishing Group.)

product of BaTiO$_{2.4}$N$_{0.4}$, via intermediate oxide–hydride–nitride [175]. The BaTi^{4+}O$_{2.4}$N$_{0.4}$ structure goes back to the tetragonal *P4mm* phase as a result of complete loss of conduction electrons. Therefore, this oxynitride exhibits room-temperature ferroelectricity. Notably, the same product can be obtainable using N$_2$ at 400 °C, indicating that the triple bond of N$_2$ molecule can be cleaved at low temperatures [176]. When BaTiO$_{2.5}$H$_{0.5}$ is exposed to HF at 150 °C, BaTiO$_{2.5}$H$_{0.25}$F$_{0.25}$ results. A further treatment in a NaOMe/H$_2$O solution at around 100 °C leads to a novel oxide–hydride–hydroxide BaTi(O^{2-},H$^-$,(OH)$^-$)$_3$, where hydride and proton coexist in the perovskite structure. The reaction scheme is summarized in Figure 12.21 [176].

An interesting mechanism for hydride diffusion has been proposed on the basis of density functional calculations [177]. When examining the diffusion of a single hydride anion in BaTiO$_3$, it is found that it is energetically favorable for the hydride at the anionic site to transfer two electrons to the neighboring titanium atoms, and then diffuse as a proton to an interstitial site, after which it converts back to a hydride on reaching another anionic site. Such proton-assisted hydride diffusion is not only nontrivial but also useful in a system where only a small amount of anion vacancy is present like ATiO$_{3-x}$H$_x$. In addition, the formation of H$_2$ molecule-like state is also proposed [178], a finding that can

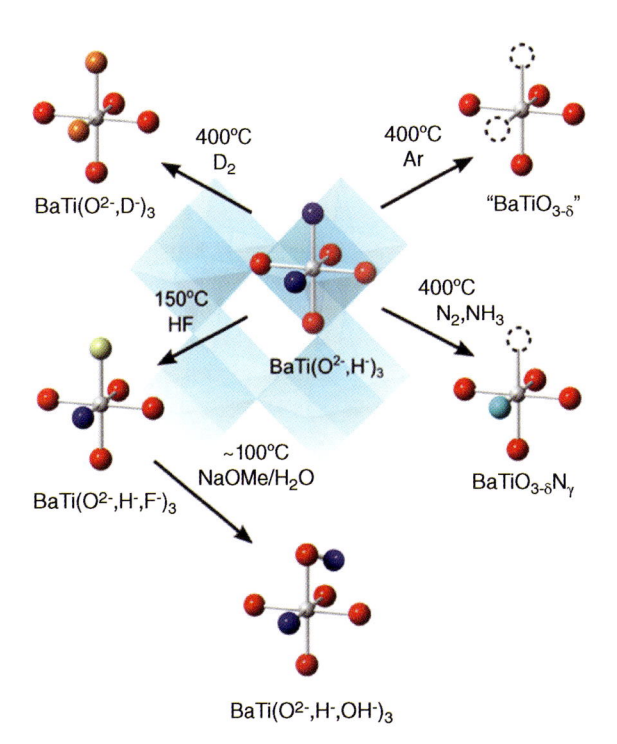

Figure 12.21 Hydride exchange reactions studied on BaTiO$_{2.5}$H$_{0.5}$ [159,174,175]. (With permission from *Journal of the American Chemical Society*.)

expand the role of the hydrogen in carrier control technology in transition metal oxides and may explain experimentally observed puzzling valence states of the oxygen vacancy in $SrTiO_3$.

12.6.2.3 Vanadium Oxyhydrides

Low-temperature hydride reduction also converts vanadium oxide perovskites to oxyhydrides. The Hayward group synthesized $Sr_{n+1}V_nO_{2n+1}H_n$ ($n = 1$, 2, and ∞; Sr_2VO_3H, $Sr_3V_2O_5H_2$, and $SrVO_2H$), topochemically converted from Ruddlesden–Popper-type perovskite oxides $Sr_{n+1}V_nO_{3n+1}$ ($Sr_3V_2O_7$, Sr_2VO_4, and $SrVO_3$) [164]. As shown in Figure 12.22, all these structures are fully anion ordered, with the hydride anions occupying the apical site of VO_4H_2 octahedra with two short V—H bonds (1.83 Å) and four long V—O bonds (1.95 Å). The resultant *trans*-VO_4H_2 octahedral coordination (or VO_4 square planar coordination) is stabilized by the d^2 electronic configurations of $(d_{xz}, d_{yz})^2$. It is noteworthy that the hydride reduction of iron oxides $Sr_{n+1}Fe_nO_{3n+1}$ ($Sr_3Fe_2O_7$, Sr_2FeO_4, and $SrFeO_3$) with Ruddlesden–Popper structures yields the isostructural oxides $Sr_{n+1}Fe_nO_{2n+1}$ (Sr_2FeO_3, $Sr_3Fe_2O_5$, and $SrFeO_2$) with hydride ions replaced by vacancies [179].

The possession of the series of vanadium oxyhydrides enables one to investigate the correlation between magnetism and dimensionality. In $Sr_{n+1}V_nO_{2n+1}H_n$, the dominant interaction should be antiferromagnetic V—O—V superexchange interaction of π-type (via V $3d_{xz/yz}$-O 2p-V $3d_{xz/yz}$), while the V—H—V interaction appears to be negligible because of the orthogonal nature of V $3d_{xz/yz}$ and H 1s orbitals, which is in marked contrast to $LaSrCoO_3H_{0.7}$ with σ-type V—H—V superexchange between Co e_g and H 1s orbitals. Sr_2VO_3H and $SrVO_2H$ may be

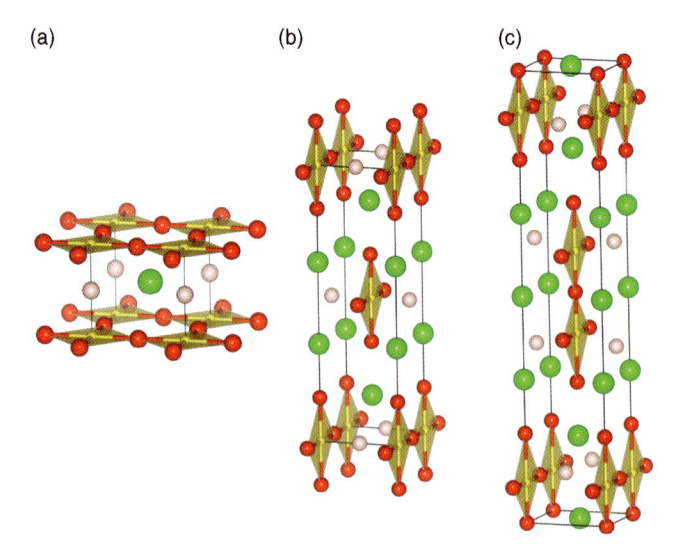

Figure 12.22 Crystal structures of $SrVO_2H$ (a), Sr_2VO_3H (b), and $Sr_3V_2O_5H_2$ (c). The red, pink, yellow, and green balls represent the oxide, hydride, vanadium, and strontium ions.

thus regarded as $S=1$ 1D and 2D systems, respectively, while $Sr_3V_2O_5H_2$ is an $S=1$ ladder with an intermediate dimension bridging 1D and 2D. The antiferromagnetic transition temperatures of $SrVO_2H$, Sr_2VO_3H, and $Sr_3V_2O_5H_2$ show a correlation with dimensionality, with T_N of 378, 240, and 170 K, respectively [162,179,180]. However, the Néel temperature of $SrVO_2H$ seems to be too high for the 2D system, suggesting a sizable out-of-plane antiferromagnetic interaction. Theoretical calculations provide several possible scenarios for the out-of-plane interaction, including direct V t_{2g}-V t_{2g} overlap and indirect exchange mediated by Sr [181,182].

12.6.3
High-Pressure Synthesis of Oxyhydrides

As already shown, topochemical hydride reductions have so far yielded perovskite-type oxyhydride with Co, Ti, and V. In the meantime, high-pressure reaction has been recognized as another approach to access transition metal oxyhydrides. High-pressure synthesis has already been widely used in the synthesis of metal hydrides that include binary systems (e.g., FeH) as well as ternary systems (e.g., $LiNiH_3$ and $CaPdH_3$). The formation of hydrides usually requires synthesis temperatures far beyond the decomposition temperatures of hydride precursors (e.g., $SrH_2 \rightarrow Sr + H_2$), but high external (and high hydrogen pressure) can avoid such decomposition and instead stabilize a solid hydride with a dense structure.

$SrCrO_2H$ is obtained by the high-pressure reaction of a mixture of SrO, Cr_2O_3, and SrH_2 at 1000 °C under 5 GPa [183]. It crystallizes in the ideal cubic perovskite structure with anion disorder. This material exhibits an antiferromagnetic order at ∼380 K, the highest T_N among Cr^{3+}-containing oxides, thus offering a new direction for controlling functionalities. In perovskite oxides ABO_3, the tolerance factor t (or $\angle$B–O–B) is known to play a key role in controlling physical properties. For example, the transition temperature of $RECr^{3+}O_3$ with the *distorted* $GdFeO_3$ structure ($t=0.85$–0.95) is linearly scaled by t ($T_N = 288$–110 K for $RE = $La $\rightarrow$ Lu) (Figure 12.23) [184]. Here, the anion composition of "O_2H" enables one to use divalent Sr^{2+} with a larger ionic radius, giving the ideal cubic structure ($t=1.0$ and $\angle$Cr–O–Cr $= 180°$) and the enhanced T_N.

We have addressed earlier that the topochemical reaction of Sr_2VO_4 yields the stoichiometric orthorhombic phase Sr_2VO_3H. However, a high-pressure reaction using a mixture of SrH_2, SrO, and VO_2 leads to a solid solution of $Sr_2VO_{4-x}H_x$ ($0 \leq x \leq 1$) [185]. An orthorhombic-to-tetragonal transition is observed at $x \sim 0.8$, below which a complete disorder between H^-/O^{2-} anions is suggested. $LaSrMnO_{3.3}H_{0.7}$ also adopts the K_2NiF_4 structure, with H^- positioned exclusively at the equatorial site [186]. This manganese oxyhydride is in striking contrast to topochemical reductions of $LaSrMnO_4$ that result in only oxygen-deficient phases down to $LaSrMnO_{3.5}$. This suggests that high (H_2) pressure condition stabilizes the oxyhydride phase, offering an opportunity to synthesize other transition metal oxyhydrides.

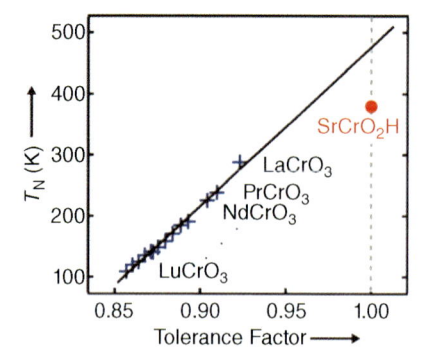

Figure 12.23 T_N of RECrO$_3$ and SrCrO$_2$H as a function of the tolerance factor t [183]. (With permission from Ref. [183].)

La$_2$LiO$_3$H prepared at 2 GPa is orthorhombic and isostructural with Sr$_2$VO$_3$H. Under a lower pressure of 1 GPa, a tetragonal phase La$_2$Li(O$_{1.21}$H$_{0.53}\square_{0.26}$)O$_2$ ($\square$ = vacancy) is formed, with disordered (H$^-$/O^{2-}/V) anions in the equatorial plane [90]. The La-site substitution results in the increase of hydride content. LaSrLiO$_2$H$_2$ is with H$^-$ partially occupying the apical site (Figure 12.24). Novel hydride anion conductivity is found in this series of oxyhydrides. The richest H$^-$ structure (Sr$_2$LiOH$_3$) exhibits the highest conductivity of 3.2×10^{-5} S/cm at 573 K. Further tuning of the structure is possible by inserting anionic vacancies, leading to the optimum composition of La$_{0.6}$Sr$_{1.4}$LiO$_2$H$_{1.6}$ with hydride anions and vacancies in the equatorial plane. This structure exhibits a hydride conductivity of 2.1×10^{-4} S/cm at 590 K with an activation energy of 68.4 kJ/mol.

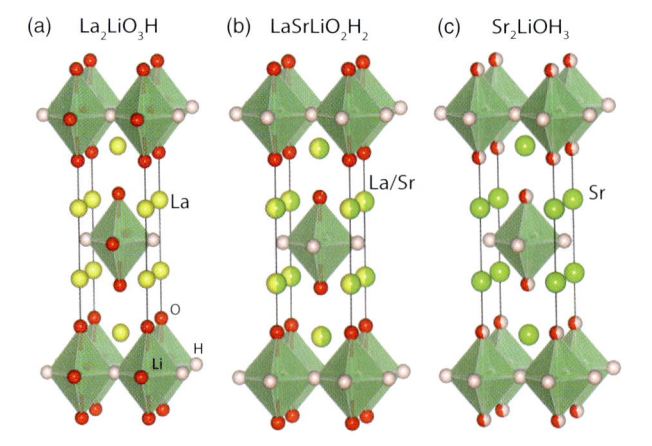

Figure 12.24 Crystal structure evolution of the La$_2$LiO$_3$H-Sr$_2$LiOH$_3$ system.

Acknowledgement

H. K. thanks JSPS Grant-in-Aid for Scientific Research on Innovative Areas "Mixed anion" (JP16H06439) from MEXT and CREST (JPMJCR1421).

References

1 Grochala, W. and Edwards, P.P. (2004) *Chem. Rev.*, **104**, 1283–1315.
2 Bergsma, J. and Loopstra, B.O. (1962) *Acta Crystallogr.*, **15**, 92–93 (short communications).
3 Brese, N.E., O'Keeffe, M., and Von Dreele, R.B. (1990) *J. Solid State Chem.*, **88**, 571–576.
4 Bronger, W., Scha, C.C., and Muller, P. (1987) *Z. Anorg. Allg. Chem.*, **545**, 69–74.
5 Shannon, R.D. (1976) *Acta Crystallogr. A*, **32**, 751–767.
6 Shriver, D.F., Atkins, P.W., and Langford, C.H. (1990) *Inorganic Chemistry*, 2nd edn, Oxford University Press, Oxford.
7 Pretzel, F.E., Rupert, G.N., Mader, C.L., Storms, E.K., Gritton, G.V., and Rushing, C.C. (1960) *J. Phys. Chem. Solids*, **16**, 10–20.
8 Kordes, E. (1962) *Z. Phys. Chem. (N F)*, **33**, 1–14.
9 Huot, J. (2010) *Handbook of Hydrogen Storage*, Wiley-VCH Verlag GmbH, Weinheim, pp. 81–116.
10 Libowitz, G.G. (1965) *The Solid State Chemistry of Binary Metal Hydrides*, 1st edn, W.A. Benjamin Inc., New York.
11 Calder, R.S., Cochran, W., Griffiths, D., and Lowde, R.D. (1962) *J. Phys. Chem. Solids*, **23**, 621–632.
12 Rodriguez, C.O. and Kunc, K. (1987) *Solid State Commun.*, **64**, 19–22.
13 Baraille, I., Pouchan, C., Causa, M., and Pisani, C. (1994) *Chem. Phys.*, **179**, 39–46.
14 Verbraeken, M.C., Suard, E., and Irvine, J.T.S. (2009) *J. Mater. Chem.*, **19**, 2766–2770.
15 Alonso, J.A., Retuerto, M., Sanchez-Benitez, J., and Fernandez-Diaz, M.T. (2010) *Z. Kristallogr.*, **225**, 225–229.
16 Verbraeken, M.C., Cheung, C., Suard, E., and Irvine, J.T.S. (2015) *Nat. Mater.*, **14**, 95–100.
17 Songster, J. and Pélton, A.D. (1993) *J. Phase Equilib.*, **14**, 373–381.
18 San-Martin, A. and Manchester, F.D. (1990) *Bull. Alloy Phase Diagr.*, **11**, 287–294.
19 Sangster, J. and Pelton, A.D. (1997) *J. Phase Equilib.*, **18**, 387–389.
20 Sangster, J. and Pelton, A.D. (1994) *J. Phase Equilib.*, **15**, 87–89.
21 Sangster, J. and Pelton, A.D. (1994) *J. Phase Equilib.*, **15**, 84–86.
22 San-Martin, A. and Manchester, F.D. (1987) *J. Phase Equilib.*, **8**, 431–437.
23 Stampfer, J.F., Holley, C.E., and Suttle, J.F. (1960) *J. Am. Chem. Soc.*, **82**, 3504–3508.
24 Friedlmeier, G.M. and Bolcich, J.C. (1988) *Int. J. Hydrogen Energy*, **13**, 467–474.
25 Curtis, R.W. and Chiotti, P. (1963) *J. Phys. Chem.*, **67**, 1061–1065.
26 Peterson, D.T. and Nelson, S.O. (1980) *J. Less-Common Met.*, **72**, 251–256.
27 Franzen, H.F., Khan, A.S., and Peterson, D.T. (1979) *J. Less-Common Met.*, **65**, 111–116.
28 Magee, C.B. (1968) *Metal Hydrides*, Academic Press, pp. 165–240.
29 Sears, V.F. (1992) *Neutron News*, **3**, 26–37.
30 Weller, M.T., Henry, P.F., Ting, V.P., and Wilson, C.C. (2009) *Chem. Commun.*, 2973–2989.
31 Ting, V.P., Henry, P.F., Kohlmann, H., Wilson, C.C., and Weller, M.T. (2010) *Phys. Chem. Chem. Phys.*, **12**, 2083–2088.
32 Zimmerman, W.B. (1972) *Phys. Rev. B*, **5**, 4704–4707.
33 Kuznetsov, V.G. and Shkrabkina, M.M. (1962) *J. Struct. Chem.*, **3**, 532–537.
34 Zintl, E. and Harder, A. (1931) *Z. Phys. Chem. Abt. B*, **14**, 265–284.
35 Ghandehari, K., Luo, H., Ruoff, A.L., Trail, S.S., and DiSalvo, F.J. (1995) *Phys. Rev. Lett.*, **74**, 2264–2267.
36 Zintl, E. and Harder, A. (1935) *Z. Elektrochem. Angew. Phys. Chem.*, **41**, 33–52.
37 Wu, H., Zhou, W., Udovic, T.J., Rush, J.J., and Yildirim, T. (2007) *J. Alloy Compd.*, **436**, 51–55.

38 Snyder, G.J., Borrmann, H., and Simon, A. (1994) *Z. Kristallogr.*, **209**, 458–458.
39 San-Martin, A. and Manchester, F.D. (1990) *Bull. Alloy. Phase Diagr.*, **11**, 287–294.
40 Veleckis, E. (1977) *J. Phys. Chem.*, **81**, 526–531.
41 Peterson, D.T. and Indig, M. (1960) *J. Am. Chem. Soc.*, **82**, 5645–5646.
42 Peterson, D.T. and Fattore, V.G. (1961) *J. Phys. Chem.*, **65**, 2062–2064.
43 Peterson, D.T. and Colburn, R.P. (1966) *J. Phys. Chem.*, **70**, 468–471.
44 Mitchell, P.C.H., Parker, S.F., Ramirez-Cuesta, A.J., and Tomkinson, J. (2005) *Vibrational Spectroscopy with Neutrons with Applications in Chemistry, Biology, Materials Science and Catalysis*, World Scientific, Singapore.
45 Mills, R.L., Dhandapani, B., Nansteel, M., He, J., and Voigt, A. (2001) *Int. J. Hydrogen. Energy*, **26**, 965–979.
46 Maeland, A.J. (1970) *J. Chem. Phys.*, **52**, 3952–3956.
47 Morris, P., Ross, D.K., Ivanov, S., Weaver, D.R., and Serot, O. (2004) *J. Alloy Compd.*, **363**, 85–89.
48 Colognesi, D., Barrera, G., Ramirez-Cuesta, A.J., and Zoppi, M. (2007) *J. Alloy Compd.*, **427**, 18–24.
49 Andresen, A.F., Maeland, A.J., and Slotfeldt-Ellingsen, D. (1977) *J. Solid State Chem.*, **20**, 93–101.
50 Nicol, A.T. and Vaughan, R.W. (1978) *J. Chem. Phys.*, **69**, 5211–5213.
51 Franco, F., Baricco, M., Chierotti, M.R., Gobetto, R., and Nervi, C. (2013) *J. Phys. Chem. C*, **117**, 9991–9998.
52 Cui, S., Feng, W., Hu, H., Feng, Z., and Wang, Y. (2008) *Solid State Commun.*, **148**, 403–405.
53 Noritake, T., Towata, S., Aoki, M., Seno, Y., Hirose, Y., Nishibori, E., Takata, M., and Sakata, M. (2003) *J. Alloy Compd.*, **356–357**, 84–86.
54 Vajeeston, P., Ravindran, P., Hauback, B.C., Fjellvåg, H., Kjekshus, A., Furuseth, S., and Hanfland, M. (2006) *Phys. Rev. B*, **73**, 224102.
55 Zachariasen, W.H., Holley, C.E., and Stamper, J.F., Jr. (1963) *Acta Crystallogr.*, **16**, 352–353.
56 Bortz, M., Bertheville, B., Böttger, G., and Yvon, K. (1999) *J. Alloy Compd.*, **287**, L4–L6.
57 Wigner, E. and Huntington, H.B. (1935) *J. Chem. Phys.*, **3**, 764–770.
58 Dalladay-Simpson, P., Howie, R.T., and Gregoryanz, E. (2016) *Nature*, **529**, 63–67.
59 Loubeyre, P., Occelli, F., and LeToullec, R. (2002) *Nature*, **416**, 613–617.
60 Pickard, C.J. and Needs, R.J. (2007) *Nat. Phys.*, **3**, 473–476.
61 Hochheimer, H.D., Strössner, K., Hönle, W., Baranowski, B., and Filipek, F. (1985) *J. Less-Common Met.*, **107**, L13–L14.
62 Duclos, S.J., Vohra, Y.K., Ruoff, A.L., Filipek, S., and Baranowski, B. (1987) *Phys. Rev. B*, **36**, 7664–7667.
63 Zurek, E., Hoffmann, R., Ashcroft, N.W., Oganov, A.R., and Lyakhov, A.O. (2009) *Proc. Natl. Acad. Sci. USA*, **106**, 17640–17643.
64 Wang, Y., Ahuja, R., and Johansson, B. (2003) *Phys. Status Solidi B*, **235**, 470–473.
65 Pépin, C., Loubeyre, P., Occelli, F., and Dumas, P. (2015) *Proc. Natl. Acad. Sci. USA*, **112**, 7673–7676.
66 Hooper, J., Baettig, P., and Zurek, E. (2012) *J. Appl. Phys.*, **111**, 112611.
67 Moriwaki, T., Akahama, Y., Kawamura, H., Nakano, S., and Takemura, K. (2006) *J. Phys. Soc. Jpn.*, **75**, 074603.
68 Bastide, J.-P., Bonnetot, B., Létoffé, J.-M., and Claudy, P. (1980) *Mater. Res. Bull.*, **15**, 1215–1224.
69 Kinoshita, K., Nishimura, M., Akahama, Y., and Kawamura, H. (2007) *Solid State Commun.*, **141**, 69–72.
70 Smith, J.S., Desgreniers, S., Klug, D.D., and Tse, J.S. (2009) *Solid State Commun.*, **149**, 830–834.
71 Smith, J.S., Desgreniers, S., Tse, J.S., and Klug, D.D. (2007) *J. Appl. Phys.*, **102**, 043520.
72 Tse, J.S., Klug, D.D., Desgreniers, S., Smith, J.S., Flacau, R., Liu, Z., Hu, J., Chen, N., and Jiang, D.T. (2007) *Phys. Rev. B*, **75**, 134108.
73 Tse, J.S., Song, Z., Yao, Y., Smith, J.S., Desgreniers, S., and Klug, D.D. (2009) *Solid State Commun.*, **149**, 1944–1946.
74 Struzhkin, V.V., Kim, D.Y., Stavrou, E., Takaki, M., Mao, H., Pickard, C.J., Needs, R.J., Prakapenka, V.B., and Goncharov, A.F. (2016) *Nat. Commun.*, **7**, 12267.

75 Hooper, J., Altintas, B., Shamp, A., and Zurek, E. (2013) *J. Phys. Chem. C*, **117**, 2982–2992.
76 Wang, H., Tse, J.S., Tanaka, K., Iitaka, T., and Ma, Y. (2012) *Proc. Natl. Acad. Sci. USA*, **109**, 6463–6466.
77 Drozdov, A.P., Eremets, M.I., Troyan, I.A., Ksenofontov, V., and Shylin, S.I. (2015) *Nature*, **525**, 73–76.
78 Stoneham, R.P.A.A.M. (1985) *J. Phys. C Solid State Phys.*, **18**, 5289.
79 Wakamori, K. and Sawaoka, A. (1982) *J. Less-Common Met.*, **88**, 217–220.
80 Gorelov, V.P. and Pal'guev, S.F. (1992) *Elektrokhimiya*, **28**, 1294–1296.
81 Verbraeken, M.C. (2009) Dope alkaline earth (nitride) hydrides. Ph.D. thesis, University of St Andrews.
82 Ure, R.W. (1957) *J. Chem. Phys.*, **26**, 1363–1373.
83 Carr, M.V., Chadwick, V.A., and Figueroa, R.D. (1976) *J. Phys. Colloques*, **37**, C7-337–C7-341.
84 Messer, C.E., Miller, R.M., and Barrante, J.R. (1966) *Inorg. Chem.*, **5**, 1814–1816.
85 Gibson, I.R. and Irvine, J.T.S. (1996) *J. Mater. Chem.*, **6**, 895–898.
86 Steele, B.C.H. (1989) *High Conductivity Solid Ionic Conductors* (ed. T. Takahashi), World Scientific Publishing Co. Pte. Ltd, Singapore, pp. 402–446.
87 Leveque, R., Zanne, M., Vergnat-Grandjean, D., and Brice, J.-F. (1980) *J. Solid State Chem.*, **33**, 233–243.
88 Villeneuve, G. (1979) *Mater. Res. Bull.*, **14**, 1231–1234.
89 Sridharan, R., Mahendran, K.H., Gnanasekaran, T., Periaswami, G., Varadaraju, U.V., and Mathews, C.K. (1995) *J. Nucl. Mater.*, **223**, 72–79.
90 Kobayashi, G., Hinuma, Y., Matsuoka, S., Watanabe, A., Iqbal, M., Hirayama, M., Yonemura, M., Kamiyama, T., Tanaka, I., and Kanno, R. (2016) *Science*, **351**, 1314–1317.
91 Bouhadda, Y., Kheloufi, N., Bentabet, A., Boudouma, Y., Fenineche, N., and Benyalloul, K. (2011) *J. Alloy Compd.*, **509**, 8994–8998.
92 Vajeeston, P., Ravindran, P., Kjekshus, A., and Fjellvåg, H. (2008) *J. Alloy Compd.*, **450**, 327–337.
93 Ghebouli, M.A., Ghebouli, B., Bouhemadou, A., Fatmi, M., and Bin-Omran, S. (2011) *Solid State Sci.*, **13**, 647–652.
94 Reshak, A.H., Shalaginov, M.Y., Saeed, Y., Kityk, I.V., and Auluck, S. (2011) *J. Phys. Chem. B*, **115**, 2836–2841.
95 Battal, G.Y., Salmankurt, B., and Duman, S. (2016) *Mater. Res. Express*, **3**, 036301.
96 Ghebouli, B., Ghebouli, M.A., and Fatmi, M. (2010) *Eur. Phys. J. Appl. Phys.*, **53**, 51.
97 Ikeda, K., Nakamori, Y., and Orimo, S. (2005) *Acta Mater.*, **53**, 3453–3457.
98 Pottmaier, D., Pinatel, E.R., Vitillo, J.G., Garroni, S., Orlova, M., Baró, M.D., Vaughan, G.B.M., Fichtner, M., Lohstroh, W., and Baricco, M. (2011) *Chem. Mater.*, **23**, 2317–2326.
99 Yvon, K. and Bertheville, B. (2006) *J. Alloy Compd.*, **425**, 101–108.
100 Ikeda, K., Kato, S., Shinzato, Y., Okuda, N., Nakamori, Y., Kitano, A., Yukawa, H., Morinaga, M., and Orimo, S. (2007) *J. Alloy Compd.*, **446–447**, 162–165.
101 Wu, H., Zhou, W., Udovic, T.J., Rush, J.J., and Yildirim, T. (2008) *Chem. Mater.*, **20**, 2335–2342.
102 Schumacher, R. and Weiss, A. (1990) *J. Less-Common Met.*, **163**, 179–183.
103 Gingl, F., Vogt, T., Akiba, E., and Yvon, K. (1999) *J. Alloy Compd.*, **282**, 125–129.
104 Wu, H., Zhou, W., Udovic, T.J., Rush, J.J., and Yildirim, T. (2009) *J. Phys. Chem. C*, **113**, 15091–15098.
105 Park, H.H., Pezat, M., and Darriet, B. (1986) *Rev. Chim. Miner.*, **23**, 323–328.
106 Messer, C.E., Eastman, J.C., Mers, R.G., and Maeland, A.J. (1964) *Inorg. Chem.*, **3**, 776–778.
107 Maeland, A.J. and Andresen, A.F. (1968) *J. Chem. Phys.*, **48**, 4660–4661.
108 Bogdanovic, B. and Schwickardi, M. (1997) *J. Alloy Compd.*, **253**, 1–9.
109 Schlapbach, L. and Zuttel, A. (2001) *Nature*, **414**, 353–358.
110 Zuttel, A., Wenger, P., Rentsch, S., Sudan, P., Mauron, P., and Emmenegger, C. (2003) *J. Power Sources*, **118**, 1–7.
111 Chen, P., Xiong, Z.T., Luo, J.Z., Lin, J.Y., and Tan, K.L. (2002) *Nature*, **420**, 302–304.
112 Sato, T., Noréus, D., Takeshita, H., and Häussermann, U. (2005) *J. Solid State Chem.*, **178**, 3381–3388.
113 Greedan, J.E. (1971) *J. Phys. Chem. Solids*, **32**, 1039–1046.

114 Libowitz, G.G. (1994) *J. Phys. Chem. Solids*, **55**, 1461–1470.
115 Shaw, B.L. (1967) *Inorganic Hydrides*, Pergamon Press.
116 Shriver, D.F., Atkins, P.W., and Langford, C.H. (2012) *Inorganic Chemistry*, 5th edn, Oxford University Press, Oxford.
117 Holley, C.E. et al. (1955) *J. Phys. Chem.*, **59**, 1226–1228.
118 Udovic, T.J., Huang, Q., and Rush, J.J. (1996) *J. Solid State Chem.*, **122**, 151–159.
119 Renaudin, G., Fischer, P., and Yvon, K. (2000) *J. Alloy Compd.*, **313**, L10–L14.
120 Udovic, T.J., Huang, Q., and Rush, J.J. (2003) *J. Alloy Compd.*, **356**, 41–44.
121 Korst, W.L. and Warf, J.C. (1966) *Inorg. Chem.*, **5**, 1719–1726.
122 Hardcastle, K.I. and Warf, J.C. (1966) *Inorg. Chem.*, **5**, 1728–1735.
123 Libowitz, G.G. and Pack, J.G. (1969) *J. Phys. Chem.*, **73**, 2352–2356.
124 Ellner, M., Reule, H., and Mittemeijer, E.J. (2000) *J. Alloy Compd.*, **309**, 127–131.
125 Huang, Q. et al. (1995) *J. Alloy Compd.*, **231**, 95–98.
126 Pebler, A. and Wallace, W.E. (1962) *J. Phys. Chem.*, **66**, 148–151.
127 Warf, J.C. and Hardcastle, K.I. (1966) *Inorg. Chem.*, **5**, 1736–1740.
128 Auffermann, G. (2002) *Z. Anorg. Allg. Chem.*, **628**, 1615–1618.
129 Antonov, V.E. et al. (2006) *Phys. Rev. B*, **73**, 054107.
130 FIZ Karlsruhe (2016) ICSD database.
131 Titcomb, C.G., Cheetham, A.K., and Fender, B.E.F. (1974) *J. Phys. C Solid State Phys.*, **7**, 2409–2416.
132 Kohlmann, H. et al. (2007) *Chem. Eur. J.*, **13**, 4178–4186.
133 Bambakidis, G. (ed.) (1981) *Metal Hydrides*, Springer, New York.
134 Hosono, H. and Matsuishi, S. (2013) *Curr. Opin. Solid State Mater. Sci.*, **17**, 49–58.
135 Hiraishi, M. et al. (2014) *Nat. Phys.*, **10**, 300–302.
136 Sun, X., Guo, Y., Wu, C., and Xie, Y. (2015) *Adv. Mater.*, **27**, 3850–3867.
137 Malaman, B. and Brice, J.F. (1984) *J. Solid State Chem.*, **53**, 44–54.
138 Brice, J.F. and Moreau, A. (1982) *Ann. Chimie Sci. Materiaux*, 7, 623–634.
139 Widerøe, M., Fjellvåg, H., Norby, T., Poulsen, F.W., and Berg, R.W. (2011) *J. Solid State Chem.*, **184**, 1890–1894.
140 Liu, X., Bjorheim, T.S., and Haugsrud, R. (2016) *RSC Adv.*, **6**, 9822–9826.
141 Zachariasen, W.H. (1951) *Acta Crystallogr.*, **4**, 231–236.
142 Huang, B. and Corbett, J.D. (1998) *J. Solid State Chem.*, **141**, 570–575.
143 Skakle, J.M.S., Fletcher, J.G., and West, A.R. (1996) *Dalton Trans.*, 2497–2501.
144 Zhang, X., Li, D., Yang, Z., Wu, H., and Pan, S. (2015) *Z. Anorg. Allg. Chem.*, **641**, 922–926.
145 Sullivan, E., Avdeev, M., Blom, D.A., Gahrs, C.J., Green, R.L., Hamaker, C.G., and Vogt, T. (2015) *J. Solid State Chem.*, **230**, 279–286.
146 Vogt, T., Woodward, P.M., Hunter, B.A., Prodjosantoso, A.K., and Kennedy, B.J. (1999) *J. Solid State Chem.*, **144**, 228–231.
147 Boss, M., Petri, D., Pickhard, F., Zönnchen, P., and Röhr, C. (2005) *Z. Anorg. Allg. Chem.*, **631**, 1181–1190.
148 Huang, B. and Corbett, J.D. (1998) *Inorg. Chem.*, **37**, 1892–1899.
149 Jehle, M., Hoffmann, A., Kohlmann, H., Scherer, H., and Röhr, C. (2015) *J. Alloy Compd.*, **623**, 164–177.
150 Kitano, M., Inoue, Y., Yamazaki, Y., Hayashi, F., Kanbara, S., Matsuishi, S., Yokoyama, T., Kim, S.-W., Hara, M., and Hosono, H. (2012) *Nat. Chem.*, **4**, 934–940.
151 Miyakawa, M., Kim, S.W., Hirano, M., Kohama, Y., Kawaji, H., Atake, T., Ikegami, H., Kono, K., and Hosono, H. (2007) *J. Am. Chem. Soc.*, **129**, 7270–7271.
152 Hayashi, K., Sushko, P.V., Hashimoto, Y., Shluger, A.L., and Hosono, H. (2014) *Nat. Commun.*, **5**, 3515.
153 Hayashi, K. (2011) *J. Solid State Chem.*, **184**, 1428–1432.
154 Hayashi, K., Matsuishi, S., Kamiya, T., Hirano, M., and Hosono, H. (2002) *Nature*, **419**, 462–465.
155 Norby, T., Wideroe, M., Glockner, R., and Larring, Y. (2004) *Dalton Trans.*, 3012–3018.
156 Steinsvik, S., Larring, Y., and Norby, T. (2001) *Solid State Ionics*, **143**, 103–116.

157 Wideroe, M., Kochetova, N., and Norby, T. (2004) *Dalton Trans.*, 3147–3151.
158 Widerøe, M., Münch, W., Larring, Y., and Norby, T. (2002) *Solid State Ionics*, **154–155**, 669–677.
159 Poulsen, F.W. (2001) *Solid State Ionics*, **145**, 387–397.
160 Bridges, C.A., Darling, G.R., Hayward, M.A., and Rosseinsky, M.J. (2005) *J. Am. Chem. Soc.*, **127**, 5996–6011.
161 Hayward, M.A., Green, M.A., Rosseinsky, M.J., and Sloan, J. (1999) *J. Am. Chem. Soc.*, **121**, 8843–8854.
162 Tsujimoto, Y., Tassel, C., Hayashi, N., Watanabe, T., Kageyama, H., Yoshimura, K., Takano, M., Ceretti, M., Ritter, C., and Paulus, W. (2007) *Nature*, **450**, 1062–1065.
163 Kobayashi, Y., Hernandez, O.J., Sakaguchi, T., Yajima, T., Roisnel, T., Tsujimoto, Y., Morita, M., Noda, Y., Mogami, Y., Kitada, A., Ohkura, M., Hosokawa, S., Li, Z., Hayashi, K., Kusano, Y., Kim, J.-E., Tsuji, N., Fujiwara, A., Matsushita, Y., Yoshimura, K., Takegoshi, K., Inoue, M., Takano, M., and Kageyama, H. (2012) *Nat. Mater.*, **11**, 507–511.
164 Romero, F.D., Leach, A., Möller, J.S., Foronda, F., Blundell, S.J., and Hayward, M.A. (2014) *Angew. Chem., Int. Ed.*, **53**, 7556–7559.
165 Hayward, M.A., Cussen, E.J., Claridge, J.B., Bieringer, M., Rosseinsky, M.J., Kiely, C.J., Blundell, S.J., Marshall, I.M., and Pratt, F.L. (2002) *Science*, **295**, 1882–1884.
166 Blundell, S.J., Marshall, I.M., Pratt, F.L., Hayward, M.A., Cussen, E.J., Claridge, J.B., Bieringer, M., Kiely, C.J., and Rosseinsky, M.J. (2003) *Physica B*, **326**, 527–529.
167 Helps, R.M., Rees, N.H., and Hayward, M.A. (2010) *Inorg. Chem.*, **49**, 11062–11068.
168 Bridges, C.A., Fernandez-Alonso, F., Goff, J.P., and Rosseinsky, M.J. (2006) *Adv. Mater.*, **18**, 3304–3308.
169 Bouilly, G., Yajima, T., Terashima, T., Kususe, K., Fujita, K., Tassel, C., Yamamoto, T., Tanaka, K., Kobayashi, Y., and Kageyama, H. (2014) *CrystEngComm*, **16**, 9669–9674.
170 Sakaguchi, T., Kobayashi, Y., Yajima, T., Ohkura, M., Tassel, C., Takeiri, F., Mitsuoka, S., Ohkubo, H., Yamamoto, T., Kim, J.E., Tsuji, N., Fujihara, A., Matsushita, Y., Hester, J., Avdeev, M., Ohoyama, K., and Kageyama, H. (2012) *Inorg. Chem.*, **51**, 11371–11376.
171 Yajima, T., Kitada, A., Kobayashi, Y., Sakaguchi, T., Bouilly, G., Kasahara, S., Terashima, T., Takano, M., and Kageyama, H. (2012) *J. Am. Chem. Soc.*, **134**, 8782–8785.
172 Page, K., Kolodiazhnyi, T., Proffen, T., Cheetham, A.K., and Seshadri, R. (2008) *Phys. Rev. Lett.*, **101**, 205502.
173 Bouilly, G., Yajima, T., Terashima, T., Yoshimune, W., Nakano, K., Tassel, C., Kususe, Y., Fujita, K., Tanaka, K., Yamamoto, T., Kobayashi, Y., and Kageyama, H. (2015) *Chem. Mater.*, **27**, 6354–6359.
174 Yamamoto, T., Yoshii, R., Bouilly, G., Kobayashi, Y., Fujita, K., Kususe, Y., Matsushita, Y., Tanaka, K., and Kageyama, H. (2015) *Inorg. Chem.*, **54**, 1501–1507.
175 Yajima, T., Takeiri, F., Aidzu, K., Akamatsu, H., Fujita, K., Yoshimune, W., Ohkura, M., Lei, S., Gopalan, V., Tanaka, K., Brown, C.M., Green, M.A., Yamamoto, T., Kobayashi, Y., and Kageyama, H. (2015) *Nat. Chem.*, **7**, 1017–1023.
176 Masuda, N., Kobayashi, Y., Hernandez, O., Bataille, T., Paofai, S., Suzuki, H., Ritter, C., Ichijo, N., Noda, Y., Takegoshi, K., Tassel, C., Yamamoto, T., and Kageyama, H. (2015) *J. Am. Chem. Soc.*, **137**, 15315–15321.
177 Iwazaki, Y., Suzuki, T., and Tsuneyuki, S. (2010) *J. Appl. Phys.*, **108**, 083705.
178 Iwazaki, Y., Gohda, Y., and Tsuneyuki, S. (2014) *APL Mater.*, **2**, 012103.
179 Tassel, C., Seinberg, L., Hayashi, N., Ganesanpotti, S., Ajiro, Y., Kobayashi, Y., and Kageyama, H. (2013) *Inorg. Chem.*, **52**, 6906–6912.
180 Kageyama, H., Watanabe, T., Tsujimoto, Y., Kitada, A., Sumida, Y., Kanamori, K., Yoshimura, K., Hayashi, N., Muranaka, S., Takano, M., Ceretti, M., Paulus, W., Ritter, C., and André, G. (2008) *Angew. Chem., Int. Ed.*, **47**, 5740–5745.

181 Yingfen, W., Hong, G., Xin, L., Zhenjie, Z., Yong-Hong, Z., and Wenhui, X. (2015) *J. Phys. Condens. Matter*, **27**, 206001.

182 Bang, J., Matsuishi, S., Maki, S., Yamaura, J.-I., Hiraishi, M., Takeshita, S., Yamauchi, I., Kojima, K.M., and Hosono, H. (2015) *Phys. Rev. B*, **92**, 064414.

183 Tassel, C., Goto, Y., Kuno, Y., Hester, J., Green, M., Kobayashi, Y., and Kageyama, H. (2014) *Angew. Chem., Int. Ed.*, **53**, 10377–10380.

184 Lei, S., Liu, L., Wang, C., Wang, C., Guo, D., Zeng, S., Cheng, B., Xiao, Y., and Zhou, L. (2013) *J. Mater. Chem. A*, **1**, 11982–11991.

185 Bang, J., Matsuishi, S., Hiraka, H., Fujisaki, F., Otomo, T., Maki, S., Yamaura, J.-I., Kumai, R., Murakami, Y., and Hosono, H. (2014) *J. Am. Chem. Soc.*, **136**, 7221–7224.

186 Tassel, C., Goto, Y., Watabe, D., Tang, Y., Lu, H., Kuno, Y., Takeiri, F., Yamamoto, T., Brown, C.M., Hester, J., Kobayashi, Y., and Kageyama, H. (2016) *Angew. Chem., Int. Ed.* doi: 10.1002/anie.201605123.

13
Local Atomic Order in Intermetallics and Alloys

Frank Haarmann[1,2]

[1]*RWTH Aachen Univertiy, Institut für Anorganische Chemie, Landoltweg 1, 52074 Aachen, Germany*
[2]*Max-Planck-Institut für Chemische Physik fester Stoffe, Chemische Metallkunde, Nöthnitzer Straße 40, 01187 Dresden, Germany*

13.1
Motivation

Phases formed by different types of metal atoms as well as those formed by metal and metalloid atoms are considered as intermetallic compounds [1]. In contrast to this, alloys can be single- or multiphase materials [2]. We will focus on single-phase materials and give a short overview about their crystal structures focusing on local ordering of the atoms.

With more than 80 elements forming intermetallic phases (IPs), they are the largest group of inorganic materials. The main interest on IPs is related with a huge potential in applications, for example, in steel as the main construction material for centuries [3] or duralumin, which is used as a lightweight construction material. Furthermore, the use as phase-change materials (PCM) in optical storage media (CD, DVD, and Blue Ray discs) is very valuable for today's life [4]. In addition, applications in thermoelectric [5,6] and magnetic materials, semi- and superconductors [7], spintronics [8], as well as active materials in catalysis [9] are known.

Until now, the various requirements for technical applications of IPs are primarily based on trial and error approaches and empiric rules since, however, our current understanding of this fascinating class of materials is still rather limited [10,11]. This is due to the complexity of the crystal structures and the chemical bonding of IPs, which is still a sought-after issue. No consistent concept for description of the chemical bonding in the various compounds exists. Local ordering of the atoms, for example, disorder, complicates the determination and description of crystal structures. Hence, new strategies and approaches are required to investigate structure–bonding–property relationships in IPs. Quantum mechanical (QM) calculations are a promising part in these new strategies,

Handbook of Solid State Chemistry, First Edition. Edited by Richard Dronskowski, Shinichi Kikkawa, and Andreas Stein.
© 2017 Wiley-VCH Verlag GmbH & Co. KGaA. Published 2017 by Wiley-VCH Verlag GmbH & Co. KGaA.

which has been shown in many publications [12–17]. Furthermore, local probes such as nuclear magnetic resonance (NMR), being sensitive on the varying atomic arrangements, will be important for structural characterization [18–21]. A detailed knowledge of structure–bonding–property relations will be the key for innovative materials' design for an ever-increasing range of technologically relevant metallic materials.

IPs are known for centuries and it is not surprising that scientist of various backgrounds, such as physicists, chemists, metallurgists, and engineers, have extensively investigated this class of materials. Therefore, it is impossible to give a comprehensive review about all crystal structures and peculiarities of IPs within this contribution. However, Pöttgen and Johrendt recently published a textbook on this subject, which gives an excellent overview of IPs with regards to synthesis, structure, and properties of stoichiometric compounds [2].

In this contribution, we will omit uncountable number of solid solutions crystallizing in variants of densely packed atoms. Instead, we start with a brief summary of prominent representatives of IPs, that is, Hume-Rothery, Laves, Zintl, and Heusler phases. Short descriptions of thermoelectric materials and PCMs are given later on. Afterwards, we discuss a strategy to investigate local ordering of the atoms in IPs by a combined application of NMR spectroscopy, QM calculations, and diffraction methods. Finally, some investigations of local ordering in IPs following this strategy will be discussed.

13.2
Synthesis

IPs are traditionally synthesized via high-temperature routes by melting appropriate mixtures of the elements and subsequent thermal annealing. Hence, phase diagrams are the most important tool for a systematic synthesis planning [22,23]. Some low temperature attempts have also succeeded by the use of ionic liquids [24–26]. X-ray diffraction, thermal analysis, and metallurgic methods are routinely used for sample characterization [27].

13.3
Hume-Rothery Phases

Hume-Rothery phases are mainly formed by transition metals and less-typical main group metals. They are often referred as electronically determined materials possessing wide homogeneity ranges [28]. The systems Cu–Zn (bras) and Cu–Sn (bronze) have been most relevant in technological applications for decades.

α-, β-, γ-, δ-, ξ-, ε-, and η-phases with defined crystal structures and valence electron concentrations (VEC) per formula unit are characteristic for Hume-Rothery phases. High coordination numbers of the atoms are usually observed

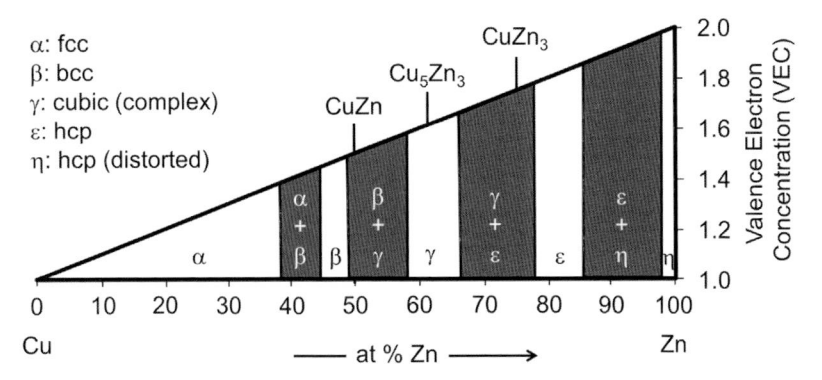

Figure 13.1 Scheme of the valence electron concentrations (VEC) per formula unit and phase sequence in the Cu–Zn system. Two phase regions are shaded in gray.

as also found in other metals. The α-phase forms a cubic closed packing while a body-centered cubic closed packing is realized in the β-phase with a statistical occupation of the atomic positions in both cases. Slow cooling of the β-phase frequently results in the formation of an ordered CsCl type structure. A complex cubic structure with 52 atoms in the unit cell is realized by the γ-phase. The δ-phase, rarely observed at high temperatures, forms large cubic unit cells. Also the ξ-phase with a hexagonal dense packing is rarely realized. A variant of the β-Mn type structure with statistical occupation of the sites is observed for the η-phase.

The VEC is characteristic for the formation of the individual phases although large homogeneity ranges are observed (Figure 13.1). Thus, only approximated values of the VEC can be given (α: 21/15, β: 21/14, γ: 21/13, ε: 21/12).

13.4
Laves Phases

Compounds with $MgCu_2$, $ZnCu_2$, or $MgNi_2$ type structures are called Laves phases (Figure 13.2). They are often realized by a typical main group element A and a less-typical transition metal B. The tetrahedron is the main motive of the crystal structures. Hence, the three structure types can be rationalized by the three-dimensional (3D) network of corner and/or face sharing B_4 tetrahedra. Corner sharing of B_4 units is observed in the cubic $MgCu_2$ type structure (Figure 13.2a). The A atoms are embedded into Frank-Kasper polyhedra with 16-fold coordination. The motive of a cubic diamond structure is realized by the A type atoms. Face and corner sharing Zn_4 tetrahedra form columns in the 3D network of the hexagonal $MgZn_2$ type (Figure 13.2b). Here, the A atoms show the motive of hexagonal diamond. Elements of both structure types are combined in the $MgNi_2$ type, in which four

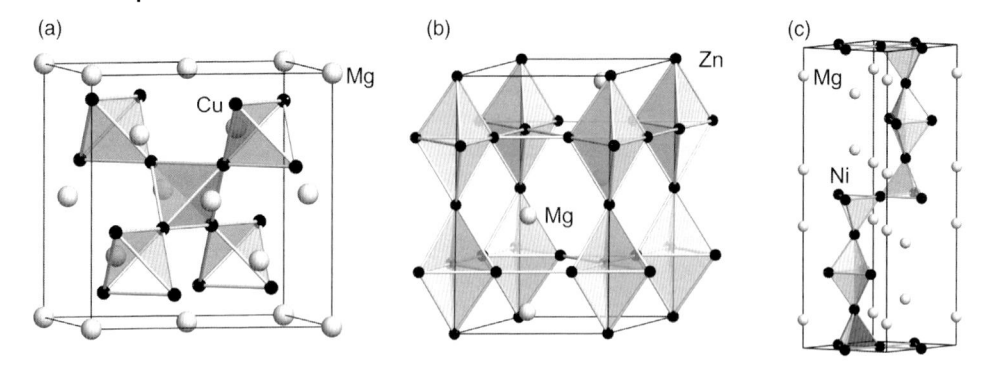

Figure 13.2 Crystal structures of Laves phases AB$_2$. MgCu$_2$ (a), MgZn$_2$ (b), and MgNi$_2$ type (c). The Mg and B atoms are represented by gray and black spheres, respectively. The B$_4$ tetrahedra are highlighted in gray.

corner and face sharing tetrahedra are connected by shared corners within the *ab* plane of the hexagonal unit cell (Figure 13.2c).

Assuming an ideal ratio of the atomic radii $r_A/r_B = 1.225$ a tetrahedral dense packing (tcp) with a space filling of 71% is formed. A variation of the atomic radii ratio $1.05 \leq r_A/r_B \leq 1.68$ is realized by nature [29,30]. Although several Laves phases possess homogeneity ranges they are often referred to as stoichiometric AB$_2$ compounds [31,32].

13.5
Zintl Phases

Zintl phases are polar intermetallics formed by electropositive and moderately electronegative elements [33]. They are realized by alkali- and alkaline-earths metals as well as rare-earth metals as electropositive elements. The more electronegative counterparts are elements on the right side of the Zintl border between group 13 and 14 of the periodic table of the elements.

The Zintl–Klemm–Busmann concept describes the bonding situation within this class of materials [34,35]. Assuming a complete electron transfer from the electropositive to the more electronegative component, a polyanionic substructure is formed, which can be described by the extended 8-N rule. The substructures often resemble those structures of the isoelectronic main group element. For example, isolated Si$_4^{4-}$ tetrahedra are observed in the monosilicides of the alkali metals M_4Si$_4$ with M = Na, K, Rb, Cs, and Ba$_2$Si$_4$ (Figure 13.3a) [36–42]. The anions are isoelectronic and isostructural to P$_4$ in white phosphorus.

Formally, Zintl phases are electron-precise, stoichiometric semiconductors. Nevertheless, NaTl, one of the most prominent prototypes for

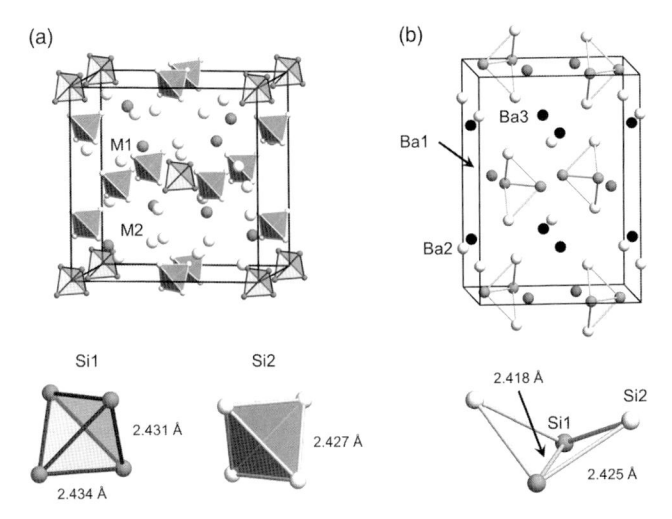

Figure 13.3 Cut out of the crystal structures of M_4Si_4 with M = K, Rb, Cs (a) and Ba_3Si_4 (b). The tetrahedral and butterfly-like anions are given below. Distances corresponding to Ba_3Si_4 and K_4Si_4 are given in Angstrom (Å).

Zintl phases, possesses a significant homogeneity range and metallic conductivity [43]. No significant homogeneity range is observed in metallic Ba_3Si_4 where the anions form butterfly-like Si_4^{6-} units (Figure 13.3b) [44,45].

Metallic conductivity and/or the formation of homogeneity ranges are/is frequently observed for compounds with group 13 elements as the more electronegative element. A Zintl-like character can be derived by the anionic sublattice obeying the extended 8-N rule by the formation of element-like substructures. $CaGa_2$ crystallizes in the $CaIn_2$ type structure with puckered six-membered Ga rings resembling a diamond like substructure with 3+1 coordination of the Ga atoms [46,47]. MGa_2 with M = Sr, Ba crystallizes in the AlB_2 type structure with honeycomb like slightly puckered 2D nets of Ga atoms resembling the graphite modification of C [47,48]. While $BaGa_2$ possesses a stoichiometric 1:2 composition without any significant homogeneity range, a complex substitution of the Sr atoms by Ga_3 units is observed in $Sr_{1-x}Ga_{2+3x}$ with $0 \leq x \leq 0.076$ at $T = 950\,°C$ (Figure 13.4) [47]. The same feature is also observed in the crystal structure of $Ca_{1-x}Ga_{2+3x}$ with $0.069 \leq x \leq 0.135$ at $T = 750\,°C$ [49]. Many other compounds with the Ga_3 unit are known in literature [50–53]. They are even observed as isolated Ga_3 groups in $Sr_2Au_6Ga_3$ [54,55].

Additional examples for Zintl phases are given in the textbooks of Kauzlarich [56] and Fässler [57], as well as in various review articles [44,58] focussing on the work of Eduard Zintl.

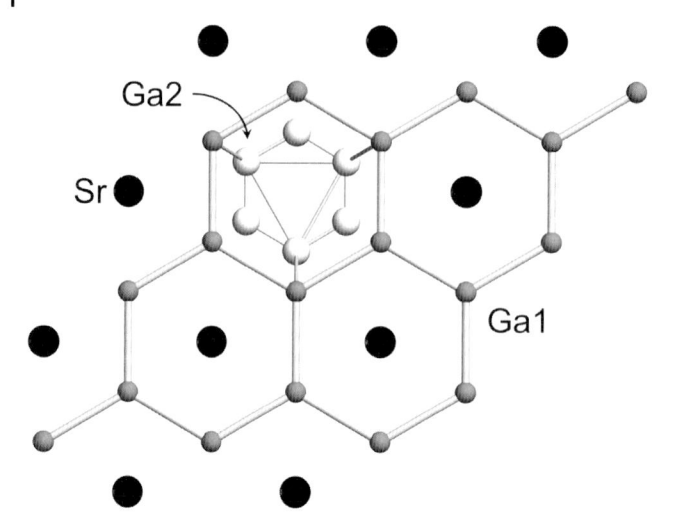

Figure 13.4 Cut out of the crystal structure of $Sr_{1-x}Ga_{2+3x}$. Sr and Ga1 atoms forming an AlB_2 type structure are represented by black and gray spheres, respectively. Additional Ga positions resulting from the Sr substitution are given in light gray. The three Ga2 atoms forming the Ga_3 unit are connected to the honeycomb network of the Ga1 atoms.

13.6
Heusler Phases

Already with the discovery of Heusler phases by the mining engineer Fritz Heusler in 1903 it was realized that this class of IPs have extraordinary physical properties. Although none of the constitution elements possesses magnetic order, a ferromagnetic coupling was detected at ambient temperature in Cu_2MnAl [59].

Since that time, many spectacular materials with various magnetic ordering variants [60–65], giant-magnetoresistance [8], shape memory alloys [66–68], topological insulators [69–71], have been reported. Furthermore, they are discussed in spintronics [8,72–74] and as potential thermoelectric materials [75]. The magnetic properties make an application of Heusler phases in data storages and as magnetic sensors challenging.

Heusler phases are typically formed by transition metals X, Y, and main group metals Z. Three structure types are related to the Heusler compounds: Cu_2MnAl, Li_2AgSb, and MgAgAs type (Figure 13.5). Their crystal structures can be described as filled variants of the NaCl type. The NaCl lattice is formed by Y and Z in so-called *full*-Heusler phases X_2YZ. All tetrahedral voids are filled by X-type atoms (Figure 13.5a). Since only half of the tetrahedral voids are filled, XYZ compounds with MgAgAs type structures are frequently designated as *half*-Heusler compounds (Figure 13.5c). Half of the X-type atoms exchange their positions with Y-type atoms in the so-called *inverse*-Heusler compounds

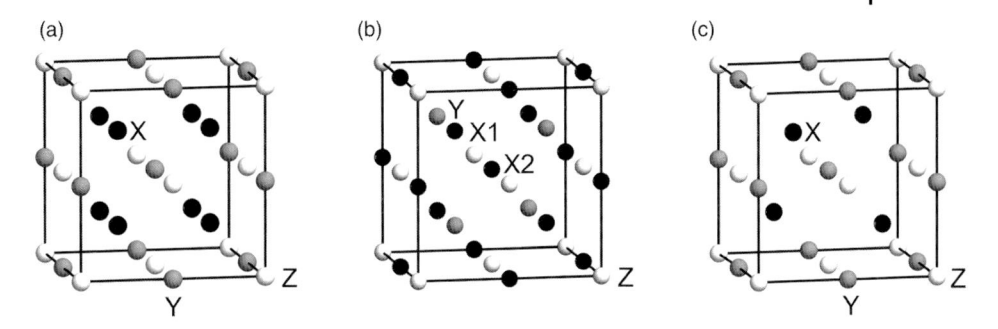

Figure 13.5 Unit cells of the so-called *full*-Heusler (a), *inverse*-Heusler (b), and *half*-Heusler compounds (c) with Cu$_2$MnAl, Li$_2$AgSb, and MgAgAs type structures, respectively. X, Y, Z atoms are represented by black, gray, and white spheres.

forming the Li$_2$AgSb type structure (Figure 13.5b). According to the *Strukturbericht* notation, *full*- and *half*-Heusler structures are also designated as L2$_1$ and C1$_b$, respectively.

13.7 Thermoelectric Materials

Thermoelectric materials are suited to convert thermal into electric energy [76]. Different types of IPs can be considered as potential thermoelectric materials such as several chalcogenide compounds, FeSb$_2$, *half*-Heusler phases, Skutterudites, Clathrates, and the Zintl phase Yb$_{14}$MnSb$_{11}$ [77–92].

A maximum of the thermoelectric figure of merit (ZT), including the power factor ($S^2\sigma$) divided by the thermal conductivity (κ), has to be obtained for a good material with high-thermoelectric performance. S is the thermoelectric power and σ the electrical conductivity. The thermal conductivity consists of the thermal lattice conductivity (κ_{latt}) and the electronic contribution (κ_{el}). In order to maximize ZT, either the thermal conductivity can be minimized and/or the electrical conductivity has to be maximized. While our current understanding to systematically manipulate the latter is rather limited, several ideas about the reduction of the thermal conductivity exist. Some are based on the shape and size of the materials, for example, thin-film techniques, nanomaterials, and nanowires [92–95]. Additionally, there are concepts to reduce the thermal conductivity by the introduction of phonon scattering. This can be achieved by a modification of the crystal structures throughout the introduction of disorder, preferably with heavier substituents or the usage of materials with complex crystal structures. The most intuitive route to achieve low thermal conductivities is based on the phonon glass electron crystal (PGEC) concept proposed by Slack [96]. Crystal structures with large cavities or tunnels filled with atoms possess low thermal conductivity since the filling atoms move–often referred to as rattling–within the cavities. The lattice vibrations are reduced due the collision (scattering) of the phonons with the independent vibrations of the filling atoms. Several intermetallic clathrates fulfill the PGEC concept due to commonly suited crystal structures.

13.8
Clathrates

Clathrates are known from hydrates that encapsulate gas molecules at low temperature and high pressure. They are often observed in deep sea as methane hydrates. The crystal structures comprise open frameworks, which are build up by sp^3-like atoms with large cavities. Group 14 elements in combination with alkali, alkaline-earth, and/or rare earth-metals are known in literature. Furthermore, the metastable empty framework of an allotrope of Ge and an almost empty framework of Si are realized [26,97].

Clathrate type I and type II structures are mainly formed by IPs (Figure 13.6). Type I with the general formula M_8X_{46} is realized by a pentagondodecahedron consisting of exclusively five-membered rings and a tetracaidecahedron formed by five- and six-membered rings. Clathrate type II having a composition $M_{24}X_{136}$ consists of pentagondodecahedra and hexakaidecahedra possessing the largest volume for the cations.

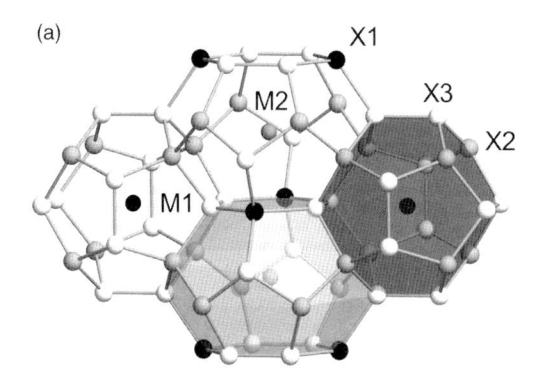

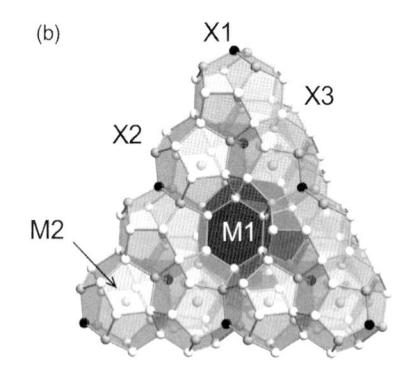

Figure 13.6 Cut out of the Clathrate type I (a) and type II (b) crystal structure. The positions of the cationic sublattice M1 and M2 are represented in black and medium gray, respectively, those of the anionic sublattice X1, X2, and X3 by black, medium gray, and white spheres, respectively. The polyhedra embedding M1 and M2 are highlighted in light and medium gray.

13.9
Phase Change Materials (PCM)

Most PCMs are found within the ternary GeSbTe phase diagram. The quasi binary section between GeTe and Sb_2Te_3 hosts the most prominent PCMs (Figure 13.7) [98]. In addition, quaternary compounds of the AgInSbTe (AIST) system are used as PCMs [99].

Those PCMs formed by Ge, Sb, and Te are often also mentioned as GST materials. Due to their physical properties being closely related to the structure of the various modifications they are used in optical and electrical data storage devices [4,100,101]. In general, an amorphous low temperature modification possesses electrically insulating or semiconducting behavior accompanied by low optical reflectivity. In contrast, a crystalline metallic modification with comparably high optical reflectivity is observed at higher temperatures [4,100,101].

In the following, the quasi-binary section of the GeTe–Sb_2Te_3 system is used to introduce the structural peculiarities of PCMs. Above 420 °C a cubic NaCl type structure is realized by GeTe [101]. Below this temperature a rhombohederal distortion along the [111] direction of the cubic NaCl type unit cell is observed [102,103]. From this, a (3+3)-fold coordination with short and long bonds of the atoms leading to the formation of Ge–Te bilayers results (Figure 13.8a). This can also be described as face sharing layers of distorted $GeTe_6$ octahedra being perpendicular to the c axis, the former [111] direction of the NaCl type cell. The main building units in crystalline GST materials are distorted $GeTe_6$ and $SbTe_6$ octahedra forming face-sharing layers.

With increasing Sb_2Te_3 content, $Ge_2Sb_2Te_5$ is formed. A double layer of $GeTe_6$ and $SbTe_6$ face-sharing units is perpendicular to the c-axis of the unit cell. Two of these blocks are related by inversion symmetry (Figure 13.8b). The Te $\cdots$ Te contacts of two Te terminated layers are often described by van der Waals type bonding. A further increase of Sb_2Te_3 results in $GeSb_2Te_4$ with building units of

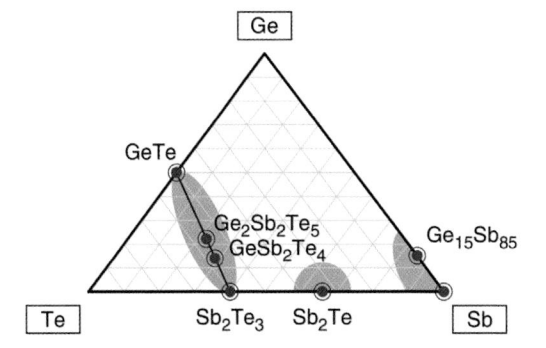

Figure 13.7 Ternary phase diagram of the GeSbTe system. The quasi binary section of GeTe and Sb_2Te_3 is highlighted together with two prominent PCMs $Sb_2Ge_2Te_5$ and $GeSb_2Te_4$. (Reproduced with permission from Ref. [98]. Copyright 2011, John Wiley & Sons.)

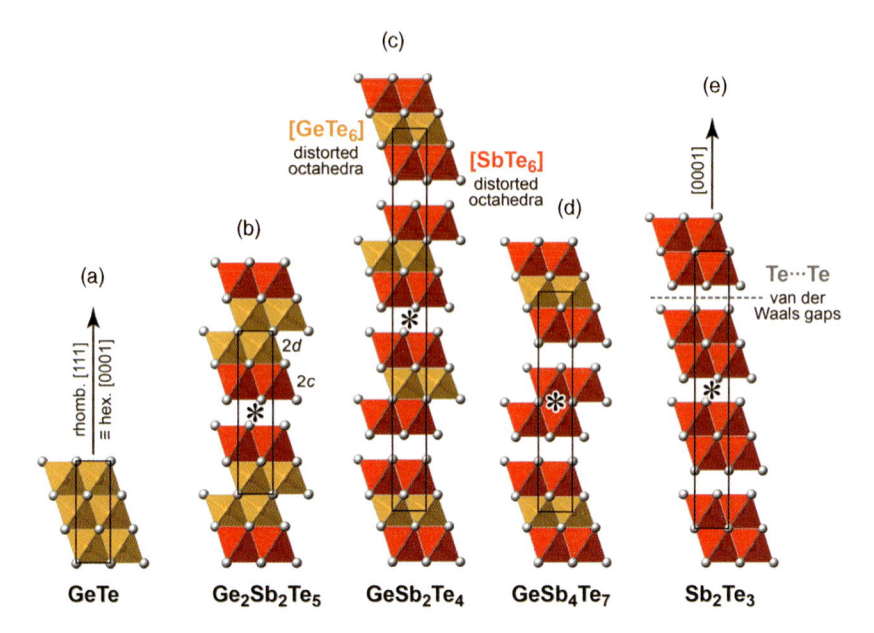

Figure 13.8 Idealized crystal structures along the GeTe–Sb$_2$Te$_3$ tieline in polyhedral representation; the Sb content increases from left to right. For Ge$_2$Sb$_2$Te$_5$, Wyckoff labels (2c/2d) are given for the two inequivalent cation sites; in reality, intermixing of Ge and Sb atoms is to be expected, as discussed in the text. All phases except GeTe take inversion-symmetric space groups, and the inversion centers have been highlighted by asterisks (*). (Reproduced with permission from Ref. [10]. Copyright 2007, John Wiley & Sons.)

distorted face sharing octahedra in the sequence SbTe$_6$, GeTe$_6$, and SbTe$_6$ (Figure 13.8c). In GeSb$_2$Te$_7$, a double layer of face-sharing distorted SbTe$_6$ octahedra separates two of the building units observed in GeSb$_2$Te$_4$ (Figure 13.8c,d). Please note, a mixed occupation of Ge and Sb sites was also observed [104]. Thus, the above mentioned PCM structure models have to be considered as idealized.

Sb$_2$Te$_3$ is finally realized by face sharing SbTe$_6$ polyhedra only. Van der Waals type Te$\cdots$Te contacts separate double layers of SbTe$_6$ octahedra (Figure 13.8e).

Another way to describe the structures is based on the NaCl type motive. This description focusses more on the distortions within the lattice than the above mentioned idealized structure models. The Te atoms of GeSb$_2$Te$_4$ occupy the Cl positions in a NaCl type lattice. The Na positions are filled for ¼ by Ge and for ½ by Sb. The remaining ¼ of positions in the cationic sublattice are vacancies. Due to the vacancies, large shifts of the atoms for relaxation of the lattice are expected. The fraction of vacancies is even reduced from 25% in GeSb$_2$Te$_4$ to 20% in Ge$_2$Sb$_2$Te$_5$, were the local atomic arrangements seem to be the origin of a metal to insulator transition [105].

The coordination of the atoms in the crystalline PCMs is rather exclusively sixfold. In contrast fourfold and sixfold coordination with homopolar Ge–Ge contacts are realized in the amorphous PCMs [106].

13.10
Disorder in IPs

In principle, three types of local ordering of the atoms have to be considered for IPs. These are the formation of vacancies, substitution, and the occupation of interstitials (Figure 13.9a–c).

Considering the tendency of local ordering of the atoms in IPs, there is high demand on methods and/or strategies to investigate disorder in detail. Since variations of the local environments have to be detected, the use of local probes such as NMR, EPR, Mössbauer, and other spectroscopic techniques is mandatory. Furthermore, the electronic structures are of great importance to understand the chemical bonding of these materials. Most of the coupling parameter of spectroscopic methods can be calculated by QM methods providing the electronic structures in addition [107–110].

In case of NMR of electronically conducting IPs, the most relevant coupling parameter is the so-called quadrupole coupling of the nuclear quadrupole moment with the electric field gradient (EFG) [111–115]. The EFG is a highly sensitive measure of the charge distribution of the atoms because it is mainly determined by the electron distribution in the vicinity of the observed nuclei [18,19]. Recently, also the Knight shift (K), the coupling of the nuclear spins with the conduction electrons, has been implemented into WIEN2K [116]. For nonconducing materials or semiconductors with a reasonable large band gap also the chemical shielding (CSA) can be calculated [108,109,117,118].

Using NMR as local probe, a strategy for structural investigations of IPs has recently been derived [18–20,42]. The triangle of methods building the basis of this strategy is briefly discussed in the following (Figure 13.10). Diffraction techniques are used to determine the average crystal structure. The result represents the starting model of the structure for the investigation of disordered materials

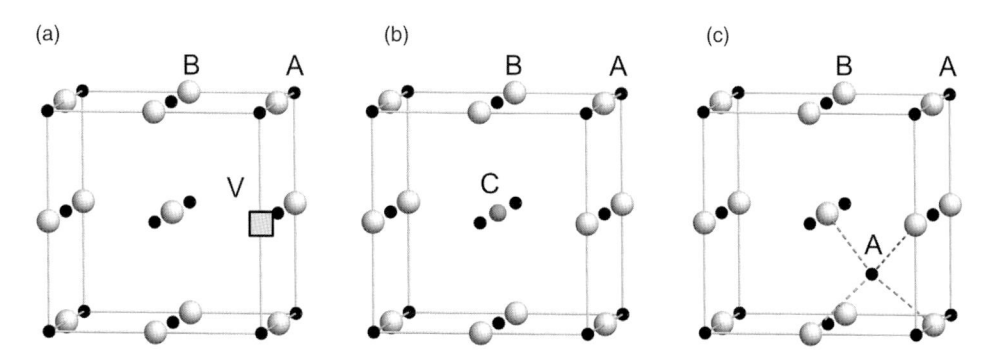

Figure 13.9 Presentation of local ordering of the atoms in binary compounds. Vacancies (V, marked by a square, (a)), substitution by foreign atoms of type C (b), and occupation of an interstitial highlighted by broken lines (c). The A, B, and C atoms are represented by black, light gray, and dark gray spheres, respectively.

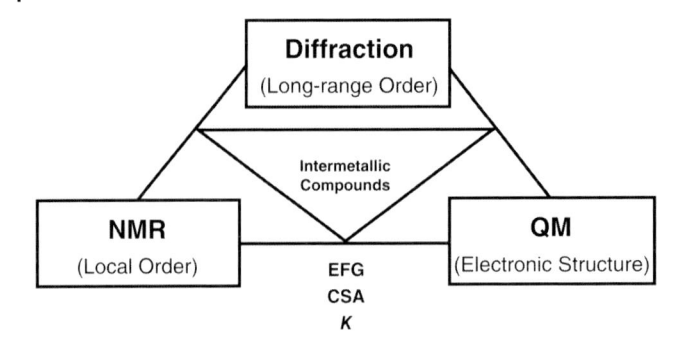

Figure 13.10 Triangle of methods for investigation of intermetallics. The relevant coupling parameters providing a comparison of measurement and calculation are given below the connection line of NMR and QM. These coupling parameters are the electric field gradient (EFG) for metallic and insulating compounds, the chemical shift anisotropy (CSA) for insulating materials, and the Knight shift (K) for metallic materials.

by combined application of diffraction, NMR, and QM calculations. It serves as the basis for QM calculations and the prediction of the NMR signals. Compared to diffraction methods providing only the periodicity of the atomic arrangements NMR is suited to investigate the local atomic arrangements. Occasionally, diffraction methods already indicate the occurrence of local ordering of the atoms in IPs. But more evidence for disorder is found by closer inspection of the NMR signal. Lineshape analysis by least-square methods provide the NMR coupling parameter [119], even for compounds with several similar atomic environments and consequently overlapping NMR signals [20,21].

To derive a structure model describing local atomic arrangements in IPs an iterative process has to be performed. The starting model of the structure is used to calculate the NMR coupling parameter for comparison with the experimental results. Superlattice structures describing the various local atomic environments being simultaneously present in the material can be obtained with the help of the Bärnighausen formalism [120]. The additional coupling parameter received by these QM calculations have to be compared with the experimental results to improve the structure model step by step.

13.11
Vacancies in $Cu_{1-x}Al_2$

Although the $CuAl_2$ type structure [121–123] is realized by many intermetallics no compound with stoichiometric composition 1:2 exists in the phase diagram of the Cu–Al system [22,124]. A Cu deficient phase $Cu_{1-x}Al_2$ is known instead [125].

Two honeycomb like interpenetrating nets of Al atoms and linear chains of Cu atoms along the tetragonal c axis of the unit cell are the main features of the

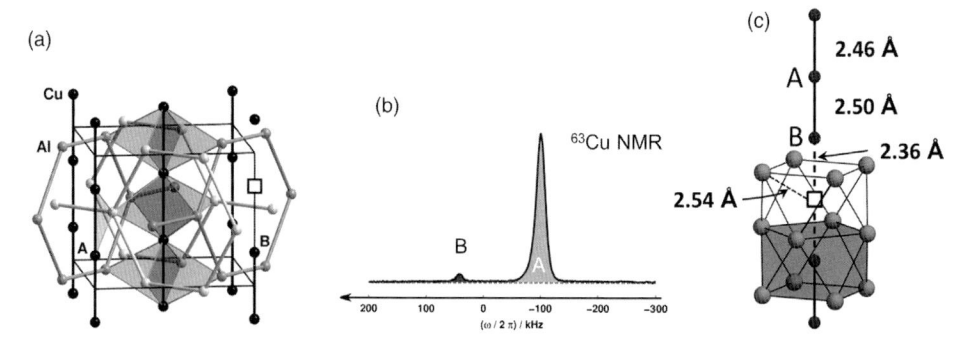

Figure 13.11 (a) Crystal structure of $Cu_{1-x}Al_2$. Cu and Al atoms are represented by black and gray spheres, respectively. Cu positions are named by A with two neighboring Cu atoms and B with one neighboring Cu atom and a vacancy, being represented by a square. (b) ^{63}Cu NMR signal of $Cu_{1-x}Al_2$. The assignment of the signal contributions to the positions of the crystal structure is given. (c) Coordination of a vacancy (square) and interatomic distances resulting from a relaxation of the lattice.

$CuAl_2$ type structure. It is realized by one site for each Cu and Al (Figure 13.11a). Various structural models for the atomic description of the homogeneity range such as the formation of vacancies, antisite order, and substitution are discussed in literature [126,127]. An unambiguous prove of the vacancy model is obtained by NMR spectroscopy on magnetically aligned crystallites [128]. Two well resolved Cu NMR signals allowing ^{27}Al–^{65}Cu and ^{63}Cu–^{65}Cu double resonance experiments are detected at ambient temperature (Figure 13.11b). Further evidence for the vacancy model and an analysis of the atomic shifts was obtained by QM calculations without symmetry constraints in space group P1. Both Al and Cu atoms shift toward the position of the vacancy by about 0.05 and 0.08 Å, respectively (Figure 13.11c).

13.12
Preferred Site Occupation

The following two sections will focus on a consequent application of the triangle of methods (Figure 13.10). The solid solutions $M(Al_{1-x}Ga_x)_4$ with $M = Sr$, Ba show full miscibility and an outstanding behavior of the lattice parameter a (Figure 13.12a) [21]. The compounds crystallize in the $BaAl_4$ type structure with one position for the alkaline-earth metal and two for the 13 group elements (Figure 13.12b). The latter are labeled according to the number of next neighbor 13 group element contacts by 4b and 5b.

NMR experiments on magnetically aligned crystallites result in a much higher resolution than measurements on randomly oriented crystallites. In addition, evidence for a preferred site occupation of the Al(4b) position is obtained (Figure 13.12c,d). This is proven by X-ray single crystal diffraction and QM

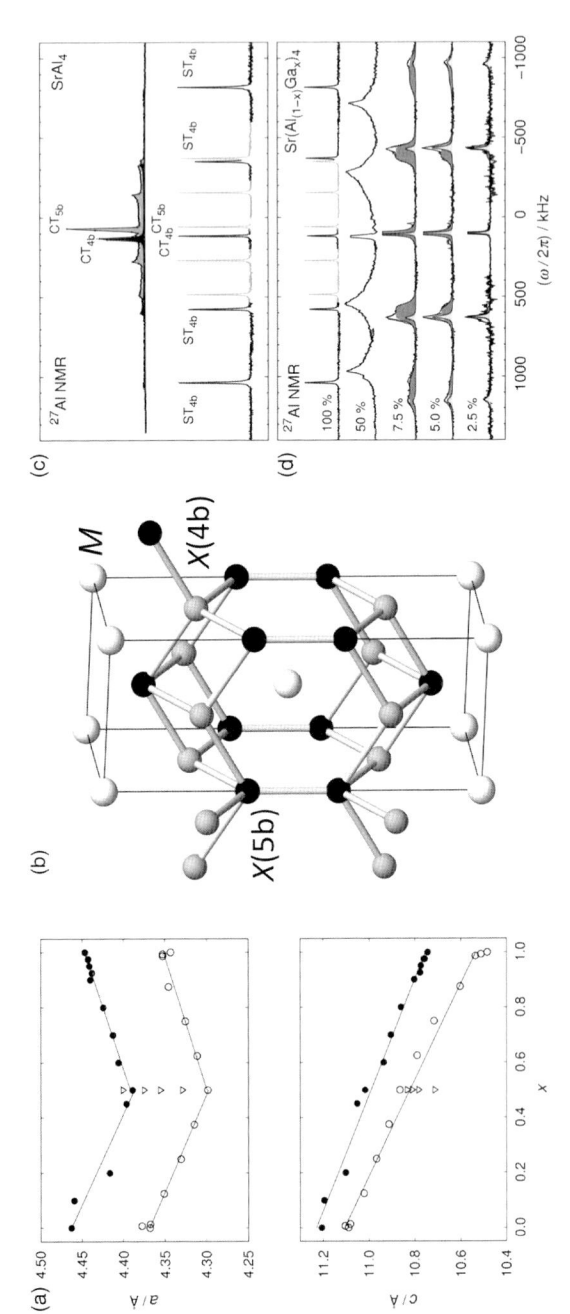

Figure 13.12 (a) Lattice parameter of $Sr(Al_{1-x}Ga_x)_4$. Open and filled symbols represent QM and experimental results, respectively. Triangles show the results for varying Al(4b) occupation in $SrAl_2Ga_2$. (b) Cut out of the crystal structure of $M(Al_{1-x}Ga_x)_4$ with $M = Sr$, Ba. White spheres illustrate the environments of the cations. Gray and black spheres show the environments of the 13 group elements. They are labeled by X(4b) and X(5b) according to their number of next neighbor 13 group elements. (c) ^{27}Al NMR signals of randomly and magnetically aligned crystallites of $SrGa_4$. The central and satellite transitions are indicated by CT and ST, respectively. (d) ^{27}Al NMR signals of magnetically aligned crystallites of $Sr(Al_{1-x}Ga_x)_4$ with varying Al concentrations.

calculation for various structure models (*vide infra*). The calculated NMR coupling parameters are in good agreement with the experimental ones. Furthermore, the lattice parameters follow the experimental trend. They are slightly smaller due to the so-called LDA over binding effect [12]. Considering of a Al (4b) preferred site occupation for MAl_2Ga_2 an decrease of the *a* parameter with an increase of the Al(4b) occupation is observed (Figure 13.12a). Additional support of the preferred site occupation is obtained by estimation of the formation energies.

A structuring of the satellite transition signals is attributed to Al–Al pairs within a single unit cell even if a statistical distribution of the Al(4b) sites should result in almost isolated Al atoms. QM calculations using superlattice models with only one substituted atom feature a decay of the atomic shifts around the substituted atoms as well as an almost exclusive influence of the substitution on the EFGs main component V_{ZZ} within the first coordination sphere [21].

13.13
Substitution by Similar Type of Atoms

The solid solution $Sr_{1-x}Ba_xGa_2$ shows complete miscibility and an increase of the lattice parameter with increasing x. Using superlattice structures this trend is well reproduced by QM calculations (Figure 13.13a) [20]. Both binary parent compounds crystallize in the AlB_2 type structure with one Ga site and the other site for the alkaline-earth metals. Both sites are of high symmetry. A reduction of the site symmetry and therefore also of the charge distribution is expected for a substitution of a cation in the vicinity of the Ga atoms (Figure 13.13b, top). QM calculations of a superlattice structure with only one substituted atom allow a prediction of the expected NMR signals for both the Ga atoms in the vicinity of the substitution center (SC) and those being in a larger distance. By this knowledge the experiments on magnetically aligned crystallites can be optimized with respect to the orientation and the measurement conditions (Figure 13.13c).

QM calculations without symmetry constraints in space group $P1$ are used to analyze the shift of the atoms and the influence of the substitution on the bonding situation of the atoms. While a substitution of Sr by Ba results in larger distances to the SC the reverse substitution causes in a shift toward the SC.

The bonding situation of the atoms is only slightly influenced as expected for substitution of the similar types of atoms (Sr versus Ba). This can be seen by inspection of the EFG as a sensitive measure for the charge distribution. While the main component (V_{ZZ}) of the EFG is almost exclusively influenced in the first coordination sphere, the asymmetry parameter (η) is affected in higher coordination spheres. η is sensitive on the symmetry of the charge distribution.

It is sufficient to describe samples with low degrees of substitution by two Ga environments being represented by their Ga NMR signal contributions. One corresponding to the Ga1 atoms in large distance to the SC and the other Ga2 in the first coordination sphere to the SC (Figure 13.13c) [20].

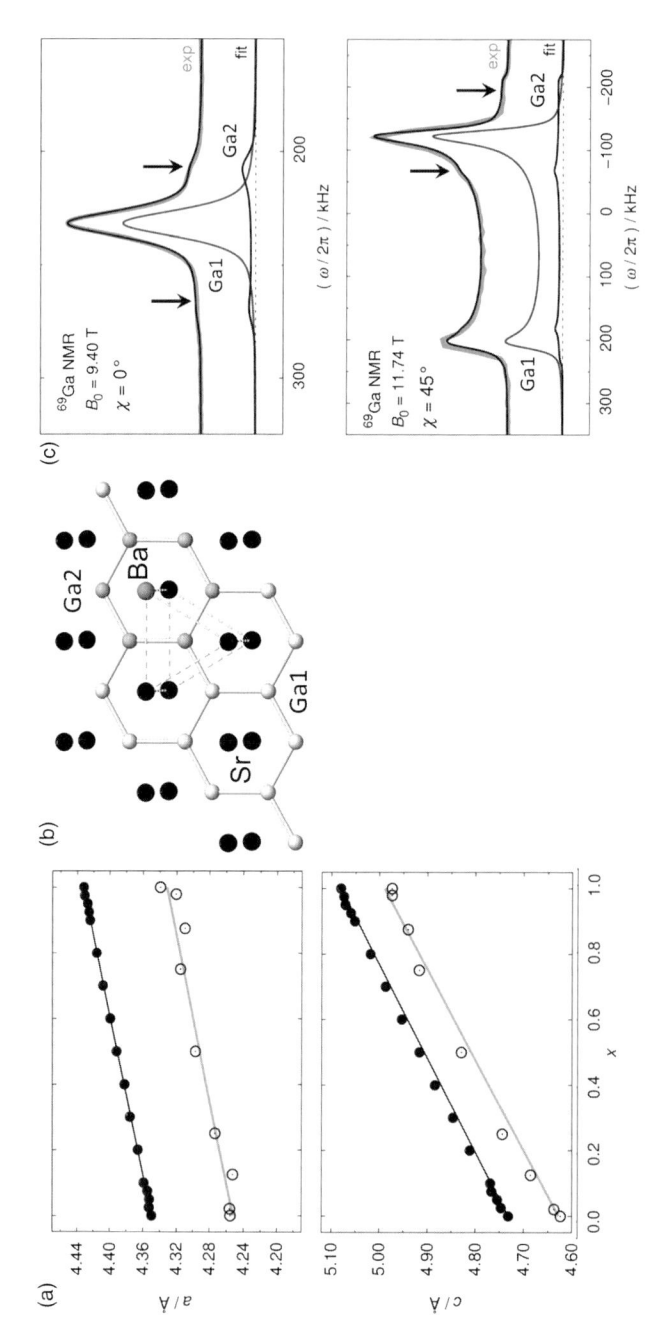

Figure 13.13 (a) Lattice parameter of $Sr_{1-x}Ba_xGa_2$ with open and filled symbols representing QM and experimental results, respectively. (b) Cut out of the crystal structure of $Sr_{1-x}Ba_xGa_2$. Black, medium gray spheres correspond to Sr and Ba. Light and medium gray spheres correspond to the environments of Ga1, and Ga2, respectively. The trigonal prismatic coordination of Ga2 is highlighted by dashed lines. (c) ^{71,69}Ga NMR signals of magnetically aligned crystallites of $Sr_{1-x}Ba_xGa_2$ with $x = 0.975$ at different magnetic fields and orientations. χ represents the angle between the magnetic fields used for alignment of the crystallites and the measurements. Frequency ranges being significantly influenced by the Ga2 atoms are indicated by arrows.

References

1. Westbrook, J.H. and Fleischer, R.L. (1995) *Intermetallic Compounds: Principles and Applications*, Wiley-VCH Verlag GmbH.
2. Pöttgen, R. and Johrendt, D. (2014) *Intermetallics: Synthesis, Structure, Function*, De Gruyter.
3. Chen, W.F. (1997) *Handbook of Structural Engineering*, CRC Press.
4. Raoux, S. and Wuttig, M. (2009) *Phase Change Materials – Science and Applications*, Springer.
5. Blatt, F.J. (2011) *Thermoelectric Power of Metals*, Springer-Verlag, Inc, New York.
6. Zeier, W.G., Zevalkink, A., Gibbs, Z.M., Hautier, G., Kanatzidis, M.G., and Snyder, G.J. (2016) *Angew. Chem., Int. Ed.*, **55**, 2.
7. Johrendt, D. (2011) *J. Mater. Chem.*, **21**, 13726.
8. Felser, C., Fecher, G., and Balke, B. (2007) *Angew. Chem., Int. Ed.*, **46**, 668.
9. Armbrüster, M., Kovnir, K., Friedrich, M., Teschner, D., Wowsnick, G., Hahne, M., Gille, P., Szentmiklósi, L., Feuerbacher, M., Heggen, M., Girgsdies, F., Rosenthal, D., Schlögl, R., and Grin, Yu. (2012) *Nat. Mater.*, **11** 690.
10. Grin, Yu., Schwarz, U., and Steurer, W. (2007) *Alloy Physics: A Comprehensive Reference*, Wiley-VCH Verlag GmbH, Weinheim.
11. Nesper, R. (1991) *Angew. Chem., Int. Ed. Engl.*, **30**, 789.
12. Dronskowski, R. (2005) *Computational Chemistry of Solid State Materials*, Wiley-VCH Verlag GmbH.
13. Grin, Yu., Wagner, F.R., Armbrüster, M., Kohout, M., Leithe-Jasper, A., Schwarz, U., Weding, U., and von Schnering, H.G. (2006) *J. Solid State Chem.*, **179**, 1707.
14. Wagner, F.R., Bezugly, V., Kohout, M., and Grin, Yu. (2007) *Chem. Eur. J.*, **13**, 5724.
15. Miller, G.J., Li, F., and Franzen, H.F. (1993) *J. Am. Chem. Soc.*, **115**, 3739.
16. Häussermann, U., Amerioun, S., Eriksson, L., Lee, Chi.-Shen., and Miller, G.J. (2002) *J. Am. Chem. Soc.*, **124**, 4371.
17. Burdett, J.K. and Miller, G.J. (1990) *Chem. Mat.*, **2**, 12.
18. Haarmann, F., Koch, K., Grüner, D., Schnelle, W., Pecher, O., Cardoso-Gil, R., Borrmann, H., Rosner, H., and Grin, Yu. (2009) *Chem. Eur. J.*, **15**, 1673.
19. Haarmann, F., Koch, K., Jeglič, P., Pecher, O., Rosner, H., and Grin, Yu. (2011) *Chem. Eur. J.*, **17**, 7560.
20. Pecher, O., Mausolf, B., Lamberts, K., Oligschläger, D., Niewieszol (née Merkens), C., Englert, U., and Haarmann, F. (2015) *Chem. Eur. J.*, **21** (40), 13971.
21. Pecher, O., Mausolf, B., Peters, V., Lamberts, K., Korthaus, A., and Haarmann, F. (2016) *Chem. Eur. J.*, **22**, 17833.
22. Massalski, T.B. (ed.) (1990) *Binary Alloy Phase Diagrams*, 2nd edn, ASM International, Materials Park, Ohio.
23. Villars, P., Prince, A., and Okamoto, H. (eds) (1995) *Handbook of Ternary Alloy Phase Diagrams*, ASM International.
24. Wolf, S., Reiter, K., Weigend, F., Klopper, W., and Feldmann, C. (2015) *Inorg. Chem.*, **54**, 3989.
25. Schütte, K., Doddi, A., Kroll, C., Meyer, H., Wiktor, C., Gemel, C., van Tendeloo, G., Fischer, R.A., and Janiak, C. (2014) *Nanoscale*, **6**, 3116.
26. Guloy, A.M., Ramlau, R., Tang, Z., Schnelle, W., Baitinger, M., and Grin, Yu. (2006) *Nature*, **443**, 320.
27. Campbell, F.C. (2008) *Elements of Metallurgy and Engineering Alloys*, ASM International.
28. Mizutani, U. (2010) *Hume-Rothery Rules for Structurally Complex Alloy Phases*, CRC Press.
29. Dwight, A.E. (1961) *Trans. ASM*, **53**, 479.
30. Ferro, R. and Saccone, A. (2008) *Intermetallic Chemistry*, 1st edn, Elsevier.
31. Kerkau, A., Grüner, D., Ormeci, A., Prots, Y., Borrmann, H., Schnelle, W., Bischoff, E., Grin, Yu., and Kreiner, G. (2009) *Z. Anorg. Allg. Chem.*, **635**, 637.
32. Riedel, E. and Janiak, C. (2011) *Anorganische Chemie*, De Gruyter.
33. Zintl, E. (1929) *Naturwissenschaften*, **17**, 782.
34. Schäfer, H., Eisenmann, B., and Müller, W. (1973) *Angew. Chem.*, **17**, 742.

35 Klemm, W. and Busmann, E. (1963) *Z. Anorg. Allg. Chem.*, **319**, 297.
36 Schäfer, R. and Klemm, W. (1961) *Z. Anorg. Allg. Chem.*, **312**, 214.
37 Witte, J., von Schnering, H.G., and Klemm, W. (1964) *Z. Anorg. Allg. Chem.*, **327**, 260.
38 Schäfer, H., Janzon, K.K., and Weiss, A. (1963) *Angew. Chem.*, **75**, 451.
39 Janzon, K.H., Schäfer, H., and Weiss, A. (1970) *Z. Anorg. Allg. Chem.*, **372**, 87.
40 Goebel, T., Prots, Yu., and Haarmann, F. (2008) *Z. Kristallogr. NCS*, **223**, 187.
41 Goebel, T., Prots, Yu., and Haarmann, F. (2009) *Z. Kristallogr. NCS*, **224**, 7.
42 Goebel, T., Ormeci, A., Pecher, O., and Haarmann., F. (2012) *Z. Anorg. Allg. Chem.*, **638**, 1437.
43 Schneider, J. (1988) *Mater. Sci. Forum*, **27/28**, 63.
44 Eisenmann, B., Janzon, K.H., Schäfer, H., and Weiss, A. (1969) *Z. Naturforsch.*, **24b**, 457.
45 Aydemir, U., Ormeci, A., Borrmann, H., Böhme, B., Zürcher, F., Uslu, B., Goebel, T., Schnelle, W., Simon, P., Carrillo-Cabrera, W., Haarmann, F., Baitinger, M., Nesper, R., von Schnering, H.G., and Grin, Yu. (2008) *Z. Anorg. Allg. Chem.*, **634** 1651.
46 Hofmann, W. and Janiche, W. (1935) *Naturwissenschaften*, **23**, 851.
47 Haarmann, F. and Prots, Yu. (2006) *Z. Anorg. Allg. Chem.*, **632**, 2135.
48 Iandelli, A. (1964) *Z. Anorg. Allg. Chem.*, **330**, 221.
49 Pecher, O. (2013) NMR-Spektroskopie an intermetallischen Phasen in den Systemen EA-Al-Ga mit EA=Ca, Sr und Ba. Dissertation. RWTH Aachen University, Aachen (Germany).
50 Sichevych, O., Raumlau, R., Giedigkeit, R., Schmidt, M., Niewa, R., and Grin, Yu. (2000) Abstracts. 13th International Conference on Solid Compounds of Transition Elements, Stresa, Italy, O-13.
51 Sichevych, O., Prots, Yu., and Grin, Yu. (2006) *Z. Kristallogr. NCS*, **221**, 265.
52 Cirafici, C. and Fornasini, M.L. (1990) *J. Less-Common Met.*, **163**, 331.
53 Haarmann, F., Prots, Yu., Göbel, S., and von Schnering, H.G. (2006) *Z. Kristallogr. NCS*, **221**, 257.
54 Kußmann, D., Hoffmann, R.-D., and Pöttgen, R. (2001) *Z. Anorg. Allg. Chem.*, **627**, 2053.
55 Gerke, B., Korthaus, A., Niehaus, O., Haarmann, F., and Pöttgen, R. (2016) *Z. Naturforsch.*, **71b**, 567.
56 Kauzlarich, S.M. (1996) *Chemistry, Structure, and Bonding of Zintl Phases and Ions: Selected Topics and Recent Advances*, 1st edn, Wiley-VCH Verlag GmbH.
57 Fässler, T.F. (ed.) (2011) *Zintl Ions – Principles and Recent Developments*, Springer.
58 Nesper, R. (1990) *Prog. Solid State Chem.*, **20**, 1.
59 Heusler, F., Starck, W., and Balke, E. (1903) *Verhandl. Dtsch. Phys. Ges.*, **5**, 219.
60 Webster, P.J. (1969) *Contemp. Phys.*, **10**, 559.
61 Dunlap, R.A. (2003) Magnetic properties of Heusler alloys. Proceedings of the 10th CF/DRDC Meeting on Naval Applications of Materials Technology, p. 600.
62 Winterlik, J., Chadov, S., Gupta, A., Alijani, V., Gasi, T., Filsinger, K., Balke, B., Fecher, G.H., Jenkins, C.A., Casper, F., Kübler, J., Liu, G., Gao, L., Parkin, S.S.P., and Felser, C. (2012) *Adv. Mater.*, **24** 6283.
63 Alijani, V., Winterlik, J., Fecher, G.H., and Felser, C. (2011) *Appl. Phys. Lett.*, **99**, 222510.
64 Ouardi, S., Fecher, G.H., Kübler, J., and Felser, C. (2013) *Phys. Rev. Lett.*, **110**, 100401.
65 Chadov, S., Kiss, J., and Felser, C. (2013) *Adv. Funct. Mater.*, **23**, 832.
66 Kainuma, R., Imano, Y., Ito, W., Sutou, Y., Morito, H., Okamoto, S., Kitakami, O., Oikawa, K., Fujita, A., Kanomata, T., and Ishida, K. (2006) *Nature*, **439**, 957.
67 Krenke, T., Duman, E., Acet, M., Wassermann, E.F., Moya, X., Mañosa, L., and Planes, A. (2005) *Nat. Mater.*, **4**, 450.
68 Takeuchi, I., Famodu, O.O., Read, J.C., Aronova, M.A., Chang, K.-S., Craciunescu, C., Lofland, S.E., Wuttig, M., Wellstood, F.C., Knauss, L., and Orozco, A. (2003) *Nat. Mater.*, **2**, 180.

69 Chadov, S., Qi, X., Kübler, J., Fecher, G.H., Felser, C., and Zhang, S.C. (2010) *Nat. Mater.*, **9**, 541.

70 Müchler, L., Zhang, H., Chadov, S., Yan, B., Casper, F., Kübler, J., Zhang, S.-C., and Felser, C. (2012) *Angew. Chem., Int. Ed.*, **51**, 1.

71 Franz, M. (2010) *Nat. Mater.*, **9**, 536.

72 de Groot, R.A., Mueller, F.M., van Engen, P.G., and Buschow, K.H.J. (1983) *Phys. Rev. Lett.*, **50**, 2024.

73 Kübler, J., Williams, A.R., and Sommers, C.B. (1983) *Phys. Rev. B*, **28**, 1745.

74 Wurmehl, S., Fecher, G.H., Kandpal, H.C., Ksenofontov, V., Felser, C., Lin, H.-J., and Morais, J. (2005) *Phys. Rev. B*, **72**, 184434.

75 Zhu, T., Fu, C., Xie, H., Liu, Y., and Zhao, X. (2015) *Adv. Energy Mater.*, **5**, 1614.

76 Sootsman, J.R., Chung, D.Y., and Kanatzidis, M.G. (2009) *Angew. Chem., Int. Ed.*, **48**, 8616.

77 Wright, D.A. (1958) *Nat. Mater.*, **181**, 834.

78 Leithe-Jasper, A., Schnelle, W., Rosner, H., Senthilkumaran, N., Rabis, A., Baenitz, M., Gippius, A., Morozova, E., Mydosh, J.A., and Grin, Yu. (2003) *Phys. Rev. Lett.*, **91**, 037208, 2004, 93, 089904 (E).

79 Bentien, A., Johnsen, S., Madsen, G.K.H., Iversen, B.B., and Steglich, F. (2007) *EPL*, **80**, 17008.

80 Uher, C., Yang, J., Hu, S., Morelli, D.T., and Meisner, G.P. (1999) *Phys. Rev. B*, **59**, 8615.

81 Xia, Y., Bhattacharya, S., Ponnambalam, V., Pope, A.L., Poon, S.J., and Tritt, T.M. (2000) *J. Appl. Phys.*, **88**, 1952.

82 Bhattacharya, S., Pope, A.L., Littleton J IV, R.T., Tritt, T.M., Ponnambalam, V., Xia, Y., and Poon, S.J. (2000) *Appl. Phys. Lett.*, **77**, 2476.

83 Hohl, H., Ramirez, A.P., Goldmann, C., Ernst, G., Wolfing, B., and Bucher, E. (1999) *J. Phys. Condens. Matter*, **11**, 1697.

84 Sakurada, S. and Shutoh, N. (2005) *Appl. Phys. Lett.*, **86**, 082105.

85 Shen, Q., Chen, L., Goto, T., Hirai, T., Yang, J., Meisner, G.P., and Uher, C. (2001) *Appl. Phys. Lett.*, **79**, 4165.

86 Condron, C.L., Kauzlarich, S.M., Ikeda, T., Snyder, G.J., Haarmann, F., and Jeglič, P. (2008) *Inorg. Chem.*, **47**, 8204.

87 Zaikina, J.V., Kovnir, K.A., Burkhardt, U., Schnelle, W., Haarmann, F., Schwarz, U., Grin, Yu., and Shevelkov, A.V. (2009) *Inorg. Chem.*, **48**, 3720.

88 Zaikina, J.V., Kovnir, K.A., Haarmann, F., Schnelle, W., Burkhardt, U., Borrmann, H., Schwarz, U., Grin, Yu., and Shevelkov, A.V. (2008) *Chem. Eur. J.*, **14**, 5414.

89 Kovnir, K.A., Shatruk, M.M., Reshetova, L.N., Presniakov, I.A., Dikarev, E.V., Baitinger, M., Haarmann, F., Schnelle, W., Baenitz, M., Grin, Yu., and Shevelkov, A.V. (2005) *Solid State Sci.*, **7**, 957.

90 Chan, J.Y., Olmstead, M.M., Kauzlarich, S.M., and Webb, D.J. (1998) *Chem. Mater.*, **10**, 3583.

91 Hu, Y., Wang, J., Kawamura, A., Kovnir, K., and Kauzlarich, S.M. (2015) *Chem. Mater.*, **27**, 343.

92 Böttner, H., Chen, G., and Venkatasubramanian, R. (2006) *MRS Bull.*, **31**, 211.

93 Chen, G. (2001) *Semicond. Semimetals*, **71**, 203.

94 Chen, G., Dresselhaus, M.S., Fleurial, J.-P., and Caillat, T. (2003) *Int. Mater. Rev.*, **48**, 45.

95 Dresselhaus, M.S., Lin, Y.M., Cronin, S.B., Rabin, O., Black, M.R., Dresselhaus, G., and Koga, T. (2001) *Semicond. Semimetals*, **71**, 1.

96 Slack, G.A. (1995) *CRC Handbook of Thermoelectrics*, (ed D.M. Rowe), CRC Press, p. 407.

97 Gryko, J., McMillan, P.F., Marzke, R.F., Ramachandran, G.K., Patton, D., Deb, S.K., and Sankey, O.F. (2000) *Phys. Rev. B*, **62**, R7707.

98 Lencer, D., Salinga, M., and Wuttig, M. (2011) *Adv. Mater.*, **23**, 2030.

99 Matsunaga, T., Akola, J., Kohara, S., Honma, T., Kobayashi, K., Ikenaga, E., Jones, R.O., Yamada, N., Takata, M., and Kojima, R. (2011) *Nat. Mater.*, **10**, 129.

100 Bensch, W. and Wuttig, M. (2010) *Chem. Unserer Zeit*, **44**, 92.

101 Deringer, V.L., Dronskowski, R., and Wuttig, M. (2015) *Adv. Funct. Mater.*, **25**, 6343.

102 Chattopadhyay, T., Boucherle, J.X., and von Schnering, H.G. (1987) *J. Phys. C Solid State Phys.*, **20**, 1431.

103 Goldak, J., Barrett, C.S., Innes, D., and Youdelis, W. (1966) *J. Chem. Phys.*, **44**, 3323.

104 Matsunaga, T., Yamada, N., and Kubota, Y. (2004) *Acta Crystallogr. Sect. B*, **60**, 685.

105 Siegrist, T., Jost, P., Volker, H., Woda, M., Merkelbach, P., Schlockermann, C., and Wuttig, M. (2011) *Nat. Mater.*, **10**, 202.

106 Deringer, V.L., Zhang, W., Lumeij, M., Maintz, S., Wuttig, M., Mazzarello, R., and Dronskowski, R. (2014) *Angew. Chem., Int. Ed.*, **53**, 10817.

107 Blaha, P., Schwarz, K., Madsen, G.K.H., Kvasnicka, D., and Luitz, J. (2001) *WIEN2k, An Augmented Plane Wave Plus Local Orbitals Program for Calculating Crystal Properties*, Vienna University of Technology, Vienna.

108 Pickard, C.J. and Mauri, F. (2001) *Phys. Rev. B*, **63**, 245101.

109 Profeta, M., Mauri, F., and Pickard, C.J. (2003) *J. Am. Chem. Soc.*, **125**, 541.

110 Clark, S.J., Segall, M.D., Pickard, C.J., Hasnip, P.J., Probert, M.J., Refson, K., and Payne, M.C. (2005) *Z. Kristallogr.*, **220**, 567.

111 Blaha, P., Schwarz, K., and Herzig, P. (1985) *Phys. Rev. Lett.*, **54**, 1192.

112 Blaha, P., Schwarz, K., and Dederichs, P.H. (1988) *Phys. Rev. B*, **37**, 2792.

113 Fojud, Z., Herzig, P., Żogał, O.J., Pietraszko, A., Dukhnenko, A., Jurga, S., and Shitsevalova, N. (2007) *Phys. Rev. B*, **75**, 184102.

114 Herzig, P., Fojud, Z., Zogał, O.J., Pietraszko, A., Dukhnenko, A., Jurga, S., and Shitsevalova, N. (2008) *J. Appl. Phys.*, **103**, 083534.

115 Haarmann, F. (2011) *eMagRes* (eds R.K. Harris and R.E. Wasylishen), John Wiley & Sons, Ltd., Chichester.

116 Laskowski, R., Khoo, K.H., Haarmann, F., and Blaha, P. (2017) *J. Phys. Chem. C*, **121**, 753.

117 Laskowski, R. and Blaha, P. (2012) *Phys. Rev. B*, **85**, 245117.

118 Laskowski, R. and Blaha, P. (2014) *Phys. Rev. B*, **89**, 014402.

119 Bak, M., Rasmussen, J.T., and Nielsen, N.C. (2000) *J. Magn. Reson.*, **147**, 296.

120 Bärnighausen, H. (1980) *J. Math. Chem.*, **9**, 139.

121 Friauf, B. (1927) *J. Am. Chem. Soc.*, **49**, 3107.

122 Meetsma, A., De Boer, J.L., and Van Smaalen, S. (1989) *J. Solid State Chem.*, **82**, 370.

123 Grin, Yu., Wagner, F.R., Armbrüster, M., Kohout, M., Leithe-Jasper, A., Schwarz, U., Weding, U., and von Schnering, H.G. (2006) *J. Solid State Chem.*, **179**, 1707.

124 Murray, J.K. (1985) *Int. Met. Rev.*, **30**, 211.

125 Gödecke, T. and Sommer, F. (1996) *Z. Metallkd.*, **87**, 581.

126 Zogg, H. (1979) *J. Appl. Crystallogr.*, **12**, 91.

127 Bastow, T.J. and Celotto, S. (2003) *Acta Mater.*, **51**, 4621.

128 Haarmann, F., Armbrüster, M., and Grin, Yu. (2007) *Chem. Mater.*, **19**, 1147.

14
Layered Double Hydroxides: Structure–Property Relationships

Shan He, Jingbin Han, Mingfei Shao, Ruizheng Liang, Min Wei, David G. Evans, and Xue Duan

Beijing University of Chemical Technology, State Key Laboratory of Chemical Resource Engineering, North Third Ring Road 15, Beijing 100029, P. R. China

14.1
Two-Dimensional Structure of Layered Double Hydroxides (LDHs)

14.1.1
Introduction of LDHs

Layered double hydroxides (LDHs) are a class of anion-intercalated functional materials that are also known as hydrotalcite-like compounds. LDHs have the general formula $[M^{II}_{1-x}M^{III}_{x}(OH)_2]^{x+}A^{n-}_{x/n} \cdot mH_2O$, where M^{II} and M^{III} are divalent and trivalent metal cations, respectively, and A^{n-} is an anion [1–7]. The host structure consists of brucite-like layers of edge-sharing $M(OH)_6$ octahedra, and the partial substitution of M^{III} for M^{II} induces positively charged host layers, which are balanced by the interlayer anions (Figure 14.1) [7]. Most interest in LDHs arises from their versatility in chemical composition and structural architecture that leads to the possibility of tailoring the properties of the material for applications in areas such as catalysis, electrochemistry, photochemistry, and biochemistry [8]. Many materials with the LDH structure have been found in nature, and these are classified as the sjögrenite-hydrotalcite group of minerals [9–13]. The formula of the parent mineral, hydrotalcite $[Mg_6Al_2(OH)_{16}]CO_3 \cdot 4H_2O$ was first determined almost 100 years ago and its crystal structure was determined in the 1960s [13–15]. Since then, considerable efforts have been made to investigate the structural details of LDHs, including ordering of metal cations within the host layers, the possible range of composition, and the distribution of interlayer anions and water molecules [8]. In this chapter, we focus on the structural properties of LDH materials and recent advances in theoretical studies of the LDHs structure.

Handbook of Solid State Chemistry, First Edition. Edited by Richard Dronskowski, Shinichi Kikkawa, and Andreas Stein.
© 2017 Wiley-VCH Verlag GmbH & Co. KGaA. Published 2017 by Wiley-VCH Verlag GmbH & Co. KGaA.

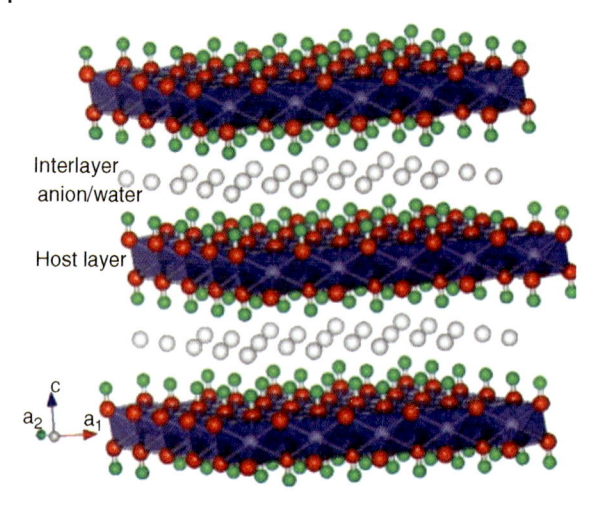

Figure 14.1 Schematic model of the LDH structure. (Reproduced with permission from Ref. [7]. Copyright 2006, The Royal Society of Chemistry.)

14.1.2
Basic Structural Features

14.1.2.1 Brucite-Like Layers

To understand the structure of LDHs, it is necessary to start from the structure of brucite [Mg(OH)$_2$], which is of the CdI$_2$ type, typically associated with small polarizing cations and polarizable anions. Brucite consists of octahedra of Mg^{2+} (sixfold coordinated to OH$^-$) sharing edges to form infinite sheets, which are stacked on top of each other to form a three-dimensional structure, held together by hydrogen bonding [16]. From the viewpoint of close packing, the structure is composed of close-packed planes of hydroxyl anions that lie on a triangular lattice. The metal cations randomly occupy the octahedral holes in the close-packed arrangement of OH$^-$ ions. In practice, the local geometry around the metal cations (D$_{3d}$) is strongly distorted away from the idealized arrangement (O$_h$) due to compression of octahedra along the stacking axis [16,17]. This leads to increases in the O···O and Mg···Mg distances parallel to the plane from 0.2973 nm (ideal O$_h$ geometry) to 0.3142 nm (experimental distance) and a decrease in thickness of the layers from 0.2427 to 0.2112 nm, as well as O—M—O bond angles of 96.7° and 83.3° rather than the value of 90° associated with a regular octahedron [17]. However, the hexagonal symmetry ($a_0 = b_0 = 0.3142$ nm, $c_0 = 0.4766$ nm, $\gamma = 120°$) is not influenced by the distortion of the brucite layers and the space-group symmetry $P\bar{3}m1$ is predetermined by the single octahedral layer. It should be noted that the value of a_0 is equal to the nearest neighbor Mg···Mg distance parallel to the plane and the O—H bonds are oriented along the threefold axes toward the vacant tetrahedral site in the adjacent layers [16,18–20].

14.1.2.2 Host Layer Structure of LDHs

The host layer structure of LDHs consists of brucite-like layers of edge-sharing $M(OH)_6$ octahedra, and the partial substitution of M^{II} and M^{III} induces positively charged host layers, balanced by the interlayer anions. A unique structural characteristic of LDH layers is that the M^{II} and M^{III} cations are distributed in a highly ordered manner in the hydroxide layers [1]. Previous abundant experiments indicated that M^{II} and M^{III} cations, which can be introduced into brucite-like layers, usually have a radius close to that of Mg^{2+} (0.072 nm) [21–23]. However, the empirical rule based on ionic radii cannot explain some experimental results. For example, the radii of Ca^{2+} and Cd^{2+} are significantly different from Mg^{2+}, but they can form stable LDHs [24,25]; while the radii of Pd^{2+} and Pt^{2+} are close to that of Mg^{2+}, but they are difficult to incorporate in LDHs [26,27]. Recently, the structural distortion angle θ of the octahedral hexahydrated metal ions was calculated by density functional theory (DFT) to quantitatively describe the construction of the LDHs layer: when θ of the octahedral hexahydrated metal ions is smaller than 1°, the metal cations are easily incorporated into LDHs layers, while those which form octahedral hexahydrated structure in heavy distortion with θ larger than 10° are difficult to incorporate in LDHs layers [28–30]. It can be found that θ of the octahedral hexahydrated metal ions could be used as a more effective criterion for the construction of the LDHs layers than ionic radii.

In addition, many studies have suggested that pure LDH phases can only be formed for stoichiometries in the range $0.20 < x < 0.33$ [4,31,32]. $M^{III}-O-M^{III}$ linkages are required when $x > 0.33$, and these are unfavorable from the viewpoint of charge repulsion (the so-called Lowenstein rule) [32,33]. However, the $Cr^{III}-O-Cr^{III}$ unit does exist in pure $Zn_7Cr_4-CO_3$-LDHs [34]. In addition, LDHs with the formula $[M^{II}Al_4(OH)_{12}](NO_3)_2 \cdot nH_2O$ (M^{II} = Zn, Ni, Co, Cu) have been prepared by reaction of activated gibbsite with $M^{II}(NO_3)_2$ [35]. Half of the cation vacancies are filled with M^{II} cations, so that although the M^{II}/Al ratio is 1: 4, there are still cation vacancies, whereas all the octahedral sites are occupied in the trioctahedral brucite-like layers of conventional LDHs.

14.1.2.3 Interlayer Structure of LDHs

The structure of the interlayer galleries contain both interlayer anions and water molecules and there is a complex network of hydrogen bonds between layer hydroxyl groups, anions, and water molecules [36]. Every anion has to balance the excess positive charges on both of the sandwiching octahedral layers, which are electrically balanced by two neighboring interlayers [37]. The bonding between octahedral layers and interlayer anions involves a combination of hydrogen bonding and electrostatic effects. Particularly, hydroxyl groups bonded to trivalent cations are strongly polarized and interact with the interlayer anions [4]. Moreover, the M^{II} and M^{III} cations distributed in a uniform manner in the hydroxide layers can lead to ordered arrangement of interlayer anions including inorganic anions [38], heteropolyacids [39], organic acids [40,41], layered compounds [42], and even polymers [43]. The XRD pattern of a

Mg$_2$Al–benzoate LDH was indexed to the (100) reflection of a superlattice with dimensions a_0, which is attributed to an ordered arrangement of benzoate anions associated with a regular arrangement of Al^{3+} cations within the layers [44]. A similar conclusion was drawn for an oxalate-intercalated LDH with Mg/Al = 2.2 [45]. Moreover, the single crystal structure of [Mg$_4$Al$_2$(OH)$_{12}$]CO$_3$·3H$_2$O shows an ordered distribution of layer cations [46]. Accordingly, the interlayer structure consists of a statistical distribution of carbonate ions and a network of water molecules, which gives a structure with space group $P\bar{6}2m$ with $a = a_0 = 0.5283$ nm and $c = 1.5150$ nm [46].

14.1.3
Theoretical Studies of LDHs Structure

With recent development in the computational hardware and software, theoretical calculations have been performed in the investigation of LDHs structure. Two kinds of computational methods including molecular dynamics (MD) simulation and density functional theory (DFT) have been widely used to calculate the electronic structure, cation ordering, guest arrangement, and some other properties of LDHs in recent decades [28–30,47–63].

14.1.3.1 Molecular Dynamics Simulations of LDHs

MD simulations, which can be used for large-scale models, have been increasingly employed to understand the microstructure and property of LDHs [47–58]. The force field (FF, potential function) in MD simulation have been developed, to provide a set of suitable parameters accounting for the interactions of all atoms in a modeled system. As far as LDH simulation, a modified Dreiding force field [47] and ClayFF force field [48] are mostly used. The former was developed by Newman *et al.*, who introduced a harmonic cosine function into Dreiding to simulate angle-bending terms in LDHs, but the structural distortions were found to be inevitable in their MD simulations. The latter (ClayFF) was originally developed by Cygan *et al.* for simulating clay substance, which is dependent on van der Waals and Coulombic interactions instead of covalent bonds to keep the coordination configurations of central metal cations. Recently, a force field for LDH materials (LDHFF) was systematically developed (Figure 14.2), which involves the parameters of all the atoms in the host sheets of LDHs including metal cations, bridging oxygen atoms, as well as hydroxyl hydrogen [49]. The potential function of the polymer consistent force field (PCFF) was used in LDHFF by introducing a double-well potential to model O—M—O bending in the host sheets. The bonded (intramolecular) parameters for modeling host sheets were originated from DFT calculations on the representative cluster models. The potential parameters are validated by employing a series of MD simulations for 24 LDHs models. The resulting structures, vibrational frequencies, as well as binding energies are well consistent with the experimental results, which provides a theoretical insight for understanding the structural property and exploiting the preparation of functional LDH. With the LDHFF

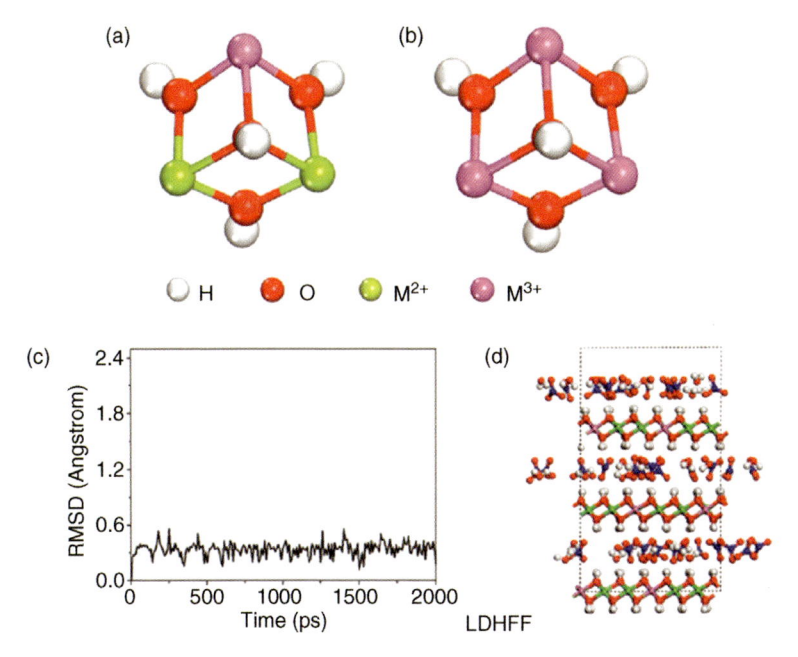

Figure 14.2 computational models of (a) $[M^{II}_2M^{III}(OH)_2)_9(OH)_4]^{3+}$ ($M^{II}_2M^{III}$ = Mg$_2$Al, Zn$_2$Al, Co$_2$Al, Ni$_2$Al, Cu$_2$Al, Mg$_2$Fe, Zn$_2$Fe, Ni$_2$Fe, Mg$_2$Cr, Zn$_2$Cr, Cu$_2$Cr, Co$_2$Cr) clusters and (b) $[M^{III}_3(OH_2)_9(OH)_4]^{3+}$ (M^{III} = Al, Fe, Cr) clusters. (c) RMSD values of the metal cations in the Mg$_2$Al–NO$_3$–LDH models monitored along MD trajectory of LDHFF. (d) The last snapshot of each simulation of LDHFF [49]. (Copyright 2010, American Chemical Society.)

force field, the interlayer behavior of guest anions in LDHs gallery can be calculated accurately. MD simulations are performed on different ZnAl-LDHs cointercalated with dye and alkylsulfonate with different alkyl chain length ($C_nH_{2n+1}SO_3$, $n = 5$, 6, 7, 10, and 12, respectively) [50]. The structure, binding energy, and the thermal motion were characterized by the diffusion coefficient of each dye, indicating that the dye–alkylsulfonate/LDH system is very effective in restraining both the thermal motion and the aggregation of the dye due to the confined microenvironment provided by the LDH layer.

14.1.3.2 Density Functional Theory Calculations of LDHs

DFT have been widely used to calculate the structural properties of LDHs, such as thermal decomposition mechanism, band structure, density of states (DOS), band edge placement [59–63]. Leitao and coworkers reported that the thermal decomposition mechanism of MgAl-CO_3^{2-}-LDH by DFT calculations [59,60]: (1) when close to 180 °C LDH loses its intercalated water molecules completely, (2) the next step involves the formation of a lamellar structure in which CO_3^{2-} anions are monodentate grafted to the layers at about 280 °C, (3) at 350 °C, a reorganization of the carbonate is observed, which leads to a bidentate bond of this anion with the layer. Recently, the electronic properties (band structure,

DOS, and band edge placement) of $M^{II}M^{III}$-LDHs (M^{II} = Mg, Co, Ni, and Zn; M^{III} = Al and Ga) have been studied in detail [61]. The calculation results of band structure indicate that MgAl-, ZnAl-, MgGa-, and ZnGa-LDHs (bandgap energies larger than 3.1 eV) are ultraviolet responsive, while CoAl-, NiAl-, and NiGa-LDHs are responsive to visible light (bandgap energies less than 3.1 eV). The DOS calculations reveal that the photogenerated hole localizes on the surface hydroxyl group of LDHs, facilitating the oxidization of a water molecule without a long transportation route. The band edge placements of $M^{II}M^{III}$-LDHs show that NiGa-, CoAl-, ZnAl-, and NiAl-LDHs have a driving force (0.965, 0.836, 0.667, and 0.426 eV, respectively) toward oxygen evolution. The dispersion of MO_6 octahedra (M = Ti, Zn, Cr) in the LDH structure, and the subsequent generation of surface defects (e.g., oxygen vacancies), has an influence on their electronic structure. Zhao and coworkers have derived the band structures of defect-rich ZnAl-LDH and NiTi-LDH based on periodic DFT calculations [64,65]. The results show that the presence of defects leads to a new defect energy level in the forbidden zone, and the calculated bandgap is much lower than that of the ideal LDH system. This can expand the light response spectrum of LDHs and enhance the efficiency of electron transfer to reactant molecules, which results in a significantly improved photocatalytic performance.

14.2
Properties and Applications of LDHs

14.2.1
Confinement Effect

The unique structure of LDHs provides a confined microenvironment for the intercalated guest molecules and thus results in controllable structures and performances for LDHs-based materials. The confinement effect of LDHs imposes several influences on the hybrid materials as follows: (1) the intercalated guest molecules exhibit well orientation, ordered alignment, and polarized fluorescence emission [40,66–68]; (2) the molecular motions are restricted, which will block the nonradiative decay and enhance luminescence efficiencies of the intercalated guests [69,70]; (3) the confined environment provides a nanoscale gallery required for the fluorescence resonance energy transfer for the intercalated photofunctional materials [71–73]. Most importantly, the confinement of LDHs imposes significant effect on the electron distribution of guest molecules, which directly determines the performances.

Multiple quantum well (MQW) structure is an important insight to study the confinement and host–guest interaction of LDHs. It has been reported [74] that an anionic poly(p-phenylene) derivative can be incorporated between LDH layers to give ultrathin films (UTFs) (Figure 14.3). For poly(p-phenylene) (APPP), the top of the valence band (TVB) and the bottom of the conducting band (BCB) are mainly dominated by the carbon atomic orbitals of APPP molecules,

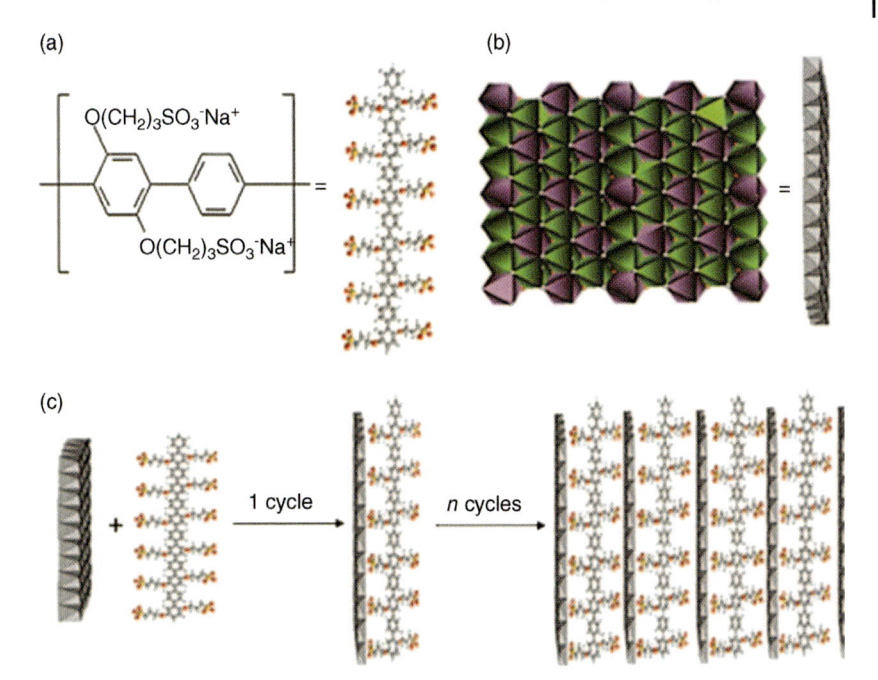

Figure 14.3 (a) The chemical formula of the anionic p-phenylene derivative APPP. (b) A representation of one sheet of Mg–Al layered double hydroxide (Mg–Al-LDH; purple: Al(OH)$_6$ octahedra; green: Mg(OH)$_6$ octahedra). (c) Assembly process for (APPP/LDH)$_n$ UTF. (Reproduced with permission from Ref. [74]. Copyright 2009, John Wiley and Sons, Inc.)

with the bandgap of 2.6 eV. The O 2p and the H 1s orbitals in OH and Mg/Al 3s orbitals from the LDH layers contribute to the total electronic densities of states below and above the TVB and BCB, with a gap of 5.7 eV. Therefore, this calculation demonstrates that the valence electrons are localized in APPV, while the LDH monolayer functions well as an energy blocking layer. The constructed MQW structure is beneficial to the improvement of light-emitting efficiency for APPP molecules. Similar LDH-based MQW structure were also observed in the APPV/LDH [75] and ZnTSPc/LDH [76] system, where the valence electrons of guest molecules can be confined within the energy wells imposed by the LDH monolayers effectively.

The confinement of LDHs plays an important role in the optimization of Au nanoclusters, especially for the emissive Au(I) content [77]. Compared with the spherical configuration of Au NCs in aqueous solution, the Au NCs localized on the LDH nanosheet exhibit a rearranged geometric configuration with the ligands attached to the LDH nanosheet due to the host–guest interaction and confined microenvironment of LDHs (Figure 14.4). Both experimental and computational studies show that the valence state of Au NCs in the confined interlayer of LDH was greatly affected, and the content of emissive Au(I)-thiol in Au NCs/LDH system was promoted from 34.4% (for pristine Au NCs) to 46.4%.

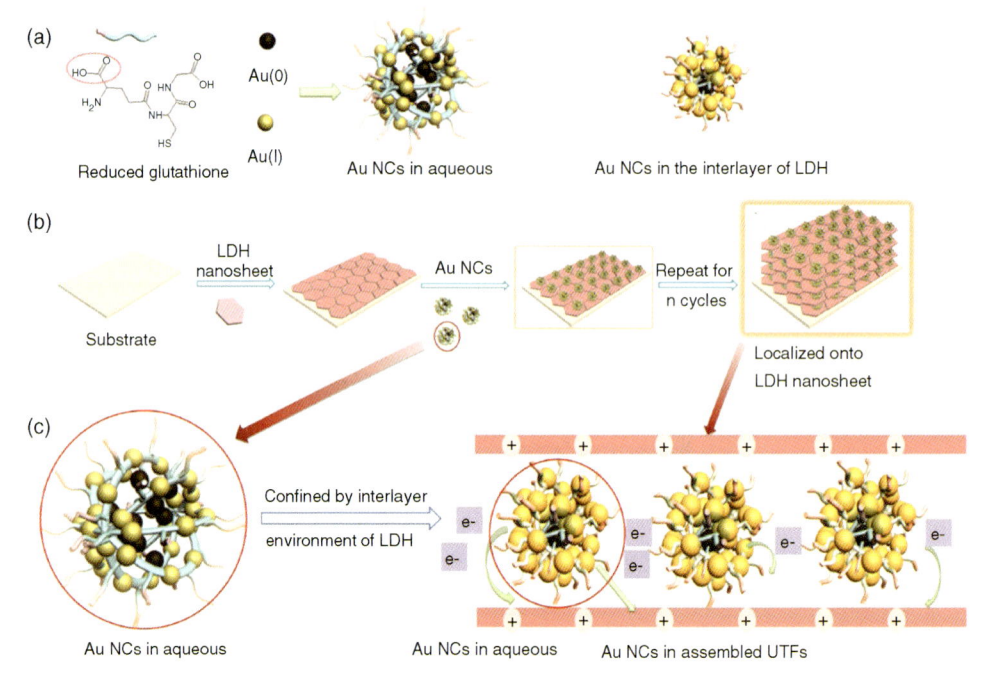

Figure 14.4 Schematic representations for (a) weak-emissive Au NCs in aqueous solution and strong-emissive Au NCs in UTFs, (b) the LbL fabrication of multilayer luminous (Au NCs/LDH) *n* UTFs, and (c) transformation of Au NCs from aqueous solution into interlayer microenvironment of LDHs [77]. (Copyright 2015, John Wiley and Sons, Inc.)

Therefore, as a result of confinement induced by LDH, the localized Au NCs undergo changes in geometric configuration and charge distribution within the 2D gallery of LDH nanosheets, resulting in the significantly promoted fluorescence performance.

Moreover, taking advantages of the confined interlayer galleries, LDHs could serve as a 2D nanoreactor for *in situ* synthesize of various multifunctional materials. Yang and coworkers. have successfully fabricated the carbon nanorings via catalytic growth in the confined space of the interlayer galleries in LDHs, and the obtained carbon nanorings exhibit enhanced lithium storage properties as a result of their unique structure [78]. In addition, graphene nanosheet [79] and graphene quantum dots [80] with precisely control over layer number or size distribution can be achieved based on the confinement of LDH.

14.2.2
Catalytic Properties of LDHs

By virtue of their versatility in composition, morphology, and architecture, LDHs display great potential as heterogeneous catalysts, precursors, and supports. The catalytic sites can be preferentially oriented, highly dispersed, and firmly anchored to afford materials with excellent catalytic performance and recyclability. The approaches to fabricate catalysts based on LDH materials include – but are not limited to – atomic-scale uniform distribution of metal cations in the brucite-like layers, lattice confinement by the brucite-like layers, and intercalation. This section summarizes the latest developments in the design and fabrication of nanocatalysts by using LDHs.

14.2.2.1 Solid Base Catalysis

By virtue of the abundance of hydroxyl groups, simple binary as-prepared materials such as MgAl LDHs are potential heterogeneous solid base catalysts for a wide variety of organic transformations [81,82]. Furthermore, after appropriate calcination at temperatures from 450 to 600 °C, LDHs can be converted into well-dispersed mixed metal oxides (MMOs) with large surface area and numerous Lewis base sites [83,84]. As noted above, through subsequent rehydration in water, the MMOs can reform the layered structure with enhanced exposure of hydroxyl anions incorporated in the interlayer region, endowing the resulting so-called activated LDHs with more abundant Brønsted-type basic sites [85]. As a result, LDHs, MMOs, or activated LDHs can be employed as environmentally benign and recyclable heterogeneous solid base catalysts and have attracted increasing attention as catalysts for important organic reactions (e.g., Knoevenagel condensations [86,87], Michael additions [88], aldol and Claisen–Schmidt condensations [89,90], and cyanoethylation of alcohols [91]).

14.2.2.2 Photocatalysis

By means of incorporating appropriate photoactive components (e.g., Zn, Ti, Fe, or Cr) into the LDH layers [92–94], a variety of LDH-based photocatalysts with a tunable bandgap can be obtained, and these materials exhibit promising applications in the elimination of organic pollutants, water reduction, and water oxidation. For instance, Wei and coworkers [64,95,96] reported the fabrication of a series of Cr- or Ti-containing LDHs by a simple and scalable coprecipitation method, and showed that the materials displayed a high photocatalytic activity for water splitting, with an H_2-production rate of 31.4 µmol/h^1. Most early investigations of LDH-based photocatalysts focused on the introduction of highly dispersed photoactive sites (e.g., CrO_6, TiO_6, or ZnO_6 octahedra) into the LDH matrix so as to obtain an effective electron–hole separation. However, recent studies have shown that an abundance of defect sites (e.g., Ti^{3+} and Zn^+) can be self-doped in LDHs themselves by the manipulation of their size, shape, and morphology via a bottom-up synthesis approach [64]. This further enhanced both the efficiency of electron–hole separation and the resulting photocatalytic performance. For example, Zhao *et al.* [65] reported the fabrication of ultrathin ZnAl-LDH nanosheets with coordinatively unsaturated Zn ions (Zn^+) by using a facile bottom-up strategy, and these materials show extraordinarily high activities for the photoreduction of CO_2 to CO in the presence of water vapor (Figure 14.5).

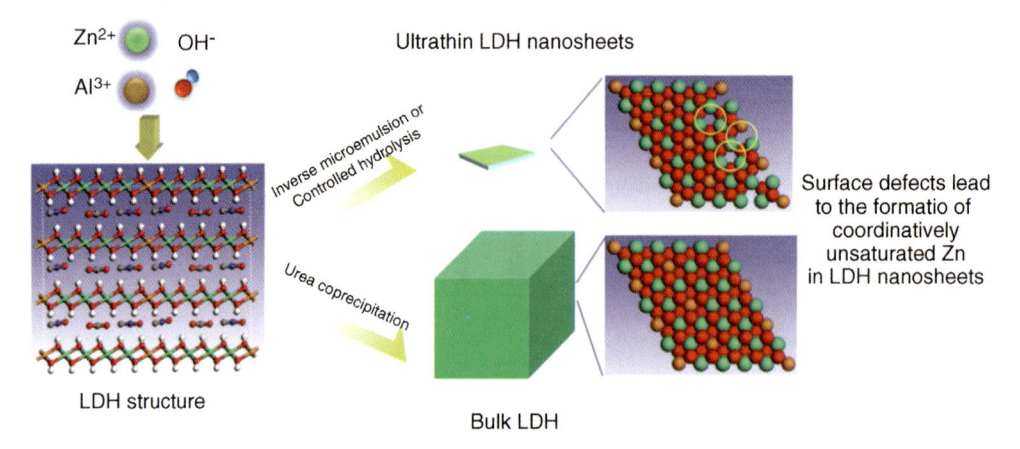

Figure 14.5 Schematic showing the formation of coordinatively unsaturated ZnAl-LDH nanosheets. (Reproduced with permission from Ref. [65]. Copyright 2015, John Wiley and Sons, Inc.)

14.2.2.3 Metal-Based Catalysts

A key structural characteristic of LDH materials is that the M^{II} and M^{III} cations are distributed uniformly within the hydroxide layers [1], without clusters of like cations. If LDHs, as precursors or templates, are converted to metal oxide-supported nanometal catalysts by heating under reducing conditions, the enhanced interactions between metal nanoparticles and the metal oxide substrate can promote high dispersion of the active species and prevent sintering of the nanocatalyst during use. Wei and coworkers [97,98] reported the preparation of highly dispersed Fe, Co, Ni metal nanocatalysts by the reductive treatment of LDH precursors. In particular, the resulting iron-based catalysts derived from MoO_4^{2-} intercalated FeMgAl-LDH precursors possess a high density of Fe nanoparticles (10^{14} to $10^{16}\,m^{-2}$) and good thermal stability (at 900 °C), and display high activity for the formation of single-walled carbon nanotube (SWNT)-array double helices (Figure 14.6).

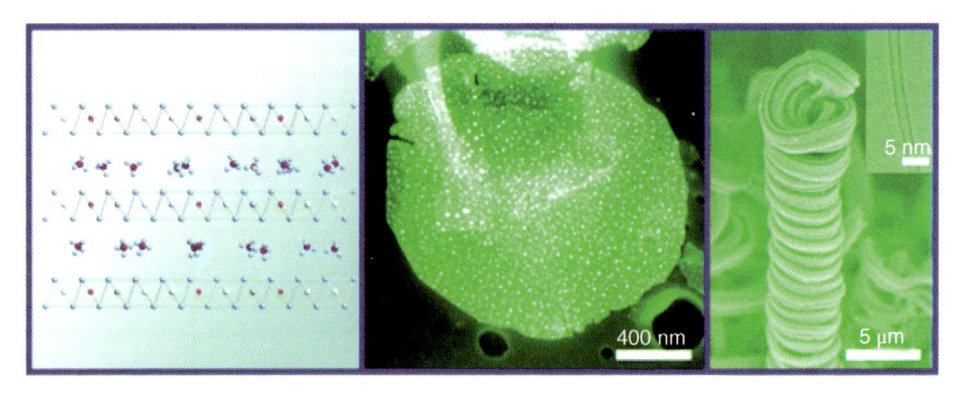

Figure 14.6 Schematic illustration of the formation of high density metal nanoparticles from a precursor with MoO_4^{2-} incorporated in FeMgAl-LDHs. (Reproduced with permission from Ref. [97]. Copyright 2010, American Chemical Society.)

In addition, bimetallic/alloy catalysts with a tunable structure of metal active sites can be obtained by reduction of LDH precursors containing two transition metal elements. Tomishige and coworkers [99] reported that uniform Ni–Fe alloy nanoparticles supported on Mg/Al MMOs with high dispersion were obtained by calcination and reduction of a Ni/Fe/Mg/Al-LDH precursor, and showed high catalytic activity for the steam reforming of toluene. Recently, Wei and coworkers [100] synthesized core–shell Cu@(CuCo-alloy) nanoparticles (NPs) embedded on a Al_2O_3 matrix via an *in situ* growth of CuCoAl-LDH nanoplatelets on aluminum substrates followed by a calcination–reduction process (Figure 14.7). The resulting Cu@(CuCo-alloy) materials exhibit efficient catalysis of CO hydrogenation to produce higher alcohols; this was attributed to the electronic and geometric interactions between Cu and Co. Although some noble metals, such as Rh or Ru, cannot be incorporated into LDH host layers in significant amounts, their bimetal/alloy nanocomposites can be fabricated via a bottom-up method. Recently, Wei *et al.* [101] reported the fabrication of Cu-decorated Ru catalysts supported on MgAl-LDH by a facile two-step procedure involving the reduction of $RuCl_3$-impregnated CuMgAl-LDH by hydrogen, followed by the rehydration of the resulting Ru_xCu_y/MgAl-MMO to afford Ru_x-Cu_y/MgAl-LDH.

As an alternative to traditional bimetal or alloy catalysts, intermetallic compounds (IMCs) have attracted extensive recent research interest as catalysts due to their unique electronic and geometric structures [102]. LDH materials are useful precursors to IMCs by virtue of the uniform distribution of metal cations in the host layers [103]. By taking advantage of the topological transformation of LDH precursors, a variety of IMCs with improved catalytic performance can be fabricated, such as PdGa IMCs for alkane dehydrogenation [104], PdIn IMCs for

Figure 14.7 Illustration of the Cu@(CuCo-alloy)/Al_2O_3 catalyst with a core–shell architecture derived from a CuCoAl-LDH film based on an *in situ* growth reaction followed by a calcination–reduction process. (Reproduced with permission from Ref. [100]. Copyright 2015, The Royal Society of Chemistry.)

ethane and propane dehydrogenation [105], and PdGa and PdZn IMCs for methanol steam reforming [103]. Wei and coworkers [106] have provided further insight into the structural features of IMCs. They first synthesized a series of NiIn IMCs with different Ni/In ratios (Ni_3In, Ni_2In, NiIn, and Ni_2In_3), by tuning the elemental ratio in the LDH layers. When used as catalysts for the hydrogenation of α,β-unsaturated aldehydes, the resulting NiIn IMCs favor nucleophilic addition at a C=O group rather than electrophilic addition at C=C, accounting for the observed highly regioselective hydrogenation activity.

14.2.2.4 Active Species-Intercalated Nanocatalysts

In the last decade, active species-intercalated nanocatalysts based on LDH materials have been widely reported, and shown to exhibit improved catalytic activity, selectivity, and stability when compared with their homogeneous counterparts. LDHs have a number of advantages as hosts for catalytically active anions: (1) the electrostatic interactions between LDH layers and catalytically active anions can induce an ordered arrangement of the anions and effectively tailor the orientation of the active site; (2) the extent of the dispersion of the active sites can be controlled by modulation of the layer charge density (which depends on the ratio of M^{II} to M^{III} cations); (3) the electrostatic interactions between LDH layers and catalytically active anions can also lead to an increase in the stability of the latter.

LDH-containing nanocatalysts have promising applications in the field of enantioselective chiral catalysis [107–110]. After incorporation of a chiral active species (e.g., chiral amino acids) into the interlayer galleries of LDHs, the resulting supported catalysts possess many advantages over traditional homogeneous catalysts in terms of long lifetime, high-thermal stability, and convenience of catalyst separation and recovery (Figure 14.8). The intercalation of other types of well-known homogeneous catalysts, such as metal complexes, into the interlayer galleries of LDHs to form functional heterogeneous catalysts has also been widely reported. For example, Wei *et al.* [111,112] reported a heterogeneous catalyst in which the rhodium complex (*trans*-RhCl(CO)(TPPTS-Na_3)$_2$) (TPPTS = tris(3-sulfophenyl) phosphine trisodium salt) was intercalated in Zn/Al-LDHs, which achieved improved selectivity, activity, and stability in the hydroformylation of higher olefins compared with the corresponding homogeneous catalyst. The excellent catalytic performance was attributed to the ordered arrangement of the intercalated rhodium complex anions in the interlayer galleries of LDHs as well as the strong electrostatic interactions between the LDH layers and the active rhodium species. Song and coworkers [113] recently reported the intercalation of a series of polyoxometalate anions into the interlayer galleries of LDHs. The resulting immobilized catalysts were used to catalyze the oximation of aromatic aldehydes with aqueous H_2O_2 as the oxidant. The results showed that a Zn_3Al–$Zn_2Mn^{III}_3WO$ heterogeneous catalyst exhibited a much higher selectivity (85%) than the corresponding homogeneous catalyst (55%). This was explained in terms of the synergistic interaction between the LDH host and intercalated POM anions.

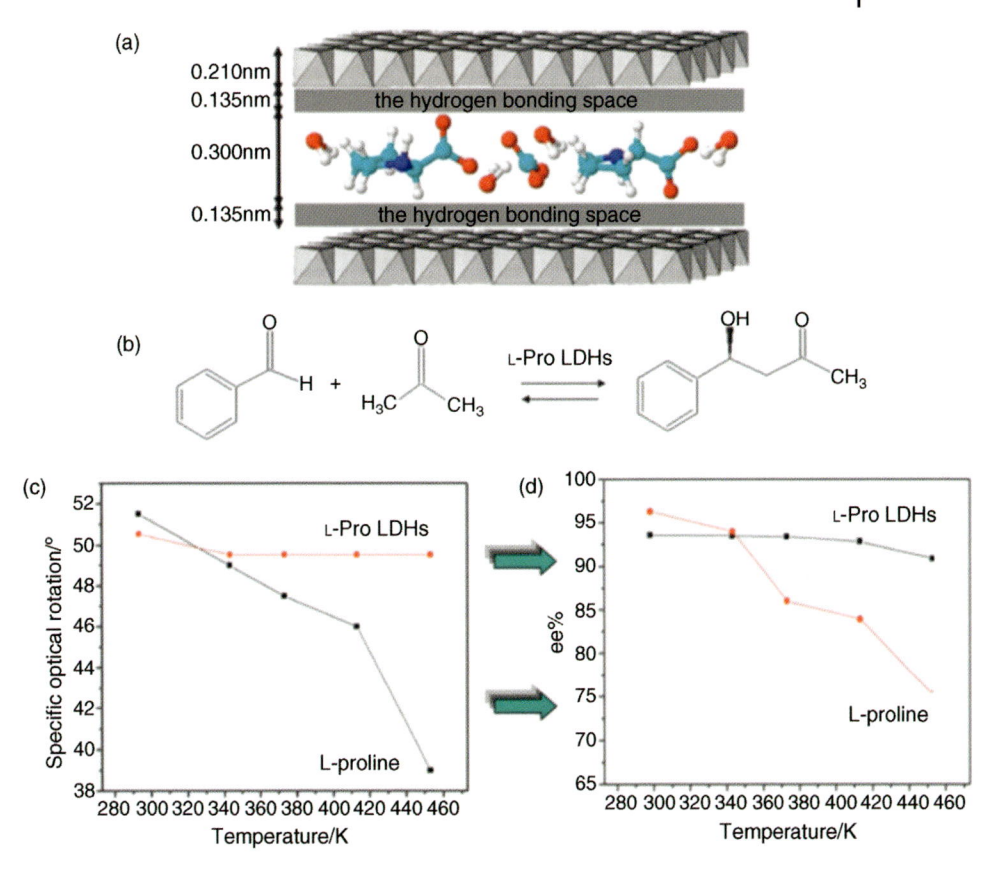

Figure 14.8 (a) A schematic model of L-proline intercalated LDHs. (b) A schematic representation of aldol reaction between benzaldehyde and acetone. Both (c) the thermal stability of specific optical rotation and (d) the enantiomeric excess (ee) value for the products of aldol reactions are significantly improved after the L-proline catalyst is intercalated in an LDH host. (Reproduced with permission from Ref. [110]. Copyright 2006, Elsevier Inc.)

14.2.3
Electrochemical Energy Storage Properties

Increasing energy demands have stimulated intensive research on alternative energy conversion and storage systems (such as batteries, supercapacitors (SCs), and water splitting) with high efficiency, low cost, and environmental benignity, which has encouraged the discovery and optimization of new high-performance materials. LDHs have been shown to be promising candidates in electrochemical energy storage/conversion owing to their tunable chemical composition, cost effectiveness, and environment-friendly nature. For instance, electrically active transition metal ions (e.g., Ni, Co, Fe, Mn) can be uniformly dispersed into the host layers of LDHs, and the resulting materials used in high-performance energy storage and conversion devices. Moreover, the facile exfoliation of LDHs

14.2.3.1 Supercapacitors

Supercapacitors (SCs) are considered as promising power sources because of their advantages of fast charging and discharging properties, high-power delivery, and excellent cycling lifespan in comparison with conventional batteries and dielectric capacitors [114–116]. Recently, LDHs have been reported as excellent candidates for use in pseudocapacitors owing to their high theoretical specific capacity and low cost [117,118]. However, LDH materials often suffer from a low-power density and life cycle because the redox kinetics is limited by the rate of mass diffusion and electron transfer. As a result, many efforts have been made to optimize the performance of LDH-based SCs by creating nanostructures (e.g., hollow microspheres [119,120], core/shell nanowires [121], nanowall films [122,123]) or incorporating conductive materials (carbon nanotubes (CNTs) [124,125], graphene [126–128], or polymers [129–131]). As shown in Figure 14.9a, LDH microspheres with tunable core–shell, yolk–shell, and hollow interior architecture were synthesized by an *in situ* growth approach [119]. It was found that the hollow NiAl-LDH microspheres exhibited excellent pseudo-capacitance performance with a high specific capacitance of 735 F/g (at 2 A/g), good cycle performance (the capacitance increases by 16.5% after 1000 cycles at a current density of 8 A/g), and remarkable rate capability (75% retention at 25 A/g), superior to the yolk–shell and core–shell structures. The fabrication of LDH nanoplatelets on conductive substrates with well-defined architectures has also been shown to be a feasible approach for enhancing ion- and charge-carrier transport. It has been reported that CoNi-LDH arrays on a nickel foam electrode display excellent supercapacitance properties with a specific capacitance of 2682 F/g, an energy density of 77.3 Wh/kg at 623 W/kg, as well as good stability (~82% of the original capacitance is retained after 5000 cycles) (Figure 14.9b) [122]. LDH/carbon nanocomposites with superior charge mobility and high surface area have also been developed for efficient electrode materials. NiMn-LDH microcrystals grafted on a CNT backbone can deliver a maximum specific capacitance of 2960 F/g (at 1.5 A/g) and excellent rate capability (79.5% retention at 30 A/g) (Figure 14.9c) [124]. CoMn-LDHs have also been grown on flexible carbon fibers (CF) via an *in situ* growth approach. The resulting CoMn-LDH/CF electrode delivers a high specific capacitance (1079 F/g at 2.1 A/g) and excellent rate capability even at a rather high current density (82.5% capacitance retention at 42.0 A/g). Furthermore, a flexible solid-state supercapacitor device fabricated using the CoMn-LDH/CFs achieves a specific energy up to 126.1 Wh/kg and a specific power of 65.6 kW/kg [123]. Moreover, graphene, as a unique 2D carbon materials, has been effectively used to support LDH materials for efficient SCs. An NiAl-LDH/graphene composite exhibits a maximum specific capacitance of 781.5 F/g and an excellent life cycle [126]. A CoNi-LDH/rGO

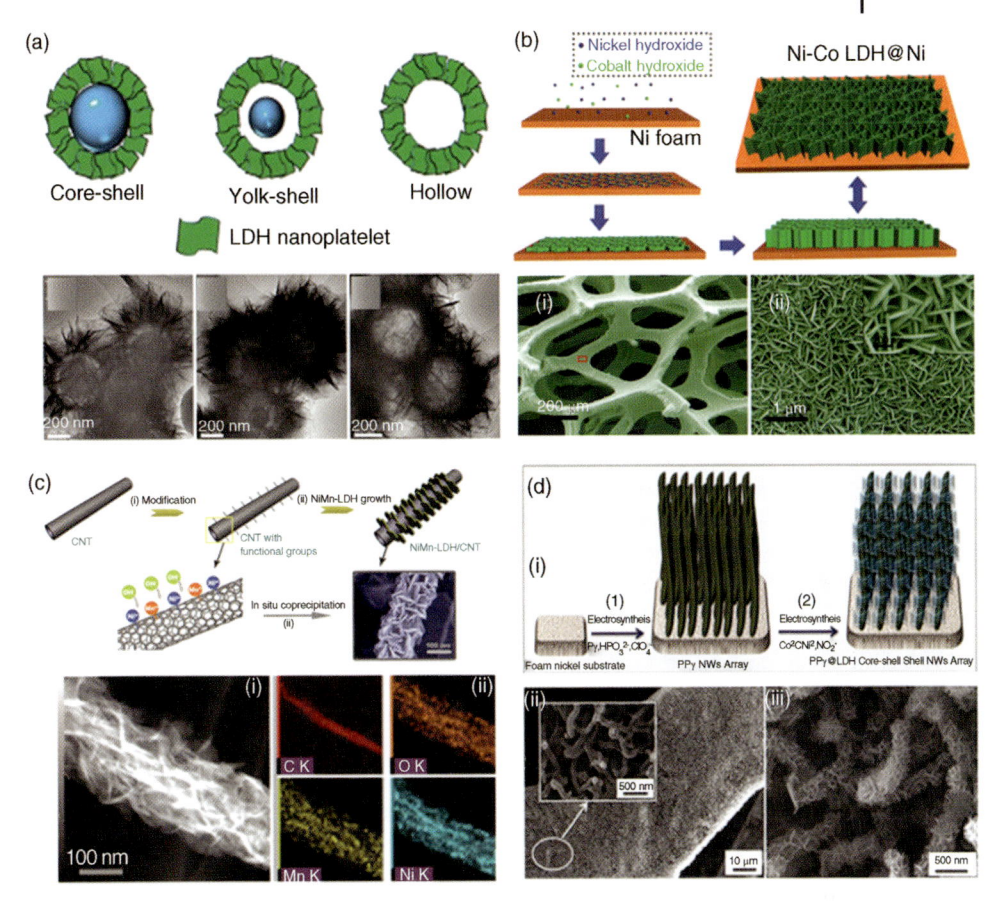

Figure 14.9 (a) LDH microspheres with tunable core–shell, yolk–shell, and hollow interior architecture [119]. (b) CoNi-LDH arrays on a nickel foam electrode [122]. (c) NiMn-LDH microcrystals grafted on a CNT backbone [124]. (d) A sophisticated nanoarray consisting of a PPy core and LDHs shell (PPy@CoNi-LDH core–shell NWs array) [129]. (Copyright 2014, 2015 John Wiley and Sons.)

(rGO = reduced graphene oxide) superlattice composite shows a high specific capacitance (∼500 F/g) at discharge currents up to 30 A/g [128]. In addition, the construction of LDH/conducting polymer nanocomposites provides another effective route to obtain excellent all round supercapacitive performance. Recently, a sophisticated hierarchical nanoarray consisting of a polypyrrole (PPy) core and LDHs shell was fabricated via a two-step electrosynthesis method (Figure 14.9d). The specific capacitance of the PPy@CoNi-LDH core–shell NWs array (2342 F/g; current density 1 A/g) was 2.1, 2.6, and 1.2 times higher than that of pristine PPy (1137 F/g), CoNi-LDH (897 F/g), and a physical mixture of these two constituents (2034 F/g), respectively [129].

14.2.3.2 Li-Ion Batteries

Considerable attention has focused on Li-ion batteries (LIBs) because of their environmental compatibility, high energy-density, high operating power, and long cycling life [116,132,133]. However, the performance of LIBs still needs to be improved because of the increasing demands for use in portable electronics and electric vehicles. The electrode is one of the important parts of LIBs, and extensive efforts have been made to develop the electrode materials with optimized structures and enhanced performance. The tunable metal ion content in the layers and exchangeable anions between the layers of LDHs as well as the ability of LDHs to be converted to the corresponding mixed metal oxides make them promising candidates for anode materials for LIBs. A stoichiometric $NiFe_2O_4$ phase was prepared from NiFe LDH via calcination at 700 °C and showed an initial discharge capacity of 1239 mAh/g and a reversible capacity of 470 mAh/g after 20 cycles [134]. The good electrochemical performance of many materials that are derived from LDH precursors can be attributed to the small particle size and uniform distribution of metal cations.

Carbon materials have been extensively utilized in LIBs due to their high surface area, extraordinary electronic and mechanical properties, and low cost. However, a high degree of dispersion and ordering are necessary when carbon materials are employed in LIBs. LDHs are considered as promising candidates for catalytic growth of carbon materials, since the metal catalyst can be dispersed in the framework of LDHs on an atomic scale. In a typical process, mixed metal oxide flakes derived from an LDH precursor were used as an effective catalyst for the growth of robust herringbone CNFs, which had a reversible capacity of 330 mAh/g and an initial coulombic efficiency of 45.2% [135]. Porous graphene microspheres (GMSs) and unstacked graphene were also obtained by a similar method, and exhibited a high reversible capacity of 1034 mAh/g at high discharge rates of 5 C and an initial discharge capacity of 1066 mAh/g at 0.14 mA/cm^2, respectively [136,137]. In addition, the confined space of LDHs can be utilized to fabricate carbon nanorings (CNRs), which show a reversible specific capacity of 1263 mAh/g after 100 cycles and the coulombic efficiency increased to nearly 100% [78].

14.2.3.3 Electrochemical Catalysis

Highly active oxygen evolution reaction (OER) catalysts play a key role in water splitting that generates clean, renewable H_2 for use as the fuel for fuel cells and other energy devices. LDHs possess a layered and relatively open structure, which makes the rapid diffusion of reactants and products, and even the fast proton-coupled electron transfer process in the water oxidation reaction, much easier. For example, ZnCo-LDH can serve as an efficient electrocatalyst for water oxidation, and exhibits a lower overpotential than monometallic cobalt-based solid-state materials [138]. The latest reports demonstrate that NiFe-based LDHs show a superb OER activity [139], which is comparable to the best noble metal catalysts for the OER. Moreover, exfoliated single-layer NiFe-LDH nanosheets exhibit significantly higher OER activity than the corresponding bulk

LDHs in alkaline conditions [140]. In the pursuit of materials with high electrocatalytic activity for water oxidation, well-aligned LDH nanoarrays with active components have also been developed [141,142]. The high porosity and large active-site density of the LDH nanoarrays enhances their OER activity and stability by providing pathways for both electrolyte ions and electrons. A variety of carbon materials (e.g., carbon nanotubes [139], graphene [143–148] and quantum dots [149]) have been mixed with LDH catalysts to improve their conductivities and inhibit particle aggregation. The excellent performance of these materials has been attributed to the synergistic effect of the nanosized LDHs catalysts and the microstructure of carbon materials. Figure 14.10 shows the different overpotentials of various LDH-based materials at current density of 10 mA/cm^2 [139,140,143–157].

LDH materials have been employed to improve the photocurrent of photoelectrochemical (PEC) splitting electrodes by facilitating electron–hole pair separation and improving O_2 evolution kinetics. A ZnO@CoNi–LDH core–shell nanoarray exhibits promising behavior in photoelectrochemical water splitting, giving a high photocurrent density as well as excellent stability [158]. The photocorrosion of Ta_3N_5 nanostructures can be dramatically reduced by using NiFe-LDH as cocatalyst [159]. The as-synthesized NiFe-LDH/Ta_3N_5 arrays exhibited solar photocurrents of 6.3 mA/cm^2 at 1.23 versus a reversible hydrogen electrode (RHE), which is the highest value reported for Ta_3N_5 nanorod arrays. A ZnCr-mixed metal oxide (MMO) has been developed for PEC water splitting by using LDHs as a precursor, and delivered a stable and high photocurrent [160]. Dopant anions were also effective when incorporated in MMO crystal structures by using LDH as a precursor. The uniformly anion-doped MMO nanostructures absorbed visible light exhibiting a significant red shift of the absorption spectra with high activity under visible light.

LDHs have also been reported as promising noble metal-free electrode materials for the electrocatalytic oxidation of other small molecules (e.g., glucose [161,162],

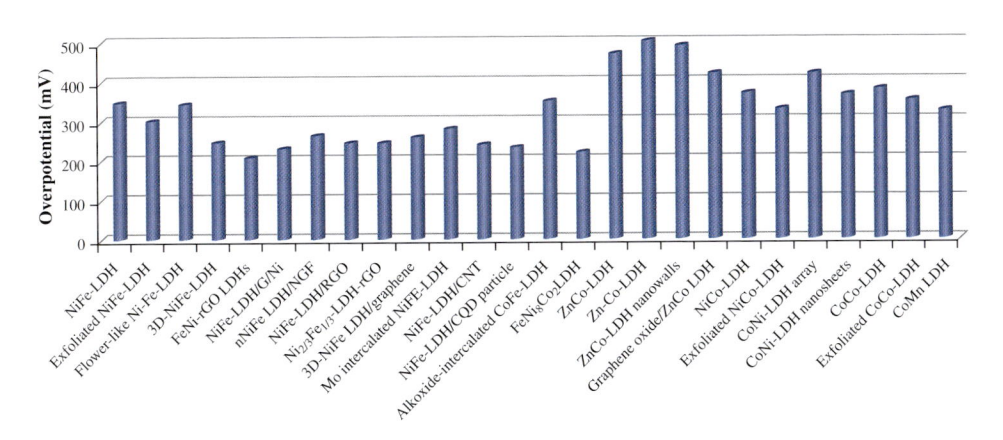

Figure 14.10 Overpotentials of various LDH-based materials for a current density of 10 mA/cm^2.

ethanol [163,164], H_2O_2 [165], and dopamine [166]) in alkaline media. For instance, ZnAl-LDH nanosheets/Au nanoparticle UTFs show high electrocatalytic activity and a broad linear response toward the oxidation of glucose [161]. With the assistance of an external magnetic field employed during their synthesis, CoFe-LDH/MnTPPS UTFs exhibit an ordered nanostructure and improved sensitivity and selectivity for the detection of glucose [162]. In addition, hierarchical LDH microspheres with tunable interior architecture were synthesized by a facile and cost-effective surfactant-templated approach, and exhibited promising ethanol oxidation behavior in alkaline solution [163].

14.2.4
Gas Barrier Properties

Polymeric packaging materials have extensive applications because of their lightweight, low cost, and good processability. However, their application as gas barrier materials is restricted due to their relatively high gas permeability [167–169]. In order to improve the barrier properties, a wide variety of hybrid materials containing polymers and two-dimensional inorganic platelets have been designed and fabricated [170–173]. Since the integration of 2D platelets into polymer matrices results in a long diffusion length and strong resistance to the permeating gas, considerable efforts have been dedicated to the construction of organic–inorganic composites as gas barrier materials. In this section, we review recent progress in the design and fabrication of LDH/polymer composite films with excellent barrier properties.

14.2.4.1 Gas Barrier Films with Brick–Mortar Structure
Although the infilling of platelets into polymer matrices enhances the gas barrier properties relative to the pristine polymer, previously employed inorganic nanoplatelets with random alignment and limited aspect ratio restrict further improvement [174,175]. Wei's group fabricated highly oriented films with a brick–mortar structure via layer-by-layer spin-coating of LDH nanoplatelets and cellulose acetate (CA) (Figure 14.11a) [176]. The oxygen transmission rate (OTR) of $(LDH/CA)_n$ multilayer films can be tailored by tuning the aspect ratio of LDH nanoplatelets from 20 to 560 (Figure 14.11b). The resulting $(LDH/CA)_{20}$ film displays excellent oxygen barrier properties with an OTR equal to or below the detection limit of commercial instrumentation ($<0.005\,cm^3/(m^2\,day)$), much better than previously reported inorganic platelet-filled barrier films (Figure 14.11c) [177–181]. The enhancement of barrier properties was attributed to the parallel orientation of the LDH nanoplatelets in the brick–mortar structure and their extremely large aspect ratio, which leads to a long diffusion length and strong resistance to the diffusion of gas molecules.

14.2.4.2 Gas Barrier Films with Brick–Mortar–Sand Structure
Although considerable efforts have been made to construct organic–inorganic composites with a brick–mortar structure, surface incompatibility between the

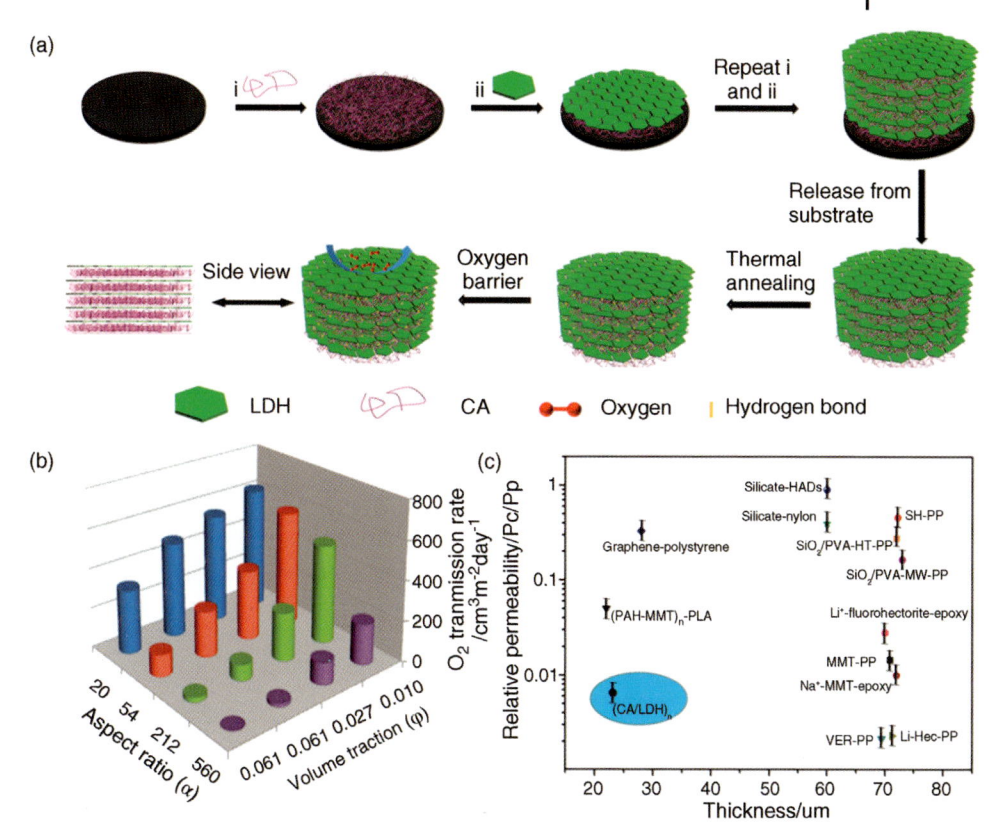

Figure 14.11 (a) Schematic representation of the (LDH/CA)$_n$ multilayer film [176]. (b) The influence of aspect ratio (20, 54, 212, and 560) and volume fraction (1.00, 2.70, 5.10, and 6.10%) of LDH nanoplatelets on the OTR of barrier film [176]; (c) Comparison study of the OTR of the (LDH/CA)$_n$ multilayer film with previously reported barrier films [177–181]. (Copyright 2014, John Wiley and Sons, Inc.)

polymeric and inorganic components is normally inevitable, which induces some free volume into the barrier films. In order to eliminate the free volume in a brick–mortar structure, a three-component hybrid structure with a brick–mortar–sand structure was fabricated by the alternate assembly of XAl-LDH nanoplatelets (X = Mg, Ni, Zn, Co) and polyacrylic acid (PAA) and subsequent CO_2 adsorption (Figure 14.12a) [182]. A CO_2 temperature programmed desorption (TPD) (Figure 14.12b) study indicated that a series of strong basic sites exist in the LDH powders, which facilitate sorption of an acidic gas by the LDH-based films, resulting in a decrease in their free volume (Figure 14.12c). The (XAl-LDH/PAA)$_n$-CO_2 films exhibit significantly enhanced gas barrier properties in comparison with brick–mortar structure films (Figure 14.12d). The ultrahigh oxygen-barrier properties of (XAl-LDH/PAA)$_n$-CO_2 films was attributed to the synergistic effects in the brick–mortar–sand structure: the 2D LDH nanoplatelets and polymer form a brick–mortar structure, which suppresses the

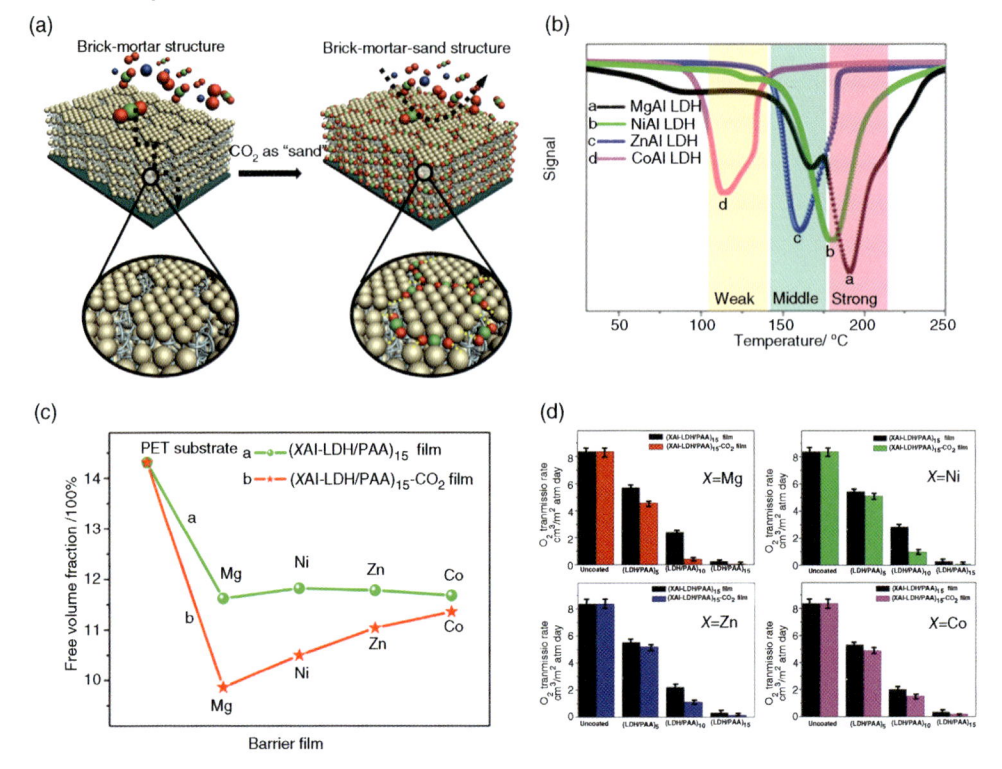

Figure 14.12 (a) Schematic illustration of (XAl-LDH/PAA)$_n$-CO$_2$ (X = Mg, Ni, Zn, Co) barrier films with a brick–mortar–sand structure. (b) CO$_2$ TPD profiles for XAl-LDH nanoplatelets. (c) Free-volume fraction for the PET substrate, (XAl-LDH/PAA)$_{15}$ and (XAl-LDH/PAA)$_{15}$-CO$_2$ films coated on PET. (d) OTR values for pristine PET (uncoated), (XAl-LDH/PAA)$_{15}$, and (XAl-LDH/PAA)$_{15}$-CO$_2$ films [182]. (Copyright 2015, John Wiley and Sons, Inc.)

permeability of gas molecules owing to the increased diffusion length and strong diffusion resistance, whereas CO$_2$ molecules as "sand" is immobilized in the brick–mortar structure, decreasing the free volume and further improving the gas barrier performance.

14.2.4.3 Gas Barrier Films with Self-Healing Properties

Long-term durability is an important issue of concern for practical application of gas barrier films. The flexing of barrier films as well as crack formation during usage allows the permeation of gas molecules, which results in an irreversible sharp decrease of the barrier properties. Designing a hybrid structure constructed from inorganic nanoplatelets and self-healing polymers is a potential way to improve the long-term durability of the gas barrier materials. Han and coworkers [183] reported the fabrication of a self-healing film with high gas barrier performance via layer-by-layer assembly of poly(sodium styrene-4-sulfonate) (PSS) and LDH nanoplatelets followed by incorporation of poly(vinyl acetate) (PVA) (Figure 14.13a). Because of the incorporation of PVA, the (LDH/PSS)$_{20}$–PVA gas barrier film is capable

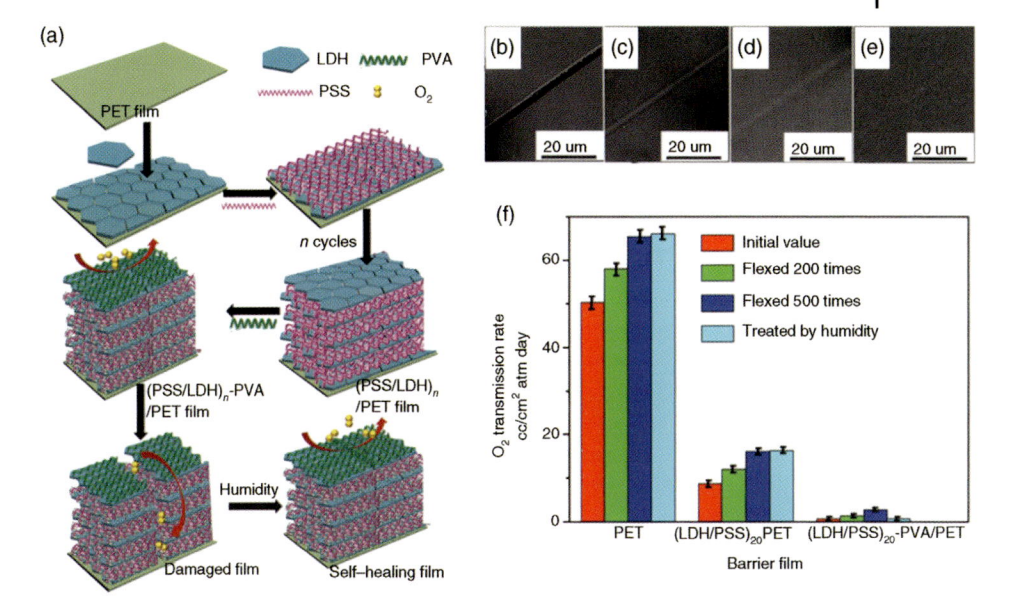

Figure 14.13 (a) Schematic illustration of the fabrication of an (LDH/PSS)$_n$–PVA film with self-healing ability. (b–e) SEM images of damaged (LDH/PSS)$_{20}$–PVA film after exposure to ~85% relative humidity for 0, 25, 50, and 75 h, respectively. (f) The variation of OTR values for pristine PET, (LDH/PSS)$_{20}$, and (LDH/PSS)$_{20}$–PVA film after flexing and humidity treatment [183]. (Copyright 2014, The Royal Society of Chemistry.)

of self-repairing the cracked areas upon exposure to humidity (Figures 14.13b–e), which can be attributed to the formation of a hydrogen-bonding network in the presence of moisture. While neither the polyethylene terephthalate (PET) substrate nor (LDH/PSS)$_{20}$ films show self-healing behavior (Figure 14.13f), it is proposed that the hydrogen bonding network formed in the (LDH/PSS)$_{20}$–PVA film with the assistance of water vapor hinders the gas diffusion and maintains the durability of the barrier film.

14.2.5
Biological and Medical Properties

By virtue of their capability to act as stable chemical storage stable materials with nontoxic metal composition, LDHs have great potential in biorelated applications, such as imaging, therapy, drug delivery, and multifunctional applications [184,185]. Due to the specific structure and morphology characteristics, LDHs can serve as a building block to provide a stable microenvironment for the guest molecules and offer high performance in bioapplications.

14.2.5.1 Synthesis of LDH-Based Biomaterials
Various biomolecules such as nucleosides, ATP, DNA, antisense RNA, vitamers, and anticancer drugs, can be intercalated into LDHs leading to performance-

improved nanohybrids [186,187]. By employing various synthetic routes such as coprecipitation, ion exchange, calcination–reconstruction, and exfoliation–reassembly, intercalated nanohybrids can be fabricated and employed in bio-LDH hybrid systems [188,189]. The size and morphology of these nanohybrids can be precisely controlled by tuning the synthesis conditions, and size distributions in the range 80–200 nm can be achieved allowing the materials to be accumulated in tumor sites.

14.2.5.2 Advantages of LDH-Based Materials in Bioapplications

As versatile carriers in drug/gene delivery and biomedicine, LDH materials possess several advantages as follows. (1) Low cytotoxicity and excellent biocompatibility–LDHs are an ideal drug nanocarrier, since it has been reported that at concentrations up to 1 mg/ml, the cytotoxicity of LDHs is rather low and no cytotoxic effects can be observed [190,191]. (2) Due to the ion-exchangeable properties of LDHs, various materials including drugs, imaging agents, DNA, or genes can be intercalated into the interlayer galleries of LDHs. Moreover, the composition of the host layers can also be changed and metal ions with specific desired functionality can be substituted for all or part of the inactive metal ions. (3) Intercalation into the LDH interlayer galleries is an effective way to improve the dispersion and stability of guests, and with the resulting better dispersion, the performance of drug molecules can be significantly improved (e.g., zinc phthalocyanine used in photodynamic therapy–see below). (4) Enhanced permeability and retention (EPR) effects have also been reported for LDHs and LDH-intercalated materials due to their controllable size distribution (Figure 14.14a) [185].

14.2.5.3 Applications of LDH-Based Materials in Drug Delivery and Treatment

The outstanding properties of LDHs and LDH-intercalated materials give them great potential in imaging, therapy, drug delivery, and multimodal applications. Imaging is a very important bioapplication, which can be used to track cells and diagnose lesions. When combined with optical, magnetic, or radioactive materials, the resulting nanohybrids can exhibit fluorescence, magnetic resonance imaging (MRI) or computed tomography (CT) performance allowing molecular imaging, disease diagnostics, and tumor visualization in a biological system [195]. Moreover, Yoon *et al.* have grafted fluorescein onto exfoliated Gd-doped LDHs, and realized simultaneous fluorescence and MRI imaging for tissues at different depths (Figure 14.14b) [192].

Porphyrin and phthalocyanine molecules (such as [tetrakis(4-carboxyphenyl)-porphyrinato] zinc(II) (ZnTPPC), Pd(II)-5,10,15,20-tetrakis (4-carboxyphenyl) porphyrin (PdTPPC), and zinc phthalocyanine (ZnPc)) intercalated LDH-based materials have also been studied for their biological and medical applications [196–198]. These materials exhibit largely improved singlet oxygen production efficiency and excellent anticancer behavior, and can therefore be used for photodynamic therapy. For example, a supramolecular photosensitizer was fabricated by intercalation of a zinc phthalocyanine (ZnPc) into LDH galleries,

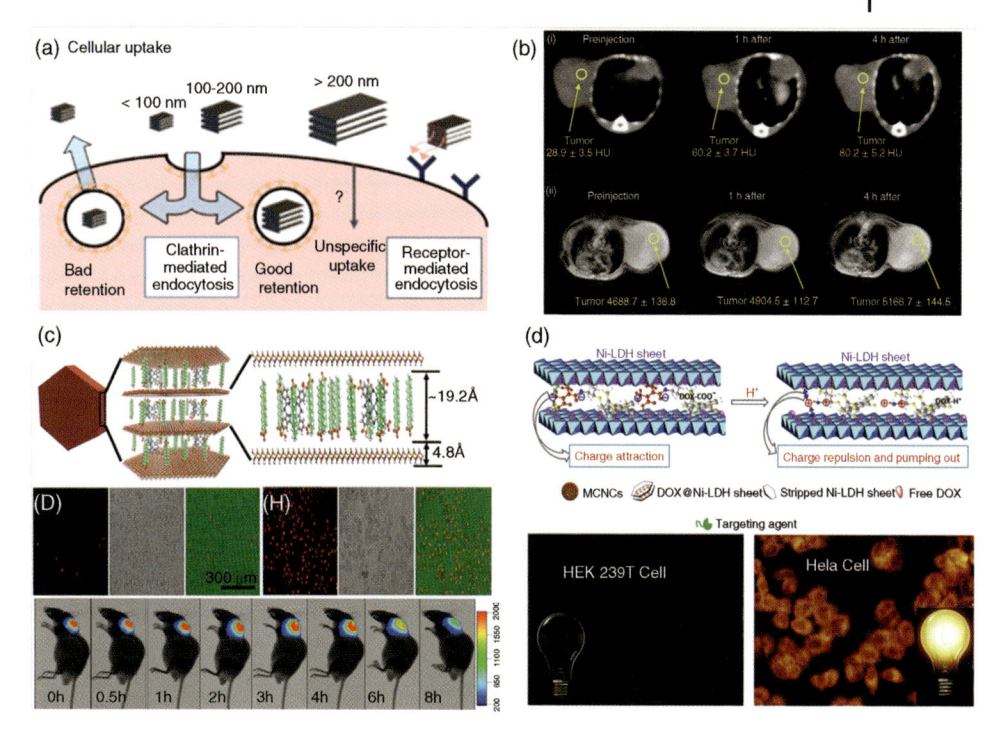

Figure 14.14 Biorelated applications of LDH-based materials in (a) cellular uptake mechanism [185], (b) CT and MRI imaging [192], (c) photodynamic therapy [193], and (d) drug delivery [194]. (Copyright 2015, Springer Science + Business Media; copyright 2009 and 2014, John Wiley and Sons; copyright 2013, Elsevier Inc.)

and the product showed excellent photodynamic therapy (PDT) anticancer behavior due to the high dispersion of ZnPc in the interlayer galleries of LDHs (Figure 14.14c) [193].

Drug-delivery performance of LDH-based materials has also been investigated, and through external stimuli, the loaded drug can be released in tumor cells [189,199]. As seen in Figure 14.14d, Wang and coworkers have synthesized DOX@MSPs/Ni-LDH-folate nanohybrids, and the loaded doxorubicin (DOX) could only be released in tumor cells, which enabled the diagnosis and imaging of cancer cells [194]. In addition, separation and purification of recombinant proteins can be realized. Shao and coworkers have designed a core–shell structure of Fe_3O_4@SiO_2@NiAl-LDH, where the NiAl-LDH shell provides abundant docking sites for the His-tagged protein via the binding between Ni^{2+} and the His-tag, and thus allowed the efficient magnetic separation of His-tagged proteins [200].

Many such bioapplications of LDHs and LDH-based materials have been developed recently, and great efforts have been paid to this fast-growing field. Further investigation of multifunctional materials with significantly enhanced performance remains an important challenge.

References

1 Sideris, P., Nielsen, U., Gan, Z., and Grey, C. (2008) *Science*, **321**, 113.
2 Evans, D.G. and Slade, R. (2006) *Struct. Bond.*, **119**, 1.
3 Khan, A., Ragavan, A., Fong, B., Markland, C., O'Brien, M., Dunbar, T., Williams, G., and O'Hare, D. (2009) *Ind. Eng. Chem. Res.*, **48**, 10196.
4 Khan, A. and O'Hare, D. (2002) *J. Mater. Chem.*, **12**, 3191.
5 He, J., Wei, M., Li, B., Kang, Y., Evans, D.G., and Duan, X. (2006) *Struct. Bond.*, **119**, 89.
6 Williams, G., Khan, A., and O'Hare, D. (2006) *Struct. Bond.*, **119**, 161.
7 Ma, R., Liu, Z., Li, L., Lyi, N., and Sasaki, T. (2006) *J. Mater. Chem.*, **16**, 3809.
8 Duan, X., Zhang, F.Z., Pu, M., Yan, H., Lu, J., Jin, L. et al. (2009) *Wuji Chaofenzi Cailiao de Chaceng Zuzhuang Huaxue*, Science Press, Beijing.
9 Frondel, C. (1941) *Am. Mineral.*, **26**, 295.
10 Brindley, G.W. (1977) *Am. Mineral.*, **62**, 458.
11 Brindley, G.W. and Kikkawa, S. (1979) *Am. Mineral.*, **64**, 836.
12 Brown, G. and Gastuche, M.C. (1967) *Clay Miner.*, **7**, 193.
13 Ingram, L. and Taylor, H.F.W. (1967) *Mineral. Mag.*, **36**, 465.
14 Taylor, H.F.W. (1969) *Mineral. Mag.*, **37**, 338.
15 Taylor, H.F.W. (1973) *Mineral. Mag.*, **39**, 377.
16 Catti, M., Ferraris, G., Hull, S., and Pavese, A. (1995) *Phys. Chem. Miner.*, **22**, 200.
17 Greaves, C. and Thomas, M. (1986) *Acta Cryst. B*, **42**, 51.
18 Parise, J., Leinenweber, K., Weidner, D., Tan, K., and von Dreele, R. (1994) *Am. Mineral.*, **79**, 193.
19 Nagai, T., Hattori, T., and Yamanaka, T. (2000) *Am. Mineral.*, **85**, 760.
20 Shinoda, K. and Aikawa, N. (1998) *Phys. Chem. Miner.*, **25**, 197.
21 Rives, V. and Ulibarri, M.A. (1999) *Coord. Chem. Rev.*, **181**, 61.
22 Williams, G.R. and O'Hare, D. (2006) *J. Mater. Chem.*, **16**, 3065.
23 Khan, A.I. and O'Hare, D. (2002) *J. Mater. Chem.*, **12**, 3191.
24 Renaudin, G. and Francois, M. (1999) *Acta Cryst. C*, **55**, 835.
25 Guo, Y., Zhang, H., Zhao, L., Li, G., Chen, J., and Xu, L. (2005) *J. Solid State Chem.*, **178**, 830.
26 Tichit, D., Lorret, O., Coq, B., Prinetto, F., and Ghiotti, G. (2005) *Microporous Mesoporous Mater.*, **80**, 213.
27 Zhou, H., Zhuo, G., and Jiang, X. (2006) *J. Mol. Catal. A Chem.*, **248**, 26.
28 Yan, H., Lu, J., Wei, M., Ma, J., Li, H., He, J., Evans, D.G., and Duan, X. (2008) *J. Mol. Struct. Theochem.*, **866**, 34.
29 Yan, H., Wei, M., Ma, J., Li, F., Evans, D.G., and Duan, X. (2009) *J. Phys. Chem. A*, **113**, 6133.
30 Yan, H., Wei, M., Ma, J., Evans, D.G., and Duan, X. (2010) *J. Phys. Chem. A*, **114**, 7369.
31 de Roy, A., Forano, C., and Besse, J. (2001) *Catal. Today*, **11**, 123.
32 Mascolo, G. and Marino, O. (1980) *Miner. Mag.*, **43**, 619.
33 Mellot-Draznieks, C., Newsam, J., Gorman, A., Freeman, C., and Férey, G. (2000) *Angew. Chem., Int. Ed.*, **39**, 2270.
34 Gutmann, N. and Müller, B. (1996) *J. Solid State Chem.*, **122**, 214.
35 Fogg, A.M., Williams, G.R., Chester, R., and O'Hare, D. (2004) *J. Mater. Chem.*, **14**, 2369.
36 Drissi, S., Refait, P., Abdelmoula, M., and Génin, J. (1995) *Corros. Sci.*, **37**, 2025.
37 Hofmeister, W. and Platen, H. (1992) *Cryst. Rev.*, **3**, 3.
38 Prasad, B., Kamath, P., and Vijayamohanan, K. (2011) *Langmuir*, **27**, 13539.
39 Chen, Y., Yan, D., and Song, Y. (2014) *Dalton Trans.*, **43**, 14570.
40 Costa, A., Gomes, A., Pillinger, M., Gonçalves, I., and de melo, J. (2015) *Langmuir*, **31**, 4769.
41 Yan, D., Lu, J., Ma, J., Wei, M., Evans, D.G., and Duan, X. (2010) *Phys. Chem. Chem. Phys.*, **12**, 15085.
42 Gunjakar, J., Kim, T., Kim, H., Kim, I., and Hwang, S. (2011) *J. Am. Chem. Soc.*, **133**, 14998.

43 Ma, S., Huang, L., Ma, L., Shim, Y., Islam, S., Wang, P., Zhao, L., Wang, S., Sun, G., Yang, X., and Kanatzidis, M. (2015) *J. Am. Chem. Soc.*, **137**, 3670.

44 Vucelic, M., Moggridge, G., and Jones, W. (1995) *J. Phys. Chem.*, **99**, 8328.

45 Roelofs, J., van Bokhoven, J., van Dillen, A., Geus, J., and de Jong, K. (2002) *Chem. Eur. J.*, **8**, 5571.

46 Arakcheeeva, A., Pushcharovskii, D., Rastsvetaeva, P., Atencio, D., and Lubman, G. (1996) *Crystallogr. Rep.*, **41**, 972.

47 Newman, S., Williams, S., Coveney, P., and Jones, W. (1998) *J. Phys. Chem. B*, **102**, 6710.

48 Thyveetil, M., Coveneny, P., Greenwell, H., and Suter, J. (2008) *J. Am. Chem. Soc.*, **130**, 12485.

49 Zhang, S., Yan, H., Wei, M., Evans, D.G., and Duan, X. (2012) *J. Phys. Chem. C*, **116**, 3421.

50 Xu, S., Zhang, S., Shi, W., Ning, F., Fu, Y., and Yan, H. (2014) *RSC Adv.*, **4**, 47472.

51 Cadars, S., Layrac, G., Gerardin, C., Deschamps, M., Yates, J., Tichit, D., and Massiot, D. (2011) *Chem. Mater.*, **23**, 2821.

52 Murthy, V., Smith, H., Zhang, H., and Smith, S. (2011) *J. Phys. Chem. A*, **115**, 13673.

53 Zhang, H., Xu, Z., Lu, G., and Smith, S. (2010) *J. Phys. Chem. C*, **114**, 12618.

54 Kafunkova, E., Taviot-Gueho, C., Bezdicka, P., Klementova, M., Kovar, P., Kubat, P., Mosinger, J., Pospisil, M., and Lang, K. (2010) *Chem. Mater.*, **22**, 2481.

55 Swadling, J., Suter, J., Greenwell, H., and Coveney, P. (2013) *Langmuir*, **29**, 1573.

56 Pisson, J., Morel-Desrosiers, N., Morel, J., de Roy, A., Leroux, F., Taviot-Gueho, C., and Malfreyt, P. (2011) *Chem. Mater.*, **23**, 1482.

57 Naik, V. and Vasudevan, S. (2011) *J. Phys. Chem. C*, **115**, 8221.

58 Dutta, D., Tummanapelli, A., and Vasudevan, S. (2013) *J. Phys. Chem. C*, **117**, 3930.

59 Costa, D., Rocha, A., Souza, W., Chiaro, S., and Leitao, A. (2012) *J. Phys. Chem. C*, **116**, 13679.

60 Costa, D., Rocha, A., Diniz, R., Souza, W., Chiaro, S., and Leitao, A. (2010) *J. Phys. Chem. C*, **114**, 14133.

61 Xu, S., Pan, T., Dou, Y., Yan, H., Zhang, S., Ning, F., Shi, W., and Wei, M. (2015) *J. Phys. Chem. C*, **119**, 18823.

62 Cunha, V., Petersen, P., Goncalves, M., Petrilli, H., Taviot-Gueho, C., Leroux, F., Temperini, M., and Constantino, V. (2012) *Chem. Mater.*, **24**, 1415.

63 Costa, D., Rocha, A., Souza, W., Chiaro, S., and Leitao, A. (2011) *J. Phys. Chem. B*, **115**, 3531.

64 Zhao, Y., Li, B., Wang, Q., Gao, W., Wang, C., Wei, M., Evans, D.G., Duan, X., and O'Hare, D. (2014) *Chem. Sci.*, **5**, 951.

65 Zhao, Y., Chen, G., Bian, T., Zhou, C., Waterhouse, G., Wu, L., Tung, C., Smith, L., O'Hare, D., and Zhang, T. (2015) *Adv. Mater.*, **27**, 7824.

66 Shi, W., Wei, M., Lu, J., Evans, D.G., and Duan, X. (2009) *J. Phys. Chem. C*, **113**, 12888.

67 Yan, D., Lu, J., Ma, J., Wei, M., Li, S., Evans, D.G., and Duan, X. (2011) *J. Phys. Chem. C*, **115**, 7939.

68 Li, S., Lu, J., Xu, J., Dang, S., Evans, D.G., and Duan, X. (2010) *J. Mater. Chem.*, **20**, 9718.

69 Guan, W., Lu, J., Zhou, W., and Lu, C. (2014) *Chem. Commun.*, **50**, 11895.

70 Li, Z., Lu, J., Qin, Y., Li, S., and Qin, S. (2013) *J. Phys. Chem. C*, **1**, 5944.

71 Li, Z., Lu, J., Li, S., Qin, S., and Qin, Y. (2012) *Adv. Mater.*, **24**, 6053.

72 Liang, R., Xu, S., Yan, D., Shi, W., Tian, R., Yan, H., Wei, M., Evans, D.G., and Duan, X. (2012) *Adv. Funct. Mater.*, **22**, 4940.

73 Qin, Y., Lu, J., Li, S., Li, Z., and Zheng, S. (2014) *J. Phys. Chem. C*, **118**, 20538.

74 Yan, D., Lu, J., Wei, M., Han, J., Ma, J., Li, F., Evans, D.G., and Duan, X. (2009) *Angew. Chem., Int. Ed.*, **48**, 3073.

75 Yan, D., Lu, J., Ma, J., Wei, M., Wang, X., Evans, D.G., and Duan, X. (2010) *Langmuir*, **26**, 7007.

76 Yan, D., Qin, S., Chen, L., Lu, J., Ma, J., Wei, M., Evans, D.G., and Duan, X. (2010) *Chem. Commun.*, **46**, 8654.

77 Tian, R., Zhang, S., Li, M., Zhou, Y., Lu, B., Yan, D., Wei, M., Evans, D.G., and Duan, X. (2015) *Adv. Funct. Mater.*, **25**, 5006.

78 Sun, J., Liu, H., Chen, X., Evans, D.G., Yang, W., and Duan, X. (2013) *Adv. Mater.*, **25**, 1125.

79 Sun, J., Liu, H., Chen, X., Evans, D.G., Yang, W., and Duan, X. (2012) *Chem. Commun.*, **48**, 8126.
80 Song, L., Shi, J., Lu, J., and Lu, C. (2015) *Chem. Sci.*, **6**, 4846.
81 Kantam, M., Choudary, B., Reddy, C., Rao, K., and Figueras, F. (1998) *Chem. Commun.*, **9**, 1033.
82 Kumbhar, P., Sanchez-Valente, J., and Figueras, F. (1998) *Chem. Commun.*, **34**, 1091.
83 Tichit, D., Lutic, D., Coq, B., Durand, R., and Teissier, R. (2003) *J. Catal.*, **219**, 167.
84 Xu, Z., Zhang, J., Adebajo, M., Zhang, H., and Zhou, C. (2011) *Appl. Clay Sci.*, **53**, 139.
85 Roelofs, J., Lensveld, D., van Dillen, A., and de Jong, K. (2001) *J. Catal.*, **203**, 184.
86 Choudary, B., Kantam, M., Neeraja, V., Rao, K., Figueras, F., and Delmotte, L. (2001) *Green Chem.*, **3**, 257.
87 Veloso, C., Pérez, C., de Souza, B., Lima, E., Dias, A., Monteiro, J., and Henriques, C. (2008) *Microporous Mesoporous Mater.*, **107**, 23.
88 Choudary, B., Kantam, M., Reddy, C., Rao, K., and Figueras, F. (1999) *J. Mol. Catal. A Chem.*, **146**, 279.
89 Abelló, S., Dhir, S., Colet, G., and Pérez-Ramírez, J. (2007) *Appl. Catal. A*, **325**, 121.
90 Abeló, S., Medina, F., Tichit, D., Pérez-Ramírez, J., Sueiras, J., Salagre, P., and Cesteros, Y. (2007) *Appl. Catal. B*, **70**, 577.
91 Valente, J., Pfeiffer, H., Lima, E., Prince, J., and Flores, J. (2011) *J. Catal.*, **279**, 196.
92 Kim, S., Lee, Y., Lee, D., Lee, J., and Kang, J. (2014) *J. Mater. Chem. A*, **2**, 4136.
93 Baliarsingh, N., Mohapatra, L., and Parida, K. (2013) *J. Mater. Chem. A*, **1**, 4236.
94 Parida, K., Satpathy, M., and Mohapatra, L. (2012) *J. Mater. Chem.*, **22**, 7350.
95 Tian, L., Zhao, Y., He, S., and Duan, X. (2012) *Chem. Eng. J.*, **184**, 261.
96 Zhao, Y., Chen, P., Zhang, B., Su, D., Zhang, S., Tian, L., Lu, J., Li, Z., Cao, X., Wang, B., Wei, M., Evans, D.G., and Duan, X. (2012) *Chem. Eur. J.*, **18**, 11949.
97 Zhao, M., Zhang, Q., Zhang, W., Huang, J., Zhang, Y., Su, D., and Wei, F. (2010) *J. Am. Chem. Soc.*, **132**, 14739.
98 Tian, G., Zhao, M., Zhang, B., Zhang, Q., Zhang, W., Huang, J., Chen, T., Qian, W., Su, D., and Wei, F. (2014) *J. Mater. Chem. A*, **2**, 1686.
99 Koike, M., Li, D., Nakagawa, Y., and Tomishige, K. (2012) *ChemSusChem*, **5**, 2312.
100 Gao, W., Zhao, Y., Chen, H., Chen, H., Li, Y., He, S., Zhang, Y., Wei, M., Evans, D.G., and Duan, X. (2015) *Green Chem.*, **17**, 1525.
101 Liu, J., Xu, S., Bing, W., Wang, F., Li, C., Wei, M., Evans, D.G., and Duan, X. (2015) *ChemCatChem*, **7**, 846.
102 Xiao, C., Wang, L., Maligal-Ganesh, R., Smetana, V., Walen, H., Thiel, P., Miller, G., Johnson, D., and Huang, W. (2013) *J. Am. Chem. Soc.*, **135**, 9592.
103 Ota, A., Kunkes, E., Kasatkin, I., Groppo, E., Ferri, D., Poceiro, B., Yerga, R., and Behrens, M. (2012) *J. Catal.*, **293**, 27.
104 Siddiqi, G., Sun, P., Galvita, V., and Bell, A. (2010) *J. Catal.*, **274**, 200.
105 Sun, P., Siddiqi, G., Vining, W., Chi, M., and Bell, A. (2011) *J. Catal.*, **282**, 165.
106 Li, C., Chen, Y., Zhang, S., Xu, S., Zhou, J., Wang, F., Wei, M., Evans, D.G., and Duan, X. (2013) *Chem. Mater.*, **25**, 3888.
107 Yuan, Q., Wei, M., Evans, D.G., and Duan, X. (2004) *J. Phys. Chem. B*, **108**, 12381.
108 Wei, M., Yuan, Q., Evans, D.G., Wang, Z., and Duan, X. (2005) *J. Mater. Chem.*, **15**, 1197.
109 Wei, M., Xu, X., He, J., Yuan, Q., Rao, G., Evans, D.G., Pu, M., and Yang, L. (2006) *J. Phys. Chem. Solids*, **241**, 319.
110 An, Z., Zhang, W., Shi, H., and He, J. (2006) *J. Catal.*, **241**, 319.
111 Wei, M., Zhang, X., Evans, D.G., Duan, X., Li, X., and Chen, H. (2007) *AIChE J.*, **53**, 2916.
112 Zhang, X., Wei, M., Pu, M., Li, X., Chen, H., Evans, D.G., and Duan, X. (2005) *J. Solid State Chem.*, **178**, 2701.
113 Zhao, S., Xu, J., Wei, M., and Song, Y. (2011) *Green Chem.*, **13**, 384.
114 Winter, M. and Brodd, R. (2008) *Chem. Rev.*, **104**, 4245.
115 Burke, A. (2000) *J. Power. Sources*, **91**, 37.
116 Simon, P. and Gogotsi, Y. (2008) *Nat. Mater.*, 7, 845.
117 Wang, Q. and O'Hare, D. (2012) *Chem. Rev.*, **112**, 4124.

118 Fan, G., Li, F., Evans, D.G., and Duan, X. (2014) *Chem. Soc. Rev.*, **43**, 7040.

119 Shao, M., Ning, F., Zhao, Y., Zhao, J., Wei, M., Evans, D.G., and Duan, X. (2012) *Chem. Mater.*, **24**, 1192.

120 Xu, J., He, F., Gai, S., Zhang, S., Li, L., and Yang, P. (2014) *Nanoscale*, **6**, 10887.

121 Ning, F., Shao, M., Zhang, C., Wei, M., and Duan, X. (2014) *Nano Energy*, **7**, 134.

122 Chen, H., Hu, L., Chen, M., Yan, Y., and Wu, L. (2014) *Adv. Funct. Mater.*, **24**, 934.

123 Zhao, J., Chen, J., Xu, S., Shao, M., Yan, D., Wei, M., Evans, D.G., and Duan, X. (2013) *J. Mater. Chem. A*, **1**, 8836.

124 Zhao, J., Chen, J., Xu, S., Shao, M., Zhang, Q., Wei, F., Ma, J., Wei, M., Evans, D.G., and Duan, X. (2014) *Adv. Funct. Mater.*, **24**, 2938.

125 Li, X., Shen, J., Sun, W., Hong, X., Wang, R., Zhao, X., and Yan, X. (2015) *J. Mater. Chem. A*, **3**, 13244.

126 Gao, Z., Wang, J., Li, Z., Yang, W., Wang, B., Hou, M., He, Y., Liu, Q., Mann, T., Yang, P., Zhang, M., and Liu, L. (2011) *Chem. Mater.*, **23**, 3509.

127 Wu, X., Jiang, L., Long, C., Wei, T., and Fan, Z. (2015) *Adv. Funct. Mater.*, **25**, 1648.

128 Ma, R., Liu, X., Liang, J., Bando, Y., and Sasaki, T. (2014) *Adv. Mater.*, **26**, 4173.

129 Shao, M., Li, Z., Zhang, R., Ning, F., Wei, M., Evans, D.G., and Duan, X. (2015) *Small*, **11**, 3529.

130 Han, J., Dou, Y., Zhao, J., Wei, M., Evans, D.G., and Duan, X. (2013) *Small*, **9**, 98.

131 Zhao, J., Xu, S., Tschulik, K., Compton, R., Wei, M., O'Hare, D., Evans, D.G., and Duan, X. (2015) *Adv. Funct. Mater.*, **25**, 2745.

132 Whittingham, M. (2004) *Chem. Rev.*, **104**, 4271.

133 Kang, B. and Ceder, G. (2009) *Nature*, **458**, 190.

134 Li, X., Yang, W., Li, F., Evans, D.G., and Duan, X. (2006) *J. Phys. Chem. Solids*, **67**, 1286.

135 Cheng, X., Tian, G., Liu, X., Nie, J., Zhao, M., Huang, J., Zhu, W., Hu, L., Zhang, Q., and Wei, F. (2013) *Carbon*, **62**, 393.

136 Shi, J., Peng, H., Zhu, L., Zhu, W., and Zhang, Q. (2015) *Carbon*, **92**, 96.

137 Zhao, M., Zhang, Q., Huang, J., Tian, G., Nie, J., Peng, H., and Wei, F. (2014) *Nat. Commun.*, **5**, 3410.

138 Qiao, C., Zhang, Y., Zhu, Y., Cao, C., Bao, X., and Xu, J. (2015) *J. Mater. Chem. A*, **3**, 6878.

139 Gong, M., Li, Y., Wang, H., Liang, Y., Wu, J., Zhou, J., Wang, J., Regier, T., Wei, F., and Dai, H. (2013) *J. Am. Chem. Soc.*, **135**, 8452.

140 Song, F. and Hu, X. (2014) *Nat. Commun.*, **5**, 4477.

141 Zhou, L., Huang, X., Chen, H., Jin, P., Li, G., and Zou, X. (2015) *Dalton Trans.*, **44**, 11592.

142 Trotochaud, L., Young, S., Ranney, J., and Boettcher, S. (2014) *J. Am. Chem. Soc.*, **136**, 6744.

143 Zou, X., Goswami, A., and Asefa, T. (2013) *J. Am. Chem. Soc.*, **135**, 17242.

144 Li, Y., Zhang, L., Xiang, X., Yan, D., and Li, F. (2014) *J. Mater. Chem. A*, **2**, 13250.

145 Tang, D., Han, Y., Ji, W., Qiao, S., Zhou, X., Liu, R., Han, X., Huang, H., Liu, Y., and Kang, Z. (2014) *Dalton Trans.*, **43**, 15119.

146 Jiang, J., Zhang, A., Li, L., and Ai, L. (2015) *J. Power Sources*, **278**, 445.

147 Liang, H., Meng, F., Acevedo, M., Li, L., Forticaux, A., Xiu, L., Wang, Z., and Jin, S. (2015) *Nano Lett.*, **15**, 1421.

148 Song, F. and Hu, X. (2014) *J. Am. Chem. Soc.*, **136**, 16481.

149 Han, N., Zhao, F., and Li, Y. (2015) *J. Mater. Chem. A*, **3**, 16348.

150 Long, X., Li, J., Xiao, S., Yan, K., Wang, Z., Chen, H., and Yang, S. (2014) *Angew. Chem., Int. Ed.*, **53**, 7584.

151 Wang, H., Tang, C., and Zhang, Q. (2015) *J. Mater. Chem. A*, **3**, 16183.

152 Tang, C., Wang, H., Wang, H., Zhang, Q., Tian, G., Nie, J., and Wei, F. (2015) *Adv. Mater.*, **27**, 4516.

153 Youn, D., Park, Y., Kim, J., and Magesh, G. (2015) *J. Power Sources*, **294**, 437.

154 Ma, W., Ma, R., Wang, C., Liang, J., Liu, X., Zhou, K., and Sasaki, T. (2015) *ACS Nano*, **9**, 1977.

155 Yu, X., Zhang, M., Yuan, W., and Shi, G. (2015) *J. Mater. Chem. A*, **3**, 6921.

156 Tang, D., Liu, J., Wu, X., Liu, R., Han, X., Han, Y., Huang, H., Liu, Y., and Kang, Z. (2014) *ACS Appl. Mater. Interfaces*, **6**, 7918.

157 Yang, Q., Li, T., Lu, Z., Sun, X., and Liu, J. (2014) *Nanoscale*, **6**, 11789.

158 Shao, M., Ning, F., Wei, M., Evans, D.G., and Duan, X. (2014) *Adv. Funct. Mater.*, **24**, 580.
159 Wang, L., Dionigi, F., Nguyen, N., Kirchgeorg, R., Gliech, M., Grigorescu, S., Strasser, P., and Schmuki, P. (2015) *Chem. Mater.*, **27**, 2360.
160 Cho, S., Jang, J., Park, Y., Kim, J., Magesh, G., Kim, J., Seol, M., Yong, K., Lee, K., and Lee, J. (2014) *Energy Environ. Sci.*, **7**, 2301.
161 Zhao, J., Kong, X., Shi, W., Shao, M., Han, J., Wei, M., Evans, D.G., and Duan, X. (2011) *J. Mater. Chem.*, **21**, 13926.
162 Shao, M., Wei, M., Evans, D.G., and Duan, X. (2011) *Chem. Commun.*, **47**, 3171.
163 Shao, M., Ning, F., Zhao, J., Wei, M., Evans, D.G., and Duan, X. (2013) *Adv. Funct. Mater.*, **23**, 3513.
164 Zhao, J., Shao, M., Yan, D., Zhang, S., Lu, Z., Li, Z., Cao, X., Wang, B., Wei, M., Evans, D.G., and Duan, X. (2013) *J. Mater. Chem. A*, **1**, 5840.
165 Shao, M., Han, J., Shi, W., Wei, M., and Duan, X. (2010) *Electrochem. Commun.*, **12**, 1077.
166 Han, J., Xu, X., Rao, X., Wei, M., Evans, D.G., and Duan, X. (2011) *J. Mater. Chem.*, **21**, 2126.
167 Yao, H., Fang, H., Tan, Z., Wu, L., and Yu, S. (2010) *Angew. Chem., Int. Ed.*, **122**, 2186.
168 Bai, H., Huang, C., Xiu, H., Zhang, Q., Deng, H., Wang, K., Chen, F., and Fu, Q. (2014) *Biomacromolecules*, **15**, 1507.
169 Yang, Y., Bolling, L., Haile, M., and Grunlan, J. (2012) *RSC Adv.*, **2**, 12355.
170 Osman, M., Mittal, V., Morbidelli, M., and Suter, U. (2004) *Macromolecules*, **37**, 7250.
171 Layek, R., Das, A., Park, M., Kim, N., and Lee, J. (2014) *J. Mater. Chem. A*, **2**, 12158.
172 Triantafyllidis, K., LeBaron, P., Park, I., and Pinnavaia, T. (2006) *Chem. Mater.*, **18**, 4393.
173 Svagan, A., Åkesson, A., Cárdenas, M., Bulut, S., Knudsen, J., Risbo, J., and Plackett, D. (2012) *Biomacromolecules*, **13**, 397.
174 Priolo, M., Holder, K., Gamboa, D., and Grunlan, J. (2011) *Langmuir*, **27**, 12106.
175 Yang, Y., Bolling, L., Priolo, M., and Grunlan, J. (2013) *Adv. Mater.*, **25**, 503.
176 Dou, Y., Xu, S., Liu, X., Han, J., Yan, H., Wei, M., Evans, D.G., and Duan, X. (2014) *Adv. Funct. Mater.*, **24**, 541.
177 Compton, O., Kim, S., Pierre, C., Torkelson, J., and Nguyen, S. (2010) *Adv. Mater.*, **22**, 4759.
178 Möller, M., Lunkenbein, T., Kalo, H., Schieder, M., Kunz, D., and Breu, J. (2010) *Adv. Mater.*, **22**, 5245.
179 Laachachi, A., Ball, V., Apaydin, K., Toniazzo, V., and Ruch, D. (2011) *Langmuir*, **27**, 13879.
180 Kuraoka, K. and Hashimoto, A. (2008) *J. Ceram. Soc. Jpn.*, **116**, 832.
181 Russo, G., Simon, G., and Incarnato, L. (2006) *Macromolecules*, **39**, 3855.
182 Dou, Y., Pan, T., Xu, S., Yan, H., Han, J., Wei, M., Evans, D.G., and Duan, X. (2015) *Angew. Chem., Int. Ed.*, **54**, 9673.
183 Dou, Y., Pan, T., Zhou, A., Xu, S., Liu, X., Han, J., Wei, M., Evans, D.G., and Duan, X. (2014) *Chem. Commun.*, **50**, 8462.
184 Park, D., Hwang, S., Oh, J., Yang, J., and Choy, J. (2013) *Prog. Polym. Sci.*, **38**, 1442.
185 Park, D., Choi, G., and Choy, J. (2015) *Struct. Bond.*, **166**, 137.
186 Oh, J., Park, D., Choi, S., and Choy, J. (2012) *Recent Pat. Nanotechnol.*, **6**, 200.
187 Choi, S. and Choy, J. (2011) *Nanomedicine*, **6**, 803.
188 Hitzky, E., Aranda, P., Darder, M., and Rytwo, G. (2010) *J. Mater. Chem.*, **20**, 9306.
189 Khan, A., Lei, L., Norquist, A., and O'Hare, D. (2001) *Chem. Commun.*, **37**, 2342.
190 Kriven, W., Kwak, S., Wallig, M., and Choy, J. (2004) *MRS Bull.*, **29**, 33.
191 Xu, Z., Walker, T., Liu, K., Cooper, H., Lu, G., and Bartlett, P. (2007) *Nanomedicine*, **2**, 163.
192 Yoon, Y., Lee, K., Im, G., Lee, B., Byeon, S., Lee, J., and Lee, I. (2009) *Adv. Funct. Mater.*, **19**, 3375.
193 Liang, R., Tian, R., Ma, L., Zhang, L., Hu, Y., Wang, J., Wei, M., Yan, D., Evans, D.G., and Duan, X. (2014) *Adv. Funct. Mater.*, **24**, 3144.

194 Li, D., Zhang, Y., Yu, M., Guo, J., Chaudhary, D., and Wang, C. (2013) *Biomaterials*, **32**, 7913.
195 Wang, X., Li, J., Zhu, Q., Li, X., Sun, X., and Sakka, Y. (2014) *J. Alloys Compd.*, **603**, 28.
196 Daniel, S. and Prabir, K. (1996) *Langmuir*, **12**, 402.
197 Eva, K., Christine, T., Petr, B., Mariana, K., Petr, K., Pavel, K., Jiří, M., Miroslav, P., and Kamil, L. (2010) *Chem. Mater.*, **22**, 2481.
198 Marie, J., Jan, D., Pavel, K., Jiří, H., František, K., and Kamil, L. (2011) *J. Phys. Chem. C*, **115**, 21700.
199 Oh, J., Choi, S., Lee, G., Han, S., and Choy, J. (2009) *Adv. Funct. Mater.*, **19**, 1617.
200 Shao, M., Ning, F., Zhao, J., Wei, M., Evans, D.G., and Duan, X. (2012) *J. Am. Chem. Soc.*, **134**, 1071.

15
Structural Diversity in Complex Layered Oxides

S. Uma

University of Delhi, Materials Chemistry Group, Department of Chemistry, Vishvavidalaya Marg, 110007 Delhi, India

15.1 Introduction

Tremendous advancement has been achieved in the area of materials research over the past few decades to understand the fundamental concepts as well as to obtain functional materials with various technological applications. The investigation of structure and properties of materials including electronic, ionic, magnetic, ferroelectric, multiferroic, optical, and nonlinear optical has resulted in successful catalysts, photocatalysts, battery cathodes, electrolytes, phosphors, and thermoelectric materials to mention a few. Among the various materials, oxides enjoy a privileged status based on their stability, and the ease with which they are available for practical applications. Consequently, oxides play a crucial role and examples are known for every single applications mentioned above. Therefore, they have become the natural choice for inorganic solid-state chemists and physicists to study the different structure–property correlations [1–3]. The fascination to work with oxides increased manifolds after the discovery of superconductivity in a variety of single-phase copper containing oxides possessing unique critical temperatures in spite of the structural and compositional complexity.

It is imperative to note that majority of the oxide-based functional materials crystallize in layered structures. The reduction of the structural three-dimensionality of the solids to a two-dimensional arrangement plays a crucial role in imparting specific properties and applications. The significance associated with layered oxides is not only reflected in the resultant anisotropy of their physical properties but also provides a paramount advantage in terms of isolating new oxide materials through chimie douce or soft chemistry approach. Thus, layered oxides provide the basic templates for targeted rational designing of materials, which may be comparable to the sequential multistep synthesis of molecular designing. Several noteworthy reviews on layered oxides published over the years

Handbook of Solid State Chemistry, First Edition. Edited by Richard Dronskowski, Shinichi Kikkawa, and Andreas Stein.
© 2017 Wiley-VCH Verlag GmbH & Co. KGaA. Published 2017 by Wiley-VCH Verlag GmbH & Co. KGaA.

highlight the technological importance of synthesizing metastable materials by carrying out reactions at low temperature (<500 °C) as compared to the traditional high temperature (>700 °C) solid-state reactions resulting only in thermodynamically stable products [4–6]. Some of the low temperature reactions are ion exchange, intercalation, deintercalation, exfoliation, and so on.

This chapter will give a review of complex layered oxides, comprising of their structures, inherent properties, and applications. The concepts related to layering will be introduced in the first section along with their classifications. This will be followed by a description of unique topochemical reactions that are characteristic of layered arrangement. The final section of the chapter will describe the properties and advantages of various layered oxide materials.

15.2
Types of Layered Oxides

15.2.1
Origin of Interest and Basic Features

The interest in layered oxides originated from the investigation of the ion-exchange chemistry in the various silicates, zeolites, including silicon and other metal ions substituted aluminum phosphates [7]. The field of ion-exchange materials based on clays and zeolites were useful to develop adsorbents, ion exchangers for water purification and catalysts in petroleum cracking. Further progress toward this subject involved the usage of hydrous oxides (e.g., $ZrO_2 \cdot xH_2O$, $(H_3O)_2Sb_2O_6$) and phosphates (e.g., $Zr(HPO_4)_2 \cdot H_2O$) meant typically for the removal of ions in nuclear effluents. The relevance of solid-state chemistry toward identifying new inorganic ion-exchanger materials resulted in fast ionic (e.g., β alumina, $Na_{1+x}Al_{11}O_{17+x/2}$) and proton conductors (e.g., $H_3PMo_{12}O_{40} \cdot 29H_2O$). Furthermore, these attempts led to the discovery of ion-exchangers possessing structures consisting of frameworks (e.g., NASICON) and tunnels (e.g., α, β, γ forms of MnO_2). The critical turning point has been the study of various layered oxides of titanium and niobium capable of ion transport and ion-exchange reactions [7].

The concept of layering among oxides and their classifications can be understood based either on their structures or on their properties or based on both. A straightforward but a broad classification is to divide the layered oxides into two types. Those are the type I layered oxides capable of exhibiting ion-exchange chemistry and the type II, those which do not exhibit any ion-exchange reactions. The former class of oxides belongs to the majority of the layered oxides mostly manifested by the segregation of alkali metals ions (Li^+, Na^+, K^+, etc.) in two-dimensions leading to their relatively weaker bonding with the adjacent structural units. As a result, the ions in the interlayer region are weakly bound and merely serve the purpose of charge compensation to balance the residual charge of the remaining structural units. Consequently, the weakly held charge compensating cations or anions as the case may be can pave way for a variety of interlayer reactions depending on their mobilities. For example, the layered nature of α-NaFeO$_2$ structure arises by the alternate arrangement of octahedrally coordinated Fe^{3+} and Na$^+$ ions (Figure 15.1a). It is feasible to obtain the corresponding lithium compounds by ion exchanging the sodium analogues with excess of LiNO$_3$ at 300 °C for 24 h [8]. The layering observed in the type II oxides without the possibility of any mobile interlayer ions and interlayer reactions is because of the different structural units carrying appropriate positive and negative charges that are held together. Sr$_4$Ti$_3$O$_{10}$ can be cited as an example and does not exhibit ion-exchange behavior despite having a layered structure consisting of (Sr$_2$Ti$_3$O$_{10}$)$^{4-}$ slabs separated by two Sr^{2+} ions (Figure 15.1b) [9]. In few special cases, the type II oxides can also be converted into type I by careful acid leaching of the appropriate structural units. Bi$_4$Ti$_3$O$_{12}$, a layered oxide with [Bi$_2$O$_2$]$^{2+}$ and [Bi$_2$Ti$_3$O$_{10}$]$^{2-}$ units, (Figure 15.1c) by itself does not contain any mobile univalent ions restrained to a specific plane [10]. However, later studies have shown that it is possible to remove the covalent [Bi$_2$O$_2$]$^{2+}$ units by acid leaching [11,12]. The various layered structures possessed by the superconducting oxides belong to the type II and the layered structure is responsible for the strong anisotropic properties as realized from their conductivity and superconductivity arising by careful electron and hole doping. These copper containing high-temperature superconducting oxides form a class of materials by themselves [13,14] and will not be included in this chapter.

The classification of layered oxides based on specific structural aspects would be to categorize them as oxides containing metal-oxygen octahedra coupled with metal-oxygen tetrahedra and those oxides which contain only metal-oxygen octahedra. Most of the oxides containing both metal-oxygen octahedra and tetrahedra (e.g., NbO(PO$_4$)·3H$_2$O) have been reviewed earlier [7] and the present discussion will focus on the subsequent progress made in the family of oxides consisting of metal-oxygen octahedra.

15.2.2
Layered Oxides Containing Metal-Oxygen Octahedra

In order to comprehend the subject of complex layered oxides, it is also beneficial to study them under the various structural families as has been the practice adopted in the literature. The various families of layered oxides and their characteristics are detailed in the following sections.

15.2.2.1 Layered Perovskites
The earliest examples of the layered perovskites are the Aurivillius oxides [10] of the general formula, Bi$_2$O$_2$[A$_{n-1}$B$_n$O$_{3n+1}$], and the Ruddlesden–Popper oxides [9] of the general formula, A$_2$[A$_{n-2}$B$_n$O$_{3n+1}$]. These oxides are named layered perovskites because their structures are derived from the three-dimensional (ABO$_3$) perovskite structure (Figure 15.1d). The perovskite unit cell consists of corner-shared BO$_{6/2}$ octahedra (B = smaller transition metal cation) and the A cation

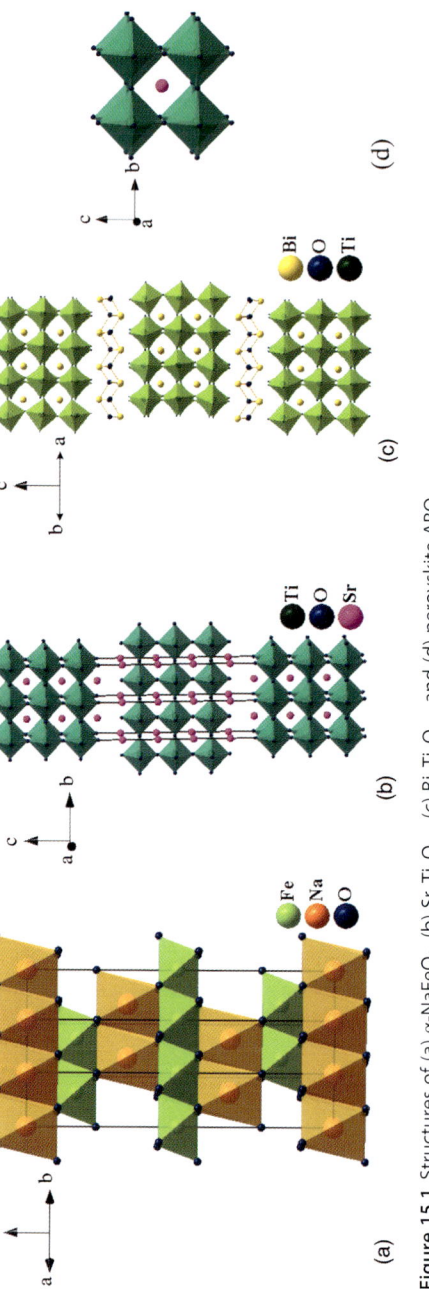

574 *15 Structural Diversity in Complex Layered Oxides*

Figure 15.1 Structures of (a) α-NaFeO$_2$, (b) Sr$_4$Ti$_3$O$_{10}$, (c) Bi$_4$Ti$_3$O$_{12}$, and (d) perovskite ABO$_3$.

(larger s, d, or f-block cation) occupies an ideally 12-coordinate cavity created by the corner-sharing of eight $BO_{6/2}$ octahedra. The layered perovskite formation can be envisaged from the cubic ABO_3 perovskite structure by cutting it along [001] direction with an insertion of an additional oxygen. Thus, the layered perovskites of Aurivillius and the Ruddlesden–Popper series of oxides contain two dimensional slabs of $[A_{n-1}B_nO_{3n+1}]^{2-}$ and $[A_{n-2}B_nO_{3n+1}]^{2-}$ separated respectively by $[Bi_2O_2]^{2+}$ layers and two divalent A cations. The thickness of the slab is determined by the number (n) of BO_6 octahedra in the direction perpendicular to the slab. Bi_2WO_6, $Bi_2SrNb_2O_9$, and $Bi_4Ti_3O_{12}$ are the $n=1$, $n=2$, and $n=3$ members of the Aurivillius phases (Figure 15.2). The different strontium titanates Sr_2TiO_4, $Sr_3Ti_2O_7$, and $Sr_4Ti_3O_{10}$ belong to the $n=1$, $n=2$, and $n=3$ members of the Ruddlesden–Popper series (Figure 15.3). The structure of the $n=1$ member is more popularly referred as the K_2NiF_4 structure, made up of the perovskite $KNiF_3$ along with a rock salt KF layer and the popular members adopting this structure are La_2CuO_4 and La_2NiO_4. As pointed out earlier, none of the strontium-based Ruddlesden–Popper oxides undergo any ion-exchange reactions. However, suitable replacement of the interlayer Sr^{2+} ions can bring the desired ion-exchange properties. $A_2Ln_2Ti_3O_{10}$ (A = K, Rb, Ln = rare earth) that are isostructural to $Sr_4Ti_3O_{10}$ undergoing ion-exchange reactions have been reported by Gopalakrishnan and Bhat [15]. The two monovalent K^+ ions occupy the interlayer region, while the Ln^{3+} ions prefer to be part of the perovskite blocks. The K^+ ions in the interlayer can be readily replaced with ions such as H^+, Ag^+, and so on. The parent and the protonated oxides undergo hydration and the structures of the anhydrous and hydrated oxides differ in the adjacent packing of layers along the c-axis. Similarly, $NaLaTiO_4$ analogous to Sr_2TiO_4 having ion-exchangeable NaO layers are useful to carry out interlayer exchange reactions [16]. Several alkali metal ions containing Ruddlesden–Popper phases in combination with other rare earth ions (e.g., $Na_2Gd_2Ti_3O_{10}$) [15,17] and Nb^{5+}, Ta^{5+} ions (e.g., $K_2CaNaTa_3O_{10}$, $K_2Ca_2Ta_2TiO_{10}$) [18] have been synthesized. Notable are the lithium-based $Li_2A_{0.5n}B_nO_{3n+1}$ (A = Ca, Sr; B = Nb, Ta) oxides reported by Fourquet and coworkers [19] belonging to the Ruddlesden–Popper type having cationic vacancies.

After the discovery of $Bi_4Ti_3O_{12}$ by Aurivillius over 60 years ago [10], followed by the report of ferroelectric properties in related members, the family of Aurivillius type oxides has grown substantially. Many of these oxides find applications as oxide ion conductors, catalysts, and photocatalysts [11,12]. The very recent report of the selective acid leaching of bismuth oxide sheets, $[Bi_2O_2]^{2+}$ from the Aurivillius phases resulting in the respective protonated forms has attracted considerable attention and has opened up intercalation and exfoliation reactions (Section 1.3.2) that were unknown in this family of oxides. A concise review of the crystal chemistry and dielectric properties of the Aurivillius family of layered oxides has been provided by Frit and Mercurio [20]. Because of the compositional flexibility of the perovskite slabs in $Bi_2O_2[A_{n-1}B_nO_{3n+1}]$, a wide range cations such as Na^+, K^+, Ca^{2+}, Sr^{2+}, Ba^{2+}, Pb^{2+}, Bi^{3+}, Ln^{3+} can be substituted at the A-site and smaller cations like Fe^{3+}, Cr^{3+}, Mn^{4+}, Ti^{4+}, Nb^{5+}, Mo^{6+},

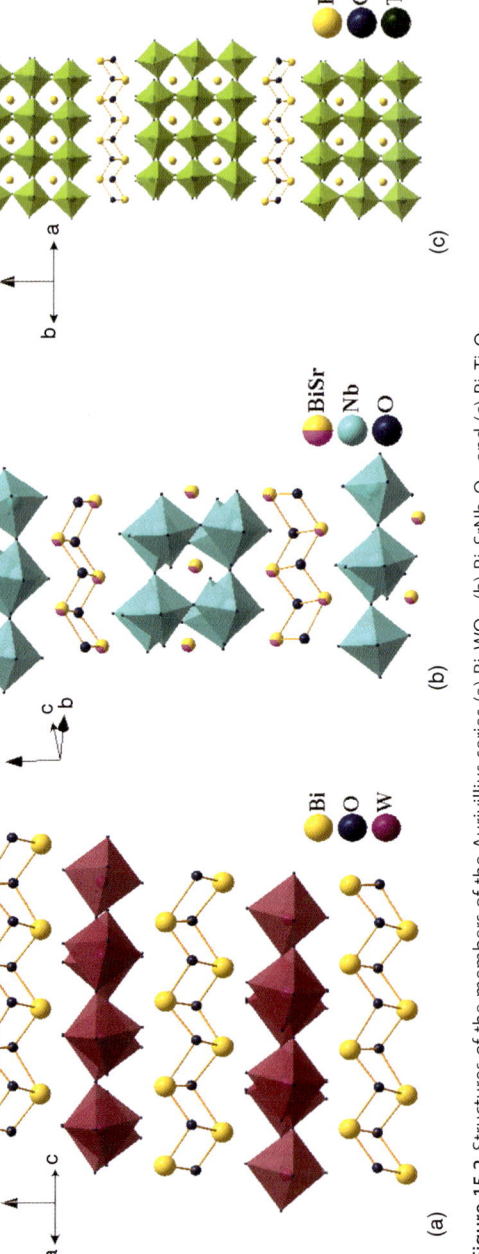

Figure 15.2 Structures of the members of the Aurivillius series (a) Bi_2WO_6, (b) $Bi_2SrNb_2O_9$, and (c) $Bi_4Ti_3O_{12}$.

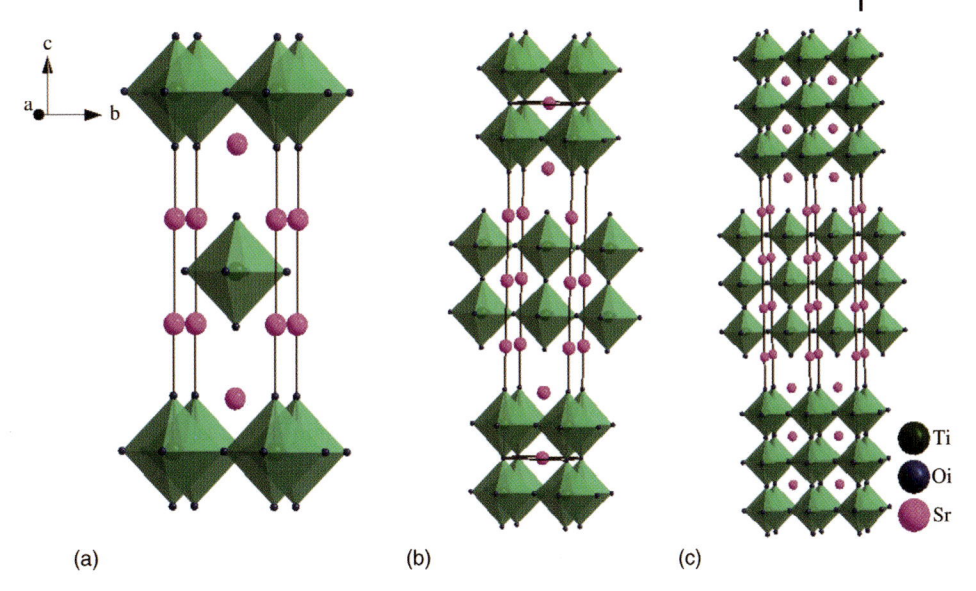

Figure 15.3 Structures of the members of the Ruddlesden–Popper series (a) Sr_2TiO_4, (b) $Sr_3Ti_2O_7$, and (c) $Sr_4Ti_3O_{10}$.

or W^{6+} can be substituted at the B-site resulting in several Aurivillius phases. Oxides, specifically $Bi_2SrTa_2O_9$ ($n=2$) and $Bi_3Ti_4O_{12}$ ($n=3$) have been reported to show large spontaneous polarizations over a wide temperature range. Specially, the disorder existing between the Bi^{3+} ions from the $[Bi_2O_2]^{2+}$ layers and the A-site cations of the perovskite block seem to influence the dielectric constants near the Curie temperature. So careful crystal structures determination for the various members have been studied by techniques such as single crystal X-ray diffraction (SXRD), powder X-ray diffraction (PXRD), and electron microscopy and others. Several mixed layer and intergrowth structures such as $Bi_9Ti_6CrO_{27}$ and $Bi_9Ti_4NbO_{21}$ [21] with general formula, $Bi_4A_{2-1}B_{2n+1}O_{6n+9}$ are also known. High resolution electron microscopy confirmed such intergrowth structures. Oxygen deficient Aurivillius-like phases $Bi_2Sr_2Nb_2GaO_{11}$ and $Bi_2Sr_2Nb_2AlO_{11.5}$ are also possible by the intergrowth between Aurivillius phases ($Bi_2A_{n-1}B_nO_{3n+1}$) and Brownmillerite-related oxides such as $Sr_2Ga_2O_5$ and $Sr_2Al_2O_5$ [22]. Additionally, aliovalent-substituted Aurivillius phases such as $Bi_2Sr_2M'_2Ti_2O_{12}$ (M'_2 = Nb, Ta), [23] $Bi_2SrNaNb_2TaO_{12}$, and $Bi_2Sr_2Nb_{3-x}M_xO_{12}$ (M = Zr, Hf, Fe, Zn) [24] have also been reported.

Another most important family of layered perovskites to be considered is the Dion–Jacobson series of layered perovskites, $A'[A_{n-1}B_nO_{3n+1}]$. The series of oxides corresponding to $n=3$ ($ACa_2Nb_3O_{10}$, A = K, Rb, Cs, Tl) originally reported by Tournoux and coworkers [25] have a lower layer charge density as compared to $Sr_4Ti_3O_{10}$. Subsequently, Jacobson et al. [26] reported the series, K[$Ca_2Na_{n-3}Nb_nO_{3n+1}$] ($3 \leq n \leq 7$), which undergoes facile ion-exchange reactions in 6 M HCl (at 60 °C, 16 h) to give the solid protonic acids. These Brønsted acids

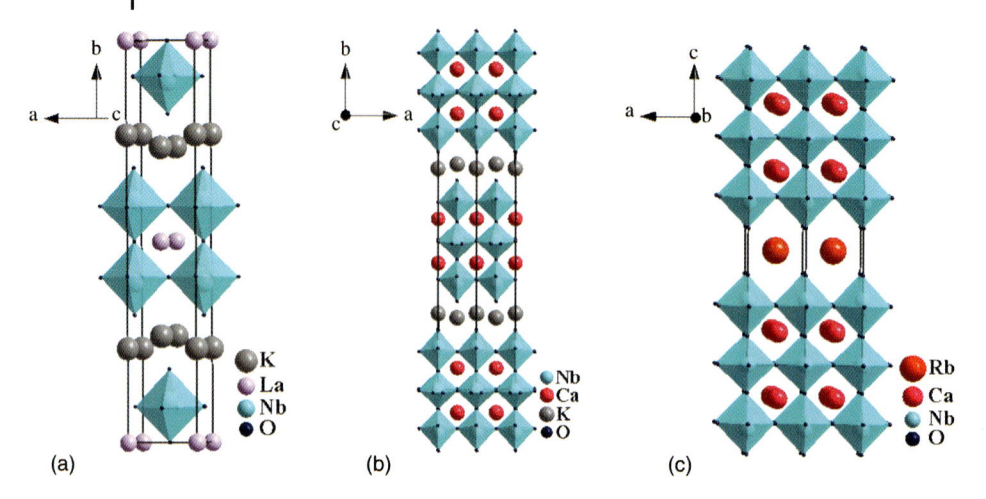

Figure 15.4 Structures of the members of the Dion–Jacobson series (a) $n=2$; $KLaNb_2O_7$ (b) $n=3$; $KCa_2Nb_3O_{10}$, and (c) $n=3$; $RbCa_2Nb_3O_{10}$.

can then be intercalated with alkylamines and exfoliated into single sheets of nano dimensions. Gopalakrishnan *et al.* [27] have contributed consistently to the development of layered perovskites starting from the report of the $n=2$ members, $ALaNb_2O_7$ (A = Li, Na, K, Rb, Cs, NH_4) and their interlayer chemical reactions. The corresponding $ABiNb_2O_7$ (A = Rb, Cs) [28] could also be obtained. Different structural types are possible and smaller sized interlayer cations such as Li^+, Na^+ have an oxygen coordination of four. While slightly bigger cations (K^+) have a prismatic six coordination and the adjacent perovskite blocks are also displaced along both the x and y directions (a shift commonly referred as $(a+b)/2$). Finally, ions such as Rb^+ and Cs^+ adopt a structure without the shifting of the adjacent layers and are stabilized by an oxygen coordination of eight (Figure 15.4). It is also worth mentioning that $BaLn_2Ti_3O_{10}$ without any alkali metal ions belongs to the Dion–Jacobson family. The Dion–Jacobson series of layered perovskites gained significance because of the possibility of designing and achieving several new layered oxides [29,30]. In addition, the B cationic sites are flexible to accommodate other metal ions [31] and inclusion of Ti^{4+} ions with the simultaneous adjustment of charges at the 12 cationic coordination site (A) within the perovskite blocks of $A'[A_{n-1}B_nO_{3n+1}]$ resulted in the formation of oxides such as $ACa_{2-x}La_xNb_{3-x}Ti_xO_{10}$ (A = K, Rb, Cs; $0 \leq x \leq 2$) [30]. Many studies have also reported the expansion of the series using divalent ions such as Sr^{2+} ($KSr_2Nb_3O_{10}$) [26], Ba^{2+} ($CsBa_2Ta_3O_{10}$) [32,33], and Pb^{2+} ($CsPb_2Nb_3O_{10}$) [34] along with pentavalent cations such as Ta^{5+}. The rare earth ion like La^{3+} can also be replaced by other Ln^{3+} (Pr, Nd, etc.) ions possessing suitable ionic sizes [31–35].

Anion deficient layered perovskites, $ACa_2Nb_{3-x}M_xO_{10-x}$ (A = Rb, Cs; M (II) = Al, Fe) for $0 < x \leq 1.0$ [36] and $ALaSrNb_2MO_9$ (A = Rb, Cs; M(II) = Mg, Co, Cu, Ni, Zn) [37] analogous to $CsCa_2Nb_3O_{10}$ have also been synthesized.

Interestingly, the presence of bigger interlayer cations (Rb^+ or Cs^+) favor the slicing of the three-dimensional perovskites leading to the formation of layered perovskites, while smaller cations (Na^+ or K^+) stabilize only in the three-dimensional perovskite structures. For example, $KCa_2Nb_2FeO_9$ and some of the $ALaSrNbMO_9$ (A = Na; M(II) = Co, Ni, Cu, Zn) oxides were known to crystallize in three-dimensional cubic perovskite structures. Anion (oxygen) vacancies are created because of the incorporation of lower valence transition metal cations at the B site. The existence of similar perovskite blocks, but with varying interlayer cation density has resulted in the formation of solid solutions, $A_{2-x}La_2Ti_{3-x}Nb_xO_{10}$ (A = K, Rb) $0 \leq x \leq 1.0$ between the $n = 3$ members of the Ruddlesden–Popper series ($A'_2A_{n-1}B_nO_{3n+1}$) and the Dion–Jacobson's series [38]. It has also been possible to vary the interlayer cation density as shown by the synthesis of the series $K_{1-x}La_xCa_{2-x}Nb_3O_{10}$ [39]. Novel bismuth containing Dion–Jacobson series, $A[Bi_3Ti_4O_{13}]$, and $A[Bi_3PbTi_5O_{16}]$ [A = K, Cs] undergoing spontaneous hydration have also been synthesized [40]. Recently, layered oxides possessing noncentrosymmetric characteristics resulting in ferroelectricity and piezoelectricity were reported in $CsBiNb_2O_7$ [41] and in $CsBi_{1-x}Eu_xNb_2O_7$ ($x = 0, 0.1, 0.2$) [42].

A complete literature on layered perovskites consisting of all the identified titanates, niobates, and tantalates together with their structural description has been extensively covered in two successive reviews by Lichtenberg et al. [31,43]. These authors for the first time synthesized layered perovskites using reduced niobates by floating zone melting method with the intention of identifying electrical conductors and/or superconductors. In addition to the layered perovskites discussed in this section that were formed by cutting the perovskite ABO_3 along [001], the review compiled two more series of oxides that were possible by cutting along [110] and [111] directions, with the insertion of one extra oxygen in all the three cases. The resulting series are $A_nB_nO_{3n+2}$ and $A_mB_{m-1}O_{3m}$, respectively, for the [110] and [111] cases. Rare earth ions (Ln^{3+} = La, Pr, Nd) are usually present in the A-site of $A_nB_nO_{3n+2}$ and a transition metal ion (e.g., Fe^{3+}) at the B site leading to the formation of paramagnetic and in some cases resulting in nonstoiochiometric oxides with the A site deficiency. $LaTaO_4$, $La_3Ti_2TaO_{11}$, and $La_{2.4}Ca_{1.6}Ti_{2.4}Nb_{1.6}O_{14}$, respectively are the $n = 2$, $n = 3$, and $n = 4$ members belonging to this series (Figure 15.5). Some of these oxides could only be synthesized under high pressure. Typical examples for the $m = 3$, $m = 4$, and $m = 5$ members of the hexagonal $A_mB_{m-1}O_{3m}$ series [31] are $Ba_3Re_2O_9$, and $LaSr_3Ta_3O_{12}$ and $Ba_5Ta_4O_{15}$ (Figure 15.6). It is also possible to synthesize Dion–Jacobson series of oxides without alkali metal ions but using divalent alkaline earth metal ions along with transition metal ions (Nb, Ta) in oxidation states lower than their common pentavalent oxidation states. Examples include $BaCaTa_2O_7$ ($n = 2$) and $BaCa_2Nb_3O_{10.07}$ ($n = 3$) and metallic behavior has been observed in the case of $BaCa_{0.6}La_{0.4}Nb_2O_{7.00}$ and $BaCa_2Nb_3O_{10.07}$ [43]. Layered perovskites of the Dion–Jacobson series with reduced tantalam were obtained by the reaction of the unreduced parent oxides with n-butyl lithium or by reacting with NaN_3 in vacuum. The intercalation of the additional Li or Na resulted in

580 | *15 Structural Diversity in Complex Layered Oxides*

⊠ = BO_6 octahedra (O located at the corners, B hidden in the center)

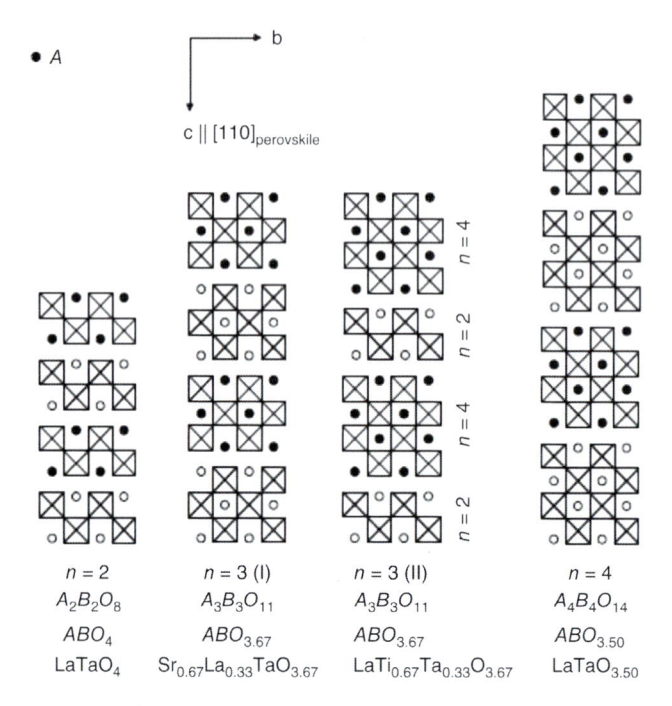

• A

c ∥ [110]$_{perovskite}$

n = 2	n = 3 (I)	n = 3 (II)	n = 4
$A_2B_2O_8$	$A_3B_3O_{11}$	$A_3B_3O_{11}$	$A_4B_4O_{14}$
ABO_4	$ABO_{3.67}$	$ABO_{3.67}$	$ABO_{3.50}$
$LaTaO_4$	$Sr_{0.67}La_{0.33}TaO_{3.67}$	$LaTi_{0.67}Ta_{0.33}O_{3.67}$	$LaTaO_{3.50}$

Figure 15.5 Sketches of the idealized crystal structures of $n = 2$, 3, and 4 members of the layered perovskites series $A_nB_nO_{3n+2}$ projected along the a-axis. In the $n = 3$ case, there are two different ordered stacking sequence. (Reproduced with permission from Ref. 43.)

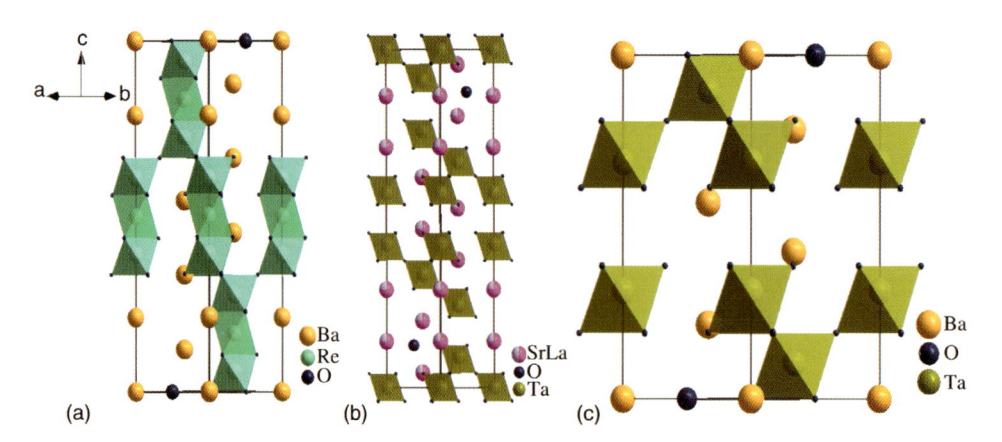

Figure 15.6 Structure of the layered perovskites $A_mB_{m-1}O_{3m}$ by cutting along [111] of the ABO_3 perovskite (a) $m = 3$, $Ba_3Re_2O_9$, (b) $m = 4$, $LaSr_3Ta_3O_{12}$, and (c) $m = 5$, $Ba_5Ta_4O_{15}$.

the formation of Ruddlesden–Popper structure type $Li_2LaTa_2O_7$, $Li_2Ca_2Ta_3O_{10}$, and $Na_2Ca_2Ta_3O_{10}$ [44].

The existence of three different families of layered perovskites has provided the solid-state chemists innumerable synthetic manipulation strategies over the years. Mallouk and coworkers demonstrated for the first time that the layered perovskites can be inter-converted by selective set of sequential topochemical reactions resembling the set of tools available in a tool box [5]. The application of layered perovskites for the rational designing of inorganic materials leading to the next generation of perovskite related compounds has been initiated and well researched by Wiley and coworkers [6]. Significant results include the unique topochemical insertion of the transition metal and alkali metal halide layers into the layered perovskites forming alternate metal–nonmetal layered structures (Section 1.3.2).

15.2.2.2 Layered Oxides Possessing Rocksalt Superstructures

The rock salt (NaCl) structure is adopted by the compounds of AX type stoichiometry. A is usually an alkali metal, alkaline earth, or transition metal cations and X is mostly an oxide, sulphide, or halide anions. Simplest are the first row divalent transition metal oxides (MO; M(II) = Mn, Fe, Co, Ni) possessing a wide range of electronic and magnetic properties [1]. A variety of new compositions and structures are possible with the introduction of more than one cation [45]. The formula type AMO_2 (A(I) = Li, Na; M (III) = Cr, Mn, Fe, Co) possesses a layered structure (α-$NaFeO_2$) because of the alternate ordering of AO_6 and MO_6 units along (111) direction (Figure 15.1a). $LiCoO_2$, currently used as a cathode in lithium ion batteries belong to this α-$NaFeO_2$ structure type. Other layered oxides crystallizing in rocksalt superstructure are listed in Table 15.1. A_2MO_3 oxides (A (I) = Li, Na; M (IV) = Ru, Sn) are formed by the alternate AO_6 and $(LiM)O_6$ layers along the (111) direction. In addition, simultaneous substitutions at both A and M sites are also possible (Figure 15.7). The common occurrence of alkali metal ions in these type of oxides are responsible for layering in such a way that the transition metal cations forming edge-shared octahedral layers are separated by alkali metal cation octahedral layers. The alkali metal cations are mobile and ion exchangeable. Further, the edge sharing (MO_6) octahedra may result in magnetically ordered honeycomb arrays in the case of metal ions with unpaired electrons. Different stacking sequences of the honeycomb layers are possible [46] and can be represented as O3, O6, P2, and P3, where O and P stands for octahedral and prismatic coordination and the numbers refer to the number of layers (Figure 15.7).

The oxides obtained by the multiple cationic substitutions in the basic rocksalt unit resulting in their ordering in distinct crystallographic sites and the respective rocksalt superstructures form equally important class of layered oxides. Common stoichiometries are $Li_3M_2XO_6$ (M(II) = Mg, Co, Ni, Cu, X(V) = Nb, Ta, Sb) [47,48] and $Na_3M_2XO_6$ (M(II) = Mg, Co, Ni, Cu; X(V) = Sb) [49]. Nalbandyan and coworkers [50] reported the series of ionic conductors, $Na_2M_2TeO_6$ (M (II) = Ni, Co, Zn, Mg). The solid solution formation in this series and the accompanying structural changes were reported subsequently [51]. Bi^{5+}

Table 15.1 Summary of layered oxides possessing rocksalt superstructures.[a]

Space group	Compound (isostructural analogues)	a (Å)	b (Å)	c (Å)	β (°)
$R\bar{3}m$	α-NaFeO$_2$ (LiMO$_2$; M(II) = Co, Mn, Ni, V, Cr)	3.027(1)	16.12(2)		
$C2/c$	Li$_2$TiO$_3$ (Na$_2$MO$_3$; M = Sn, Ir)	5.069(2)	8.799(7)	9.7599(9)	100.1
	Na$_3$LiFeSbO$_6$	5.3274(2)	9.2049(2)	11.377(3)	108.47(1)
	Li$_4$CrSbO$_6$ (Li$_4$MSbO$_6$; (M(III) = Mn, Al, Ga))	5.136(1)	8.890(2)	9.809 (3)	99.93(2)
$C2/m$	Li$_2$MnO$_3$ (Li$_2$RuO$_3$)	4.9246(1)	8.5216(1)	5.0245(1)	109.398(1)
	Na$_2$Cu$_2$TeO$_6$ (Li$_2$Cu$_2$TeO$_6$)	5.7059(6)	8.6751(9)	5.9380(6)	113.740(2)
	Li$_3$Ni$_2$SbO$_6$ (Li$_3$M$_2$SbO$_6$; M(II) = Zn, Cu) Li$_3$M$_2$BiO$_6$ (M(II) = Ni, Zn) Li$_3$NiMgBiO$_6$ (Li$_3$NiM'BiO$_6$; M'(II) = Cu, Zn)	5.1828(2)	8.9677(3)	5.1577(2)	109.69(2)
	Na$_3$Zn$_2$SbO$_6$ (Na$_3$M$_2$SbO$_6$; M(II) = Mg, Ni, Cu, Co) Na$_3$M$_2$BiO$_6$ (M(II) = Zn, Ni)	5.3636(1)	9.2722(1)	5.6645(1)	108.47(1)
	Li$_4$FeSbO$_6$ (Li$_4$MTeO$_6$; M(II) = Co, Ni, Cu, Zn, Mg)	5.1706(5)	8.9382(5)	5.1635(2)	109.49(1)
	NaNi$_2$BiO$_{6-\delta}$·1.7 H$_2$O	5.2667(1)	9.1042(3)	7.3398(2)	104.041(4)
$P\bar{3}1m$	NaNi$_2$BiO$_{6-\delta}$	5.220(3)	5.705(5)		
$P6_3/mcm$	Na$_2$Ni$_2$TeO$_6$	5.2042(5)		11.1383(5)	
$P6_322$	Na$_2$Mg$_2$TeO$_6$ (Na$_2$M$_2$TeO$_6$; M(II) = Co, Zn)	5.253(1)		11.2603(2)	
$P3_112$	Na$_4$FeSbO$_6$	5.4217(7)		16.2715(1)	

[a] Relevant references are included in the text.

15.2 Types of Layered Oxides

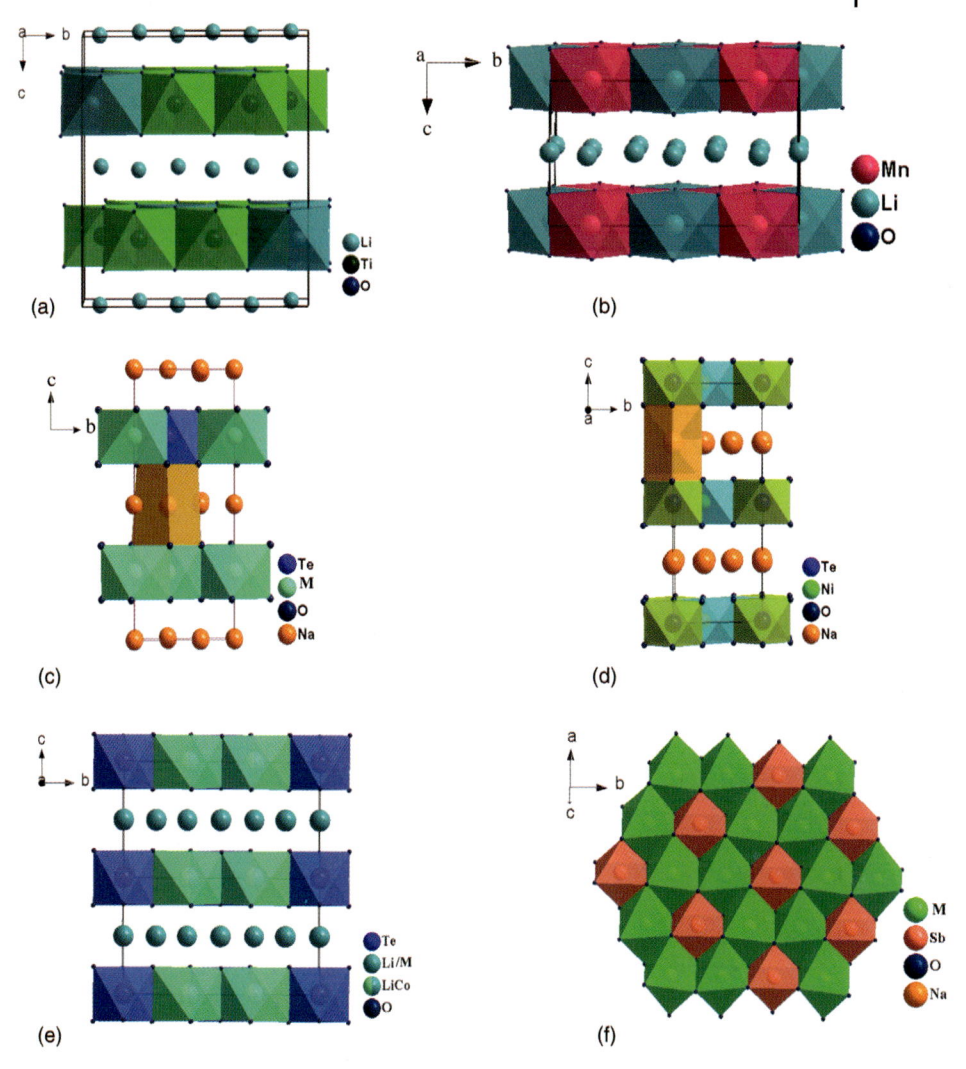

Figure 15.7 Structures of layered oxides belonging to rocksalt superstructure family (a) Li$_2$TiO$_3$ (S.G. C2/m, O3), (b) Li$_2$MnO$_3$ (S.G. C2/c, O3), (c) Na$_2$M$_2$TeO$_6$ (M (II) = Co, Zn) (S.G. P6$_3$22, P2), (d) Na$_2$M$_2$TeO$_6$ (M (II) = Ni) (S.G. P6$_3$/mcm, P2), (e) Li$_4$MTeO$_6$ (M (II) = Co, Ni, Zn) (S.G. C2/m, O3), and (f) honeycomb array in Na$_3$M$_2$SbO$_6$ (M (II) = Cu) (S. G. C2/m, O3). The space groups along with the type of structural stacking types are given in brackets.

containing oxides, Li$_3$Ni$_2$BiO$_6$ and Li$_3$NiMBiO$_6$ (M = Mg, Cu, Zn) were synthesized and characterized by Subramanian and coworkers [52]. Li$_2$RuO$_3$ [53] and Na$_2$Cu$_2$TeO$_6$ [54] are rocksalt-ordered structures having interesting magnetic spin-gap behavior along with Na$_2$Co$_2$TeO$_6$ and Na$_3$Co$_2$SbO$_6$ oxides known to have low-temperature magnetic ordering [55]. Additional series of oxides, Li$_4$M$_2^{II}$TeO$_6$ (M (II) = Co, Ni, Cu, Zn) [56] and Li$_4$M$_2^{III}$SbO$_6$ (M (III) = Cr, Fe,

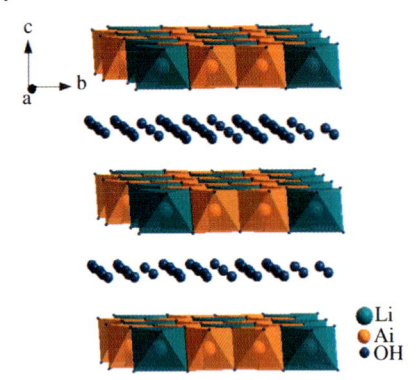

Figure 15.8 General structure of LDH.

Mn, Al, Ga) [57,58] containing excess of lithium with their complete structural characterization have been reported recently. The interlayer chemistry to produce novel metastable oxides has also been reported by the formation of $Li_2Cu_2TeO_6$ and the two different polymorphs of $Li_2Ni_2TeO_6$ [59]. The structure and magnetism of the Fe^{3+} oxides such as Na_4FeSbO_6 [60] and $Na_3LiFeSbO_6$ [61] have also been carried out. The flexibility of the honeycomb arrays to accommodate different metal ions has been demonstrated [62] by varying the cationic proportions in $Li_3(Li_{1.5x}Fe_{3-(x+1.5x)}Te_x)O_6$ ($0.1 \leq x \leq 1.0$) and $Li_{4.5}M_{0.5}TeO_6$ (M(III) = Cr, Mn, Al, Ga) Among several potential application possibilities, these oxides are primarily relevant to identify new cathode material for batteries (e.g., Li_4NiTeO_6 [63], $Na_3Ni_2SbO_6$ [64] and ionic conductors (e.g., $Na_2M_2TeO_6$ (M = Co, Ni, Zn, Mg)) [50].

Layered double hydroxides (LDH) are another distinct class of layered oxides [65] possessing ion-exchangeable anions separated by edge-shared metal hydroxide octahedra (Figure 15.8). LDH are derived from the brucite ($Mg(OH)_2$) structure by the partial substitution of Mg(II) by other cations (e.g., hydrotalcite, $Mg_6Al_2(OH)_{16}[CO_3]\cdot 4H_2O$). The general composition is represented by the formula, $[M^{z+}_{1-x}M^{3+}_x(OH)_2]^{y+}(X^{n-}_{y/n})\cdot mH_2O$ when $z = 2$, $M^{2+} = Mg^{2+}$, Ca^{2+}, Zn^{2+}, Ni^{2+}, Mn^{2+}, Co^{2+}, Fe^{2+} with $M^{3+} = Al^{3+}$, Cr^{3+}, Mn^{3+}, Fe^{3+}, Ga^{3+}, Co^{3+}, Ni^{3+}, such that $y = x$. When $z = 1$ with $M^+ = Li^+$ and $M^{3+} = Al^{3+}$ leading to $y = 2x-1$ and x can vary between $0.1 \leq x \leq 0.5$. Various members of LDH have been extensively investigated for a wide variety of anion-exchange intercalation reactions and materials based on LDH have been the focus of developing ion-exchange materials, catalysts, photocatalysts, sorbents, halogen absorbers, and as materials for drug delivery [66].

15.2.2.3 Layered Oxides Made up of Edge- and/or Corner-Shared Metal-Oxygen Octahedra

The layered oxides discussed in Sections 15.2.2.1 and 15.2.2.2 are composed respectively of corner- and edge-shared metal-oxygen octahedra. There are

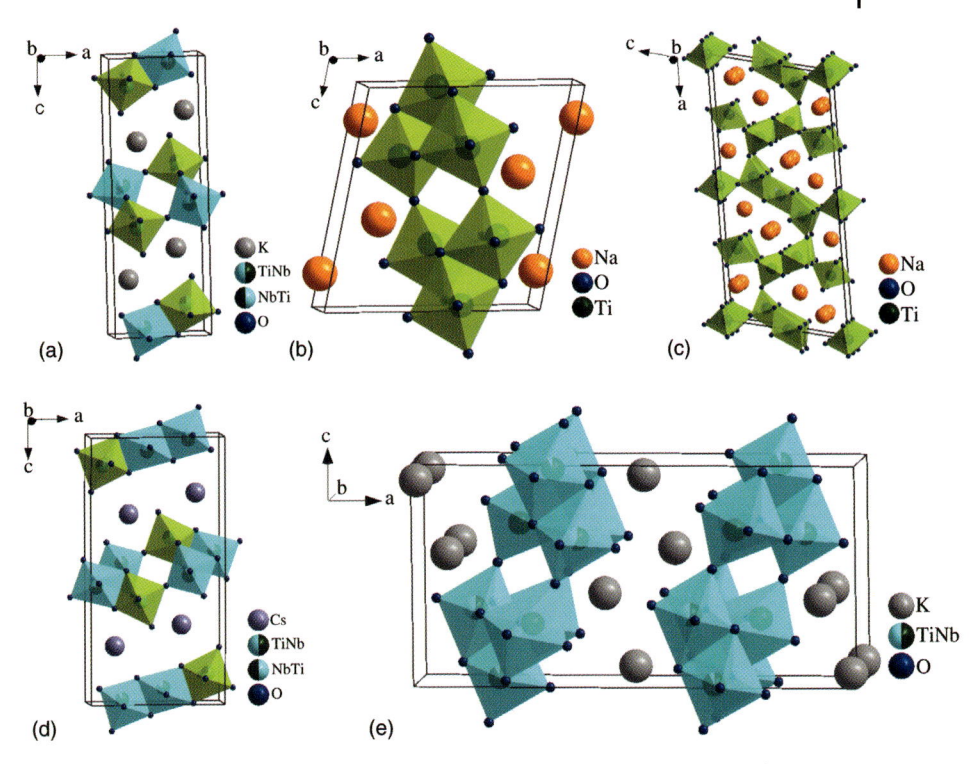

Figure 15.9 Structures of (a) $Na_2Ti_3O_7$, (b) $Na_2Ti_4O_9$, (c) $K_3Ti_5NbO_{14}$, (d) $CsTi_2NbO_7$, and (e) $KTiNbO_5$.

many layered titanates, niobates, tantalates, and mixed titanium niobates or tantalates that are also known to containing metal–oxygen octahedra sharing both edges and corners (Figure 15.9). The series with the formula, $A_mM_nO_{2n+1}$ (A = interlayer (mostly alkali) metal ions, M = Ti, Nb) and n denotes the number of edge-shared octahedra in each chain [67]. $K_2Ti_2O_5$, $Na_2Ti_3O_7$, $Na_2Ti_4O_9$, and $Na_4Ti_5O_{12}$ are some of the known members of this series capable of undergoing ion-exchange reaction [7]. The corresponding protonated oxides can be obtained by treating with hydrochloric acid. The dehydration of $H_2Ti_4O_9$ resulted in the formation of a new form of titanium dioxide TiO_2 (B) [68]. Some of the mixed niobium titanates and tantalum titanates are $ATiNbO_5$ [69–71], $CsTi_2NbO_7$ [72], $A'_3Ti_5NbO_{14}$ [73], and KTi_3NbO_9 [69] (Figure 15.10). For $ATiNbO_5$ (A = K, Rb, Tl, Cs) type, Nb(V) can also be replaced with Ta(V). In $KTiNbO_5$, alkali metal ions are present in between the layers obtained by zig-zag chains of edge-sharing MO_6 octahedra. Recently, cations (Mg^{2+}, Fe^{2+}) substitutions leading to the formation of $Rb(Mg_{0.34}Nb_{1.66})O_5$ and $K(Fe_{0.43}Nb_{1.57})O_5$ have been reported [74]. It is also possible to reduce the amount of interlayer cations by carefully adjusting the ratio between Ti and Nb leading to $A_{1-x}(Ti_{1-x}Nb_{1+x})O_5$ series [70]. Divalent substitution is also feasible in $K_2Ti_6O_{13}$ structural type to form

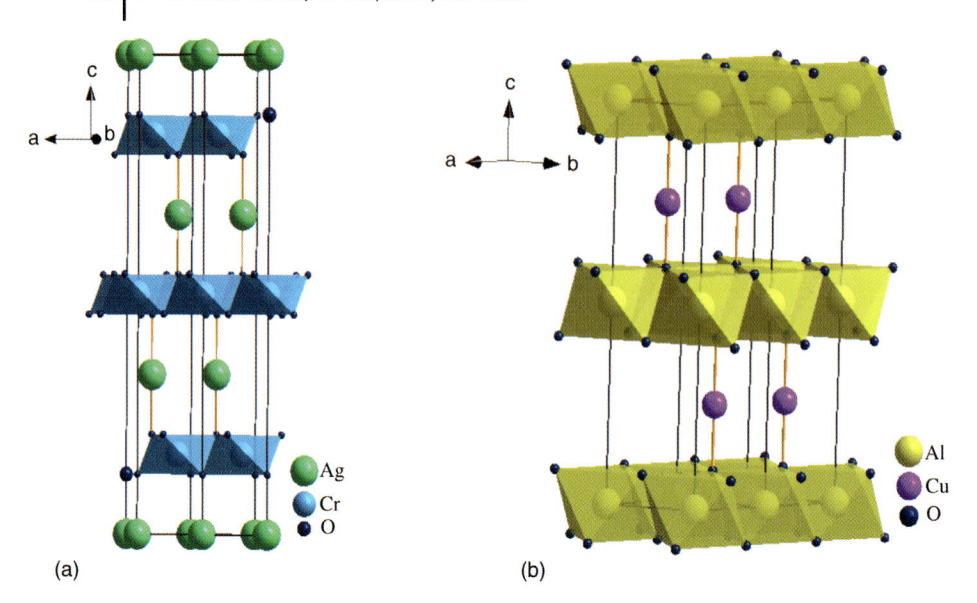

Figure 15.10 Structures of delafossite (a) 3R ($AgCrO_2$) and (b) 2H ($CuAlO_2$).

$K_2M_2Nb_4O_{13}$ (M= Mg, Fe) [75]. $K_4Nb_6O_{17}$ and KNb_3O_8 are some of the examples for the layered niobates [76]. $K_4Nb_6O_{17}$ layers consist of a double ReO_3 like chain possessing two types of interlayer regions with distinctly different intercalation properties. The structure of KNb_3O_8 is similar to $K_4Nb_6O_{17}$ consisting of NbO_6 octahedra but contains only one type of interlayer region. $LiNbWO_6$ and $LiTaWO_6$ are the ordered trirutile structures possessing layering [77]. Lithium and Nb(Ta)/W atoms are ordered in the octahedral sites to give layers of M/W (M = Nb, Ta) separated by lithium atoms perpendicular to the c-axis. Lithium ions can be exchanged with protons that in turn can undergo ion-exchange and intercalation reactions.

15.2.2.4 **Miscellaneous Layered Oxides**
Apart from the structural chemistry of the above mentioned oxides, the category of complex layered oxides certainly could include many more two-dimensional members. For example, the oxides with delafossite (ABO_2) structure (A(I) = Cu, Ag, Pd, Pt; B(III) = Fe, Co, Mn, Ga including Ln^{3+} ions) (Figure 15.10) made up of brucite layers separated by monovalent ions in a linear coordination [78]. Specifically, the Cu^+ and Ag^+ ion-based oxides are investigated as potential p-type transparent conductors because of their wide optical band gaps [79]. Here, again several combinations of metal ions are possible to add new members to the delafossite family of oxides [80]. Oxides derived from ilmenite structure such as AMO_3 (A (I) = Na, Ag; M(V) = Sb, Bi) and $PbSb_2O_6$ structure type also are made up of layered structures separated by monovalent and divalent metal ions. Layered oxides of Bi^{3+}

containing different anions having the $(Bi_2O_2)^{2+}$ layers are also a common occurrence. They are the sillen-phases [81–83] such as BiOCl, MBiO$_2$Cl (M = Ca, Pb, etc.), and the sillen-Aurivillius intergrowths like Bi$_6$Ti$_2$MO$_{14}$Cl (M^{3+} = Cr^{3+}, Mn^{3+}, Fe^{3+}) [84], the recently reported BiO(IO$_3$) [85], Bi$_2$O$_2$[B$_3$O$_5$(OH)] [86] and Bi$_2$O$_2$[BO$_2$(OH)] [86], Bi$_2$O$_2$(OH)(NO$_3$) [87], and [Bi$_6$O$_{4.5}$(OH)$_{3.5}$]$_2$(NO$_3$)$_{11}$ [88]. It is also possible to add newer members of layered oxyhalides, oxysulphides, and oxynitrides by varying the type of anions.

15.3
Interlayer Reactions

The significance associated with layered materials is their ability to undergo topochemical reactions at low temperatures (<500 °C). This particular approach commonly referred to as the chimie douce referring to the soft chemistry methods suggests a definite structural relationship between the reactants and the products. As noted earlier, topochemical reactions are useful to arrive at metastable inorganic solids that could not be stabilized by the conventional high temperature routes. Almost all of the topochemical reaction strategies reviewed with respect to the layered perovskites in general can be considered to be applicable for the majority of the remaining type of layered oxides. Only very few of the topochemical reactions involving redox chemistry are characteristics of specific structural types.

15.3.1
Ion Exchange

Ion exchange is the simplest and the most developed method over the past few decades. The earlier examples include exchange of the monovalent ions including H$^+$ ions. The synthesis of proton-based oxides like the exchange of K$^+$ ions in K$_2$Ti$_4$O$_9$ with protons (H$^+$), the exchange of H$^+$ for Li$^+$ in LiAlO$_2$ by reacting with molten benzoic acid, and the preparation of several Brønsted acids with varying strengths using many of the layered perovskites and other types of layered oxides that are available in the literature. The dehydration of the protonated oxides to form novel three dimensional metastable oxides such as TiNbO$_{4.5}$ (from HTiNbO$_5$) [70], La$_2$Ti$_3$O$_9$ (from H$_2$La$_2$Ti$_3$O$_{10}$) [15], SrTa$_2$O$_6$ (from H$_2$SrTa$_2$O$_7$) [89], CaNaTa$_3$O$_9$ (from H$_2$CaNaTa$_3$O$_{10}$) [18], Ca$_2$Ta$_2$TiO$_9$ (from H$_2$Ca$_2$Ta$_2$Ti$_2$O$_{10}$) [18], SrLaTi$_2$TaO$_9$ (from H$_2$SrLaTi$_2$TaO$_{10}$) [18], and Ln$_2$□Ti$_2$O$_7$ (Ln = La,Nd, Sm; □ = vacancy) (from HLnTiO$_4$) [90] are some of the few interesting topochemical dehydration reactions. For the first time, Sugahara and coworkers reported the formation of H$_{1.8}$Bi$_{0.2}$Sr$_{0.8}$NaNb$_3$O$_{12}$ starting from Bi$_2$SrNaNb$_3$O$_{12}$ [11]. Subsequently, other oxides Bi$_2$SrTa$_2$O$_9$ [12], Bi$_2$W$_2$O$_9$ [91], and Bi$_2$CaNaNb$_3$O$_{12}$ [92] were reported to form the respective protonated oxides.

Uni(iso)valent ion-exchange reaction replacing the interlayer cations such as K^+, Rb^+, Cs^+ with other smaller cations like Li^+, Na^+, or by NH_4^+, Tl^+, Ag^+ using the respective molten salts (around 300 °C) has become somewhat a routinely adopted method. This route has been particularly useful to stabilize structures that crystallize based on the sizes of the interlayer cations. It is possible to exchange Rb^+ ions in $RbCa_2Nb_3O_{10}$ with NH_4^+, Li^+, Na^+, and K^+ by the reaction in the temperature range 200–400 °C for 4 days with the corresponding nitrates in excess (in ratios of 1:10 or 1:20) [93]. The PXRD patterns of the products (after washing using distilled water) showed the presence of ion-exchanged layered perovskites. The thermal stability of the ion (H^+ and NH_4^+) exchanged products was studied. In these two cases, the oxides decomposed when heated to 365 °C, while $LiCa_2Nb_3O_{10}$ and β $NaCa_2Nb_3O_{10}$ underwent irreversible transition around 600–650 °C and finally decomposed at higher temperatures (>700 °C). A reversible ferroelastic transition was noticed for $KCa_2Nb_3O_{10}$ at 1000 °C. Various alkali metal ions containing Ruddlesden–Popper and Dio-Jacobson phases also showed the ability to replace their interlayer ions. $RbLaNb_2O_7$, $CsCa_2Nb_3O_{10}$, $RbCa_2Nb_3O_{10}$, $NaLaTiO_4$, $RbLa_2Ti_2NbO_{10}$, $K_2Sr_2Ta_2O_7$, $K_2La_2Ti_3O_{10}$, and $Rb_2La_2Ti_3O_{10}$ have been studied in detail [4–6,15]. Several Ag^+ containing oxides can only be accessed by the exchange reaction with $AgNO_3$ around 250–300 °C because of inherent reduction possibility of Ag^+ salts to Ag^0 in air at molten elevated temperatures (>700 °C) [94].

In addition to the divalent metal ions such as alkaline earths (Ca, Sr, Ba), transition metal ions such as Co, Ni, Cu, Zn have replaced the univalent ions both in Dion–Jacobson and Ruddlesden–popper phases [95]. Special mention is made to the conversion of $K_2Eu_2Ti_3O_{10}$ to form A^{II}-$Eu_2Ti_3O_{10}$ (A(II) = Ca, Sr, Ni, Cu, Zn) by reaction in aqueous dinitrate solutions [96]. Interestingly, metal–nonmetal arrays could be introduced within the interlayer galleries as shown by Wiley and coworkers in which copper halide layers by the reaction of $RbLaNb_2O_7$ and $CuCl_2$ in vacuum [97]. Many of the layered perovskites were demonstrated to undergo this type of copper-halide exchange chemistry. Additionally, the reaction was extended to include transition-metal chloride layers $(MCl)LaNb_2O_7$ (M = V, Cr, Mn, Fe, Co) [6]. The formation of $(CuCl)LaNb_2O_7$ makes the oxide a frustrated magnetic system showing spin-gap behavior. The structure consists of CuO_2Cl_4 octahedra connecting the perovskite blocks and the edge-shared motif resembles that of $HgCl_6$ octahedra observed in NH_4HgCl_3. $(FeCl)LaNb_2O_7$ show two magnetic transitions, one antiferromagnetic ordering below 80 K and another resulting from the realignment of spins below 10 K. $(MnCl)LaNb_2O_7$ also showed antiferromagnetic behavior that can be explained as a Heisenberg square-planar system. These copper halides-based oxide can further be treated with n-butyl lithium in order to obtain lithium chloride-incorporated layered perovskites. The efficiency of simple ion-exchange chemistry to provide unusual oxides has been shown from time to time with many examples.

Within the layered rock salt super structure family of oxides, most of the lithium salts ($LiMO_2$) can be synthesized from the corresponding $NaMO_2$ (M(III) =

Co, Cr, In, etc.) oxides using excess molten excess $LiNO_3$ at 300 °C [8]. Simple reaction of sodium cobalt oxide (Na_xCoO_2, $0.5 \leq x \leq 1.0$) with calcium nitrate in vacuum around 350 °C resulted in Ca_xCoO_2 ($x \leq 0.5$) oxides [98]. Oxides with delafossite structure such as $CuMO_2$, $AgMO_2$, $PdMO_2$ (M(II) = Fe, Co, Cr, Rh) can be synthesized by the reaction of $LiMO_2$ oxides (possessing α-$NaFeO_2$ structure) with the corresponding CuCl, $AgNO_3$, or $PdCl_2$ (along with Pd) reagents [78–80]. Ion-exchange reactions using $AgNO_3$ have become the standard mode of synthesis for the Ag^+-based delafossites (Ag_2TiO_3 [99], Ag_2SnO_3 [99], $Ag_3Co_2SbO_6$ [100], $Ag_3LiMTeO_6$ (M(II) = Co, Ni, Zn), etc.) [56,57]. $Li_2M_2TeO_6$ isostructural with $Na_2Cu_2TeO_6$ could only be synthesized by treating the latter with excess of lithium nitrate around 300 °C [59]. Similarly, $Li_2Ni_2TeO_6$ obtained by ion exchange from $Na_2Ni_2TeO_6$ is different from the disordered structure obtained by the reaction at high temperature [59].

15.3.2
Intercalation

The simplest intercalation could be the incorporation of the neutral water molecules within the layered perovskites. Many oxides are known to reversibly take water molecules to form the respective hydrated and anhydrous oxides. Some of the examples include $HLaNb_2O_7$, $K_2La_2Ti_3O_{10}$, $NaLa_2Ti_2TaO_{10}$, and $K_2SrTa_2O_7$. Majority of the proton-based oxides take up water molecules because of the stability associated through hydrogen bonding between water, the proton and the oxygen atoms of the perovskites blocks. The most widely studied reactions are the intercalation of long chain amines by the reaction of the Brønsted acidic proton oxides with the basic amines. The amine chains stabilize in the form of self-assembled monolayer and the amount of amine incorporated estimated from the increase in the interlayer spacings coupled with mass loss obtained by thermal analysis. The layered perovskites belonging to Dion–Jacobson and Ruddlesden–Popper structures have been known to intercalate organic amine to form the organic/inorganic hybrid solids based on the Brønsted acidity and the interlayer flexibility to accommodate the organic species [4–7]. Amine intercalation has also been possible in the protonated forms of Aurivillius phases. Amine intercalated $H_2W_2O_7$ has been found by Wang et al. [101]. Recently, functionalization of the Dion–Jacobson, Ruddlesden–Popper and Aurivillius phases has been pushed to greater heights by carrying out the acid extraction followed by the functionalization of organic amines under microwave irradiation [102–104]. This approach has improved the kinetics manifolds and the reactions could be completed in few hours in $Bi_2SrTa_2O_9$ instead of weeks [104].

The lattice expansion observed by incorporating long-chain alkyl amines is proportional to the number of carbon atoms present in the organic bases. This trend has also been observed in the other titanium, titanium niobium containing oxides such as $HNbWO_6$ or $HTaWO_6$. Intercalation of amines with short alkyl chains and bulkier bases such as $(C_4H_9)_4NOH$ (TBAOH) does not result in layer expansion because of their inability to pack tightly in the interlayer region.

However, solid delamination occurs leading to colloidal suspension of sheets [105]. Similar exfoliation is also possible by other organics such as n-butylamine and polymeric surfactants. Both the Dion–Jacobson and Ruddlesden–Popper types of protonated oxides can be exfoliated into colloidal sheets. The exfoliated sheets can then be assembled layer by layer on cationic substrates [106].

15.4
Potential Applications and Future Possibilities

This chapter is intended to present an overview of the materials that can be categorized as complex layered oxides with the possibility of identifying many more oxides possessing similar characteristics. It is important to understand the structure type and the possible consequences in the physical properties brought by layering. Novel synthetic ideas are being applied to maximize the variety of metastable materials that can be discovered by low-temperature topochemical reactions. Specific example has been the extraction of K_2O from the Ruddlesden–Popper oxide, $K_2La_2Ti_3O_{10}$ leading to $KLa_2Ti_3O_{9.5}$ and $La_2Ti_3O_9$ [107]. PPh_4Br and PBu_4Br (Ph = phenyl; Bu = n-butyl) were used to eliminate KBr under vacuum showing that simultaneous and reversible extraction of both cations and anions is possible. Similarly, unique organic–inorganic hybrids have also been synthesized by reacting layered perovskites with different organophosphonic acids [108] and triethylphospine ($(C_2H_5)_3P=O$, TEPO) [109]. It is appropriate to look for synthesis of unusual metastable oxides possessing interesting electronic and magnetic properties. Superconductivity has been reported in the $Li_xACa_2Nb_3O_{10}$ (A = K, Rb) with a transition temperature of 6 K via reductive intercalation [110]. It has also been shown that nanosheets ($LaNb_2O_7$ and $Ca_2Nb_3O_{10}$) obtained by delamination were used to form superlattices giving rise to ferroelectricity [111]. Functionalization of various organic species in the interlayer will benefit particularly to form inorganic nanosheets that are being viewed to play a crucial role as dielectric alternatives in nanoelectronics [112]. Layered oxides in the form of bulk, nanosheets incorporated with catalytically active metal ions [113] and molecules [114] are seriously looked upon for catalytic [115] and photocatalytic [116,117] applications.

Acknowledgment

Author thanks DST (SR/S1/PC-07/2011), DU-DST Purse Grant phase-II and University of Delhi under the "Scheme to Strengthen R&D Doctoral Research Program." The author wishes to thank Professor R. Nagarajan for encouragement and support and the help rendered by research students Ms. Akanksha Gupta, Meenakshi Pokhriyal, and Ms. Aanchal Sethi during the preparation of the manuscript.

References

1 Gopalakrishnan, J. and Rao, C.N.R. (1997) *New Directions in Solid State Chemistry*, 2nd edn, Cambridge University Press.
2 Rao, C.N.R. (2013) *Readings in Solid-State and Materials Chemistry*, World Scientific Publishing Co. Pvt. Ltd.
3 Bruce, D.W., Hare, D.O., and Walton, R.I. (2010) *Functional Oxides, Inorganic Materials Series*, John Wiley & Sons, Ltd.
4 Gopalakrishnan, J. (1995) *Chem. Mater.*, **7**, 1265–1275 and the references listed therein.
5 Schaak, R.E. and Mallouk, T. (2002) *Chem. Mater.*, **14**, 1455–1471 and the references listed therein.
6 Ranmohotti, K.G.S., Josepha, E., Choi, J., Zhang, J., and Wiley, J.B. (2011) *Adv. Mater.*, **23**, 442–460 and the references listed therein.
7 Clearfield, A. (1988) *Chem. Rev.*, **88**, 125–148 and the references listed therein.
8 England, W.A., Goodenough, J.B., and Wiseman, P.J. (1983) *J. Solid State Chem.*, **49**, 289–299.
9 (a) Ruddlesden, S.N. and Popper, P. (1958) *Acta Cryst.*, **10**, 538–539; (b) Ruddlesden, S.N. and Popper, P. (1958) *Acta Cryst.*, **11**, 54–55.
10 (a) Aurivillius, B. (1949) *Ark. Kemi.*, **1**, 463–480; (b) Aurivillius, B. (1949) *Ark. Kemi.*, **1**, 499–512.
11 Sugimoto, W., Shirata, M., Sugahara, Y., and Kuroda, K. (1999) *J. Am. Chem. Soc.*, **121**, 11601–11602.
12 Tsunoda, Y., Shirata, M., Sugimoto, W., Liu, Z., Terasaki, O., Kuroda, K. and Sugahara, Y. (2001) *Inorg. Chem.*, **40**, 5768–5771.
13 Sleight, A.W. (1988) *Science*, **242**, 1519–1527.
14 Cava, R.J. (1990) *Sci. Am.*, **263**, 24–31.
15 Gopalakrishnan, J. and Bhat, V. (1987) *Inorg. Chem.*, **26**, 4299–4301.
16 (a) Blasse, G. (1968) *J. Inorg. Nucl. Chem.*, **30**, 656–658; (b) Toda, K., Kameo, Y., Kurita, S., and Sato, M. (1996) *J. Alloys Compd.*, **234**, 19–25; (c) Byeon, S.-H., Park, K. and Itoh, M. (1996) *J. Solid State Chem.*, **121**, 430–436.
17 Gondrand, M. and Joubert, J.C. (1987) *Rev. Chim. Miner.*, **24**, 33–41.
18 Schaak, R.E. and Mallouk, T.E. (2000) *J. Solid State Chem.*, **155**, 46–54.
19 (a) Bhuvanesh, N.S.P., Lopez, M.P.C., Bohnke, O., Emery, J., and Fourquet, J.L. (1999) *Chem. Mater.*, **11**, 634–641; (b) Bhuvanesh, N.S.P., Lopez, M.P.C., Duroy, H., and Fourquet, J.L. (1999) *J. Mater. Chem.*, **9**, 3093–3100; (c) Bhuvanesh, N.S.P., Lopez, M.P.C., Duroy, H., and Fourquet, J.L. (2000) *J. Mater. Chem.*, **10**, 1685–1692.
20 Frit, B. and Mercurio, J.P. (1992) *J. Alloys Compd.*, **188**, 27–35.
21 (a) Kikuchi, T., Watanabe, A. and Uchida, K. (1977) *Mat. Res. Bull.*, **12**, 299–304; (b) Gopalakrishnan, J., Ramanan, A., Rao, C.N.R., Jefferson, D.A., and Smith, D.J. (1984) *J. Solid State Chem.*, **55**, 101–105.
22 Kendall, K.R., Thomas, J.K., and zur Loye, H.C. (1994) *Solid State Ion.*, **70–71**, 221–224.
23 Kendall, K.R., Thomas, J.K., and zur Loye, H.C. (1995) *Chem. Mater.*, **7**, 50–57.
24 Mandal, T.K., Sivakumar, T., Augustine, S., and Gopalakrishnan, J. (2005) *Mat. Sci. Eng. B*, **121**, 112–119.
25 (a) Dion, M., Ganne, M. and Tournoux, M. (1981) *Mat. Res. Bull.*, **16**, 1429–1435; (b) Dion, M., Ganne, M., Tournoux, M., and Ravez, J. (1984) *Rev. Chim. Miner.*, **21**, 92–103.
26 (a) Jacobson, A.J., Johnson, J.W., and Lewandowski, J.T. (1985) *Inorg. Chem.*, **24**, 3729–3733; (b) Jacobson, A.J., Lewandowski, J.T., and Johnson, J.W. (1986) *J. Less-Common Met.*, **116**, 137–146; (c) Treacy, M.M.J., Rice, S.B., Jacobson, A.J., and Lewandowski, J.T. (1990) *Chem. Mater.*, **2**, 279–286.
27 Gopalakrishnan, J., Bhat, V., and Raveau, B. (1987) *Mat. Res. Bull.*, **22**, 413–417.
28 Subramanian, M.A., Gopalakrishnan, J., and Sleight, A.W. (1988) *Mater. Res. Bull.*, **23**, 837–842.
29 MohanRam, R.A. and Clearfield, A. (1991) *J. Solid State Chem.*, **94**, 45–51.
30 Gopalakrishnan, J., Uma, S., and Bhat, V. (1993) *Chem. Mater.*, **5**, 132–136.

31 Lichtenberg, F., Herrnberger, A., and Wiedenmann, K. (2008) *Prog. Solid State Chem.*, **36**, 253–387.
32 Hong, Y.S. and Kim, S.J. (1996) *Bull. Korean Chem. Soc.*, **17**, 730–735.
33 (a) Thangadurai, V., Beurmann, P.S. and Weppner, W. (2001) *J. Solid State Chem.*, **158**, 279–289; (b) Thangadurai, V. and Weppner, W. (2001) *J. Mater Chem.*, **11**, 636–639.
34 Hojamberdiev, M., Bekheet, M.F., Zahedi, E., Wagata, H., Kamei, Y., Yubuta, K., Gurlo, A., Matsushita, N., Domen, K., and Teshima, K. (2016) *Cryst. Growth Des.*, **16**, 2302–2308.
35 Yoshimura, J., Ebina, Y., Kondo, J., Domen, K., and Tanaka, A. (1993) *J. Phys. Chem.*, **97**, 1970–1973.
36 Uma, S. and Gopalakrishnan, J. (1994) *Chem. Mater.*, **6**, 907–912.
37 Gopalakrishnan, J., Uma, S., Vasanthacharya, N.Y., and Subbanna, G.N. (1995) *J. Am. Chem. Soc.*, **117**, 2353–2354.
38 Uma, S., Raju, A.R., and Gopalakrishnan, J. (1993) *J. Mater Chem.*, **3**, 709–713.
39 Uma, S. and Gopalakrishnan, J. (1993) *J. Solid State Chem.*, **102**, 332–339.
40 Gopalakrishnan, J., Sivakumar, T., Thangadurai, V., and Subbanna, G.N. (1999) *Inorg. Chem.*, **38**, 2802–2806.
41 (a) Snedden, A., Knight, K.S. and Lightfoot, P. (2003) *J. Solid State Chem.*, **173**, 309–313; (b) Jin Oh, S., Shin, Y., Tran, T.T., Lee, D.W., Yoon, A., Halasyamani, P.S., and Ok, K.M. (2012) *Inorg. Chem.*, **51**, 10402–10407.
42 Kim, H.G., Yoo, J.S., and Ok, K.M. (2015) *J. Mater. Chem. C*, **3**, 5625–5630.
43 Lichtenberg, F., Herrnberger, A., Wiedenmann, K., and Mannhart, J. (2001) *Prog. Solid State Chem.*, **29**, 1–70.
44 (a) Toda, K., Takahashi, M., Teranishi, T., Ye, Z.G., Sato, M., and Hinatsu, Y. (1999) *J. Mater. Chem.*, **9**, 799–803; (b) Toda, K., Teranishi, T., Ye, Z.G., Sato, M., and Hinatsu, Y. (1999) *Mat. Res. Bull.*, **34**, 971–982.
45 Mather, G.C., Dussarrat, C., Etourneau, J., and West, A.R. (2000) *J. Mater. Chem.*, **10**, 2219–2230 and the references listed therein.
46 Delmas, C., Fouassier, C., and Hagenmuller, P. (1980) *Phys. B*, **99**, 81–85.
47 (a) Mather, G.C., Smith, R.I., Skakle, J.M.S., Fletcher, J.G., Castellanos, M.A., Gutierrez, R.M.P., and West, A.R. (1995) *J. Mater. Chem.*, **5**, 1177–1182; (b) Greaves, C. and Katib, S.M.A. (1990) *Mater. Res. Bull.*, **25**, 1175–1182.
48 (a) Skakle, J.M.S., Castellanos, M.A.R., Tovar, S.T., and West, A.R. (1997) *J. Solid State Chem.*, **131**, 115–120; (b) Mather, G.C. and West, A.R. (1996) *J. Solid State Chem.*, **124**, 214–219.
49 (a) Smirnova, O.A., Nalbandyan, V.B., Petrenko, A.A., and Avdeev, M. (2005) *J. Solid State Chem.*, **178**, 1165–1170; (b) Politaev, V.V., Nalbandyan, V.B., Petrenko, A.A., Shukaev, I.L., Volotchaev, V.A., and Medvedev, B.S. (2010) *J. Solid State Chem.*, **183**, 684–691.
50 Evstigneeva, M.A., Nalbandyan, V.B., Petrenko, A.A., Medvedev, B.S., and Kataev, A.A. (2011) *Chem. Mater.*, **23**, 1174–1181.
51 Berthelot, R., Schmidt, W., Sleight, A.W., and Subramanian, M.A. (2012) *J. Solid State Chem.*, **196**, 225–231.
52 Berthelot, R., Schmidt, W., Muir, S., Eilertsen, J., Etienne, L., Sleight, A.W., and Subramanian, M.A. (2012) *Inorg. Chem.*, **51**, 5377–5385.
53 (a) James, A.C.W.P. and Goodenough, J.B. (1988) *J. Solid State Chem.*, **74**, 287–294; (b) Miura, Y., Yasui, Y., Sato, M., Igawa, N., and Kakurai, K. (2007) *J. Phys. Soc. Jpn.*, **76**, 033705.
54 Xu, J., Assoud, A., Soheilnia, N., Derakhshan, S., Cuthbert, H.L., Greedan, J.E., Whangbo, M.H., and Kleinke, H. (2005) *Inorg. Chem.*, **44**, 5042–5046.
55 Viciu, L., Huang, Q., Morosan, E., Zandbergen, H.W., Greenbaum, N.I., McQueen, T., and Cava, R.J. (2007) *J. Solid State Chem.*, **180**, 1060–1067.
56 Kumar, V., Bhardwaj, N., Tomar, N., Thakral, V., and Uma, S. (2012) *Inorg. Chem.*, **51**, 10471–10473.
57 Bhardwaj, N., Gupta, A., and Uma, S. (2014) *Dalton Trans.*, **43**, 12050–12057.
58 (a) Nalbandyan, V.B., Avdeev, M. and Evstigneeva, M.A. (2013) *J. Solid State Chem.*, **199**, 62–65; (b) Zvereva, E.A.,

58 Savelieva, O.A., Titov, Y.D., Evstigneeva, M.A., Nalbandyan, V.B., Kao, C.N., Lin, J.-Y., Presniakov, I.A., Sobolev, A.V., Ibragimov, S.A., Hafiez, M.A., Krupskaya, Y., Jahne, C., Tan, G., Klingeler, R., Buchner, B., and Vasiliev, A.N. (2013) *Dalton Trans.*, **42**, 1550–1566.
59 Kumar, V., Gupta, A., and Uma, S. (2013) *Dalton Trans.*, **42**, 14992–14998.
60 Politaev, V.V. and Nalbandyan, V.B. (2009) *Solid State Sci.*, **11**, 144–150.
61 Schmidt, W., Berthelot, R., Etienne, L., Wattiaux, A., and Subramanian, M.A. (2014) *Mater. Res. Bull.*, **50**, 292–296.
62 (a) Gupta, A., Kumar, V. and Uma, S. (2015) *J. Chem. Sci.*, **127**, 225–233; (b) Uma, S. and Gupta, A. (2016) *Mat. Res. Bull.*, **76**, 118–123.
63 Sathiya, M., Ramesha, K., Rousse, G., Foix, D., Gonbeau, D., Guruprakash, K., Prakash, A.S., Doublet, M.L., and Tarascon, J.M. (2013) *Chem. Commun.*, **49**, 11376–11378.
64 Yuan, D., Liang, X., Wu, L., Cao, Y., Ai, X., Feng, J., and Yang, H. (2014) *Adv. Mater.*, **26**, 6301–6306.
65 (a) Williams, G.R. and O'Hare, D. (2006) *J. Mater. Chem.*, **16**, 3065–3074; (b) Khan, A.I. and O'Hare, D. (2002) *J. Mater. Chem.*, **12**, 3191–3198.
66 (a) Kuang, Y., Zhao, L., Zhang, S., Zhang, F., Dong, M., and Xu, S. (2010) *Materials*, **3**, 5220–5235; (b) Mohapatra, L. and Parida, K. (2016) *J. Mater. Chem. A*, **4**, 10744–10766.
67 Raveau, B. (1984) *Rev. Chim. Miner.*, **21**, 391–406.
68 Tournoux, M., Marchand, R., and Brohan, L. (1986) *Prog. Solid State Chem.*, **17**, 33–52.
69 (a) Wadsley, A.D. (1964) *Acta Cryst.*, **17**, 623–628; (b) Rebbah, H., Desgardin, G. and Raveau, B. (1980) *J. Solid State Chem.*, **31**, 321–328.
70 Rebbah, H., Desgardin, G., and Raveau, B. (1979) *Mat. Res. Bull.*, **14**, 1125–1131.
71 Kikkawa, S. and Koizumi, M. (1980) *Mat. Res. Bull.*, **15**, 533–539.
72 Hervieu, M. and Raveau, B. (1980) *J. Solid State Chem.*, **32**, 161–165.
73 Grandin, A., Borel, M.M., Desgardin, G., and Raveau, B. (1981) *Revue de Chimie Minerale*, **18**, 322–332.
74 Kumada, N., Iwase, E., and Kinomura, N. (1998) *Mat. Res. Bull.*, **33**, 1729–1738.
75 Kumada, N. and Kinomura, N. (1997) *Mat. Res. Bull.*, **32**, 559–537.
76 Nassau, K., Shiever, J.W., and Bernstein, J.L. (1969) *J. Electrochem. Soc.*, **116**, 348–353.
77 Bhat, V. and Gopalakrishnan, J. (1988) *Solid State Ion.*, **26**, 25–32.
78 Prewitt, C.T., Shannon, R.D., and Rogers, D.B. (1971) *Inorg. Chem.*, **10**, 719–723.
79 Marquardt, M.A., Ashmore, N.A., and Cann, D.P. (2006) *Thin Solid Films*, **496**, 146–156.
80 Nagarajan, R., Uma, S., Jayaraj, M.K., Tate, J., and Sleight, A.W. (2002) *Solid State Sci.*, **4**, 787–792.
81 Sillen, L.G. (1942) *Naturwissenschaften.*, **30**, 318–324.
82 Charkin, D.O., Berdonosov, P.S., Moisejev, A.M., Shagiakhmetov, R.R., Dolgikh, V.A., and Lightfoot, P. (1999) *J. Solid State Chem.*, **147**, 527–535.
83 Charkin, D.O., Dytyatiev, O.A., Dolgikh, V.A., and Lightfoot, P. (2003) *J. Solid State Chem.*, **173**, 83–90.
84 Liu, S., Miiller, W., Liu, Y., Avdeev, M., and Ling, C.D. (2012) *Chem. Mater.*, **24**, 3932–3942.
85 Nguyen, S.D., Yeon, J., Kim, S.-H., and Halasyamani, P.S. (2011) *J. Am. Chem. Soc.*, **133**, 12422–12425.
86 Cong, R., Sun, J., Yang, T., Li, M., Liao, F., Wang, Y., and Lin, J. (2011) *Inorg. Chem.*, **50**, 5098–5104.
87 (a) Wang, C., Liu, Q. and Li, Z. (2011) *Cryst. Res. Technol.*, **46**, 655–658; (b) Henry, N., Evain, M., Deniard, P., Jobic, S., Abraham, F., and Mentre, O. (2005) *Z. Naturforsch.*, **60b**, 322–327.
88 Henry, N., Evain, M., Deniard, P., Jobic, S., Mentré, O., and Abraham, F. (2003) *J. Solid State Chem.*, **176**, 127–136.
89 Ollivier, P.J. and Mallouk, T.E. (1998) *Chem. Mater.*, **10**, 2585–2587.
90 Thangadurai, V., Gopalakrishnan, J., and Subbanna, G.N. (1998) *Chem. Commun.*, 1299–1300.
91 Kudo, M., Ohkawa, H., Sugimoto, W., Kumada, N., Liu, Z., Terasaki, O., and Sugahara, Y. (2003) *Inorg. Chem.*, **42**, 4479–4484.

92 Sugimoto, W., Shirata, M., Kuroda, K., and Sugahara, Y. (2002) *Chem. Mater.*, **14**, 2946–2952.
93 Guertin, S.L., Josepha, E.A., Montasserasadi, D., and Wiley, J.B. (2015) *J. Alloys Compd.*, **647**, 370–374.
94 (a) Toda, K., Suzuki, T. and Sato, Mineo. (1997) *Solid State Ion.*, **93**, 177–181; (b) Sato, M., Watanabe, J. and Uematsu, K. (1993) *J. Solid State Chem.*, **107**, 460–470. (c) Toda, K., Watanabe, J. and Sato, M. (1990) *Solid State Ion.*, **90**, 15–19.
95 Hyeon, K.A. and Byeon, S.H. (1999) *Chem. Mater.*, **11**, 352–357.
96 Schaak, R.E. and Mallouk, T.E. (2000) *J. Am. Chem. Soc.*, **122**, 2798–2803.
97 Kodenkandath, T.A., Lalena, J.N., Zhou, W.L., Carpenter, E.E., Sangregorio, C., Falster, A.U. Jr., Simmons, W.B., O'Connor, C.J., and Wiley, J.B. (1999) *J. Am. Chem. Soc.*, **121**, 10743–10746.
98 Cushing, B.L. and Wiley, J.B. (1998) *J. Solid State Chem.*, **141**, 385–391.
99 Hosogi, Y., Kato, H., and Kudo, A. (2008) *J. Mater. Chem.*, **18**, 647–653.
100 Zvereva, E.A., Stratan, M.I., Ushakov, A.V., Nalbandyan, V.B., Shukaev, I.L., Silhanek, cA.V., -Hafiez, M.A., Streltsovb, S.V., and Vasiliev, A.N. (2016) *Dalton Trans.*, **45**, 7373–7384.
101 Wang, C., Tang, K., Wang, D., Liu, Z., Wang, L., Zhu, Y., and Qian, Y. (2012) *J. Mater. Chem.*, **22**, 11086–11092.
102 Boykin, J.R. and Smith, L.J. (2015) *Inorg. Chem.*, **54**, 4177–4179.
103 Tefaghi, S.A., Veiga, E.T., Amand, G., and Wiley, J.B. (2016) *Inorg. Chem.*, **55**, 1604–1612.
104 Wang, Y., Delahaye, E., Leuvrey, C., Leroux, F., Rabu, P., and Rogez, G. (2016) *Inorg. Chem.*, **55**, 4039–4046.
105 (a) Schaak, R.E. and Mallouk, T.E. (2000) *Chem. Mater.*, **12**, 3427–3434; (b) Fang, M., Kim, C.H., Saupe, G.B., Kim, H.N., Waraksa, C.C., Miwa, T., Fujishima, A., and Mallouk, T.E. (1999) *Chem. Mater.*, **11**, 1526–1532; (c) Schaak, R.E. and Mallouk, T.E. (2000) *Chem. Mater.*, **12**, 2513–2516.
106 (a) Bizeto, M.A., Shiguihara, A.L., and Constantino, V.R.L. (2009) *J. Mater. Chem.*, **19**, 2512–2525; (b) Osada, M. and Sasaki, T. (2009) *J. Mater. Chem.*, **19**, 2503–2511.
107 Gönen, Z.S., Paluchowski, D., Zavalij, P., Eichhorn, B.W., and Gopalakrishnan, J. (2006) *Inorg. Chem.*, **45**, 8736–8742.
108 Shimada, A., Yoneyama, Y., Tahara, S., Mutin, P.H., and Sugahara, Y. (2009) *Chem. Mater.*, **21**, 4155–4162.
109 Toihara, N., Yoneyama, Y., Shimada, A., Tahara, S., and Sugahara, Y. (2015) *Dalton Trans.*, **44**, 3002–3008.
110 (a) Takano, Y., Taketomi H., Tsurumi, H., Yamadaya, T., and Mori, N. (1997) *Physica B* **68**, 237–238; (b) Takano, Y., Takayanagi, S., Ogawa, S., Yamadaya, T., and Mori, N. (1997) *Solid State Commun.*, **103**, 215–217.
111 Li, B.-W., Osada, M., Ozawa, T.C., Ebina, Y., Akatsuka, K., Ma, R., Funakubo, H., and Sasaki, T. (2010) *ACS Nano*, **4**, 6673–6680.
112 Osada, M. and Sasaki, T. (2012) *Adv. Mater.*, **24**, 210–228.
113 (a) Boltersdorf, J. and Maggard, P.A. (2013) *ACS Catal.*, **3**, 2547–2555; (b) Hata, H., Kobayashi, Y., Bojan, V., Youngblood, W.J., and Mallouk, T.E. (2008) *Nano Lett.*, **8**, 794–799.
114 Maeda, K., Sahara, G., Eguchi, M., and Ishitani, O. (2015) *ACS Catal.*, **5**, 1700–1707.
115 Centi, G. and Perathoner, S. (2008) *Microporous Mesoporous Mater*, **107**, 3–15.
116 Kudo, A. and Miseki, Y. (2009) *Chem. Soc. Rev.*, **38**, 253–278.
117 Osterloh, F.E. (2008) *Chem. Mater.*, **20**, 35–54.

16
Magnetoresistance Materials

Ichiro Terasaki

Nagoya University, Department of Physics, Furo-cho, Chikusa-ku, 464-8602 Nagoya, Japan

The magnetoresistance (MR) is a most fundamental transport parameter for metals and semiconductors [1]. It is defined as difference in resistance ($\Delta R = R(H) - R(0)$) induced by external field $\mu_0 H$ normalized by the zero-field resistance $R(0)$, given by

$$\text{MR} = \Delta R / R(0). \tag{16.1}$$

By definition, it is a dimensionless quantity, and independent of sample dimensions unlike the resistance or the Hall voltage. It can be either negative or positive, and can be quite large in magnitude. To emphasize the large, negative MR, another definition written as

$$\text{MR} = -\Delta R / R(H) \tag{16.2}$$

is often used.

The magnetoresistance has a long-time research history over century. It was a characterization tool for metals and semiconductors at the initial stage, and later, together with development of low-temperature and superconductor technologies, it has become a decisive tool for the determination of the Fermi surface. From the viewpoint of modern technology, spin-dependent transport is used as a magnetic recording of hard disks.

In this chapter, we will briefly review various classes of bulk materials showing characteristic magnetoresistance, by focusing mainly on transition metal oxides. We will also explain the mechanisms of their magnetoresistance intuitively. For further reading, we recommend good reviews for magnetoresistance in general [1], quantum transport [2], spintronics [3], giant magnetoresistance [4,5], colossal magnetoresistance [6–8], dilute magnetic semiconductors [9–11], and graphene [12].

16.1
Classical Magnetoresistance

Hall was the first person who tried to measure magnetoresistance in metals [13]. He thought that external fields should cause additional resistance by bending the

Handbook of Solid State Chemistry, First Edition. Edited by Richard Dronskowski, Shinichi Kikkawa, and Andreas Stein.
© 2017 Wiley-VCH Verlag GmbH & Co. KGaA. Published 2017 by Wiley-VCH Verlag GmbH & Co. KGaA.

trajectory of the electrons through the Lorentz force, but he failed in detecting a measurable change in the resistance. The absence of the magnetoresistance is understood within the framework of the Drude model, in which an electron in metals is treated as a classical particle with a mass of m and a density of n. The electron feels a viscous resistance through scattering with a rate of $1/\tau$, and the friction force balances the acceleration by the electric field to hold Ohm's law. In this situation, the equation of motion for the electron is written as

$$m\frac{d^2}{dt^2}\vec{r} = q(\vec{E} + \vec{v} \times \vec{B}) - \frac{m}{\tau}\vec{v}, \tag{16.3}$$

where q is the charge of the electron ($q < 0$). Just for the sake of simplicity, let the external field $\vec{B} = \mu_0\vec{H}$ be parallel to the z axis and the external electrical current be parallel to the x axis. In a steady state, we equalize the right-hand side with zero, and immediately get

$$-mv_x/\tau + qE_x = 0, \tag{16.4}$$

$$q(E_y - v_x B) = 0. \tag{16.5}$$

From Eq. (16.4), we arrive at $v_x = q\tau E_x/m$. Since the current density is given by $j = nqv_x = nq^2\tau E_x/m$, the conductivity σ, the ratio of the current density to the electric field is given by $\sigma = nq^2\tau/m$, which is independent of B, or equivalently, magnetoresistance is exactly zero. Instead, a finite transverse voltage E_y arises in proportion to the magnetic field given as $E_y = v_x B = j_x B/nq = \rho_{xy} j_x$, where we define the Hall resistivity ρ_{xy} as E_y/j_x. It is convenient here to define the Hall angle θ_H as $\theta_H \equiv \rho_{xy}/\rho$. By using $\rho = m/nq^2\tau$ and $\rho_{xy} = B/nq$, we rewrite θ_H as

$$\theta_H = qB\tau/m = \omega_c\tau, \tag{16.6}$$

where $\omega_c \equiv qB/m$ is the cyclotron frequency.

Finite values of magnetoresistance are, however, often observed in conventional semiconductors and metals. This can be understood by introducing the frequency and wavenumber dependence of τ in the Boltzmann equations. In this framework, we identify the magnetoresistance to the variance of the Hall angle $\langle(\Delta\theta_H)^2\rangle = \langle(\theta_H - \langle\theta_H\rangle)^2\rangle = \langle\theta_H^2\rangle - \langle\theta_H\rangle^2$, where $\langle\cdots\rangle$ denotes an average in the momentum space [14]. Accordingly, the magnetoresistance is always positive, and usually small in conventional metals, for their Hall angle is usually small in magnitude. This is a second-order process against magnetic fields, and involves various conceivable intermediate processes. Since the Drude model assumes $\theta_H = \langle\theta_H\rangle$, it leads the variance of the Hall angle and thus the magnetoresistance to be zero. In another approximation of $\langle(\Delta\theta_H)^2\rangle = \langle\theta_H\rangle^2$, we get

$$\mathrm{MR} \sim \langle\theta_H\rangle^2 \sim \left(\frac{\rho_{xy}\mu_0 H}{\rho}\right)^2 \propto \left(\frac{\mu_0 H}{\rho}\right)^2. \tag{16.7}$$

This relationship of MR $\propto (H/\rho)^2$ is known as Kohler's rule. Figure 16.1 shows an example of classical magnetoresistance, in which all the data fall into a single curve as a function of $(\mu_0 H/\rho)^2$ [15]. A high-mobility semimetal of WTe$_2$ shows a

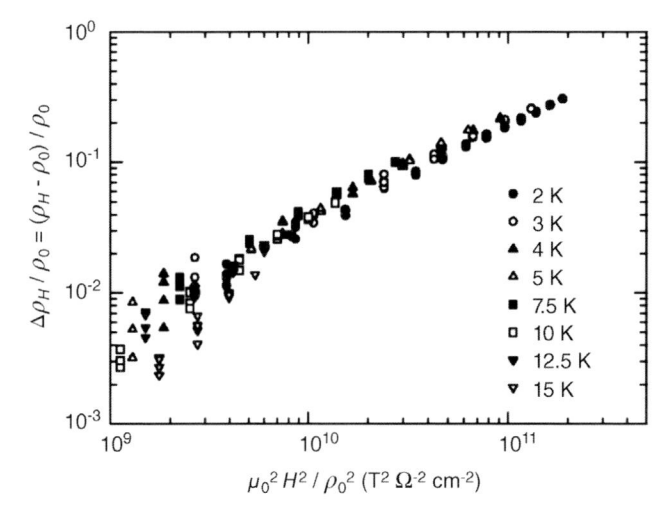

Figure 16.1 Magnetoresistance of $Na_{0.6}CoO_2$ [15] plotted as a function of $(\mu_0 H/\rho)^2$. All the data fall into a single curve to show the validity of Kohler's law.

nonsaturating quadratic magnetoresistance up to 60 T [16]. If the electron and hole conductions are identical, $\theta_H = 0$ and $\langle \theta_H \rangle^2$ can be very large and nonsaturating.

The resistivity at 300 K is often proportional to temperature owing to the electron–phonon interaction, and thus the magnetoresistance is proportional to $(H/T)^2$. If one employ a classical value for the mobility, the classical magnetoresistance is of the order of $(\mu_B \mu_0 H/k_B T)^2$, where μ_B is the Bohr magneton. This is as small as only 0.1% at room temperature in 10 T, but can be of order of unity at 1 K in 1 T. As such, the study of the magnetoresistance has been often done at low temperatures.

Here, we briefly clarify the difference between $\vec{B}$ and $\vec{H}$. The two quantities are associated with the relation of $\vec{B} = \mu_0(\vec{H} + \vec{M})$, where $\vec{M}$ is the magnetization. The quantity $\vec{B}$ is the source of the Lorentz field, and is made of the magnetic fields induced from various current sources. In contrast, the quantity $\vec{H}$ is ususally associated with the external field $\vec{B}_{ext} = \mu_0 \vec{H}$ that is determined by the electrical current flowing in the solenoid on the MR measurement apparatus employed. Although the quantity $\vec{B}$ predominantly determines the magnetoresistance, this quantity is a complicated field including the materials' response $\vec{M}$. Thus, we refer $\mu_0 \vec{H}$ as a control parameter used in experiments in this chapter, unless noted.

16.2
Classification of Magnetoresistance

The Boltzmann equation predicts that the classical magnetoresistance is always positive and small, and the field dependence is quadratic at low fields. When one takes quantum natures into account, however, the magnetoresistance can be negative, large, or nonmonotonic with fields.

The interaction between the conduction electron and the external field is classified into two: the orbital part and the spin part. The orbital part comes from a change in the trajectories of moving electrons through the Lorentz force, only appearing in the configuration of $\vec{I} \perp \vec{H}$ (transverse configuration). The classical magnetoresistance mentioned in Section 16.1 is a prime example for this.

In a pure sample, the Hall angle θ_H or $\omega_c \tau$ can be larger than 2π at low temperatures in high magnetic fields. In such situations, the electron makes a round to come back to the original point. Similarly to the electron bound in an atom, the classically circulating electron is subject to quantization, and its kinetic energy is quantized into discrete energy levels, known as the Landau levels. Electrons in the Landau levels exhibit resistivity oscillation with a frequency proportional to $1/H$; this phenomenon is known as the Shubnikov–de Haas (SdH) oscillation [2].

In the opposite situation where the mobility is extremely low, the electrons tend to localize around the impurity potential at low temperatures, known as "weak (Anderson) localization." The essence of this phenomenon is that the moving electron forms a standing wave through interference with its scattered wave. Magnetic field suppresses the interference, and it makes the system more conductive, that is, causes *negative* magnetoresistance.

The spin part stems from the interaction between the conduction electrons and the localized moments in magnetic materials, and the magnetoresistance is independent of the magnetic field direction. Depending on the type of the interaction, the magnetoresistance can take either a positive or negative value, and it often shows complicated temperature and magnetic field dependence. External field aligns the localized moment parallel to the field direction through the Zeeman effect, and hence it helps the ferromagnetic order, while affecting adversely the antiferromagnetic arrangement. Thus, the magnetic field causes negative magnetoresistance in itinerant ferromagnet [17,18], whereas it causes positive magnetoresistance in spin-density-wave materials [19].

Table 16.1 summarizes various types of magnetoresistance. Unfortunately, the spin and orbital contributions often appear in the same material, and the fraction changes with temperature and magnetic field. Since there is no universal panacea for specifying each component, one should measure the temperature,

Table 16.1 Summary of various types of magnetoresistance.

Contribution	Name	Sign	Note
Orbital	Classical MR	Positive	$\propto (H/T)^2$, Kohler's rule
	Shubnikov–de Haas	Positive	Oscillation with $1/H$
	Weak localization	Negative	$\propto \ln H$ in 2D
Spin	Kondo effect	Negative	$\propto \ln H$
	Double exchange	Negative	$\propto M^2$, GMR

GMR denotes giant magnetoresistance. $\mu_0 H$ is an external field.

magnetic field, orientation dependence of the resistivity carefully and quantitatively. Basically, one should measure magnetoresistance for both the longitudinal ($\vec{I} \parallel \vec{H}$) and transverse ($\vec{I} \perp \vec{H}$) configurations. When the magnetoresistance is identical for the two configurations, it will come from the spin contribution. Otherwise, the subtraction of the longitudinal one from the transverse one corresponds to the orbital contribution.

16.3
Giant Magnetoresistance Materials

As was mentioned in Section 16.1, the magnetoresistance is usually as small as $(\mu_B \mu_0 H / k_B T)^2$. It is only 10^{-5} at 300 K in an external field of 1 T. Thus substantial magnetoresistance much larger than $(\mu_B \mu_0 H / k_B T)^2$ is likely to arise from a nonclassical, unconventional origin. Such magnetoresistance called giant magnetoresistance (GMR) has been originally recognized in magnetic multilayers, and its discovery has been awarded by the Nobel Prize in 2007 [20,21]. Here we will skip the details of the magnetic multilayers, but introduce bulk GMR materials with peculiar interactions between localized spins and conduction electrons.

Manganese oxides with the perovskite structure (Figure 16.2) often exhibit a ferromagnetic phase transition accompanying a metallic conduction [22]. At high temperatures, the resistivity increases with decreasing temperature, and it turns to decrease after showing a broad peak around the Curie temperature T_c, below which saturation magnetization grows. GMR was first observed in $La_{1-x}Pb_xMnO_3$ single crystals [23]. Since the ground state is metallic and

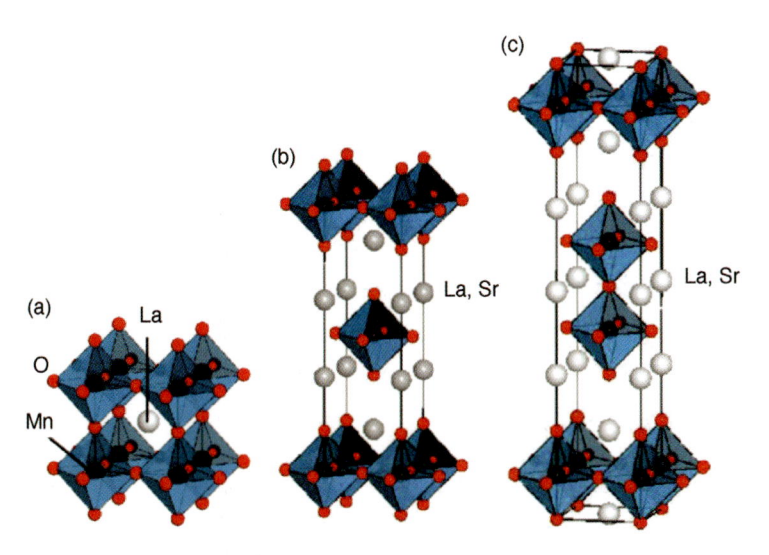

Figure 16.2 Crystal structure of the Ruddlesden–Popper-type manganese oxides with Mn^{3+} ions. (a) $LaMnO_3$, (b) $LaSrMnO_4$, and (c) $La_2SrMn_2O_7$.

ferromagnetic, external fields help the ferromagnetic order, and hence they lower the resistivity to cause the large negative magnetoresistance.

This behavior has been explained in terms of double-exchange mechanism [24,25], which is closely related to the electron configuration of the Mn ions in hole-doped LaMnO$_3$. Figure 16.3a shows a schematic of the electronic states of trivalent and tetravalent Mn ions. In the trivalent Mn ion (Mn^{3+}), four electrons occupy the 3d energy levels. As is well known, the d energy levels, being fivefold degenerate in vacuum, are split into the doubly degenerate e_g levels and triply degenerate t_{2g} levels in six oxygen anions octahedrally coordinated. Owing to strong Hund coupling, the four electrons occupy four energy levels with aligning their spins parallel, and they take the electron configuration of $(e_g)^1(t_{2g})^3$. Similarly, in the tetravalent Mn ion (Mn^{4+}), three electrons occupy the t_{2g} levels with parallel spins to take the electron configuration of $(t_{2g})^3$.

The three t_{2g} electrons tightly bound through the Hund coupling behave as a classical spin of $S = 3/2$, whereas the one e_g electron in the Mn^{3+} ion (or equivalently one hole in the Mn^{4+} ion) acts as a mobile carrier. The Hund coupling also tends to align the e_g spin parallel to the t_{2g} spin, and thus a ferromagnetic interaction will act between the neighboring t_{2g} spins in order that the e_g electron lowers the kinetic energy by hopping from one site to another. This situation is schematically drawn in Figure 16.3c; the e_g electron in the center can move to the right, but not to the left. In this way, the ferromagnetic order happens together with the metallic conduction below T_c. This e_g-electron-mediated magnetic interaction is referred to as the double-exchange interaction [24].

Figure 16.4 shows the resistivity of a single-crystal sample of La$_{1-x}$Sr$_x$MnO$_3$ ($x = 0.175$) in external fields [26,27]. The zero-field resistivity is nonmetallic above $T_c = 300$ K, whereas metallic below T_c. The external fields lower the resistivity, and shift the broad peak to higher temperatures. As a result, the magnetoresistance given by Eq. (16.1) exceeds 80% at 284 K in 15 T, indicating that the resistance decreases by a factor of 5. A quantitative relationship between the magnetization M and the magnetoresistance $\Delta R/R$ has been calculated on the basis of dynamical mean-field approximation given by [28]

$$\frac{R(H)}{R(0)} = 1 - C\left(\frac{M}{M_s}\right)^2, \tag{16.8}$$

where M_s is the saturation magnetization, and C is a constant of the order of unity.

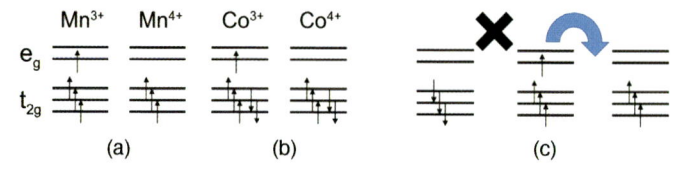

Figure 16.3 Electronic states of (a) Mn^{3+} and Mn^{4+}, and (b) Co^{3+} and Co^{4+}. (c) Schematic picture of the double-exchange mechanism.

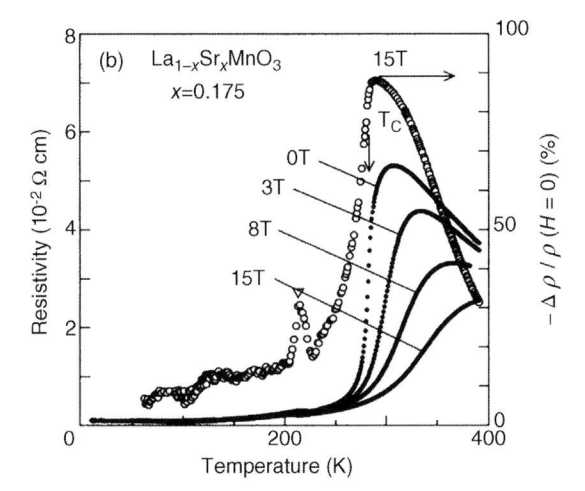

Figure 16.4 Resistivity in various magnetic fields and magnetoresistance in 15 T in a single-crystal sample of $La_{0.825}Sr_{0.175}MnO_3$ [27].

Although the double-exchange mechanism explains the ferromagnetic state quantitatively, it fails in explaining the paramagnetic insulating state above T_c shown in Figure 16.4. This requires us to consider one additional degree of freedom – the orbital. In the Mn^{3+} ion, the one e_g electron can occupy either one of the two orbitals (say, $d_{3x^2-r^2}$ and $d_{3y^2-r^2}$); this situation is often referred to as the orbital degree of freedom. Kugel and Khomskii [29] have treated the orbital degrees of freedom as an Ising pseudo-spin, and found an effective interaction between the orbital–orbital interaction. Accordingly, one can naturally extend the idea of magnetically ordered states to the idea of *orbital-ordered states*. A resonant X-ray diffraction experiment has revealed that the e_g orbital is ordered at the Néel temperature in $LaMnO_3$ [30]. The MnO_6 octahedra elongated along the x direction form NaCl-type order with those elongated along the y direction. This distortion couples tightly with the e_g electron, and it moves together when the electron hops to the e_g-empty Mn^{4+} site. Such an electron as accompanies the lattice distortion is referred to as small polaron, the mobility of which shows activation-type temperature dependence [31]. The e_g electron in the perovskite manganese oxides is a typical small polaron with an activation energy between the e_g levels split by the distortion of the octahedron, and it exhibits nonmetallic resistivity with the activation energy.

Similar double-exchange interaction acts in the perovskite cobalt oxide $La_{1-x}Sr_xCoO_3$ [32], in which Co^{3+} and Co^{4+} ions are responsible for the electric and magnetic properties. Figure 16.3b shows a schematic of the electronic states of trivalent and tetravalent Co ions. In this case, the electronic configuration of the Co^{3+} and Co^{4+} ions are $(e_g)^1(t_{2g})^5$ and $(t_{2g})^5$, respectively. Aside from the magnitude of the t_{2g} spin, one can find the close resemblance to the manganese oxides. While the metallic ferromagnetism emerges below T_c, the anomaly at T_c is milder in this oxide than in $La_{1-x}Sr_xMnO_3$, possibly because of the smaller t_{2g}

spin of $S = 1/2$. Concomitantly, the magnetoresistance is smaller in magnitude than that in $La_{1-x}Sr_xMnO_3$ [33].

The layered manganese oxide $La_{2-2x}Sr_{1+2x}Mn_2O_7$ belongs to another class of GMR materials. This particular oxide is one member ($p = 2$) of the Ruddlesden–Popper series $A_{p+1}B_pO_{3p+1}$, and it consists of the MnO_2-(La,Sr)O-MnO_2 double layer and the rock-salt $(La,Sr)_2O_2$ layer (Figure 16.2c). As shown in Figure 16.5a, the cross-layer resistivity ρ_c is nonmetallic, while the in-layer resistivity ρ_{ab} is metallic from 260 down to 90 K [34]. This implies that the in-layer ferromagnetic order develops below around 260 K, while the three-dimensional magnetic order grows to complete a long-range order below $T_c = 90$ K.

The cross-layer magnetoresistance below T_c is found to be large at low fields as shown in Figure 16.5b. The neutron diffraction reveals that in the absence of external field, the ferromagnetic in-layer spins align antiferromagnetically along the cross-layer direction. External field aligns all the ferromagnetic in-layer spins parallel, and concomitantly lets the conduction electron hop from layer to layer more frequently. This behavior is identical to a *spin valve* of magnetic multilayer, where the paramagnetic metal layer (Cr) is sandwiched by the two magnetic metal layer (Fe) [20,21]. Comparing this magnetoresistance with that in Figure 16.4, one can find the giant response in a small value of external magnetic field (less than 1 T). Whereas this high sensitivity is advantageous for sensor applications a drawback is the low working temperature below 90 K; This is inevitable because the nature-made spin valve works in the two-dimensional materials that always have T_c lower than the three-dimensional relatives.

Aside from the perovskite manganese oxides, there are many variations for the GMR effects. The pyrochlore manganese oxide $Tl_2Mn_2O_7$ [35] and the spinel chromium sulfide $FeCr_2S_4$ [36] show similar GMR effects, but the mechanism is definitely different from the double-exchange interaction, because the Mn and Cr ions are not in a mixed valence state. An ordered perovskite oxide Sr_2FeMoO_6 shows a significant magnetoresistance at room temperature, where the

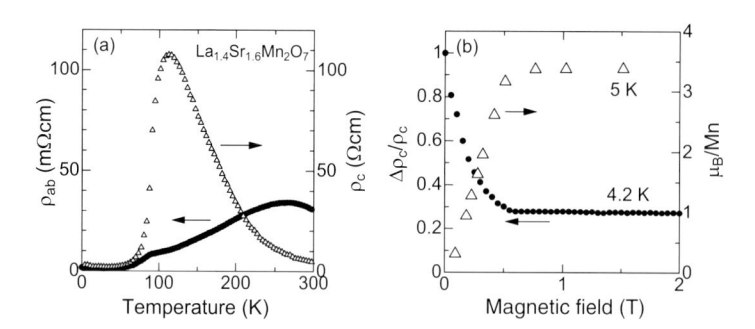

Figure 16.5 (a) Resistivity of $La_{1.4}Sr_{1.6}Mn_2O_7$ plotted as a function of temperature. (b) Magnetoresistance of the cross-layer resistivity at 5 K and the magnetization at 4 K plotted as a function of magnetic field applied in parallel to the cross-layer direction. (The data are taken from Ref. [34].)

Fe^{3+} ions act as local moments, and the Mo^{5+} ions act as fully polarized carriers [37].

There are also many reports on the nature-made spin-valve materials. The cross-layer resistivity of $GdBaCo_2O_{5+\delta}$ shows a step-like negative magnetoresistance (40% in 3 T at 240 K) [38], which is explained in terms of the field-induced spin orientation of the intermediate-spin state of Co^{3+}. The layered cobalt oxides of Sr_2CoO_4 and $Sr_3(Fe,Co)_2O_7$ show spin-valve-type magnetoresistance [39,40] as is similar to the layered manganese oxides. As an odd example, $SrCo_6O_{11}$ exhibits a two-step spin valve behavior together with a two-step magnetic plateau at 4.2 K [41].

Other than the double-exchange interaction, widely known are various forms of interaction between the conduction electrons and the localized moments. In rare-earth intermetallic compounds, the 4f/5f electrons act as magnetic moments whereas s (or p,d) conduction electrons are responsible for the electrical conduction. Through the interaction between these (s − f interaction), the s electrons largely respond to the external field. One significant difference from the manganese oxides is that the f electrons show truly quantum characters, and tend to play a dual role of itinerant and localized electrons; when the Kondo effect is dominant, the f electron is itinerant to show the Fermi liquid behavior with a significantly enhanced mass (so-called heavy fermion) [42]. When the Ruderman–Kittel–Kasuya–Yoshida (RKKY) interaction is dominant, they are localized to show a magnetic order. Magnetic field can modify these ground states to cause large magnetoresistance. A typical heavy fermion compound $CeCu_6$ shows nonmetallic conduction above around 1 K, where the conduction electron starts to form a singlet pair with the f electron on the Ce site. External fields prevent the singlet-pair formation, and cause a large negative magnetoresistance of 50% at 1 K in 6 T [43].

The layered cobalt oxides with the CdI_2-type CoO_2 layer have been investigated as thermoelectric materials [44–47]. The high performance of the thermoelectricity is due to the spin degrees of freedom [48], and the resistivity depends on magnetic field. The magnetoresistance is negative and large at low temperatures [49,50]. Figure 16.6 shows the magnetoresistance of a single-crystal $Ca_3Co_4O_9$ [51]. The magnetoresistance follows the scaling relation proposed for GMR manganese oxides given by

$$\rho(T,H) = \rho(T,0)\exp(-\alpha M^\mu/T), \quad (16.9)$$

where α and μ are some constants. On the contrary, another layered cobalt oxide Bi–Sr–Co–O shows anisotropic large negative magnetoresistance, which is due to the itinerant ferromagnetism below 2 K [52]. A similarity to heavy fermion compounds is discussed in terms of pseudogap [53], and the magnetic field suppresses the pseudogap to cause negative magnetoresistance [54].

Some organic conductors with transition metal elements also show similar magnetoresistance effect. In this case, the conductive π electrons interact with the localized d electrons (π − d interaction), and respond to external fields. Fe-phthalocyanine is a typical example, where an external field of 10 T causes a

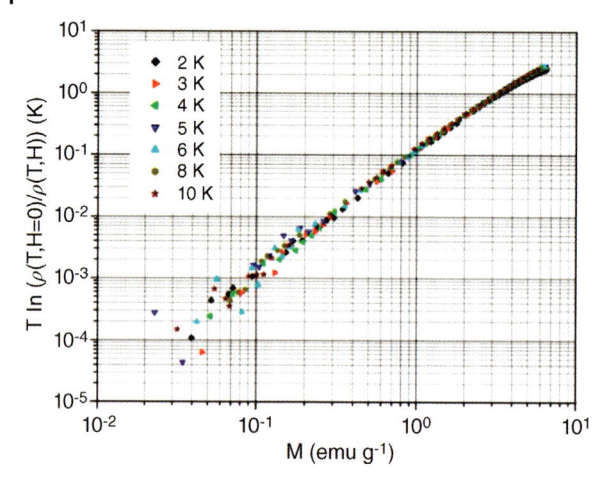

Figure 16.6 Scaling of the magnetoresistance of the layered cobalt oxide $Ca_3Co_4O_9$ [51].

large negative magnetoresistance of 1000% defined by Eq. (16.2) at 20 K [55]. Another type of large magnetoresistance is reported in θ-(BEDT-TTF)$_2$CsZn (SCN)$_4$, where the charge ordering grows at low temperature. In this particular organic conductor, the large *positive* magnetoresistance appears at low temperatures (2000% defined by Eq. (16.1) at 0.14 K around 3 T) [56]. This magnetoresistance is identical between the longitudinal and transverse directions, and thus this comes from the spin contribution.

16.4
Colossal Magnetoresistance Materials

Some materials exhibit a metal–insulator transition driven by external fields. In such materials, the resistivity can change by many orders of magnitude with magnetic fields, and the magnetoresistance can be much larger than GMR. Such magnetoresistance is called colossal magnetoresistance (CMR). In this section, we will see why and how external magnetic fields cause a metal–insulator transition in such materials.

Figure 16.7a shows the temperature dependence of the magnetic diffraction peaks of $Nd_{0.5}Sr_{0.5}MnO_3$, in which a ferromagnetic order occurs around 250 K (T_c), but it suddenly quenches around 160 K (T_N^{CE}), below which a CE-type antiferromagnetic order grows instead [57]. In the CE order, ferromagnetic zig-zag chains are aligned in antiparallel with each other owing to the orbital–orbital and spin–spin interactions (For a schematic figure, See Ref. [4]). At the same temperature of T_N^{CE}, the lattice parameter discontinuously changes to show a first-order transition (not shown). A resonant X-ray diffraction experiment has revealed that this transition is due to a charge ordering, in which the Mn^{4+} and Mn^{3+} ions form NaCl-type order owing to the long-range Coulomb

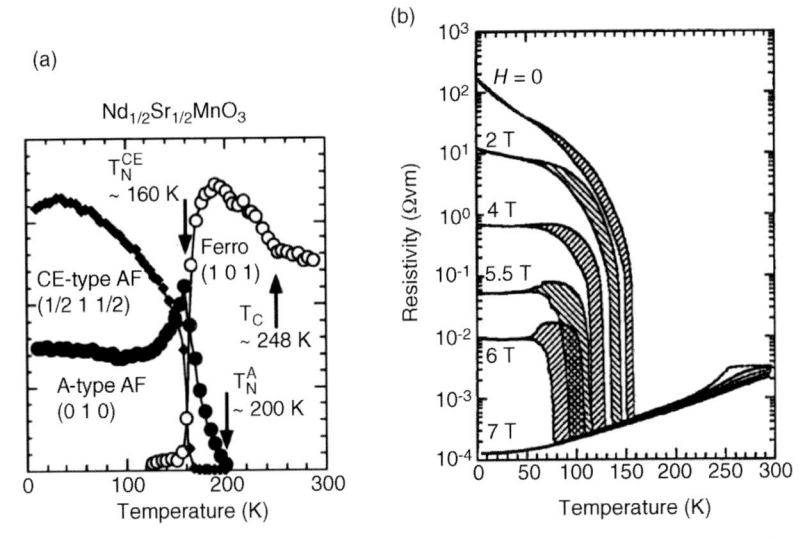

Figure 16.7 (a) Magnetic diffraction [57] and (b) resistivity of a single-crystal sample of $Nd_{0.5}Sr_{0.5}MnO_3$ [60]. Colossally large magnetoresistance is observed. The hatched areas represent temperature hysteresis.

repulsion [58]. Concomitantly, the e_g electrons in the Mn^{3+} ions exhibit orbital ordering, while the Mn^{4+} ion lacks the orbital degrees of freedom.

As shown in Figure 16.7b, the zero-field resistivity changes with decreasing temperature from nonmetallic to metallic around $T_c = 250$ K thanks to the double-exchange mechanism, and it then shows a metal–insulator transition characterized by a steep rise at $T_N^{CE} = 160$ K [59,60]. This indicates a strong competition between the ferromagnetic metallic phase and the charge-ordered, antiferromagnetic, insulating phase.

This competition is understood in terms of an interplay among the electron–electron, electron–spin, and spin–spin interactions. The t_{2g} spins interact antiferromagnetically through the super-exchange interaction, while they interact with conduction e_g electrons to favor ferromagnetic alignment. Thus the t_{2g} spins are frustrated between the ferro- and antiferromagnetic interactions. When the ratio of the Mn^{3+} to Mn^{4+} fraction takes a particular rational number, the charge ordered phase becomes stable to favor an insulating state and eventually the corresponding antiferromagnetic arrangement of the t_{2g} spins. The insulating nature of the charge-ordered state prefers narrow-band systems. In the perovskite oxides, the bandwidth is known to be narrower for a smaller lattice volume, where the oxygen octahedron is more canted to make the transfer energy smaller. Accordingly, smaller ions such as Pr^{3+}, Nd^{3+}, Ca^{2+} together with a particular ratio of $Mn^{3+}:Mn^{4+} = 0.5:0.5$ stabilize the charge-ordered state. Since the Sr^{2+} ion is fairly large, $Nd_{0.5}Sr_{0.5}MnO_3$ is on the verge of competing phases. A combination of small ions of Pr^{3+} and Ca^{2+} gives a stable charge-ordered state in the wide range of x ($0.3 < x < 0.5$) [61].

External fields can control the phase competition mentioned above; They help growth of the ferromagnetic itinerant state, and suppress the charge-ordered insulating state. In the absence of magnetic field, the resistivity reaches a high value of 10^2 Ωcm at 4 K, and it dramatically decreases down to 10^{-4} Ωcm in $\mu_0 H = 7$ T at the same temperature. Furthermore, the temperature dependence below 160 K changes from nonmetallic in 0 T to metallic in 7 T; This is a typical example of field-induced metal–insulator transition. In the case of (Pr,Ca)MnO$_3$, external fields reduce the resistivity by a factor of 10^{10} [61].

One might regard this metal–insulator transition as a homogeneous transition [62]. Microscopic probes such as transmission-electron microscope and scanning probe spectroscopy experiments have, however, revealed that the electronic states consist of highly inhomogeneous mixture of the metal and insulator phases [63,64]. External fields help to grow a volume fraction of the ferromagnetic phases, but even in a completely metallic sample, the two-phase mixture remains. The temperature hysteresis indicated by the hatched areas in Figure 16.7b implies the two-phase mixture. In this sense, electron tunneling across charge-ordered insulating domains determines the bulk resistance of such materials, which can be largely affected by the domain distributions modified by the external fields. Such a type of the magnetoresistance has been called tunneling magnetoresistance (TMR) in the field of spintronics [65]. The CMR phenomenon in Nd$_{0.5}$Sr$_{0.5}$MnO$_3$ is TMR in a disordered magnet, and a similar magnetoresistance is often observed in organic polymers [66]. A theoretical study of a competing Ising magnet reveals that competing phases can induce a colossal response with the corresponding conjugate external field [67].

Further studies have clarified that the solid solution of Nd and Sr ions significantly affect the physical properties. A nominal composition of Nd$_{0.5}$Ba$_{0.5}$MnO$_3$ can take an A-site-ordered phase of NdBaMn$_2$O$_6$ and an A-site-disordered phase of Nd$_{0.5}$Ba$_{0.5}$MnO$_3$, depending on the preparation conditions [68]. The phase diagram is completely different between the two, and the charge ordering is more stable in the A-site-ordered phase [69]. Theoretically, the charge ordered phase is more susceptible against impurities than the double-exchange ferromagnetic phase, and the A-site disorder enhances the metallic phase [70].

A similar CMR phenomenon is seen in the SrCo$_{12}$O$_{19}$ below 50 K, where the charge ordered state likely competes with the itinerant ferromagnetic state [71]. In this oxide, the Co ions responsible for the conduction align in a Kagomé lattice, and possibly owing to this geometrical frustration, the metal–insulator transition seems of second order, as shown in Figure 16.8. The resistivities are nonmetallic for the both directions in zero field, and the external fields dramatically suppress the nonmetallic behavior to decrease the resistivity by a factor of 10^3 at low temperatures. Most remarkably, the resistivity shows a metallic temperature dependence below 4 K, meaning the field-induced metallic nature.

Some organic salts also exhibit a field-induced metal–insulator transition. λ–(BETS)FeCl$_4$ shows a field-induced metal–insulator transition at 8 K in 14 T, and concomitantly the resistivity changes colossally with magnetic fields [72]. When the magnetic field is parallel to the conducting BETS layer, the external

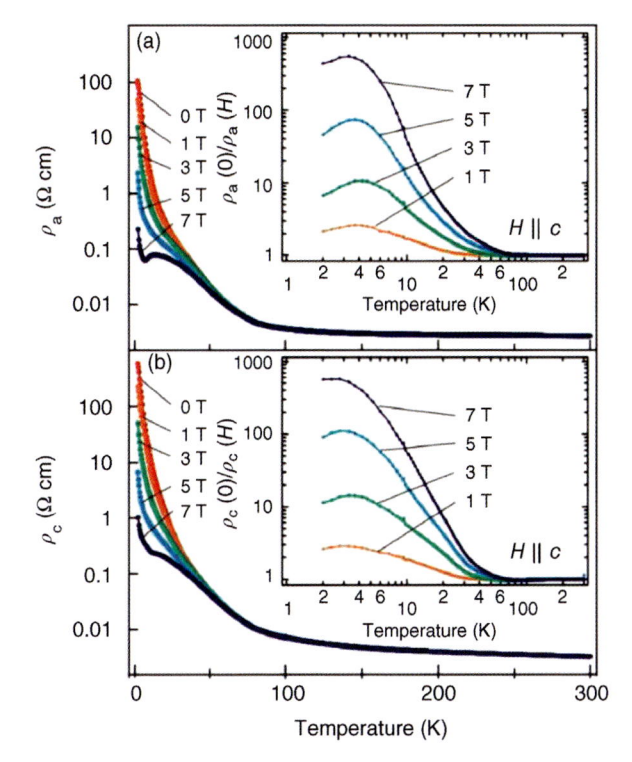

Figure 16.8 The resistivity of a single-crystal sample of SrCo$_{12}$O$_{19}$ [71]. (a) In-layer resistivity and (b) cross-layer resistivity. The insets expand the low-temperature data, where the resistivity shows a metallic conduction below 4 K in external fields.

field induces the superconductivity by canceling the magnetization of the FeCl$_4$ layer [73]. However, the zero-field ground state and the field-driven mechanism are yet to be explored.

16.5 Dilute Magnetic Semiconductors

Semiconductor technology has made full use of the charge degree of freedom of carriers. One can naturally expect that a complete control of the spin degree of freedom will be a next step, which is indeed a most important target in the field of spintronics. For this purpose, a ferromagnetic semiconductor with fully polarized spins is indispensable.

The Stoner ferromagnetism is a simplest explanation for metallic ferromagnetic materials [74]. The electron–electron repulsion is reduced when all the spins are polarized, because such electrons keep away from each other in order to satisfy the Pauli principle. On the contrary, the Fermi wavenumber increases by a factor of $2^{1/3}$ and the corresponding Fermi energy (or kinetic energy) increases by a factor of $2^{2/3}$. Thus ferromagnetism occurs when the energy gain for the electron–electron repulsion overcomes the increase in the Fermi energy. Stoner discussed this situation with a Hartree–Fock-type mean-field approximation, and found the necessary condition of $UD(E_F) > 1$, where U is a measure of electron–electron repulsion and $D(E_F)$ is the density of states at the Fermi energy E_F. According to this condition, ferromagnetism is unstable for a small $D(E_F)$, and semiconductors cannot be ferromagnetic. Actually metallic ferromagnetic materials were element metals or metallic alloys. At present, however, many ferromagnetic semiconductors are discovered as shown in Figure 16.9 [75], where the materials cover a wide range of the conductivity from semiconductors to metals. This is because the Stoner theory is based on nearly free-electron systems with a moderate electron–electron interaction. Completely different systems like the perovskite manganese oxides (Sections 16.3 and 16.4) can show ferromagnetism with a low carrier concentration.

EuX (X = O, S, Se, Te) is a time-honored magnetoresistive semiconductor, where the 4f and 6s bands of Eu contribute to the electrical conduction. A delicate overlap of the two bands is susceptible against magnetic field, and causes large negative magnetoresistance [76,77]. However, the physical properties are sensitive to nonstoichiometry, and thus sample-dependent.

At present, the ferromagnetism has been discovered in magnetic-ion-doped semiconductors. In such materials, we can control the carrier density and the magnetic-ion density independently, and we can also use various techniques developed in semiconductor engineering. This class of materials has been called

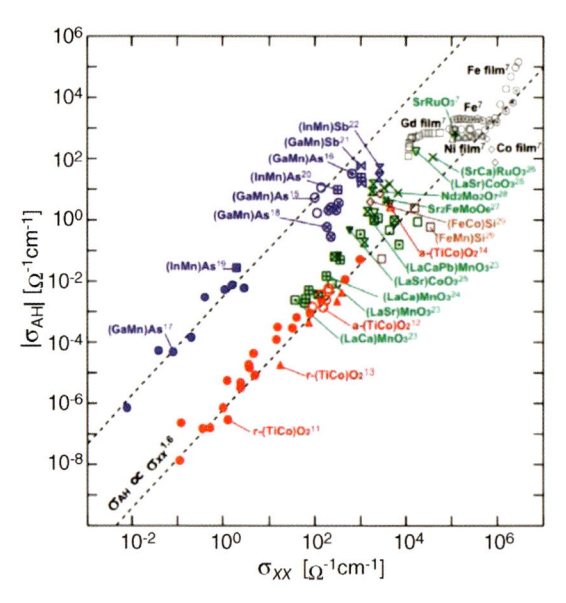

Figure 16.9 Scaling of the anomalous Hall conductivity σ_{AH} to the magnetoconductivity σ_{xx} for various ferromagnetic magnetic semiconductors [75]. (© 2007 The Japan Society of Applied Physics.)

dilute magnetic semiconductor (DMS), which has quickly become a central issue in the field of spintronics [10].

The DMS first discovered is Mn-doped GaAs. The Mn ion substituted for the Ga site acts as a localized moment, which interacts with carriers doped from donor/acceptor impurities. The highest Curie temperature is 200 K, which is unusually high in such a low carrier density system [9]. The magnetoresistance is negative and large, and its mechanism is understood in terms of the RKKY interaction [78].

One serious drawback of (Mn,Ga)As is the low Curie temperature. Many researchers have tried to synthesize a room temperature DMS, and a most extensively investigated material is Co-doped TiO_2 [79]. The nature of the ferromagnetism has been controversial. In spite of clear evidence for the interaction between the doped carriers and the ferromagnetism [80], a possible precipitation of Co metal into the TiO_2 matrix tends to explain the high Curie temperature of 600 K and the small saturation magnetization [81]. Very recently, x-ray fluorescent holography experiment has proposed that a cluster of $CoTi_4O_2$ is formed in ferromagnetic samples [82]. According to the local-density approximation, the phase of $CoTi_4O_2$ is unstable as bulk, but seems to be stable in the TiO_2 matrix. Although it is yet to be explored whether this cluster is responsible for high-temperature ferromagnetism, the precipitation scenario seems too simple to explain microscopic nature of this oxide.

16.6
Quantum Transport Phenomena

The magnetoresistance due to the Lorentz force is not always classical, but also of quantum nature, when the Hall angle $\theta_H = \omega_c \tau$ given by Eq. (16.6) is much larger than 2π. This occurs in high-mobility conductors at low temperature, often called "quantum transport."

When $\omega_c \tau \gg 1$, the electrons are no longer in the Bloch states, but in the Landau levels discretely separated by $\hbar \omega_c$. Thus the chemical potential is pinned at one Landau level for a particular range of magnetic fields, and it discontinuously jumps into the adjacent level in sweeping magnetic fields. Since the energy levels are descrete at intervals of $\hbar \omega_c$, the chemical potential and eventually all the thermodynamic quantities oscillate with a period proportional to $1/H$. The oscillating magnetization and specific heat are known as de Haas–van Alphen effect, while the oscillating resistivity is called Shubnikov–de Haas effect. Figure 16.10 shows an example of the SdH oscillation of an organic superconductor [83]. The resistivities along the a- and b-axis directions show the superconducting zero-resistance state below 3–4 T, and shows transition to the normal state in higher fields. Clearly, the resistivity oscillation is seen above around 6 T, and the period of the oscillation looks to be expanded with increasing fields. The periodicities correspond to the extrema of the cross-sectional area of the Fermi surface, and accordingly, the SdH oscillation has now been a powerful and decisive tool for the Fermi surface of metals.

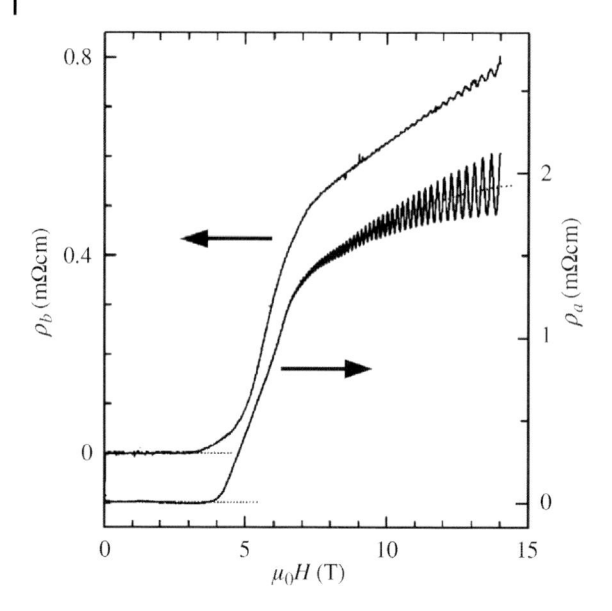

Figure 16.10 Shubnikov–de Haas oscillation in κ-(BEDT-TTF)$_2$Cu(NCS)$_2$ [83].

Angular magnetoresistance oscillation (AMRO) is another type of quantum transport observed in many low-dimensional materials. In a quasi-two-dimensional conductor, the Fermi surface is basically cylindrical, but is slightly warped with a finite cross-plane transfer energy. The cyclotron motion of the conduction electrons occurs in the momentum space perpendicular to the field direction, and it changes with the direction of external fields [84]. Thus the magnetoresistance shows oscillation as a function of the angle between the cross-plane direction and the field direction [85]. From this angular dependence, one can reconstruct the Fermi surface along the in-layer direction [86], and also can evaluate the cross-plane transfer energy [87,88].

Graphene is a single-layer graphite, which has attracted much interests from both fundamental scientists and electronic engineers. This exhibits an ideally high mobility coming from its linear dispersion relation know as "Dirac cone," and also exhibits an ideally strong hardness like diamond. This material is a promising candidate for electrode of next-generation electronic devices. From the fundamental side, the high mobility and the zero band gap generates the zero-energy Landau levels and the quantum Hall states even at room temperature [89], and the fractional quantum Hall states at low temperatures [90].

16.7
Concluding Remarks

In this chapter, we have reviewed various magnetic materials showing large magnetoresistance, mainly forcusing on transition metal oxides with strong electron correlations. Strongly correlated electron systems have attracted a renewed interest since the discovery of the high-temperature superconducting copper oxide in 1986, and vast numbers of researches have been conducted on various transition metal compounds. As the strong correlation is the source of magnetism, many of them exhibit characteristic responses against external fields. Independently around the same time, rapid development of nanotechnologies has driven materials hunting for magnetic semiconductors. The materials introduced here are some of such materials. The materials development is actively ongoing, from which many new materials and phenomena are being discovered one after another even now. The author is confident that in the course of such studies, a new class of magnetoresistance materials will be found in near future.

References

1 Pippard, A.B. (1989) *Magnetoresistance in Metals*, Cambridge University Press.
2 Singleton, J. (2000) Studies of quasi-two-dimensional organic conductors based on BEDT-TTF using high magnetic fields. *Rep. Prog. Phys.*, **63**, 1111.
3 Žutić, I., Fabian, J., and Das Sarma, S. (2004) Spintronics: fundamentals and applications. *Rev. Mod. Phys.*, **76**, 323–410.
4 Maekawa, S., Tohyama, T., Barnes, S., Ishihara, S., Koshibae, W., and Khaliullin, G. (2004) *Physics of Transition Metal Oxides*, Springer Series in Solid-State Sciences, Springer-Verlag, Berlin.
5 Thompson, S.M. (2008) The discovery, development and future of GMR: the Nobel Prize 2007. *J. Phys. D Appl. Phys.*, **41**, 093001.
6 Nagaev, E. (2001) Colossal-magnetoresistance materials: manganites and conventional ferromagnetic semiconductors. *Phys. Rep.*, **346**, 387–531.
7 Ramirez, A.P. (1997) Colossal magnetoresistance. *J. Phys. Condens. Matter*, **9**, 8171.
8 Dagotto, E. (2003) *Nanoscale Phase Separation and Colossal Magnetoresistance*, Springer Series in Solid-State Sciences, Springer-Verlag, Berlin.
9 Dietl, T. and Ohno, H. (2014) Dilute ferromagnetic semiconductors: physics and spintronic structures. *Rev. Mod. Phys.*, **86**, 187–251.
10 Pearton, S.J., Heo, W.H., Ivill, M., Norton, D.P., and Steiner, T. (2004) Dilute magnetic semiconducting oxides. *Semicond. Sci. Tech.*, **19**, R59.
11 MacDonald, A.H., Schiffer, P., and Samarth, N. (2005) Ferromagnetic semiconductors: moving beyond (Ga,Mn)As. *Nat. Mater.*, **4**, 195–202.
12 Castro Neto, A.H., Guinea, F., Peres, N.M.R., Novoselov, K.S., and Geim, A.K. (2009) The electronic properties of graphene. *Rev. Mod. Phys.*, **81**, 109–162.
13 Ashcroft, N. and Mermin, N. (1976) *Solid State Physics*, Saunders College, Philadelphia.
14 Ong, N.P. (1991) Geometric interpretation of the weak-field Hall conductivity in two-dimensional metals with arbitrary Fermi surface. *Phys. Rev. B*, **43**, 193–201.
15 Motohashi, T., Ueda, R., Naujalis, E., Tojo, T., Terasaki, I., Atake, T., Karppinen, M., and Yamauchi, H. (2003) Unconventional magnetic transition and transport behavior in $Na_{0.75}CoO_2$. *Phys. Rev. B*, **67**, 064–406.
16 Ali, M.N., Xiong, J., Flynn, S., Tao, J., Gibson, Q.D., Schoop, L.M., Liang, T., Haldolaarachchige, N., Hirschberger, M., Ong, N.P., and Cava, R.J. (2014) Large, non-saturating magnetoresistance in WTe_2. *Nature*, **514**, 205–208.
17 Smit, J. (1951) Magnetoresistance of ferromagnetic metals and alloys at low temperatures. *Physica*, **17**, 612–627.

18 McGuire, T. and Potter, R. (1975) Anisotropic magnetoresistance in ferromagnetic 3d alloys. *Mag. IEEE Trans.*, **11**, 1018–1038.

19 Audouard, A. and Askenazy, S. (1995) Spin-density-wave transition and resistivity minimum of the Bechgaard salt (TMTSF)$_2$NO$_3$ at high magnetic field, where TMTSF is tetramethyltetraselenafulvalene. *Phys. Rev. B*, **52**, R700–R703.

20 Baibich, M.N., Broto, J.M., Fert, A., Van Dau, F.N., Petroff, F., Etienne, P., Creuzet, G., Friederich, A., and Chazelas, J. (1988) Giant magnetoresistance of (001)Fe/(001) Cr magnetic superlattices. *Phys. Rev. Lett.*, **61**, 2472–2475.

21 Binasch, G., Grünberg, P., Saurenbach, F., and Zinn, W. (1989) Enhanced magnetoresistance in layered magnetic structures with antiferromagnetic interlayer exchange. *Phys. Rev. B*, **39**, 4828–4830.

22 Santen, J.V. and Jonker, G. (1950) Electrical conductivity of ferromagnetic compounds of manganese with perovskite structure. *Physica*, **16**, 599–600.

23 Searle, C.W. and Wang, S.T. (1970) Studies of the ionic ferromagnet (LaPb) MnO$_3$. V. Electric transport and ferromagnetic properties. *Can. J. Phys.*, **48**, 2023–2031.

24 Zener, C. (1951) Interaction between the d-shells in the transition metals. II. Ferromagnetic compounds of manganese with perovskite structure. *Phys. Rev.*, **82**, 403–405.

25 Anderson, P.W. and Hasegawa, H. (1955) Considerations on double exchange. *Phys. Rev.*, **100**, 675–681.

26 Tokura, Y., Urushibara, A., Moritomo, Y., Arima, T., Asamitsu, A., Kido, G., and Furukawa, N. (1994) Giant magnetotransport phenomena in filling-controlled kondo lattice system: La$_{1-x}$Sr$_x$MnO$_3$. *J. Phys. Soc. Jpn.*, **63**, 3931–3935.

27 Urushibara, A., Moritomo, Y., Arima, T., Asamitsu, A., Kido, G., and Tokura, Y. (1995) Insulator-metal transition and giant magnetoresistance in La$_{1-x}$Sr$_x$MnO$_3$. *Phys. Rev. B*, **51**, 14103–14109.

28 Furukawa, N. (1995) Magnetoresistance of the double-exchange model in infinite dimension. *J. Phys. Soc. Jpn.*, **64**, 2734–2737.

29 Kugel, K.I. and Khomskii, D.I. (1982) The Jahn-Teller effect and magnetism: transition metal compounds. *Sov. Phys. Usp.*, **25**, 231–256.

30 Murakami, Y., Hill, J.P., Gibbs, D., Blume, M., Koyama, I., Tanaka, M., Kawata, H., Arima, T., Tokura, Y., Hirota, K., and Endoh, Y. (1998) Resonant x-ray scattering from orbital ordering in LaMnO$_3$. *Phys. Rev. Lett.*, **81**, 582–585.

31 Palstra, T.T.M., Ramirez, A.P., Cheong, S.W., Zegarski, B.R., Schiffer, P., and Zaanen, J. (1997) Transport mechanisms in doped LaMnO$_3$: evidence for polaron formation. *Phys. Rev. B*, **56**, 5104–5107.

32 Jonker, G. and Santen, J.V. (1953) Magnetic compounds wtth perovskite structure III. Ferromagnetic compounds of cobalt. *Physica*, **19**, 120–130.

33 Yamaguchi, S., Taniguchi, H., Takagi, H., Arima, T., and Tokura, Y. (1995) Magnetoresistance in metallic crystals of La$_{1-x}$Sr$_x$CoO$_3$. *J. Phys. Soc. Jpn.*, **64**, 1885–1888.

34 Kimura, T., Tomioka, Y., Kuwahara, H., Asamitsu, A., Tamura, M., and Tokura, Y. (1996) Interplane tunneling magnetoresistance in a layered manganite crystal. *Science*, **274**, 1698–1701.

35 Shimakawa, Y., Kubo, Y., and Manako, T. (1996) Giant magnetoresistance in Ti$_2$Mn$_2$O$_7$ with the pyrochlore structure. *Nature*, **379**, 53–55.

36 Ramirez, A.P., Cava, R.J., and Krajewski, J. (1997) Colossal magnetoresistance in Cr-based chalcogenide spinels. *Nature*, **386**, 156–159.

37 Kobayashi, K.-I., Kimura, T., Sawada, H., Terakura, K., and Tokura, Y. (1998) Room-temperature magnetoresistance in an oxide material with an ordered double-perovskite structure. *Nature*, **395**, 677–680.

38 Taskin, A.A., Lavrov, A.N., and Ando, Y. (2003) Ising-like spin anisotropy and competing antiferromagnetic-ferromagnetic orders in GdBaCo$_2$O$_{5.5}$ single Crystals. *Phys. Rev. Lett.*, **90**, 227201.

39 Matsuno, J., Okimoto, Y., Fang, Z., Yu, X.Z., Matsui, Y., Nagaosa, N., Kawasaki,

M., and Tokura, Y. (2004) Metallic ferromagnet with square-lattice CoO_2 sheets. *Phys. Rev. Lett.*, **93**, 167202.

40 Nakamura, R., Tozawa, J., Akaki, M., Akahoshi, D., Itatani, K., and Kuwahara, H. (2012) Anisotropic magneto-transport properties of layered perovskite $Sr_3Fe_{2-x}Co_xO_{7-\delta}$ crystals. *J. Phys. Conf. Ser.*, **400**, 032–060.

41 Ishiwata, S., Terasaki, I., Ishii, F., Nagaosa, N., Mukuda, H., Kitaoka, Y., Saito, T., and Takano, M. (2007) Two-staged magnetoresistance driven by the ising-like spin sublattice in $SrCo_6O_{19}$. *Phys. Rev. Lett.*, **98**, 217201.

42 Onuki, Y., Settai, R., Sugiyama, K., Takeuchi, T., Kobayashi, T.C., Haga, Y., and Yamamoto, E. (2004) Recent advances in the magnetism and superconductivity of heavy fermion systems. *J. Phys. Soc. Jpn.*, **73**, 769–787.

43 Sumiyama, A., Oda, Y., Nagano, H., Onuki, Y., Shibutani, K., and Komatsubara., T. (1986) Coherent kondo state in a dense kondo substance: $Ce_xLa_{1-x}Cu_6$. *J. Phys. Soc. Jpn.*, **55**, 1294–1304.

44 Terasaki, I., Sasago, Y., and Uchinokura, K. (1997) Large thermoelectric power in $NaCo_2O_4$ single crystals. *Phys. Rev. B*, **56**, R12685–R12687.

45 Maignan, A., Klein, Y., Hébert, S., Pelloquin, D., Hervieu, M., and Raveau, B. (2005) Oxides as new thermoelectric materials. *Mater. Integr.*, **18**, 37–40.

46 Koumoto, K., Terasaki, I., and Funahashi, R. (2006) Complex oxide materials for potential thermoelectric applications. *MRS Bull.*, **31**, 206–210.

47 He, J., Liu, Y., and Funahashi, R. (2011) Oxide thermoelectrics: the challenges, progress, and outlook. *J. Mater. Res.*, **26**, 1762–1772.

48 Koshibae, W., Tsutsui, K., and Maekawa, S. (2000) Thermopower in cobalt oxides. *Phys. Rev. B*, **62**, 6869–6872.

49 Masset, A.C., Michel, C., Maignan, A., Hervieu, M., Toulemonde, O., Studer, F., Raveau, B., and Hejtmanek, J. (2000) Misfit-layered cobaltite with an anisotropic giant magnetoresistance: $Ca_3Co_4O_9$. *Phys. Rev. B*, **62**, 166–175.

50 Maignan, A., Hébert, S., Hervieu, M., Michel, C., Pelloquin, D., and Khomskii, D. (2003) Magnetoresistance and magnetothermopower properties of Bi/Ca/Co/O and Bi(Pb)/Ca/Co/O misfit layer cobaltites. *J. Phys. Condens. Matter*, **15**, 2711.

51 Limelette, P., Soret, J.C., Muguerra, H., and Grebille, D. (2008) Magnetoresistance scaling in the layered cobaltate $Ca_3Co_4O_9$. *Phys. Rev. B*, **77**, 245123.

52 Tsukada, I., Yamamoto, T., Takagi, M., Tsubone, T., Konno, S., and Uchinokura, K. (2001) Ferromagnetism and large negative magnetoresistance in Pb doped Bi–Sr–Co–O misfit-layer compound. *J. Phys. Soc. Jpn.*, **70**, 834–840.

53 Terasaki, I. (2001) Cobalt oxides and Kondo semiconductors: a pseudogap system as a thermoelectric materials. *Mater. Trans.*, **42**, 951–955.

54 Terasaki, I. and Fujii, T. (2003) Magneto-thermoelectric effects of the layered cobalt oxides, in Proceedings of The 22nd International Conference on Thermoelectrics (ICT2003), IEEE, Piscataway, pp. 207–210.

55 Hanasaki, N., Tajima, H., Matsuda, M., Naito, T., and Inabe, T. (2000) Giant negative magnetoresistance in quasi-one-dimensional conductor $TPP[Fe(Pc)(CN_2]_2$: interplay between local moments and one-dimensional conduction electrons. *Phys. Rev. B*, **62**, 5839–5842.

56 Takahide, Y., Konoike, T., Enomoto, K., Nishimura, M., Terashima, T., Uji, S., and Yamamoto, H.M. (2007) Large positive magnetoresistance of insulating organic crystals in the non-ohmic region. *Phys. Rev. Lett.*, **98**, 116602.

57 Kawano-Furukawa, H., Kajimoto, R., Yoshizawa, H., Tomioka, Y., Kuwahara, H., and Tokura, Y. (2003) Orbital order and a canted phase in the paramagnetic and ferromagnetic states of 50% hole-doped colossal magnetoresistance manganites. *Phys. Rev. B*, **67**, 174422.

58 Nakamura, K., Arima, T., Nakazawa, A., Wakabayashi, Y., and Murakami, Y. (1999) Polarization-dependent resonant-x-ray diffraction in charge- and orbital-ordering phase of $Nd_{1/2}Sr_{1/2}MnO_3$. *Phys. Rev. B*, **60**, 2425–2428.

59 Kuwahara, H., Tomioka, Y., Asamitsu, A., Moritomo, Y., and Tokura, Y. (1995) A first-order phase transition induced by a magnetic field. *Science*, **270**, 961–963.

60 Tokura, Y., Kuwahara, H., Moritomo, Y., Tomioka, Y., and Asamitsu, A. (1996) Competing instabilities and metastable states in $(Nd,Sm)_{1/2}Sr_{1/2}MnO_3$. *Phys. Rev. Lett.*, **76**, 3184–3187.

61 Tomioka, Y., Asamitsu, A., Moritomo, Y., and Tokura., Y. (1995) Anomalous magnetotransport properties of $Pr_{1-x}Ca_xMnO_3$. *J. Phys. Soc. Jpn.*, **64**, 3626–3630.

62 Dagotto, E., Hotta, T., and Moreo, A. (2001) Colossal magnetoresistant materials: the key role of phase separation. *Phys. Rep.*, **344**, 1–153.

63 Uehara, M., Mori, S., Chen, C.H., and Cheong, S.-W. (1999) Percolative phase separation underlies colossal magnetoresistance in mixed-valent manganites. *Nature*, **399**, 560–563.

64 Fäth, M., Freisem, S., Menovsky, A.A., Tomioka, Y., Aarts, J., and Mydosh, J.A. (1999) Spatially inhomogeneous metal-insulator transition in doped manganites. *Science*, **285**, 1540–1542.

65 Julliere, M. (1975) Tunneling between ferromagnetic films. *Phys. Lett. A*, **54**, 225–226.

66 Hu, B. and Wu, Y. (2007) Tuning magnetoresistance between positive and negative values in organic semiconductors. *Nat. Mater.*, **6**, 985–991.

67 Burgy, J., Mayr, M., Martin-Mayor, V., Moreo, A., and Dagotto, E. (2001) Colossal effects in transition metal oxides caused by intrinsic inhomogeneities. *Phys. Rev. Lett.*, **87**, 277202.

68 Kageyama, H., Nakajima, T., Ichihara, M., Ueda, Y., Yoshizawa, H., and Ohoyama, K. (2003) New stacking variations of the charge and orbital ordering in the metal-ordered manganite $YBaMn_2O_6$. *J. Phys. Soc. Jpn.*, **72**, 241–244.

69 Nakajima, T., Kageyama, H., Yoshizawa, H., and Ueda, Y. (2002) Structures and electromagnetic properties of new metal-ordered manganites: $RBaMn_2O_6$ (R=Y and rare-earth elements). *J. Phys. Soc. Jpn.*, **71**, 2843–2846.

70 Motome, Y., Furukawa, N., and Nagaosa, N. (2003) Competing orders and disorder-induced insulator to metal transition in manganites. *Phys. Rev. Lett.*, **91**, 167204.

71 Ishiwata, S., Nakano, T., Terasaki, I., Nakao, H., Murakami, Y., Uwatoko, Y., and Takano, M. (2011) Uniaxial colossal magnetoresistance in the Ising magnet $SrCo_{12}O_{19}$. *Phys. Rev. B*, **83**, 020401.

72 Brossard, L., Clerac, R., Coulon, C., Tokumoto, M., Ziman, T., Petrov, D., Laukhin, V., Naughton, M., Audouard, A., Goze, F., Kobayashi, A., Kobayashi, H., and Cassoux, P. (1998) Interplay between chains of $S = 5/2$ localised spins and two-dimensional sheets of organic donors in the synthetically built magnetic multilayer $\lambda\text{-}(BETS)_2FeCl_4$. *Eur. Phys. J. B*, **1**, 439–452.

73 Uji, S., Shinagawa, H., Terashima, T., Yakabe, T., Terai, Y., Tokumoto, M., Kobayashi, A., Tanaka, H., and Kobayashi, H. (2001) Magnetic-field-induced superconductivity in a two-dimensional organic conductor. *Nature*, **410**, 908–910.

74 Blundell, S. (2001) *Magnetism in Condensed Matter*, Oxford University Press.

75 Fukumura, T., Toyosaki, H., Ueno, K., Nakano, M., Yamasaki, T., and Kawasaki, M. (2007) A scaling relation of anomalous Hall effect in ferromagnetic semiconductors and metals. *Jpn. J. Appl. Phys.*, **46**, L642.

76 Shapira, Y. and Reed, T.B. (1972) Resistivity and Hall effect of EuS in fields up to 140 kOe. *Phys. Rev. B*, **5**, 4877–4890.

77 Shapira, Y., Foner, S., Oliveira, N.F., and Reed, T.B. (1972) EuTe. II. resistivity and Hall effect. *Phys. Rev. B*, **5**, 2647–2657.

78 Matsukura, F., Ohno, H., Shen, A., and Sugawara, Y. (1998) Transport properties and origin of ferromagnetism in (Ga,Mn)As. *Phys. Rev. B*, **57**, R2037–R2040.

79 Matsumoto, Y., Takahashi, R., Murakami, M., Koida, T., Fan, X.J., Hasegawa, T., Fukumura, T., Kawasaki, M., Koshihara, S.Y., and Koinuma, H. (2001) Ferromagnetism in Co-doped TiO_2 rutile thin films grown by laser molecular beam epitaxy. *Jpn. J. Appl. Phys.*, **40**, L1204.

80 Griffin, K.A., Pakhomov, A.B., Wang, C.M., Heald, S.M., and Krishnan, K.M.

(2005) Intrinsic ferromagnetism in insulating cobalt doped anatase TiO_2. *Phys. Rev. Lett.*, **94**, 157204.

81 Kim, J.-Y., Park, J.-H., Park, B.-G., Noh, H.-J., Oh, S.-J., Yang, J.S., Kim, D.-H., Bu, S.D., Noh, T.-W., Lin, H.-J., Hsieh, H.-H., and Chen, C.T. (2003) Ferromagnetism induced by clustered Co in co-doped anatase TiO_2 thin films. *Phys. Rev. Lett.*, **90**, 017401.

82 Hu, W., Hayashi, K., Fukumura, T., Akagi, K., Tsukada, M., Happo, N., Hosokawa, S., Ohwada, K., Takahasi, M., Suzuki, M., and Kawasaki, M. (2015) Spontaneous formation of suboxidic coordination around Co in ferromagnetic rutile $Ti_{0.95}Co_{0.95}O_2$ film. *Appl. Phys. Lett.*, **106**.

83 Sasaki, T., Fukuda, T., Yoneyama, N., and Kobayashi, N. (2003) Shubnikov–de haas effect in the quantum vortex liquid state of the organic superconductor κ-BEDT-$TTF_2Cu(NCS)_2$. *Phys. Rev. B*, **67**, 144521.

84 Yamaji, K. (1989) On the angle dependence of the magnetoresistance in quasi-two-dimensional organic superconductors. *J. Phys. Soc. Jpn.*, **58**, 1520–1523.

85 Kajita, K., Nishio, Y., Takahashi, T., Sasaki, W., Kato, R., Kobayashi, H., Kobayashi, A., and Iye, Y. (1989) A new type oscillatory phenomenon in the magnetotransport of θ-$(BEDT-TTF)_2I_3$. *Solid State Commun.*, **70**, 1189–1193.

86 Kartsovnik, M.V., Laukhin, V.N., Pesotskii, S.I., Schegolev, I.F., and Yakovenko., V.M. (1992) Angular magnetoresistance oscillations and the shape of the fermi surface in $\beta(ET)_2IBr_2$. *J. Phys. I France*, **2**, 89–99.

87 Ohmichi, E., Adachi, H., Mori, Y., Maeno, Y., Ishiguro, T., and Oguchi, T. (1999) Angle-dependent magnetoresistance oscillation in the layered perovskite Sr_2RuO_4. *Phys. Rev. B*, **59**, 7263–7265.

88 Hussey, N.E., Abdel-Jawad, M., Carrington, A., Mackenzie, A.P., and Balicas, L. (2003) A coherent three-dimensional Fermi surface in a high-transition-temperature superconductor. *Nature*, **425**, 814–817.

89 Novoselov, K.S., Jiang, Z., Zhang, Y., Morozov, S.V., Stormer, H.L., Zeitler, U., Maan, J.C., Boebinger, G.S., Kim, P., and Geim, A.K. (2007) Room-temperature quantum hall effect in graphene. *Science*, **315**, 1379.

90 Novoselov, K.S., Geim, A.K., Morozov, S.V., Jiang, D., Katsnelson, M.I., Grigorieva, I.V., Dubonos, S.V., and Firsov, A.A. (2005) Two-dimensional gas of massless Dirac fermions in graphene. *Nature*, **438**, 197–200.

17
Magnetic Frustration in Spinels, Spin Ice Compounds, $A_3B_5O_{12}$ Garnet, and Multiferroic Materials

Hongyang Zhao,[1] Hideo Kimura,[2] Zhenxiang Cheng,[3] and Tingting Jia[2]

[1]Wuhan Institute of Technology, School of Materials Science and Engineering, Hubei Key Laboratory of Plasma Chemistry and Advanced Materials, 206 Guanggu first Road, Wuhan 430205, China
[2]National Institute for Materials Science, Magnetoelectric Crystal Group, Research Center for Functional Materials, 1-2-1 Sengen, Tsukuba 305-0047, Japan
[3]University of Wollongong, Institute for Superconducting and Electronics Materials, Innovation Campus, North Wollongong, NSW 2500, Australia

17.1
Frustration: An Introduction

Sometimes, a little frustration can make life interesting. In physics, frustration refers to the presence of competing forces that cannot be simultaneously satisfied [1]. Frustrated magnetism has become an extremely active field of research. After undergoing a revival in the 1990s in the context of applying Anderson's resonating-valence-bond (RVB) theory to high-temperature superconductors, the subject has experienced a tremendous burst of theoretical and experimental activity in the last decade [2]. In some sense, frustrated systems are excellent candidates to test approximations and improve theories. The study of order–disorder phenomena is a fundamental task of equilibrium statistical mechanics. The concept has been applied broadly from magnetism, which we discuss here, to negative thermal expansion of solids [3] and to soft materials [4]. Great efforts have been made to understand the basic mechanisms responsible for spontaneous ordering as well as the nature of the phase transition in many kinds of systems, especially in physics, where frustration often leads to exotic properties of materials. When interactions between magnetic degrees of freedom in a lattice are incompatible with the underlying crystal geometry, exotic phenomena such as spin ice and spin liquid phases can emerge [5]. The word "frustration" has been introduced [6] to describe the situation where a spin (or a number of spins) in the system cannot find an orientation to fully satisfy all the interactions with its neighboring spins. This definition can be applied to Ising spins, Potts models, and Vector spins. In general, the frustration is caused either by competing

Handbook of Solid State Chemistry, First Edition. Edited by Richard Dronskowski, Shinichi Kikkawa, and Andreas Stein.
© 2017 Wiley-VCH Verlag GmbH & Co. KGaA. Published 2017 by Wiley-VCH Verlag GmbH & Co. KGaA.

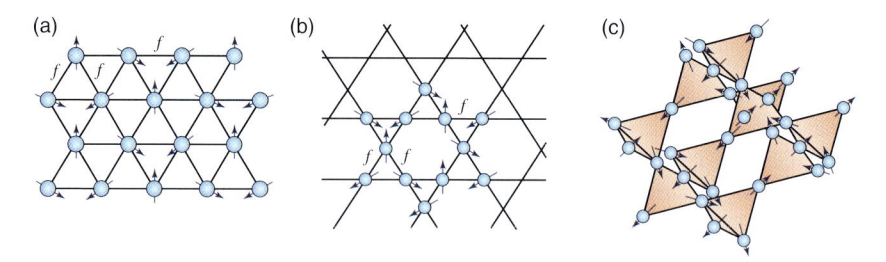

Figure 17.1 Frustrated magnetism on 2D and 3D lattices. Two types of 2D lattice are depicted: a triangular lattice (a) and a kagome lattice (b). The 3D lattice depicted is a pyrochlore lattice (c). Blue circles denote magnetic ions, arrows indicate the direction of spin and black lines indicate the shape of the lattice. In (b), ions and spins are depicted on only part of the illustrated lattice. (Reproduced with permission from Ref. [1]. Copyright 2010, Nature Publishing Group.)

interactions (such as the Villain model [7]) or by lattice structure, as in the triangular, face-centered cubic (fcc), and hexagonal-close-packed (hcp) lattices, with antiferromagnetic nearest-neighbor (nn) interaction. Such geometric frustration occurs in systems of spins on lattices that involve triangular motifs (Figure 17.1), in which the nearest-neighbor interactions favor antialigned spins (Figure 17.2). On a triangular lattice, all the three spins cannot be antiparallel. Instead, depending on the circumstances, the spins fluctuate, or order, in a less obvious manner. The effects of frustration are rich and often unexpected, and the field of frustrated magnetism has emerged naturally over the years: classical versus quantum, two-dimensional versus three-dimensional, rare earth versus transition metal ions, corner-sharing versus edge-sharing lattices, Ising versus Heisenberg interactions, and so on. It would be dangerous to neglect these differences,

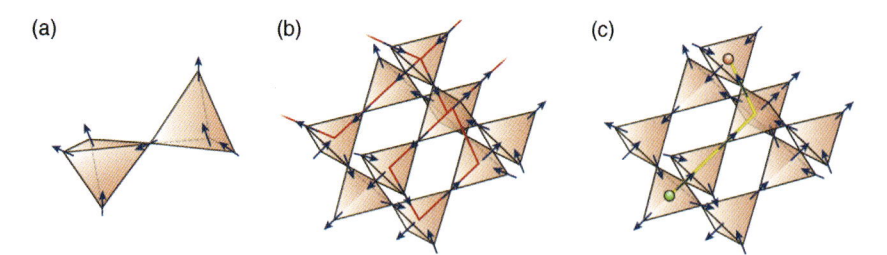

Figure 17.2 Spins, artificial magnetic fields and monopoles in spin ice. (a) A ground-state configuration of spins is shown in a pyrochlore lattice. The spin obeys the constraints of the ice rules that mandate two inward-pointing spins and two outward-pointing ones on each tetrahedron. (b) For the same lattice type, some of the loops of "magnetic flux" are shown (red lines), as defined by the mapping of spins to an artificial magnetic field. (c) For the same lattice type, a monopole (green) and antimonopole (red) are shown. These are created by flipping the "string" of spins connected by the yellow line (compare with the spins in (b)). (Reproduced with permission from Ref. [1]. Copyright 2010, Nature Publishing Group.)

because this is a field where details matter. Indeed, small interactions are often responsible for the ultimate selection from among several candidate ground states. This is why, with the exception of a small number of more general chapters, a significant fraction of the contributions to this volume concentrate on specific aspects or materials rather than on broader principles [2]. Many of them are not understood yet.

Frustrated magnets are materials in which localized magnetic moments, or spin, interact through competing exchange interactions that cannot be simultaneously satisfied, giving rise to a large degeneracy of the system ground state. Studies of frustration began with antiferromagnets, in which frustration usually has a simple geometric origin [1]. In addition to the fact that real magnetic materials are often frustrated due to several kinds of interactions, frustrated spin systems have their own interest in statistical mechanics. Recent studies show that many established statistical methods and theories have encountered many difficulties in dealing with frustrated systems. Since the mechanisms of many phenomena are not understood in real systems (disordered systems, systems with long-range interaction, three-dimensional systems, etc.), it is worth searching for the origins of these phenomena in exactly solved systems. These exact results will help us to understand qualitatively the behavior of real systems, which are, in general, much more complicated.

In this review, we survey how the discovery of new materials and improved experimental probes, together with complementary advances in theory, have reinvigorated the study of spin liquids and frustrated magnetism in general. We begin with several concepts in magnetic frustration: Geometric Frustration in Simple Atomic Systems, Fluctuations of the Spins in a Spin Liquid–Classical or Quantum, Spin Vice. After describing their basic physics, we discuss the experimental situation and Interesting Magnetic Frustrated Materials: Highly Frustrated Magnetism in Spinels, Spin Ice Compounds: $Ho_2Ti_2O_7$ and $Dy_2Ti_2O_7$, $A_3B_5O_{12}$ Garnet Structure, and Magnetic Frustration in Multiferroic Materials.

17.2
Several Concepts in Frustrated Magnets

Frustrated magnets are very important in condensed matter physics. On the fundamental side is the search for principles that help to organize the variety of behavior that we observe around us. On the applied side, the instabilities exhibited by frustrated magnets open a window on the richness of nature realized in different materials [5]. An antiferromagnet consisting of a two-dimensional triangular lattice of Ising spins (which point either upward or downward along a single axis) provides one of the prototypes of frustration. Wannier showed in 1950 that this model has very large ground state degeneracy [8]. Now, such degeneracy is considered to be characteristic of frustration. The antiferromagnetic triangle is the simplest case in which a conflict arises between the geometry of the space inhabited by a set of degrees of freedom and the local correlations

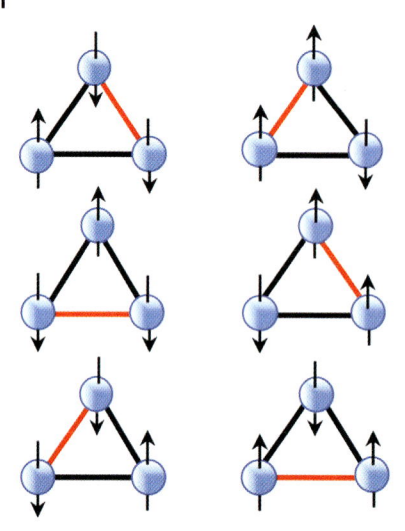

Figure 17.3 Elements of frustration. A triangle of antiferromagnetically interacting Ising spins, which must point upward or downward, is the simplest example of frustration. All three spins cannot be antiparallel. Instead of the two ground states mandated by the Ising symmetry (up and down), there are six ground states (blue circles denote magnetic ions, arrows indicate the direction of spin, and black and red lines indicate the shape of the triangular lattice, with red lines denoting the axis on which the spins are parallel). (Reproduced with permission from Ref. [5]. Copyright 2006, American Institute of Physics.)

favored by their interactions. This phenomenon is a powerful paradigm for discovery over the past few decades. Two other particularly well-studied aspects are low effective dimensionality for electronic systems and tunable optical lattices for systems of cold atoms [5]. At low temperatures, the spins continue to fluctuate thermally, although in a correlated manner, because they are restricted to the ground states of the Ising antiferromagnet. By analogy to an ordinary liquid, in which the molecules form a dense, highly correlated state that has no static order, the spins in the triangular Ising antiferromagnet form a "spin liquid," or cooperative paramagnet. The "frustration" parameter, f (Figure 17.1), provides a quantitative measure of the depth of the spin-liquid regime; $f = \infty$ if the spins remain liquid down to a temperature, T, of absolute zero [1]. Figure 17.3 shows the elements of frustration on 2D and 3D lattices; such degeneracies can persist, fluctuations are enhanced, and ordering is suppressed. Ramirez introduced a simple empirical measure of frustration that has become widely used [1,9].

17.2.1
Geometric Frustration in Simple Atomic Systems: The Curved-Space Approach

Toulouse introduced spin models in frustration [6]. These describe the situations in which one cannot minimize the energy of the system by merely minimizing all local interactions. The three antiferromagnetic interactions along the bonds forming the plaquette can never be satisfied simultaneously [10]. One

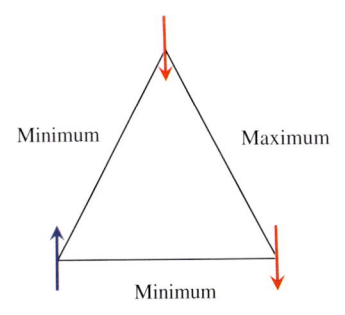

Figure 17.4 Elementary plaquette of a triangular lattice. When interactions between nearest neighbors at the neighboring Ising spins are antiferromagnetic, one can never satisfy (i.e., anti-align spins) more than two bonds at the same time.

simple illustration is shown in Figure 17.4, which is provided by an Ising spin model on a triangular lattice with antiferromagnetic interactions between nearest-neighbor spins: an elementary "plaquette" of the lattice made by one triangle of nearest-neighbor sites is frustrated. The classical frustrated systems have Hamiltonians with competing interactions that make contributions to the energy that cannot simultaneously be minimized. The concept was originally discussed in relation to spin glasses. As shown in Figure 17.5, frustration that destabilizes the Néel order has been investigated extensively in models with competing nearest-neighbor and further-neighbor interactions [11]. The J_1–J_2 model with the relation of $J_1 = J_2$ on the square lattice follows the classical ground state [2]. Spin glasses are a well-known example of frustrated systems; however, in this case, frustration is induced by the presence of quenched disorder, due, for instance, to frozen-in impurities [12]. This inhomogeneous and externally imposed frustration is not relevant for supercooled liquids, in which glassiness and heterogeneous behavior are generated by themselves. To emphasize the difference,

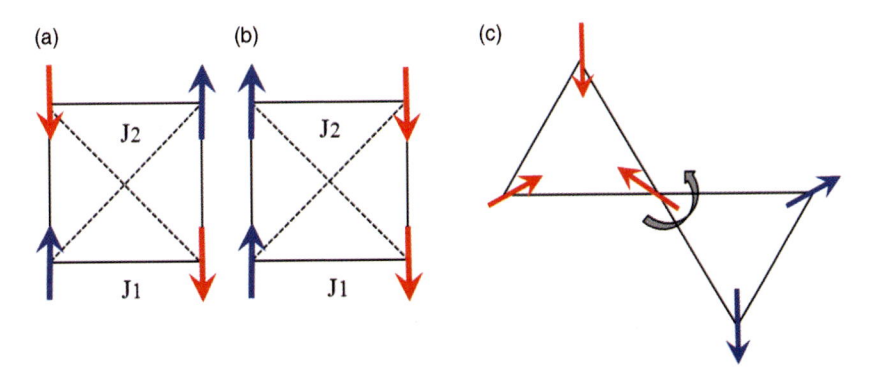

Figure 17.5 (a and b): J_1 – J_2 model, showing ground state spin configurations for: $2J_1 > J_2$ (a) and $J_2 > 2J_1$ (b). (c) Ground states of classical Heisenberg spins at vertices of two corner-sharing triangles, with degeneracy arising from rotations about the common spin, as indicated.

frustration in liquids has been given several detailed definitions, for example, "uniform," "geometric," "topological," or "structural" [10]. In icosahedral order in simple one-component liquids, the atoms interact through spherically symmetric pair potentials. The ground state of four atoms is a perfect tetrahedron, with the atoms on the vertices, and 20 tetrahedra are combined to form a regular icosahedron with 13 atoms. The importance of local icosahedral order in simple atomic liquids and liquid metals was first proposed by Frank [13]. The most stable cluster made of a central atom and a shell of 12 neighbors is indeed an icosahedron, and not an arrangement associated with the actual crystalline phases, body-centered cubic (bcc), face-centered cubic (fcc), or hexagonal close packed (hcp). In three-dimensional Euclidean space, however, the locally preferred structure (tetrahedral or icosahedra) cannot propagate freely to tile the whole space: this is *geometric frustration*. The global ground state of the system is instead an fcc or hcp crystal [10].

The concept of frustration is more easily grasped by contrasting the case of spherical particles in 3 dimensions with other situations [14,15]. The geometrical frustration in classical magnets can lead to macroscopic ground state degeneracy and the suppression of long-range order. Low temperature states in model systems, although disordered, are very different from those of a noninteracting paramagnet: correlations are power-law in space and decay in time at a rate set by the temperature alone. Many experimental systems display these features within the temperature window $T_c < T < |\Theta_{CW}|$ where the Curie–Weiss temperature Θ_{CW} characterizes the sign and strength of interactions. The behavior is dominated by nearest-neighbor exchange and is well summed up in the term coined by Jacques Villain [16], a pioneer in the field of cooperative paramagnetism [2]. Systems of spherical particles in two-dimensional Euclidean space, that is, discs on a plane, are not subject to frustration; the locally preferred structure is a regular hexagon, with one atom at the center and six vertices, and this structure can be periodically repeated to form a triangular lattice. Similarly, aligned hard cubes in three-dimensional Euclidean space form simple cubic arrangements that are both locally and globally preferred, thereby excluding any frustration effect [17].

17.2.2
Fluctuations of the Spins in a Spin Liquid can be Classical or Quantum [1]

17.2.2.1 Classical Spin Liquid
We start with a simple example, in which the spins can be regarded as classical: spin-ice materials [18–20]. In spin ices such as $Dy_2Ti_2O_7$, $Ho_2Ti_2O_7$, and $Ho_2Sn_2O_7$, the rare-earth atoms (Dy and Ho) are magnetic and sit on a pyrochlore lattice with a network of corner-sharing tetrahedra, as shown in Figure 17.1c. Their f-electron spins are large and classical, and they behave as Ising doublets aligned with the local <111> axis, which connects the centers of the two tetrahedra shared by that spin. Their interactions are predominantly long range and dipolar, but much of their physics can be understood solely from

an effective nearest-neighbor exchange energy, J_{eff}, which is ferromagnetic [21,22]. This ferromagnetic interaction is frustrated, owing to the different Ising axes of the spins. As illustrated in Figure 17.2a, the states that minimize the energy of a single tetrahedron are highly degenerate, consisting of all six configurations in which there are two spins in and two spins out from the center of the tetrahedron.

The system fluctuates almost entirely within the two-in and two-out manifold of states when $k_B T$, where k_B is the Boltzmann constant, has a far smaller value than J_{eff}. Thus, the number of such states is exponentially large, and within this limit, the low-temperature entropy is even. This entropy was measured in spin ice [23–25]. The spins remain paramagnetic in this regime, the "ice rules" imply strong correlations. The correlated paramagnet is a simple example of a classical spin liquid. If two spins on a tetrahedron point out, then the other two spins must point in. The spin liquid is qualitatively distinguishable from an ordinary paramagnet. The physics of frustrated magnetism and spin liquids has intriguing connections to diverse exotic phenomena. Now connetion between the resonating valence bond (RVB) states and high-temperature copper-oxide-based superconductivity attracts much interest in spin liquids [26]. In an RVB state, electrons bound into valence bonds are paired, even though the state is nonsuperconducting. Despite the elegance of this suggestion and decades of theoretical work, RVB superconductivity remains an experimentally unresolved question [27].

17.2.2.2 Quantum Spin Liquid (QSL)

In some systems, "quantum paramagnets" down to zero temperature, states are conveniently represented in terms of spins paired into rotationally invariant singlets, or "valence bonds" (VBs). We should at least distinguish two very different physical phases, valence bond crystals (or solids) and resonating-valence-bond (RVB) liquids. The properties of these two types of state are completely different at long distances; a VB crystal wave function is "localized" in the vicinity of one (or a small number of) simple "parent" VB configuration(s), which is (are) spatially regular; the wave function of an RVB liquid spreads over macroscopically many VB configurations which are very different, and distant from each other in configuration space [2]. It is difficult to decide whether a given spin model has a valence bond configuration (VBC) or a short-range RVB-liquid ground state, or something else. These states differ in their long-distance properties: confinement versus deconfinement of spinons and long- versus short-range bond–bond correlation functions. It is instructive to formulate the spin model in terms of spinons interacting with gage fields. Its application may be extended to other types of spin-liquid wave functions based on states with long-range VBs and possibly also possessing gapless spinons [2,28].

Valence band solid (VBS) states are interesting, however, they do not represent a true quantum spin liquid (QSL). It is a challenge for theorists to find the correct QSL among the all-possible RVB phases. In quantum mechanics, the magnitude of a spin is quantized in half integer units of $\hbar$ (where $\hbar$

denotes Planck's constant divided by 2π), the quantum of angular momentum. Classical fluctuations dominate for large spins (those with a size, S, much larger than the minimum size of ½) and are driven by thermal energy. Spins can be thought of as reorienting randomly with time, cycling through different microstates. When the energy $k_B T$ becomes too small, classical fluctuations cease and the spins either freeze or order. For small spins, with S comparable to ½, the quantum mechanical uncertainty principle produces zero-point motions comparable to the size of the spin itself, which persist down to $T=0$ K. Although they are similar to thermal fluctuations in some ways, quantum fluctuations can be phase coherent. If they are strong enough, the result is a QSL, which is a superposition state in which the spins simultaneously point in many different directions. In a QSL, the spins are highly entangled with one another in a subtle way that involves entanglement between widely separated spins. QSLs are more elusive than their classical counterparts, but offer a much greater conceptual pay-off. They may also be associated with exotic forms of superconductivity. Researchers are seeking QSL states in solid-state materials known as Mott insulators (in which electrons are localized to individual atomic or molecular orbitals but maintain their spin degree of freedom) and recently, in artificial frustrated lattices of ultracold atoms that have been created optically [1,26,29].

17.2.3
Spin Ice

Since the discovery of spin-ice materials over 10 years ago, their experimental and theoretical investigation has raised a range of profound questions concerning several aspects of the physics of frustrated magnetic systems. While these studies have led to an enhanced global understanding of many fundamental issues pertaining to frustration in condensed matter systems, some unanswered questions remain, particularly in the context of low-temperature spin dynamics, random disorder, and the properties of metallic Ising or pyrochlore systems. These questions can be expected to form the focus of spin-ice research for several years to come [2]. For materials with Ising spins, geometrical frustration takes a different form. An Ising spin has only two discrete orientations – up or down – because it is constrained to point along one axis. The spin cannot exhibit small deviations from those directions, which is different from the continuous case.

As shown in Figure 17.6, in common water ice the location of oxygen atoms is strictly periodic, however, the locations of the hydrogen ions are not so perfectly periodic. The ice forms a hexagonal structure that preserves the H–O distance found in water molecules in atmosphere. That distance is far less than half the distance between oxygen atoms. Therefore, each hydrogen atom can occupy one of two sites: one "ice rule" states that around each oxygen atom are two hydrogens that are close by, as in an H_2O molecule, and two that are far away. The approximate equivalence of pyrochlore spin ice to water ice follows from

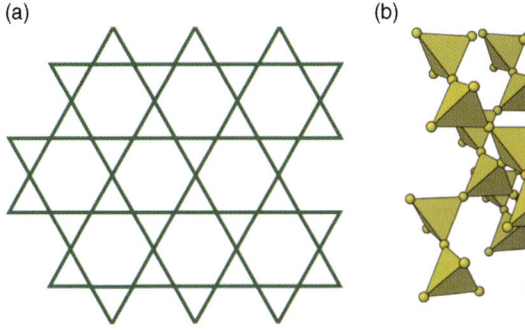

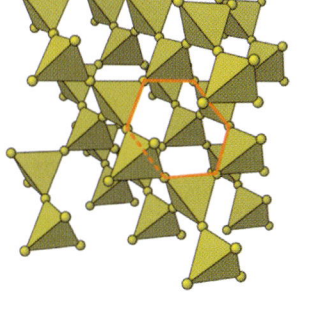

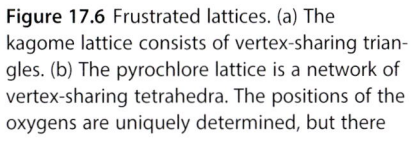

Figure 17.6 Frustrated lattices. (a) The kagome lattice consists of vertex-sharing triangles. (b) The pyrochlore lattice is a network of vertex-sharing tetrahedra. The positions of the oxygens are uniquely determined, but there are exponentially many allowed proton configurations. (Reproduced with permission from Ref. [5]. Copyright 2006, American Institute of Physics.)

identifying the centers of the pyrochlore tetrahedra with the locations of the oxygen atoms. The spins are then located at the midpoint of the bond between a pair of neighboring oxygens. The ground-state entropy of pyrochlore spin ice is thus expected to be approximately the same as it is in water ice [5]. Ramirez et al. tested that exception [30] by measuring the specific heat of dysprosium titanate, $Dy_2Ti_2O_7$, an Ising pyrochlore similar to $Ho_2Ti_2O_7$. Indeed, the resulting entropy agreed with the theoretical Pauling value for ice to within a few percent. In the absence of any structural disorder in the host compound, the nonzero entropy indicates that the spin ice represents a new state of magnetism [5].

17.3
Interesting Magnetic Frustrated Materials

More generally, the problem of frustration and spin-liquid physics in systems with mobile charges is relatively unstudied. There are several interesting materials in which this type of physics can be explored experimentally. The ancient Greeks were aware of the phenomenon of magnetic order in lodestone, a type of rock containing the ferromagnet magnetite Fe_3O_4 [5]. Magnetic moments in a ferromagnet tend to align and thereby sum to an easily observed macroscopic magnetic moment. The origin of this behavior can be simply illustrated by as few as three spins on a triangular lattice, as illustrated in Figure 17.7. On reorienting a single spin by δS, the total spin of the two tetrahedra would change by δS. If two of the spins on an elementary triangle are antialigned to satisfy their antiferromagnetic interation, the third one can no longer point in a direction opposite to both other spins. Not all the antiferromagnetic interactions can exist

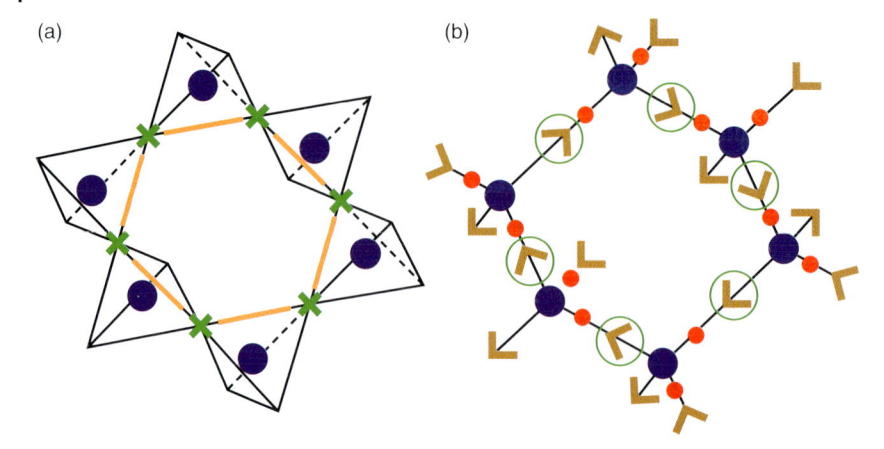

Figure 17.7 Local modes. (a) A hexagonal loop (orange), consisting of edges of a group of tetrahedra in the pyrochlore lattice, can support a zeroenergy mode. The mode involves reorienting neighboring spins (green crosses) by equal and opposite amounts δS. (b) Ice representation of such a loop. Oxygen atoms (blue) reside in the centers of the tetrahedra, while the spins (brown) sit on the midpoints of the bonds and point in the direction of the hydrogen atoms (red). (Reproduced with permission from Ref. [5]. Copyright 2006, American Institute of Physics.)

in their lowest energy state, and thus, they are incompatible with triangular lattice symmetry, a situation known as geometrical frustration [5].

17.3.1
Highly Frustrated Magnetism in Spinels [2]

The spinel structure, which has the chemical formula AB_2X_4 ($X = O$, S, Se), is one of the most frequently stabilized structural categories in complex transition metal oxides and chalcogenides. Materials with the spinel structure have provided physicists with a surprisingly rich variety of phenomena, including ferromagnetism, ferrimagnetism, and Jahn–Teller transitions [31]. As shown in Figure 17.8a, there are two basic structural units: AO_4 tetrahedra and BO_6 octahedra, which form the spinel structure. The BO_6 octahedra are connected by their edges, while the A ions form the AO_4 units with the coordinated O ions. B ions are located at every intersection of the chains, as shown in Figure 17.8b, and this network is known as the pyrochlore lattice, which is considered to give rise to very strong geometrical frustration effects. In Figure 17.8c, the pyrochlore structure is viewed from the <111> direction. There are two types of planes with alternate stacking: one is a two-dimensional triangular lattice, consisting of the apical B ions of the tetrahedra; the other is a two-dimensional kagomé lattice consisting of the B triangles at the bases of the tetrahedra [2].

Spinels have attracted increasing interest because they exemplify the physics of frustration. Here, we focus on frustrated magnetism and the phenomena that are

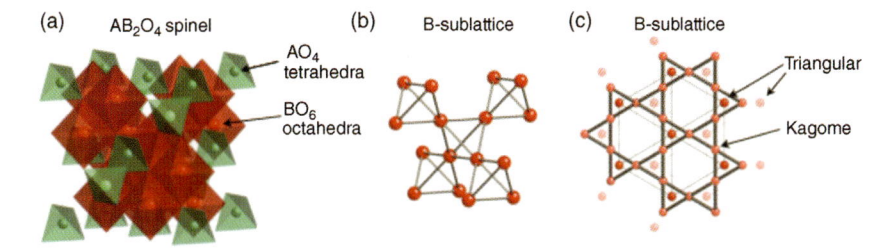

Figure 17.8 (a) Spinel structure, emphasizing the two basic structural units of AO_4 tetrahedra and BO_6 octahedra. (b) B-sublattice of the spinel structure, which defines a pyrochlore lattice. (c) B-sublattice structure (pyrochlore lattice) viewed from the <111> direction. (Reproduced with permission from Ref. [2]. Copyright 2011, Springer.)

induced by the oxide spinel structure. First, we discuss the structure and electronic properties of spinel oxides and propose a unique spinel structure. We then provide an overview of different spinel materials, emphasizing systems with spin and charge frustration on the B-sites and with spin frustration on the A-sites, to explain the physics of geometrical frustration in these compounds. How can long-range magnetic order emerge from such a spin-liquid state? A spin–lattice coupling is responsible for the appearance of magnetic ordering [2]. Usually, the distortion in the lattice makes magnetic coupling nonuniform, which inherently suppresses the geometrical frustration and lifts the spin degeneracy. This may induce a long-range magnetic order, which can be considered as a spin Jahn–Teller transition or as a three-dimensional analog of the spin-Peierls transition.

We selected normal spinels with magnetic B ions and nonmagnetic A ions: ACr_2O_4 (A = Zn, Cd, and Hg); the Cr in this spinel is trivalent. We selected $ZnCr_2O_4$ to discuss the highly frustrated magnetism in spinels. $ZnCr_2O_4$ shows geometrical frustration associated with the pyrochlore geometry of the B-sublattice. In the triply degenerate t_{2g} orbitals of Cr^{3+}, three electrons are accommodated and $S = 3/2$, indicating a Mott insulator with no orbital degrees of freedom. The strong frustration is clear from the temperature-dependence of the magnetic susceptibility χ_T, shown in Figure 17.9a. Curie–Weiss behavior is observed at high temperatures and an estimate of the effective moment gives $p_{eff} \approx 3.7\ \mu_B$, consistent with the expected value ($S = 3/2$). The antiferromagnetic Curie–Weiss temperature is −390 K, as obtained from the data. Despite the large value for the Curie–Weiss temperature, $|\Theta_{CW}| \geq 100$ K, no evidence for ordering was observed down to $T \approx 10$ K. The decrease in χ_T is discontinuous, with the Néel temperature, $T_N = 12.5$ K, which corresponds to long-range antiferromagnetic order [32,33]. The spin-liquid phase can be seen from the paramagnetic phase well below the Curie–Weiss temperature and above T_N. The magnetic entropy $S_m(T)$ for $ZnCr_2O_4$ as estimated from the magnetic specific heat $C_m(T)$, recovers to a value of ∼40% of the total spin entropy $S = R \cdot \ln 4$, where R is the gas

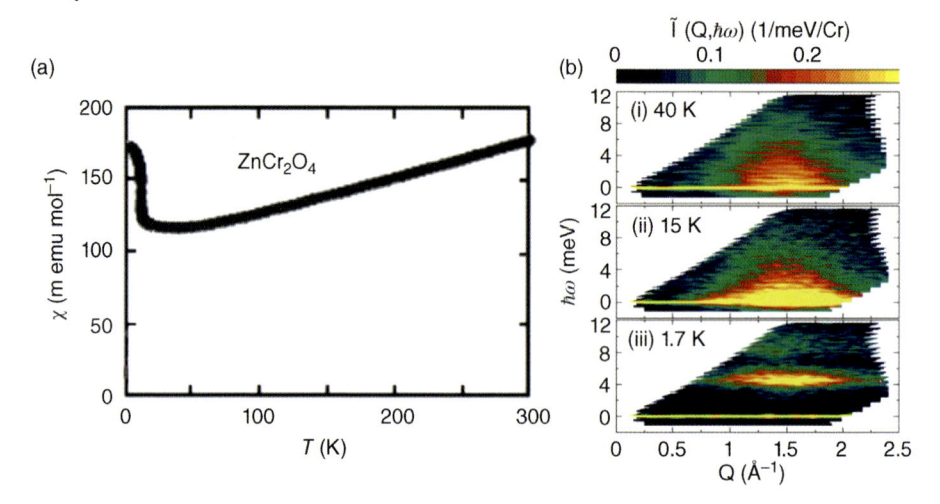

Figure 17.9 (a) Magnetic susceptibility χ_T for the $S = 3/2$ pyrochlore antiferromagnet ZnCr$_2$O$_4$, shown as $1/\chi$ versus T. (Reproduced with permission from Ref. [35]. Copyright 2006, Elsevier B.V. All rights reserved.) (b) Contour plots of magnetic neutron scattering intensity in the plane with energy transfer $\hbar\omega$ and wave number Q for ZnCr$_2$O$_4$. (Reproduced with permission from Ref. [32]. Copyright 2000, American Physical Society.)

constant, even at only 30 K, one order of magnitude lower in temperature than Θ_{CW} [34]. This large entropy should signal the presence of low-lying, strongly degenerate spin excitations produced by geometrical frustration. As shown in Figure 17.9b, these low excitations were detected by inelastic neutron scattering measurements with a very intense signal ($\omega = 0$ above $T_N = 12.5$ K) [32]. The presence of hexagonal antiferromagnetic spin clusters was revealed through the inelastic neutron scattering at low energy (1.5 meV) with $T = 15$ K ($T > T_N$) [34]. The antiferromagnetic ordering processes occurring at low temperatures are accompanied by structural phase transitions from cubic to tetragonal. This could be understood in terms of a spin Jahn–Teller picture. In ZnCr$_2$O$_4$, the spin structure is collinear and commensurate, but the lattice distortion is not uniform. If we use the magnetic field to tune the frustration, it may create a new phase of the spin–lattice complex in spinel materials [2].

As shown in Figure 17.10, the pyrochlore lattice contains hexagons in its kagomé planes. Magnetic fluctuations select configurations in which the spins have local antiferromagnetic order, and the ground-state manifold of states has zero spin on each tetrahedron. This gives rise to local excitations that are related to the spin orientation for each antiferromagnetic hexagon. The 1/2 plateau corresponds to configurations with three up and one down spin(s) in each tetrahedron. Based on model calculations, it has been noted that stabilizing the 1/2-plateau state over such an extended field range is possible only through the introduction of further terms in the magnetic Hamiltonian, most probably involving a spin–lattice coupling [36]. Indeed, a relatively large magnetostriction at the transition to the plateau state has been observed, although the spin–orbit

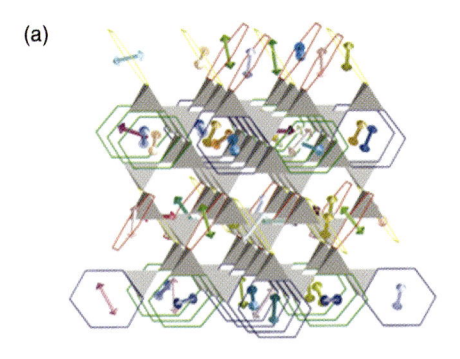

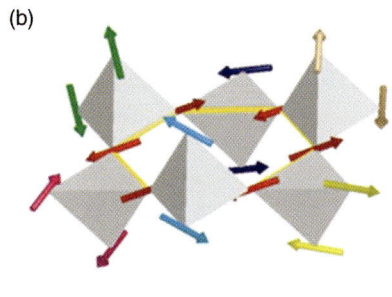

Figure 17.10 Antiferromagnetic hexagonal spin cluster formed in the spin-liquid phase of $ZnCr_2O_4$. (a) The lattice of corner-sharing tetrahedra formed by the octahedrally coordinated B-sites in a spinel structure with chemical formula AB_2O_4. A periodic assignment of all spins in the pyrochlore lattice is made to four different types of nonoverlapping hexagons, represented by the colors blue, green, red, and gold. Every spin belongs to just one hexagon, and each such hexagon carries a six-spin director. The resulting tetragonal structure of these hexagons has a unit cell of $2a \times 2a \times 3c$, and can be described by a stacking of two different types of three-layer slabs along the c-axis. The hexagon coverage on consecutive slabs is in fact uncorrelated, so that a macroscopic number of random slab-sequences can be generated. (b) Possible spin fluctuations in the classical ground-state manifold a spin cluster surrounding a hexagon (shown in yellow) in the pyrochlore lattice of (a). (Reproduced with permission from Ref. [34]. Copyright 2002, Nature Publishing Group.)

coupling is believed to be negligibly small here. The magnetizations measured at high magnetic fields revealed two other transitions [37]. It was found that the spin structure in the 1/2-plateau state has a rather complicated ordering pattern of "3-up, 1-down" tetrahedra with symmetry $P4_332$, which is reflected in the structural distortion [38].

17.3.2
Spin Ice Compounds: $Ho_2Ti_2O_7$ and $Dy_2Ti_2O_7$

The name "spin ice" refers to a class of magnetic compounds that may be described by spins on a lattice obeying a local "ice-rule" constraint. The specific examples of spin ice compounds that we will consider are $Ho_2Ti_2O_7$ and $Dy_2Ti_2O_7$. For a review on spin ice, see Ref. [39]. The dynamical objects in models of these compounds are the large spins of the rare-earth ions (e.g., angular momentum $J_{Ho} = 8$, $J_{Dy} = 15/2$), which reside on the sites of a pyrochlore lattice, as shown in Figure 17.11. As the figure shows, each pyrochlore site features corner sharing by two tetrahedra. Figure 17.12 shows a single tetrahedron inscribed in a cube with some important crystallographic directions labeled.

An important effect of the neighboring Ti and O atoms is that they cause a crystal field anisotropy which strongly favors maximizing the component of the Ho/Dy spin pointing along its "easy-axis," which is the local [111] direction. As

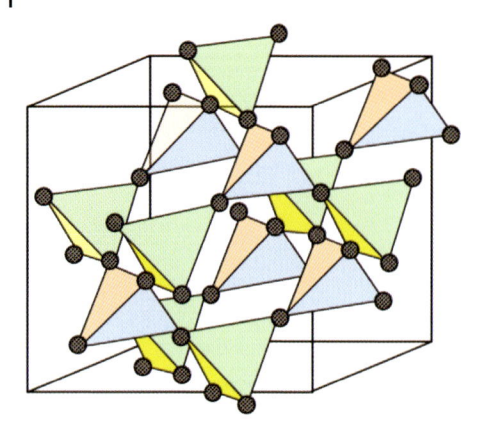

Figure 17.11 The pyrochlore lattice composed of corner-sharing tetrahedra.

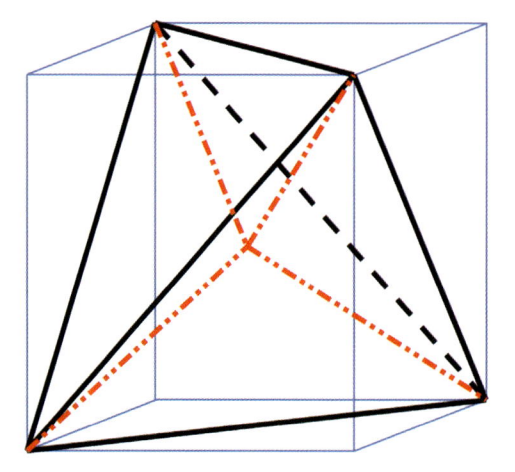

Figure 17.12 A single tetrahedron inscribed in a cube. The edges of the cube denote the [100] direction, and the edges of the tetrahedron are in the [110] direction. The [111] directions of the pyrochlore lattice are the body diagonals of the cube, and indicated by the dashed lines.

shown in Figures 17.11 and 17.12, this axis is the line joining the centers of the two tetrahedra sharing the corner where the spin resides. Here, the anisotropy energy is infinite, so that with respect to one of its tetrahedra, the spin either points "in" toward the center of the tetrahedron or "out" away from the center. The configuration space becomes more constrained when the interaction between spins is considered. The net interaction may be modeled by a ferromagnetic coupling between nearest-neighbor spins. With reference to a single tetrahedron, this implies that if a particular spin points "in," then its interaction is optimized if the other spins of the tetrahedron point "out." This is also correct for the other spins, so that the ferromagnetic nearest-neighbor coupling, combined with the pyrochlore geometry and easy-axis constraint, gives rise to frustration. An optimal configuration for a given tetrahedron is one where two of its spins point inwards and two outwards; for each tetrahedron, there are six such configurations. An optimal configuration for the whole system is one where the spins obey what is called an "ice rule": on every tetrahedron, two spins point in and two out. As is commonly the case in frustrated systems, the number of such optimal configurations grows exponentially with system size [40].

17.3.3
$A_3B_5O_{12}$ Garnet Structure [41]

Ho garnet is of interest due to its unique frustrated magnetic interactions in the low temperature region below 1 K [42–45]. These frustrated properties have been studied using muon spin rotation (μSR) spectroscopy [46], Mössbauer effects [43], infrared spectroscopy [44], and specific heat [45]. The magnetic interaction has not been studied, however, such as by determining the magnetic spin states of single crystals by neutron diffraction. This is because large-size single crystals, more than 10 mm in diameter, are required, although single crystals 3–5 mm in diameter are enough for such a study over a long period. The single crystals of $Ho_3Ga_5O_{12}$ and $Ho_3Al_5O_{12}$ garnets were grown to 3–5 mm in diameter using the floating zone (Fz) technique. The growth conditions, such as seed crystals used, deviation from stoichiometry (congruent melting composition), post-annealing, growth interface shape, and optimum growth direction to prevent cracking, are important. $HoAlO_3$ single crystals were grown to 15–20 mm in diameter by the Czochralski (Cz) technique. Then, we could demonstrate the magnetic properties for $HoAlO_3$ single crystals in the liquid helium temperature range and at magnetic fields varying between 0 and 50 kOe. The anisotropic magnetothermal properties of $HoAlO_3$ single crystals were evaluated from the dependence of the paramagnetic properties on distance along the crystal axis. Finally, we could show the frustrated properties on $Ho_3Al_5O_{12}$ garnet single crystals.

In the $A_3B_5O_{12}$ garnet structure, the A-site forms a network of corner-sharing triangles. Hence, when the A-site is occupied by antiferromagnetically coupled isotropic (Heisenberg) spins (such as Gd^{3+}), geometrical frustration prevents the system from selecting a unique ground state, even at zero temperature, resulting in novel highly degenerate ground states with remaining entropy [47]. In contrast, when the A-site is occupied by other rare-earth elements besides Gd^{3+}, the strong crystalline-electric field (CEF) introduces uniaxial (Ising) anisotropy. This anisotropy removes the ground state degeneracy, and thus, the system is expected to behave as a prototypical antiferromagnet [48].

For the non-Kramers ions, however, such as Ho and Tb, a different situation may be realized. The low-point symmetry D_2 of the A-site lifts the degeneracy of the 4f J-multiplet completely, and thus, the ground state (along with any other excited states) becomes a nonmagnetic singlet. Nonetheless, many non-Kramers garnets order magnetically at low temperature (e.g., $T_N = 0.839$ K for

Ho$_3$Al$_5$O$_{12}$ [48]), due to inter-spin interactions, which induce magnetic moment by mixing ground state and excited state wave functions. Even then, owing to the competition between CEF (favoring the nonmagnetic singlet) and the inter-spin interactions (favoring the long-range-magnetic order), the Néel temperature and ordered moment size are significantly reduced. In this way, the non-Kramers garnets may provide a new playground for studying the competition between singlet and ordered ground states. For a detailed understanding of the competition effect, the CEF level scheme has to be experimentally determined, as well as the inter-spin interactions. Therefore, we have performed both inelastic and elastic neutron scattering experiments on the non-Kramers Ho$_3$Al$_5$O$_{12}$ garnet. The inelastic scattering provides information on the CEF level scheme, whereas the elastic diffuse scattering pattern may be used to discuss details of the inter-spin interactions.

The neutron scattering experiments were performed using the thermal-neutron and the cold-neutron-triple-axis spectrometers, ISSP-GPTAS, and ISSP-HER, installed at the JRR-3 research reactor (Tokai, Japan). For the thermal neutron inelastic experiment, the final neutron energy was fixed at the neutron Fermi energy, $E_F = 14.7$ meV, and the collimations 20'-40'-40'-40' are employed. This configuration yields the energy resolution $\Delta E = 0.72$ meV at the elastic position. The diffuse scattering pattern is collected using the (energy-integrating) double-axis mode, with the incident energy $E_i = 14.7$ meV, and with the collimations 40'-80'-40'. The cold neutron experiment was carried out with the final energy fixed to $E_F = 3.0$ meV, using a doubly focusing analyzer to increase the counting efficiency. Energy resolution at the elastic position is estimated as $\Delta E = 0.082$ meV. The samples were loaded into an aluminum sample can, and then placed inside a closed-cycle 3He-refrigerator ($T_{min} < 0.7$ K) or a one-shot-type 3He-cryostat ($T_{min} < 0.4$ K.)

Figure 17.13a shows the inelastic scattering spectrum from the powder sample of Ho$_3$Al$_5$O$_{12}$ crystals at $Q = 1$ Å^{-1} and $T = 6$ K, measured at ISSP-GPTAS in the

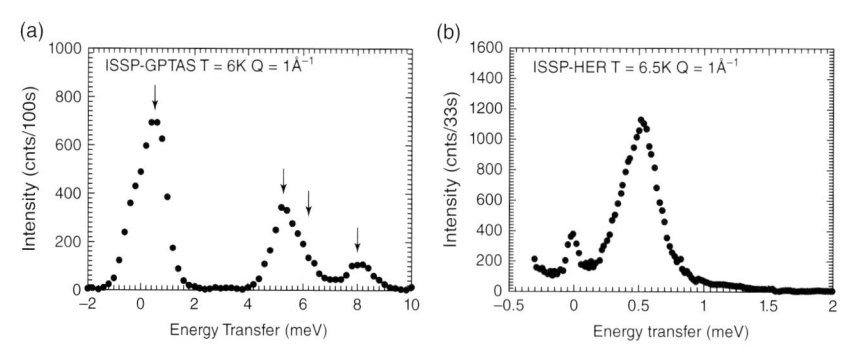

Figure 17.13 Neutron inelastic scattering spectra of Ho$_3$Al$_5$O$_{12}$ garnet measured at (a) $T = 6$ K using ISSP-GPTAS, and (b) $T = 6.5$ K using ISSP-HER. A powder sample is used in the measurements. Arrows in (a) indicate the peak positions [41].

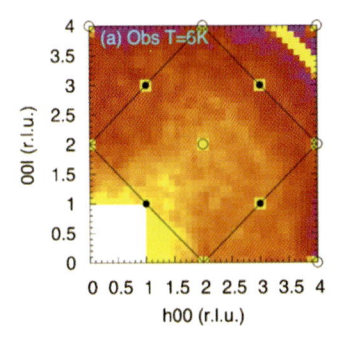

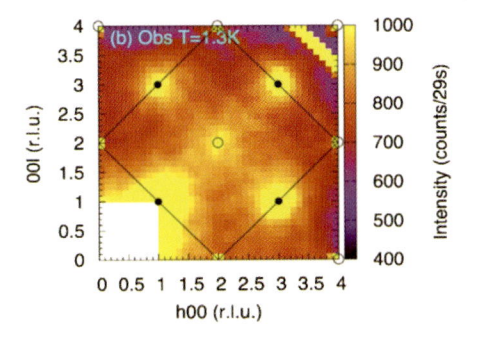

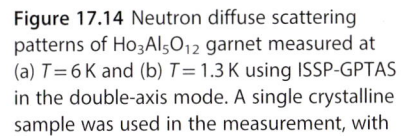

Figure 17.14 Neutron diffuse scattering patterns of Ho$_3$Al$_5$O$_{12}$ garnet measured at (a) $T=6$ K and (b) $T=1.3$ K using ISSP-GPTAS in the double-axis mode. A single crystalline sample was used in the measurement, with the (h0l) planes set as the scattering planes. The filled circles denote the positions where the antiferromagnetic Bragg reflections appear below T_N, whereas the open circles denote the nuclear Bragg positions [41].

energy range of $0 < \hbar\omega < 10$ meV. There are four peaks at $\hbar\omega = 0.5$, 5.3, 6.2, and 8.0 meV in the spectrum. From their Q-dependence, we conclude that there are excitation peaks between the CEF splitting levels. For the lower energy region, a high-energy-resolution spectrum was measured at $Q = 1$ Å^{-1} and $T = 6.5$ K using ISSP-HER, which provides the very detailed spectrum shown in Figure 17.13b. A clear inelastic peak is found at $\hbar\omega = 0.5$ meV, which is separate from the elastic ($\hbar\omega = 0$) peak. The elastic peak is very weak compared to the inelastic ones, indicating that the ground state is indeed singlet. (The small remaining intensity may be due to Ho incoherent scattering.) We note that the second excited state, observed at 5.3 meV, is roughly 10 times higher than the first excited state (0.5 meV). This CEF level scheme greatly simplifies the problem; only the two low-energy states need to be included in the discussion on the low-temperature magnetic properties of Ho$_3$Al$_5$O$_{12}$ garnet. Systems with such energy level schemes are called singlet–singlet systems [49], and have been studied intensively because of their possible proximity to the quantum critical point [50].

Figure 17.14 shows the magnetic diffuse scattering pattern in the (h0l) planes, obtained using Ho$_3$Al$_5$O$_{12}$ single crystal. At the paramagnetic temperature $T = 6$ K, only featureless scattering appears in the figure, whereas at lower temperature $T = 1.3$ K, sufficiently close to T_N, strong diffuse scattering appears around the (101), (301), (103), (303), and (202) positions. We note that (101) and other (h0l) reflections (with h and l being odd numbers) characterize the positions where the magnetic Bragg reflections appear below T_N, indicating that the diffuse scattering corresponds to the short-range antiferromagnetic correlations. In addition, a very prominent feature may be seen in the diffuse scattering pattern; at $T = 1.3$ K, in that a diffuse scattering peak also appears around the (202) position, which indeed corresponds to the nuclear Bragg reflection position. This indicates that, in addition to the antiferromagnetic correlations, Ho$_3$Al$_5$O$_{12}$ definitely has coexisting ferromagnetic correlations just above T_N. Therefore, we can conclude that $Ho_3Al_5O_{12}$ garnet is a rare and intriguing example of (1) a singlet–singlet ground state system with (2) competing ferromagnetic and antiferromagnetic interactions. Further study is highly desirable to search for novel quantum phenomena originating from this competition between the local quantum singlet and the coexisting ferromagnetic and antiferromagnetic correlations. In conclusion, $Ho_3Al_5O_{12}$ garnet is a rare and intriguing example of a singlet–singlet ground state system with competing ferromagnetic and antiferromagnetic interactions.

17.3.4
Magnetic Frustration in Multiferroic Materials – Aurivillius Bilayered Multiferroic: $Bi_5Ti_3FeO_{15}$ [51]

Multiferroic materials, such as those characterized by the coexistence of ferromagnetic and ferroelectric orders, have recently attracted ever-increasing interest and provoked a great number of research projects, driven by the intriguing physics and novel functionality resulting from the coupling between ferroelectric and magnetic orders [52–58]. The key feature of the multiferroic materials, the magnetoelectric (ME) effect, the coupling interaction between two ferroic orders, provides an additional degree of freedom in the design of novel multifunctional devices. It would be ideal if there were a single-phase multiferroic material with an ME effect at room temperature, but unfortunately, such materials are very rare [59]. Large magnetoelectric coupling in a single-phase multiferroic material can be achieved by a spin-orbit-coupling-driven inverse Dyaloshinski–Moriya (DM) interaction between two neighboring spins in spin-driven ferroelectric systems, as in helimagnets [53]. The materials with the general formula $(Bi_2O_2)^{2+}(A_{m-1}B_mO_{3m+1})^{2-}$ (m: number of pseudoperovskite layers) and an Aurivillius layer structure is naturally ferroelectric at room temperature, with very high ferroelectric transition temperatures [60–62]. Importantly, the B-sites of these Aurivillius compounds can accommodate heterovalent elements [63–68], making it possible to achieve the general formula of $Bi_4Ti_3O_{12} \cdot n\text{-}BiMO_3$ (M = Fe, Mn; n = 1, 2), which creates the possibility of introducing novel properties not possessed by the parent compound, especially the capability for magnetic ion doping to introduce magnetic ordering and make the material multiferroic. Magnetic ions in $Bi_4Ti_3O_{12} \cdot nBiMO_3$ would be responsible for the observation of magnetism. The critical parameters determining the type of magnetism in terms of the exchange interaction are the local environment of the magnetic ions and the distance between neighboring ions. The ordered transition metal atoms in the lattice will lead to the formation of a natural superlattice with the layers in single-crystal unit-cell thickness in the form of a single $BiMO_3$ layer sandwiched by $Bi_4Ti_3O_{12}$ layers. This would make $Bi_4Ti_3O_{12} \cdot nBiFeO_3$ (BTFO) a possible multiferroic, considering that $BiFeO_3$ is a well-known room-temperature multiferroic material.

The magnetic properties of BTFO film were examined, as shown in Figures 17.15–17.18. The Figure 17.15 inset shows the zero-field-cooled and field-cooled

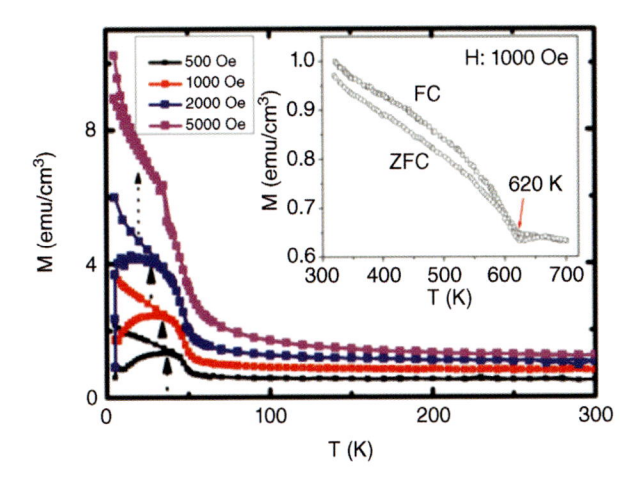

Figure 17.15 ZFC–FC curves of BTFO films measured under different magnetic fields; inset shows high-temperature ZFC–FC curves at 1000 Oe [51].

(ZFC–FC) magnetization curves above room temperature, where a weak ferromagnetic transition around 620 K is observed, which is close to the paramagnetic to spiral antiferromagnetic transition temperature of BiFeO$_3$ and BTFO [69,70]. Weak ferromagnetism is observed at room temperature, which is evidenced by the Figure 17.16 inset, a well-saturated M–H loop with a saturated magnetic moment of ~8 emu/cm^3 (0.066 μ_B per unit cell). To confirm that the abnormality in the magnetization is caused by the existence of long-range magnetic ordering,

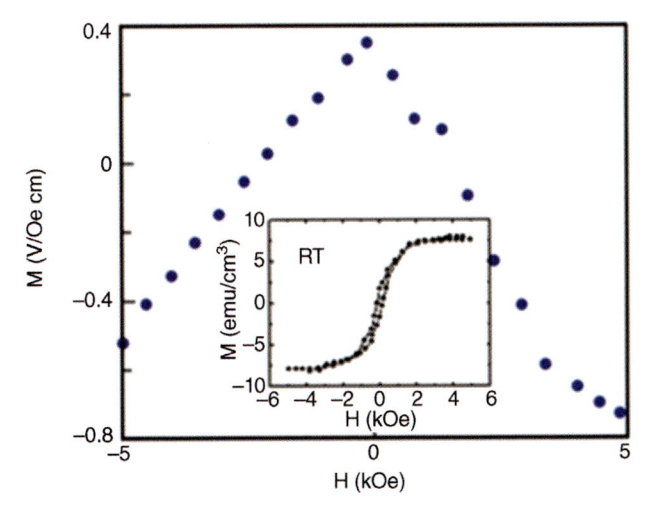

Figure 17.16 ME coupling versus magnetic field of BTFO15 film; inset shows the magnetic hysteresis loop measured at room temperature. The magnetic field is normal to the thin film surface [51].

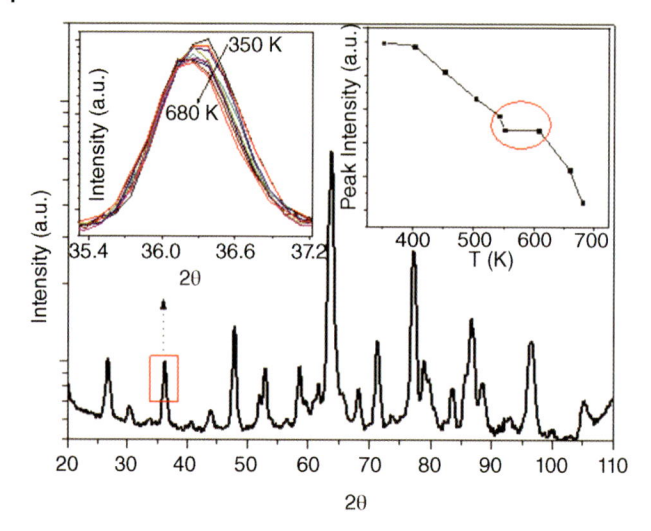

Figure 17.17 Neutron powder diffraction of BTFO ceramic sample and analysis of a specific peak (insets). The diffraction peak intensity decreases with increasing temperature. An abnormal feature was observed around ~600 K (right inset), which is regarded as due to the short-range magnetic ordering caused by spin–phonon interaction. Neutron diffraction was carried out on the powder samples using the high intensity diffractometer Wombat ($\lambda = 2.410$ Å) of the OPAL Research Reactor, Lucas Heights. Over the whole temperature range (5–680 K), no magnetic peak was detected [51].

neutron powder diffraction (NPD) was carried out across the temperature of the magnetic abnormality. As shown in Figure 17.17, the NPD of a ceramic sample shows that there is no new peak that appears around the magnetic transition temperature of ~620 K, but a clear abnormality in the peak intensity at around ~620 K is observed. This indicates that, although there is no obvious long-range magnetic ordering, some kind of magnetic ordering does occur at that temperature. The Mössbauer spectra of BTFO ceramic sample at room temperature contain features consistent with quadrupolar effects (can be fitted by one doublet), which indicate the absence of long-range magnetic ordering (as shown in Figure 17.18). There might be short-range magnetic ordering, however, from the Fe-rich areas in both the thin film and the bulk samples, where the Fe-rich areas mimic $BiFeO_3$, considering their very similar magnetic ordering temperature. In addition, the very small magnetic moment per unit cell of BTFO indicates a canted antiferromagnetic structure in those Fe-rich nanoregions, which mimic $BiFeO_3$ to some extent. At low temperature, an enhancement of the magnetic moment on both the FC and the ZFC magnetization curves, accompanied by bifurcations between the curves, were observed for all cooling fields of 500 Oe, 1 kOe, and 2 kOe, as indicated in Figure 17.14, but the bifurcation feature is largely suppressed at 5 kOe, and the temperature for this feature gradually decreases with increasing applied magnetic field. These observations indicate a spin-glass-like behavior, which is similar to the case of $BiFeO_3$ [71]. It arises

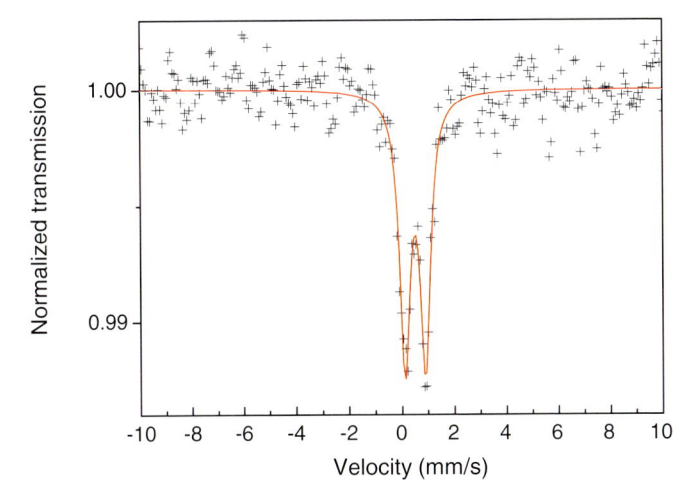

Figure 17.18 Mössbauer spectrum of BTFO15 ceramic sample obtained at room temperature using a standard constant-acceleration spectrometer and a ^{57}CoRh source. The spectra were calibrated at room temperature with a piece of α-Fe foil. Mössbauer spectra of this ceramic sample at room temperature only displayed the quadrupolar effect and can be fitted by one doublet, which indicates that there is no long-range magnetic ordering [51].

from weak ferromagnetism, from the competition between the ferromagnetic and antiferromagnetic transitions in the thin film or from the spin frustration caused by strain effects at the film/substrate interface [72,73].

The magnetoelectric coupling in BTFO film is regarded as driven by the inverse Dyaloshinski–Moriya (DM) interaction in the form of $\vec{P} = \alpha \sum \sum_{ij} \vec{e_{ij}} \times (\vec{S_i} \times \vec{S_j})$, where e_{ij} is the unit vector connecting the neighboring spins S_i and S_j, and the proportionality constant is determined by the spin–orbit and spin-exchange interactions, as well as the possible spin-lattice coupling term [53]. Although the exact spin-ordering structure of the Fe-rich nanoregions in BTFO is not clear, the weak magnetic moment indicates a canted antiferromagnetic ordering. Such a canted antiferromagnetic structure can induce an electric polarization based on the above inverse DM interaction, which will be added to the dominant ion-displacement driven ferroelectric polarization. Therefore, at low and room temperature, only part of the polarization will show a response to external magnetic field. This part of the polarization, which is driven by spin (canted antiferromagnetism), can be flipped by magnetic field (as evidenced by direct ME coupling measurements). The polarization may also be enhanced by magnetic field though complete spin-orbit-lattice coupling when the magnetic field works on spin (as evidenced by magnetopolarization measurements).

The introduction of iron is not only the reason for the observed room temperature ME coupling, but also the reason for the enhancement in ferroelectric polarization. Short-range magnetic ordering is proposed for the first time as one source of the magnetoelectric coupling in Aurivillius layer-structured bismuth titanate. This novel concept will broaden the number of single-phase material candidates that can show a magnetoelectric coupling effect for room temperature application. For example, by simply doping some magnetic ions into a ferroelectric host material, which has very large chemical solubility, a single phase magnetoelectric material could be designed if there is no impurity formation.

17.4
Conclusions

We have only touched the surface of the deep well of phenomena to be explored in frustrated magnets, most of which are highly amenable to laboratory studies. In several frustrated magnets, researchers seem finally to have uncovered the phenomena, but the detailed physics have not yet been elucidated. Explaining them will be a theoretical and experimental challenge. Numerous examples have demonstrated that frustrated magnetism presents an excellent proving ground in which to discover new states and new properties. For theorists, this conviction comes from the simple observation that long-range magnetic order, the standard low-temperature instability, cannot be achieved due to the proliferation of possible ground states. This is true both for classical systems, where averaging over the various ground states often leads to decaying correlations at large distances, and for quantum systems, where this proliferation translates into a very soft spectrum and diverging fluctuations. In the process of searching for QSLs, examples of dimensional reduction and quantum criticality were unexpectedly found.

Now, the definition of frustration is ubiquitous in physics, which was first used in spin glasses to describe systems that reach their true ground state when both disorder and frustration are at play. In these systems, disorder has an important effect on the properties of the materials. The intrinsic effect of "geometrical" frustration is the competition between interaction pathways that arises in clean and periodic, but frustrated systems. Frustration and one-dimensionality share many common features: they both lead to diverging fluctuations, to exotic excitations, and to reduced critical temperatures due to small additional interactions. In frustrated systems, however, the presence of several competing states leads to a very large number of low-lying excitations, or more generally, to a redistribution of spectral weight into narrow bands.

Frustrated magnetism is a vast research field, and dichotomies have emerged naturally over the years: classical versus quantum, 2D versus 3D, rare-earth versus transition-metal ions, corner-sharing versus edge-sharing lattices, Ising versus Heisenberg interactions, and so on. [2]. It would be dangerous to neglect any of them. Here, we have only chosen the aspects, which seemed most interesting to us. In the future, what will be the leading issues and challenges in frustrated magnetism? Each reader will probably have her or his own opinion. We hope that this review may serve both as a reference and as a springboard that can contribute to the solution of these fascinating problems in frustrated magnets.

Acknowledgments

Author H.Y. Zhao is grateful for support from the Chinese National Natural Science Foundation (Grant no. 51402327). Author Z.X. Cheng thanks the Australian Research Council (ARC) for support through a Future Fellowship. Author T.T. Jia acknowledges support from the Japan Society for the Promotion of Science (JSPS). The authors wish to thank Mrs. M. Nakabayashi at the Green Network of Excellence (GRENE) for thin film preparation. Part of this work was conducted in GRENE, sponsored by the Ministry of Education, Culture, Sports, Science and Technology (MEXT), Japan.

References

1 Balents, L. (2010) Spin liquids in frustrated magnets. *Nature*, **464**, 199–208.
2 Claudine, L., Philippe, M., and Frédéric, M. (eds) (2011) *Introduction to Frustrated Magnetism: Materials, Experiments, Theory*, Springer, Heidelberg.
3 Ramirez, A.P., Broholm, C.L., Cava, R.J., and Kowach, G.R. (2000) Geometrical frustration, spin ice and negative thermal expansion – the physics of underconstraint. *Physica B Condens. Matter*, **280**, 290–295.
4 Kleman, M. and Lavrentovich, O.D. (2003) *Soft Matter Physics: An Introduction*, Springer, NY.
5 Moessner, R. and Ramirez, A.P. (2006) Geometrical frustration. *Phys. Today*, **59**, 24–29.
6 Toulouse, G. (1977) Theory of frustration effect in spin-glass. I. *Commun. Phys.*, **2**, 115–119.
7 Villain, J. and Toulouse, G. (1977) Theory of the frustration effect. II. Ising spins on a square lattice. *J. Phys. C*, **10**, L537–L544.
8 Wannier, G.H. (1950) Antiferromagnetism: the triangular Ising net. *Phys. Rev.*, **79**, 357–364.
9 Ramirez, A.P. (1994) Strongly geometrically frustrated magnets. *Annu. Rev. Mater. Sci.*, **24**, 453–480.
10 Tarjus, G., Akivelson, S., Nussinov, Z., and Viot, P. (2005) The frustration-based approach of supercooled liquids and the glass transition: a review and critical assessment. *J. Phys. Condens. Matter*, **17**, R1143–R1182.
11 Chandra, P. and Doucot, B. (1988) Possible spin-liquid state at large S for the frustrated square Heisenberg lattice. *Phys. Rev. B*, **38**, 9335–9338.
12 Young, A.P. (ed.) (1998) *Spin Glasses and Random Fields*, World Scientific, Singapore.
13 Frank, F.C. (1952) Supercooling liquids. *Proc. R. Soc. A*, **215**, 43–46.
14 Nelson, D.R. (2002) *Defects and Geometry in Condensed Matter*, Cambridge University Press, Cambridge.
15 Sadoc, J.F. and Mosseri, R. (1999) *Geometrical Frustration*, Cambridge University Press, Cambridge.
16 Villain, J. (1979) Insulating spin glasses. *Z. Phys. B*, **33**, 31–42.
17 Jagla, E.A. (1997) Statistical physics, plasmas, fluids, and related interdisciplinary topics. *Phys. Rev. E*, **58** (4), 4701–4705.
18 Bramwell, S.T. and Gingras, M.J.P. (2001) Spin ice state in frustrated magnetic pyrochlore materials. *Science*, **294**, 1495–1501.
19 Gingras, M.J.P. (2012) Spin ice – a magnetic analogue of common water ice and a classical spin liquid. *Phys. Can.*, **68**, 89–94.
20 Harris, M.J., Bramwell, S.T., McMorrow, D.F., Zeiske, T., and Godfrey, K.W. (1997) Geometrical frustration in the ferromagnetic pyrochlore $Ho_2Ti_2O_7$. *Phys. Rev. Lett.*, **79**, 2554–2557.

21 Isakov, S.V., Moessner, R., and Sondhi, S.L. (2005) Why spin ice obeys the ice rules. *Phys. Rev. Lett.*, **95**, 217201.

22 den Hertog, B.C. and Gingras, M.J.P. (2000) Dipolar interactions and origin of spin ice in Ising pyrochlore magnets. *Phys. Rev. Lett.*, **84**, 3430–3433.

23 Ramirez, A.P., Hayashi, A., Cava, R.J., Siddharthan, R., and Shastry, B.S. (1999) Zero-point entropy in 'spin ice'. *Nature*, **399**, 333–334.

24 Pauling, L. (1935) The structure and entropy of ice and of other crystals with some randomness of atomic arrangement. *J. Am. Chem. Soc.*, **57**, 2680–2684.

25 Tchernyshyov, O., Moessner, R., and Sondhi, S.L. (2002) Order by distortion and string modes in pyrochlore antiferromagnets. *Phys. Rev. Lett.*, **88**, 067203.

26 Anderson, P.W. (1987) The resonating valence bond state in La_2CuO_4 and superconductivity. *Science*, **235**, 1196–1198.

27 Lee, P.A., Nagaosa, N., and Wen, X.G. (2006) Doping a Mott insulator: physics of high-temperature superconductivity. *Rev. Mod. Phys.*, **78**, 17–85.

28 Misguich, G. (2011) Quantum spin liquids and fractionalization, in *Introduction to Frustrated Magnetism Materials, Experiments, Theory* (eds L. Claudine, M. Philippe, and Frédéric M.), Springer, Heidelberg, pp. 407–475.

29 Liang, S., Doucot, B., and Anderson, P.W. (1988) Some new variational resonating-valence-bond-type wave functions for the spin-½ antiferromagnetic Heisenberg model on a square lattice. *Phys. Rev. Lett.*, **61**, 365–368.

30 Ramirez, A.P., Hayashi, A., Cava, R.I., Siddharthan, R., and Shastry, B.S. (1999) Zero-point entropy in 'spin ice'. *Nature*, **399**, 333–335.

31 Goodenough, J.B. (1963) *Magnetism and the Chemical Bond*, Interscience Publishers, NY.

32 Lee, S.H., Broholm, C., Kim, T.H., Ratcliff, W. II, and Cheong, S.W. (2000) Local spin resonance and spin-Peierls-like phase transition in a geometrically frustrated antiferromagnet. *Phys. Rev. Lett.*, **84**, 3718–3721.

33 Ueda, H., Aruga-Katori, H., Mitamura, H., Goto, T., and Takagi, H. (2005) Magnetic-field induced transition to the ½ magnetization plateau state in the geometrically frustrated magnet $CdCr_2O_4$. *Phys. Rev. Lett.*, **94**, 047202.

34 Lee, S.H., Broholm, C., Ratcliff, W., Gasparovic, G., Huang, Q., Kim, T.H., and Cheong, S.-W. (2002) Emergent excitations in a geometrically frustrated magnet. *Nature*, **418**, 856–858.

35 Ueda, H., Yamaura, J., Mitamura, H., Goto, T., Aruga-Katori, H., Takagi, H., and Ueda, Y. (2007) S = 3/2 spin systems on frustrated pyrochlore lattice. *J. Magn. Magn. Mater.*, **310**, 1275–1279.

36 Penc, K., Shannon, N., and Shiba, H. (2004) Half-magnetization plateau stabilized by structural distortion in the antiferromagnetic Heisenberg model on a pyrochlore lattice. *Phys. Rev. Lett.*, **93**, 197203.

37 Mitamura, H., Ueda, H., Aruga-Katori, H., Takeyama, S., Sakakibara, T., Ueda, Y., and Takagi, H. (2007) Phase transitions of a geometrically frustrated spin system $CdCr_2O_4$ in very high magnetic fields. *J. Phys. Soc. Jpn.*, **76**, 085001.

38 Matsuda, M., Ueda, H., Kikkawa, A., Tanaka, Y., Katsumata, K., Narumi, Y., Inami, T., Ueda, Y., and Lee, S.-H. (2007) Spin-lattice instability to a fractional magnetization state in the spinel $HgCr_2O_4$. *Nat. Phys.*, **3**, 397–400.

39 Bramwell, S.T. and Gingras, M.J.P. (2001) Spin ice state in frustrated magnetic pyrochlore materials. *Science*, **294**, 1495–1501.

40 Raman, Kumar S. (2005) Geometry, frustration, and exotic order in magnetic systems, Ph.D. thesis, Princeton University. Available at https://www.princeton.edu/physics/graduate-program/theses/theses-from-2005/K.Ramanthesis.pdf.

41 Kimura, H., Numazawa, T., and Sato, T.J. (2011) Holmium oxide single crystals and their properties, in *Advances in Chemistry Research*, vol. **6** (ed. J.C. Taylor), Nova Science Publishers, NY, pp. 185–199.

42 Dalmas de Réoter, P., Yaouanc, A., Gubbens, P.C.M., Kaiser, C.T., Baines, C., and King, P.J.C. (2003) Absence of

magnetic order in $Yb_3Ga_5O_{12}$: relation between phase transition and entropy in geometrically frustrated materials. *Phys. Rev. Lett.*, **91**, 167201.

43 Hodges, J.A., Bonville, P., Rams, M., and Królas, K. (2003) Low temperature spin fluctuations in geometrically frustrated $Yb_3Ga_5O_{12}$. *J. Phys. Condens. Matter*, **15**, 4631–4639.

44 Papagelis, K., Kanellis, G., Zorba, T., Ves, S., and Kourouklis, G.A. (2002) Infrared lattice spectra of $Tm_3Al_5O_{12}$ and $Yb_3Al_5O_{12}$ single crystals. *J. Phys. Condens. Matter*, **14**, 915–923.

45 Nagata, S., Sasaki, H., Suzuki, K., Kikuchi, J., and Wada, N. (2001) Specific heat anomaly of the holmium garnet $Ho_3Al_5O_{12}$ at low temperature. *J. Phys. Chem. Solids*, **62**, 1123–1130.

46 Numazawa, T., Kimura, H., Sato, M., and Maeda, H. (1993) Carnot magnetic refrigerator operating between 1.4 and 10 K. *Cryogenics*, **33**, 547–554.

47 Petrenko, O.A., Ritter, C., Yethiraj, M., and Paul, M.c.K.D. (1998) Investigation of the low-temperature spin-liquid behavior of the frustrated magnet gadolinium gallium garnet. *Phys. Rev. Lett.*, **80**, 4570–4573.

48 Wolf, W.P. (2000) The ising model and real magnetic materials. *Braz. J. Phys.*, **30** (4), 794–810.

49 Birgeneau, R.J. (1973) Singlet ground state dynamics in magnetism and magnetic materials. *AIP Conf. Proc.*, **10**, 1664–1668.

50 Rønnow, H.M., Jensen, J., Parthasarathy, R., Aeppli, G., Rosenbaum, T.F., McMorrow, D.F., and Kraemer, C. (2007) Magnetic excitations near the quantum phase transition in the Ising ferromagnet $LiHoF_4$. *Phys. Rev. B*, **75**, 054426.

51 Zhao, H.Y., Kimura, H., Cheng, Z.X., Osada, M., Wang, J.L., Wang, X.L., Dou, S.X., Liu, Y., Yu, J.D., Matsumoto, T., Tohei, T., Shibata, N., and Ikuhara, Y. (2014) Large magnetoelectric coupling in magnetically short-range ordered $Bi_5Ti_3FeO_{15}$ film. *Sci. Rep.*, **4**, 5255.

52 Hur, N., Park, S., Sharma, P.A., Ahn, J.S., Guha, S., and Cheong, S.W. (2004) Electric polarization reversal and memory in a multiferroic material induced by magnetic fields. *Nature*, **429**, 392–395.

53 Kimura, T., Goto, T., Shintani, H., Ishizaka, K., Arima, T., and Tokura, Y. (2003) Magnetic control of ferroelectric polarization. *Nature*, **426**, 55–58.

54 Alexe, M., Ziese, M., Hesse, D., Esquinazi, P., Yamauchi, K., Fukushima, T., Picozzi, S., and Gosele, U. (2009) Ferroelectric switching in multiferroic magnetite (Fe_3O_4) thin films. *Adv. Mater.*, **21**, 4452–4455.

55 Catalan, G. and Scott, J.F. (2009) Physics and applications of bismuth ferrite. *Adv. Mater.*, **21**, 2643–2485.

56 Wang, J.B., Neaton, J., Zheng, H., Nagarajan, V., Ogale, S.B., Liu, B., Viehland, D., Vaithyanathan, V., Schlom, D.G., Waghmare, U.V., Spaldin, N.A., Rabe, K.M., Wutting, M., and Ramesh, R. (2003) Epitaxial $BiFeO_3$ multiferroic thin film heterostructures. *Science*, **299**, 1719–1722.

57 Nan, C.W. (1994) Magnetoelectric effect in composites of piezoelectric and piezomagnetic phases. *Phys. Rev. B*, **50**, 6082–6088.

58 Spaldin, N.A., Cheong, S.W., and Ramesh, R. (2010) Multiferroics: past, present and future. *Phys. Today*, **63**, 38–43.

59 Hill, N.A. (2000) Why are there so few magnetic ferroelectrics? *J. Phys. Chem. B*, **104**, 6694–6709.

60 Nakashima, S., Fujisawa, H., Ichikawa, S., Park, J.M., Kanashima, T., Okuyama, M., and Shimizu, M. (2010) Structural and ferroelectric properties of epitaxial $Bi_5Ti_3FeO_{15}$ and natural-superlattice-structured $Bi_4Ti_3O_{12}$–$Bi_5Ti_3FeO_{15}$ thin films. *J. Appl. Phys.*, **108**, 074106.

61 Li, J.B., Huang, Y.P., Rao, G.H., Liu, G.Y., Luo, J., Chen, J.R., and Liang, J.K. (2010) Ferroelectric transition of Aurivillius compounds $Bi_5Ti_3FeO_{15}$ and $Bi_6Ti_3Fe_2O_{18}$. *Appl. Phys. Lett.*, **96**, 222903.

62 Suzuki, M., Nagata, H., Ohara, J., Funakubo, H., and Kakenaka, T. (2003) $Bi_{3-x}M_xTiTaO_9$ (M = La or Nd) ceramics with high mechanical quality factor Q_m. *Jpn. J. Appl. Phys.*, **42** (Part 1), 6090–6093.

63 Aurivillius, B. (1949) Mixed bismuth oxides with layer lattices. *Ark. Kemi*, **1**, 463–480.

64 Noguchi, Y., Miyayama, M., and Kudo, T. (2001) Direct evidence of A-site-deficient strontium bismuth tantalate and its enhanced ferroelectric properties. *Phys. Rev. B*, **63**, 214102.

65 Wang, C.M. and Wang, J.F. (2006) High performance Aurivillius phase sodium–potassium bismuth titanate lead-free piezoelectric ceramics with lithium and cerium modification. *Appl. Phys. Lett.*, **89**, 202905.

66 Noguchi, Y., Murata, K., and Miyayama, M. (2007) Effects of defect control on the polarization properties in Bi_2WO_6-based single crystals. *Ferroelectrics*, **355**, 55–60.

67 Noguchi, Y., Miyayama, M., and Kudo, T. (2000) Ferroelectric properties of intergrowth $Bi_4Ti_3O_{12}$–$SrBi_4Ti_4O_{15}$ ceramics. *Appl. Phys. Lett.*, **77**, 3639–3641.

68 Srinivas, A., Suryanarayana, S.V., Kumar, G.S., and Kumar, M.M. (1999) Magnetoelectric measurements on $Bi_5FeTi_3O_{15}$ and $Bi_6Fe_2Ti_3O_{18}$. *J. Phys. Condens. Matter*, **11**, 3335–3340.

69 Nakashima, S., Nakamura, Y., Yun, K.Y., and Okuyama, M. (2007) Preparation and characterization of Bi-layer-structured multiferroic $Bi_5Ti_3FeO_{15}$ thin films prepared by pulsed laser deposition. *Jpn. J. Appl. Phys.*, **46** (10B), 6952–6955.

70 Cheng, Z.X., Wang, X.L., Dou, S.X., Ozawa, K., and Kimura, H. (2008) Improved ferroelectric properties in multiferroic $BiFeO_3$ thin films through La and Nb codoping. *Phys. Rev. B*, 77, 092101.

71 Singh, M.K., Katiyar, R.S., Prelier, W., and Scott, J.F. (2009) Almeida-Thouless line in $BiFeO_3$: is bismuth ferrite a mean-field spin glass? *J. Phys. Condens. Matter*, **21**, 042202.

72 Venkatesan, S., Daumont, C., Kooi, B.J., Noheda, B., and De Hosson, Jeff Th. M. (2009) Nanoscale domain evolution in thin films of multiferroic $TbMnO_3$. *Phys. Rev. B*, **80**, 214111.

73 Venkatesan, S., Vlooswijk, A., Kooi, B.J., Morelli, A., Palasantzas, G., Jeff Th. M., DeHosson., and Noheda, B. (2008) Monodomain strained ferroelectric $PbTiO_3$ thin films: phase transition and critical thickness study. *Phys. Rev. B*, **78**, 104112.

18
Structures and Properties of Dielectrics and Ferroelectrics

Mitsuru Itoh

Tokyo Institute of Technology, Laboratory for Materials and Structures, 4259 Nagatsuta, 226-8503 Midori-kuYokohama, Japan

18.1
Crystal Symmetry and the Properties

The concept of "ferroelectric (ferroelectrische)" was first proposed by Schrödinger in 1912. Ferroelectric and ferromagnetics occupy an important position in materials science. The word "dielectrics" is almost a synonym for "insulators" in materials science. An atom is composed of an atomic core and electron cloud, and a crystal is composed of a framework built up of atomic cores. The definition of insulators may be vague if one considers that the conductivity of materials is defined by the strength of binding electrons to the atomic cores. When a conductor is put in an electric field gradient, free electrons are distributed over its surface to cancel the internal electric field. In the case of insulators, electrons localize around the atomic cores because the electrons cannot overwhelm the potential barrier to be free. However, overlapping charge centers of positive atomic cores and negative electrons under a zero field can shift each other, and exude positive and negative charges at the surface of the material in order to weaken the internal electric field.

Dielectrics are classified broadly into two groups (Figure 18.1). One is material belonging to centrosymmetric group, and the other to noncentrosymmetric one. Positive and negative charge centers overlap each other in dielectric materials such as diamond, sodium chloride, and sapphire. However, the external electric field induces off-centering of the positive and negative charges in the atoms to exhibit electric polarization. After releasing the external electric field, the induced electric polarization disappears. Such a materials group is classified into paraelectric. As in case of sodium nitrate $NaNaO_2$, which has a permanent electric dipole of NO_2^-, certain dielectrics have permanent dipoles that are thermally agitated at high temperature and aligned to one direction collectively below a certain temperature [1]. Such materials are called "order–disorder-type ferroelectrics." Usually, they have ferroelectric domains where each electric dipole is

Handbook of Solid State Chemistry, First Edition. Edited by Richard Dronskowski, Shinichi Kikkawa, and Andreas Stein.
© 2017 Wiley-VCH Verlag GmbH & Co. KGaA. Published 2017 by Wiley-VCH Verlag GmbH & Co. KGaA.

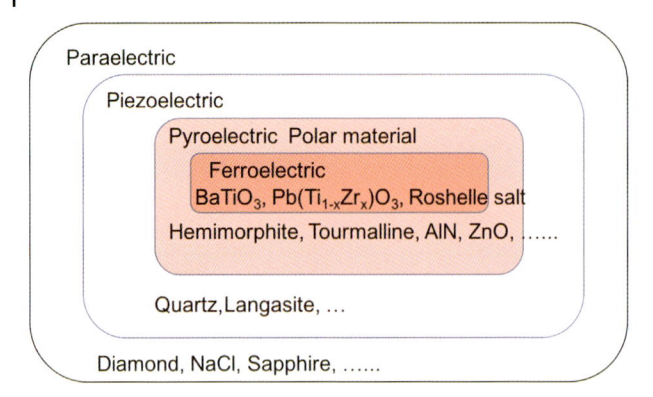

Figure 18.1 Classification of materials in insulators.

Paraelectric
Piezoelectric
Pyroelectric Polar material
Ferroelectric
$BaTiO_3$, $Pb(Ti_{1-x}Zr_x)O_3$, Roshelle salt
Hemimorphite, Tourmalline, AlN, ZnO, ……
Quartz, Langasite, …
Diamond, NaCl, Sapphire, ……

aligned in one direction to decrease the electrostatic energy. Another example of ferroelectrics is perovskite-type $BaTiO_3$. In $BaTiO_3$, there is an optically soft mode whose frequency shows a strong temperature dependence above T_c and becomes zero at T_c. This optically soft mode is called the ferroelectric optical soft mode. The three types of soft modes for perovskite-type oxides are shown in Figure 18.2. The ferroelectric soft mode frequency decreases with temperature and freezes at T_c (freezing of the soft mode at the zone center in an inverse lattice). At and below T_c, the structure of $BaTiO_3$ and its related compounds decrease in symmetry from centrosymmetric and nonpolar ($Pm\bar{3}m$) to noncentrosymmetric and polar structures ($P4mm$). In the fields of materials science and physics, ferroelectrics are included in the polar material group. In practice, however, ferroelectrics are defined as a material group with polarization that is switchable by the electric field, although there is no distinct difference between ferroelectric and polar materials in physical expression.

Many complex oxide materials are known to be ferroelectric, especially perovskite-type oxides. It is generally judged from the point group or space group of the material, whether it possesses ferroelectricity or not. As shown in Figure 18.1, dielectric (insulate) materials can be classified as (1) paraelectric, (2) piezoelectric, (3) polar material, or (4) ferroelectric. As already described, materials with an electric polarization whose direction can be switched into opposite

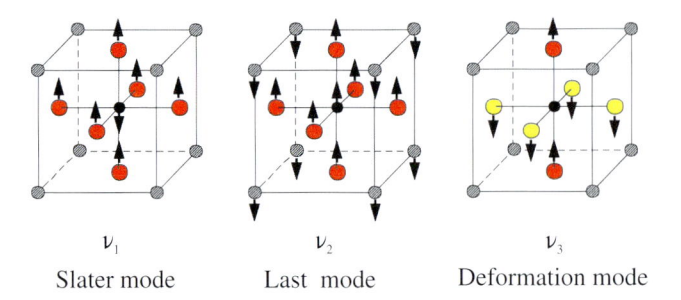

ν_1 Slater mode ν_2 Last mode ν_3 Deformation mode

Figure 18.2 Ferroelectric soft modes in perovskite.

directions by the external electric field are defined as ferroelectric. Thus, materials with nonswitchable polarization can be conveniently classified into pyroelectric and polar materials group. This definition is not a scientifically meaningful one, however, and consequently, depends on the quality of the sample (e.g., number of defects, sintering density), measurement instruments, and conditions. In true ferroelectric materials, the typical polarization (P) versus electric field (E) loop is difficult to measure at room temperature due to the following two reasons: (1) When the T_c of the material is much higher than room temperature, coercive field of the ferroelectric becomes too large to switch the polarization. (2) Leak current originated from the defects and/or carriers coming from the transition metal ions prevents the sample from the high insulating state. However, in the scientific field, measurement of the P–E loop and its hysteretic behavior are indispensable to manifest the ferroelectricity of a material.

Crystal symmetry and dielectric properties are related to each other. There are 32 point group classifications for the crystals, 20 of these ($\bar{4}3m, 23, 422, \bar{4}2m, 4mm, \bar{4}, 4, 622, 6m2, 6mm, \bar{6}, 6, 32, 3m, 3, 222, mm2, m, 2$ and $\bar{1}$) are crystallographically classified into noncentrosymmetric and piezoelectric groups, and 11 ($m3m, m3, 4/mmm, 4/m, 6/mmm, 6/m, 3m, \bar{3}, mmm, 2/m, 1$) into centrosymmetric ones. The crystal belonging to the point group 432 does not have a centrosymmetry, and it does not show a piezoelectricity because the symmetry operation along the polarization direction cancels the polarization. Positive "(normal) piezoelectricity" (C/N) in materials is defined as an appearance of an electric charge on the crystal surface that is proportional to the magnitude of the uniaxial stress. In contrast, "inverse piezoelectricity" (m/V) is defined by the manifestation of distortion in the crystal by application of an external electric field. Although the apparent units are different between normal and inverse piezoelectric coefficients, the absolute values of their coefficients are the same. As mentioned below, the magnitude of the distortion induced by the application of the external electric field should change linearly. The materials belonging to 10 point groups with polar vector elements are called pyroelectric or polar crystals that are structurally noncentrosymmetric and have a spontaneous polarization in the crystal. Usually, in ambient air, moisture or gases in the air adhere to the surface of the polar material to neutralize the positive and negative charges at the surface. When the temperature of such a material is increased, electric charge is gradually released from the material because the polarization decreases its value toward T_c, and eventually an electric charge appears at the surface when the T_c of the material is higher than room temperature. Taking advantage of this pyroelectricity, pyroelectric materials are used as a component of motion sensors in the industry.

As already mentioned, polar crystals, whose polarization direction can be switched by the external electric field, are defined as "ferroelectric" and show a hysteretic P–E loop. Figure 18.3a shows a P–E loop and Figure 18.3b shows the temperature dependence of a dielectric constant for ferroelectric triglycine sulfate (TGS) (Fu, D. and Itoh, M., unpublished work). TGS is a ferroelectric and has a paraelectric–ferroelectric phase transition of the second-order type at

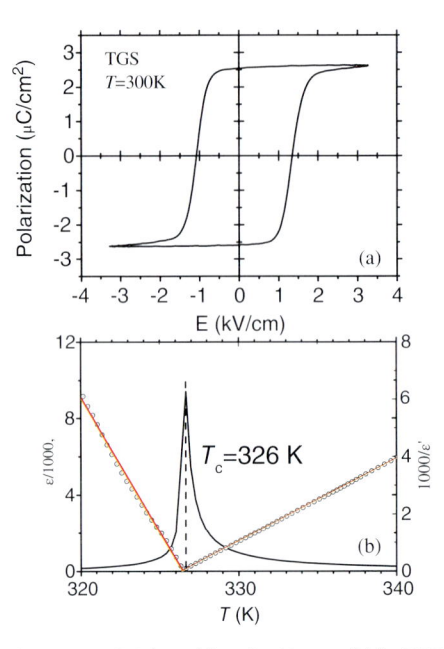

Figure 18.3 P–E loop (a) and ε–T curve (b) for TGS.

326 K (53 °C). A decrease in the polarization for TGS proportional to $\sqrt{T_c - T}$, where $T_c = 53$ °C, is suitable for the application to motion sensors. The square-like P–E hysteresis loop shown in Figure 18.3a is usually difficult to acquire for normal ferroelectrics due to experimental instrument problems or leak currents. Instead, generally a rounded P–E loop is obtained due to the superposition of the leak current. Practically important ferroelectric materials belonging to 20 piezoelectric groups are perovskite and its related materials. Among perovskite-related materials, $BaTiO_3$, $PrTiO_3$, and $PbZrO_3$ are key materials. $BaTiO_3$ was discovered in 1943, and its ferroelectricity was revealed a short time later. $BaTiO_3$ shows phase transitions from $Pm\bar{3}m$ (O_h) to $P4mm$ (C_{4v})(130 °C), $Amm2$ (C_{2v})(~5 °C), and finally to $R3m$ (C_{3v})(−90 °C) [2]. $PbTiO_3$ changes from paraelectric $Pm\bar{3}m$(O_h) to $P4mm$ (C_{4v}) at 490 °C [3], while $PbZrO_3$ is antiferroelectric. The high-temperature phase of $PbZrO_3$ changes from paraelectric $Pm\bar{3}m$ (O_h) to antiferroelectric $Pbam$ (D2h) at 230 °C [4]. The solid solution of ferroelectric $PbTiO_3$ and an antiferroelectric $PbZrO_3$, with a composition of 42–47 mol% ferroelectric $PbTiO_3$, forms an almost vertical phase boundary between rhombohedral and tetragonal phases. This peculiar phase boundary is called as a morphotropic phase boundary (MPB) [5]. The electromechanical coupling coefficient shows very large values in the vicinity of the MPB. This solid solution $PbZrO_3$–$PbTiO_3$ (PZT) is used for piezoelectric devices. Evolution of ferroelctricty and antiferroelectricity is closely related to the covalent bonding between Pb6s and O2p orbitals in lead perovskites.

18.2
Second-Order Phase Transition in Ferroelectric

The Landau theory [6] for the second-order phase transition expresses the free energy by even power series of electric polarization P. Temperature dependence is included in the coefficient of the quadratic term of P^2. In the Landau theory, the parameter P is called an order parameter, given by

$$F(P, T) = F(0, T) + \frac{\alpha(T - T_c)}{2}P^2 + \frac{\beta}{4}P^4 + \frac{\gamma}{6}P^6, \quad (18.1)$$

where α, β, and γ are constants.

Starting from the uniaxial ferroelectric, the free energy is given by

$$F(P, T) = F(0, T) + \frac{\alpha(T - T_c)}{2}P^2 + \frac{\beta}{4}P^4, \quad (18.2)$$

where $\beta > 0$.

Partial differentiation of Eq. (18.2) by P gives

$$\alpha(T - T_c)P + \beta P^2 = E. \quad (18.3)$$

Putting $E = 0$ in Eq. (18.3) and the solution gives a spontaneous polarization of P_s below T_c. Partial differentiation of Eq. (18.3) by E gives the temperature dependence of the electric susceptibility by

$$\chi = \frac{C}{T - T_c}, \quad T \geq T_c, \quad (18.4)$$

and

$$\chi = \frac{C}{2(T_c - T)}, \quad T \leq T_c, \quad (18.5)$$

where $C = 1/\alpha$. The susceptibility measured substantially at $E = 0$ is called a primary susceptibility. Equations (18.4) and (18.5) are the so-called Curie–Weiss law, and the C is the Curie constant. It is clear that the slope of $1/\chi$ is equal to 2:1 above and below T_c, as shown in Figure 18.4a. Free energy for the paraelectric phase shows a single minimum at $E = 0$ and $T > T_c$, and two minimums below T_c. When the electric field is a finite value, the energy term PE is added to the free energy of Eq. (18.2), and the relation P versus E curve can be drawn. The compulsory first-order phase transition is the most important property of ferroelectric materials. The electric field where the polarization jumps to the opposite value is called the coercive field (E_c). With T approaching to T_c, the coercive field approaches zero.

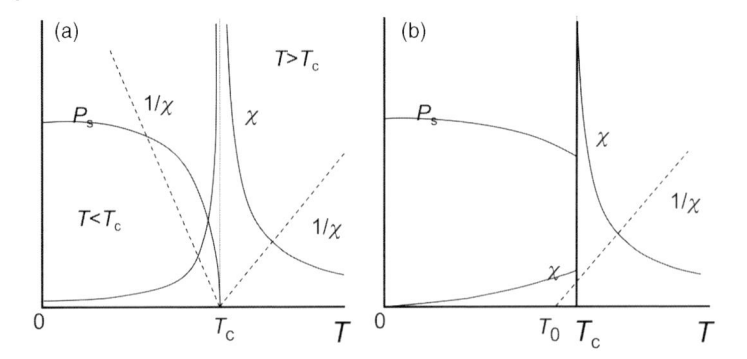

Figure 18.4 Schematic explanation for the first- and second-order phase transitions in ferroelectric. (a) Temperature dependences of P and χ beyond the phase transition in the second-order phase transition. (b) Temperature dependences of P and χ beyond the phase transition in the first-order phase transition.

18.3
First-Order Phase Transition in Ferrelectric

Polarization suddenly appears below a certain temperature in the first-order phase transition in ferroelectric materials. In phenomenological theory, free energy is expressed by including up to sixth order term of P by

$$F(P, T) = F(0, T) + \frac{\alpha(T - T_0)}{2}P^2 + \frac{\beta}{4}P^4 + \frac{\gamma}{6}P^6, \tag{18.6}$$

where $\beta < 0$ and $\gamma > 0$. It should be noted that T_0 is different from T_c.

Through Eq. (18.6), we get

$$\chi = \frac{C}{T - T_0}, \quad T \geq T_c. \tag{18.7}$$

χ for the ferroelectric phase is approximately given by

$$\chi = \frac{C}{-4(T - T_0) - 2\beta C P_s^2}, \quad T \leq T_c. \tag{18.8}$$

At $T = T_c$, we get

$$\chi = \frac{C}{4(T_c - T_0)}. \tag{18.9}$$

The temperature dependences of polarization and electric susceptibility are depicted schematically in Figure 18.4b.

18.4
Successive Phase Transition and Free Energy of Ferroelectric

A simple example is given for BaTiO$_3$ that undergoes a phase transition from paraelectric cubic phase to ferroelectric tetragonal phase. Helmholtz free energy of a material with polarizations of P_1, P_2, and P_3 along x, y, and z directions, respectively, is given by

$$F(P_1, P_2, P_3, T) = \alpha_0 + \frac{\alpha}{2}\left(P_1^2 + P_2^2 + P_3^2\right) + \frac{\beta}{4}\left(P_1^4 + P_2^4 + P_3^4\right)$$
$$+ \frac{\beta'}{2}\left(P_1^2 P_2^2 + P_2^2 P_3^2 + P_3^2 P_1^2\right) + \frac{\gamma}{6}\left(P_1^6 + P_2^6 + P_3^6\right). \tag{18.10}$$

To simplify the equation, we generally replace the terms $F(0,0,0,T) = \alpha_0$ and $\alpha \equiv \alpha(T-T_0)$. The features of Eq. (18.10) are explained by the fact that P_1, P_2, and P_3 are contained in a symmetric form, and that each term is included as an invariant form (for point group O$_h$). Required conditions are that a product of odd-order terms cannot be included, and that the coefficient of the highest-order term must be positive, and so on. The solution of P_1, P_2, and P_3 can be obtained through the derivative of Eq. (18.10) to bring them to zero.

Figure 18.5 shows the polarization direction of each phase. In case of BaTiO$_3$, the coefficient β is negative because the cubic–tetragonal phase transition is of the first order. Each coefficient in Eq. (18.10) can be determined from the temperature dependence of electric susceptibility and the discontinuity of the spontaneous polarization at T_c. Using such coefficients, and consequently derived free energy function, we can depict the behavior of the phase transition schematically. This is an example of successive phase transition occurring in one material of BaTiO$_3$. The Landau phenomenological equation is effective in depicting the behavior of the phase transitions of ferroelectrics.

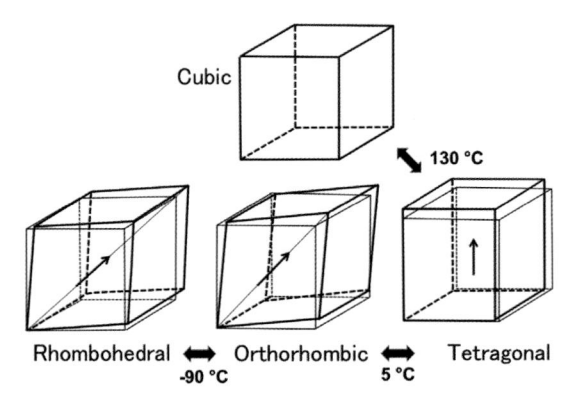

Figure 18.5 Temperature dependent polarization directions in the phases in BaTiO$_3$.

18.5
Generalization of the Formula to a Real Ferroelectric

Equation (18.10) is free energy expression using polarizations as an order parameter. Generally, electrically polar ferroelectric material accompanies a strain in the crystal. Here, two prototypical cases are introduced, the second-order phase transition and the uniaxial ferroelectric.

When a paraelectric phase is nonpiezoelectric, the free energy of the crystal can be given by

$$F(P, T) = F(0, T) + \frac{\alpha(T - T_c)}{2}P^2 + \frac{\beta}{4}P^4 + \frac{c_0}{2}x^2 - gxP^2, \qquad (18.11)$$

where x, c_0, and gxP^2 are strain induced in the crystal, a stiffness coefficient, and a coupling energy between the strain and the polarization, respectively. Paraelectricity and nonpiezoelectricity of the materials should give an invariance of the energy against the point group symmetry operation.

Spontaneous polarization can be given by

$$P_s = \sqrt{-\frac{\alpha(T - T_c)}{\beta - (2g^2/c_0)}}, \quad T \leq T_c. \qquad (18.12)$$

Strain and piezoelectric coefficients can be derived from the above equations. Spontaneous strain can be given by

$$x_s = \frac{g}{c_0}P_s^2, \quad T \leq T_c, \qquad (18.13)$$

where the g value can be both positive and negative. In cases where g is positive, the crystal expands below T_c and vice versa. This effect is called an electrostriction effect, and eventually the magnitude becomes proportional to E^2. For this reason, the term gxP^2 is termed an electrostriction coupling.

One important point is that electric susceptibilities measured under conditions of free deformation and non-free deformation of the crystal are different. The electric susceptibility measured under the free deformation condition is defined as a susceptibility of free crystal χ^X, where X is a stress. In contrast, susceptibility measured under nondeformation conditions (i.e., a restricted condition) is a restricted susceptibility χ^x.

Using the spontaneous polarization P_s and spontaneous strain x_s, χ^x becomes

$$\chi^x = \frac{1}{2\beta P_s^2} = \frac{C}{2(T_c - T)} \frac{\beta - (2g^2/c)}{\beta}, \quad T < T_c. \qquad (18.14)$$

Thus, depending on the magnitude of g, χ^x becomes smaller than χ^X by the amount of the last term of Eq. (18.14). This is the summary of the contribution from the piezoelectric deformation of the crystal.

When the paraelectric phase is piezoelectric, free energy, including lattice strain, is given by

$$F(P, T) = F(0, T) + \frac{\alpha(T - T_0)}{2}P^2 + \frac{\beta}{4}P^4 + \frac{c_0}{2}x^2 - hxP. \quad (18.15)$$

The addition of the term hxP in Eq. (18.15) leads to a higher phase transition temperature than T_0. Free electric susceptibility χ^X and restricted susceptibility χ^x at $T \geq T_c$ are given by, respectively,

$$\chi^X = \frac{C}{T - T_C} \quad (18.16)$$

and

$$\chi^x = \frac{C}{T - (T_C - (h^2/c_0))}. \quad (18.17)$$

Those for ferroelectric phase at $T \leq T_c$ are, respectively,

$$\chi^X = \frac{C}{2(T - T_C)} \quad (18.18)$$

and

$$\chi^x = \frac{C}{2(T - T_C) + (h^2/c_0)}. \quad (18.19)$$

18.6 Antiferroelectric and Ferrielectric

PbZrO$_3$ is known as an antiferroelectric that shows a phase transition at 230 °C from cubic $Fm\bar{3}m$ (O$_h$) to antiferroelectric and orthorhombic $Pbam$ (D$_{2h}$). In the antiferroelectric phase, unit cell dimension expands to $\sqrt{2} \times \sqrt{2} \times 2\,a$ of the high-temperature cubic lattice a, where Pb ions are shifted along ±[110] directions. Figure 18.6a shows the displacement of Pb^{2+} ions in the antiferroelectric phase [7]. Such an antiparallel arrangement of dipole moments is called an antipolar structure. When an external electric field is applied to such antiferroelectric material, the antipolar state is forced to change into a polar state. This transition occurs discontinuously, like polarization switching in a ferroelectric material. This kind of enforced first-order-like phase transition yields a double hysteresis loop, as shown in Figure 18.6b.

Kittel [8] has provided a phenomenological explanation for the anitiferroelectric as follows:

$$F(P_a, P_b, T) = F(0, T) + \frac{\alpha(T - T_0)}{2}(P_a^2 + P_b^2) + \alpha' P_a P_b + \frac{\beta}{4}(P_a^4 + P_b^4), \quad (18.20)$$

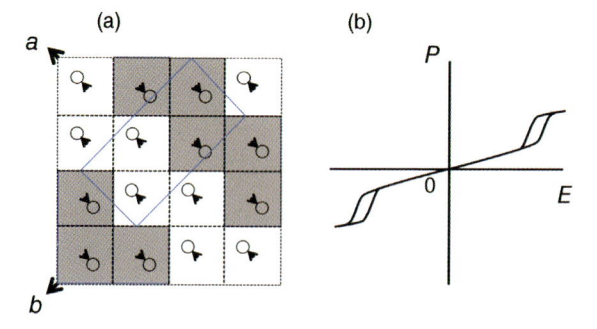

Figure 18.6 (a) Schematic illustration of the displacement pattern of Pb^{2+} ions in antiferroelectric PbZrO$_3$ at room temperature [7]. Lattice parameters of PbZrO$_3$ at room temperature are $a = 5.8883$ Å, $b = 11.7571$ Å, and $c = 8.2222$ Å. White and shaded areas show the opposite directions of ion shifts each other in the primitive perovskite cell. (b) Double hysteresis loop observed for antiferroelectric.

where P_a and P_b are sublattice polarizations. $P_a = -P_b$ and $\alpha' > 0$ are the necessary conditions.

Through Eq. (18.20), we finally obtain

$$P_a^2 = \frac{\alpha' - \alpha(T - T_0)}{\beta}. \tag{18.21}$$

The phase transition temperature, where the real solution is obtained, is given by

$$T'_c = T_0 + \frac{\alpha'}{\alpha}. \tag{18.22}$$

Below this temperature, finite values of P_a and P_b evolves and satisfies the condition $P_a = -P_b$.

Electric susceptibility above and below T'_c are derived as follows:

$$\chi = \frac{(2/\alpha)}{T - (T'_c - (2\alpha'/\alpha))}, \quad T \geq T'_c, \tag{18.23}$$

and

$$\chi = \frac{(1/\alpha)}{T'_c + (3\alpha'/2\alpha) - T}, \quad T \leq T'_c, \tag{18.24}$$

where E should be small to satisfy P_a–P_b. When the electric field is large, an enforced flipping of polarization may occur in one sublattice. A double hysteresis loop is observed in such an antiferroelectric, as shown in Figure 18.6b. When some antiferroelectric has a polarization of $P_a \neq P_b$, as explained in the later section, spontaneous polarization emerges in the crystal. It is defined as ferrielectric.

18.7
Order–Disorder-Type Ferroelectric: Rochelle Salt – Potassium Sodium Tartrate (NaKC$_4$H$_4$O$_6$·4H$_2$O)

Ferroelectricity in Rochelle salt was discovered relatively recently in 1921 by J. Valasek [9] in the United States. Rochelle salt gets its name from the French city, La Rochelle. Marie and Pierre Curie discovered the pyroelectricity and piezoelectricity of Rochelle salts in 1880. The strong inverse piezoelectricity effect of this salt has attracted researchers' attention from the viewpoint of its potential as a vibrator. Langevin invented a resonator using Rochelle salt. This material is very rare in that ferroelectricity appears only at temperatures between two phase transitions at 21 and −19 °C [10]. Both phase transitions are second order, and the structures of both the highest and lowest phases are the same. Such behavior is called a reentrant phase transition. Point groups for the highest-temperature phase and mid-temperature phase are orthorhombic and piezoelectric D_2 and monoclinic C_2, respectively. Phenomenological considerations of the ferroelectricity and piezoelectricity of Rochelle salt was first established mainly by Takahashi and Mitsui [11,12]. The potassium ions in Rochelle salt can be replaced by ammonium ions. Ammonium Rochelle salt NaK$_{1-x}$(NH$_4$)$_x$C$_4$H$_4$O$_6$·4H$_2$O [13] is also a ferroelectric and shows a complicated manner of manifestation of ferroelectricity in terms of change in temperature and composition. Rubidium Rochelle salt, NaRbC$_4$H$_4$O$_6$·4H$_2$O, does not show ferroelectricity down to 0 K [13]. Deuteron replacement for hydrogen expands the stability range of ferroelectricity temperatures between 35 and −22 °C [10] that manifests the importance of hydrogen bonding in the ferroelectricity evolution mechanism of this compound. Rochelle salt and its related compounds are considered to be order–disorder-type ferroelectrics.

18.8
Hydrogen-Bonded System: Potassium Dihydrogen Phosphor (KDP)

Based on the results with Rochelle salt, compounds with hydrogen bonding were targeted in the search for new ferroelectrics, including water itself. Potassium dihydrogen phosphate (KH$_2$PO$_4$) (KDP) is prepared from an aqueous solution. The ferroelectric transition temperature is 123 K [14], and the transition is second order mixed with a first-order component. The symmetry changes from tetragonal D_{2h} to orthorhombic C_{2v}, and the paraelectric high temperature phase is piezoelectric. Figure 18.7 shows the crystal structure of paraelectric KDP at room temperature [15]. The PO$_4$ network is connected by hydrogen bonding, and potassium ions occupy the opening. One PO$_4$ tetrahedron has four hydrogen bonds. Each proton in the hydrogen bond has two stable points, that is, two protons occupy a closer position and the other two a distant position. Since the hydrogen bonds are located inside the plane perpendicular to the c-axis, the

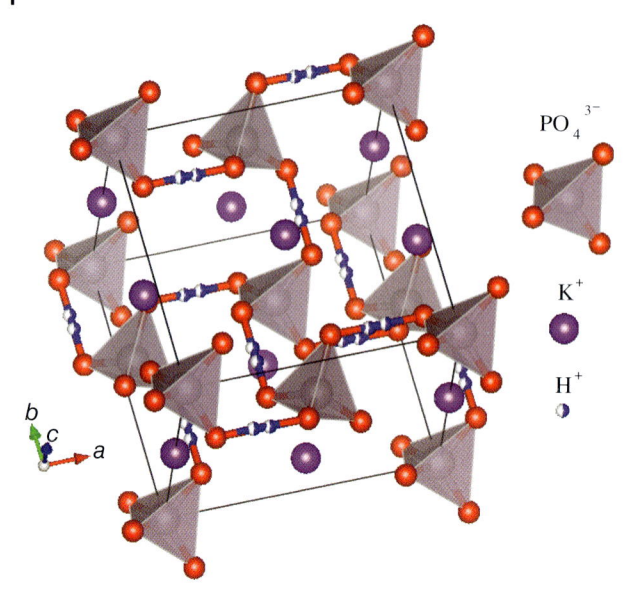

Figure 18.7 Crystal structure of paraelectric KDP.

proton arrangement does not contribute to the polarization along the c-axis. However, the deformation of PO_4 tetrahedron contributes to the polarization. Slater developed a statistical treatment model for such an order–disorder-type phase transitions [16]. The weak point of the Slater model has to do with the isotope effect. Deuterium-substituted KD_2PO_4 (DKDP) shows a Curie temperature at 218 K [17], higher by 95 K than KDP, but such a drastic change cannot be explained within the framework of the Slater model. In order to explain the experimental result of the isotope effect, several physical models were proposed. The proton-tunneling model is one example, in which protons occupying bistable positions tunnel to other position accompanied by the displacement of potassium and phosphor ions. There are many kinds of KDP related compounds [18].

18.9
Perovskites and Related Compounds

During the period of the Second World War, researchers in Japan (Waku and Ogawa), the United States (Wainer and Salomon), and the Soviet Union (Wu) independently discovered the ferroelectric $BaTiO_3$. Arthur von Hippel confirmed its ferroelectricity in 1946 [19]. Concurrently, ABO_3 oxides such as perovskite-type ferroelectric $PbTiO_3$ [20,21] and $KNbO_3$ and $LiNbO_3$-type $LiNbO_3$ and $LiTaO_3$ were successfully synthesized [22]. Meanwhile, antiferroelectric $PbZrO_3$

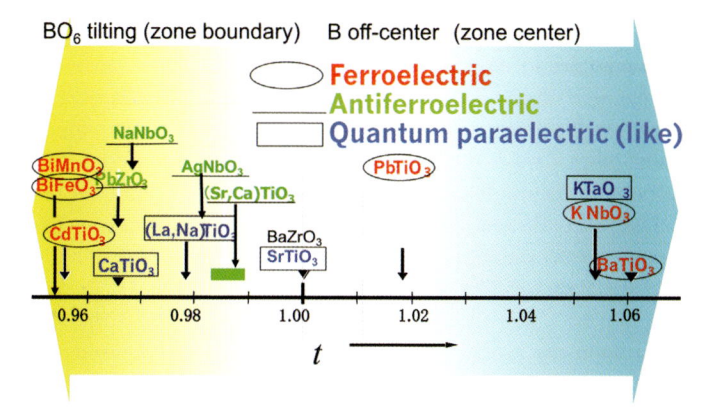

Figure 18.8 Tolerance factor dependence of dielectric and ferroelectric properties for various perovskite-type oxides.

was discovered in 1951 [7]. Figure 18.8 summarizes the dielectric and ferroelectric properties of typical perovskite-type oxides plotted against a tolerance factor of perovskite structure,

$$t = \frac{r_A + r_O}{\sqrt{2}(r_B + r_O)}, \tag{18.25}$$

where, r_A, r_B, and r_O are the ionic radius of A, B, and O ions, respectively. Using the ionic radius of cations at given coordination numbers (A-site = 12 and B-site = 6), we can calculate t for ABO$_3$. We can see that only SrTiO$_3$ and BaZrO$_3$ are located at $t = 1.00$, which means the ideal packing of Sr^{2+} or Ba^{2+}, Ti^{4+} or Zr^{4+}, and O^{2-} ions. Although $t = 1.06$ for BaTiO$_3$ ($P_s = 28\,\mu C/cm^2$) is the largest compound among perovskite-type oxides, its polarization is not the largest among the ABO$_3$. This means that covalency between cations and anions largely contributes to the evolution of the ferroelectricity. When $t > 1.06$, hexagonal-type oxides with face-shared BO$_6$ octahedra appear instead of corner-shared BO$_6$. The face-shared hexagonal phase is seen as a high-temperature form of BaTiO$_3$ above 1430 °C [23]. There are three kinds of ferroelectric BaO-TiO$_2$ compounds [24]. It should be noted that tetragonal BaTiO$_3$, hexagonal BaTiO$_3$, and BaTi$_2$O$_5$ show T_c's at 140, −193, and ~1000 °C, respectively. Three perovskite-type oxides – BaTiO$_3$, KNbO$_3$, and PbTiO$_3$ – occupy the region $t > 1$. However, PbTiO$_3$ has the largest value of polarization ($P_s = 80\,\mu C/cm^2$) in this region. In the soft mode theory, three kinds of ferroelectric soft modes are proposed, as shown in Figure 18.2. The relative vibration of O$_6$ and B, shown in Figure 18.2a, is named the Slater mode. The relative vibration of BO$_6$ versus A is called the Last mode (Figure 18.2b). The other is a deformation mode (Figure 18.2c). It has been recognized experimentally that the Slater mode is the predominant soft mode in BaTiO$_3$ and KNbO$_3$, while the Last mode is predominant in PbTiO$_3$. Structural investigations of BaTiO$_3$ and PbTiO$_3$ revealed that relative displacement of Pb and O in PbTiO$_3$ are larger than in BaTiO$_3$, which manifests as the freezing of the Last mode in PbTiO$_3$.

When $t > 1$, displacement of cations relative to oxygen ions is predominant in the phase transition mechanism. Condensation of such a vibrational mode induces a phase transition to lower the symmetry of the crystal. In physics, this is called an optical soft mode condensation at the zone center (Γ point) in the Brillouin zone. A model was developed to explain the phase transition mechanism of BaTiO$_3$ and other piezoelectric crystals mainly by Cochran [25,26] and J.F. Scott [27]. Such soft modes can be observed by neutron scattering or Raman spectral study. It should be noted that Raman observation of ferroelectric soft modes are possible only in the region where $T < T_c$ by the normal method. So, generally speaking, neutron and Raman scattering experiments are complementary to each other. In Raman scattering studies, observation of the softening of the ferroelectric soft mode (E mode) is difficult in BaTiO$_3$ due to the collision of the soft mode with acoustic phonons (overdamping). Observation of the ferroelectric soft mode manifests displacive-type ferroelectricity in the material, and it has been typically reported for PbTiO$_3$ [28], CdTiO$_3$ [29,30], (Ba,Ca)TiO$_3$ [31], and perovskite-related Bi$_2$SiO$_5$ [32].

In the case $t < 1$, it is generally recognized that perovskite-type oxides tend to have an orthorhombic symmetry, typically a GdFeO$_3$-type distortion. It is thought that this distortion is caused by the rotation of BO$_6$ octahedra to reduce the electrostatic repulsion. From the viewpoint of the definition for displacive-type ferroelectricity, where central cations move to one direction in the oxygen cage, it is hard to imagine the evolution of ferroelectricity in materials for $t < 1$. However, there are many ferroelectric materials in the region $t < 1$, including BiFeO$_3$, BiMnO$_3$, CdTiO$_3$, (Bi,Na)TiO$_3$, (Ag,Li)NbO$_3$, and (Ag,Li)TaO$_3$. This is when the concept of the conventional displacive-type ferroelectricity (i.e., larger space for the displacement of cation is profitable for the evolution of ferroelectricity) breaks down. Manifestation of ferroelectricity in $t < 1$ means the covalency between A and O in the perovskite plays an important role in the evolution of the ferroelectricity, as corroborated by the enhancement of ferroelectricity in PbTiO$_3$ (compared with BaTiO$_3$), whose T_c is higher and polarization larger than those of BaTiO$_3$, although the tolerance factor is smaller than BaTiO$_3$. It is generally recognized that the covalency between Pb6s or Bi6s and O induces and enhances the ferroelectricity of displacive-type ferroelectrics.

The role of the covalency between A and O is demonstrated in two perovskite-types CdTiO$_3$ and CaTiO$_3$ [30]. Both CdTiO$_3$ and CaTiO$_3$ have a GdFeO$_3$-type orthorhombic structure (*Pnma*) at room temperature, and no piezoelectricity. Due to the smaller ionic radius of Cd, the tolerance factor for CdTiO$_3$ ($t = 0.957$) is a little bit smaller than of CaTiO$_3$ ($t = 0.963$). However, only the smaller CdTiO$_3$ turns ferroelectric below 80 K. Ferroelectric soft mode has been observed for CdTiO$_3$ below T_c, confirming the displacive-type ferroelectricity. The only reasonable explanation for this difference is the covalency between Ca and O and between Cd and O. The covalency between Cd4s and O2p induces

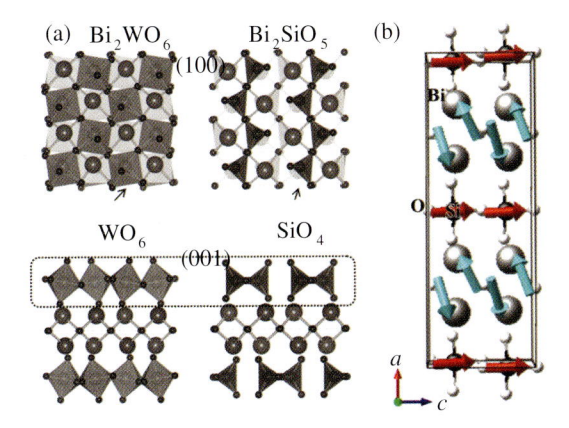

Figure 18.9 (a) Structure of Bi_2SiO_5 compared with Bi_2WO_6 having Aurivillius-type structure ($n=1$). WO_6 two-dimensional layer in Bi_2WO_6 is replaced by SiO_4 one-dimensional chain in Bi_2SiO_5 [32]. (b) Ferrielectric-like arrangement of dipole moments in Bi_2SiO_5 [33].

the anisotropy of the vibration of Cd versus O, and finally enhances the displacement of the Ti ion against O. A similar mechanism is at work in a perovskite-related silicate ferroelectric, Bi_2SiO_5 [33]. Bi_2SiO_5 is a relative of Bi_2WO_6 that belongs to an Aurivillius group of $(Bi_2O_2)(A_{n-1}B_nO_{3n+1})$ ($n=1$ $B=W^{6+}$), where W^{6+} is replaced by Si^{4+}. Crystal structures for Bi_2SiO_5 and Bi_2WO_6 are similar, except that the $(SiO_4)_n$ chain runs in one direction, as shown in Figure 18.9a. Precise structural investigation revealed that Bi ions move in one direction to form a stronger covalent bonding with O. This disproportionation of the chemical bonds in Bi_2O_2 layer, that is, a change from an ionic to covalent-bonded Bi_2O_2 layer also induces the disproportionation of the Si—O bond to yield a shift of the Si ion to one side inside SiO_4 tetrahedra. As depicted in Figure 18.9b, synthesized dipole moments in the Bi_2O_2 layer and $(SiO_4)_n$ chain are aligned in antiparallel directions to each other; consequently, Bi_2SiO_5 can be defined a ferrielectric [33].

18.10
Quantum Paraelectricity

When a ferroelectric transition temperature of displacive-type ferroelectricity is near 0 K, the magnitude of the quantum fluctuation (zero-point vibration) becomes comparable to that of the dipoles, and consequently leads to a saturation of the susceptibility down to 0 K. In such cases, the Barret formula is often used to explain the behavior. The Barret formula [34] is given by

$$\varepsilon = A + \frac{C}{((T_1/2)) \coth ((T_1/2T)) - T_0}, \quad (18.26)$$

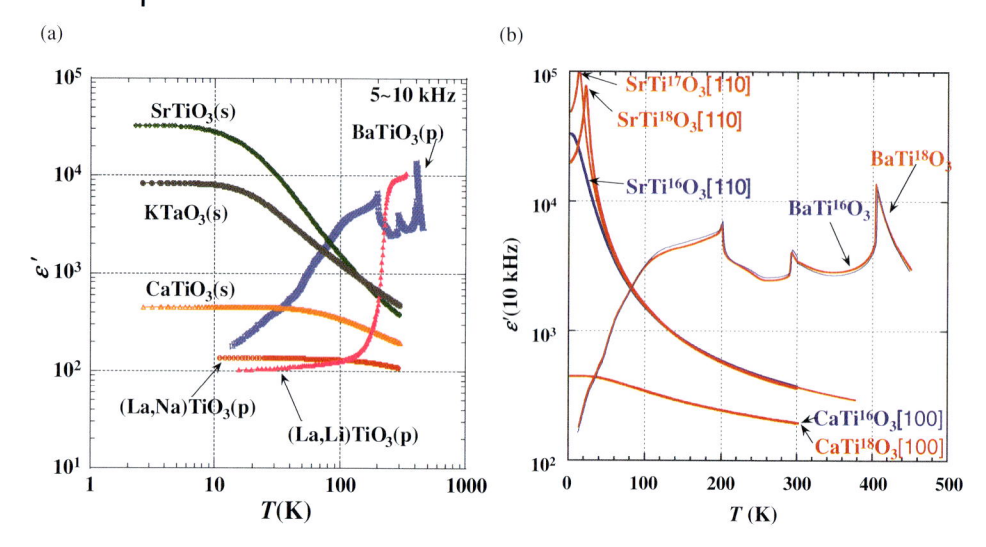

Figure 18.10 (a) Temperature dependence of dielectric constant in various titanate perovskites and KTaO$_3$ and (b) effect of oxygen isotope exchange on the dielectric constants for CaTiO$_3$, BaTiO$_3$, and BaTiO$_3$. s and p in round brackets in (a) represent single crystal and polycrystal, respectively.

where T_0 represents the Curie–Weiss temperature in the classical (high T) limit and T_1 is the characteristic cross over temperature dividing the quantum–mechanical and classical region. Figure 18.10a compares the temperature dependence of the dielectric constant for the perovskites, CaTiO$_3$, SrTiO$_3$, BaTiO$_3$, KTaO$_3$, (La,Na)TiO$_3$, and (La,Li)TiO$_3$. Titanate oxides generally tend to show a saturation of dielectric constant or susceptibility at low temperatures. SrTiO$_3$ is known as a representative quantum paraelectric. In addition, CaTiO$_3$ shows a behavior similar to SrTiO$_3$. KTaO$_3$ also shows a saturation behavior like that of SrTiO$_3$. The Barret formula has a drawback from the physical perspective, although it is quite often used to describe the saturation behavior of the susceptibility or a dielectric constant. The fluctuation spectrum is assumed to be independent of the wave vector in the Barret formula, but the ferroelectric soft mode (optical polar mode) freezes in displacive-type ferroelectrics, although the formula is quite often used to evaluate the Curie–Weiss temperature T_0. In order to solve this problem, a formula describing quantitatively the dielectric behavior of the materials was recently proposed [35]. Using this theory, the inverse of the dielectric constant should be proportional to the square of the temperature in the vicinity of quantum critical limit, as shown in Figure 18.11a. At the quantum critical point, the magnitude of the intercept on the $10^5/\varepsilon$-axis should be zero. The $10^5/\varepsilon$ intercept is positive in the quantum paraelectric region and zero at the quantum critical point. Figure 18.11b shows the phase diagram for the paraelectric near the quantum critical limit. As may be seen in the figure, quantum paraelectric SrTiO$_3$ steps over the ferroelectric critical point and turns into a

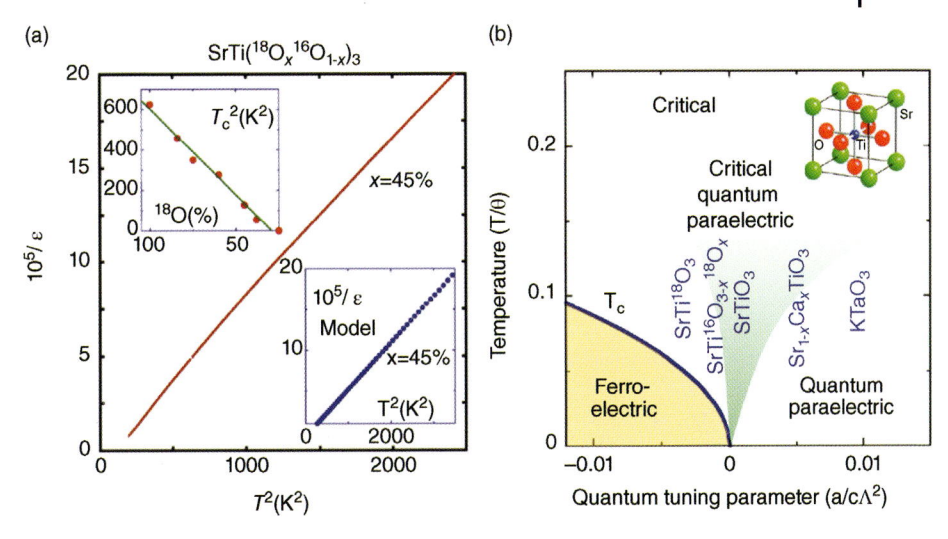

Figure 18.11 (a) $10^5/\varepsilon$ versus T^2 plot in oxygen isotope exchanged SrTiO$_3$ [34]. Theoretically calculated values are plotted in the lower insert. Upper column shows the square of T_c versus magnitude of oxygen isotope substitution. (b) Phase diagram for materials near the ferroelectric quantum critical point. θ is the Debye temperature. Quantum tuning parameter is composed of the inverse static susceptibility a, the mode stiffness parameter c, and Debye wave vector Λ. Usually, the quantum tuning parameter can be varied, for example, by the application of hydrostatic pressure, by isotopic substitution as in the case of SrTi(^{18}O$_x$^{16}O$_{1-x1-x}$)$_3$, or by chemical substitution as in the case of Sr$_{1-x}$Ca$_x$TiO$_3$.

ferroelectric by the oxygen isotope substitution through the quantum critical point (33% ^{18}O substitution), as shown in Figure 18.10b. Other examples of quantum paraelectric KTaO$_3$ and small cation-substituted (Sr,Ca)TiO$_3$ are farther than SrTiO$_3$ from the quantum critical point. Crossovers from quantum paraelectrics to ferroelectrics through the quantum critical point other than oxygen-isotope exchanged SrTiO$_3$ have been reported in nonoxide ferroelectric of tetrathiafulvalene (TTF) under hydrostatic pressure [36].

18.11
Relaxor Ferroelectric

One of the most intriguing properties exhibited by relaxors is the huge dielectric response with strong frequency dispersion in the large temperature range, as shown in Figure 18.12 for a prototypical relaxor Pb(Mg$_{1/3}$Nb$_{2/3}$)O$_3$ (PMN) crystal. However, in contrast to normal ferroelectrics such as BaTiO$_3$, as shown in Figure 18.12, the PMN crystal shows no sharp phase transition in the crystal, despite continuous increases in the dielectric constant around room temperature. Such an anomalous dielectric behavior has puzzled solid-state physicists for the past 50 years. Two models have mainly been proposed for the origin of

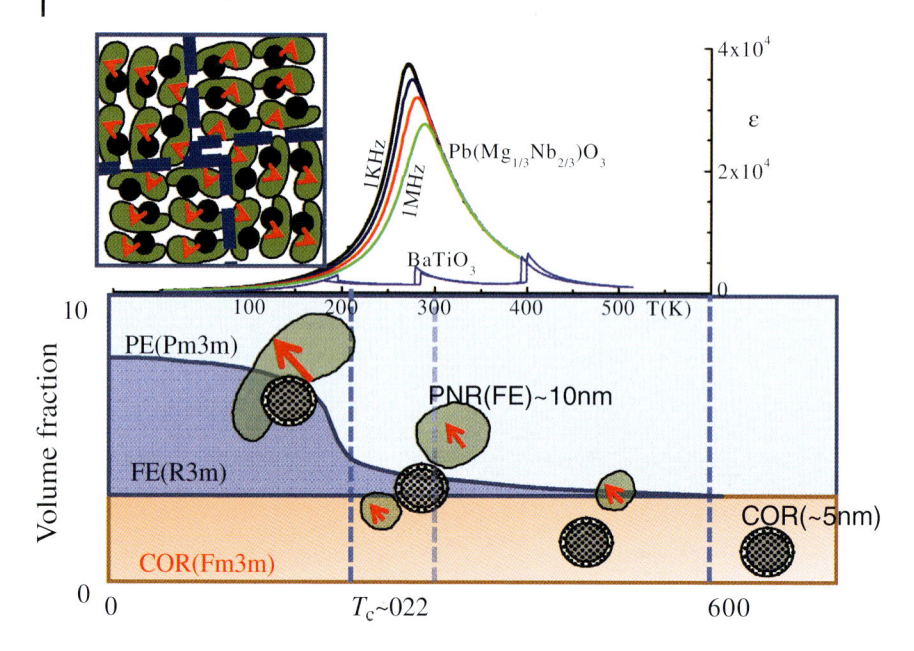

Figure 18.12 Temperature dependence of dielectric constant for PMN and the possible microscopic model for the structure [36]. Dielectric constant for BaTiO$_3$ is shown for a comparison. Diagram in the lower column schematically represents the proportion and morphology of each phase based on the TEM observation and Raman spectra. Figure at the top left represents the microscopic model below T_c based on the nanosized ferroelectric model.

the strong frequency dispersion: the first is nanosized ferroelectric and the second is dipole glass, as shown in Figure 18.12. It is generally known that there are multiple inhomogeneities in PMN: (1) The chemically ordered region (COR), where Mg and Nb is 1:1 ordered and different from the average cation composition, persists at all temperatures. (2) The ferroelectric region grows with a decrease in temperature. (3) The paraelectric region, which persists at all temperature. Regions (2) and (3) mutually alter their volume fractions with a temperature change. Presence of the ferroelectric soft mode and ferroelectric domain manifest the ferroelectricity of PMN; however, the large frequency dispersion of the susceptibility reflects different sensitivities of each region in the crystal. Figure 18.12 is an image of the temperature evolution of the volume change of the three regions in PMN [37]. Although there have been many reports on the relaxor behavior of the materials, most report only the temperature variation of the dielectric constant. Dipole interaction in dielectric is of long-range order, in contrast to the magnetic system, so one should take care about the sample status, single crystal or polycrystal, and the quality of the sample like defect concentration, nonstoichiometry of component ions, and multiple phases. Also, one should inspect the microscopic structure of the material at least by TEM before discussing the relaxor behavior of the material against the external electric field. Quite often, Vogel Fulcher's formula [38,39] given by

$$f_m = f_0 \left[-\frac{E_a}{k(T'_m - T_f)} \right], \tag{18.27}$$

where f_0, E_a, T_f, k, and T'_m, respectively, are Debye frequency, activation energy, freezing temperature, Boltzmann constant, and the peak temperature of dielectric constant. Equation (18.27) is used to analyze the dielectric behavior of the "relaxor." However, detailed investigations on the microscopic structure and dynamical motion are needed before discussing the real nature of the dielectric dispersion of the material.

18.12
Piezoelectric: Pb(Zr$_{1-x}$Ti$_x$)O$_3$ (PZT)

PbZrO$_3$ is an antiferroelectric showing the phase transition from cubic to orthorhombic at 230 °C. When ferroelectric PbTiO$_3$ is alloyed, the antiferroelectric phase is replaced by ferroelectric rhombohedral and tetrahedral phases. A phase diagram is presented in Figure 18.13 [40]. The most important phase boundary appears between $x(\text{PbTiO}_3) = 0.4$ and 0.5. This is called a morphotropic phase boundary (MPB), where the electromechanical response is

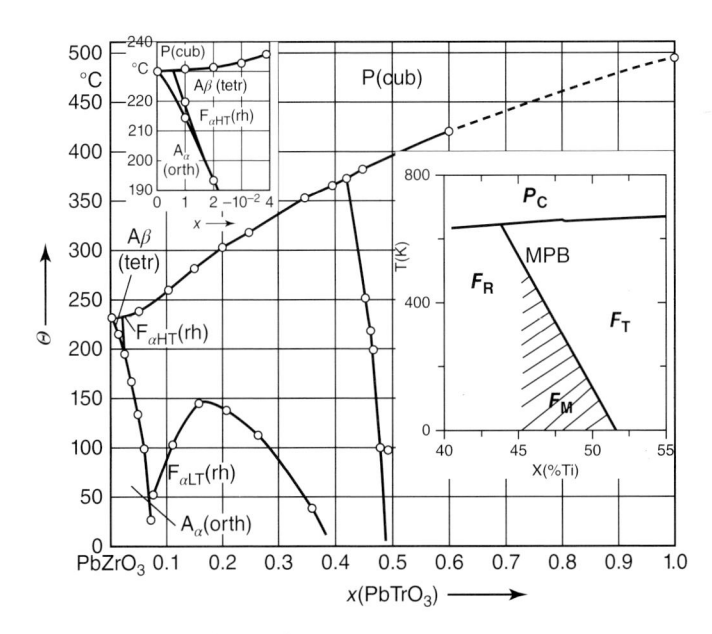

Figure 18.13 Phase diagram for Pb(Zr$_{1-x}$Ti$_x$)O$_3$ system [39]. Recent investigations have revealed the detailed phase diagram near the morphotropic phase boundary by the detailed neutron scattering experiments (*inset*) [41].

maximum. Pb(Zr$_{1-x}$Ti$_x$)O$_3$ (PZT) near this composition has been utilized as a high performance piezoelectric material. Free energy for Eq. (18.10) is simplified including up to fourth term by

$$F = \frac{\alpha}{2}\left(P_1^2 + P_2^2 + P_3^2\right) + \frac{\beta}{4}\left(P_1^4 + P_2^4 + P_3^4\right) + \frac{\beta'}{2}\left(P_1^2 P_2^2 + P_2^2 P_3^2 + P_3^2 P_1^2\right). \quad (18.28)$$

Equation (18.28) means the free energy is isotropic against the direction of polarization; that is, phase transition accompanying polarization direction change rarely occurs. In the vicinity of MPB, the polarization direction can be changed by a small external electric or elastic field [41]. This clearly explains that elastic and piezoelectric constants of shear strain diverge at MPB. Recently, detailed structural analysis revealed that monoclinic phase appears in the MPB region that facilitates the continuous rotation of polarization direction against the external electric field [42].

18.13
Organic Ferroelectric

There are many organic ferroelectrics but most of them have low ferroelectric transition temperatures, and consequently, low temperature T_c and small polarization. However, quite recently, croconic acid, which has a hydrogen-bonded polar structure in its crystalline state, was found to have a ferroelectricity at room temperature with a magnitude of spontaneous polarization 21–22 µC/cm^2 [43]. Figure 18.14 shows the structure and P–E loop. Protons in OH groups

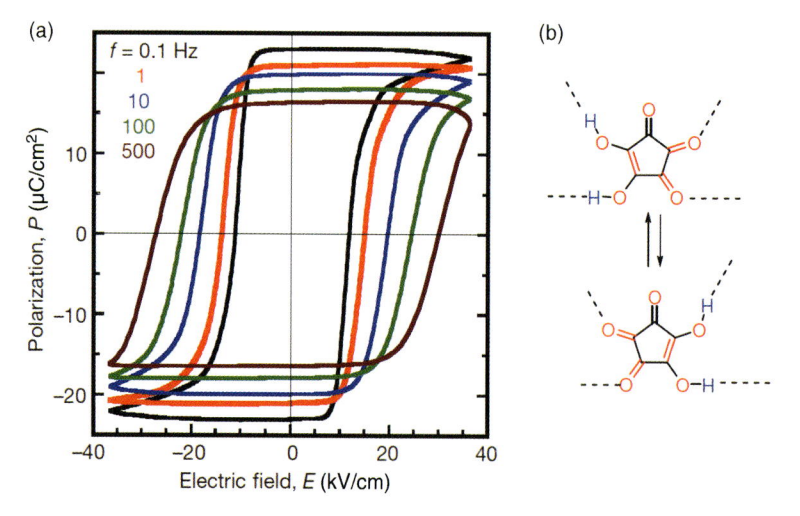

Figure 18.14 P–E loops for crocon at room temperature (a) and the structural change before and after the polarization switching (b). Note that the position of the C=C double bond is changed after the switching [42].

move all together to C=O groups in the next molecule via the hydrogen bonding, yielding a change in the position of C=C double bonds, and consequently causing polarization switching. Such a breakthrough in molecular materials – ferroelectricity comparable to that of inorganic materials and high-temperature stability up to at least 150 °C – may create new possibilities in the field of materials science for future application.

References

1. Sawada, S., Nomura, S., Fujii, S., and Yoshida, I. (1958) Ferroelectricity in $NaNO_2$. *Phys. Rev. Lett.*, **1**, 320–321.
2. Kay, H.F. and Vousden, P. (1949) Symmetry change in Barium titanate at low temperatures and their relation to its ferroelectric properties. *Philos. Mag.*, **40**, 1019.
3. Shirane, G. and Suzuki, K. (1951) On the phase transition in Barium–Lead Titanate. *J. Phys. Soc. Jpn.*, **6**, 274–278.
4. Shirane, G., Sawaguchi, E., and Takagi, Y. (1951) Dielectric properties of lead zirconate. *Phys. Rev.*, **84**, 476.
5. Sawaguchi, E. (1953) Ferroelectricity versus antiferroelectricity in the solid solutions of $PbZrO_3$ and $PbTiO_3$. *J. Phys. Soc. Jpn.*, **8**, 615–629.
6. Landau, L.D. and Lifshitz, E.M. (1969) *Statistical Physics: Course of Theoretical Physics*, Pegamon Press.
7. Sawaguchi, E., Maniwa, H., and Hoshino, S. (1951) Antiferroelectric structure of lead zirconate. *Phys. Rev.*, **83**, 1078.
8. Kittel, C. (1951) Theory of antiferroelectric crystals. *Phys Rev.*, **82**, 729–732.
9. Valasek, J. (1921) Piezo-electric and allied phenomena in Rochelle salt. *Phys. Rev.*, **17**, 475–481.
10. Hablützel, J. (1939) Schweres Seignettesaltz: Dielectrische Untersuchungen an $KNaC_4H_2D_2O_6$-$4H_2O$-Kristallen. *Helv. Phys. Acta*, **12**, 488–510.
11. Takahashi, H. (1959) On the polarization mechanism in Rochelle salt (in Japanese) *Busseironkenkyuu*, 22, 1–2.
12. Mitsui, T. (1958) Theory of the ferroelectric effect in Rochelle salt. *Phys. Rev.*, **111**, 1259–1267.
13. Makita, Y. and Takagi, Y. (1958) Ferroelectric and optical properties of Na (K-NH_4)-tartrate mixed crystals. *J. Phys. Soc. Jpn.*, **13**, 367–377.
14. Busch, G. and Scherrer, O. (1935) Eine neue seinette-elektrische Substanz. *Kurze Originalmittelkungen*, **23**, 737.
15. Nelmes, R.J., Meyer, G.M., and Tibballs, J.E. (1982) The crystal structure of tetragonal KH_2PO_4 and KD_2PO_4 as a function of temperature. *J. Phys. C Solid State Phys.*, **15**, 59–75.
16. Slater, J.C. (1941) Theory of the transition in KH_2PO_4. *J. Chem. Phys.*, **9**, 16–33.
17. Zwicker, B. and Scerrer, P. (1944) Electrooptische der seignette-electrischen Kristalle KH_2PO_4 and KD_2PO_4. *Helv. Phys. Acta*, **17**, 346–345.
18. Blinc, R. (1960) On the isotopic effect in the ferroelectric behavior of crystals with short hydrogen bonds. *J. Phys. Chem. Solids*, **13**, 204–211.
19. von Hippel, A. (1950) Ferroelectricity, domain structure, and phase transitions of barium titanate. *Rev. Mod. Phys*, **22**, 221–237.
20. Shirane, G., Hoshino, S., and Suzuki, K. (1950) X-ray study of the phase transition in lead titanete. *Phys. Rev.*, **80**, 1105–1106.
21. Smolenslii, G.A. (1950) *Dokl. Akad. Nauk SSSR*, **70**, 405.
22. Matthias, B.T. and Remeika, J.P. (1949) Ferroelectricity in the ilmenite structure. *Phys. Rev.*, **76**, 1886–1887.
23. Zhu, N. and West, A.R. (1955) Formation and stability of ferroelectric $BaTi_2O_5$. *J. Am. Ceram. Soc.*, **38**, 102–113.
24. Kimura, T., Goto, T., Yamane, H., Iwata, H., Kajiwara, T., and Akashi, T. (2003) A ferroelectric barium titanate $BaTi_2O_5$. *Acta Crystallogr.*, **C59**, i128–i130.

25 Cochran, W. (1960) Crystal stability and the theory of ferroelectricity. *Rev. Adv. Phys.*, **9**, 387–423.
26 Cochran, W. (1961) Crystal stability and the theory of ferroelectricity part II: piezoelectric crystals. *Rev. Adv. Phys.*, **10**, 401–420.
27 Scott, J.F. (1974) Soft-mode spectroscopy: experimental studies of structural phase transition. *Rev. Adv. Phys.*, **46**, 83–128.
28 Burns, G. and Scott, B.A. (1973) Lattice modes in ferroelectric perovskite: $PbTiO_3$. *Phys. Rev.*, **7**, 3068–3101.
29 Taniguchi, H., Shan, Y.-J., Mori, H., and Itoh, M. (2007) Critical soft-mode dynamics and unusual anticrossing in $CdTiO_3$ studied by Raman scattering. *Phys. Rev. B*, **76** (212103), 1–4.
30 Taniguchi, H., Soon, H.-P., Shimizu, T., Moriwake, H., Shan, Y.-J., and Itoh, M. (2011) Mechanism for suppression of ferroelectricity in $Cd_{1-x}Ca_xTiO_3$. *Phys. Rev. B*, **84** (174106), 1–5.
31 Shimizu, T., Fu, D., Taniguchi, H., Taniyama, T., and Itoh, M. (2012) Origin of the dielectric response in $Ba_{0.767}Ca_{0.233}TiO_3$. *Appl. Phys. Lett.*, **100** (102908), 1–4.
32 Taniguchi, H., Kuwabara, A., Kim, J., Kim, Y., Moriwake, H., Kim, S., Hoshiyama, T., Koyama, T., Mori, S., Takata, M., Hosono, H., Inaguma, Y., and Itoh, M. (2013) Ferroelectricity driven by twisting of silicate tetrahedral chains. *Angew. Chem., Int. Ed.*, **52**, 8088–8092.
33 Kim, Y., Kim, J., Fujiwara, A., Taniguchi, H., Kim, S., Tanaka, H., Sugimoto, K., Kato, K., Itoh, M., Hosono, H., and Takata, M. (2014) Hierarchical dielectric orders in layered ferroelectrics Bi_2SiO_5. *IUCrJ*, **1**, 160–164.

34 Barret, J.H. (1952) Dielectric constant in perovskite-type crystals. *Phys. Rev.*, **86**, 118–120.
35 Rowley, S.E. et al. (2014) Ferroelectric quantum ferroelectricity. *Nat. Phys.*, **10**, 367–372.
36 Horiuchi, S., Kobayashi, K., Kumai, R., Minami, N., Kagawa, F., and Tokura, Y. (2015) Quantum ferroelectricity in charge transfer complex crystals. *Nat. Commun.*, **6**, 7469.
37 Fu, D., Taniguchi, H., Itoh, M., Koshihara, S., Yamamoto, N., and Mori, S. (2009) Relaxor $Pb(Mg_{1/3}Nb_{2/3})O_3$: a ferroelectric with multiple inhomogeneities. *Phys. Rev. Lett.*, **103** (207601), 1–4.
38 Vogel, H. (1921) Das temperaturabhängigkeitsgesetz der viskosität von flüssigkeiten. *Phys. Zeitschrift*, **22**, 645–636.
39 Fulcher, G.S. (1925) Analysis of recent measurements of the viscosity of glasses. *J. Am. Ceram. Soc.*, **8**, 339–355.
40 Berlincourt, D. (1966) Transducers using forced transitions between ferroelectric and antiferroelectrc states. *IEEE Trans. Sonics Ultrason.*, **SU-13**, 116.
41 Uesu, Y. (2002) Colossal piezoelectric effect in complex perovskite oxides. *J. Phys. Soc. Jpn.*, **57**, 646–653.
42 Guo, R., Cross, L.E., Park, S.-E., Noheda, B., Cox, D.E., and Shirane, G. (2000) Origin of the high piezoelectric response in $PbZr_{1-x}Ti_xO_3$. *Phys. Rev. Lett.*, **84**, 5423–5426.
43 Horiuchi, S., Tokunaga, Y., Giovannetti, G., Picozzi, S., Itoh, H., Shimano, R., Kumai, R., and Tokura, Y. (2010) Above room temperature ferroelectricity in a single-component molecular crystal. *Nature*, **463**, 489–793.

19
Defect Chemistry and Its Relevance for Ionic Conduction and Reactivity

Joachim Maier

Max-Planck-Institute for Solid State Research, Heisenbergstr. 1, 70569 Stuttgart, Germany

19.1
Introduction: On the Relevance of Defect Chemistry

This contribution deals with the fundamental role of defect chemistry for solid-state chemistry in general and for electrochemical energy applications in particular. Casually formulated, solid-state chemistry is chemistry of the perfect state plus defect chemistry, not in the sense that defect chemistry adds nonidealities or pathological properties to the perfect solid state rather in the sense of Figure 19.1 highlighting that point defects are the active particles. They are for solid-state chemistry what H^+ and OH^- are for aqueous chemistry, namely, constituting the ionic excitations and enabling the "inner chemical life" [1,2]. Only after having realized this, one is able to understand solid materials and tune functionalities in the same way as one is able to do with H_2O.

Nonetheless, defect chemistry is equally disliked in the chemists' community as it is in the physicists' community. That it has not become a natural part of freshmen textbooks is particularly embarrassing for the first as it is a truly chemical issue. It is even more embarrassing as the beginning of defect chemistry dates almost 100 years back [3,4]. However, instead of having triggered a gold rush in general chemistry, defect chemistry has been mainly limited to physical chemistry and materials science ("solid state ionics").

Figure 19.1 is instructive in this context. It shows the two best understood materials of our world, the "chemists' water" and the "physicists' silicon." In both cases, it is not so much the perfect structure, rather the dissociated particles H^+ and OH^- that are of relevance for chemistry in the case of water and electrons and holes in the case of silicon [2,5]. Obviously, it is crucial to identify ionic (generally vacancies and interstitials) and electronic (conduction electrons and holes) imperfections when professionally dealing with ionic crystals.

Handbook of Solid State Chemistry, First Edition. Edited by Richard Dronskowski, Shinichi Kikkawa, and Andreas Stein.
© 2017 Wiley-VCH Verlag GmbH & Co. KGaA. Published 2017 by Wiley-VCH Verlag GmbH & Co. KGaA.

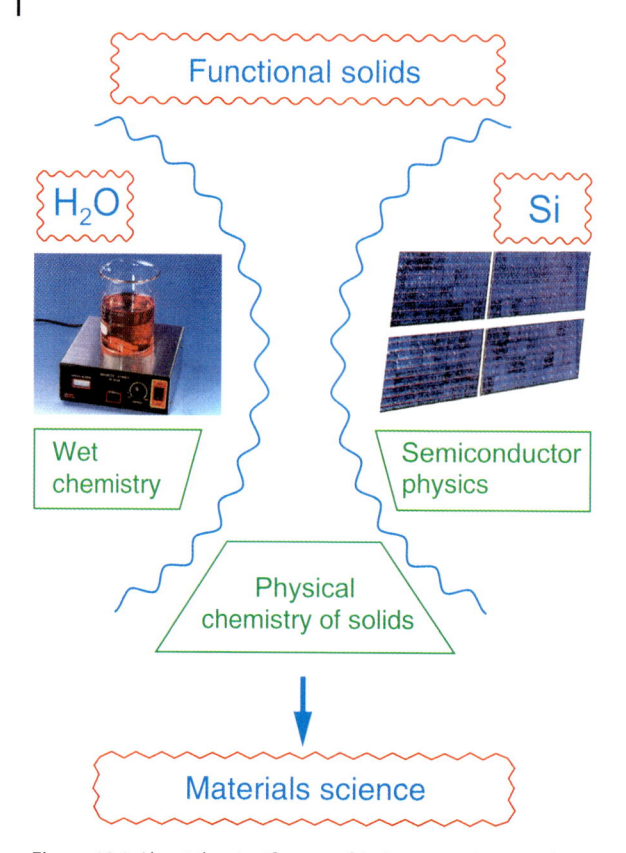

Figure 19.1 About the significance of "relevant, active particles.".

Let us dwell on this for a moment (Figure 19.2a). Dissociation of water is usually written as

$$H_2O + H_2O \rightleftharpoons H_3O^+ + OH^- \qquad (19.1)$$

Structurally more correct would be the involvement of a greater number of H_2O molecules (cf. $H_9O_4^+$, etc.). Thermodynamically, however, it suffices to even subtract one H_2O molecule and to write

$$H_2O \rightleftharpoons H^+ + OH^- \qquad (19.2)$$

or even to subtract one more H_2O to yield

$$\text{Nil} \rightleftharpoons \text{excess proton} + \text{lacking proton} \qquad (19.3)$$

The two species on the right-hand side are the point defects in the phase of water, directly reflecting the relevance of excess and lacking particles for the solid state (cf. Figure 19.2a). Let us consider AgCl as example where such excitations are mainly restricted to the Ag-sublattice (see Figures 19.2b and 19.3a). AgCl is dissociated in terms of being constituted of ions. However, these are not

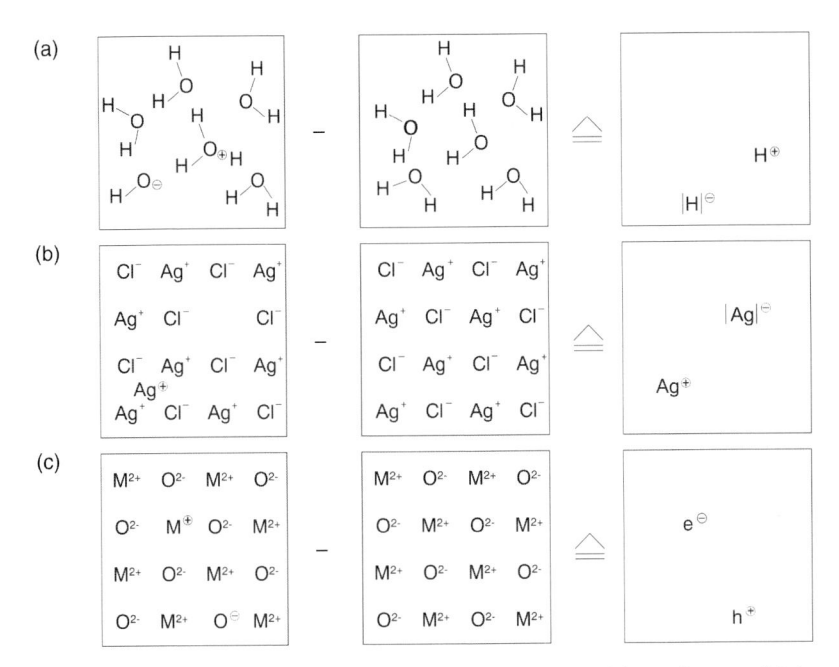

Figure 19.2 "Defect chemistry = real chemistry minus chemistry of the perfect state." Point defect chemistry for water (a), AgCl (b) and a purely electronically disordered oxide (c). (Figure reproduced with permission from [6]. Copyright 2004, John Wiley & Sons Inc.)

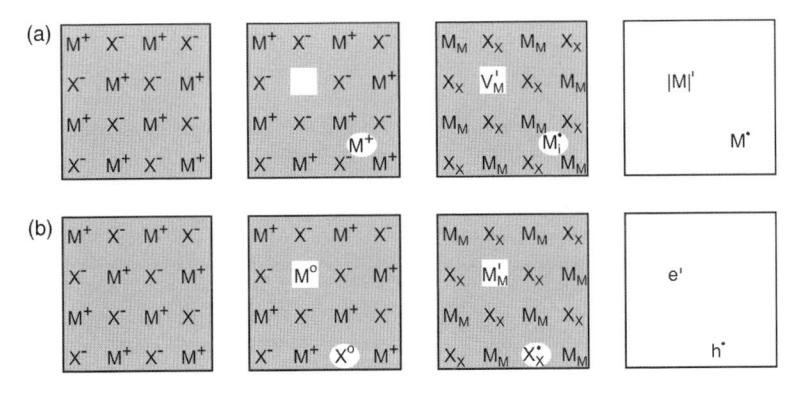

Figure 19.3 Perfect (left) and defective crystal situations for the compound M^+X^-. A specific example may be Ag^+Cl^-. (a) Ionic defects. (b) Electronic defects. First and second columns: structure elements in absolute notation of perfect and real state. Third column: structure elements in relative notation as far as charge is concerned (V_i, that is, the vacant interstitial site is not designated). Fourth column: building elements (second column "minus" first column) as relative constituents as far as charge and matter is concerned. (Figure reproduced with permission from [7]. Copyright 2010, Taylor & Francis.)

at all free, rather they are strongly bound by Coulomb forces (Figure 19.3 first column). The decisive dissociation (superionic dissociation) out of the Coulomb bonding (Figure 19.3a, second column) can be written as

$$Ag^+Cl^- + Ag^+Cl^- \rightleftharpoons Ag_2^+Cl^- + \square Cl^- \tag{19.4}$$

Structurally, more adequate would be to involve more AgCl entities (e.g., $Ag_5Cl_4^+$, $Ag_5Cl_6^-$). Thermodynamically, however, it suffices to write

$$\text{Nil} \rightleftharpoons \text{excess silver ion} + \text{lacking silver ion} \tag{19.5}$$

which is arrived at by subtracting 2AgCl from Eq. (19.4) (cf. Figure 19.3a, fourth column).

As far as the electronic disorder is concerned (Figures 19.3b and 19.2c), the situation is analogous as well. For AgCl, the transition from valence to conduction band can be described as

$$Ag^+Cl^- + Ag^+Cl^- \rightleftharpoons Ag^\circ Cl^- + Ag^+Cl^\circ \tag{19.6}$$

or in minimal language (subtracting 2AgCl) as

$$\text{Nil} \rightleftharpoons \text{excess electron} + \text{lacking electron} \tag{19.7}$$

(cf. Figure 19.3b, fourth column). For the case of ionic disorder, it turns out that the detailed notation of Eq. (19.4) is too clumsy and the minimal notation of Eq. (19.5) too abstract. Thus, Kröger–Vink nomenclature [8] as a relative notation is generally used. This nomenclature is not a completely relative notation (unlike building elements, in fourth column of Figure 19.3 and in Eq. (19.5)), rather it uses real structural elements but with relative charges (third column in Figure 19.3). Hence, Eq. (19.4) is rewritten as

$$Ag_{Ag} + V_i \rightleftharpoons Ag_i^\bullet + V'_{Ag} \tag{19.8}$$

(V_i stands for a vacant interstitial site). A rigorous treatment of the difference between building elements and structure elements as far as statistics is concerned is given in Ref. [9]. Luckily the nomenclature is not really important, rather the question which concentrations enter equilibrium conditions and rate equations.

Also, as far as acid–base and redox chemistry is concerned, defect chemistry is the key. It can even be shown [5] that if one wants to construct an acid–base concept that is completely analogous to the Brønsted concept for H_2O, it has to be based on point defect chemistry (note that acidity and basicity of the ions are both in bulk and on the surface thermodynamically coupled to the concentration of the point defects). Equally, redox effects enter by including the electronic defects.

Most important are point defects as centers enabling chemical kinetics in the bulk but also at the surface (cf. heterogeneous catalysis). Neglecting them is the reason why chemical kinetics of the solid state is still in its infancy (see Section 19.5). Notwithstanding the complexity, a solid-state reaction involves interfacial reaction steps for which point defects are crucial but also the comparatively sluggish transport steps for which again the point defects are decisive. They enable

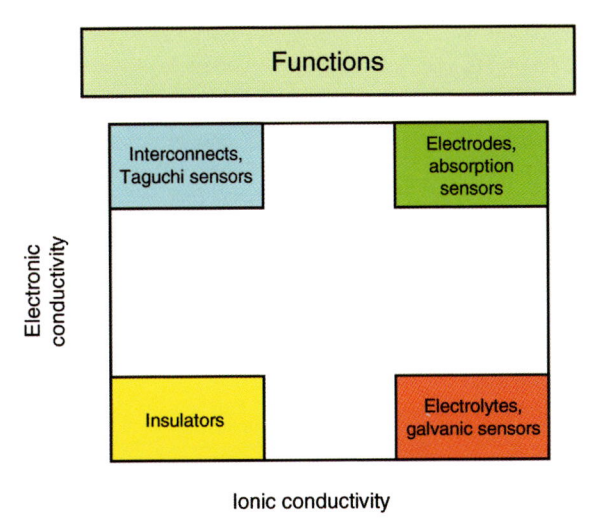

Figure 19.4 Functional materials for electrochemistry can be distinguished according to ionic and electronic conductivity contributions. (Figure reproduced with permission from [10]. Copyright 2003, Taylor & Francis.)

ionic and electronic transport and hence, finally device components such as solid electrolytes and solid electrodes that are the key elements for batteries, fuel cells, and other devices of electrochemical energy technology (see Figure 19.4) [6].

So far we concentrated on point defects that are necessary part of the ionic crystals in equilibrium. Very often point defects are present in higher concentrations than equilibrium would demand. This is typically the case when high-temperature materials are quenched. Higher dimensional defects such as dislocations or grain boundaries are necessarily nonequilibrium structure elements of the solid state. (The presence of heterophase boundaries – including surfaces – may be demanded by equilibrium, however, their number, area, and form are typically out of equilibrium.) As they are typically immobile, they will enter our considerations in the form of fixed frame conditions. It is also their presence together with the manifold of different structures that make solid-state chemistry more colorful than it is possible with fluid phases.

19.2
Ionic and Electronic Defects as Acid–Base and Redox Particles in Equilibrium

Figures 19.2 and 19.3 illustrate graphically the defect generation as "subtraction" of the perfect state from the real state. Note that in all cases it is (configurational) entropy that is responsible for a finite defect concentration as it costs energy to dissociate H_2O, to disorder Ag^+, or to transfer e^- from O^{2-} to M^{2+}. It is very revealing to use a common language for describing the energetics.

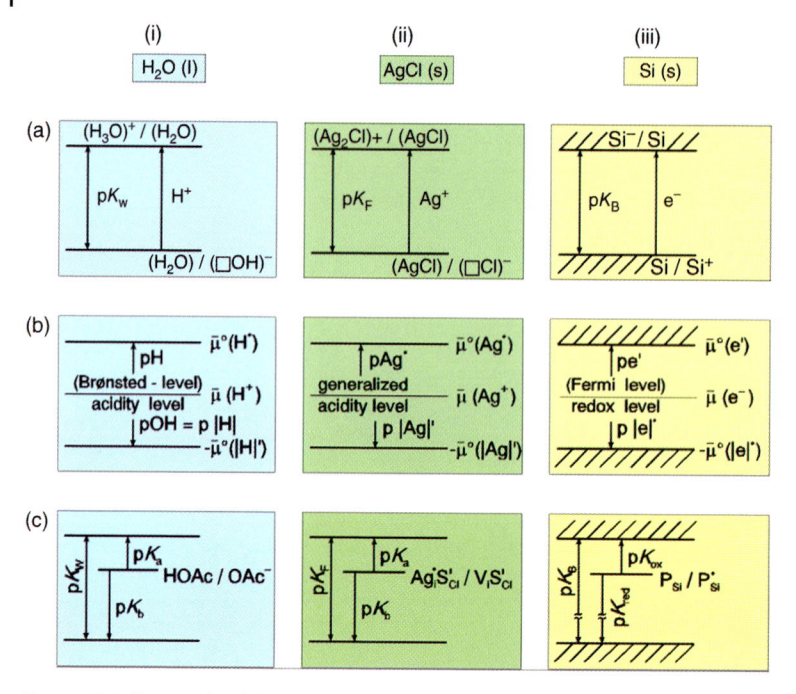

Figure 19.5 "Energy" level diagrams for water, AgCl, and Si ("physical picture"). (Figure reproduced with permission from [5]. Copyright 2001, John Wiley & Sons Inc.)

Let us first use the "physical picture" of level diagrams that is very illustrative for a simple defect chemistry (see Ref. [11] for a precise definition of the terms used) [6,12,13]. (This picture refers to a rigid level diagram, where excitation does not affect the levels, as it is the case for dilute defect chemistry.)

Figure 19.5a(iii) depicts the well-known semiconductor level diagram for electrons (here in silicon). The electron transfer (in a nonlocalized) picture is the transition from the valence to the conduction band by bridging the electronic gap. If this gap shrinks to zero, one speaks of a metal. If we use the analogous construction (by employing standard electrochemical potentials; in the case of interactions energetic activity coefficients are also included) for water (Figure 19.5a(i)), we bridge the ionic gap by bringing a H^+ from one H_2O to another one, forming H_3O^+ and leaving OH^- (H^+-vacancy). When this ionic gap is zero, one speaks of a strong electrolyte.

Figure 19.5a(ii) refers to AgCl, where now the ionic gap is bridged by exciting one Ag^+ from a regular site to an energetically higher lying interstitial site leaving the Ag^+-vacancy. If the gap shrinks to zero, we speak of a superionic conductor where all the Ag^+ are disordered as is the case in $\alpha - AgI$ (see further).

Figure 19.5b includes the full electrochemical potential that now contains concentration terms (from the configurational entropy) such that the distance to the levels bears information on the realized defect concentrations. Figure 19.5b(iii) is

19.2 Ionic and Electronic Defects as Acid–Base and Redox Particles in Equilibrium

well known and the inserted level is called Fermi level (electrochemical potential of the electron), in the case of H_2O (Figure 19.5b(i)) this level (electrochemical potential of the proton) may be termed Brønsted level and in the solid-state case (electrochemical potential of the silver ion) let us refer to it as Frenkel level (Figure 19.5b(ii)). As far as the underlying statistics is concerned, see Refs. [9–11].

The comparison makes it immediately clear that the – above mentioned – exact analogue to a Brønsted acid–base picture for a solid has to be based on counting point defects. This striking but thermodynamically precise correspondence is set out rigorously in Ref. [5]. (It also enables a generalized treatment of bulk and surface acidity as well as definition and comparison of different acidity scales in different solids.) In this sense, V'_{Ag} is a basic and $Ag_i^{\bullet}$ an acidic center that is immediately obvious if one refers to H^+ or O^{2-} instead of Ag^+ (see Section 19.5).

Also, the role of impurities is analogous in all cases (Figure 19.5c). P in Si is a donor exciting e^- to the upper level, in H_2O the impurity HOAc is termed an acid exciting H^+ to the upper level, and in AgCl S'_{Cl} (S^{2-} substituting for Cl^- as acidic center) forms associates with $Ag_i^{\bullet}$ out of which Ag^+ can be excited into free (i.e., unassociated) interstitial sites.

To fully describe the defect chemistry in a solid, one also needs to consider the coupling of ionic and electronic diagrams. Clearly, AgCl does not only show ionic disorder, it also has a finite electronic band gap. The coupling of ionic and electronic disorder is given in Figure 19.6 (see Ref. [14]). As there the electron excitation is plotted upside down, the distance of the two electrochemical

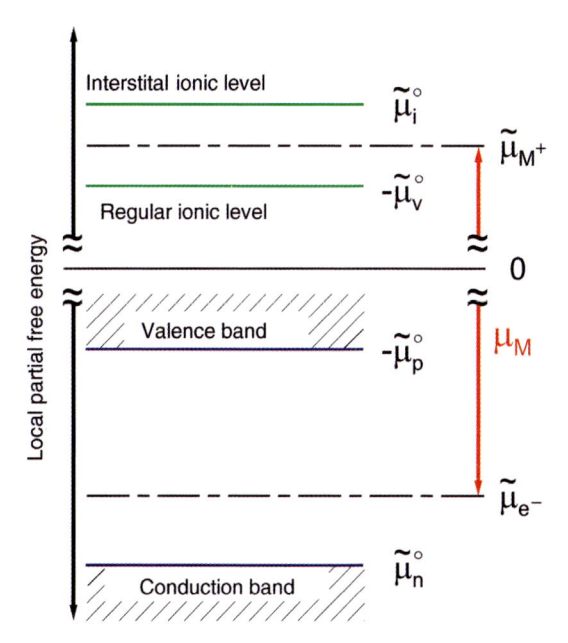

Figure 19.6 Coupling of ionic excitation and electronic excitation in the "physical picture." (Figure reproduced with permission from [14]. Copyright 2003, Elsevier.)

potentials (for Ag$^+$ and e$^-$), refers to their sum that is the chemical potential of Ag. This quantity can be tuned by varying the Ag-partial pressure of the gas phase or more generally the Ag-activity of the neighboring phase. It measures the increase in Gibbs energy on increasing the Ag content at constant Cl content and is hence, determined by the nonstoichiometry δ in Ag$_{1+\delta}$Cl (δ can be positive or negative).

Obviously, the major control parameters that are decisive for the variation of defect chemistry are (i) component activity or partial pressure (P), (ii) temperature (T), and (iii) doping content (C) [6] as they directly influence the electrochemical potentials. It is the P-impact that directly involves the neighboring phase.

Now let us turn to the commonly used "chemical picture" that describes the disorder processes in terms of defect chemical reactions. While the "physical picture" may be more illustrative in terms of energetics, the chemical picture is much more appropriate in complex situation. For dilute situations, Boltzmann statistics is valid directly leading to ideal mass action laws, as it is known in aqueous chemistry as well as in semiconductor physics (cf. Ref. [9]). For our AgCl example these are

$$K_F = [Ag_i^\bullet][V_{Ag}'] ,$$

$$K_B = [e'][h^\bullet], \quad (19.9)$$

$$K_{Ag} = \frac{[Ag_i^\bullet][e']}{P_{Ag}} .$$

The last equation is the mass action law for the interaction with the neighboring phase. If it is a gas phase (g), the corresponding reaction is

$$Ag(g) + V_i \rightleftharpoons Ag_i^\bullet + e' \quad (19.10)$$

expressing the variation of $\delta = [Ag_i^\bullet] - [V_{Ag}']$. Note that P_{Ag} can also be tuned by varying the partial pressure of the anionic component according to $P_{Cl_2}^{1/2} = K_f/P_{Ag}$, where K_f is the formation constant of AgCl (formation from the elements Ag(s) and Cl$_2$(g)).

The wanted solutions for $[Ag_i^\bullet]$, $[V_{Ag}']$, $[h^\bullet]$, and $[e']$ follow from Eq. (19.9) once the electroneutrality equation

$$[Ag_i^\bullet] + [h^\bullet] + zC = [V_{Ag}'] + [e'] \quad (19.11)$$

is taken account of.

In Eq. (19.11), C is the impurity content and z the effective charge of the dopant, for example, given by $[Mn_{Ag}^\bullet]$ (with $z = +1$) if AgCl is doped by manganese, or by $[S_{Cl}']$ (with $z = -1$) if it is sulfur-doped. For a highly pure AgCl, C can be neglected. Note however, that at low enough intrinsic defect concentrations and hence, at low enough T any material will be impurity dominated (or dominated by frozen-in disorder).

Even though not a problem to solve this set of equations strictly, the solution becomes particularly simple if Eq. (19.11) can be approximated by single terms on both sides each (Brouwer approximation) [4,15]. It is well known for AgCl that the ionic disorder is much larger than the electronic one, hence

$$[Ag_i^\bullet] + C \simeq [V_{Ag}'] . \tag{19.12}$$

For pure AgCl ($C = 0$, more accurately $C \ll \sqrt{K_F}$) it follows that

$$[Ag_i^\bullet] = [V_{Ag}'] = K_F^{1/2},$$
$$[e'] = PK_{Ag}K_F^{-1/2}, \tag{19.13}$$
$$[h^\bullet] = P^{-1}K_B K_F^{1/2} K_{Ag}^{-1}.$$

Here we drop the index Ag in P_{Ag} to simplify notation.

If $C = [Mn_{Ag}^\bullet] \gg \sqrt{K_F}$, then $[Ag_i^\bullet]$ sinks and $[V_{Ag}']$ increases such that

$$[Mn_{Ag}^\bullet] \simeq [V_{Ag}'] \tag{19.14}$$

hence,

$$[V_{Ag}'] = C,$$
$$[Ag_i^\bullet] = C^{-1} K_F,$$
$$[e'] = PC K_{Ag} K_F^{-1}, \tag{19.15}$$
$$[h^\bullet] = P^{-1} C^{-1} K_B K_F K_{Ag}^{-1}.$$

If $C = [S_{Cl}']$ the situation is opposite:

$$[Ag_i^\bullet] = C,$$
$$[V_{Ag}'] = C^{-1} K_F,$$
$$[e'] = PC^{-1} K_{Ag}, \tag{19.16}$$
$$[h^\bullet] = P^{-1} C K_{Ag}^{-1} K_B.$$

Obviously, these equations describe the defect chemistry as functions of P, C, and T. The T-dependence occurs via the K_r's. As $K_r \propto \exp - \Delta_r G^\circ / RT$ the $1/T$ dependence for the mass action constant of the reaction r is governed by the reaction enthalpy $\Delta_r H^\circ$.

The above results can be generalized for Brouwer (i.e., low complexity) and Boltzmann conditions (i.e., low concentrations) to

$$c_j = \alpha_j P^{N_j} C^{M_j} \Pi_r K_r^{\gamma_{rj}} \tag{19.17}$$

with α, N, M, γ being simple rational numbers (cf. Eqs. (19.13)–(19.16)).

This leads to straight lines in all three presentations (i) $\log c_j$ versus $\log P$ (Kröger–Vink diagram) [8], (ii) $\log c$ versus $\log C$ [6,16], and (iii) $\log c$ versus $1/T$ (van't Hoff diagram) [3]. Equation (19.17) is also valid if other so far neglected disorder types and point defects are present, such as anion vacancies, anion interstitials, or antisite defects. It is the great advantage of the "chemical picture" that an increased complexity can be easily taken account of by adding further disorder equations, that is, mass action laws, as long as defect concentrations are small (cf. Refs. [4,6]). Should there additional but frozen native point defects be present, they will not lead to further K's, rather these concentrations would be included in C.

The simple rational numbers N, M, γ in Eq. (19.17) characterize the P, T, and C dependencies. The respective trends are given by the signs of these numbers. The qualitative behavior can be summarized as P–T–C rules [6,17] that presuppose simple defect chemistry:

P-rule: Increase of P (e.g., P_{Ag}) increases (decreases) those ionic concentrations that increase the stoichiometry of the element in the compound, for example, of $Ag_i^{\bullet}$ ($V_{Ag}^{|}$). If P refers to the electropositive (electronegative) element, an increase of P increases (decreases) $[e^{|}]$ and decreases (increases) $[h^{\bullet}]$.

T-rule: Temperature increase typically increases defect concentrations, as the formation enthalpies of the defects are positive.

C-rule: If the dopant is effectively positively (negatively) charged, for example, $Mn_{Ag}^{\bullet}$ ($S_{Cl}^{|}$), its concentration increases (decreases) all the negatively charged carriers ($V_{Ag}^{|}$, $e^{|}$) and decreases (increases) all the positively charged carriers ($Ag_i^{\bullet}$, $h^{\bullet}$). (This rule of homogeneous doping will be complemented later by the rule of heterogeneous doping, $\sum$-rule.).

Without going into further detail, it should just be stated that there is a manifold of association reactions between these simple defects, namely, association between ionic defects (e.g., interstitial-vacancy pairs, atomistic pores i.e., pair of anion and cation vacancies), between electronic defects (e.g., excitons, cooper pairs), and also between ionic and electronic defects (e.g., color centers). An example of the latter form is given by

$$Ag_i^{\bullet} + e^{|} \rightleftharpoons Ag_i^{x} \qquad (19.18)$$

describing the formation of a neutral silver interstitial. Note again (i) that the argument that such valencies are rather unusual, only means that their equilibrium concentrations are small and (ii) that Eq. (19.18) corresponds to an association of an already "unusual" and hence dilute species ($e^{|}$ in Eq. (19.18) corresponds in a localized picture to a neutral regular silver ion). Equation (19.18) is interesting, as it reflects both a redox and an acid–base reaction and – in the physical picture – finds its place in the gaps of both diagrams (cf. Figure 19.6).

The qualitative rules formulated above are of extreme importance for designing functional materials. The quantitative behavior depends on the magnitude of the exponents as well as on the K-values and hence, the defect formation energetics.

Let us now address the three decisive control parameters step-by-step and concentrate mostly on the important class of the oxides with predominant disorder in the oxygen sublattice. (Table 19.1 shows how to obtain the defect concentrations from mass action considerations for this regime.)

Figure 19.7 displays the variation of the defect chemistry within the solubility range of such an oxide when the oxygen partial pressure is varied. Even though the solubility range may be very tiny, the variations in the defect chemistry are typically spectacular. Stoichiometric variations are usually of secondary importance for properties such as mass, volume, or phase energy, yet of first order for the charge carrier concentrations and hence, for chemical kinetics and electrochemistry. The usually large changes of carrier concentrations (that follow the above P-rule) reflect the usually huge P_{O_2}-window irrespective of the absolute phase width. A typical example is the range spanned from nominally 10^{-18} to 1 bar for CoO at 1000 K. The extremely small partial pressure value corresponds to the coexistence of CoO with Co and the high value to the coexistence with pure O_2.

All properties that decisively depend on defect concentrations are hence, not definable for a given compound unless the defect chemistry is specified. Questions such as what is the electronic conductivity of lead(II)oxide, what is T_c of yttrium–barium–copper oxide, what is the color of rock salt (violet at the reducing side of the phase diagram!), what is the reactivity of ZnO (even if the reaction is specified), and so on are all not good questions even if the materials are ideally pure. Also, the modern computational materials screening approaches are not very helpful in this respect, unless one specifically deals with defect chemical parameters such as defect energies.

The defect chemistry is most directly reflected by the electrical conductivity (σ), as σ_j is proportional to c_j. The proportionality factor is (besides molar charge) given by the mobility u_j that will be described further in greater detail. Here it suffices to say that the mobility depends on temperature (typically according to an Arrhenius law, activated by the positive migration threshold $\Delta H_i^{\neq}$), but naturally not on P or C as long as the defect concentrations are small. Hence, we can apply the above rules for the conductivities also. Usually the mobility of electronic defects is much larger than that of ionic defects. This leads to the conclusion that the basic Kröger–Vink diagram in Figure 19.7 describes the transition from an n- to p-type behavior. Around the p–n minimum, an (P_{O_2} independent) ionic conductivity may appear (for good solid electrolytes, this "electrolytic domain" is rather extended), provided the mobility ratio of ions to electrons is not too small. Typically not all branches are realized, since upper and lower thermodynamic bounds for P_{O_2} (see above) limit the accessible range. So, Kröger–Vink diagrams for SnO_2 show only oxygen deficiency and hence n-type conductivity ($SnO_{2-|\delta|}$) [18], while La_2CuO_4 only shows O-excess ($La_2CuO_{4+|\delta|}$) and p-type conductivity (see Figure 19.8) [16]. PbO is an example of a situation of primarily ionic disorder [19], with the electronic conductivity

Table 19.1 Typical defect chemical treatment shown for an oxide with dominant anti-Frenkel-disorder in the oxygen sublattice.

Conditions	Equations	Defect chemical treatment of the regime of predominant ionic disorder (I-regime)		
		Solutions for pure material ($C = 0$)	Solutions for strong acceptor doping ($z = +1, C \gg 0$)	Solutions for strong donor doping ($z = -1, C \gg 0$)
$\frac{1}{2}O_2 + V_O^{\cdot\cdot} \rightleftharpoons O_O + 2h^{\cdot}$	$K_O = [h^{\cdot}]^2/(p^{1/2}[V_O^{\cdot\cdot}])$	$[V_O^{\cdot\cdot}] = [O_i''] = K_F^{1/2}$	$[V_O^{\cdot\cdot}] = C/2$	$[V_O^{\cdot\cdot}] = 2K_F(T)/C$
$O_O + V_i \rightleftharpoons O_i'' + V_O^{\cdot\cdot}$	$K_F = [O_i''][V_O^{\cdot\cdot}]$		$[O_i''] = 2K_F(T)/C$	$[O_i''] = C/2$
$\phi \rightleftharpoons e' + h^{\cdot}$	$K_B = [e'][h^{\cdot}]$	$[h^{\cdot}] =$ $= p^{1/4}K_F^{1/4}(T)K_O^{1/2}(T)$	$[h^{\cdot}] =$ $= 2^{-1/2}p^{1/4}C^{1/2}K_O^{1/2}(T)$	$[h^{\cdot}] =$ $= 2^{1/2}p^{1/4}C^{-1/2}K_O^{1/2}(T)K_F^{1/2}(T)$
Electroneutrality	$zC + 2[O_i''](+[e']) =$ $= 2[V_O^{\cdot\cdot}](+[h^{\cdot}])$	$[e'] =$ $= p^{-1/4}K_F^{-1/4}(T)K_B(T)K_O^{-1/2}(T)$	$[e'] =$ $= 2^{1/2}p^{-1/4}C^{-1/2}K_B(T)K_O^{-1/2}(T)$	$[e'] =$ $= 2^{-1/2}p^{-1/4}C^{1/2}K_B(T)K_O^{-1/2}(T)K_F^{-1/2}(T)$

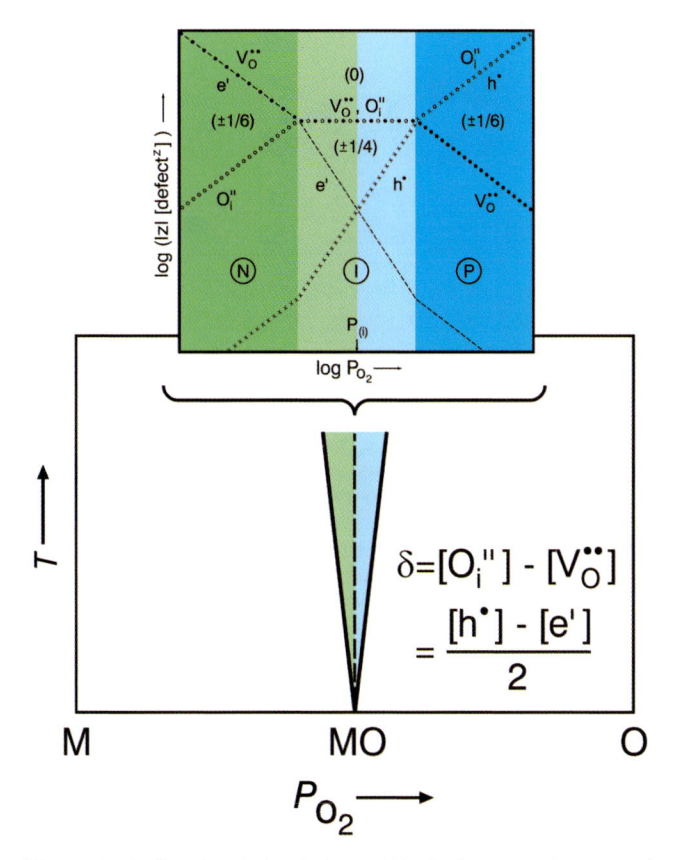

Figure 19.7 Defect chemical variations within the homogeneity range of an oxide. (Figure reproduced with permission from [17]. Copyright 2005, Springer, New York.)

showing transition from n- to p-type, and the ionic conductivity being independent of P_{O_2}.

The just considered La_2CuO_4 is a ternary, however, we treated it as a quasibinary. This approximation is naturally allowed if cation defects can be neglected. But also if their concentrations are significant but immobile, they only enter the quasibinary electroneutrality equation as dopants. This point is taken up again later.

If however, more components are mobile, the number of P's in Eq. (19.19) has to be increased

$$c_j = \alpha_j \Pi_k P_k^{N_{kj}} C^{M_j} \Pi_r K_r^{\gamma_{rj}}. \tag{19.19}$$

This is, for instance, the case for the important class of H_2O absorbing oxides that hereupon become proton conductors. Then for reversible conditions besides T, C, P_{O_2}, P_{H_2O}, or alternatively P_{H_2} must be fixed to define the c_j's.

A simpler example for which ionic and electronic defects depend on P_{O_2} and P_{H_2O} are hydroxides [20]. Point defects of relevance in KOH are hydroxide

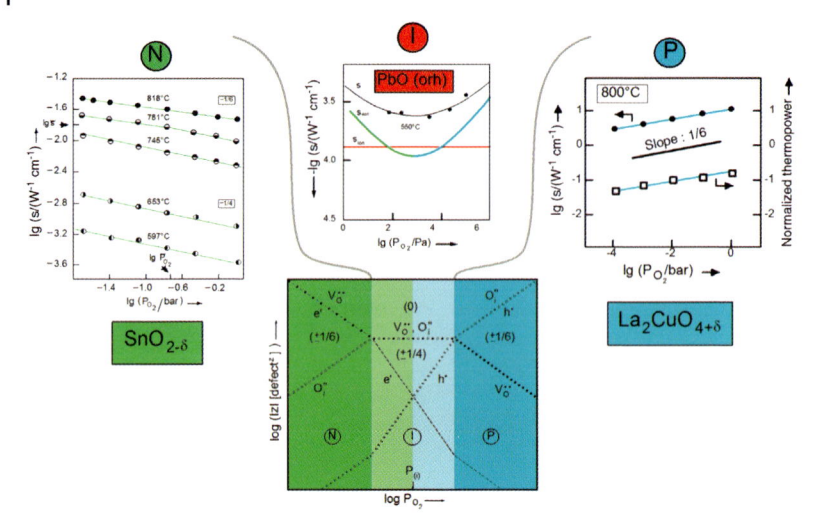

Figure 19.8 Defect chemical variations in the homogeneity ranges of SnO_2, PbO, and La_2CuO_4. (Figure reproduced with permission from [10]. Copyright 2003, Taylor & Francis.)

vacancies and interstitials but also protonated OH^- ions (H_2O at OH^- site) and deprotonated OH^- ions (O^{2-} at OH^- site) both enabling internal proton conductivity. A similar degree of complexity applies for the defect chemistry of peroxides or superoxides [21].

Let us now discuss the C-dependence of simple oxides. The calculation for the regime of predominant anti-Frenkel disorder (i.e., O_i'' and $V_O^{\bullet\bullet}$ are in the majority) is given in Table 19.1 and the results displayed in Figure 19.9. Analogous relations hold for a primarily Schottky-disordered material such as $SrTiO_3$ [22–24]. Here, O_i'' has to be replaced by V_{Sr}''. For this diagram it does not matter whether or not, the cationic sublattice is in equilibrium with the outer gas phase as well. Obviously, intentional doping (e.g., by an acceptor A') is very helpful to define

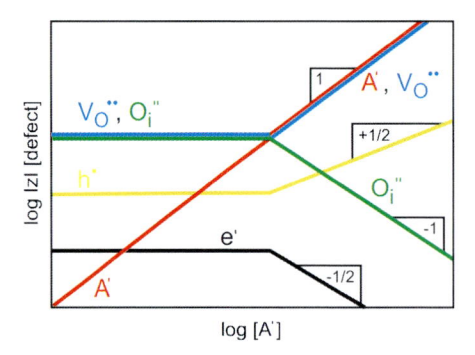

Figure 19.9 Defect chemical modification of an anti-Frenkel disordered oxide on acceptor (A') doping. (Figure reproduced with permission from [7]. Copyright 2010, Taylor & Francis.)

the defect chemistry and to avoid that Sr- (and Ti-defects) whose concentrations depend on prehistory, become relevant in the electroneutrality equation (see right-hand side of Figure 19.9). The right-hand side of Figure 19.9 also applies nominally to Y-doped ZrO_2. However, in the practically used solid electrolyte doping contents of several percent are realized and interactions cannot be neglected [25].

As partial pressures and temperature are often not freely controllable, the doping content is the most decisive materials design parameter. Owing to the importance of the rule of homogeneous doping (C-rule), let us rephrase it as $z_j \delta c_j / z \delta C < 0$ (for all j), where z_j and z are the charge numbers of j and dopant, respectively, or, more concisely, in terms of the exponent M defined in Eqs. (19.17) and (19.19) in the form of $(z_j/z)M_j < 0$.

It is important that this inequality affects all (mobile) carriers in equilibrium; so on Y-doping of ZrO_2 not only the $V_O^{\bullet\bullet}$ concentration but also the $h^{\bullet}$ concentration is increased and in Sr-doped La_2CuO_4 not only [$h^{\bullet}$] but also $[V_O^{\bullet\bullet}]$. Owing to the double charge of $V_O^{\bullet\bullet}$, the ratio $[V_O^{\bullet\bullet}]/[h^{\bullet}]$ increases on A'-doping. In the solid electrolyte ZrO_2 this is desired, while in the HTSC material La_2CuO_4 this is detrimental. The latter is one possible reason for the so-called overdoping.

Among the control parameters, C plays a special role as it does, by definition, not refer to total equilibrium. Total equilibrium is only established at those temperatures where the dopant can be reversibly in-/excorporated, that is, where the dopant behaves as a component rather than a frozen structure element. In $Zr(Y)O_2$, this requires extremely high temperatures. (At these temperatures Eq. (19.19) applies with the two component partial pressures P_{O_2} and P_Y (while $C = 1$).) At temperatures of operation, the Y-content is frozen. Then Eq. (19.17) applies with $C = [Y]$ (while P_Y disappears as control parameter). In Ref. [26], such a frozen parameter was termed ex-situ parameter, as it always requires a new preparation/quenching procedure to vary it. The transition between total reversible and partially quenched situations is not trivial (see Ref. [26–28]) but of great importance for materials science. Moreover, the choice of quenching conditions introduces further valuable control parameters (P, T conditions at quenching temperature) and hence, further practical degrees of freedom.

A further instructive example is CO_2 contamination of hydroxides (Figure 19.10). Absorbed CO_2 herein acts as an acid by increasing the concentration of the internal acidic defects ($V_{OH}^{\bullet}$, $HOH_{OH}^{\bullet}$) and decreasing the concentration of the basic defects (OH_i', O_{OH}') (note that the regular OH^- is neutral as far as the internal scale for hydroxides is concerned). At high temperatures the incorporation reaction of CO_2 is in equilibrium and P_{CO_2} enters as a control parameter in addition to P_{H_2O} and P_{O_2} (Figure 19.10a). At low temperatures, the CO_2 content is fixed and the pertinent treatment is based on $C = [CO_2]$ (Figure 19.10b). The latter is the doped situation [20].

This transition from reversible to partially frozen-in conditions is highly relevant for oxides, as typically the interaction with the oxygen atmosphere is only active at elevated temperature, whereas the solid behaves as "sealed" at room

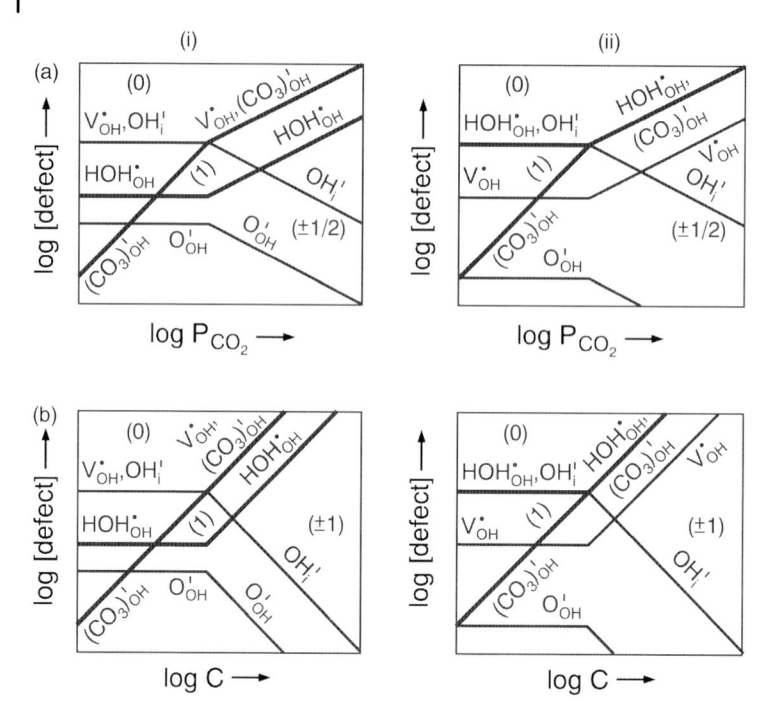

Figure 19.10 CO_2 effect on the defect chemistry in hydroxides. (a) Reversible CO_2 exchange. (b) Frozen CO_2 content (termed C). (i) and (ii) columns refer to different native situations. (Figure reproduced with permission from [20]. Copyright 1999, Elsevier.)

temperature. This freezing-in is often not primarily caused by a low diffusion coefficient (as the decrease of sample size would show) but typically by too sluggish a surface reaction. This is treated in more detail in Section 19.5.

Measurements of T-dependencies are very helpful in determining defect formation energies. As they are positive, a T increase usually increases defect concentrations (if not fixed by dopant) and even more so the conductivities. Contributions from concentrations and contributions from mobilities can be typically separated using doping experiments. A well-investigated example is again AgCl. While the T-dependence of the ion conductivity of pure AgCl gives the sum of half of the Frenkel enthalpy (via the appearance of $\sqrt{K_F}$, see Eq. (19.13)) plus the migration energy of the most mobile defect ($Ag_i^\bullet$), S-doping fixes $[Ag_i^\bullet]$ and only the migration term remains (see Eq. (19.16)) [29].

In this contribution, consideration of nonidealities coming along with high defect concentrations is beyond scope. The exception is the next example, as it refers to the fundamental "thermal destiny of an ionic crystal" giving deep insight into the coupling of perfect solid, defective solid and molten state, and providing a quantification of Tamann's rule [30].

Figure 19.11 sketches Frenkel disorder in, say, the Ag-sublattice of the "normal" ion conductor β – AgI at lower temperatures. At $T = 0$ we refer to the

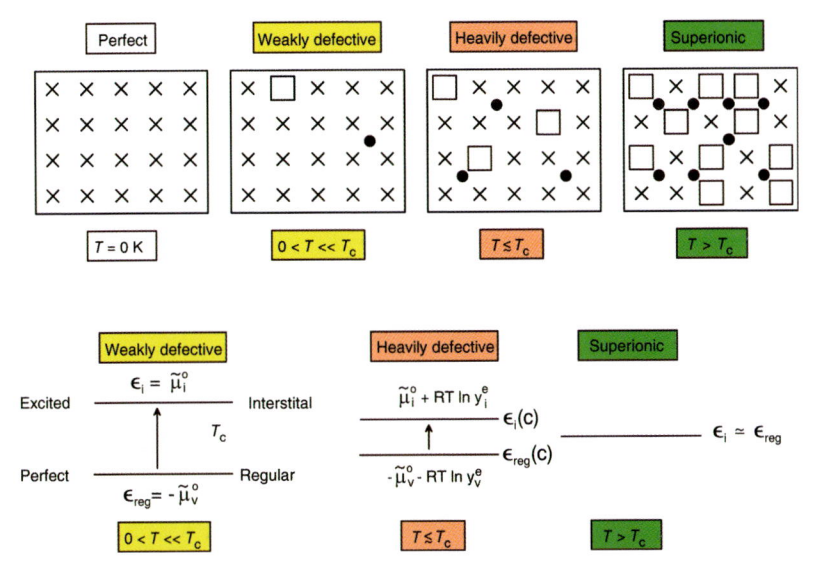

19.2 Ionic and Electronic Defects as Acid–Base and Redox Particles in Equilibrium

Figure 19.11 Thermal destiny of an ionic crystal. (Figure reproduced with permission from [33]. Copyright 2000, John Wiley & Sons Inc.)

perfect structure; the ionic gap is insurmountable. At finite T (still far from sublattice melting point), we refer to the above considered regime of dilute defect chemistry. At even higher temperatures, the defect concentration is already so large that the attractive Coulomb interaction of effectively positively charged $Ag_i^{\bullet}$ and negatively charged V'_{Ag} are perceptible. This leads to an effectively facilitated defect formation via varied activity coefficients and to a level narrowing in the "physical picture": The increased defect concentration now facilitates further carrier formation. This avalanche effect leads to a transition into a fully disordered sublattice corresponding to a single degenerate defect level [31,32]. Such a superionic transition occurs in β – AgI at 146 °C, leading to the superionic conductor α – AgI. The fact that the crystal structure changes means that T_c assessed in Figure 19.12 is somewhat overestimated. However, the structural effect is often not very pronounced so that even the melting process for AgCl and AgBr can be referred to as superionic transitions in the above sense. Obviously, in these phases the molten state is energetically not very different from the estimated virtual solid superionic state. The conceptual significance of the above picture is that the transition to the superionic or molten state is already predetermined by the defect formation in the weakly defective solid that itself is the excitation of the perfect structure. This connection between the various states directly leads to a deeper understanding of empirical relationships between defect energies and melting points as well as between melting point and "reaction temperature" (Tamann's rule) [30].

It is also instructive to have a look at the inserts of Figure 19.12, showing at different temperatures the Gibbs energy as a function of defect concentration c.

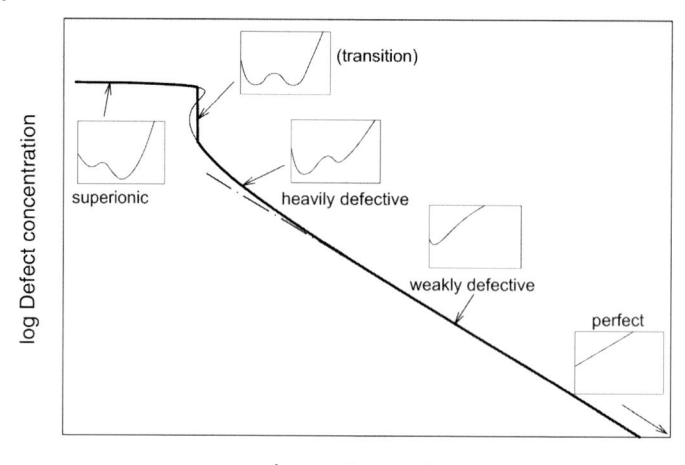

Figure 19.12 Defect concentration as a function of temperature: from dilute defect chemistry up to superionic transition. (Figure reproduced with permission from [33]. Copyright 2000, John Wiley & Sons Inc.)

At $T = 0$, the minimum is at $c = 0$, that is, at the perfect state. At finite temperature, the concentration of the minimum is different from zero, owing to the configurational entropy that defect formation brings in. When interactions become important, two comparable relative minima can be observed, one for small defect concentrations and one for a rather complete disorder (large c).

Above the phase transformation the superionic situation refers to the latter state that becomes more stable. The instability of the situation in the premelting regime expressed by the nonsingle-valued behavior is evaded by a first-order or higher order "phase" transformation [31].

19.3
Transport and Transfer

As already mentioned, the transition from carrier concentration to carrier conductivity involves the consideration of its mobility. In a "chemical picture," ion transport of, for example, Ag^+ (i.e., forward/backward hops between 2 equivalent sites) can be seen as a pseudochemical reaction with a symmetrical threshold (i.e., zero standard reaction free energy), for which we can write a bimolecular rate law (Figure 19.13) [6,34,35]. Let us address the hopping process with an applied electric potential.

For a vacancy mechanism

$$V'_{Ag}(x) + Ag_{Ag}(x + \Delta x) \rightleftharpoons Ag_{Ag}(x) + V'_{Ag}(x + \Delta x), \tag{19.20}$$

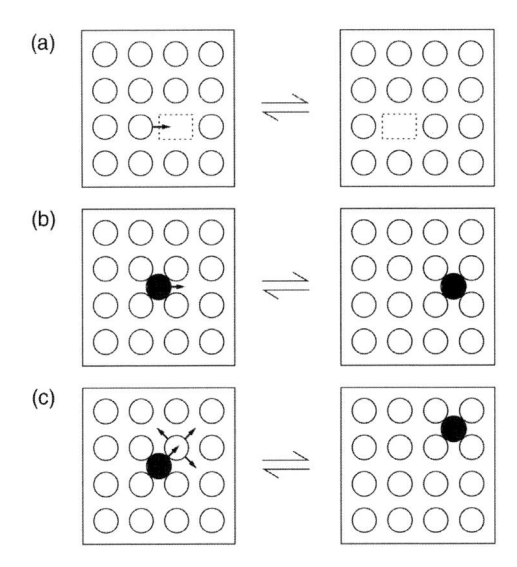

Figure 19.13 Mechanism of ion conduction. (a) Vacancy, (b) interstitial, and (c) interstitialcy (indirect interstitial). (Figure reproduced with permission from [36]. Copyright 2005, Elsevier.)

the forward reaction rate, that is, the rate of moving the vacancy from x to $x + \Delta x$, reads (Figure 19.13a)

$$\vec{\Re} = \vec{k}\left[V'_{Ag}(x)\right]\left[Ag_{Ag}(x+\Delta x)\right] \propto \exp\frac{-(\Delta G^{\neq} + F\Delta\phi/2)}{RT}\left[V'_{Ag}(x)\right]\left[Ag_{Ag}(x+\Delta x)\right]. \tag{19.21}$$

Equation (19.21) contains the migration barrier $\Delta G^{\neq}$ as well as half of the electrical potential drop from x to $x + \Delta x$. An analogous equation holds for the back reaction rate $\overleftarrow{\Re}$, the difference $\vec{\Re} - \overleftarrow{\Re}$ being the net reaction rate.

For constant concentrations (steady-state electrical transport), Eq. (19.21) simplifies to a transport equation that only involves the electrical potential in the exponential term as variable. For small driving forces ($F\Delta\phi \ll RT$) and dilute conditions ($\left[Ag_{Ag}\right] = \text{const}$) one obtains

$$\Re \propto -k_{chem}\left[V'_{Ag}\right]\Delta\phi, \tag{19.22}$$

k_{chem} is the purely chemical part of the rate constant (given by $\exp - (\Delta G^{\neq}/RT)$) that is the same for forward and backward reaction. Equation (19.22) is identical to Ohm's law ($j \propto -\sigma\nabla\phi$) showing that the conductivity σ is proportional to k_{chem} as well as to defect concentrations. In electrical terms k_{chem} yields the mobility u.

If on the other hand, the defect concentrations vary from site to site, that is, a concentration gradient is present, but electrical gradients are absent ($\Delta\phi = 0$), pure diffusion is referred to, and the analogous consideration leads to Fick's law ($j \propto -D\nabla\phi$) whereby the diffusion coefficient D is obviously directly proportional to k_{chem} and thus to u (but not to concentrations). These approximations show that diffusion and conduction are mechanistically coupled (Nernst–Einstein equation).

A more thorough treatment of mobility, conductivity, particle diffusion, tracer diffusion, and chemical diffusion in the light of chemical reaction kinetics is given in Ref. [6].

Let us refer to the electrical conductivity and address a few relevant points:

i) Electronic defects are typically more mobile than ionic point defects. As a consequence, various materials in which ionic defects prevail are electronic conductors nonetheless. Conversely, in order for materials to show predominant ionic conduction, the electronic carrier concentrations must be very small.

ii) Vacancy and interstitial defects have different mobilities. In Ag—halides, for example, the mobility of silver interstitials is greater than that of vacancies. As a consequence, the total conductivity of the pure halide is predominantly of interstitial type ($\sigma_i \gg \sigma_v$ though $c_i = c_v$). As Cd-doping ($Cd^{\bullet}_{Ag}$) leads to an increase of $[V'_{Ag}]$ and a decrease of $[Ag^{\bullet}_i]$, a characteristic minimum is traversed before the ion conductivity monotonically increases on further doping [29].

iii) As ionic mobilities do not exceed $0.1\ cm^2/Vs$ and carrier concentrations cannot be higher than the inverse molar volume, ion conductivities are not expected to exceed 10^2 S/cm [37,38]. Figure 19.14 gives ionic conductivities for very good solid electrolytes, confirming this statement.

As far as the search for ion conductors is concerned, the use of structural arguments based on the perfect structure are often misleading. In the case of defective ion conductors, it is the mobility of a few point defects that is decisive and that can only be addressed by considering the defective state. But also as far as superionic conductors (i.e., complete disorder) are concerned, structural arguments based on the perfect structure can be misleading. Many structures show open channels or open 2D spaces that are not necessarily good conduction pathways, as the energetic situation for ions may not be favorable if the pathways are not adjusted to the ions size. A good example is $\beta - Al_2O_3$ that shows very high Na^+ conductivity along the conduction planes. However, these planes turn out to be too roomy for Li^+, leading to unfavorable chemical environments during transport [37,38,41,42]. A guideline that very often applies is softness: Polarizability of the moving ion (e.g., Ag^+ in silver halides) or of the environment (N^{3-} in Li_3N) enables site exchange under the conservation of favorable bonding situations [6]. Examples of recent interest are Li–Ge–thiophosphates [39] or the even slightly better conducting Si-variants [40] showing unprecedented Li-ion conductivities.

iv) The last example already refers to the very interesting class of mixed conductors, that is, materials that exhibit both significant ionic and electronic conduction. While purely ionic conductors are natural candidates for solid

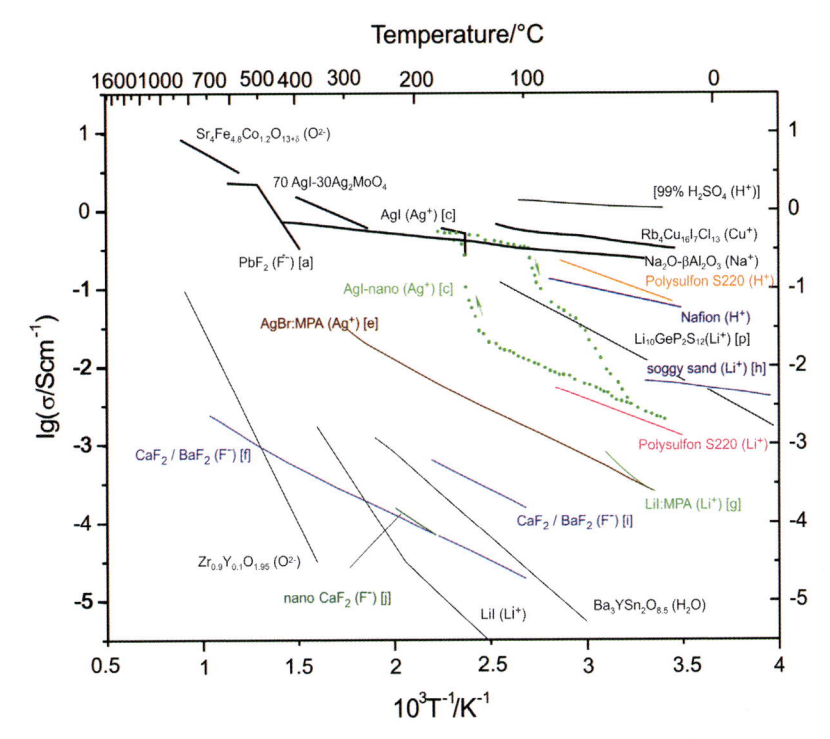

Figure 19.14 Ion conductivities of selected solid electrolytes as a function of temperature. Note that values of ~1 S/cm are clearly below the theoretical limit. The Li-conductivity of the $Li_{10}GeP_2S_{12}$ [39] is slightly exceeded by the Si-analog [40]. (Sources are given in Refs. [6,38].)

electrolytes, these mixed conductors are appropriate for electrodes, catalysts, permeation filters, or electrochemical reactors (cf. last section). In such cases, the joint transport of ionic and electronic carriers is key that is characterized by ambipolar conductivity (steady state) and chemical diffusion coefficient D^δ (transient). Interesting recent examples refer to methyl–ammonium–lead iodide and their "anomalies" during operation as photo-electrodes [43], as well as to very fast interfacially controlled artificial mixed conductors [44]. Figure 19.15 gives an overview on materials with high D^δ-values. In this context, it should be emphasized that the chemical diffusion coefficient is a most (if not the most) important kinetic parameter in solid-state chemistry (cf. Section 19.5).

Let us now again consider our basic reaction, Eq. (19.21) and consider transport along asymmetric chemical thresholds, that is, $\overleftarrow{k}_{chem} \neq \overrightarrow{k}_{chem}$. Such extensions can describe charge transfer processes at electrode/electrolyte interfaces where the asymmetric threshold is modified by an electrical field (Butler–Volmer equation) as well as simple chemical reactions as they occur, for example, at the gas/solid interface (cf. Section 19.5). Ref. [45] treats cases where both chemical and electrical driving forces are present and large.

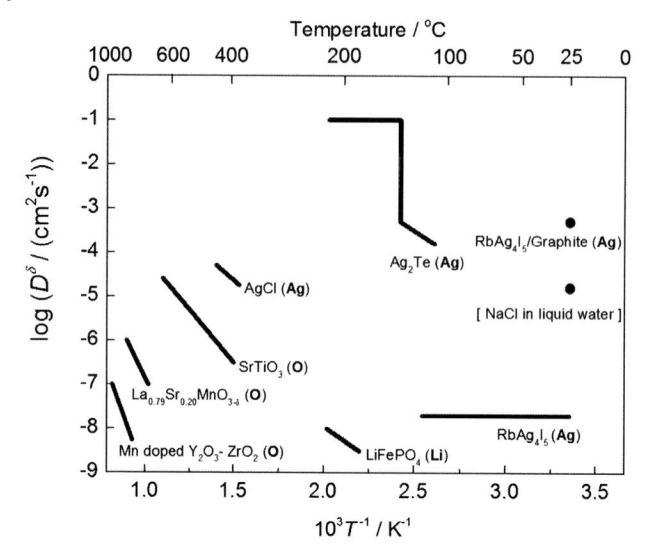

Figure 19.15 Chemical diffusion coefficients of selected mixed conductors as a function of temperature. Note the enormously high value of the RbAg$_4$I$_5$/C composite at room temperature that even exceeds NaCl diffusion in water [44]. (Sources are given in Refs. [6,44].)

19.4
Defect Chemistry at Boundaries and Heterogeneous Electrolytes

A very attractive scientific field from a fundamental as well as application point of view refers to the variation of defect chemistry at boundaries [46]. Such considerations not only lead to the explanation of various anomalies but also to novel functional materials.

If the density of interfaces is extremely high, such effects become very prominent and, in the extreme, even characterized by mesoscopic effects (overlap of boundary effects) [13,46,47]. This field called "nanoionics" might become as promising for energy research as nanoelectronics has already become for information technology. For details the reader is referred to the literature. Here, only the basic principles are set out. While bulk defect chemistry is determined (and the formation of easily formable bulk defects is limited) by electroneutrality, this is not the case at interfaces where positive and negative carries are rather decoupled. When referring to the boundary of two phases, these two phases can be appropriately chosen such that ion conductivities can be giantly increased or depressed, such that insulators can be turned into conductors, electronic conductors can be turned into ion conductors and vice versa, or such that the dominant defect-type can be varied. Figure 19.16 shows the effect of boundary charging on the distribution of the mobile carriers [14]. A rule analogous to the rule of homogeneous doping (C-rule) applies, namely, the rule of heterogeneous doping (Σ-rule) in which the charge of the dopant (αzC) is replaced by the interfacial charge [6].

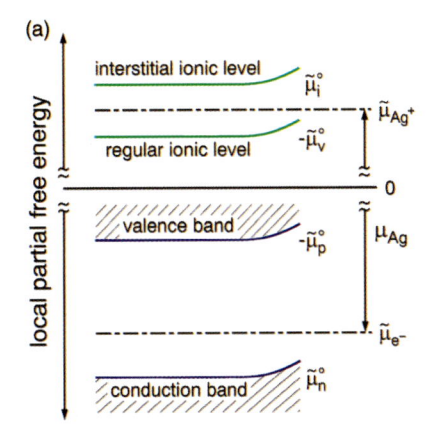

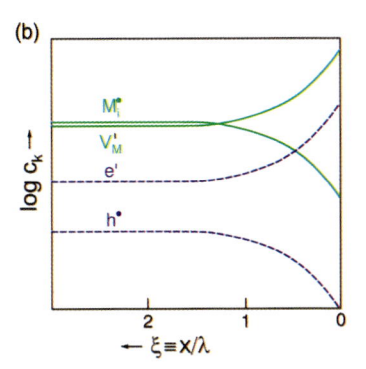

Figure 19.16 (a) Physical picture (energy level diagram) for ionic and electronic disorder at interfaces. (Figure reproduced with permission from [14]. Copyright 2003, Elsevier.) (b) Corresponding concentration changes. (Figure reproduced with permission from [48]. Copyright 2002, Elsevier.)

Σ-*rule:* If the interfacial charge is positive (negative), all mobile negative (positive) charge carriers are accumulated and all positive (negative) ones depleted $\left(\frac{z_j \delta c_j}{\delta \Sigma} < 0\right)$.

In the chemical picture, the electroneutrality equation has to be replaced by Poisson's equation and the necessary boundary conditions. In the physical picture (Figure 19.16), the occurrence of a space charge field leads to bending of the levels not only for the electronic but also for the ionic levels because all carriers perceive the same field. This is also highly relevant for electronic materials in which the ionic defects may be still in majority and their interaction determine the field that the electrons have to follow (fellow traveler effect) [49].

Figure 19.17 gives an overview on phenomena based on ion redistribution at abrupt junctions. In heterogeneous electrolytes [6,46], where ion stabilizing second insulating phases are dispersed, the ion conductivity in the matrix is increased by vacancies induced by internal adsorption of the ions. This works for cationic conductors (e.g., AgX, LiX) as well as for anionic conductors (e.g., CaF_2) [50–52]. This is particularly interesting for Li-halides as they cannot be doped homogeneously to a large extent by higher valent cations. In CaF_2/BaF_2 heterostructures, F-redistribution affects both space charge zones (transition from F^- from BaF_2 to CaF_2); these heterostructures can serve as master examples as here the transition from isolated space charge zones to overlapped space charge zones (artificial ion conductors) can be studied [51]. In nanocrystalline ceramics, various anomalies due to space charge zones occurring at grain boundaries are observed. In many oxides, the grain boundaries are O^{2-} deficient [53–58]. This positive charging leads to $h^{\bullet}$ depression (e.g., $SrTiO_3$

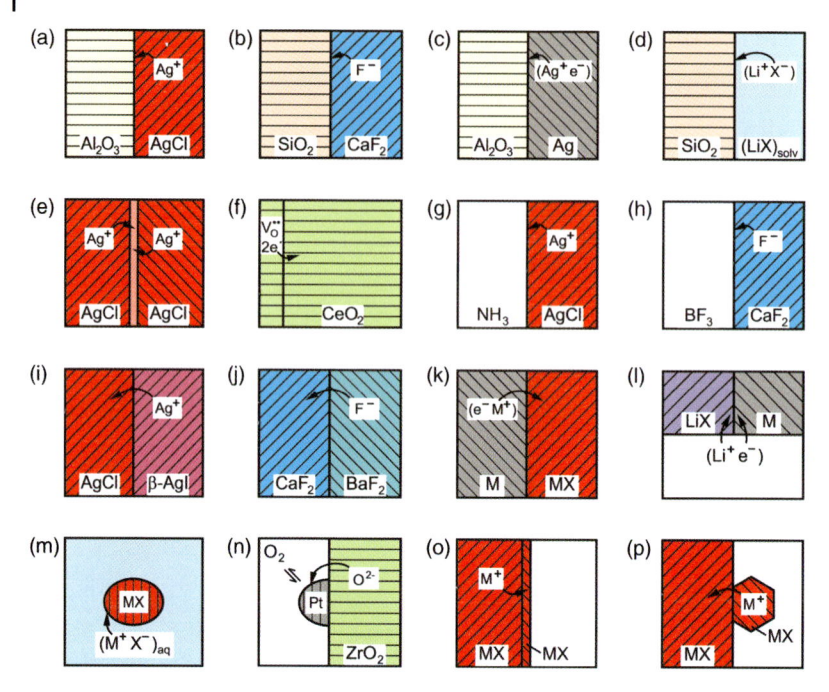

Figure 19.17 Ionic redistribution effects at contacts. The figure includes conductivity and storage anomalies, homo- and heterophase boundaries, and effects involving solids and fluids. (Figure reproduced with permission from [59]. Copyright 2004, Elsevier.)

bicrystal), $OH^{\bullet}$ depression (BaZrO$_3$ ceramics), $V_O^{\bullet\bullet}$ depression (SrTiO$_3$ bicrystal), and e' enrichment (nano CeO$_2$).

Such effects can be observed in nanocrystalline SrTiO$_3$ [60] as a function of P_{O_2}. Here, we refer to the previous section where the defect chemistry of oxides was discussed. The expected and experimentally realized conductivity isotherms of SrTiO$_3$ bulk are shown in Figure 19.18 by the black line along with the situation of nanosized mesoscopic ceramics (red line). The changes of n-, p- and oxygen vacancy conductivity introduced by "downsizing" SrTiO$_3$ are spectacular but conform very well with the space charge concept.

Apart from space charge effects, elastic effects can also play a certain role (a nice example is given by ZrO$_2$/CeO$_2$ heterostructures) [61] but have not been shown to reach the enormous variations that are caused by charge redistribution.

Very recent works on composite electrodes show analogous effects on storage [62–64]. An extreme anomaly is that two phases, α and β can store a compound A^+B^- at the contact dissociatively (job-sharing) even though none of the phases might do the job alone (Figure 19.17l). A relevant example is $\alpha/\beta =$ Li$_2$O/Ru or LiF/Ni in the context of Li-storage, that is, $A^+ \equiv $ Li$^+$, $B^- = e^-$. Moreover, job-sharing diffusion can be enormously fast [44]. For the thermodynamics and kinetics in such more delicate cases, see Refs. [44,65].

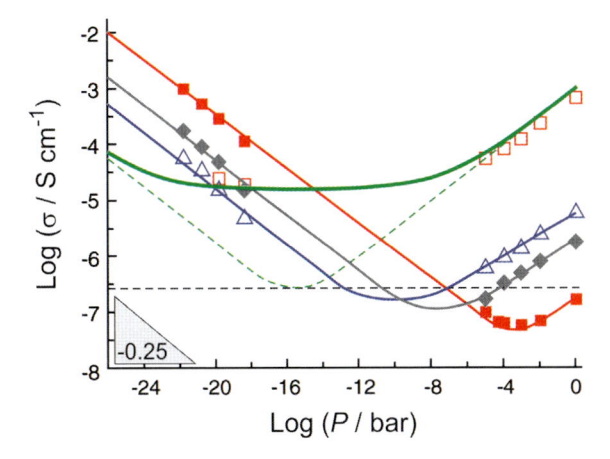

Figure 19.18 Conductivity isotherm for single crystalline and mesoscopic nanocrystalline SrTiO$_3$. (Figure reproduced with permission from [60]. Copyright 2010, John Wiley & Sons Inc.)

19.5
Reaction and Catalysis

In order to show the fundamental relevance of point defects for a solid-state reaction, it is sensible to consider a most simple reaction, namely, the kinetics of stoichiometry change in an oxide:

$$\text{MO}_{1+\delta} + \frac{\Delta\delta}{2}\text{O}_2 \rightleftharpoons \text{MO}_{1+\delta+\Delta\delta}. \tag{19.23}$$

Here the structural perturbations are marginal, and we do not have to involve solid–solid interfaces, nucleation, or growth. Nonetheless, the details of the kinetics are already complex [66].

Notwithstanding the complexity of the reaction scheme, we can distinguish at least two major lumped steps. The first is the chemical surface reaction and the second is the chemical transport process:

oxygen in the gas phase ⇌ oxygen in the surface ⇌ oxygen in the bulk.

Figure 19.19 shows the measured evolution of the oxygen stoichiometry [55,66–68] in time and space, if diffusion is rate determining (Figure 19.19a) and if surface reaction is rate determining (Figure 19.19b). Figure 19.19 also includes the fact that transfer across internal grain boundaries is rate determining (Figure 19.19c). The reasons for this blocking lie in the depletion of both decisive charge carriers (h$^{\bullet}$ and V$_{\text{O}}^{\bullet\bullet}$), as discussed in the previous section.

In the first case, such experiments nicely allow the determination of the chemical diffusion coefficient of oxygen as decisive rate parameters and in the second case of effective rate constants (k^δ). Even in the case of mixed kinetics, a safe deconvolution of k^δ and D^δ is possible by this spatially resolved

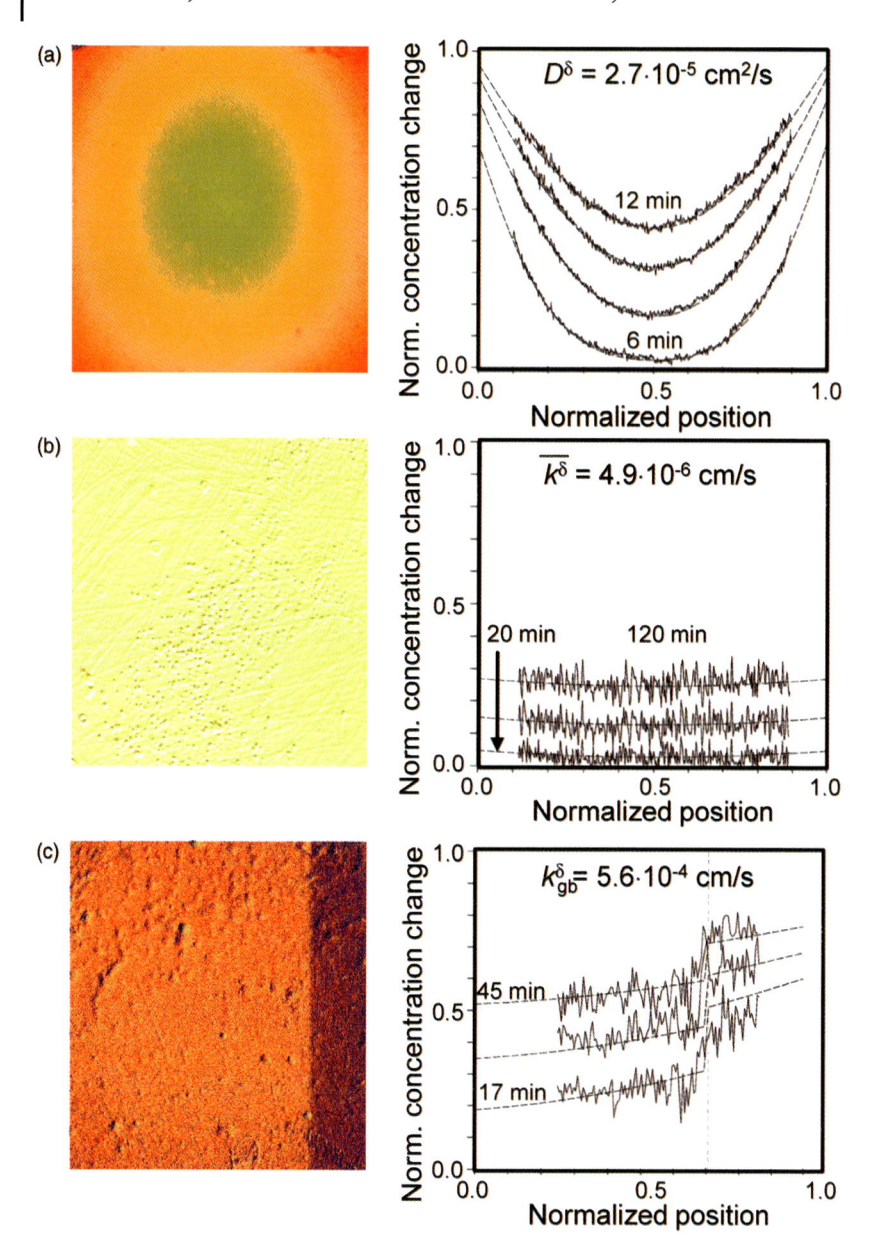

Figure 19.19 Oxygen stoichiometry profiles for Fe-doped $SrTiO_3$ developing as a consequence of an oxygen partial pressure change. Measured by time and space resolved optical absorption spectroscopy under different conditions. (a) Diffusion control. (b) Surface reaction control. (c) grain boundary control. (Figure reproduced with permission from [69]. Copyright 2000, Elsevier.)

technique [66–68]. The defect chemical analysis enables a precise prediction of such D^δ values for SrTiO$_3$. The surface reaction is rather a conventional chemical reaction step (involving an asymmetric threshold with nonidentical educts/products). However, as already mentioned, the surface reaction scheme is complex in that it comprises a manifold of elementary steps involving point defects. Again for SrTiO$_3$, a detailed deconvolution of the surface steps into individual contributions (adsorption, dissociation, ionization, charge transfer) was possible by applying a combination of experimental techniques [66]. Figure 19.20 indicates the way of how to conceptually approach the evaluation of the various methods [35]. (Note the similarity to the procedure for evaluating transport coefficients (Eq. (19.21), cf. also Ref. [35]).)

Needless to say that similar treatments give a proper insight into electrocatalysis, for example, of fuel cell cathodes or into heterogeneous catalysis [70]. Figure 19.21 illustrates the mechanistic situation of oxygen interaction for typical fuel cell cathode materials based on a combination of experimental and theoretical studies [70].

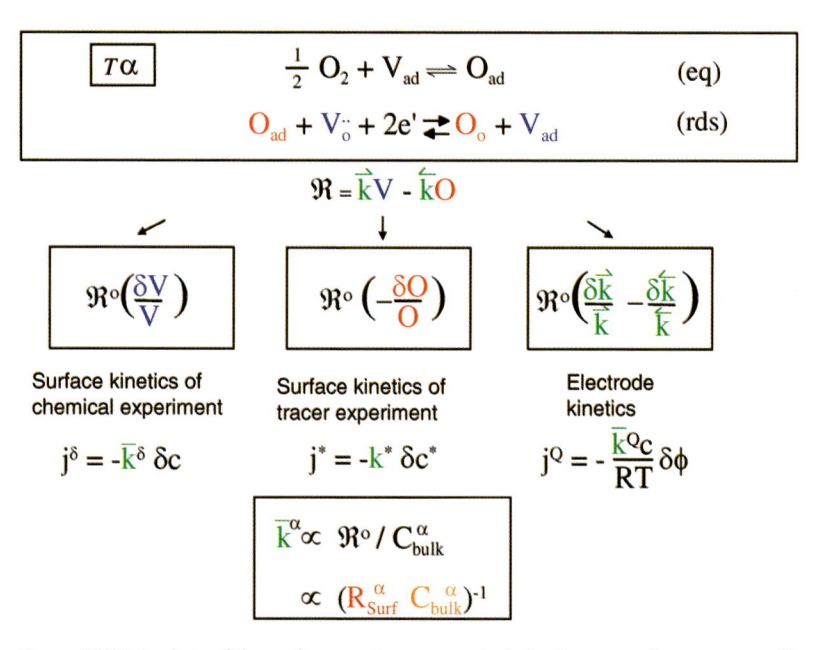

$$T\alpha \quad \tfrac{1}{2}O_2 + V_{ad} \rightleftharpoons O_{ad} \quad \text{(eq)}$$
$$O_{ad} + V_o^{\cdot\cdot} + 2e' \rightleftharpoons O_o + V_{ad} \quad \text{(rds)}$$

$$\mathfrak{R} = \vec{k}V - \overleftarrow{k}O$$

$$\mathfrak{R}^\circ\left(\tfrac{\delta V}{V}\right) \qquad \mathfrak{R}^\circ\left(-\tfrac{\delta O}{O}\right) \qquad \mathfrak{R}^\circ\left(\tfrac{\delta \vec{k}}{\vec{k}} - \tfrac{\delta \overleftarrow{k}}{\overleftarrow{k}}\right)$$

Surface kinetics of chemical experiment
$$j^\delta = -\overline{k}^\delta\,\delta c$$

Surface kinetics of tracer experiment
$$j^* = -\overline{k}^*\,\delta c^*$$

Electrode kinetics
$$j^Q = -\tfrac{\overline{k}^Q c}{RT}\,\delta\phi$$

$$\overline{k}^\alpha \propto \mathfrak{R}^\circ/C^\alpha_{bulk}$$
$$\propto (R^\alpha_{Surf}\, C^\alpha_{bulk})^{-1}$$

Figure 19.20 Analysis of the surface reaction kinetics [35]. Generalized treatment yielding the special cases for the rates of chemical changes, isotope changes, and electrical potential changes exemplified for a simple transfer reaction (Tα). The $\overline{k}$'s are relaxation constants for the respective processes. They are the reaction analogues to the diffusion constants. The R's and C's denote effective resistances and capacitances ($\mathfrak{R}^\circ$ = exchange rate).

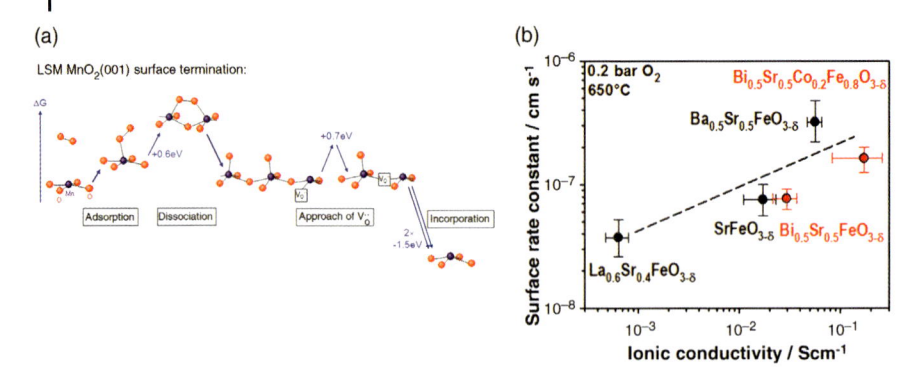

Figure 19.21 (a) Gibbs free energy diagram for oxygen interaction at the Lanthanum manganite fuel cell cathode (MnO_2 (001) termination). (Figure reproduced with permission from [70]. Copyright 2013, Royal Society of Chemistry.) (b) Correlation of effective rate constant of oxygen incorporation with bulk ionic conductivity in (Bi, La, Sr, Ba)(Fe, Co)$O_{3-\delta}$ perovskite SOFC cathode materials. (Figure reproduced with permission from [71]. Copyright 2014, The Electrochemical Society.)

As outlined in the beginning, point defects are acid–base and redox active centers and hence, can also strongly affect catalysis. This is most important for heterogeneous catalysis that takes place at the surface. The surface itself is a region of enhanced energy (2D-defect) but point defects in the surface layer are even more reactive. Consider, for example, a metal vacancy in a metal oxide. This site has a highly basic character (metal ion missing in the center of a pure O^{2-} environment). Surprisingly, outside electrocatalysis catalytic processes are not systematically investigated in this respect (e.g., with respect to doping) [72].

While this is state-of-the-art in the transport case, for the surface reaction systematic doping experiments are very rare. It can be foreseen that taking defect chemistry in catalysis seriously, will be of great help in analyzing or optimizing catalytic functions.

A higher degree of complexity is reached if interfaces are formed or annihilated [73]. A still very simple solid-state reaction where this is the case, is the oxidation of a metal film to a metal oxide [73–75]. If the molar volumes of educt and product are very dissimilar, there will be enough metal/gas contacts present until the reaction has come to an end. In many cases, however, the oxide scale grows densely and the mass transport of M or O through the oxide becomes decisive. Let us focus on such a case. At small thicknesses, typically the surface reaction is rate limiting, then the growth velocity is independent of thickness. At larger thicknesses the diffusion is likely to become rate determining, the growth velocity will then be proportional to the thickness giving rise to a parabolic rate law. Such parabolic rate constants are determined by the conductivities of the decisive ionic and of the decisive electronic defects. The slowest one determines the overall rate, and simple doping can accelerate or decelerate corrosion.

Zn corrosion is a good example [73]. ZnO exhibits an oxygen deficiency making it n-type conducting. The conduction electrons must be compensated by oxygen vacancies or zinc interstitials. As $\sigma_n \gg \sigma_{ion}$, the ion conductivity acts limiting that – according to our C-rule – can be increased by doping Zn with a lower valent cation, while doping with a higher valent cation, corrosion rate is decreased. Indeed such effects are observed and could be quantitatively analyzed [73]. (As in these cases it is Zn that is doped, but nevertheless the doping affects the oxide in the above sense, the conclusion may be drawn that the oxide grows into the metal. This implies that oxygen vacancies are the dominant ionic mobile species.)

A more typical chemical reaction is a powder reaction of BaO and TiO_2 for $BaTiO_3$ [76,77]. Already this simple example is highly involved in view of phase and interface distribution topology and also in view of orientation, nucleation, and growth kinetics. In such cases, an accurate description is extremely difficult and may not even be meaningful.

19.6
Electrochemical Applications

Owing to the fact that conductivity and storage is enabled by the point defects, defect chemistry is the decisive field for electrochemical applications and devices, as far as the solid state is concerned. Most important applications refer to energy storage and conversion (see Figure 19.22) as well as to sensing. In particular, the tuning of σ_{ion} and σ_{eon} is crucial for materials optimization in this respect (cf. also Figure 19.4). Let us start with solid electrolytes for which σ_{ion} should be as large and σ_{eon} as small as possible.

19.6.1
Ion Conductors as Solid Electrolytes

19.6.1.1 Fuel Cells
Owing to stability and high concentrations of mobile oxygen vacancies, the aforementioned Y-doped ZrO_2 is the classic oxygen ion conductor for high temperature fuel cells. It works very well at high temperatures [25,78]. At less high temperatures, doped CeO_2 appears to be a better choice, yet n-type contributions stemming from the reducibility of Ce^{4+} relevant at the anode side (hydrogen side) cause problems [79]. These electronic contributions are particularly pronounced at grain boundaries. Here, space charge effects depress vacancy concentrations and enhance n-type concentrations (cf. Σ-rule) [54].

Ternary oxidic systems are also investigated, yet have not been able to replace these two master candidates for a multitude of reasons. Note that the demands for a solid electrolyte in high-temperature fuel cells are severe, the most important ones being chemical stability in the large redox window, negligible electronic conductivity therein, mechanical strength, stable contacts, and matching of thermal expansion.

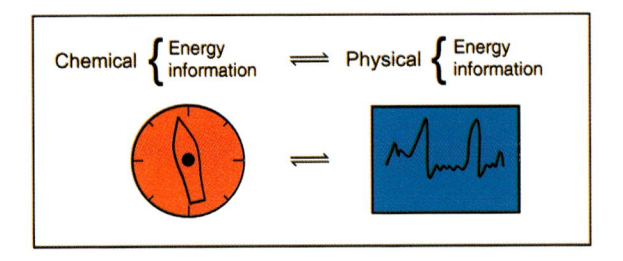

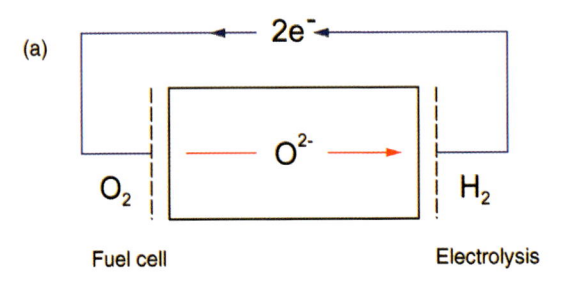

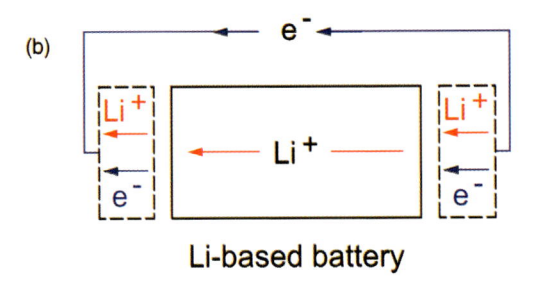

694 *19 Defect Chemistry and Its Relevance for Ionic Conduction and Reactivity*

Figure 19.22 Electrochemical devices for energy research relying on the presence of ionic and ionic-electronic conductors. (Figure reproduced with permission from [7]. Copyright 2010, Taylor & Francis.)

Low-temperature fuel cells use proton conducting electrolytes such as Nafion, PBI-based systems, or polysulfones [80,81]. In polyelectrolytes, proton conductivity occurs in the space charge zones of internal pore channels. The protons stem from acidic functional groups of the organic backbone, and are injected into the liquid phase in the pores. Here stability problems of the membrane (together with the nontrivial electrode catalysis) are major drawbacks.

Intermediate-temperature fuel cells promise a good match between electrochemical kinetics and materials problems. Envisaged electrolytes are partially hydroxylated oxides enabling proton conductivity by phonon-assisted proton hopping. Typical examples are $BaCeO_3$ and $BaZrO_3$. Grain boundary effects are very detrimental in these systems, for which again space charge effects are accounted for [82].

19.6.1.2 Batteries

As far as batteries are concerned, solid electrolytes are conductors of choice for thin film batteries of high safety and low power density. High–performance batteries such as lithium- or sodium-based batteries use nanoparticulate electrodes and it is hard to replace liquid electrolytes that offer optimum contact behavior. (A counter example are the classic Na–S cells, here the electrolyte is solid, yet the electrode phase liquid.) As to the latter, solid electrolytes are discussed as passivation membranes attached to Li or Na as to avoid solid electrolyte dendrite formation. Glass or polymer electrolytes are envisaged as alternatives, even more promising are semisolid electrolytes such as solid–liquid composites, for example, soggy sand electrolytes or ion-exchanged polyelectrolytes [83–85].

19.6.1.3 Chemical Sensors

Solid electrolytes can also be used in potentiometric cells that are designed such that the open circuit voltage is an unambiguous measure of the activity of a component to be detected. The classical example is P_{O_2} determination by the above-mentioned ZrO_2 electrolytes [86]. Here, P_{O_2} on the reference is known and the measured cell voltage allows for P_{O_2} determination on the measuring side. A more modern example is the thermodynamically exact determination of CO_2-partial pressure by the cell [87]

$$(Au, O_2), CO_2, Na_2CO_3 \left| \beta'' - Al_2O_3(Na_2O) \right| TiO_2, Na_2Ti_6O_{13}(O_2Au)$$

In these cases, the amount of ion conductivity is not very crucial, as only very small measurement currents need to be drawn. It is more important that σ_{eon} be negligible and that the electrochemical kinetics of the gas–solid interaction be fast enough, issues for which again ion point defects are decisive. Another interesting sensor application of pure ion conductors as acid–base active surface sensors is briefly touched upon at the very end of this chapter (19.6.3.2).

19.6.2
Mixed Conductors as Electrodes or Absorbers

19.6.2.1 Fuel Cells

If σ_{ion} and σ_{eon} are equally substantial, one speaks of mixed conductors. In fuel cells, it may suffice that electrodes only work as electronic conductors provided the catalytic activity is sufficient and enough phase contacts are present. It is however, highly beneficial if these electrodes are mixed conductors as then the active zones for the gas/solid interactions involve the electrode bulk, and are thus largely increased. In fact the best cathode materials for SOFC are pronounced mixed electron–oxygen ion conductors. It is striking, but in view of defect chemistry not surprising, that many good heterogeneous catalysts are mixed conductors [72,88].

Complex mixed conducting materials may also be good catalysts for room temperature fuel cells with the potential to replace the previous noble metal catalysts presently used [89].

Not much research has been devoted to electrodes for intermediate fuel cells. Here again mixed conductors are beneficial. In various cases, the conductivity of three carriers (e^-, O^{2-}, H^+) is to be considered the treatment of which is rather complex [90].

19.6.2.2 Batteries

As far as battery electrodes are concerned, mixed conductivity is a must as here the electroactive component should be stored rather than only converted. As at room temperature, the diffusion coefficients are not very large, size reduction is the most important tool to increase performance.

Since all these nano particles have to be connected with electronic and ionic "leads" efficiently, the following conclusion is crucial: The best solution is to use a bicontinuous network of electron collector and electrolyte that is heterogeneous on a scale comparable with the size of the electroactive particles, whereby this size is chosen such that diffusion problems become just negligible [91]. Implementing electronically conducting phases within the nanostructured electrodes (electrochemical integrated circuits) while simultaneously guaranteeing good electrolyte contact down to the nanoscale is a sophisticated problem of materials research.

19.6.2.3 Sensors, Chemical Storage and Permeation Devices

A pertinent selective chemical sensor can be based on a solid mixed conductor with single-ion conduction. An example is $SrTiO_3$ as it can absorb oxygen, and the resulting equilibrium stoichiometry of which can be easily detected by conductivity measurements [92]. Selectivity increases on increasing the sample thickness, as the surface effects then become negligible. Simultaneously, however, the response time (chemical diffusion) suffers. The purely ionic analogue would be to, for example, detect H_2O by incorporating it into a H^+/O^{2-} conducting material and to record the evolving proton conductivity [93].

Mixed conductivity is also decisive for storing H_2 or other gases dissociatively in ionic materials. Ceramic membranes that allow for permeation of gases are technologically relevant for gas purification or catalysis [94].

19.6.3
Electronic Conductors as Current Collectors or Taguchi Sensors

19.6.3.1 Fuel Cells and Batteries

Needless to say that pure electron conductors are important in order to connect the cell components with the outer circuit. As ionic contributions are irrelevant here, this is not a typical (electro)chemical problem apart from the topological implementation during synthesis. As to the latter, much exciting work is devoted to generate sophisticated electrochemical integrated circuits.

19.6.3.2 Sensors

A significant issue refers to the important class of Taguchi sensors where the electronic surface conductivity is measured that is caused by electron uptake (or injection) induced by an oxidizing (or reducing) gas. Here the fast response is due to the fact that the gas/solid interaction only comprises the surface reaction and subsurface response. Owing to the negligible ionic contributions, there is no significant stoichiometric change that penetrates into the bulk. This enhances the response time yet at the expense of selectivity. Selectivity may be achieved by coupling such Taguchi sensors to selective prestepts, by using the temperature dependence of adsorption equilibria or by using an array of different sensing materials (pattern recognition, chemical nose) [95].

More recently, it was pointed out that the ion conduction analogue (e.g., detection of NH_3 via ionic surface conductivities) to a Taguchi sensor allows rapidly detecting acid–base active gases with potentially high selectivity [96].

These few examples that only give a taste of the exciting field of electrochemical devices, show that major progress relies on progress of defect chemistry.

19.7
Conclusions

Owing to their mobility and relative charge, point defects are key constituents of the solid state as far as physical (energy application) and chemical functionality (catalysis) is concerned. Even more important than the sheer application point of view is their fundamental, conceptual role in solid-state chemistry. Solid-state chemistry without them is like aqueous chemistry without H^+ and OH^-. It is the presence of point defects that allows for an "inner chemical life" of solids. Less casually formulated: Whenever the kinetics of the crystalline solid state is considered, point defects as active particles are the major players. This refers to both reaction and transport.

In addition to these phenomena, immobile point defects or higher dimensional defects (mostly interfaces) contribute to the overall properties in form of setting local boundary conditions. At high enough temperature, they can be mobile and become active as such.

A most exciting aspect of defect chemistry is its generalizing power combined with the great potential to cross-fertilize neighboring disciplines.

References

1 Maier, J. (2006) Inner chemical life of solids. *Chem. Int.*, **28**, 4–7.
2 (a) Maier, J. (1993) Defect chemistry: composition, transport, and reactions in the solid state; Part I – thermodynamics. *Angew. Chem., Int. Ed. Engl.*, **32**, 313–335 (b) Maier, J. (1993) Defect chemistry: composition, transport, and reactions in the solid state; Part II – kinetics. *Angew. Chem. Int. Ed. Engl.*, **32**, 528–542.
3 (a) Wagner, C. and Schottky, W. (1930) Theory of controlled mixed phases. *Z. Phys. Chem.*, **11**, 163–210; (b) Wagner, C. (1936) The theory of the warm-up process. II. *Z. Phys. Chem.*, **32**, 447–462.
4 Kröger, F.A. (1964) *Chemistry of Imperfect Crystals*, North-Holland, Amsterdam.
5 Maier, J. (2001) Acid-base centers and acid-base scales in ionic solids. *Chem. Eur. J.*, **7**, 4762–4770.

6 Maier, J. (2004) *Physical Chemistry of Ionic Materials: Ions and Electrons in Solids*, John Wiley & Sons, Ltd, Chichester.

7 Maier, J. (2010) Nanoionics, in *Handbook of Nanophysics, Vol. 1: Principles and Methods* (ed. K.D. Sattler), RC Press, Taylor & Francis Group, Boca Raton, FL., pp. 8-1–8-11.

8 Kröger, F.A. and Vink, H.J. (1956) Relations between the concentrations of imperfections in crystalline solids, in *Solid State Physics*, vol. **2** (eds F. Seitz and D. Turnbull), Academic Press, New York, pp. 307–435.

9 Maier, J. (2012) Building versus structure elements: ionic charge carriers in solids. *Z. Phys. Chem.*, **226**, 863–870.

10 Maier, J. (2003) Ionic and mixed conductors for electrochemical devices. *Radiat. Eff. Defect. S.*, **158**, 1–10.

11 Maier, J. (2005) Chemical potential of charge carriers in solids. *Z. Phys. Chem.*, **219**, 35–46.

12 Jamnik, J., Pejovnik, S., and Maier, J. (1993) A new approach for the computation of the frequency response of space-charge containing interfaces. *Electrochim. Acta*, **14**, 1975–1978.

13 Maier, J. (2005) Nanoionics: ion transport and electrochemical storage in confined systems. *Nat. Mater.*, **4**, 805–815.

14 Maier, J. (2003) Defect chemistry and ion transport in nano-structured materials. Part II. Aspects of nanoionics. *Solid State Ion.*, **157**, 327–334.

15 Brouwer, G. (1954) A general asymptotic solution of reaction equations common in solid state chemistry. *Philips Res. Rep.*, **9** (5), 366–376.

16 Maier, J. and Pfundtner, G. (1991) Defect chemistry of the high-T_c-superconductors. *Adv. Mater.*, **3** (6), 292–297.

17 Maier, J. (2005) Solid state electrochemistry I: thermodynamics and kinetics of charge carriers in solids, in *Modern Aspects of Electrochemistry*, vol. **38** (eds B.E. Conway, C.G. Vayenas, and R.E. White), Kluwer Academic, New York, pp. 1–173.

18 Maier, J. and Göpel, W. (1988) Investigations of the bulk defect chemistry of polycrystalline Tin(IV) oxide. *J. Solid State Chem.*, **72**, 293–302.

19 Maier, J. and Schwitzgebel, G. (1983) Electrochemical polarization of orthorhombic PbO. *Mater. Res. Bull.*, **18**, 601–608.

20 Spaeth, M., Kreuer, K.D., Cramer, C., and Maier, J. (1999) Giant haven ratio for proton transport in sodium hydroxide. *J. Solid State Chem.*, **148**, 169–177.

21 (a) Gerbig, O., Merkle, R. and Maier, J. (2013) Electron and ion transport in Li_2O_2. *Adv. Mater.*, **25**, 3129–3133 (b) Gerbig, O., Merkle, R. and Maier, J. (2015) Electrical transport and oxygen exchange in the superoxides of potassium, rubidium, and cesium. *Adv. Funct. Mater.*, **25**, 2552–2563.

22 Balachandran, U. and Eror, N.E. (1981) Electrical conductivity in strontium titanate. *J. Solid State Chem.*, **39**, 351–359.

23 Choi, G.M. and Tuller, H.L. (1988) Defect structure and electrical properties of single-crystal $Ba_{0.03}Sr_{0.97}TiO_3$. *J. Am. Ceram. Soc.*, **71** (4), 201–205.

24 Denk, I., Münch, W., and Maier, J. (1995) Partial conductivities in $SrTiO_3$: bulk polarization experiments, oxygen concentration cell measurements, and defect-chemical modeling. *J. Am. Ceram. Soc.*, **78**, 3265–3272.

25 (a) Steele, B.C.H. (1995) Interfacial reactions associated with ceramic ion transport membranes. *Solid State Ion.*, **75**, 157–165 (b) Sasaki, K. and Maier, J. (2000) Re-analysis of defect equilibria and transport parameters in Y_2O_3-stabilized ZrO_2 using {EPR} and optical relaxation. *Solid State Ion.*, **134**, 303–321.

26 Maier, J. (2003) Complex oxides: high temperature defect chemistry versus low temperature defect chemistry. *Phys. Chem. Chem. Phys.*, **5**, 2164–2173.

27 Waser, R. (1991) Bulk conductivity and defect chemistry of acceptor-doped strontium-titanate in the quenches state. *J. Am. Ceram. Soc.*, **74**, 1934–1940.

28 Sasaki, K. and Maier, J. (1999) Low-temperature defect chemistry of oxides. I. General aspects and numerical calculations. *J. Appl. Phys.*, **86** (10), 5422–5433.

29 Teltow, J. (1950) On the ionic and electronic conductivity and lattice defects of silver bromide with impurities of silver,

cadmium, and lead sulphide. *Z. Phys. Chem.*, **195**, 213–224.

30 Merkle, R. and Maier, J. (2005) On the Tammann-Rule. *Z. Anorg. Allg. Chem.*, **631**, 1163–1166.

31 Hainovsky, N. and Maier, J. (1995) Simple phenomenological approach to premelting and sublattice melting in Frenkel disordered ionic crystals. *Phys. Rev. B*, **51**, 15789–15797.

32 Huberman, R.A. (1974) Cooperative phenomena in solid electrolytes. *Phys. Rev. Lett.*, **32**, 1000–1002.

33 Maier, J. and Münch, W. (2000) Thermal destiny of an ionic crystal. *Z. Anorg. Allg. Chem.*, **626**, 264–269.

34 Maier, J. (2005) Utility of simple rate equations for solid state reactions. *Z. Anorg. Allg. Chem.*, **631**, 433–442.

35 Maier, J. (1998) On the correlation of macroscopic and microscopic rate constants in solid state chemistry. *Solid State Ion.*, **112**, 197–228.

36 Maier, J. (2005) Ionic and mixed conductivity in condensed phases, in *Encyclopedia of Condensed Matter Physics* (eds G. Bassani, G. Liedl, and P. Wyder), Elsevier/Academic Press, pp. 9–21.

37 Geller, S. (ed.) (1977) *Solid Electrolytes*, vol. **21**, Springer, Berlin.

38 Maier, J. (2014) Pushing nanoionics to the limits: charge carrier chemistry in extremely small systems. *Chem. Mater.*, **26**, 348–360.

39 Kayama, N., Homma, K., Yamakawa, Y., Kanno, R., Yonemura, M., Kamiyama, T., Kato, Y., Hama, S., Kawamoto, K., and Mitsui, A. (2011) A lithium superionic conductor. *Nat. Mater.*, **10**, 682–686.

40 Kuhn, A., Gerbig, O., Zhu, C., Falkenberg, F., Maier, J., and Lotsch, B.V. (2014) A new ultrafast superionic Li-conductor: ion dynamics in $Li_{11}SiPS_2$ and comparison with other tetragonal LGPS-type electrolytes. *Phys. Chem. Chem. Phys.*, **16**, 14669–14674.

41 Kennedy, J.H. (1977) The β-Aluminas, in *Solid Electrolytes*, vol. **21** (ed. S. Geller), Springer, Berlin, pp. 105–141.

42 Adams, S. and Rao, R.P. (2009) Transport pathways for mobile ions in disordered solids from the analysis of energy-scaled bond-valence mismatch landscapes. *Phys. Chem. Chem. Phys.*, **11**, 3210–3216.

43 Yang, T.-Y., Gregori, G., Pellet, N., Grätzel, M., and Maier, J. (2015) The significance of ion conduction in a hybrid organic-inorganic lead-iodide-based perovskite photosensitizer. *Angew. Chem., Int. Ed.*, **54**, 7905–7910.

44 Chen, C., Fu, L., and Maier, J. (2016 Synergistic, ultrafast mass storage and removal in artificial mixed conductors. *Nature*, **536**, 159–164

45 Riess, I. and Maier, J. (2008) Symmetrized general hopping current equation. *Phys. Rev. Lett.*, **100** (1–4), 205901.

46 Maier, J. (1995) Ionic conduction in space charge regions. *Prog. Solid State Chem.*, **23**, 171–263.

47 Maier, J. (1987) Space charge regions in solid two phase systems and their conduction contribution – III: defect chemistry and ionic conductivity in thin films. *Solid State Ion.*, **23**, 59–67.

48 Maier, J. (2002) Nanosized mixed conductors (aspects of nanoionics. Part III). *Solid State Ion.*, **148**, 367–374.

49 Maier, J. (1989) Space charge regions in solid two phase systems and their conduction contribution. IV. The behavior of minority charge carriers. Part A: concentration profiles, conductivity contribution, determination by generalized wagner–Hebb procedure. *Ber. Bunsenges. Phys. Chem.*, **93**, 1468–1473.

50 (a) Liang, C.C. (1973) Conduction characteristics of lithium iodide aluminium oxide solid electrolytes. *J. Electrochem. Soc.*, **120**, 1289–1292 (b) Shahi, K. and Wagner, J.B. (1982) Enhanced ionic conduction in dispersed solid electrolyte systems (DSES) and/or multiphase systems: $AgI - Al_2O_3$, $AgI - SiO_2$, AgI–Fly ash, and AgI–AgBr. *J. Solid State Chem.*, **42**, 107–119.

51 Sata, N., Eberman, K., Eberl, K., and Maier, J. (2000) Mesoscopic fast ion conduction in nanometre-scale planar heterostructures. *Nature*, **408**, 946–949.

52 Adams, S., Hariharan, K., and Maier, J. (1994) Crystallization in fast ionic glassy silver oxysalt systems. *Solid State Phenom.*, **39–40**, 285–288.

53 Waser, R. (1991) Bulk conductivity and defect chemsitry of acceptor-doped strontium-titanate in the quenched state. *J. Am. Ceram. Soc.*, **74**, 1934–1940.

54 Kim, S. and Maier, J. (2002) On the conductivity mechanism of nanocrystalline ceria. *J. Electrochem. Soc.*, **149**, J73–J83.

55 Leonhardt, M., Jamnik, J., and Maier, J. (1999) In situ monitoring and quantitative analysis of oxygen diffusion through schottky-barriers in $SrTiO_3$ bicrystals. *Electrochem. Solid State Lett.*, **2**, 333–335.

56 Babilo, P. and Haile, S.M. (2005) Enhanced sintering of yttrium-doped barium zirconate by addition of ZnO. *J. Am. Ceram. Soc.*, **88**, 2362–2368.

57 Iguchi, F., Sata, N., Tsurui, T., and Yugami, H. (2007) Microstructures and grain boundary conductivity of $BaZr_{1-x}Y_xO_3$ (x – 0.05, 0.10, 0.15) ceramics. *Solid State Ion.*, **178**, 691–695.

58 Shirpour, M., Merkle, R., and Maier, J. (2012) Space charge depletion in grain boundaries of $BaZrO_3$ proton conductors. *Solid State Ion.*, **225**, 304–307.

59 Maier, J. (2004) Ionic transport in nanosized systems. *Solid State Ion.*, **175**, 7–12.

60 Lupetin, P., Gregori, G., and Maier, J. (2010) Mesoscopic charge carriers chemistry in nanocrystalline $SrTiO_3$. *Angew. Chem., Int. Ed.*, **49**, 10123–10126.

61 Korte, C., Peters, A., Janek, J., Hesse, D., and Zakharov, N. (2008) Ionic conductivity and activation energy for oxygen ion transport in superlattices – the semicoherent multilayer system YSZ ((ZrO_2 + 9.5 mol% Y_2O_3)/Y_2O_3). *Phys. Chem. Chem. Phys.*, **10**, 4623–4635.

62 Jamnik, J. and Maier, J. (2003) Nanocrystallinity effects in lithium battery materials. Aspects of nano-ionics. Part IV. *Phys. Chem. Chem. Phys.*, **5**, 5215–5220.

63 Zhukovskii, Y.F., Balaya, P., Dolle, M., Kotomin, E.A., and Maier, J. (2007) Enhanced lithium storage and chemical diffusion in metal-LiF nanocomposites: experimental and theoretical results. *Phys. Rev. B*, **76** (1–6), 235414.

64 Fu, L.J., Chen, C.C., Samuelis, D., and Maier, J. (2014) Thermodynamics of lithium storage at abrupt junctions: modeling and experimental evidence. *Phys. Rev. Lett.*, **112** (1–5), 208301.

65 Maier, J. (2013) Thermodynamics of electrochemical lithium storage. *Angew. Chem., Int. Ed.*, **52**, 4998–5026.

66 Merkle, R. and Maier, J. (2008) How is oxygen incorporated into oxides? A comprehensive kinetic study of a simple solid-state reaction with $SrTiO_3$ as a model material. *Angew. Chem., Int. Ed.*, **47**, 3874–3894.

67 Denk, I., Noll, F., and Maier, J. (1997) In-situ profiles of oxygen diffusion in $SrTiO_3$: bulk behavior and boundary effects. *J. Am. Ceram. Soc.*, **80**, 279–285.

68 De Souza, R.A., Fleig, J., Merkle, R., and Maier, J. (2003) $SrTiO_3$: a model electroceramic. *Z. Metallkd.*, **94**, 218–225.

69 Maier, J. (2000) Interaction of oxygen with oxides: how to interpret measured effective rate constants? *Solid State Ion.*, **135**, 575–588.

70 Kuklja, M.M., Kotomin, E.A., Merkle, R., Mastrikov, Yu.A., and Maier, J. (2013) Combined theoretical and experimental analysis of processes determining cathode performance in solid oxide fuel cells. *Phys. Chem. Chem. Phys.*, **15**, 5443–5471.

71 Wedig, A., Merkle, R., and Maier, J. (2014) Oxygen exchange kinetics of (Bi, Sr)(Co, Fe)$O_{3-\delta}$ thin-film microelectrodes. *J. Electrochem. Soc.*, **161**, F23–F32.

72 Merkle, R. and Maier, J. (2006) The significance of defect chemistry for the rate of gas–solid reactions: three examples. *Top. Catal.*, **38**, 141–145.

73 Gensch, C. and Hauffe, K. (1951) On the speed of oxidation of zinc alloys. *Z. Phys. Chem.*, **196**, 427–437.

74 (a) Wagner, C. (1969) Distribution of cations in metal oxide and metal sulphide solid solutions formed during oxidation of alloys. *Corros. Sci.*, **9**, 91–109 (b) Wagner, C. (1933) The theory of the warm-up process. *Z. Phys. Chem.*, **21**, 25–41.

75 Fromhold, A.T. (1980) *Theory of Metal Oxidation*, North-Holland, Amsterdam.

76 Templeton, L.K. and Pask, J.A. (1959) Formation of $BaTiO_3$ from $BaCO_3$ and TiO_2 in air and in CO_2. *J. Am. Ceram. Soc.*, **42**, 212–216.

77 Niepce, J.C. and Thomas, G. (1990) About the mechanism of the solid-way synthesis of barium metatitanate. Industrial consequences. *Solid State Ion.*, **43**, 69–76.

78 Kilner, J.A. and Steele, B.C.H. (1981) Mass transport in anion-deficient fluorite oxides, in *Nonstoichiometric Oxides* (ed. O.T. Sørensen), Academic Press, New York, pp. 233–269.

79 Steele, B.C.H. (2000) Appraisal of $Ce_{1-y}Gd_yO_{2-y/2}$ electrolytes for IT-SOFC operation at 500 °C. *Solid State Ion.*, **129**, 95–110.

80 Kreuer, K.D. (2014) Ion conducting membranes for fuel cells and other electrochemical devices. *Chem. Mater.*, **26**, 361–380.

81 Schuster, M., Kreuer, K.D., Andersen, H.T., and Maier, J. (2007) Ion conducting membranes for fuel cells and other electrochemical devices. *Macromolecules*, **40**, 598–607.

82 Norby, T. (2009) Proton conduction in solids: bulk and interfaces. *MRS Bull.*, **34**, 923–928.

83 Knauth, P. (2009) Inorganic solid Li ion conductors: an overview. *Solid State Ion.*, **180**, 911–916.

84 Pfaffenhuber, C., Göbel, M., Popovic, J., and Maier, J. (2013) Soggy-sand electrolytes: status and perspectives. *Phys. Chem. Chem. Phys.*, **15**, 18318–18335.

85 Kreuer, K.D., Wohlfarth, A., de Araujo, C.C., Fuchs, A., and Maier, J. (2011) Single alkaline-ion (Li^+, Na^+) conductors by ion exchange of proton-conducting ionomers and polyelectrolytes. *Chemphyschem.*, **12**, 2558–2560.

86 Fischer, W.A. and Janke, D. (eds) (1975) *Metallurgische Elektrochemie*, Springer, Berlin.

87 Maier, J., Holzinger, M., and Sitte, W. (1994) Fast potentiometric CO_2 sensors with open reference electrodes. *Solid State Ion.*, **74**, 5–9.

88 Fleig, J. and Maier, J. (2004) The polarization of mixed conducting SOFC cathodes: effects of surface reaction coefficient, ionic conductivity and geometry. *J. Eur. Ceram. Soc.*, **24**, 1343–1347.

89 Suntivich, J., Gasteiger, H., Yabuuchi, N., Nakanishi, H., Goodenough, J., and Shao-Horn, Y. (2011) Design principles for oxygen-reduction activity on perovskite oxide catalysts for fuel cells and metal-air batteries. *Nat. Chem.*, **3**, 546–550.

90 Poetzsch, D., Merkle, R., and Maier, J. (2015) Stoichiometry variation in materials with three mobile carriers – thermodynamics and transport kinetics exemplified for protons, oxygen vacancies, and holes. *Adv. Funct. Mater.*, **25**, 1542–1557.

91 Maier, J. (2014) Control parameters for electrochemically relevant materials: the significance of size and complexity. *Faraday Discuss.*, **176**, 17–29.

92 Menesklou, W., Schreiner, H.J., Härdtl, K.H., and Ivers-Tiffee, E. (1999) High temperature oxygen sensors based on doped $SrTiO_3$. *Sens. Actuators B*, **59**, 184–189.

93 Maier, J. (2000) Electrochemical sensor principles for redox-active and acid-base-active gases. *Sens. Actuators B*, **65**, 199–203.

94 Zhou, M., Deng, H., and Abeles, B. (1997) Transport in mixed ionic electronic conductor solid oxide membranes with porous electrodes: large pressure gradient. *Solid State Ion.*, **93**, 133–138.

95 Göpel, W. (1985) Chemisorption and charge transfer at ionic semiconductor surfaces: implications in designing gas sensors. *Prog. Surf. Sci.*, **20**, 9–103.

96 Holzinger, M., Fleig, J., Maier, J., and Sitte, W. (1995) Chemical sensors for acid–base active gases: applications to CO_2 and NH_3. *Ber. Bunsenges. Phys. Chem.*, **99**, 1427–1432.

20
Molecular Magnets

J.V. Yakhmi

Homi Bhabha National Institute, Anushakti Nagar, Mumbai 400085, India

20.1
Introduction

Most molecular materials are organic in nature, and molecular crystals are built from well-defined molecules, which do not change their geometries appreciably upon entering the crystal lattice. This is because intermolecular interactions are non-covalent, such as hydrogen bonding, Van der Waals interactions, donor–acceptor charge transfer, etc. that are much weaker than the energies of typical chemical bonds, ionic or covalent. This makes it possible to modify the properties of a molecular solid in a predetermined way by attaching a function to the building block (i.e., the molecule), and thus engineer a bulk "designer" molecular material, and synthesize it at or close to ambient temperature. Molecular materials have been produced which exhibit several properties originally attributed to the metallic lattices, such as, high electric conductivity/photoactivity in polymers, and superconductivity in charge-transfer organic complexes/fullerenes. No wonder that there has been a desire to develop ferro(ferri)magnetic molecular materials, too, by designing new combinations of interactions between magnetic centers in organic (or polymeric) materials, preferably with *p*-orbital-based spins. Theoretical models do predict the possibility of attaining long-range magnetic order in molecular materials. However, organic compounds are mostly diamagnetic, with closed-shell structures, and even if one (or more) unpaired electron is maintained stable per organic molecule, stabilization of a triplet state (parallel alignment of spins) satisfying the orthogonality conditions (Hund's rules) is difficult, and more often than not they get antiferromagnetically coupled with no spontaneous magnetic moment. In the conventional magnetic materials used in present-day technology, such as, Fe, Fe_2O_3, Cr_2O_3, $SmCo_5$, $Nd_2Fe_{14}B$, and so on, magnetic order arises from co-operative spin–spin interactions between unpaired electrons located in *d*-orbitals or *f*-orbitals. But their synthesis

Handbook of Solid State Chemistry, First Edition. Edited by Richard Dronskowski, Shinichi Kikkawa, and Andreas Stein.
© 2017 Wiley-VCH Verlag GmbH & Co. KGaA. Published 2017 by Wiley-VCH Verlag GmbH & Co. KGaA.

typically depends on solid-state chemistry or high temperature metallurgical routes.

20.2
Strategies of Design and Synthesis of Molecular Magnets

First ever ferro(ferri)magnetic molecular compounds exhibiting a spontaneous magnetization below a certain temperature, T_c, were reported in 1986 [1,2]. Subsequently, molecular magnets of many different categories have been proposed and synthesized, with an ever-growing interest in the field of molecule-based magnets. Typical synthetic approach to design molecule-based magnets consists of selecting molecular precursors, each with an unpaired spin (i.e., the function, as shown in Figure 20.1), and assembling them in a manner that there is no compensation of spins at the scale of the crystal lattice [3–7]. Magnetism being a co-operative effect, the spin–spin interaction must extend to all the three dimensions. And if this interaction between spin carriers occurs through space, then we have a genuine molecular lattice. But the spin–spin interaction may also occur through bond in which case we would have a polymeric or extended structure. In the latter case, the interactions are usually much stronger, particularly so if there is conjugation among the bridging ligands. In short, the design of a molecular magnet requires that: (a) all the molecules in the lattice have unpaired electrons; and (b) the unpaired electrons should have their spins aligned parallel

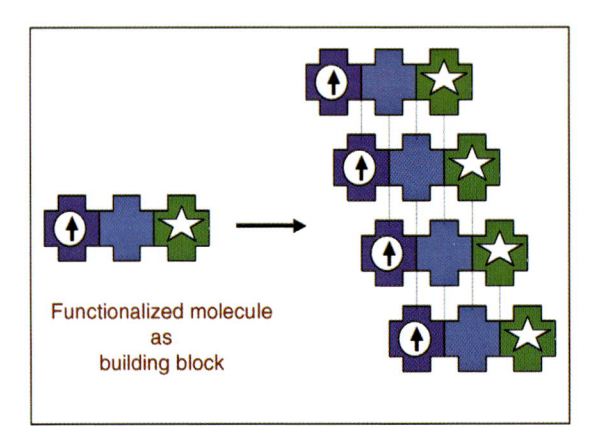

Figure 20.1 A cartoon depicting the self-assembly of a molecule-based magnet from functionalized molecules, each bearing an unpaired spin (the function, indicated by an arrow). The remainder of this functionalized molecule – the body and the tail marked with a "star" serves as an orientation controlling site designed to steer the organization of these building blocks, using noncovalent forces, in such a way that there is no compensation of spins at the scale of the crystal lattice, yielding a three-dimensional lattice with a long-range magnetic order.

along a given direction. Tools of supramolecular chemistry can also be employed to design molecular magnets [8].

20.3
Purely Organic Magnets

Magnetism in metal-free compounds must involve electrons from p-orbitals, which was considered impossible not long ago because the origin of magnetism was thought to be the regime of d- or f-electrons, exclusively. Therefore, the discovery of ferromagnetism involving p-electrons in iron-containing organic-based material in 1986 [1] was path-breaking. Over the years, there has been sporadic progress in the preparation of π-conjugated oligomers and polymers with large values of spin quantum number, S, but success has eluded thus far in the design of polymer magnets. Whenever this happens, it would have a very substantial impact on their applications in industry [4,9,10]. Possibility of band ferromagnetism in purely organic polymeric structures, such as, PAT [poly(4-amino 1,2,4 triazole], which are chains of five-membered rings, was suggested by Arita *et al.* [11] from their spin density functional calculations, which predicted that the ground state can become ferromagnetic when the flat band of such materials is made half-filled, with suitable dopings. Interestingly, a long-range ferromagnetic order (Figure 20.2a) has indeed been shown to exist in compounds made from purely organic constituents (only C, H, N, and O elements), because unlike in the case of ionic/metallic compounds, the occurrence of spin delocalization and spin polarization in molecular lattices, can lead to intermolecular exchange interactions conducive for a long-range ferromagnetic order. The delocalization of spin density in certain molecules makes it possible for magnetic interactions to take place across extended bridges between magnetic centers far apart from each other, propagating through conjugated bond

(a) Ferromagnet   (b) Ferrimagnet

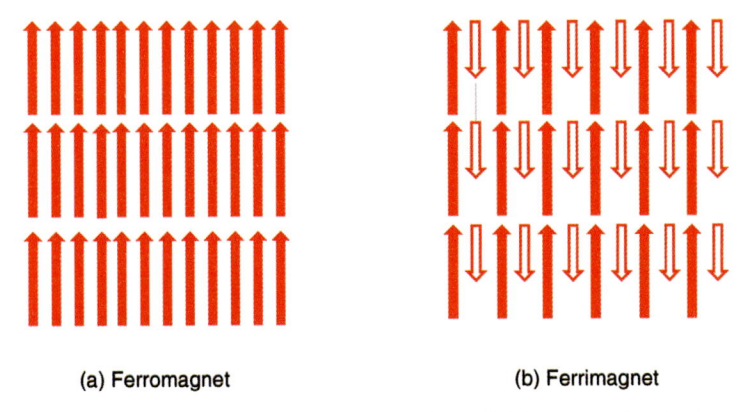

Figure 20.2 (a) Schematic of a ferromagnet with parallel alignment of 'like' spins; and (b) a ferrimagnet with antiparallel alignment of unequal spins, yielding a net aggregate spin along a given direction, for the bulk sample.

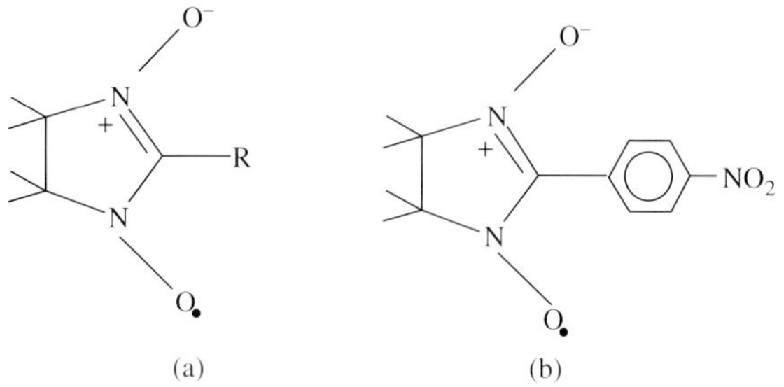

Figure 20.3 The chemical structure of (a) NITR and (b) p-NPNN (dark dot indicates an unpaired electron).

linkages. Spin polarization, that is, the simultaneous existence of positive and negative spin densities at different locations within a given radical is crucial for intermolecular exchange interactions to bring about ferromagnetic interaction between organic radicals, as per McConnell's model [12]. Spin density across different regions of the nitronyl nitroxide radical NITR (R = alkyl), a versatile building block with spin $S = 1/2$ ground state (Figure 20.3a), for instance, shows positive values, equally delocalized between N and O within each N–O group, and a small negative value on the bridging sp^2 carbon, due to spin polarization. By substituting different alkyl groups (like R = benzyl, isopropyl, methyl, ethyl, phenyl, etc.) in NITR, one can tune the single-radical ground state to establish new exchange pathways through varied coordination sites. For instance, ferromagnetism at 0.6 K arises solely from p-orbital spins in the β-phase of R = phenyl compound (4-nitrophenylnitronyl nitroxide) (p-NPNN, with formula $C_{13}H_{16}N_3O_4$, shown in Figure 20.3b), a metal-free organic magnet which contains only C, H, N, and O elements [13].

Ferromagnetism has also been obtained for the purely organic fullerene-based charge-transfer material, [tetrakis(dimethylamino)ethylene][C_{60}] with T_c of 16.1 K [14]. Fullerene has no intrinsic magnetic moment. For a magnetic moment to exist, an electron must be transferred to C_{60} from a donor molecule. Another example of a fullerene-based ferromagnet was the cobaltocene-doped derivative, which has a T_c of 19 K [15]. The highest ordering temperature reported to date for an organic magnet has been for the β-phase of the 4'-cyano-tetrafluorophenyldithiadiazolyl, a sulfur-based free radical, which was found to be a weak ferromagnet below 35.5 K [16,17]. Under applied pressure of 16 kilobars, its T_c value could be raised to 65 K [18].

20.4
Ferrimagnetic Building-Blocks

Kahn [3,19] proposed a novel strategy on how to build a molecule-based magnet based on the use of ferrimagnetic chains containing alternating spins of unequal

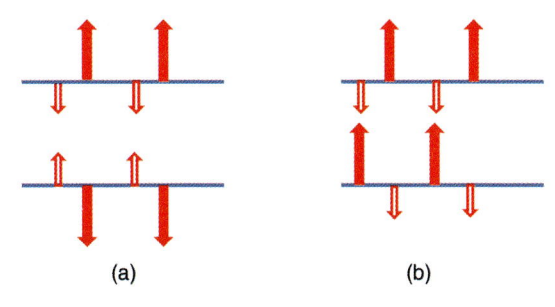

Figure 20.4 Assembly of ferrimagnetic chains leading to net (a) zero, or (b) nonzero magnetic moment in bulk.

magnitude, $S_A \neq S_B$ (Figure 20.2b), and assembling them in such a way that there is a net spin (Figure 20.4b), leading to a long-range magnetic order in the bulk lattice. Here, S_A stands for the large spin and S_B for the small spin on two different spin carriers, A and B, such as Mn^{II} ions ($S = 5/2$) and Cu^{II} ions ($S = 1/2$), respectively, within the same molecular precursor. Using this strategy, a large number of molecular (ferro)ferrimagnets have been assembled, the spin carriers in them being either two different metal ions [20,21] or a metal ion and an organic radical [22–25], with intervening ligands which serve as effective exchange pathways.

Heterobimetallic species, in which two different metal ions are bridged by extended *bis*-bidentate ligands such as oxamato [26,27], oxamido [20,28], or oxalato [21], in particular, allow a variety of spin topologies. Kahn's group synthesized several Mn(II)Cu(II) molecular magnets in which the ferrimagnetic interactions are propagated through *bis*-bidentate ligands, namely, MnCu(*opba*)· 0.7DMSO, which exhibits spontaneous magnetization below $T_c = 6.5$ K, and is synthesized by reacting the Cu(II) precursor Cu(*opba*)$^{2-}$ (Figure 20.5), where *opba* stands for *ortho*-phenylene*bis*(oxamato), with a divalent ion, Mn(II) in a 1 : 1 stoichiometry. With a view to raise the Curie temperatures of this class of compounds, the 2 : 3 Mn(II)Cu(II) compounds $A_2M_2[Cu(opba)]_3 \cdot n\text{Solv}$ were synthesized, by employing [Cu(*opba*)]$^{2-}$ to cross-link the chains in a two-dimensional network, using the cation A^+. Among them is (NBu$_4$)$_2$Mn$_2$[Cu(*opba*)]$_3$·6DMSO·1 H$_2$O, exhibiting a ferromagnetic behavior with $T_c = 15$ K [27], which increased to 22.5 K when all the solvent molecules are removed [27,29]. The structure of 1 : 2 Mn(II)Cu(II) compound (NBu$_4$)$_2$Mn[Cu(*opba*)]$_2$ in the class of "*opba*" molecular magnets revealed a new crystallographic arrangement [30], unlike a one-dimensional chain where the ratio is 1 : 1, or planar graphite-like sheets for the 2 : 3 ratio.

An effective process for assembling spin-bearing precursors is when polymerization is associated with dehydration. When the precursors are linked to each other after polymerization, the interaction involving a spin carrier A from a given unit and a spin carrier B from the adjacent unit becomes large, and the resulting ground state spin of the polymer [AB]$_n$, where n is the number of units, becomes $n(S_A - S_B)$. If n is infinite and the system is three-dimensional, a

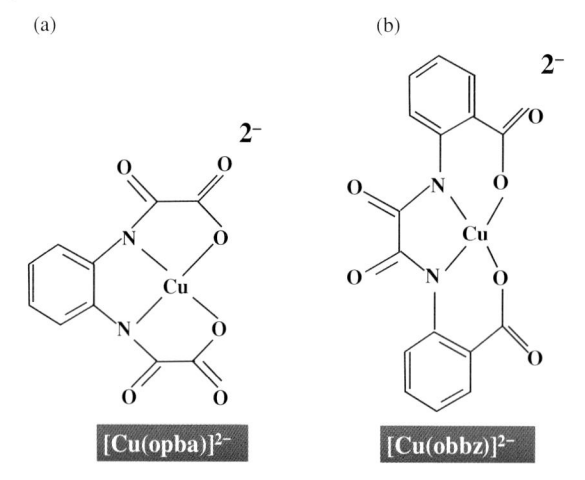

Figure 20.5 A schematic of the copper dianion precursors (a) [Cu(*opba*)]$^{2-}$ and (b) [Cu(*obbz*)]$^{2-}$, where "*opba*" and "*obbz*" stands for *ortho*-phenylenebis(oxamato) and oxamido bis(benzoato), respectively.

long-range ferrimagnetic ordering may occur, which can also be considered as a ferromagnetic coupling of the combined spin/S_A–S_B/of the AB units, which in the case of Mn(II)Cu(II)-based compounds is, $S = 2$. The compound MnCu(*obbz*)·5 H$_2$O [20], obtained by the polymerization of ferrimagnetic molecular precursor units prepared by reacting the copper dianion [Cu(*obbz*)]$^{2-}$ (Figure 20.5), where *obbz* is the ligand oxamido bis(benzoato), with Mn(II) ions. It exhibits a minimum at 44 K in its $\chi_M T$ versus T plot, χ_M being the molar magnetic susceptibility and T the temperature – a signature of a one-dimensional ferrimagnet, and a maximum at 2.3 K due to a 3-dimensional antiferromagnetic ordering. Upon dehydration, it yields the monohydrate, MnCu(*obbz*)·1 H$_2$O, which orders ferromagnetically below the critical temperature (T_c) of 14 K, due to the noncompensation of the magnetic moments on Mn(II) and Cu(II) [20,31]. Subsequently, Ni(II)-, Fe(II)-, and Co(II)-based bimetallic chain compounds were also synthesized using the [Cu(*obbz*)]$^{2-}$ precursor [20,31–34].

Using the mixed metal ion spin–organic radical spin approach, a 46 K magnet was synthesized by the reaction of a trinitroxide radical, with three parallel spins ($S = 3/2$), with *bis*(hexafluoroacetylacetonato) Mn(II), [Mn(II)(hfac)$_2$] [25]. However, Mn(II) ion, being magnetically isotropic cannot prevent the domains from rotating freely under an applied field. Hence, the Mn(II)Cu(II)-based magnets, though exhibit a high value of T_c (12–30 K), are soft ferromagnets, exhibiting rather narrow magnetic hysteresis loops below T_c with rather weak coercive field values ($H_c < 50$ Oe at 4.2 K) [27,35–37]. The same is generally true for Fe(II)Cu(II)-based magnets. It is the coercivity of a magnet that confers a memory effect on it (Figure 20.6). Stronger coercive fields are expected for Co^{2+}-based molecular magnets where Co^{2+} ion in distorted octahedral environment, being magnetically anisotropic, can assume preferred orientations. Replacing Mn(II), with an

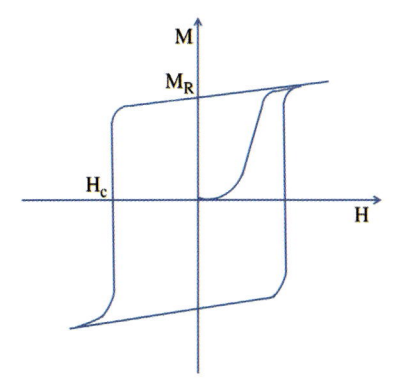

Figure 20.6 Memory effect is conferred on a sample by its Hysteresis loop (coercivity). Within $-H_C$ to $+H_C$, the sample can be considered to be bistable.

orbital singlet state ($^6A_1$), by Co(II) with an orbital triplet ground state ($^4T_1$) in $(cat)_2Mn_2[Cu(opba)]_3 \cdot S$ [27,35,36] resulted in a dramatic rise in coercivity. For instance, $H_c = 3000$ Oe for $(NBu_4)_2Co_2[Cu(opba)]_3 \cdot 3DMSO \cdot 3 H_2O$ and 3100 Oe for $(rad)_2Co_2[Cu(opba)]_3 \cdot 0.5DMSO \cdot 3 H_2O$ [37], where "rad" stands for the radical cation 2-(4-N-methylpyridinium)-4,4,5,5-tetramethylimidazolin-1-oxyl 3-oxide and "opba" denotes orthophenylene-*bis*(oxamato). [(*Etrad*)$_2$Co$_2${Cu(*opba*)}3(DMSO)$_{1.5}$]·0.25 H$_2$O exhibits high coercivity, up to 24 kOe at 6 K [38]. These values are much higher than that for the commercial atom-based materials, Fe_2O_3 or CrO_2.

20.5
Molecular Magnetic Sponges and Hydrogels

The synthetic procedures employed to obtain molecular materials are different from those employed in solid-state chemistry and the molecular crystal lattice is characteristically soft, as compared to the ionic/metallic lattices. We have exploited these attributes to demonstrate that for certain molecular magnets, assembled from Co(II)Cu(II)-based ferrimagnetic chains, it is possible to modify the magnetic properties dramatically and reversibly through a mild dehydration–rehydration process, and have named this class of compounds as molecular magnetic sponges [32,33,39,40]. This is because they show "sponge"-like characteristics, namely, a reversible cross-over under dehydration to a polymerized long range magnetically ordered state with spontaneous magnetization, and transforming back into the isolated units underlying the initial nonmagnetic phase by reabsorbing water, i.e., rehydration of both noncoordinated and coordinated water molecules. Coercivity values for these sponges are high and for some of them a color change, too, occurs reversibly and simultaneously with the change in magnetic properties at the transition temperature corresponding to the dehydration–rehydration process. In the case of CoCu(*obbz*)·nH$_2$O, we confirmed that the Co—O bonds could be broken and created without destroying the essence of the molecular architecture. The main features of the four Co(II)Cu(II)-based molecular magnetic sponges synthesized by our group, namely, CoCu(*pbaOH*)(H$_2$O)$_3$·2 H$_2$O, CoCu(*pba*)(H$_2$O)$_3$·2 H$_2$O, CoCu(*obbz*)(H$_2$O)$_4$·2 H$_2$O, and CoCu(*obze*)(H$_2$O)$_4$·2 H$_2$O are high values of T_c (38, 33, 25, and 25 K, respectively) and H_c (5.66, 3, 1·3, and 1 kOe, respectively).

Magnetic sponge-like behavior has also been observed in the case of 3D ferrimagnetic octacyanate, $\{[Mn^{II}(imH)]_2[Nb^{IV}(CN)_8]\}_n$ with $T_c = 62$ K [41]. Main discussion on octacyanates follows in Section 20.6.4.

Hydrogels find widespread applications in biomedical engineering due to their hydrated environment and tunable properties (e.g., mechanical, chemical, biocompatible) similar to the native extracellular matrix (ECM). Recently, significant advances have been achieved in the development of magnetic hydrogels (i.e., the combination of hydrogels with micro- and/or nanomagnetic particles that can quickly respond to an external magnetic field (MF), under which, the mechanical behavior of magnetic hydrogels, including shape deformation and the swelling/deswelling process can be controlled through noncontact triggering using the external MF. The applications of magnetic hydrogels in biomedical engineering are in the areas of tissue engineering, drug delivery and release, enzyme immobilization, cancer therapy, and soft actuators [42].

20.6
Polycyanometallates

20.6.1
Prussian Blue and its Hexacyanate Analogues

A unique feature of the molecular magnets is that they are usually weakly colored unlike the opaque classical magnets. Design of molecular ferromagnets of low density that are transparent and have a tunable high T_c is a cherished goal and photomagnetic switching has been reported in molecular magnets, especially where hexacyanometalates, $[M(CN)_6]^{n-}$, are used as molecular building blocks. Most metal hexacyanometalates have a cubic (*fcc*) structure and are analogues of the well-known Prussian blue salt, $Fe^{III}_4[Fe^{II}(CN)_6]_3 \cdot 15\ H_2O$, whose structure was described by Ludi and Gudel [43] as highly disordered cubic cell consisting of alternating ferrocyanide and ferric ions, with linear Fe^{III}–N–C–Fe^{II} bridges.

Prussian blue analogues (PBA), are represented by the general formula $A_x[B(CN)_6]_y \cdot zH_2O$, where A and B are 3d or 4d transition metal magnetic ions, which occupy the corners of the cube in the face-centered cubic structure (Figure 20.7). The cyanide groups bridges the metal ions, along the cube edges, the metal ions A and B being octahedrally coordinated by N and C, respectively. Depending on the charged-state of the N- and C-coordinated metal ions, a certain number of cations can occupy interstitial positions to satisfy charge compensation. These

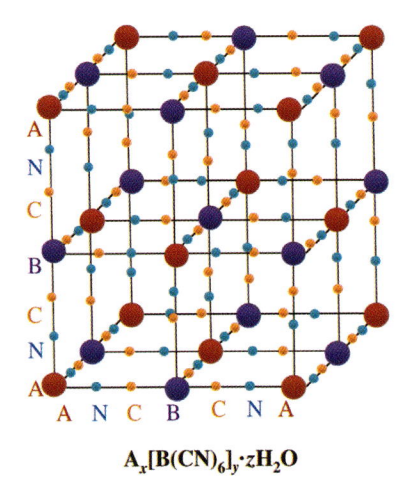

Figure 20.7 FCC structure of Prussian blue analogues, A[B(CN)$_6$]·zH$_2$O. For clarity, H and O atoms are not shown.

$$A_x[B(CN)_6]_y \cdot zH_2O$$

interstitial metal ions may be the same as the N- and C-coordinated metal ions, or they may be alkali metal ions. When $x/y = 1$, the first coordination of A and B is {A(NC)$_6$} and {B(NC)$_6$}, respectively. For a specific case of Fe[Fe(CN)$_6$]·4H$_2$O, the four water molecules are noncoordinated and occupy the 8c (1/4; 1/4; 1/4) interstitial positions. But for $x > y$, some of the {B(NC)$_6$} sites remain vacant and the first coordination of A and B will be {A(NC)$_{6-n}$·(H$_2$O)$_n$} ($n = 1, 6$) and {B(NC)$_6$}, respectively. The [B(CN)$_6$] vacancies, which maintain the required charge neutrality, and partial occupancies of the noncoordinated water molecules constitute the main structural defects. For example, for M$_3$[Fe(CN)$_6$]$_2$·14–16H$_2$O, 33% of the [Fe(CN)$_6$] sites remains vacant yielding a network of pores, which are generally filled by water molecules. PBAs can be synthesized very easily in the chemical lab, and therefore, are much studied for the variety of characteristics they offer. We have shown that the [Fe(CN)$_6$] vacancies, alongwith the presence of both interstitial water molecules and alkali ions give rise to structural disorder in Ba$_x$Mn[Fe(CN)$_6$]$_{2(x+1)/3}$·zH$_2$O ($x = 0.0$ and 0.3) compounds, and the number of both coordinated and noncoordinated water molecules decreased with Ba substitution [44].

The cyano bridge is known to mediate strong antiferromagnetic or ferromagnetic interactions [45]. T_c values above room temperatures, namely, 376 K [46], have been reported for hexacyanometalates but their high *fcc* symmetry is often accompanied by inherent disorder among different cationic sites, making it difficult to grow single crystals or to study any magnetic anisotropy.

There exists a correlation between the structural disorder and the observed magnetic properties of PBA compounds. The structural defects/disorders in a 3d–3d ions-based PBA, Ni$_3$[Cr(CN)$_6$]$_2$·zH$_2$O, have been studied by us [47], by substitution of 4d Ru^{3+} ions at Ni^{2+} sites. For the series of compounds

Ru$_x$Ni$_{3-3x/2}$[Cr(CN)$_6$]$_2 \cdot z$H$_2$O ($x = 0.0$ to 0.5), gradual increase in x first leads to an increase in T_c, from 57 to 62 K, when $x = 0.2$, but decreased for a higher ($x > 0.2$) composition [48]. This results from the change in effective number of the magnetic nearest neighbors and their average spin values. The initial increase in T_c is due to a decrease in the [Cr(CN)$_6$] vacancies, which for Ru content when increased beyond ($x > 0.2$), falls due to a decrease in the average spin of the Ru^{3+}/Ni^{2+} site.

We have demonstrated that one can grow crystals (size ~200 nm) of Prussian blue analogues (NiHCF, CoHCF) at the air–water interface [49–51], which when transferred as a monolayer showed magnetic order below 20 K. Nickel hexacyanoferrate (NiHCF) nanoparticles could be extracted into an organic phase using cetyltrimethylammonium bromide (CTAB) as the surfactant [52]. Potassium ion sensing action was demonstrated with measurements on NiHCF–DODA films (DODA: dioctadecyl dimethylammonium bromide) [53]. We have also used the double-stranded CT–DNA as a template to self-assemble crystals of nickel hexacyanoferrate, Ni$_3$[Fe(CN)$_6$]$_2 \cdot 14$ H$_2$O, with an average diameter of 400 nm [54].

Thin films and nanoparticles of PBAs are useful and can be prepared by different routes. An increase in the surface-to-volume ratio, due to a reduction in particle size, and inter-particle interactions play an important role in deciding the magnetic properties of nanoparticle systems. We have compared the observed magnetic behavior of Fe[Fe(CN)$_6$]$\cdot x$H$_2$O nanoparticles (mean diameter ~50 nm) prepared using a water-in-oil microemulsion technique, with that of its bulk counterpart [55]. A magnetic ordering temperature, T_c of 13.0 K was estimated from the magnetization data, which is low compared to that for the bulk polycrystalline sample, $T_c = 17.4$ K, and suggests the presence of a weak surface spin disorder in the ferromagnetically ordered cores of nanoparticles, the surface spin-disorder causing a spin glass behavior. Nanoparticles of Cu$_{1.5}$[Cr(CN)$_6$]$\cdot z$H$_2$O, were prepared in our lab by using PVP (polyvinylpyrrolidone) as a capping agent [56]. The size of these *fcc* particles could be controlled by varying the PVP to Cu ion feed ratio during the precipitation reaction, and specifically were of sizes 18, 9, and 5 nm, as estimated by TEM measurements. A ferromagnetic ordering of Cu and Cr ionic moments was observed with the magnetic ordering temperature decreasing from 65 to 52 K with decreasing particle size. Monodisperse nanoparticles of PBAs can be prepared easily in the lab, as has been demonstrated for the case of Cobalt(II) hexacyanoferrate(III), which were prepared using Co-ions released from a citrate complex [57]. As a typical orally administered drug in clinics, Prussian blue nanoparticles (PB NPs) have approved biosafety in the human body, and have, therefore been explored recently as a new generation of near-infrared (NIR) laser-driven photothermal ablation agents for cancer phototherapy [58].

Crystalline thin films of K$_j$Fe$_k^{II}$[Cr^{III}(CN)$_6$]$_l \cdot m$H$_2$O were prepared electrochemically with variation in film thickness and composition [59]. With an increase in film thickness from 1 to 5 µm, the Curie temperature increased from 11 to 21 K. However, for the films prepared with higher electrode voltage, a maximum T_C of 65 K was obtained for K^{I}Fe^{II}[Cr^{III}(CN)$_6$]$\cdot 2$ H$_2$O, and the enhancement in T_C was attributed to the change in the Fe^{II}/Cr^{III} ratio due to incorporation of K^+ ions.

20.6.2
Negative Magnetization and Pole Reversal

A crossover of the field-cooled magnetization from positive to negative below the magnetic ordering temperature is a unique feature exhibited by molecular multication ferrimagnets [29,60] and by Prussian blue analogues [61]. We have provided the first neutron magnetic structure evidence toward the microscopic understanding of the negative magnetization phenomenon in the PBA, $Cu_{0.73}Mn_{0.77}[Fe(CN)_6] \cdot zH_2O$ ($T_c = 17.9$ K). The Reverse Monte Carlo analysis, combined with the Rietveld refinement technique shows an antiferromagnetic ordering of Mn moments with respect to the Cu as well as the Fe moments [62]. This bistabilty, i.e., a temperature dependent pole reversal exhibited by this compound, can be used for volatile memory applications.

20.6.3
Magnetocaloric Effect

Magnetocaloric effect (MCE) deals with the change in thermal state of a material upon the application of a magnetic field, and has potential in the area of refrigeration. For a gainful cryocooling, one looks for materials with a large change in entropy upon application of even small magnetic fields. As per Maxwell equations, the change in entropy ΔS_m that occurs for different magnitudes of applied fields at initial and final temperatures, can be written as

$$\Delta S_m(T)_{\Delta H} = \int_{H_i}^{H_f} \left[\frac{\partial M(T,H)}{\partial T}\right]_H dH.$$

The hexacyanate molecular magnet, $Cu_{0.73}Mn_{0.77}[Fe(CN)_6] \cdot zH_2O$ exhibited a reversible and bipolar switching of magnetization under low values of applied magnetic field, which leads to the reversal of magnetic entropy change (ΔS_m), causing a bipolar magnetocaloric effect [63]. The existence of both positive and negative MCEs below the magnetic ordering temperature, i.e., a pole reversal of magnetic entropy change ($-\Delta S_m$) can be exploited for thermomagnetic switching or for magnetic cooling/heating-based constant temperature bath applications.

20.6.4
Hepta/Octacyanates

An alternate strategy adopted by us to grow single crystals of PBA's was to employ the heptacyanate anion $[Mo^{III}(CN)_7]^{4-}$ as a precursor because its pentagonal bipyramidal coordination sphere is incompatible with a cubic lattice.

Reaction of Mn^{II} ions with the heptacyanate $[Mo^{III}(CN)_7]^{4-}$ precursor led to lowering of symmetry and high values of Curie temperatures ($T_c = 51$ K) arising from a ferromagnetic interaction between the low-spin Mo^{3+} and the high-spin Mn^{2+} through the Mo^{III}–C– N–Mn^{II} bridges. Additional use of a macrocycle reduced the symmetry further by imposing heptacoordination on Mn^{II} ion, too. But, the crystals of the compound obtained thus, $[Mn^{II}L]_6[Mo^{III}(CN)_7][Mo^{IV}(CN)_8]_2 \cdot 19.5 H_2O$, where L is the macrocycle, had a low value of T_c (3 K), which was attributed to the existence of diamagnetic Mo^{4+} along certain –CN– bridges, in addition to the paramagnetic Mo^{3+} [64]. We also demonstrated the use of octacyanometalates as versatile building blocks [65]. Ohkoshi et al. [66] reported reversible switching between paramagnetic and ferromagnetic states for $Cu^{II}_2[Mo^{IV}(CN)_8] \cdot 8 H_2O$ solid by irradiating it with visible light, which is very encouraging. However, its T_c is low (25 K), and to be practically useful, the T_c value must be raised above room temperature.

20.7
High Spin (HS)–Low Spin (LS) transitions. Spin Crossover (SCO)

The phenomenon of spin crossover (SCO) deals with bistability at the molecular level, and is found among the $3d^4$–$3d^7$ coordination compounds, particularly for the Fe^{II} compounds which are diamagnetic ($S=0$ for LS) and paramagnetic ($S=2$ for HS state), transitions between which marks not only a magnetic switching as a function of temperature but also is often accompanied by profound optical changes. Consequently, magnetic and optical measurements under external stimuli such as temperature or light irradiation are the common techniques to investigate these spin transitions, though at times the entropy driven SCO can be addressed under the application of pressure, as well. Typically, a SCO molecule consists of a transition metal center surrounded by organic ligands that provide structural stability [67]. The all-important bistability in it stems from a competition between the ligand field favoring an electronic occupation of the levels lowest in energy, the LS state and the spin-pairing energy favoring a parallel alignment of the spins of the d electrons, the HS state [68,69].

Triazole and tetrazole ligands possess the needed ligand-field strengths around the Fe^{II} ion to bring about a SCO, at or even above ambient temperature, which has led to a number of proposed applications, such as bulk sensors and optoelectronic displays and data storage [70]. The SCO polymer $[Fe(Htrz)_2(trz)](BF_4) \cdot H_2O$, with Htrz = 1,2,4-4$H$-triazole and trz = the deprotonated triazolato ligand, displays a spin transition around 385 K in the warming mode and 345 K in the cooling mode. Even particles of this material coated with gold show quite good hysterisis, needed for switching applications [71].

Tiny molecular magnetic switches that can be controlled optically, is an exciting proposition. To learn about the mechanism of a SCO transition, ultrafast phototransformations are commonly studied as a phenomenon now called, light induced excited-spin state trapping (LIESST) [68,72]. Time-resolved studies of ultrafast light-induced spin-state switching, triggered by a femtosecond laser flash, and the following out-of-equilibrium dynamics in Fe^{III} spin-crossover crystals has been reviewed recently [73]. It has been established that the response of a SCO material to femtosecond excitation is a multistep process [74].

Interesting new findings among the light-switchable materials are: a bistable CoFe-PBA@CrCr-PBA hexacynaometallate heterostructure, designed from Prussian blue anologue-based coordination polymers, which shows reversible photomagnetic switching upto well above liquid nitrogen [75]; and an Fe-free polymer chain copper nitroxide-based molecular magnet $[Cu(hfac)_2L^{Pr}]$ (hfac = hexafluoroacetylacetonate, L^{Pr} = nitronyl nitroxide material) that exhibits ultrafast (<50 fs) spin-state photoswitching [76].

20.8
Design of Molecular Magnets Using AZIDO Bridge to Mediate Spin–Spin Couplings

For versatility in coordination, the azide (N^{3-}) ligand is used a lot to mediate magnetic couplings, since an azido bridge can link two or more metal ions in end-to-end (EE), or end-on (EO), equally effectively. An end-to-end coordination of the azido group results in moderate to strong antiferromagnetic coupling whereas an end-on coordination gives rise to ferromagnetic coupling between paramagnetic centers (Figure 20.8).

Nearly, all Mn(II) azide polymetallic complexes reported in literature had the bridging of two metal ions by two azido groups, whether in EE or EO modes. We could synthesize a *trans* coordinated azido-Mn(II) chain compound, $[Mn(L)(N_3)PF_6]_n$, where L = 2,13-dimethyl-3,6,9,12,18-pentaazabicyclo{12.3.1}octadeca-1,(18),2,12,14,16-pentaene), the crystal structure of which has 1-D arrays of Mn(II) units linked by single end-to-end azido bridges which remain isolated from the neighboring chains, and the compound exhibits long-range antiferromagnetic order below 30 K [77].

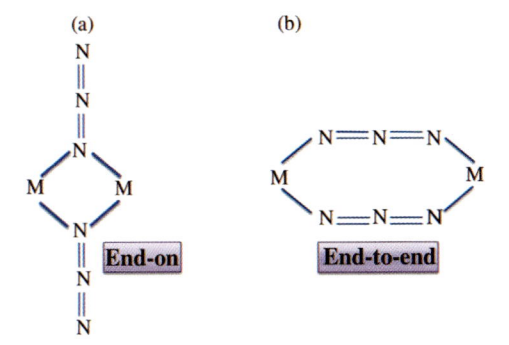

Figure 20.8 The azide-ligand, N_3, which can act as an (a) end-on bridge; or (b) an end-to-end bridge, between two spin-bearing metal-ions.

20.9
Single Molecule Magnets (SMM) and Single Chain Magnets (SCM)

In the early 1990s, the molecular cluster $Mn_{12}O_{12}(CH_3COO)_{16}(H_2O)_4$ was shown to display slow magnetic relaxation at low temperatures, similar to the behavior of a superparamagnet below its blocking temperature [78,79]. Molecules of this type, referred to as single molecule magnets (SMM), exhibit slow relaxation because of an energy barrier to magnetization reversal with a spin ground state, which is large (e.g., $S=10$ in the case of Mn_{12}-acetate), and retain their magnetization at the molecular level, after removal of an applied field, as single-domain nanoparticles exhibiting slow relaxation of the magnetization and magnetic hysteresis below a blocking temperature (T_B).

After Mn_{12}, other SMMs were discovered, such as $Fe_8O_2(OH)_{12}(C_6N_3H_{15})_6 Br_7(H_2O)]Br(H_2O)_8$, called Fe_8 [80], and also the 3d–4f single molecule magnets [81]. All SMMs were found to behave as "typical" magnets at a "molecule" level, being able to retain magnetization after removal of an applied field below the blocking temperature, T_B. The hysteresis loop M versus H always shows the existence of slow magnetic relaxations, that is, the slow dynamics of magnetization, and therefore, these class of materials are called "magnets". Owing to the small size of these samples, there can be no domain walls present as in the case of bulk ferro- or ferri-magnets, and one finds a single magnetic domain. Owing to their magnetic anisotropy, the slow relaxation of the magnetization comes from the existence of an energy barrier.

Their magnetization relaxation time being unusually long below T_B, a memory device can be constructed using SMMs, where the information is contained in the magnetization direction of SMMs, and the on/off states are achieved by switching their magnetization direction. Crystals of SMMs function as collections of monodisperse nanomagnets, unlike other nanoscale magnetic materials that have size distribution. As magnetic aggregates, the SMMs are examples of "zero-dimensional" systems, though essentially mesoscopic, and therefore, lie on the border between the classical and quantum regimes, serving as new sources of magnetic phenomena including quantum tunneling, quantum coherence, QuBits, and magnetocaloric effect, with potential applications in high-density information storage devices, quantum computing, and in spin-valves, etc. [82–85]. SMMs provide a frontier to store information at a molecular level. They possess well-defined quantum states and long coherence time, both suited for this job. Besides, the spin-orbit and hyperfine interactions of organic molecules are weak which are conducive to preserve electron-spin polarization over a much greater distance than found in conventional semiconductor systems. A spin-polarized current, passing through a SMM, should allow the reading and reversing of its magnetization, hence are suited to fabrication of spintronic devices, with switching possibilities using light, electric field, and so on.

More recently, there is also much interest in what is known as single-chain magnets (SCM), which are ferro- or ferrimagnetic chains, and mostly Ising systems. One-dimensional chains cannot, in principle, exhibit long-range magnetic order. However, polymer chains comprising anisotropic TM-ions with alignment of unpaired spins ferro- or ferrimagnetically are known to show slow magnetic relaxations and single-chain magnet behavior [86]. A typical example of SCMs is the cobalt(II)-organic radical polymer Co(*hfac*)$_2$NITPhOMe, nicknamed "CoPhOMe" [87]. 1D coordination polymers composed of Mn porphyrin radical molecule pairs have been studied for their interesting properties [88,89].

Molecular chains of antiferrimagnetically coupled Mn^{III}-ion ($S=2$) in the center of the porphyrin disc and TCNE (tetracyanoethylene) radical moments ($S=\frac{1}{2}$) exhibit a variety of behavior decided by the site where the R=F is substituted to TPP (tetraphenylporphyrin) at the anisotropic chain 1D [MnR$_4$TPP][TCNE], or if the substituted group is bulky. The compound manganese^{III}tetra(*ortho*-fluorophenyl)porphyrin-tetracyanoethylene with F in *ortho* position is a single-chain magnet showing slow relaxations with blocking temperature of 6.6 K [90], while that with R=F in *meta* position shows not only a blocking below 5.4 K but also a long-range magnetic ordering below 10 K [91]. For bulky groups R=OC$_n$H$_{2n+1}$, a state with magnetic order appears below $T_c \approx 22$ K, and slow relaxation is observed below 8 K, and hysterisis at 2.3 K points to a SCM-like behavior [92].

There have been extensive efforts to look for large enough magnetocaloric effect in molecular magnets [93]. While the giant magnetocaloric effect (MCE) in intermetallic compounds is related to the interplay between long-range magnetic and lattice order [94], molecular nanomagnets have recently shown superior cooling performances at cryogenic temperatures. The molecular cage, Fe$_{14}$(bta)$_6$ (bta = benzotriazole) [95], was one of the first examples for which enhanced MCE was experimentally observed in bulk samples [96,97]. Corradini et al. [98] have reported that a large MCE for Fe$_{14}$(bta)$_6$ molecular nanomagnet is a property inherent at the single molecule level.

Magnetic molecules with high-spin ground states and small anisotropies have emerged as good candidates for cryocooling. Using 3d/4f chemistry to construct [Cu$^{II}_5$Gd$^{III}_4$] clusters, a magnetic refrigerant, [Cu$^{II}_5$Gd$^{III}_4$O$_2$(OMe)$_4$(teaH)$_4$(O$_2$CC(CH$_3$)$_3$)$_2$(NO$_3$)$_4$].2MeOH.2Et$_2$O(1.2MeOH.2Et$_2$O), where teaH$_3$ stands for triethanolamine was fabricated for low temperature applications [99]. It shows a huge MCE with $-\Delta S_m$ reaching record values of 30J/(kgK). Peng et al. [100] have also proposed several high-nuclearity 3d–4f clusters as potential magnetic coolers.

The difference between the SMMs and SCMs is that for SMMs the slow relaxation of magnetization results from isolated anisotropic clusters, the intrinsic magnetic anisotropy of which cannot be easily tuned. On the other hand, for the SCM materials, the origin of the slow relaxation of the magnetization lies in the magnetic interactions prevailing between anisotropic repeating units along a single chain. The magnetic dynamics of such chains was first described by Glauber in the framework of the Ising model [101], and the SCM behavior arises from the presence of a short-range order along the chains, which brings about a slowing down of the spin dynamics as a function of temperature. And since intrachain interaction can be tuned through clever synthesis strategies involving substitution and so on, SCM are emerging as potential candidates for information storage. An excellent account of the properties of SCMs is provided in review articles [102,103].

20.10
Molecular Spintronics

At present, the charge of electrons is employed for transporting, storing, or processing information in communications and information technologies. But the electron has a charge as well as a spin, which too can be used to store and process information, because a spin can get influenced if transported through a magnetic material, or by the application of a magnetic field. The spin-based electronics, that is, spintronics, therefore, is based on the detection of a response of spins to an external stimulus, which allows one to implement logic, memory, and sensing capabilities.

Nobel prize winning discovery of giant magnetoresistance (GMR) by Albert Fert and Peter Gruenberg [104,105] brought about the fulfilments of increased data storage density, and read-head devices for computer disk drives. Spintronics-based applications already exist now, such as, the read-head used in a modern hard disk, and, the magnetic random access memory (MRAM), which obviates the need to have a voltage applied to preserve the orientation of a magnet, unlike in a conventional RAM.

A long-spin lifetime is expected from the very small spin-orbit coupling in organic semiconductors. The development of the subject of molecular spintronics, which implies manipulating spins and charges in molecular electronic devices is, therefore, a natural extension to molecular electronics by combining it with the principles of spintronics. Continued progress in the synthesis of molecular magnetic materials with high Curie temperatures provides a further impetus to the development of molecular spintronics, with a goal to achieve powerless nonvolatility in devices made from molecular magnets.

The first ever organic spintronic device was reported in 2002 and employed a planar structure of two LSMO ($La_{2/3}Sr_{1/3}MnO_3$) electrodes separated by an ≈ 100 nm long channel of α-sexithiophene, which showed a spin-diffusion length of $\lambda s \sim 70$ nm, thus demonstrating the phenomenon of spin injection into organic semiconductors [106]. Molecular spintronics research picked up momentum with further publications on devices like an organic spin valve using organic molecules by Vardeny's group [107] as well as magnetic tunnel junctions using organic semiconducting materials as the spacer layer [108]. The simplest OSV consists of a diamagnetic molecule in between two magnetic leads, which can be metallic or semiconducting, namely, a C_{60} (fullerence) sandwiched between Ni electrodes, showing a very large negative magnetoresistance effect [109]. Whereas, the long spin relaxation times [110] in organics are promising, their weakness is low mobility, often less than 0.1 $cm^2/(Vs)$.

Using LSMO/organic semiconductor-based heterojuctions, Epstein's group has employed the room temperature organic-based magnet V(TCNE)$_2$ as an electron spin polarizer in the standard spintronics device geometry to fabricate magnetic tunnel junctions, and showed that molecule/organic-based magnetic material can function as a spin injector/detector for spintronics [111], and have also observed giant magnetoresistance (GMR) effects via injection and transport of spin polarized carriers and transport through the rubrene (C$_{42}$H$_{28}$) as organic semiconductor layer [112]. Spin polarization has also been reported for transition-metal phthalocyanines [113], and magnetotransport investigated for polyaniline nanofiber networks [114].

Interestingly, the SMMs show long spin coherence and spin relaxation times [115], making it desirable to incorporate them into electric circuits [116]. Spin-valve devices using TbPc$_2$ single-molecule magnets (Pc = phthalocyanine) have been fabricated [117,118].

Molecular spintronics is an emerging multidisciplinary subject located at the cross-section of spintronics with molecular electronics and molecular magnetism, and excellent reviews/popular articles are now available on a variety of topics related to molecular spintronics, such as, on graphene spintronics [119]; on organic spin valve devices [120]; organic spintronics [121,122]; polymer spintronics [123]; single-molecule spintronics [124]; magnetoresistance in molecular wires [125]; molecular spintronics with SAMs [126]; and general reviews [127–130].

20.11
Multifunctional Molecular Magnetic Materials

Apart from the fact that a molecule is the ultimate unit for data-storage, the design of molecular magnets has also opened the doors for the unique possibility of designing poly-functional materials, at the molecular level such as those exhibiting ferromagnetic behavior and second-order optical nonlinearity [131], or, liquid crystallinity and magnetism [132]. A three-dimensional chiral, transparent ferrimagnet, K$_{0.4}$[Cr(CN)$_6$][Mn(S)-*pn*](S)-*pn*H$_{0.6}$ with $T_c = 53$ K, where (S)–*pn* stands for (S)-1,2-diaminopropane has been reported [133], and since then there have been many additions to the general class of bifunctional molecular magnets.

Magnetic zeolitic structures, too, offer excellent conditions to encapsulate different functional systems with conducting optical, chiral, and NLO properties, among others. It is desirable, therefore, to evolve a synergism of magnetic and nanoporous behavior, together with the molecular characteristics of coordination polymers, as a new route to the development of multifunctional molecular materials. Veciana and co-workers [134,135] have fabricated a nanoporous molecular magnet, Cu$_3$(PTMTC)$_2$(*py*)$_6$(CH$_3$CH$_2$OH)$_2$(H$_2$O), (PTMTC being a radical, namely, polychlorinated triphenylmethyl with three carboxylic groups) with reversible solvent-induced mechanical and magnetic properties. This compound has an open framework structure, with very large pores (2.8–3.1 nm) and bulk magnetic ordering. Reversible behavior has only been observed for ethanol and methanol, showing the selectivity of this sponge-like magnetic sensor.

Wang et al. [136] have reported a family of porous magnets of $[M_3(HCOO)_6]$ where M = TM ion, which exhibit a wide spectrum of guest inclusion behavior as well as 3D long-range magnetic ordering.

Technology of nonlinear properties and complex electrodynamics can lead to new paradigms of applications if electronic ferroelectricity can be developed in low-dimensional (1D or 2D) molecular solids [137,138]. Coexistence of ferroelectricity and ferromagnetism ($T_c = 11$ K) was reported in $Rb^I_{0.82}Mn^{II}_{0.20}Mn^{III}_{0.80}[Fe^{II}(CN)_6]_{0.80}[Fe^{III}(CN)_6]_{0.14} \cdot H_2O$ [139], with ferroelectricity arising from a mixing of Fe^{II}, Fe^{III}, Fe vacancies, Mn^{II}, and Jahn–Teller distorted Mn^{III} centers, and ferromagnetism from a parallel ordering of the magnetic spins on the Mn^{III} centers.

Magnetic spin ordering and magnetoelectric coupling, i.e., multiferroicity, simultaneously at room temperature has been reported recently in three-dimensional charge transfer compounds [140]. These were designed from crystallized thiophene nanowire donor and fullerene acceptor, and demonstrate charge-driven magnetism, with anisotropy. Similarly, supramolecular chemistry using self-assembly methods and noncovalent interactions has been shown to have potential to produce useful organic ferroelectric phases with high figures of merit, which may even operate at room temperature [141].

20.12
Biogenic Magnetism

Most organisms do not respond to magnetic fields, and are diamagnetic. Magnetotactic bacteria and migratory animals are among organisms that exploit magnetism. Both can sense geomagnetic fields and alter their behavior accordingly. Apart from academic interest in it, biogenic magnetization has industrial potential, too. Because of the contactless, remote, and permeable nature of magnetic interactions, they can be integrated into biological systems, and thus lead to a new dimension to biomedical engineering and therapy, and provide a unique interface between cells.

20.12.1
Magnetotactic Bacteria

Aquatic magnetotactic bacteria (MTB) evolved on Earth 2 billion years ago. Self-assembly processes of nature are much evident since monodisperse nanoscale single-domain magnetite crystals are synthesized by magnetotactic bacteria, which collectively behave as bar magnets, imparting the bacteria an ability to orient and migrate along geomagnetic field lines. Magnetotactic bacteria are able to mineralize, with precise biological control, pure magnetic nanocrystals of magnetite (Fe_3O_4), or greigite (Fe_3S_4), with specific crystal morphology, and narrow size distribution. The magnetic nanoparticle is enveloped by a 3–4 nm thick phospholipid bilayer membrane with embedded proteins. The nanocrystal and its enveloping membrane are together known as a magnetosome [142]. Magnetosomes are organelles produced by a variety of magnetotactic bacteria. Since these magnetosomes self-assemble into a chain-like structure within a cell, they act like dipole magnets and confer a magnetic moment to the cells, thereby helping them to align along Earth's magnetic field as they swim and navigate, the reason why they are called magnetotactic. Magnetotaxis is thought to function in conjunction with chemotaxis to aid MTB in locating and maintaining an optimal position at the oxic–anoxic interface in chemically stratified water columns or sediments, presumably in search of food. Interest in MTB has recently increased because of the potential biomedical applications of the magnetosomes.

In nature, the synthesis of magnetosomes and the production of monodisperse single magnetic domain nanoparticles in them at ambient temperatures involves strict genetic control over intracellular differentiation, biomineralization, and their assembly into highly ordered chains. Several specific genes are known to be participating in different steps of formation of magnetosomes [143].

Inadequate knowledge about the biosynthetic functions involved in the magnetosomes, and their formation, and the structural and genetic complexity of these organelles have so far limited the applications of MTB to diverse nanotechnological and biomedical areas [144]. But this could become possible if the expression of the underlying biosynthetic pathway from these unique microorganisms within other organisms could be established. In an important development, it has been recently shown that the ability to biomineralize highly ordered magnetic nanostructures can be transferred to a foreign recipient [145]. Synthesis of artificial magnetic bacteria has also been tried by attempting to organize superparamagnetic maghemite nanoparticles on the external surfaces of nonmagnetic probiotic bacteria *Lactobacillus fermentum* or *Bifidobacteria breve*, using them as bioplatforms [146].

20.12.2
Magnetoreception

The avian magnetic compass developed over 90 million years ago, enabling pigeons to detect magnetic field changes of around 20 nT, or even lower. The ability of animals to detect geomagnetic fields is used by many invertebrates and vertebrates for compass orientation purposes, including to maintain a given path during long-distance migration [147], using biogenic magnetite or other iron-mineral particles used for navigation or "homing" action [143,148]. Elaborate iron mineral containing dendrites in the upper beak seem to be a common feature of birds, and these provide possible link with avian magnetoreception in chickens, homing pigeons, European robin, garden warbler [149]. Magnetoreception in birds [150] needs chemical detection of the direction of Earth's magnetic field (~50 μT). Besides, the ferromagnetic nanoparticles must be mechanically coupled to nerve fibres or sensory organelles in order that they can transmit magnetic information to the brain of the avian species. Ability of birds to use magnetic field information under different conditions, light intensity and color, magnetic field strength, and presence of oscillating fields, suggests a complex sensory system of multiple receptors.

In birds, two models of magnetoreception have been proposed: a magnetite-based process in which the magnetic particles are located in a bird's beak [151,152]; and chemical-based reactions, a model under which the magnetic information is transmitted to the nervous system through the light-induced product of magnetically sensitive radical-pair reactions in specialized photoreceptors in the bird's eye [153]. Until now, the only molecules found to have such photoreceptor characteristics are the cryptochromes (CRY) [154,155]. The first genetic evidence for a cryptochrome-based magnetosensitive system in animals has been obtained in the fruit fly *Drosophila melanogaster* [156].

What is the radical pair mechanism (RPM)? Molecules that have an odd number of electrons have consequently an unpaired electron spin. A radical pair is a short-lived reaction intermediate comprising two radicals formed in tandem whose unpaired electron spins may be either antiparallel (a singlet state, S) or parallel (a triplet state, T). Each electron spin has an associated magnetic moment, the minimum requirement for a radical pair reaction to be sensitive to an external magnetic field is that at least one of the S and T states undergoes a reaction that is not open to the other. Luckily, the phenomenon of the initial charge separation steps of bacterial photosynthetic energy conversion does provide a precedence for a magnetically sensitive radical pair chemistry of this kind. The limited information currently available suggests that cryptochrome radical pairs could, in principle, form the basis of a compass magnetoreceptor. What is not yet understood is the mechanism of how the magnetic field modulated signaling of cryptochrome enters into a bird's sensory perception and further into its orientation behavior [157].

To date, neither physiological nor molecular mechanisms for the formation of such magnetic particles in nerve tissue are understood [158]. However, several avian species, such as, European robin, the loggerhead sea turtle, the brown bat, the Caribbean spiny lobster, and the red-spotted newt have a light-dependent compass in the eye, with evidence that it is based on the RPM. In certain cases, including turtles, lobsters, and newts, this is in addition to magnetite, whereas in salamanders, frogs, lizards, and some fish the light-dependent MF compass is housed in the pineal gland.

Whereas, specific genes are crucial for the formation of magnetic particles in magnetotactic bacteria, no such genes have yet been characterized in migratory animals. In humans, formation of magnetic particles can be observed in the neuronal tissue in neurodegenerative diseases.

20.13
Other Recent Developments, Current scenario, and Future Perspectives

A high-efficiency thermally regenerative electrochemical cycle (TREC) has been presented [159] for harvesting low grade heat energy by employing solid copper hexacyanoferrate (CuHCF) as a positive electrode and Cu/Cu^{2+} as a negative electrode in an aqueous electrolyte.

^{137}Cs is among the main fission product in radioactive wastes. Nanoparticle film of copper hexacyanoferrate (CuHCF^{III}) has been developed for cesium separation from wastewater, electrochemically [160].

First rechargeable Ca-ion battery concept has been demonstrated recently by Lipson et al. [161], utilizing manganese hexacyanoferrate (MFCN) as the cathode to intercalate Ca reversibly in a dry nonaqueous electrolyte. MFCN maintains its crystallinity during cycling, with only minor structural changes associated with expansion and contraction.

Because of the presence of noncoordinated water molecules in PBAs, which can escape from the lattice upon slight warming, creating voids, hydrogen storage characteristics have been exhibited by core–shell nanoparticles (av. size ~25 nm) formed from $Mn_{1.5}[Cr(CN)_6]\cdot 7.5\,H_2O$ as the core, surrounded by a shell of $Ni_{1.5}[Cr(CN)_6]\cdot 7.5\,H_2O$ [162].

Magnets fabricated from molecules are unique to magnetism. They exhibit all the phenomena observed in conventional transition metal and rare earth-based magnets, and more, by exhibiting liquid crystallinity, chirality, low density, solubility, spin-crossover [163], negative magnetization [29,62,164], photoinduced magnetization/switching due to their colored nature [165–167], and possibility of modulation of their properties electrochemically, all of which are arising from their characteristic molecular nature. In this context, an exciting development is the fabrication of a trifunctional phenalenyl-based neutral radical, by Itkis et al. [168], which exhibits magnetooptoelectronic bistability with hysteresis loops centered near 335 K, thus opening the possibility for new type of electronic devices, where multiple physical channels can be used for writing, reading, and transferring information.

Some groups are attempting to utilize the high degree of directionality of hydrogen bonding between open-shell molecules to obtain supramolecular self-organization aimed at achieving new molecular ferromagnets. There is also a substantial activity and progress in the development of spin-crossover compounds that exhibit magnetic bistability [163,169], which is at times accompanied by a change of color, too. Then, there is a continuing quest for discovering new molecular materials which can exhibit long-range magnetic order at room temperature (or even above), the earliest example of which was $V[TCNE]_x \cdot y CH_2Cl_2$ ($x \sim 2$; $y \sim 1/2$), a disordered, amorphous magnet depicting a T_c of ~ 400 K, but extremely water/air sensitive [170]. Jain et al. [171] recently reported the synthesis of three air-stable Ni_2A-based metal-organic magnets (T_c above room temperature), by reacting a metal (M) precursor complex bis(1,5-cyclooctadiene)nickel with organics A = TCNE (tetracyanoethylene), TCNQ (7,7,8,8-tetracyanoquinodimethane) or DDQ (2,3-dichloro-5,6-dicyano-1,4-benzoquinone). A structural model for these materials is not available, though spectroscopic features suggested that the organic ligands are σ-bound to nickel. However, subsequent study by Miller and Pokhodnya [172] has negated these findings, and proposed that the black powdery magnetic material made upon dissolution of $Ni(COD)_2$ in CH_2Cl_2 presumably consists of "nano"- or bigger-sized particles of nickel metal, and lacks the "organic" species which may entitle it to be called a molecular magnet.

A high T_C value of 210 K has been reported [173] for a vanadium octacyanoniobate ferrimagnet with a high coordination number, $K_{0.59}V^{II}_{1.59}V^{III}_{0.41}[Nb^{IV}(CN)_8]\cdot(SO_4)_{0.50}\cdot 6.9 H_2O$, owing to a strong superexchange interaction between V^{II} ($S = 3/2$) and Nb^{IV} ($S = 1/2$) through the CN groups.

Miller's group [174] have shown that $Mn^{II}(TCNE)[C_4(CN)_8]_{1/2}$ (TCNE = tetracyanoethylene) exhibits a reversible pressure-induced piezomagnetic transition from a low magnetization antiferromagnetic state to a high magnetization ferrimagnetic state above 0.50 ± 0.15 kbar. In the ferrimagnetic state, the critical temperature, T_c, increases with increasing hydrostatic pressure and is ~97 K at 12.6 kbar, the magnetization increases by three orders of magnitude (1000-fold), and the material becomes a hard magnet with a significant remnant magnetization.

Unique combination of tunability, processability, and other attributes of molecular lattices hold promise for molecule-based magnets for future applications, but much progress has to be made before that happens because of certain vexing issues like their air stability, degradability (in some cases), and processability, which are some phenomena unique to molecular lattices.

Acknowledgments

The author has benefitted enormously by his collaborations with the late Prof. Olivier Kahn, a pioneer in the area of molecular magnetism. The author has also benefitted from interaction with Prof. K Inoue and Dr. M.D. Sastry, and would, in addition like to place on record his deep appreciation for the contributions of his colleagues and past students Drs. S.M. Yusuf, Sipra Choudhury, G.K. Dey, M. D. Mukadam, Amit Kumar, S.A. Chavan, Aman K. Sra, Prasanna Ghalsasi, Nitin Bagkar, Nidhi Sharma on which this chapter is based. Expert help of Dr. Ajay Singh is acknowledged toward designing some of the figures used in this chapter.

References

1 Miller, J.S., Calabrese, J.C., Epstein, A.J., Bigelow, R.B., Zang, J.H., and Reiff, W.M. (1986) Ferromagnetic properties of one-dimensional decamethylferrocenium tetracyanoethylenide (1 : 1):

[Fe(η^5-C$_5$Me$_5$)$_2$]$^{\bullet+}$[TCNE]$^{\bullet-}$. *J. Chem. Soc. Chem. Commun.*, 1026–1028.

2 Pei, Y., Verdaguer, M., Kahn, O., Sletten, J., and Renard, J.P. (1986) Ferromagnetic transition in a bimetallic molecular system. *J. Am. Chem. Soc.*, **108**, 7428–7430.

3 Kahn, O. (1993) *Molecular Magnetism*, Wiley-VCH Verlag GmbH, New York.

4 Itoh, K. and Kinoshita, M. (eds) (2000) *Molecular Magnetism*, Gordon and Breach Science Publishers, Amsterdam.

5 Miller, J.S. (2000) Organometallic- and organic-based magnets: new chemistry and new materials for the new millennium. *Inorg. Chem.*, **39**, 4392–4408.

6 Yakhmi, J.V. (2003) Self-assembly of nano-scale magnets: molecule upwards, in *Proc. of INAE Conference on Nanotechnology* (ed. M.J. Zarabi), Indian National Academy of Engineering, (ICON-2003), pp. 78–87.

7 Yakhmi, J.V. (2009) Molecule-based magnets. *Bull. Mater. Sci.*, **32**, 217–225.

8 Kahn, O. (ed.) (1996) Magnetism: a supramolecular function, in *NATO ASI Series C*, vol. **484**, Kluwer, Dordrecht, Netherlands.

9 Ovchinnikov, A.A. (1978) Multiplicity of the ground state of large alternant organic molecules with conjugated bonds. *Theor. Chim. Acta*, **47**, 297–304.

10 Rajca, S., Wongsriratanakul, J., and Rajca, S. (2001) Magnetic ordering in an organic polymer. *Science*, **294**, 1503–1505.

11 Arita, R., Suwa, Y., Kuroki, K., and Aoki, H. (2002) Gate-induced band ferromagnetism in an organic polymer. *Phys. Rev. Lett.*, **88**, 127202-4.

12 McConnell, H.M. (1963) Ferromagnetism in solid free radicals. *J. Chem. Phys.*, **39**, 1910–1910.

13 Tamura, M., Nakazawa, Y., Shiomi, D., Nozawa, K., Hosokoshi, Y., Ishikawa, M., Takahashi, M., and Kinoshita, M. (1991) Bulk ferromagnetism in the β-phase crystal of the *p*-nitrophenyl nitronyl nitroxide radical. *Chem. Phys. Lett.*, **186**, 401–404.

14 Allemand, P., Khemani, K., Koch, K., Wudl, F., Holczer, K., Donovan, S., Gruner, G., and Thompson, J.D. (1991) Organic molecular soft ferromagnetism in a fullerene C$_{60}$. *Science*, **253**, 301–302.

15 Mrzel, A., Omerzu, A., Umek, P., Mihailovic, D., Jaglicic, Z., and Trontelj, Z. (1998) Ferromagnetism in a cobaltocene-doped fullerene derivative below 19 K due to unpaired spins on fullerene molecules. *Chem. Phys. Lett.*, **298**, 329–334.

16 Banister, A.J., Bricklebank, N., Lavender, I., Rawson, J.M., Gregory, C.I., Tanner, B.K., Clegg, W., Elsegood, M.R.J., and Palacio, F. (1996) Spontaneous magnetization in a sulfur–nitrogen radical at 36 K. *Angew. Chem., Int. Ed.*, **35**, 2533–2535.

17 Palacio, F., Antorrena, G., Castro, M., Burriel, R., Rawson, J., Smith, J.N., Bricklebank, N., Novoa, J., and Ritter, C. (1997) High-temperature magnetic ordering in a new organic magnet. *Phys. Rev. Lett.*, **79**, 2336–2339.

18 Mito, M., Kawae, T., Takeda, K., Takagi, S., Matsushita, Y., Deguchi, H., Rawson, J.M., and Palacio, F. (2001) Pressure-induced enhancement of the transition temperature of a genuine organic weak-ferromagnet up to 65 K. *Polyhedron*, **20**, 1509–1512.

19 Kahn, O. (1995) Magnetism of heterobimetallics: towards molecule-based magnets. *Adv. Inorg. Chem.*, **43**, 179–259.

20 Nakatani, K., Carriat, J.V., Journaux, Y., Kahn, O., Lloret, F., Renard, J.P., Pei, Y., Sletten, J., and Verdaguer, M. (1989) Chemistry and physics of the novel molecular-based compound exhibiting a spontaneous magnetization below $T_c = 14$K, MnCu(obbz)·1H$_2$O (obbz = oxamidobis(benzoato)). Comparison with the antiferromagnet MnCu(obbz)·5H$_2$O. Crystal structure and magnetic properties of NiCu(obbz)·6H$_2$O. *J. Am. Chem. Soc.*, **111**, 5739–5748.

21 Okawa, H., Mitsumi, M., Ohba, M., Kodera, M., and Matsumoto, N. (1994) Dithiooxalato(dto)-bridged bimetallic assemblies {NPr$_4$[MCr(dto)$_3$]}$_x$ (M = Fe, Co, Ni, Zn; NPr$_4$ = tetrapropylammonium ion): new

complex-based ferromagnets. *Bull. Chem. Soc. Jpn.*, **67**, 2139–2144.

22 Caneschi, A., Gatteschi, D., Sessoli, R., and Rey, P. (1989) Toward molecular magnets: the metal radical approach. *Acc. Chem. Res.*, **22**, 392–398.

23 Broderick, W.E., Thompson, J.A., Day, P., and Hoffman, B.M. (1990) A Molecular ferromagnet with a curie temperature of 6.2 kelvin: $[Mn(C_5(CH_3)_5)_2]^+[TCNQ]^-$. *Science*, **249**, 401–403.

24 Inoue, K. and Iwamura, H. (1994) Ferro- and ferrimagnetic ordering in a two-dimensional network formed by manganese(II) and 1,3,5-tris[p-(N-tert-butyl-N-oxyamino)phenyl]benzene. *J. Am. Chem. Soc.*, **116**, 3173–3174.

25 Inoue, K., Hayamizu, T., Iwamura, H., Hashizume, D., and Ohashi, Y. (1996) Assemblage and alignment of the spins of the organic trinitroxide radical with a quartet ground state by means of complexation with magnetic metal ions. A molecule-based magnet with three-dimensional structure and high T_C of 46 K. *J. Am. Chem. Soc.*, **118**, 1803–1804.

26 Kahn, O., Pei, Y., Verdaguer, M., Renard, J.P., and Sletten, J. (1988) Magnetic ordering of manganese(II) copper(II) bimetallic chains; design of a molecular based ferromagnet. *J. Am. Chem. Soc.*, **110**, 782–789.

27 Stumpf, H.O., Pei, Y., Kahn, O., Sletten, J., and Renard, J.P. (1993) Dimensionality of manganese(II)-copper(II) bimetallic compounds and design of molecular-based magnets. *J. Am. Chem. Soc.*, **115**, 6738–6745.

28 Pei, Y., Kahn, O., Nakatani, K., Codjovi, E., Mathoniere, C., and Sletten, J. (1991) Design of a molecular-based ferromagnet through polymerization reaction in the solid state of manganeseII copperII molecular units. Crystal structure of $MnCu(obze)(H_2O)_4 \cdot 2H_2O$ (obze = oxamido-N-benzoato-N'-ethanoato). *J. Am. Chem. Soc.*, **113**, 6558–6564.

29 Chavan, S.A., Ganguly, R., Jain, V.K., and Yakhmi, J.V. (1996) Magnetization behavior of $(NBu_4)_2Mn_2[Cu(opba)]_3$ and related solvated ferromagnets. *J. Appl. Phys.*, **79**, 5260–5262.

30 Neels, A., Stoeckli-Evans, H., Chavan, S.A., and Yakhmi, J.V. (2001) $\{(NBu_4)_2Mn[Cu(opba)]_2\}n$: a new structural class among 'opba' bimetallic magnets. *Inorg. Chim. Acta*, **326**, 106–110.

31 Chavan, S.A., Yakhmi, J.V., and Gopalakrishnan, I.K. (1995) Magnetism in binuclear compounds MCu[(obbz)·nH_2O with M = Mn or Co. *Mol. Cryst. Liq. Cryst.*, **274**, 11–16.

32 Larionova, J., Chavan, S.A., Yakhmi, J.V., Froystein, A.G., Sletten, J., Sourisseau, C., and Kahn, O. (1997) Dramatic modifications of magnetic properties through dehydration-rehydration processes of the molecular magnetic sponges $CoCu(obbz)(H_2O)_4 \cdot 2H_2O$ and $CoCu(obze)(H_2O)_4 \cdot 2H_2O$, with obbz = N,N'-Bis(2-carboxyphenyl)oxamido and obze = N-(2-Carboxyphenyl)-N'-(carboxymethyl)oxamido. *Inorg. Chem.*, **36**, 6374.

33 Kahn, O., Larionova, J., and Yakhmi, J.V. (1999) Molecular magnetic sponges. *Chem. Eur. J.*, **5**, 3443–3449.

34 Sra, A.K., Tokumoto, M., and Yakhmi, J.V. (2001) Magnetic studies on FeIICu (obbz)·1H2O, where obbz = N,N'-oxamido bis(benzoato). *Philos. Mag. B*, **81**, 477–487.

35 Stumpf, H.O., Pei, Y., Kahn, O., Ouahab, L., and Grandjean, D. (1993) A Molecular-based magnet with a fully interlocked three-dimensional structure. *Science*, **261**, 447–449.

36 Stumpf, H.O., Ouahab, L., Pei, Y., Bergerat, P., and Kahn, O. (1994) Chemistry and physics of a molecular-based magnet containing three spin carriers, with a fully interlocked structure. *J. Am. Chem. Soc.*, **116**, 3866–3874.

37 Stumpf, H.O., Pei, Y., Michaut, C., Kahn, O., Renard, J.P., and Ouahab, L. (1994) Bimetallic molecular-based magnets with large coercive fields. *Chem. Mater.*, **6**, 257–259.

38 Vaz, M.G.F., Pinheiro, L.M.M., Stumpf, H.O., Alcantara, A.F.C., Golhen, S., Ouahab, L., Cador, O., Mathoniere, C., and Kahn, O. (1999) Soft and hard molecule-based magnets of formula

[(Etrad)$_2$M$_2${Cu(opba)}$_3$]·S [Etrad$^+$ = radical cation, M^{II} = Mn^{II} or Co^{II,} opba = *Ortho*-phenylenebis(oxamato), S = solvent molecules], with a fully interlocked structure. *Chem. Eur. J.*, **5**, 1486–1495.

39 Turner, S., Kahn, O., and Rabardel, L. (1996) Crossover between three-dimensional antiferromagnetic and ferromagnetic states in Co(II)Cu(II) ferrimagnetic chain compounds. A new molecular-based magnet with $T_c = 38$ K and a coercive field of 5.66×10^3 Oe. *J. Am. Chem. Soc.*, **118**, 6428–6432.

40 Chavan, S.A., Larionova, J., Kahn, O., and Yakhmi, J.V. (1998) Magnetic behaviour of the molecular sponge CoCu(pba)(H$_2$O)$_3$·2H$_2$O, where pba = 1,3-propylenebis(oxamato). *Philos. Mag. B*, **77**, 1657–1668.

41 Pinkowicz, D., Podgajny, R., Bałanda, M., Makarewicz, M., Gaweł, B., Łasocha, W., and Sieklucka, B. (2008) Magnetic spongelike behavior of 3D ferrimagnetic {[Mn^{II}(imH)]$_2$[Nb^{IV}(CN)$_8$]}$_n$ with $T_c = 62$ K. *Inorg. Chem.*, **47**, 9745–9747.

42 Li, Y., Huang, G., Zhang, X., Li, B., Chen, Y., Lu, T., Lu, T.J., and Xu, F. (2013) Magnetic hydrogels and their potential biomedical applications. *Adv. Funct. Mater.*, **23**, 660–672.

43 Ludi, A. and Gudel, H.U. (1973) Structural chemistry of polynuclear transition metal cyanides. *Struct. Bond.*, **14**, 1–21.

44 Thakur, N., Yusuf, S.M., and Yakhmi, J.V. (2010) Structural disorder in alkaline earth metal doped Ba$_x$Mn[Fe(CN)$_6$]$_{2(x+1)/3}$·zH$_2$O molecular magnets: a reverse Monte Carlo study. *Phys. Chem. Chem. Phys.*, **12**, 12208–12216.

45 Entley, W.R. and Girolami, G.S. (1995) High-temperature molecular magnets based on cyanovanadate building blocks: spontaneous magnetization at 230 K. *Science*, **268**, 397.

46 Holmes, S.M. and Girolami, G.S. (1999) Sol–gel synthesis of KV^{II}[Cr^{III}(CN)$_6$]·2H$_2$O: a crystalline molecule-based magnet with a magnetic ordering temperature above 100 °C. *J. Am. Chem. Soc.*, **121**, 5593–5594.

47 Kumar, A., Yusuf, S.M., Keller, L., Yakhmi, J.V., Srivastava, J.K., and Paulose, P.L. (2007) Variation of structural and magnetic properties with composition in the (Co$_x$Ni$_{1-x}$)$_{1.5}$[Fe(CN)$_6$]·zH$_2$O series. *Phys. Rev. B*, **75**, 224419 (11 pp).

48 Yusuf, S.M., Thakur, N., Kumar, A., and Yakhmi, J.V. (2010) Cyanide-bridged Ru$_x$Ni$_{3(3x/2)}$[Cr(CN)$_6$]$_2$·zH$_2$O molecular magnets: controlling structural disorder and magnetic properties by a 4d ion (Ruthenium) susbtitution. *J. Appl. Phys.*, **107**, 053902 (9 pp).

49 Choudhury, S., Bagkar, N., Dey, G.K., Subramanian, H., and Yakhmi, J.V. (2002) Crystallization of prussian blue analogues at the air-water interface using an octadecylamine monolayer as a template. *Langmuir*, **18**, 7409–7414.

50 Choudhury, S., Dey, G.K., and Yakhmi, J.V. (2003) Growth of cubic crystals of cobalt-hexacyanoferrate under the octadecyl amine monolayer. *J. Cryst. Growth*, **258**, 197–203.

51 Bagkar, N., Choudhury, S., Kim, K.H., Chowdhury, P., Lee, S.I., and Yakhmi, J.V. (2006) Crystalline thin films of transition metal hexacyanochromates grown under Langmuir monolayer. *Thin Solid Films*, **513**, 325–330.

52 Bagkar, N., Ganguly, R., Choudhury, S., Hassan, P.A., and Yakhmi, J.V. (2004) Synthesis of surfactant encapsulated nickel hexacyanoferrate nanoparticles and deposition of their Langmuir–Blodgett film. *J. Mater. Chem.*, **14**, 1430–1436.

53 Bagkar, N., Betty, C.A., Hassan, P.A., Kahali, K., Bellare, J.R., and Yakhmi, J.V. (2006b) Self-assembled films of nickel hexacyanoferrate: electrochemical properties and application in potassium ion sensing. *Thin Solid Films*, **497**, 259–266.

54 Bagkar, N., Choudhury, S., Bhattacharya, S., and Yakhmi, J.V. (2008) DNA-templated assemblies of nickel hexacyanoferrate crystals. *J. Phys. Chem. C*, **112**, 6467–6472.

55 Mukadam, M.D., Kumar, K., Yusuf, S.M., Yakhmi, J.V., Tewari, R., and Dey, G.K. (2008) Spin-glass behaviour in

ferromagnetic Fe[Fe(CN)$_6$]·xH$_2$O nanoparticles. *J. Appl. Phys.*, **103**, 123902.

56 Kumar, A., Yusuf, S.M., and Yakhmi, J.V. (2010) Cu$_{1.5}$[Cr(CN)$_6$]·6.5H$_2$O Nanoparticles: synthesis, characterization and magnetic properties. *Appl. Phys. A*, **99**, 79–83.

57 Shiba, F., Fujishiro, R., Kojima, T., and Okawa, Y. (2012) Preparation of monodisperse cobalt(II) hexacyanoferrate (III) nanoparticles using cobalt ions released from a citrate complex. *J. Phys. Chem. C*, **116**, 3394–3399.

58 Fu, G., Liu, W., Feng, S., and Yue, X. (2012) Prussian blue nanoparticles operate as a new generation of photothermal ablation agents for cancer therapy. *Chem. Commun.*, **48**, 11567–11569.

59 Bhatt, P., Yusuf, S.M., Mukadam, M.D., and Yakhmi, J.V. (2010) Enhancement of Curie temperature in electrochemically prepared crystalline thin films of Prussian blue analogues K$_j$Fe$^{II}_k$[Cr^{III}(CN)$_6$]$_l$·mH$_2$O. *J. Appl. Phys.*, **108**, 023916 (6 pp).

60 Mathoniere, C., Nuttal, C.J., Carling, S.G., and Day, P. (1996) Ferrimagnetic mixed-valency and mixed-metal tris(oxalato) iron(III) compounds: synthesis, structure, and magnetism. *Inorg. Chem.*, **35**, 1201–1206.

61 Ohkoshi, S.I., Abe, Y., Fujishima, A., and Hashimoto, K. (1999) Design and preparation of a novel magnet exhibiting two compensation temperatures based on molecular field theory. *Phys. Rev. Lett.*, **82**, 1285–1288.

62 Kumar, A., Yusuf, S.M., Keller, L., and Yakhmi, J.V. (2008) Microscopic understanding of negative magnetization in Cu, Mn, and Fe based prussian blue analogues. *Phys. Rev. Lett.*, **101**, 207206 (4 pp).

63 Yusuf, S.M., Kumar, A., and Yakhmi, J.V. (2009) Temperature- and magnetic-field-controlled magnetic pole reversal in a molecular magnetic compound. *Appl. Phys. Lett.*, **95**, 182506 (3 pp).

64 Sra, A.K., Andruh, M., Kahn, O., Golhen, S., Ouahab, L., and Yakhmi, J.V. (1999) A mixed-valence and mixed-spin molecular magnetic material: [Mn^{II}L]$_6$[Mo^{III}(CN)$_7$][Mo^{IV}(CN)$_8$]$_2$·19.5H$_2$O. *Angew. Chem., Int. Ed.*, **38**, 2606–2609.

65 Sra, A.K., Rombaut, G., Lahitete, F., Golhen, S., Ouahab, L., Mathoniere, C., Yakhmi, J.V., and Kahn, O. (2000) Hepta/octa cyanomolybdates with Fe^{2+}: influence of the valence state of Mo on the magnetic behavior. *New J. Chem.*, **24**, 871–876.

66 Ohkoshi, S.I., Tokoro, H., Hozumi, T., Zhang, Y., Hashimoto, K., Mathoniere, C., Bord, I., Rombaut, G., Verelst, M., dit Moulin, C.C., and Villain, F. (2006) Photoinduced magnetization in copper octacyanomolybdate. *J. Am. Chem. Soc.*, **128**, 270–277.

67 Guetlich, P. and Goodwin, H.A. (eds) (2004) Spin crossover in transition metal compounds, in *Topics in Current Chemistry*, Springer, Berlin.

68 Halcrow, M.A. (ed.) (2013) *Spin-Crossover Materials: Properties and Applications*, John Wiley & Sons, Ltd, Chichester, UK.

69 Halcrow, M.A. (2013) The foundation of modern spin-crossover. *Chem. Commun.*, **49**, 10890–10892.

70 Roubeau, O. (2012) Triazole-based one-dimensional spin-crossover coordination polymers. *Chem. Eur. J.*, **18**, 15230–15244.

71 Tobon, Y.A., Etrillard, C., Nguyen, O., Létard, J.-F., Faramarzi, V., Dayen, J.-F., Doudin, B., Bassani, D.M., and Guillaume, F. (2012) ResonanceRaman study of spin-crossover [Fe(Htrz)$_2$(trz)](BF$_4$)·H$_2$O particles coated with gold. *Eur. J. Inorg. Chem.*, **2012** (Issue 35), 5837–5842.

72 Bertoni, R., Lorenc, M., Tissot, A., Servol, M., Boillot, M.-L., and Collet, E. (2012) Femtosecond spin-state photoswitching of molecular nanocrystals evidenced by optical spectroscopy, *Angew. Chem., Int. Ed.*, **51**, 7485.

73 Bertoni, R., Lorenca, M., Tissot, A., Boillot, M.-L., and Collet, E. (2015) Femtosecond photoswitching dynamics and microsecond thermal conversion driven by laser heating in FeIII spin-crossover solids. *Coord. Chem. Rev.*, **282–283**, 66–76.

74 Lorenc, M., Hébert, J., Moisan, N., Trzop, E., Servol, M., Buron-Le Cointe, M., Cailleau, H., Boillot, M.L., Pontecorvo, E., Wulff, M., Koshihara, S., and Collet, E. (2009) Successive dynamical steps of photoinduced switching of a molecular Fe(III) spin-crossover material by time-resolved X-ray diffraction. *Phys. Rev. Lett.*, **103**, 028301 (4 pp).

75 Risset, O.N., Brinzari, T.V., Meisel, M.W., and Talham, D.R. (2015) Light switchable magnetism in a coordination polymer heterostructure combining the magnetic potassium chromiumhexacyanochromate with the light-responsive rubidium cobalthexacyanoferrate. *Chem. Mater.*, **27**, 6185–6188.

76 Kaszub, W., Marino, A., Lorenc, M., Collet, E., Bagryanskaya, E.G., Tretyakov, E.V., Ovcharenko, V.I., and Fedin, M.V. (2014) Ultrafast photoswitching in a copper-nitroxide-based molecular magnet, *Angew. Chem. Int. Ed.*, **53**, 10636–10640.

77 Sra, A.K., Sutter, Jean-Pascal, Guionneau, P., Chasseau, D., Yakhmi, J.V., and Kahn, O. (2000) Example of a single *trans*-azido-bridged Mn(II) chain: synthesis, structural and magnetic characteristics. *Inorganica Chim. Acta*, **300–302**, 778–782.

78 Sessoli, R., Tsai, H.-L., Schake, A.R., Wang, S., Vincent, J.B., Folting, K., Gatteschi, D., Christou, G., and Hendrickson, D.N. (1993) High-spin molecules: [$Mn_{12}O_{12}(O_2CR)_{16}(H_2O)_4$]. *J. Am. Chem. Soc.*, **115**, 1804–1816.

79 Sessoli, R., Gatteschi, D., Caneschi, A., and Novak, M.A. (1993) Magnetic bistability in a metal-ion cluster. *Nature*, **365**, 141–143.

80 Delfs, C., Gatteschi, D., Pardi, L., Sessoli, R., Wieghardt, K., and Hade, D. (1993) Magnetic properties of an octanuclear iron(III) cation. *Inorg. Chem.*, **32**, 3099–3103.

81 Ledezma-Gairaud, M., Grangel, L., Aromí, G., Fujisawa, T., Yamaguchi, A., Sumiyama, A., and Sañudo, E.C. (2014) From serendipitous assembly to controlled synthesis of 3d–4f single-molecule magnets. *Inorg. Chem.*, **53**, 5878–5880.

82 Friedman, J.R., Sarachik, M.P., Tejada, J., and Ziolo, R. (1996) Macroscopic measurement of resonant magnetization tunneling in high-spin molecules. *Phys. Rev. Lett.*, **76**, 3830–3833.

83 Barbara, B. and Gunther, L. (1999) Magnets, molecules and quantum mechanics. *Phys. World*, **12**, 35–40.

84 Leuenberger, M.N. and Loss, D. (2001) Quantum computing in molecular magnets. *Nature*, **410**, 789–793.

85 Caneschi, A., Gatteschi, D., and Totti, F. (2015) Molecular magnets and surfaces: a promising marriage. A DFT insight. *Coord. Chem. Rev.*, **289–290**, 357–378.

86 Caneschi, A., Gatteschi, D., Lalioti, N., Sangregorio, C., Sessoli, R., Venturi, G., Vindigni, A., Rettori, A., Pini, M.G., and Novak, M.A. (2001) Cobalt(II)-nitronyl nitroxide chains as molecular magnetic nanowires. *Angew. Chem.*, **40**, 1760–1763.

87 Bogani, L., Sessoli, R., Pini, M.G., Rettori, A., Novak, M.A., Rosa, P., Massi, M., Fedi, M.E., Giuntini, L., Caneschi, A., and Gatteschi, D. (2005) Finite-size effects on the static properties of a single-chain magnet. *Phys. Rev. B*, **72**, 064406 (10 pp).

88 Brandon, E.J., Rittenberg, D.K., Arif, A.M., and Miller, J.S. (1998) Ferrimagnetic behavior of multiple phases and solvates of (*meso*-tetrakis(4-chlorophenyl)porphinato)manganese(III) tetracyanoethenide, [MnTClPP]$^+$[TCNE]•$^-$. Enhancement of magnetic coupling by thermal annealing. *Inorg. Chem.*, **37**, 3376–3384.

89 Fardis, M., Diamantopoulos, G., Papavassiliou, G., Pokhodnya, K., Miller, J.S., Rittenberg, D.K., and Christides, C. (2002) ^{1}H NMR investigation of the magnetic spin configuration in the molecule-based ferrimagnet [MnTFPP][TCNE]. *Phys. Rev., B*, **66**, 064422 (6 pp).

90 Bałanda, M., Rams, M., Nayak, S.K., Tomkowicz, Z., Haase, W., Tomala, K., and Yakhmi, J.V. (2006) Slow magnetic relaxations in the anisotropic Heisenberg chain compound Mn(III) tetra(*ortho*-fluorophenyl)porphyrin-tetracyanoethylene. *Phys. Rev. B*, **74**, 224421-9.

91 Tomkowicz, Z., Rams, M., Bałanda, M., Foro, S., Nojiri, H., Krupskaya, Y., Kataev, V., Büchner, B., Nayak, S.K., Yakhmi, J.V., and Haase, W. (2012) Slow magnetic relaxations in manganese(III) tetra(meta-fluorophenyl)-porphyrin-tetracyanoethenide. Comparison with the relative single chain magnet ortho compound. *Inorg. Chem.*, **51**, 9983–9994.

92 Bałanda, M., Tomkowicz, Z., Haase, W., and Rams, M. (2011) Single-chain magnet features in 1D [MnR$_4$TPP][TCNE] compounds. *J. Phys.*, **303**, 012036 (14 pp).

93 Pełka, R., Konieczny, P., Zieliński, P.M., Wasiutyński, T., Miyazaki, Y., Inaba, A., Pinkowicz, D., and Sieklucka, B. (2014) Magnetocaloric effect in {[Fe(pyrazole)$_4$]$_2$[Nb(CN)$_8$]·4H$_2$O}$_n$ molecular magnet. *J. Magn. Magn. Mater.*, **354**, 359–362.

94 Liu, J., Gottschall, T., Skokov, K.P., Moore, J.D., and Gutfleisch, O. (2012) Giant magnetocaloric effect driven by structural transitions. *Nat. Mater.*, **11**, 620–626.

95 Low, D.M., Jones, L.F., Bell, A., Brechin, E.K., Mallah, T., Rivière, E., Teat, S.J., and McInnes, E.J.L. (2003) Solvothermal synthesis of a tetradecametallic Fe^{III}Cluster. *Angew. Chem., Int. Ed.*, **42**, 3781–3784.

96 Evangelisti, M., Candini, A., Ghirri, A., Affronte, M., Brechin, E.K., and McInnes, E.J.L. (2005) Spin-enhanced magnetocaloric effect in molecular nanomagnets. *Appl. Phys. Lett.*, **87**, 072504 (3 pp).

97 Affronte, M., Ghirri, A., Carretta, S., Amoretti, G., Piligkos, S., Timco, G., and Winpenny, R.E.P. (2004) Engineering molecular rings for magnetocaloric effect. *Appl. Phys. Lett.*, **84**, 3468–3470.

98 Corradini, V., Ghirri, A., Candini, A., Biagi, R., del Pennino, U., Dotti, G., Otero, E., Choueikani, F., Blagg, R.J., McInnes, E.J.L., and Affronte, M. (2013) Magnetic cooling at a single molecule level: a spectroscopic investigation of isolated molecules on a surface. *Adv. Mater.*, **25**, 2816–2820.

99 Langley, S.K., Chilton, N.F., Moubaraki, B., Hooper, T., Brechin, E.K., Evangelisti, M., and Murray, K.S. (2011) Molecular coolers: the case for [Cu$^{II}_5$Gd$^{III}_4$]. *Chem. Sci.*, **2**, 1166–1169.

100 Peng, J.B., Zhang, Q.C., Kong, X.J., Zheng, Y.Z., Ren, Y.P., Long, L.S., Huang, R.B., Zheng, L.S., and Zheng, Z.P. (2012) High-nuclearity 3d–4f clusters as enhanced magnetic coolers and molecular magnets. *J. Am. Chem. Soc.*, **134**, 3314–3317.

101 Glauber, R.J. (1963) Time-dependent statistics of the ising model. *J. Math. Phys.*, **4**, 294–307.

102 Gatteschi, D. and Vindigni, A. (2014) Chapter 8, Single-chain magnets, in *Molecular Magnets: Physics of Applications* (eds J. Bartolomé, F. Luis, and J.F. Fernández), Springer-Verlag, Berlin, Heidelberg, pp. 191–220.

103 Coulon, C., Pianet, V., Urdampilleta, M., and Clerac, R. (2015) Single-chain magnets and related systems. *Struct. Bond.*, **164**, 143–184.

104 Baibich, M.N., Broto, J.M., Fert, A., Dau, F.N.V., Petro, F., Eitenne, P., Creuzet, G., Friederich, A., and Chazelas, J. (1988) Giant magnetoresistance of (001)Fe/(001)Cr magnetic superlattices. *Phys. Rev. Lett.*, **61**, 2472–2475.

105 Binasch, G., Gruenberg, P., Saurenbach, F., and Zinn, W. (1989) Enhanced magnetoresistance in layered magnetic structures with antiferromagnetic interlayer exchange. *Phys. Rev. B*, **39**, 4828–4830.

106 Dediu, V., Murgia, M., Matacotta, F.C., Taliani, C., and Barbanera, S. (2002) Room temperature spin-polarized injection in organic semiconductor. *Solid State Commun.*, **122**, 181–184.

107 Xiong, Z.H., Wu, D., Vardeny, Z.V., and Shi, J. (2004) Giant magnetoresistance in organic spin-valves. *Nature*, **427**, 821–824.

108 Santos, T.S., Lee, J.S., Migdal, P., Lekshmi, I.C., Satpati, B., and Moodera, J.S. (2007) Room-temperature tunnel magnetoresistance and spin-polarized tunneling through an organic semiconductor barrier. *Phys. Rev. Lett.*, **98**, 016601-4.

109 Pasupathy, A.N., Bialczak, R.C., Martinek, J., Grose, J.E., Donev, L.A.K., McEuen,

P.L., and Ralph, D.C. (2004) The kondo effect in the presence of ferromagnetism. *Science*, **306**, 86–89.

110 Dediu, V.A., Hueso, L.E., Bergenti, I., and Taliani, C. (2009) Spin routes in organic semiconductors. *Nat. Mater.*, **8**, 707–716.

111 Yoo, J.-W., Chen, C.-Y., Jang, H.W., Bark, C.W., Prigodin, V.N., Eom, C.B., and Epstein, A.J. (2010) Spin injection/detection using an organic-based magnetic semiconductor. *Nat. Mater.*, **9**, 638–642.

112 Yoo, J.-W., Jang, H.W., Prigodin, V.N., Kao, C., Eom, C.B., and Epstein, A.J. (2010) Tunneling vs. giant magnetoresistance in organic spin valve. *Synth. Met.*, **160**, 216–222.

113 Miyamoto, K., Iori, K., Sakamoto, K., Kimura, A., Qiao, S., Shimada, K., Namatame, H., and Taniguchi, M. (2008) Spin-dependent electronic band structure of Co/Cu(001) with different film thicknesses. *J. Phys. Cond. Matter*, **20**, 225001 (5 pp).

114 Bozdag, K.D., Chiou, N.-R., Prigodin, V.N., and Epstein, A.J. (2010) Magnetic field, temperature and electric field dependence of magneto-transport for polyaniline nanofiber networks. *Synth. Met.*, **160**, 271–274.

115 Ardavan, A., Rival, O., Morton, J.J.L., Blundell, S.J., Tyryshkin, A.M., Timco, G.A., and Winpenny, R.E.P. (2007) Will spin-relaxation times in molecular magnets permit quantum information processing? *Phys. Rev. Lett.*, **98**, 057201-4.

116 Bogani, L. and Wernsdorfer, W. (2008) Molecular spintronics using single-molecule magnets. *Nat. Mater.*, **7**, 179–186.

117 Urdampilleta, M., Klyatskaya, S., Cleuziou, J.-P., Ruben, M., and Wernsdorfer, W. (2011) Supramolecular spin valves. *Nat. Mater.*, **10**, 502–506.

118 Urdampilleta, M., Klyatskaya, S., Ruben, M., and Wernsdorfer, W. (2015) Magnetic interaction between a radical spin and a single-molecule magnet in a molecular spin-valve. *ACS Nano*, **9** (4), 4458–4464.

119 Han, W., Kawakami, R.K., Gmitra, M., and Fabian, J. (2014) Graphene spintronics. *Nat. Nanotechnol.*, **9**, 794–807.

120 Nguyen, T.D., Ehrenfreund, E., and Valy Vardeny, Z. (2014) The development of organic spin valves from unipolar to bipolar operation. *MRS Bull.*, **39**, 585–589.

121 Boehme, C. and Lupton, J.M. (2013) Challenges for organic spintronics. *Nat. Nanotechnol.*, **8**, 612–615.

122 Moodera, J.S., Koopmans, B., and Oppeneer, P.M. (2014) On the path toward organic spintronics. *MRS Bull.*, **39**, 578–581.

123 Koopmans, B. (2014) Organic spintronics: pumping spins through polymers. *Nat. Phys.*, **10**, 249–250.

124 Wagner, S., Kisslinger, F., Ballmann, S., Schramm, F., Chandrasekar, R., Bodenstein, T., Fuhr, O., Secker, D., Fink, K., Ruben, M., and Weber, H.B. (2013) Switching of a coupled spin pair in a single-molecule junction. *Nat. Nanotechnol.*, **8**, 575–579.

125 Mahato, R.N., Lülf, H., Siekman, M.H., Kersten, S.P., Bobbert, P.A., de Jong, M.P., De Cola, L., and van der Wiel, W.G. (2013) Ultrahigh magnetoresistance at room temperature in molecular wires. *Science*, **341**, 257–260.

126 Galbiati, M., Barraud, C., Tatay, S., Bouzehouane, K., Deranlot, C., Jacquet, E., Fert, A., Seneor, P., Mattana, R., and Petroff, F. (2012) Unveiling self-assembled monolayers' potential for molecular spintronics: spin transport at high voltage. *Adv. Mater.*, **24**, 6429–6432.

127 Mooser, S., Cooper, J.F.K., Banger, K.K., Wunderlich, J., and Sirringhaus, H. (2012) Spin injection and transport in a solution-processed organic semiconductor at room temperature. *Phys. Rev. B*, **85**, 235202 (7 pp).

128 Ehrenfreund, E. and Valy Vardeny, Z. (2012) Effects of magnetic field on conductance and electroluminescence in organic devices. *Isr. J. Chem.*, **52**, 552–562.

129 Shiraishi, M. and Ikoma, T. (2011) Molecular spintronics. *Phys. E*, **43**, 1295–1317.

130 Yakhmi, J.V. and Bambole, V.A. (2012) Molecular spintronics, in *Ferroics and*

Multiferroics vol. **189** (eds. H.S. Virk and W. Kleemann), Solid State Phenomena. ISBN: 13: 978-3-03785-431-0, Trans Tech Publications, Ltd., Zurich-Durnten, Switzerland, pp. 95–127.

131 Benard, S., Yu, P., Audiere, J.P., Riviere, E., Clement, R., Guilhem, J., Tchertanov, L., and Nakatani, K. (2000) Structure and NLO properties of layered bimetallic oxalato-bridged ferromagnetic networks containing stilbazolium-shaped chromophores. *J. Am. Chem. Soc.*, **122**, 9444–9454.

132 Binnemans, K., Galyametdinov, Y.G., Deun, R.K., Bruce, D.W., Collinson, S.R., Polishchumk, A.P., Bikchantaev, I., Haase, W., Prosvirin, A.V., Tinchurina, L., Litvinov, I., Gubajdullin, A., Rakhmatullin, A., Uytterhoeven, K., and Meervelt, L.V. (2000) Rare-earth-containing magnetic liquid crystals. *J. Am. Chem. Soc.*, **122**, 4335–4344.

133 Inoue, K., Imai, H., Ghalsasi, P.S., Ohba, M., Okawa, H., and Yakhmi, J.V. (2001) A three-dimensional ferrimagnet with a high magnetic transition temperature (T_C) of 53 K based on a chiral molecule. *Angew. Chem., Int. Ed.*, **40**, 4242–4245.

134 Maspoch, D., Ruiz-Molina, D., Wurst, K., Domingo, N., Cavallini, M., Biscarini, F., Tejada, J., Rovira, C., and Veciana, J. (2003) A nanoporous molecular magnet with reversible solvent-induced mechanical and magnetic properties. *Nat. Mater.*, **2**, 190–195.

135 Maspoch, D., Domingo, N., Ruiz-Molina, D., Wurst, K., Vaughan, G., Tejada, J., Rovira, C., and Veciana, J. (2004) A robust purely organic nanoporous magnet. *Angew. Chem., Int. Ed.*, **43**, 1828–1832.

136 Wang, Z., Zhang, B., Zhang, Y., Kurmoo, M., Liu, T., Gao, S., and Kobayashi, H. (2007) A damily of porous magnets, [$M_3(HCOO)_6$] (M = Mn, Fe, Co and Ni). *Polyhedron*, **26**, 2207–2215.

137 Tomić, S. and Dressel, M. (2015) Ferroelectricity in molecular solids: a review of electrodynamic properties. *Rep. Prog. Phys.*, **78**, 096501 (26 pp).

138 Martins, P. and Lanceros-Méndez, S. (2013) Polymer-based magnetoelectric materials. *Adv. Funct. Mater.*, **23**, 3371–3385.

139 Ohkoshi, S., Tokoro, H., Matsuda, T., Takahashi, H., Irie, H., and Hashimoto, K. (2007) Coexistence of ferroelectricity and ferromagnetism in a rubidium manganese hexacyanoferrate. *Angew. Chem.*, **119**, 3302–3305.

140 Qin, W., Chen, X., Li, H., Gong, M., Yuan, G., Grossman, J.C., Wuttig, M., and Ren, S. (2015) Room temperature multiferroicity of charge transfer crystals. *ACS Nano*, **9** (9), 9373–9379.

141 Tayi, A.S., Kaeser, A., Matsumoto, M., Aida, T., and Stupp, S.I. (2015) Supramolecular ferroelectrics. *Nat. Chem.*, **7**, 281–294.

142 Philipse, A.P. and Mass, D. (2002) Magnetic colloids from magnetotactic bacteria: chain formation and colloidal stability. *Langmuir*, **18**, 9977–9984.

143 Jogler, C. and Schueler, D. (2009) Genomics, genetics, and cell biology of magnetosome formation. *Annu. Rev. Microbiol.*, **63**, 501–521.

144 Lang, C. and Schueler, D. (2006) Biogenic nanoparticles: production, characterization, and application of bacterial magnetosomes. *J. Phys. Condens. Matter*, **18**, S2815–S2828.

145 Kolinko, I., Lohße, A., Borg, S., Raschdorf, O., Jogler, C., Tu, Q., Posfai, M., Tompa, E., Plitzko, J.M., Brachmann, A., Wanner, G., Mueller, R., Zhang, Y., and Schueler, D. (2014) *Nat. Nanotechnol.*, **9**, 193–197.

146 Martín, M., Carmona, F., Cuesta, R., Rondón, D., Gálvez, N., and Domínguez-Vera, J.M. (2014) Artificial magnetic bacteria: living magnets at room temperature. *Adv. Funct. Mater.*, **24**, 3489–3493.

147 Wiltschko, W. and Wiltschko, R. (2005) Magnetic orientation and magnetoreception in birds and other animals. *J. Comp. Physiol. A*, **191**, 675–693.

148 Lohmann, K.J. and Johnsen, S. (2000) The neurobiology of magnetoreception in vertebrate animals. *Trends Neurosci.*, **23**, 153–159.

149 Falkenberg, G., Fleissner, G., Schuchardt, K., Kuehbacher, M., Thalau, P. *et al.*

(2010) Avian magnetoreception: elaborate iron mineral containing dendrites in the upper beak seem to be a common feature of birds. *PLoS One*, **5** (2), e9231 (9 pp).

150 Kishkinev, D.A. and Chernetsov, N.S. (2015) *Biol. Bull. Rev.*, **5**, 46–62.

151 Eder, S.H.K., Cadiou, H., Muhamad, A., McNaughton, P.A., Kirschvink, J.L., and Winklhofer, M. (2012) Magnetic characterization of isolated candidate vertebrate magnetoreceptor cells. *Proc. Natl. Acad. Sci. USA*, **109**, 12022–12027.

152 Shaw, J., Boyd, A., House, M., Woodward, R., Mathes, F., Cowin, G., Saunders, M., and Baer, B. (2015) Magnetic particle-mediated magnetoreception. *J. Roy. Soc. Interface*, **12**, 20150499.

153 Ritz, T., Adem, S., and Schulten, K. (2000) A model for photoreceptor-based magnetoreception in birds. *Biophys. J.*, **78**, 707–718.

154 Schulten, K., Swenberg, C.E., and Weller, A. (1978) A biomagnetic sensory mechanism based on magnetic field modulated coherent electron spin motion. *Z. Phys. Chem.*, **111**, 1–5.

155 Fusani, L., Bertolucci, C., Frigato, E., and Foà, A. (2014) Cryptochrome expression in the eye of migratory birds depends on their migratory status. *J. Exp. Biol.*, **217**, 918–923.

156 Gegear, R.J., Casselman, A., Waddell, S., and Reppert, S.M. (2008) Cryptochrome mediates light-dependent magnetosensitivity in Drosophila. *Nature*, **454**, 1014–1018.

157 Rodgers, C.T. and Horea, P.J. (2009) Chemical magnetoreception in birds: the radical pair mechanism. *Proc. of Natl. Acad. Sci.*, **106**, 353–360.

158 Winklhofer, M. (2010) Theme supplement on 'Magnetoreception'. *J. R. Soc. Interface*, **7** (Suppl. 2), S131–S134.

159 Lee, S.W., Yang, Y., Lee, H.W., Ghasemi, H., Kraemer, D., Chen, G., and Cui, Y. (2014) An electrochemical system for efficiently harvesting low-grade heat energy. *Nat. Commun.*, **5**. Article number: 3942. doi: 10.1038/ncomms4942

160 Chen, R., Tanaka, H., Kawamoto, T., Asai, M., Fukushima, C., Kurihara, M., Ishizaki, M., Watanabe, M., Arisaka, M., and Nankawa, T. (2013) Thermodynamics and mechanism studies on electrochemical removal of cesium ions from aqueous solution using a nanoparticle film of copper hexacyanoferrate. *ACS Appl. Mater. Interfaces*, **5**, 12984–12990.

161 Lipson, A.L., Pan, B., Lapidus, S.H., Liao, C., Vaughey, J.T., and Ingram, B.J. (2015) Rechargeable ca-ion batteries: a new energy storage system. *Chem. Mater.*, **27**, 8442–8447.

162 Bhatt, P., Banerjee, S., Anwar, S., Mukadam, Mayuresh D., Meena, S.S., and Yusuf, S.M. (2014) Core–shell prussian blue analogue molecular magnet $Mn_{1.5}[Cr(CN)_6]\cdot mH_2O@Ni_{1.5}[Cr(CN)_6]\cdot nH_2O$ for hydrogen storage. *ACS Appl. Mater. Interfaces*, **6**, 17579–17588.

163 Krober, J., Codjovi, E., Kahn, O., Groliere, F., and Jay, C. (1993) A spin transition system with a thermal hysteresis at room temperature. *J. Am. Chem. Soc.*, **115**, 9810–9811.

164 Mathoniere, C., Carling, S.G., Yusheng, D., and Day, P. (1994) Molecular-based mixed valency ferrimagnets $(XR_4)Fe^{II}Fe^{III}(C_2O_4)_3 (X = N, P; R = n\text{-propyl}, n\text{-butyl}, \text{phenyl})$: anomalous negative magnetisation in the tetra-n-butylammonium derivative. *J. Chem. Soc. Chem. Commun.*, 1551–1552.

165 Hashimoto, K. and Ohkoshi, S. (1999) Design of novel magnets using Prussian blue analogues. *Philos. Trans. R. Soc. London A*, **357**, 2977–3003.

166 Sastry, M.D., Bhide, M.K., Kadam, R.M., Chavan, S.A., Yakhmi, J.V., and Kahn, O. (1999) Photo-induced changes in magnetic order in the molecular magnet $(NBu_4)_2Mn_2[Cu(opba)]_3\cdot 6DMSO\cdot 1H_2O$. *Chem. Phys. Lett.*, **301**, 385–388.

167 Pejakovic, D.A., Kitamura, C., Miller, J.S., and Epstein, A.E. (2002) Photoinduced magnetization in the organic-based magnet $Mn(TCNE)_x\cdot y(CH_2Cl_2)$. *Phys. Rev. Lett.*, **88**, 057202 (4 pp).

168 Itkis, M.E., Chi, X., Cordes, A.W., and Haddon, R.C. (2002) Magneto-optoelectronic bistability in a phenalenyl-based neutral radical. *Science*, **296**, 1443–1445.

169 Fujita, W. and Awaga, K. (1999) Room-temperature magnetic bistability in organic radical crystals. *Science*, **286**, 261–262.
170 Zhang, J., Miller, J.S., Vazquez, C., Zhou, P., Brinckerhoff, W.B., Epstein, A.J., and McLean, R.S. (1996) Improved synthesis of the V(tetracyanoethylene)$_x \cdot y$ (solvent) room-temperature magnet: doubling of the magnetization at room temperature. *ACS Symp. Ser.*, **644**, 311–318.
171 Jain, R., Kabir, K., Gilroy, J.B., Mitchell, K.A.R., Wong, K.C., and Hicks, R.G. (2007) High-temperature metal-organic magnets. *Nature*, **445**, 291–294.
172 Miller, J.S. and Pokhodnya, K.I. (2007) Formation of $Ni[C_4(CN)_8]$ from the reaction of $Ni(COD)_2$ (COD = 1,5-cyclooctadiene) with TCNE in THF. *J. Mater. Chem.*, **17**, 3585–3587.
173 Imoto, K., Takemura, M., Tokoro, H., and Ohkoshi, S.I. (2012) A cyano-bridged vanadium–niobium bimetal assembly exhibiting a high curie temperature of 210 K. *Eur. J. Inorg. Chem.*, **2012** (Issue 16), 2649–2652.
174 McConnell, A.C., Bell, J.D., and Miller, J.S. (2012) Pressure-induced transition from an antiferromagnet to a ferrimagnet for $Mn^{II}(TCNE)[C_4(CN)_8]_{1/2}$ (TCNE = tetracyanoethylene). *Inorg. Chem.*, **51**, 9978–9982.

21
Ge–Sb–Te Phase-Change Materials

Volker L. Deringer[1] and Matthias Wuttig[2,3]

[1] RWTH Aachen University, Institute of Inorganic Chemistry, Landoltweg 1, 52056 Aachen, Germany
[2] RWTH Aachen University, Institute of Physics IA, 52056 Aachen, Germany
[3] RWTH Aachen University, Jülich–Aachen Research Alliance (JARA-FIT), 52056 Aachen, Germany

21.1
Introduction

Encoded in "ones" and "zeroes," digital data are being created and stored in ever-increasing amounts. This makes it necessary to develop better, faster, and more reliable technologies for data processing and data storage. In fact, these technologies are often enabled by very basic research in the solid-state sciences that must continue to supply newly synthesized materials with desirable properties [1].

Phase-change materials (PCMs) from the ternary Ge–Sb–Te system are among the most promising candidates for data storage applications [2]; they are widely used in mass-market devices and at the same time in most of the current prototypes. PCMs can be switched between markedly different solid-state phases, typically an amorphous and a crystalline one. Both exhibit a large contrast in their physical properties allowing one to encode digital ones and zeroes. Switching proceeds rapidly at high temperatures, but is extremely slow at ambient conditions (this is crucial for data retention); it is reversible and in practice can be induced either by laser beams or by electrical current pulses of different intensity (Figure 21.1) [2].

The functional principle of PCMs was already put forward half a century ago in the visionary works of Ovshinsky [4]. Finding materials that are practically suitable, however, has been a task far from trivial, as it requires an entire portfolio of particular properties to be present in a single compound (cf. Table 21.1). In the 1990s, Yamada *et al.* discovered such suitable materials in the Ge–Sb–Te system [5–7] that were initially proposed for use in optical media such as the DVD-RAM but later found viable for electronic memories as well [8] (note that the latter type of application had been envisioned by Ovshinsky in the very

Handbook of Solid State Chemistry, First Edition. Edited by Richard Dronskowski, Shinichi Kikkawa, and Andreas Stein.
© 2017 Wiley-VCH Verlag GmbH & Co. KGaA. Published 2017 by Wiley-VCH Verlag GmbH & Co. KGaA.

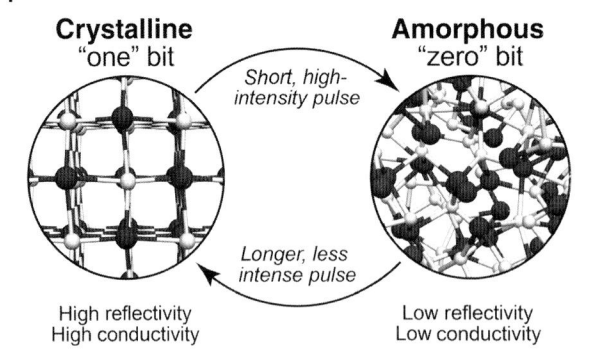

Figure 21.1 The functional principle of phase-change data storage. Ones and zeroes are encoded by small bits of a PCM that differ strongly in their physical properties. In optical disks, the materials are switched by laser pulses and the contrast in reflectivity is read out to retrieve the digital information. In electronic media, current pulses (and thus Joule heating) are employed; in this case, the different conductivities of the crystalline and amorphous phases encode ones and zeroes. (Reprinted from Ref. [3] with permission. Copyright 2015, Wiley-VCH Verlag GmbH & Co. KGaA, Weinheim.)

beginning [4]). It is these Ge–Sb–Te materials with which this chapter is concerned.

This text is part of a handbook on solid-state *chemistry*, and so we have tried to tailor it to complement previous reviews – which have mainly been taking perspectives of solid-state physics [9] or industrial device fabrication [10,11]. We must focus on a few fundamental aspects, for simple reasons of brevity; interested readers who would like to delve deeper into the field are referred to introductory texts [2,12,13] as well as to comprehensive textbooks [14,15].

In the following sections, we will argue that PCMs are intriguingly (and sometimes unexpectedly) complex on the atomic scale, and that it is this complexity that gives rise to both the diverse property portfolio and exciting opportunities for new applications, as noted in a previous review by the present authors [3].

Table 21.1 Requirements that a material must fulfill to be suitable for phase-change data storage. This is the case, largely, in Ge-Sb-Te alloys.

Application requirement	Material requirement
Encoding of distinct "ones" and "zeroes"	Optical and/or electrical contrast between different solid-state phases (typically crystalline *versus* amorphous)
Fast data transfer	Fast crystallization at high temperature
Long-term retention	No spontaneous crystallization at room temperature
Scalability	Simple and robust chemical composition; functionality must be retained when downsizing devices
Cyclability	Long-term chemical stability

Source: Adapted and simplified from Ref. [9].

Particular emphasis will be on the structural chemistry of Ge–Sb–Te compounds, which we will discuss in sequence for the crystalline (Sections 21.2) and amorphous phases (Section 21.3). From there, the way leads on to the chemical-bonding nature, and to the nature of structural vacancies and their effect on properties (Section 21.4). Finally, we will give an outlook onto emerging and possible future applications of Ge–Sb–Te PCMs (Section 21.5).

21.2
Crystalline Ge–Sb–Te Phases: Stable and Metastable

21.2.1
GeTe and Sb_2Te_3

Before looking at the ternary title compounds such as the "Blu-ray material" $Ge_8Sb_2Te_{11}$, let us consider the binary parent compositions, germanium telluride (GeTe) and antimony telluride (Sb_2Te_3). Both of them are very interesting materials in themselves, notwithstanding their formal simplicity, and many structural properties and principles of more complex PCMs can best be understood by starting with the binaries.

The first reported synthesis of GeTe dates back to the 1930s and was achieved by fusing the constituent elements [16]. GeTe is the only stable composition in the binary system, at variance with the lighter germanium chalcogenides: crystalline GeS_2 and $GeSe_2$ exist, but $GeTe_2$ does not. Subsequently, the crystal structure of GeTe was carefully characterized by single-crystal X-ray [17] and neutron diffraction [18]. At ambient conditions, GeTe takes the rhombohedral space group $R3m$ and a slightly distorted rock salt like structure: While both cations and anions are octahedrally coordinated in a rock salt lattice, there is a small distortion along [111] in GeTe that leads to alternating shorter and longer Ge–Te bonds; the coordination environment is thus "3 + 3"-fold. Due to this Peierls-like distortion [19], GeTe exhibits a noncentrosymmetric structure, other than all stable ternary Ge–Sb–Te alloys derived from it (see below). Furthermore, the material contains a significant amount of cation vacancies [20], and its sum formula would thus more precisely be written as $Ge_{(1-x)}Te$; we will discuss the nature and effects of these vacancies in Section 21.4.2.

The second binary parent compound, Sb_2Te_3, takes a layered structure in the space group $R\bar{3}m$ (note the presence of inversion symmetry in this case) [21]. The cations reside in distorted octahedra once more, but between these building blocks one finds Te $\cdots$ Te bilayers that are held together by much weaker van der Waals forces.

Compared to the Ge–Te binary phase diagram, that of Sb–Te is almost beautifully complex and hosts several structures beyond Sb_2Te_3 with well-defined stacking sequences [22]. One example is Sb_2Te, which is formed of "Sb_2Te_3"-like building blocks and of layers that resemble pure Sb; both motifs alternate along the [0001] direction [23]. We note in passing that Sb_2Te is parent to another important PCM family, but in this case, the pure amorphous material crystallizes *too* rapidly to be suitable for application (cf. Table 21.1). The crystallization speed of Sb–Te compounds can be reduced by alloying with a few percent of Ag and In each. This way, one arrives at the "AIST" material [24] that is likewise used in data storage and has been structurally characterized using several complementary techniques [25–27].

21.2.2
The Stable Quasibinary Phases

We now turn to the stable crystalline Ge–Sb–Te compounds that form a range of quasibinary alloys according to

$$n\,GeTe + m\,Sb_2Te_3 \rightarrow Ge_nSb_{2m}Te_{(n+3m)}$$

as shown in Figure 21.2. The close relation to the binary parent compounds is immediately apparent: all stable ternary Ge–Sb–Te alloys exhibit Te⋯Te bilayers and can be rationalized as put together from "building blocks" derived from the binaries [28].

Interestingly, the initial structural reports on many layered ternary Ge–Sb–Te alloys were based on electron diffraction in the 1960s [29,30], and were

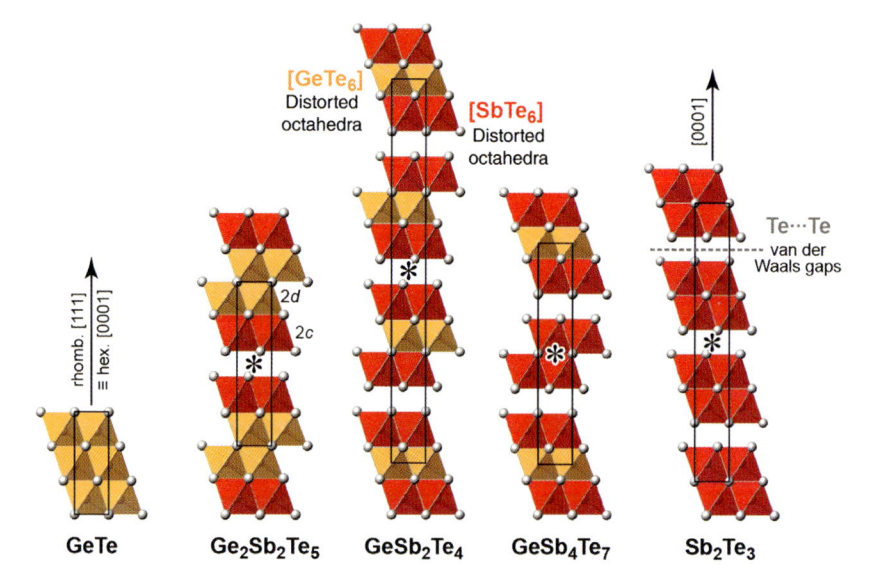

Figure 21.2 Crystal structures of quasibinary GeTe–Sb_2Te_3 compounds, all visualized in the (pseudo)hexagonal cell setup of the corresponding rhombohedral space groups. All structures except that of GeTe exhibit an inversion center (indicated by asterisks) and also a varying amount of "van der Waals" gaps with Te⋯Te contacts. Color code: Ge, yellow; Sb, red; and Te, gray. (Reprinted from Ref. [3] with permission. Copyright 2015, Wiley-VCH Verlag GmbH & Co. KGaA, Weinheim.)

later supplemented and partially revised using synchrotron-quality X-ray diffraction [31,32]. Furthermore, transmission electron microscopy (TEM) has played a key role in clarifying the atomic structures, in particular regarding the cation ordering [33] and the formation of stable phases during thermal annealing [34,35].

Finally, it is important to stress that Figure 21.2 is merely an approximation regarding the distribution of Ge/Sb on the cationic lattices: germanium (yellow) and antimony (red) are both octahedrally coordinated, and they can intermix, which has indeed been observed in structural refinements [31].

21.2.3
Metastable (Rock Salt-Type) Phases

Compared to the aforementioned, highly ordered Ge–Sb–Te structures, their metastable (as-crystallized) counterparts appear very simple at first. Reminiscent of binary GeTe, the structure of the metastable ternaries has been derived from the rock salt-type early on [31,36–38]. Conceptually, one starts with an anionic fcc sublattice fully occupied by Te, and then fills the cationic, interpenetrating sublattice with an appropriate share of Ge, Sb, and vacancies. It is these rock salt-like structures that are relevant for many applications and devices in which the compounds do not have time to anneal and reach the layered equilibrium structure.

However, the picture of fully random occupations is again an oversimplification: There is typically some order, and more importantly, a gradual transition between rock salt-like structures and the thermodynamically stable, well-ordered ones; this transition can be traced by TEM (Figure 21.3) [34,35]. In a freshly quenched sample, vacancies are disordered but not fully at random, rather forming canted vacancy layers [34]; given sufficient time to anneal at high temperature, these layers will order into extended planes, and ultimately lead from the metastable to the stable layered compounds as shown in Figure 21.2 [35]. How vacancies and their ordering determine macroscopic properties of Ge–Sb–Te alloys will be a central theme in Section 21.4.

21.3
The Amorphous Phases

21.3.1
What Makes Amorphous PCMs So Special?

Amorphous solids lack long-range order, but the local atomic environments within them often resemble those in the crystalline counterparts; this is the classical Zachariasen model [39]. For example, both silica glass (amorphous SiO_2) and α-quartz (crystalline SiO_2) contain $[SiO_4]^{4-}$ tetrahedra throughout, and both are transparent and colorless solids [40].

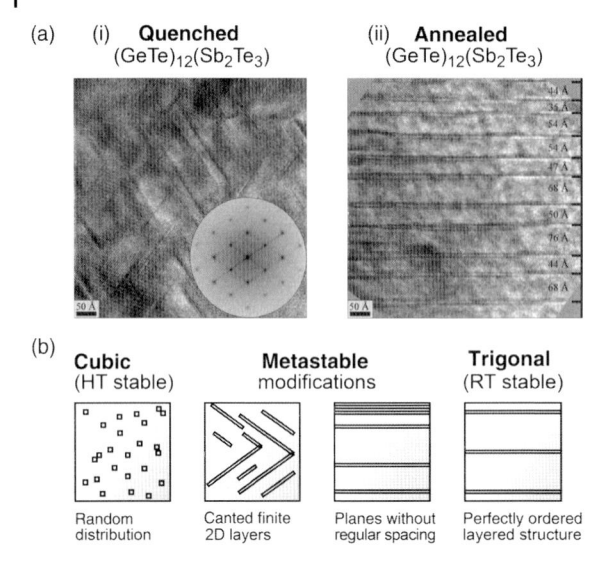

Figure 21.3 (a) High-resolution transmission electron microscopy (HRTEM) images of metastable crystallized $(GeTe)_{12}(Sb_2Te_3)$ samples; (i) quenched from 500 °C (with Fourier transformation in the inset) and (ii) annealed at 400 °C for 20 h and subsequently quenched, with the spacing of vacancy layers indicated. (Reprinted figure with permission from Ref. [34]. Copyright 2010 by the American Physical Society.) (b) Schematic sketch of different vacancy distributions in these GeTe-rich phases. (Adapted (as in Ref. [3]) from Ref. [35] with permission by the Royal Society of Chemistry. Copyright 2012.)

Amorphous PCMs are much different in this regard, and amorphizing even the binary GeTe gives rise to a very complex solid-state phase. Figure 21.4a shows a representative fragment from an atomistic simulation, featuring a tetrahedron in which furthermore two homopolar Ge–Ge bonds are formed; none of these structural characteristics occur in crystalline GeTe. (We stress that terms like "tetrahedra" are necessarily approximations here, and that the strongly distorted coordination polyhedra in amorphous PCMs can often only be classified using numerical order parameters [41]). Such unusual coexistence of local fragments has been predicted by simulations [41] and later verified by Raman spectroscopy [42]. Finally, the amorphous phase evolves over time; it *ages*, that is, it changes its microscopic structure and thereby its electrical resistivity [43–45]. This is a very important (and problematic) issue for the long-term stability of digital "zero bits."

The structural complexity of amorphous solids extends well beyond the first coordination shell [49], and this is true for Ge–Sb–Te alloys as well. On the medium-range scale, four-membered rings (with alternating cations and anions) constitute an important structural fragment in amorphous PCMs (Figure 21.4b) [47,50]; to first approximation, these rings resemble the ordering in the rock salt phase. Vacancy defects are uniquely defined for

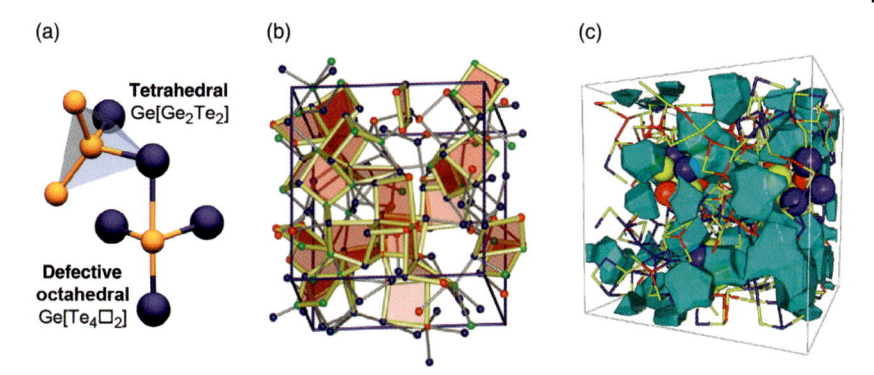

Figure 21.4 Microscopic complexity in amorphous (a-) phase-change materials, as observed on three increasingly larger length scales. All structural models have been generated by atomistic simulations (see text). (a) *Short-range order*: Fragment from an *ab initio* MD simulation of a-GeTe, highlighting coexisting local environments. (Adapted with permission. [46] Copyright 2014, Wiley-VCH Verlag GmbH & Co. KGaA, Weinheim.) (b) *Medium-range order*: Structural model of a-$Ge_2Sb_2Te_5$, emphasizing the presence of fourfold rings (shaded). (Reprinted by permission from Macmillan Publishers Ltd: Nature Materials, Ref. [47], copyright 2008.) (c) The role of *void volumes*, with teal isosurfaces encasing "empty" areas in the amorphous phase. (Reprinted figure with permission from Ref. [48]. Copyright 2009 by the American Physical Society.)

crystalline PCMs but not for their amorphous phases; in the latter case, one may look at "void volumes" instead (Figure 21.4c) [48].

Note a certain vagueness of language and definitions here, which is exactly what makes amorphous PCMs so challenging (and interesting) to study: One cannot directly visualize the atoms but must rely on indirect observations; more often than not, only several techniques combined can answer a specific question.

21.3.2
Experimental Probes

The layered crystalline phases had already posed challenges to experimental structure determination (Section 21.2.2); this is even more so for their amorphous counterparts. To probe local coordination environments, extended X-ray fine structure absorption spectroscopy (EXAFS) is a key technique that led to the initial (and unexpected) observation of a contrast in bond distances: Ge–Te bonds in amorphous $Ge_2Sb_2Te_5$ are *shorter* than in the crystalline phase, and presumably stronger [51]. This is in line with vibrational properties reported later on, which likewise point to a "stiffer," more rigid bonding network in amorphous $Ge_2Sb_2Te_5$ [52]. EXAFS experiments also initially pointed toward the existence of homopolar Ge–Ge bonds in the amorphous phase [53].

Further insight can be gained from reverse Monte Carlo (RMC) techniques [54], by adjusting an initial structural model to the available experimental evidence (e.g., scattering data). RMC has proven particularly powerful when combined with first-principles molecular dynamics (MD) simulations of atomic structure [48]. Indeed, the latter have become a cornerstone of structural research on amorphous PCMs, and we will hence discuss them here in some detail.

21.3.3
First-Principles Simulations

Throughout materials science, and regarding amorphous Ge–Sb–Te alloys in particular, atomistic simulations using density-functional theory (DFT) and *ab initio* MD nowadays complement experiments [55]. Initial theoretical work has been concerned with the very structure of amorphous Ge–Sb–Te phases [41,56], and not much later the amorphous–crystalline transition (that is, the switching dynamics) has been under study [47] with models of increasing size and computational sophistication [57,58].

The interplay between experiments and theory in studying amorphous PCMs is highly successful, to a large share, because both techniques go to their extremes regarding length scales, and thereby approach one another: A memory cell on the order of a few nanometers is extremely small and challenging to manufacture, but it is large to the theorist, and constitutes the upper limit of what (today's) simulations are capable of. Thus, a goal for theory is to make larger time and length scales accessible: for example, novel machine learning based potentials speed up simulations while retaining close-to-DFT accuracy [59], and metadynamics allows one to sample the (rare) nucleation events that induce crystallization of PCMs [60]. Ultimately, these theoretical endeavors require carefully chosen models, sturdy computational techniques [55], but most importantly, an ongoing exchange with experimentalists to further unravel the microscopic nature of amorphous Ge–Sb–Te materials.

21.4
Materials Properties and Property Contrast

21.4.1
Resonant Bonding in PCMs

Bonding between atoms is a concept most central to chemistry, but again Ge–Sb–Te alloys have defied "simple" classification schemes; they are not textbook semiconductors, nor salts, and their bonding nature has been vividly discussed. Albeit often intrinsically vague, bonding models have proven useful and predictive all throughout chemistry [61], and we here introduce one such concept that has been suggested to play a key role in crystalline PCMs.

The "resonant" electronic structure of the benzene molecule, with more than one ground state superimposed, is taught in every first year chemistry course. Interestingly, a somewhat reminiscent concept has been proposed for solids in the 1970s, arguing for "resonant" long-range interactions of aligned p-orbitals in IV–VI semiconductors [62]. More recently, such "resonant bonding" has been identified in crystalline PCMs [63]. In their amorphous counterparts, by contrast, the long-range order (and thus the alignment of p-orbitals) is lost, and so there is no resonant but "classical" covalent bonding in amorphous PCMs.

Resonant bonding gives rise to a portfolio of peculiar properties [63]. Crystalline Ge–Sb–Te alloys show unusually large Born effective charges (in other words, they are sensitive toward lattice distortions that would decrease the alignment of p-orbitals); at the same time, they exhibit high dielectric constants ϵ_∞ [63] which is the origin of the optical contrast between crystalline and amorphous PCMs, and thus the functional principle of optical storage media [65].

Further, it has been suggested that resonance bonding can serve as a guideline to find new materials *beyond* Ge–Sb–Te. To survey and classify candidate compositions, the creation of "structure maps" with suitable coordinates has a long history in solid-state physics [66,67]. Precisely, such a scheme has been applied to PCMs, evidencing that those are found in a particular region where low lattice distortions and sufficiently low ionicity prevail (Figure 21.5) [64]. If one of these

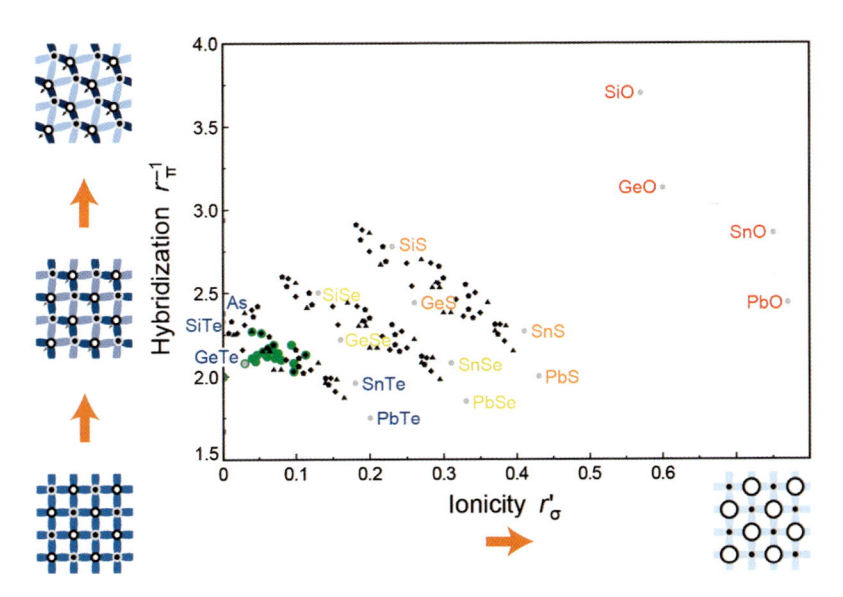

Figure 21.5 A structure map for group V elements and binary IV–VI compounds. The ionicity and hybridization (the latter measuring the tendency to form "sp^3" networks such as in zinc blende lattices) are determined according to orbital radii. Successful PCMs have been highlighted by green bullets and cluster in a small region on the lower left-hand side of the map. (Reprinted by permission from Macmillan Publishers Ltd: Nature Materials, Ref. [64], copyright 2008.)

conditions is not fulfilled, resonance bonding cannot occur and no viable PCMs are found throughout the corresponding regions of the map.

21.4.2
Origin and Nature of Vacancies

Judging from their composition alone, quasibinary alloys of GeTe and Sb_2Te_3 should be semiconductors and exhibit a bandgap. The crystalline phases, however, are electrically conducting, with quite significant concentrations of hole (p-type) charge carriers [68]. This is because a large number of cation vacancies form both in GeTe (cf. Section 21.2.1) and the ternary alloys; these move the Fermi level downward into the valence band [69]. Importantly, p-type conductivity in Ge–Sb–Te alloys is achieved without external doping (that is, without changing the chemical composition, unlike in typical semiconductors like Si or GaAs).

At this point, it is important to distinguish two types of vacancies. The first is the stoichiometric ones – which occur on the metastable rock salt-like Ge–Sb–Te lattices due to the dissimilar number of cations and anions (and thus not in GeTe). Their existence has been rationalized by quantum-chemical analyses that revealed antibonding (destabilizing) cation–anion interactions in the highest occupied crystal orbitals, and these electronic levels become depopulated when forming cation vacancies (38). Second, there are "excess" vacancies beyond the stoichiometric composition that give rise to p-type conductivity as the "missing" cations generate hole carriers. No matter how created, cation vacancies have pronounced effects on the macroscopic properties of crystalline Ge–Sb–Te alloys, as will be discussed later.

21.4.3
How Disorder Determines Electronic Properties

A typical switching experiment to characterize a candidate PCM proceeds as follows: Heat a freshly prepared amorphous sample until it crystallizes, then cool down to room temperature and measure the electronic conductivity during this cycle. An interesting effect was now observed by annealing $GeSb_2Te_4$ films up to different temperatures (Figure 21.6a): Depending on the final annealing temperature chosen, the crystallized materials span different resistance values, so widely spread that one observes a metal–insulator transition [70]. Structurally, this can be traced back to the aforementioned progressive ordering of vacancies (Section 21.2.3), and large-scale DFT computations helped to rationalize the observed effects: Vacancy clusters localize the electronic wavefunction (Figure 21.6b), whereas in the ordered phase, the highest occupied bands are evenly spread out and thus conducting (Figure 21.6c) [71].

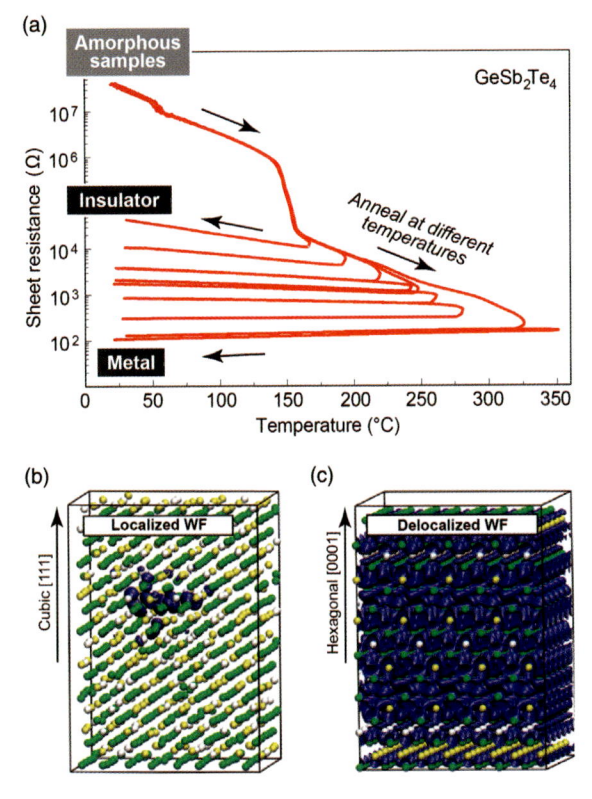

Figure 21.6 Metal–insulator transition in GeSb$_2$Te$_4$ and an atomistic explanation. (a) Electrical properties of GeSb$_2$Te$_4$ films during heating–cooling cycles (arrows). The final (room temperature) resistance values of crystallized GeSb$_2$Te$_4$ span several orders of magnitude, depending on the annealing temperature that may result either in cubic or in (pseudo) hexagonal modifications. (Adapted from Ref. [70].) (b,c) Isosurfaces encasing the highest occupied crystal orbital in large models of crystalline GeSb$_2$Te$_4$, either with fully randomly distributed vacancies (b) or fully ordered layers (c). (Adapted from Ref. [71].)

21.5
From Phase-Change Data Storage to a Versatile Materials Platform

Ge–Sb–Te alloys are almost synonymous with phase-change data storage nowadays, but there is reason to believe they could be valuable for new and diverse applications in the future [3]. We will hence conclude this chapter by mentioning three such examples.

Besides controlling electronic properties (Section 21.4.3), vacancy disorder also impacts thermal transport – that is, the heat conduction in crystalline PCMs. The latter is again unconventional in that it is very low [72], and being dependent on vacancy ordering, it can be tuned via both the sample composition and the degree of annealing [73]. The combination of low thermal and high electronic conductivity makes Ge–Sb–Te materials interesting for thermoelectric applications, in which waste heat is converted to energy [74,75]. Microscopically, a link between resonant bonding and low thermal conductivity has been established recently [76], and indeed a number of widely used thermoelectrics can be localized in the vicinity of PCMs on a "treasure map" similar to that in Figure 21.6 [40].

Likewise, the optical properties – which originally paved the way for the success of Ge–Sb–Te based optical memory disks [12]– can be put to new and creative uses. For example, tunable optoelectronic frameworks have been reported that could find application in flexible displays and "smart paper," again exploiting the contrast in optical properties between crystalline and amorphous Ge–Sb–Te phases [77].

Finally, these materials promise to play a pivotal role in the development of entirely new computing concepts. "Neuromorphic" architectures are inspired by the brain, relying on artificial synapses and neurons and could enable faster and more energy-efficient computing by orders of magnitude. Ovshinsky already envisioned such applications of PCMs in 2004 [78], and a decade later prototypes based on $Ge_2Sb_2Te_5$ have indeed been reported [79,80].

Whether in commercial optical disks, in novel memory chips, or in the very early prototypes just mentioned, Ge–Sb–Te alloys continue to be intriguing materials that challenge fundamental research and enable diverse applications. In all these endeavors, a good deal of solid-state *chemistry* (synthesis, structure determination, bonding theory, etc.) has been vital – and is likely to be in the future.

Acknowledgments

V.L.D. is grateful to Professor Richard Dronskowski for guidance and support, and to the Studienstiftung des deutschen Volkes for a doctoral fellowship. V.L.D. and M.W. acknowledge financial support by Deutsche Forschungsgemeinschaft (DFG) through SFB 917 "Nanoswitches". M.W. also acknowledges financial aid by the European Research Council (ERC) through the ERC Advanced Grant "Disorder Control".

References

1 Dronskowski, R. (2001) *Adv. Funct. Mater.*, **11**, 27–29.
2 Wuttig, M. and Yamada, N. (2007) *Nat. Mater.*, **6**, 824–832.
3 Deringer, V.L., Dronskowski, R., and Wuttig, M. (2015) *Adv. Funct. Mater.*, **25**, 6343–6359.
4 Ovshinsky, S.R. (1968) *Phys. Rev. Lett.*, **21**, 1450–1453.
5 Yamada, N., Ohno, E., Akahira, N., Nishiuchi, K., Nagata, K., and Takao, M. (1987) *Jpn. J. Appl. Phys.*, **26** (26-4), 61–66.

6 Yamada, N., Ohno, E., Nishiuchi, K., Akahira, N., and Takao, M. (1991) *J. Appl. Phys.*, **69**, 2849–2856.

7 Yamada, N. and Matsunaga, T. (2000) *J. Appl. Phys.*, **88**, 7020–7028.

8 Lankhorst, M.H.R., Ketelaars, B.W.S.M.M., and Wolters, R.A.M. (2005) *Nat. Mater.*, **4**, 347–352.

9 Lencer, D., Salinga, M., and Wuttig, M. (2011) *Adv. Mater.*, **23**, 2030–2058.

10 Burr, G.W., Breitwisch, M.J., Franceschini, M., Garetto, D., Gopalakrishnan, K., Jackson, B., Kurdi, B., Lam, C., Lastras, L.A., Padilla, A., Rajendran, B., Raoux, S., and Shenoy, R.S. (2010) *J. Vac. Sci. Technol. B*, **28**, 223–262.

11 Raoux, S., Wełnic, W., and Ielmini, D. (2010) *Chem. Rev.*, **110**, 240–267.

12 Yamada, N. (2012) *Phys. Status Solidi B*, **249**, 1837–1842.

13 Raoux, S., Xiong, F., Wuttig, M., and Pop, E. (2014) *MRS Bull.*, **39**, 703–710.

14 Raoux, S. and Wuttig, M., (Eds.) (2009) *Phase Change Materials: Science and Applications*, Springer, New York.

15 Kolobov, A.V. and Tominaga, J. (2012) *Chalcogenides: Metastability and Phase Change Phenomena*, Springer, Heidelberg.

16 Klemm, W. and Frischmuth, G. (1934) *Z. Anorg. Allg. Chem.*, **218**, 249–251.

17 Goldak, J., Barrett, C.S., Innes, D., and Youdelis, W. (1966) *J. Chem. Phys.*, **44**, 3323–3325.

18 Chattopadhyay, T., Boucherle, J.X., and von Schnering, H.G. (1987) *J. Phys. C Solid State Phys.*, **20**, 1431–1440.

19 Waghmare, U.V., Spaldin, N.A., Kandpal, H.C., and Seshadri, R. (2003) *Phys. Rev. B*, **67**, 125111.

20 Kolobov, A.V., Tominaga, J., Fons, P., and Uruga, T. (2003) *Appl. Phys. Lett.*, **82**, 382–384.

21 Semiletov, S.A. (1956) *Sov. Phys. Crystallogr.*, **1**, 317–319.

22 Schneider, M.N., Seibald, M., Lagally, P., and Oeckler, O. (2010) *J. Appl. Crystallogr.*, **43**, 1012–1020.

23 Agafonov, V., Rodier, N., Céolin, R., Bellissent, R., Bergman, C., and Gaspard, J.P. (1991) *Acta Crystallogr. C*, **47**, 1141–1143.

24 Iwasaki, H., Ide, Y., Harigaya, M., Kageyama, Y., and Fujimura, I. (1992) *Jpn. J. Appl. Phys. Pt. 1*, **31**, 461–465.

25 Matsunaga, T., Umetani, Y., and Yamada, N. (2001) *Phys. Rev. B*, **64**, 184116.

26 Matsunaga, T. and Yamada, N. (2004) *Jpn. J. Appl. Phys.*, **43**, 4704–4712.

27 Matsunaga, T., Akola, J., Kohara, S., Honma, T., Kobayashi, K., Ikenaga, E., Jones, R.O., Yamada, N., Takata, M., and Kojima, R. (2011) *Nat. Mater.*, **10**, 129–134.

28 Da Silva, J.L.F., Walsh, A., and Lee, H. (2008) *Phys. Rev. B*, **78**, 224111.

29 Agaev, K.A. and Talybov, A.G. (1968) *Sov. Phys. Crystallogr.*, **11**, 400–402.

30 Petrov, I.I., Imamov, R.M., and Pinsker, Z.G. (1968) *Sov. Phys. Crystallogr.*, **13**, 339–342.

31 Matsunaga, T., Yamada, N., and Kubota, Y. (2004) *Acta Crystallogr. B*, **60**, 685–691.

32 Matsunaga, T., Morita, H., Kojima, R., Yamada, N., Kifune, K., Kubota, Y., Tabata, Y., Kim, J.-J., Kobata, M., Ikenaga, E., and Kobayashi, K. (2008) *J. Appl. Phys.*, **103**, 093511.

33 Kooi, B.J. and De Hosson, J.T.M. (2002) *J. Appl. Phys.*, **92**, 3584–3590.

34 Schneider, M.N., Urban, P., Leineweber, A., Döblinger, M., and Oeckler, O. (2010) *Phys. Rev. B*, **81**, 184102.

35 Schneider, M.N., Biquard, X., Stiewe, C., Schröder, T., Urban, P., and Oeckler, O. (2012) *Chem. Commun.*, **48**, 2192–2194.

36 Matsunaga, T. and Yamada, N. (2004) *Phys. Rev. B*, **69**, 104111.

37 Matsunaga, T., Kojima, R., Yamada, N., Kifune, K., Kubota, Y., Tabata, Y., and Takata, M. (2006) *Inorg. Chem.*, **45**, 2235–2241.

38 Wuttig, M., Lüsebrink, D., Wamwangi, D., Wełnic, W., Gilleßen, M., and Dronskowski, R. (2007) *Nat. Mater.*, **6**, 122–128.

39 Zachariasen, W. (1932) *J. Am. Chem. Soc.*, **54**, 3841–3851.

40 Wuttig, M. and Raoux, S. (2012) *Z. Anorg. Allg. Chem.*, **638**, 2455–2465.

41 Caravati, S., Bernasconi, M., Kühne, T.D., Krack, M., and Parrinello, M. (2007) *Appl. Phys. Lett.*, **91**, 171906.

42 Mazzarello, R., Caravati, S., Angioletti-Uberti, S., Bernasconi, M., and Parrinello, M. (2010) *Phys. Rev. Lett.*, **104**, 085503.

43 Ielmini, D., Lacaita, A.L., and Mantegazza, D. (2007) *IEEE Trans. Electron Devices*, **54**, 308–315.

44 Fantini, P., Brazzelli, S., Cazzini, E., and Mani, A. (2012) *Appl. Phys. Lett.*, **100**, 013505.

45 Raty, J.-Y., Zhang, W., Luckas, J., Chen, C., Mazzarello, R., Bichara, C., and Wuttig, M. (2015) *Nat. Commun.*, **6**, 7467.

46 Deringer, V.L., Zhang, W., Lumeij, M., Maintz, S., Wuttig, M., Mazzarello, R., and Dronskowski, R. (2014) *Angew. Chem., Int. Ed.*, **53**, 10817–10820.

47 Hegedüs, J. and Elliott, S.R. (2008) *Nat. Mater.*, **7**, 399–405.

48 Akola, J., Jones, R.O., Kohara, S., Kimura, S., Kobayashi, K., Takata, M., Matsunaga, T., Kojima, R., and Yamada, N. (2009) *Phys. Rev. B*, **80**, 020201(R).

49 Elliott, S.R. (1991) *Nature*, **354**, 445–452.

50 Akola, J. and Jones, R.O. (2008) *J. Phys. Condens. Matter*, **20**, 465103.

51 Kolobov, A.V., Fons, P., Frenkel, A.I., Ankudinov, A.L., Tominaga, J., and Uruga, T. (2004) *Nat. Mater.*, **3**, 703–708.

52 Matsunaga, T., Yamada, N., Kojima, R., Shamoto, S., Sato, M., Tanida, H., Uruga, T., Kohara, S., Takata, M., Zalden, P., Bruns, G., Sergueev, I., Wille, H.C., Hermann, R.P., and Wuttig, M. (2011) *Adv. Funct. Mater.*, **21**, 2232–2239.

53 Baker, D.A., Paesler, M.A., Lucovsky, G., Agarwal, S.C., and Taylor, P.C. (2006) *Phys. Rev. Lett.*, **96**, 255501.

54 Ohara, K., Temleitner, L., Sugimoto, K., Kohara, S., Matsunaga, T., Pusztai, L., Itou, M., Ohsumi, H., Kojima, R., Yamada, N., Usuki, T., Fujiwara, A., and Takata, M. (2012) *Adv. Funct. Mater.*, **22**, 2251–2257.

55 Zhang, W., Deringer, V.L., Dronskowski, R., Mazzarello, R., Ma, E., and Wuttig, M. (2015) *MRS Bull.*, **40**, 856–865.

56 Akola, J. and Jones, R.O. (2007) *Phys. Rev. B*, **76**, 235201.

57 Kalikka, J., Akola, J., and Jones, R.O. (2014) *Phys. Rev. B*, **90**, 184109.

58 Zhang, W., Ronneberger, I., Zalden, P., Xu, M., Salinga, M., Wuttig, M., and Mazzarello, R. (2014) *Sci. Rep.*, **4**, 6529.

59 Sosso, G.C., Miceli, G., Caravati, S., Giberti, F., Behler, J., and Bernasconi, M. (2013) *J. Phys. Chem. Lett.*, **4**, 4241–4246.

60 Ronneberger, I., Zhang, H., Eshet, H., and Mazzarello, R. (2015) *Adv. Funct. Mater.*, **25**, 6407–6413.

61 Dronskowski, R. (2005) *Computational Chemistry of Solid State Materials*, Wiley-VCH Verlag GmbH, Weinheim.

62 Lucovsky, G. and White, R.M. (1973) *Phys. Rev. B*, **8**, 660–667.

63 Shportko, K., Kremers, S., Woda, M., Lencer, D., Robertson, J., and Wuttig, M. (2008) *Nat. Mater.*, **7**, 653–658.

64 Lencer, D., Salinga, M., Grabowski, B., Hickel, T., Neugebauer, J., and Wuttig, M. (2008) *Nat. Mater.*, **7**, 972–977.

65 Wełnic, W., Botti, S., Reining, L., and Wuttig, M. (2007) *Phys. Rev. Lett.*, **98**, 236403.

66 Phillips, J.C. and van Vechten, J.A. (1969) *Phys. Rev. Lett.*, **22**, 705–708.

67 St. John, J. and Bloch, A.N. (1974) *Phys. Rev. Lett.*, **33**, 1095–1098.

68 Yáñez-Limón, J.M., González-Hernández, J., Alvarado-Gil, J.J., Delgadillo, I., and Vargas, H. (1995) *Phys. Rev. B*, **52**, 16321–16324.

69 Edwards, A.H., Pineda, A.C., Schultz, P.A., Martin, M.G., Thompson, A.P., Hjalmarson, H.P., and Umrigar, C.J. (2006) *Phys. Rev. B*, **73**, 045210.

70 Siegrist, T., Jost, P., Volker, H., Woda, M., Merkelbach, P., Schlockermann, C., and Wuttig, M. (2011) *Nat. Mater.*, **10**, 202–208.

71 Zhang, W., Thiess, A., Zalden, P., Zeller, R., Dederichs, P.H., Raty, J.-Y., Wuttig, M., Blügel, S., and Mazzarello, R. (2012) *Nat. Mater.*, **11**, 952–956.

72 Risk, W.P., Rettner, C.T., and Raoux, S. (2009) *Appl. Phys. Lett.*, **94**, 101906.

73 Siegert, K.S., Lange, F.R.L., Sittner, E.R., Volker, H., Schlockermann, C., Siegrist, T., and Wuttig, M. (2015) *Rep. Prog. Phys.*, **78**, 013001.

74 Schneider, M.N., Rosenthal, T., Stiewe, C., and Oeckler, O. (2010) *Z. Kristallogr.*, **225**, 463–470.

75 Sittner, E.-R., Siegert, K.S., Jost, P., Schlockermann, C., Lange, F.R.L., and Wuttig, M. (2013) *Phys. Status Solidi A*, **210**, 147–152.

76 Lee, S., Esfarjani, K., Luo, T., Zhou, J., Tian, Z., and Chen, G. (2014) *Nat. Commun.*, **5**, 3525.
77 Hosseini, P., Wright, C.D., and Bhaskaran, H. (2014) *Nature*, **511**, 206–211.
78 Ovshinsky, S.R. (2004) *Jpn. J. Appl. Phys.*, **43**, 4695–4699.
79 Wright, C.D., Liu, Y., Kohary, K.I., Aziz, M.M., and Hicken, R.J. (2011) *Adv. Mater.*, **23**, 3408–3413.
80 Kuzum, D., Jeyasingh, R.G.D., Lee, B., and Wong, H.-S.P. (2012) *Nano Lett.*, **12**, 2179–2186.

Index

a

$A_2BB'O_6$ double perovskites
– B-site cation order patterns 229
ab initio calculations 6, 111, 272
ab initio methods 297
$A^{3+}B^{3+}O^{2-}_3$ compounds
– classification of 226
ABX_3 perovskite
– cubic crystal structure of 224
– schematic drawing of 225
acid–base reactions 62, 674
acoustic phonons (overdamping) 656
actinide hydrides 500
activation energies 494, 496, 514, 601
adiabatic temperature 400
AeH_2 cotunnite structure 486
AeH_x polyhedra 486
Ae–H coordination 486
$AE_2[ZnN_2]$
– crystal structures of 291
Ag–Ag interactions 43
$AgPb_{18}SbTe_{20}$ 456
$AgTaN_2$ 293
A- and B-site cation
– ferrimagnetic spin structure of 237
A- and B-site cation-ordered perovskite
– crystal structure of 236
A-type-La_2O_3 structure 262
A-site cation ordered perovskite
– crystal structure of 234
A-site deficient perovskites 206
A-site lanthanide ions 227
AlB_2-structure type 444
AlB_2-type borides 441
$Al_{13}(Co,Ni)_4$ structure 77
Al-Cu-Fe-Si system 83
Al-Fe-Si system 83
alkali borate glasses 122
– Mayenite with 505
alkali halide crystals 487
alkali hydride, cubic lattice parameters 485
alkali hydrides 487, 488
alkali-metal based 1/1 Bergman phases 47
alkali metal antimonides
– photoemissivity of 457
alkali metal halide layers 581
alkali metal hydrides 484
alkali metals 3, 139, 140, 141, 525
alkali metal suboxides 140, 147, 149, 150, 157
alkaline earth hydride crystal structures 486
alkaline earth hydrides 479, 485, 486, 487, 488, 493
alkaline earth ions 486
alkaline earth metal containing compounds 317
alkaline earth metal halides 257
alkaline earth metal-rich systems 291
alkaline earth metal nitrides, of transition metals 326
alkaline earth metal nitridocuprates 292
alkaline earth metal nitridometalates 300
alkaline earth metals 140, 158, 443, 477, 525, 535, 537
– subnitrides of 150
alkaline earth subnitrides 157
alkali silicate 122
allotropes 436
– of Ge 529
Al-Mn-Fe-Si system 83
Al-Mn-Si system 83
Al–Cu network
Al–O–P linkages 117
Al-Pd-Mn-Si system 83
α-Mg_3Sb_2
– crystal structure of 457
α-$NaFeO_2$- type structure 325, 574

α-PbO_2 structure 269
$AlPO_4$ segregation scenario 117
Al-Re-Si system 83
Al-Rh-Si system 83
aluminum 440, 479
aluminum-cobalt system 76
ambient temperature VO_2 structure 183
ambipolar conductivity 685
amine intercalation 589
ammonia catalysis 505
ammonium Rochelle salt 653
ammonolysis 370
– preparation of 371
– reactions 255, 371
amorphous–crystalline transition 742
amorphous phase-change materials (PCMs)
– characterestics 740
– microscopic complexity 741
amorphous phases 740
– experimental probes 742
– first-principles simulations 742
AMRO. See angular magnetoresistance oscillation (AMRO)
angular dependence 610
angular magnetoresistance oscillation (AMRO) 610
angular momentum 4
anionic conduction 364
anionic formal charge 361
anionic lattice 362
anionic p-phenylene derivative
– chemical formula of 547
anionic poly(p-phenylene) derivative 546
anionic tetrahedral networks
– dimensionalities of 317
anion vacancy orde 233
anisotropic magnetic properties 447
anisotropic magnetothermal properties, of $HoAlO_3$ single crystals 631
anisotropy 571, 630
anomalous Hall conductivity
– scaling 608
antibonding (destabilizing) cation–anion interactions 744
antibonding interactions 41
antibonding orbitals 20, 62
antibonding peak 37
antibonding states 37
antiferroelectric $PbZrO_3$
– displacement pattern of Pb^{2+} ions 652
antiferromagnet 412, 415, 619
antiferromagnetic
– arrangement 598

– correlations 633
– coupling 716
– hexagonal spin cluster 629
– interaction 414, 513
antiferromagnetic order 269, 628, 708
antiferromagnetic spin clusters 628
antiferromagnetic state 39, 501
antiferromagnetic transition temperature 502
antiferromagnetic triangle 619
antiferromagnetic V–O–V superexchange interaction 512
antiferromagnetism 231
antiferromagnets 619
anti-Frenkel disorder 678
anti-2H-$BaNiO_3$ 274
antimonides 456, 471
– semiconductance of 465
– for thermoelectric energy conversion 471
antimony telluride (Sb_2Te_3) 737
antimony, toxicity of 472
antiperovskite 243, 503
antipolar structure 651
antiprisms 460
aperiodic crystals 73
2/1 approximant structure 84
aqueous chemistry 672
aristotype Li_4SrN_2 294
arrangement of truncated icosahedra in the 1/1 Tsai-type approximant 88
Arrhenius law 675
arsenides 456
arsenopyrite 463
artificial ion conductors 687
artificial synapses 746
atom, composition 643
atomic hydrogen 478, 479
atomic orbital (AO) 2, 3, 5, 6, 7, 52
atomic positions and symmetry information for the CCP lattice 165
atomic positions and symmetry information for the HCP lattice 166
atomic radii ratio 525
atomic size effects 57
atomistic simulation 741
Au–Sn and K–Sn COHP curves 16
Au_2O_3, structure 178
aurivillius compounds 634
aurivillius-type structure 657
aurivillius series members, structures 576
avalanche effect 681
average magnetic moment per Fe ion 415
avian magnetic compass 722

avian magnetoreception 722
A$_2$X$_3$ bixbyite
– structures of 366
AX$_2$ fluorite
– structures of 366
azide-ligand 715
– end-on bridge 715
– end-to-end bridge 715
azide (N^{3-}) ligand 715
azido bridge 715

b

Ba$_3$AlO$_4$H crystal structure 504
BaAl$_4$ type structure 535
Ba$_{14}$(Ca,Sr)N$_6$ clusters 153
back reaction rate 683
Ba$_{16}$[CuN]$_8$[Cu$_2$N$_3$][Cu$_3$N$_4$] 297
baddeleyite (P2$_1$/c) unit cell for ZrO$_2$ and HfO$_2$ 188
Ba$_3$[FeN$_3$] 301
BaGa$_2$Sb$_2$ 467
– crystal structure of 467
Ba^{2+} ions 180
Ba$_{14}$LiN$_6$ cluster 153
ball milling 385
Ba$_4$[Mn$_3$N$_6$] 320
Ba$_6$Mn$_{20}$N$_{21.5-d}$ 307
Ba$_5$Na$_{12}$ polyhedra 156
Ba$_{16}$[NbN$_4$]$_3$[Nb$_2$N$_7$] 319
band ferromagnetism 705
bandgap 8, 16, 546, 744
band insulators 426
Ba$_2$[Ni$_3$N$_2$] 298
– crystal structure of 298
Ba$_2$[Ni$_2$(Ni$_{1-x}$Li$_x$)N$_2$] 298
Ba$_8$[NiN]$_6$N
– crystal structure 297
barium cations 504
barium hydride 493
bar magnets 721
Barret formula 657
Ba$_8$Sb$_4$OH$_2$ crystal structures 505
BaTiO$_3$ 644, 649
– temperature dependent polarization directions 649
BaTi^{4+}O$_{2.4}$N$_{0.4}$ structure 511
Bärnighausen formalism 533
B–O–Al linkages 118
B–O(1) sublattice consists of corner sharing B–O(1) octahedra 194
bcc Jones zone 30
bcc solid solution 267
Bergman-type clusters 81

Bergman 1/1 quasi-crystalline approximants 45
Berthollet'sches Knallsilber 272
β-Mn nitrides
– Al$_2$Mo$_3$C type 280
β-Mn structure 76
β-rhombohedral boron 435
– crystal structure 436
– type borides 449
β-Zn$_4$Sb$_3$
– crystal structure of 464
B-site cation
– crystal structure B-site cation in perovskite of 233, 238
B-site cation orders 230
B-site-deficient perovskites 207
B-sublattice of spinel structure 627
B-sublattice structure (pyrochlore lattice) 627
Bifidobacteria breve 721
bifunctional molecular magnets 720
bimetallic/alloy catalysts 551
bimolecular rate law 682
binary antimonides 456–464
– early transition metal antimonides 459
– late transition metal antimonides 462
– main group antimonides 457
binary borides
– group elements 437
– lanthanides 437
– main group metals 436
binary hydrogen compounds 497
binary intermetallic phases 56
binary IV–VI compounds
– structure map 743
binary model glasses, medium-range order effects in 118
– cation next-nearest neighbor environments and spatial distributions 118
– ring-size distributions and "super-structural units 124
binary nitrides
– chemical bonding 252
– crystal structures 264, 268
– enthalpies of 259
– idealized crystal structures 263
binary phase diagrams 24
binary P$_2$O$_5$–GeO$_2$ glass system 115
binary system B$_2$O$_3$–SiO$_2$ 114
binary TaN 285
binary TMO with ideal rutile structure 182
binary TM oxides
– MO, with NaCl structure 170
binary transition metal nitrides 251

binary transition metals 252, 464
– nitrides, synthesis of 251, 252, 258, 260
binary Zn_3N_2 272
biogenic magnetism 720–723
– magnetotactic bacteria 721
biogenic magnetite 722
biomineralization 721
bioplatforms 721
$Bi_{95}Sb_5$ 458
Bi_2SiO_5 657
– ferrielectric-like arrangement of dipole moments 657
– vs. Bi_2WO_6 657
bismuth Aurivillius phases 244
bismuthides 456
Bi_2Te_3
– superlattices of 456
$Bi_4Ti_3O_{12}$ structure 574
Bi_2WO_6 657
bixbyites 190, 365
Bloch states 609
boat-like conformation 445
body-centered tetragonal structure 38
Bohr magneton 597
Boltzmann conditions 673
Boltzmann constant 661
Boltzmann equations 596, 598
Boltzmann statistics 672
bond angle distribution 112
bond angles 27
bond valence sums (BVS) 488
borate glasses 95, 120
boric acid 438
boride
– synthesis 439
boride crystals 440
boride materials 439
borides 435, 442
– icosahedra 448
– synthesis 439
– – nanomaterials 440
– – powder preparation 439
– – single-crystal growth 439
Born-Mayer potentials 146
boron 435
– allotropes 435
– atom arrangement, in metallic borides 438
– icosahedron atom 436
– structural arrangement of atoms 441
boron halides 439
boron hydrides 479
boron-rich borides 449
boron-rich materials 438

boron-rich metal borides 436
boron-rich solids 446
boron–oxygen compounds 438
boron nitride 443
boron oxide 113, 439
bottom of conducting band (BCB) 546
Bragg reflections 633
Bravais lattice 30, 32
bridging ligands 704
bridging oxygen atoms 544
Bridgman methods 256
Brillouin zone 656
brittleness 500
brookite (TiO_2) showing channels 186
Brønsted acid–base picture 671
Brønsted acids 577, 587
Brønsted concept 668
Brønsted-type basic sites 549
Brønsted level 671
Brouwer approximation 673
brownmillerite, $A_2BB'O_5$, structure 209
brownmillerite-type $A_2B_2O_5$
– crystal structure of 239
brownmillerites 207
– structure 210
– type 239
brucite 584
– distortion of 542
– structure of 542
B/Si ratio 116
bulk defects 686
bulk sensors 715
Butler–Volmer equation 685

c
$Ca_7Ag_{2.72}N_4$ 330
Ca@Ag_{18} polyhedra
– chemical pressure 59
CaAgSb 466
– crystal structures of 466
Ca_3AlSb_3 468
– crystal structures of 468
$Ca_5Al_2Sb_6$
– crystal structures of 470
$Ca_{14}AlSb_{11}$ 470
$CaAl_2Si_2$ 466
Ca_2AuN 330
$Ca_2B_2C_2$ 17
$Ca_4[Cr_2N_6]$
– edge-sharing double tetrahedra 319
$Ca_6[Cr_2N_6]H$ 304
$CaCu_5$-type phase $SrAg_5$ 26
$CaCu_5$-type structure 283

CaCu$_5$ structure 59
Cadmium 86
CaF$_2$ fluorite structure 187
CaIn$_2$ type structure 526
calcination 369, 551
calcination–reduction process 551
Ca$_2$[Li(Fe$_{1-x}$Li$_x$)N$_2$] 293
Ca$_2$[(Li$_{1-y}$Cu$_y$)(Cu$_{1-x}$Li$_x$)N$_2$] 293
Ca(Li[Mn$_{1-x}$Li$_x$N])$_2$
– crystal structures of 295
Ca$_9$Mn$_4$Bi$_9$ 468
Ca$_3$[MnN$_3$]
– structural relationship 302
Ca$_{12}$[Mn$_{19}$N$_{23}$]
– crystal structures 307
Ca[MnN]·Li$_2$[MnN] 295
cancer phototherapy 713
cancer therapy 710
Ca–Ag phase diagram 59
Ca–Ag system 59
(Ca$_3$N)$_2$[MN$_3$]
– isolated trigonal-planar nitridometalate anions 301
capping agent 712
carbon-containing nanocomposites 462
carbon-like frameworks of boron atoms, in borides 449
carbon nanorings (CNRs) 548, 556
carbon nanotubes 462
carbothermal reaction 365
carlsbergite 267
catalytic sites 548
CaTiO$_3$ 656
– structure parameters of 227
cation clustering 120
cation coordination polyhedra, for 8a site in bixbyite 190
cation defects 677
cationic conductors 687
cation ion radii 166
cation next-nearest environments and spatial distributions studied by dipolar NMR 120
cation ordering 739
cation vacancies 744
Ca$_2$[VN$_3$]
– one-dimensional chains 320
Ca$_5$[VN$_4$]N
– crystal structures of 318
CaWO$_4$
– Ca- and W-based polyhedra in
– – scheelite arrangement of 367
– luminescent properties of 369

CCP lattice 165
Cd$_{58}$Ce$_{13}$ and Zn$_{58+\delta}$Ho$_{13,}$ complex modulated behavior 89
Cd-doping 684
CdSe and ZnS, mechanically induced self-propagating reactions 399
CdTiO$_3$ 656
Ce$_6$Cr$_{20-x}$N$_{22}$ 306
cellulose acetate (CA) 558
Ce$_2$[MnN$_3$]
– crystal structure 322
cesium 140, 141, 143
– atom 149
– suboxides 148
cetyltrimethylammonium bromide (CTAB) 712
chalcogenide compounds 528
chalcogenides 13, 383
chalcogens 383
charge compensation 711
charge density wave 417
charge disproportionation
– schematic drawings of 235
charge-transfer material
– purely organic fullerene-based 706
charge-transfer organic complexes 703
chemical bonding 532
chemical diffusion 684
– coefficient 685, 689
chemical hydrides 477
chemical-bonding 737
chemical kinetics 668
chemically ordered region 660
chemical potential of Ag 672
chemical pressure 58, 225
chemical reaction 693
chemical reactions, simple 685
chemical shielding 532
chemical shift anisotropy (CSA) 532
chemical transport process 689
chemical vapor deposition (CVD) 253
chemotaxis 721
chimie douce-type process 371
chiral organic fluids 427
chiral proline 430
chocolate-brown Hg$_3$N$_2$ 272
chromium-containing steels
– steel-hardening process 267
chromium ions 167
chromium oxides 167
citrate complexation 369
classical magnetoresistance 595
classical spin liquid 622

clathrates 529
– type I and type II crystal structure, cut out 528
ClayFF force field 544
close packing (CP) 163
clustering behavior 124
clusters in primitive Mackay 1/1 approximant 85
cobalt-rich disordered phase $FeCo_3N$ orders 275
cobalt(II) hexacyanoferrate(III) 713
cobaltocene-doped derivative 706
cobalt oxyhydrides 507
$Co_{23}B_6$ crystal structure 442
CO_2 effect on the defect chemistry in hydroxides 680
coercive field 647
Co(II)Cu(II)-based ferrimagnetic chains 710
coloring problem 17
colossal magnetoresistance (CMJ) 595, 604
columbite structure of $CoNb_2O_6$ 184
compass magnetoreceptor 722
complex anions $[Fe_2N_4]^{8-}$, 1D $[Fe_2N_3]^{5-}$ 303
Complex Layered Oxides, structural diversity 571
compositional complexity 571
compositional gradient 397
composition dependence 414
compound $(Au_{12}Sn_9)^{23-}$ clusters 15
computational models 545
computed tomography (CT) 562
Co_4N crystallizes 271
– in perovskite structure 271
conduction band 668
conduction electrons 598, 610
conductivities 478, 492, 493, 506, 514, 528, 680, 684, 693
– isotherm for single crystalline and mesoscopic nanocrystalline 689
connectivities
– cation distributions, in glasses with multiple network formers 129
– and cation distributions, in glasses with multiple network formers 126, 127
connectivity distribution of the bridging oxygen atoms in $(K_2O)_{0.33}((B_2O_3)_x(P_2O_5)_{1-x})_{0.67}$ glasses derived from NMR data and comparison with 128
connectivity distribution of the bridging oxygen atoms in $(Na_2O)_{0.33}((Ge_2O_4)_x(P_2O_5)_{1-x})_{0.67}$ glasses derived from NMR data and comparison with 128

connectivity distribution of the bridging oxygen atoms in $(Na_2O)_{0.33}((Te_2O_4)_x(P_2O_5)_{1-x})_{0.67}$ glasses derived from NMR data and comparison with 129
consecutive shells of the Bergman cluster in the icosahedral 1/1 approximant 82
continuous random network model 93
contour plots of magnetic neutron scattering intensity 628
cooling 406
cooperative paramagnet 620
cooperative paramagnetism 622
coordination number (CN) 166
coordination polyhedron of the A-site in pyrochlore 192
coordination sphere of boron atoms 441
coordinatively unsaturated ZnAl-LDH nanosheets 550
copper 440
copper chalcogenides, synthesis of
– by ball milling 385
copper dianion 708
copper dianion precursors 708
copper halides-based oxide 588
copper hexacyanoferrate (CuHCF) 723
copper-red crystals Rb_9O_2 141
copper metal 385
corrosion 692
CORSee chemically ordered region 660
corundum or Al_2O_3 177
– binary oxides 178
– ilmentite and other ordered corundum-like structures 180
– structure 177
$CoSb_3$
– thermoelectric properties of 463
Coulomb bonding 668
Coulomb forces 668
Coulombic interactions 544
Coulomb interaction 681
Coulomb repulsion 148, 605
counterintuitive orbital mixing 6
coupling of ionic excitation and electronic excitation 671
covalent bond 477, 478
covalent bonding 384
covalent network glasses 93
covalent network glasses, medium range order in 109
– glassy silica and the binary system 109
– network connectivity in 113
– ring size distributions 112

covalent nitrides 331
covalent oxide glasses 96
Coxeter-Boerdijk helices 76
CrB_4 structure 450
$CrGa_4$
– 18-electron configurations 52
$CrGa_4$'s stoichiometry 53
Cr_3GaN
– electronic structures 278
$CrGa_4$ structure 51
crocon
– P–E loops 662
croconic acid 662
crocon polarization switching
– structural change before and after 662
cross-layer magnetoresistance 602
cross-linking density 361
cross-plane transfer energy 610
Cr_3Se_4, magnetic structure of 412
crucible materials 254
cryocooling 713, 717
cryogenic temperatures 717
cryptochromes (CRY) 722
crystal
– orbitals 7, 8, 49
– piezoelectric deformation 650
– properties 643
– structures 27, 362, 483, 681
– – and compositions 498
– – $La_2Ni_2O_5$ 240
– – TMO, representation of 169
crystal chemistry 478
– TMO 163
– – close packing and hole filling 163
crystalline Ge–Sb–Te phases 737
– metastable 737, 739
– stable 737
– – quasibinary phases 739
crystalline-electric field (CEF) 631
crystallinity 385
crystallization 440
crystallized thiophene nanowire 720
Crystallographic Information File 163
crystal orbital Hamilton populations (COHP) 1, 7, 9, 11
– analysis 10, 15
– curves 10
– data 16
crystal orbital overlap population (COOP) 1, 7, 9, 10
– analysis 11, 13, 15
– Hoffmann and Zheng's application 12

crystal symmetry 643
CsCl-type high-pressure modification 261
CsCl structures 17, 64
Cs/O ratio 141
$CsTi_2NbO_7$ structure 585
CTAB. See cetyltrimethylammonium bromide
Cu_2AlMn-type 306
– metal substructure 308
$CuAl_2$ type structure 535
cube-octahedra 446
cubic Al_2Mo_3C structure type 279
cubic $Ca_{19}Ag_8N_7$ 329
cubic crystal structures 33
cubic Cu_3N with ReO_3-type structure 277
cubic η-carbide Fe_3W_3C structure 280
cubic face-centered lattice 362
cubic γ(-Fe_4N 270
– structure 273
cubic high-temperature phase γ-Mo_2N 267
cubic perovskite
– Cartesian axes of 224
cubic (Pm-3m) perovskite structure 197
cubic $SrTiO_3$
– crystal structures of 226
Cu ions 397
$Cu_{1-x}Al_2$ crystal structure 533
Cu–Al system 535
Cu–Cd and Au–Cd binary phase diagrams 33
Cu–Se system, partial phase diagram of 395
Cu–Ti
– binary phase diagrams 65
Cu–Zn system 523
Cu NMR signals 535
Cu_3PdN 277
Curie constant 413, 647
Curie–Weiss behavior 300, 627
– magnetic susceptibility 300
Curie–Weiss law 6, 647
Curie–Weiss temperature 622, 627, 658
Curie temperatures 276, 283, 600, 609, 654
Cu_3Se_2 by making α-CuSe and α-Cu_2Se, formation of 394
$Cu_{15}Si_4$ 31
CuTi structures 17, 65
cyano bridge 712
cyanogen, $(CN)_2$ 331
cyano-nitrido-metalates 255
cyanometallates 714
cyclotron frequency 596
cyclotron motion 610
Czochralski (Cz) technique 631

d

data processing 735
data storage 735
– applications 735
– density 718
Debye frequency 661
Debye–Waller factors 30
Debye temperature 659
Debye wave vector 659
decagonal approximants 76
defect chemical treatment, shown for an oxide with dominant anti-Frenkel-disorder in oxygen sublattice 676
defect chemical variations
– within homogeneity range of oxide 677
– in homogeneity ranges of, PbO and 678
defect chemistry 665, 667, 668, 671, 675
– at boundaries and heterogeneous electrolytes 686
defect concentration, as function of temperature 682
defect fluorite-type
– structures of 364
defect fluorites
– vacancies, role of 378
defect-fluorite solid solutions
– R–Ta–O–N system 376
– R–W–O–N system 373
degeneracy 619
degree of anisotropy 488
de Haas–van Alphen effect 610
dehydration–rehydration process 710
Delafossite-type nitrides $CuNbN_2$ 293, 328
deltahedra 446
δ-MoN, hexagonal modifications 267
δ-NbN in nitrogen atmosphere 265
dendrites 722
density functional theory (DFT) 4, 105, 543, 544, 742
– calculations 15, 26
– calibrated Hückel model 63
– calibrated simple Hückel DOS 8
density of states (DOS) 7
– curves 9, 13
– distribution 62
desintercalation 364
deuterides 485
deuterium 477, 484
DFT. See density functional theory
3d-transition metal chalcogenides, structural relation between 403
diamond 46
diantimonides 462

diatomic covalent bond 3
diborides 442, 443
dielectrics 643
– classification 643
– constants, effect of oxygen isotope exchange 658
– materials, classification 644
– properties 232
diffraction 73
– techniques 533
diffuse scattering pattern 632, 633
diffusion barriers 251
diffusion coefficient 545, 683
dilute magnetic semiconductor (DMS) 595, 608, 609
3+1 dimensional direct space 74
Dion–Jacobson phases 243, 579
– crystal structures of 244
Dion–Jacobson series members, structures of 578
dipolar second moments in sodium ultraphosphate glasses 125
dipolar second moment values, measured via spin echo decay method 121
dipole glass 660
dipole interaction 660
dipole magnets 721
dipole–dipole interactions 114, 120
Dirac cone 611
disproportionation, between light Li and heavy Ag sites 42
DMS. See dilute magnetic semiconductor (DMS)
$3d^4$–$3d^7$ coordination 714
dodecaborides, icosahedra 448
dodecagonal approximants 80
donor/acceptor impurities 609
dopant anions 557
doping effect 223, 493, 638, 680, 692, 705
double-exchange ferromagnetic phase 606
double-exchange interaction 602, 603
double-exchange mechanism 600, 601
double-layer Ruddlesden–Popper perovskite structure 508
double hysteresis loop 651
double perovskite $A_2BB'O_6$
– spin glass magnetic structure of 232
double perovskites
– charge differences of 230
– group–subgroup relationships of space groups 231
doxorubicin (DOX) 563
Dreiding force field 544

Drosophila melanogaster 722
Drude model 596
3D ThCr$_2$Si$_2$ structure 13
3d transition metal chalcogenides 403
duralumin 521
DVD-RAM 735
Dyaloshinski–Moriya (DM) interaction 634, 637
dye–alkylsulfonate/LDH system 545

e
effective Bohr magnetons 413
eH orbitals 41
eHtuner 6
elastic effects 688
electrical conductivity 438, 443, 444, 528, 603, 609, 675, 684
electrical data storage devices 530
electrical driving forces 685
electrical gradients 683
electrical potential 683
electrical resistivity 385, 418, 420, 741
– β-Cu$_{2-x}$Se 386
– In$_x$Nb$_3$Te$_4$ 421
– measurements 288
– temperature variations of 423, 424
electric dipole 643
electric energy 528
electric field gradient (EFG) 104, 114, 532, 643
electric polarization 643, 644
electric potential 682
electric quadrupole moment eQ 104
electric susceptibilities 650
electrocatalysis 691, 692
electrocatalytic oxidation 557
electrochemical applications 693
electrochemical devices, for energy research 694
electrochemical energy applications 665
electrochemical energy technology 669
electrochemical potentials 479, 672
electrochemical reactors 685
electrochemistry 494
– functional materials for 669
electrode/electrolyte interfaces 685
electromotive force (EMF) 398
– α-Cu$_2$Se | β-CuSe system 399
electron affinity 6
electron beam lithography 448
electron density 2, 479, 483
– antimony atom 74
electron diffraction 421, 739

electron doping 510
electronegative elements 18
electronegative metals 61
electronegativity differences 60
electron energy DOS 32
electroneutrality 365
electroneutrality equation 672
electron-mediated magnetic interaction 600
electronic carriers 685
electronic charge density 417
electronic conduction 500, 506, 675, 684
electronic conductors 686
– current collectors or taguchi sensors 696
– – fuel cells and batteries 696
– – sensors 696
electronic defects 668, 684
electronic disorder 668
electronic DOS pseudogaps 51
electronic energy 20
electronic influences on site symmetry 168
electronic materials 687
electronic occupation 714
electronic properties determination
– disorder, role of 745
electronic stability 13
electronic structure 7
electronic wavefunction 746
electron localization function (ELF) 315
electron microscopy 77, 389
electron–electron repulsion 608
electron–hole separation 549
electron probe microanalysis (EPMA) 390
electron spin polarizer 719
electron tunneling 606
electrophilicity 1
electropositive elements 525
electropositive metal 14, 61
electrostatic interactions 552
electrostatic lattice energy 230
electrostriction coupling 650
electrostriction effect 650
elemental mixture in methanol 391
elementary plaquette of a triangular lattice 621
elements of frustration 620
elements that accommodated on A-site and B-site of perovskite structure oxides 198
EMF. See electromotive force
enantiomeric secondary alcohols 428
enantioselective crystallization 428, 430
– amorphous Se 427
energy-band structure 420

energy level diagrams for water, AgCl, and Si) 670
energy resolution 632
enhanced permeability and retention (EPR) 562
– measurements 306
enthalpy 115, 478, 482, 483, 673
entries in Shannon Tables of Effective Ionic Radii (Å), example 167
entropies 483
entropy 682
enzyme immobilization 710
EPMA. *See* electron probe microanalysis electron probe microanalysis (EPMA)
ε-Fe_2CoN 279
ε-Fe_2NiN 279
ε-Fe_3N_{1+x} 269
ε-Mn_4N crystallizes 268
ε-phase 270
ε-TaN 266
ε-TaN transforms 265
ε-Ti_2N 262
equiatomic antimonide 466
equilibrium conditions 668
essentially discrete diffraction diagram 73
EuB_6 crystals 447
$(Eu_3N)In$ 278
$Eu_4[TaMgN_5]$
– crystal structures 321
EXAFS. *See* extended X- ray fine structure absorption spectroscopy (EXAFS)
extended Hückel (eH) 5
– methods 18, 297
extended X-ray absorption fine structure 94
extended X-ray fine structure absorption spectroscopy (EXAFS) 742
extracellular matrix (ECM) 710

f

fabrication of an $(LDH/PSS)_n$–PVA film
– schematic illustration of 561
face-centered cubic (fcc) compounds 39
face-centered cubic (fcc) symmetry 712
face-centered cubic (fcc) 164, 618
$Fe_{3.1}Ga_{0.9}N$ 275
Fe-containing phase 293
Fe-phthalocyanine 604
Fe(II)Cu(II)-based magnets 709
$[Fe^{III}N_3]^{6-}$ anions 303
$FeMoN_2$
– crystal structures of 327
Fe_3Nb_3N 281

Fe–Fe bond distances 37
Fe–N phase diagram 270
FeNiN 285
Fe^{3+} oxides 584
fergusonite 377
fermi energy 8, 52
fermi level 671
fermi liquid behavior 603
fermi sphere 31
fermi surface 422, 595
– Nb_3Te_4 420
fermi wavenumber 608
fermi wave vectors 417
ferrelectric
– first-order phase transition 648
ferrielectric 651
ferrimagnet 412, 705, 720
ferrimagnetic building-blocks 706
ferrimagnetic chains
– assembly of 707
ferrimagnetic interactions 707
ferrimagnetic octacyanate 710
ferrimagnetic ordering 708
ferrimagnetism 415
ferroaxial coupling 236
ferroelectric 579, 720
– definition 644
– displacive-type 656
– first-order phase transitions 648
– free energy 649
– optical soft mode 644
– phenomenological considerations 653
– polarization 236, 637
– second-order phase transition 647
– soft mode frequency 644
– successive phase transition 649
ferroelectric Aurivillius phases
– crystal structure of 246
ferroelectric transition temperatures 653, 657, 662
ferromagnet 705
ferromagnetic and antiferromagnetic interactions 634
ferromagnetic coupling 527, 630
ferromagnetic interaction 235, 414, 714
ferromagnetic interactions 712
ferromagnetic itinerant state 606
ferromagnetic magnetic semiconductors 608
ferromagnetic metal 509
ferromagnetic nanoparticles 722
ferromagnetic phases 606
ferromagnetic phase transition 600
ferromagnetic semiconductor 608

ferromagnetism 35, 232, 509, 608, 635, 720
- Fe, Co, and Ni 37
Fe_3Se_4 and $FeTi_2Se_4$, magnetic structures of 415
$FeSe_x$, phase diagram of 406
Fick's law 683
field-cooled magnetization 713
filiation 362
filled skutterudites 463
- semiconductance in 463
- thermoelectric properties of 463
first-order J–T (FOJT) 168
first-order or higher order phase transformation 682
fluctuations of the spins in a spin liquid– classical or quantum, spin vice 619
fluoride 501
fluoride ion, ionic radius 479
fluorite 362
- composition 363
- structure of 363
- structures, other defect 196
fluorite, CaF_2, and related structures 188
- binary oxides 188
- ordered defect fluorite strutures; bixbyite, pyrochlore, and weberite 189
fluorite-type structures
- defect fluorites 365
- – bixbyite and antibixbyite 366
- – pyrochlore 369
- – scheelite 367
- – zirconium-based oxynitrides 365
- from diamond to fluorite; antifluorite 364
- structural filiation to 362
fluorspar 363
Frank-Kasper polyhedra 524
free crystal
- susceptibility 650
free electric susceptibility 651
free electron concepts 45
free electron model 32
free energy 647
Frenkel disorder 680
Frenkel enthalpy 680
Frenkel level 671
frustrated lattices 625
frustrated magnetism 617, 618, 619
frustrated magnetism on 2D and 3D lattices 618
frustrated magnets 619
frustrated materials 619
frustrated systems 631
frustration 617

fullerene acceptor 720
fullerenes 703
fulminating silver 272

g

Ga -hybrid orbitals 10
γ-AgCuS + Cu → Ag + Cu_2S 388
γ-AgCuS, decomposition of 388
γ'-Fe_4N 270
γ'-Fe_4N-type perovskite structure 268
γ'-Fe_4N structure 272
γ'-M_4N phases 275
γ-Mo_2N 267
Ga–Ga bonding interactions 9
Ga–Ga bonds 8
Ga–Ga curve 10
Ga–Ga interactions 9, 10
Ga NMR signal 537
Ga 4p orbitals 9
GaSb 458
gas barrier films
- with brick–mortar–sand structure 558
- with brick–mortar structure 558
- with self-healing properties 560
gas barrier properties 558
gas-phase interstitial modification technique 253
$GdFeO_3$-type distortion 656
$GdFeO_3$-type orthorhombic structure 656
Ge–Sb–Te alloys 739, 744
Ge–Sb–Te based optical memory disks 746
Ge–Sb–Te system 735
Ge–Te binary phase diagram 738
GeO_2 glass 113
geometrical frustration 618, 622
- in simple atomic systems 619
- – curved-space approach 620
germanate 95
germanium chalcogenides 737
germanium telluride (GeTe) 737
$GeSb_2Te_4$
- atomistic explanation 745
- metal–insulator transition 745
$GeSb_2Te_4$ films 745
- electrical properties 745
GeSbTe system
- ternary phase diagram 529
$GeTe$–Sb_2Te_3 compounds
- crystal structures 738
$GeTe$–Sb_2Te_3 system 530
GGA-DFT electronic structure of $ScNi_2$ 7
giant magnetoresistance 595
giant magnetoresistance (GMR) 599, 718

giant magnetoresistance materials 599
Gibbs free energy 503, 672, 681, 692
glassy magnetic state 413
Glazer tilt systems 199
golden metallic rock salt-type δ-TiN$_x$ 262
Golden yellow metallic rock salt-type ZrN 262
Goodenough–Kanamori rules 163
grain boundaries 669
graphene 595
– microspheres (GMSs) 556
– nanosheet 548
– quantum dots 548
– spintronics 719
graphite 443
greigite (Fe$_3$S$_4$) 721
group V elements, structure map 743
growth kinetics 693

h

half-Heusler compounds 2, 527
half-Heusler phases 51
half-metallic electronic structure 232
Hall angle 596, 609
Hall resistivity 596
Hall voltage 595
Hamiltonian matrix 5
hardness 448
H$_2$/Ar gas mixtures 280
Hartree–Fock-type mean-field approximation 608
HCP lattice 165
HDDR process (hydrogenation–decomposition–desorption–recombination) 253
heat conduction 746
heat-resistant alloys 252
heavy alkali metals 141
heavy alkaline earth (Ca, Sr, and Ba) hydrides 485
heavy fermion 603
heavy rare earth trihydrides 500
HeI photoelectron spectra
– of Cs, Cs$_{11}$O$_3$·Cs$_{10}$, and Cs$_{11}$O$_3$ 146
Helical chains 297
Helmholtz free energy 649
heptacyanate anion 714
hepta/octacyanates 714
heterogeneous catalysis 455
heterogeneous electrolytes 687
Heusler-phase 306, 322
Heusler phases 527
hexaborides 447

hexacyanometalates 710
hexagonal bipyramidal 289
hexagonal ε-NbN 265
hexagonal-close-packed 618
hexagonal lattice 403, 405
hexagonal Mn$_5$Si$_3$ structure 285
hexagonal Nb$_5$N$_6$ 265
hexagonal perovskite compounds 241
– tolerance factor 241
hexagonal perovskite structures 212, 213, 241
– projections, of octahedra network 241
hexagonal single crystals 256
hexagonal symmetry 542
hexakaidecahedra 529
Hf d orbitals 18
Hf–Hf and Hf–P bond strengths 18
Hg–In
– binary phase diagrams 61
[Hg$_2$N]$^+$ in mercury-containing nitride phases 273
4H-BaRuO$_3$-type structure 277
high density metal nanoparticles
– schematic illustration of 550
higher borides 450
higher moments 25
highest occupied crystal orbital
– isosurfaces encasing 745
high-density information storage devices 717
high-energy-resolution spectrum 633
high-pressure behavior 489
high-pressure phase 387
– pyrite-type CuSe$_2$, 387
– tetragonal Cu$_2$−xS 385
high-pressure phase transitions 490, 491
high-resolution transmission electron microscopy (HRTEM) 740
high-spin state 169
high-T$_c$ superconducting copper oxides
– crystal structure of 240, 246
high-temperature centrifugation-aided filtration (HTCAF) 288
highly frustrated magnetism in spinels 626
high spin –low spin transitions 714
H$_2$ molecular orbital diagram 3
Ho$_3$Al$_5$O$_{12}$ garnet 634
hollandite structure exemplified by Ba$_2$Mn$_8$O$_{16}$ 186
homing 722
homogeneity 371
homogeneous catalysts 552
homogeneous glasses 117
homogeneous transition 606
homonuclear pnictogen–pnictogen bond 462

homopolar Ge–Ge bonds 742
Hume-Rothery electron concentration
    phases 1, 32, 35, 522
– tight-binding energy 24
Hund's rules 499, 703
Hund coupling 600
Hückel approaches 6
Hückel calculations 25, 300
Hückel energy 25
Hückel-based model 7
Hückel methods 4
Hückel model 6
Hückel theory 4
Hückel tight-binding method 426
hybridization 500
hydride 507, 511
hydride anion 479, 504, 506, 510, 512
– in oxides 501, 506
hydride compounds, thermodynamics of
    480
hydride diffusion 511
hydride exchange reactions 511
hydride ions 488
– mobility, in saline hydrides 478, 491
hydride phase 481
hydride species 477
hydroformylation 552
hydrogen 477, 481
hydrogen AO electron clouds, change in 2
hydrogenation reaction 502
hydrogen atom
– electron affinity 479
hydrogen bond 477
hydrogen bonded system
– potassium dihydrogen phosphor 653
hydrogen bonding 703
hydrogen chemistry 478
– hydride compounds 479
– hydrogen species 478
hydrogen impurities 255
hydrogen partial pressure 481
hydrogen storage 455, 478, 481
hydrogen sulfide 501
hydrotalcite-like compounds 541
hydrous oxides 572
hydroxyl hydrogen 544
18-electron rule 51
μ_2-Hückel chemical pressure analysis 58
-Hückel model of $SrAg_5$ 26
-scaled sH 27
μ_2-scaling 25
hypothetical $CaCu_5$-type $CaAg_5$ phase 60
hysteresis loop 709, 716

i
icosahedra 446
icosahedral approximants 80
icosidodecahedral shells 85
icosidodecahedron 85
Imma orthorhombic 239
Imma structure 210
impurity phases 253
indium 440
industrial device fabrication 737
inelastic neutron scattering 628
inelastic scattering spectrum 632
infinite zigzag chain anions 296
infrared absorption 94
infrared spectroscopy 631
inorganic crystal structure database
    163
insulator–metal transition 225
insulators 643
– classification of materials 644
interatomic distances 25, 500
interatomic interactions 25, 59
intercalation 364, 589
interfacial charge 686
Intergrowths 211
inter-particle interactions 712
inter-spin interactions 632
interlayer reactions 587
intermediate-range order 94
intermetallic bonding 1, 2
intermetallic community 13
intermetallic compound (IMC) 8, 16, 17, 31,
    51, 55, 521, 551
– characteristic of 17
intermetallic phases 6, 33, 521
intermetallics, investigation of 532
intermetallic structures 27, 59
intermetallic systems 6
intermolecular exchange interactions
    705
intermolecular interactions 703
interpolyhedral interaction 446
interstitial compounds 440
interstitial nitrides 331
Interstitial nitrogen 284
intrinsic magnetic anisotropy
    718
inverse-Heusler compounds 527
inverse piezoelectricity 645
inverse piezoelectricity effect 653
iodine 440
ion conduction, mechanism of
    683

ion conductivities 684
– solid electrolytes, as function of temperature 685
ion conductors 684
– solid electrolytes 693, 695
ion exchange 587
ion-exchange materials 584
ionic and electronic defects, as acid–base and redox particles in equilibrium 669
– C-rule 674
– P-rule 674
– T-rule 674
ionic and electronic diagrams 671
ionic bonding 477
ionic bonds 477, 478
ionic compounds 478
ionic conductivity for Cu ions 398
ionic conductors 584, 684
ionic crystal 680
ionic crystals 665, 669
– concept 222
ionic defects 687
ionic disorder 668, 671, 673
ionic gap 670
ionic liquids 522
ionic materials 479
ionic mobile species 693
ionic model 488
ionic nitrides 330
ionic radius 479
– of large halides 479
ionic redistribution effects at contacts 688
ionization energy 6, 479
Ir DOS distribution 64
Iron (Fe) deficiency 327
iron-based nitrides 275
iron-based superconductor 406
Ising systems 717
isolated double tetrahedra $[M_2N_6]^{8-}$ 319
isolated dumbbell anions 290
isolated nonplanar anions 304
isolated planar anions 300
isolated tetrahedra $[MN_4]$ 318
isolobal T–T bonds 55
isostructural compounds
– semiconductance of 462
isostructural oxides 512
isotropic chemical shift 104
isotypic $Ba[Mg_{3.33}Nb_{0.67}N_4]$ 321
isotypic high-temperature phases $\beta\text{-}Ba_3[MoN_4]$ 319
isotypic phase $Ba_2[NNiCN]$ 291
isotypic phase $Sr_2[Ni_2(Ni_{1-x}Li_x)N_2]$ 298
isotypic quaternary nitridorhenates 315
isotypic $Sr_3[CrN_4]$ 318
isotypic structures $Li_3Ba_2[NbN_4]$ 314
itinerant ferromagnetic state 606

j

Jahn–Teller distortion 223, 231, 300
Jahn–Teller effect 27
Jahn–Teller J–T phenomena 168
Jahn–Teller transition 627
Jahn–Teller type, symmetry breaking 40
$J_1 - J_2$ model, showing ground state spin configurations for 621
Jones model 33
Jones reflections 35
Jones theory 32, 41
Jones zones 31, 34
– for γ-brass (Cu_5Zn_8-type) Cu_5Cd_8 and Cd_3Cu_4 32

k

Kanamori–Goodenough rule 232
κ-space 33
$K_{23}Au_{12}Sn_9$
– electronic structure analysis 16
$K[CoO_2]$-type structure 323
K-edge XANES 290
k-space 27
K-values 675
kinetic energies 2, 3
$KNbO_3$ 654
K–Au interactions 16
Knight shift 532
Kohler's law 597
Kohler's rule 596
Kohn–Sham Ansatz 4
Kohn–Sham equation 4, 6
Kondo effect 603
$KOsO_3N$ monocrystals 368
K_5Pb_{24}
– eH bonding analysis 15
Kröger–Vink diagram 675
Kröger–Vink nomenclature 668
Kröger–Vink notation 480
$KTaO_3$
– temperature dependence of dielectric constant 658
$K_3Ti_5NbO_{14}$ structure 585
$KTiNbO_5$ structure 585

l

La$_3$[Cr$_2$N$_6$]
– crystal structures 324
Lactobacillus fermentum 721
LaHO crystal structure 503
La$_2$LiO$_3$H-Sr$_2$LiOH$_3$ system 514
Landau levels 598, 609
Landau phenomenological equation 649
Landau theory 227, 647
lanthanide 4f orbitals 8
lanthanides 448
– Ce–Lu 259
lanthanoides 140, 435
lanthanum hexaboride 448
laser-heated diamond-anvil cells 262
LaSrCoO$_3$H$_{0.7}$ crystal structures 508
LaSrCoO$_3$H$_{0.7}$ QENS data 509
La$_{1.4}$Sr$_{1.6}$Mn$_2$O$_7$
– resistivity of 602
latent heat 426
lattice constants 225
lattice thermal conductivity 462, 468
Laves phases 2, 523
– AB$_2$, MgCu$_2$, crystal structures 524
layered cobalt oxide Ca$_3$Co$_4$O$_9$
– magnetoresistance scaling 604
layered double hydroxide (LDH) 541, 584
– active species-intercalated
  nanocatalysts 552
– based materials
– – biorelated applications of 563
– – overpotentials of 557
– basic structural features 542
– biological and medical properties 561
– – in bioapplications 562
– – in drug delivery and treatment 562
– – drug-delivery performance 563
– – synthesis of LDH-Based biomaterials 561
– brucite-like layers 542
– catalytic properties of 548
– confinement effect 546
– – on electron distribution of guest
  molecules 546
– – on hybrid materials 546
– construction of 543
– density functional theory calculations of 545
– electrochemical catalysis 556
– electrochemical energy storage
  properties 553
– electronic properties 545
– fabrication of 554
– facile exfoliation of 553
– formula 541

– general structure 584
– host layer structure of 543
– host–guest interaction of 546
– interlayer galleries of 552
– interlayer structure of 543
– metal-based catalysts 550
– microspheres 555
– microstructure and property of 544
– molecular dynamics simulations of 544
– photocatalysis of 549
– properties and applications of 546
– rearranged geometric configuration 547
– solid base catalysis of 549
– for stoichiometries 543
– structural characteristic of 543
– structural properties of 545
– supercapacitors (SC) 554
– theoretical studies of 544
layered oxides 571
– containing metal-oxygen octahedra 573
– made up of edge- and/or corner-shared
  metal-oxygen octahedra 584
– miscellaneous 586
– possessing rocksalt superstructures 581,
  582
– rocksalt superstructure family, structures
  of 583
– types 572
layered oxyhalides 587
layered oxynitrides 587
layered oxysulphides 587
layered perovskites 573
layered perovskites A$_m$B$_{m-1}$O$_{3m}$,
  structure 580
LDA-DFT
– eH-fitted band structure of γ-brass
  42
LDA–DFT band calculations 41
(LDH/CA)$_n$ multilayer film
– schematic representation of 559
"spintronicmaterials 161
leak currents 646
LED industries 331
Lennard-Jones 6–ä12 potential 25
Li$_{52.0}$Al$_{88.7}$Cu$_{19.3}$ 46
Lewis acid 479
Lewis acid–base reaction 62
Lewis bases 63, 478, 479
Lewis notation 139
Lewis structure 1
Lewis theory 62
L-proline intercalated LDHs
– schematic model of 553

Li$_8$Ag$_5$
– γ-brass structure 44
Li$_{32}$Ag$_{20}$
– DOS and COOP for Ag framework 44
Li$_{33}$Ag$_{19}$ 43
Li$_{52.0}$Al$_{88.7}$Cu$_{19.3}$
– DOS and COOP 48
– Jones zones 45
Li$_{54.0}$Al$_{89.2}$Cu$_{19.0}$ powder 47
Li atoms 44
Li$_6$C a metal-rich species 139
Li$_6$(Ca$_{1-x}$Sr$_x$)$_2$[Mn$_2$N$_6$] 304
Li$_6$Ca$_2$[Mn$_2$N$_6$]
– crystal structures 306
Li$_{15}$[Cr^{VI}N$_4$]$_2$N
– crystal structures of 313
LIESST. See light induced excited state trapping (LIESST)
ligand field favoring 714
ligand holes, localization of 235
light-switchable materials 715
light induced excited-spin state trapping 715
light irradiation 714
Li-ion batteries
– transition metal as anode material in 364
Li-ion batteries (LIB) 556
– coulombic efficiency 556
– degree of dispersion 556
– performance of 556
– use of, carbon materials 556
Li-ion conduction 684
– lithium nitrides for 455
Li-Mg-Zn-Al Bergman phases
– using Mulliken populations 19
Li-storage 688
Li$^+$ ions 19
Li$_2$[(Li$_{0.5}$Co$_{0.5}$)N]
– DC magnetization 288
Li$_2$[(Li$_{1-x}$Cu$_x$)]N 290
Li$_2$[(Li$_{1-x}$Fe$_x$)N] 288
Li$_5$[(Li$_{1-x}$Mn$_x$)N]$_3$ 295
Li$_2$[(Li$_{1-x}$M$_x$)]N 290
Li$_5$[(Li$_{1-x}$M$_x$)N]$_3$
– crystal structures 289
Li(2)$_2$[Li(1)N] 287
Li$_2$[LiN] crystal structure 289
Li$_2$[LiN]-type structure 288
Li$_{24}$[Mn^{III}N$_3$]$_3$N$_2$ 305
[LiN] chains 295
Li–Ga curve 10
Li–Ga interactions 9, 10
Li–Ga overlap 10

Li–Ge–thiophosphates 684
Li–Hg and In–Hg phase diagrams 61
linear combination of atomic orbitals (LCAO) 4, 42
linearly coordinated Li 287
linear magnetoresistance (LMR) 393
Linear Muffin Tin Orbital-Atomic Sphere Approximation-Tight-Binding (LMTO-ASA-TB) 15
Li$_2$N layers 289
Li$_3$N substitution variants 290
Li$_3$N superstructures 287
Li$_2$O-type defect and order variants 304
Li$_2$O structure 310
liquid crystallinity 719
Li$_3$[ScN$_2$] 310
Li$_2$Sr$_5$[MoN$_4$]$_2$ 314
Li$_3$Sr$_3$[NiN]$_4$
– predominant structural features 295
Li$_2$Ta$_3$N$_5$ 326
Li$_{16}$[TaN$_4$]$_2$O
LiTaO$_3$ 654
lithium alkaline earth metal nitridometalates
– linear chains 294
lithium battery cathodes 163
lithium clustering 122
lithium extraction 313
lithium ion conductor 287
lithium ions 120, 586
lithium salts (LiMO$_2$) 588
lithium silicate glasses 120, 122
Li$_7$[VN$_4$] 312
Li$_x$[Mn$_{2-x}$N] 304
LiZrN$_2$ 328
LMR. See linear magnetoresistance
localized bonding models 48
local modes 626
Lorentz field 597
Lorentz force 596, 609
LT-SrCeN$_2$ 325

m
MA. See mechanical alloying (MA)
2/1 Mackay approximant, shells of cluster 86
magnesium borides 443
magnesium hydride (MgH$_2$) 488
magnetic abnormality 636
magnetic anisotropy 716
magnetic centers 703
magnetic diffuse scattering pattern 633
magnetic dipole–dipole coupling 115
magnetic dynamics 718
magnetic entropy change 713

magnetic field 598, 710
magnetic field (MF) 710
magnetic frustrated materials 625
magnetic frustration 619
– in multiferroic materials 634
magnetic hydrogels 710
magnetic-ion density 609
magnetic hysteresis 716
magnetic information 722
magnetic interaction 413, 631
magnetic ion doping 634
magnetic MAX phases 282
magnetic moment 703
magnetic ordering 720
– temperature 712, 713
magnetic particles 722
magnetic properties 412
– of BTFO film 634
magnetic random access memory (MRAM) 718
magnetic relaxations 716, 717
magnetic resonance imaging (MRI) 562
magnetic sensor 720
magnetic spin ordering 720
magnetic susceptibility 413, 425, 628
– data 306
magnetic tunnel junctions 719
magnetic zeolitic structures 720
magnetism 512, 617, 704
magnetite (Fe_3O_4) 161, 721
– crystals 721
magnetization 704
– relaxation time 716
magnetocaloric effect 713, 717
magnetocaloric effect (MCE) 713, 717
magnetocaloric refrigeration 438
magnetocrystalline anisotropy 443
magnetoelectric coupling 637, 720
– in BTFO film 637
magnetoelectric effect 634
magnetopolarization measurements 637
magnetoreception 722
magnetoresistance (MR) 595, 719
– classification of 598
– effect 223, 719
– types of 599
magnetoresistive semiconductor 609
magnetosensitive system
– cryptochrome-based 722
magnetosome 721
magnetotactic bacteria 720, 721
magnetotactic bacteria (MTB) 721
magnetotaxis 721

magnets 716
– molecule-based 704
– rare earth-based 723
manganese borides 438
manganese hexacyanoferrate (MFCN) 723
manganese-rich $Ca_{12}[Mn_{19}N_{23}]$ 305
manganese oxides 602
MAS-NMR spectrum of glassy B_2O_3 104
mass action constant 673
mass action laws 674
mass-market devices 735
mass spectrometry 510
MAX-phases $M_{n+1}AN_n$
– crystal structures 282
maximally localized Wannier function 49
MAX phases 272, 278, 281, 282
Maxwell equations 713
mayenite 505
MB_4 variants 447
McConnell's model 706
MCE. See magnetocaloric effect
MD. See molecular dynamics
MD simulations 113, 120
mechanical alloying (MA) 384
mechanical energy 401
mechanochemical reactions 384
mechanochemical synthesis 384
ME coupling versus magnetic field of BTFO15 film 635
medium-range order 94
– experimental characterization of 99
medium-range order, experimental characterization of
– solid-state NMR 103
– – dipolar NMR methods 105
– – magic-angle spinning NMR 104
– vibrational spectroscopy 101
– X-ray and neutron diffraction techniques 99
memory effect 709
metal atom coordination, in diborides 443
metal atom matrix 139
metal atoms 403
metal borides 435, 438, 439, 441
metal cations 544
metal chalcogenides 383, 384, 402
– preparation methods of 383
metal diborides 445
metal film 692
metal fluxes 257
metal halides 439
metal hydrides 478, 501
metal-like electrical resistivities 326
metal-organic magnets

– Ni$_2$A-based 723
metal-oxygen octahedra 573
metal-rich borides 440
metal-rich compounds
– multinary derivatives of 272
– – β-Mn derivatives 279
– – ε-Fe$_3$N derivatives 278
– – η-carbide derivatives 280
– – γ'-Fe$_4$N derivatives 273
metal-rich Rb$_6$O 143
metal ions 572, 586, 588, 590
metallic bond 478
metallic compounds La$_3$[Cr$_2$N$_6$] 324
metallic conductivity 509, 526
metallic η-carbide-type compounds 280
metallic glasses 93
metallic hydrides 477, 479, 487, 496, 500
– structural information for 498
metallic lattice 501
metallide nitrides 258, 328
metal ligand 507
metallization 490
metal-metal bonds 139, 140, 159
metal-metal-bonded clusters 159, 160
metal–boron interactions 445
metal–insulator transition 604, 605, 745
– field-induced 606
metal–metal hydride phase diagram 481
metal–metal multibond 450
metal–nonmetal arrays 588
metal–nonmetal layered structures 581
metal oxides 93, 383, 439, 692
metal vacancies 405
metastable oxides 587, 590
metastable semiconducting Cu$_3$N 271
metastable ternaries 739
metathesis product 439
methylammonium cation 223
methylammonium lead halides 223
methyl–ammonium–lead iodide 685
MF. See magnetic field (MF)
MFCN. See manganese hexacyanoferrate (MFCN)
Mg$_2$Al–benzoate LDH 544
Mg$_{32}$(Al,Zn)$_{49}$ phase 19
MgCu$_2$
– tight-binding energies of 23
MgCu$_2$ structure 57
MgEu$_4$[TaN$_4$]N 320
MgH$_2$ rutile structure 488
Mg/Mg^{2+} couple 479
Mg$_3$Sb$_2$ 457
– thermoelectric properties of 457

m-Co$_4$Al$_{11}$ structure of two layers 78
microcrystalline samples 271
microscopic probes 606
microwave communication devices 233
microwave irradiation 589
migration energy 680
migration of Cu$^+$ ions 398
milling 384
Millon's base [Hg$_2$N]N$_3$ 272
missing cations 744
mixed conductors
– as electrodes or absorbers 695
– – batteries 696
– – fuel cells 695
– – sensors, chemical storage and permeation devices 696
mixed metal oxide (MMO) 549, 557
MM'$_3$N phases
– crystal structure types 274
Mn-based perovskite nitrides 276
Mn(II)Cu(II) molecular magnets 707
Mn$_4$N
– Curie temperature 276
Mn$_3$SbN 276
Mn$_3$ZnN exhibits antiferromagnetic order 276
mobile carrier 600
mobility 675, 684
Mo–Mo bond 460
modified oxide glasses 124
– medium range order in 118
modified random network (MRN) 119
molecular building blocks 711
molecular cluster 716
molecular crystal lattice 709
molecular dynamics (MD) 544, 742
– simulations 95
molecular magnet
– design 704
molecular magnetic sponges 709, 710
molecular magnetic switches 715
molecular magnets
– design and synthesis strategies 704
molecular magnets design
– using azido bridge to mediate spin–spin couplings 715
molecular materials 703
molecular multication ferrimagnets 713
molecular orbital theory 6
molecular spintronics 718
molecular wires 719
molecule-based magnet
– self-assembly 704

molybdenum diboride 445
moment analyses 20
– higher moments 21
monosilicides 525
morphotropic phase boundary 646, 661
morphotropic phase boundary (MPB) 646, 661
Mo_3Sb_7 462
– crystal structure of 461
– density of states of 461
Mößbauer spectra 290
– for Fe atoms 288
MoS_2-type structure 327
Mössbauer effects 631
Mössbauer spectra
– of BTFO ceramic sample 636
– of BTFO15 ceramic sample 637
motion sensors 645
motivation 521
Mo_3 triangles
– in MoS_2 50
Mott insulator 627
Mott–Hubbard model 163
MPB. See morphotropic phase boundary (MPB)
MR. See magnetoresistance (MR)
MRAM. See magnetic random access memory (MRAM)
MRN hypothesis 120
M_4Si_4 crystal structures, cut out 525
MTB. See magnetotactic bacteria (MTB)
multielectron wave function 49
multiferroicity 236
multiferroic materials 634
multifunctional molecular magnetic materials 719
multiple inhomogeneities 660
multiple quantum well (MQW) 546
muon spin rotation (μSR) spectroscopy 631
M_2 values 120, 122, 123

n

Na_2B_{29} structure 449
Na_3B_{20} structure 448
NaCl crystal structures of binary nitrides 261
Na^+ conductivity 684
$Na_3Co_2SbO_6$ oxide 583
$Na_2Co_2TeO_6$ oxide 583
$Na_2[HgO_2]$-type 291
nanocrystalline 688
nanodomains 456, 464
nanoionics 686
nanomagnets 717
nanomaterials 528
nanoparticle systems
– magnetic properties 712
nanoporous molecular magnet 720
nanostructuring 456, 466
nanowires 528
$Na_4[Re^VN_3]$ contains pyramidal anions $[ReN_3]^{5-}$ 304
narrow-gap semiconductors 458, 462, 467
narrow magnetic hysteresis loops 709
$Na_2Ti_3O_7$ structure 585
$Na_2Ti_4O_9$ structure 585
N_2 atmosphere 252
Nb–O bonds 184
Nb–O octahedron in $CoNb_2O_6$ 184
$NbSb_2$ 460
– crystal structures of 459
Nd–O coordination polyhedron with 201
$Nd_{0.5}Sr_{0.5}MnO_3$
– magnetic diffraction 605
– resistivity 605
Néel temperature 269, 508, 632
nearest-neighbor 618
near-infrared 713
near-infrared (NIR) 713
nearly free electron model 29
negative magnetization 713
neodymium iron boride 438
neodynium phase 369
Nernst–Einstein equation 684
nerve fibres 722
nesting 422
net reaction rate 683
network formers 93
neurodegenerative diseases
– magnetic particles formation 723
neuromorphic 746
neutron diffractions 369, 377, 408, 445, 510, 602, 636, 737
– data 446
– experiments 120
neutron diffuse scattering patterns of $Ho_3Al_5O_{12}$ garnet measured at 633
neutron inelastic scattering spectra of $Ho_3Al_5O_{12}$ garnet measured at 632
neutron powder diffraction 636
– of BTFO ceramic sample and analysis of specific peak 636
neutron scattering 656
– experiments 632
next-nearest neighbor correlations in tetrahedral network glasses 97
N_2/H_2 gas mixture 280

N-poor phase boundary 267
Ni$_2$Al$_3$ structure 57
NiAs-type structure 265
nickel hexacyanoferrate (NiHCF)
  nanoparticles 712
Ni···Ni contacts 297
NiHCF–DODA films 712
Ni-based perovskites Ni$_3$MN 277
Ni membrane 255
Ni$_2$Mo$_3$N antiferromagnetism 280
niobium chalcogenides 418
NITR, chemical structure 706
nitridation 366, 370, 373, 377
nitrided R$_2$WO$_6$ precursors
  – composition and unit cell parameter of 373
nitride ions 327
nitride–metalide phases 329
nitride oxides 251
nitrides 251
nitrides, non-main group elements
  – preparative aspects 252
nitridoferrates(II) Ca$_2$[FeN$_2$] 303
nitridometalate-azide-nitride Ba$_9$[(Ta, Nb)
  N$_4$]$_2$[N$_3$]N 255
nitridometalate oxide Ba$_3$[ZnN$_2$]O 291
nitridometalates 251, 254, 286, 300, 309, 310,
  323, 331
  – binary transition metal nitrides 253
  – chemistry 309
  – containing heavier alkali metals 306, 316
  – crystal structures of 305
  – planar zigzag chains 296
  – prototypes 325
  – synthesis 254
  – tetrahedra 316
nitridosilicates 251
nitridotungstates 257
nitrogen 253, 259
  – sensitizing effect 366
nitrogen/europium concentration ratio 366
nitrogen-rich tantalum nitride 266
NMR chemical shifts 507
NMR coupling parameters 533, 535
NMR crystallography 132
NMR methodology 94
NMR signals 533, 537
NMR spectroscopy 120, 535
noble metal chemical bond 42
noble metal γ-brass e$^-$/a ranges 33
noncoherently magnetic 1 : 3 transition metal
  species 29
nonconducting materials 532
non-Kramers garnets 632

non-spin polarized bcc Cr band structure 41
noninteracting paramagnet 622
nonmagnetic Cr Fermi energy 39
nonmagnetic Fe
  – DOS and COHP 37
nonmagnetic metals 412, 413
nonmagnetic probiotic bacteria 721
nonmagnetic transition metal compounds 38
nonmetallic impurities 254
Nowotny chimney ladder (NCL) 53
  – phases 2, 53, 54
n = 1 RP phase, Sr$_2$TiO$_4$, as an intergrowth of
  one perovskite 211
nuclear Bragg reflection position 633
nuclear magnetic resonance (NMR) 94, 506,
  522
nuclear quadrupole moment 532
nucleation 481, 689, 693
nucleophilicity 1

O
occupancy rate (p$_o$) 404
octagonal approximants 75
octahedral 446
  – distortion 223
  – linkage pattern in anatase (TiO$_2$) 186
  – rotations 224
octahedral rotations
  – lattice connectivity 224
  – magnitude and phase of 224
octahedral site preference energies
  (OPSE) 168, 169
octahedral sites 543
octahedral tilting distortion 228
Ohm's law 596, 683
O holes 165
o-Al$_{13}$Co$_4$ structure of two layers 78
olefins 552
oligomeric anions 303
one-electron Schrödinger equation 5
opaque classical magnets 710
o′-Al$_{13}$Co$_4$ structure of two layers 79
optical data storage devices 530
optical micrographs of ZnSe 402
optical rotation 430
optical storage media 744
optimization process 228
optoelectronic displays 715
orange-red Ta$_3$N$_5$ 266
orbital-ordered states 601
orbitals 5
  – degrees of freedom 601
ordered perovskite structures

– involving combinations of A- and B-site and oxygen vacancy ordering 211
order–disorder-type ferroelectrics 643, 653
order–disorder phenomena 617
order–disorder transition 493
organic chemistry 1
organic ferroelectric 662
organic-based material
– iron-containing 705
organic–inorganic hybrids 590
organic polymeric structures 705
organic salts 607
organic spintronic device 719
organometallic compound 51
organophosphonic acids 590
orientation 693
orthorhombic approximant 81
orthorhombic $Ba_2[Ni_3N_2]$ 297
orthorhombic $CaTiO_3$ crystal structures 226
orthorhombic compounds 282
orthorhombic phase o-Co_4Al_{13} 77
osmium atoms 446
O_2 system
– darker contrast 36
overdoping 679
oxic–anoxic interface 721
oxidation 140, 148, 163
– state 161, 166
– – of Mg atoms 444
oxide-based functional materials 571
oxides 581
– ions 192
– materials 572
oxonitridometalates 251
oxyfluorides 229, 503, 504
oxygen 140
– atoms 142
– crystal structure of 238
– defect perovskites 207
– evolution reaction (OER) 556
– interaction 691
– stoichiometry 689
– – profiles for Fe-doped 690
– transmission rate (OTR) 558
– vacancies 238, 693
oxygen isotope exchanged $SrTiO_3$
– ε versus T^2 plot 659
oxyhydrides 477, 501, 502, 503, 514
– high-pressure synthesis of 513
oxynitrides 229, 361, 511
– crystal chemistry of 361
– environmental applications 361
– optical properties of 361

– and oxides, comparison between 361
– powders 375
– preparation of
– – solid/gas reactions 370
– with pyrochlore structure
– – preparation of 370
– scheelite structure type 368
– solid solutions
– – characterization of 376
– synthesis 370
oxynitride tungstates
– spectral selectivity of 376

p

packing of characteristic, Li_{26} and Ba_5Na_{12} units in $Li_{13}Na_{29}Ba_{19}$ and central Li_4N unit 156
paraelectric KDP 643
– crystal structure 654
paramagnetic centers 716
partial amorphization 401
partial electron transfer 27
particle diffusion 684
Pauling bond order function 18
Pauli paramagnets 412
PBA. See prussian blue analogues (PBA)
p block states 14
Pb–Pb interactions 14, 15
PB NPs. See Prussian blue nanoparticles
$PbTiO_3$ 654
$Pb(Zr_{1-x}Ti_x)O_3$ system
– phase diagram 661
$PbZrO_3$ 651, 654, 661
$PbZrO_3$–$PbTiO_3$ (PZT) 646
PCM. See phase-change material (PCM)
PD_2 Al-Co-Ni structure of two layers 79
Pearson Crystal Database 75
Peierls-like distortion 737
Peierls-type distortion 450
Peltier effect 456
pentagondodecahedron 529
perfect and defective crystal situations
– for compound M^+X^- 667
periodic table 139
– with the transition elements identified 162
perovskite
– X anion vacancy order 237
perovskite-related layered structure compounds 243
perovskites 221
– ABO_3 structure 574
– A-site cation order 233
– A-site cations and B-site cations 236
– as ionic crystals 222

- B-site cation order 230
- Ca_3AuN 329
- chemical formula 221
- chemical variety of 222
- compositional flexibility 229
- crystal structure of 223
- electronic structure of 222
- ferroelectric soft modes 644
- hydrides, structural information for 495
- -type oxides 655
- – tolerance factor dependence of dielectric and ferroelectric properties 655
- ion order in 229
- mineral, of calcium titanium oxide 221
- nitrides 278
- phase 368, 369
- physical and chemical properties 221
- related compounds 654–656
- related compounds r R1_Start1 654
- related structures 202
- and related structures 197
- rock salt-type order 230
- solar cells 223
- structure 511
- – ideal 495
phase
- boundary 661
- crystallize 323
- separation 411
- stability/flexibility 494
- transformation 682
- transition
- – temperature 652
- transitions 500
phase-change material (PCM) 530, 735
- origin and nature of vacancies 744
- resonant bonding 743
phase competition 606
phase diagram 606
- $A_2BB'O_6$ double perovskites showing existence range for ordered "rock salt" structures 203
- $ANiO_3$ 227
- $(Co_xFe_{1-x})_3Se_4$ 410
- $CrSe_x$ 406
- Cs/O 140
- $Cs_{11}O_3$/Rb 142
- $FeSe_x$ 407
- $(Ni_xV_{1-x})_3Se_4$ 411
- Rb/O 142
- $TiSe_x$ 404
phase-change data storage 746
- functional principle 736

- material requirements 736
phase-change materials (PCM) 735
- materials properties 743
- property contrast 743
phase-pure nitridometalates 255
phenomenological theory 648
phonon glass electron crystal (PGEC) 529
phosphates 95, 572
phosphides 18
photoactive components 549
photoactivity 378, 703
photocatalysis 373, 455, 549
photocatalysts 584
photocatalytic
- activity, for water splitting 549
- performance 546
photocathodes 457
- Cs_3Sb as 457
- Na_2KSb as 457
- Na_3Sb as 457
photoconductivity 427
photodynamic therapy (PDT) 563
photoelectrochemical (PEC) 557
photoelectrochemical water splitting 557
photoelectrodes 685
photoelectron spectroscopy 146
photoinduced magnetization 723
photoluminescence 377
photomagnetic switching 710
photothermal ablation agents 713
phototransformations 715
phthalocyanine molecules 562
p- and f-block oxyhydrides 503
p-and f-block oxyhydrides
- rare earth oxyhydride 503
p-NPNN
- chemical structure 706
p-orbital-based spins 703
P-rule 675
p-type conductivity 744
physical picture (energy level diagram)
- for ionic and electronic disorder at interfaces 687
piezoelectric (PZT) 579, 645, 661
PMN structure
- possible microscopic model 660
- temperature dependence of dielectric constant 660
pnictides 455
pnictogens (Pn) 455
Poisson's equation 687
polarizability 684
polarization 637

– switching 663
polarization (P) versus electric field (E) loop 645
pole reversal 713
polyacrylic acid (PAA) 559
polyantimonides
– in lithium batteries 455
polycyanometallates 710–714
– prussian blue and its hexacyanate analogues 710
polyethylene terephthalate (PET) 561
polyhedra 446
polyhedra in boron-rich borides 446
polyhedral connectivity 182
polyhedral distortion 166
polyhedral morphology 429
polymer consistent force field (PCFF) 544
polymeric anions
– linear arrangement 293
polymer magnets
– design 705
polymer spintronics 719
poly(p-phenylene) (APPP) 546
polypyrrole (PPy) core 555
poly(sodium styrene-4-sulfonate) (PSS) 560
poly(vinyl acetate) (PVA) 560
polyvinyl chloride (PVC) 140
polyvinylpyrrolidone (PVP) 712
porphyrin 562
post-perovskite
– crystal structure of 242
post-perovskite compounds 243
– crystal structure of 243
– in geoscience research 243
postulated charge redistribution
– occurring in alkali borophosphate glasses 130
potassium ion sensing 712
potential energies 2
Potts models 617
powder X-ray diffraction (PXRD) 577
premilling process 401
primary susceptibility 647
pristine polymer 558
(100) projections of CaF_2 and AA'_2BO_7 weberite unit cells 195
projections of crystal structures
– $Ba_{14}CaN_6 Na_7$ 152
– $Ba_{14}CaN_6 Na_8$ 152
– $Ba_{14}CaN_6 Na_{14}$ 152
– $Ba_{14}CaN_6 Na_{17}$ 152
– $Ba_{14}CaN_6 Na_{21}$ 152
– $Ba_{14}CaN_6 Na_{22}$ 152

– $Cs_{11}O_3$ 145
– $Cs_{11}O_3$ Cs 145
– $Cs_{11}O_3$ Cs_{10} 145
– $Cs_{11}O_3$ Rb 145
– $Cs_{11}O_3$ Rb_2 145
– $Cs_{11}O_3$ Rb_7 145
– Rb_9O_2 144
– $Rb_9O_2Rb_3$ 144
projections of rod structures
– Ba_2N 150
– Ba_3N 150
– Ba_3N Na 150
– Ba_3N Na_5 150
protonated oxides 590
proton conductors 572
proton-assisted hydride diffusion 511
prototypical relaxor 659
Prussian blue analogues (PBA) 710, 711
– fcc structure 711
Prussian blue nanoparticles (PB NPs) 713
Prussian blue salt 711
pseudobinary M_3X_4–M'_3X_4 system 408
pseudobinary M_3X_4–X_4 system, phase diagram and site preference in 406
pseudobinary systems 410
pseudobrookite (Fe_2TiO_5)-type structure 266
pseudocapacitance performance 554
pseudocapacitors 554
pseudogap 8, 13, 14, 32, 46, 604
pseudopotentials 1, 33–35
– theory 33–35
PtN synthesis 271
purely organic magnets 705
pyramidal anion $[Re^VN_3]^{4-}$ 304
pyrochlore lattice 626
– composed of corner-sharing tetrahedra 630
pyrochlores 191, 365
– phase 370
– structures 192, 369

q

quadratic–pyramidal coordinated transition metal sandwich 324
quadrupolar coupling
– parameters 111
quadrupole coupling 532
qualitative treatments 29
quantization 598
quantum critical point 659
quantum mechanics 58
quantum paraelectricity 657
quantum spin liquid 623
quantum theory 1

quantum theory of atoms in molecules (QTAIM) method 332
quantum transport 595, 609
– phenomena 609
quasibinary electroneutrality equation 677
quasi-crystalline approximants 45, 75
– $Li_{52.0}Al_{88.7}Cu_{19.3}$ 46
quasi-crystals 2, 45
quasi-inelastic neutron scattering (QENS) 508
quasi-one-dimensional compound NbSe, charge density wave instability in 417
quaternary compounds 530
quaternary hexanitridodichromate(V) $Li_4Sr_2[Cr_2N_6]$ 315
quaternary nitrides $Li_3Sr_2[NbN_4]$ 313
quaternary nitridometalates
– crystal structures of 315
quaternary nitridometalates containing Li
– crystal structures of 314
quaternary system Li-Sr-M-N 294
quenching 141, 440

r

radical pair mechanism 722
radical pair mechanism (RPM) 722
radius ratios 166
Raman scattering 94
Raman spectra 113
– $(Na_2O)_x(B_2O_3)_{1-x}$ glasses 102
Raman spectroscopy 741
raMO analysis 54, 55
rare earth elements 326, 368, 376, 435
rare earth metal nitridometalates 317
rare earth metals 140, 477, 478, 479, 499, 525
rare earth–tungsten–oxygen system
– stoichiometries in 372
rare earth oxyhydrides 503
rare earth tantalates 370
rate constant 683
rate equations 668
rattling effect 463
$Rb_2Ba_6Sb_5OH$ crystal structures 505
RbH_5
– interpretation 50
Rb_{13} icosahedra 144
Rb^+ ions 588
Rb_9O_2 and $Cs_{11}O_3$ clusters
– in alkali metal suboxides 143
Rb_9O_2 clusters 142, 143, 144
Rb_6O octahedra 144
Rb/O system 147
reaction temperature 681
reactive molecular nitrogen 253

read-head devices 718
real ferroelectric
– formula generalization 650
ReB_2-structure type 444
ReB_2-type crystallizing compounds 445
rechargeable Ca-ion battery concept 723
REDOR experiment
– on a borophosphate glass 109
– principle 106
REDOR pulse sequence timing and corresponding evolution of dipolar Hamiltonian 107
reentrant phase transition 653
refractory materials 254
regioselective hydrogenation 552
REHO compounds 503
relaxors 659
– dielectric behavior 661
– ferroelectric 659
$R_{E_2}M_{17}N_x$
– magnetic properties 284
$R_{E_2}M_{17}N_x$ compounds 284
$R_EM_{12}N_x$ crystal structure 284
repulsive forces 166
residual acidities 64
resistivity 388, 601
resonant bonding 743
– low thermal conductivity, relation with 746
resonant X-ray diffraction 605
resonating-valence-bond (RVB) 617
Restricted Hartree–Fock calculations on $(OH)_3Si-O-Si(OH)_3$ clusters 111
reversed approximation molecular orbital (raMO) 51
reverse Monte Carlo (RMC) 742
– analysis 713
– fitting 119
– modeling (RMC) 110
reversible hydrogen electrode (RHE) 557
rhenium arsenide 462
– isostructural with Mo3Sb7 462
rhenium nitrides ReN_x 269
rhodium 552
rhombicosidodecahedron 85
R-phase 45
R-value 147
Rietveld refined Ag-rich phase 45
Rietveld refinement technique 713
rigid band model 20, 21
ring size distributions
– in glassy silica 112
– in glassy SiO_2, and binary sodium silicate glasses 126

ring structures
– motifs in tetrahedral network glasses 98
– postulated for modified borate glasses 126
RKKY. See Ruderman Kasuya Yoshida (RKKY)
Rochelle salt
– potassium sodium tartrate 653
rocksalt 484
– -type group 3, 285
– -type phase $Mo_{1-x}Nb_xN$ 285
– -type UN 262
– phase 741
– structure 581
3R ($AgCrO_2$) structure 586
$R_2Ti_2O_7$ 370
rubidium 141, 143
rubidium Rochelle salt 653
rubidium suboxides 141, 143
Ruddlesden-Popper phases
– crystal structures of 245
Ruddlesden–Popper-type manganese oxides 599
Ruddlesden–Popper-type perovskite oxides 512
Ruddlesden–Popper oxide 590
Ruddlesden–Popper phases 244, 588
Ruddlesden–Popper phase $Sr_3Ti_2O_7$ 324
Ruddlesden–Popper series 602
– members, structures of 577
Ruddlesden–Popper structures 512
Ruderman–Kittel–Kasuya–Yoshida (RKKY) interaction 603
Ru sublattice 176
rutile and related structures 181
– binary oxides 181
– other forms of TiO_2 and MnO_2 187
– site-ordered rutiles, trirutile structure and columbite 183
– structures compositionally related to rutile, hollandites, and others 184
rutile (TiO_2) structure 182
R_6WO_{12}
– thermal ammonolysis reaction of 374
$R_{14}W_4O_{33}$
– thermal ammonolysis reaction of 374
R_6WO_{12} tungstates
– preparation of
– – amorphous citrate method 372

s
saline hydrides 477, 479, 480
sample size 680
satellite positions 74
saturated magnetization 415

Sb_2–Sb_2 bonds 460
Sb_2Te_3
– superlattices of 456
scanning probe spectroscopy 606
Sc DOS curve 64
scheelites 365, 368
schematic decomposition
– of $Li_{80}Ba_{39}N_9$ structure into Li_{13}, Ba_4, Ba_6N, and $Ba_5N_6Li_{12}$ fragments 154
Schrödinger equation 2, 4
Sc_3InN 278
ScIr formation 63
$ScNbN_2$ 327
Sc–Cu system 61
$ScTaN_2$ 327
secondary alcohols, enantiomeric configuration 428
second-order Jahn–Teller (SOJT) effect 509
second-order optical nonlinearity 719
second order J–T (SOJT) 168
Seebeck coefficients 425, 438, 448, 456
Seebeck effect 456
– used in spacecrafts 456
selenide systems 408
self-propagating high-temperature synthesis 399
self-propagating high-temperature synthesis (SHS) 399
semiconducting behavior 530
semiconductive-to-metallic transition 424
semiconductors 455, 458, 532, 670
– technology 608
semiempirical parameters 4
semi-empirical relation 111
SEM images of Cu particles 392
sensory organelles 722
shell of centered 1/1 Mackay-type approximant 84
Shubnikov–de Haas effect 610
Shubnikov–de Haas oscillation 598, 610
S-doping 680
Si–O–B linkages 115
Si–O–Si bond angle distribution 111
– functions 110
signal-to-noise ratio 484
Si-variants 684
silica glass (amorphous SiO_2) 740
silicate 95
simple Hückel (sH) 5
simple perovskite oxides
– typical charge combinations of 222
simple perovskites 230

– group–subgroup relationships of space
  groups 228
single chain magnets (SCM) 716
single crystal X-ray diffraction (SXRD) 577
single-chain magnet behavior 717
single-chain magnets 717
single-crystal sample of $SrCo_{12}O_{19}$
– resistivity 607
single-crystal X-ray 737
single-molecule spintronics 719
single-walled carbon nanotube (SWNT)-array
  double helices 550
single molecule magnets 716
single molecule magnets (SMM) 716
single tetrahedron inscribed in a cube 630
singlet–singlet ground state system 634
singlet–singlet systems 633
sintering 440
sinter metal boride powders 439
site-ordered double perovskites 203
site-ordered perovskites 205
site-selective values extracted from 124
site preference and magnetic properties 412
sketches of different bonding situations, with
  metal-rich compounds 158
skewness 20
Slater determinants 49
Slater-type orbital (STO) 5
Slater model 654
smart paper 746
sodium ion distributions 120
sodium oxotungstate 268
sodium silicate glasses 123
soft actuators 710
soft mode theory 655
solid electrodes 669
solid electrolyte 364
– for fuel cells 365
– for gas sensors 365
solid electrolytes 669, 684
solid-state chemistry 362, 665, 669, 685, 709
solid-state metathes 252
solid-state metathesis 257
solid-state NMR experiments 120
Solid-state NMR results on B_2O_3–SiO_2 glasses
  as a function of glass composition 116
solid-state NMR spectroscopies 132
solid-state reactions 392, 398, 440, 502, 572,
  668, 689, 692
solid-state sciences 735
solid-state structures 6
solid–solid interfaces 689
solid solution system 410

solid state ionics 665
sol–gel system 117
sol–gel technique 117
solubility 437, 494, 675
sonochemical reactions 389
sonochemical synthesis 385
– of Cu and Ag chalcogenides 389
sonochemistry 389
space charge effects 688
space group "family tree" for double perovskites
  $A_2BB'O_6$ 204
space group "family tree" for perovskites with a
  single B-site ion based on Glazer's octahedral
  tilt systems 200
space group symmetry, dependence, of
  Brownmillerites on 208
specific capacitance 554
specific heat 631
spectroscopic techniques 532
sphalerite 362
spin crossover (SCO) 714
spin delocalization 705
spin echo double resonance (SEDOR) 108
spinel AB_2O_4
– showing edge-sharing octahedra of B-sites
  and A-site tetrahedra 213
spinels 213, 626
spinel structure 627
spin frustration 637
spin-based electronics 718
spin-glass-like behavior 636
spin-liquid phase of $ZnCr_2O_4$ 629
spin-orbit-lattice coupling 637
spin-Peierls transition 627
spin-polarized conduction electrons 232
spin-state switching
– ultrafast light-induced 715
spin-valve devices 719
spin-valve-type magnetoresistance 603
spin ice 624
spin ice compounds 629
– $Ho_2Ti_2O_7$ and $Dy_2Ti_2O_7$ 629
spin injection 719
spin liquid 620
spin–lattice complex 628
spin–lattice coupling 627, 628
spin–spin interaction 704
spinodal decomposition 406
spin polarization 705
spin quantum number 705
spins, artificial magnetic fields and monopoles
  in spin ice 618
spin transitions 714

spintronics 595, 609
spontaneous polarization 650
spontaneous strain 650
square-planar coordination 322
$Sr(Al_{1−x}Ga_x)_4$ lattice parameter 534
$Sr_{0.53}Ba_{0.47}[CuN]$ 296
$Sr_6[CuN_2][Cu_2N_3]$
– crystal structures of 292
$Sr_8[Fe^{III}N_3]_2[Fe^{II}N_2]$ 299
$Sr_{10}[FeN_2][Fe_2N_4]_2$
– crystal structures 299
Sr_3GaSb_3 468
– crystal structures of 468
$Sr_2[Li(M_{1−x}Li_x)N_2]$ 294
$Sr[LiN]$ 293
$Sr_2[LiNCoN]$ 293
$Sr_2(Li_4N_2[Li_{1−x}Ni_xN])$ 295
$Sr_{1−x}Ba_xGa_2$ lattice parameter 536
$Sr_{1−x}Ga_{2+3x}$, crystal structure, cut out 526
$Sr_8[Mn^{III}N_3]_2[Mn^{II}N_2]$ 302
$Sr_8[MN_3]_2[MN_2]$ 299
$Sr_8[MnN_3]_2[FeN_2]$ 299, 300
Sr–Fe–N system 299
SrTe nanocrystals 456
$Sr[TiN_2]$
– crystal structures 324
$SrTiO_3$ 658
– structure parameters of 227
$SrTiO_3$-based oxides 507
$Sr_4Ti_3O_{10}$ structure 574
$SrVO_2H$ crystal structures 512
stacking sequence for Li_2ReO_3 176
stacking sequence of O−Li−O−Li/Sn−O layers in β-Li_2SnO_3 structure 175
stacking sequences 738
steady-state electrical transport 683
steric strain 57
stoichiometric compounds 141
stoichiometry 366, 373
Stoner ferromagnetism 608
Stoner theory 609
STP transition metal element crystal structure type 22
strain effects 637
strong-emissive Au NCs 548
strontium hydride 493
structural complexity 571
structural distortion angle 543
structural energetic stabilization energy 34
structural energy difference theorem 25
structural families, derived from ABO_3 perovskites 202

structural order in glasses, general concepts of 95
structural transition temperature 502
structure field map for brownmillerites 209
structure field map for stability of pyrochlore structure
– TMO with the A^{3+}/B^{4+} combination 191
structure-dependent energy of atomic arrangement 34
structure maps 744
structure–bonding–property relations 522
structure prediction diagnostic software (SPuDS) 228
structure types for TMO 169
– binary oxides 169
– NaCl or rock salt 169
– ternary and more complex oxides, with ordered-rock salt structure 172
$μ_3$ acidity model 62
$μ_3$-neutralization 64
suboxide hydrides 504
suboxides, and other low-valent species 139
suboxides of lithium and sodium 141
substitution center (SC) 537
successive phase transitions
– in quasi-one-dimensional sulfides ACu_7S_4 (A= T l, K, Rb) 423
successive shells
– of Tsai-type approximants 87
sulfides 18
sulfur 13
– content 13
supercapacitor (SC) 553
superconducting copper oxides 161
superconducting region 501
superconducting temperature 420
superconducting transition 420
– temperature 502
superconductivity 384, 385, 388, 420, 423, 501, 505, 571, 573, 590, 607
– in quasi-one-dimensional compound $InxNb_3Te_4$ 418
superconductor 438, 443, 501
super-exchange interaction 605
superionic conductors 670, 681, 684
superionic dissociation 668
superionic situation 682
superionic transitions 681
superlattice reflections 421
superlattice structures 533, 537
superparamagnet 716
superparamagnetic
– maghemite nanoparticles 721

surface reaction kinetics, analysis of 691
surface spin-disorder 712
switching 735
symmetry-breaking 38
synchrotron-quality X-ray diffraction 739
synchrotron X-ray diffraction studies
– on glassy P_2O_5 115
synergism 40
– between the nearly free electron and tight-binding models 42
– Cu_5Zn_8 40
– Li_8Ag_5 42
system Mn-Si-V 80

t
Ta–O octahedron in $CoTa_2O_6$ 184
Tamann's rule 680, 681
Ta_3N_5
– crystal structure 266
Ta–B system 440
Ta–S compound 13
Ta–S phases 13
Ta–Ta antibonding states 13
Ta–Ta bonding 13
Ta–Ta interactions 13
Ta_3N_5. $Na_xTa_3N_5$ 326
tantalum-centered tantalum hexagonal antiprisms 80
tantalum oxynitride 372, 376
Ta_6S_5 compound 13
Ta_6S_5 phase 14
Ta/V-Te quasicrystals 80
T–O–T bond angles 113
tellurides 456
tellurium
– toxicity of 472
temperature dependences of electrical resistivity 424
temperature hysteresis 605, 606
temperature programmed desorption (TPD) 559
ternary antimonides 464–471
– without Sb —Sb bonds 465
– with Sb–Sb bonds 469
ternary azidometalates 251
ternary B_2O_3–SiO_2–GeO_2 glasses 115
ternary ionic hydrides 494
ternary nitridometalates
– tetrahedral anions 318
ternary systems 87, 95
ternary transition metal 18
tetragonal Nb_4N_5 265

tetragonal phases 283
tetrahedral coordination 309
tetrahedral voids
– order patterns of 312
tetrahydridoborates 440
tetrathiafulvalene (TTF) 659
ThB_4 (UB_4) structure 446
$ThCr_2Si_2$ structure 11
thermal ammonolysis 369
thermal analysis 140, 522, 589
thermal annealing 522, 739
thermal conductivities 438, 458, 466, 471, 528, 529
thermal destiny 680
thermal destiny, of an ionic crystal 681
thermal lattice conductivity 528
thermally regenerative electrochemical cycle 723
thermally regenerative electrochemical cycle (TREC) 723
thermal transport
– vacancy disorder, effect of 746
thermodynamic stability 478, 481
thermoelectric materials 522, 528, 603
– alkali metal antimonides 457
thermoelectrics 456
– efficiency of 456
– thermal conductivity of 456
θ-Mn_6N_5 269
thin films 456
2 H ($CuAlO_2$) structure 586
T holes 165
thorium 262
three-cation coordination polyhedra in AA'_2BO_7 weberites 196
Ti, binary perovskite nitride 278
TiCoSb
– crystal structure of 465
tight binding-based structure maps 27
tight-binding bonding models 40
tight-binding energy eigenvalues 41
tight-binding-based energy 41
tight-binding theory 27
tight-binding theory, emerging directions in 48
tilt systems among known perovskites with a single B-site ion, distribution of 201
$Ti_{21}Mn_{25}$ structures 65
$TiSb_2$ 459
– crystal structures of 459
$TiSe_x$, phase diagram of 404
titanate oxides 658

titanate perovskites
- temperature dependence of dielectric constant 658
titanium 438
- diboride 443
titanium niobium containing oxides 589
titanium oxyhydrides 509
TlCu$_7$S$_4$, schematic crystal structure 423
TM ions 167
TMO catalysts 163
TMO with bixbyite structure 190
TMO with columbite structure 185
TMO with ilmenite structure 179
TMO with trirutile, $P4_2/mnm$ structure 185
T–E bonding 54, 55
T–E compounds 55
T–E intermetallic compounds 51
T–T bonding 54
T–T contacts 54
tolerance factor 224, 514
toluene 551
top of the valence band (TVB) 546
topotactic oxidation 231
total energies 18, 25, 26, 58, 62
total number of P–O–Al linkages n_{P-O-Al} as extracted from 117
tracer diffusion 684
transition elements 45, 161
transition-metal phthalocyanines 719
transition metal 13, 54, 462, 502, 581
- interact with the filled low-lying d orbitals 62
- oxidation states and valence electron configurations 49
transition metal (T) 51
transition metal atoms 463
transition metal borides 449
transition metal cations 234, 581
transition metal chemistry 140
transition metal compounds
- tight-binding derived energies 29
transition metal d orbitals 8
- band 13
transition metal elements 13, 27, 39
transition metal (T) elements 61
transition metal-based CsCl-type phases 27
transition metal intermetallic phase formation 63
transition metal ions 166, 588, 645
transition metal MAX nitrides 281
transition metal oxide (TMO) 161, 506, 512
transition metal oxyhydrides 502, 507, 513
transition metal positions 77

transition metals 63, 326, 361, 435, 440
- nitridometalates of 286, 332
- structures 63
- tight-binding energies 22
transition/rare earth metals
- binary nitrides of 258
transition temperature 408, 420, 501, 590
transmission electron microscopy (TEM) 419, 606, 739
transparency 6
transparent phases Ca$_2$[ZnN$_2$] 290
transport and transfer 682
treasure map 746
triclinic Sr$_{10}$[FeN$_2$][Fe$_2$N$_4$]$_2$ 299
triethylphospine 590
triglycine sulfate 645
triglycine sulfate (TGS)
- P–E loop 646
trigonal Li$_{24}$[Mn^{III}N$_3$]$_3$N$_2$ 304
Triple-quantum (TQ-) MAS-NMR spectrum of glassy B$_2$O$_3$ 105
triple perovskites 239
triplet state 703
trirutile structure, AB$_2$O$_6$ 184
tritium 477
truncated icosahedron 82, 87
Tsai-type approximant 87
2/1 Tsai-type approximant, arrangement of truncated icosahedra in 88
Tsai-type cluster 86
T sublattice orbitals 55
tubular furnace
- nitridation of 370
tunable metal ion 556
tungstates
- thermal nitridation of 374
tunneling magnetoresistance (TMR) 606
two-phase mixture 606
two-step electrosynthesis method 555

u

UB$_4$ structure 447
UB$_{12}$ structure 448
UB$_{12}$ with boron atom cube-octahedra structure 449
ultrathin film (UTF) 546
uniaxial ferroelectric 647
unit cell 527
- 2/1 approximant 83
- of Ba$_6$N·Na$_{10}$ 149
- constants and cell volumes for binary TMO
- - with Al$_2$O$_3$ structure 178

- constants, for binary TMO oxides with the CaF_2 structure 188
- of doubly ordered double perovskite $NaLaMnWO_6$ 205
- of $FeTiO_3$ 179
- of γ-$LiFeO_2$ 173
- of $La_{0.33}NbO_3$ 206
- of $LiBa_2N$, increasing sizes for N, Li, and Ba 153
- $Li_8Ba_{12}N_6 \cdot Na_{15}$ with $Li_8Ba_{12}N_6$ clusters 155
- Li_4MgReO_6 176
- $Li_3Mg_2RuO_6$ 176
- $Li_2Ti_2O_4$ 173
- NbO unexpected SQP coordination of $Nb^{2+}(4d^3)$ 171
- $NdTiO_3$ (Pnma) $a^-b^+a^-$ 201
- suboxometalate Cs_9InO_4 and coordination of InO_4^{5-} ion by Cs atoms 148
urea route
- using halides $(M_xCl_y+(NH_2)_2CO)$/oxides $(M_xM'_yO_z+(NH_2)_2CO)$ 253

v
vacancy 684
- defects 741
vacancy-ordered rock salt-type Ta_4N_5 266
vacancy-ordered structures 406
- in binary MX MX_2 system 402
vacancy-ordering 404
vacancy ordered brownmillerite structure (I2mb) 207
valence electron concentration (VEC) 18, 23, 51, 522, 523
valence electrons 547
valence states 222
vanadium octacyanoniobate ferrimagnet 724
vanadium oxyhydrides 512
Van't Hoff plot 483
van der Waals forces 738
van der Waals gap 268
van der Waals interactions 459, 544, 703
vaporization 140
V–V bond distances 183
Vector spins 617
versatile materials platform 746
vertex-sharing tetrahedra 309
- infinite chains 316
vertex-sharing tetrahedral 317
vibrational spectroscopy 94
Villain model 618
virial theorem 3
Vogel Fulcher's formula 661

void volumes 741
V^{2+}/V^{3+} ions 180

w
Wannier analysis
- of H-based bands 50
Wannier function 1, 48–50
water
- photooxidation 378
- photoreduction 378
water-in-oil microemulsion technique 712
wave function 4
wave vector 421
WC-type structure 265
weak-emissive Au NCs 548
weak (Anderson) localization 598
weberites 195
Weiss temperature 413
$[WN_2N_{2/2}]$ tetrahedra build six-membered rings 316
Wolfsberg–Helmholz approximation 6
Woodward–Hoffmann rules 1
Wurtzite-type structures 285
Wyckoff symbols 189

x
$(XAl-LDH/PAA)_n$-CO_2:
- schematic illustration of 560
X-ray and neutron diffraction 94
x-ray diffraction 404, 483, 522
- analysis 426
- data 259, 446
x-ray fluorescent holography experiment 609
X-ray machine collection factors 30
X-ray/neutron diffraction 132
X-ray photoelectron spectroscopy 94
X-ray scattering factors
- of 1/1 quasi-crystalline Bergman phases and Jones zones 46
XMCD measurements 275
XRD method. 395
XRD pattern of
- $CuInSe_2$ 388
- $Cu_{1.96}S$ 387
- $CuSe_2$ 387
- tetragonal $Cu_{1.96}S$ 386

y
YBa_2MnO_5 and $YBa_2Cu_3O_7$ structures 210
$Yb_5In_2Sb_6$
- crystal structures of 470
$Yb_9Mn_{4+x}Sb_9$
- crystal structure of 469

YCoC-type structure 293
ytterbium dihydride 499
yttrium–barium–copper oxide 675
$Y_6WO_{12-3x}N_{2x}$ oxynitride powders
– diffuse reflectance spectra of 375

z
Zeeman effect 598
zeolites 572
zero band gap 611
zero bits 741
zero-energy Landau levels 611
zero-field cooling (ZFC) 288
zero-field resistance 595
zero standard reaction free energy
    682
ζ-phase 270
$\zeta(\zeta')$-Mn_2N 268

ZFC–FC curves of BTFO films measured under
    different magnetic fields 635
zigzag Au chains 330
zinc phthalocyanine (ZnPc) 562
Zintl concept 8
Zintl electropositive/electronegative atom
    ratios 27
Zintl phases 2, 14, 525, 526
Zintl ZE_3 compounds 28
zirconium 159
– metal 500
– oxides 371
Zn corrosion 692
$ZnCr_2O_4$, spin structure 628
Zr d orbitals 18
$Zr_3Er_4O_{12}$
– nitridation 366
Zr_2ON_2 366